SENSORS HANDBOOK

SENSORS HANDBOOK

Sabrie Soloman
Chairman and CEO
American SensoRx, Inc.

McGRAW-HILL

New York   San Francisco   Washington, D.C.   Auckland   Bogotá
Caracas   Lisbon   London   Madrid   Mexico City   Milan
Montreal   New Delhi   San Juan   Singapore
Sydney   Tokyo   Toronto

Library of Congress Cataloging-in-Publication Data

Soloman, Sabrie.
   Sensors handbook / Sabrie Soloman.
      p.   cm.
   Includes index.
   ISBN 0-07-059630-1 (alk. paper)
   1. Detectors—Handbooks, manuals, etc. I. Title.
TA165.S729 1998
681′.25—dc21   97-48542
                                                CIP

McGraw-Hill
A Division of The **McGraw·Hill** Companies

Copyright © 1999 by The McGraw-Hill Companies, Inc. All rights reserved. Printed in the United States of America. Except as permitted under the United States Copyright Act of 1976, no part of this publication may be reproduced or distributed in any form or by any means, or stored in a data base or retrieval system, without the prior written permission of the publisher.

1 2 3 4 5 6 7 8 9 0   DOC/DOC   9 0 3 2 1 0 9 8

ISBN 0-07-059630-1

The sponsoring editor for this book was Harold B. Crawford, the editing supervisor was Marc Campbell, and the production supervisor was Clare Stanley. It was set in Times Roman by North Market Street Graphics.

Printed and bound by R. R. Donnelley & Sons Company.

 This book is printed on recycled, acid-free paper containing a minimum of 50% recycled de-inked fiber.

McGraw-Hill books are available at special quantity discounts to use as premiums and sales promotions, or for use in corporate training programs. For more information, please write to the Director of Special Sales, McGraw-Hill, Inc. 11 West 19th Street, New York, NY 10011. Or contact your local bookstore.

Information contained in this work has been obtained by The McGraw-Hill Companies, Inc. ("McGraw-Hill") from sources believed to be reliable. However, neither McGraw-Hill nor its authors guarantee the accuracy or completeness of any information published herein, and neither McGraw-Hill nor its authors shall be responsible for any errors, omissions, or damages arising out of use of this information. This work is published with the understanding that McGraw-Hill and its authors are supplying information but are not attempting to render engineering or other professional services. If such services are required, the assistance of an appropriate professional should be sought.

The humble work of this Handbook *is dedicated to the most cherished individuals in my life: my late grade-school private tutor, Mr. Kamel Reyad, and my articulate student and methodologist, Miss Cerlinde Chahin.*

Mr. Reyad departed this earth in perfect glory and full honor. He selflessly gave of himself unceasingly to my modest beginning, and his memory never failed to seize the uppermost of my intellect. The echo of his voice still resonates through my soul, remembering his words of wisdom and courage. Although a great deal of his teachings were dedicated toward acquiring the knowledge for science and innovation, yet his few divine words about Christ, the founder of science and the creator of the universe, made me embrace the ideology of creation and accept the Christian faith—eternally.

Miss Chahin has permeated in me the spirit of permutation and brought forth the hidden wealth of knowledge that I unknowingly possessed. While Mr. Reyad had instilled in me the desire to follow in his footsteps for a short period only to teach me how to bypass him, in a like manner, and in the fullness of time, I found my student Miss Chahin performing the same professional role that I portrayed at one time.

Indeed, the loving living example of Mr. Reyad will always live on, through me, my students, and my students' students.

Dr. Sabrie Soloman
June 21, 1998

ACKNOWLEDGMENTS

Mere thanks is insufficient to Dr. Tamer Wasfy for his immeasurable efforts in assisting me in developing the most advanced technologies ever known in the fields of spectroscopy and vision: the SpectR and the InspectR. Both technologies are described in this book. Many thanks also is insufficient to North Market Street Graphics, Lancaster, PA, for their boundless efforts in reviewing and preparing the graphics and index for this book. This book also was made possible by the efforts of my colleagues and friends in various universities and industries, and by the encouragement of the staff of Columbia University, New York.

Dr. Sabrie Soloman
June 21, 1998

CONTENTS

Foreword xxxvii
Preface xxxix

Chapter 1. Introduction   1.1

Establishing an Automation Program / *1.2*
Understanding Workstations, Work Cells, and Work Centers / *1.3*
Classification of Control Processes / *1.8*
Open- and Closed-Loop Control Systems / *1.9*
Understanding Photoelectric Sensors / *1.11*
    Principles of Operation / *1.11*
    Manufacturing Applications of Photodetectors / *1.12*
Detection Methods / *1.17*
    Through-Beam Detection Method / *1.17*
    Reflex Detection Method / *1.18*
    Proximity Detection Method / *1.18*
Proximity Sensors / *1.21*
    Typical Applications of Inductive Proximity Sensors / *1.21*
    Typical Applications of Capacitive Proximity Sensors / *1.22*
Understanding Inductive Proximity Sensors / *1.23*
    Principles of Operation / *1.23*
    Inductive Proximity Sensing Range / *1.24*
    Sensing Distance / *1.26*
    Target Material and Size / *1.27*
    Target Shape / *1.29*
    Variation Between Devices / *1.30*
    Surrounding Conditions / *1.31*
Understanding Capacitive Proximity Sensors / *1.33*
    Principles of Operation / *1.33*
    Features of Capacitive Sensors / *1.35*
    Sensing Range / *1.35*
    Target Material and Size / *1.36*
    Surrounding Conditions / *1.36*
Understanding Limit Switches / *1.37*
Inductive and Capacitive Sensors in Manufacturing / *1.37*
    Relays / *1.39*
    Triac Devices / *1.39*
    Transistor dc Switches / *1.41*
    Inductive and Capacitive Control/Output Circuits / *1.42*
    Accessories for Sensor Circuits / *1.44*
    Inductive and Capacitive Switching Logic / *1.45*
    Inductive and Capacitive Sensor Response Time—Speed of Operation / *1.49*
Understanding Microwave Sensing Applications / *1.53*
    Characteristics of Microwave Sensors / *1.54*
    Principles of Operation / *1.54*

Detecting Motion with Microwave Sensors / 1.56
Detecting Presence with Microwave Sensors / 1.58
Measuring Velocity with Microwave Sensors / 1.59
Detecting Direction of Motion with Microwave Sensors / 1.59
Detecting Range with Microwave Sensors / 1.60
Microwave Technology Advancement / 1.62
Understanding Laser Sensors / 1.63
Properties of Laser Light / 1.64
Essential Laser Components / 1.64
Semiconductor Displacement Laser Sensors / 1.67
Industrial Applications of Laser Sensors / 1.68
References / 1.80

Chapter 2. Fiber Optics in Sensors and Control Systems   2.1

Introduction / 2.1
Photoelectric Sensors—Long-Distance Detection / 2.1
    Light-Emitting Diodes / 2.2
    Through-Beam Sensors / 2.3
    Reflex Photoelectric Controls / 2.5
    Polarized Reflex Detection / 2.5
    Proximity (Diffuse-Reflection) Detection / 2.7
    Automated Guided Vehicle System / 2.8
Fiber Optics / 2.9
    Individual Fiber Optics / 2.10
    Bifurcated Fiber Optics / 2.10
Optical Fiber Parameters / 2.12
    Excess Gain / 2.12
    Background Suppression / 2.14
    Contrast / 2.14
    Polarization / 2.14
Inductive Proximity Sensors—Noncontact Metal Detection / 2.15
Limit Switches—Traditional Reliability / 2.16
Factors Affecting the Selection of Position Sensors / 2.17
Wavelengths of Commonly Used Light-Emitting Diodes / 2.18
Sensor Alignment Techniques / 2.18
    Opposing Sensing Mode / 2.18
    Retroreflective Sensing Mode / 2.18
    Proximity (Diffuse) Sensing Mode / 2.19
    Divergent Sensing Mode / 2.19
    Convergent Sensing Mode / 2.20
    Mechanical Convergence / 2.20
Fiber Optics in Industrial Communication and Control / 2.21
Principles of Fiber Optics in Communications / 2.21
Fiber-Optic Information Link / 2.22
Configurations of Fiber Optics / 2.23
    Optical Power Budget / 2.23
    Digital Links—Pulsed / 2.24
    Digital Links—Carrier-Based / 2.25
    Analog Links / 2.26
    Video Links / 2.26
    Data Bus Networks / 2.27
Configurations of Fiber Optics for Sensors / 2.30
    Fiber-Optic Bundle / 2.30
    Bundle Design Considerations / 2.32
    Fiber Pairs for Remote Sensing / 2.33
    Fiber-Optic Liquid Level Sensing / 2.34

Flexibility of Fiber Optics / 2.34
    Fiber-Optic Terminations / 2.34
Testing of Fiber Optics / 2.36
    Test Light Sources / 2.36
    Power Meters / 2.36
    Dual Laser Test Sets / 2.37
    Test Sets/Talk Sets / 2.38
    Attenuators / 2.40
    Fault Finders / 2.40
    Fiber Identifiers / 2.41
Networking with Electrooptic Links / 2.42
    Hybrid Wire/Fiber Network / 2.43
    Daisy Chain Network / 2.44
    Active Star Network / 2.44
    Hybrid Fiber Network / 2.44
    Fiber-Optic Sensory Link for Minicell Controller / 2.46
Versatility of Fiber Optics in Industrial Applications / 2.47
    High-Clad Fiber-Optic Cables / 2.48
    Fiber-Optic Ammeter / 2.50
References / 2.52

Chapter 3. Networking of Sensors and Control Systems in Manufacturing   3.1

Introduction / 3.1
Number of Products in a Flexible System / 3.2
Sensors Tracking the Mean Time Between Operator Interventions / 3.3
Sensors Tracking the Mean Time of Intervention / 3.3
Sensors Tracking Yield / 3.3
Sensors Tracking the Mean Processing Time / 3.4
Network of Sensors Detecting Machinery Faults / 3.5
    Diagnostic Systems / 3.5
    Resonance and Vibration Analysis / 3.6
    Sensing Motor Current for Signature Analysis / 3.6
    Acoustics / 3.7
    Temperature / 3.7
    Sensors for Diagnostic Systems / 3.7
    Quantifying the Quality of a Workpiece / 3.7
    Evaluation of an Existing Flexible Manufacturing Cell Using a Sensing Network / 3.8
Understanding Computer Communications and Sensors' Role / 3.14
    Application Layer Communication / 3.16
    Presentation Layer Communication / 3.16
    Session Layer Communication / 3.17
    Transport Layer Communication / 3.17
    Network Layer Communication / 3.17
    Data Link Layer Communication by Fiber Optics or Coaxial Cable / 3.17
    Physical Layer Communication / 3.17
    Adding and Removing Information in Computer Networks Based on Open System
        Interconnect (OSI) / 3.18
Understanding Networks in Manufacturing / 3.19
    RS-232–Based Networks / 3.20
    Ethernet / 3.21
    Transmission Control Protocol (TCP)/Internet Protocol (IP) / 3.22
Manufacturing Automation Protocol / 3.23
    Broadband System for MAP Protocol / 3.23
    Carrier-Band System for MAP Protocol / 3.25
    Bridges MAP Protocol / 3.25

Token System for MAP Protocol / 3.26
Multiple-Ring Digital Communication Network—AbNET / 3.27
Universal Memory Network / 3.28
References / 3.30

Chapter 4. The Role of Sensors and Control Technology in Computer-Integrated Manufacturing 4.1

Introduction / 4.1
CIM Plan / 4.2
    CIM Plan in Manufacturing / 4.2
    CIM Plan in Engineering and Research / 4.2
    CIM Plan in Production Planning / 4.2
    CIM Plan in Physical Distribution / 4.2
    CIM Plan for Business Management / 4.3
    CIM Plan for the Enterprise / 4.3
Manufacturing Enterprise Model / 4.3
    Marketing / 4.5
    Engineering and Research / 4.6
    Production Planning / 4.8
    Plant Operations / 4.9
    Physical Distribution / 4.12
    Business Management / 4.13
Design of CIM with Sensors and Control Systems / 4.14
    Components of CIM with Sensors and Control Systems / 4.16
    CIM with Sensors and Control Systems at the Plant Level / 4.16
Decision Support System for CIM with Sensors and Control Systems / 4.19
    Computer-Integrated Manufacturing Database (CIM DB) / 4.20
    Structure of Multiobjective Support Decision Systems / 4.20
Analysis and Design of CIM with Sensors and Control Systems / 4.21
    Structured Analysis and Design Technique (SADT) / 4.21
    A Multiobjective Approach for Selection of Sensors in Manufacturing / 4.23
Data Acquisition for Sensors and Control Systems in CIM Environment / 4.23
    Real-World Phenomena / 4.24
    Sensors and Actuators / 4.24
    Signal Conditioning / 4.24
    Data Acquisition for Sensors and Control Hardware / 4.24
    Computer System / 4.27
    Communication Interfaces / 4.28
    Software / 4.28
Developing CIM Strategy with Emphasis on Sensors' Role in Manufacturing / 4.28
    CIM and Building Blocks / 4.29
    CIM Communications / 4.30
    Plant Floor Communications / 4.30
    Managing Data in the CIM Environment / 4.31
    CIM Environment Presentation / 4.32
    The Requirement for Integration / 4.33
References / 4.38

Chapter 5. Advanced Sensor Technology in Precision Manufacturing Applications 5.1

Identification of Manufactured Components / 5.1
    Bar-Code Identification Systems / 5.1
    Transponders / 5.3
    Electromagnetic Identification of Manufactured Components / 5.3

CONTENTS   xi

  Surface Acoustic Waves / 5.3
  Optical Character Recognition / 5.3
Digital Encoder Sensors / 5.4
  Position Encoder Sensors in Manufacturing / 5.6
Fuzzy Logic for Optoelectronic Color Sensors in Manufacturing / 5.7
  Sensing Principle / 5.8
  Color Theory / 5.8
  Units of Color Measurement / 5.10
  Color Comparators and True Color Measuring Instruments / 5.10
  Color Sensor Algorithms / 5.12
  Design Considerations in Fuzzy Logic Color Sensors / 5.12
  Fuzzy Logic Controller Flowchart / 5.13
Sensors Detecting Faults in Dynamic Machine Parts (Bearings) / 5.15
Sensors for Vibration Measurement of a Structure / 5.17
Optoelectronic Sensor Tracking Targets on a Structure / 5.18
Optoelectronic Feedback Signals for Servomotors Through Fiber Optics / 5.19
Acoustooptical/Electronic Sensor for Synthetic-Aperture Radar Using Vision
  Technology / 5.20
The Use of Optoelectronic/Vision Associative Memory for High-Precision Image Display
  and Measurement / 5.23
Sensors for Hand-Eye Coordination of Microrobotic Motion Utilizing Vision
  Technology / 5.24
Force and Optical Sensors Controlling Robotic Gripper for Agriculture and Manufacturing
  Applications / 5.26
Ultrasonic Stress Sensor Measuring Dynamic Changes in Materials / 5.27
Predictive Monitoring Sensors Serving CIM Strategy / 5.29
Reflective Strip Imaging Camera Sensor—Measuring a 180°-Wide Angle / 5.30
Optical Sensor Quantifying Acidity of Solutions / 5.31
Sensors for Biomedical Technology / 5.32
  Sensor for Detecting Minute Quantities of Biological Materials / 5.33
  Sensors for Early Detection and Treatment of Lung Tumors / 5.33
  Ultrasensitive Sensor for Single-Molecule Detection / 5.34
References / 5.36

Chapter 6. Industrial Sensors and Control   6.1

Introduction / 6.1
Sensors in Manufacturing / 6.3
Temperature Sensors in Process Control / 6.5
  Semiconductor Absorption Sensors / 6.5
  Semiconductor Temperature Detector Using Photoluminescence / 6.6
  Temperature Detector Using Point-Contact Sensors in Process Manufacturing
    Plant / 6.8
  Noncontact Sensors—Pyrometers / 6.8
Pressure Sensors / 6.10
  Piezoelectric Crystals / 6.11
  Strain Gages / 6.11
Fiber-Optic Pressure Sensors / 6.12
Displacement Sensors for Robotic Applications / 6.13
Process Control Sensors Measuring and Monitoring Liquid Flow / 6.15
  Flow Sensor Detecting Small Air Bubbles for Process Control in Manufacturing / 6.16
  Liquid Level Sensors in Manufacturing Process Control for Petroleum and Chemical
    Plants / 6.17
  On-line Measuring and Monitoring of Gas by Spectroscopy / 6.20
Crack Detection Sensors for Commercial, Military, and Space Industry Use / 6.22
Control of Input/Output Speed of Continuous Web Fabrication Using Laser Doppler
  Velocity Sensor / 6.24
Ultrasonic/Laser Nondestructive Evaluation Sensor / 6.25

xii CONTENTS

Process Control Sensor for Acceleration / 6.26
An Endoscope as Image Transmission Sensor / 6.27
Sensor Network Architecture in Manufacturing / 6.28
Power Line Fault-Detection System for Power Generation and Distribution Industry / 6.30
References / 6.31

Chapter 7. Sensors in Flexible Manufacturing Systems   7.1

Introduction / 7.1
The Role of Sensors in FMS / 7.1
    Current Available Sensor Technology for FMS / 7.2
Robot Control Through Vision Sensors / 7.4
    Image Transformation / 7.4
    Robot Vision and Human Vision / 7.5
    Robot Vision and Visual Tasks / 7.5
    Robot Visual Sensing Tasks / 7.6
    Robots Utilizing Vision Systems to Recognize Objects / 7.7
Robot Vision Locating Position / 7.8
Robot Guidance with Vision System / 7.9
    Robot Vision Performing Inspection Tasks / 7.9
    Components of Robot Vision / 7.10
End Effector Camera Sensor for Edge Detection and Extraction / 7.11
    Shape and Size / 7.11
    Position and Orientation / 7.12
    Multiple Objects / 7.12
End Effector Camera Sensor Detecting Partially Visible Objects / 7.15
    Run-Time Phase / 7.19
Ultrasonic End Effector / 7.19
End Effector Sound-Vision Recognition Sensor / 7.19
    Standoff / 7.21
    Large Surface Measurements / 7.21
    Sensitivity of Measurements / 7.21
    Small Surfaces / 7.21
    Positioning / 7.24
End Effector Linear Variable-Displacement Transformer Sensor / 7.27
    Extreme Environments / 7.27
    Cryogenic Manufacturing Applications / 7.29
    Measurement at High Temperatures in Manufacturing / 7.30
Robot Control Through Sensors / 7.30
Multisensor-Controlled Robot Assembly / 7.31
    Control Computer / 7.34
    Vision Sensor Modules / 7.34
    Software Structure / 7.34
    Vision Sensor Software / 7.35
    Robot Programming / 7.35
    Handling / 7.36
    Gripper and Gripping Methods / 7.36
    Accuracy / 7.38
References / 7.38

Chapter 8. SPECTR$_x$: An Online Production Analytical Sensor Developed for Pharmaceutical, Food, Petroleum, Agriculture, Beef, Pork, and Poultry Industries   8.1

Principle of Operation / 8.1
Industrial Applications / 8.1

The Purpose of the Development / 8.2
  Applicable Industries / 8.2
Technical Description / 8.3
  The Software / 8.4
  Presentation of Data / 8.4
Theory of Operation / 8.4
  Configuration / 8.5
  The Spectrometer / 8.6
  Fiber Optics / 8.6
  The Detector / 8.6
  SPECTR Software / 8.7
  Data Acquisition / 8.8
  Dimensions and Specs / 8.8
Plug-and-Play / 8.8
  Power-Up / 8.9
  Start the SPECTR / 8.9
  SPECTR Alignment Procedure / 8.9
  SPECTR Reference / 8.10
  Optimum Object Distance / 8.10
  Setting Standard Scan / 8.10
  Setting Production Run / 8.11
  Setting Production Parameters / 8.11
  Calibrating Sabrie's Index for Hardness Monitor / 8.11
  Option for Synchronous Production / 8.12
  Applications / 8.12
  Saving Current Scan Plus Settings / 8.12
  Load Scan Plus Settings / 8.12
  Setting User Passwords / 8.12
  Changing Internal Parameters / 8.13
  SPECTR Sensor Commands / 8.14
  Text Boxes / 8.15
  Data Storage / 8.15
Scan Parameters / 8.15
Internal Parameters / 8.17
The Spectrum / 8.18
  Function Commands / 8.19
  Types of Spectra / 8.21
  3D Spectrum Screen / 8.21
Production / 8.22
  Production Function Commands / 8.22
  Production Parameters / 8.25
Database / 8.26
  Database Fields / 8.27
Quantitative Analysis / 8.27
  Adding Spectra / 8.28
  Quantitative Analysis Parameters / 8.28
  Use of Quantitative Spectral Analysis (Optimization of Weights) / 8.29
Neural Networks / 8.29
  Manual Neural Network Generation Screen / 8.31
  Train Manual Neural Screen / 8.31
  Run Network Generation Screen / 8.32
  Use of Neural Network Analysis / 8.33

Chapter 9. Communications   9.1

Introduction / 9.1
Single-Board Computer / 9.1

Sensors for Input Control / 9.2
Microcomputer Interactive Development System / 9.4
Personal Computer as a Single-Board Computer / 9.6
    Role of Sensors in Programmable Logic Controllers / 9.7
    Central Control Unit / 9.9
    Process Computer / 9.10
The NC Controller / 9.10
    Manufacturing Procedure and Control / 9.11
    Machining Program / 9.12
    Absolute Control / 9.16
    NC Software / 9.18
    Operation of an NC System / 9.19
    Computer Numerical Control System / 9.27
Industrial Handling / 9.27
Packaging Technology / 9.31
Linear Indexing for Manufacturing Applications / 9.32
Synchronous Indexing for Manufacturing Applications / 9.35
Parallel Data Transmission / 9.35
Serial Data Transmission / 9.36
Collection and Generation of Process Signals in Decentralized Manufacturing Systems / 9.41
References / 9.44

Chapter 10. Microelectromechanical Systems Applications in Energy Management   10.1

Introduction / 10.1
Towards Improved Efficiency / 10.1
The Role of MEMS in Improved Efficiency / 10.2
A Low-Pressure Solution / 10.4
Summary / 10.9
References / 10.9

Chapter 11. The MEMS Program   11.1

Introduction / 11.1
Sensor Programs / 11.1
    SOI Sensors / 11.1
    High-Temperature Sensors / 11.2
    Capacitive Pressure Sensor Process / 11.3
    Material Properties of ZMR SOI / 11.3
    Piezoresistance of ZMR / 11.5
Bulk Micromachined Accelerometer / 11.6
    Proof Mass Die / 11.10
    Force Mass Die / 11.11
    Assembly / 11.12
    Results / 11.12
Surface Micromachined Microspectrometer / 11.12
    Basic Configuration / 11.14
    Theory and Considerations / 11.14
    Process Development / 11.16
    Results / 11.18
Conclusions / 11.18
Reference / 11.19

Chapter 12. MEMS in the Medical Industry   12.1

Introduction / *12.1*
History / *12.1*
    Revisable Blood Pressure Monitoring Transducers / *12.1*
    Disposable Blood Pressure Monitoring Transducers / *12.2*
Current Uses for MEMS Devices in the Medical Industry / *12.3*
    Infusion Pumps / *12.3*
    Kidney Dialysis / *12.4*
    Respirators / *12.4*
    Other Applications / *12.4*
Future Applications / *12.4*
    Neural Interface / *12.4*
    Clinical Diagnostics / *12.7*
Hurdles/Enablers / *12.8*
    Retrofits vs. Enablers / *12.8*
    Technical Hurdles / *12.9*
    Regulatory Hurdles / *12.10*
    Economic Hurdles / *12.10*
Conclusion/Summary / *12.10*
References / *12.11*

Chapter 13. MEMS: A Future Technology?   13.1

Introduction / *13.1*
MEMS: A Current or Future Technology? / *13.1*
    What Are MEMS? / *13.1*
    Current Market / *13.2*
    Market Projections / *13.2*
    Comparison to Semiconductors / *13.3*
What Are the Obstacles? / *13.3*
Concluding Remarks / *13.4*
References / *13.5*

Chapter 14. MEMS Advanced Research and Development   14.1

Introduction / *14.1*
CMOS Compatible Surface Micromachining / *14.1*
Microinstrumentation / *14.2*
Biomedical Applications / *14.2*
    New Process Concepts (DRIE/SFB) / *14.3*
Stanford CIS and the National Nanofabrication Users Network / *14.3*
Summary / *14.3*
References / *14.4*

Chapter 15. Functional Integration of Microsystems in Silicon   15.1

Introduction / *15.1*
The Challenge / *15.1*
The Appeal of On-Chip Integration / *15.2*
The Technical Problems and the Economic Limitations / *15.2*
Wafer Bonding as a Compromise / *15.5*
The Multichip Module on Silicon as the Optimum Solution / *15.6*
Conclusions / *15.7*

Chapter 16. Automotive Applications of Microelectromechanical Systems (MEMS)   16.1

Introduction / *16.1*
Automotive Requirements / *16.2*
Unique MEMS Features / *16.2*
System Applications / *16.2*
    Safety / *16.3*
    Comfort, Convenience, and Security / *16.4*
    Engine/Drive Train / *16.5*
    Vehicle Diagnostics/Monitoring / *16.7*
Market Figures / *16.8*
Conclusions / *16.11*
References / *16.11*

Chapter 17. A Brief Study of Magnetism and Magnetic Sensors   17.1

Introduction: Qualitative Description of Magnetic Fields / *17.1*
The Si and Gaussian Units / *17.2*
Field Sources / *17.3*
AC Fields and DC Fields / *17.8*
Magnetometers and Applications / *17.8*
Conclusion / *17.10*

Chapter 18. The Fundamentals and Value of Infrared Thermometry   18.1

Introduction / *18.1*
Fundamentals of Infrared Thermometry / *18.3*
The Selection Process / *18.7*
Getting Started / *18.12*

Chapter 19. GMR: The Next Generation of Magnetic Field Sensors   19.1

Introduction / *19.1*
GMR Materials / *19.1*
    Physics / *19.1*
    GMR and Saturation Field / *19.3*
    Hysteresis and Linearity / *19.4*
    Resistivity / *19.5*
    Temperature Coefficient of Resistivity (TCR) / *19.6*
    High-Temperature Endurance / *19.6*
    Noise / *19.7*
    Magnetic Field Sensors / *19.7*
GMR Sensor Element / *19.7*
Integrated GMR Sensor / *19.10*
Potential of GMR Sensor Technology / *19.12*
    High Field Sensors / *19.13*
    Low Field Sensors / *19.15*
    Derivative Products / *19.15*
References / *19.16*

Chapter 20. Smart Civil Structures, Intelligent Structural Systems   20.1

Introductions / 20.1
Smart Structure? / 20.2
Fiber-Optic Sensing / 20.3
A Few Fiber Optics Smart Structure Results / 20.3
Summary / 20.4
References / 20.5

Chapter 21. A New Approach to Structural Health Monitoring for Bridges and Buildings   21.1

Introduction / 21.1
Savannah River Bridge Project / 21.6
Arpa Bridge Monitoring Project / 21.8
Embedment Applications / 21.9
References / 21.11

Chapter 22. True Online Color Sensing and Recognition   22.1

Introduction / 22.1
Sensing Light and Color / 22.1
Definition of Color / 22.1
Light/Energy Spectrum Distribution / 22.2
Light Distribution / 22.3
Metamerism / 22.5
Background / 22.5
System Description / 22.6
Advantages of Online Color Sensors / 22.6
Color Theory / 22.7
Principle of Operation / 22.7
Examples of Applications / 22.8
Conclusion / 22.8

Chapter 23. Fundamentals of Solid-State Presence-Sensing Technologies   23.1

Presence Detection / 23.1
Presence Sensors / 23.1
   Noncontact Sensors versus Limit Switches / 23.1
Magnetic-Actuated Switches Applications / 23.5
   Magnetic-Actuated Switch Characteristics / 23.6
   General Terminology for Sensing Distance / 23.6
Components of a Solid-State Sensor / 23.7
   General Terminology / 23.7
   Discrete Sensing Requires Differential / 23.7
   Differential / 23.7
   Repeatability / 23.8
Inductive Principles / 23.8
Shielded and Nonshielded Inductive Sensors / 23.8
Capacitive Principles / 23.9
General Photoelectric Terminology / 23.10

Photoelectric Principles / 23.10
Thru-Beam Scanning / 23.11
Retroreflective Scanning / 23.13
Retroreflective Polarized Scanning / 23.14
Proximity (Diffuse) Scanning / 23.14
Proximity (Diffuse) Background Suppression / 23.15
Color Registration / 23.16
Fiber-Optic Sensors / 23.16
    Thru-Beam and Proximity (Diffuse) Scanning with Extension Cords / 23.16
    Bending Light around Corners / 23.17
    Theory of Operation / 23.17
    Fiber Optics and Sensing / 23.17
    Fiber-Optic Thru-Beam Scanning / 23.17
    Fiber Optics Applications / 23.17
Solid-State Sensor Technologies / 23.18
    Electromechanical Contact Advantages / 23.18
    Electromechanical Contact Drawbacks / 23.19
    Solid-State Advantages / 23.20
    Solid-State Drawbacks / 23.20
    Transistor Switching for DC / 23.20
    Sourcing and Sinking / 23.20
Three-Wire Technology / 23.21
Two-Wire Technology / 23.22
    AC/DCV Two-Wire Technology / 23.23
    Matching Sensors with PLC Input Thresholds / 23.23
Radio Frequency Immunity / 23.24
Weld Field Immunity / 23.24
Response Time: Inertia / 23.25
Power-up Delay Protection / 23.25
    On Delay / 23.26
    Off Delay / 23.26
Response Time / 23.27
Standard Operating Frequency / 23.28

Chapter 24. Design and Application of Robust Instrumentation Sensors in Extreme Environments   24.1

Introduction / 24.1
Design Challenges / 24.4
Extreme Environmental Conditions / 24.5
    Extreme Temperatures / 24.5
    Humidity / 24.5
    Icing / 24.5
    High Wind / 24.5
Power Disturbances / 24.6
Electromagnetic Interference / 24.7
Lightning and Static Discharge / 24.7
Reliability and Maintenance / 24.8
Case Histories / 24.8
Summary / 24.9

Chapter 25. Color Machine Vision   25.1

Why Color Vision? / 25.1
Principles of Color Sensing and Vision / 25.1

Lighting for Machine Vision / 25.5
   Color CCD Cameras / 25.6
   Traditional Color-Based Classification / 25.7
Apples and Oranges: A Classification Challenge / 25.9
Minimum Description: Classification by Distribution Matching / 25.11
Typical Industrial Applications / 25.13
References / 25.14

Chapter 26. Monolithic Integrated Physical and Chemical Sensors in CMOS Technology   26.1

Introduction / 26.1
Physical Sensors / 26.2
   Surfaced-Micromachined Capacitive Pressure Sensor / 26.2
   Integrated Flow Sensor / 26.5
Chemical and Biochemical Sensors / 26.7
   Conductivity Sensor / 26.7
   Hydrogen-Sensitive MOSFET / 26.9
   Sensor Matrix for Two-Dimensional Measurement of Concentrations / 26.10
Summary / 26.13
References / 26.13

Chapter 27. A Research Prototype of a Networked Smart Sensor System   27.1

Introduction / 27.1
Background / 27.1
Overview of Distributed Methods / 27.2
   Transducer-Related Properties of Distributed Measurement Nodes / 27.2
   Measurement-Related Properties of Distributed Measurement Nodes / 27.3
   Sensor- or Application-Related Properties of Distributed Measurement Nodes / 27.3
   Communication Protocol Issues / 27.4
   Data Management Issues / 27.4
   Control Protocol and Real-Time Issues / 27.5
Prototype System / 27.6
   Design Objectives and Specifications / 27.6
   General Description of an Application Using the Prototype System / 27.6
   Smart Node Architecture / 27.7
   Operational Aspects of Smart Nodes / 27.9
Interface Definitions / 27.9
   Transducer Interface / 27.10
   Network Interface / 27.10
Experience Using the Prototype System / 27.10
   Printer Circuit Board Manufacturing / 27.10
   Laboratory Ambient Condition Monitoring / 27.11
   Miscellaneous Systems / 27.11
   Observations / 27.11
Topics for Future Research / 27.12
Conclusions / 27.12
Appendix: Detailed Description of System Models / 27.13
   Network Interface / 27.13
   Behavioral Models / 27.14
   Transducer Interface / 27.15
References / 27.17

Chapter 28. Sensors and Transmitters Powered by Fiber Optics — 28.1

Introduction / *28.1*
Fiber-Optic Power Interface / *28.2*
Advantages of Fiber-Optic Power / *28.3*
Practical Considerations of Fiber-Optic Power / *28.4*
System Configurations and Applications / *28.4*
Conclusions / *28.5*
References / *28.6*

Chapter 29. A Process for Selecting a Commercial Sensor Actuator Bus as an Industry Interoperable Standard — 29.1

Introduction / *29.1*
Background and Related Work / *29.2*
  Sensor/Actuator Bus Evaluation Efforts / *29.2*
  Sensor/Actuator Bus Candidates / *29.3*
The Process of Evaluation and Selection / *29.4*
Sensor/Actuator Bus Survey / *29.6*
Selection Criteria / *29.7*
Candidate Presentation and Review / *29.11*
SAB Interoperability Standard Selection / *29.13*
Conclusions / *29.13*
  Lessons Learned / *29.13*
  Summary / *29.14*
Appendix: Listing of Acronyms / *29.14*
References / *29.15*

Chapter 30. A Portable Object-Oriented Environment Model (POEM) for Smart Sensors — 30.1

Introduction / *30.1*
  Smart Sensor System Integration Issues / *30.2*
  An Outline of the Approach / *30.2*
An Illustrative Example of OO Technology for Smart Sensors / *30.6*
  Example of Programming Model / *30.7*
The Object Model in Detail / *30.8*
  Active Objects / *30.8*
  Reactive Objects / *30.10*
Programming Support / *30.11*
  Active Object Classes for Supporting Smart Sensors / *30.11*
  Reactive Object Classes / *30.13*
The Example Revisited / *30.14*
  Active Objects / *30.15*
  The Reactive Object / *30.16*
Related Work / *30.17*
Conclusions / *30.18*
References / *30.18*

Chapter 31. New Generation of High-Temperature Fiber-Optic Pressure Sensors — 31.1

Introduction / *31.1*

Sensor System Descriptions / *31.2*
Sensor Head Design / *31.3*
Autoreferencing Technique / *31.4*
Sensor Calibration and Laboratory Tests / *31.6*
Engine Test Results / *31.7*
Conclusions / *31.8*
References / *31.8*

Chapter 32. Peer-to-Peer Intelligent Transducer Networking   32.1

Introduction / *32.1*
Why Peer-to-Peer Transducers Are Better / *32.2*
    Form Follows Function / *32.2*
    Easier to Build Small or Large Systems . . . and Expand Them / *32.4*
    Better Loop Performance or Lower Cost or Both / *32.5*
    Better Flexibility from à la Carte Computing to à la Carte Controls / *32.5*
    "But We Don't Think We Can Completely Replace Our Controller" / *32.6*
    The Bottom Line: A Matter of Pure Economics / *32.6*
Implementing Intelligent Transducers / *32.7*
    Implementing the Hardware: It's All in the IC, Network Transceiver, and I/O Objects / *32.7*
    But What About the Software Development and System Integration? / *32.7*
    Interoperability / *32.8*
    Developing Software for an Individual Transducer / *32.10*
    Toolboxes for Verifying Multidevice Operation / *32.10*
    Systems Integration and Maintenance / *32.10*
Intelligent Transducers and Self-Documentation / *32.11*
Problems and Diagnosis / *32.12*
Summary / *32.12*

Chapter 33. Principles and Applications of Acoustic Sensors Used for Gas Temperature and Flow Measurement   33.1

Introduction / *33.1*
Historical Review of Temperature and Flow Measurements / *33.1*
High-Temperature Gas Measurements / *33.6*
    Thermocouples / *33.6*
    Optical Pyrometers and Radiation Pyrometers / *33.8*
Acoustic Pyrometers / *33.11*
    Background Information / *33.11*
    Applications / *33.14*
    Slagging Measurements and Control / *33.16*
    Emission Reduction Using Sorbent Injection / *33.16*
    Refuse Fired Boilers / *33.17*
    Online Measurement of Boiler Performance and Unit Heat Rate / *33.17*
The Measurement of Gas Flow in Large Ducts and Stacks / *33.18*
Instruments Used to Measure Gas Flow in Ducts and Stacks / *33.19*
    Thermal Dispersion / *33.20*
    Differential Pressure Sensors / *33.20*
    Ultrasonic / *33.20*
    A Practical Method for Obtaining Accurate and Reliable Measurements of Volumetric Flow Rates in Large Ducts and Stacks / *33.26*
Conclusions / *33.28*
References / *33.30*

Chapter 34. Portable PC-Based Data Acquisition: An Overview   34.1

Introduction / *34.1*
Portable Applications / *34.2*
A Lack of Slots / *34.2*
Alternatives to Plug-in DAQ / *34.2*
Serial-Port DAQ Devices / *34.3*
Parallel-Port DAQ Devices / *34.4*
PCMCIA DAQ Cards / *34.5*
Power Considerations / *34.6*

Chapter 35. Understanding and Applying Intrinsic Safety   35.1

Introduction / *35.1*
Where Can Intrinsic Safety Be Used? / *35.1*
Methods to Prevent Explosions / *35.2*
Limiting the Energy to the Hazardous Area / *35.2*
Which Sensors and Instruments Can Be Made Intrinsically Safe? / *35.4*
Make Sure the Circuit Works / *35.5*
    Temperature Sensors: Thermocouples and RTDs / *35.5*
Barrier Types / *35.6*
Rated Voltage / *35.7*
Internal Resistance / *35.7*

Chapter 36. Application of Acoustic, Strain, and Optical Sensors to NDE of Steel Highway Bridges   36.1

Introduction / *36.1*
    WIDOT Structure B-5-158, Green Bay, Wisconsin / *36.2*
    Caltrans Structure B-28-153, Benicia Martinez, California / *36.3*
    Caltrans Structure B-22-26 (Bryte Bend), Sacramento, California / *36.5*
    WIDOT Structure B-47-40, Prescott, Wisconsin / *36.6*
Acoustic Emission Testing / *36.7*
Strain Gage Testing / *36.8*
Laser Displacement Gage Testing / *36.9*
Summary and Conclusions / *36.9*

Chapter 37. Long-Term Monitoring of Bridge Pier Integrity with Time Domain Reflectometry Cables   37.1

Introduction / *37.1*
Background / *37.2*
    Bridge Column Failure / *37.2*
    Time Domain Reflectometry (TDR) / *37.2*
TDR Cable Installation in New Column Construction / *37.4*
    Project Description / *37.4*
    Cable Selection / *37.4*
    Cable Installation / *37.4*
TDR Cable Installation in Existing Columns / *37.6*
    Project Description / *37.6*
    Proposed Cable Installation / *37.7*
Summary / *37.10*
References / *37.10*

Chapter 38. Nondestructive Evaluation (NDE) Sensor Research, Federal Highway Administration (FHWA) — 38.1

NDE for Bridge Management / *38.1*
    Objectives / *38.1*
    Background / *38.1*
    Current NDE Research Program / *38.6*
    Future NDE Research Program / *38.6*
    Conclusion / *38.6*

Chapter 39. Sensors and Instrumentation for the Detection and Measurement of Humidity — 39.1

Introduction / *39.1*
Definition of Humidity / *39.1*
    What Is Humidity? / *39.1*
    What Is Its Importance? / *39.2*
Sensor Types / *39.2*
    Relative Humidity / *39.2*
    Bulk Polymer-Humidity Sensor / *39.3*
    Resistive Polymer Sensor / *39.3*
    Capacitive Polymer Sensor / *39.5*
    Displacement Sensor / *39.7*
    Aluminum Oxide / *39.7*
    Electrolytic Hygrometer / *39.8*
    Chilled Mirror Hygrometer / *39.9*
    Continuous Balance—Dual-Mirror Twin-Beam Sensor / *39.14*
    Cycling Chilled Mirror Dew Point Hygrometer (CCM) / *39.15*
    Chilled Mirror Dew Point Transmitters / *39.20*
Summary of Balancing Methods / *39.21*
    Manual Balance / *39.21*
    Automatic Balance Control (ABC) / *39.21*
    PACER Cycle / *39.22*
    Continuous Balance / *39.22*
    Cycled Mirror (CCM) Technique / *39.22*
    CCM with Sapphire Mirror and Wiper / *39.23*
Other Types of Dew Point Hygrometers / *39.23*
    Dew Cup / *39.23*
    Fog Chamber / *39.23*
    Piezoelectric Hygrometer / *39.24*
    Wet Bulb/Dry Bulb Psychrometer / *39.24*
    Saturated Salt (Lithium Chloride) / *39.25*
Calibration / *39.26*
    National Calibration Laboratories / *39.26*
    The NBS Standard Hygrometer / *39.26*
    Precision Humidity Generators / *39.27*
    Two-Flow Method / *39.28*
    Two-Temperature Method / *39.28*
    Two-Pressure Method / *39.28*
    Secondary Standards / *39.28*
Applications / *39.29*
    Automobile Emissions / *39.29*
    Computer Rooms / *39.30*
    Nuclear Power Stations / *39.30*
    Petrochemical Gases / *39.30*
    Natural Gas / *39.30*

Semiconductor Manufacturing / *39.30*
Pharmaceutical Applications / *39.30*
Energy Management / *39.30*
Heat Treat Applications / *39.31*
Meteorological Applications / *39.31*
Laboratory Calibration Standards / *39.31*
Engine Testing / *39.31*
Museums / *39.31*

Chapter 40. Thermal Imaging for Process and Quality Control   40.1

Introduction / *40.1*
Cameras / *40.1*
Processors / *40.1*
System Development / *40.4*
Conclusions / *40.5*

Chapter 41. The Detection of ppb Levels of Hydrazine Using Fluorescence and Chemiluminescence Techniques   41.1

Introduction / *41.1*
The Experiment / *41.2*
    Apparatus / *41.2*
    Chemicals and Stock Solutions / *41.3*
    Procedure / *41.4*
Conclusion / *41.12*
References / *41.12*

Chapter 42. Sensitive and Selective Toxic Gas Detection Achieved with a Metal-Doped Phthalocyanine Semiconductor and the Interdigitated Gate Electrode Field-Effect Transistor (IGEFET)   42.1

Introduction / *42.1*
Sensor Concept / *42.4*
Sensor Fabrication / *42.9*
Sensor Operation / *42.10*
Sensor Performance / *42.14*
Conclusion / *42.20*
References / *42.22*

Chapter 43. Molecular Relaxation Rate Spectrometer Detection Theory   43.1

References / *43.10*

Chapter 44. Current State of the Art in Hydrazine Sensing   44.1

Introduction / *44.1*
Hydrazine Detection Infrared Spectrometer / *44.2*
Electromechanical Sensors / *44.2*
Colorimetric Detectors / *44.2*

Colorimetric Dosimetry / *44.3*
Ion Mobility Spectrometry / *44.4*
Hydrazine Area Monitor / *44.5*
Fluorescence Detection / *44.5*
Conductive Polymer Hydrazine Sensor / *44.6*
Conclusions / *44.6*
References / *44.7*

Chapter 45. Microfabricated Sensors: Taking Blood Testing out of the Laboratory   45.1

Biosensor for Automated Immunoanalysis / *45.1*

Chapter 46. Closed-Loop Control of Flow Rate for Dry Bulk Solids   46.1

Introduction / *46.1*
Structure and Nature of Closed-Loop Control / *46.1*
Weigh Belt Feeder and Its Flow Rate Control Loop / *46.3*
Loss-in-Weight Feeder and Its Flow Rate Control Loop / *46.4*
Closure / *46.5*
References / *46.5*

Chapter 47. Weigh Belt Feeders and Scales: The Gravimetric Weigh Belt Feeder   47.1

Overview / *47.1*
   Introduction / *47.1*
   Why Feeders? What Are They? / *47.1*
   Definitions / *47.2*
The Basics / *47.2*
   Technology Triangle / *47.2*
   Basics of Feeding / *47.4*
   Controlling Mass Flow / *47.4*
Principles of Weigh Belt Feeder Operation / *47.5*
   Basic Function of the Weigh Belt Feeder / *47.5*
   Mechanical Design Strategies for Weigh Belt Feeders / *47.6*
   Sensors and Controls / *47.13*
Applications of Weigh Belt Feeders / *47.20*
   Introduction / *47.20*
   Weigh Belt Feeder Calibration Issues / *47.22*
   Basics of Belt Scales / *47.22*
Multiingredient Proportioning for Dry Bulk Solids / *47.31*
References / *47.33*

Chapter 48. Low-Cost Infrared Spin Gyro for Car Navigation and Display Cursor Control Applications   48.1

Introduction / *48.1*
Theory of Operation / *48.1*
Cursor Control Applications / *48.2*
Car Navigation Applications / *48.3*
The Effect of the Pendulum on Performance / *48.4*

Software Compensation / 48.4
Navigation System Configuration / 48.5
Road Test Results / 48.5
Conclusion / 48.6

Chapter 49. Quartz Rotation Rate Sensor: Theory of Operation, Construction, and Applications — 49.1

Theory of Operation / 49.1
Construction / 49.3
Applications / 49.4
    Instrumentation / 49.4
    Control / 49.6

Chapter 50. Fiber Optic Rate Gyro for Land Navigation and Platform Stabilization — 50.1

Introduction / 50.1
Gyro Design / 50.2
Performance / 50.5
Conclusion / 50.7
References / 50.7

Chapter 51. A Micromachined Comb Drive Tuning Fork Gyroscope for Commercial Applications — 51.1

Introduction / 51.1
Theory of Operation / 51.2
Fabrication / 51.3
Electronics / 51.5
Test Results / 51.6
Conclusions / 51.8
References / 51.9

Chapter 52. Automotive Applications of Low-G Accelerometers and Angular Rate Sensors — 52.1

Introduction / 52.1
Sensing Technologies / 52.2
Acceleration Sensor / 52.3
Angular Rate Gyroscope / 52.4
Circuit Technology / 52.4
Low-G Accelerometer Applications / 52.7
Angular Rate Gyroscope Applications / 52.9
Conclusion / 52.9
References / 52.9

Chapter 53. Microfabricated Solid-State Secondary Batteries for Microsensors — 53.1

Introduction / 53.1

Experimental / *53.2*
Results and Discussion / *53.3*
    Contacts / *53.3*
    Cathode / *53.3*
    Electrolyte / *53.4*
    LiI Layer / *53.5*
    Anode / *53.5*
    Microbattery / *53.5*
Summary / *53.11*
References / *53.11*

Chapter 54. High-Temperature Ceramic Sensors   54.1

Introduction / *54.1*
Ceramic Gas Sensors / *54.2*
Ceramic Thermistors / *54.6*
References / *54.9*

Chapter 55. Microfabricated and Micromachined Chemical and Gas Sensor Developments   55.1

Introduction / *55.1*
Tin-Oxide-Based Sensors / *55.2*
Schottky-Diode-Type Sensors / *55.2*
Solid Electrolyte Electrochemical Sensors / *55.3*
Calorimetric Sensors / *55.4*
References / *55.5*

Chapter 56. Electro-formed Thin-Film Silica Device as Oxygen Sensor   56.1

Introduction / *56.1*
Device Preparation / *56.2*
Precursor Chemistry / *56.2*
Device Structure / *56.2*
    Electrical Measurements / *56.3*
    Device Characteristics / *56.4*
Sensor Operation / *56.6*
Discussion / *56.7*
Summary / *56.8*
References / *56.8*

Chapter 57. Using Leg-Mounted Bolt-On Strain Sensors to Turn a Tank into a Load Cell   57.1

Introduction / *57.1*
Bolt-On Weight Sensing / *57.2*
    The Bolt-On Microcell® Sensor / *57.2*
    Two-Axis Strain Sensors / *57.3*
Bolt-On Weight Sensors versus Load Cells / *57.3*
Vessel Leg and Brace Temperature-Induced Stresses and the Cure / *57.5*
    H-Beam Leg Effects / *57.5*
    X-Brace Effects / *57.5*

Load Cells Using Microcell Strain Sensors / 57.6
　　Physical Description of a Load Stand® Transducer / 57.6
　　Electrical Characterization / 57.6
Calibration without Moving Premeasured Live Material / 57.7
Summary and Conclusions / 57.7
References / 57.8

Chapter 58. Five New Technologies for Weighing Instrumentation　　58.1

Introduction / 58.1
Sigma Delta A/D Conversion / 58.1
Dynamic Digital Filtering / 58.2
Multichannel Synchronous A/D Control / 58.3
Expert System Diagnostics / 58.3
Digital Communication Networks / 58.6
　　Model / 58.6
Synergy / 58.7
References / 58.7

Chapter 59. Multielement Microelectrode Array Sensors and Compact Instrumentation Development at Lawrence Livermore National Laboratory　　59.1

Introduction / 59.1
Results and Discussion / 59.1
Conclusions / 59.5
References / 59.6

Chapter 60. Enabling Technologies for Low-Cost, High-Volume Pressure Sensors　　60.1

Introduction / 60.1
Medical Disposable Pressure Sensors / 60.1
　　Market Overview / 60.1
　　Disposable Sensor Technology Overview / 60.2
Miniature Pressure Sensors / 60.5
　　Sensor Die / 60.6
　　Leadframe Packaging / 60.8
　　IsoSensor / 60.8
Smart Sensor Technology / 60.9
Sensor Communication / 60.10
Conclusions / 60.11
References / 60.12

Chapter 61. A Two-Chip Approach to Smart Sensing　　61.1

Background / 61.1
Approaches to Solving Problems / 61.1
　　Discrete Component Approach / 61.1
　　Single-Chip Approach / 61.2
　　Two-Chip Approach / 61.2
Product Examples / 61.3

Accelerometer / 61.3
    Pressure Sensor / 61.4
Conclusions / 61.4

Chapter 62. Specifying and Selecting Semiconductor Pressure Transducers — 62.1

General Factors / 62.1
    Details / 62.1
    Physical/Mechanical / 62.2
    Electrical / 62.4
    Performance / 62.4

Chapter 63. Introduction to Silicon Sensor Terminology — 63.1

Introduction / 63.1
General Definitions / 63.1
Performance-Related Definitions / 63.5

Chapter 64. Silicon Sensors and Microstructures: Integrating an Interdisciplinary Body of Material on Silicon Sensors — 64.1

Introduction / 64.1
Markets and Applications / 64.2
    Introduction / 64.2
    Characteristics of Sensors and Transducers / 64.2
    Classification of Silicon Sensors / 64.3
    Generic Sensor Classification / 64.3
    Radiant Signal Domain / 64.5
    Mechanical Signal Domain / 64.5
    Thermal Signal Domain / 64.6
    Magnetic Signal Domain / 64.6
    Chemical Signal Domain / 64.7
    Evolution and Growth of Silicon Sensor Technology / 64.8
Silicon Micromechanics: Advantages and Obstacles / 64.13
    Educated Technologists / 64.14
    Deep Silicon Etch / 64.15
    Chip Stress Isolation / 64.15
    Dimensional Control of Silicon Structures / 64.17
    Stability of Silicon Resistors / 64.17
    Wafer Lamination / 64.18
    High-Volume, Low-Cost Pressure/Acceleration/Temperature Testing / 64.18
    Packaging / 64.18
Sensor Market Definition / 64.18
The World's Market Size and Growth / 64.19
Characterization of Emerging Markets / 64.23
    Automotive Market / 64.23
    Medical Market / 64.24
    Process/Industrial Controls Markets / 64.25
    Elevator Vibration Monitoring / 64.25
    Consumer Market / 64.25
    HVAC Market / 64.26
    Aerospace / 64.26
    Micromachining Market / 64.27

Technology Trend / 64.29
    Electronic On-Chip Integration / 64.30
    Mechanical On-Chip Integration / 64.31
    Application-Specific Sensors (ASS) / 64.32
    Smart Sensor / 64.32
Market Trends / 64.32
References / 64.33

Chapter 65. Understanding Silicon Processing and Micromachining   65.1

What Is Silicon? / 65.1
    Electrical Properties / 65.2
    Mechanical Performance of Silicon / 65.3
Basic Sensor Materials and Processing Techniques / 65.5
    Photolithography / 65.5
    Diffusion and Ion Implantation / 65.5
    Metal Layers and Interconnects / 65.7
    Insulators / 65.7
    Additive Processes / 65.8
    Subtractive Processes / 65.9
Basic Pressure Sensor Process / 65.10
Conclusions / 65.15
References / 65.15

Chapter 66. Universal Sensors Technology: Basic Characteristics of Silicon Pressure Sensors   66.1

Silicon Piezoresistive Pressure Sensors / 66.1
    Piezoresistance in Silicon / 66.1
    Typical Integrated Sensor Geometries / 66.5
    Low-Pressure Sensors / 66.10
    High-Pressure Sensors / 66.11
    Effect of Dopant Concentration on Piezoresistive Sensors / 66.12
    Piezoresistive Temperature Coefficients versus Doping Level / 66.14
    Piezoresistive Sensors for Extended Temperature Operation / 66.16
    Piezoresistive Sensors for Use at High Temperatures / 66.16
    Total Error Band Concept / 66.18
    Overpressure Capabilities of Piezoresistive Sensors / 66.20
    Linearity Performance / 66.22
    High-Stability Piezoresistive Sensors / 66.22
    Voltage Sensitivity of Semiconductor Devices / 66.24
    Dynamic Performance of Piezoresistive Sensors / 66.25
Silicon Capacitive Pressure Sensor / 66.26
    Principle of Operation / 66.28
    Relative Advantages of Capacitance versus Piezoresistance / 66.28
    Silicon Accelerometers / 66.29
    Summary / 66.31
    References / 66.31

Chapter 67. Advanced Sensor Designs   67.1

Introduction / 67.1
Fully On-Chip Compensated, Calibrated Pressure Sensors / 67.2

Introduction / *67.2*
Sensor Design / *67.3*
Sensor Characteristics / *67.4*
Packaging / *67.4*
Conclusion / *67.5*
Pressure Sensors Using Si/Si Bonding / *67.5*
Introduction / *67.5*
1 × 1-mm Pressure Sensor for High-Volume Applications / *67.5*
Catheter Tip Pressure Sensors / *67.6*
High-Pressure Sensors / *67.8*
Low-Pressure Sensor with High Overpressure Protection / *67.10*
High-Temperature Pressure Sensors / *67.11*
Capacitive Absolute Pressure Sensors / *67.12*
Silicon Accelerometers / *67.13*
Conclusion / *67.17*
Very Low Pressure Sensor / *67.18*
Conclusions / *67.19*
References / *67.19*

Chapter 68. Silicon Microstructures 68.1

Introduction / *68.1*
Microplumbing / *68.2*
    Synoptic View / *68.2*
    Nozzles / *68.2*
    Channel and Flow Resistors / *68.2*
    Valves / *68.5*
Thermally Isolated Silicon Microstructures / *68.7*
    Introduction / *68.7*
    Sensors for Thermal Conductivity and Flow / *68.7*
    Radiation Detectors / *68.9*
    Vacuum Sensors / *68.9*
    Heated Filaments / *68.9*
    Gas Sensors Using Suspended Structures / *68.10*
    Microwave Power Sensing / *68.11*
Electrical Switches / *68.11*
Light Modulators and Deflectors / *68.15*
Micromotors / *68.17*
Resonant Structures for Measurement and Actuation / *68.18*
Applications in Microbiology / *68.19*
    Cell Manipulation / *68.19*
    Atomic Force Microscopes / *68.20*
Conclusion / *68.20*
References / *68.21*

Chapter 69. Computer Design Tools 69.1

Introduction / *69.1*
Computer Modeling / *69.2*
    Analytical Modeling / *69.3*
    Finite Element Modeling (FEM) / *69.4*
    Finite Difference Modeling / *69.8*
Process Modeling / *69.9*
Computer-Aided Layout of Sensors and Microstructures / *69.10*
Electrical Modeling for Silicon Sensors / *69.11*
    Digital Circuit Simulators for Analog Circuits / *69.12*

Digital Simulation and Design Aids for Digital Circuits / 69.12
Software Development Aids / 69.13

Chapter 70. Signal Conditioning for Sensors   70.1

Introduction / 70.1
Characteristics of Pressure Sensors / 70.2
Constant Current versus Constant Voltage Excitation / 70.4
Analog Electrical Models of Piezoresistive Pressure Sensors / 70.4
    Basic Model / 70.6
    High-Performance Analog Model / 70.10
Basic Constant Current Compensation / 70.13
    Compensation of Offset Voltage / 70.13
    Compensation of Full-Scale Output / 70.15
    Calculation of Compensating Resistor Values / 70.16
    Required Performance of Compensating Resistors / 70.18
Constant Voltage FSO Compensation / 70.20
    Resistor Compensation / 70.20
    Thermistor Compensation / 70.20
    Diode Compensation / 70.20
Gain Programming for Normalization / 70.24
    Basic Circuit / 70.24
Measurement of Differential Pressure Using Two Pressure Sensors / 70.25
Digital Compensation and Normalization / 70.28
    Linear Approximation of Pressure and Temperature Characteristics / 70.28
    Linear Approximation of Pressure Characteristics and Parabolic Approximation
        of Temperature Characteristics / 70.31
    Parabolic Approximation of Both Pressure and Temperature Characteristics / 70.32
    Third-Order Polynomial Distribution / 70.32
Current Sources for Sensor Excitation / 70.33
Instrumentation Amplifiers / 70.35
    Amplifier Performance Requirements / 70.36
    Three-Amplifier Configuration / 70.37
    Two-Amplifier Configuration / 70.38
    Switched Capacitor Instrumentation Amplifier / 70.38
Autozeroing Circuit with Eight-Bit Resolution / 70.40
Smart Sensors / 70.44
References / 70.44

Chapter 71. Sensor Packaging Technology   71.1

Introduction / 71.1
    The Design Process / 71.1
    Functions of the Sensor Package / 71.1
Evolution of Silicon Sensor Packaging / 71.3
The Application-Driven Nature of the Silicon Package / 71.4
Wafer-Level Operations / 71.10
    The Concept of the Micropackage / 71.10
    Wafer Lamination Techniques / 71.11
Generic Die Operations Common to All Packages / 71.12
    Wafer Sawing / 71.13
    Die Characterization / 71.13
    Die Down / 71.13
    Wire Bond / 71.15
    Die Protection / 71.15

Packaging Options for Silicon Sensors / 71.16
    The Design Process / 71.16
    Unpackaged Die / 71.16
    TO Series Packages / 71.16
    Metal Diaphragm, Oil-Isolated, All-Media Package / 71.18
    Ultra Low Cost Package / 71.19
References / 71.22

Chapter 72. Advances in Surface Micromachined Force Sensors   72.1

Introduction / 72.1
Surface Micromachined Absolute Pressure Transducers / 72.2
Resonant Integrated Microsensor (RIM) / 72.5
Conclusions / 72.6
References / 72.7

Chapter 73. Peer-to-Peer Distributed Control for Discrete and Analog Systems   73.1

Introduction: What Is Peer-to-Peer Distributed Control? / 73.1
Why Is Peer-to-Peer Distributed Control Useful? / 73.2
    Reliability / 73.2
    Flexibility / 73.3
    Expandability / 73.3
    Interoperability / 73.4
What Characteristics Are Needed of a Technology Used to Implement Peer-to-Peer Distributed Control? / 73.4
    A Low-Cost, Standardized Controller Element / 73.5
    A Fully Featured Network, Appropriate for Control / 73.5
    A Migration Path from Existing Products / 73.6
What Does LONWORKS Include That Makes It a Fit for Peer-to-Peer Distributed Control Systems? / 73.7
    Range of Applications / 73.7
    NEURON IC—Node Controller / 73.7
    I/O Structure / 73.8
    A Full OSI Seven-layer Protocol Definition / 73.8
    Operating System Integral to Distributed Control / 73.10
Users Move Toward Peer-to-Peer, Distributed Control Networks / 73.11

Chapter 74. Distributed, Intelligent I/O for Industrial Control and Data Acquisition: The Seriplex Sensor/Actuator Bus   74.1

Introduction / 74.1
System Description / 74.5
How the System Works / 74.8
ASIC General Description / 74.9
Communication System—Master/Slave Mode / 74.10
    Throughput Time for Master/Slave System / 74.11
Communication System—Peer-to-Peer Mode / 74.12
    Throughput Time for Peer-to-Peer System / 74.12
The CPU Interfaces / 74.13
I/O Devices / 74.19
Open Architecture / 74.20

Chapter 75. Thin/Thick Film Ceramic Sensors   75.1

Introduction / 75.1
Thin Film Process / 75.2
Thick Film Process / 75.2
Process for Electrode Contacts of Thin/Thick Ceramic Sensors / 75.4
Why Thin/Thick Films for Ceramic Sensors / 75.5
References / 75.7

Chapter 76. Low-Noise Cable Testing and Qualification for Sensor Applications   76.1

Introduction / 76.1
Overview / 76.1
High-Frequency (AC) Cable Measurements / 76.7
    Measuring High-Frequency Cable Characteristics / 76.7
    Time Domain Measurements / 76.7
    Frequency Domain Measurements / 76.10
Low-Frequency (DC) Cable Measurements / 76.13
    Cable Capacitance Testing / 76.13
    Insulation Resistance and Leakage Current Measurements / 76.14
    Series Resistance and Continuity Testing / 76.16
    Dielectric Withstand Voltage / 76.18
    Low-Level Systems Considerations / 76.19
Mechanical Testing for Low-Noise Coaxial Cables / 76.21
    Drop Tests / 76.21
    Bowstring Excitation Test Method / 76.22
    Electrical Response to the Bowstring Test Method / 76.22
    Flex Degradation/Flex Life / 76.24
    Tick Tock Testing / 76.24
    Rolling Flex Test / 76.24
Cable Comparisons / 76.25
References / 76.26

Chapter 77. Resonant Microbeam Technology for Precision Pressure Transducer Applications   77.1

Introduction / 77.1
Resonant Microbeam Technology / 77.4
    Principle of Operation / 77.4
    Advantages and Features / 77.6
Interface Considerations / 77.7
Conclusions and Summary / 77.10
References / 77.10

Chapter 78. Two-Chip Smart Accelerometer   78.1

Introduction / 78.1
Sensor Element / 78.1
Signal-Conditioning IC / 78.2
    Signal Processing / 78.3
    Error Detection Functions / 78.4
    Addressing / 78.4

Electrical Trimming / 78.6
Package / 78.6
Customization / 78.7
References / 78.8

Chapter 79. Quartz Resonator Fluid Monitors for Vehicle Applications 79.1

Introduction / 79.1
Quartz Resonator Sensors / 79.2
Oscillator Electronics / 79.8
Lubricating Oil Monitor / 79.10
Battery State-of-Charge Monitor / 79.13
Coolant Capacity Monitor / 79.16
Conclusion / 79.18
References / 79.19

Chapter 80. Overview of the Emerging Control and Communication Algorithms Suitable for Embedding into Smart Sensors 80.1

Introduction / 80.1
Generic Model of a Control System / 80.2
Computers and Communication in Control / 80.3
    Hub-Based Control Configuration / 80.3
    Bus-Based Control Configuration / 80.4
    Distributed Control Configuration / 80.4
    Smart Sensor Model / 80.5
Plug-and-Play Communication Requirements / 80.5
Modern Computation Techniques for Smart Sensors / 80.7
    Fuzzy Representation / 80.7
    Rough Representation / 80.9
    Sample Application / 80.11
    Plug-and-Play Approach to Software Development / 80.13
Flexible Architecture for Smart Sensors / 80.15
Summary and Conclusions / 80.16
References / 80.18

Chapter 81. Automotive Applications of Conductive Polymer-Based Chemical Sensors 81.1

Introduction / 81.1
Experimental / 81.1
Results and Discussion / 81.2
    Methanol Content in Hexane / 81.2
    Acid and Water Detection in Nonpolar Media / 81.3
    Degradation of Automatic Transmission Fluid / 81.5
Summary / 81.6
References / 81.6

Chapter 82. Modeling Sensor Performance for Smart Transducers 82.1

Introduction / 82.1
Compensating Sensor Errors / 82.1

Statistical Compensation / 82.3
    Zero / 82.3
    Full-Scale Output / 82.4
Digital Compensation and Normalization / 82.5
    Look-up Tables / 82.5
    Compensating Algorithm / 82.6
    Modeling of Zero and FSO / 82.7
    Single-Function Model / 82.9
Conclusions / 82.10
References / 82.11

Chapter 83. Infrared Gas and Liquid Analyzers: A Review of Theory and Applications   83.1

Introduction / 83.1
The Source / 83.2
The Sample Cell / 83.3
Sample Cell Window Materials / 83.3
Optical Filter(s) / 83.4
Detectors / 83.5
Applications / 83.5

Chapter 84. Infrared Noncontact Temperature Measurement: An Overview   84.1

Introduction / 84.1
Hardware Requirements / 84.4
Target / 84.4
Detectors / 84.4
Optical Materials / 84.5
Optical Filters / 84.5
Two-Color Analysis / 84.5
Applications / 84.6

Chapter 85. Quality Control Considerations   85.1

Design Assurance / 85.2

Chapter 86. Microsystem Technologies   86.1

Introduction / 86.1
    Surface Micromachined Pressure / 86.1
Monolithic Magnetic Field-Sensor with Adaptive Offset Reduction / 86.2
    Sensing Element / 86.4
    Signal Processing Electronics / 86.5
A Planar Fluxgate-Sensor with CMOS-Readout Circuitry / 86.5
A Thermoelectric Infrared Radiation Sensor / 86.7
Conclusion / 86.8
References / 86.8

Index   I.1

FOREWORD

There is not much time left until the beginning of the third millennium.

Relations between nations and institutions, and also between individuals, are increasingly characterized by a comprehensive and rapid exchange of information. Communication between scientists and engineers, bankers and brokers, manufacturers and consumers proceeds at an ever quickening pace. Exchanging ideas and thoughts with others is no longer a matter of weeks, months, or years. Humankind has now reached the point of being able to distribute large volumes of information to any number of addresses within the blink of an eye.

Human intelligence is thus nourished from many different sources producing globally useful scientific, technical, and economic improvements within only a short time. The question of whether the globalization of our thoughts is an advantage or a disadvantage for humankind still remains to be answered, however.

Dr. Sabrie Soloman devotes his scientific work to the challenging area of translating scientific ideas into technically applicable solutions. He consolidates thoughts and knowledge within controlled systems and participates to ensure, prior to the end of the second millennium, that technical progress is not conceived of as a risk but rather as an opportunity.

Dr. Soloman's new handbook introduces us into the world of advanced sensor technology, starting with information in a simple form, and, with increasing complexity, offers sophisticated solutions for problems that to date were considered insoluble.

Profound knowledge and understanding of all phases of the processes within complex systems are the prerequisites to securing control and quality. This is clearly illustrated by a large number of applications and examples from chemistry, physics, medicine, and other allied subjects covered in Dr. Soloman's book.

Adaptive systems are increasingly being called for in the processing of complex information. These systems must react quickly and adequately to expected, and even unforeseeable, developments.

Uhlmann, as a producer of highly sophisticated packaging systems for the pharmaceutical industry worldwide, stands for a high level of reliability thanks to mechanical, electrical, electronic, and optical modules permitting quality and efficiency in pharmaceutical packaging to be combined.

Reflecting on almost fifty years of tradition as a supplier to the pharmaceutical industry, forward thinking and the ambition to combine advanced science with state-of-the-art technology in our packaging systems have been the cornerstone of our success. Developments for the future include the application of advanced sensor technology and fiber-optic transmission techniques, as well as adaptive and decision-making computers on our machinery. Thus, we will contribute our share to ensure that medication, whether in the form of liquids, tablets, or coated tablets, is supplied to the consumer in perfect quality as far as Uhlmann is responsible for the packaging process.

Dr. Soloman's handbook offers the user a large variety of references in regard to target-oriented sensor technology—the analysis of the subsequent process information transformed into signals that secure function and quality. *Sensors Handbook* will certainly provide impulses for the development of highly sophisticated systems, continuing to challenge us in the future.

Hedwig Uhlmann
Chairman and CEO
Uhlmann Pac-Systeme GmbH & Co. KG

PREFACE

MANUFACTURING OF ARTIFICIAL ORGANS

A month ago my closest, most cherished friend Rochelle Good donated part of herself to her sister by giving one of her kidneys. The control of diabetes with insulin shots had failed to maintain barely adequate function of both Rochelle's sister's kidneys. The bond between the two women was always deep and strong; however, the concept of one's organs living in another's body is rarely realized, except by a few who are brave and noble. In the third-century legend of Saints Cosmos and Damian, the leg of a recently deceased Moorish servant is transplanted onto a Roman cleric whose own limb has just been amputated. The cleric's life hangs in the balance, but the transplant takes, and the cleric lives. The miraculous cure is attributed to the intervention of the saintly brothers, both physicians, who were martyred in A.D. 295.

What was considered miraculous in one era may become merely remarkable in another. Surgeons have been performing reimplantation of severed appendages for almost three decades now, and transplants of organs such as the heart, liver, and kidney are common—so common, in fact, that the main obstacle to transplantation lies not in surgical technique but in an ever worsening shortage of the donated organs themselves. In the next three decades, medical science will move beyond the practice of transplantation and into the era of fabrication. The idea is to make organs rather than simply to move them.

SENSORS AND CONTROL SYSTEMS IN MANUFACTURING

The advent of advanced sensor and control technology,* described in Chaps. 1–8, has caused an advancement in cell biology and plastic manufacture. These have already enabled researchers to construct artificial tissues that look and function like their natural counterparts. Genetic engineering may produce universal donor cells—cells that do not provoke rejection by the immune system—for use in these engineered tissues. "Bridging" technologies of sensors and medicine may serve as intermediate steps before such fabrication becomes commonplace. Transplantation of organs from animals, for example, may help alleviate the problem of organ shortage. Several approaches under investigation involve either breeding animals, such as the genetic hogs produced by Swift & Co.,† whose tissues will be immunologically accepted in humans, or developing drugs to allow the acceptance of these tissues.

---

* Dr. Sabrie Soloman, *Sensors and Control Systems in Manufacturing* (New York: McGraw-Hill Publishing Company, 1995).
† Swift & Co. is located in Greeley, Colorado. It is a subsidiary of ConAgra corporation.

Alternatively, microelectronics may help bridge the gap between the new technologies and the old. The results will bring radical changes in the treatment of a host of devastating conditions. Engineering artificial tissue is the natural successor to treatments for injury and disease.

Millions of people suffer organ and tissue loss every year from accidents, birth defects, and diseases such as cancer and diabetes. In the last half of this century, innovative drugs, surgical procedures, and medical devices have greatly improved the care of these patients. Immunosuppressive drugs such as cyclosporine and tacrolimus (Prograf) prevent rejection of transplanted tissue; minimally invasive surgical techniques such as laparoscopy have reduced trauma; dialysis and heart-lung machines sustain patients whose conditions would otherwise be fatal.

Yet these treatments are imperfect and often impair the quality of life. The control of diabetes with insulin shots, for example, is only partly successful. Injection of the hormone insulin once or several times a day helps the cells of diabetics to take up the sugar glucose (a critical source of energy) from the blood. But the appropriate insulin dosage for each patient may vary widely from day to day and even hour to hour. Often amounts cannot be determined precisely enough to maintain blood sugar levels in the normal range and thus prevent complications of diabetes—such as blindness, kidney failure, and heart disease—later in life.

Innovative research in biosensor design and drug delivery, described in Chaps. 8, 12, 14, 39, 41, 45, 60, and 64, may someday make insulin injections obsolete. In many diabetics, the disease is caused by the destruction in the pancreas of so-called islet tissue, which produces insulin. In other people, the pancreas makes insulin, but not enough to meet the body's demands. It is possible to envision a sensor-controlled device that would function like the pancreas, continuously monitoring glucose levels and releasing the appropriate amount of insulin in response. The device could be implanted or worn externally.

THE SPECTR$_x$™

Much of the technology for an external glucose sensor that might be worn like a watch already exists. Recent studies at the Massachusetts Institute of Technology, the University of California at San Francisco, and elsewhere have shown that the permeability of the skin can temporarily be increased by electric fields or low-frequency ultrasonic waves, allowing molecules such as glucose to be drawn from the body. The amount of glucose extracted in this way can be measured by reaction with an enzyme such as glucose oxidase; or infrared sensors, such as the SpectR$_x$,* described in Chap. 8, could detect the level of glucose in the blood.

These sensing devices could be coupled via microprocessors to a power unit that would pass insulin through the skin and into the bloodstream by the same means that the sugar was drawn out as described in Chaps. 9, 12, 13, 14, 15, 26, and 27. The instrument would release insulin in proportion to the amount of glucose detected.

An implantable device made of a semipermeable plastic could also be made. The implant, which could be inserted at any of several different sites in the body, would have the form of a matrix carrying reservoirs of insulin and glucose oxidase. As a patient's glucose level rose, the sugar would diffuse into the matrix and react with the enzyme, generating an acidic breakdown product. The increase in acidity would

---

* Invented by Dr. Sabrie Soloman, Patent Number 5,679,954 and patent pending PCT/US97/06624— Based on US-SN 08/635,773

alter either the permeability of the plastic or the solubility of the hormone stored within it, resulting in a release of insulin proportional to the rise in glucose. Such an implant could last a lifetime, but its stores of glucose oxidase and insulin would have to be replenished.

The ideal implant would be one made of healthy islet cells that would manufacture insulin themselves.* Investigators are working on methods to improve the survival of the tissue, but supply remains a problem. As is the case with all transplantable organs, the demand for human pancreas tissue far out strips the availability. Consequently, researchers are exploring ways to use islets from animals. They are also attempting to create islet tissue, not quite from scratch, but from cells taken from the patient, a close relative, or a bank of universal donor cells. The cells could be multiplied outside the body and then returned to patient.

SPINNING PLASTIC INTO TISSUE

Many strategies in the field of tissue engineering depend on the manipulation of ultrapure, biodegradable plastics or polymers suitable to be used as substrates for cell culture and implementation. These polymers possess both considerable mechanical strength and a high surface-to-volume ratio. Many are descendants of the degradable sutures introduced two decades ago. Using computer-aided manufacturing methods, researchers design and manipulate plastics into intricate scaffolding beds that mimic the structure of specific tissues and even organs. The scaffolds are treated with compounds that help cells adhere and multiply, then "seeded" with cells. As the cells divide and assemble, the plastic degrades. Finally, only coherent tissue remains. The new, permanent tissue can then be implanted in the patient.

This approach has already been demonstrated in animals, most recently in engineered heart valves in lambs; these valves were created from cells derived from the animals' blood vessels. During the past several years, human skin grown on polymer substrates has been grafted onto burn patients and foot ulcers of diabetic patients with some success. The epidermal layer of the skin may be rejected in certain cases, but the development of universal donor epidermal cells will eliminate that problem.

Eventually, whole organs such as kidneys and livers will be designed, fabricated, and transferred to patients. Although it may seem unlikely that a fully functional organ could grow from a few cells on a polymer frame, research with heart valves suggests that cells are remarkably adept at organizing the regeneration of their tissue of origin. They are able to communicate in three-dimensional culture using the same extracellular signals that guide the development of organs in utero. We have good reason to believe that, given the appropriate initial conditions, the cells themselves will carry out the subtler details of organ reconstruction. Surgeons will need only to orchestrate the organs' connections with patients' nerves, blood vessels, and lymph channels.

Similarly, engineered structural tissue will replace the plastic and metal prostheses used today to repair damage to bones and joints. These living implants will merge seamlessly with the surrounding tissue, eliminating problems such as infection and loosening at the joint that plague contemporary prostheses. Complex, customized shapes such as noses and ears can be generated by constructed computer-aided contour mapping and the loading of cartilage cells onto polymer constructs; indeed,

---

* Paul E. Lacy, "Treating Diabetes With Transplanted Cells," *Scientific American,* July 1995.

these forms have been made and implanted in laboratory animals. Other structural tissues, ranging from urethral tubes to breast tissue, can be fabricated according to the same principle. After mastectomy, cells that are grown on biodegradable polymers would be able to provide a completely natural replacement for the breast.

Ultimately, tissue engineering will produce complex body parts such as hands and arms. The structure of these parts can already be duplicated in polymer scaffolding, and most of the relevant tissue types—muscle, bone, cartilage, tendon, ligaments, and skin—grow readily in culture. A mechanical bioreactor system could be designed to provide nutrients, exchange gases, remove waste, and modulate temperature while the tissue matures. The only remaining obstacle to such an accomplishment is the resistance of nervous tissue to regeneration. So far no one has succeeded in growing human nerve cells. But a great deal of research is being devoted to this problem, and many investigators are confident that it will be overcome.

INNOVATIVE ELECTRONICS

In the meantime, innovative microelectronic devices, described in Chaps. 10, 12, 13, and 14, may substitute for implants of engineered nervous tissue. For example, a microchip implant may someday be able to restore some vision to people who have been blinded by diseases of the retina, the sensory membrane that lines the eye. In two of the more common retinal diseases, retinitis pigmentosa and macular degeneration, the light-receiving ganglion cells of the retina are destroyed, but the underlying nerves that transmit images from those cells to the brain remain intact and functional.

An ultrathin chip, described in Chaps. 24, 25, 26, 27, and 28, placed surgically at the back of the eye, could work in conjunction with a miniature camera to stimulate the nerves that transmit images. The camera would fit on a pair of eyeglasses; a laser attached to the camera would both power the chip and send it visual information via an infrared beam. The microchip would then excite the retinal nerve endings much as healthy cells do, producing the sensation of sight. At MIT and the Massachusetts Eye and Ear Infirmary, recent experiments in rabbits with a prototype of this "vision chip" have shown that such a device can stimulate the ganglion cells, which then send signals to the brain. Researchers will have to wait until the chip has been implanted in humans to know whether those signals approximate the experience of sight. Mechanical devices will also continue to play a part in the design of artificial organs, as they have in this century. They will be critical components in, say, construction of the so-called artificial womb. In the past few decades, medical science has made considerable progress in the care of premature infants. Current life support systems can sustain babies at 24 weeks of gestation; their nutritional needs are met through intravenous feeding, and ventilators help them to breathe.

Younger infants cannot survive, primarily because their immature lungs are unable to breathe air. A sterile, fluid-filled artificial womb would improve survival rates for these newborns. The babies would breathe liquids called perfluorocarbons, which carry oxygen and carbon dioxide in high concentrations. Perfluorocarbons can be inhaled and exhaled just as air is. A pump would maintain continuous circulation of the fluid, allowing for gas exchange. The uterine environment is more closely approximated by liquid breathing than by traditional ventilators, and liquid breathing is much easier on the respiratory tract. Indeed, new work on using liquid ventilation in adults with injured lungs is under way. Liquid ventilation systems for older babies are currently in clinical trials. Within a decade or so, such systems will be used to sustain younger fetuses.

In addition to a gas exchange apparatus, the artificial womb would be equipped with filtering devices to remove toxins from the liquid. Nutrition would be delivered intravenously, as it is now. The womb would provide a self-contained system in which development and growth could proceed normally until the baby's second "birth." For most premature babies, such support would be enough to ensure survival. The developing child is, after all, the ultimate tissue engineer.

SELF-ASSEMBLY MIMICS CREATION?

Nature abounds with examples of self-assembly. Consider a raindrop on a leaf. The liquid drop has a smooth, curved surface of just the kind required for optical lenses. Grinding a lens of that shape would be a major undertaking. Yet the liquid assumes this shape spontaneously, because molecules at the interface between liquid and air are less stable than those in the interior. The laws of thermodynamics require that a raindrop take the form that maximizes its energetic stability. The smooth, curved shape does so by minimizing the area of the unstable surface.

This type of self-assembly, known as thermodynamic self-assembly, works to construct only the simplest structures. Living organisms, on the other hand, represent the extreme in complexity. They, too, are self-assembling: cells reproduce themselves each time they divide. Complex molecules inside a cell direct its function. Complex subcomponents help to sustain cells. The construction of a cell's complexity is balanced thermodynamically by energy-dissipating structures within the cell and requires complex molecules such as ATP. An embryo, and eventually new life, can arise from the union of two cells, whether or not human beings attend to the development.

The kind of self-assembly embodied by life is called coded self-assembly because instructions for the design of the system are built into its components. The idea of designing materials with a built-in set of instructions that will enable them to mimic the complexity of life, as described in Chaps. 15, 27, 66, and 67, is immensely attractive. Researchers are only beginning to understand the kinds of structures and tasks that could exploit this approach. Coded self-assembly is truly a concept for the next century.

SUPER-INTELLIGENT MATERIALS

Imagine, for a moment, music in your room or car that emanates from the doors, floor, or ceiling; ladders that alert us when they are overburdened and may soon collapse under the strain; buildings and bridges that reinforce themselves during earthquakes and seal cracks of their own accord, as described in Chaps. 20, 21, and 38. Like living beings, these systems would alter their structure, account for damage, effect repairs, and retire—gracefully, one hopes—when age takes its toll.

Such structures may seem far-fetched. But, in fact, many researchers have demonstrated the feasibility of such "living" materials. To animate an otherwise inert substance, modern-day alchemists enlist a variety of devices: actuators and motors that behave like muscles; sensors that serve as nerves and memory; and communications and computational networks that represent the brain and spinal column. In some respects, the systems have features that can be considered superior to biological functions—some substances can be hard and strong one moment but made to act like Jell-O the next.

These so-called intelligent materials systems have substantial advantages over traditionally engineered constructs. Henry Petroski, in his book *To Engineer Is Human*, perhaps best articulated the traditional principles. A skilled designer always considers the worst-case scenario. As a result, the design contains large margins of safety, such as numerous reinforcements, redundant subunits, backup subsystems, and added mass. This approach, of course, demands more natural resources than are generally required and consumes more energy to produce and maintain. It also requires more human effort to predict those circumstances under which an engineered artifact will be used and abused.

Trying to anticipate the worst case has a much more serious and obvious flaw, one we read about in the newspapers and hear about on the evening news from time to time: that of being unable to foresee all possible contingencies. Adding insult to injury is the costly litigation that often ensues.

Intelligent materials systems, in contrast, would avoid most of these problems. Made for a given purpose, they would also be able to modify their behavior under dire circumstances. As an example, a ladder that is overloaded with weight could use electrical energy to stiffen and alert the user of the problem. The overload response would be based on the actual life experience of the ladder, to account for aging or damage. As a result, the ladder would be able to evaluate its current health; when it could no longer perform even minimal tasks, the ladder would announce its retirement. In a way, then, the ladder resembles living bone, which remodels itself under changing loads. But unlike bone, which begins to respond within minutes of an impetus but may take months to complete its growth, an intelligent ladder needs less than a second to change.

ARTIFICIAL MUSCLES FOR INTELLIGENT SYSTEMS

Materials that allow structures such as ladders to adapt to their environment are known as *actuators*, and are described in Chap. 29. Such substances can change shape, stiffness, position, natural frequency, and other mechanical characteristics in response to temperature or electromagnetic fields. The four most common actuator materials being used today are shape-memory alloys, piezoelectric ceramics, magnetostrictive materials, and electrorheological and magnetorheological fluids. Although none of these categories stands as the perfect artificial muscle, each can nonetheless fulfill particular requirements of many tasks.

Shape-memory alloys are metals that at a certain temperature revert back to their original shape after being strained. In the process of returning to their "remembered" shape, the alloys can generate a large force useful for actuation. Most prominent among them, perhaps, is the family of the nickel-titanium alloys developed at the Naval Ordnance Laboratory (now the Naval Surface Warfare Center). The material, known as Nitinol (Ni for nickel, Ti for titanium, and NOL for Naval Ordnance Lab), exhibits substantial resistance to corrosion and fatigue and recovers well from large deformations. Strains that elongate up to 8 percent of the alloy's length can be reversed by heating the alloy, typically with electric current.

JAPANESE NITINOL

Japanese engineers are using Nitinol in micromanipulators and robotics actuators to mimic the smooth motions of human muscles. The controlled force exerted when the Nitinol recovers its shape allows these devices to grasp delicate paper cups filled with water. Nitinol wires embedded in composite materials have also been used to modify vibrational characteristics. They do so by altering the rigidity or state of stress in the structure, thereby shifting the natural frequency of the composite. Thus, the structure would be unlikely to resonate with any external vibrations; this process is known to be powerful enough to prevent the collapse of a bridge. Experiments have shown that embedded Nitinol can apply compensating compression to reduce stress in a structure. Other applications for these actuators include engine mounts and suspensions that control vibration.

The main drawback of shape-memory alloys is their slow rate of change. Because actuation depends on heating and cooling, they respond only as fast as the temperature can shift.

THE FRENCH—PIERRE AND JACQUES CURIE

A second kind of actuator, one that addresses the sluggishness of the shape-memory alloys, is based on piezoelectrics. This type of material, discovered in 1880 by French physicists Pierre and Jacques Curie, expands and contracts in response to an applied voltage. Piezoelectric devices do not exert nearly so potent a force as shape-memory alloys; the best of them recover only from less than 1 percent strain. But they act much more quickly, in thousandths of a second. Hence, they are indispensable for precise, high-speed actuation. Optical tracking devices, magnetic heads and adaptive optical systems for robots, ink-jet printers, and speakers are some examples of systems that rely on piezoelectrics. Lead zirconate titanate (PZT) is the most widely used type.

Recent research has focused on using PZT actuators to attenuate sound, dampen structural vibrations, and control stress. At Virginia Polytechnic Institute and State University, piezoelectric actuators were used in bonded joints to resist the tension near locations that have a high concentration of strain. The experiments extended the fatigue life of some components by more than an order of magnitude.

A third family of actuators is derived from magnetostrictive materials. This group is similar to piezoelectrics except that it responds to magnetic, rather than electric, fields. The magnetic domains in the substance rotate until they line up with an external field. In this way, the domains can expand the material. Terfenol-D, which contains the rare earth element terbium, expands by more than 0.1 percent. This relatively new material has been used in low-frequency, high-power sonar transducers, motors, and hydraulic actuators. Like Nitinol, Terfenol-D is being investigated for use in the active damping of vibrations.

The fourth kind of actuator for intelligent systems is made of special liquids called electrorheological and magnetorheological fluids. These substances contain micron-size particles that form chains when placed in an electric or magnetic field, resulting in increases in apparent viscosity of up to several orders of magnitude in milliseconds. Applications that have been demonstrated with these fluids include tunable dampers, vibration-isolation systems, joints for robotic arms, and frictional devices such as clutches, brakes, and resistance controls on exercise equipment. Still, several problems such as abrasiveness and chemical instability plague these fluids, and much recent work to improve these conditions is aimed at the magnetic substances.

AMERICAN NERVES OF GLASS

Providing the actuators with information are the sensors, which describe the physical state of the materials system. Advances in micromachining, contributed largely by American electronic industries and research institutes and described in Chaps. 1, 2, 5, 6, 10, 13, and 23, have created a wealth of promising electromechanical devices that can serve as sensors. The main focus is on two types that are well developed now and are the most likely to be incorporated in intelligent systems: optical fibers and piezoelectric materials.

Optical fibers embedded in a "smart" material can provide data in two ways. First, they can simply provide a steady light signal to a sensor; breaks in the light beam indicate a structural flaw that has snapped the fiber. The second, more subtle, approach involves looking at key characteristics of the light intensity, phase, polarization, or similar feature. The National Aeronautics and Space Administration and other research centers have used such a fiber-optic system to measure the strain in composite materials. Fiber-optic sensors can also measure magnetic fields, deformations, vibrations, and acceleration. Resistance to adverse environments and immunity to electrical or magnetic noise are among the advantages of optical sensors.

In addition to serving as actuators, piezoelectric materials make good sensors. Piezoelectric polymers, such as polyvinylidene fluoride (PVDF), are commonly exploited for sensing because they can be formed in thin films and bonded to many kinds of surfaces. The sensitivity of PVDF to pressure has proved suitable for sensors that can read braille and distinguish grades of sandpaper. Ultrathin PVDF films, perhaps 200 to 300 µm thick, have been proposed for use in robotics. Such a sensor might be used to replicate the capability of human skin, detecting temperature and geometric features such as edges and corners, or distinguishing between different fabrics.

Actuators and sensors are crucial elements in an intelligent materials system, as described in Chaps. 7 and 29, but the essence of this new design philosophy in the manifestation of the most critical of life functions, intelligence—the extent to which the material should be smart or merely adaptive—is debatable. At a minimum, there must be an ability to learn about the environment and live within it.

The thinking features that the intelligent materials community is trying to create have constraints that the engineering world has never experienced before. Specifically, the vast number of sensors and actuators and their associated power sources would argue against feeding all these devices into a central processor. Instead designers have taken clues from nature. Neurons are not nearly so fast as modern-day silicon chips, but they can nonetheless perform complex tasks with amazing speed because they are networked efficiently.

The key appears to be hierarchical architecture. Signal processing and the resulting action can take place at levels below and far removed from the brain. The reflex of moving your hand away from a hot stove, for example, is organized entirely within the spinal cord. Less automatic behaviors are organized by successively higher centers within the brain. Besides being efficient, such an organization is fault-tolerant: unless there is some underlying organic reason, we rarely experience a burning sensation when holding an iced drink.

The brains behind an intelligent materials system follow a similar organization. In fact, investigators take their cue from research into artificial life, an outgrowth of the cybernetics field. Among the trendiest control concepts is the artificial neural network, which is computer programming that mimics the functions of real neurons. Such software can learn, change in response to contingencies, anticipate needs, and correct mistakes—more than adequate functions for intelligent materials systems. Ultimately, computational hardware and the processing algorithms will determine how complex these systems can become—that is, how many sensors and actuators we can use.

ENGINEERING MICROSCOPIC MACHINES

The electronics industry relies on its ability to double the number of transistors on a microchip every 18 months, as described in Chaps. 27, 28, and 29, a trend that drives the dramatic revolution in electronics. Manufacturing millions of microscopic elements in an area no larger than a postage stamp has now begun to inspire technology that reaches beyond the field that produced the pocket telephone and the personal computer.

Using the materials and processes of microelectronics, researchers have fashioned microscopic beams, pits, gears, membranes, and even motors that can be deployed to move atoms or to open and close valves that pump microliters of liquid. The size of these mechanical elements is measured in microns—a fraction of the width of a human hair. And, like transistors, millions of these elements can be fabricated at one time.

In the next 50 years, this structural engineering of silicon may have as profound an impact on society as did the miniaturization of electronics in preceding decades. Electronic computing and memory circuits, as powerful as they are, do nothing more than switch electrons and route them on their way over tiny wires. Micromechanical devices will supply electronic systems with a much-needed window to the physical world, allowing them to sense and control motion, light, sound, heat, and other physical forces.

The coupling of micromechanical and electronic systems will produce dramatic technical advances across diverse scientific and engineering disciplines. Thousands of beams with cross sections of less than a micron will move tiny electrical scanning heads that will read and write enough data to store a small library of information on an area the size of a microchip. Arrays of valves will release drug dosages into the bloodstream at precisely timed intervals. Inertial guidance systems on chips will aid in locating the positions of military combatants and direct munitions precisely at targets.

Microelectromechanical systems (MEMS) is the name given to the practice of making and combining miniaturized mechanical and electronic components. MEMS devices are made using manufacturing processes that are similar, and in some cases identical, to those for electronic components.

SURFACE MICROMACHINING

One technique, called *surface micromachining,* parallels electronics fabrication so closely that it is essentially a series of steps added to the making of a microchip, as described in Chaps. 10, 11, 12, 13, and 14. Surface micromachining acquired its name because the small mechanical structures are "machined" onto the surface of a silicon disk known as a *wafer*. The technique relies on photolithography as well as other staples of the electronic manufacturing process that deposit or etch away small amounts of material on the chip.

Photolithography creates a pattern on the surface of a wafer, marking off an area that is subsequently etched away to build up micromechanical structures such as a motor or a freestanding beam. Manufacturers start by patterning and etching a hole in a layer of silicon dioxide deposited on the wafer. A gaseous vapor reaction then deposits a layer of polycrystalline silicon, which coats both the hole and the remaining silicon dioxide material. The silicon deposited into the hole becomes the base of the beam, and the same material that overlays the silicon dioxide forms the suspended part of the beam structure. In the final step, the remaining silicon dioxide is etched away, leaving the polycrystalline silicon beam free and suspended above the surface of the wafer.

Such miniaturized structures exhibit useful mechanical properties. When stimulated with an electrical voltage, a beam with a small mass will vibrate more rapidly than a heavier device, making it a more sensitive detector of motion, pressure, or even chemical properties. For instance, a beam could adsorb a certain chemical (adsorption occurs when thin layers of a molecule adhere to a surface). As more of the chemical is adsorbed, the weight of the beam changes, altering the frequency at which it would vibrate when electrically excited. This chemical sensor could therefore operate by detecting such changes in vibrational frequency. Another type of sensor that employs beams manufactured with surface micromachining functions on a slightly different principle. It changes the position of suspended parallel beams that make up an electrical capacitor—and thus alters the amount of stored electrical charge—when an automobile goes through the rapid deceleration of a crash. Analog Devices, a Massachusetts-based semiconductor company, manufactures this acceleration sensor to trigger the release of an air bag. The company has sold more than half a million of these sensors to automobile makers over the past two years.

This air bag sensor may one day be looked back on as the microelectromechanical equivalent of the early integrated electronics chips. The fabrication of beams and other elements of the motion sensor on the surface of a silicon microchip has made it possible to produce this device on a standard integrated circuit fabrication line.

The codependence link of machines and sensors demonstrates that integrating more of these devices with electronic circuits will yield a window to the world of motion, sound, heat, and other physical forces.

The structures that serve as part of an acceleration sensor for triggering air bags are made by first depositing layers of silicon nitride (an insulating material) and silicon dioxide on the surface of a silicon substrate. Holes are lithographically patterned and etched into the silicon dioxide to form anchor points for the beams. A layer of polycrystalline silicon is then deposited. Lithography and etching form the pattern of the beams. Finally, the silicon dioxide is etched away to leave the freestanding beams.

In microelectronics the ability to augment continually the number of transistors that can be wired together has produced truly revolutionary developments: the microprocessors and memory chips that made possible small, affordable computing devices such as the personal computer. Similarly, the worth of MEMS may become apparent only when thousands or millions of mechanical structures are manufactured and integrated with electronic elements.

The first examples of mass production of microelectromechanical devices have begun to appear, and many others are being contemplated in research laboratories all over the world. An early prototype demonstrates how MEMS may affect the way millions of people spend their leisure time in front of the television set. Texas Instruments has built an electronic display in which the picture elements, or pixels, that make up the image are controlled by microelectromechanical structures. Each pixel consists of a 16-micron-wide aluminum mirror that can reflect pulses of colored light onto a screen. The pixels are turned off or on when an electric field causes the mirrors to tilt 10 degrees to one side or the other. In one direction, a light beam is reflected onto the screen to illuminate the pixel. In the other, it scatters away from the screen, and the pixel remains dark.

This micromirror display could project the images required for a large-screen television with a high degree of brightness and resolution of picture detail. The mirrors could compensate for the inadequacies encountered with other technologies. Display designers, for instance, have run into difficulty in making liquid-crystal screens large enough for a wall-size television display.

The future of MEMS can be glimpsed by examining projects that have been funded during the past three years under a program sponsored by the U.S. Department of Defense's Advanced Research Projects Agency. This research is directed toward building a number of prototype microelectromechanical devices and systems that could transform not only weapons but also consumer products.

A team of engineers at the University of California at Los Angeles and the California Institute of Technology wants to show how MEMS may eventually influence aerodynamic design. The group has outlined its ideas for a technology that might replace the relatively large moving surfaces of a wing—the flaps, slats, and ailerons—that control both turning and ascent and descent. It plans to line the surface of a wing with thousands of 150-µm-long plates that, in their resting position, remain flat on the wing surface. When an electrical voltage is applied, the plates rise from the surface at up to a 90° angle. Thus activated, they can control the vortices of air that form across selected areas of the wing. Sensors can monitor the currents of air rushing over the wing and send a signal to adjust the position of the plates.

These movable plates, or *actuators,* function similarly to a microscopic version of the huge flaps on conventional aircraft. Fine-tuning the control of the wing surfaces would enable an airplane to turn more quickly, stabilize against turbulence, or burn less fuel because of greater flying efficiency. The additional aerodynamic control achieved with this "smart skin" could lead to radically new aircraft designs that move beyond the cylinder-with-wings appearance that has prevailed for 70 years. Aerospace engineers might dispense entirely with flaps, rudders, and even the wing surface, called a vertical stabilizer. The aircraft would become a kind of "flying wing," similar to the U.S. Air Force's Stealth bomber. An aircraft without a vertical stabilizer would exhibit greater maneuverability—a boon for fighter aircraft and perhaps also one day for high-speed commercial airliners that must be capable of changing direction quickly to avoid collisions.

MICRO-MICROSCOPES

The engineering of small machines and sensors allows new uses for old ideas. For a decade, scientists have routinely worked with scanning probe microscopes that can manipulate and form images with individual atoms. The most well known of these devices is the scanning tunneling microscope (STM).

The STM, an invention for which Gerd Binnig and Heinrich Rohrer of IBM won the Nobel Prize in Physics in 1986, caught the attention of micromechanical specialists in the early 1980s. The fascination of the engineering community stems from calculations of how much information could be stored if STMs were used to read and write digital data. A trillion bits of information—equal to the text of 500 *Encyclopedia Britannicas*—might be fit into a square centimeter on a chip by deploying an assembly of multiple STMs.

The STM is a needle-shaped probe, the tip of which consists of a single atom. A current that "tunnels" from the tip to a nearby conductive surface can move small groups of atoms, either to create holes or to pile up tiny mounds on the silicon chip. Holes and mounds correspond to the zeros and ones required to store digital data. A sensor, perhaps one constructed from a different type of scanning probe microscope, would "read" the data by detecting whether a nanometer-size plot of silicon represents a zero or a one.

Only beams and motors a few microns in size, and with a commensurately small mass, will be able to move an STM quickly and precisely enough make terabit (trillion-bit) data storage on a chip practicable. With MEMS, thousands of STMs could be suspended from movable beams built on the surface of a chip, each one reading or writing data in an area of a few square microns. The storage medium, moreover, could remain stationary, which would eliminate the need for today's spinning media disk drives.

Noel C. MacDonald, an electrical engineering professor at Cornell University, has taken a step toward fulfilling the vision of the pocket research library. He has built an STM-equipped microbeam that can be moved in the vertical and horizontal axes or even at an oblique angle. The beam hangs on a suspended frame attached to four motors, each of which measures only 200 µm (two hair widths) across. These engines push or pull on each side of the tip at speeds as high a million times a second. MacDonald's next plan is to build an array of STMS.

THE PERSONAL SPECTROPHOTOMETER

The Lilliputian infrastructure afforded by MEMS might let chemists and biologists perform their experiments with instruments that fit in the palm of the hand. Westinghouse Science and Technology Center is in the process of reducing to the size of a calculator a 50-pound benchtop spectrometer used for measuring the mass of atoms or molecules. A miniaturized mass spectrometer presages an era of inexpensive chemical detectors for do-it-yourself toxic monitoring.

In the same vein, Richard M. White, a professor at the University of California at Berkeley, contemplates a chemical factory on a chip. White has begun to fashion millimeter-diameter wells each of which holds a different chemical, in a silicon chip. An electrical voltage causes liquids or powders to move from the wells down a series of channels into a reaction chamber. These materials are pushed there by micropumps made of piezoelectric materials that constrict and then immediately release sections of the channel. The snakelike undulations create a pumping motion. Once the chemicals are in the chamber, a heating plate causes them to react. An outlet channel from the chamber then pumps out what is produced in the reaction.

A pocket-calculator-size chemical factory could thus reconstitute freeze-dried drugs, perform DNA testing to detect waterborne pathogens, or mix chemicals that can then be converted into electrical energy more efficiently than can conventional batteries. MEMS gives microelectronics an opening to the world beyond simply processing and storing information. Automobiles, scientific laboratories, televisions, airplanes, and even the home medicine cabinet will never be the same.

The handbook previews the use of several technologies fundamental to the development of sensor applications in many fields. It provides also valuable yet simple understanding of sensor implementations in the fields of manufacturing, engineering, engineering design, aerospace, military science, pharmaceuticals, medicine, agriculture, manufacturing control, environmental applications, and the work of students and research organizations in medicine and engineering. The book will be useful for various types of management.

The sensor technology described in this handbook will provide the scientist, the engineer, and the system implementer with a very powerful tool to implement the latest in flexible manufacturing control using the sensors to monitor and control productivity quantitatively and qualitatively. Additionally, it serves the advanced manufacturing organizations in providing a clear understanding of the role of sensors and control systems in the computer-integrated manufacturing strategies.

Information regarding sensors has been limited and difficult to find. This handbook is also tailored for those who design, operate, and/or manage operating plants. To improve the quality of an operation, one must understand what is happening within the operation and the product itself. Sensors are the keys to communication between the operation and those who operate/manage it.

Sensors detect deviations and provide for continuous correction. This handbook contributes the knowledge and the understanding of effective use of sensors to advance the manufacturing technology and medical applications. It gives manufacturers, research scientists, and readers hands-on techniques and methods to ensure an error-free environment. It will help any manufacturing organization to monitor and improve the productivity of production lines with cost effective sensors and simple control devices.

Undoubtedly, this handbook will play a key role in the information system. However, sensors and control technology alone can not shorten lead time, reduce inventories, and minimize excess capacity to the extent required by today's manufacturing operation. This can be accomplished by integrating various sensors with appropriate control means throughout the manufacturing operation. The result is that individual manufacturing processes will be able to flow, communicate, and respond together as a unified cell, well structured for its functions.

COST AND COMPETITION

Rising cost. Shorter lead time. Complex customer specifications. Competition from across the street . . . and around the world. Today's businesses face an ever increasing number of challenges. The manufacturers that meet these challenges will be those that develop more effective and efficient forms of production, development, and marketing.

Advanced sensors and control technology can make a fundamental commitment to manufacturing solutions based on simple and affordable integration. With sensors and control technology, one can integrate manufacturing processes, react to rapidly changing production conditions, help people to react more effectively to complex qualitative decisions, lower costs, and improve product quality throughout the manufacturing enterprise.

The first step in achieving such flexibility is establishing an information system that can be reshaped whenever necessary. This will enable it to respond to the changing requirements of the enterprise and the environment. This reshaping must be accomplished with minimal cost and disruption to the operation.

In order to develop a sensory and control information system that will achieve these objectives, the enterprise must start with a long-range architectural strategy, one that provides a foundation that accommodates today's needs as well as taking tomorrow's into account. These needs include supporting new manufacturing processes, incorporating new data functions, and establishing new data bases and distributed channels. The tools for this control and integration are available today.

Advanced sensory and control technology, discussed in this handbook, is more than an implementation of new sensing technologies. It is a long-range strategy that allows the entire manufacturing and research operation to work together to achieve the business qualitative and quantitative goals. It must have the top management commitment. It may entail changing the mind-set of people in the organization and managing the change. The major success of this manufacturing strategy is largely credited to the success of implementing the advanced technology of sensory and control systems.

SETTING THE STAGE FOR ADVANCED SENSORS AND CONTROL SYSTEMS

This handbook deals with setting up relatively small devices—often called *sensors*—designed to sense and measure an object's physical characteristics such as size, speed, acceleration, color, temperature, pressure, volume, flow rate, altitude, latitude, shape, orientation, quantity, deformation, homogeneity, topography, viscosity, electric voltage, electric current, electric resistance, surface textures, microcracks, vibrations, noise, acidity, contamination, active ingredient, assay concentration, chemical composition of pharmaceutical drugs, and blood viruses.

"THERE IS NO NEW THING UNDER THE SUN!"

>   All the rivers run into the sea; yet the sea is not full; unto the place from whence the rivers come, thither they return again. All things are full of labour, man cannot utter it: the eye is not satisfied with seeing, nor the ear filled with hearing. The thing that hath been, it is that which shall be; and that which is done is that which shall be done: and there is no new thing under the sun. Is there any thing whereof it may be said, See, this is new? It hath been already of old time, which was before us. . . . I looked on all the works that my hands had wrought, and on the labour that I had laboured to do: and, behold, all was vanity and vexation of spirit, and there was no profit under the sun. My son, be admonished: of making many books there is no end; and much study is a weariness of the flesh. Let us hear the conclusion of the whole matter: Fear God, and keep his commandments: for this is the whole duty of man.
>   —Ecclesiastes 1:3–10, 2:11–13, 12:12–13

November 25, 1997

Dr. Sabrie Soloman
Chairman & CEO
American SensoRx, Inc.
Professor, Advanced Manufacturing Technology
Columbia University

CHAPTER 1
INTRODUCTION

Integrated sensors and control systems are the way of the future. In times of disaster, even the most isolated outposts can be linked directly into the public telephone network by portable versions of satellite earth stations called *very small aperture terminals* (VSATs). They play a vital role in relief efforts such as those for the eruption of Mount Pinatubo in the Philippines, the massive oil spill in Valdez, Alaska, the 90,000-acre fire in the Idaho forest, and Hurricane Andrew's destruction in south Florida and the coast of Louisiana.

VSATs are unique types of sensors and control systems. They can be shipped and assembled quickly and facilitate communications by using more powerful antennas that are much smaller than conventional satellite dishes. These types of sensors and control systems provide excellent alternatives to complicated conventional communication systems, which in disasters often experience serious degradation because of damage or overload.

Multispectral sensors and control systems will play an expanding role to help offset the increasing congestion on America's roads by creating "smart" highways. At a moment's notice, they can gather data to help police, tow trucks, and ambulances respond to emergency crises. Understanding flow patterns and traffic composition would also help traffic engineers plan future traffic control strategies. The result of less congestion will be billions of vehicle hours saved each year.

The spacecraft Magellan, Fig. 1.1, is close to completing its third cycle of mapping the surface of planet Venus. The key to gathering data is the development of a synthetic aperture radar as a sensor and information-gathering control system, the sole scientific instrument aboard Magellan. Even before the first cycle ended, in mid-1991, Magellan had mapped 84 percent of Venus' surface, returning more digital data than all previous U.S. planetary missions combined, with resolutions 10 times better than those provided by earlier missions. To optimize radar performance, a unique and simple computer software program was developed, capable of handling the nearly 950 commands per cycle. Each cycle takes one venusian day, the equivalent of 243 Earth days.

Manufacturing organizations in the United States are under intense competitive pressure. Major changes are being experienced with respect to resources, markets, manufacturing processes, and product strategies. As a result of international competition, only the most productive and cost-effective industries will survive.

Today's sensors and control systems have explosively expanded beyond their traditional production base into far-ranging commercial ventures. They will play an important role in the survival of innovative industries. Their role in information assimilation, and control of operations to maintain an error-free production environment, will help enterprises to stay effective on their competitive course.

FIGURE 1.1 The Spacecraft Magellan. (*Courtesy Hughes Corp.*)

ESTABLISHING AN AUTOMATION PROGRAM

Manufacturers and vendors have learned the hard way that technology alone does not solve problems. A prime example is the gap between the information and the control worlds, which caused production planners to set their goals according to dubious assumptions concerning plant-floor activities, and plant supervisors then could not isolate production problems until well after they had arisen.

The problem of creating effective automation for an error-free production environment has drawn a long list of solutions. Some are as old as the term *computer-*

integrated manufacturing (CIM) itself. However, in many cases, the problem turned out to be not technology, but the ability to integrate equipment, information, and people.

The debate over the value of computer-integrated manufacturing technology has been put to rest, although executives at every level in almost every industry are still questioning the cost of implementing CIM solutions. Recent economic belt tightening has forced industry to justify every capital expense, and CIM has drawn fire from budget-bound business people in all fields.

Too often, the implementations of CIM have created a compatibility nightmare in today's multivendor factory-floor environments. Too many end users have been forced to discard previous automation investments and/or spend huge sums on new equipment, hardware, software, and networks in order to effectively link together data from distinctly dissimilar sources. The expense of compatible equipment and the associated labor cost for elaborate networking are often prohibitive.

The claims of CIM open systems are often misleading. This is largely due to proprietary concerns, a limited-access database, and operating system compatibility restrictions. The systems fail to provide the transparent integration of process data and plant business information that makes CIM work.

In order to solve this problem, it is necessary to establish a clearly defined automation program. A common approach is to limit the problem description to a workable scope, eliminating the features that are not amenable to consideration. The problem is examined in terms of a simpler, workable model. A solution can then be based on model predictions.

The danger associated with this strategy is obvious: if the simplified model is not a good approximation of the actual problem, the solution will be inappropriate and may even worsen the problem.

Robust automation programs can be a valuable asset in deciding how to solve production problems. Advances in sensor technology have provided the means to make rapid, large-scale improvements in problem solving and have contributed in essential ways to today's manufacturing technology.

The infrastructure of an automation program must be closely linked with the use and implementation of sensors and control systems, within the framework of the organization. The problem becomes more difficult whenever it is extended to include the organizational setting. Organization theory is based on a fragmented and partially developed body of knowledge, and can provide only limited guidance in the formation of problem models. Managers commonly use their experience and instinct in dealing with complex production problems that include organizational aspects. As a result, creating a competitive manufacturing enterprise—one involving advanced automation technology utilizing sensors and control systems and organizational aspects—is a task that requires an understanding of both how to establish an automation program and how to integrate it with a dynamic organization.

In order to meet the goals of integrated sensory and control systems, an automated manufacturing system has to be built from compatible and intelligent subsystems. Ideally, a manufacturing system should be computer-controlled and should communicate with controllers and materials-handling systems at higher levels of the hierarchy as shown in Fig. 1.2.

UNDERSTANDING WORKSTATIONS, WORK CELLS, AND WORK CENTERS

Workstations, work cells, and work centers represent a coordinated cluster of a production system. A production machine with several processes is considered a work-

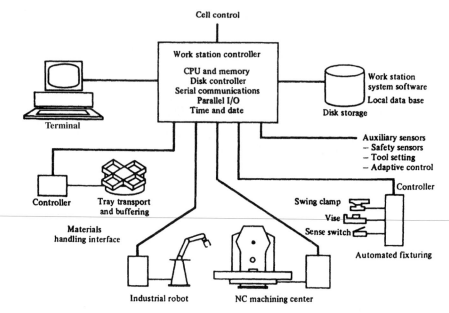

FIGURE 1.2 Computer-controlled manufacturing system.

station. A machine tool is also considered a workstation. Integrated workstations form a work cell. Several complementary workstations may be grouped together to construct a work cell. Similarly, integrated work cells may form a work center. This structure is the basic concept in modeling a flexible manufacturing system. The flexible manufacturing system is also the corner stone of the computer-integrated manufacturing strategy (Fig. 1.3).

The goal is to provide the management and project development team with an overview of major tasks to be solved during the planning, design, implementation, and operation phases of computer-integrated machining, inspection, and assembly systems. Financial and technical disasters can be avoided if a clear understanding of the role of sensors and control systems in the computer-integrated manufacturing strategy is asserted.

Sensors are largely applied within the workstations. Sensors are the only practical means of operating a manufacturing system and tracking its performance continuously.

Sensors and control systems in manufacturing provide the means of integrating different, properly defined processes as input to create the expected output. Input may be raw material and/or data which have to be processed by various auxiliary components such as tools, fixtures, and clamping devices. Sensors provide the feedback data to describe the status of each process. The output may also be data and/or materials which can be processed by further cells of the manufacturing system. A flexible manufacturing system, which contains workstations, work cells, and work centers and is equipped with appropriate sensors and control systems, is a distributed management information system, linking together subsystems of machining, packaging, welding, painting, flame cutting, sheet-metal manufacturing, inspection, and assembly with material-handling and storage processes.

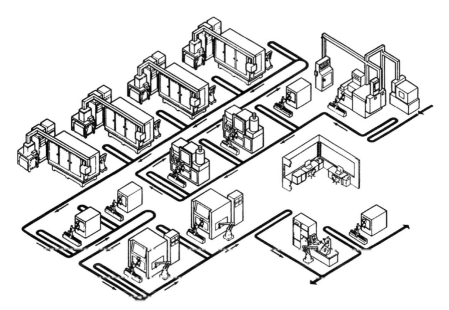

FIGURE 1.3 Workstation, work cell, and work center.

In designing various workstations, work cells, and work centers in a flexible manufacturing system, within the computer-integrated manufacturing strategy, the basic task is to create a variety of sensors interconnecting different material-handling systems, such as robots, automated guided-vehicle systems, conveyers, and pallet loading and unloading carts, to allow them to communicate with data processing networks for successful integration with the system.

Figure 1.4 illustrates a cell consisting of several workstations with its input and output, and indicates its basic functions in performing the conversion process, storing workpieces, linking material-handling systems to other cells, and providing data communication to the control system.

The data processing links enable communication with the databases containing part programs, inspection programs, robot programs, packaging programs, machining data, and real-time control data through suitable sensors. The data processing links also enable communication of the feedback data to the upper level of the control hierarchy. Accordingly, the entire work-cell facility is equipped with current data for real-time analysis and for fault recovery.

A cluster of manufacturing cells grouped together for particular production operations is called a *work center*. Various work centers can be linked together via satellite communication links irrespective of the location of each center. Manufacturing centers can be located several hundred feet apart or several thousand miles apart. Adequate sensors and control systems together with effective communication links will provide practical real-time data analysis for further determination.

The output of the cell is the product of the module of the flexible manufacturing system. It consists of a finished or semifinished part as well as data in a computer-readable format that will instruct the next cell how to achieve its output requirement. The data are conveyed through the distributed communication networks. If,

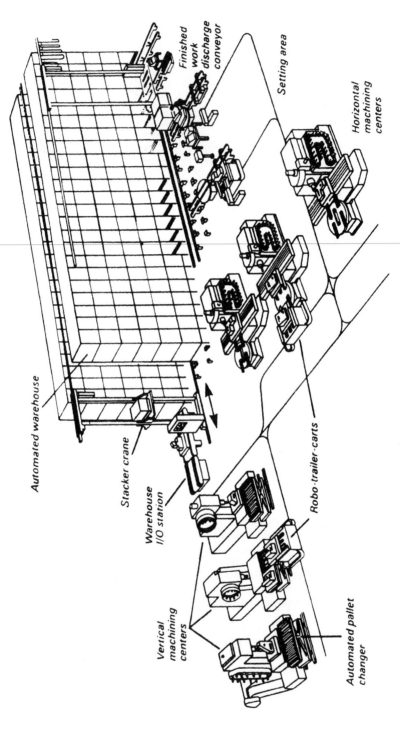

FIGURE 1.4 Conversion process in a manufacturing cell.

for example, a part is required to be surfaced to a specific datum in a particular cell, sensors will be adjusted to read the required acceptable datum during the surfacing process. Once the operation is successfully completed, the part must once again be transferred to another cell for further machining or inspection processes. The next cell is not necessarily physically adjacent; it may be the previous cell, as programmed for the required conversion process.

The primary reason for the emphasis on integrating sensors and control systems into every manufacturing operation is the worldwide exponentially increasing demand for error-free production operations. Sensors and control technology can achieve impressive results only if effectively integrated with corporate manufacturing strategy.

The following benefits can be achieved:

1. *Productivity.* A greater output and a lower unit cost.
2. *Quality.* Product is more uniform and consistent.
3. *Production reliability.* The intelligent, self-correcting sensory and feedback system increases the overall reliability of production.
4. *Lead time.* Parts can be randomly produced in batches of one or in reasonably high numbers, and the lead time can be reduced by 50 to 75 percent.
5. *Expenses.* Overall capital expenses are 5 to 10 percent lower. The cost of integrating sensors and feedback control systems into the manufacturing source is less than that of stand-alone sensors and feedback system.
6. *Greater utilization.* Integration is the only available technology with which a machine tool can be utilized as much as 85 percent of the time—and the time spent cutting can also be over 90 percent.

In contrast, a part, from stock to finished item, spends only 5 percent of its time on the machine tool, and actual productive work takes only 30 percent of this 5 percent. The time for useful work on stand-alone machines without integrated sensory and control systems is as little as 1 to 1.5 percent of the time available (see Tables 1.1 and 1.2).

To achieve the impressive results indicated in Table 1.1, the integrated manufacturing system carrying the sensory and control feedback systems must maintain a high degree of flexibility. If any cell breaks down for any reason, the production planning and control system can reroute and reschedule the production or, in other words, reassign the system environment. This can be achieved only if both the processes and the routing of parts are programmable. The sensory and control systems will provide instantaneous descriptions of the status of parts to the production and planning system.

If different processes are rigidly integrated into a special-purpose, highly productive system such as a transfer line for large batch production, then neither modular development nor flexible operation is possible.

However, if the cells and their communication links to the outside world are programmable, much useful feedback data may be gained. Data on tool life, measured dimensions of machined surfaces by in-process gaging and production control, and fault recovery derived from sensors and control systems can enable the manufacturing system to increase its own productivity, learn its own limits, and inform the part programmers of them. The data may also be very useful to the flexible manufacturing system designers for further analysis. In non-real-time control systems, the data cannot usually be collected, except by manual methods, which are time-consuming and unreliable.

TABLE 1.1 Time Utilization of Integrated Manufacturing Center Carrying Sensory and Control Systems

	Active, %	Idle, %
Tool positioning and tool changing	25	
Machining process	5	
Loading and inspection	15	
Maintenance	20	
Setup	15	
Idle time		15
Total	85	15

TABLE 1.2 Productivity Losses of Stand-alone Manufacturing Center Excluding Sensory and Control Systems

	Active, %	Idle, %
Machine tool in wait mode		35
Labor control		35
Support services		15
Machining process	15	
Total	15	85

CLASSIFICATION OF CONTROL PROCESSES

An *engineering integrated system* can be defined as a machine responsible for certain production output, a controller to execute certain commands, and sensors to determine the status of the production processes. The machine is expected to provide a certain product as an output, such as computer numerical control (CNC) machines, packaging machines, and high-speed press machine. The controller provides certain commands arranged in a specific sequence designed for a particular operation. The controller sends its commands in the form of signals, usually electric pulses. The machine is equipped with various devices, such as solenoid valves and step motors, that receive the signals and respond according to their functions. The sensors provide a clear description of the status of the machine performance. They give detailed accounts of every process in the production operation (Fig. 1.5).

Once a process is executed successfully, according to a specific sequence of operations, the controller can send additional commands for further processes until all processes are executed. This completes one cycle. At the end of each cycle a command is sent to begin a new loop until the production demand is met.

In an automatic process, the machine, the controller, and the sensors interact with one another to exchange information. Mainly, there are two types of interaction between the controller and the rest of the system: through either an open-loop control system or a closed-loop control system.

An *open-loop control system* (Fig. 1.6) can be defined as a system in which there is no feedback. Motor motion is expected to faithfully follow the input command. Stepping motors are an example of open-loop control.

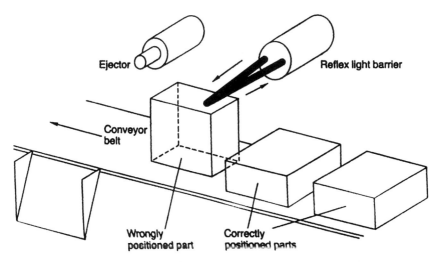

FIGURE 1.5 Sensors providing machine status.

A *closed-loop control system* (Fig. 1.7) can be defined as a system in which the output is compared to the command, with the result being used to force the output to follow the command. Servo systems are an example of closed-loop control.

OPEN- AND CLOSED-LOOP CONTROL SYSTEMS

In an open-loop control system, the actual value in Fig. 1.6 may differ from the reference value in the system. In a closed-loop system, the actual value is constantly monitored against the reference value described in Fig. 1.7.

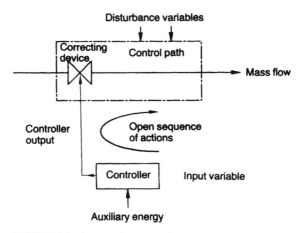

FIGURE 1.6 An open-loop control system.

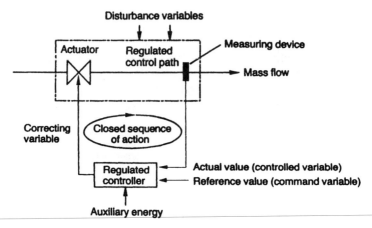

FIGURE 1.7 A closed-loop control system.

The mass flow illustrated in Fig. 1.8 describes the amount of matter per unit time flowing through a pipeline that must be regulated. The current flow rate can be recorded by a measuring device, and a correcting device such as a valve may be set to a specific flow rate. The system, if left on its own, may suffer fluctuations and disturbances which will change the flow rate. In such an open-loop system, the reading of the current flow rate is the *actual value,* and the *reference value* is the desired value of the flow rate. The reference value may differ from the actual value, which then remains unaltered.

If the flow rate falls below the reference value because of a drop in pressure, as illustrated in Fig. 1.9, the valve must be opened further to maintain the desired actual value. Where disturbances occur, the course of the actual value must be continuously observed. When adjustment is made to continuously regulate the actual value, the loop of action governing measurement, comparison, adjustment, and reaction within the process is called a *closed loop.*

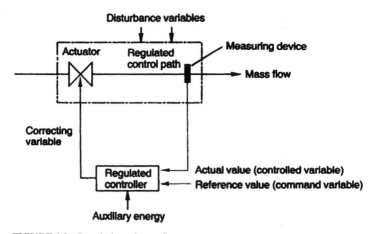

FIGURE 1.8 Regulation of mass flow.

INTRODUCTION 1.11

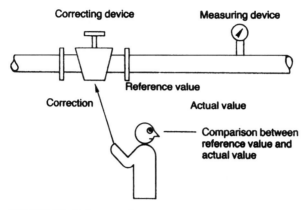

FIGURE 1.9 Reference value.

UNDERSTANDING PHOTOELECTRIC SENSORS

In order to successfully automate a process, it is necessary to obtain information about its status. The sensors are the part of the control system which is responsible for collecting and preparing process status data and for passing it onto a processor (Fig. 1.10).

Principles of Operation

Photoelectric controls use light to detect the presence or absence of an object. All photoelectric controls consist of a sensor, a control unit, and an output device. A logic module or other accessories can be added to the basic control to add versatility. The sensor consists of a source and a detector. The source is a light-emitting

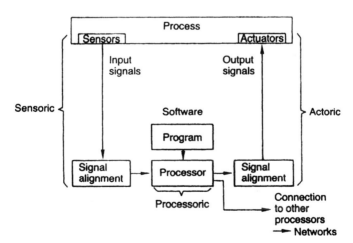

FIGURE 1.10 Components of controlled process.

diode (LED) that emits a powerful beam of light either in the infrared or visible light spectrum. The detector is typically a photodiode that senses the presence or absence of light. The detection amplifier in all photoelectric controls is designed so that it responds to the light emitted by the source; ambient light, including sunlight up to 3100 metercandles, does not affect operation.

The source and detector may be separated or may be mounted in the same sensor head, depending on the particular series and application (Fig. 1.11).

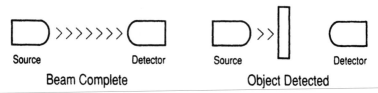

FIGURE 1.11 Components of photoelectric control.

The control unit modulates and demodulates the light sent and received by the source and detector. This assures that the photoelectric control responds only to its light source. The control unit also controls the output device in *self-contained* photoelectric controls; the control unit and sensor are built into an integral unit.

Controls can be configured to operate as light-actuated devices. The output is triggered when the detector sees light. They can also be dark-actuated devices, where the output is triggered when the detector does not see light.

Output devices may include relays such as *double pole, double throw* (DPDT) and *single pole, double throw* (SPDT). Output devices may also include a triac or other high-current device and may be programmable-controller–compatible.

Logic modules are optional devices that allow addition of logic functions to a photoelectric control. For example, instead of providing a simple ON/OFF signal, a photoelectric control can (with a logic module) provide time-delay, one-shot, retriggerable one-shot, motion-detection, and counting functions.

Manufacturing Applications of Photodetectors

The following applications of photoelectric sensors are based on normal practices at home, at the workplace, and in various industries. The effective employment of photoelectric sensors can lead to successful integration of data in manufacturing operations to maintain an error-free environment and assist in obtaining instantaneous information for dynamic interaction.

A photoelectric sensor is a semiconductor component that reacts to light or emits light. The light may be either in the visible range or the invisible infrared range. These characteristics of photoelectric components have led to the development of a wide range of photoelectric sensors.

A photoelectric reflex sensor equipped with a time-delay module set for *delay dark* ignores momentary beam breaks. If the beam is blocked longer than the predetermined delay period, the output energizes to sound an alarm or stop the conveyer (Fig. 1.12).

A set of photoelectric through-beam sensors can determine the height of a scissor lift as illustrated in Fig. 1.13. For example, when the control is set for *dark-to-light*

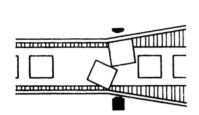

FIGURE 1.12 Jam detection with photoelectric sensor.

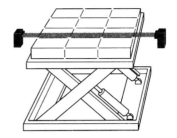

FIGURE 1.13 Stack height measurement with photoelectric sensor.

energizing, the lift rises after a layer has been removed and stops when the next layer breaks the beam again.

Cans on a conveyer are diverted to two other conveyers controlled by a polarized photoelectric reflex sensor with a divider module (Fig. 1.14). Items can be counted and diverted in groups of 2, 6, 12, or 24. A polarized sensor is used so that shiny surfaces may not falsely trigger the sensor control.

Two photoelectric control sensors can work together to inspect a fill level in cartons on a conveyer (Fig. 1.15) A reflex photoelectric sensor detects the position of the carton and energizes another synchronized photoelectric sensor located above the contents. If the photoelectric sensor located above the carton does not "see" the fill level, the carton does not pass inspection.

A single reflex photoelectric sensor detects boxes anywhere across the width of a conveyer. Interfacing the sensor with a programmable controller provides totals at specific time intervals (Fig. 1.16).

High-temperature environments are accommodated by the use of fiber optics. The conveyer motion in a 450°F cookie oven can be detected as shown in Fig. 1.17. If the motion stops, the one-shot logic module detects light or dark that lasts too long, and the output device shuts the oven down.

Placing the photoelectric sensor to detect a saw tooth (Fig. 1.18) enables the programmable controller to receive an input signal which rotates the blade into position for sharpening of the next tooth.

A through-beam photoelectric sensor is used to time the toll gate in Fig. 1.19. To eliminate toll cheating, the gate lowers the instant the rear of the paid car passes the control. The rugged sensor can handle harsh weather, abuse, and 24-h operation.

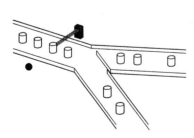

FIGURE 1.14 Batch counting and diverting with photoelectric sensor.

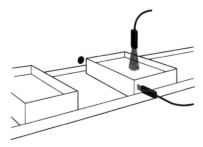

FIGURE 1.15 Measuring carton fill with photoelectric sensor.

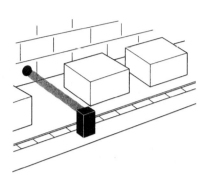

FIGURE 1.16 Box counting with photoelectric sensor.

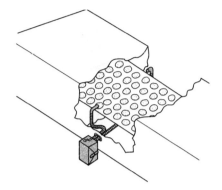

FIGURE 1.17 Detecting proper cookie arrangement.

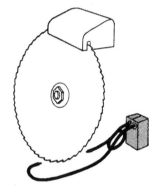

FIGURE 1.18 Sawtooth inspection.

FIGURE 1.19 Toll-booth control with photoelectrical sensor.

A safe and secure garage is achieved through the use of a through-beam photoelectric sensor interfaced to the door controller. The door shuts automatically after a car leaves, and if the beam is broken while the door is lowering, the motor reverses direction and raises the door again (Fig. 1.20).

A photoelectric sensor that generates a "curtain of light" detects the length of a loop on a web drive system by measuring the amount of light returned from an array of retroreflectors. With this information, the analog control unit instructs a motor controller to speed up or slow down the web drive (Fig. 1.21).

Small objects moving through a curtain of light are counted by a change in reflected light. A low contrast logic module inside the photoelectric sensor unit responds to slight but abrupt signal variations while ignoring slow changes such as those caused by dust buildup (Fig. 1.22).

A pair of through-beam photoelectric sensors scan over and under multiple strands of thread. If a thread breaks and passes through one of the beams, the low-contrast logic module detects the sudden changes in signal strength and energizes

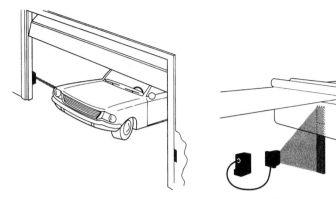

FIGURE 1.20 Garage door control with photoelectric control.

FIGURE 1.21 Web loop control.

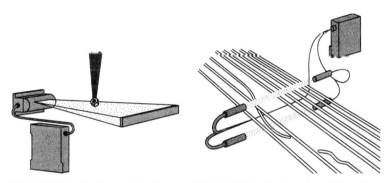

FIGURE 1.22 Small parts detection.

FIGURE 1.23 Broken-thread detection.

the output. As this logic module does not react to slow changes in signal strength, it can operate in a dusty environment with little maintenance (Fig. 1.23).

A remote photoelectric source and detector pair inspects for passage of light through a hypodermic needle (Fig. 1.24). The small, waterproof stainless-steel housing is appropriate for crowded machinery spaces and frequent wash-downs. High signal strength allows quality inspection of hole sizes down to 0.015 mm.

Index marks on the edge of a paper are detected by a fiber-optic photoelectric source/detector sensor to control a cutting shear down line (Fig. 1.25).

Liquids are monitored in a clear tank through beam sensors and an analog control. Because the control produces a voltage signal proportional to the amount of detected light, liquid mixtures and densities can be controlled (Fig. 1.26).

Remote photoelectric sensors inspect for the presence of holes in a metal casting (Fig. 1.27). Because each hole is inspected, accurate information is recorded. A rugged sensor housing and extremely high signal strength handle dirt and grease with minimum maintenance. The modular control unit allows for dense packaging in small enclosures.

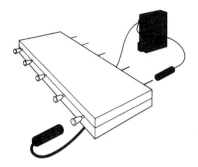

FIGURE 1.24 Hypodermic needle quality assurance.

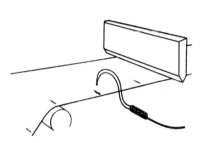

FIGURE 1.25 Indexing mark detection.

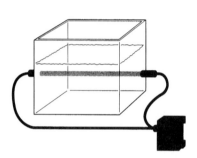

FIGURE 1.26 Liquid clarity control.

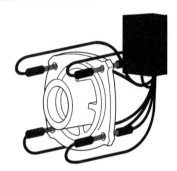

FIGURE 1.27 Multihole casting inspection.

In a web flaw detection application, a web passes over an array of retroreflectors (Fig. 1.28). When light is returned to the sensor head, the output is energized and the web shuts down. High web speeds can be maintained because of the superior response time of the control unit.

A reflex photoelectric sensor with a motion control module counts the revolutions of a wheel to monitor over/underspeed of a rotating object. Speed is controlled by a programmable controller. The rate ranges from 2.4 to 12,000 counts per minute (Fig. 1.29).

When the two through-beam photoelectric sensors in Fig. 1.30 observe the same signal strength, the output is zero. When the capacity of the web changes, as in a splice, the signal strengths are thrown out of balance and the output is energized. This system can be used on webs of different colors and opacities with no system reconfiguration.

Understanding the environment is important to effective implementation of an error-free environment. An awareness of the characteristics of photoelectric controls and the different ways in which they can be used will establish a strong foundation. This understanding also will allow the user to obtain a descriptive picture of the condition of each manufacturing process in the production environment.

Table 1.3 highlights key questions the user must consider.

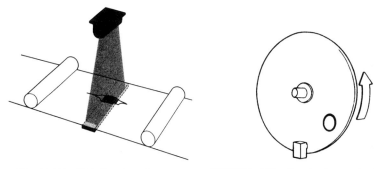

FIGURE 1.28 Web flaw detection.

FIGURE 1.29 Over/underspeed of rotating disk.

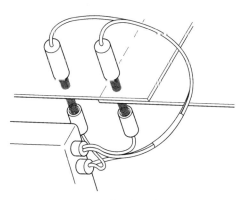

FIGURE 1.30 Web splice detection.

DETECTION METHODS

There are three modes of detection used by photoelectric sensors:

1. Through-beam detection
2. Reflex detection
3. Proximity detection

Through-Beam Detection Method

The through-beam method requires that the source and detector are positioned opposite each other and the light beam is sent directly from source to detector (Fig. 1.31). When an object passes between the source and detector, the beam is broken, signaling detection of the object.

Through-beam detection generally provides the longest range of the three operating modes and provides high power at shorter range to penetrate steam, dirt, or

TABLE 1.3 Key Characteristics of Sensors

Key point	Consideration
1. Range	How far is the object to be detected?
2. Environment	How dirty or dark is the environment?
3. Accessibility	What accessibility is there to both sides of the object to be detected?
4. Wiring	Is wiring possible to one or both sides of the object?
5. Size	What size is the object?
6. Consistency	Is the object consistent in size, shape, and reflectivity?
7. Requirements	What are the mechanical and electrical requirements?
8. Output Signal	What kind of output is needed?
9. Logic functions	Are logic functions needed at the sensing point?
10. Integration	Is the system required to be integrated?

other contaminants between the source and detector. Alignment of the source and detector must be accurate.

Reflex Detection Method

The reflex method requires that the source and detector are installed at the same side of the object to be detected (Fig. 1.32). The light beam is transmitted from the source to a retroreflector that returns the light to the detector. When an object breaks a reflected beam, the object is detected.

The reflex method is widely used because it is flexible and easy to install and provides the best cost-performance ratio of the three methods. The object to be detected must be less reflective than the retroreflector.

Proximity Detection Method

The proximity method requires that the source and detector are installed on the same side of the object to be detected and aimed at a point in front of the sensor (Fig. 1.33). When an object passes in front of the source and detector, light from the

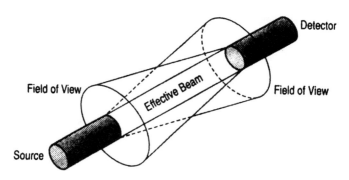

FIGURE 1.31 Through-beam-detection.

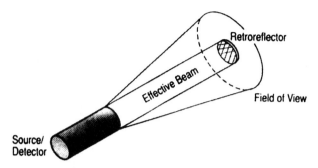

FIGURE 1.32 Reflex detection.

source is reflected from the object's surface back to the detector, and the object is detected.

Each sensor type has a specific operating range. In general, through-beam sensors offer the greatest range, followed by reflex sensors, then by proximity sensors.

The maximum range for through-beam sensors is of primary importance. At any distance less than the maximum range, the sensor has more than enough power to detect an object.

The optimum range for the proximity and reflex sensors is more significant than the maximum range. The optimum range is the range at which the sensor has the most power available to detect objects. The optimum range is best shown by an excess gain chart (Fig. 1.34).

Excess gain is a measure of sensing power available in excess of that required to detect an object. An excess gain of 1 means there is just enough power to detect an object, under the best conditions without obstacles placed in the light beam. The distance at which the excess gain equals 1 is the maximum range. An excess gain of 100 means there is 100 times the power required to detect an object. Generally, the more excess gain available at the required range, the more consistently the control will operate.

For each distance within the range of sensor, there is a specific excess gain. Through-beam controls generally provide the most excess gain, followed by reflex and then proximity sensors.

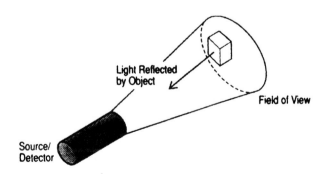

FIGURE 1.33 Proximity detection.

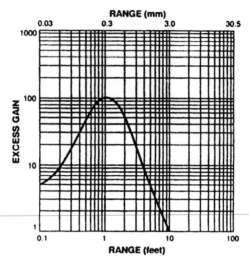

FIGURE 1.34 Photoelectric excess gain and range.

General guidelines can be provided for the amount of excess gain required for the amount of contamination in an environment. Environments can be relatively clean, lightly dirty, dirty, very dirty, and extremely dirty. Table 1.4 illustrates the excess gain recommended for these types of environments for each sensing mode.

Example. If, in a through-beam setup, the source is in a lightly dirty environment where excess gain is 1.8, and the detector is in a very dirty environment where excess gain is 25, the recommended excess gain is 1.8 × 25 = 45, from Table 1.4.

TABLE 1.4 Excess Gain Chart

Environment	Through beam	Reflex	Proximity
Relatively clean	1.25 per side	1.6 per side	
Office clean	1.6 total	2.6 total	2.6 total
Lightly dirty	1.8 per side	3.2 per side	
Warehouse, post office	3.2 total	10.5 total	3.2 total
Dirty	8 per side	64 per side	
Steel mill, saw mill	64 total		64 total
Very dirty	25 per side		
Steam tunnel, painting, rubber or grinding, cutting with coolant, paper plant	626 total		
Extremely dirty	100 per side		
Coal bins or areas where thick layers build quickly	10,000 total		

PROXIMITY SENSORS

Proximity sensing is the technique of detecting the presence or absence of an object with an electronic noncontact sensor.

Mechanical limit switches were the first devices to detect objects in industrial applications. A mechanical arm touching the target object moves a plunger or rotates a shaft which causes an electrical contact to close or open. Subsequent signals will produce other control functions through the connecting system. The switch may be activating a simple control relay, or a sophisticated programmable logic control device, or a direct interface to a computer network. This simple activity, once done successfully, will enable varieties of manufacturing operations to direct a combination of production plans according to the computer-integrated manufacturing strategy.

Inductive proximity sensors are used in place of limit switches for noncontact sensing of metallic objects. Capacitive proximity switches are used on the same basis as inductive proximity sensors; however, capacitive sensors can also detect nonmetallic objects. Both inductive and capacitive sensors are limit switches with ranges up to 100 mm.

The distinct advantage of photoelectric sensors over inductive or capacitive sensors is their increased range. However, dirt, oil mist, and other environmental factors will hinder operation of photoelectric sensors during the vital operation of reporting the status of a manufacturing process. This may lead to significant waste and buildup of false data.

Typical Applications of Inductive Proximity Sensors

Motion position detection (Fig. 1.35)

1. Detection of rotating motion
2. Zero-speed indication
3. Speed regulation

Motion control (Fig. 1.36)

FIGURE 1.35 Motion/position detection with inductive proximity sensor.

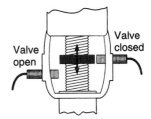

FIGURE 1.36 Motion control, inductive proximity sensor.

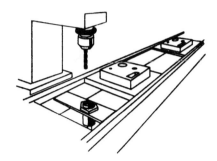

FIGURE 1.37 Conveyer system control, inductive proximity sensor.

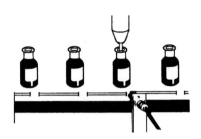

FIGURE 1.38 Process control, inductive proximity sensor.

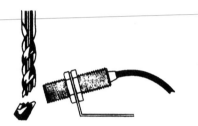

FIGURE 1.39 Machine control, inductive proximity sensor.

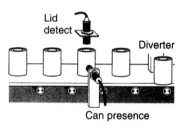

FIGURE 1.40 Verification and counting, inductive proximity sensor.

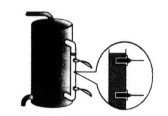

FIGURE 1.41 Liquid level detection, capacitive proximity sensor.

1. Shaft travel limiting
2. Movement indication
3. Valve open/closed

   Conveyer system control (Fig. 1.37)

1. Transfer lines
2. Assembly line control
3. Packaging machine control

   Process control (Fig 1.38)

1. Product complete
2. Automatic filling
3. Product selection

   Machine control (Fig. 1.39)

1. Fault condition indication
2. Broken tool indication
3. Sequence control

   Verification and counting (Fig. 1.40)

1. Product selection
2. Return loop control
3. Product count

Typical Applications of Capacitive Proximity Sensors

Liquid level detection (Fig. 1.41)

1. Tube high/low liquid level
2. Overflow limit
3. Dry tank

FIGURE 1.42 Bulk material level control, capacitive proximity sensor.

FIGURE 1.43 Process control, capacitive proximity sensor.

Bulk material level control (Fig. 1.42)

1. Low level limit
2. Overflow limit
3. Material present

Process control (Fig. 1.43)

1. Product present
2. Bottle fill level
3. Product count

UNDERSTANDING INDUCTIVE PROXIMITY SENSORS

Principles of Operation

An inductive proximity sensor consists of four basic elements (Fig. 1.44).

1. Sensor coil and ferrite core
2. Oscillator circuit
3. Detector circuit
4. Solid-state output circuit

The oscillator circuit generates a radio-frequency electromagnetic field that radiates from the ferrite core and coil assembly. The field is centered around the axis of the ferrite core, which shapes the field and directs it at the sensor face. When a metal target approaches and enters the field, *eddy currents* are induced into the surfaces of the target. This results in a loading effect, or "damping," that causes a reduction in amplitude of the oscillator signal (Fig. 1.45).

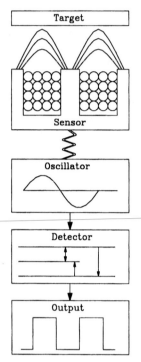

FIGURE 1.44 Operating principle of inductive proximity sensor.

The detector circuit detects the change in oscillator amplitude (Fig. 1.46). The detector circuit will switch ON at a specific operate amplitude. This ON signal generates a signal to turn ON the solid-state output. This is often referred to as the *damped* condition. As the target leaves the sensing field, the oscillator responds with an increase in amplitude. As the amplitude increases above a specific value, it is detected by the detector circuit, which switches OFF, causing the output signal to return the normal or OFF (*undamped*) state.

The difference between the operate and the release amplitude in the oscillator and corresponding detector circuit is referred to as *hysteresis* (H) of the sensor. It corresponds to a difference in point of target detection and release distance between the sensor face and the target surface (Fig. 1.47).

Inductive Proximity Sensing Range

The *sensing range* of an inductive proximity sensor refers to the distance between the sensor face and the target. It also includes the shape of the sensing field generated through the coil and core. There are several mechanical and environmental factors that affect the sensing range:

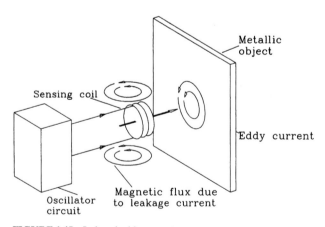

FIGURE 1.45 Induced eddy current.

INTRODUCTION

Mechanical factors	Environmental factors
Core size	Ambient temperature
Core shield	Surrounding electrical conditions
Target material	Surrounding mechanical conditions
Target size	Variation between devices
	Target shape

The geometry of the sensing field can be determined by the construction factor of the core and coil. An open coil with no core produces an omnidirectional field. The geometry of an air-core is a toroid. Such sensors could be actuated by a target

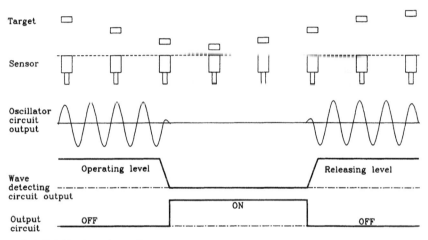

FIGURE 1.46 Detection cycle.

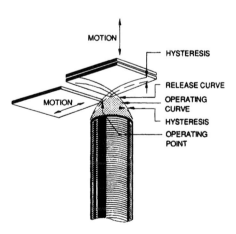

FIGURE 1.47 Core assembly.

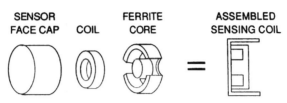

FIGURE 1.48 Open coil without core.

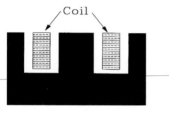

FIGURE 1.49 Cup-shaped coil/core assembly.

approaching from any direction, making them undesirable for practical industrial applications (Fig. 1.48).

Ferrite material in the shape of a cup core is used to shape the sensing field. The ferrite material absorbs the magnetic field, but enhances the field intensity and directs the field out of the open end of the core (Fig. 1.49).

A standard field range sensor is illustrated in Fig. 1.50. It is often referred to as *shielded* sensing coil. The ferrite contains the field so that it emanates straight from the sensing face. Figure 1.51 shows the typical standard-range sensing-field plot.

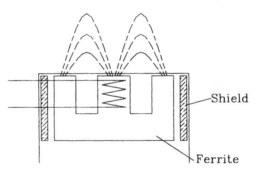

FIGURE 1.50 Standard range core coil.

An extended range coil and core assembly does not have the ferrite around the perimeter of the coil (Fig. 1.52). This unshielded device accordingly has an extended range. Figure 1.53 illustrates a typical extended sensing-field plot.

Sensing Distance

The electromagnetic field emanates from the coil and core at the face of the sensor and is centered around the axis of the core. The nominal sensing range is a function of the coil diameter and the power that is available to operate the electromagnetic field.

INTRODUCTION 1.27

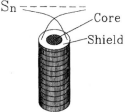

FIGURE 1.51 Standard range field plot.

The sensing range is subject to manufacturing tolerances and circuit variations. Typically it varies by 10 percent. Similarly, temperature drift can affect the sensing range by 10 percent. Applied to the nominal sensing switch, these variations mean that sensing range can be as much as 120 percent or as little as 81 percent of the nominal stated range (Fig. 1.54).

$$S_r = 0.9 < S_n < 1.1$$

$$S_n = 0.9 < S_r < 1.1$$

where S_n = nominal sensing range
S_r = effective sensing range
S = usable sensing range

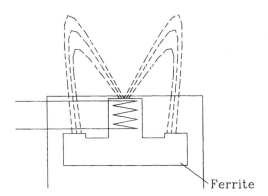

FIGURE 1.52 Extended range core and coil.

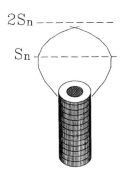

FIGURE 1.53 Extended range field plot.

Target Material and Size

As a target approaches the sensing field, eddy currents are induced in the target. In order to ensure that the target has the desired damping affect on the sensor, the target has to be of appropriate size and material. Metallic targets can be defined as:

- *Ferrous.*   Containing iron, nickel, or cobalt
- *Nonferrous.*   All other metallic materials, such as aluminum, copper, and brass. Eddy currents induced in ferrous targets are stronger than in nonferrous targets, as illustrated in Table 1.5.

An increase in target size will not produce an increase in sensing range. However, a decrease in target size will produce a decrease in sensing range, and may also

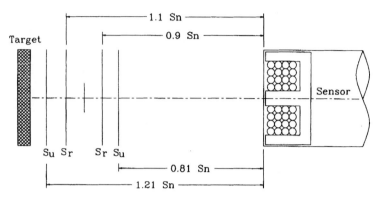

FIGURE 1.54 Sensing distance tolerances.

increase response time. Figure 1.55 illustrates the relationship of target size and target material to the sensing range of a limit-switch-type sensor with nominal 13-mm sensing range.

Table 1.6, shows correction factors by which the rated nominal sensing range of most inductive proximity sensors can be multiplied. This will determine the effective sensing range for a device sensing a stated target material of standard target size.

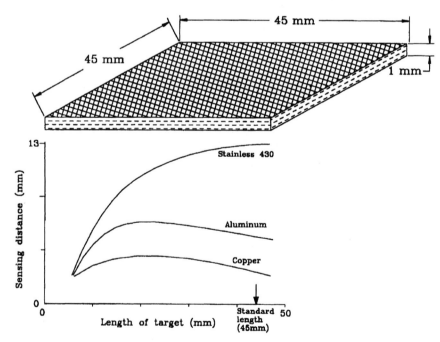

FIGURE 1.55 Sensing range correction factor.

INTRODUCTION

TABLE 1.5 Standard Target

Device	Standard target dimensions	Standard target material
Modular limit switch type	45 mm square × 1 mm thick	Mild steel
8 mm tubular	8 mm square × 1 mm thick	Mild steel
12 mm tubular	12 mm square × 1 mm thick	Mild steel
18 mm tubular	18 mm square × 1 mm thick	Mild steel
30 mm tubular	30 mm square × 1 mm thick	Mild steel

TABLE 1.6 Target Material Correction

Target material	Limit-switch type	Pancake type	Tubular, mm			
			8	12	18	30
Steel (1020)	1.0	1.0	1.0	1.0	1.0	1.0
Stainless steel (400)	1.03	0.90	0.90	0.90	1.0	1.0
Stainless steel (300)	0.85	0.70	0.60	0.70	0.70	0.65
Brass	0.50	0.54	0.35	0.45	0.45	0.45
Aluminum	0.47	0.50	0.35	0.40	0.45	0.40
Copper	0.40	0.46	0.30	0.25	0.35	0.30

Example	18-mm tubular extended range	8 mm
	Copper target correction factor	× 0.35
	Sensing range detecting copper standard target	2.80 mm

Target Shape

Standard targets are assumed to be of a flat, square shape with the stated dimensions. Targets of round shape or with a pocketed surface have to be of adequate dimensions to cause the necessary dampening effect on the sensor. Allowing the sensor-to-target distance less than the nominal range will help to assure proper sensing. Also using the next larger size or an extended-range sensor will also minimize problems with other than standard target dimensions or shape. Figure 1.56 illustrates the axial (head-on) approach, indicating that the target approaches the face of the sensor on the axis of the coil core. When the target approaches axially, the sensor should not be located such that it becomes an end stop. If axial operation is considered, good application practice is to allow for 25 percent overtravel.

Lateral (*side-by*) *approach* means the target approaches the face of the sensor perpendicular to the axis of the coil core (Fig. 1.57). Good application practice (GAP), a terminology often used in "world-class" manufacturing strategies, dictates that the tip of the sensing field envelope should not be used. That is

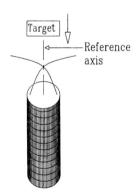

FIGURE 1.56 Axial approach.

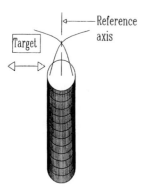

FIGURE 1.57 Lateral approach.

the point where sensing range variations start to occur. Therefore, it is recommended that the target pass not more than 75 percent of the sensing distance D from the sensor face. Also, the target should not pass any closer than the basic tolerance incorporated in the machine design, to prevent damage to the sensor. Hysteresis can be greater for an axial approach (Fig. 1.58).

Variation Between Devices

Variations of sensing range between sensors of the same type often occur. With modern manufacturing technologies and techniques, variations are held to a minimum. The variations can be attributed to collective tolerance variations of the electrical components in the sensor circuit and to subtle differences in the manufacturing process from one device to the next; 5 percent variation is typical (Fig. 1.59).

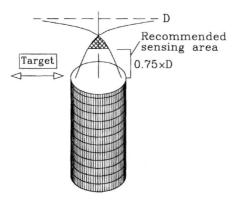

FIGURE 1.58 Lateral approach—recommended sensing distance.

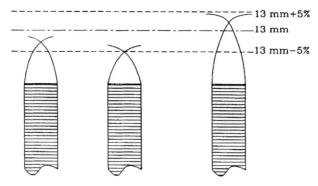

FIGURE 1.59 Sensing range variation.

INTRODUCTION

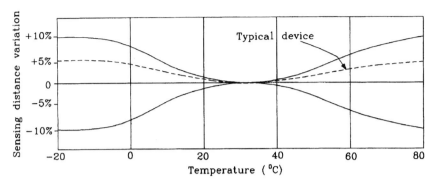

FIGURE 1.60 Sensing range variation—with temperature.

Sensing distance also will vary from one temperature extreme to the other because of the effect of temperature change on the components of the sensor. Typical temperature ranges are −25°C (−3°F) to +70°C (+180°F). Figure 1.60 illustrates sensing range variation with temperature.

Surrounding Conditions

There are several environmental factors that must also be considered in order to obtain reliable information from inductive proximity sensors. These surrounding factors are

1. *Embeddable mounting.* The shielded sensor in Fig. 1.61 is often referred to as a *flush-mounted* sensor. Shielded sensors are not affected by the surrounding metal.
2. *Flying metal chips.* A chip removed from metal during milling and drilling operations may affect the sensor performance depending on the size of the chip, its location on the sensing face, and type of material. In these applications, the sensor face should be oriented so that gravity will prevent chips from accumulating on the sensor face. If this is not possible, then coolant fluid should wash the sensor face to remove the chips. Generally, a chip does not have sufficient surface area to cause a sensor turn on. If a chip lands on the center of the sensor face, it

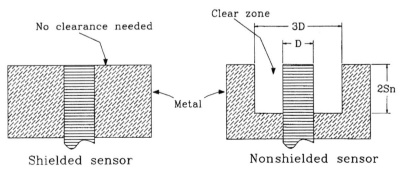

FIGURE 1.61 Embeddable and nonembeddable sensors.

will have a negligible effect, but elsewhere on the sensor face, it will extend the range of the sensor.

3. *Adjacent sensors.* When two similar sensors are located adjacent to or opposite each other, the interaction of their fields can affect operation. Figure 1.62 provides the guidelines for placing two similar sensors adjacent to each other. *Alternate-frequency* heads will allow adjacent mounting of sensors without interaction of their sensing fields.

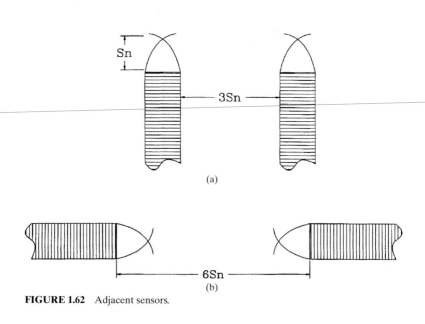

FIGURE 1.62 Adjacent sensors.

4. *Magnetic fields.* Electrical wiring in the vicinity of the sensor face may affect sensor operation. If the magnetic field around the electrical wiring reaches an intensity that would saturate the sensor ferrite or coil, the sensor will not function properly. Use of inductive sensors in the presence of high-frequency radiation can also unexpectedly affect their operation. Sensors specially designed for welding application can be used with programmable logic control (PLC). The PLC can be programmed to ignore the signal from the sensor for the period that the high-frequency welder is operated. A slight OFF-time delay assures proper operation of the sensor.

5. *Radio-frequency interference (RFI).* Radio transceivers, often called *walkie-talkie* devices, can produce a signal that can cause an inductive proximity sensor to operate falsely. The radio transceiver produces a radio-frequency signal similar to the signal produced by the oscillator circuit of the sensor. The effect that RFI has on an inductive proximity switch can vary. The factors that determine this variation are as follows:

   a. *Distance between RFI source and the sensor.* Typically, inductive proximity switches are not affected by RFI when a transceiver is 1 ft away from the inductive switch. However, if closer than 1 ft, the switch may operate without a target present.

b. *Signal frequency.* The signal frequency may be the determining factor that will cause a particular device to false-operate.
   c. *Signal intensity.* Radio-frequency transceivers usually are portable devices with power rating of 5 W maximum.
   d. *Inductive proximity package.* The sensor package construction may determine how well the device resists RFI.
   e. *Approach to the sensor.* A transceiver approaching the connecting cable of a switch may affect it at a greater distance than if it was brought closer to the sensing face. As RFI protection varies from device to device and manufacturer to manufacturer, most manufacturers have taken steps to provide the maximum protection against false operation due to RFI.
6. *Showering arc.* Showering arc is the term applied to induced line current/voltage spikes. The spike is produced by the electrical arc on an electromechanical switch or contactor closure. The current spike is induced from lines connected to the electromechanical switch to the lines connected to the inductive proximity switch, if the lines are adjacent and parallel to one another. The result can be false operation of the inductive proximity switch. The spike intensity is determined by the level of induced voltage and the duration of the spike. Avoiding running cables for control devices in the same wiring channel as those for the contactor or similar leads may eliminate spikes. Most electrical code specifications require separation of control device leads from electromechanical switch and contactor leads.

UNDERSTANDING CAPACITIVE PROXIMITY SENSORS

Principles of Operation

A capacitive proximity sensor operates much like an inductive proximity sensor. However, the means of sensing is considerably different. Capacitive sensing is based on dielectric capacitance. *Capacitance* is the property of insulators to store an electric charge. A capacitor consists of two plates separated by an insulator, usually called a *dielectric.* When the switch is closed (Fig. 1.63) a charge is stored on the two plates.

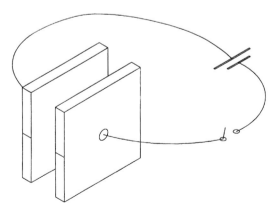

FIGURE 1.63 Capacitive principle.

The distance between the plates determines the ability of a capacitor to store a charge and can be calibrated as a function of stored charge to determine discrete ON and OFF switching status.

Figure 1.64 illustrates the principle as it applies to the capacitive sensor. One capacitive plate is part of the switch, the sensor face (the enclosure) is the insulator, and the target is the other plate. Ground is the common path.

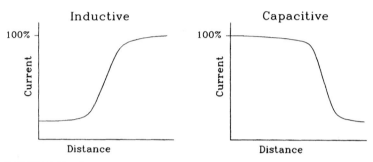

FIGURE 1.64 Capacitive sensor.

The capacitive proximity sensor has the same four basic elements as an inductive proximity sensor:

1. Sensor (the dielectric plate)
2. Oscillator circuit
3. Detector circuit
4. Solid-state output circuit

The oscillator circuit in a capacitive switch operates like one in an inductive proximity switch. The oscillator circuit includes feedback capacitance from the external target plate and the internal plate. In a capacitive switch, the oscillator starts oscillating when sufficient feedback capacitance is detected. In an inductive proximity switch, the oscillation is damped when the target is present (Fig. 1.65).

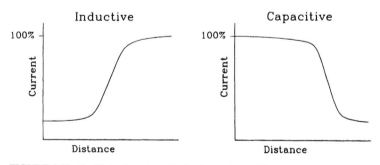

FIGURE 1.65 Oscillator damping of inductive and capacitive sensors.

In both capacitive and inductive switch types, the difference between the operate and the release amplitude in the oscillator and corresponding detector circuit is referred to as *hysteresis* of the sensor. It corresponds to the difference between target detection and release distances from the sensor face.

Features of Capacitive Sensors

The major characteristics of capacitive proximity sensors are

1. They can detect nonmetallic targets.
2. They can detect lightweight or small objects that cannot be detected by mechanical limit switches.
3. They provide a high switching rate for rapid response in object counting applications.
4. They can detect liquid targets through nonmetallic barriers, (glass, plastic, etc.).
5. They have long operational life with a virtually unlimited number of operating cycles.
6. The solid-state output provides a bounce-free contact signal.

Capacitive proximity sensors have two major limitations:

1. They are affected by moisture and humidity.
2. They must have an extended range for effective sensing.

Sensing Range

Capacitive proximity sensors have a greater sensing range than inductive proximity sensors as illustrated below.

Tubular diameter, mm	Inductive extended range, mm	Capacitive extended range, mm
18	8	10
30	15	20
34	—	40

Sensing distance for capacitive proximity sensors is a matter of plate diameter, as coil size is for inductive proximity sensors. Capacitive sensors basically measure a dielectric gap. Accordingly, it is desirable to be able to compensate for target and application conditions with a sensitivity adjustment for the sensing range. Most capacitive proximity switches are equipped with a sensitivity adjustment potentiometer (Fig. 1.66).

FIGURE 1.66 Sensitivity adjustment.

Target Material and Size

The sensing range of capacitive sensors, like that of inductive proximity sensors, is determined by the type of material. Table 1.7 lists the sensing-range derating factors which apply to capacitive proximity sensors. Capacitive sensors can be used to detect a target material through a nonmetallic interposing material like glass or plastic. This is beneficial in detecting a liquid through the wall of a plastic tank or through a glass sight tube. The transparent interposing material has no effect on sensing. For all practical purposes, the target size can be determined in the same manner as for inductive proximity sensors.

Surrounding Conditions

Capacitive proximity devices are affected by component tolerances and temperature variations. As with inductive devices, capacitive proximity devices are affected by the following surrounding conditions:

1. *Embeddable mounting.* Capacitive sensors are generally treated as nonshielded, nonembeddable devices.
2. *Flying chips.* Capacitive devices are more sensitive to metallic and nonmetallic chips.
3. *Adjacent sensors.* Allow more space than with inductive proximity devices because of the greater sensing range of capacitive devices.

TABLE 1.7 Target Material Correction

Material	Factor
Mild steel	1.0
Cast iron	1.0
Aluminum and copper	1.0
Stainless steel	1.0
Brass	1.0
Water	1.0
Polyvinylchloride (PVC)	0.5
Glass	0.5
Ceramic	0.4
Wood	≥0.2
Lubrication oil	0.1

4. *Target background.* Relative humidity may cause a capacitive device to operate even when a target is not present. Also, the greater sensing range and ability to sense nonmetallic target materials dictate greater care in applying capacitive devices with target background conditions.
5. *Magnetic fields.* Capacitive devices are not usually applied in welding environment.
6. *Radio-frequency interference.* Capacitive sensor circuitry can be affected by RFI in the same way that an inductive device can.
7. *Showering arc.* An induced electrical noise will affect the circuitry of a capacitive device in the same way that it does an inductive device.

UNDERSTANDING LIMIT SWITCHES

A limit switch is constructed much like the ordinary light switch used in home and office. It has the same ON/OFF characteristics. The limit switch usually has a pressure-sensitive mechanical arm. When an object applies pressure on the mechanical arm, the switch circuit is energized. An object might have a magnet attached that causes a contact to rise and close when the object passes over the arm.

Limit switches can be either *normally open* (NO) or *normally closed* (NC) and may have multiple poles (Fig. 1.67). A normally open switch has continuity when pressure is applied and a contact is made, while a normally closed switch opens when pressure is applied and a contact is separated. A single-pole switch allows one circuit to be opened or closed upon switch contact, whereas a multiple-pole switch allows multiple circuits to be opened or closed.

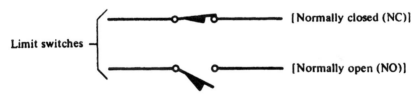

FIGURE 1.67 Normally open—normally closed microswitches.

Limit switches are mechanical devices. They have three potential problems:

1. They are subject to mechanical failure.
2. Their mean time between failures (MTBF) is low compared to noncontact sensors.
3. Their speed of operation is relatively slow; the switching speed of photoelectric microsensors is up to 3000 times faster.

INDUCTIVE AND CAPACITIVE SENSORS IN MANUFACTURING

Inductive and capacitive proximity sensors interface to control circuits through an output circuit, for manufacturing applications. Also, the control circuit type is a determining factor in choosing an output circuit. Control circuits, whether powered by AC, DC, or AC/DC, can be categorized as either *load powered* or *line powered*.

The load-powered devices are similar to limit switches. They are connected in series with the controlled load. These devices have two connection points and are often referred to as *two-wire switches*. Operating current is drawn through the load. When the switch is not operated, the switch must draw a minimum operating current referred to as *residual current*. When the switch is operated or damped (i.e., a target is present), the current required to keep the sensor operating is the minimum holding current (Fig. 1.68). The residual current is not a consideration for low-impedance loads such as relays and motor starters. However, high-impedance loads, most commonly programmable logic controllers, require residual current of less than 2 mA. Most sensors offer 1.7 mA or less.

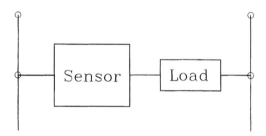

FIGURE 1.68 Load-powered residual current.

In some manufacturing applications, a particular type of PLC will require less than 1.7 mA residual current. In such applications a loading resistor is added in parallel to the input to the PLC load. Then minimum holding current may range up to 20 mA, depending on the sensor specification. If the load impedance is too high, there will not be enough load current level to sustain the switch state.

Inductive proximity sensors with holding current of 4 mA or less can be considered low-holding-current sensors. These devices can be used with PLCs without concern for minimum holding current.

Line-powered devices derive current, usually called *burden current*, from the line and not through the controller load. These devices are called three-wire switches because they have three connections (Fig. 1.69).

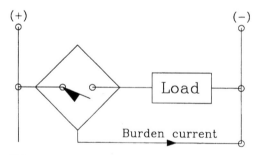

FIGURE 1.69 Line-powered burden current.

The operating current for a three-wire sensor is burden current, and is typically 20 mA. Since the operating current does not pass through the load, it is not a major concern for the circuit design.

Relays

An output circuit relay is a mechanical switch available in a variety of contact configurations. Relays can handle load currents at high voltages, allowing the sensor to directly interface with motors, large solenoids, and other inductive loads. They can switch either AC or DC loads. Contact life depends on the load current and frequency of operation. Relays are subject to contact wear and resistance buildup. Because of contact bounce, they can produce erratic results with counters and programmable controllers unless the input is filtered. They can add 10 to 25 ms to an inductive or capacitive switch response time because of their mechanical nature (Fig. 1.70).

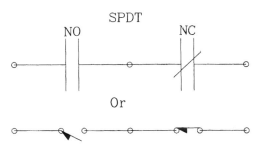

FIGURE 1.70 Relay output.

Relays are familiar to manufacturing personnel. They are often used with inductive or capacitive proximity sensors, as they provide multiple contacts. The good and bad features of a relay are summarized below.

Relay advantages	Relay disadvantages
Switches high currents/loads	Slow response time
Multiple contacts	Mechanical wear
Switches AC or DC voltages	Contact bounce
Tolerant of inrush current	Affected by shocks and vibration

Triac Devices

A triac is a solid-state device designed to control AC current (Fig. 1.71). Triac switches turn ON in less than a microsecond when the gate (control leg) is energized, and shut OFF at the zero crossing of the AC power cycle.

Because a triac is a solid-state device, it is not subject to the mechanical limitations of a relay such as mechanical bounce, pitting, corrosion of contacts, and shock

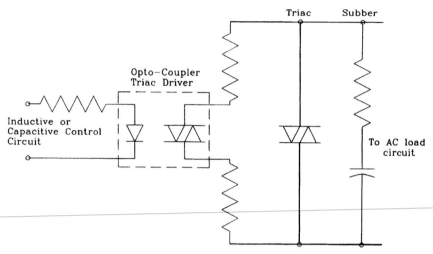

FIGURE 1.71 Triac circuit.

and vibration. Switching response time is limited only by the time it takes the 60-Hz AC power to go through one-half cycle (8.33 ms) (Fig. 1.72).

As long as a triac is used within its rated maximum current and voltage specifications, life expectancy is virtually infinite. Triac devices used with inductive or capacitive sensors generally are rated at 2-A loads or less. Triac limitations can be summarized as follows: (1) shorting the load will destroy a triac and (2) directly connected inductive loads or large voltage spikes from other sources can false-trigger a triac.

To reduce the effect of these spikes, a snubber circuit composed of a resistor and capacitor in series is connected across the device. Depending on the maximum switching load, an appropriate snubber network for switch protection is used. The snubber network contributes to the OFF state leakage to the load. The leakage must be considered when loads requiring little current, such as PLCs, are switched. In the ON state, a drop of about 1.7 V rms is common (Fig. 1.73). Good and bad features of triacs are listed below.

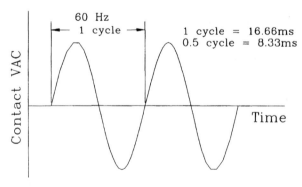

FIGURE 1.72 AC power cycle.

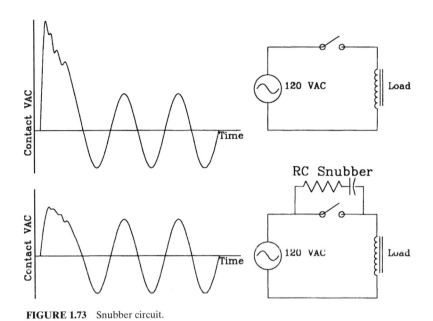

FIGURE 1.73 Snubber circuit.

Triac advantages	Triac disadvantages
Fast response time (8.33 ms)	Can be falsely triggered by large inductive current
Tolerant of large inrush currents	Snubber contributes to OFF state leakage current
Can be directly interfaced with programmable controllers	Can be destroyed by short circuits
Infinite life when operated within rated voltage/current limits	

Transistor DC Switches

Transistors are solid-state DC switching devices. They are most commonly used with low-voltage DC-powered inductive and capacitive sensors as the output switch. Two types are employed, depending on the function (Fig. 1.74).

In an NPN transistor, the current source provides a contact closure to the DC positive rail. The NPN current sink provides a contact to the DC common. The transistor can be thought of as a single-pole switch that must be operated within its voltage and maximum current ratings (Fig. 1.75).

Any short circuit on the load will immediately destroy a transistor that is not short-circuit protected. Switching inductive loads creates voltage spikes that exceed many times the maximum rating of the transistor. Peak voltage clamps such as zener diodes or transorbs are utilized to protect the output device. Transistor outputs are typically rated to switch loads of 250 mA at 30 V DC maximum (Fig. 1.76).

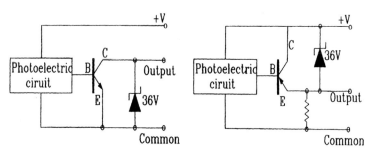

FIGURE 1.74 DC circuit logic.

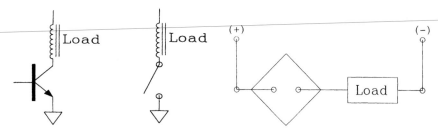

FIGURE 1.75 Transistor switch. **FIGURE 1.76** Voltage clamp.

Transistor advantages	Transistor disadvantages
Virtually instantaneous response	Low current handling capacity
Low OFF state leakage and voltage drop	Cannot handle inrush current unless clamped
Infinite life when operated within rated current/voltage	Can be destroyed by short circuit unless protected
Not affected by shock and vibration	

Output Configuration. Output configurations are categorized as follows:

1. Single output—normally open (NO)
2. Single output—normally closed (NC)
3. Programmable output—NO or NC
4. Complementary output—NO and NC

The functions of normally open and normally closed output are defined in Table 1.8.

Inductive and Capacitive Control/Output Circuits

A single output sensor has either an NO or an NC configuration and cannot be changed to the other configuration (Fig. 1.77).

A programmable output sensor has one output, NO or NC, depending on how the output is wired when installed. These sensors are exclusively two-wire AC or DC (Fig. 1.78).

TABLE 1.8 Output Logic

Output configuration	Target state	Oscillator state	Output
NO	Absent	Undamped	Nonconducting (OFF)
	Present	Damped	Conducting (ON)
NC	Absent	Undamped	Conducting (ON)
	Present	Damped	Nonconducting (OFF)

A complementary output sensor has two outputs, one NO and one NC. Both outputs change state simultaneously when the target enters or leaves the sensing field. These sensors are exclusively three-wire AC or DC (Fig. 1.79).

The choice of control circuit and output logic plays an important part in determining the reliability of data collection. The choice of control circuit and output logic depends on the following parameters:

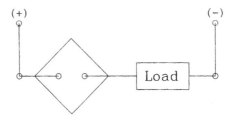

FIGURE 1.77 Single output.

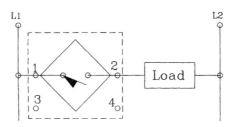

FIGURE 1.78 Programmable output.

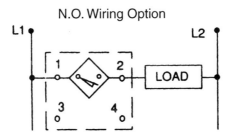

FIGURE 1.79 Complementary output.

1. *AC or DC control voltage.* Use of AC control may seem to require the use of an AC-configured sensor. However, interface circuitry can allow for DC sensors even if the main control voltage source is AC.
2. *Control circuit current requirements.* Usually control circuits operating in the 200- to 300-mA range can use either AC or DC sensors. Circuits with 0.5-A and higher current will dictate the type of sensor to be used.
3. *Application output requirements.* NO output is the most commonly used output type. Controlled circuit configurations may dictate use of NC or complementary-type configured sensors.
4. *Switching speed requirements.* AC circuits are limited in their operations per second. DC circuits may be required for applications involving counting or high speed.
5. *Connecting logic device.* The device to which the sensor is connected—such as programmable controller, relay, solenoid, or timer/counter—is usually the most important factor in sensor circuit and output configuration.

Accessories for Sensor Circuits

Sensor circuits and their output configurations must have various types of indicators and protection devices, such as:

1. Light-emitting diode (LED) indicators
2. Short-circuit protectors
3. Reverse-polarity protectors—DC three-wire
4. Wire terminators—color-coded wire
5. Pin connector type and pin-out designator

LED Indicators. LED indicators provide diagnostic information on the status of sensors, e.g., operated or not operated, that is vital in computer-integrated manufacturing. Two LEDs also indicate the status of complementary-type sensor switches and power ON/OFF status, and short-circuit condition.

Short-Circuit Protection. Short-circuit protection is intended to protect the switch circuit from excessive current caused by wiring short circuits, line power spikes from high inrush sources, or lightning strikes. This option involves special circuitry which either limits the current through the output device or turns the switch OFF. The turn-off–type switch remains inoperative until the short circuit has been cleared—with power disconnected. Then power is reapplied to the sensor. A second LED is usually furnished with this type of device to indicate the shorted condition.

Reverse-Polarity Protection. Reverse-polarity protection is special circuitry that prevents damage in a three-wire DC sourcing (PNP) or sinking (NPN) device when it is connected to control circuitry incorrectly. Although reverse polarity is relatively common, not all switches are equipped with this option.

Wire Termination. Wire terminals are common on limit-switch enclosure-type sensors. The terminal designations are numbered and correspond to the device wiring diagram (Fig. 1.80). Cable/wire stub terminations are most common on tubu-

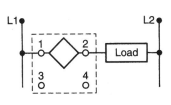

FIGURE 1.80 Wire terminal.

lar sensors. Color-coded conductors are essential for correct wiring. Most sensor wires are color-coded to comply with industry wire color-code standards.

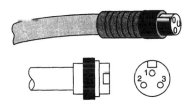

Pin Connectors. Pin-connector-terminal sensors feature a male pin connector receptacle on the switch or at the end of the wire/cable stub. The female receptacle is at the end of the matching cable cord. Most industry-standard pin connectors are either the mini type—approximately 18 mm in diameter—or the micro type—approximately 12 mm in diameter (Fig. 1.81).

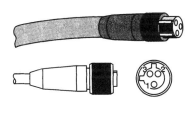

Inductive and Capacitive Switching Logic

The outputs of two or more inductive or capacitive proximity sensors can be wired together in series or parallel to perform logic functions. The ON, or activated, switch function can be either a normally open or a normally closed output function, depending on the desired control logic.

Although sensors are the most effective means for the data acquisition role in manufacturing, care must be exercised when sensors are integrated with various production operations. The following factors will affect the performance of the switch logic circuit:

FIGURE 1.81 Pin connector.

1. Excessive leakage current in parallel-connected load-powered devices
2. Excessive voltage drop in series-connected devices
3. Inductive feedback with line-powered sensors with parallel connections

TABLE 1.9 Binary Logic Chart—
Parallel Connection

A	B	C	OUT
0	0	0	0
0	0	1	1
0	1	0	1
1	0	0	1

0 = OFF
1 = ON

Parallel-Connection Logic—OR Function. The binary OR logic in Table 1.9 indicates that the circuit output is ON (1) if one or more of the sensors in parallel connection is ON.

It is important to note that, in two-wire devices, the OFF state residual current is additive (Fig. 1.82). If the circuit is affected by the total leakage applied, a shunt (loading) resistor may have to be applied. This is a problem in switching to a programmable controller or other high-impedance device.

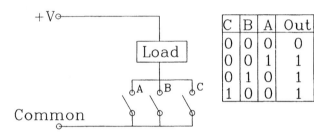

FIGURE 1.82 Parallel sensor arrangement.

Example $\qquad I_a + I_b + I_c = I_t$

$1.7 + 1.7 + 1.7 = 5.1 \text{ mA}$

Three-wire 10 to 30 V can also be connected in parallel for a logic OR circuit configuration. Figure 1.83 shows a current *sourcing* (PNP) parallel connection.

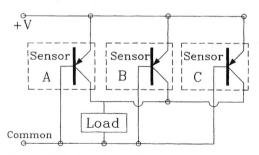

FIGURE 1.83 Sourcing (PNP) parallel sensor arrangement.

Figure 1.84 shows a current *sinking* (NPN) parallel connection. It may be necessary to utilize blocking diodes to prevent inductive feedback (or reverse polarity) when one of the sensors in parallel is damped while the other is undamped. Figure 1.85 demonstrates the use of blocking diodes in this type of parallel connection.

Series-Connection Logic—AND Function. Figure 1.86 shows AND function logic indicating that the series-connected devices must be ON (1) in order for the series-connected circuit to be ON.

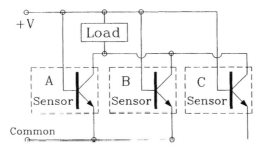

FIGURE 1.84 Sinking (NPN) parallel sensor arrangement.

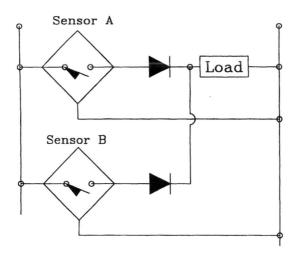

FIGURE 1.85 Blocking diodes.

B	A	Out
0	0	0
0	1	0
1	0	0
1	1	1

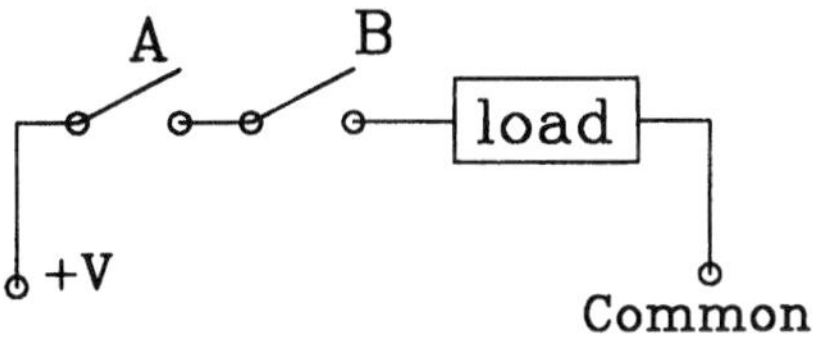

FIGURE 1.86 Series AND logic.

The voltage drop across each device in series will reduce the available voltage the load will receive. Sensors, as a general rule, have a 7- to 9-V drop per device. The minimum operating voltage of the circuit and the sum of the voltage drop per sensor will determine the number of devices in a series-connected circuit. Figure 1.87 shows a typical two-wire AC series-connected circuit.

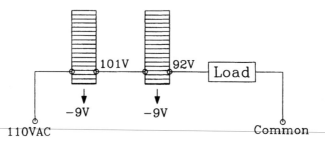

FIGURE 1.87 Series connected, load powered.

Series connection is generally applied to two-wire devices, most commonly two-wire AC. 10- to 30-V DC two-wire connections are not usually practical for series connection because of the voltage drop per device and minimum operating voltage. Three-wire devices are generally not used for series connection. However, the following characteristics should be considered for three-wire series-connected circuits (Fig. 1.88):

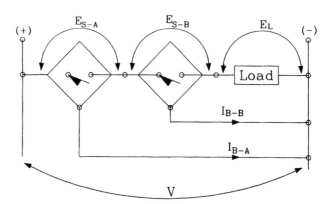

FIGURE 1.88 Series connected, line powered.

1. Each sensor must carry the load current and the burden current for all the downstream sensors (Fig. 1.88).
2. When conducting, each sensor will have a voltage drop in series with the load, reducing the available voltage to the load. As with two-wire devices, this and the minimum operating voltage will limit the number of devices wired in series.
3. When upstream sensors are not conducting, the downstream sensors are disconnected from their power source and are incapable of responding to a target until

the upstream sensors are activated (damped). Time before availability will be increased due to the response in series.

Series and parallel connections that perform logic functions with connection to a PLC are not common practice. Utilizing sensors this way involves the above considerations. It is usually easier to connect directly to the PLC inputs and perform the desired logic function through the PLC program.

Inductive and Capacitive Sensor Response Time—Speed of Operation

When a sensor receives initial power on system power-up, the sensor cannot operate. The sensor operates only after a delay called *time delay before availability* (Fig. 1.89).

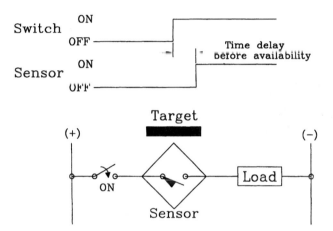

FIGURE 1.89 Time delay prior to availability.

In AC sensors, this delay is typically 35 ms. It can be as high as 100 ms in AC circuits with very low residual current and high noise immunity. In DC sensors, the time delay is typically 30 ms.

Response and Release Time. A target entering the sensing field of either an inductive or a capacitive sensor will cause the detector circuit to change state and initiate an output. This process takes a certain amount of time, called *response time* (Fig. 1.90).

Response time for an AC sensor is typically less than 10 ms. DC devices respond in microseconds. Similarly, when a target leaves the sensing field, there is a slight delay before the switch restores to the OFF state. This is the *release time*. Release time for an AC device is typically one cycle (16.66 ms). The DC device release time is typically 3 ms.

High-Speed Operation. Mechanical devices such as limit switches and relays do not operate at speeds suitable for high-speed counting or other fast-operating-circuit needs. Solid-state devices, however, can operate at speeds of 10, 15, or more operations per second. DC devices can operate at speeds of 500, 1000, or more operations per second.

1.50 CHAPTER ONE

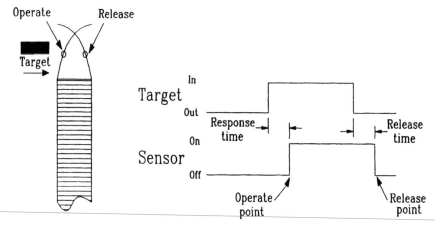

FIGURE 1.90 Response/release times.

In order to properly achieve high-speed operation, there are some basic principles that need be applied.

Maximum Target Length. There is a response delay when a sensor has a target entering the sensing field, as previously stated. There is a similar delay for the respective load to operate. The time from when the sensor conducts and the load operates is the *load response time*. Together these delays make up the *system response time* T_o.

Similarly, there are delays when the target reaches the release point in the sensing field caused by the sensor release time and the corresponding *load release time*. In order to ensure that the sensor will operate reliably and repeatedly, the target must stay in the field long enough to allow the load to respond. This is called *dwell time*. Figure 1.91 illustrates the time functions for reliable, repeatable sensor operation. Figure 1.92 illustrates the dwell range.

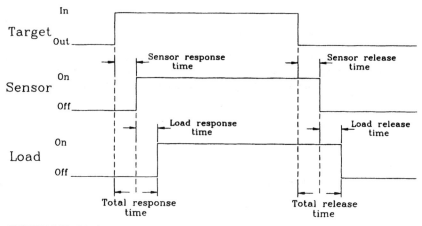

FIGURE 1.91 Maximum target length.

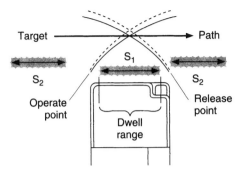

FIGURE 1.92 Dwell range.

Target Duty Cycle. Response (turn-on) times for the sensor and the controlled load may be considerably different from the release (turn-off) times for the same devices. Conditions for the target duty cycles are illustrated by Fig. 1.93. Note that the target is not out of the sensing field long enough to allow the sensor to turn off the controlled load. The application must be arranged so that both sensor and load turn ON and OFF reliably and repeatedly.

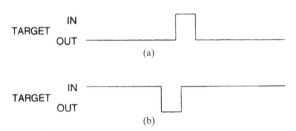

FIGURE 1.93 Target duty cycle. (*a*) Critical response time (turn-on). (*b*) Critical release time (turn-off).

Timing Functions. When an inductive control is operating a logic function, an output is generated for the length of time an object is detected (Fig. 1.94).

ON Delay Logic. ON delay logic allows the output signal to turn on only after the object has been detected for a predetermined period of time. The output will turn off immediately after the object is no longer detected. This logic is useful if a sensor

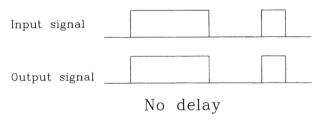

FIGURE 1.94 No delay.

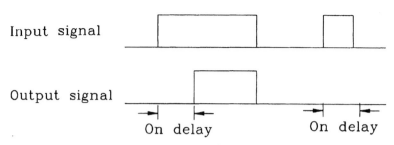

FIGURE 1.95 ON delay.

must avoid false interruption from a small object. ON delay is useful in bin fill or jam detection, since it will not false-trigger in the normal flow of objects going past (Fig. 1.95).

OFF Delay Logic. OFF delay logic holds the output on for a predetermined period of time after an object is no longer detected. The output is turned on as soon as the object is detected. OFF delay ensures that the output will not drop out despite a short period of signal loss. If an object is once again detected before the output times out, the signal will remain ON. OFF delay logic is useful in applications susceptible to periodic signal loss (Fig. 1.96).

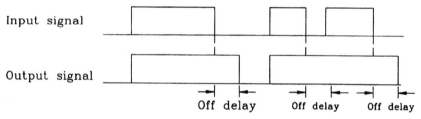

FIGURE 1.96 OFF delay.

ON/OFF Delay Logic. ON/OFF delay logic combines ON and OFF delay so that the output will be generated only after the object has been detected for a predetermined period of time, and will drop out only after the object is no longer detected for a predetermined period of time. Combining ON and OFF delay smoothes the output of the inductive proximity control (Fig. 1.97).

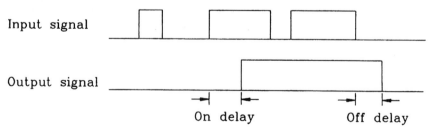

FIGURE 1.97 ON/OFF delay.

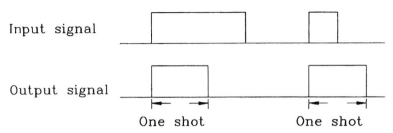

FIGURE 1.98 One-shot.

One-Shot Delay. One-shot logic generates an output of predetermined length no matter how long an object is detected. A standard one-shot must time out before it can be retriggered. One-shot logic is useful in applications that require an output of specified length (Fig. 1.98).

UNDERSTANDING MICROWAVE SENSING APPLICATIONS

Microwave sensors are valuable tools in the industrial environment for measuring motion, velocity, direction of movement, and range. They are rugged devices capable of operating in hostile environments. They are intrinsically safe, since they have no moving parts, and require low power. They will not cause any harm to operators and function effectively in explosive environments. They can successfully measure large military and industrial objects over large distances and can provide a great deal of information about the objects, as observed during the Persian Gulf War in 1991.

Microwave technology has long been an effective method of measuring the parameters of motion and presence. Applications range from simple intrusion alarms which merely indicate that an object has entered its field of view to complex military radar systems which define the existence, location, and direction of motion.

Microwave sensing technology can be classified into five categories:

1. *Motion sensing.* Sensing a moving object in a defined domain, for example, detecting an intruder in a prohibited area.
2. *Presence sensing.* Sensing that an object exists in a defined domain at a given time. This concept is vital in industrial control systems where the arrival of an object may not be noticed.
3. *Velocity sensing.* Sensing the linear speed of an object in a specified direction. This concept is used by police to detect speeding cars.
4. *Direction-of-motion sensing.* Determining whether a target is moving away from or toward the microwave sensor device. This concept is particularly important for manufacturers of automated guided vehicle systems for obstacle avoidance. It is also used to detect whether objects or personnel are approaching or departing from automatic doors.
5. *Range sensing.* Measuring the distance from the sensor to an object of interest. Applications include sensing the level of oil or chemical solutions in tanks and containers.

Characteristics of Microwave Sensors

Microwave sensor general characteristics important in industrial and commercial applications are

1. *No contact.* Microwave sensors operate without actually contacting the object. This is particularly important if the object is in a hostile environment or sensitive to wear. They can monitor the speed of power-plant generator shafts, continuously monitoring acceleration and deceleration in order to maintain a constant rotational speed. Microwave sensors can effectively penetrate nonmetallic surfaces, such as fiberglass tanks, to detect liquid levels. They can also detect objects in packaged cartons.
2. *Rugged.* Microwave sensors have no moving parts and have proven their reliability in extensive military use. They are packaged in sealed industrial enclosures to endure the rigors of the production environment.
3. *Environmental reliability.* Microwave sensors operate reliably in harsh, inhospitable environments. They can operate from −55°C to +125°C in dusty, dirty, gusty, polluted, and poisonous areas.
4. *Intrinsically safe.* Industrial microwave sensors can be operated in an explosive atmosphere because they do not generate sparks due to friction or electrostatic discharge. Microwave energy is so low that it presents no concern about hazard in industrial applications.
5. *Long range.* Microwave sensors are capable of detecting objects at distances of 25 to 45,000 mm or greater, depending on the target size, microwave power available, and the antenna design.
6. *Size of microwave sensors.* Microwave sensors are larger than inductive, capacitive, and limit switch sensors. However, use of higher microwave frequencies and advances in microwave circuit development allow the overall package to be significantly smaller and less costly.
7. *Target size.* Microwave sensors are better suited to detect large objects than smaller ones such as a single grain of sand.

Principles of Operation

Microwave sensors consist of three major parts: (1) transmission source, (2) focusing antenna, and (3) signal processing receiver.

Usually the transmission and receiver are combined together in one module, which is called a *transceiver*. A typical module of this type is used by intrusion alarm manufacturers for an indoor alarm system. The transceiver contains a Gunn diode mounted in a small precession cavity which, upon application of power, oscillates at microwave frequencies. A special cavity design will cause this oscillation to occur at 10.525 GHz, which is one of the few frequencies that the U.S. Federal Communications Commission (FCC) has set aside for motion detectors. Some of this energy is coupled through an iris into an adjoining waveguide. Power output is in the 10- to 20-mW range. The DC input power for this stage (8 V at 150 mA) should be well-regulated, since the oscillator is voltage-sensitive. The sensitivity of the system can be significantly reduced by noise (interference).

At the end of the waveguide assembly, a flange is fastened to the antenna. The antenna focuses the microwave energy into a beam, the characteristics of which are determined by the application. Antennas are specified by beam width or gain. The higher the gain, the longer the range and the narrower the beam. An intrusion alarm protecting a certain domain would require a wide-beam antenna to cover the area, while a traffic control microwave sensor would require a narrow-beam, high-gain antenna to focus down the road.

Regardless of the antenna selection, when the beam of microwave energy strikes an object, some of the microwave energy is reflected back to the module. The amount of energy will depend on the composition and shape of the target. Metallic surfaces will reflect a great deal, while styrofoam and plastic will be virtually transparent. A large target area will also reflect more than a small one.

The reflected power measured at the receiver decreases by the fourth power of the distance to the target. This relationship must be taken into consideration when choosing the transmitted power, antenna gain, and signal processing circuitry for a specific application.

When the reflected energy returns to the transceiver, the mixer diode will combine it with a portion of the transmitted signal. If the target is moving toward or away from the module, the phase relationships of these two signals will change and the signal out of the mixer will be an audio frequency proportional to the speed of the target. This is called the *Doppler frequency*. This is of primary concern in measuring velocity and direction of motion. If the target is moving across in front of the module, there will not be a Doppler frequency, but there will be sufficient change in the mixer output to allow the signal processing circuitry to detect it as unqualified motion in the field.

The signal from the mixer will be in the microvolt to millivolt range so that amplification will be needed to provide a useful level. This amplification should also include 60-Hz and 120-Hz notch filters to eliminate interference from power lines and fluorescent light fixtures, respectively. The remaining bandwidth should be tailored to the application.

Besides amplification, a comparator and output circuitry relays are added to suit the application (Fig. 1.99).

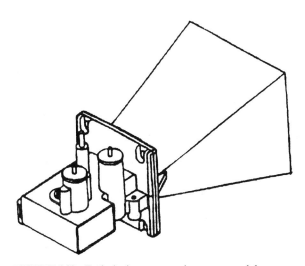

FIGURE 1.99 Typical microwave motion sensor module.

Detecting Motion with Microwave Sensors

The presence of an object in the microwave field disturbs the radiated field. There may be a Doppler frequency associated with the disturbance. The signal from the mixer to the signal processing circuitry may vary with a large amplitude and long duration so that it can be detected. The amplitude gain and the delay period are of specific importance in tailoring the device for particular application, such as motion detection. These sensors are primarily used in intrusion alarm applications where it is only necessary to detect the movement rather than derive further information about the intruder. The sensitivity would be set to the minimum necessary level to detect a person-sized object moving in the domain to be protected to prevent pets or other nonhostile moving objects from causing a false alarm. In addition, some response delay would be introduced for the same reason, requiring continuous movement for some short period of time.

Other applications include parts counting on conveyer belts; serial object counting in general; mold ejection monitoring, particularly in hostile environments; obstacle avoidance in automated guided vehicle systems; fill indication in tanks; and invisible protection screens. In general, this type of sensor is useful where the objects to be sensed are moving in the field of interest (Fig. 1.100).

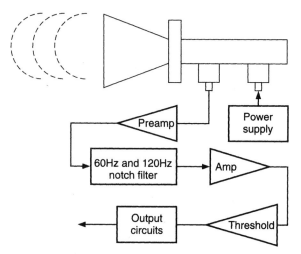

FIGURE 1.100 Microwave motion sensor.

Other devices which compete for the same applications are ultrasonic, photoelectric, and infrared sensors.

In the intrusion alarm manufacturing industry, microwave sensors have the advantages of longer range and insensitivity to certain environmental conditions. Ultrasonic sensors are sensitive to drafts and high-frequency ambient noise caused by bells and steam escaping from radiators. Infrared sensors are sensitive to thermal gradients caused by lights turning on and off. The effectiveness of infrared sensors is severely reduced at high ambient temperatures. However, utilizing dual technologies is recommended to minimize false alarms—combining microwave technology with infrared technology, for example. It is necessary for the intruder to be sensed by both technologies before an alarm is given. In other applications, microwave sensors can show advantages over photoelectric sensors in the areas of longer range, increased area of coverage, operation in hostile environments, and in applications where it is necessary to see through one medium (such as a cardboard box or the side of a nonmetallic tank) to sense the object on the other side.

If the target is moving toward or away from the transceiver there will be an audio-frequency (Doppler) signal out of the mixer diode which is proportional to the velocity of the target. The frequency of this signal is given by the formula:

$$F_d = 2V\,(F_t/c)$$

where F_d = Doppler frequency
V = velocity of the target
F_t = transmitted microwave frequency
c = speed of light

If the transmitted frequency is 10.525 GHz (the motion detector frequency), this equation simplifies to:

$$F_d = 31{,}366 \text{ Hz} \times V \quad \text{in miles/hour}$$

or

$$F_d = 19.490 \text{ kHz} \times V \quad \text{in kilometers/hour}$$

or

$$F_d = 84.313 \text{ kHz} \times V \quad \text{in furlongs/fortnight}$$

This assumes that the target is traveling directly at or away from the transceiver. If there is an angle involved, then the equation becomes

$$F_d = 2V(F_t/c)\cos\Theta$$

where Θ is the angle between the transceiver and the line of movement of the target. Evidently, as the target is moving across the face of the transceiver, $\cos\Theta = 0$, and the frequency is 0. If the angle is kept below 18°, however, the measured frequency will be within 5 percent of the center frequency (Fig. 1.101).

Signal processing for this module must include amplification, a comparison network to shape the signal into logic levels, and a timing and counting circuit to either drive a display device or compare the frequency to certain limits. If more than one moving object is in the microwave field, it may be necessary to discriminate on the basis of amplitude or frequency bandwidth, limiting to exclude unwanted frequencies. Velocities near 3 km/h and 6 km/h are also difficult to measure with this system

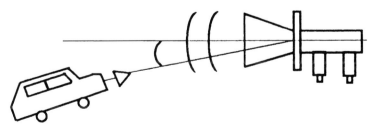

FIGURE 1.101 Angular velocity measurement.

since the corresponding Doppler frequencies are 60 and 120 Hz, which are prime interference frequencies from power lines and fluorescent fixtures. Extra shielding or isolation will be necessary in this case. False alarm rate may also be reduced by counting a specific number of cycles before triggering an output. This will actually correspond to the target moving a defined distance.

Microwave sensors are well-suited for measuring velocity of objects, which most other sensors cannot do directly. Inductive, photoelectric, and other sensors can measure radial velocity. For example, inductive photoelectric sensors measure radial velocity when configured as a tachometer, and if the rotating element is configured as a trailing wheel, then linear velocity can be defined. Photoelectric sensors can also be set up with appropriate signal processing to measure the time that a moving object takes to break two consecutive beams. This restricts the measurement to a specific location. Multiple beams would be needed to measure velocity over a distance, whereas a single microwave sensor could accomplish the same result.

Aside from their use in police radars, microwave sensors can measure the speed of baseball pitches. There are many industrial applications for these sensors as well. Microwave sensors are an excellent means of closed-loop speed control of a relatively high-speed rotating shaft (3600 r/min). Other applications include autonomous-vehicle speed monitoring and independent safety monitoring equipment for heavy and high-speed machine tools. Also, a microwave sensor will detect an overvelocity condition (Fig. 1.102).

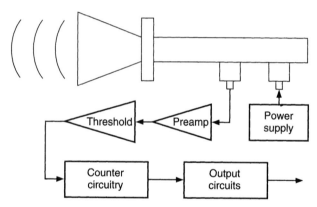

FIGURE 1.102 Velocity sensing module.

A microwave sensor, mounted on a tractor or other farm equipment to measure ground speed, will play an important role in reducing excessive distribution of seeds and fertilizer per acre. The ordinary wheel driven-speedometer is not sufficiently accurate because of wheel slippage. Accurate speed measurement is necessary in these vehicles, so that seeds and fertilizer are spread at a specific rate by the accessory equipment; an over- or underestimate of the speed will result in the wrong density per acre.

Detecting Presence with Microwave Sensors

A static object can be detected in the field of a microwave sensor. The purpose of this detection is to determine that the object is still in the field of interest and has not departed. This is particularly desirous in control systems where the controller is performing other tasks and then accesses the sensor to determine whether there is a sensed object at that particular time. In this situation, presence sensing is especially advantageous since the output can be verified by further interrogations to eliminate false sensing.

To detect the presence of an object, a microwave sensor with separate transmitter and receiver must be used. A transceiver in this application is not adequate, although the transmitter and the receiver can be mounted in the same enclosure. The receiver must not sense any energy unless the object is present in the field. A means to modulate the transmitter is needed, and the receiver should be narrow-band to amplify and detect the modulated reflection. The sensitivity of the receiver must be adjustable to allow for ambient reflections.

Microwave sensors have been extensively and successfully tested at various fast-food drive-through vending locations. Other types of sensors such as ultrasonic and photoelectric sensors, were also tested, less successfully. They were sensitive to the environment. It was discovered that frost heaving of the ground would eventually cause their buried loop to fail, and the cost of underground excavation to replace the loop was exorbitant.

Another application of the microwave sensor is the door-opening market. The microwave sensor will check for safety reasons the area behind a swinging door to detect whether there is an individual or an object in the path way. Ultrasonic sensors may perform the same task, yet range and environmental conditions often make a microwave sensor more desirable.

A microwave sensor can check boxes to verify that objects actually have been packed therein. The sensor has the ability to see through the box itself and triggers only if an object is contained in the box. This technology relies on the sensed object being more reflective than the package, a condition that is often met.

Measuring Velocity with Microwave Sensors

Microwave sensors are ideally suited to measuring linear velocity. Police radar is a simple example of a Doppler-frequency-based velocity sensor. This technology can be applied wherever it is necessary to determine velocity in a noncontact manner.

Detecting Direction of Motion with Microwave Sensors

Direction of motion—whether a target is moving toward or away from the microwave sensor—can be determined by the use of the Doppler-frequency concept (Fig. 1.103) by adding an extra mixer diode to the module. A discriminating effect is generated by the additional diode, which is located in the waveguide such that the Doppler outputs from the two mixers differ in phase by one-quarter wavelength, or 90°. These outputs will be separately amplified and converted into logic levels. The resulting signals can then be fed into a digital phase-discrimination circuit to determine the direction of motion. Such circuits are commonly found in motion control applications in conjunction with optical encoders. Figure 1.104 shows the phase relationships of the different directions.

Outputs from this module can vary widely to suit the application. The simplest is two outputs, one for motion and the other for direction (toward or away). These outputs can be added to a third, which provides the velocity of the target. The combination of signals could be analyzed to provide a final output when specific amplitude, direction, distance, and velocity criteria are met (Fig. 1.103).

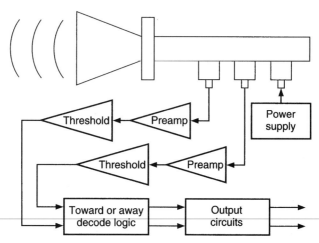

FIGURE 1.103 Direction of motion sensor schematic.

In the door-opening field, using the amplitude, direction, distance, and velocity information reduces the number of false openings. This extends the life of the door mechanism, besides saving heat if the door is an entrance to a heated building.

In this case, the measurements by circuitry indicate the following characteristics:

Characteristic	Measurement
Person-sized object	Amplitude of return
Moving at walking pace	Velocity
Toward or away	Direction
Specific time before opening	Distance

Detecting Range with Microwave Sensors

An early-warning military radar system depends on costly microwave sensors. A small yacht may use a microwave sensor selling for less than $1000 to detect targets at ranges up to 5 mi.

Regardless of their cost, microwave range sensors for commercial, industrial, and military applications employ essentially the same measuring technique. They transmit a narrow pulse of energy and measure the time required for the return from the target. Since microwave energy propagates at the speed of light, the time for the pulse to reach the target and return is 2 ns per foot of range. If the range to the target is 1 mi, the time required is 10.56 µs.

Although the microwave power needed is sufficient to raise the sensor temperature to 500°F, the design of the signal processing circuitry to measure the response is not difficult. However, if the target is very close to the transmitter, then the short response time may pose a real problem. At 3 ft, the time response is 6 ns. For 1-in resolution, the circuitry must be able to resolve 167 ps. This may pose a significant problem.

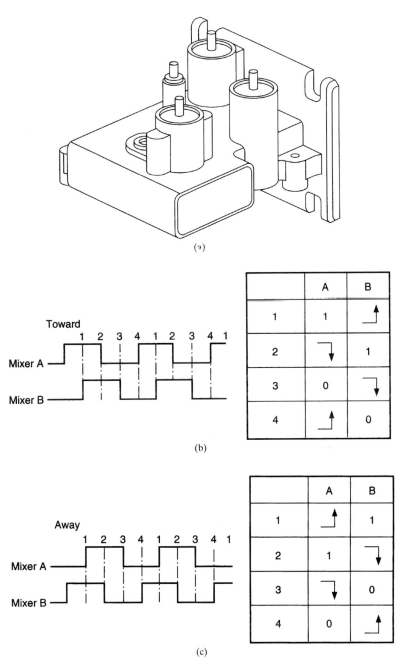

FIGURE 1.104 (*a*) Direction of motion sensor device. (*b*) Motion logic "away." (*c*) Direction logic "toward."

The alternative method to resolve a target at a short range involves changing the frequency of continuous oscillator. This method is better-suited to industrial applications. An oscillator starting at 10.525 GHz and sweeping at 50 MHz in 10 ms in the 6 ns mentioned above will have changed its frequency by:

$$(6 \text{ ns} \times 50 \text{ MHz}/0.01 \text{ s}) = 30 \text{ Hz}$$

The returning wave will be still at 10.525 GHz. The output from the mixer diode as the sweep continues will be the 30-Hz difference. If this frequency is averaged over time, it is not difficult to resolve a range to 0.001 in.

The above calculation indicates that the frequency is high for faraway objects and low for targets that are close. This leads to two conclusions:

1. The closer the object is, the lower the frequency, and therefore the longer the measurement will take.
2. The signal processing amplifier should have a gain which increases with frequency.

Problems can arise when the target is moving or there are multiple targets in the area of interest. Movement can be detected by comparing consecutive readings and can be used as a discrimination technique. Multiple targets can be defined by narrow-beam antennas to reduce the width of the area of interest. Gain adjustments are also required to eliminate all but the largest target. Audio-bandwidth filters may be used to divide the range into sectors for greater accuracy.

Other sensor types, such as photoelectric and inductive sensors, may be utilized to measure distances. Inductive sensors are used in tank level measurements. They must be coupled with a moving component which floats on the surface of the substance to be measured. Photoelectric sensors measure position by focusing a beam on a point in space and measuring the reflection on a linear array. This can give very precise measurements over a limited range but is subject to adverse environmental conditions. Photoelectric sensors can focus a camera on a target for a better picture. Ultrasonic sensors may perform the same function, but their range is limited and they can be defeated by a hostile environment.

Microwave sensors for measuring range have an impressive array of applications, including measurement of the level of liquid or solid in a tank, sophisticated intrusion alarms, autonomous guided vehicle industrial systems, and noncontact limit switching. In tank level sensing in the chemical industry (Fig. 1.105) the microwave sensor is mounted at the top of the tank and measures the distance from that position to the surface of the contents. Since the electronic circuitries can be isolated from the tank contents by a sealed window, it is intrinsically safe. It has the advantage of being a noncontact system, which means that there are no moving parts to break or be cleaned. This allows the microwave sensor to be used on aggressive chemicals, liquids, liquefied gases, highly viscous substances, and solids such as grain and coal.

Microwave Technology Advancement

Advances in technology have opened up important new applications for microwave sensors. The expected governmental permission to utilize higher frequencies and the decreasing size of signal processing circuitry will significantly reduce the cost of the sensors and will enable them to detect even smaller targets at a higher

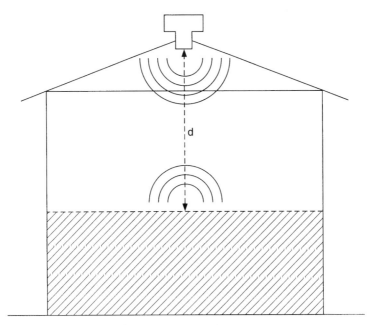

FIGURE 1.105 Tank level sensor.

resolution. Microwave integrated circuit technology (MICT), presently developed for the Military Microwave Integrated Circuit (MIMIC) program will overflow into the industrial market, causing increases in performance and in analysis capabilities. Consequently, the utilization of computer-integrated manufacturing technology will be broadened.

UNDERSTANDING LASER SENSORS

Two theories about the nature of light have been recognized. The particle theory was the first presented to explain the phenomena that were observed concerning light. According to this theory, light is a particle with mass, producing reflected beams. It was believed that light sources actually generated large quantities of these particles. Through the years, however, many phenomena of light could not be explained by the particle theory, such as reflection of light as it passes through optically transparent materials.

The second theory considered that light was a wave, traveling with characteristics similar to those of water waves. Many, but not all, phenomena of light can be explained by this theory.

A dual theory of light has been proposed, and is presently considered to be the true explanation of light propagation. This theory suggests that light travels in small packets of wave energy called *photons*. Even though photons are *bundles* of wave energy, they have momentum like particles of mass. Thus, light is wave energy traveling with some of the characteristics of a moving particle. The total transmitted as light is the sum of energies of all the individual photons emitted.

Velocity, frequency, and wavelength are related by the equation:

$$c = f\lambda$$

where c = velocity of light, km/s
 f = frequency, Hz
 λ = wavelength, m

This equation shows that the frequency of a wave is inversely proportional to the wavelength; that is, higher-frequency waves have shorter wavelengths.

Properties of Laser Light

Laser stands for *light amplification by stimulated emission of radiation.* Laser light is monochromatic, whereas standard white light consists of all the colors in the spectrum and is broken into its component colors when it passes through a standard glass prism (Figs. 1.106, 1.107).

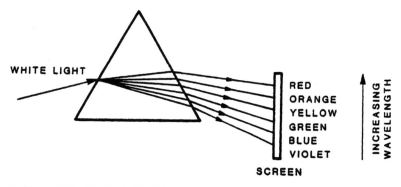

FIGURE 1.106   Standard white light.

Essential Laser Components

Laser systems consist of four essential components:

1. The active medium
2. The excitation mechanism
3. The feedback mechanism
4. The output coupler

The Active Medium.   The active medium is the collection of atoms, ions, or molecules in which stimulated emission occurs. It is in this medium that laser light is produced. The active medium can be a solid, liquid, gas, or semiconductor material. Often the laser takes its name from that of the active medium. For example, the ruby laser has a crystal of ruby as its active medium while the CO_2 laser has carbon dioxide gas.

The wavelength emitted by a laser is a function of the active medium. This is because the atoms within the active medium have their own characteristic energy levels at which they release photons. It will be shown later that only certain energy

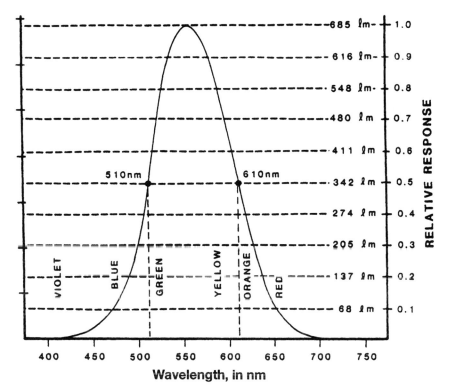

FIGURE 1.107 Spectrum of standard white light.

levels within the atom can be used to enhance stimulated emission. Therefore a given active medium can produce a limited number of laser wavelengths, and two different active media cannot produce the same wavelengths. Table 1.10 contains a list of materials commonly used in lasers and the corresponding wavelengths that these materials produce.

The active medium is the substance that actually lases. In the helium-neon laser, only the helium lases.

Excitation Mechanism. The excitation mechanism is the device used to put energy into the active medium. There are three primary types of excitation mechanisms: optical, electrical, and chemical. All three provide the energy necessary to raise the energy state of the atom, ion, or molecule of the active medium to an excited state. The process of imparting energy to the active medium is called *pumping the laser.*

Optical Excitation. An optical excitation mechanism uses light energy of the proper wavelength to excite the active medium. The light may come from any of several sources, including a flash lamp, a continuous arc lamp, another laser, or even the sun. Although most of these use an electric power supply to produce the light, it is not the electrical energy that is used directly to excite the atoms of the active medium but rather the light energy produced by the excitation mechanism.

Optical excitation is generally used with active media that do not conduct electricity—solid lasers like the ruby. Fig. 1.108 is a schematic drawing of a solid laser with an optical pumping source.

TABLE 1.10 Wavelengths of Laser Materials

Type of active medium	Common material	Wavelength produced, nm
Solid	Ruby	694
	Nd:YAG	1,060
	Nd:glass	1,060
	Erbium	1,612
Liquid	Organic dyes	360–650
Gas	Argon (ionized)	488
	Helium-neon	632.8
	Krypton (ionized)	647
	CO_2	10,600
Semiconductor	Gallium arsenide	850
	Gallium antimonide	1,600
	Indium arsenide	3,200

The sun is considered a possible optical pumping source for lasers in space. The optical energy from the sun could be focused by curved mirrors onto the laser's active medium. Since the size and weight of an electric power supply is of concern in space travel, solar pumping of lasers is an interesting alternative.

Electrical Excitation. Electrical excitation is most commonly used when the active medium will support an electric current. This is usually the case with gases and semiconductor materials.

When a high voltage is applied to a gas, current-carrying electrons or ions move through the active medium. As they collide with the atoms, ions, or molecules of the active medium, their energy is transferred and excitation occurs. The atoms, ions, and electrons within the active medium are called *plasma*.

Figure 1.109 is a schematic drawing of a gas laser system with electrical excitation. The gas mixture is held in a gas plasma tube and the power supply is connected to the ends of the plasma tube. When the power supply is turned on, electron movement within the tube is from the negative to the positive terminal.

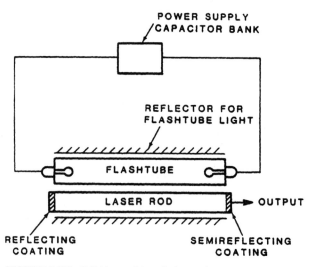

FIGURE 1.108 Solid laser with optical pumping source.

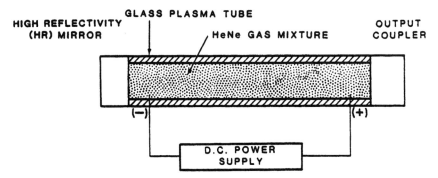

FIGURE 1.109 Gas laser with electrical excitation.

Chemical Excitation. Chemical excitation is used in a number of lasers. When certain chemicals are mixed, energy is released as chemical bonds are made or broken. This energy can be used as a pumping source. It is most commonly used in hydrogen-fluoride lasers, which are extremely high-powered devices used primarily in military weapons and research. These lasers are attractive for military applications because of the large power-to-weight ratio.

Feedback Mechanism. Mirrors at each end of the active medium are used as a *feedback mechanism.* The mirrors reflect the light produced in the active medium back into the medium along its longitudinal axis. When the mirrors are aligned parallel to each other, they form a resonant cavity for the light waves produced within the laser. They reflect the light waves back and forth through the active medium.

In order to keep stimulated emission at a maximum, light must be kept within the amplifying medium for the greatest possible distance. In effect, mirrors increase the distance traveled by the light through the active medium. The path that the light takes through the active medium is determined by the shape of the mirrors. Figure 1.110 shows some of the possible mirror combinations. Curved mirrors are often used to alter the direction in which the reflected light moves.

Output Coupler. The feedback mechanism keeps the light inside the laser cavity. In order to produce an output beam, a portion of the light in the cavity must be allowed to escape. However, this escape must be controlled. This is most commonly accomplished by using a partially reflective mirror in the feedback mechanism. The amount of reflectance varies with the type of laser. A high-power laser may reflect as little as 35 percent, with the remaining 65 percent being transmitted through the mirror to become the output laser beam. A low-power laser may require an output mirror reflectivity as high as 98 percent, leaving only 2 percent to be transmitted. The output mirror that is designed to transmit a given percentage of the laser light in the cavity between the feedback mirrors is called the *output coupler.*

Semiconductor Displacement Laser Sensors

Semiconductor displacement laser sensors, consisting of a light-metering element and a position-sensitive detector (PSD), detect targets by using triangulation. A light-emitting diode or semiconductor laser is used as the light source. A semiconductor laser beam is focused on the target by the lens. The target reflects the beam, which is then focused on the PSD, forming a beam spot. The beam spot moves on the

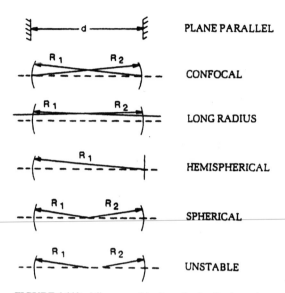

FIGURE 1.110 Mirror combinations for feedback mechanism.

PSD as the target moves. The displacement of the workpiece can then be determined by detecting the movement of the beam spot.

Industrial Applications of Semiconductor Displacement Lasers. The laser beam emitted from laser diode in the transmitter is converged into a parallel beam by the lens unit. The laser beam is then directed through the slit on the receiver and focused on the light-receiving element. As the target moves through the parallel laser beam, the change in the size of the shadow is translated into the change in received light quantity (voltage). The resulting voltage is used as a comparator to generate an analog output voltage.

Industrial Applications of Laser Sensors*

Electrical and electronics industries:

1. *Warpage and pitch of IC leads.* The visible beam spot facilitates the positioning of the sensor head for small workpieces. Warpage and pitch can be measured by scanning IC leads with the sensor head (Fig. 1.111).
2. *Measurement of lead pitch of electronic components.* The sensor performs precise noncontact measurement of pitch using a laser beam (Fig. 1.112).
3. *Measurement of disk head movement.* The laser sensor is connected to a computer in order to compare the pulse input to the disk head drive unit with actual movement. The measurement is done on-line, thus increasing productivity (Fig. 1.113).

---

* A few nonlaser optical sensors are included, as indicated.

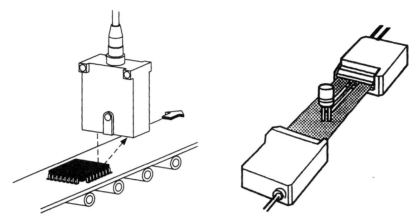

FIGURE 1.111 Warpage and pitch of IC lead.

FIGURE 1.112 Measurement of lead pitch of electronic components.

4. *Detection of presence/absence of resin coating.* The laser displacement sensor determines whether a resin coating was formed after wire bonding (Fig. 1.114).
5. *Detection of double-fed or mispositioned resistors prior to taping.* Through-beam-type sensor heads are positioned above and below the resistors traveling on a transfer line. A variation on the line changes the quantity of light in the laser beam, thus signaling a defect (Fig. 1.115).
6. *Detection of defective shrink wrapping of videocassette.* Defective film may wrap or tear during shrink wrapping. The laser sensor detects defective wrapping by detecting a change in the light quantity on the surface of the videocassette (Fig. 1.116).
7. *Measurement of gap between roller and doctor blade.* Measures the gap between the roller and the doctor blade in submicrometer units. The sensor's automatic measurement operation eliminates reading errors (Fig. 1.117).

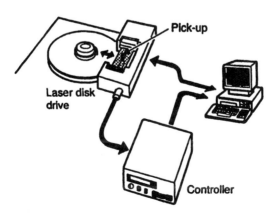

FIGURE 1.113 Measurement of disk head movement.

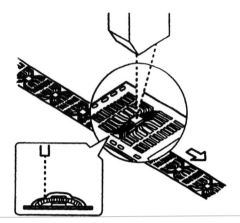

FIGURE 1.114 Detection of presence/absence of resin coating.

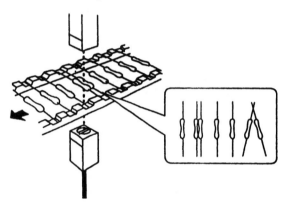

FIGURE 1.115 Detection of double-fed or mispositioned resistors.

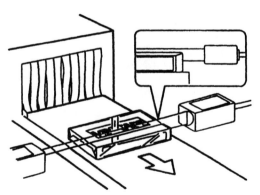

FIGURE 1.116 Detection of defective shrink wrapping of videocassette.

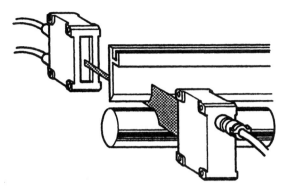

FIGURE 1.117 Measurement of gap between roller and doctor blade.

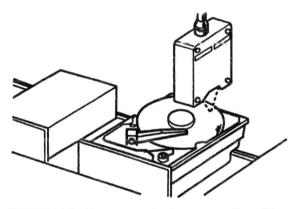

FIGURE 1.118 Measurement of surface run-out of laser disk.

8. *Measurement of surface run-out of laser disk.* The surface run-out of a laser disk is measured at a precision of 0.5 μm. The sensor head enables measurement on a mirror-surface object (Fig. 1.118).

9. *Displacement of printer impact pins.* The visible beam spot facilitates positioning of the head of a pin-shaped workpiece, enabling measurement of the vertical displacement of impact pins (Fig. 1.119).

Automotive manufacturing industries:

1. *Measurement of thickness of connecting rod.* Measures the thickness of the connecting rod by processing the analog inputs in the digital meter relay (Fig. 1.120).
2. *Measurement of depth of valve recesses in piston head.* Measures the depth of the valve recesses in the piston head so that chamber capacity can be measured. Iron jigs are mounted in front of the sensor head, and the sensor measures the distance the jigs travel when they are pressed onto the piston head (Fig. 1.121).*

---

* Nonlaser sensor.

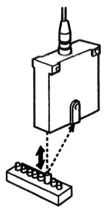

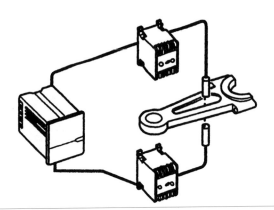

FIGURE 1.119 Displacement of printer impact pins.

FIGURE 1.120 Measurement of thickness of connecting rod.

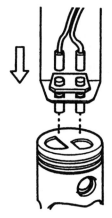

FIGURE 1.121 Measurement of depth of valve recesses in piston head.

3. *Measurement of height of radiator fin.* Detects improper radiator fin height by comparing the bottom value of the analog output with a stored pair of tolerances (Fig. 1.122).

4. *Measurement of outer diameter of engine valve.* The laser scan micrometer allows on-line measurement of the outer diameter of engine valves simply by positioning a separate sensor head on either side of the conveyer (Fig. 1.123).

5. *Positioning of robot arm.* The laser displacement sensor is used to maintain a specific distance between the robot arm and target. The sensor outputs a plus or minus voltage if the distance becomes greater or less, respectively, than the 100-mm reference distance (Fig. 1.124).

6. *Detection of damage on microdiameter tool.* Detects a break, chip, or excess swarf from the variation of light quantity received (Fig. 1.125).

Metal/steel/nonferrous industries:

1. *Detection of misfeeding in high-speed press.* The noncontact laser sensor, timed by a cam in the press, confirms the material feed by monitoring the pilot holes and outputs the result to an external digital meter relay (Fig. 1.126).

2. *Simultaneous measurement of outer diameter and eccentricity of ferrite core.* Simultaneously measures the outer diameter and eccentricity of a ferrite core with a single sensor system. The two measured values can then be simultaneously displayed on a single controller unit (Fig. 1.127).

INTRODUCTION

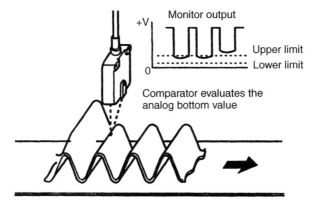

FIGURE 1.122 Measurement of height of radiator fin.

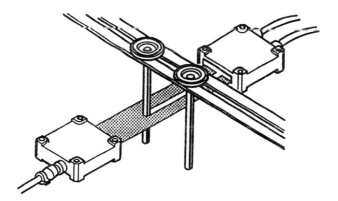

FIGURE 1.123 Measurement of outer diameter of engine valve.

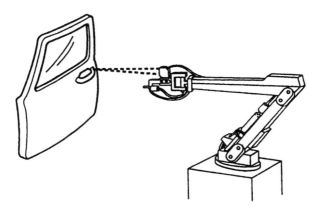

FIGURE 1.124 Positioning of robot arm.

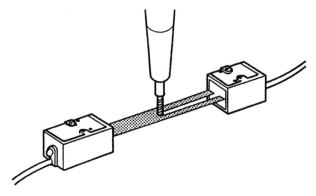

FIGURE 1.125 Detection of damage on microdiameter tool.

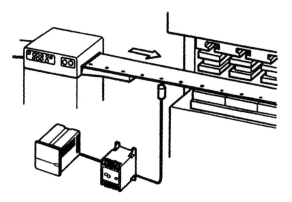

FIGURE 1.126 Detection of misfeeding in high-speed press.

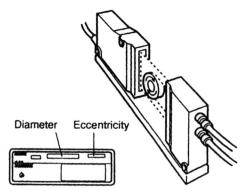

FIGURE 1.127 Simultaneous measurement of outer diameter and eccentricity of ferrite core.

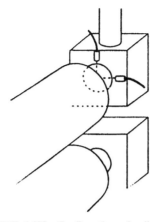

FIGURE 1.128 Confirmation of roller centering.

3. *Confirmation of roller centering.* The analog output of the inductive displacement sensor is displayed as a digital value, thus allowing a numerical reading of the shaft position (Fig. 1.128).
4. *Measurement of the height and inner diameter of sintered metal ring.* Determines the height and inner diameter of the metal ring by measuring the interrupted areas of the parallel laser beam (Fig. 1.129).
5. *Measurement of outer diameter of wire in two axes.* Simultaneously measures in x axis to determine the average value of the outer diameter, thereby increasing dimensional stability (Fig. 1.130).
6. *Measurement of outer diameter after centerless grinding.* The scanning head allows continuous noncontact measurement of a metal shaft immediately after the grinding process (Fig. 1.131).

Food processing and packaging:

1. *Detection of material caught during heat sealing.* Detects material caught in the distance between rollers (Fig. 1.132).
2. *Detection of missing or doubled packing ring in cap.* Detects a missing or doubled rubber ring in caps by using the comparator to evaluate sensor signals (Fig. 1.133).
3. *Detection of incorrectly positioned small objects.* The transmitter and the receiver are installed to allow a parallel light beam to scan slightly above the

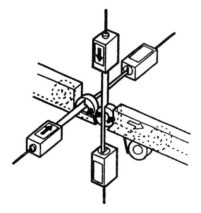

FIGURE 1.129 Measurement of the height and inner diameter of sintered metal ring.

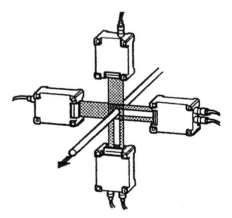

FIGURE 1.130 Measurement of outer diameter of wire in two axes.

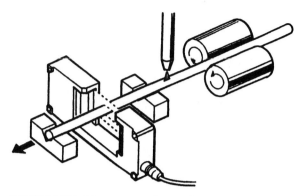

FIGURE 1.131 Measurement of outer diameter after centerless grinding.

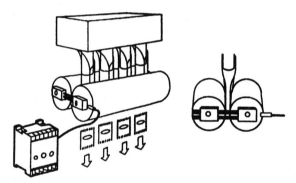

FIGURE 1.132 Detection of material caught during heat sealing.

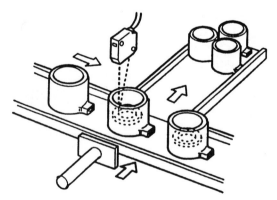

FIGURE 1.133 Detection of missing or doubled packing ring in cap.

tablet sheets. When a single tablet projects from a line of tablets, the optical axis is interrupted and the light quantity changes (Fig. 1.134).
4. *Measurement of tape width.* Measures the width of running tape to the submicrometer level; 100 percent inspection improves product quality (Fig. 1.135).
5. *Measurement of sheet thickness.* Use of two controllers enables thickness measurement by determining the distance in input values. Thus, thickness measurement is not affected by roller eccentricity (Fig. 1.136).

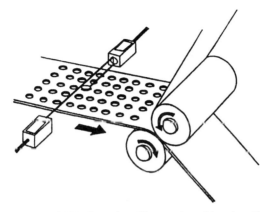

FIGURE 1.134 Detection of incorrectly positioned small objects.

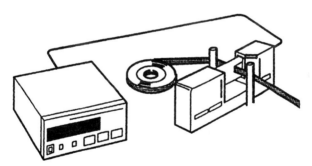

FIGURE 1.135 Measurement of tape width.

Automatic machinery:

1. *Detection of surface run-out caused by clamped error.* Improper clamping due to trapped chips will change the rotational speed of the workpiece. The multifractional digital meter relay sensor calculates the surface run-out of the rotating workpiece, compares it with the stored value, and outputs a detection signal (Fig. 1.137).
2. *Detection of residual resin in injection molding machine.* When the sensor heads are such that the optical axis covers the surface of the die, any residual resin will interface with this axis (Fig. 1.138).

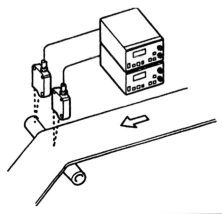

FIGURE 1.136 Measurement of sheet thickness.

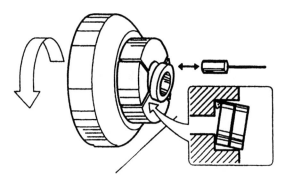

FIGURE 1.137 Detection of surface run-out caused by clamped error.

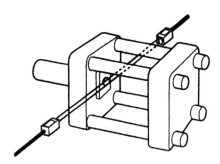

FIGURE 1.138 Detection of residual resin.

3. *Measurement of travel of camera lens.* A separate sensor can be installed without interfering with the camera body, thus assuring a highly reliable reading of lens travel (Fig. 1.139).

4. *Measurement of rubber sheet thickness.* With the segment function that allows the selection of measuring points, the thickness of a rubber sheet (i.e., the distance between the rollers) can be easily measured (Fig. 1.140).

5. *Measurement of stroke of precision table.* Detects even minute strokes

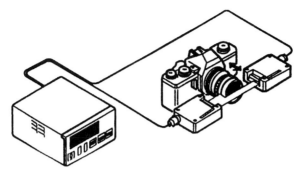

FIGURE 1.139 Measurement of travel of camera lens.

FIGURE 1.140 Measurement of rubber sheet thickness.

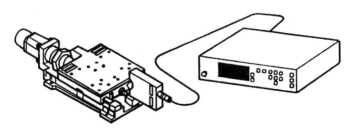

FIGURE 1.141 Measurement of stroke of precision table.

at a resolution of 0.05 µm. In addition, the AUTO ZERO function allows indication of relative movement (Fig. 1.141).

6. *Measurement of plasterboard thickness.* A sensor head is placed above and below the plasterboard, and its analog outputs are fed into a digital meter relay. The meter relay indicates the absolute thickness value (Fig. 1.142).

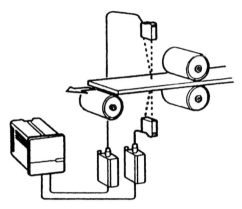

FIGURE 1.142 Measurement of plasterboard thickness.

REFERENCES

1. Chappel, A. (ed.), *Optoelectronics: Theory and Practice,* McGraw-Hill, New York, 1978.
2. Doebelin, E. O., *Measurement Systems: Application and Design,* 4th ed., McGraw-Hill, New York, 1990.
3. Holliday, D., and R. Resnick, *Physics,* Wiley, New York, 1975.
4. International Organization for Standardization, "Statistical Interpretation of Data: Comparison of Two Means in the Case of Paird Observations," ISO 3301-1975.
5. Lion, K. L., *Elements of Electrical and Electronic Instrumentation,* McGraw-Hill, New York, 1975.
6. Neubert, H. K. P., *Instrument Transducers,* 2d ed., Clarendon Press, Oxford, 1975.
7. Ogata, K., *Modern Control Engineering,* 2d ed., Prentice-Hall, Englewood Cliffs, N.J, 1990.
8. Rock, I., *Lightness Constancy, Perception,* W. H. Freeman, New York, 1984.
9. Seippel, R. G., *Optoelectronics,* Reston Publishing Co., Reston, Va., 1981.
10. Shortley, G., and D. Williams, *Quantum Property of Radiation,* Prentice-Hall, Englewood Cliffs, N.J., 1971.
11. Todd, C. D. (Bourns Inc.), *The Potentiometer Handbook,* McGraw-Hill, New York, 1975.
12. White, R. M., "A Sensor Classification Scheme," *IEEE Trans. Ultrasonics, Ferroelectrics, and Frequency Control,* March, 1987.

CHAPTER 2
FIBER OPTICS IN SENSORS AND CONTROL SYSTEMS

INTRODUCTION

Accurate position sensing is crucial to automated motion control systems in manufacturing. The most common components used for position sensing are photoelectric sensors, inductive proximity sensors, and limit switches. They offer a variety of options for manufacturing implementation from which highly accurate and reliable systems can be created. Each option has its features, strengths, and weaknesses that manufacturing personnel should understand for proper application. There are three types of sensors used in manufacturing applications:

1. *Photoelectric sensors.*   Long-distance detection
2. *Inductive proximity sensors.*   Noncontact metal detection
3. *Limit switches.*   Detection with traditional reliability

PHOTOELECTRIC SENSORS—LONG-DISTANCE DETECTION

A photoelectric sensor is a switch that is turned on and off by the presence or absence of receiving light (Fig. 2.1). The basic components of a photoelectric sensor are a power supply, a light source, a photodetector, and an output device. The key is the photodetector, which is made of silicon, a semiconductor material that conducts current in the presence of light. This property is used to control a variety of output devices vital for manufacturing operation and control, such as mechanical relays, triacs, and transistors, which in turn control machinery.

Early industrial photoelectric controls used focused light from incandescent bulbs to activate a cadmium sulfide photocell (Fig. 2.2). Since they were not modulated, ambient light such as that from arc welders, sunlight, or fluorescent light fixtures could easily false-trigger these devices. Also, the delicate filament in the incandescent bulbs had a relatively short life span, and did not hold up well under high vibration and the kind of shock loads normally found in an industrial environment. Switching speed was also limited by the slow response of the photocell to light/dark changes (Fig. 2.1).

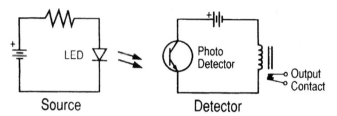

FIGURE 2.1 Photoelectric sensor.

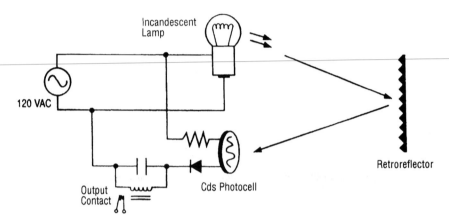

FIGURE 2.2 Early photoelectric control.

Light-Emitting Diodes

Photoelectric sensors use an effective light source, light-emitting diodes (LEDs), which were developed in the early 1960s. LEDs are solid-state devices that emit light when current is applied (Fig. 2.3). This is the exact opposite of the photodetector, which emits current when light is received.

LEDs have several advantages over incandescent bulbs and other light sources. LEDs can be turned on and off very rapidly, are extremely small, consume little

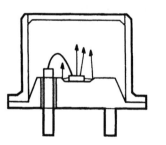

Light Emitting Diodes (LED's)
- Fast turn-on and turn-off
- No warm-up
- Small
- Rugged
- Low power consumption
- High radiant efficiency
- Long life

FIGURE 2.3 Light-emitting diode.

power, and last as long as 100,000 continuous hours. Also, since LEDs are solid-state devices, they are much more immune to vibration than incandescent bulbs.

LEDs emit light energy over a narrow wavelength (Fig. 2.4a). Infrared (IR) gallium arsenide LEDs emit energy only at 940 nm (Fig. 2.4b). As this wavelength is at the peak of a silicon photodiode's response, maximum energy transfer between source and detector is achieved.

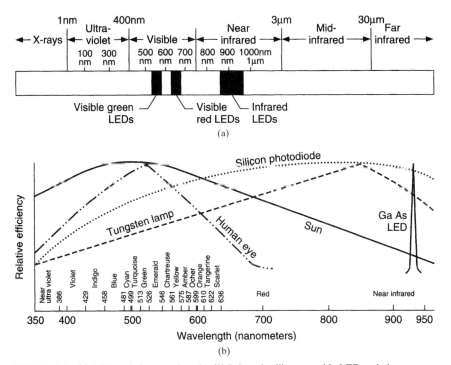

FIGURE 2.4 (a) LED emission wavelengths. (b) Infrared gallium arsenide LED emission.

A silicon photodetector's sensitivity to light energy also peaks in the infrared light spectrum. This contributes to the high efficiency and long range possible when silicon photodetectors are used in conjunction with gallium arsenide LEDs.

In recent years, visible LEDs have been introduced as light sources in photoelectric controls. Because the beam is visible to the naked eye, the principle advantage of visible LEDs is ease of alignment. Visible beam photoelectric controls usually have lower optical performance than IR LEDs.

The modes of detection for optical sensors are (1) through-beam and (2) reflection (diffuse reflection and reflex detection).

Through-Beam Sensors

Through-beam sensors have separate source and detector elements aligned opposite each other, with the beam of light crossing the path that an object must cross (Fig. 2.5).

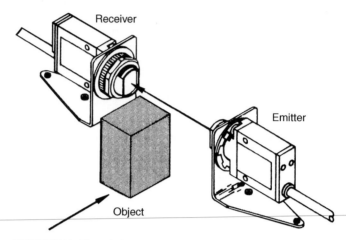

FIGURE 2.5 Through-beam sensor.

The effective beam area is that of the column of light that travels straight between the lenses (Fig. 2.6).

Because the light from the source is transmitted directly to the photodetector, through-beam sensors offer the following benefits:

1. Longest range for sensing
2. Highest possible signal strength
3. Greatest light/dark contrast ratio
4. Best trip point repeatability

The limitations of through-beam sensors are as follows:

1. They require wiring of the two components across the detection zone.
2. It may be difficult to align the source and the detector.

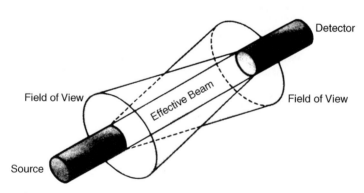

FIGURE 2.6 Effective beam area.

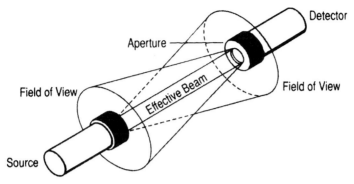

FIGURE 2.7 Sensor with aperture over lens for detecting small objects.

3. If the object to be detected is smaller than the effective beam diameter, an aperture over the lens may be required (Fig. 2.7).

Reflex Photoelectric Controls

Reflex photoelectric controls (Fig. 2.8) position the source and detector parallel to each other on the same side of the target. The light is directed to a retroreflector and returns to the detector. The switching and output occur when an object breaks the beam.

Since the light travels in two directions (hence twice the distance), reflex controls will not sense as far as through-beam sensors. However, reflex controls offer a powerful sensing system that is easy to mount and does not require that electrical wire be run on both sides of the sensing area. The main limitation of these sensors is that a shiny surface on the target object can trigger false detection.

Polarized Reflex Detection

Polarized reflection controls use a polarizing filter over the source and detector that conditions the light such that the photoelectric control sees only light returned from

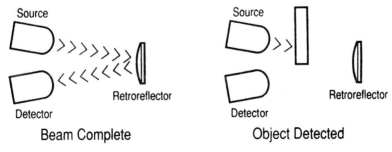

FIGURE 2.8 Reflex photoelectric controls.

the reflector (Fig. 2.9). A polarized reflex sensor is used in applications where shiny surfaces such as metal or shrink-wrapped boxes may false-trigger the control.

Polarized reflex sensing is achieved by combining some unique properties of polarizers and retroreflectors. These properties are (1) polarizers pass light that is aligned along only one plane and (2) corner-cube reflectors depolarize light as it travels through the face of the retroreflector (Fig. 2.10).

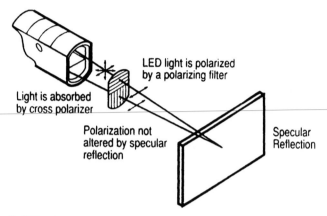

FIGURE 2.9 Polarization reflection controls.

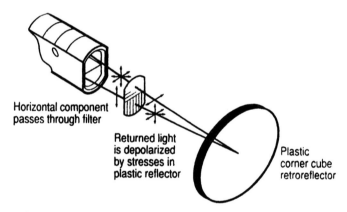

FIGURE 2.10 Corner-cube reflector.

Light from the source is aligned by a polarizer. When this light reflects off the retroreflector, it is depolarized. The returning light passes through another polarizing filter in front of the detector. The detector's polarizer is oriented at 90° to the source's polarizer. Only the light which has been rotated by the corner cube retroreflector can pass through the detector's polarizer. Light that bounces off other shiny objects, and has not been rotated 90°, cannot pass through the detector's polarizer, and will not trigger the control.

Polarized reflex sensors will not work with reflective tape containing glass beads. Also, shiny objects wrapped with clear plastic shrink-wrap will potentially false-trigger a polarized reflex control, since under certain conditions these act as a corner-cube reflector.

The polarized reflex detection sensor has the following advantages:

1. It is not confused by the first surface reflections from target objects.
2. It has a high dark/light contrast ratio.
3. It is easily installed and aligned. One side of the sensing zone only need be wired.

It also has certain limitations:

1. Operating range is half that of a nonpolarized sensor since much of the signal is lost in the polarizing filters.
2. The sensor can be fooled by shiny objects wrapped with shrink-wrap material.

Proximity (Diffuse-Reflection) Detection

Proximity detection is similar to reflex detection, because the light source and detector elements are mounted on the same side (Fig. 2.11). In this application, the sensors detect light that is bounced off the target object, rather than the breaking of the beam. The detection zone is controlled by the type, texture, and composition of the target object's surface.

Focused proximity sensors are a special type of proximity sensor where the source and the detector are focused to a point in front of the sensor (Fig. 2.12).

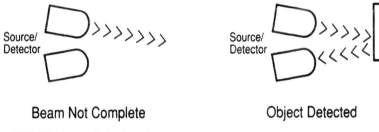

FIGURE 2.11 Proximity detection.

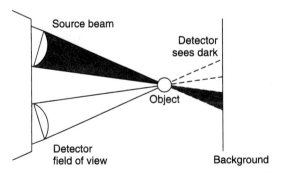

FIGURE 2.12 Focused proximity sensor.

Focused proximity sensors can detect extremely small objects, or look into holes or cavities in special applications. Background objects will not false-trigger a focused proximity sensor since they are "cross-eyed" and cannot see past a certain point.

Advantages of the focused proximity sensor are

1. Installation and alignment are simple. The control circuit can be wired through only one side of the sensing zone.
2. It can detect differences in surface reflectivity.

It also has certain limitations:

1. It has limited sensing range.
2. The light/dark contrast and sensing range depend on the target object's surface reflectivity.

Automated Guided Vehicle System

The wide sensing field of diffuse reflective photoelectric sensors makes them ideally suited for use as obstacle detection sensors. Sensing distance and field width are adjustable to form the obstacle detection zone vital for manufacturing operations.

Two outputs, *near* and *far*, provide switching at two separate sensing distances which are set by corresponding potentiometers. The sensing distance of the far output is adjustable up to 3 m maximum. The sensing distance of the near output is adjustable from 30 to 80 percent of the far output. Indicators include a red LED which glows with the near output ON and a yellow LED which glows with the far output ON.

An ideal application for this family of sensors is the automated guided vehicle (AGV), which requires both slow-down and stop controls to avoid collisions when obstacles enter its path (Fig. 2.13).

A modulated infrared light source provides immunity to random operation caused by ambient light. Additionally, unwanted sensor operation caused by adjacent sensor interference (crosstalk) is also eliminated through the use of multiple-position modulated frequency adjustments.

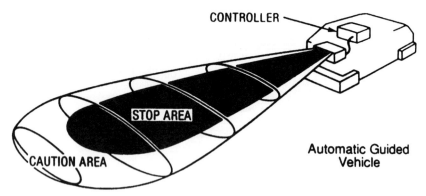

FIGURE 2.13 Application of diffuse reflective photoelectric sensor in automated guided vehicle system.

FIBER OPTICS

Fiber optics has greatly expanded the applications of photoelectric sensors. Fiber optics uses bundles of thin plastic or glass fibers that operate on a principle discovered in 1854 by John Tyndahl. When Tyndahl shined a beam of light through a stream of water, instead of emerging straight from the stream of water as might be expected, the light tended to bend with the water as it arced towards the floor. Tyndahl discovered that the light was transmitted along the stream of water. The light rays inside the water bounced off the internal walls of the water and were thereby contained (Fig. 2.14). This principle has come to be known as *total internal reflection*.

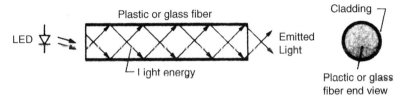

FIGURE 2.14 Total internal reflection.

Industry has since discovered that the principle of total internal reflection also applies to small-diameter glass and plastic fibers, and this has lead to rapid growth of applications throughout the industry. Because optical fibers are small in diameter and flexible, they can bend and twist into confined places. Also, because they contain no electronics, they can operate in much higher temperatures—as high as 400°F—and in areas of high vibration. They are limited by sensing distances, which typically are 80 mm in the proximity mode and 400 mm in the through-beam mode. Also, because of their small sensing area, optical fibers can be fooled by a small drop of water or dirt over the sensing area.

Fiber optics is used to transmit data in the communication field and to transmit images or light in medicine and industry. Photoelectric controls use fiber optics to bend the light from the LED source and return it to the detector so sensors can be placed in locations where common photoelectric sensors could not be applied.

Fiber optics used with photoelectric controls consists of a large number of individual glass or plastic fibers which are sheathed in suitable material for protection. The optical fibers used with photoelectric controls are usually covered by either PVC or stainless-steel jackets. Both protect the fibers from excessive flexing and the environment (Fig. 2.15).

Optical fibers are transparent fibers of glass or plastic used for conducting and guiding light energy. They are used in photoelectric sensors as "light pipes" to conduct sensing light into and out of a sensing area.

Glass optical-fiber assemblies consist of a bundle of 0.05-mm-diameter discrete glass optical fibers housed within a flexible sheath. Glass optical fibers are also able to withstand hostile sensing environments. Plastic optical-fiber assemblies consist of one or two acrylic monofilaments in a flexible sheath.

There are two basic styles of fiber-optic assemblies: (1) individual fiber optics (Fig. 2.16) and (2) bifurcated fiber optics (Fig. 2.17).

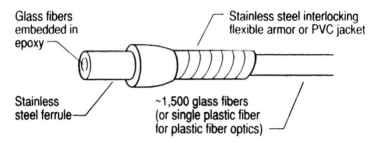

FIGURE 2.15 Jacketed glass fibers.

FIGURE 2.16 Individual fiber-optic assembly.

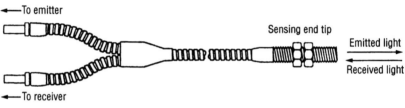

FIGURE 2.17 Bifurcated fiber-optic assembly.

Individual fiber-optic assemblies guide light from an emitter to a sensing location, or to a receiver from a sensing location. Bifurcated fibers use half their fiber area to transmit light and the other half to receive light.

Individual Fiber Optics

A fiber-optic assembly having one control end and one sensing end is used for piping photoelectric light from an emitter to the sensing location or from the sensing location back to a receiver. It is usually used in pairs in the opposed sensing mode, but can also be used side by side in the diffuse proximity mode or angled for the specular reflection or mechanical convergent mode.

Bifurcated Fiber Optics

A bifurcated fiber-optic assembly is branched to combine emitted light with received light in the same assembly. Bifurcated fibers are used for diffused (divergent) proximity sensing, or they may be equipped with a lens for use in the retro-reflective mode.

There are three types of sensing modes used in positioning of a sensor so that the maximum amount of emitted energy reaches the receiver sensing element:

1. Opposed sensing mode (Fig. 2.18)

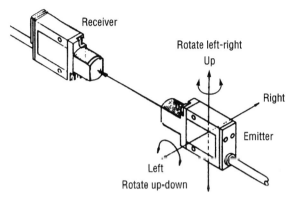

FIGURE 2.18 Opposed sensing mode. For alignment, move emitter or receiver up-down, left-right, and rotate.

2. Retroreflective sensing mode (Fig. 2.19)
3. Proximity (diffused) sensing mode (Fig. 2.20)

Opposed sensing is the most efficient photoelectric sensing mode and offers the highest level of optical energy to overcome lens contamination, sensor misalignment, and long scanning ranges. It is also often referred to as *direct scanning* and sometimes called the *beam break* mode.

The addition of fiber optics to photoelectric sensing has greatly expanded the application of photoelectric devices. Because they are small in diameter and flexible, optical fibers can bend and twist into tiny places formerly inaccessible to bulky electronic devices.

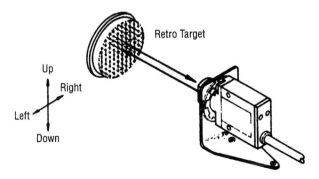

FIGURE 2.19 Retroreflective sensing mode. For alignment, move target up-down, left-right.

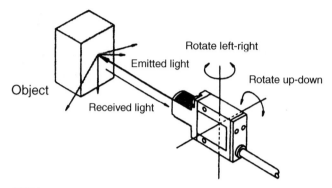

FIGURE 2.20 Proximity sensing mode. For alignment, rotate up-down, left-right.

Optical fibers operate in the same sensing modes as standard photoelectric controls—through-beam, proximity, and reflex. The sizes and shapes of sensing tips have been developed to accommodate many applications.

Optical fibers have a few drawbacks:

1. Limited sensing distance. Typical sensing distance in the proximity mode is 80 mm; 380 mm for the through-beam mode.
2. Typically more expensive than other photoelectric sensing controls.
3. Easily fooled by a small drop of water or dirt over the sensing surface.

Optical fibers' advantages are

1. Sensing in confined places
2. Ability to bend around corners
3. No electronics at sensing point
4. Operation at high temperatures (glass)
5. Total immunity from electrical noise and interference
6. Easily cut to desired lengths (plastic)

OPTICAL FIBER PARAMETERS

The most important parameters affecting optical-fiber performance are

1. Excess gain
2. Background suppression
3. Contrast
4. Polarization

Excess Gain

Excess gain is the measure of energy available between the source and the detector to overcome signal loss due to dirt or contamination. Excess gain is the single most

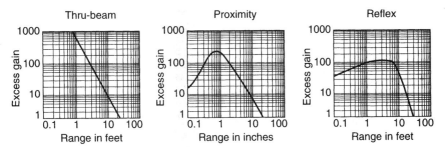

FIGURE 2.21 Excess gain curves.

important consideration in choosing a photoelectric control in manufacturing. It is the extra punch that the sensor has available within its detecting region.

By definition, excess gain is the ratio of the amount of light the detector sees to the minimum amount of light required to trip the sensor. This ratio is depicted graphically for all photoelectric sensors; in Fig. 2.21, excess gain is plotted along the vertical logarithmic axis, starting at 1, the minimum amount of light required to trigger the detector. Every point above 1 represents the amount of light above that required to trigger the photoelectric control—the excess gain.

Often, the standard of comparison for choosing between different photoelectric sensors is range. Actually, more important to most applications is the excess gain. For a typical application, the higher the excess gain within the sensing region, the more likely the application will work. It is the extra margin that will determine whether the photoelectric control will continue to operate despite the buildup of dirt on the lens or the presence of contamination in the air.

Example An application requires detecting boxes on a conveyer in a filthy industrial environment (Fig. 2.22). The boxes will pass about 2 to 5 mm from the sensor as they move along the conveyer at the sensing location. Given a choice between the two proximity sensors whose excess gain curves appear in Figs. 2.23 and 2.24, which photoelectric control should be selected for this application?

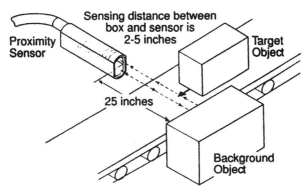

FIGURE 2.22 Box detection.

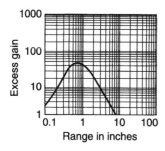
FIGURE 2.23 Excess gain curve for sensor 1.

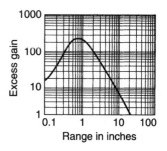
FIGURE 2.24 Excess gain curve for sensor 2.

If the decision were based solely on specified range, the unit described in Fig. 2.23 would be selected. However, if units were installed in this application, it might fail after a short time in operation. Over time, contaminants from the environment would settle on the lens, decreasing the amount of light the sensor sees. Eventually, enough lens contamination would accumulate that the photoelectric control would not have enough excess gain to overcome the signal loss created by the coating, and the application would fail.

A better choice for this application would be the unit represented in Fig. 2.24. It delivers much more excess gain in the operating region required for this application and will therefore work much more successfully than the other unit.

Background Suppression

Background suppression enables a diffuse photoelectric sensor to have high excess gain to a predetermined limit and insufficient excess gain beyond that range, where it might pick up objects in motion and yield a false detection. By using triangular ranging, sensor developers have created a sensor that emits light that reflects on the detector from two different target positions. The signal received from the more distant target is subtracted from that of the closer target, providing high excess gain for the closer target.

Contrast

Contrast measures the ability of a photoelectric control to detect an object; it is the ratio of the excess gain under illumination to the excess gain in the dark. All other things being equal, the sensor that provides the greatest contrast ratio should be selected. For reliable operation, a ratio 10:1 is recommended.

Polarization

Polarization is used in reflection sensors in applications where shiny surfaces, such as metal or shrink-wrapped boxes, may trigger the control falsely. The polarizer passes light along only one plane (Fig. 2.25), and the corner-cube reflectors depolarize the light as it passes through the plastic face of the retroreflector (Fig. 2.10). Only light

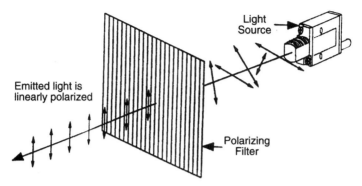

FIGURE 2.25 Polarization.

that has been rotated by the corner-cube retroreflector can pass through the polarizer, whereas light that bounces off other shiny objects cannot.

Like regular reflex photoelectric sensors, polarized sensors have a high light/dark contrast ratio and are simple to install and align. However, the polarizers do limit the sensor's operating range because light is lost passing through them.

INDUCTIVE PROXIMITY SENSORS—NONCONTACT METAL DETECTION

Inductive proximity sensors are another common choice for position sensing. An inductive proximity sensor consists of four basic elements:

1. The sensor, which comprises a coil and ferrous core
2. An oscillator circuit
3. A detector circuit
4. A solid-state output

In the circuitry in Fig. 2.26, the oscillator generates an electromagnetic field that radiates from the sensor's face. This field is centered around the axis and detected by the ferrite core to the front of the sensor assembly. When a metal object enters the electromagnetic field, eddy currents are induced in the surface of the target. This loads in the oscillator circuit, which reduces its amplitude.

The detector circuit detects the change in the oscillator amplitude and, depending on its programming, switches ON and OFF at a specific oscillator amplitude. The sensing circuit returns to its normal state when the target leaves the sensing area and the oscillator circuit regenerates.

The nominal sensing range of inductive proximity sensors is a function of the diameter of the sensor and the power that is available to generate the electromagnetic field. This is subject to a manufacturing tolerance of ±10 percent, as well as a temperature drift tolerance of ±10 percent. The target size, shape, and material will have an effect on the sensing range. Smaller targets will reduce the sensing range, as will targets that are not flat or are made of nonferrous material.

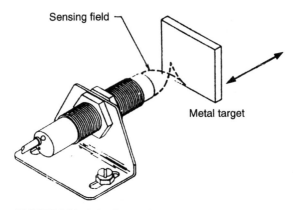

FIGURE 2.26 Inductive proximity sensor.

There are basically two types of inductive proximity sensors: (1) shielded and (2) nonshielded. The shielded version has a metal cover around the ferrite core and coil assembly. This focuses the electromagnetic field to the front of the sensor and allows it to be imbedded in metal without influencing the sensing range. The nonshielded sensor can sense on the side as well as in front of a sensor. It requires a nonmetallic area around the sensor to operate correctly.

Inductive proximity sensors have several benefits:

1. High repeatability. Visibility of the environment is not an issue, since inductive proximity sensors can sense only electromagnetic fields. Therefore, environments from dirt to sunlight pose no problem for inductive proximity sensors. Also, because they are non-contact sensors, nothing wears.
2. Shock and vibration resistance.
3. Versatile connections. They can connect two or three wires with an ac, ac/dc, or dc power supply and up to four-wire dc connections.
4. Wide operating temperature range. They operate between −20 to +70°C, ±10 percent.
5. Very fast response, particularly in the dc models. These sensors can detect presence, send a signal, and reset in 50 μs (2000 times per second) in dc models.

Inductive proximity sensors are generally limited by their sensing distances and the material they sense. The effective range is limited to about 25 mm for most models and can be extended to only about 100 mm with the large models.

LIMIT SWITCHES—TRADITIONAL RELIABILITY

Limit switches are mechanical position-sensing devices that offer simplicity, robustness, and repeatability to processes. Mechanical limit switches are the oldest and simplest of all presence- or position-sensing devices: contact is made and a switch is engaged. This simplicity contributes generally to the cost advantage of limit switches. Yet, they can provide the control capabilities and versatility demanded in today's error-free manufacturing environment. The key to their versatility is the various forms they can take in the switch, actuating head, and lever operator. There are two-step, dual-pole limit switches which can detect and count two products of different sizes and can provide direct power control to segregate or process the items differently. The lever operator will rotate 10° to activate one set of contacts and 20° to activate another set. Because of the high amperage they can handle, limit switches can control up to 10 contacts from the movement of a single lever.

They are easy to maintain because the operator can hear the operation of the switch and can align it easily to fit the application. They are also robust. Limit switches are capable of handling an inrush current 10 times that of their steady-state current rating. They have rugged enclosures and they have prewiring that uses suitable strain-relief bushings to enable the limit switch to retain cables with 500 to 600 pounds of force on them. Limit switches can also handle direct medium-power switching for varying power factors and inrush stresses. For example, they can control a multihorsepower motor without any interposing starter, relay, or contactor.

Reliability is another benefit. Published claims for repeat accuracy for standard limit switches vary from within 0.03 mm to within 0.001 mm over the temperature range of −4 to +200°F. Limit switches dissipate energy spikes and rarely break down under normal mode surges. They will not be affected by electromagnetic interferences (EMI); there are no premature responses in the face of EMI. However, because they are mechanical devices, limit switches face physical limitations that can shorten their service life even though they are capable of several million operations. Also, heavy sludge, chips, or coolant can interfere with their operation.

FACTORS AFFECTING THE SELECTION OF POSITION SENSORS

In selecting a position sensor, there are several key factors that should be considered:

1. *Cost.* Both initial purchase price and life-cycle cost must be considered.
2. *Sensing distance.* Photoelectric sensors are often the best selection when sensing distances are longer than 25 mm. Photoelectric sensors can have sensing ranges as long as 300,000 mm for outdoor or extremely dirty applications, down to 25 mm for extremely small parts or for ignoring background. Inductive proximity sensors and limit switches, on the other hand, have short sensing distances. The inductive proximity sensors are limited by the distance of the electromagnetic field—less than 25 mm for most models—and limit switches can sense only as far as the lever operator reaches.
3. *Type of material.* Inductive proximity sensors can sense only ferrous and non-ferrous materials, whereas photoelectric and limit switches can detect the presence of any solid material. Photoelectric sensors, however, may require polarizer if the target's surface is shiny.
4. *Speed.* Electronic devices using dc power are the fastest—as fast as 2000 cycles per second for inductive proximity models. The fastest-acting limit switches can sense and reset in 4 ms or about 300 times per second.
5. *Environment.* Proximity sensors can best handle dirty, gritty environments, but they can be fooled by metal chips and other metallic debris. Photoelectric sensors will also be fooled or left inoperable if they are fogged or blinded by debris.

6. *Types of voltages, connections, and requirements of the device's housing.* All three types can accommodate varying requirements, but the proper selection must be made in light of the power supplies, wiring schemes, and environments.
7. *Third-party certification.* Underwriters Laboratories (UL), National Electrical Manufacturers Association (NEMA), International Electrotechnical Commission (IEC), Factory Mutual, Canadian Standards Association (CSA), and other organizations impose requirements for safety, often based on the type of application. The certification will ensure the device has been tested and approved for certain uses.
8. *Intangibles.* These can include the availability of application support and service, the supplier's reputation, local availability, and quality testing statements from the manufacturer.

WAVELENGTHS OF COMMONLY USED LIGHT-EMITTING DIODES

An LED is a semiconductor that emits a small amount of light when current flows through it in the forward direction. In most photoelectric sensors, LEDs are used both as emitters for sensing beams and as visual indicators of alignment or output status for a manufacturing process. Most sensor manufacturers use visible red, visible green, or infrared (invisible) LEDs (Fig. 2.4b). This simple device plays a significant part in industrial automation. It provides instantaneous information regarding an object during the manufacturing operation. LEDs, together with fiber optics, allow a controller to direct a multitude of tasks, simultaneously or sequentially.

SENSOR ALIGNMENT TECHNIQUES

A sensor should be positioned so that the maximum amount of emitted energy reaches the receiver element in one of three different modes:

1. Opposed sensing mode
2. Retroreflective sensing mode
3. Proximity (diffuse) sensing mode

Opposed Sensing Mode

In this photoelectric sensing mode, the emitter and receiver are positioned opposite each other so that the light from the emitter shines directly at the receiver. An object then breaks the light beam that is established between the two. Opposed sensing is always the most reliable mode.

Retroreflective Sensing Mode

Retroreflective sensing is also called the *reflex* mode or simply the *retro* mode. A retroreflective photoelectric sensor contains both emitter and receiver. A light beam is established between the sensor and a special retroreflective target. As in opposed sensing, an object is sensed when it interrupts this beam.

Retro is the most popular mode for conveyer applications where the objects are large (boxes, cartons, etc.), where the sensing environment is relatively clean, and where scanning ranges are typically a few meters in length. Retro is also used for code-reading applications. Automatic storage and retrieval systems and automatic conveyer routing systems use retroreflective code plates to identify locations and/or products.

Proximity (Diffuse) Sensing Mode

In the proximity (diffuse) sensing mode, light from the emitter strikes a surface of an object at some arbitrary angle and is diffused from the surface at all angles. The object is detected when the receiver captures some small percentage of the diffused light. Also called the *direct reflection* mode or simply the photoelectric *proximity* mode, this method provides direct sensing of an object by its presence in front of a sensor. A variation is the ultrasonic proximity sensor, in which an object is sensed when its surface reflects a sound wave back to an acoustic sensor.

Divergent Sensing Mode

This is a variation of the diffuse photoelectric sensing mode in which the emitted beam and the receiver's field of view are both very wide. Divergent mode sensors (Fig. 2.27), have loose alignment requirements, but have a shorter sensing range than diffuse mode sensors of the same basic design. Divergent sensors are particularly useful for sensing transparent or translucent materials or for sensing objects with irregular surfaces (e.g., webs that flutter). They are also used effectively to sense objects with very small profiles, such as small-diameter thread or wire, at close range.

All unlensed bifurcated optical fibers are divergent. The divergent mode is sometimes called the *wide-beam diffuse* (or *proximity*) mode.

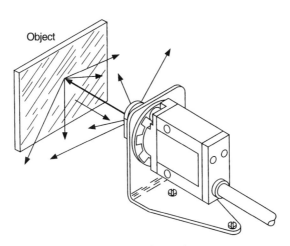

FIGURE 2.27 Divergent sensing mode.

Convergent Sensing Mode

This is a special variation of diffuse mode photoelectric proximity sensing which uses additional optics to create a small, intense, and well-defined image at a fixed distance from the front surface of the sensor lens (Fig. 2.28). Convergent beam sensing is the first choice for photoelectric sensing of transparent materials that remain within a sensor's depth of field. It is also called the *fixed-focus proximity* mode.

Mechanical Convergence

In mechanical convergence (Fig. 2.29), an emitter and a receiver are simply angled toward a common point ahead of the sensor. Although less precise than the optical

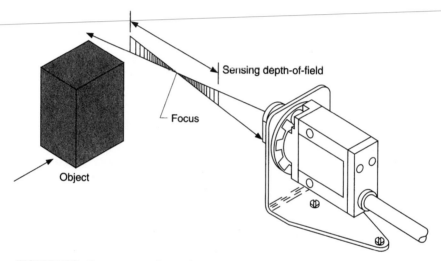

FIGURE 2.28 Convergent sensing mode.

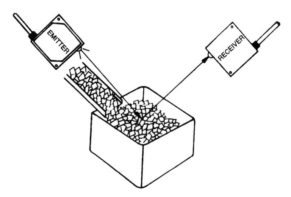

FIGURE 2.29 Mechanical convergence.

convergent-beam sensing mode, this approach to reflective sensing uses light more efficiently than diffuse sensing and gives a greater depth of field than true optical convergence.

Mechanical convergence may be customized for an application by mounting the emitter and the receiver to converge at the desired distance. Depth of field is controlled by adjusting the angle between the emitter and the receiver.

FIBER OPTICS IN INDUSTRIAL COMMUNICATION AND CONTROL

The application of fiber optics to industrial information transfer is a natural extension of the current commercial uses of this technology in high-data-rate communications. While the primary advantage of fiber optics in traditional application areas has been extremely reliable communication at high rates, exceeding 1 Gbit/s over distances exceeding 100 km, other intrinsic features of the technology are more important than data rate and distance capability in industrial uses.

The physical mechanism of light propagating through a glass fiber has significant advantages that enable sensors to carry data and plant communications successfully and in a timely manner—a fundamental condition that must be constantly maintained in a computer-integrated manufacturing environment:

1. The light signal is completely undisturbed by electrical noise. This means that the fiber-optic cables can be laid wherever convenient without special shielding. Fiber-optic cables and sensors are unaffected by electrical noise when placed near arc welders, rotating machinery, electrical generators, etc., whereas in similar wired applications, even the best conventional shielding methods are often inadequate.

2. Fiber-optic communication is devoid of any electrical arcing or sparking, and thus can be used successfully in hazardous areas without danger of causing an explosion.

3. The use of a fiber link provides total electrical isolation between terminal points on the link. Over long plant distances, this can avoid troublesome voltage or ground differentials and ground loops.

4. A fiber-optic system can be flexibly configured to provide additional utility in existing hardware.

PRINCIPLES OF FIBER OPTICS IN COMMUNICATIONS

An optical fiber (Fig. 2.30) is a thin strand composed of two layers, an inner core and an outer cladding. The core is usually constructed of glass, and the cladding structure, of glass or plastic. Each layer has a different index of refraction, the core being higher. The difference between the index of refraction of the core material n_1 and that of the surrounding cladding material n_2 causes rays of light injected into the core to continuously reflect back into the core as they propagate down the fiber.

The light-gathering capability of the fiber is expressed in terms of its numerical aperture NA, the sine of the half angle of the acceptance cone for that fiber. Simply stated, the larger the NA, the easier it is for the fiber to accept light (Fig. 2.31).

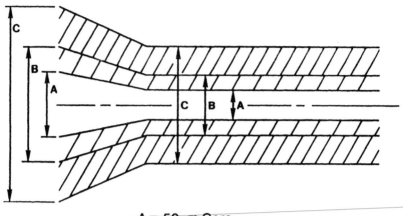

A = 50μm Core
B = 100μm Core
C = 200μm Core

FIGURE 2.30 Structure of optical fiber.

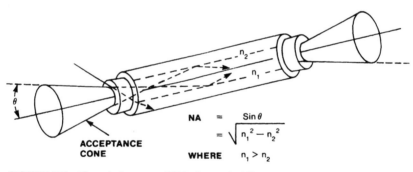

FIGURE 2.31 Numerical aperture (NA) of an optical fiber.

FIBER-OPTIC INFORMATION LINK

A fiber-optic communication system (Fig. 2.32) consists of:

1. A light source (LED or laser diode) pulsed by interface circuitry and capable of handling data rates and voltage levels of a given magnitude.
2. A detector (photodiode) which converts light signals to electrical signals and feeds interface circuitry to recreate the original electrical signal.
3. Fiber-optic cables between the light source and the detector (called the *transmitter* and the *receiver,* respectively).

It is usual practice for two-way communication to build a transmitter and receiver into one module and use a duplex cable to communicate to an identical module at the

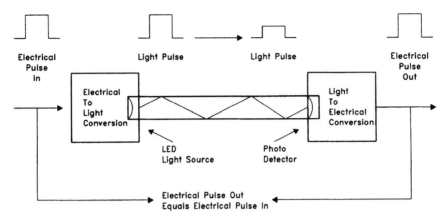

FIGURE 2.32 Fiber-optic communication system.

end of the link. The 820- to 850-nm range is the most frequent for low-data-rate communication, but other wavelengths (1300 and 1550 nm) are used in long-distance systems. Fiber selection must always take transmission wavelength into consideration.

CONFIGURATIONS OF FIBER OPTICS

The selection of optical fiber plays a significant part in sensor performance and information flow. Increased fiber diameter results in higher delivered power, but supports a lower bandwidth. A fiber with a 200-μm core diameter (glass core, silicone plastic cladding), by virtue of its larger diameter and acceptance angle, can transmit 5 to 7 times more light than a 100-μm core fiber, or up to 30 times more light than a 40-μm fiber, the historical telecommunication industry standard. Computer data communications systems commonly employ 62.5-μm fiber because of its ability to support very high data rates (up to 100 Mbaud) while offering increased power and ease of handling. Factory bandwidth requirements for most links are typically one or more orders of magnitude less. Therefore, fiber core size may be increased to 200 μm to gain the benefits of enhanced power and handling ease, the decrease in bandwidth being of no consequence (Fig. 2.31).

Optical Power Budget

An optical power budget examines the available optical power and how it is used and dissipated in a fiber-optic system. It is important to employ the highest possible optical power budget for maximum power margin over the detector requirement. A budget involves four factors:

1. Types of light source
2. Optical fiber acceptance cone
3. Receiver sensitivity
4. Splice, coupling, and connector losses

Laser-diode sources are generally not economically feasible or necessary in industrial systems. Light-emitting diodes are recommended for industrial applications. Such systems are frequently specified with transmitted power at 50 µW or greater if 200-µm-core fiber is used.

Successful communication in industry and commercial applications is determined by the amount of light energy required at the receiver, specified as *receiver sensitivity*. The higher the sensitivity, the less light required from the fiber. High-quality systems require power only in the hundreds of nanowatts to low microwatts range.

Splice losses must be low so that as little light as possible is removed from the optical-fiber system. Splice technology to repair broken cable is readily available, permitting repairs in several locations within a system in a short time (minutes) and causing negligible losses. Couplers and taps are generally formed through the process of glass fusion and operate on the principle of splitting from one fiber to several fibers. New active electronic couplers replenish light as well as distribute the optical signal.

An example of an optical power budget follows:

1. The optical power injected into a 200-µm-core fiber is 200 µW (the same light source would inject approximately 40 µW into a 100-µm-core fiber)
2. The receiver sensitivity is 2 µW
3. The receiver budget (dB) is calculated as:

    $$dB = 10 \log [(\text{available light input})/(\text{required light output})]$$

    $$= 10 \log [(200 \ \mu W)/(2 \ \mu W)]$$

    $$= 10 \log 100$$

    $$= 20 \ dB$$

4. Three major sources of loss are estimated as:
    a. 2 to 3 dB loss for each end connector
    b. 1 to 2 dB loss for each splice
    c. 1 dB/150 m loss for fiber of 200 µm diameter

Most manufacturers specify the optical power budget and translate this into a recommended distance.

Digital Links—Pulsed

The one-for-one creation of a light pulse for an electrical pulse is shown in Fig. 2.32. This represents the simplest form of data link. It does not matter what format and signal level the electrical data takes (e.g., whether IEEE RS-232 or RS-422 standard format or CMOS or TTL logic level), as long as the interface circuitry is designed to accept them at its input or reproduce them at the output. Voltage conversion may be achieved from one end of the link to the other, if desired, through appropriate interface selection.

The light pulses racing down the fiber are independent of electrical protocol. Several design factors are relevant to these and other types of data links as follows:

1. *Minimum output power.* The amount of light, typically measured in microwatts, provided to a specific fiber size from the data link's light source
2. *Fiber size.* Determined from the data link's light source

3. *Receiver sensitivity.* The amount of light, typically measured in microwatts or nanowatts, required to activate the data link's light detector
4. *Data rate.* The maximum rate at which data can be accurately transmitted
5. *Bit error rate (BER).* The frequency with which a light pulse is erroneously interpreted (for example, 10^{-9} BER means no more than one of 10^9 pulses will be incorrect)
6. *Pulse-width distortion.* The time-based disparity between input and output pulse widths

The simple pulse link is also the basic building block for more complex links. Figure 2.33 provides an example of a 5-Mbit three-channel link used to transmit encoder signals from a servomotor to a remote destination.

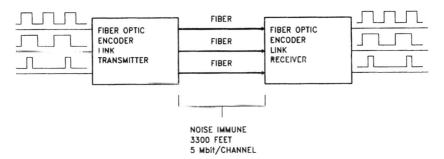

FIGURE 2.33   Three-channel optical link.

Digital Links—Carrier-Based

A carrier-based digital link is a system in which the frequency of the optical carrier is varied by a technique known as *frequency-shift keying* (FSK). Figure 2.34 illustrates the modulation concept; two frequencies are employed to create the logic 0 and 1 states. This scheme is especially useful in systems where electrical "handshaking" (confirmation of reception and acceptance) is employed. Presence of the optical carrier is the equivalent of the handshake signal, with the data signal presented by frequency.

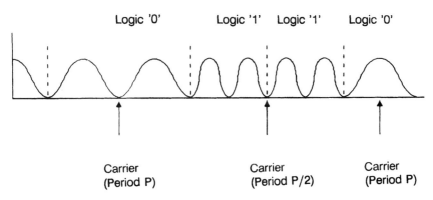

FIGURE 2.34   Modulation concept.

Figure 2.35 illustrates a system where the logic of the fiber-optic line driver recognizes the optical carrier to create a handshake between terminal and processor.

Additionally, since the processor is capable of recognizing only one terminal, the carrier is controlled to deny the handshake to all other terminals once one terminal is actively on line to the processor.

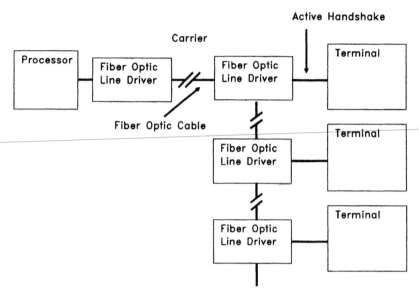

FIGURE 2.35 System employing optical carrier.

Analog Links

It is well-recognized that, in motion control and process measurement and control, transmitting analog information without distortion is important. There are several ways of treating analog information with fiber optics.

Analog data cannot easily be transmitted through light intensity variation. A number of external factors, such as light source variation, bending losses in cable, and connector expansion with temperature, can affect the amount of raw light energy reaching the detector. It is not practical to compensate for all such factors and deliver accurate analog data. A viable method of transmitting data is to use an unmodulated carrier whose frequency depends on the analog signal level. A more advanced means is to convert the analog data to digital data, where accuracy also is determined by the number of bits used, multiplex the digital bits into one stream, and use the pulsed digital link approach.

Figure 2.36 illustrates a link in which this last approach is used to produce both digital and analog forms of the data at the output.

Video Links

Long-distance video transmission in industrial situations is easily disrupted by radiated noise and lighting. Repeaters and large-diameter coaxial cables are often used

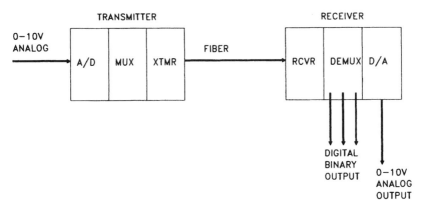

FIGURE 2.36 Analog and digital data transmission.

for particularly long runs. The use of fiber optics as a substitute for coaxial cable allows propagation of noise-free video over long distances. Either an intensity- or frequency-modulated optical carrier signal is utilized as the transmission means over fiber. With intensity-modulated signals, it is mandatory that some sort of automatic gain control be employed to compensate for light degradation due to varying cable losses, splices, etc. Figure 2.37 illustrates a typical fiber-optic video link in a machine-vision application.

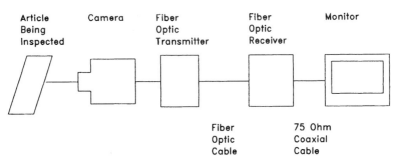

FIGURE 2.37 Fiber-optic video link.

Data Bus Networks

Wiring a system often causes serious problems for designers and communication system integrators regarding the choice of topology. The basic difference between fiber and wiring is that one normally does not splice or tap into fiber as one would with coaxial or twin axial cable to create a drop point.

Daisy Chain Data Bus. The simplest extension of a point-to-point data link is described in Fig. 2.38. It extends continuously from one drop point (node) to the next by using each node as a repeater. The fiber-optic line driver illustrated in Fig. 2.35 is such a system, providing multiple access points from remote terminals to a programmable controller processor. A system with several repeater points is vulnerable to the loss of any repeater, and with it all downstream points, unless some optical bypass scheme is utilized. Figures 2.38 and 2.39 exhibit such a scheme.

Ring Coupler. A preferred choice among several current fiber-optic system designs is the token-passing ring structure.

Signals are passed around the ring, with each node serving to amplify and retransmit. Care must be taken to provide for a node becoming nonoperational. This is usually handled by using some type of bypass switching technique, given that the system provides sufficient optical power to tolerate a bypassed repeater. Another contingency method is to provide for the transmitting node to read its own data coming around the ring, and to retransmit in the other direction if necessary, as illus-

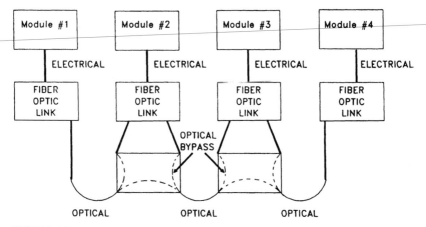

FIGURE 2.38 Point-to-point data link.

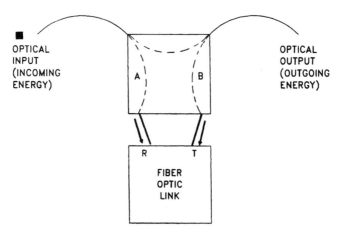

FIGURE 2.39 Daisy chain data bus.

FIBER OPTICS IN SENSORS AND CONTROL SYSTEMS   2.29

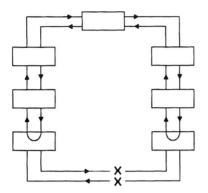

FIGURE 2.40 Ring coupler.

trated in Fig. 2.40. Yet another is to provide for a second pair of fibers paralleling the first, but routed on a physically different path.

Passive Star Coupler. Certain systems have attempted to utilize a fiber-optic coupling technology offered from the telecommunications and data communications applications areas. When successful, this technique allows tapping into fiber-optic trunk lines, a direct parallel with coaxial or twin axial systems. Light entering into the tap or coupler is split into a given number of output channels. The amount of light in any output channel is determined by the total amount of light input, less system losses, divided by the number of output channels. Additional losses are incurred at the junction of the main data-carrying fibers with the fiber leads from the tap or star. As such, passive couplers are limited to systems with few drops and moderate distances. Also, it is important to minimize termination losses at the coupler caused by the already diminished light output from the coupler. A partial solution is an active in-line repeater, but a superior solution, the *active star coupler,* is described below.

The Active Star Coupler. The basic principle of the active star coupler is that any light signal received as an input is converted to an electrical signal, amplified, and reconverted to optical signals on all other output channels. Figure 2.41 illustrates an eight-port active star coupler, containing eight sets of fiber-optic input/output (I/O) ports. A signal received on the channel 1 input will be transmitted on the channel 2

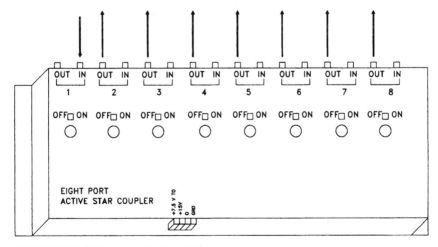

FIGURE 2.41 Eight-port active star coupler.

to 8 output ports. One may visualize the use of the active star coupler as aggregating a number of taps into one box. Should the number of required taps exceed the number of available I/O ports, or should it be desirable to place these tap boxes at several locations in the system, the active star couplers may be jumpered together optically by tying a pair of I/O ports on one coupler to that on another in a hub-and-spoke system.

With the active star coupler serving as the hub of the data bus network, any message broadcast by a unit on the network is retransmitted to all other units on the network. A response of these other units is broadcast back to the rest of the network through the star, as in an electrical wired data bus network.

CONFIGURATIONS OF FIBER OPTICS FOR SENSORS

Fiber-optic sensors for general industrial use have largely been restricted to applications in which their small size has made them convenient replacements for conventional photoelectric sensors. Until recently, fiber-optic sensors have almost exclusively employed standard bundle technology, whereby thin glass fibers are bundled together to form flexible conduits for light.

Recently, however, the advances in fiber optics for data communications have introduced an entirely new dimension into optical sensing technology. Combined with novel but effective transducing technology, they set the stage for a powerful class of fiber-optic sensors.

Fiber-Optic Bundle

A typical fiber-optic sensor probe, often referred to as a *bundle* (Fig. 2.42), is 1.25 to 3.15 mm in diameter and made of individual fiber elements approximately 0.05 mm in diameter. An average bundle will contain up to several thousand fiber elements, each working on the conventional fiber-optic principle of total internal reflection.

Composite bundles of fibers have an acceptance cone of the light based on the numerical aperture of the individual fiber elements.

$$NA = \sin \Theta$$
$$= \sqrt{n_1^2 - n_2^2}$$

where $n_1 > n_2$ and Θ = half the cone angle.

Bundles normally have NA values in excess of 0.5 (acceptance cone full angle greater than 60°), contrasted with individual fibers for long-distance, high-data-rate

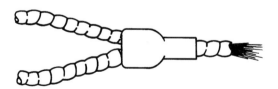

FIGURE 2.42 Fiber-optic sensor probe.

applications, which have NA values approaching 0.2 (acceptance cone full angle approximately 20°).

The ability of fiber-optic bundles to readily accept light, as well as their large total cross-sectional surface area, have made them an acceptable choice for guiding light to a remote target and from the target area back to a detector element. This has been successfully accomplished by using the pipe as an appendage to conventional photoelectric sensors, proven devices conveniently prepackaged with adequate light source and detector elements.

Bundles are most often used in either opposed beam or reflective mode. In the opposed beam mode, one fiber bundle pipes light from the light source and illuminates a second bundle—placed on the same axis at some distance away—which carries light back to the detector. An object passing between the bundles prevents light from reaching the detector.

In the reflective mode, all fibers are usually contained in one probe but divided into two legs at some junction point in an arrangement known as *bifurcate*. One bifurcate leg is then tied to the light source and the other to the detector (Fig. 2.43). Reflection from a target provides a return path to the detector for the light. The target may be fixed so that it breaks the beam, or it may be moving so that, when present in the probe's field of view, it reflects the beam.

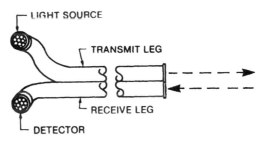

FIGURE 2.43 Reflective mode bifurcate fiber optics.

Typical bundle construction in a bifurcate employs one of two arrangements of individual fibers. The sending and receiving fibers are arranged either randomly or hemispherically (Fig. 2.44). As a practical matter, there is little, if any, noticeable impact on the performance of a photoelectric system in any of the key parameters such as sensitivity and scan range.

Application areas for bundle probes include counting, break detection, shaft rotation, and displacement/proximity sensing.

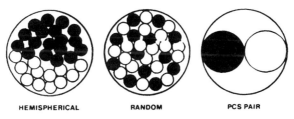

FIGURE 2.44 Bundle construction.

Bundle Design Considerations

Microscopic fiber flaws (Fig. 2.45) such as impurities, bubbles, voids, material absorption centers, and material density variations all diminish the ability of rays of light to propagate down the fiber, causing a net loss of light from one end of the fiber to the other. All these effects combine to produce a characteristic absorption curve, which graphically expresses a wavelength-dependent loss relationship for a given fiber (Fig. 2.46). The fiber loss parameter is expressed as attenuation in dB/km as follows:

$$\text{Loss} = -10 \log (p_2/p_1)$$

where p_2 = light power output and p_1 = light power input.

Therefore, a 10-dB/km fiber would produce only 10 percent of the input light at a distance of 1 km.

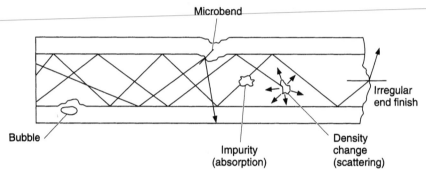

FIGURE 2.45 Microscopic fiber flaws.

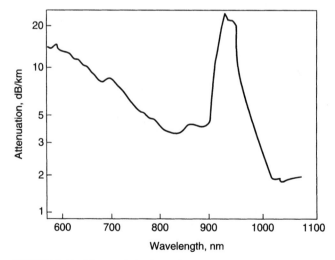

FIGURE 2.46 Characteristic absorption curve.

Because of their inexpensive lead silicate glass composition and relatively simple processing techniques, bundles exhibit losses in the 500-dB/km range. This is several orders of magnitude greater than a communications-grade fiber, which has a loss of 10 dB/km. The maximum practical length for a bundle is thus only about 3 m. Further, the absence of coating on individual fibers and their small diameter make them susceptible to breakage, especially in vibratory environments.

Also, because of fiber microflaws, it is especially important to shield fibers from moisture and contaminants. A fiber exposed to water will gradually erode to the point of failure. This is true of any optical fiber, but is especially true of uncoated fibers in bundles.

Fiber Pairs for Remote Sensing

A viable solution to those design problems that exist with fiber bundles is to use large-core glass fibers that have losses on a par with telecommunication fibers. Although its ability to accept light is less than that of a bundle, a 200- or 400-μm core diameter plastic clad silica (PCS) fiber provides the ability to place sensing points hundreds of meters away from corresponding electronics. The fiber and the cable construction in Fig. 2.47 lend themselves particularly well to conduit pulling, vibratory environments, and general physical abuse. These fibers are typically proof-tested for tensile strength to levels in excess of 50,000 lb/in². A pair of fibers (Fig. 2.44) is used much like a bundle, where one fiber is used to send light to the sensing point and the other to return light to the detector. The performance limitation of a fiber pair compared to a bundle is reduced scan range; however, lenses may be used to extend the range. A fiber pair may be used in one of two configurations: (1) a single continuous probe, i.e., an unbroken length of cable from electronics to sensing point, or (2) a fiber-optic extension cord to which a standard probe in either a bundle or a fiber pair is coupled mechanically. This allows the economical replacement, if necessary, of the standard probe, leaving the extension cord intact.

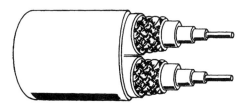

FIGURE 2.47 Pair of fibers.

The typical application for a fiber pair is object detection in explosive or highly corrosive environments, e.g., ammunition plants. In such cases, electronics must be remote by necessity. Fiber pairs also allow the construction of very small probes for use in such areas as robotics, small object detection, thread break detection, and small target rotation.

Fiber-Optic Liquid Level Sensing

Another technique for interfacing with fiber-optic probes involves the use of a prism tip for liquid sensing (Fig. 2.48). Light traveling down one leg of the probe is totally internally reflected at the prism-air interface. The index of refraction of air is 1. Air acts as a cladding material around the prism. When the prism contacts the surface of a liquid, light is stripped from the prism, resulting in a loss of energy at the detector. A properly configured system can discriminate between liquid types, such as gasoline and water, by the amount of light lost from the system, a function of the index of refraction of the liquid.

This type of sensor is ideal for setpoint use in explosive liquids, in areas where electronics must be remote from the liquid by tens or hundreds of meters, and where foam or liquid turbulence make other level-sensing techniques unusable.

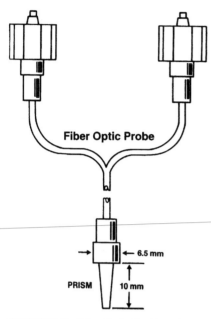

FIGURE 2.48 Prism tip for liquid sensing.

FLEXIBILITY OF FIBER OPTICS

The power of fiber optics is further shown in the flexibility of its system configurations. A master industrial terminal (Fig. 2.49) can access any of a number of remote processors. The flexibility of switching, distance capability, and noise immunity of such a system are its primary advantages.

Figure 2.50 illustrates a passive optical coupler with a two-way fiber-optic link communicating over a single fiber through an on-axis rotary joint. Such a system allows a simple, uninterrupted communication link through rotary tables or other rotating machinery. This is a true challenge for high-data-rate communication in a wire system.

Fiber-Optic Terminations

Optical fibers are becoming increasingly easier to terminate as rapid advances in termination technology continue to be made. Several manufacturers have connector systems that require no polishing of the fiber end, long a major objection in fiber optics. Products that eliminate epoxy adhesives are also being developed. Field installation times now typically average less than 10 min for large-core fibers (100 and 200 μm) with losses in the 1- to 3-dB range. Further, power budgets for well-designed industrial links normally provide a much greater latitude in making a connection. A 5- to 6-dB-loss connection, while potentially catastrophic in other

FIBER OPTICS IN SENSORS AND CONTROL SYSTEMS

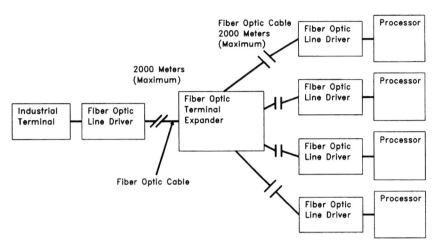

FIGURE 2.49 Master industrial terminal and remote processors.

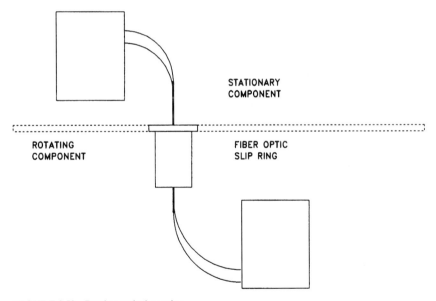

FIGURE 2.50 Passive optical coupler.

types of systems, may be quite acceptable in short-haul systems with ample power budgets.

The most popular connector style for industrial communications is the SMA style connector, distinguished by its nose dimensions and configuration, as well as the thread size on the coupling nut. The coupling nut is employed to mechanically join

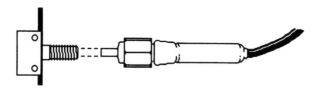

FIGURE 2.51 SMA connection to an information link.

the connector to a mating device on the data link or to a thread splice bushing. Figure 2.51 illustrates an SMA connection to an information link.

TESTING OF FIBER OPTICS

Optical measurements, perhaps among the most difficult of all physical measurements, are fundamental to the progress and development of fiber-optic technology. Recently various manufacturers have offered lines of fiber-optic test equipment for use in field and laboratory. Typical field measurement equipment determines the average optical power emitted from the system source, the component and overall system loss, the bit error rate, and the location of breaks in the fiber. Laboratory equipment measures loss through connectors and splicing, characterizes transmitters and receivers, and establishes bit error rate.

Testing of fiber-optic cables or systems is normally done with a calibrated light source and companion power meter. The light source is adjusted to provide a 0-dB reading on the power meter with a short length of jumper cable. The cable assembly under test is then coupled between the jumper and the power meter to provide a reading on the meter, in decibels, that corresponds to the actual loss in the cable assembly.

Alternatively, the power through the cable from the system's transmitter can be read directly and compared with the system's receiver sensitivity specification. In the event of a cable break in a long span, a more sophisticated piece of test equipment, an optical time-domain reflectometer (OTDR), can be employed to determine the exact position of the break.

Test Light Sources

The Photodyne 9XT optical source driver (Fig. 2.52) is a hand-held unit for driving LED and laser fiber-optic light sources. The test equipment is designed to take the shock and hard wear of the typical work-crew environment. The unit is powered from two rechargeable nicad batteries or from line voltage. The LED series is suited for measurement applications where moderate dynamic range is required or coherent light should be avoided. The laser modules are used for attenuation measurements requiring extreme dynamic range or where narrow spectral width and coherent light are required. The laser modules are the most powerful source modules available.

Power Meters

The Photodyne 2285XQ fiber-optic power meter (Fig. 2.53) is a low-cost optical power meter for general-purpose average-power measurement, but particularly for

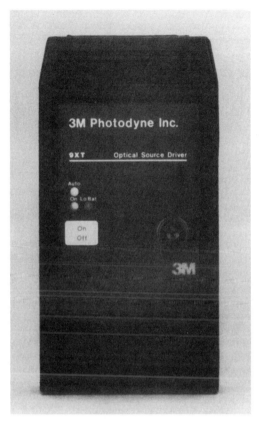

FIGURE 2.52 Photodyne 9XT optical source driver. (*Courtesy 3M Corporation.*)

fiber-optic applications, in both manual and computer-controlled test setups. A unique and powerful feature of the 2285XSQ is its ratio function. This allows the user to make ratio (A/B) measurements by stacking several instruments together via the interface without the need for a controller.

Another very powerful feature is the built-in data logger. With this function, data may be taken at intervals from 1 to 9999 s. This feature is useful in testing optical devices for short- and long-term operation.

At the heart of the instrument is a built-in large-area high-sensitivity indium gallium arsenide (InGaAs) sensor. All industry connectors may be interfaced to the sensor using any of the Photodyne series 2000 connector adapters. The sensor is calibrated for the all the fiber-optic windows: 820, 850, 1300, and 1550 nm.

Dual Laser Test Sets

The Photodyne 2260XF and 2260XFA are switchable dual laser test sets, with a transmit and receive section in one unit. They measure and display loss in fiber links

FIGURE 2.53 Photodyne 2285XQ fiber-optic power meter. (*Courtesy 3M Corporation.*)

at both 1300 and 1550 nm simultaneously in one pass. They are designed for use in installing, maintaining, and troubleshooting single-mode wavelength-division multiplexed (WDM) fiber-optic links operating at 1300- and 1550-nm wavelengths. They may also be used for conventional links operating at either 1300 or 1550 nm.

The essential differences between the two models are that the XF version (Fig. 2.54) has a full complement of measurement units, log and linear (dBm, dB, mW, µW, nW), whereas the XFA version has log units only (dBm, dB). In the XF version, laser power output is adjustable over the range 10 to 20 dBm (100 µW to 21 mW). In the XFA version, it is automatically set to a fixed value of 10 dBm. The XF version allows the user to access wavelengths over the ranges 1250 to 1350 nm and 1500 to 1600 nm in 10-nm steps. The XFA version is fixed at 1300 nm to 1550 nm. To control all functions, the XF version has six keys; the XFA has only two. Although both instruments perform identical tasks equally well, the XF may be seen as the more flexible version and the XFA as the simpler-to-use version.

Test Sets/Talk Sets

The hand-held Photodyne 21XTL fiber-optic test set/talk set (Fig. 2.55) is for use in installation and maintenance of fiber cables. This instrument functions as a power meter and test set, as well as a talk set. For maintenance purposes, the user may establish voice communication over the same fiber pair that is being measured.

The 21XTL as a power meter covers an exceptionally wide dynamic range (−80 to +3 dBm). As an option, the receiver may include a silicon or an enhanced InGaAs photodiode. With the InGaAs version, the user can perform measurements and voice communication at short and long wavelengths. The silicon version achieves a superior dynamic range at short wavelengths only.

The highly stabilized LED ensures repeatable and accurate measurements. Precision optics couples the surface-emitter LED to the fiber core. With this technique the fiber end will not wear out or scratch. The transmitter is interchangeable, providing complete flexibility of wavelengths and connector types.

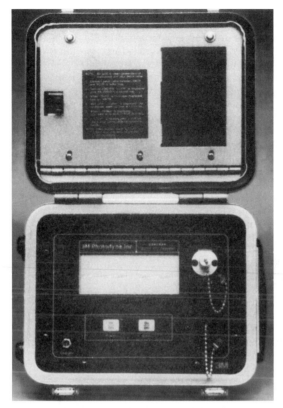

FIGURE 2.54 Photodyne 2260XF switchable dual laser test set. (*Courtesy 3M Corporation.*)

FIGURE 2.55 Photodyne 21XTL fiber-optic test set/talk set. (*Courtesy 3M Corporation.*)

Attenuators

The 19XT optical attenuator (Fig. 2.56) is a hand-held automatic optical attenuator. It provides continuous attenuation of both short- and long-wave optical signals in single-mode and multimode applications. The calibrated wavelengths are 850 nm/1300 nm multimode and/or 1300 nm/1550 nm single mode. An attenuation range of 70 dB is offered with 0.1-dB resolution and an accuracy of ±0.2 dB typical (±0.5 dB maximum). Unique features allow scanning between two preset attenuation values and including the insertion loss in the reading.

FIGURE 2.56 Photodyne 19XT optical attenuator. (*Courtesy 3M Corporation.*)

The 19XT allows simple front panel operation or external control of attenuation through analog input and output connections.

Fault Finders

The Photodyne 5200 series optical fault finders (Fig. 2.57) offer flexible alternatives for localizing faults or trouble areas on any fiber-optic network. The 5200 series fault

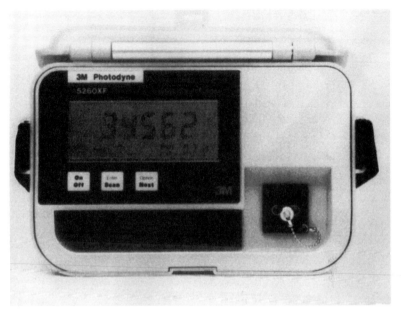

FIGURE 2.57 Photodyne 5200 series optical fault finder. (*Courtesy 3M Corporation.*)

finders can easily be integrated into the troubleshooting routines of fiber-optic crews. They offer a variety of features, such as:

1. Fast, accurate analysis of faults
2. Autoranging for greatest accuracy
3. Reflective and nonreflective fault detection
4. Multiple-fault detection capabilities
5. Go/no-go splice qualification
6. Variable fault threshold setting (0.5 to 6.0 dB)
7. Fault location up to 82 km
8. Automatic self-test and performance check
9. AC or rechargeable battery operation
10. Large, easy-to-read liquid-crystal display (LCD)

Fiber Identifiers

The Photodyne 8000XG fiber identifier (Fig. 2.58) is designed for fast, accurate identification and traffic testing of fiber-optic lines without cutting the fiber line or interrupting normal service. Ideal for use during routine maintenance and line modification, this small, hand-held unit can be used to locate any particular fiber line, identify live fibers, and determine whether or not traffic is present. Features of the 8000XG are:

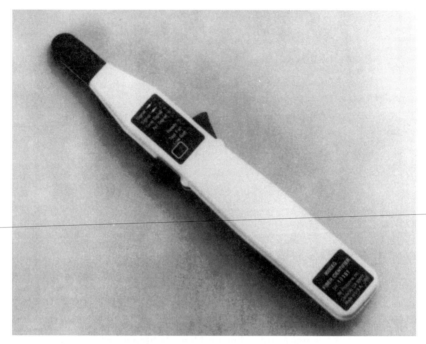

FIGURE 2.58 Photodyne 8000XG fiber identifier. (*Courtesy 3M Corporation.*)

1. Light weight, portability, and battery operation
2. Automatic self-test after each fiber insertion
3. Mechanically damped fiber action
4. Operation over 850- to 1550-nm range
5. Transmission direction indicators
6. 1- and 2-kHz tone detection
7. Low insertion loss at 1300 and 1550 nm
8. Completely self-contained operation

NETWORKING WITH ELECTROOPTIC LINKS

The following examples describe a number of products utilizing communication through fiber electrooptic modules. The function of the fiber is to replace wire. This can be achieved by interconnecting the electrooptic modules in a variety of ways. Figure 2.59 shows a programmable controller communication network through-coaxial-cable bus branched to four remote input/output stations. The programmable controller polls each of the input/output stations in sequence. All input/output stations hear the programmable controller communication, but only the one currently being addressed responds.

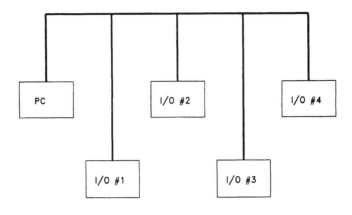

─── WIRE

FIGURE 2.59 Programmable controller communication network.

Hybrid Wire/Fiber Network

Figure 2.60 shows an electrooptic module used to replace a troublesome section of coaxial cable subject to noise, grounding problems, lightning, or hazardous environment. When fiber is used in this mode, the electrooptic module should be placed as close to the input/output drop as possible in order to minimize the effect of potential electrical noise problems over the section of coax cable connecting them.

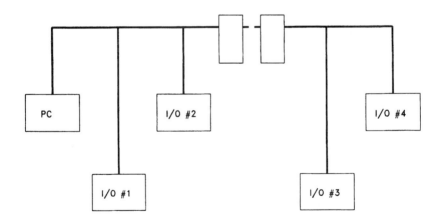

─── WIRE
- - - FIBER

FIGURE 2.60 Hybrid wire/fiber network.

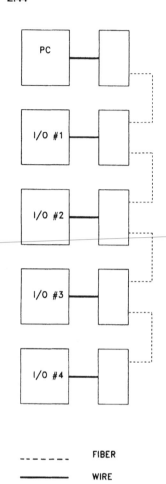

FIGURE 2.61 Daisy chain network.

Daisy Chain Network

The daisy chain configuration (Fig. 2.61) is an economical choice for long, straight-line installations (e.g., conveyer or mine shaft applications). The signal generated at the programmable controller is converted to light and transmitted outward. At each transmitted section, the electrical signal is reconstructed and a light signal is regenerated and sent down the chain.

Active Star Network

The electrooptical programmable controller links may be joined to an electrooptic module with four-port and eight-port active star couplers for a hub-and-spoke-type system. Figure 2.62 shows two active star couplers joined via fiber, a configuration which might be appropriate where clusters of input/output racks reside in several locations separated by some distance. The star coupler then becomes the distributor of the light signals to the racks in each cluster as well as a potential repeater to a star in another cluster.

Hybrid Fiber Network

Star and daisy chain network structures can be combined to minimize overall cabling costs (Fig. 2.63). The fiber network configuration exactly duplicates the coax network. The final decision on exactly which configuration to choose depends on the following criteria:

1. Economical cabling layout
2. Length of cable runs versus cost of electronics
3. Location of end devices and power availability
4. Power-out considerations

Other considerations specific to the programmable controller being used can be summarized as follows:

- *Pulse width.* Total acceptable pulse-width distortion, which may limit the number of allowable repeater sites, is

    Allowable distortion (ns) = allowable percent distortion × period of signal (ns)

- *Signal propagation delay.* Allowable system signal propagation delay may limit the overall distance of the fiber network.

FIBER OPTICS IN SENSORS AND CONTROL SYSTEMS

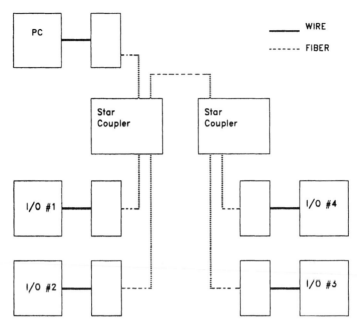

FIGURE 2.62 Two active star couplers.

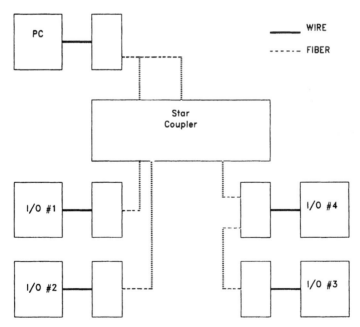

FIGURE 2.63 Hybrid fiber network.

Example:

Electrooptic module distortion = 50% allowable distortion × 1 Mbaud transmission

$$= 50 \text{ ns}$$

Calculate:

$$\text{Allowable distortion} = 50\% \text{ allowable distortion} \times 10^{-6}$$

$$= 500 \text{ ns}$$

$$\text{Maximum number of repeater sites} = 500 \text{ ns}/50 \text{ ns}$$

$$= 10$$

(Note: A star coupler counts as a repeater site.)

Distance Calculation (Signal Propagation Delay)	
Delay of light in fiber-optic regeneration	1.5 ns/ft
Delay in module	50 ns

Manufacturers of programmable controllers will provide the value of the system delay. This value must be compared with the calculated allowable delay. If the overall fiber system is longer than published maximum length of the wired system, the system must be reconfigured.

Fiber-Optic Sensory Link for Minicell Controller

A minicell controller is typically used to coordinate and manage the operation of a manufacturing cell, consisting of a group of automated programmable machine controls (programmable controllers, robots, machine tools, etc.) designed to work together and perform a complete manufacturing or process-related task. A key benefit of a minicell controller is its ability to adjust for changing products and conditions. The minicell controller is instructed to change data or complete programs within the automation work area. A minicell controller is designed to perform a wide variety of functions such as executing programs and data, uploading/downloading from programmable controllers, monitoring, data analysis, tracking trends, generating color graphics, and communicating in the demanding plant floor environment. Successful minicell controllers use fiber-optic links that can interface with a variety of peripheral devices. A minicell controller can be used in a variety of configurations, depending on the optical-fiber lengths, to provide substantial system design and functional flexibility (Fig. 2.64).

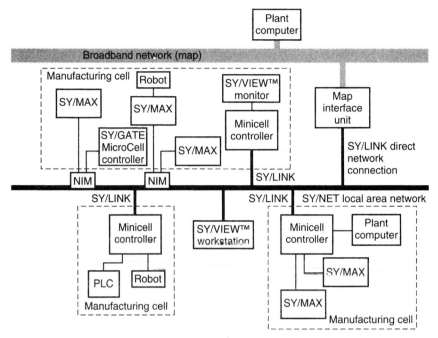

FIGURE 2.64 Minicell controller.

VERSATILITY OF FIBER OPTICS IN INDUSTRIAL APPLICATIONS

A constant concern in communication is the ever-increasing amount of information that must be sent with greater efficiency over a medium requiring less space and less susceptibility to outside interferences. As speed and transmission distance increase, the problems caused by electromagnetic interference, radio-frequency interference, crosstalk, and signal distortion become more troublesome. In terms of signal integrity, just as in computer-integrated manufacturing data acquisition and information-carrying capacity, fiber optics offers many advantages over copper cables. Furthermore, optical fibers emit no radiation and are safe from sparking and shock. These features make fiber optics the ideal choice for many processing applications where safe operation in hazardous or flammable environments is a requirement.

Accordingly, fiber-optic cables have these advantages in industrial applications:

1. Wide bandwidth
2. Low attenuation
3. Electromagnetic immunity
4. No radio-frequency interference
5. Small size

6. Light weight
7. Security
8. Safety in hazardous environment

High-Clad Fiber-Optic Cables

Large-core, multimode, step-index high-clad silica fiber-optic cables make fiber-optic technology user-friendly and help designers of sensors, controls, and communications realize substantial cost saving. The coupling efficiency of high-clad silica fibers allows the use of less expensive transmitters and receivers. High-clad silica polymer technology permits direct crimping onto the fiber cladding. Field terminations can be performed in a few minutes or less, with minimal training. The following lists describe the structure and characteristics of several fiber-optic cables used in industry:

Simplex fiber-optic cables (Fig. 2.65) are used in:

FIGURE 2.65 Simplex fiber-optic cable.

- Light-duty indoor applications
- Cable trays
- Short conduits
- Loose tie wrapping
- Subchannels for breakout cables

Zipcord fiber-optic cables (Fig. 2.66) are used in:

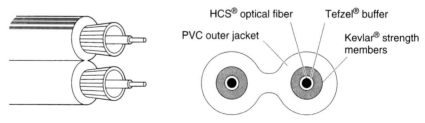

FIGURE 2.66 Zipcord fiber-optic cable.

- Light-duty (two-way) transmission
- Indoor runs in cable trays
- Short conduits
- Tie wrapping

Multichannel fiber-optic cables (Fig. 2.67) are used in:

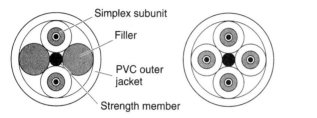

FIGURE 2.67 Multichannel fiber-optic cable.

- Outdoor environments
- Multifiber runs where each channel is connectorized and routed separately
- Aerial runs
- Long conduit pulls
- Two, four, and six channels (standard)
- Eight to 18 channels

Heavy-duty duplex fiber-optic cables (Fig. 2.68) are used in:

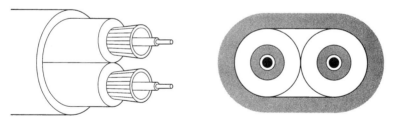

FIGURE 2.68 Heavy-duty duplex fiber-optic cable.

- Rugged applications
- Wide-temperature-range environments
- Direct burial
- Loose tube subchannel design
- High-tensile-stress applications

Table 2.1 summarizes fiber-optic cable characteristics.

TABLE 2.1 Fiber-Optic Cable Characteristics

Characteristics	Simplex	Zipcord	Duplex	Channels 2	4	6
Cable diameter, mm	2.5	2.5 × 5.4	3.5 × 6	8	8	10
Weight, kg/km	6.5	11	20	41	41	61
Jacket type	PVC	PVC	PVC	PE	PE	PE
Jacket color	Orange	Orange	Orange	Black	Black	Black
Pull tension	330 N	490 N	670 N	890 N	1150 N	2490 N
Max. long-term tension	200 N	310 N	400 N	525 N	870 N	1425 N
Max. break strength	890 N	1340 N	1780 N	2370 N	4000 N	11,000 N
Impact strength at 1.6 N · m	100	100	150	200	200	200
Cyclic flexing, cycles	>5000	>5000	>5000	>2000	>2000	>2000
Minimum bend radius, mm	25	25	25	50	50	75
Cable attenuation	0.6 dB/km at 820 nm					
Operating temperature	−40 to +85°C					
Storage temperature	−40 to +85°C					

Fiber-Optic Ammeter

In many applications, including the fusion reactors, radio-frequency systems, and telemetry systems, it is often necessary to measure the magnitude and frequency of current flowing through a circuit in which high dc voltages are present. A fiber-optic current monitor (Fig. 2.69) has been developed at the Princeton Plasma Physics Laboratory (PPPL) in response to a transient voltage breakdown problem that caused failures of Hall-effect devices used in the Tokamak fusion test reactor's natural-beam heating systems.

The fiber-optic current monitor measures low current in a conductor at very high voltage. Typical voltages range between tens of kilovolts and several hundred kilovolts. With a dead band of approximately 3 mA, the circuit derives its power from the conductor being measured and couples information to a (safe) area by means of fiber optics. The frequency response is normally from direct current to 100 kHz, and a typical magnitude range is between 5 and 600 mA.

The system is composed of an inverting amplifier, a current regulator, transorbs, diodes, resistors, and a fiber-optic cable. Around an inverting amplifier, a light-emitting diode and a photodiode form an optical closed feedback loop. A fraction of the light emitted by the LED is coupled to the fiber-optic cable.

As the current flows through the first diode, it splits between the 1.5-mA current regulator and the sampling resistor. The voltage across the sampling resistor causes a small current to flow into the inverting amplifier summing junction and is proportional to the current in the sampling resistor. Since photodiodes are quite linear, the light power from the LED is proportional to the current through the sampling resistor. The light is split between the local photodiode and the fiber cable. A photodiode, located in a remote safe area, receives light that is linearly proportional to the conductor current (for current greater than 5 mA and less than 600 mA).

To protect against fault conditions, the design utilizes two back-to-back transorbs in parallel with the monitor circuit. The transorbs are rated for 400 A for 1 ms. The fiber-optic ammeter is an effective tool for fusion research and other applications where high voltage is present.

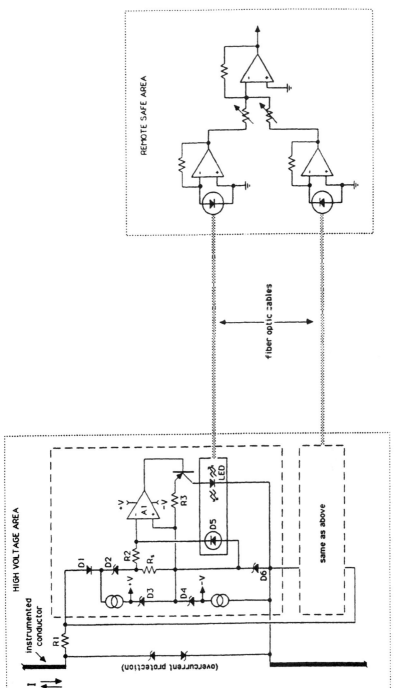

FIGURE 2.69 Fiber-optic current monitor.

REFERENCES

1. Berwick, M., J. D. C. Jones, and D. A. Jackson, "Alternating Current Measurement and Non-Invasive Data Ring Utilizing the Faraday Effect in a Closed Loop Fiber Magnetometer," *Optics Lett.,* **12**(294) (1987).
2. Cole, J. H., B. A. Danver, and J. A. Bucaro, "Synthetic Heterodyne Interferometric Demodulation," *IEEE J. Quant. Electron.,* **QE-18**(684) (1982).
3. Dandridge, A., and A. B. Tveten, "Phase Compensation in Interferometric Fiber Optic Sensors," *Optics Lett.,* **7**(279) (1982).
4. Desforges, F. X., L. B. Jeunhomme, Ph. Graindorge, and G. L. Baudec, "Fiber Optic Microswitch for Industrial Use," presented at SPIE O-E Fiber Conf., San Diego, no. 838–41 (1987).
5. Favre, F., and D. LeGuen, "High Frequency Stability in Laser Diode for Heterodyne Communication Systems," *Electron. Lett.* **16**(179) (1980).
6. Giallorenzi, T. G., "Optical Fiber Interferometer Technology and Hydrophones in Optical Fiber Sensors," NATO ASI Series E, No. 132, Martinus Nijhoff Dordrecht, 35–50 (1987).
7. Hocker, G. B., "Fiber Optic Sensing for Pressure and Temperature," *Appl. Optics,* **18**(1445) (1979).
8. Jackson, D. A., A. D. Kersey, and A. C. Lewin, "Fiber Gyroscope with Passive Quadrature Demodulation," *Electron. Lett.,* **20**(399) (1984).
9. Optical Society of America, *Optical Fiber Optics.* Summaries of papers presented at the Optical Fiber Sensors Topical Meeting, January 27–29, 1988, New Orleans, La. (IEEE, Catalog No. 88CH2524–7).
10. Popovic, R. S., "Hall Effect Devices," *Sensors and Actuators,* **17**, 39–53 (1989).
11. Saxena, S. C., and S. B. Lal Seksena, "A Self-Compensated Smart LVDT Transducer," *IEEE Trans. Instrum. Meas.,* **38**, 748–783 (1989).
12. Wong, Y. L., and W. E. Ott, *Function Circuits and Applications,* McGraw-Hill, New York, 1976.

CHAPTER 3
NETWORKING OF SENSORS AND CONTROL SYSTEMS IN MANUFACTURING

INTRODUCTION

Central to the development of any computer-integrated manufacturing facility is the selection of the appropriate automated manufacturing system and the sensors and control systems to implement it. The degree to which a CIM configuration can be realized depends on the capabilities and cost of available equipment and the simplicity of information flow.

When designing an error-free manufacturing system, the manufacturing design group must have an appreciation for the functional limits of the automated manufacturing equipment of interest and the ability of the sensors to provide effective information flow, since these parameters will constrain the range of possible design configurations. Obviously, it is not useful to design a manufacturing facility that cannot be implemented because it exceeds the equipment's capabilities. It is desirable to match automated manufacturing equipment to the application. Although sensors and control systems are—by far—less costly than the automated manufacturing equipment, it is neither useful nor cost-effective to apply the most sophisticated sensors and controls, with the highest performance, to every possible application. Rather, it is important that the design process determines the preferred parameter values.

The preferred values must be compatible with available equipment and sensors and control systems, and should be those appropriate for the particular factory. The parameters associated with the available equipment and sensors and control systems drive a functional process of modeling the manufacturing operation and facility. The parameters determine how the real-world equipment constraints will be incorporated into the functional design process. In turn, as many different functional configurations are considered, the cost-benefit relations of these alternatives can be evaluated and preferred parameter values determined. So long as these preferred values are within the limits of available automated manufacturing equipment and sensory and control systems, the design group is assured that the automated manufacturing equipment can meet its requirements. To the degree that optimum design configurations exceed present equipment capabilities, original equipment manufacturers (OEMs) are motivated to develop new equipment designs and advanced sensors and control systems.

Sensors and control systems, actuators/effectors, controllers, and control loops must be considered in order to appreciate the fundamental limitations associated with manufacturing equipment for error-free manufacturing. Many levels of factory automation are associated with manufacturing equipment; the objective at all times should be to choose the levels of automation and information flow that are appropriate for the facility being designed, as revealed through cost-benefit studies. Manufacturing facilities can be designed by describing each manufacturing system—and the sensors and controls to be used in it—by a set of functional parameters. These parameters are:

1. The number of product categories for which the automated manufacturing equipment, sensors, and control systems can be used (with software downloaded for each product type)
2. The mean time between operator interventions (MTOI)
3. The mean time of intervention (MTI)
4. The percentage yield of product of acceptable quality
5. The mean processing time per product

An *ideal equipment unit* would be infinitely flexible so that it could handle any number of categories desired, would require no operator intervention between setup times, would produce only product of acceptable quality, and would have unbounded production capabilities.

The degree to which real equipment containing sensors and control systems can approach this ideal depends on the physical constraints associated with the design and operation of the equipment and the ability to obtain instantaneous information about equipment performance through sensors and control systems. The performance of the equipment in each of the five parameters above is related to details of the equipment's operation in an error-free environment. Relationships must be developed between the physical description of the equipment's operation and the functional parameters that will be associated with this operation. The objective is to link together the physical design of the equipment and its functional performance through sensory and control systems in the factory setting.

This concept provides insight into an area in which future manufacturing system improvements would be advantageous, and also suggests the magnitude of the cost-benefit payoffs that might be associated with various equipment designs. It also reveals the operational efficiency of such systems.

An understanding of the relationships between the equipment characteristics and the performance parameters based on sensors and control systems can be used to select the best equipment for the parameter requirements associated with a given factory configuration. In this way, the manufacturing design team can survey alternative types of available equipment and select the units that are most appropriate for each potential configuration.

NUMBER OF PRODUCTS IN A FLEXIBLE SYSTEM

The first parameter listed above, the number of product categories for which the manufacturing system can be used, represents the key concern in flexible manufacturing. A unit of automated manufacturing equipment is described in terms of the number of product categories for which it can be used with only a software download to distinguish among product types. A completely fixed automated manufacturing system that cannot respond to computer control might be able to accommodate only one product category without a manual setup. On the other hand, a very flexible manufacturing system would be able to accommodate a wide range of product categories with the aid of effective sensors and control systems. This parameter will thus be defined by the breadth of the processes that can be performed by an automated manufacturing equipment unit and the ability of the unit to respond to external control data to shift among these operations.

The most effective solution will depend on the factory configuration that is of interest. Thus OEMs are always concerned with anticipating future types of factories in order to ensure that their equipment will be an optimum match to the intended configuration. This also will ensure that the *concept of error-free manufacturing can be implemented with a high degree of spontaneity.* There is a continual trade-off between flexibility and cost. In general, more flexible and "smarter" manufacturing equipment will cost more. Therefore, the objective in a particular setting will be to achieve just the required amount of flexibility, without any extra capability built into the equipment unit.

SENSORS TRACKING THE MEAN TIME BETWEEN OPERATOR INTERVENTIONS

The MTOI value should be matched to the factory configuration in use. In a highly manual operation, it may be acceptable to have an operator intervene frequently. On the other hand, if the objective is to achieve operator-independent manufacturing between manual setups, then the equipment must be designed so that the MTOI is longer than the planned duration between manual setups. The manufacturer of automated equipment with adequate sensors and control systems must try to assess the ways in which factories will be configured and produce equipment that can satisfy manufacturing needs without incurring any extra cost due to needed features.

SENSORS TRACKING THE MEAN TIME OF INTERVENTION

Each time an intervention is required, it is desirable to compare the intervention interval with that for which the system was designed. If the intervention time becomes large with respect to the planned mean time between operator interventions, then the efficiency of the automated manufacturing equipment drops rapidly in terms of the fraction of time it is available to manufacture the desired product.

SENSORS TRACKING YIELD

In a competitive environment, it is essential that all automated manufacturing equipment emphasize the production of quality product. If the automated manufacturing equipment produces a large quantity of product that must be either discarded or reworked, then the operation of the factory is strongly affected, and costs will increase rapidly. The objective, then, is to determine the product yields that are required for given configurations and to design automated manufacturing equipment containing sensors and control systems that can achieve these levels of yield. Achieving higher yield levels will, in general, require additional sensing and adaptability features for the equipment. These features will enable the equipment to adjust and monitor itself and, if it gets out of alignment, to discontinue operation.

SENSORS TRACKING THE MEAN PROCESSING TIME

If more product units can be completed in a given time, the cost of automated manufacturing equipment with sensors and control systems can be more widely amortized. As the mean processing time is reduced, the equipment can produce more product units in a given time, reducing the manufacturing cost per unit. Again, automated manufacturing equipment containing sensory and control systems generally becomes more expensive as the processing time is reduced. Tradeoffs are generally necessary among improvements in the five parameters and the cost of equipment. If high-performance equipment is to be employed, the factory configuration must make effective use of the equipment's capabilities to justify its higher cost. On the other hand, if the factory configuration does not require the highest parameters, then it is far more cost-effective to choose equipment units that are less sophisticated but adequate for the purposes of the facility. This interplay between parameter values and equipment design and cost is an essential aspect of system design.

Table 3.1 illustrates the difference between available parameter values and optimum parameter values, where the subscripts for equipment E represent increasing levels of complexity. The table shows the type of data that can be collected to evaluate cost and benefits. These data have significant impact on system design and perfor-

TABLE 3.1 Values of Available Parameters

Equipment	MTOI, min	MTI, min	Yield, %	Process, min	R&D expense, thousands of dollars	Equipment cost, thousands of dollars
Production function A							
E_1	0.1	0.1	90	12		50
E_2	1.0	0.1	85	8		75
E_3	10	1.0	80	10		85
E_4	18	1.0	90	8	280	155
Production function B							
E_1	1.0	0.1	95	10		150
E_2	10	0.5	90	2		300
Production function C							
E_1	0.1	0.1	98	3		125
E_2	5.0	1.0	98	2		250
E_3	8.0	2.0	96	1		300
E_4	20	2.0	96	1	540	400

mance which, in turn, has direct impact on product cost. Given the type of information in Table 3.1, system designers can evaluate the effects of utilizing various levels of sensors and control systems on new equipment and whether they improve performance enough to be worth the research and development and production investment.

One of the difficulties associated with manufacturing strategy in the United States is that many companies procure manufacturing equipment only from commercial vendors and do not consider modifying it to suit their own needs. Custom modification can produce pivotal manufacturing advantage, but also requires the company to expand both its planning scope and product development skills. The type of analysis indicated in Table 3.1 may enable an organization to determine the value and return on investment of customizing manufacturing equipment to incorporate advanced sensors and control systems. Alternatively, enterprises with limited research and development resources may decide to contract for development of the optimum equipment in such a way that the sponsor retains proprietary rights for a period of time.

NETWORK OF SENSORS DETECTING MACHINERY FAULTS

A comprehensive detection system for automated manufacturing equipment must be seriously considered as part of the manufacturing strategy. A major component of any effort to develop an intelligent and flexible automatic manufacturing system is the concurrent development of automated diagnostic systems, with a network of sensors, to handle machinery maintenance and process control functions. This will undoubtedly lead to significant gains in productivity and product quality. Sensors and control systems are one of the enabling technologies for the "lights-out" factory of the future.

A flexible manufacturing system often contains a variety of manufacturing work cells. Each work cell in turn consists of various workstations. The flexible manufacturing cell may consist of a CNC lathe or mill whose capabilities are extended by a robotic handling device, thus creating a highly flexible machining cell whose functions are coordinated by its own computer. In most cases, the cell robot exchanges workpieces, tools (including chucks), and even its own gripping jaws in the cell (Fig. 3.1).

Diagnostic Systems

A diagnostic system generally relies on copious amounts of *a priori* and *a posteriori* information. *A priori* information is any previously established fact or relationship that the system can exploit in making a diagnosis. *A posteriori* information is the information concerning the problem at hand for which the diagnosis will be made. The first step in collecting data is to use sensors and transducers to convert physical states into electrical signals. After processing, a signal will be in an appropriate form for analysis (perhaps as a table of values, a time-domain waveform, or a frequency spectrum). Then the analysis, including correlations with other data and trending, can proceed.

After the data have been distilled into information, the deductive process begins, leading finally to the fault diagnosis. Expert systems have been used effectively for diagnostic efforts, with the diagnostic system presenting either a single diagnosis or a set of possibilities with their respective probabilities, based on the *a priori* and *a posteriori* information.

FIGURE 3.1 Flexible machining cell.

Resonance and Vibration Analysis

Resonance and vibration analysis is a proven method for diagnosing deteriorating machine elements in steady-state process equipment such as turbomachinery and fans. The effectiveness of resonance and vibration analysis in diagnosing faults in machinery operating at variable speed is not proved, but additional study has indicated good potential for its application in robots. One difficulty with resonance and vibration analysis is the attenuation of the signal as it travels through a structure on the way to the sensors and transducers. Moreover, all motion of the machine contributes to the motion measured by the sensors and transducers, so sensors and transducers must be located as close as possible to the component of concern to maximize the signal-to-noise ratio.

Sensing Motor Current for Signature Analysis

Electric motors generate back electromotive force (emf) when subjected to mechanical load. This property makes a motor a transducer for measuring load vibrations via current fluctuations. Motor current signature analysis uses many of the same techniques as vibration analysis for interpreting the signals. But motor current signature analysis is nonintrusive because motor current can be measured anywhere along the motor power cables, whereas a vibration sensor or transducer must be mounted close to the machine element of interest. The limited bandwidth of the signals associated with motor drive signature analysis, however, may restrict its applicability.

Acoustics

A good operator can tell from the noise that a machine makes whether a fault is developing or not. It is natural to extend this concept to automatic diagnosis. The operator, obviously, has access to subtle, innate pattern recognition techniques, and thus is able to discern sounds within a myriad of background noises. Any diagnostic system based on sound would have to be able to identify damage-related sounds and separate the information from the ambient noise. Acoustic sensing (looking for sounds that indicate faults) is a nonlocal, noncontact inspection method. Any acoustic technique is subject to outside disturbances, but is potentially a very powerful tool, provided that operating conditions are acoustically repeatable and that the diagnostic system can effectively recognize acoustic patterns.

Temperature

Using temperature as a measurement parameter is common, particularly for equipment running at high speed, where faults cause enough waste heat to raise temperature significantly. This method is generally best for indicating that a fault has occurred, rather than the precise nature of the fault.

Sensors for Diagnostic Systems

Assuming an automated diagnostic system is required, the necessary sensors are normally mounted permanently at their monitoring sites. This works well if data are required continuously, or if there are only a few monitoring locations. However, for those cases where many sites must be monitored and the data need not be continuously received during operation of the flexible manufacturing cell, it may be possible to use the same sensor or transducer, sequentially, in the many locations.

The robot is well-suited to gathering data at multiple points with a limited number of sensors and transducers. This would extend the mandate of the robot from simply moving workpieces and tools within the flexible manufacturing cell (for production) to include moving sensors (for diagnostic inspection).

Within the flexible manufacturing cell, a robot can make measurements at sites inside its work space by taking a sensor or transducer from a tool magazine, delivering the sensor or transducer to a sensing location, detaching it during data collection, and then retrieving it before moving to the next sensing position.

Sensor mobility does, however, add some problems. First, the robot will not be able to reach all points within the flexible manufacturing cell because its work space is only a subspace of the volume taken up by the flexible manufacturing cell. The manipulator may be unable to assume an orientation desired for a measurement even inside its work space. Also, the inspection procedure must limit the robot's influence on the measurement as much as possible. Finally, sensors require connectors on the robot end effectors for signals and power. The end effector would have to be able to accommodate all the types of sensors to be mounted on it.

Quantifying the Quality of a Workpiece

If workpiece quality can be quantified, then quality can become a process variable. Any system using product quality as a measure of its performance needs tight error checks so as not to discard product unnecessarily while the flexible manufacturing cell adjusts its operating parameters. Such a system would depend heavily, at first, on the continued supervision of an operator who remains in the loop to assess product quality. Since it is forbidden for the operator to influence the process while it is under automatic control, it is more realistic for the operator to look for damage to product after each stage of manufacture within the cell. In that way, the flexible manufacturing cell receives diagnostic information about product deficiencies close to the time that improper manufacture occurred.

In the future, these quality assessments will be handled by the flexible manufacturing cell itself, using sensors and diagnostic information for process control. Robots, too, will be used for maintenance and physical inspection as part of the regular operation of the flexible manufacturing cell. In the near term, the flexible manufacturing cell robot may be used as a sensor-transfer device, replacing inspectors who would otherwise apply sensors to collect data.

Evaluation of an Existing Flexible Manufacturing Cell Using a Sensing Network

A study was conducted at the Mi-TNO in the Netherlands of flexible manufacturing cells for low-volume orders (often called job production, ranging from 1 to 100 parts per order). The automated manufacturing equipment used in the study consisted of two free-standing flexible manufacturing cells. The first cell was a turning-machine cell; the second, a milling-machine cell. The turning cell contained two armed gantry robots for material handling. The study was mainly conducted to assess the diagnostics for flexible manufacturing systems (FMS). In considering the approach to setting up diagnostics for an FMS, it was decided to divide development of the diagnostics program into three major blocks (Fig. 3.2):

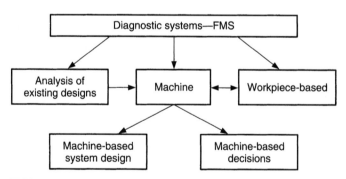

FIGURE 3.2 Diagnostics for an FMS.

1. Analyzing the existing design
2. Setting up diagnostics that are machine-based
    a. Choosing important points in the flexible manufacturing cells where critical failure can occur and where sensors are mounted
    b. Setting up a diagnostic decision system for the hardware system

3. Establishing a workpiece-based diagnostic system that is actually part of quality control

The flexible manufacturing cells were divided into control volumes (as shown in Fig. 3.3 for the turning cell). For the turning cell, for example, the hardware control volumes were denoted *a, b, c, d,* and *e,* and software control volumes *p1, p2, p3,* and *p4.* The failures within each of these control volumes were further categorized as:

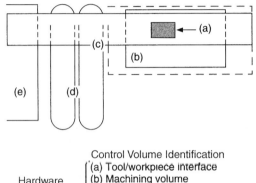

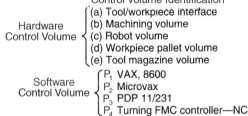

FIGURE 3.3 Hardware control volumes in a manufacturing cell.

1. Fault detected by an existing sensor
2. Fault that could have been detected if a sensor had been present
3. Operator learning phase problem
4. Failure due to manufacturer problem
5. Software logic problem
6. Repeat of a problem
7. System down for an extended period

Software Problems. It is often assumed that disturbances in the cell software control system can be detected and evaluated relatively easily. Software diagnostics are common in most turnkey operations; however, it has been shown that software diagnostics are far from perfect. Indeed, software problems are of particular concern when a revised program is introduced into a cell which has been running smoothly (existing bugs having been ironed out). The availability of the cell plummets in these situations, with considerable loss of production capabilities and concomitant higher cost. The frequency of software faults increases dramatically when revised packages are introduced to upgrade the system (Fig. 3.4). This is mainly due to human error on the part of either the vendor or the operator. Two possible approaches to software defect prevention and detection are:

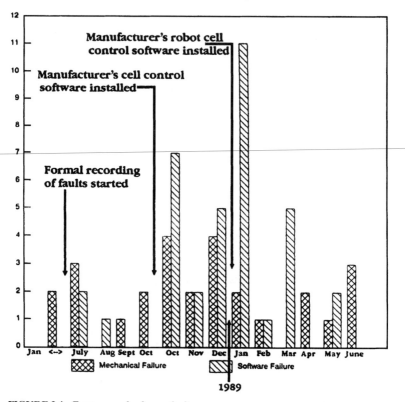

FIGURE 3.4 Frequency of software faults.

1. Investigate software methodologies and procedures and recommend alternative languages or more tools as defect prevention measurements. This is tedious and uses ambiguous results because such investigations are not based on data.
2. Analyze the problems that result from the current design and develop a solution for each class of problem. This produces less ambiguous solutions and is typically used to solve only immediate problems, thereby producing only short-term solutions.

To identify the types of faults that occur in programs, it is necessary to know what caused the problem and what remedial actions were taken. Program faults can be subdivided into the categories shown below and restated in Figure 3.5:

- Wrong names of global variables or constants
- Wrong type of structure or module arguments

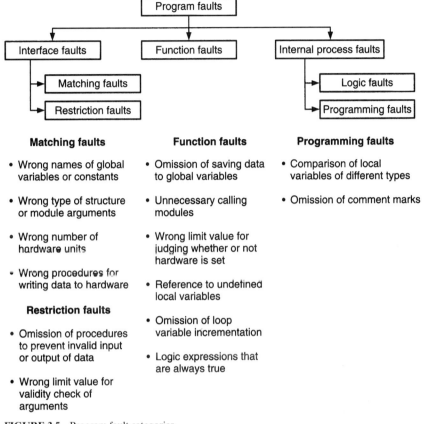

FIGURE 3.5 Program fault categories.

- Wrong number of hardware units
- Wrong procedures for writing data to hardware

  Restriction faults

- Omission of procedures to prevent invalid input or output of data
- Wrong limit value for validity check of arguments

  Function faults

- Omission of saving data to global variables
- Unnecessary calling modules
- Wrong limit value for judging whether or not hardware is set
- Reference to undefined local variables
- Omission of loop variable incrementation
- Logic expressions that are always true

Programming faults

- Comparison of local variables of different types
- Omission of comment marks

This categorization provides significant insight into the location of fault conditions, the reasons for their occurrence, and their severity. If the faults are in either a hardware (mechanical) or software category, then the frequency of failure by month can be summarized as indicated in Fig. 3.4. Two major and unexpected milestones in the program represented in Fig. 3.4 are the routine introduction of revised cell control software and revised robot control software. In both cases there was a substantial increase in the downtime of the flexible manufacturing cell. In an industrial environment this would have been very costly.

In this study, it was found that interface faults (mismatched data transfer between modules and hardware) were the major cause of downtime. Also, in the new machine software, it was found faults occurred because the software had not been properly matched to the number of tool positions physically present on the tool magazine. Once, such a fault actually caused a major collision within the machining volume.

Detecting Tool Failure. An important element in automated process control is real-time detection of cutting tool failure, including both wear and fracture mechanisms. The ability to detect such failures on line allows remedial action to be undertaken in a timely fashion, thus ensuring consistently high product quality and preventing potential damage to the process machinery. The preliminary results from a study to investigate the possibility of using vibration signals generated during face milling to detect both progressive (wear) and catastrophic (breakage) tool failure are discussed below.

Experimental Technique. The experimental studies were carried out using a 3-hp vertical milling machine. The cutting tool was a 381-mm-diameter face mill employing three Carboloy TPC-322E grade 370 tungsten carbide cutting inserts. The standard workpiece was a mild steel plate with a length of 305 mm, a height of 152 mm, and a width of 13 mm. While cutting, the mill traversed the length of the workpiece, performing an interrupted symmetric cut. The sensor sensed the vibration generated during the milling process on the workpiece clamp. The vibration signals were recorded for analysis. Inserts with various magnitudes of wear and fracture (ranging from 0.13 mm to 0.78 mm) were used in the experiments.

Manufacturing Status of Parts. Figure 3.6 shows typical acceleration versus time histories. Figure 3.7a is the acceleration for three sharp inserts. Note that the engagement of each insert in the workpiece is clearly evident and that all engagements share similar characteristics, although they are by no means identical.

Figure 3.7b shows the acceleration for the combination of two sharp inserts and one insert with a 0.39-mm fracture. The sharp inserts produce signals consistent with those shown in Fig. 3.6, while the fractured insert produces a significantly different output.

The reduced output level for the fractured insert is a result of the much smaller depth of cut associated with it. It would seem from the time-domain data that use of either an envelope detection or a threshold crossing scheme would provide the ability to automate the detection of tool fracture in a multi-insert milling operation.

Figure 3.7 shows typical frequency spectra for various tool conditions. It is immediately apparent that, in general, fracture phenomena are indicated by an increase in the level of spectra components within the range of 10,000 to 17,000 Hz. A comparison of Fig. 3.7a and b indicates a notable increase in the spectra around 11 kHz

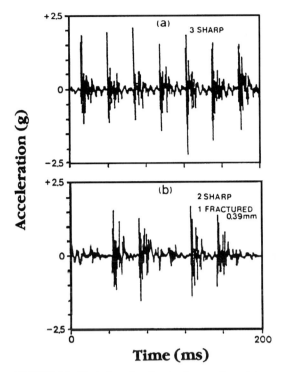

FIGURE 3.6 Typical acceleration level versus time plot.

when a single insert fracture of 0.39 mm is present. For two fractured inserts (Fig. 3.7c), the peak shifts to around 13.5 kHz. For three fractured inserts (Fig. 3.7d), both the 13.5-kHz peak and an additional peak at about 17 kHz are apparent.

Comparing Figs. 3.7e and c, it is seen that, in general, increasing the size of insert fracture results in an increase in the spectral peak associated with the failure condition. Usually, the assemblies used to obtain the spectral data are not synchronously averaged. Therefore, it may be possible to improve the spectral data by utilizing a combination of synchronous averaging and delayed triggering to ensure that data representative of each insert in the cutter is obtained and processed.

In general, the acceleration-time histories for the worn inserts do not produce the noticeably different engagement signals evident in a case of fracture. However, by processing the data in a slightly different manner, it is possible to detect evidence of tool wear.

Figure 3.8 shows the amplitude probability density (APD) as a function of time for several tool conditions. The data are averaged for eight assemblies. It thus seems possible that insert wear could be detected using such features as the location of the peak in the APD, the magnitude of peak, and the area under specific segments of the distribution.

As with fracture, the presence of insert wear resulted in a significant increase in the spectral components within the 10- to 13-kHz band. Although this would seem to indicate that the presence of flank wear could be detected by simple spectral analysis, it is not yet clear if this method would be sufficiently discriminating to permit reliable determination of the magnitude of flank wear.

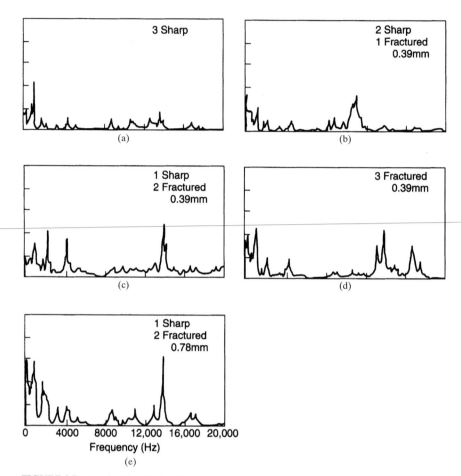

FIGURE 3.7 Acceleration level versus time for various inserts.

UNDERSTANDING COMPUTER COMMUNICATIONS AND SENSORS' ROLE

The evolution in computer software and hardware has had a major impact on the capability of computer-integrated manufacturing concepts. The development of smart manufacturing equipment and sensors and control systems, as well as methods of networking computers, has made it feasible to consider cost-effective computer applications that enhance manufacturing. In addition, this growth has changed approaches to design of manufacturing facilities.

Messages are exchanged among computers according to various protocols. The open system interconnect (OSI) model developed by the International Standards Organization (ISO) provides such a framework. Figure 3.9 illustrates how two users might employ a computer network. As illustrated, the transfer of information takes

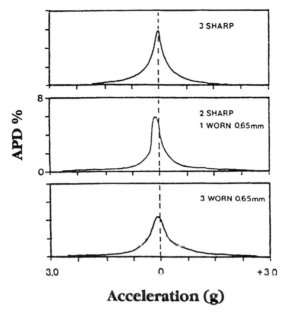

FIGURE 3.8 Amplitude probability density (APD) for several tool conditions.

place from User 1 to User 2. Each message passes through a series of layers that are associated with message processing:

Sequence of layers—User 1	Sequence of layers—User 2
(PhL)1 Physical layer	(PhL)1 Physical layer
(DL)2 Data link layer	(DL)2 Data link layer
(NL)3 Network layer	(NL)3 Network layer
(TL)4 Transport layer	(TL)4 Transport layer
(SL)5 Session layer	(SL)5 Session layer
(PL)6 Presentation layer	(PL)6 Presentation layer
(AL)7 Application layer	(AL)7 Application layer

A message is sent from User 1 to User 2, with message acknowledgment sent back from User 2 to User 1. The objective of this communication system is to transfer a message from User 1 to User 2 and to confirm message receipt. The message is developed at application layer (AL)7, and passes from there to presentation layer (PL)6, from there to session layer (SL)5, and so forth until the message is actually transmitted over the communication path. The message arrives at physical layer (PhL)1 for User 2 and then proceeds from physical layer (PhL)1 to data link layer (DL)2, to network layer (NL)3, and so forth, until User 2 has received the message. In order for Users 1 and 2 to communicate with one another, every message must pass though all the layers.

The layered approach provides a structure for the messaging procedure. Each time a message moves from User 1 [application layer (AL)7 to physical layer (PhL)1] additional processing and addressing information is added to the beginning or end of the

3.16 CHAPTER THREE

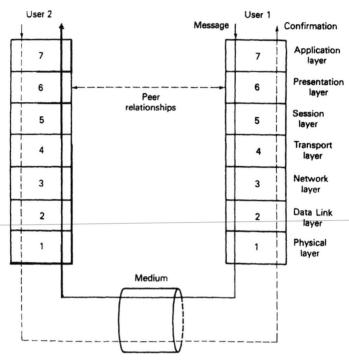

FIGURE 3.9 Message movement within a computer network.

message. As the original message moves on, new information is added to ensure correct communication. Then, as the message moves to User 2 [from physical layer (PhL)1 to application layer (AL)7], this additional information is removed as illustrated in Fig. 3.9.

The layers work together to achieve "peer" communication. Any information or operation instruction that is added at session layer (SL)5 for User 1 will be addressing session layer (SL)5 for User 2. The layers thus work as peers; each layer has a given operational or addressing task to make sure the message is correctly communicated from User 1 to User 2. Each layer associates only with the layers above and below itself. The layer receives messages from one direction, processes the messages, and then passes them on to the next layer.

Application Layer Communication

Communication begins when User 1 requests that a message be transferred from location 1 to location 2. In developing this message, it may be necessary to have application software that will provide supporting services to the users. This type of service is provided by application layer (AL)7. Common software application tools enable, for example, the transfer of files or the arranging of messages in standard formats.

Presentation Layer Communication

The initial message is passed down from the application layer (AL)7 to presentation layer (PL)6, where any necessary translation is performed to develop a common message syntax that User 2 will understand. If the two different users apply different computer or equipment languages, it will be necessary to define the differences in such a way that the users can communicate with one another. The basic message that began with User 1 is translated to a common syntax that will result in understanding by User 2. Additional information is added to the message at presentation layer (PL)6 to explain to User 2 the nature of the communication that is taking place. An extended message begins to form.

Session Layer Communication

The message now passes from presentation layer (PL)6 to session layer (SL)5. The objective of the session layer is to set up the ability for the two users to converse, instead of just passing unrelated messages back and forth. The session layer will remember that there is an ongoing dialog and will provide the necessary linkages between the individual messages so that an extended transfer of information can take place.

Transport Layer Communication

The message then passes from session layer (SL)5 to transport layer (TL)4, which controls the individual messages as part of the communication sequence. The purpose of the transport layer is to make sure that individual messages are transferred from User 1 to User 2 as part of the overall communication session that is defined by session layer (SL)5. Additional information is added to the message so that the transport of this particular portion of the communication exchange results.

Network Layer Communication

The message is then passed to network layer (NL)3, divided into packets, and guided to the correct destination. The network layer operates so that the message (and all accompanying information) is tracked and routed correctly so that it will end up at User 2. This includes making sure that addresses are correct and that any intermediate stops between User 1 and User 2 are completely defined.

Data Link Layer Communication by Fiber Optics or Coaxial Cable

The system now passes a message to data link layer (DL)2, which directs each frame of the message routing in transit. Each frame is loaded into the communication system in preparation for transmission. The message is prepared to leave the User 1 environment and to move into the communication medium. The next step is from data link layer (DL)2 to physical layer (PhL)1. At this point, the frame is converted into series of digital or analog electronic signals that can be placed on the communication medium itself—wire, fiber optics, coaxial cables, or other means—to achieve the transmission of the message from User 1 to User 2.

Physical Layer Communication

The electronic signal arrives at the correctly addressed location for User 2 and is received by physical layer (PhL)1. The physical layer then converts the electronic signal back to the original frame that was placed on the medium. This extended message is passed up to the data link layer, which confirms that error-free communication has taken place and that the frame has been received at the correct location. Data link layer (DL)2 directs the frame routing in transit. When the full frame is received at data link layer (DL)2, the routing information is removed and the remaining information is transferred to network layer (NL)3. Network layer (NL)3 confirms the appropriate routing and assembles the packets. Then the routing information is stripped from the message. The message is then passed to transport layer (TL)4, which controls the transport of the individual messages. Transport layer (TL)4 confirms that the correct connection has been achieved between User 1 and User 2 and receives the information necessary to achieve the appropriate connection for this particular exchange.

The remaining information is often passed to session layer (SL)5, which interprets whether the message is part of a continuing communication and, if so, identifies it as part of an ongoing conversation. The information is then passed to presentation layer (PL)6, which performs any necessary translation to make sure that User 2 can understand the message as it has been presented. The message is then passed to application layer (AL)7, which identifies the necessary support programs and software that are necessary for interpretation of the message by User 2. Finally, User 2 receives the message and understands its meaning.

Adding and Removing Information in Computer Networks Based on Open System Interconnect (OSI)

This step-by-step passing of the message from User 1 "down" to the actual communication medium and "up" to User 2 involves adding information to the original message prior to the transfer of information and removing this extra information on arrival (Fig. 3.10). At User 1, additional information is added step by step until the communication medium is reached, forming an extended message. When this information arrives at User 2, the additional information is removed step by step until User 2 receives the original message. As noted, a peer relationship exists between the various levels. Each level communicates only with the level above or below it, and the levels perform a matching service. Whatever information is added to a message by a given level at User 1 is removed from the message by the matching level associated with User 2. Communication is made possible by having User 1's physical layer (PhL)1 communicate with User 2's physical layer (PhL)1, data link layer (DL)2 communicate with data link (DL)2, network layer (NL)3 communicate with network layer (NL)3, and so forth to achieve an exchange between User 1 and User 2.

The two users see only the message that is originated and delivered; they do not see all of the intermediate steps. This is analogous to the steps involved in making a telephone call or sending a letter, in which the two users know only that they have communicated with one another, but do not have any specific knowledge of the details involved in passing a message from one location to another. This orderly and structured approach to communications is useful because it separates the various tasks that must take place. It provides a means for assuring that the methods for processing and addressing messages are always the same at every node. Whether a message is being sent or is being received, a sequential processing activity always takes place.

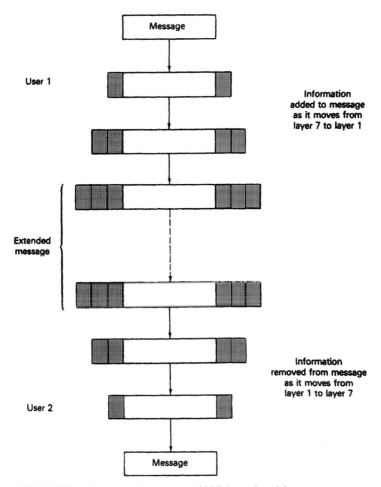

FIGURE 3.10 Open system interconnect (OSI) layered model.

UNDERSTANDING NETWORKS IN MANUFACTURING

A seemingly simple problem in the development of computer networks is to establish the ability to interconnect *any* two computer system elements. This might involve a computer terminal, a modem, a bar-code scanner, a printer, sensors, and other system elements that must exchange information in manufacturing. It might seem reasonable that such one-to-one interconnection would follow a well-defined strategy and make maximum use of standards. Unfortunately, for historical reasons and because of the wide diversity of the equipment units that are available today, this situation does not hold. In fact, achieving interconnection between typical system elements can be a disconcerting experience.

RS-232–Based Networks

One of the most common approaches used to interconnect computer system elements in manufacturing is associated with a strategy that was never really intended for this purpose. As noted by Campbell (1984), "In 1969, the EIA (Electronic Industries Association), Bell Laboratories, and manufacturers of communication equipment cooperatively formulated and issued 'EIA RS-232,' which almost immediately underwent minor revisions to become 'RS-232-C.' The RS-232 interface was developed to allow data equipment terminals to be connected to modems so that data could be transmitted over the telephone network. The entire purpose of this standard was to assure that the use of telephone lines for computer communications would be handled in a way acceptable to the telephone company.

Thus, in its general application today, RS-232 is not a standard. It is more a guideline for addressing some of the issues involved in interfacing equipment. Many issues must be resolved on an individual basis, which leads to the potential for difficulties. Essentially, a vendor's statement that a computer system element is RS-232-compatible provides a starting point to consider how the equipment unit might be interconnected. However, the detailed aspects of the interconnection require further understanding of the ways in which equipment units are intended to communicate. Campbell (1984) is a helpful introduction to applying RS-232 concepts.

In a sense, the history of the RS-232 interface illustrates the difficulties associated with creating a well-defined means for allowing the various elements of a computer network to interface. Past experience also indicates how difficult it is to develop standards that will apply in the future to all the difficult situations that will be encountered. As it has evolved, the RS-232 approach to the communications interface is an improvement over the "total anarchy" at position A in Fig. 3.11, but it still leads to a wide range of problems.

An entire computer network can be configured by using combinations of point-to-point RS-232 connections. In fact, a number of networks of this type are in common use. Such networks require that multiple RS-232 interfaces be present on each equipment unit, as is often the case. Each particular interconnection must be customized for the two equipment units being considered. Thus, a system integrator not only must decide on the elements of the system and how they should perform in a functional sense, but also must develop a detailed understanding of the ways in which RS-232 concepts have been applied to the particular equipment units that are used for the network. The system integrator incurs a substantial expense in achieving the required interconnections.

It is essential to realize that the RS-232 pseudo-standard only addresses the ability to transfer serially information bit by bit from one system element to another. The higher level communications protocol in Fig. 3.11 is not considered. RS-232 provides the means for running wires or fiber-optic cables from one element to another in order to allow digital signals to be conveyed between system elements. The meaning associated with these bits of information is completely dependent on the hardware and software that is implemented in the system elements. RS-232 is a widely used approach for computer elements to transfer information. Using RS-232 is certainly much better than using no guidelines at all. However, because RS-232 does not completely define all of the relationships that must exist in communication links, it falls far short of being a true standard or protocol.

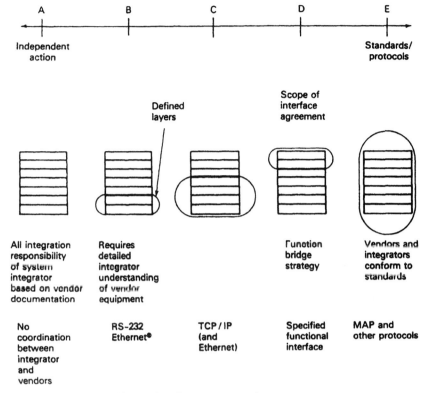

FIGURE 3.11 Levels of integration of computer networks.

Ethernet*

As illustrated in Fig. 3.11, one approach to local-area networks (LANs) is to define a protocol for the first two layers of a communication strategy and then allow individual users to define the upper layers. This approach has been widely applied using a method referred to as *Ethernet* [Metcalfe and Boggs (1976), Shock and Hupp (1980), Tanenbaum (1988)].

In every computer communication system, there must be a means of scheduling for each node to transmit onto the network and listen to receive messages. This may be done on a statistical basis. For example, when a unit needs to transmit over the network, it makes an effort to transmit. If another node tries to transmit at the same time, both nodes become aware of the conflict, wait for a random length of time, and try again. It might seem that this would be an inefficient means of controlling a network, since the various nodes are randomly trying to claim the network for their own use, and many collisions may occur. As it turns out, for lower communication volumes, this method works very well. As the number of nodes on the system and the number of messages being exchanged increases, however, the number of collisions between active nodes goes up and reduces the effectiveness of the system (Fig. 3.12).

---

* Ethernet is a trademark of Xerox Corp.

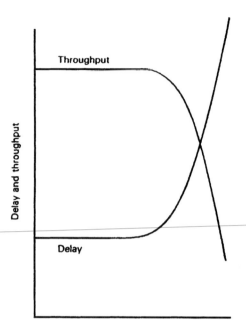

FIGURE 3.12 Delay and throughput versus network load (number of users).

This type of access control for a computer network is referred to as *carrier-sense multiple-access with collision detection* (CSMA/CD). Ethernet and similar solutions are widely applied to create CSMA/CD networks, particularly in settings in which a maximum waiting time for access to the network does not have to be guaranteed. This type of network is simple to install, and a wide range of hardware and software products are available for support. On the other hand, as indicated in Fig. 3.12, network performance can degrade significantly under high load; therefore, the utility of an Ethernet-oriented network will depend on the particular configuration and loads that are expected for the network.

Transmission Control Protocol (TCP)/Internet Protocol (IP)

TCP/IP applies to the transport and network layers indicated in Fig. 3.9. TCP/IP thus provides a means for addressing intermediate protocol levels, and in fact is often combined with Ethernet in a communication approach that defines both the lower and middle aspects of the system. TCP/IP functions by dividing any message provided to these middle layers into *packets* of 64 kbytes and then sending packets one at a time to the communication network. TCP/IP must also reassemble the packets in the correct order at the receiving user.

TCP/IP provides a common strategy to use for networking. It allows extension of the Ethernet lower layers to a midlayer protocol on which the final application and presentation layers may be constructed.

MANUFACTURING AUTOMATION PROTOCOL

The manufacturing automation protocol (MAP) is one of the protocols that has been developed for computer communication systems. MAP was developed specifically for use in a factory environment. General Motors Corp. has been the leading advocate of this particular protocol. When faced with a need for networking many types of equipment in its factory environment, General Motors decided that a new type of protocol was required. Beginning in 1980, General Motors began to develop a protocol that could accommodate the high data rate expected in its future factories and provide the necessary noise immunity expected for this environment. In addition, the intent was to work within a mature communications technology and to develop a protocol that could be used for all types of equipment in General Motors factories. MAP was developed to meet these needs. The General Motors effort has drawn on a combination of Institute of Electrical and Electronics Engineers (IEEE) and ISO standards, and is based on the open system interconnect (OSI) layered model as illustrated in Fig. 3.10.

Several versions of MAP have been developed. One difficulty among many has been obtaining agreement among many different countries and vendor groups on a specific standard. Another problem is that the resulting standards are so broad that they have become very complex, making it difficult to develop the hardware and software to implement the system and consequently driving up related costs. The early version of MAP addressed some of the OSI layers to a limited degree, and made provision for users to individualize the application layer for a particular use. The latest version of MAP makes an effort to define more completely all the application layer software support as well as the other layers. This has led to continuing disagreements and struggles to produce a protocol that can be adopted by every vendor group in every country to achieve MAP goals.

Because of its complexity, MAP compatibility among equipment units, or *interoperability*, has been a continuing difficulty. MAP has not been applied as rapidly as was initially hoped for by its proponents because of the complexity, costs, and disagreements on how it should be implemented. Assembling a complete set of documentation for MAP is a difficult activity that requires compiling a file of standards organization reports, a number of industry organization reports, and documentation from all the working committees associated with ISO.

Broadband System for MAP Protocol

The MAP protocol was developed with several alternatives for physical layer (PhL)1. MAP can be implemented through what is called a *broadband* system (Fig. 3.13). So that the manufacturing units can talk to one another, transmitted messages are placed on the cable; a *head-end remodulator* retransmits these messages and directs them to the receiving station. The broadband version of MAP has the highest capabilities because it allows several types of communications to take place on the same cabling at the same time. On the other hand, because of its greater flexibility, a broadband system is more complex and more expensive to install. It requires modems and MAP interface equipment for each item of equipment and a head-end remodulator to serve the entire network. The main cable used for broadband is unwieldy (approximately 25 mm in diameter) and is appropriate only for wiring very large factories. Multiple drop cables can be branched off the main cable for the different MAP nodes. Broadband communication can achieve a high data rate of 10 Mb/s and can split the frequency spectrum to allow several different communications to take place simultaneously. As indicated in Fig. 3.14, the three transmit fre-

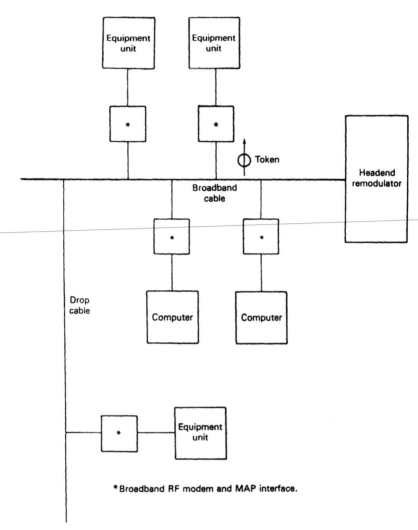

FIGURE 3.13 Broadband system.

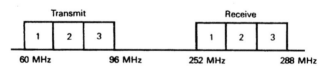

FIGURE 3.14 Broadband frequencies.

quencies and the three receive frequencies are separated from one another in the frequency domain. Three different channels can coexist on the MAP network. The head-end remodulator transfers messages from the low frequencies to the high frequencies.

Carrier-Band System for MAP Protocol

Another type of MAP network makes use of a carrier-band approach, which uses somewhat less expensive modems and interface units and does not require heavy duty cable (Fig. 3.15). For a small factory, a carrier-band version of MAP can be much more cost-effective. The carrier-band communication can also achieve a high data rate of 5 to 10 Mb/s, but only one channel can operate on the cable at a given time. A single channel is used for both transmission and reception.

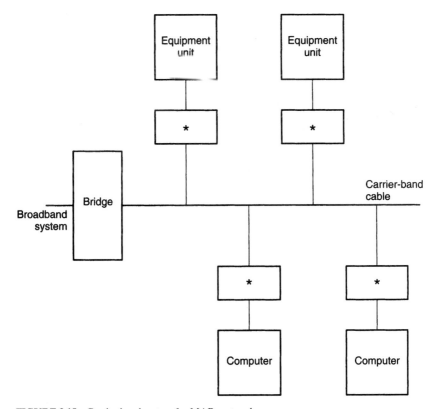

FIGURE 3.15 Carrier-band system for MAP protocol.

Bridges MAP Protocol

It is possible to use devices called *bridges* to link a broadband factory-wide communication network to local carrier-band networks (Fig. 3.16). The bridge transforms the message format provided on one side to the message formats that are required on the other. In this sense, a bridge transforms one operating protocol to another.

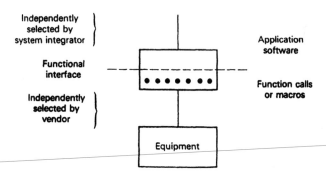

FIGURE 3.16 Bridges system for MAP protocol.

Token System for MAP Protocol

In developing MAP, General Motors was concerned with assuring that every MAP mode would be able to claim control of the network and communicate with other nodes within a certain maximum waiting time. Within this waiting time, every node would be able to initiate its required communications. To do this, MAP implements a *token passing* system (Fig. 3.17). The token in this case is merely a digital word that is recognized by the computer. The token is rotated from node address to node address; a node can claim control of the network and transmit a message only when it holds the token. The token prevents message collisions and also ensures that, for a given system configuration, the maximum waiting time is completely defined (so long as no failures occur). The token is passed around a logic ring defined by the sequence of node addresses, not necessarily by the physical relationships.

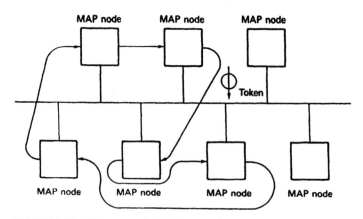

FIGURE 3.17 Token system for MAP protocol.

The token is a control word, and each MAP node can initiate communication only if it passes the token. It is interesting to note that MAP nodes that are not a part of the logic ring will never possess the token, but they may still respond to a token holder if a message is addressed to them. Token management is handled at data link layer (DL)2 of the OSI model. This layer controls the features of the token application with respect to how long a token can be held, the sequence of addresses that is to take place, and the amount of time that is allowed for retrying communications before failure is assumed. If the logic ring is broken at some point—for example, if one equipment unit is no longer able to operate—the other nodes will wait a certain length of time and then will reform the token passing scheme. They will do this by an algorithm through which the token is awarded to the highest station address that contends. The rest of the stations on the ring are then determined by the highest-address successors. This process is repeated until the token ring is re-formed.

Physical layer (PhL)1 involves encoding and modulation of the message so that the digital data are transferred into analog and digital communication signals. Each MAP application requires a modem for this purpose. The modem takes the extended message that has been developed at higher layers and modifies it so that it can be used to provide an electronic signal to the communications medium. The medium itself provides the means for transferring the signal from User 1 to User 2.

MAP continues to be an important protocol approach for sensors and control systems in computer integrated manufacturing. For very large factories the broadband option is available, and for a smaller factories a carrier band system is also available. A number of vendors now produce the hardware and software necessary to establish a MAP network. However, such networks typically are quite high in cost and, because of the complexity of the protocol, can also be difficult to develop and maintain. Thus MAP is only one of several solutions that are available to planning and implementation teams.

MULTIPLE-RING DIGITAL COMMUNICATION NETWORK—AbNET

An optical-fiber digital communication network has been proposed to support the data acquisition and control functions of electric power distribution networks. The optical-fiber links would follow the power distribution routes. Since the fiber can cross open power switches, the communication network would include multiple interconnected loops with occasional spurs (Fig. 3.18). At each intersection a node is needed. The nodes of the communication network would also include power distribution substations and power controlling units. In addition to serving data acquisition and control functions, each node would act as a repeater, passing on messages to the next node.

Network topology is arbitrary, governed by the power system. The token ring protocols that are used in single-ring digital communication networks are not adequate. The multiple-ring communication network would operate on the new AbNET protocol, which has already been developed, and would feature fiber optics, for this more complicated network.

Initially, a message inserted anywhere in the network would pass from node to node throughout the network, eventually reaching all connected nodes. On the first reception of a message, each node would record an identifying number and transmit the message to the next node. On second reception of the message, each node would recognize the identifying number and refrain from retransmitting the message. This

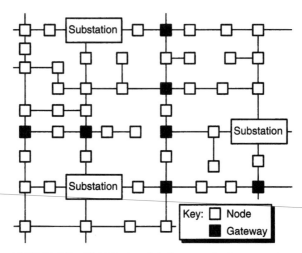

FIGURE 3.18 Multiple-ring digital communication network—AbNET.

would prevent the endless repetition and recirculating of messages. This aspect of the protocol resembles the behavior of cells in the immune system, which learn to recognize invading organisms on first exposure and kill them with antibodies when they encounter the organisms again. For this reason, the protocol is called *AbNET* after the microbiologists' abbreviation *Ab* for *antibodies*. The AbNET protocols include features designed to maximize the efficiency and fault-tolerant nature of the approach. Multiple service territories can be accommodated, interconnected by *gateway* nodes (Fig. 3.18).

The AbNET protocol is expected to enable a network to operate as economically as a single ring that includes an *active monitor* node to prevent the recirculation of messages. With AbNET the performance of the proposed network would probably exceed that of a network that relies on a central unit to route messages. Communications would automatically be maintained in the remaining intact parts of the network even if fibers were broken.

For the power system application, the advantages of optical-fiber communications include electrical isolation and immunity to electrical noise. The AbNET protocols augment these advantages by allowing an economical system to be built with topology-independent and fault-tolerant features.

UNIVERSAL MEMORY NETWORK

The universal memory network (UMN) is a modular, digital data communication system that enables computers with differing bus architectures to share 32-bit-wide data between locations up to 3 km apart with less than 1 ms of latency (Fig. 3.19). This network makes it possible to design sophisticated real-time and near-real-time data processing systems without the data transfer bottlenecks that now exist when

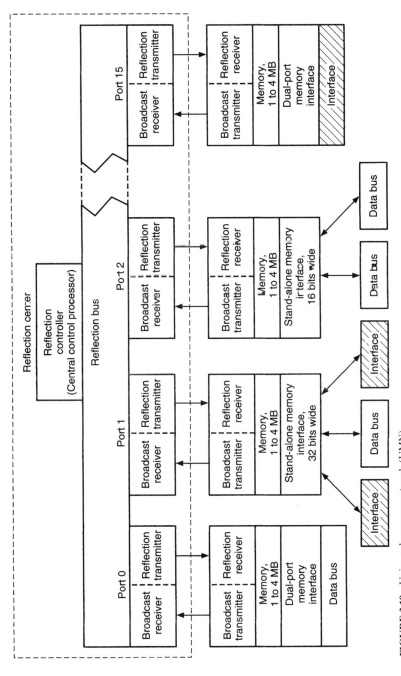

FIGURE 3.19 Universal memory network (UMN).

computers use the usual communication protocols. This enterprise network permits the transmission of the volume of data equivalent to an average encyclopedia each second (40 Mbyte/s). Examples of facilities that can benefit from the universal memory network include telemetry stations, real-time-monitoring through laser sensors, simulation facilities, power plants, and large laboratories (e.g., particle accelerators), or any facility that shares very large volumes of data. The main hub of the universal memory network uses a *reflection center*—a subsystem containing a central control processor (the reflection controller) and a data bus (the reflection bus) equipped with 16 dual memory parts. Various configurations of host computers, workstations, file servers, and small networks or subnetworks of computers can be interconnected by providing memory speed-bandwidth connectivity. The reflection center provides full duplex communication between the ports, thereby effectively combining all the memories in the network into dual-ported, random-access memory. This dual-port characteristic eliminates the CPU overhead on each computer that is incurred with Ethernet.

The reflection bus carries write transfers only and operates at a sustained data rate of 40 Mbyte/s. This does not include address, error correction, and coordination information, which makes actual universal memory network bus traffic approach 100 Mbyte/s. The universal memory network can be implemented in copper cables for distances up to 15 m and in fiber optics for distances up to 3 km. A combination of both for media can be used in the same network. Multiple reflection centers can be interconnected to obtain configurations requiring more ports.

In addition to the reflection center of main hub, the universal memory network includes smaller subsystems called *shared memory interfaces* (SMIs), which make it possible for computers based on different bus architectures (e.g., SELBus, DEC BI, Multi Bus, and VME, or other selected buses) to communicate via the reflection bus. Each host computer is attached to the reflection center by a bus-interface circuit card, which translates the read and write transfers of the host computer to and from the reflection-bus standard. This translation centers around the ordering of bits and conversation used by various vendor architectures to a common strategy required by the 100-ns cycle time of the reflection bus.

The standard memory interface enhances the modular nature of the network. It provides computer memory access to processors of lower cost and enables a large number of workstations to be supported from one reflection center. For example, one reflection center can support up to 12 SMI memory interfaces, each with capacity to support between 8 and 16 workstations, depending on local hardware configurations. Multiple reflection centers can be interconnected to support even more workstations.

REFERENCES

1. Ames, J. G., "Which Network Is the Right One," *Manufacturing Engineering*, **56,** May 1988.
2. Borggraaf, P., "Semiconductor Factory Automation: Current Theories," *Semiconductor International*, **88,** October 1985.
3. Bux, W., "Local Area Subnetworks: A Performance Comparison," *IEEE Trans. Communications*, **COM-29**(10):1465 (1981).
4. Campbell, J., *The RS-232 Solution,* Alameda, California: Sybex, Inc. (1984).
5. Kaminski, A. M., Jr., "Protocols for Communicating in the Factory," *IEEE Spectrum*, **65,** April 1986.

6. Kleinrock, L., and S. S. Lam, "Packet Switching in a Multiaccess Broadcast Channel: Performance Evaluation," *IEEE Trans. Communications,* **COM-23**(4):410 (1975).
7. Metcalfe, R. M., and D. R. Boggs, "Ehternet: Distributed Packet-Switching for Local Computer Networks," *Communications of the ACM* **19:**395 (1976).
8. Shock, J. F., and J. A. Hupp, "Measured Performance of an Ethernet Local Network," *Communications of the ACM,* **23:**711 (1980).
9. Tanenbaum, A. S., *Computer Networks,* 2d ed., Englewood Cliffs, N.J.: Prentice-Hall (1988).
10. Talvage, J., and R. G. Hannam, *Flexible Manufacturing Systems in Practice: Application Design and Simulation,* New York: Marcel Decker (1988).
11. Voelcker, J., "Helping Computers Communicate," *IEEE Spectrum,* **61,** March 1986.
12. Warndorf, P. R., and M. E. Merchant, "Development and Future Trends in Computer Integrated Manufacturing in the USA," *International Journal in Technology Management,* **1**(1–2):162 (1986).
13. Wick, C., "Advances in Machining Centers," *Manufacturing Engineering,* **24,** October 1987.
14. Wittry, E. J., *Managing Information Systems: An International Approach,* Dearborn, Mich.: Society of Manufacturing Engineers (1987).

CHAPTER 4
THE ROLE OF SENSORS AND CONTROL TECHNOLOGY IN COMPUTER-INTEGRATED MANUFACTURING

INTRODUCTION

According to various studies conducted in the United States, nearly 50 percent of the productivity increase during the period 1950–1990 was due to technological innovation. That is, the increase was due to the introduction of high-value-added products and more efficient manufacturing processes, which in turn have caused the United States to enjoy one of the highest living standards in the world. However, unless the United States continues to lead in technological innovation, the relative living standard of the country will decline over the long term.

This clearly means that the United States has to invest more in research and development, promote scientific education, and create incentives for technological innovation. In the R&D arena, the United States has been lagging behind other nations: about 1.9 percent of the U.S. gross national product (GNP) (versus about 2.6 percent of the GNP in Japan and West Germany) goes for R&D. The introduction of computer-integrated manufacturing (CIM) strategy in U.S. industry has begun to provide a successful flow of communication which may well lead to a turnaround. *Sensors and control systems in manufacturing* are a powerful tool for implementing CIM. Current world business leaders view CIM as justifying automation to save our standard of living. Successful implementation of CIM depends largely on creative information gathering through sensors and control systems, with information flow as feedback response. Information gathering through sensors and control systems is *imbedded* in CIM. CIM provides manufacturers with the ability to react more quickly to market demands and to achieve levels of productivity previously unattainable.

Effective implementation of sensors and control subsystems within the CIM manufacturing environment will enable the entire manufacturing enterprise to work together to achieve new business goals.

CIM PLAN

This chapter will address implementation of a CIM plan through the technique of modeling.

A model can be defined as a tentative description of a system or theory; the model accounts for many of the system's known properties. An *enterprise model* can be defined (in terms of its functions) as the function of each area, the performance of each area, and the performance of these areas interactively. The creation of a model requires an accurate description of the needs of an enterprise.

In any manufacturing enterprise there is a unique set of business processes that are performed in order to design, produce, and market the enterprise's products. Regardless of how unique an enterprise or its set of processes is, it shares with others the same set of high-level objectives. To attain the objectives, the following criteria must be met:

1. Management of manufacturing finances and accounting
2. Development of enterprise directives and financial plans
3. Development and design of product and manufacturing processes utilizing adequate and economical sensors and control systems
4. Management of manufacturing operations
5. Management of external demands

CIM Plan in Manufacturing

In manufacturing, CIM promotes customer satisfaction by allowing order entry from customers, faster response to customer enquiries and changes, via electronic sensors, and more accurate sales projections.

CIM Plan in Engineering and Research

In engineering and research, CIM benefits include quicker design, development, prototyping, and testing; faster access to current and historical product information; and paperless release of products, processes, and engineering changes to manufacturing.

CIM Plan in Production Planning

In production planning, CIM offers more accurate, realistic production scheduling that requires less expediting, canceling, and rescheduling of production and purchase orders.

In plant operations, CIM helps to control processes, optimize inventory, improve yields, manage changes to product and processes, and reduce scrap and rework. CIM also helps in utilizing people and equipment more effectively, eliminating production crises, and reducing lead time and product costs.

CIM Plan in Physical Distribution

In physical distribution, where external demands are satisfied with products shipped to the customer, CIM helps in planning requirements; managing the flow of products; improving efficiency of shipping, vehicle, and service scheduling; allocating supplies to distribution centers; and expediting processing of returned goods.

CIM Plan for Business Management

For business management activities such as managing manufacturing, finance, and accounting, and developing enterprise directives and financial plans, CIM offers better product cost tracking, more accuracy in financial projections, and improved cash flow.

CIM Plan for the Enterprise

For the enterprise as a whole, these advantages add up to faster release of new products, shorter delivery times, optimized finished goods inventory, shorter production planning and development cycles, reduced production lead time, improved product quality, reliability and serviceability, increased responsiveness, and greater competitiveness. In effect, CIM replaces an enterprise's short-term technical improvements with a long-term strategic solution.

The advantages of CIM with sensors and control systems are not just limited to the four walls of an enterprise. It can also deliver real productivity gains in the outside world. For example, suppliers will be able to plan production, schedule deliveries, and track shipments more efficiently. Customers will benefit from shorter order-to-delivery times, on-time deliveries, and less expensive, higher-quality products.

MANUFACTURING ENTERPRISE MODEL

The integration and productivity gains made possible by CIM with sensors and control systems are the key to maintaining a competitive edge in today's manufacturing environments. The enterprise model defines an enterprise in terms of its functions. In a traditional enterprise that relies on a complex organization structure, operations and functional management are divided into separate departments, each with its own objectives, responsibilities, resources, and productivity tools.

Yet, for the enterprise to operate profitably, these departments must perform in concert. Sensors and control systems that improve one operation at the expense of another, and tie up the enterprise's resources, are counterproductive. New sensors and control systems in CIM can create a systematic network out of these insulated pockets of productivity. But to understand how, one must examine the elements of an enterprise model and see how its various functional areas work—independently and with each other.

Creating a model of the enterprise can help to expose operations that are redundant, unnecessary, or even missing. It can also help determine which information is critical to a successful implementation once effective sensors and control systems are incorporated.

Obviously, this model is a general description. There are many industry-unique variations to the model. Some enterprises may not require all of the functions described, while others may require more than are listed. Still other enterprises may use the same types of functions, but group them differently.

For example, in the aerospace industry, life-cycle maintenance of products is an essential requirement and may require extensions to the model. In the process industry, real-time monitoring and control of the process must be included in the model.

Computer integrated manufacturing harnesses sensors and control system technology to integrate these manufacturing and business objectives. When implemented properly, CIM can deliver increased productivity, cost-efficiency, and responsiveness throughout the enterprise. CIM can accomplish these objectives by addressing each of the major functional areas of the manufacturing enterprise:

1. Marketing
2. Engineering research and development of sensors in CIM strategy
3. Production planning
4. Plant operations incorporating sensors and shop floor control systems
5. Physical distribution
6. Business management

Integrating these functions and their resources requires the ability to share and exchange information about the many events that occur during the various phases of production; manufacturing systems must be able to communicate with the other information systems throughout the enterprise (Fig. 4.1). There must also be the

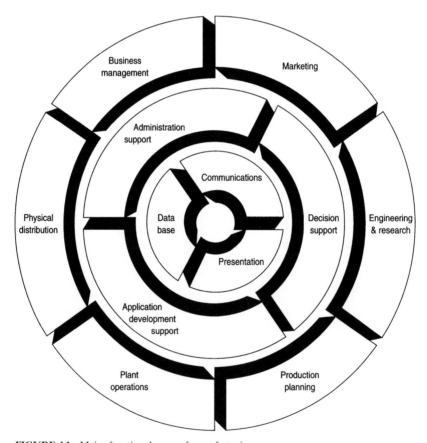

FIGURE 4.1 Major functional areas of manufacturing.

means to capture data close to the source, then distribute the data at the division or corporate level, as well as to external suppliers, subcontractors, and even customers.

To meet these needs, the CIM environment requires a dynamic network of distributed functions. These functions may reside on independent system platforms and require data from various sources. Some may be general-purpose platforms, while others are tailored to specific environments. But the result is an environment that encompasses the total information requirements of the enterprise, from developing its business plans to shipping its products.

With this enterprise-wide purpose, CIM can deliver its benefits to all types of manufacturing operations, from plants that operate one shift per day to processes that must flow continuously, from unit fabrication and assembly to lots with by-products and coproducts. These benefits can also be realized in those enterprises where flexible manufacturing systems are being used to produce diversified products over shorter runs, as well as in those that use CIM with sensors and control systems to maintain an error-free environment.

By creating more efficient, more comprehensive management information systems through sensors and control systems, CIM supports management efforts to meet the challenges of competing effectively in today's world markets.

Marketing

Marketing acts as an enterprise's primary contact with its customer (Fig. 4.2). To help meet the key objectives of increasing product sales, a number of functions are performed within marketing. These include market research; forecasting demand and sales; analyzing sales; tracking performance of products, marketing segments, sales personnel, and advertising campaigns; developing and managing marketing

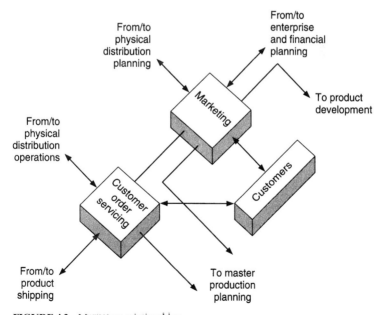

FIGURE 4.2 Marketing relationships.

channels; controlling profits and revenues; and managing sales personnel, sales plans, and promotion. Input comes from business management and customers. Output goes to customers, product development, customer order servicing, and master production planning.

The information handling requirements of marketing include monumental texts and graphics as well as queries and analysis of internal and external data. The internal data is gathered through a variety of software routines.

Customer order servicing involves entering and tracking customer orders. This can be for standard products or custom-designed products. Other customer order servicing activities include providing product quotations, checking customer credit, pricing product, allocating order quantities, and selecting shipments from distribution centers.

Input to this area includes order and forecast data from marketing or directly from customers as well as available-to-promise data from production planning. Output can include allocation of all orders, quotations for custom products, communication with production engineering regarding custom products, order consideration, and shipping releases. Customer service will significantly improve through electronic data interchange (EDI).

Engineering and Research

The engineering and research areas of an enterprise can be broken down into separate activities (Fig. 4.3). Each of these has its own special needs, tools, and relationships to other areas.

The research activities include investigating and developing new materials, products, and process technology. Information processing needs include complex analyses, extensive texts, imaging, graphics, and videos.

Research input comes from such outside research sources as universities, trade journals, and laboratory reports. Then research must communicate its findings to three other functional areas—product development, process development, and facilities engineering.

Product Development. In today's increasingly competitive manufacturing markets, creation of new material and production technologies is essential for the successful development of products. The product development area uses these materials and production technologies to design, model, simulate, and analyze new products.

Product development activities include preparing product specifications and processing requirements, drawings, materials or parts lists, and bills of material for new products or engineering changes.

In this area, laboratory analysis tools, computer-aided design/computer-aided engineering (CAD/CAE) tools, and group technology (GT) applications are helping to reduce product development time, increase productivity, and improve product quality.

Product development comes from marketing, research, and plant operations. Its output—product specifications, manufacturing control requirements, drawings, text, and messages—is directed to process development and engineering release control.

Process Development. This functional area creates process control specifications, manufacturing routings, quality test and statistical quality control specifications, and numerical control (NC) programming. It also validates the manufacturability of

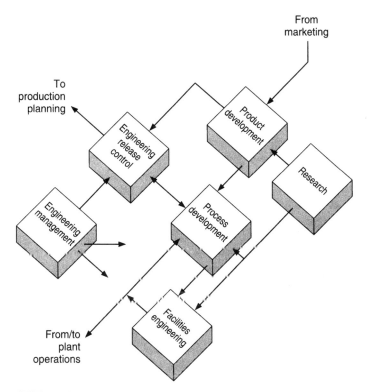

FIGURE 4.3 Engineering and research.

product designs. Computer-aided process planning (CAPP) programs and group technology applications for routing similar parts have helped streamline these functions. Expert systems have also been used to supplement traditional product testing and defect-analysis processes. This area is also responsible for the application of new manufacturing technologies such as work cells, conveyer systems, and robotics. Sensors and control systems play a fundamental role within this work-cell structure.

Process development receives input from research and product development as well as statistical process data from plant operations. Output includes providing routings, process control algorithms, and work-cell programming to plant operations by way of engineering release control.

Facilities Engineering. The chief responsibility of facilities engineering is the installation of plant automation incorporating new equipment with sensors and control systems, material flow, inventory staging space, and tools. Tools may include driverless material handling equipment, conveyers, and automated storage and retrieval systems. This area is also responsible for plant layout and implementation of such plant services and utilities as electricity, piping, heat, refrigeration, and light.

Input to facilities engineering is from research and process development. Output, such as plant layouts, changes of schedule, and forecasts of new equipment availability, goes to plant operations.

Engineering Release Control. This function involves the coordination of the release of new products, processes, tools, and engineering changes to manufacturing. A major check point in the product cycle is obtaining assurance that all necessary documentation is available, after which the information is released to manufacturing.

Input is received from product and process development activities. Specific output, including product and tool drawings, process and quality specifications, NC programs, and bills of material, is transferred to production planning and plant operations.

Engineering Management. Among the activities of engineering management are introducing engineering releases, controlling product definition data, estimating cost, entering and controlling process data, and defining production resources.

Input and output for engineering management include exchanging designs and descriptions with engineering design; reviewing product data, routings, and schedules with production planning; and accepting engineering change requests from plant operations.

Production Planning

Production planning can be viewed as five related functional areas (Fig. 4.4).

Master Production Planning. In master production planning, information is consolidated from customer order forecasts, distribution centers, and multiple plans in order to anticipate and satisfy demands for the enterprise's products. Output includes time-phased requirements sent to the material planning function as well as the final assembly schedule.

Material Planning and Resource Planning. These two areas require timely and accurate information—demand schedules, production commitment, inventory and work-in-progress status, scrap, actual versus planning receipts, shortages, and equipment breakdowns—in order to keep planning up to date with product demands.

Product and process definition data come from the engineering areas. Output is to plant operation and procurement and includes production schedules, order releases, and plans for manufactured and purchased items.

Procurement. Procurement involves selecting suppliers and handling purchase requirements and purchase orders for parts and materials. Among the input is material requirements from material planning and just-in-time delivery requests from plant operations. Other input includes shipping notices, invoices, and freight bills.

Output to suppliers includes contracts, schedules, drawings, purchase orders, acknowledgments, requests for quotations, release of vendor payments, and part and process specifications. In order to streamline this output, as well as support just-in-time concepts, many enterprises rely on sensors for electronic data interchange with vendors.

Plant Release. The functions of this area can vary, depending on the type of manufacturing environment. In continuous-flow environments, for example, this area produces schedules, recipes to optimize use of capacity, specifications, and process routings.

For job-shop fabrication and assembly environments, this area prefers electronic or paperless-shop documents consisting of engineering change levels; part, assembly,

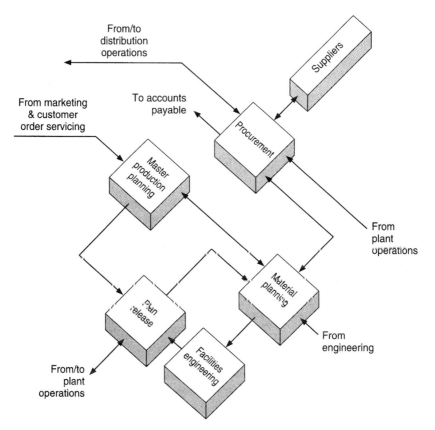

FIGURE 4.4 Production planning.

setup, and test drawings and specifications; manufacturing routings; order and project control numbers; and bar codes, tags, or order-identification documents.

However, large-volume industries are evolving from typical job-shop operations to continuous-flow operations, even for fabrication and assembly-type processes.

Input—typically an exploded production plan detailing required manufacturing items—comes from material planning. Output—schedules, recipes or shop packets—is used in plant operations for scheduling.

Plant Operations

Plant operations can be described in terms of nine functions (Fig. 4.5).

Production Management. This area provides dynamic scheduling functions for the plant floor by assigning priorities, personnel, and machines. Other activities include sending material and tool requests for just-in-time delivery.

Input for production management includes new orders and schedules from production planning and real-time feedback from plant operation. Output includes

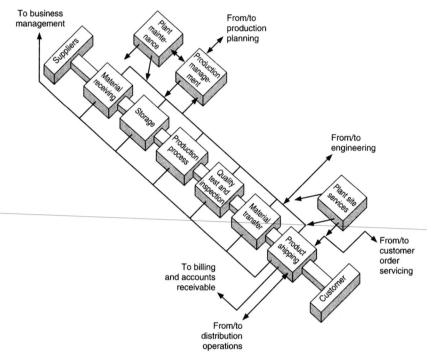

FIGURE 4.5 Plant operations.

dynamic schedules and priorities that are used to manage operations on the plant floor.

Material Receiving.   The material receiving function includes accepting and tracking goods, materials, supplies, and equipment from outside suppliers or other locations within the enterprise. Input to this area includes receiving reports and purchase order notices.

Output includes identifying receipts with appropriate documentation, then routing materials to the proper destination. Data are also sent to accounting, procurement, and production management. Production management can also direct materials based on more immediate needs.

Storage.   This represents inventory management—where materials are stored and are accessible to the proper production locations. The materials include finished goods, raw materials, parts, supplies, work-in-progress, and tools, as well as nonproduction materials and equipment.

Storage functions include preparing item identification and storage tags, managing storage locations, processing pick requests, reporting picks and kit activities, and planning physical inventory cycles and reporting counts.

Storage input includes storage and picking requests from production management scheduling functions. Output includes receiving and disbursement reports for use in production management and accounting.

Production Process. Production process functions include managing the production process, processing materials, fabricating parts, grading or reworking components, assembling final products, and packaging for distribution.

One of today's trends in fabrication and assembly processes is toward continuous processing, such as continuous movement with minimal intermediate inventories. This can be described with such terms as continuous-flow manufacturing, flexible manufacturing cells, just-in-time logistics, and group technology. Unfortunately, in many instances, these automation efforts are autonomous, without regard to the other functions of the enterprise.

The information handling needs of the production process can include analog and digital data, text, graphics, geometries, applications programs—even images, video, and voice. Processing this information may require subsecond access and response time.

Input to this area includes shop documents, operator instructions, recipes, and schedules from production management as well as NC programs from process development. Output consists of material and tool requests, machine maintenance requests, material transfer requests, production, and interruption reports for production management, production and labor reports for cost accounting and payroll, and statistical process reports for production management and process development.

Quality Test and Inspection. Testing items and products to assure the conformity of specifications is the main activity in quality test and inspection. This includes analyzing and reporting results quickly by means of metrological sensors and control systems, in order to reduce scrap and rework costs.

Quality test and product specifications are input from engineering. Chief output includes purchased-item inspection results to procurement, manufactured-item inspection and product test results to production process and production management, quality test and inspection activity reports to cost accounting, and rejected part and product dispositions to material handling.

Material Transfer. Material transfer involves the movement of materials, tools, parts, and products among the functional areas of the plant. These activities may be manual. Or they may be semiautomated, using control panels, forklifts, trucks, and conveyors. Or they may be fully automated, relying on programmable logical controllers, distribution control systems, stacker cranes, programmed conveyors, automated guided vehicles, and pipelines.

Input in this area may be manual requests or those generated by the system. Output includes reporting completed moves to production management.

Product Shipping. This area supports the movement of products to customers, distributors, warehouses, and other plants. Among the activities are selecting shipment and routing needs, consolidating products for a customer or carrier order, preparing shipping lists and bills of lading, reporting shipments, and returning goods to vendors.

The primary input is from customer order servicing, and it includes the type of product and the method and date of shipment. Output includes reporting shipment dates to customer order servicing, billing, and accounts receivable.

Plant Maintenance. Plant maintenance includes those functions that ensure the availability of production equipment and facilities. Maintenance categories include routine, emergency, preventive, and inspection services. In addition, many of today's advanced users of systems are moving toward diagnostic tools based on expert systems, which reduce equipment downtime. Input (maintenance requests) can be initiated by plant personnel, a preventive maintenance and inspection system, or a process and equipment monitoring system. Output includes requests for purchase of maintenance items, schedules for maintenance for use in production management, requests for equipment from facilities engineering, and maintenance work order costs to cost accounting.

Plant Site Services. The final area of plant operation is plant site services. Input received and output provided cover such functions as energy supply and utilities management, security, environmental control, grounds maintenance, and computer and communications installations.

Physical Distribution

Physical distribution can be viewed as having two functional areas (Fig. 4.6).

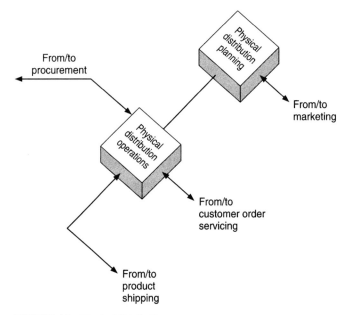

FIGURE 4.6 Physical distribution.

Physical Distribution Planning. This involves planning and control of the external flow of parts and products through warehouses, distribution centers, other manufacturing locations, and points of sale. These functions may also include allocating demand, planning finished goods inventory, and scheduling vehicles. Some industries require another major set of functions that relate to logistics reports. This includes spare parts, maintenance, training, and technical documentation.

Input and output are exchanged with marketing and physical distribution operations. The information handling requirements of physical distribution planning are usually medium to heavy, especially if multiple distribution centers exist.

Physical Distribution Operations. This area includes receiving, storing, and shipping finished goods at the distribution center or warehouse. Receipts arrive from plants or other suppliers, and shipments are made to customers and dealers. Other functions can include scheduling and dispatching vehicles, processing returned goods, and servicing warranties and repairs.

Input is received from plant site product shipping, procurement, physical distribution planning, and customer order servicing. Output includes acknowledgments to plant site product shipping and procurement, as well as data for updates to physical distribution planning and customer order servicing.

Business Management

Within the enterprise model, a business management function may be composed of seven areas (Fig. 4.7).

Financial Planning and Management. In financial planning and management, financial resource plans are developed and enterprise goals are established. Among the functions are planning costs, budgets, revenues, expenses, cash flow, and investments.

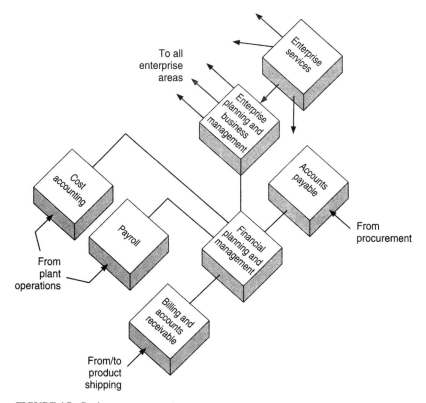

FIGURE 4.7 Business management.

Input includes financial goals and objectives established by management as well as summarized financial data received from all areas of the enterprise. Output includes financial reports to stockholders, departmental budgets, and general ledger accounting.

Accounts Payable. Accounts payable primarily involves paying vendors. Input includes vendor invoices and goods-received reports. The output includes discount calculations to vendors and payment checks. This last function lends itself to the use of electronic data interchange to electronically transfer funds to vendors.

Billing and Accounts Receivable. This area prepares invoices to customers and manages customer account collections. Input to this area consists of product shipping data and cash received. Output includes customer invoices, delinquent account reports, and credit ratings. Transferring funds electronically through EDI can simplify this area significantly.

Cost Accounting. Cost accounting activity supports product pricing and financial planning by establishing product costs. These costs can include those of materials, direct and indirect labor, fixed production elements (machinery and equipment), variable production elements (electricity, fuels, or chemicals), and overhead. Other functions can include establishing standard costs, reporting variances to standards, costing job orders, and determining accrued costs.

Cost accounting input is acquired primarily from plan operations. Output is sent to financial management and planning.

Payroll. This area computes payments, taxes, and other deductions for employees. It also includes reporting and paying employee tax withholdings to government agencies. Input includes time and attendance information and production data from plant operations. Output includes payroll checks, labor distribution, and government reports and payments.

Enterprise Planning and Business Management. These functions include establishing goals and strategies for marketing, finance, engineering and research, plant automation, sensors and control systems, and information systems. Input and output are exchanged through sensors and control systems with virtually every other area of the enterprise.

Enterprise Services. Enterprise services include such support services as office personnel, management information services, personnel resources, and public relations. These services require extensive administrative support tools, such as text processing, decision support, and graphic tools. But since input and output will be exchanged throughout the entire enterprise, it is imperative that these tools be integrated with the enterprise's other systems.

DESIGN OF CIM WITH SENSORS AND CONTROL SYSTEMS

With the advent of low-priced computers and sensors and control systems, there have been a number of technological developments related to manufacturing that can be used to make production more efficient and competitive. The primary purpose is to develop several computer-concepts as related to the overall manufacturing plan of CIM systems.

In order for the manufacturing enterprise to succeed in the future, it is imperative that it adopt a manufacturing strategy that integrates its various functions. CIM systems have the potential to accomplish this task. The implementation of CIM with sensory and control systems on the shop floor represents a formidable, albeit obtainable, objective. To accomplish this goal, enterprises must have access to information on what is available in CIM. A secondary purpose of obtaining access to information is to provide a framework that can aid in the search for information. Once the information is obtained, it becomes necessary to look at the current system objectively and decide how to approach the problem of implementing CIM with sensors and control systems.

While many of the ideas associated with CIM are new and untried, progressive enterprises, with the realization that old methods are ineffective, are doing their part in implementing this new technology. Some of the concepts currently being implemented are flexible manufacturing systems, decision support systems (DSS), artificial intelligence (AI), just-in-time (JIT) inventory management, and group technology. While all of these concepts are intended to improve efficiency, each one alone can only accomplish so much. For example, an FMS may reduce work-in-process (WIP) inventory while little is accomplished in the area of decision support systems and artificial intelligence to relate all aspects of manufacturing management and technology to each other for FMS. The advent of inexpensive sensors and control systems enables the concept of CIM to be implemented with greater confidence.

Recent advances in computer technology and sensors in terms of speed, memory, and physical space have enabled small, powerful, personal computers to revolutionize the manufacturing sector and become an essential part of design, engineering, and manufacturing, through, for example, database management systems (DBMSs) and local-area networks (LANs). The coordination of the various aspects of a manufacturing environment means that complex systems inherently interact with one another. Due to a lack of standards and poor communication between departments, many components and databases are currently incompatible.

Table 4.1 describes some benefits of CIM.

The potential for CIM, according to Table 4.1, is overwhelming, but the main issue is how to analyze and design CIM that incorporates sensors, control systems, and decision support so that it is utilized effectively.

Manufacturing problems are inherently multiobjective. For example, improving quality usually increases cost and/or reduces productivity. Furthermore, one cannot maximize quality and productivity simultaneously; there is a tradeoff among these objectives. These conflicting objectives are treated differently by different levels and/or units of production and management. Obviously, without a clear understanding of objectives and their interdependence at different levels, one cannot successfully achieve CIM with sensors and control systems.

TABLE 4.1 CIM Benefits

Application	Improvement with CIM, %
Manufacturing productivity	120
Product quality	140
Lead time (design to sale)	60
Lead time (order to shipment)	45
Increase in capital equipment utilization	340
Decrease in WIP inventory	75

Components of CIM with Sensors and Control Systems

The decision making in design of CIM with effective sensors and control systems can be classified into three stages.

1. *Strategic level.* The strategic level concerns those decisions typically made by the chief executive officer (CEO) and the board of directors. Upper management decisions of this type are characterized by a relatively long planning horizon, lasting anywhere from 1 to 10 years. Implementing CIM with sensors and control systems has to begin at this level. Even though small enterprises may not have as many resources at their disposal, they have the added advantage of fewer levels of management to work through while constructing CIM.
2. *Tactical level.* At the tactical level, decisions are made that specify how and when to perform particular manufacturing activities. The planning horizon for these decisions typically spans a period from 1 to 24 months. Activities at this level include such intermediate functions as purchasing and inventory control. They affect the amount of material on the shop floor but do not control the use of the material within the manufacturing process.
3. *Operational level.* Day-to-day tasks, such as scheduling, are performed at the operational level. The primary responsibility at this level is the effective utilization of the resources made available through the decisions made on the strategic and tactical levels. Because of the variability in demand or machine down time, the planning horizon at this level must be relatively short, normally 1 to 15 days.

While each of these levels has certain responsibilities in a manufacturing plant, the objectives are often conflicting. This can be attributed to inherent differences between departments, e.g., sales and marketing may require a large variety of products to serve every customer's needs while the production department finds its job easier if there is little product variation. One of the main causes of conflicting decisions is a lack of communication due to ineffective sensors and control systems between levels and departments. CIM with adequate sensors and control systems provides the ability to link together technological advances, eliminate much of the communication gap between levels, and bring all elements into a coherent production system.

CIM with Sensors and Control Systems at the Plant Level

Some of the important emerging concepts related to CIM with effective sensors and control systems are flexible manufacturing systems, material handling systems, automated storage and retrieval systems (AS/RS), computer-aided design (CAD), computer-aided engineering (CAE), computer-aided manufacturing (CAM), and microcomputers. These components of CIM can be classified into three major groups (Fig. 4.8).

Flexible Manufacturing Systems Incorporating Sensors and Control Systems. An FMS can link several elements on the shop floor through sensors in order to coordinate those elements. While CIM can be applied to any manufacturing industry, FMSs find their niche in the realm of discrete production systems such as job shops.

The most important FMS elements are numerical control machines and an automated material handling network to transport the product from raw material inventory, through the NC operations, and finally to the finished goods inventory.

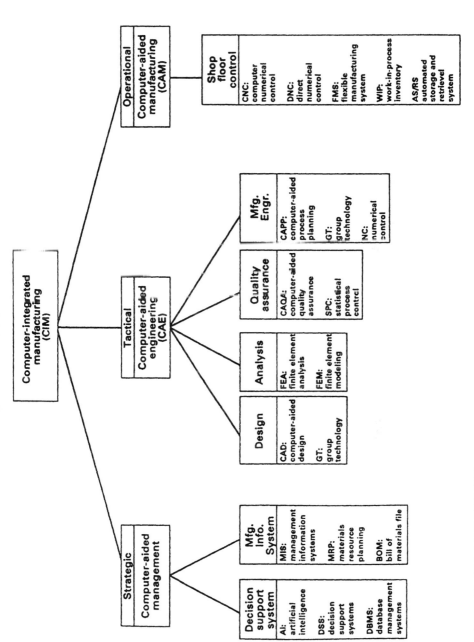

FIGURE 4.8 Components within CIM.

Numerical control technology has made major advances with the advent of computer numerical control and direct numerical control (CNC/DNC). Microprocessors and sensors located on the machine itself can now provide the codes necessary for the parts to be operated on.

Material Handling. Material handling is the means of loading, unloading, and transporting workpieces among different machines and departments. There are several ways in which material handling is accomplished:

1. *Transfer lines* consist of fixed automation machinery such as conveyer belts. Their advantages are high speed and low cost. Their major disadvantage is their lack of flexibility. Dedicated transfer lines can handle only a limited number of parts and cannot be easily changed once in place, thus defeating the goals of an FMS.
2. *Robots* provide another alternative for moving workpieces. Generally robots can be made very flexible because of their programmability, but they are limited to their region of access.
3. *Automated guided vehicles* can move workpieces a great distance, but they lack the speed found in both robot and transfer lines. Yet, because of their ability to be programmed to different routes, they are more flexible than transfer lines.

Automated Storage and Retrieval Systems. By means of AGVs, raw materials can be taken from the loading dock and placed in a designated place in inventory. By means of an AS/RS, inventory can be tracked throughout the manufacturing process and optimized for strategically locating items in storage. Because the process is computerized, data on what exactly is in inventory assist planners in determining order and production schedules.

Inventories consist of raw materials, work in process, and finished goods. Inventories should be controlled by keeping track of inventory locations and amounts. An AS/RS can accomplish this task.

Computer-Aided Engineering/Design/Manufacturing (CAE/CAD/CAM). Computer-aided design helps the engineer in many ways during the design stage. Simply drawing the part on a computer increases the productivity of designers, but CAD is more than just automated drafting. CAD can facilitate group technology and the construction of a bill of materials (BOM) file.

Computer-aided engineering consists of the many computerized facets of engineering that go into a particular product. When a part has been designed, the CAE subgroup is responsible for generating the NC code that can be used by the NC machines on the floor.

By using GT, similar parts can be classified by similar attributes and placed in part families. By grouping parts this way, much redundancy in design and manufacturing is eliminated.

In *computer-aided manufacturing* (CAM), computers are used to plan and conduct production. They can correct and process data, control the manufacturing process, and provide information that can be used in decision making. CAM can involve distributed quality control and product testing and inspection, which are built into the manufacturing process to support the larger functional relationships.

Microcomputers for CIM. The integration of different elements of CIM and sensor systems can be accomplished only with the aid of a computer. Because of the magnitude of the CIM and sensor database, a mainframe (or a minicomputer for a small enterprise) is necessary for storage and retrieval of information. The power of microcomputers has increased so dramatically, however, that they are suitable for most other applications, such as:

1. *Programmable logic controllers (PLCs).* On the factory floor, microcomputers are subject to harsh conditions. Dust, high temperature, and high humidity can quickly destroy an ordinary personal computer. Traditionally the shop floor belongs to the programmable logic controller. The components can operate in extreme conditions, e.g., 55°C and 90 percent humidity. In addition, PLCs are geared to real-time control of factory operations. Through advances in microprocessor technology, the functions that were once controlled by relays and mechanical switching mechanisms can now be performed by PLCs. The ladder diagrams used to notate logic circuits can now be programmed directly into the memory of the programmable controller. Because of microelectronic circuitry, PLCs can process control information quickly and shut down automatically in case of emergencies. Whereas PLCs have become commonplace on the floor of discrete product operations, process control computers have become indispensable in the control of process plants where conditions must be monitored constantly. They are also used in the control of office and factory environments. For example, a PLC can turn furnaces and air-conditioning units ON and OFF to provide suitable working conditions while optimizing energy use.

2. *Industrial personal computers (PCs).* Until recently, PCs were commonly found in the protected environments of offices. Now, manufacturers have introduced rugged versions of popular personal computers. For example, IBM has developed the 5531 and 7531/32, industrialized versions of PC/XT and PC/AT, respectively, to withstand the environment of the factory floor. They have the advantage of being able to run any PC-DOS–based software, but are unable to perform real-time control. This problem has been remedied with the advent of the industrial computer disk operating system (IC-DOS), a real-time operating system that is compatible with other IBM software. This allows a shop floor computer to provide real-time control while using software packages previously found only on office models.

3. *Microsupercomputers.* The hardware of microsupercomputers has increased computing power significantly. Offering high performance, transportability, and low price, the new microsupercomputers compare favorably to mainframes for many applications.

DECISION SUPPORT SYSTEM FOR CIM WITH SENSORS AND CONTROL SYSTEMS

With an increase in production volume and efficiency comes a need to have a more effective method of scheduling and controlling resources. Herein lies a connection between CAE and computer-aided management. The long-range plans of a company must include forecasts of what the demand will be for various products in the future. Through these forecasts, the enterprise determines what strategy it will take to ensure survival and growth.

For the enterprise to make intelligent decisions, reliable information must be available. In regard to the three levels of decision making, it is also important that the information be consistent throughout each level. The best way to assure this availability and consistency is to make the same database available to all individuals involved in the production process. Because of lack of good communication between levels and sometimes the reluctance of upper-level managers to commit themselves to CIM, constructing a centralized database represents one of the most difficult problems in the implementation of CIM.

Computer-Integrated Manufacturing Database (CIM DB)

The creation of a CIM DB is at the heart of the effective functioning of CIM. Most manufacturers have separate databases set up for nearly every application. Since data from one segment of an enterprise may not be structured for access by other segments' software and hardware, a serious problem for meeting the CIM goal of having readily available data for all levels occurs. Another problem with multiple databases is in the redundancy of data. Both the strategic and tactical decision makers, for example, may need information concerning a bill of material file. Even with the assumption that the databases contain consistent data (i.e. the same information in each), maintaining them both represents inefficient use of computer time and storage and labor. To install CIM, these databases must be consolidated. Unfortunately, bringing multiple databases into one CIM DB that remains available to everyone and consistent in all levels presents a significant obstacle because of the large investment needed in time and computer hardware and software.

Structure of Multiobjective Support Decision Systems

The success of CIM also depends largely on the ability to incorporate sensor technology with a database. The database is utilized in making decisions on all levels—decisions that are used to update to the database. Decision support systems can provide a framework for efficient database utilization by allowing storage and retrieval of information and problem solving through easy communications.

Decision making problems in manufacturing can be grouped into two general classes:

- *Structured decisions* are those that are constrained by physical or practical limitations and can be made almost automatically with the correct input. An example is generating a group technology part code given the geometry of the part.
- *Unstructured decisions* typically are those that contain a large degree of uncertainty. Decisions considered by strategic planners are almost always unstructured. Deciding whether or not to expand a certain product line, for example, may be based on demand forecasting and on the expected growth of competitors.

Due to the predictive nature of these decisions, they inherently contain more uncertainty than structured decisions. Long-range planning consists primarily of unstructured information.

Decision support systems mainly consist of three separate parts:

1. *Language systems.* The function of a language system (LS) is to provide a means for the user to communicate with the DSS. Some considerations for the choice of a language are that the formulation should be easily understood, implementable, and modifiable. Moreover, processing the language should be possible on a separate level or on the problem processing system (PPS) level. An obvious

choice for a language would be the spoken language of the user. This would require little or no training for the user to interact with a computer, but the complexity of sentences and the use of words that have multiple meanings present difficult problems that, when solved, would introduce unwanted inefficiency into the language system. An alternative would be to use a more formal language based on logic, e.g., PROLOG. The advantage here is that a language can be used at all levels of the DSS. In the design and use of an LS for the user interface, one can consider objectives such as ease of communication, the level of complexity that can be presented by the LS, and the time needed for the user to learn it.

2. *Knowledge systems.* The basic function of a knowledge system (KS) is the representation and organization of the "knowledge" in the system. Two possible approaches are storing in the information in database form or representing the data as a base for artificial intelligence using methods from, for example, predicate calculus. The objective of KS is to ease accessibility of data for the DSS. The KS should be able to organize and classify databases and problem domains according to objectives that are sensible and convenient for the user. Some of the objectives in the design of the KS are to reduce the amount of computer memory required, increase the speed with which the data can be retrieved or stored, and increase the number of classifications of data and problem domains possible.

3. *Problem processing systems.* The problem processing system of a DSS provides an interface between the LS and the KS. The primary function is to receive the problem from the user via the LS and use the knowledge and data from the KS to determine a solution. Once a solution is found, the PPS sends it through the KS to be translated into a form the user can recognize. More importantly, in the model formulation, analysis, and solution procedure of PPS, the conflicting objectives of stated problems must be considered. The PPS should provide methodology that can optimize all conflicting objectives and generate a compromise solution acceptable to the user. Some of the objectives in the development of such multiobjective approaches are to reduce the amount of time that the user must spend to solve the problem, increase the degree of interactiveness (e.g., how many questions the user should answer), reduce the difficulty of questions posed to the user, and increase the robustness of the underlying assumptions and procedures.

ANALYSIS AND DESIGN OF CIM WITH SENSORS AND CONTROL SYSTEMS

Many manufacturing systems are complex, and finding a place to begin a system description is often difficult. Breaking down each function of the system into its lowest possible level and specifying objectives for each level and their interactions will be an effective step. The objectives and decision variables, as related to elements or units for all possible levels, are outlined in the following sections.

Structured Analysis and Design Technique (SADT)

The structured analysis and design technique is a structured methodology. It combines the graphical diagramming language of structured analysis (SA) with the formal thought discipline of a design technique (DT). The advantage of SADT is that it contains a formalized notation and procedure for defining system functions.

Psychological studies have shown that the human mind has difficulty grasping more than five to seven concepts at one time. Based on this observation, SADT follows the structured analysis maxim: "Everything worth saying must be expressed in six or fewer pieces." Limiting each part to six elements ensures that individual parts are not too difficult to understand. Even complex manufacturing systems can be subjected to top-down decomposition without becoming overwhelming.

The basic unit for top-down decomposition is the structural analysis box (Fig. 4.9). Each of the four sides represents a specific action for the SA box. The four actions implement:

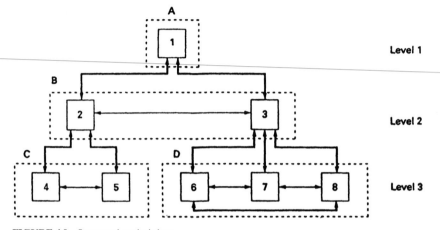

FIGURE 4.9 Structural analysis box.

1. *Input,* measured in terms of different decision variables.
2. *Control,* to represent constraints and limitations.
3. *Output,* measured in the form of a set of objectives.
4. *Mechanism for translation,* which performs the translations (for mapping) of input into output as constrained by the control action.

Since each box represents an idea in the system, each box contains a detailed diagram. A parent-child relationship exists between the box being detailed and those boxes under it. The same relationship holds for the parent diagram as the diagram directly preceding the child diagrams.

The interfaces linking boxes through their neighbors at the same level are the input, output, and control actions. In graphical terms, these interfaces are designated by arrows. Methodically proceeding through a given network, the entire system can be modeled in terms of boxes and arrows.

While SADT provides a realistic approach to modeling any system, it cannot provide the solution to any problem. The integrated computer-aided manufacturing definition method comes one step closer to the realization of a functional CIM system. It utilizes teamwork and demands that all correspondence and analysis be in written form so that others can obtain a grasp of the situation and errors can be more readily located. Because the written word is required during the implementation phases, the documentation that usually is done at the end of most major projects can be nearly eliminated. Keeping accurate records also plays an important role in debugging the system in the future.

A Multiobjective Approach for Selection of Sensors in Manufacturing

There are six criteria for the evaluation of sensors in manufacturing:

1. *Cost* is simply the price for the sensor and its integrated circuitry if it should be purchased.
2. *Integrability* is the degree to which the sensor can be used in conjunction with the manufacturing system it serves. This can be usually be measured in terms of compatibility with existing hardware control circuits and software.
3. *Reliability* is the quality of the sensors as indicated by the *mean time between failures* (MTBF), and can be measured by performing a simple stress test on the sensor under severe limits of operation. If the sensor operates under a certain high temperature for a certain period of time, it will assure the user that the system will perform satisfactorily under normal operating conditions. It will also indicate that the electronic control circuits are reliable, according to the *burn-in philosophy*.
4. *Maintenance* involves the total cost to update and maintain the sensor and how often the sensor needs to be serviced.
5. *Expandability* is how readily the sensor can be modified or expanded as new needs arise because of a changing environment.
6. *User friendliness* indicates the ease of using and understanding the unit. It may include the quality of documentation in terms of simplicity, completeness, and step-by-step descriptions of procedures.

DATA ACQUISITION FOR SENSORS AND CONTROL SYSTEMS IN CIM ENVIRONMENT

The input signals generated by sensors can be fed into an interface board, called an *I/O board*. This board can be placed inside a PC-based system. As personal computers for CIM have become more affordable, and I/O boards have become increasingly reliable and readily available, PC-based CIM data acquisition has been widely implemented in laboratory automation, industrial monitoring and control, and automatic test and measurement.

To create a data acquisition system for sensors and control systems that really meets the engineering requirements, some knowledge of electrical and computer engineering is required. The following key areas are fundamental in understanding the concept of data acquisition for sensors and control systems:

1. Real-world phenomena
2. Sensors and actuators
3. Signal conditioning
4. Data acquisition for sensors and control hardware
5. Computer systems

6. Communication interfaces
7. Software

Real-World Phenomena

Data acquisition and process control systems measure real-world phenomena, such as temperature, pressure, and flow rate. These phenomena are sensed by sensors, and are then converted into analog signals which are eventually sent to the computer as digital signals.

Some real-world events, such as contact monitoring and event counting, can be detected and transmitted as digital signals directly. The computer then records and analyzes this digital data to interpret real-world phenomena as useful information.

The real world can also be controlled by devices or equipment operated by analog or digital signals which are generated by the computer (Fig. 4.10).

Sensors and Actuators

A sensor converts a physical phenomenon such as temperature, pressure, level, length, position, or presence or absence, into a voltage, current, frequency, pulses, etc.

For temperature measurements, some of the most common sensors include thermocouples, thermistors, and resistance temperature detectors (RTDs). Other types of sensors include flow sensors, pressure sensors, strain gages, load cells, and optical sensors.

An actuator is a device that activates process control equipment by using pneumatic, hydraulic, electromechanical, or electronic signals. For example, a valve actuator is used to control fluid rate for opening and closing a valve.

Signal Conditioning

A signal conditioner is a circuit module specifically intended to provide signal scaling, amplification, linearization, cold junction compensation, filtering, attenuation, excitation, common mode rejection, etc. Signal conditioning improves the quality of the sensor signals that will be converted into digital signals by the PC's data acquisition hardware.

One of the most common functions of signal conditioning is amplification. Amplifying a sensor signal provides an analog-to-digital (A/D) converter with a much stronger signal and thereby increases resolution. To acquire the highest resolution during A/D conversion, the amplified signal should be equal to approximately the maximum input range of the A/D converter.

Data Acquisition for Sensors and Control Hardware

In general, data acquisition for sensors and control hardware performs one or more of the following functions:

1. Analog input
2. Analog output
3. Digital input

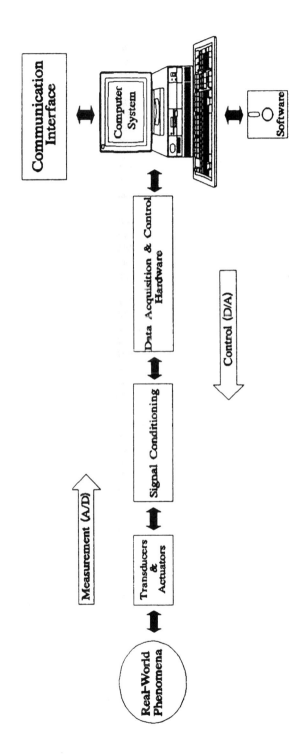

FIGURE 4.10 Integration of computer-controlled devices.

4. Digital output
5. Counter/timer

Analog Input (A/D). An analog-to-digital converter produces digital output directly proportional to an analog signal input, so that it can be digitally read by the computer. This conversion is imperative for CIM (Fig. 4.11).

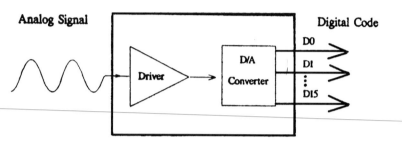

FIGURE 4.11   Analog-to-digital converter.

The most significant aspects of selecting A/D hardware are

1. Number of input channels
2. Single-ended or differential input
3. Sampling rate (in samples per second)
4. Resolution (in bits)
5. Input range (specified as full-scale volts)
6. Noise and nonlinearity

Analog Output (D/A). A digital-to-analog (D/A) converter changes digital information into a corresponding analog voltage or current. This conversion allows the computer to control real-world events.

Analog output may directly control equipment in a process that is then measured as an analog input. It is possible to perform a closed loop or proportional integral-differential (PID) control with this function. Analog output can also generate waveforms in a function generator (Fig. 4.12).

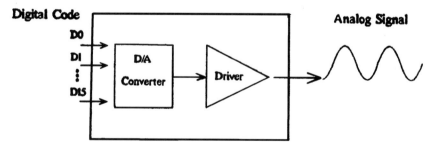

FIGURE 4.12   Analog output.

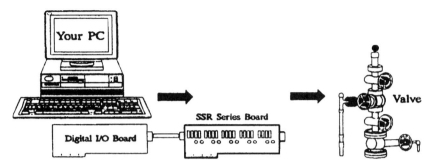

FIGURE 4.13 Application of digital input/output.

Digital Input and Output. Digital input and output are useful in many applications, such as contact closure and switch status monitoring, industrial ON/OFF control, and digital communication (Fig. 4.13).

Counter/Timer. A counter/timer can be used to perform event counting, flow meter monitoring, frequency counting, pulse width and time period measurement, etc.

Most data acquisition and control hardware is designed with the multiplicity of functions described above on a single card for maximum performance and flexibility. Multifunction data acquisition for high-performance hardware can be obtained through PC boards specially designed by various manufacturers for data acquisition systems.

Computer System

Today's rapidly growing PC market offers a great selection of PC hardware and software in a wide price range. Thus, a CIM strategy can be economically implemented.

Hardware Considerations. Different applications require different system performance levels. Currently, there are Pentium, Pentium Pro, and Pentium II CPUs which will allow a PC to run at benchmark speeds from 300 up to 800 MHz. Measurements and process control applications usually require 80286 systems. But for applications that require high speed, real-time data analysis, an 80386 or 80486 system will be much more suitable.

Industrial PCs. An *industrial PC* (IPC) is designed specifically to protect the system hardware in harsh operating environments. IPCs are rugged chasses that protect system hardware against excessive heat, dust, moisture, shock, and vibration. Some IPCs are even equipped with power supplies that can withstand temperatures from −20 to +85°C for added reliability in harsh environments.

Passive Backplane and CPU Card. More and more industrial data acquisition for sensors and control systems are using passive backplane and CPU card configurations. The advantages of these configurations are reduced mean time to repair (MTTR), ease of upgrading the system, and increased PC-bus expansion slot capacity.

A passive backplane allows the user to plug in and unplug a CPU card without the effort of removing an entire motherboard in case of damage or repair.

Communication Interfaces

The most common types of communication interfaces used in PC-based data acquisition for sensor and control system applications are RS-232, RS-422/485, and the IEEE-488 general-purpose interface bus (GPIB).

The RS-232 interface is the most widely used interface in data acquisition for sensors and control systems. However, it is not always suitable for distances longer than 50 m or for multidrop network interfaces. The RS-422 protocol has been designed for long distances (up to 1200 m) and high-speed (usually up to 56,000 bit/s) serial data communication. The RS-485 interface can support multidrop data communication networks.

Software

The driving force behind any data acquisition for sensors and control systems is its software control. Programming the data acquisition for sensors and control systems can be accomplished by one of three methods:

1. *Hardware-level programming* is used to directly program the data acquisition hardware's data registers. In order to achieve this, the control code values must determine what will be written to the hardware's registers. This requires that the programmer use a language that can write or read data from the data acquisition hardware connected to the PC. Hardware-level programming is complex, and requires significant time—time that might be prohibitive to spend. This is the reason that most manufacturers of data acquisition hardware supply their customers with either driver-level or package-level programs.
2. *Driver-level programming* uses function calls with popular programming languages such as C, PASCAL, C++, Visual C++, and BASIC, thereby simplifying data register programming.
3. *Package-level programming* is the most convenient technique of programming the entire data acquisition system. It integrates data analysis, presentation, and instrument control capabilities into a single software package. These programs offer a multitude of features, such as pull-down menus and icons, data logging and analysis, and real-time graphic displays.

DEVELOPING CIM STRATEGY WITH EMPHASIS ON SENSORS' ROLE IN MANUFACTURING

To develop a comprehensive CIM strategy incorporating sensors and control systems, an enterprise must begin with solid foundation, such as a CIM architecture. A CIM architecture is an information system structure that enables the industrial enterprise to integrate information and business processes. It accomplishes this by (1) establishing the direction integration will take and (2) defining the interfaces between the users and the providers of this integration function.

Figure 4.14 shows how CIM architecture answers the enterprise's integration means. A CIM architecture provides a core of common services. These services support every other area in the enterprise, from its common support function to its highly specialized business processes.

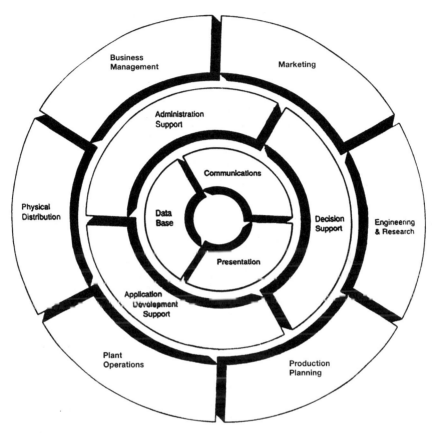

FIGURE 4.14 CIM architecture.

CIM and Building Blocks

The information environment of an industrial enterprise is subject to frequent changes in system configuration and technologies. A CIM architecture incorporating sensors and control systems can offer a flexible structure that enables it to react to the changes. This structure relies on a number of modular elements that allow systems to change easily to grow with the enterprise's needs.

Figure 4.17 shows a modular structure that gives CIM flexibility. It is based on three key building blocks:

1. *Communications.* The communication and distribution of data
2. *Data management.* The definition, storage, and use of data
3. *Presentation.* The presentation of data to people and devices throughout the enterprise

Utilizing the building blocks, CIM can provide a base for integrating the enterprise's products, processes, and business data. It can define the structure of the hardware, software, and services required to support the enterprise's complex requirements. It can also translate information into a form that can be used by the enterprise's people, devices, and applications.

CIM Communications

Communications—the delivery of enterprise data to people, systems, and devices—is a critical aspect of CIM architecture. This is because today's industrial environment brings together a wide range of computer systems, data acquisition systems, technologies, system architectures, operating systems, and applications. This range makes it increasingly difficult for people and machines to communicate with each other, especially when they describe and format data differently.

Various enterprises, in particular IBM, have long recognized the need to communicate data across multiple environments. IBM's response was to develop *systems network architecture* (SNA) in the 1970s. SNA supports communication among different IBM systems, and over the years it has become the standard for host communications in many industrial companies.

However, the CIM environment with sensor communications must be even more integrated. It must expand beyond individual areas, throughout the entire enterprise, and beyond—to customers, to vendors, and to subcontractors.

Communications in the CIM environment involves a wide range of data transfer, from large batches of engineering or planning data to single-bit messages from a plant floor device. Many connectivity types and protocols must be supported so that the enterprise's people, systems, and devices can communicate. This is especially true in cases where response time is critical, such as during process alerts.

Plant Floor Communications

Plant floor communications can be the most challenging aspect of factory control. This is due to the wide range of manufacturing and computer equipment that have been used in production tasks over the decades.

IBM's solution for communicating across the systems is the IBM plant floor series, a set of software products. One of these products, *Distributed Automation Edition* (DAE), is a systems enabler designed to provide communication functions that can be utilized by plant floor applications. These functions include:

1. Defining and managing networks
2. Making logical device assignments
3. Managing a program library for queueing and routing messages
4. Establishing alert procedures
5. Monitoring work-cell status

With these functions, Distributed Automation Edition can (1) help manufacturing engineers select or develop application programs to control work-cell operations and (2) provide communication capabilities between area- and plant-level systems (Fig. 4.15).

DAE supports several communication protocols to meet the needs of a variety of enterprises. For example, it supports SNA for connections to plant host systems and the IBM PC network, as well as the IBM token-ring protocol and manufacturing auto-

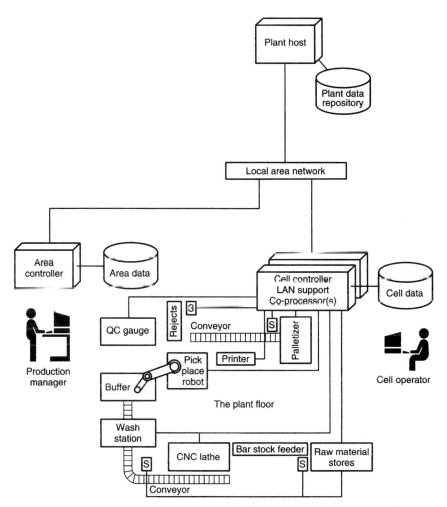

FIGURE 4.15 Distributed Automation Edition (DAE).

mated protocol (MAP) for plant floor communications. MAP is the evolving plant floor communication industry standard, adopted by the International Standards Organization for communications among systems provided by different vendors.

Managing Data in the CIM Environment

The second building block of a CIM architecture incorporating sensors and control technology is data management. This includes how data are defined, how different data elements are related, where data are stored, and who has access to that data. Data management is particularly critical in today's industrial environment, since there are very many different databases, formats, and storage and access techniques.

Standards are evolving. For example, *Structured Query Language* (*SQL*) provides a medium for *relational database* applications and for users to access a database. Unfortunately, there is a significant amount of data that exists today in other database technologies that are not accessible by current standards.

Data management defines and records the location of the data created and used by the enterprise's business functions. Data management also means enabling users to obtain the data needed without having to know where the data are located.

Relationships among several data elements must be known if data are to be shared by users and applications. In addition, other data attributes are important in sharing data. These include the type of data (text, graphic, image), their status (working, review, completed), and their source (person, application, or machine).

In CIM with sensory architecture, data management can be accomplished through three individual storage functions: (1) the data repository, (2) the enterprise data storage, and (3) the local data files.

Some of the key data management functions—the repository, for example—are already being implemented by the *consolidated design file* (CDF) established through the IBM *Data Communication Service* (DCS).

The consolidated design file operates on a relational database and is built on SQL. One example of its use is as an engineering database to integrate CAD/CAM applications with the business needs of the engineering management function. This environment, IBM's DCS/CDF, provides the following repository functions:

1. Transforming data to a user-selected format
2. Storing CAD/CAM data
3. Adding attributes to CAD/CAM data
4. Enabling users to query data and attributes

DCS/CDF also provides communications functions to transfer data between the repository and CAD/CAM applications (Fig. 4.16).

CIM Environment Presentation

Presentation in the CIM environment means providing data to and accepting data from people and devices. Obviously this data must assume appropriate data definitions and screen formats to be usable.

Because today's industrial enterprise contains such a wide array of devices and information needs, it must have a consistent way to distribute and present information to people, terminals, workstations, machine tools, robots, sensors, bar-code readers, automated guided vehicles, and part storage and retrieval systems. The range of this information covers everything from simple messages between people to large data arrays for engineering design and applications (Fig. 4.17). It may originate from a CIM user in one functional area of the enterprise and be delivered to a CIM user or device in another area.

In today's environments, presentation occurs on displays that utilize various technologies. Some are nonprogrammable terminals, some are programmable workstations, and some are uniquely implemented for each application. As a result, the same information is often treated differently by individual applications.

For example, the same manufactured part may be referred to as a part number in a bill of material in production planning, as a drawing in engineering's CAD application, and as a routing in a paperless shop order from plant operations.

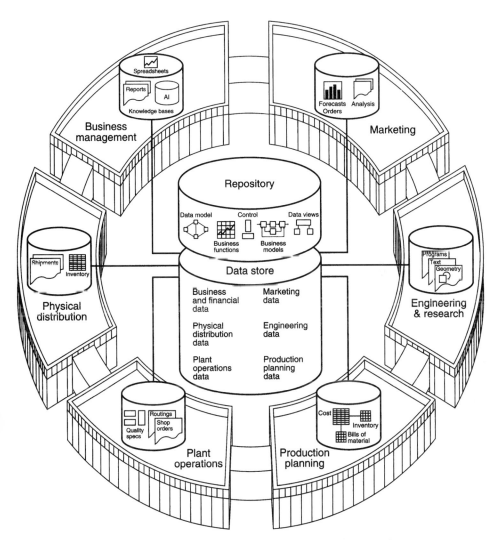

FIGURE 4.16 Data Communication Service/consolidated design file (DCS/CDF).

As data are shared across the enterprise, they must be transformed into definitions and formats that support the need of individual users and applications. Applications must be able to access shared data, collect the required information, then format that information for delivery.

The Requirement for Integration

Communication, data management, and presentation each have their own set of technical requirements. In addition, before these three building blocks can be inte-

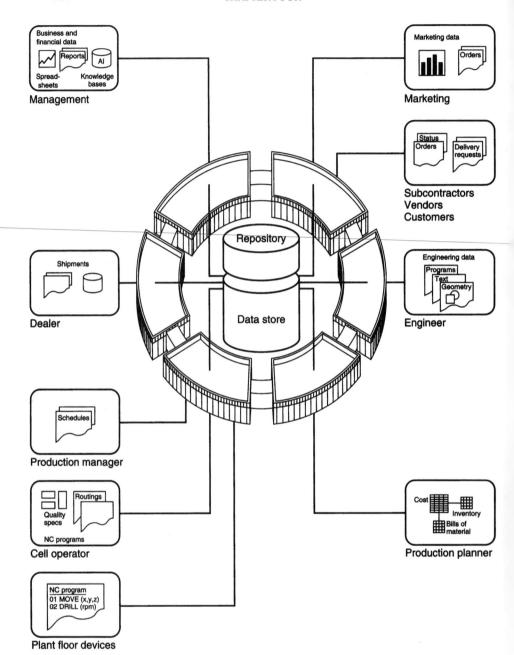

FIGURE 4.17 Presentation of data.

grated, a CIM architecture must also address a number of enterprise-wide constraints. For example, a CIM architecture should be able to:

1. Utilize standard platforms
2. Integrate data
3. Protect the installed base investment
4. Work with heterogeneous systems
5. Utilize industry-standard operator interfaces
6. Reduce application support cost
7. Provide a customizable solution
8. Offer phased implementation
9. Deliver selectable functions
10. Improve the business process

Standard Platforms. Utilizing standard computing platforms is one step industrial enterprises can take toward integration. Today, there are many products available that utilize standard platforms. These include processors, operating systems, and enablers for communications, database management, and presentation. In addition, platforms such as IBM's *systems application architecture* (SAA) and *advance interactive executive* (AIX) help make application consistency a reality across strategic IBM and UNIX operating system environments.

SAA, for example, is a comprehensive IBM blueprint for consistency and compatibility of software products. SAA begins the process by establishing definitions for four key application aspects: common user access, common programming interface, common communication support, and common applications. Through these definitions, SAA will support the development of new applications across major operating environments.

AIX is IBM's version of the UNIX operating system. AIX combines consistent user and application interfaces to aid in the development of a integrated application across UNIX environments. AIX consists of six related system enablers:

1. Base systems
2. Programming
3. Interface
4. User interface
5. Communication support
6. Distributed processing and applications

Data Integration. Integration requirements are often met by creating bridges between individual applications. Bridges usually copy a collection of data between two applications. A bridge between engineering and production planning allows these two functions to share a bill of material. Another bridge permits an engineering CAD/CAM application to download an NC program to a plant floor personal computer. Or a bridge between production planning and plant operations may be used to provide a copy of the production schedule to the plant floor system.

However, a problem with bridges is that changes made to the original set of data are not immediately incorporated into the copy of the data. This results in out-of-date information. Another problem is that bridges become difficult to maintain when more than two applications must work together.

As enterprises begin to integrate their operations, it will be imperative that the latest information is shared among multiple applications and across business functions. For example, engineering, marketing, cost accounting, production planning, and plant operations may all need access to inventory status information. At other times, the enterprise's various business functions may need information about product specifications, order status, operating cost, and more.

A CIM architecture must be able to simplify and accelerate this integration. It must provide the facilities to integrate data across the various applications of the business functions—facilities such as data query, data communication, controlled access and editing, and consistent data definitions.

Installed Base Investment. Today's industrial enterprises have made considerable investments in their installed bases, including systems, data, and even training. In the United States alone, manufacturers spend billions of dollars per annum on information systems hardware, software, and integration services for production planning, engineering, and plant operations. CIM with sensory technology must help protect this investment by permitting the integration of existing systems, applications, and data.

Heterogeneous Systems. In today's heterogeneous environment, data are located on different systems and in different formats. Applications have different needs, which are answered by processors, communications, and displays utilizing different technologies and architectures.

In an enterprise model, production planning may automate its operations on a single mainframe using an interactive database. Engineering may store drawings in an engineering database, then design and analyze products on a network of graphics workstations. Plant operations and sensors and control systems may be automated with personal computers and specialized machine controllers connected by both standard and proprietary networks. The data needed to operate the enterprise are scattered across all these diverse systems.

The heterogeneous environment is also characterized by an installed system base provided by multiple computer system suppliers, software vendors, and systems integrators. A CIM architecture must allow the integration of these varied system solutions and operating platforms.

Industry Standards and Open Interfaces. As integration technologies mature, there will be the need to support an expanding set of industry standards. Today these standards include communication protocols such as MAP, token ring, and Ethernet; data exchange formats such as the *initial graphics exchange specifications* (IGES) for engineering drawings; data access methods such as SQL; and programming interfaces such as *programmer's hierarchical interactive graphics standard* (PHIGS). A CIM architecture must be able to accommodate these and other evolving standards. One framework for accomplishing this has already been established, the open systems architecture for CIM (CIM-OSA). CIM-OSA is being defined in the *Esprit* program by a consortium of European manufacturers, universities, and information system suppliers, including IBM. Data exchange formats are also being extended to accommodate product definition in the *product definition exchange specification* (PDES). In addition, a CIM architecture must be able to support well-established solutions, such as IBM's SNA, which have become de facto standards.

In this competitive marketplace, manufacturers must also be able to extend operations as needed and support these new technologies and standards as they become available. These needs may include adding storage to a mainframe system, replacing engineering work-stations, installing a new machine tool, upgrading an operating system, and utilizing new software development tools. A CIM architecture with open interfaces will allow enterprises to extend the integration implementation over time to meet changing business needs.

Reduced Application Support Cost. A CIM architecture incorporating sensors must also yield application solutions at a lower cost than traditional stand-alone computing. This includes reducing the time and labor required to develop integrated applications and data. It also means reducing the time and effort required to keep applications up to speed with the changes in the enterprise's systems environment, technology, and integration needs.

Customizable Solutions. Every enterprise has its own business objectives, shared data, system resources and applications. For example, one enterprise may choose to address CIM requirements by reducing the cycle time of product development. It does this by focusing on the data shared between the product development and process development functions in accelerating the product release process.

Another enterprise's aim is to reduce the cycle time required to deliver an order to a customer. It addresses this by exchanging data between production planning and operations functions and automating the order servicing and production processes. It must also be able to customize operations to individual and changing needs over time.

Phased Implementation. Implementing enterprise-wide integration will take place in many small steps instead of through a single installation. This is because integration technology is still evolving, implementation priorities are different, installed bases mature at different rates, new users must be trained, and lessons will be learned in pilot installations.

As enterprises begin their implementation efforts in phases, they will be integrating past, present, and future systems and applications. A CIM architecture must be able to support the integration of this diverse installed base.

Selectable Functions. Most enterprises will want to weigh the benefits of integration against the impact this change will bring to each application and set of users. For example, an emphasis on product quality may require that production management gain greater insight into the quality of individual plant operations activities by implementing advanced sensors and control systems developed for particular applications. When an existing shop floor control application adequately manages schedules and support shop personnel with operating instructions, some additional information, such as that on quality, may be added, but rewriting the existing application may not be justified.

However, the plant manager may plan to develop a new production monitoring application to operate at each workstation. This application will make use of various sensors, share data with shop floor control, and utilize software building blocks for communications, database management, and presentation.

As is evident, the existing application requires only a data sharing capability, while the new application can benefit from both data sharing and the architecture building blocks. A CIM architecture with selectable functions will provide more options that can support the variety of needs within an enterprise.

Improved Business Process. Obviously, an enterprise will not implement integration on the basis of its technical merits alone. A CIM architecture must provide the necessary business benefits to justify change and investment.

The integration of information systems must support interaction between business functions and the automation of business processes. This is a key function if corporate goals, such as improved responsiveness to customer demands and reduced operating cost, are to be met. A CIM architecture must provide the means by which an entire enterprise can reduce the cycle times of business processes required for order processing, custom offerings, and new products. It must also reduce the impact of changes in business objectives and those business processes.

REFERENCES

1. Bucker, D. W., "10 Principles to JIT Advancement," *Manufacturing Systems,* March 1988, p. 55.
2. Campbell, J., *The RS-232 Solution,* Sybex, Inc., Alameda, Calif., 1984.
3. Clark, K. E., "Cell Control, The Missing Link to Factory Integration," International Industry Conference Proceedings, Toronto, May 1989, pp. 641–646.
4. Datapro Research Corporation, "How U.S. Manufacturing Can Thrive," *Management and Planning Industry Briefs,* March 1987, pp. 39–51.
5. Groover, M. P., and E. W. Zimmer, Jr., *CAD/CAM: Computer Aided Design and Manufacturing,* Prentice Hall, Englewood Cliffs, N.J., 1984.
6. IBM Corp., *Introducing Advanced Manufacturing Applications,* IBM, Atlanta, Ga., 1985.
7. "APICS: Tweaks and Distributed Systems Are the Main Focus—MRP No Longer Looks to the Future for Finite Capacity Scheduling," *Managing Automation,* January 1992, pp. 31–36.
8. Manufacturing Studies Board, National Research Council, *Toward a New Era in U.S. Manufacturing: The Need for a National Vision,* National Academy Press, Washington, D.C., 1986.
9. Orlicky, J., *Material Requirements Planning,* McGraw-Hill, New York, 1985.
10. Schonberger, R. J., "Frugal Manufacturing," *Harvard Business Review,* September–October 1987, pp. 95–100.
11. Skinner, W., *Manufacturing: The Formidable Competitive Weapon,* Wiley, New York, 1985.
12. "Support for The Manufacturing Floor," *Manufacturing Engineering,* March 1989, pp. 29–30.
13. Vollmann, T. E., W. L. Berry, and D. C. Whyback, *Manufacturing Planning and Control Systems,* Richard D. Irwin, Homewood, Ill., 1984.

CHAPTER 5
ADVANCED SENSOR TECHNOLOGY IN PRECISION MANUFACTURING APPLICATIONS

IDENTIFICATION OF MANUFACTURED COMPONENTS

In an automated manufacturing operation one should be able to monitor the identification of moving parts. The most common means of automatic identification is *bar-code technology*. There are other approaches that offer advantages under certain conditions.

Bar-Code Identification Systems

The *universal product code* (UPC) used in retail stores is a standard 12-digit code. Five of the digits represent the manufacturer and five the item being scanned. The first digit identifies the type of number system being decoded (a standard supermarket item, for example) and the second is a parity digit to determine the correctness of the reading. The first six digits are represented by code in an alternating pattern of light and dark bars. Figure 5.1 shows two encodings of the binary string 100111000. In both cases the minimum printed width is the same. The delta code requires nine such widths (the number of bits), while the width code requires 13 such widths (if a wide element is twice the width of the narrow element). Different bar widths allow for many character combinations. The remaining six digits are formed by dark alternating with light bars reversing the sequence of the first six digits. This allows backward scanning detection (Fig. 5.2).

A bar-code reader can handle several different bar-code standards, decoding the stripes without knowing in advance the particular standard. The military standard, code 1189, specifies the type of coding to be used by the Department of Defense, which is a modification of code 39. Code 39 consists of 44 characters, including the letters A through Z. Because of its alphanumeric capabilities, code 39 is very effective for manufacturing applications. Code 39 is structured as follows: three of nine bars (light and dark) form wide characters; the rest are narrow.

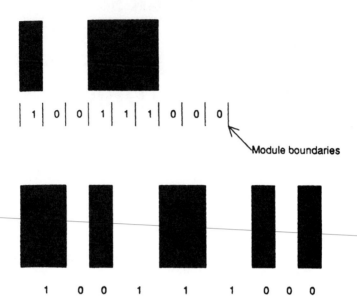

FIGURE 5.1 Encoding of the binary string 10011100 by delta code (top) and width code (bottom).

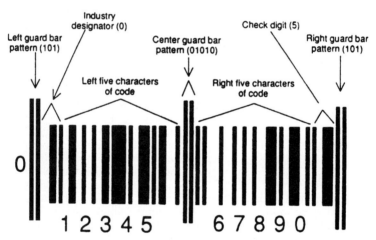

FIGURE 5.2 Specifications of the UPC symbol. The readable characters are normally printed in OCR-B font.

Bar-code labels are simple to produce. Code 39, for example, can be generated by a personal computer. Such labels are ideal for inventory identifications and other types of fixed-information gathering. Bar codes are not necessarily placed on labels. Tools, for example, have had the code etched on their surfaces to allow for tool tracking. Techniques have been developed for molding bar codes onto rubber tires.

Holographic scanners allow reading around corners so that parts need not be oriented perpendicular to the reader as they feed down a processing line.

A difficulty with bar coding has been the fact that it cannot be read if the bars become obscured by dirt, grease, or other substances. Infrared scanners are used to read codes that are coated with black substances to prevent secrecy violations through reproduction of the codes. One way to generally offset the problem of a dirty environment is to use magnetic-stripe-encoded information.

Transponders

While bar-code labels and magnetic stripes are very effective on the shop floor, shop circumstances may require more information to be gathered about a product than can be realistically handled with encoded media. For instance, with automobiles being assembled to order in many plants, significant amounts of information are necessary to indicate the options for a particular assembly. Radio-frequency (RF) devices are used in many cases. An RF device, often called a transponder, is fixed to the chassis of a car during assembly. It contains a chip that can store a great amount of information. A radio signal at specific assembly stations causes the transponder to emit information which can be understood by a local receiver. The transponder can be coated with grease and still function. Its potential in any assembly operation is readily apparent. There are several advanced transponders that have read/write capability, thus, supporting local decision making.

Electromagnetic Identification of Manufactured Components

There are many other possible electronic schemes to identify manufactured parts in motion. Information can be coded on a magnetic stripe in much the same way that bars represent information on a bar code label, since the light and dark bars are just a form of binary coding.

Operator identification data are often coded on magnetic stripes that are imprinted on the operators' badges. Magnetic stripe information can be fed into a computer. Such information might include the following: (1) the task is complete, (2) x number of units have been produced, (3) the unit part numbers, (4) the operator's identification number, etc. This same scanning station can also be set up using bar-code information; however, with magnetic striping, the information can be read even if the stripe becomes coated with dirt or grease. A disadvantage of magnetic striping is that the reader has to contact the stripe in order to recall the information.

Surface Acoustic Waves

A process similar to RF identification is *surface acoustic waves* (SAW). With this process, part identification is triggered by a radar-type signal which can be transmitted over greater distances than in RF systems.

Optical Character Recognition

Another form of automatic identification is *optical character recognition* (OCR). Alphanumeric characters form the information, which the OCR reader can "read." In mail processing centers, high-speed sorting by the U.S. Postal Service is accom-

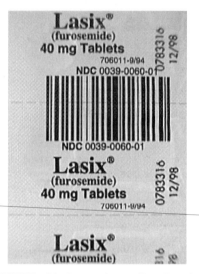

FIGURE 5.3 Bar code reading by the InspectRx Vision System. (*Courtesy of American SensoRx, Inc.*)

plished using OCR. The potential application to manufacturing information determination is obvious.

There are also many other means for part identification, such as vision systems and voice recognition systems. *Vision systems* utilize TV cameras to read alphanumeric data and transmit the information to a digital converter. OCR data can be read with such devices, as can conventionally typed characters. *Voice recognition systems* have potential where an individual's arms and hands are utilized in some function that is not conducive to reporting information. Such an application might be the inspection of parts by an operator who has to make physical measurements on the same parts.

In *laser scanning applications,* a laser beam scans and identifies objects at a constant speed. The object being scanned interrupts the beam for a time proportional to its diameter or thickness. Resolutions of less than 1 mm are possible.

In *linear array applications,* parallel light beams are emitted from one side of the object to be measured to a photooptical diode array on the opposite side. Diameters are measured by the number of array elements that are blocked. Resolutions of 5 mm or greater are possible.

In *TV camera applications,* a TV camera is used in the digitizing of the image of an object and the result is compared to the stored image (Fig. 5.3). Dimensions can be measured, part orientation can be determined, and feature presence can be checked. Some exploratory work is being accomplished with cameras that can fit in a tool changer mechanism. The camera can be brought to the part like a tool and verify part characteristics.

DIGITAL ENCODER SENSORS

Digital encoder sensors provide directly an output in a digital form and thus require only simple signal conditioning. They are also less susceptible to electromagnetic interference, and are therefore useful for information processing and display in measurement and control systems. Their ability to rapidly scan a series of patterns provides additional manufacturing automation opportunities when light and dark patterns are placed in concentric rings in a disk. Figure 5.4 illustrates a portion of such a disk that can be rigidly attached to a shaft or an object and housed in an assembly containing optical sensors for each ring (Fig. 5.5). The assembly, called an *optical encoder,* automatically detects rotation of a shaft or an object. The shaft rotation information can be fed back into a computer or controller mechanism for controlling velocity or position of the shaft. Such a device has application in robots and numerical control machine tools and for precision measurements of strip advancement to generate a closed-loop feedback actuation for displacement compensation.

There are two classes of digital encoder sensors:

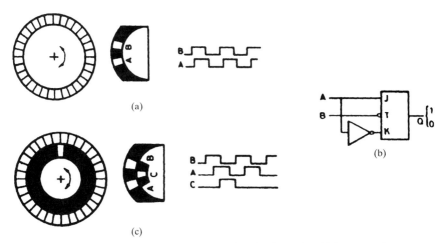

FIGURE 5.4 Detection of movement direction in directional encoders: (a) by means of two outputs with 90° phase shift; (b) output electronic circuit; (c) additional marker for absolute positioning.

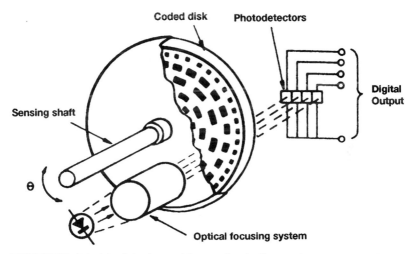

FIGURE 5.5 Principle of absolute position encoders for linear and rotary movements.

1. Encoder sensors yielding at the output a digital version of the applied analog input. This class of encoder sensors includes position encoders.
2. Encoder sensors that rely on some physical oscillatory phenomenon that is transduced by a conventional modulating sensor. This class of sensors may require an electronic circuit acting as a digital counter in order to yield a desired digital output signal.

There are no sensors where the transduction process directly yields a digital output. The usual process is to convert an analog input quantity into a digital signal by means of a sensor without the requirement to convert an analog voltage into its digital equivalent.

TABLE 5.1 Absolute Optical Encoder

Encoder ring	Angular displacement, degrees	Observed pattern	Computed value, degrees
1 (innermost)	180	1	180
2	90	0	
3	45	0	
4	22.5	1	22.5
5	11.25	0	
6	5.625	1	5.625
7	2.8125	1	2.8125
8	1.40625	0	
			210.94

Position Encoder Sensors in Manufacturing

Position encoder sensors can be categorized as linear and angular position encoder sensors. The optical encoder sensor can be either incremental or absolute. The incremental types transmit a series of voltages proportional to the angle of rotation of the shaft or object. The control computer must know the previous position of the shaft or object in order to calculate the new position. Absolute encoders transmit a pattern of voltages that describes the position of the shaft at any given time. The innermost ring reaches from dark to light every 180°, next ring every 90°, the next 45°, and so on, depending on the number of rings on the disk. The resulting bit pattern output by the encoder reveals the exact angular position of the shaft or object. For an absolute optical encoder disk that has eight rings and eight LED sensors, and in turn provides 8-bit outputs, (10010110). Table 5.1 shows how the angular position of the shaft or object can be determined.

The incremental position encoder sensor suffers three major weaknesses:

1. The information about the position is lost whenever the electric supply fails or the system is disconnected, and when there are strong perturbations.

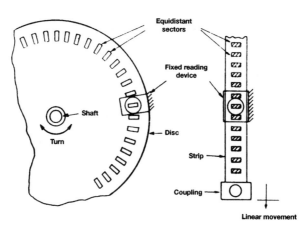

FIGURE 5.6 Principle of linear and rotary incremental position encoders.

2. The digital output, to be compatible with the input/output peripherals of a computer, requires an up/down counter.
3. The incremental position encoder does not detect the movement direction unless elements are added to the system (Fig. 5.6).

Physical properties used to define the disk pattern can be magnetic, electrical, or optical. The basic output generated by the physical property is a pulse train. By differentiating the signal, an impulse is obtained for each rising or falling edge, increasing by two the number of counts obtained for a given displacement (Fig. 5.7).

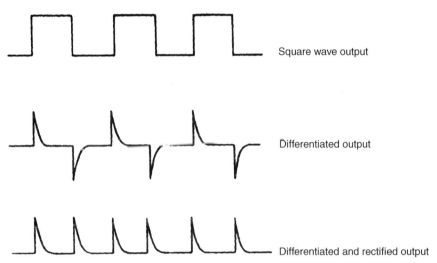

FIGURE 5.7 Improving output resolution of an incremental encoder by differentiation and rectification.

FUZZY LOGIC FOR OPTOELECTRONIC COLOR SENSORS IN MANUFACTURING

Fuzzy logic will most likely be the wave of the future in practical and economical solutions to control problems in manufacturing. Fuzzy logic is simply a technique that mimics human reasoning. This technology is now being explored throughout various industries. Fuzzy logic color sensors can relay information to microprocessors to determine color variance within an acceptable range of colors. A conventional sensor could not perform this function because it could choose only a specific color and reject all other shades—it uses a very precise set of rules to eliminate environmental interference.

The research and development activities for fuzzy logic technology began in mid-1990. The research has lead to the creation of a fuzzy logic color sensor that can learn a desired color and compare it with observed colors. The sensor can distinguish between acceptable and unacceptable colors for objects on a conveyer belt. The development of new light source technology allows the color sensor to produce more accurate color measurement (Fig. 5.8). Also, the integration of the fuzzy logic sensor with a microprocessor enables the data to be collected and interpreted accurately.

Sensing Principle

FIGURE 5.8 Integration of fuzzy logic sensor in a sensing module.

The sensor is designed with a broad-spectrum solid-state light source utilizing a light-emitting diode cluster. The LED-based light source provides stable, long-lasting, high-speed target illumination capabilities. The LED cluster is made up of three representative hues of *red, green, and blue* which provide a triple bell-shaped spectral power distribution for the light source (Fig. 5.9). The light incident on the target is reflected with varying intensities, depending on the particular target color under analysis.

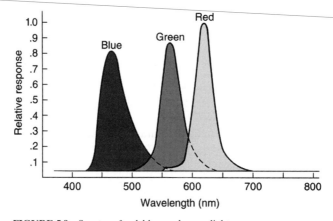

FIGURE 5.9 Spectra of red, blue, and green light.

The reflected light is received by a semiconductor receiver in the center of the LED cluster. The amount of light reflected back onto the receiver is transduced to a voltage and converted to digital format immediately via an analog-to-digital converter. The internal processing of the converted red, green, and blue (RGB) values offers variable sample size and averaging to compensate for signal noise (Fig. 5.9).

Ambient light is sampled between every component pulse and immediately subtracted from the sampled signal so that the effects of factory ambient light are suppressed. Thus, hooding the sensor is not totally necessary. In an area of a very bright or high-frequency lighting, it may be beneficial to provide some degree of hooding to at least limit the brightness. The sensor's electronics also employs temperature compensation circuitry to stabilize readings over temperature ranges.

Color Theory

Color science defines color in a space, with coordinates of *hue, saturation,* and *intensity* (HSI). These three general components uniquely define any color within HSI color space. Hue is related to the reflected wavelength of a color when a white light is shined

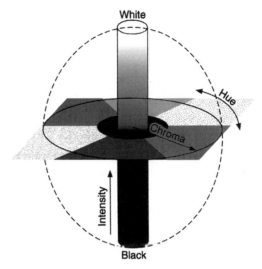

FIGURE 5.10 Coordinates of hue, saturation, and intensity of color in space.

on it. Intensity (lightness) measures the degree of whiteness, or gray scale, of a given color. Saturation is a measure of the vividness of a given hue. The term *chromaticity* primarily includes elements of the hue and saturation components. Researchers depict color in space using hue as the angle of a vector, saturation as the length of it, and intensity as a plus or minus height from a center point (Fig. 5.10).

The concepts of hue, saturation, and intensity can be further clarified by a simplified pictorial presentation. Consider Fig. 5.11, where a color is depicted at a molecular level. Color is created when light interacts with pigment molecules. Color is generated by the way pigment molecules return (bend) incoming light. For example, a red pigment causes a measurable hue component of the color. The relative density of the pigment molecules leads to the formation of the saturation component. There are some molecules present which return almost all wavelengths, and appear white as a result, leading to the intensity (lightness) component.

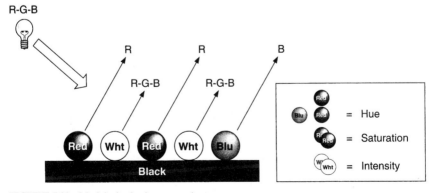

FIGURE 5.11 Model of color interpretation.

Units of Color Measurement

If color description depends on measuring the interaction of a target color with a given white light source, it is clear that in order to have the system of measurement standardized, both the light source and means of detection must be well-defined. One very popular set of standardization rules has been set up by the Commission International de l'Eclairage (CIE), a color standardization organization. From color theory, it is known that the response to a color stimulus (its determination), depends on the spectral power distribution of the light source (illuminant), times the spectral reflectance of the target (color) surface, times the spectral response of the detector (observer) (Fig. 5.12).

With this principle in mind, the CIE presented a detailed description of the standard light source and a standard observer (photodetector). The result of the study was the popular CIE diagram, which creates a two-dimensional mapping of a color space (Fig. 5.13).

Further manipulation of the CIE observations has lead to another color coordinate system, the so-called L.a.b numbers for describing a color. The L.a.b. numbering system is fairly prevalent in industrial applications. The machines that measure color according to this theory are referred to as *color spectrophotometers* or *color meters*. These machines are typically expensive, bulky, and not well-suited for distributed on-line color sensing.

The fuzzy color sensor does not offer CIE-based color measurement; however, it is a very high resolution color comparator. The sensor learns a color with its own standard light source (trio-stimulus LEDs) and its own observer (semiconductor photoreceiver). It thereby sets up its own unique color space with the three dimensional coordinates being the red, blue, and green readings (Fig. 5.13).

Theoretically there would be 256^3, or 16,777,216, unique positions in its color space for defining color (Fig. 5.14). In reality, the actual number of colors that a sensor can reliably distinguish is much less because of optical noise and practical limitations of the design.

The device compares the RGB colors it observes to the internally learned standard. Each time a standard color is relearned, it essentially recalibrates the sensor.

Color Comparators and True Color Measuring Instruments

If the learned color is initially defined by a color spectrometer, the fuzzy logic color sensor (comparator) can be installed in conjunction with it. The color sensor can learn the same area that the spectrometer has read, thereby equating its learned standard

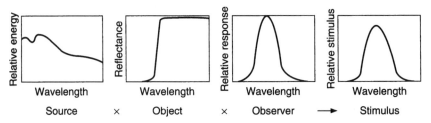

FIGURE 5.12 Stimulus response to a color (detector/determination) = illuminant × target × observer.

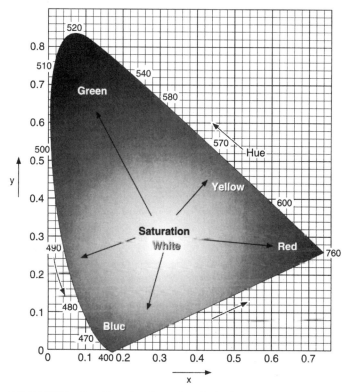

FIGURE 5.13 Two-dimensional mapping of color in a space.

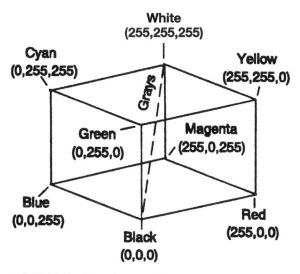

FIGURE 5.14 Three-dimensional coordinates of color in a space.

to the absolute color. By using the color sensor in this manner, a temporary correlation to an absolute color standard can be established (Fig. 5.15). This permits the relative drift from the standard to be monitored. The advantage is that more color sensing can be economically distributed across the target area. When a significant relative deviation is detected, an alert signal can flag the absolute color sensor to take a reading in the suspect area. If enough storage space is available in a central processing computer, lookup tables can be constructed to relate the serially communicated color sensor readings to a standard color coordinate system like the CIE system.

Color Sensor Algorithms

There are two internal software algorithms, or sets of rules, for analyzing fuzzy logic color sensor data:

1. The absolute algorithm: compares color on the basis of absolute voltages
2. The relative algorithm: compares color on the basis of relative percentages of each RGB component voltage

The choice of algorithm depends on sensing distance variation and the type of color distinction one needs. If the outputs vary excessively with distance when the absolute algorithm is used, a relative (ratio) algorithm must be considered. While a relative algorithm does not retain the lightness information, it greatly reduces unwanted distance-related variations. The relative algorithm shows changes in chromaticity (hue and chroma) which exist in most color differences. If the sensing distances can be held constant, the absolute algorithm works well at detecting subtle changes in density (shades) of a single color.

Design Considerations in Fuzzy Logic Color Sensors

The design of fuzzy logic color sensors aims to achieve maximum color sensing ability, while maintaining the expected simplicity of operation and durability of typical discrete industrial sensors. There are several other key goals that must be considered

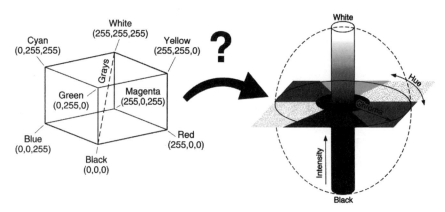

FIGURE 5.15 Correlation to an absolute color standard.

in system design such as high speed, small size, configurability, high repeatability, and long light source life.

The choice of a solid-state light source satisfies the majority of the criteria for a good industrialized color sensor design. The fuzzy logic color sensor utilizes two sets of three different LED photodiodes as its illumination source. The three LED colors, red, green, and blue, were chosen essentially for their coverage of the visible light spectrum (Fig. 5.16).

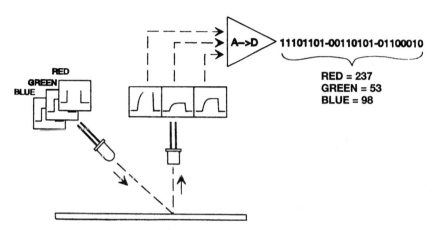

FIGURE 5.16 Conversion of red, green, and blue LED outputs from analog to digital.

The light from each LED is sequentially pulsed onto the target and its reflected energy is collected by a silicon photoreceiver chip in the LED cluster. Ambient light compensation circuitry is continually refreshed between each LED pulse, so the reported signals are almost entirely due to the LED light pulses. The LED sources offer a very fast (microsecond response) and stable (low spectral drift, steady power) source of a given wavelength band, without resorting to filters. Emerging blue LEDs, in combination with the more common red and green LEDs, have made it possible to use three solid-state spectra to define a hue. The choice of the specific LED bandwidth (or spectral distribution) is made so as to obtain the best color distinction through broader coverage of the illumination spectrum.

Fuzzy Logic Controller Flowchart

The entire sensor operation is illustrated in Fig. 5.17. An internal microcontroller governs the device operations. It directs signals in and out of the sensor head, to maintain local and remote communications, and provides color discrimination algorithms to produce the appropriate signal output at the control pin. As the device proceeds out of reset, it checks the locally (or remotely) set configuration dip-switches, which in turn define the path, or operating menu, to proceed through. There is permanent storage of learned or remotely loaded values of RGB readings, tolerance, number of readings to average, and the white-card calibration value, so these settings can be available at reset or at power-up.

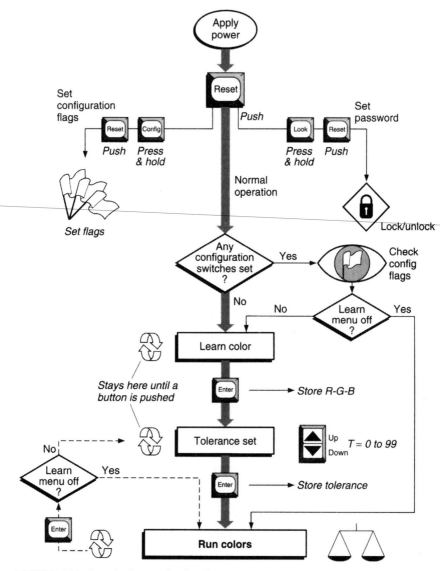

FIGURE 5.17 Fuzzy-logic controller flowchart.

One or more of three optional menus can be selected before entering into the main operation of learning and running colors. The three alternative menus are (1) white-card gain set, (2) number of reads to average set, and (3) the observed stored reading menus.

If none of the alternative menus is activated, the sensor will proceed directly to the primary modes, which are learn, tolerance set, and run modes (Fig. 5.18). By pressing

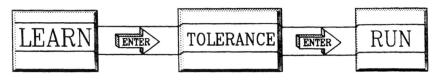

FIGURE 5.18 Simple path of operation.

and holding the appropriate buttons while pushing the reset button, two other programming menus can be entered. These are the set configuration flags (dip-switch) menu and the set password menu.

SENSORS DETECTING FAULTS IN DYNAMIC MACHINE PARTS (BEARINGS)

A system consisting of analog and digital signal processing equipment, computers, and computer programs would detect faults in ball bearings in turbomachines and predict the remaining operating time until failure. The system would operate in real time, extracting the diagnostic and prognostic information from vibrations sensed by accelerometers, strain gages, and acoustical sensors, and from the speed of the machine as measured by a tachometer.

The vibrations that one seeks to identify are those caused by impact that occurs when pits in balls make contact with races and pits in races make contact with balls. These vibrations have patterns that are unique to bearings and repeat at known rates that are related to ball-rotation, ball-pass, and cage-rotation frequencies. These vibrations have a wide spectrum that extends up to hundreds of kilohertz, where the noise component is relatively low.

The system in Fig. 5.19 would accept input from one of two sensors. Each input signal would be amplified, bandpass-filtered, and digitized. The digitized signal would be processed in two channels: one to compute the keratosis of the distribution of the amplitudes, the other to calculate the frequency content of the envelope of the signal. The *keratosis* is the fourth statistical moment and is known, from theory and experiment, to be indicative of vibrations caused by impact on faults. The keratosis would be calculated as a moving average for each consecutive digitized sample of the signal by using a number of samples specified by the technician. The trend of a keratosis moving average would be computed several times per second, and the changes in the keratosis value that are deemed to be statistically significant would be reported.

In the other signal processing channel, the amplitude envelope of the filtered, digitized signal would be calculated by squaring the signal. Optionally, the high-frequency sample data would be shifted to a lower frequency band to simplify processing by use of a Fourier transformation. This transformation would then be applied to compute the power spectrum.

The output of the tachometer would be processed in parallel with the spectral calculations so that the frequency bins of the power spectrum could be normalized on the basis of the speed of rotation of the machine. The power spectrum would be averaged with a selected number of previous spectra and presented graphically as a

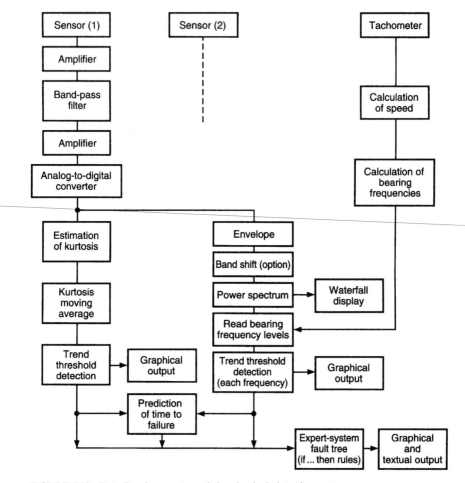

FIGURE 5.19 Data flow for an automatic bearing fault detection system.

"waterfall" display; this is similar to a sonar display with which technicians can detect a discrete frequency before an automatic system can.

The bearing frequencies would be calculated from the measured speed and the known parameters of the bearings, with allowances for slip. The power spectrum levels would be read for each bearing frequency; a moving average of the amplitude at each bearing frequency and harmonic would be maintained, and trends representing statistically significant increases would be identified by threshold detection and indicated graphically.

By using algorithms based partly on analyses of data from prior tests, the results of both keratosis and power spectrum calculations would be processed onto predictions of the remaining operating time until failure. All the results would then be processed by an expert system. The final output would be a graphical display and text that would describe the condition of the bearings.

SENSORS FOR VIBRATION MEASUREMENT OF A STRUCTURE

An advanced sensor was developed to gage structure excitations and measurements that yield data for design of robust stabilizing control systems (Fig. 5.20).

An automated method for characterizing the dynamic properties of a large flexible structure estimates model parameters that can be used by a robust control system to stabilize the structure and minimize undesired motions. Although it was developed for the control of large, flexible structures in outer space, the method is also applicable to terrestrial structures in which vibrations are important—especially aircraft, buildings, bridges, cranes, and drill rigs.

The method was developed for use under the following practical constraints:

1. The structure cannot be characterized in advance with enough accuracy for purposes of control.
2. The dynamics of the structure can change in service.
3. The numbers, types, placements, and frequency responses of sensors that measure the motions and actuators that control them are limited.
4. Time available during service for characterization of the dynamics is limited.
5. The dynamics are dominated by a resonant mode at low frequency.
6. In-service measurements of the dynamics are supervised by a digital computer and are taken at a low rate of sampling, consistent with the low characteristic frequencies of the control system.
7. The system must operate under little or no human supervision.

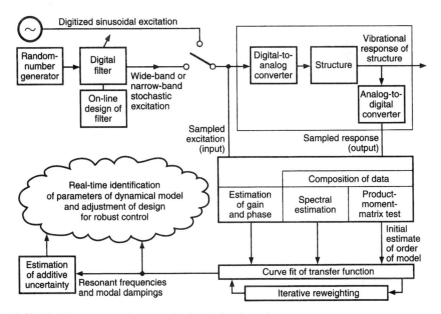

FIGURE 5.20 Automated characterization of vibrations of a structure.

The method is based on extracting the desired model and control-design data from the response of the structure to known vibrational excitations (Fig. 5.20). Initially, wideband stochastic excitations are used to obtain the general characteristics of the structure. Narrow-band stochastic and piece-wise-constant (consistent with sample-and-hold discretizations) approximations to sinusoidal excitations are used to investigate specific frequency bands in more detail.

The relationships between the responses and excitations are first computed nonparametrically—by spectral estimation in the case of stochastic excitations and by estimation of gains and phases in the case of approximately sinusoidal excitations. In anticipation of the parametric curve fitting to follow, the order of a mathematical model of the dynamics of the structure is estimated by use of a *product moment matrix* (PMM). Next, the parameters of this model are identified by a least-squares fit of transfer-function coefficients to the nonparametric data. The fit is performed by an iterative reweighting technique to remove high-frequency emphasis and assure minimum-variance estimation of the transfer-function coefficient. The order of the model starts at the PMM estimate and is determined more precisely thereafter by successively adjusting a number of modes in the fit at each iteration until an adequately small output-error profile is observed.

In the analysis of the output error, the additive uncertainty is estimated to characterize the quality of the parametric estimate of the transfer function and for later use in the analysis and design of robust control. It can be shown that if the additive uncertainty is smaller than a certain calculable quantity, then a conceptual control system could stabilize the model structure and could also stabilize the real structure. This criterion can be incorporated into an iterative design procedure. In this procedure, each controller in a sequence of controllers for the model structure would be designed to perform better than the previous one did, until the condition for robust capability was violated. Once the violation occurred, one could accept the penultimate design (if its performances were satisfactory) or continue the design process by increasing a robustness weighting (if available). In principle, convergence of this iterative process guarantees a control design that provides high performance for the model structure while guaranteeing robustness of stability to all perturbations of the structure within the additive uncertainty.

OPTOELECTRONIC SENSOR TRACKING TARGETS ON A STRUCTURE

The location and exact position of a target can be accurately sensed through optoelectronic sensors for tracking a retroreflective target on a structure. An optoelectronic system simultaneously measures the positions of as many as 50 retroreflective targets within 35° of view with an accuracy of 0.1 mm. The system repeats the measurement 10 times per second. The system provides an unambiguous indication of the distance to each target that is not more than 75 m away from its sensor module. The system is called *spatial high-accuracy position-encoding sensor* (SHAPES).

SHAPES fills current needs in the areas of system identification and control of large flexible structures, such as large space- and ground-based antennas and elements of earth-orbiting observational platforms. It is also well-suited to applications in rendezvous and docking systems. Ground-based applications include boresight determination and precise pointing of 70-m deep-space-network antennas.

SHAPES illuminates the retroreflective targets by means of a set of lasers in its sensor module. In a typical application (Fig. 5.21) a laser diode illuminates each tar-

ADVANCED SENSOR TECHNOLOGY IN PRECISION MANUFACTURING   5.19

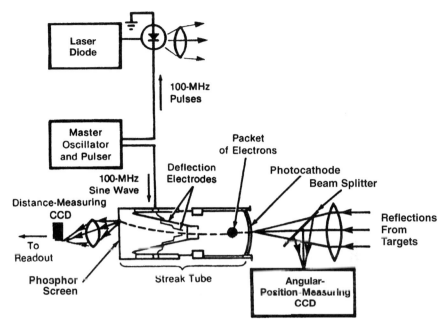

FIGURE 5.21 Beam splitter diverts reflections from a continuous-wave laser into a CCD camera for measurement of angles of reflection.

get with 30-ps pulses at a repetition rate of 100 MHz. Light reflected from the target is focused by a lens and passed through a beam splitter to form images on a charge-coupled device (CCD) and on the photocathode of a streak tube. The angular position of the target is determined simply from the position of its reflection on the charge-coupled device.

The measurement of the distance to the target is based on the round-trip time of the optical pulses. The round-trip distance can be measured in terms of the difference between the phase of the train of return pulses incident on the photocathode and the phase of a reference sine wave that drives the deflection plate of the streak tube. This difference, in turn, manifests itself as a displacement between the swept and unswept positions, at the output end of the streak tube, of the spot of light that represents the reflection from the target. The output of the streak tube is focused on a CCD for measurement and processing of the position of this spot. Three microprocessors control the operation of SHAPES and convert the raw data required from the angular-position and distance-measuring CCDs into position of the target in three dimensions.

OPTOELECTRONIC FEEDBACK SIGNALS FOR SERVOMOTORS THROUGH FIBER OPTICS

In what is believed to be among its first uses to close a digital motor-control loop, fiber-optics transmission provides immunity to noise and rapid transmission of data.

An optoelectronic system effects closed-loop control of the shaft angle of four servomotors and could be expanded to control as many as 16. The system includes a full-duplex fiber-optic link (Fig. 5.22) that carries feedforward and feedback digital signals over a distance of many meters, between commercial digital motor-control circuits that execute a PID control algorithm with programmable gain (one such control circuit dedicated to each servomotor) and modules that contain the motor-power switching circuits, digital-to analog buffer circuits for the feedforward control signals, and analog-to-digital buffer circuits for the feedback signals from the shaft-angle encoders (one such module located near, and dedicated to, each servomotor).

Besides being immune to noise, optical fibers are compact and flexible. These features are particularly advantageous in robots, which must often function in electromagnetically noisy environments and in which it would otherwise be necessary to use many stiff, bulky wires (which could interfere with movement) to accommodate the required data rates.

Figure 5.22 shows schematically the fiber-optic link and major subsystems of the control loop of one servomotor. Each digital motor-control circuit is connected to a central control computer, which programs the controller gains and provides the high-level position commands. The other inputs to the motor-control circuit include the sign of the commanded motor current and pulse-width modulation representing the magnitude of the command motor current.

The fiber-optic link includes two optical fibers—one for feedforward, one for feedback. The ends of the fibers are connected to identical bidirectional interface circuit boards, each containing a transmitter and a receiver. The fiber-optic link has a throughput rate of 175 MHz; at this high rate, it functions as though it were a 32-bit parallel link (8 bits for each motor control loop), even though the data are multiplexed into a serial bit stream for transmission. In the receiver, the bit stream is decoded to reconstruct the 8-bit pattern and a programmable logic sequencer expands the 8-bit pattern to 32 bits and checks for errors by using synchronizing bits.

ACOUSTOOPTICAL/ELECTRONIC SENSOR FOR SYNTHETIC-APERTURE RADAR UTILIZING VISION TECHNOLOGY

An acoustooptical sensor operates in conjunction with analog and digital electronic circuits to process frequency-modulated *synthetic-aperture radar* (SAR) return signals in real time. The acoustooptical SAR processor will provide real-time SAR imagery aboard moving aircraft or space SAR platforms. The acoustooptical SAR processor has the potential to replace the present all-electronic SAR processors that are currently so large and heavy and consume so much power that they are restricted to use on the ground in the postprocessing of the SAR in-flight data recorder.

The acoustooptical SAR processor uses the range delay to resolve the range coordinates of a target. The history of the phase of the train of radar pulses as the radar platform flies past a target is used to obtain the azimuth (cross-range) coordinate by processing it coherently over several returns. The range-compression signal processing involves integration in space, while the azimuth-compression signal processing involves integration in time.

Figure 5.23 shows the optical and electronic subsystems that perform the space and time integrations. The radar return signal is heterodyned to the middle frequency of an acoustooptical sensor and added electronically to a reference sinusoid to capture the history of the phase of the return signal interferometrically for com-

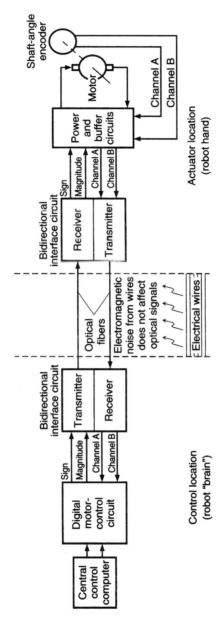

FIGURE 5.22 Full-duplex fiber-optic transmission link.

5.21

pression in azimuth. The resulting signal is applied to the acoustooptical sensor via a piezoelectric transducer. The acoustooptical sensor thus becomes a cell that encodes the evolving SAR return.

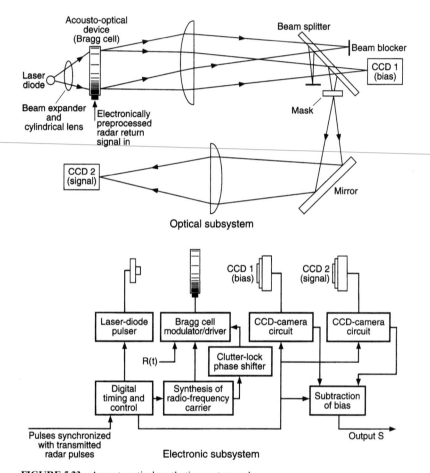

FIGURE 5.23 Acoustooptical synthetic-aperture radar.

Meanwhile, pulses of light a few tens of nanoseconds long are generated by a laser diode in synchronism with the transmitted pulses and are used to sample and process the return signal. Lenses shape that laser light into a plane wave incident upon the acoustooptical sensor. The integration in space is effected at the moment of sampling by the focusing action. The position of the focal point in the cell depends on the range delay of the corresponding target, and light is brought to focus on two CCD imaging arrays at positions that depend on the range.

The sinusoidal reference signal component of the cell interacts with laser radiation to generate a plane wave of light that interferes with the light focused by the cell. This produces interference fringes that encode the phase information in the range-compressed optical signal. These fringes are correlated with a mask that has a predetermined spatial distribution of density and that is placed in front of, or on, one of the CCD arrays. This CCD array is operated in a delay-and-integrate mode to obtain the desired correlation and integration in time for the azimuth compression. The output image is continuously taken from the bottom picture element of the CCD array.

Two CCDs are used to alleviate a large undesired bias of the image that occurs at the output as a result of optical processing. CCD_1 is used to compute this bias, which is then subtracted from the image of CCD_2 to obtain a better image.

THE USE OF OPTOELECTRONIC/VISION ASSOCIATIVE MEMORY FOR HIGH-PRECISION IMAGE DISPLAY AND MEASUREMENT

Storing an image of an object often requires large memory capacity and a high-speed interactive controller. Figure 5.24 shows schematically an optoelectronic associative memory that responds to an input image by displaying one of M remembered images. The decision about which if any of the remembered images to display is made by an optoelectronic analog computation of an inner-product-like measure of resemblance between the input image and each of the remembered images. Unlike associative memories implemented as all-electronic neural networks, this memory does not rely on the precomputation and storage of an outer-product synapse matrix. Instead, the optoelectronic equivalent of this matrix is realized by storing remembered images in two separate spatial light modulators placed in tandem. This scheme reduces the required size of the memory by an order of magnitude.

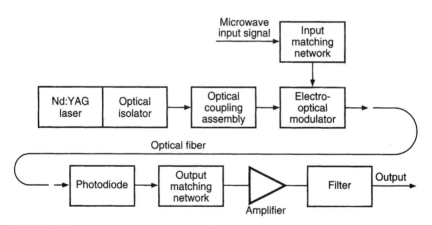

FIGURE 5.24 A developmental optoelectronic associative memory.

A partial input image is binarized and displayed on a liquid-crystal light valve spatial modulator which reprocesses the image in real time by operating in an edge-enhancement mode. This preprocessing increases the orthogonality (with respect to the inner product) between the input image and each of the remembered images, thereby increasing the ability of the memory to discriminate among different images.

The light from the input image is passed through a polarizing beam splitter, a lens, a binary diffraction grating, and another lens, to focus an array of M replicas of the input image on one face of a liquid-crystal-television spatial light modulator that is displaying the M remembered images. The position of each replica of the input image coincides with that of one of the remembered images. Light from the array of pairs of overlapping input and remembered images is focused by a corresponding array of lenslets onto a corresponding array of photodetectors. The intensity of light falling on each photodetector is proportional to the inner product between the input image and the corresponding remembered image.

The outputs of the photodetectors are processed through operational amplifiers that respond nonlinearly to inner-product level (in effect executing analog threshold functions). The outputs of the amplifiers drive point sources of white light, and an array of lenslets concentrates the light from each source onto the spot occupied by one of M remembered images displayed on another liquid-crystal-television spatial light modulator. The light that passes through this array is reflected by a pivoted ray of mirrors through a lens, which focuses the output image onto a CCD television camera. The output image consists of superpositioned remembered images, the brightest of which are those that represent the greatest inner products (the greatest resemblance to the input image). The television camera feeds the output image to a control computer, which performs a threshold computation, then feeds the images through a cathode-ray tube back to the input liquid-crystal light valve. This completes the associative recall loop. The loop operates iteratively until one (if any) of the remembered images is the sole output image.

SENSORS FOR HAND-EYE COORDINATION OF MICROROBOTIC MOTION UTILIZING VISION TECHNOLOGY

The micro motion of a robotic manipulator can be controlled with the help of dual feedback by a new method that reduces position errors by an order of magnitude. The errors—typically of the order of centimeters—are differences between real positions on the one hand and measured and computed positions on the other; these errors arise from several sources in the robotic actuators and sensors and in the kinematic model used in control computations. In comparison with current manufacturing methods of controlling the motion of a robot with visual feedback (the robotic equivalent of hand-eye coordination), the novel method requires neither calibration over the entire work space nor the use of an absolute reference coordinate frame for computing transformations between field of view and robot joint coordinates.

The robotic vision subsystem includes five cameras: three stationary ones that provide wide-angle views of the work space and two mounted on the wrist of an auxiliary robot arm to provide stereoscopic close-up views of the work space near the manipulator (Fig. 5.25). The vision subsystem is assumed to be able to recognize the objects to be avoided and manipulated and to generate data on the coordinates of the objects from sent positions in the field-of-view reference frame.

The new method can be implemented in two steps:

1. The close-up stereoscopic cameras are set initially to view a small region that contains an object of interest. The end effector is commanded to move to a nom-

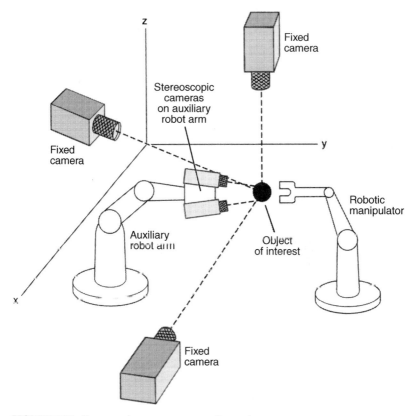

FIGURE 5.25 Stereoscopic cameras on an auxiliary robot arm.

inal position near the object and within the field of view. Typically, the manipulator stops at a slightly different position, which is measured by the cameras. Then the measured error in position is used to compute a small corrective motion. This procedure is designed to exploit the fact that small errors in relative position can be measured accurately and small relative motions can be commanded accurately.

2. The approximate direct mapping between the visual coordinates and the manipulator joint-angle coordinates can be designed without intermediate transformation to and from absolute coordinates. This is, in effect, a calibration, but it requires fewer points than does a conventional calibration in an absolute reference frame over the entire work space. The calibration is performed by measuring the position of a target (in field-of-view coordinates) when the target is held rigidly by the manipulator at various commanded positions (in manipulator joint-angle coordinates) and when the cameras are placed at various commanded positions. Interpolations and extrapolations to positions near the calibration points are thereafter performed by use of the nonlinear kinematic transformations.

FORCE AND OPTICAL SENSORS CONTROLLING ROBOTIC GRIPPER FOR AGRICULTURE AND MANUFACTURING APPLICATIONS

A robotic gripper operates in several modes to locate, measure, recognize (in a primitive sense), and manipulate objects in an assembly subsystem of a robotic cell that is intended to handle geranium cuttings in a commercial greenhouse. The basic concept and design of the gripper could be modified for handling other objects, for example, rods or nuts, including sorting the objects according to size. The concept is also applicable to real-time measurement of the size of an expanding or contracting part gripped by a constant force and to measurement of the size of a compliant part as a function of the applied gripping force.

The gripper is mounted on an industrial robot. The robot positions the gripper at a fixed distance above the cutting to be processed. A vision system locates the cutting in the x-y plane lying on a conveyer belt (Fig. 5.26).

The robot uses fiber-optic sensors in the fingertip of the gripper to locate the cutting along the axis. The gripper grasps the cutting under closed-loop digital servo force control. The size (that is, the diameter of the stem) of the cutting is determined

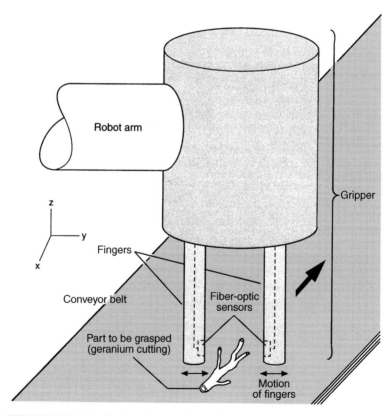

FIGURE 5.26 Robotic gripper for geranium cutting.

from the finger position feedback, while the cutting is being grasped under force control. The robot transports the cutting to a scale for weighing, to a trimming station, and finally to a potting station. In this manner, cuttings are sorted according to weight, length, and diameter.

The control subsystem includes a 32-bit minicomputer that processes the vision information and collects system grating data. The minicomputer communicates with a robot controller and the gripper node control. The gripper node control communicates with the scale. The robot controller communicates with the gripper node controller via discrete input/output triggering of each of the gripper functions.

The gripper subsystem includes a PC/AT-compatible industrial computer; a gripper mechanism, actuated by two dc servomotors, with an integrated load cell; discrete input/output components; and two fiber-optic analog-output distance sensors. The computer includes a discrete input/output circuit card, an 8-channel A/D converter circuit card, a motor-control circuit card with A/D components, two serial ports, and a 286 processor with coprocessor.

A geranium cutting comprises a main stem and several petioles (tiny stem leaves). The individual outputs from the fiber-optic sensors can be processed into an indication of whether a stem or a petiole is coming into view as the gripper encounters the cutting. Consequently the gripper can be commanded to grasp a stem but not a petiole. The axial centerline of a stem can also be recognized from the outputs of the five optic sensors. Upon recognition of a centerline, the gripper signals the robot, and the robot commands the gripper to close.

The motor-controller circuit card supplies the command signals to the amplifier that drives the gripper motors. This card can be operated as a position control with digital position feedback or as a force control with analog force feedback from the load cell mounted in the gripper. A microprocessor is located on the motor control card. Buffered command programs are downloaded from the computer to this card for independent execution by the card.

Prior to a controlled force closure, the motor-control card controls the gripper in position-servo mode until a specified force threshold is sensed, indicating contact with the cutting. Thereafter, the position-servo loop is opened, and the command signal to the amplifier is calculated as the difference between the force set point and the force feedback from the load cell. This distance is multiplied by a programmable gain value, then pulse-width-modulated with a programmable duty cycle of typically 200 percent. This technique provides integral stability to the force-control loop. The force-control loop is bidirectional in the sense that, if the cutting expands between the fingertips, the fingers are made to separate and, if the cutting contracts, the fingers are made to approach each other.

ULTRASONIC STRESS SENSOR MEASURING DYNAMIC CHANGES IN MATERIALS

An *ultrasonic dynamic vector stress sensor* (UDVSS) has recently been developed to measure the changes in dynamic directional stress that occur in materials or structures at the location touched by the device when the material or structure is subjected to cyclic load. A strain gage device previously used for the measurement of such a stress measured strain in itself, not in the part being stressed, and thus provided a secondary measurement. Other techniques, such as those that involve thermoelasticity and shearography, have been expensive and placed demands on the measured material. The optical measurement of stress required the application of a phase coat to the object under test. The laser diffraction method required notching or sharp marking of the specimen.

A UDVSS is the first simple portable device able to determine stress directly in the specimen itself rather than in a bonded gage attached to the specimen. As illustrated in Fig. 5.27, a typical material testing machine applies cyclic stress to a specimen. The UDVSS includes a probe, which is placed in contact with the specimen; an electronic system connected to the probe; and a source of a reference signal. The probe assembly includes a probe handle that holds the probe, a transducer mount that contains active ultrasonic driver and receiver, and ultrasonic waveguide transmitter and ultrasonic waveguide receiver that convert the electrical signals to mechanical motion and the inverse, and a cable that connects the probe of the electronics. When in contact with the specimen, the ultrasonic waveguide transmitter causes acoustic waves to travel across the specimen to the ultrasonic waveguide receiver, wherein the wave is converted to an electrical signal.

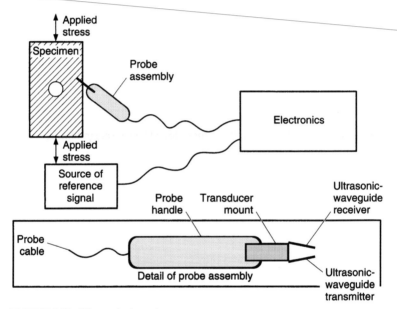

FIGURE 5.27 Ultrasonic dynamic stress sensor.

The operation of the UDVSS is based on the physical phenomenon that the propagation of sound in the specimen changes when the stress in the specimen changes. A pulse phase-locked loop reacts to a change in propagation of sound and therefore in stress by changing its operational frequency. The component of that signal represents that change in voltage needed to keep the system at quadrature to follow the system change in stress. That signal provides the information on changing stress.

The UDVSS can be moved around on the specimen to map out the stress field, and by rotating the probe, one can determine the direction of a stress. In addition, the probe is easily calibrated. The UDVSS should find wide acceptance among manufacturers of aerospace and automotive structures for stress testing and evaluation of designs.

PREDICTIVE MONITORING SENSORS SERVING CIM STRATEGY

Computer-integrated manufacturing technology can be well-served by a predictive monitoring system that would prevent a large number of sensors from overwhelming the electronic data monitoring system or a human operator. The essence of the method is to select only a few of the many sensors in the system for monitoring at a given time and to set alarm levels of the selected sensor outputs to reflect the limit of expected normal operation at the given time. The method is intended for use in a highly instrumented system that includes many interfacing components and subsystems, for example, an advanced aircraft, an environmental chamber, a chemical processing plant, or a machining work cell.

Several considerations motivate the expanding effort in implementing the concept of predictive monitoring. Typically, the timely detection of anomalous behavior of a system and the ability of the operator or electronic monitor to react quickly are necessary for the continuous safe operation of the system.

In the absence of a sensor-planning method, an operator may be overwhelmed with alarm data resulting from interactions among sensors rather than data directly resulting from anomalous behavior of the system. In addition, much raw sensor data presented to the operator may by irrelevant to an anomalous condition. The operator is thus presented with a great deal of unfocused sensor information, from which it may be impossible to form a global picture of events and conditions in the system. The predictive monitoring method would be implemented in a computer system running artificial intelligence software, tentatively named *PREMON*. The predictive monitoring system would include three modules: (1) a causal simulator, (2) a sensor planner, and (3) a sensor interpreter (Fig. 5.28).

The word *event* in Fig. 5.28 denotes a discontinuous change in the value of a given quantity (sensor output) at a given time. The inputs to the causal simulator would include a causal mathematical model of the system to be monitored, a set of events

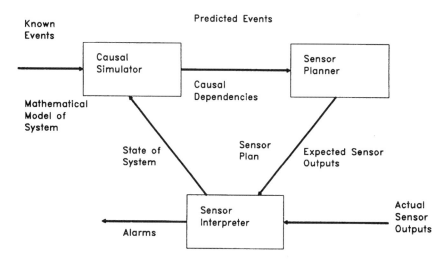

FIGURE 5.28 Predictive monitoring system would concentrate on the few sensor outputs that are most important at the moment.

that describe the initial state of the system, and perhaps some future scheduled events. The outputs of the causal simulator would include a set of predicted events and a graph of causal dependency among events.

The sensor planner would use the causal dependency graph generated by the causal simulator to determine which few of all the predicted events are important enough that they should be verified. In many cases, the most important events would be taken to be those that either caused, or are caused by, the greatest number of other events. This notion of causal importance would serve as the basis for the election of those sensors, the outputs of which should be used to verify the expected behavior of the system.

The sensor interpreter would compare the actual outputs of the selected sensors with the values of those outputs predicted by the causal simulator. Alarms would be raised where the discrepancies between predicted and actual values were significant.

REFLECTIVE STRIP IMAGING CAMERA SENSOR— MEASURING A 180°-WIDE ANGLE

A proposed camera sensor would image a thin striplike portion of a field of view 180° wide. For example, it could be oriented to look at the horizon in an easterly direction from north to south or it could be rotated about a horizontal north/south axis to make a "pushbroom" scan of the entire sky proceeding from the easterly toward the westerly half of the horizon. Potential uses for the camera sensor include surveillance of clouds, coarse mapping of terrain, measurements of the bidirectional reflectance distribution functions of aerosols, imaging spectrometry, oceanography, and exploration of the planets.

The imaging optics would be a segment of concave hemispherical reflecting surfaces placed slightly off center (Fig. 5.29). Like other reflecting optics, it would be achromatic. The unique optical configuration would practically eliminate geometric distortion of the image. The optical structure could be fabricated and athermalized fairly easily in that it could be machined out of one or a few pieces of metal, and the spherical reflecting surface could be finished by diamond turning. In comparison, a camera sensor with a fish-eye lens, which provides a nearly hemispherical field of view, exhibits distortion, chromatism, and poor athermalization. The image would be formed on a thin semicircular strip at half the radius of the sphere. A coherent bundle of optical fibers would collect the light from this strip and transfer the image to a linear or rectangular array of photodetectors or to the entrance slit of an image spectrograph. Provided that the input ends of the fibers were properly aimed, the cones of acceptance of the fibers would act as aperture stops; typically, the resulting width of the effective aperture of the camera sensor would be about one-third the focal length ($f/3$).

The camera sensor would operate at wavelengths from 500 to 1100 nm. The angular resolution would be about 0.5°. In the case of an effective aperture of $f/3$, the camera would provide an unvignetted view over the middle 161° of the strip, with up to 50 percent vignetting in the outermost 9.5° on each end.

The decentration of the spherical reflecting surface is necessary to make room for the optical fibers and the structure that would support them. On the other hand, the decentration distance must not exceed the amount beyond which the coma that results from decentration would become unacceptably large. In the case of an effective aperture of $f/3$, the coma would be only slightly in excess of the spherical aberration if the decentration were limited to about $f/6$. This would be enough to accommodate the fibers and supporting structure.

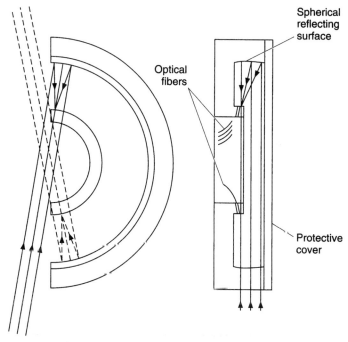

FIGURE 5.29 A thin segment of a hemispherical concave reflector would form an image from a 180° strip field of view onto optical fibers.

OPTICAL SENSOR QUANTIFYING ACIDITY OF SOLUTIONS

With environmental concerns increasing, a method for taking effective measurements of acidity will minimize waste and reduce both the cost and the environmental impact of processing chemicals. Scientists at Los Alamos National Laboratory (LANL) have developed an optical sensor that measures acid concentration at a higher level than does any current device. The optical high-acidity sensor, reduces the wear generated from acidity measurements and makes possible the economic recycling of acid waste.

The high-acidity sensor (Fig. 5.30) consists of a flow cell (about 87.5 mm across) in which two fused silica lenses are tightly mounted across from one another. Fiber-optic cables connect the two lenses to a spectrophotometer. One lens is coated with a sensing material

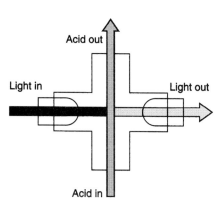

FIGURE 5.30 Optical high-acidity sensor.

consisting of a polymer that is chemically bound to the lens and an indicator that is physically entrapped with the polymer. Acidic solutions flow vertically up through the flow cell. The light from a fixed source in the spectrophotometer is collimated by the coated lens, passes through the acidic solution, and then is focused by the second lens.

The amount of light absorbed by the indicator depends on the acidity of the solution. The absorption spectrum of the indicator reflects the concentration of hydrogen ions in the solution, which is how acidity is measured.

The LANL sensor's sensitivity range—acid concentrations of 4 to 12 molar—is unmatched by any other acidity measurement technique in current manufacturing. Present techniques are unsuitable for measuring solutions with a negative pH or for measuring chemical processes. With the LANL sensor, high acid concentrations can be measured in 3 min or less, which is 50 times faster than measuring with manual sampling and titration. The sensing material can function in highly acidic solutions for 4 to 6 months with calibration less than once a week, and needs to be replaced only once or twice a year. A sensor using similar principles has recently been developed from measuring lower acidity concentrations.

The sensor is selective in its measurements because the small pores of the polymer allow only hydrogen ions to pass through the indicator. Metal ions do not form a complex with the indicator under acidic solutions. The sensor is reusable, that is, chemically reversible. In short, no comparable product for measuring high acidity is reversible or as sensitive, fast, accurate, and selective as the LANL optical high-acidity sensor.

The prime use for the sensor is monitoring highly acidic chemical processes and waste solutions. High-acidity processes are common in preparing metals from ore-mining operations, treating fuel elements from nuclear power plants, manufacturing bulk acid, and metal finishing including passivation and electroplating.

For highly acidic processor applications, the LANL sensor will save companies thousands of dollars by improving efficiency and decreasing the time devoted to acid measurements. The sensor can be used on manufacturing lines, allowing control and waste management adjustment to be made before acidity fluctuations become a problem. The sensor will improve efficiency at least 25 percent by eliminating the need to reprocess material processed incorrectly on the first try because its true acidity was not known and controlled. Higher efficiency will mean lower cost and minimal waste products generated; the sensor itself generates no caustic waste. Finally, the sensor's waste monitoring capabilities will help ensure that any discharged waste is environmentally benign.

For the 500 acidity measurements done at LANL in 1991, the sensor saved $99,500, a 99.5 percent savings in labor costs in the first year. And, because the sensor generates no waste, 20 L a year of caustic waste was avoided, a 100 percent reduction.

SENSORS FOR BIOMEDICAL TECHNOLOGY

In recent years, advanced imaging and other computer-related technology have greatly expanded the horizons of basic biological and biochemical research. Currently, such drivers as the growing needs for environmental information and increased understanding of genetic systems have provided impetus to biotechnology development. This is a relatively new specialty in the marketplace; nevertheless, the intensity of worldwide competition is escalating. Collaborative research and development projects among the U.S. government, industry, and academia constitute a major thrust for

Sensor for Detecting Minute Quantities of Biological Materials

A new device based on laser-excited fluorescence provides unparalleled detection of biological materials for which only minuscule samples may be available. This device, invented at the Ames Laboratories, received a 1991 R&D 100 award.

The Ames Microfluor detector was developed to meet a need for an improved detection technique, driven by important new studies of the human genome, abuse substances, toxins, DNA adduct formation, and amino acids, all of which may be available only in minute amounts. Although powerful and efficient methods have been developed for separating biological mixtures in small volume (i.e., capillary electrophoresis), equally powerful techniques for subsequent detection and identification of these mixtures have been lacking.

The Microfluor detector combines very high sensitivities with the ability to analyze very small volumes. The instrument design is based on the principle that many important biomaterials are fluorescent, while many other biomaterials, such as peptides and oligonucleotides, can be made to fluoresce by adding a fluorescent tag.

When a sample-filled capillary tube is inserted into the Microfluor detector and is irradiated by a laser beam, the sample will fluoresce. The detector detects, monitors, and quantifies the contents by sensing the intensity of the fluorescent light emitted. The signal is proportional to the concentration of the materials. The proportionality constant is characteristic of the material itself.

Analyses can be performed with sample sizes 50 times smaller than those required by other methods, and concentrations as low as 10^{-11} molar (1 part per trillion) can be measured. Often, the critical components in a sample are present at these minute concentrations. These two features make the Microfluor detector uniquely compatible with capillary electrophoresis. In addition, the Ames-developed detector is distinct from other laser-excited detectors in that it is not seriously affected by stray light from the laser itself; it also allows simple alignment and operation in full room light. The Microfluor detector has already been used to determine the extent of adduct formation and base modification in DNA so that the effects of carcinogens on living cells can be studied. Future uses of the sensor will include DNA sequencing and protein sequencing. With direct or indirect fluorescence detection, researchers are using this technique to study chemical contents of individual living cells. This capability may allow pharmaceutical products to be tested on single cells rather than a whole organism, with improved speed and safety.

Sensors for Early Detection and Treatment of Lung Tumors

A quick, accurate method for early detection of lung cancer would raise chances of patient survival from less than 50 percent to 80 or 90 percent. Until now, small cancer cells deep in the lung have been impossible to detect before they form tumors large enough to show up in x-rays. Researchers at Los Alamos National Laboratory in collaboration with other institutions, including the Johns Hopkins school of medicine and St. Mary's Hospital in Grand Junction, Colorado, are developing methods for finding and treating lung cancer in its earliest stages.

A detection sensor involves a porphyrin, one of an unusual family of chemicals that is found naturally in the body and concentrates in cancer cells. The chemical is added to a sample of sputum coughed up from the lung. When exposed to ultraviolet or laser light, cells in the porphyrin-treated sputum glow a bright red. When the sample is viewed under the microscope, the amount and intensity of fluorescence in the cells determines the presence of cancer.

The first clinical test of a detection technique using porphyrin was done by LANL and St. Mary's Hospital in 1988. Four different porphyrins were tested on sputum samples from two former miners, one known to have lung cancer and one with no detectable cancer. One of the porphyrins was concentrated in certain cells only in the sputum of the miner with lung cancer. Other tests concluded that these were cancer cells. Later, a blind study of sputum samples from 12 patients, 8 of whom had lung cancer in various stages of development, identified all the cancer patients as well as a ninth originally thought to be free of cancer. Further tests showed that this patient also had lung cancer.

Identifying the ninth patient prompted a new study, in which the procedure was evaluated for its ability to detect precancerous cells in the lung. In this study, historical sputum samples obtained from Johns Hopkins were treated with the porphyrin. Precancerous conditions that chest x-rays had not detected were identified in samples from patients who later developed the disease. Although further testing is needed, researchers now feel confident that the technique and the detection sensor can detect precancerous conditions 3 to 4 years before onset of the disease.

Working in collaboration with industry, researchers expect to develop an instrument in several years for rapid automated computerized screening of larger populations of smokers, miners, and other people at risk of developing lung cancer. Because lung cancer is the leading cancer killer in both men and women, a successful screening procedure would dramatically lower the nation's mortality rate from the disease. A similar procedure has potential for the analysis of Pap smears, now done by technicians who screen slides one at a time and can misread them.

In addition to a successful screening program, LANL hopes to develop an effective treatment for early lung cancer. The National Cancer Institute will help investigate and implement both diagnostic and therapeutic programs. Researchers are performing basic cell studies on animals to investigate the effect of porphyrin on lung cancer cells. They are also exploring the use of the LANL-designed porphyrin attached to radioactive copper to kill early cancer cells in the lung in a single search-and-destroy mission. The porphyrin not only seeks out cancer cells but also has a molecular structure, similar to a hollow golf ball, that can take certain metals into its center. A small amount of radioactive copper 67 placed inside the porphyrin should destroy tumors the size of pinhead, as well as function as a tracer.

Ultrasensitive Sensor for Single-Molecule Detection

Enhancing the sensitivity of research instruments has long been a goal of physicists, biologists, and chemists. With the single-molecule detector, researchers have achieved the ultimate in this pursuit: detection of a single fluorescent molecule in a liquid. The new instrument—a thousand times more sensitive than existing commercial detectors—brings new capabilities to areas of science and technology that affect lives in many ways, from DNA sequencing, biochemical analysis, and virus studies to identifying environmental pollutants.

For some time, scientists have observed individual molecules trapped in a vacuum, where they are isolated and relatively easy to find. However, many important biological and chemical processes occur in a liquid environment, where many billions of other molecules surround the molecules of interest. Developing a single-molecule detector that operates in such an environment presented a difficult challenge.

The observation technique involves attaching fluorescent dye molecules to the molecules of interest and then exciting the dye molecule by passing them in solution to a rapidly pulsed laser beam and detecting the subsequent faint light, or photons, they emit. The fluorescent lifetimes of the molecules are much shorter than the time the molecules spend in the laser beam; therefore each molecule is reexcited many times and yields many fluorescent photons. The signature of the passing molecule is the burst of photons that occurs when the molecule is passing the laser beam (Fig. 5.31).

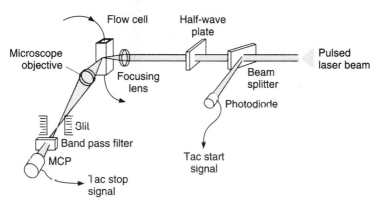

FIGURE 5.31 Single-molecule detector for individual fluorescent molecules in solution.

A lens, or microscope objective, and a slit are arranged to image the photons from a small region around the laser beam waist onto a microchannel plate photomultiplier (MCPP) that counts individual photons. The intense excitation light from the laser is blocked from reaching the MCPP by a bandpass spectral filter, which is centered near the peak fluorescent wavelength. The excitation light consists of extremely short pulses, each about 70 trillionth of a second. The dye molecule does not emit light until a billionth of a second after excitation, so the flash of laser light fades before the feeble molecular glow occurs. For reliable identification of individual molecules, the technique maximizes the number of the detected photons and minimizes the number of background photons. Although some background remains, the technique registers over 85 percent of the fluorescent molecules.

Developed to aid in sequencing chromosomes for the Human Genome Project, the sensor detector's high activity would allow DNA sequencing rates hundreds of times faster than those obtainable with present techniques. It would also eliminate the need for radioactive materials, gels, and electrophoresis solutions, which often create disposal problems, and it is expected to help make DNA sequencing a routine diagnostic and clinical tool. One eventual benefit of the DNA research may be rapid screening for any of 3500 known genetic diseases such as diabetes, cystic fibrosis, and Alzheimer's disease.

The ultrasensitive detector can be used to find and quantify minute amounts of chemicals, enzymes, and viruses in the blood and monitor the dispersal of extremely low concentrations of environmental pollutants. The device may also be useful in studying the interaction of viruses and their binding sites. It may be possible to develop a procedure for rapidly evaluating the efficiency of vaccines; such a procedure could quickly expedite the search for an effective vaccine against the AIDS virus.

REFERENCES

1. Barks, R. E., "Optical High Acidity Sensor," Los Alamos National Laboratory, June 1991.
2. Bicknel, T. J., and W. H. Farr, "Acousto Optical/Electronic Processor for SAR," *NASA Tech Briefs,* **16,** May 1992.
3. Bonnert, R., "Design of High Performance Digital Tachometer with Digital Microcontroller," *IEEE Trans. Instrum.,* **38,** 1104–1108 (1989).
4. Buser, R. A., and N. F. Rooij, "Resonant Silicon Structures," *Sensors and Actuators,* **17,** 145–154 (1989).
5. D'Amico A., and E. Verona, "SAW Sensors," *Sensors and Actuators,* **17,** 66 (1989).
6. Fleming Dias, J., "Physics Sensors Using SAW Devices," *Hewlett-Packard J.,* 18–20, December 1981.
7. Gast, T., "Sensors with Oscillating Elements," *J. Phys. E. Sci. Instrum.,* **18,** 783–789 (1985).
8. Hayman, J. S., and M. Frogatt, "Ultrasonic Dynamic Vector Stress Sensor (UDVSS)," Langley Research Center, 1992.
9. Hewlett-Packard, "Design and Operational Considerations for the HELDS-5000 Incremental Shaft Encoder," Application Note 1011, Palo Alto, 1981.
10. Higbie, B. N., "Automatic Detection of Faults in Turbomachinery Bearings," *NASA Tech Briefs,* **16,** May 1992.
11. Higbie, N., Technology Integration and Development Group Inc., *Technical Report,* May 1992.
12. Huner, B., P. Klinkhachorn, and E. B. Everton, "Hybrid Clock Oscillator Modules as Deposition Monitors," *Rev. Sci. Instrum.,* **59,** 983–986 (1988).
13. Ito, H., "Balanced Absorption Quartz Hygrometer," *IEEE Trans. Ultrasonics, Ferroelectrics, and Frequency Control,* **34,** 136–141 (1987).
14. Lokshin, A. M., "Hand/Eye Coordination for Microrobotic Motion—Utilizing Vision Technology," Caltech, 1991.
15. Los Alamos National Laboratory, Research Team, "Single Molecule Detection," January 1991.
16. Montgomery, J. L., "Force and Optical Sensors Controlling Robotic Gripper for Agriculture," Martin Marietta Corp., 1992.
17. Noble, M. N., D. N. Mark, and R. Blue, "Tracking Retroreflective Targets on a Structure," *NASA Tech Briefs,* **16,** May 1992.
18. "Optical Sensor to Quantify Highly Acidic Solution—Sensing High Acidity without Generating Waste," LANL, *Technology '91,* 57–58 (1991).
19. Tien-Hsin Chao, "Experimental Optoelectronic Associative Memory," *NASA Tech Briefs,* **16,** May 1992.
20. "Ultrasensitive Detection for Medical and Environmental Analysis—Single Molecule Detector," LANL, *Technology '91,* 80–81 (1991).
21. Vaughan, A. H., "Reflective Strip Imaging Camera Sensor—Measuring a 180° Wide Angle," Caltech, NASA's Jet Propulsion Laboratory, 1992.
22. Williams, D. E., "Laser Sensor Detecting Microfluor," Ames Laboratory, July 1991.

CHAPTER 6
INDUSTRIAL SENSORS AND CONTROL

INTRODUCTION

Current manufacturing strategy defines manufacturing systems in terms of sensors, actuators, effectors, controllers, and control loops. Sensors provide a means for gathering information on manufacturing operations and processes being performed. In many instances, sensors are used to transform a physical stimulus into an electrical signal that may be analyzed by the manufacturing system and used for making decisions about the operations being conducted. Actuators convert an electrical signal into a mechanical motion. An actuator acts on the product and equipment through an effector. Effectors serve as the "hand" that achieves the desired mechanical action. Controllers are computers of some type that receive information from sensors and from internal programming, and use this information to operate the manufacturing equipment (to the extent available, depending on the degree of automation and control). Controllers provide electronic commands that convert an electrical signal into a mechanical action. Sensors, actuators, effectors, and controllers are linked to a control loop.

In limited-capability control loops, little information is gathered, little decision making can take place, and limited action results. In other settings, "smart" manufacturing equipment with a wide range of sensor types can apply numerous actuators and effectors to achieve a wide range of automated actions.

The purpose of sensors is to inspect work in progress, to monitor the work-in-progress interface with the manufacturing equipment, and to allow self-monitoring of manufacturing by the manufacturing system's own computer. The purpose of the actuator and effector is to transform the work in progress according to the defined processes of the manufacturing system. The function of the controller is to allow for varying degrees of manual, semiautomated, or fully automated control over the processes. In a fully automated case, such as in computer-integrated manufacturing, the controller is completely adaptive and functions in a closed-loop manner to produce automatic system operation. In other cases, human activity is involved in the control loop.

In order to understand the ways in which the physical properties of a manufacturing system affect the functional parameters associated with the manufacturing system, and in order to determine the types of physical manufacturing system properties that are necessary to implement the various desired functional parameters, it is necessary to understand the technologies that are available for manufacturing systems that use automation and integration to varying degrees.

The least automated equipment makes use of detailed operator control over all equipment functions. Further, each action performed by the equipment is individually directed by the operator. Manual equipment thus makes the maximum use of human capability and adaptability. Visual observations can be enhanced by the use of microscopes and cameras, and the actions that are undertaken can be enhanced by the use of simple effectors. The linkages between the sensory information (from microscopes or through cameras) and the resulting actions are obtained by placing the operator in the loop.

This type of system is clearly limited by the types of sensors used and their relationship to the human operator, the types of the effectors that can be used in conjunction with the human operator, and the capabilities of the operator. The manufacturing equipment that is designed for a manual strategy must be matched to human capabilities. The human-manufacturing equipment interface is extremely important in many manufacturing applications. Unfortunately, equipment design is often not optimized as a sensor-operator-actuator/effector control loop.

A manufacturing system may be semiautomated, with some portion of the control loop replaced by a computer. This approach will serve the new demands on manufacturing system design requirements. Specifically, sensors now must provide continuous input data for both the operator and computer. The appropriate types of data must be provided in a timely manner to each of these control loops. Semiautomated manufacturing systems must have the capability for a limited degree of self-monitoring and control associated with the computer portion of the decision making loop. An obvious difficulty in designing such equipment is to manage the computer- and operator-controlled activities in an optimum manner. The computer must be able to recognize when it needs operator support, and the operator must be able to recognize which functions may appropriately be left to computer control. A continuing machine-operator interaction is part of normal operations.

Another manufacturing concept involves fully automated manufacturing systems. The processing within the manufacturing system itself is fully computer-controlled. Closed-loop operations must exist between sensors and actuators/effectors in the manufacturing system. The manufacturing system must be able to monitor its own performance and decision making for all required operations. For effective automated operation, the mean time between operator interventions must be large when compared with the times between manufacturing setups.

$$\text{MTOI} = \left(\sum_{i}^{n} \tau_1 + \tau_2 + \tau_3 + \tau_4 + \cdots + \tau_4 \right) \Big/ n \tag{6.1}$$

where τ = setup time
i = initial setup
n = number of setups

The processes in use must rarely fail; the operator will intervene only when such failures occur. In such a setting, the operator's function is to ensure the adequate flow of work in progress and respond to system failure.

Several types of work cells are designed according to the concept of total manufacturing integration. The most sophisticated cell design involves fully automated processing and materials handling. Computers control the feeding of work in progress, the performance of the manufacturing process, and the removal of the work in progress. Manufacturing systems of this type provide the opportunity for the most advanced automated and integrated operations. The manufacturing system must be modified to achieve closed-loop operations for all of these functions.

Most manufacturing systems in use today are not very resourceful. They do not make use of external sensors that enable them to monitor their own performance. Rather, they depend on internal conditioning sensors to feed back (to the control system) information regarding manipulator positions and actions. To be effective, this type of manufacturing system must have a rigid structure and be able to determine its own position based on internal data (largely independent of the load that is applied). This leads to large, heavy, and rigid structures.

The more intelligent manufacturing systems use sensors that enable them to observe work in progress and a control loop that allows corrective action to be taken. Thus, such manufacturing systems do not have to be as rigid because they can adapt.

The evolution toward more intelligent and adaptive manufacturing systems has been slow, partly because the required technologies have evolved only in recent years and partly because it is difficult to design work cells that effectively use the adaptive capabilities. Enterprises are not sure whether such features are cost-effective and wonder how to integrate smart manufacturing systems into the overall strategy.

The emphasis must be on the building-block elements that are necessary for many types of processing. If the most advanced sensors are combined with the most advanced manufacturing systems, concepts, and state-of-the-art controllers and control loops, very sophisticated manufacturing systems can result. On the other hand, much more rudimentary sensors, effectors, and controllers can produce simple types of actions.

In many instances today sensors are analog (they involve a continuously changing output property), and control loops make use of digital computers. Therefore, an analog-to-digital converter between the preprocessor and the digital control loop is often required.

The sensor may operate either passively or actively. In the passive case, the physical stimulus is available in the environment and does not have to be provided. For an active case, the particular physical stimulus must be provided. Machine vision and color identification sensors are an active means of sensing, because visible light must be used to illuminate the object before a physical stimulus can be received by the sensor. Laser sensors are also active-type sensors. Passive sensors include infrared devices (the physical stimulus being generated from infrared radiation that is associated with the temperature of a body) and sensors to measure pressure, flow, temperature, displacement, proximity, humidity, and other physical parameters.

SENSORS IN MANUFACTURING

Many types of sensors have been developed during the past several years, especially those for industrial process control, military uses, medicine, automotive applications, and avionics. Several types of sensors are already being manufactured by commercial companies.

Process control sensors in manufacturing will play a significant role in improving productivity, qualitatively and quantitatively, throughout the coming decades. The main parameters to be measured and controlled in industrial plants are temperature, displacement, force, pressure, fluid level, and flow. In addition, detectors for leakage of explosives or combustible gases and oils are important for accident prevention.

Optical-fiber sensors may be conveniently divided into two groups: (1) intrinsic sensors and (2) extrinsic sensors.

Although intrinsic sensors have, in many cases, an advantage of higher sensitivity, almost all sensors used in process control at present belong to the extrinsic type. Extrinsic-type sensors employ light sources such as LEDs, which have higher reliability, longer life, and lower cost than semiconductor lasers. They also are compatible with multimode fibers, which provide higher efficiency when coupled to light sources and are less sensitive to external mechanical and thermal disturbances.

As described in Chap. 1, objects can be detected by interrupting the sensor beam. Optical-fiber interrupters are sensors for which the principal function is the detection of moving objects. They may be classified into two types: reflection and transmission.

In the reflection-type sensor, the light beam emitted from the fiber is reflected back into the same fiber if the object is situated in front of the sensor.

In the transmission-type sensor, the emitted light from the input fiber is interrupted by the object, resulting in no received light in the output fiber located at the opposite side. Typical obstacle interrupters employ low-cost large-core plastic fibers because of the short transmission distance. The minimum detectable size of the object is typically limited to 1 mm by the fiber core diameter and the optical beam. The operating temperature range of commercially available sensors is typically -40 to $+70°C$. Optical-fiber sensors have been utilized in industry in many ways, such as:

1. Detection of lot number and expiration dates, for example, in the pharmaceutical and food industries (Fig. 6.1)
2. Color difference recognition, for example, colored objects on a conveyor
3. Defect detection, for example, missing wire leads in electronic components
4. Counting discrete components, for example, bottles or cans
5. Detecting absence or presence of labels, for example, packaging in the pharmaceutical and food industries

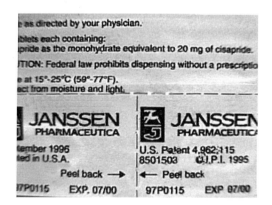

FIGURE 6.1 Lot number and expiration date by the InspectR Vision System. (*Courtesy of American SensoR, Inc.*).

Fiber-optic sensors for monitoring process variables such as temperature, pressure, flow, and liquid level are also classified into two types: (1) normally OFF type in which the shutter is inserted between the fibers in the unactivated state. Thus, this type of sensor provides high and low levels as the light output corresponds to ON and OFF states, respectively, and (2) normally ON type where the shutter is retracted from the gap in the unactivated state.

In both types, the shutter is adjusted so that it does not intercept the light beam completely but allows a small amount of light to be transmitted, even when fully closed. This transmitted light is used to monitor the cable (fiber) continuity for faults and provides an intermediate state. Commercially available sensors employ fibers of 200-μm core diameter. The typical differential attenuation which determines the ON-OFF contrast ratio is about 20 dB. According to manufacturers' specifications, these sensors operate well over the temperature range −40 to +80°C with 2-dB variation in light output.

TEMPERATURE SENSORS IN PROCESS CONTROL

Temperature is one of the most important parameters to be controlled in almost all industrial plants, since it directly affects material properties and thus product quality. During the past few years, several temperature sensors have been developed for use in electrically or chemically hostile environments. Among these, the practical temperature sensors, which are now commercially available, are classified into two groups: (1) low-temperature sensors with a range of −100 to +400°C using specific sensing materials such as phosphors, semiconductors, and liquid crystals and (2) high-temperature sensors with a range of 500 to 2000°C based on blackbody radiation.

Semiconductor Absorption Sensors

Many of these sensors can be located up to 1500 m away from the optoelectronic instruments. The operation of semiconductor temperature sensors is based on the temperature-dependent absorption of semiconductor materials. Because the energy and gap of most semiconductors decrease almost linearly with increasing temperature T, the band-edge wavelength $\lambda_g(T)$ corresponding to the fundamental optical absorption shifts toward longer wavelengths at a rate of about 3 Å/°C [for gallium arsenide (GaAs)] with T. As illustrated in Fig. 6.2, when a light-emitting diode with a radiation spectrum covering the wavelength $\lambda_g(T)$ is used as a light source, the light intensity transmitted through a semiconductor decreases with T.

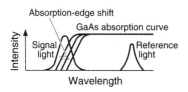

FIGURE 6.2 Operating principle of optical-fiber thermometer based on temperature-dependent GaAs light absorption.

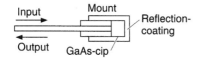

FIGURE 6.3 Sensing element of the optical-fiber thermometer with GaAs light absorber.

Figure 6.3 shows the reflection-type sensing element. A polished thin GaAs chip is attached to the fiber end and mounted in a stainless-steel capillary tube of 2-mm diameter. The front face of the GaAs is antireflection-coated, while the back face is gold-coated to return the light into the fiber.

The system configuration of the thermometer is illustrated in Fig. 6.4. In order to reduce the measuring errors caused by variations in parasitic losses, such as optical fiber loss and connector loss, this thermosensor employs two LED sources [one aluminum gallium arsenide (AlGaAs), the other indium gallium arsenide (InGaAs)] with different wavelengths. A pair of optical pulses with different wavelengths $\lambda_s = 0.88$ μm and $\lambda_r = 1.3$ μm are guided from the AlGaAs LED and the InGaAs LED to

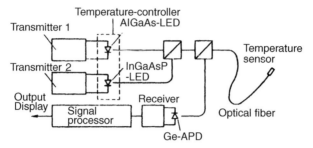

FIGURE 6.4 System configuration of the optical-fiber thermometer with GaAs light absorber.

the sensing element along the fiber. The light of λ_s is intensity-modulated by temperature. On the other hand, GaAs is transparent for the light of λ_r, which is then utilized as a reference light. After detection by a germanium avalanche photodiode (GeAPD), the temperature-dependent signal λ_s is normalized by the reference signal λ_r in a microprocessor.

The performance of the thermometer is summarized in Table 6.1. An accuracy of better than ±2°C is obtained within a range of −20 to +150°C. The principle of operation for this temperature sensor is based on the temperature-dependent direct fluorescent emission from phosphors.

Semiconductor Temperature Detector Using Photoluminescence

The sensing element of this semiconductor photoluminescence sensor is a double-heterostructure GaAs epitaxial layer surrounded by two $Al_xGa_{1-x}As$ layers. When the GaAs absorbs the incoming exciting light, the electron-hole pairs are generated in the GaAs layer. The electron-hole pairs combine and reemit the photons with a wavelength determined by temperature. As illustrated in Fig. 6.5, the luminescent wavelength shifts monotonically toward longer wavelengths as the temperature T increases. This is a result of the decrease in the energy gap E_g with T. Therefore, analysis of the luminescent spectrum yields the required temperature information. The

TABLE 6.1 Characteristics of Semiconductor Sensors (Thermometer Performance)

Property	Semiconductor absorption sensor (Mitsubishi, Japan)	Semiconductor photoluminescence sensor (ASEA Innovation, Sweden)	Phosphor A sensor (Luxton, U.S.)	Phosphor B sensor (Luxton, U.S.)
Range, °C	−20 to +150	0 to 200	20 to 240	20 to 400
Accuracy, °C	±2.0	±1.0	±2.0	±2.0
Diameter, m	2	From 0.6	0.7	1.6
Time constant, s	0.5	From 0.3	From 0.25	
Fiber type	100-μm silica core	100-μm silica core	400-μm polymer clad	Fiber silica
Fiber length, m	300	500	100	300
Light source	AlGaAs LED	AlGaAs LED	Halogen lamp	Halogen lamp

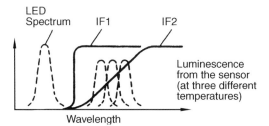

FIGURE 6.5 Operating principle of optical-fiber thermometer based on temperature-dependent photoluminescence from a GaAs epitaxial film.

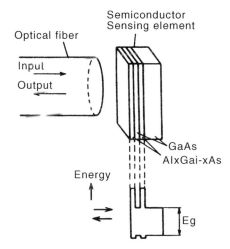

FIGURE 6.6 Sensing element of optical-fiber thermometer based on temperature-dependent photoluminescence.

double heterostructure of the sensing element provides excellent quantum efficiency for the luminescence because the generated electron-hole pairs are confined between the two potential barriers (Fig. 6.6).

The system is configured as shown in Fig. 6.7. The sensing element is attached to the end of the silica fiber (100-μm core diameter). The excitation light from an LED, with a peak wavelength of about 750 nm, is coupled into the fiber and guided to a special GRIN lens mounted to a block of glass. A first optical inference filter IF_1, located between the GRIN lens and the glass block, reflects the excitation light which is guided to the sensing element along the fiber. However, this optical filter is transparent to the returned photoluminescent light. The reflectivity of the second interference filter IF_2 changes at about 900 nm. Because the peak wavelength of the luminescence shifts toward longer wavelength with temperature, the ratio between the transmitted and the reflected light intensifies if IF_2 changes. However, the ratio is independent of any variation in the excitation light intensity and parasitic losses. The two lights separated by IF_2 are detected by photodiodes 1 and 2. The detector module is kept at a constant temperature in order to eliminate any influence of the thermal drift of IF_2.

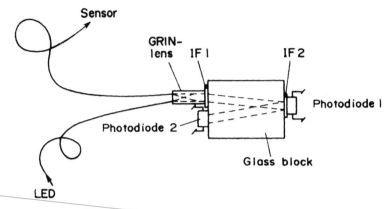

FIGURE 6.7 Optical system of optical-fiber thermometer based on temperature-dependent photoluminescence.

The measuring temperature range is 0 to 200°C, and the accuracy is ±1°C. According to the manufacturer's report, good long-term stability, with a temperature drift of less than 1°C over a period of 9 months, has been obtained.

Temperature Detector Using Point-Contact Sensors in Process Manufacturing Plant

Electrical sensors are sensitive to microwave radiation and corrosion. The needs for contact-type temperature sensors have lead to the development of point-contact sensors which are immune to microwave radiation, for use in: (1) electric power plants using transformers, generators, surge arresters, cables, and bus bars; (2) industrial plants utilizing microwave processes; and (3) chemical plants utilizing electrolytic processes.

The uses of microwaves include drying powder and wood; curing glues, resins, and plastics; heating processes for food, rubber, and oil; device fabrication in semiconductor manufacturing; and joint welding of plastic packages, for example.

Semiconductor device fabrication is currently receiving strong attention. Most semiconductor device fabrication processes are now performed in vacuum chambers; they include plasma etching and stripping, ion implantation, plasma-assisted chemical vapor deposition, radio-frequency sputtering, and microwave-induced photoresist baking. These processes alter the temperature of the semiconductors being processed. However, the monitoring and controlling of temperature in such hostile environments is difficult with conventional electrical temperature sensors. These problems can be overcome by the contact-type optical-fiber thermometer.

Noncontact Sensors—Pyrometers

Because they are noncontact sensors, pyrometers do not affect the temperature of the object they are measuring. The operation of the pyrometer is based on the spectral distribution of blackbody radiation, which is illustrated in Fig. 6.8 for several different temperatures. According to the Stefan-Boltzmann law, the rate of the total radiated energy from a blackbody is proportional to the fourth power of absolute temperature and is expressed as:

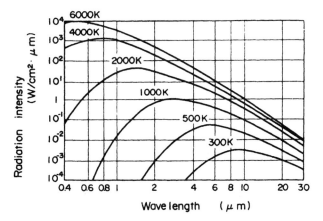

FIGURE 6.8 Spectral distribution of blackbody radiation

$$W_t = \sigma T^4 \tag{6.2}$$

where σ is the Stefan-Boltzmann constant and has the value of 5.6697×10^{-8} W/m$^2\cdot$K^4.

The wavelength at which the radiated energy has its highest value is given by Wien's displacement law,

$$\lambda_m T = 2.8978 \times 10^{-3} \text{ m}\cdot\text{K} \tag{6.3}$$

Thus, the absolute temperature can be measured by analyzing the intensity of the spectrum of the radiated energy from a blackbody. A source of measurement error is the emissivity of the object, which depends on the material and its surface condition. Other causes of error are deviation from the required measurement distance and the presence of any absorbing medium between the object and the detector.

Use of optical fibers as signal transmission lines in pyrometers allows remote sensing over long distances, easy installation, and accurate determination of the position to be measured by observation of a focused beam of visible light from the fiber end to the object. The sensing head consists of a flexible bundle with a large number of single fibers and lens optics to pick up the radiated energy (Fig. 6.9).

The use of a single silica fiber instead of a bundle is advantageous for measuring small objects and longer distance transmission of the picked-up radiated light. The lowest measurable temperature is 500°C, because of the unavoidable optical loss in silica fibers at wavelengths longer than 2 µm. Air cooling of the sensing head is usually necessary when the temperature exceeds 1000°C.

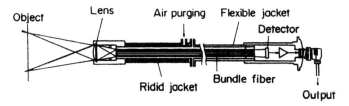

FIGURE 6.9 Schematic diagram of an optical-fiber pyrometer.

Optical-fiber pyrometers are one of the most successful optical-fiber sensors in the field of process control in manufacturing. Typical applications are

1. Casting and rolling lines in steel and other metal plants
2. Electric welding and annealing
3. Furnaces in chemical and metal plants
4. Fusion, epitaxial growth, and sputtering processes in the semiconductor industry
5. Food processing, paper manufacturing, and plastic processing

Figure 6.10 is a block diagram of the typical application of optical-fiber pyrometers for casting lines in a steel plant, where the temperature distribution of the steel slab is measured. The sensing element consists of a linear array of fused-silica optical rods, thermally protected by air-purge cooling. Light radiated from the heated slabs is collected by the optical rods and coupled into a 15-m-long bundle of fibers, which transmits light to the optical processing unit. In this system, each fiber in the bundle carries the signal from a separate lens which provides the temperature information at the designated spot of the slabs. An optical scanner in the processing unit scans the bundle and the selected light signal is analyzed in two wavelength bands by using two optical interference filters.

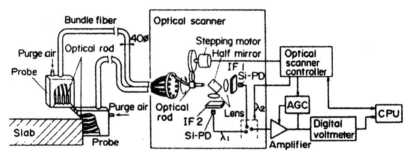

FIGURE 6.10 Temperature distribution measurement of steel slabs by an optical-fiber pyrometer using two-wavelength method.

PRESSURE SENSORS

If a pressure P acting on a diaphragm compresses a spring until an equilibrium is produced, the pressure can be represented as

$$F \text{ (kg)} = A \text{ (m}^2\text{)} \times P \text{ (kg/m}^2\text{)} \tag{6.4}$$

In this equation F represents the force of the spring and A represents a surface area of the diaphragm. The movement of the spring is transferred via a system of levers to a pointer whose deflection is a direct indication of the pressure (Fig. 6.11). If the measured value of the pressure has to be transmitted across a long distance, the mechanical movement of the pointer can be connected to a variable electrical resistance (potentiometer). A change in the resistance results in a change in the measured voltage, which can then easily be evaluated by an electronic circuit or further processed. This example illustrates the fact that a physical quantity is often subject to many transformations before it is finally evaluated.

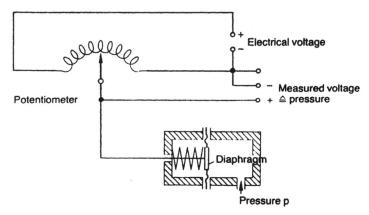

FIGURE 6.11 Deflection as a direct indication of pressure.

Piezoelectric Crystals

Piezoelectric crystals may be utilized to measure pressure. Electrical charges are produced on the opposite surfaces of some crystals when they are mechanically loaded by deflection, pressure, or tension. The electrical charge which is produced in the process is proportional to the effective force. This change in the charge is very small. Therefore, electrical amplifiers are used to make it possible to process the signals (Fig. 6.12).

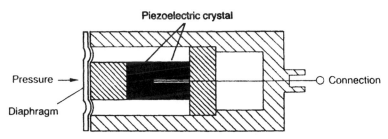

FIGURE 6.12 Electrical amplifiers are connected to a piezoelectric crystal.

Pressure in this situation is measured by transforming it into a force. If the force produced by pressure on a diaphragm acts on a piezoelectric crystal, a signal which is proportional to the pressure measured can be produced by using suitable amplifiers.

Strain Gages

Strain gages can also measure pressure. The electrical resistance of a wire-type conductor is dependent, to a certain extent, on its cross-sectional area. The smaller the cross section, i.e., the thinner the wire, the greater the resistance of the wire. A strain gage is a wire which conducts electricity and stretches as a result of the mechanical influence (tension, pressure, or torsion) and thus changes its resistance in a manner which is detectable. The wire is attached to a carrier which in turn is attached to the object to be measured. Conversely, for linear compression, which enlarges the cross-sectional area of a strain gage, resistance is reduced. If a strain gage is attached to a diaphragm (Fig. 6.13), it will follow the movement of the diaphragm. It is either pulled or compressed, depending on the flexure of the diaphragm.

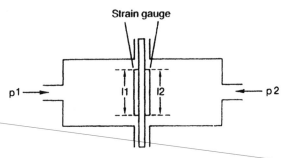

FIGURE 6.13 Strain gage for measurement of pressure.

FIBER-OPTIC PRESSURE SENSORS

A Y-guide probe can be used as a pressure sensor in process control if a reflective diaphragm, moving in response to pressure, is attached to the end of the fiber (Fig. 6.14). This type of pressure sensor has a significant advantage over piezoelectric transducers, since it works as a noncontact sensor and has a high frequency response. The pressure signal is transferred from the sealed diaphragm to the sensing diaphragm, which is attached to the end of the fiber. With a stainless-steel diaphragm about 100 μm thick, hysteresis of less than 0.5 percent and linearity within ±0.5 percent are obtained up to the pressure level of 3×10^5 kg/m^2 (2.94 MPa) in the temperature range of −10 to +60°C.

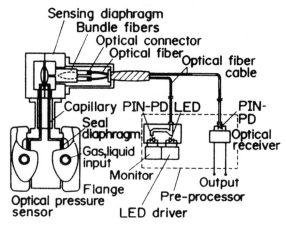

FIGURE 6.14 Schematic diagram of a fiber-optic pressure sensor using Y-guide probe with a diaphragm attached.

The material selection and structural design of the diaphragm are important to minimize drift. Optical-fiber pressure sensors are expected to be used under severe environments in process control. For example, process slurries are frequently highly corrosive, and the temperature may be as high as 500°C in coal plants. The conventional metal diaphragm exhibits creep at these high temperatures. In order to eliminate these problems, an all-fused-silica pressure sensor based on the microbending effect in optical fiber has been developed (Fig. 6.15). This sensor converts the pressure applied to the fused silica diaphragm into an optical intensity modulation in the fiber.

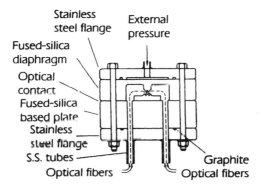

FIGURE 6.15 Fiber-optic microbend sensor.

A pressure sensor based on the wavelength filtering method has been developed. The sensor employs a zone plate consisting of a reflective surface, with a series of concentric grooves at predetermined spacing. This zone plate works as a spherical concave mirror whose effective radius of curvature is inversely proportional to the wavelength. At the focal point of the concave mirror, a second fiber is placed which transmits the returned light to two photodiodes with different wavelength sensitivities. When broadband light is emitted from the first fiber to the zone plate, and the zone plate moves back and forth relative to the optical fibers in response to the applied pressure, the wavelength of the light received by the second fiber is varied, causing a change in the ratio of outputs from the two photodiodes. The ratio is then converted into an electrical signal which is relatively unaffected by any variations in parasitic losses.

DISPLACEMENT SENSORS FOR ROBOTIC APPLICATIONS

The operating principle of a displacement sensor using Y-guide probes is illustrated in Fig. 6.16. The most common Y-guide probe is a bifurcated fiber bundle. The light emitted from one bundle is back-reflected by the object to be measured and collected by another bundle (receiving fibers). As a result, the returned light at the detector is intensity-modulated to a degree dependent on the distance between the end of the fiber bundle and the object. The sensitivity and the dynamic range are determined by the geometrical arrangement of the array of fiber bundles and by both the number and type of the fibers. Figure 6.17 shows the relative intensity of the returned light as a function of distance for three typical arrangements: random, hemispherical, and concentric circle arrays. The intensities increase with distance and reach a peak at a certain discrete distance. After that, the intensities fall off very slowly. Most sensors use the high-sensitivity regions in these curves. Among the three arrangements, the random array has the highest sensitivity but the narrowest dynamic range. The displacement sensor using the Y-guide probe provides resolution of 0.1 μm, linearity within 5 percent, and dynamic range of 100 μm displacement. Y-guide probe displacement sensors are well-suited for robotics applications as position sensors and for gaging and surface assessment, since they have high sensitivity to small distances.

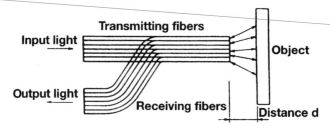

FIGURE 6.16 Principle of operation of fiber-optic mechanical sensor using a Y-guide probe.

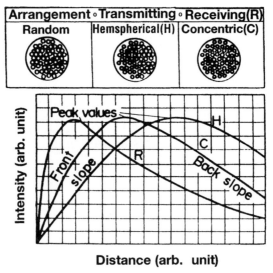

FIGURE 6.17 Relative intensity of returned light for three fiber-optic arrangements.

One profound problem of this type of displacement sensor is the measuring error arising from the variation in parasitic losses along the optical transmission line. Recalibration is required if the optical path is interrupted, which limits the range of possible applications. In order to overcome this problem, a line-loss-independent displacement sensor with electrical subcarrier phase encoder has been implemented. In this sensor, the light from an LED modulated at 160 MHz is coupled into the fiber bundle and divided into two optical paths. One of the paths is provided with a fixed retroreflector at its end. The light through the other is reflected by the object. The two beams are returned to the two photodiodes separately. Each signal, converted into an electric voltage, is electrically heterodyned into an intermediate frequency at 455 kHz. Then the two signals are fed to a digital phase comparator, the output of which is proportional to the path distance. The resolution of the optical path difference is about 0.3 mm, but improvement of the receiver electronics will provide a higher resolution.

PROCESS CONTROL SENSORS MEASURING AND MONITORING LIQUID FLOW

According to the laws of fluid mechanics, an obstruction inserted in a flow stream creates a periodic turbulence behind it. The frequency of shedding the turbulent vortices is directly proportional to the flow velocity. The flow sensor in Fig. 6.18 has a sensing element consisting of a thin metallic obstruction and a downstream metallic bar attached to a multimode fiber-microbend sensor. As illustrated in Fig. 6.19, the vortex pressure produced at the metallic bar is transferred, through a diaphragm at the pipe wall that serves as both a seal and a pivot for the bar, to the microbend sensor located outside the process line pipe. The microbend sensor converts the time-varying mechanical force caused by the vortex shedding into a corresponding intensity modulation of the light. Therefore, the frequency of the signal converted into the electric voltage at the detector provides the flow-velocity information. This flow sensor has the advantage that the measuring accuracy is essentially independent of any changes in the fluid temperature, viscosity, or density, and in the light source intensity. According to the specifications for typical optical vortex-shedding flow sensors, flow rate can be measured over a Reynolds number range from 5×10^3 to 6000×10^3 at temperatures from -100 to $+600°C$. This range is high compared to that of conventional flow meters. In addition, an accuracy of ± 0.4 and ± 0.7 percent, respectively, is obtained for liquids and gases with Reynolds numbers above 10,000.

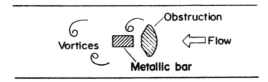

FIGURE 6.18 Principle of operation of a vortex-shedding flow sensor.

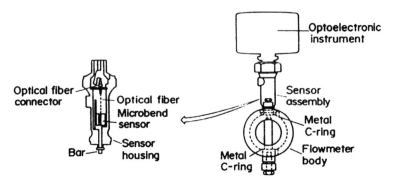

FIGURE 6.19 Schematic diagram of a vortex-shedding flow sensor.

Flow Sensor Detecting Small Air Bubbles for Process Control in Manufacturing

Another optical-fiber flow sensor employed in manufacturing process control monitors a two-fluid mixture (Fig. 6.20). The sensor can distinguish between moving bubbles and liquid in the flow stream and display the void fraction, namely, the ratio of gas volume to the total volume. The principle of operation is quite simple. The light from the LED is guided by the optical fiber to the sensing element, in which the end portion of the fiber is mounted in a stainless steel needle of 2.8-mm outer diameter. When liquid is in contact with the end of the fiber, light enters the fluid efficiently and very little light is returned. However, when a gas bubble is present, a significant fraction of light is reflected back. With this technique, bubbles as small as 50 µm may be detected with an accuracy of better than 5 percent and a response time of only 10 µs.

Potential applications of this flow sensor for the control of processes in manufacturing systems are widespread—for example, detection of gas plugs in production

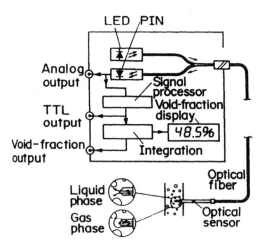

FIGURE 6.20 Flow sensor for two-phase mixtures.

wells in the oil industry and detection of fermenters and distillers in the blood-processing and pharmaceutical industries.

An optical-fiber flow sensor for a two-phase mixture based on Y-guide probes is shown in Fig. 6.21. Two Y-guide probes are placed at different points along the flow stream to emit the input light and to pick up the retroreflected light from moving solid particles in the flow. The delay time between the signals of the two probes is determined by the average velocity of the moving particles. Therefore, measurement of the delay time by a conventional correlation technique provides the flow velocity. An accuracy of better than ±1 percent and a dynamic range of 20:1 are obtained for flow velocities up to 10 m/s. A potential problem of such flow sensors for two-phase mixtures is poor long-term stability, because the optical fibers are inserted into the process fluid pipes.

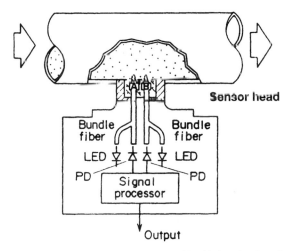

FIGURE 6.21 Flow sensor using two Y-guided probes based on a correlation technique.

Liquid Level Sensors in Manufacturing Process Control for Petroleum and Chemical Plants

Several optical-fiber liquid level sensors developed in recent years have been based on direct interaction between the light and liquid. The most common method in commercial products employs a prism attached to the ends of two single optical fibers (Fig. 6.22). The input light from an LED is totally internally reflected and returns to the output fiber when the prism is in air. However, when the prism is immersed in liquid, the light refracts into the fluid with low reflection, resulting in negligible returned light. Thus, this device works as a liquid level switch. The sensitivity of the sensor is determined by the contrast ratio, which depends on the refractive index of the liquid. Typical examples of signal output change for liquids with different refractive indices are indicated in Table 6.2.

The output loss stays at a constant value of 33 dB for refractive indices higher than 1.40. The signal output of a well-designed sensor can be switched for a change in liquid level of only 0.1 mm.

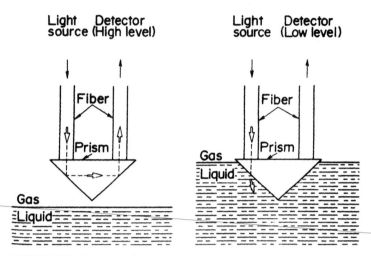

FIGURE 6.22 Principle of operation of a liquid level sensor with a prism attached to two optical fibers.

Problems to be solved for this sensor are dirt contamination on the prism surface and bubbles in the liquid. Enclosing the sensing element with a fine filter helps keep it clean and simultaneously reduces level fluctuations caused by bubbles. Since optical-fiber liquid level sensors have the advantages of low cost and electrical isolation, their use is widespread in petroleum and chemical plants, where the hazardous environment causes difficulties with conventional sensors. They are used, for example, to monitor storage tanks in a petroleum plant.

Another optical-fiber liquid level sensor, developed for the measurement of boiler-drum water level, employs a triangularly shaped gage through which red and green light beams pass. The beams are deflected as it fills with water, so that the green light passes through an aperture. In the absence of water, only red light passes through. Optical fibers transmit red or green light from individual gages to a plant control room located up to 150 m from the boiler drum (Fig. 6.23). The water level in the drum is displayed digitally.

This liquid level sensor operates at temperatures up to 170°C and pressures up to 3200 lb/in^2 gage. Many sensor units are installed in the boiler drum, and most have been operating for 7 years. This sensor is maintenance-free, fail-safe, and highly reliable.

TABLE 6.2 Refractive Index versus Output

Refractive index, n	Loss change, dB
1.333	2.1
1.366	4.6
1.380	6.0
1.395	31.0

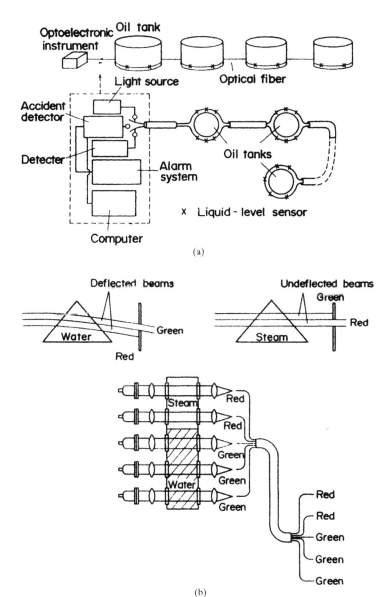

FIGURE 6.23 (*a*) Principle of operation of liquid level sensor for the measurement of boiler-drum water. (*b*) Liquid level sensor for the measurement of boiler-drum water with five-port sensors.

On-line Measuring and Monitoring of Gas by Spectroscopy

An optical spectrometer or optical filtering unit is often required for chemical sensors because the spectral characteristics of absorbed, fluorescent, or reflected light indicate the presence, absence, or precise concentration of a particular chemical species (Fig. 6.24).

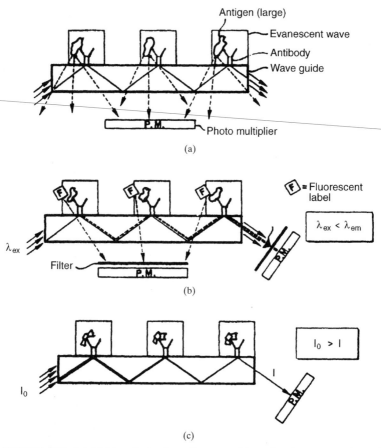

FIGURE 6.24 (a) Light scattering from the waveguide surface increases as the antigen-antibody complexes are formed, and is detected by a side-mounted photodetector. (b) Fluorescence is excited in a fluorescent marker, attached to an antigen molecule, by light passing through the waveguide. The fluorescent light can be collected either sideways from the guide, or as light which is retrapped by the guide and directed to a photodetector. (c) Absorption of light by antibody-antigen complexes on the surface attenuates light traveling down the waveguide.

Sensing of chemical parameters via fibers is usually done by monitoring changes in a suitably selected optical property—absorbance, reflectance, scattering (turbidity), or luminescence (fluorescence or phosphorescence), depending on the particular device. Changes in parameters such as the refractive index may also be employed for sensing purposes. The change in light intensity due to absorption is determined by the number of absorbing species in the optical path, and is related to the concentration C of the absorbing species by the Beer-Lambert relationship. This law describes an exponential reduction of light intensity with distance (and also concentration) along the optical path. Expressed logarithmically,

$$A = \log I_0/I = \eta l C$$

where A is the optical absorbance, l is the path length of the light, η is the molar absorptivity, and I_0 and I are the incident and transmitted light, respectively. For absorption measurements via optical fibers, the medium normally must be optically transparent.

An accurate method for the detection of leakage of flammable gases such as methane (CH_4), propane (C_3H_8), and ethylene (C_2H_4) is vital in gas and petrochemical plants in order to avoid serious accidents. The recent introduction of low-loss fiber into spectroscopic measurements of these gases offers many advantages for process control in manufacturing:

1. Long-distance remote sensing
2. On-line measurement and monitoring
3. Low cost
4. High reliability

The most commonly used method at present is to carry the sample back to the measuring laboratory for analysis; alternatively, numerous spectrometers may be used at various points around the factory. The new advances in spectroscopic measurements allow even CH_4 to be observed at a distance of 10 km with a detection sensitivity as low as 5 percent of the lower explosion limit (LEL) concentration. The optical-fiber gas measuring system employs an absorption spectroscopy technique, with the light passing through a gas-detection cell for analysis. The overtone absorption bands of a number of flammable gases are located in the near-infrared range (Fig. 6.25).

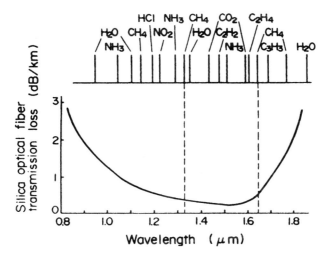

FIGURE 6.25 Absorption lines of typical flammable gases in the near-infrared and the transmission loss of silica fiber.

The optical gas sensing system can deal with a maximum of 30 detection cells (Fig. 6.26). The species to be measured are CH_4, C_3H_8, and C_2H_4 molecules. Light from a halogen lamp (an infrared light source) is distributed into a bundle of 30 single optical fibers. Each of the distributed beams is transmitted through a 1-km length of fiber to a corresponding gas detection cell. The receiving unit is constructed of three optical switches, a rotating sector with four optical interference filters, and three Ge photodiodes. Each optical switch can select any 10 returned beams by specifying the number of the cell. The peak transmission wavelength of the optical filter incorporated in the sensor is 1.666 μm for CH_4, 1.690 μm for C_3H_8, 1.625 μm for C_2H_2, and 1.600 μm for a reference beam. After conversion to electrical signals, the signal amplitudes for the three gases are normalized by the reference amplitude. Then the concentration of each gas is obtained from a known absorption-concentration calibration curve stored in a computer.

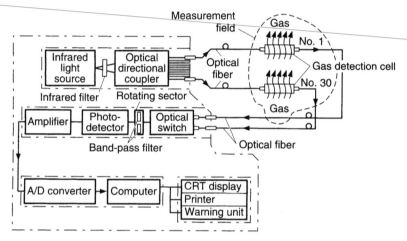

FIGURE 6.26 Gas detection system with 30 detection cells.

An intrinsic distributed optical-fiber gas sensor for detecting the leakage of cryogenically stored gases such as CH_4, C_2H_4, and N_2 has also been developed. The sensor's operation is based on the temperature-dependent transmission loss of optical fiber. That is, the optical fiber is specially designed so that the transmission loss increases with decreasing temperature by choosing the appropriate core and cladding materials. Below the critical temperature, in the region of −55°C, most of the light has transferred to the cladding layer, and the light in the core is cut off. By connecting this temperature-sensitive fiber between a light source and a detector and monitoring the output light level, the loss of light resulting from a cryogenic liquid in contact with the fiber can be detected directly.

CRACK DETECTION SENSORS FOR COMMERCIAL, MILITARY, AND SPACE INDUSTRY USE

Accurate and precise detection of crack propagation in aircraft components is of vital interest for commercial and military aviation and the space industry. A system has been recently developed to detect cracks and crack propagation in aircraft components. This system uses optical fibers of small diameter (20 to 100 μm) which can be etched to increase their sensitivity. The fibers are placed on perforated adhesive foil to facilitate attachment to the desired component for testing. The fiber is in direct contact with the component (Fig. 6.27). The foil is removed after curing of the adhesive. Alternatively, in glass-fiber-reinforced plastic (GFRP) or carbon-fiber-reinforced plastic (CFRP), materials which are used more and more in aircraft design, the fiber can be easily inserted in the laminate without disturbing the normal fabrication process. For these applications, bare single fiber or prefabricated tape with integrated bundles of fibers is used. The system initially has been developed for fatigue testing of aircraft components such as frames, stringers, and rivets. In monitoring mode, the system is configured to automatically interrupt the fatigue test. The system has also been applied to the inspection of the steel rotor blades of a 2-MW wind turbine. A surveillance system has been developed for the centralized inspection of all critical components of the Airbus commercial jetliner during its lifetime. This fiber nervous system is designed for in-flight monitoring and currently is accessible to flight and maintenance personnel.

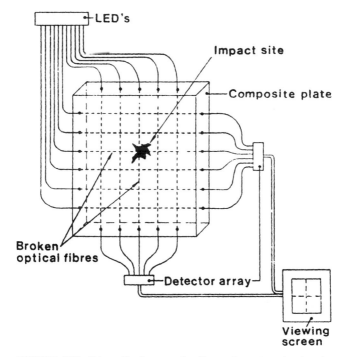

FIGURE 6.27 Schematic diagram of a fiber-optic system showing the location of impact damage in a composite structure.

An optical-fiber mesh has been tested for a damage assessment system for a GFRP submarine sonar dome. Two sets of orthogonally oriented fibers are nested in the laminate during the fabrication process. When the fibers of the mesh are properly connected to LEDs and the detectors, the system can be configured to visualize the location of a damaged area.

As an alternative, a video camera and image processing are applied to determine the position of the damaged area. The fiber end faces at the detection side of the mesh are bundled and imaged into the camera tube. Two images are subtracted: the initial image before the occurrence of damage and the subsequent image. If fibers are broken, their location is highlighted as a result of this image subtraction.

CONTROL OF INPUT/OUTPUT SPEED OF CONTINUOUS WEB FABRICATION USING LASER DOPPLER VELOCITY SENSOR

A laser Doppler velocimeter (LDV) can be configured to measure any desired component velocity, perpendicular or parallel to the direction of the optical axis. An LDV system has been constructed with a semiconductor laser and optical fibers and couplers to conduct the optical power. Frequency modulation of the semiconductor laser (or, alternatively, an external fiber-optic frequency modulator) is used to introduce an offset frequency. Some commercial laser Doppler velocimeters are available with optical-fiber leads and small sensing heads. However, these commercial systems still use bulk optical components such as acoustooptic modulators or rotating gratings to introduce the offset frequency.

With an LDV system the velocity can be measured with high precision in a short period of time. This means that the method can be applied for real-time measurements to monitor and control the velocity of objects as well as to measure their vibration. Because the laser light can be focused to a very small spot, the velocity of very small objects can be measured, or if scanning techniques are applied, high spatial resolution can be achieved. This method is used for various applications in manufacturing, medicine, and research. The demands on system performance with respect to sensitivity, measuring range, and temporal resolution are different for each of these applications.

In manufacturing processes, for example, LDV systems are used to control continuous roll milling of metal (Fig. 6.28), to control the rolling speed of paper and films, and to monitor fluid velocity and turbulence in mixing processes. Another industrial application is vibration analysis. With a noncontact vibrometer, vibration of machines, machine tools, and other structures can be analyzed without disturbing the vibrational behavior of the structure.

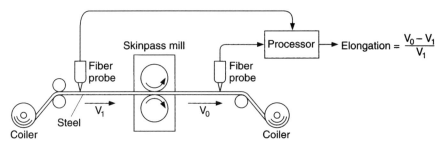

FIGURE 6.28 Fiber-optic laser Doppler velocimeter at a rolling mill controls pressure by measuring input speeds.

Interestingly, the LDV system proved useful in the measurement of arterial blood velocity (Fig. 6.29), thereby providing valuable medical information. Another application in medical research is the study of motion of the tympanic membrane in the ear.

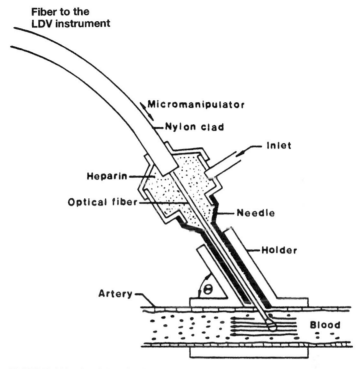

FIGURE 6.29 Special probe for measurement of blood velocity.

ULTRASONIC/LASER NONDESTRUCTIVE EVALUATION SENSOR

Ultrasonic/laser optical inspection is a relatively new noncontact technique. A laser system for generating ultrasound pulses without distortion of the object surface is shown in Fig. 6.30. A laser pulse incident on a surface will be partly absorbed by the material and will thus generate a sudden rise in temperature in the surface layer of the material. This thermal shock causes expansion of a small volume at the surface, which generates thermoelastic strains. Bulk optical systems have been used previously to generate the laser pulse energy. However, the omnidirectionality of bulk sources is completely different from other well-known sources, and is regarded as a serious handicap to laser generation.

To control the beamwidth and beam direction of the optically generated ultrasonic waves, a fiber phased array has been developed. In this way the generated ultrasonic beam can be focused and directed to a particular inspection point below the surface of an object (Fig. 6.30). This system has been optimized for the detection of fatigue cracks at rivet holes in aircraft structures.

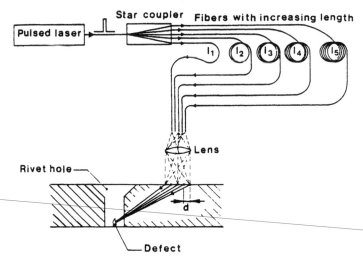

FIGURE 6.30 Setup for beam steering of laser-generated ultrasound by fiber-optic phased array.

The combination of laser-generated ultrasound and an optical-fiber interferometer for the detection of the resultant surface displacement has led to a technique which is useful for a wide variety of inspection tasks in manufacturing, including areas which are difficult to access and objects at high temperature, as well as more routine inspection and quality control in various industrial environments. Such a system can be applied to the measurement of thickness, velocity, flaws, defects, and grain size in a production process.

PROCESS CONTROL SENSOR FOR ACCELERATION

The principle of operation of the process control acceleration sensor is illustrated in Fig. 6.31. The sensor element, consisting of a small cantilever and a photoluminescent material, is attached to the end of a single multimode fiber. The input light of wavelength λ_s is transmitted along the fiber from a near-infrared LED source to the sensor element. The sensor element returns light at two different wavelengths, one of which serves as a signal light and the other as a reference light, into the same fiber. The signal light at wavelength λ_s is generated by reflection from a small cantilever. Since the relative angle of the reflected light is changed by the acceleration, the returned light is intensity-modulated. The reference light

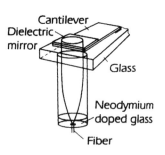

FIGURE 6.31 Cantilever-type acceleration sensor.

of wavelength λ_r is generated by photoluminescence of a neodymium-doped glass element placed close to the sensor end of the fiber.

The optoelectronic detector module has two optical filters to separate the signals λ_s and λ_r and two photodiodes to convert the signal and the reference light into separate analog voltages. The signal processing for compensation is then merely a matter of electrical division. A measuring range of 0.1 to 700 m/s^2 and a resolution of 0.1 m/s^2 is obtained over the frequency range of 5 to 800 Hz.

AN ENDOSCOPE AS IMAGE TRANSMISSION SENSOR

An imaging cable consists of numerous optical fibers, typically 3000 to 100,000, each of which has a diameter of 10 μm and constitutes a picture element (pixel). The principle of image transmission through the fibers is shown in Fig. 6.32. The optical fibers are aligned regularly and identically at both ends of the fibers. When an image is projected on one end of the image fiber, it is split into multiple picture elements. The image is then transmitted as a group of light dots with different intensities and colors, and the original picture is reduced at the far end. The image fibers developed for industrial use are made of silica glass with low transmission loss over a wide wavelength band from visible to near infrared, and can therefore transmit images over distances in excess of 100 m without significant color changes. The basic structure of the practical optical-fiber image sensing system (endoscope) is illustrated in Fig. 6.33. It consists of the image fiber, an objective lens to project the image on one end, an eyepiece to magnify the received image on the other end, a fiber protection tube, and additional fibers for illumination of the object.

Many examples have been reported of the application of image fibers in process control. Image fibers are widely employed to observe the interior of blast fur-

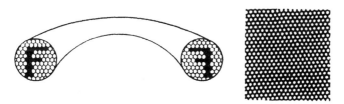

FIGURE 6.32 Image transmission through an image fiber.

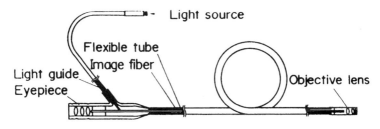

FIGURE 6.33 Basic structure of fiber scope.

naces and the burner flames of boilers, thereby facilitating supervisory control. Image fibers can operate at temperatures up to 1000°C, when provided with a cooling attachment for the objective lens and its associated equipment. Another important application of the image fiber bundles is observation, control, and inspection of nuclear power plants and their facilities. Conventional image fibers cannot be used within an ionizing radiation environment because ordinary glass becomes colored when exposed to radiation, causing increasing light transmission loss. A high-purity silica core fiber is well-known as a radiation-resistant fiber for nuclear applications.

The endoscope has demonstrated its vital importance in medical and biochemical fields such as:

1. Angioplasty
2. Laser surgery
3. Gastroscopy
4. Cystoscopy
5. Bronchoscopy
6. Cardioscopy

SENSOR NETWORK ARCHITECTURE IN MANUFACTURING

In fiber-optic sensor networks, the common technological base with communication is exploited by combining the signal generating ability of sensors and the signal transmitting capability of fiber optics. This combination needs to be realized by a suitable network topology in various manufacturing implementations. The basic topologies for sensor networking are illustrated in Fig. 6.34. The basic network topologies are classified into seven categories:

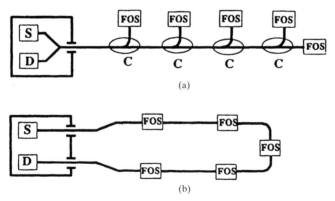

FIGURE 6.34 Basic network topologies: (*a*) linear array, (*b*) ring.

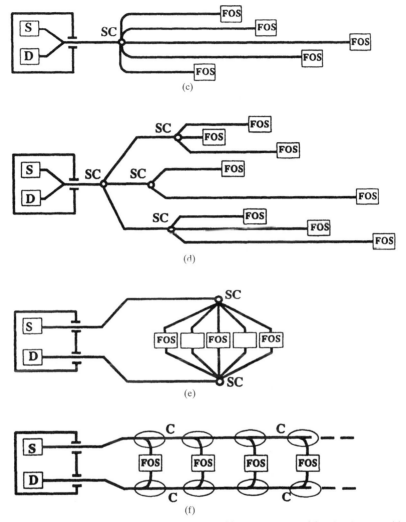

FIGURE 6.34 (*Continued*) Basic network topologies: (*c*) reflective star, (*d*) reflective tree, (*e*) transmissive star and (*f*), ladder network.

1. Linear array network with access-coupled reflective sensors (Fig. 6.34*a*).
2. Ring network with in-line transmissive sensors (Fig. 6.34*b*).
3. Star network with reflective sensors (Fig. 6.34*c*).
4. Star network with reflective sensors; one or more sensors can be replaced by a separate star network, in order to obtain a tree network (Fig. 6.34*d*).
5. Star network that can also be operated with transmissive sensors (Fig. 6.34*e*).
6. Ladder network with two star couplers. A star coupler is replaced by several access couplers, the number required being equal to the number of sensors (Fig. 6.34*f*).

Topological modifications, especially of sensor array and ladder network, may be desirable in order to incorporate reference paths of transmissive (dummy sensors) or reflective sensors (splices, open fiber end). The transmit and return fibers, or fiber highway, generally share a common single-fiber path in networks using reflective sensors.

When a suitable fiber-optic network topology is required, various criteria must be considered:

1. The sensor type, encoding principle, and topology to be used
2. The proposed multiplexing scheme, required number of sensors, and power budget
3. The allowable cross-communication level
4. The system cost and complexity constraints
5. The reliability (i.e., the effect of component failure on system performance)

POWER LINE FAULT-DETECTION SYSTEM FOR POWER GENERATION AND DISTRIBUTION INDUSTRY

In power distribution lines, faults such as short circuits, ground faults, and lightning strikes on the conductors must be detected in a very short time to prevent damage to equipment and power failure, and to enable quick repair. If the transmission line is divided in sections and a current or magnetic-field sensor is mounted in each section, a faulty section can be determined by detection of a change of the level and phase of the current on the power line. A system was developed as a hybrid optical approach to a fault-locating system that detects the phase and current difference between two current transformers on a composite fiber-optic ground wire (OPGW) wherein, due to induction, current is constantly passing (Fig. 6.35). The signal from a local electrical sensor, powered by solar cells and batteries, is transmitted over a conventional optical-fiber communication link. By three-wavelength multiplexing, three sensor signals can be transmitted over a single fiber. Seven sensors, three at each side of a substation and one at the substation itself, can monitor one substation on the power line, using one fiber in the OPGW.

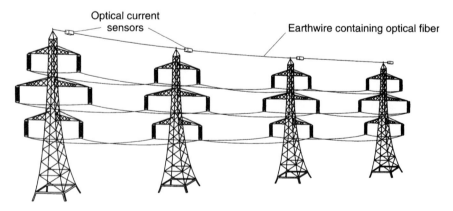

FIGURE 6.35 Fault locating system based on current and phase in ground wire.

Another system uses current transformers to pick up lightning current and thus detect lightning strikes. The signal is transmitted to a central detection point using the OPGW. Every sensor has its own OPGW fiber. This system is on a 273-kV power line in Japan.

The OPGW opens the possibility for using all kinds of sensors along the transmission line. These sensors may not only be for locating faults, but also for monitoring structural integrity. The use of optical time-domain reflectometry (OTDR) combined with passive intrinsic (distributed) sensors along the OPGW has future potential for providing a convenient and powerful monitoring method for power lines.

REFERENCES

1. Bailey Control Systems, Wickliffe, Ohio.
2. Bartman, R. K., B. R. Youmans, and N. M. Nerheim. "Integrated Optics Implementation of a Fiber Optic Rotation Sensor: Analysis and Development," *Proc. SPIE,* **719**, 122–134.
3. Berthold, J. W., "Industrial Applications of Optical Fiber Sensors," Fiber Optic and Laser Sensors III, *Proc. SPIE,* **566**, 37–44.
4. Carrol, R., C. D. Coccoli, D. Cardelli, and G. T. Coate, "The Passive Resonator Fiber Optic Gyro and Comparison to the Interferometer Fiber Gyro," *Proc. SPIE,* **719**, 169–177 (1986).
5. Chappel, A. (ed.), *Optoelectronics—Theory and Practice,* McGraw-Hill, New York, 1978.
6. Crane, R. M., A. B. Macander, D. W. Taylor, and J. Gagorik, "Fiber Optics for a Damage Assessment System for Fiber Reinforced Plastic Composite Structures," *Rev. Progress in Quantitative NDE,* 2B, Plenum Press, New York, 1419–1430, 1982.
7. Doeblin, E. O., *Measurement Systems—Application and Design,* 4th ed., McGraw-Hill, New York, 1990.
8. Fields, J. N., C. K. Asawa, O. G. Ramer, and M. K. Barnoski, "Fiber Optic Pressure Sensor," *J. Acoust. Soc. Am.,* **67**, p. 816 (1980).
9. Finkelstein, L., and R. D. Watts, "Fundamental of Transducers—Description by Mathematical Models," *Handbook of Measurement Science,* vol. 2, P. H. Sydenham (ed.), Wiley, New York, 1983.
10. Friebele, E. L. and M. E. Gingerich, "Radiation-Induced Optical Absorption Bands in Low Loss Optical Fiber Waveguides," *J. Non-Crust. Solids,* **38**(39), 245–250 (1980).
11. Henze, M., "Fiber Optics Temperature and Vibration Measurements in Hostile Environments," Technical Material, *ASEA Research and Innovation,* CF23-1071E (1987).
12. Hofer, B., "Fiber Optic Damage Detection in Composite Structure," *Proc. 15th Congress. Int. Council Aeronautical Science,* ICAS-86-4.1.2, 135–143 (1986).
13. Kapany, N. S., *Fiber Optics, Principles and Applications,* Academic Press, London, 1976.
14. Lagakos, N., et al., "Multimode Optical Fiber Displacement Sensor," *Appl. Opt.,* **20**, 167 (1981).
15. Liu, K., "Optical Fiber Displacement Sensor Using a Diode Transceiver," Fiber Optic Sensors II, A. M. Sheggi (ed.), *Proc. SPIE,* **798**, 337–341 (1987).
16. Mizuno, Y., and T. Nagai, "Lighting Observation System on Aerial Power Transmission Lines by Long Wavelength Optical Transmission," *Applications of Fiber Optics in Electrical Power Systems in Japan,* C.E.R.L. Letterhead, paper 5.
17. Mori, S., et al., "Development of a Fault-Locating System Using OPGW," *Simitomo Electric Tech. Rev.* (25), 35–47.
18. Norton, H. N., *Sensors and Analyzer Handbook,* Prentice-Hall, Englewood Cliffs, N.J., 1982.
19. Neubert, H. K. P., *Instrument Transducers,* 2d ed., Clarendon Press, Oxford, 1975.

20. Ogeta, K., *Modern Control Engineering*, 2d ed. Prentice-Hall, Englewood Cliffs, N.J., 1990.
21. Petrie, G. R., K. W. Jones, and R. Joones, "Optical Fiber Sensors in Process Control," *4th Int. Conf. Optical Fiber Sensors, OFS'86*, Informal Workshop at Tsukuba Science City, VIII 1–VIII 19, 1986.
22. Place, J. D., "A Fiber Optic Pressure Transducer Using A Wavelength Modulation Sensor," *Proc. Conf. Fiber Optics '85 (Sira)*, London, 1985.
23. Ramakrishnan, S., L. Unger, and R. Kist, "Line Loss Independent Fiberoptic Displacement Sensor with Electrical Subcarrier Phase Encoding," *5th Int. Conf. Optical Fiber Sensors, OFS '88*, New Orleans, pp. 133–136, 1988.
24. Sandborn, V. A., *Resistance Temperature Transducers*, Metrology Press, Fort Collins, Colo., 1972.
25. Scruby, C. B., R. J. Dewhurst, D. A. Hutchins, and S. B. Palmer, "Laser Generation of Ultrasound in Metals," *Research Techniques in Nondestructive Testing*, vol. 15, R. S. Sharpe (ed.), Academic Press, London, 1982.
26. Tsumanuma, T., et al., "Picture Image Transmission-System by Fiberscope," *Fujikura Technical Review* (15), 1–10 (1986).
27. Vogel, J. A., and A. J. A. Bruinsma, "Contactless Ultrasonic Inspection with Fiber Optics," *Conf. Proc. 4th European Conf. Non-Destructive Testing*, Pergamon Press, London, 1987.
28. Yasahura, T., and W. J. Duncan, "An Intelligent Field Instrumentation System Employing Fiber Optic Transmission," *Advances in Instrumentation*, ISA, Wiley, London, 1985.

CHAPTER 7
SENSORS IN FLEXIBLE MANUFACTURING SYSTEMS

INTRODUCTION

Flexibility has become a key goal in manufacturing, hence the trend toward flexible manufacturing systems. These are designed to produce a variety of products from standard machinery with a minimum of workers. In the ultimate system, raw material in the form of bars, plates, and powder would be used to produce any assembly required without manual intervention in manufacture. Clearly, this is a good breeding ground for robots.

But it should be emphasized that the early FMSs are in fact direct numerical control (DNC) systems for machining. And it must be acknowledged that an NC machine tool is really a special-purpose robot. It manipulates a tool in much the same way as a robot handles a tool or welding gun. Then, with no more than a change in programming, it can produce a wide range of products. Moreover, the controllers for robots and NC machines are almost the same. But for an NC machine to be converted into a self-supporting flexible system, it needs some extra equipment, including a handling device. It then forms an important element in an FMS.

The principle of flexible manufacturing for machining operations is that the NC machining cells are equipped with sensors to monitor tool wear and tool breakage. Such cells are able to operate unmanned so long as they can be loaded by a robot or similar device, since the sensors will detect any fault and shut the operation down if necessary. The requirements can be summarized as:

1. CNC system with sufficient memory to store many different machining programs
2. Automatic handling at the machining tool either by robot or other material handling system
3. Workpieces stored near the machine to allow unmanned operation for several hours. A guided vehicle system may be employed if workpieces are placed at a designated storage and retrieval system away from the machine
4. Various sensors to monitor, locate, and/or diagnose any malfunction

THE ROLE OF SENSORS IN FMS

The monitoring sensor devices are generally situated at the location of the machining process, measuring workpiece surface textures, cutting-tool vibrations, contact temperature between cutting tool and workpiece, flow rate of cooling fluid, electrical current fluctuations, etc. Data in the normal operating parameters are stored in memory with data on acceptable manufacturing limits. As the tool wears, the tool changer is actuated. If the current rises significantly, along with other critical signals from sensors, a tool failure is indicated. Hence, the machine is stopped. Thus, with the combination of an NC machine, parts storage and retrieval, handling devices, and sensors, the unmanned cell becomes a reality. Since the control system of the NC machine, robot, and unmanned guided vehicles are similar, central computer control can be used effectively.

Systems based on these principles have been developed. In Japan, Fanauc makes numerical controllers, small NC and EDM machines, and robots; Murata makes robot trailers as well as a variety of machinery including automated textile equipment; and Yamazaki makes NC machines. In France, Renault and Citroen use FMS to machine gear boxes for commercial vehicles and prototype engine components respectively, while smaller systems have been set up in many other countries.

The central element in establishing an error-free production environment is the availability of suitable sensors in manufacturing. The following represents a summary of sensing requirements in manufacturing applications:

1. Part identification
2. Part presence or absence
3. Range of object for handling
4. Single-axis displacement of measurement
5. Two-dimensional location measurement
6. Three-dimensional location measurement

Current Available Sensor Technology for FMS

The currently available sensors for manufacturing applications can be classified into four categories:

1. Vision sensors
   a. Photodetector
   b. Linear array
   c. TV camera
   d. Laser triangulation
   e. Laser optical time-domain reflectometry
   f. Optical fiber
2. Tactile sensors
   a. Probe
   b. Strain gages
   c. Piezoelectric
   d. Carbon material
   e. Discrete arrays
   f. Integrated arrays

3. Acoustic sensors
   a. Ultrasonic detectors and emitters
   b. Ultrasonic arrays
   c. Microphones (voice control)
4. Passive sensors
   a. Infrared
   b. Magnetic proximity
   c. Ionizing radiation
   d. Microwave radar

Integrating vision sensors and robotics manipulators in flexible manufacturing systems presents a serious challenge in production. Locating sensors on the manipulator itself, or near the end effector, provides a satisfactory solution to the position of sensors within the FMS. Locating an image sensor above the work area of a robot may cause the manipulator to obscure its own work area. Measurement of the displacement of the end effector also may suffer distortion, since the destination will be measured in relative terms, not absolute. Placing the sensor on the end effector allows absolute measurement to be taken, reducing considerably the need for calibration of mechanical position and for imaging linearity. Image sensory feedback in this situation can be reduced to the simplicity of range finding in some applications.

Extensive research and development activities were conducted recently to find ways to integrate various sensors close to the gripper jaws of robots. The promise of solid-state arrays for this particular application has not entirely materialized, primarily because of diversion of effort resulting from the commercial incentives associated with the television industry. It might be accurate to predict that, over the next decade, imaging devices manufactured primarily for the television market will be both small and affordable enough to be useful for robotics applications. However, at present, array cameras are expensive and, while smaller than most thermionic tube cameras, are far too large to be installed in the region of a gripper. Most of the early prototype arrays of modest resolution (developed during the mid-1970s) have been abandoned.

Some researchers have attacked the problem of size reduction by using coherent fiber optics to retrieve the image from the gripper array, which imposes a cost penalty on the total system. This approach can, however, exploit a fundamental property of optical fiber in that a bundle of coherent fibers can be subdivided to allow a single high-resolution imaging device to be used to retrieve and combine a number of lower-resolution images from various paths of the work area including the gripper with each subdivided bundle associated with its own optical arrangement.

Linear arrays have been used for parts moving on a conveyer in such a way that mechanical motion is used to generate one axis of a two-dimensional image. The same technique can be applied to a robot manipulator by using the motion of the end effector to generate a two-dimensional image.

Tactile sensing is required in situations involving placement. Both active and passive compliant sensors have been successfully applied in the field. This is not the case for tactile array sensors because they are essentially discrete in design, are inevitably cumbersome, and have very low resolution.

Acoustic sensors, optical sensors, and laser sensors are well developed for effective use in manufacturing applications. Although laser range-finding sensors are well developed, they are significantly underused in FMS, especially in robotic applications. Laser probes placed at the end effector of an industrial robot will form a natural automated inspection system in manufacturing.

Sensing for robot applications does not depend on a relentless pursuit for devices with higher resolution; rather, the fundamental consideration is selecting the optimum resolution for the task to be executed. There is a tendency to assume that the higher the resolution, the greater the application range for the system. However, considerable success has been achieved with a resolution as low as 50 × 50 picture elements. With serial processing architectures, this resolution will generate sufficient gray-scale data to test the ingenuity of image processing algorithms. Should its processing time fall below 0.5 s, an algorithm can be used for robots associated with handling. However, in welding applications, the image processing time must be faster.

ROBOT CONTROL THROUGH VISION SENSORS

An increasing number of manufacturing processes rely on machine-vision sensors for automation. The tasks for which vision sensors are used vary widely in scope and difficulty. Robotic applications in which vision sensing has been used successfully include inspection, alignment, object identification, and character recognition.

Human vision involves transformation, analysis, and interpretation of images. Machine-vision sensing can be explained in terms of the same functions: image transformation, image analysis, and image interpretation.

Image Transformation

Image transformation involves acquiring camera images and converting them to electrical signals that can be used by a vision computer (Fig. 7.1). After a camera image is transformed into an electronic (digitized) image, it can be analyzed to extract useful information in the image such as object edges, alignment, regions, boundaries, colors, and absence or presence of vital components.

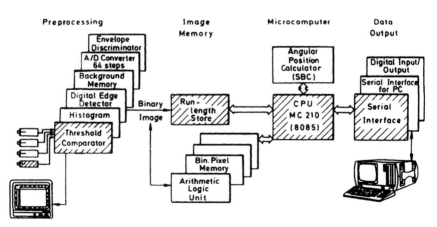

FIGURE 7.1 Image transformation involves the acquisition and conversion of camera images to electrical signals that can be used by a vision computer.

Once the image is analyzed, the vision sensing system can interpret what the image represents so that the robot can continue its task. In robot vision execution, design considerations entail cost, speed, accuracy, and reliability.

Robot Vision and Human Vision

Given that robot vision systems typically execute only part of what is normally called *seeing*, some similarities with human vision nevertheless arise. Other similarities arise in the "hardware" of human and robot visual systems.

The similarities in hardware could include an analogy between eyes and video cameras—both have lenses to focus an image on a sensitive "retina" that produces a visual signal interpreted elsewhere. In both human and robot vision this signal is passed to a device that can remember important aspects of the image for a time, perform specialized image processing functions to extract important information from the raw visual signal, and analyze the image in a more general way.

Some similarities in performance follow. Human and robot vision work well only where lighting is good. Both can be confused by shadows, glare, and cryptic color patterns. Both combine size and distance judgments, tending to underestimate size when they underestimate distance, for instance.

However, humans and machines have far more differences than similarities. A human retina contains several million receptors, constantly sending visual signals to the brain. Even the more limited video camera gathers over 7 Mbytes of visual information per second. Many of the surprising aspects of machine vision arise from the need to reduce this massive flow of data so that it can be analyzed by a computer system.

Machine-vision systems normally only detect, identify, and locate objects, ignoring many of the other visual functions. However, they perform this restricted set of functions very well, locating and even measuring objects in a field of view more accurately than any human can.

Robot Vision and Visual Tasks

Several standard visual tasks are performed by robot-vision systems. These tasks include recognizing when certain objects are in the field of view, determining the location of visible objects, assisting a robot hand with pickup and placement, and inspecting known objects for the presence of certain characteristics (usually specific flaws in manufacture) (Fig. 7.2).

A robot-vision system must exercise some judgment in performing *visual tasks*— those for which the input is a visual image (normally obtained from an ordinary video camera). Which visual tasks are relatively easy for machine vision, and which are hard? The distinction is not so much that some tasks are hard and others easy; rather, it is the detail within a task that distinguishes easy problems from hard ones.

What makes a problem hard? Some of the contributing factors are

1. Objects that vary widely in detail. (Examining stamped or milled product may be easy, while molded or sculpted items may be more difficult. Natural objects are by far the hardest with which to deal.)
2. Lighting variations, including reflections and shadows, as well as fluctuations in brightness (as found in natural sunlight). These variations may go unnoticed by human inspectors, but they can make otherwise easy problems difficult or impossible for robot vision.

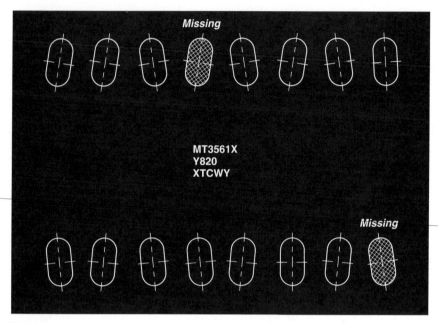

FIGURE 7.2   Image of a flaw. (*Courtesy of American SensoRx, Inc.*)

3. In general, ignoring "unimportant" variations in an image while responding to "significant" ones is very hard. (Most hard problems can be placed in this category.)

Robot Visual Sensing Tasks

Robots will work in unpleasant locations. Health hazards are of no concern to them. Special safety equipment is not required for a robot spraying paint, welding, or handling radioactive materials or chemicals. All this adds up to reduced production costs. As the day progresses, the tired worker has a tendency to pay less attention to details, and the quality of the finished product may suffer. This is especially noticeable in automobiles where spray paint can run or sag and weld joints may not be made perfectly. The panels of the car may not be aligned, and the finished product may not operate properly, with predictable customer dissatisfaction. In pharmaceutical production, too, an operator inspecting and verifying lot number and expiration date may become fatigued and fail to inspect for sensitive information.

Robots, on the other hand, do not tire or change their work habits unless they are programmed do so. They maintain the same level of operation throughout the day. With vision-sensing systems and robots, it is possible for American manufacturers to compete against lower labor costs in foreign countries. The initial investment is the only problem. After the initial investment, the overall operation costs of the production line are reduced or held constant. Small educational robots can be used to retrain humans to operate and maintain robots.

The roles of robots with machine vision can be summarized as follows:

1. *Handling and assembly:* recognizing position/orientation of objects to be handled or assembled, determining presence or absence of parts, and detecting parts not meeting required specifications
2. *Part classification:* identifying objects and recognizing characters
3. *Inspection:* checking for assembly and processing, surface defects, and dimensions
4. *Fabrication:* making investment castings, grinding, deburring, water-jet cutting, assembling wire harness, gluing, sealing, puttying, drilling, fitting, and routing
5. *Welding:* automobiles, furniture, and steel structures
6. *Spray painting:* automobiles, furniture, and other objects

Robots Utilizing Vision Systems to Recognize Objects

A basic use of vision is recognizing familiar objects. It is easy to see that this ability should be an important one for robot vision. It can be a task in itself, as in counting the number of each kind of bolt in a mixed lot on a conveyer belt. It can also be an adjunct to other tasks, for example, recognizing a particular object before trying to locate it precisely, or before inspecting it for defects.

It is important to note that this task actually has two distinct parts: first, object familiarization, that is learning what an object looks like, then object recognition (Fig. 7.3).

There are many ways of learning to recognize objects. Humans can learn from verbal descriptions of the objects, or they can be shown one or more typical items. A brief description of a *pencil* is enough to help someone identify many unfamiliar items as *pencil*. Shown a few *pencils,* humans can recognize different types of *pencils,* whether they look exactly like the samples or not.

Robot-vision systems are not so powerful, but both these approaches to training still apply to them. Robots can be given descriptions of what they are to recognize, perhaps derived from CAD data to guide a machine tool (Fig. 7.4) or they can be shown samples, then be expected to recognize objects more or less like the samples.

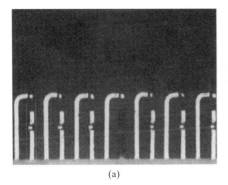
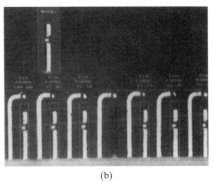

(a)   (b)

FIGURE 7.3 Recognition of a learned object. (*Courtesy of American SensoRx, Inc.*)

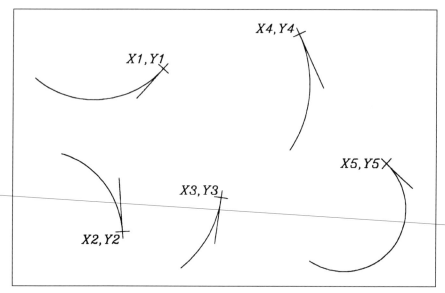

FIGURE 7.4 Machine guidance.

Recognizing objects once they have been learned is the second, and more difficult, part of the task. Several basic questions arise in virtually every recognition task. Among them are "What are the choices?" and "What can change?"

Specifying the actual task completely is normally the hardest part of the application. When this has been accomplished, the basic question is, for what particular vision features should the search begin to recognize an object, and, when the features are found, how should they be analyzed to decide which object (if any) has been found?

ROBOT VISION LOCATING POSITION

Humans use several techniques for gaging distances, especially triangulation on the left- and right-eye views, feedback from the eye's focusing mechanism, and the apparent motion produced by small head movements. This kind of object location may make use of knowledge about the object being located. By knowing how large the object really is, one can judge how far away it is from the size of its retinal image.

Few robot-vision systems use binocular vision, autofocus feedback, or moving cameras to estimate distances. However, with a rigidly mounted camera, it is possible to interpret each visible point as lying on a particular line of sight from the camera. Accurate information about the true distances between visible points on an object allows the robot-vision system to accurately calculate its distance. Similarly, if an object is resting on a platform at an accurately known distance, the robot vision system can interpret distances in the camera image as accurate distances on the object.

However locations are determined, the visual location task for a robot-vision system normally includes calibration as well as object location. Calibration normally is carried out by providing the system with a view that has easily identifiable points at known spatial locations. Once calibrated, the system can then locate objects in its own coordinate system (pixels) and translate the position into work cell coordinates (inches, millimeters, etc.).

ROBOT GUIDANCE WITH VISION SYSTEM

Another use of machine vision is in robot guidance—helping a robot to handle and place parts and providing it with the visual configuration of an assembly after successive tasks. This can involve a series of identification and location tasks. The camera can be attached to a mobile arm, making the location task seem somewhat more like normal vision. However, the camera typically is mounted on a fixed location to reduce system complexity.

While each image can give the location of certain features with respect to the camera, this information must be combined with information about the current location and orientation of the camera to give an absolute location of the object. However, the ability to move the camera for a second look at an object allows unambiguous location of visible features by triangulation. Recognition is a useful tool for flexible manufacturing systems within a CIM environment. Any of several parts may be presented to a station where a vision system determines the type of part and its exact location. While it may be economical and simpler to send the robot a signal giving the part type when it arrives, the ability to detect what is actually present at the assembly point, not just what is supposed to be there, is of real value for totally flexible manufacturing.

Robot Vision Performing Inspection Tasks

Visual inspection can mean any of a wide variety of tasks, many of which can be successfully automated. Successful inspection tasks are those in which a small number of reliable visual cues (features) are to be checked and a relatively simple procedure is used to make the required evaluation from those cues.

The differences between human and robot capabilities are most evident in this kind of task, where the requirement is not simply to distinguish between good parts and anything else—a hard enough task—but usually to distinguish between good parts or parts with harmless blemishes, and bad parts. Nevertheless, when inspection can be done by robot vision, it can be done very predictably.

Many inspection tasks are well suited to robot vision. A robot-vision system can dependably determine the presence or absence of particular items in an assembly (Fig. 7.5), providing

FIGURE 7.5 Presence or absence of items in an assembly. (*Courtesy of American SensoRx, Inc.*)

accurate information on each of them. It can quickly gage the approximate area of each item passing before it, as long as the item appears somewhere in its field of view (Fig. 7.6).

Components of Robot Vision

Figure 7.7 is a schematic diagram of the main components of a robot in a typical vision process for manufacturing. A fixed camera surveys a small, carefully lighted area where the objects to be located or inspected are placed. When visual information is needed (as signaled by some external switch or sensors), a digitizer in a robot vision system converts the camera image into a "snapshot": a significant array of integer brightness values (called gray levels). This array is sorted in a large random-access memory (RAM) array in the robot-vision system called an *image buffer* or a *frame buffer*. Once sorted, the image can be displayed on a monitor at any time. More importantly, the image can be analyzed or manipulated by a vision computer, which can be programmed to solve robot vision problems. The vision computer is often connected to a separate general-purpose (host) computer, which can be used to load programs or to perform tasks not directly related to vision.

Once an image is acquired, vision processing operations follow a systematic path. Portions of the image buffer may first be manipulated to suppress information that will not be valuable to the task at hand and to enhance the information that will be useful. Next, the vision program extracts a small number of cues from the image—perhaps allowing the region of interest to be reduced to exclude even more extraneous data.

At this stage, the vision program calculates, from the selected image region, the cues (features) of direct importance to the task at hand and makes a decision about

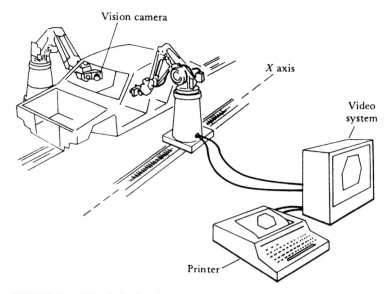

FIGURE 7.6 Object in field of view.

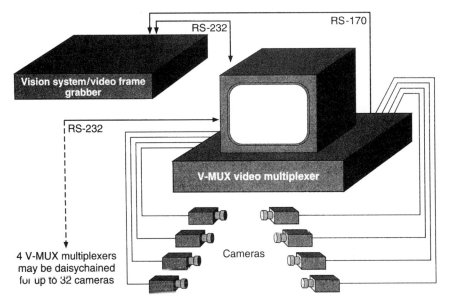

FIGURE 7.7 Schematic diagram of typical vision process.

the presence of a known part or its location in the field of view, or perhaps about the presence of specific defects in the object being inspected. Finally, the robot-vision system activates control lines based on the decisions, and (perhaps) transfers a summary of the conclusion to a data storage device or another computer.

END EFFECTOR CAMERA SENSOR FOR EDGE DETECTION AND EXTRACTION

A considerable amount of development of synchronized dual camera sensors at a strategic location on a robot end effector has been conducted for processing two-dimensional images stored as binary matrices. A large part of this work has been directed toward solving problems of character recognition. While many of these techniques are potentially useful in the present context, it is valuable to note some important differences between the requirements of character recognition and those associated with visual feedback for mechanical assembly.

Shape and Size

All objects presented to the assembly machine are assumed to match an exact template of the reference object. The object may have an arbitrary geometric shape, and the number of possible different objects is essentially unlimited. Any deviation in shape or size, allowing for errors introduced by the visual input system, is a ground for rejection of the object (though this does not imply the intention to perform 100 percent inspection of components). The derived description must therefore contain all the shape and size information originally presented as a stored image. A character recognition system has in general to tolerate considerable distortion, or style, in the characters to be recognized, the most extreme example being handwritten characters. The basic set of characters, however, is limited. The closest approach to a template-matching situation is achieved with the use of a type font specially designed for machine reading, such as optical character recognition (Fig. 7.8).

Position and Orientation

A component may be presented to the assembly machine in any orientation and any position in the field of view. Though a position- and orientation-invariant description is required in order to recognize the component, the measurement of these parameters is also an important function of the visual system to enable subsequent manipulation. While a line character may sometimes be skewed or bowed, individual characters are normally presented to the recognition system in a relatively constrained orientation, a measurement of which not required.

Multiple Objects

It is a natural requirement that the visual system for an assembly machine should be able to accommodate a number of components randomly positioned in the field of view. The corresponding problem of segmentation in character recognition is eased (for printed characters) by *a priori* knowledge of character size and pitch. Such information has fostered techniques for the segmentation of touching characters. No attempt is made to distinguish between touching objects. Their combined image will be treated by the identification procedures as that of a single, supposedly unknown object.

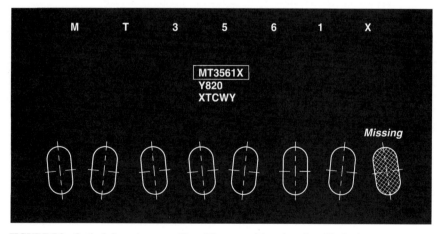

FIGURE 7.8 Optical character recognition. (*Courtesy of American SensoRx, Inc.*)

The essentially unlimited sizes of the set of objects that must be accommodated by the recognition system demands that a detailed description of shapes be extracted for each image. There are, however, a number of basic parameters which may be derived from an arbitrary shape to provide valuable classification and position information. These include:

1. Area
2. Perimeter
3. Minimum enclosing rectangle
4. Center of the area
5. Minimum radius vector (length and direction)
6. Maximum radius vector (length and direction)
7. Holes (number, size, position)

Measurements of area and perimeter provide simple classification criteria which are both position- and orientation-invariant. The dimensionless shape factor area/perimeter has been used as a parameter in object recognition. The coordinates of the minimum enclosing rectangle provide some information about the size and shape of the object, but this information is orientation dependent. The center of area is a point that may be readily determined for any object, independent of orientation, and is thus of considerable importance for recognition and location purposes. It provides the origin for the radius vector, defined as a line in the center of the area to a point on the edge of an object. The radius vectors of maximum and minimum length are potentially useful parameters for determining both identification and orientation. Holes are common features of engineering components, and the number present in a part is a further suitable parameter. The holes themselves may also be treated as objects, having shape, size, and position relative to the object in which they are found.

The requirements for the establishment of connectivity in the image and the derivation of detailed descriptions of arbitrary geometric shapes are most appropriately met by an edge-following technique. The technique starts with the location of an arbitrary point on the black/white edge of an object in the image (usually by a raster scan). An algorithm is then applied which locates successive connected points on the edge until the complete circumference has been traced and the starting point is reached. If the direction of each edge point relative to the previous point is recorded, a one-dimensional description of the object is built up which contains all the information present in the original shape. Such chains of directions have been extensively studied by Freeman. Measurements of area, perimeter, center of area, and enclosing rectangle may be produced while the edge is being traced, and the resulting edge description is in a form convenient for the calculation of radius vectors.

Edge following establishes connectivity for the object being traced. Continuing the raster scan in search of further objects in the stored image then presents the problem of the rediscovery of an already traced edge.

A computer plot of the contents of the frame with the camera viewing a square and a disk is illustrated in Fig. 7.9, and the result of applying the edge-extracting operation is illustrated in Fig. 7.10. The edge-following procedure may now be applied to the image in the same way it was to the solid object. The procedure is arranged, however, to reset each edge point as it is traced. The tracing of a complete object thus removes it from the frame and ensures that it will not be subsequently retraced.

FIGURE 7.9 Computer plot of the contents of a frame.

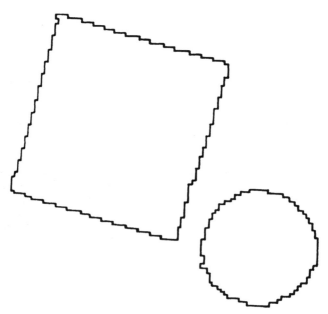

FIGURE 7.10 Result of applying edge extraction operation.

END EFFECTOR CAMERA SENSOR DETECTING PARTIALLY VISIBLE OBJECTS

A new method of locating partially visible two-dimensional objects has been developed. The method is applicable to complex industrial parts that may contain several occurrences of local features, such as holes and corners. The matching process utilizes clusters of mutually consistent features to hypothesize objects and uses templates of the objects to verify these hypotheses. The technique is fast because it concentrates on key features that are automatically selected on the basis on the detailed analysis of CAD-type models of the objects. The automatic analysis applies general-purpose routines for building and analyzing representations of clusters of local features that could be used in procedures to select features for other locational strategies. These routines include algorithms to compute the rotational and mirror symmetries of objects in terms of their local features. The class of tasks that involve the location of the partially visible object ranges from relatively easy tasks, such as locating a single two-dimensional object, to the extremely difficult task of locating three-dimensional objects jumbled together in a pallet. In two-dimensional tasks, the uncertainty is in the location of an object in a plane parallel to the image plane of the camera sensor. This restriction implies a simple one-to-one correspondence between sizes and orientations in the image, on the one hand, and sizes and orientations in the plane of the object, on the other.

This class of two-dimensional tasks can be partitioned into four subclasses that are defined in terms of the complexity of the scene:

1. A portion of one of the objects
2. Two or more objects that may touch one another
3. Two or more objects that may overlap one another
4. One or more objects that may be defective

This list is ordered roughly by the increasing amount of effort required to recognize and locate the object.

Figure 7.11 illustrates a portion of an aircraft frame member. A typical task might be to locate the pattern of holes for mounting purposes. Since only one frame member is visible at a time, each feature appears at most once, which simplifies feature identification. If several objects can be in view simultaneously and can touch one another, as in Fig. 7.12, the features may appear several times. Boundary features such as corners may not be recognizable, even though they are in the picture, because the objects are in mutual contact. If the objects can lie on one another (Fig. 7.13), even some of the internal holes may be unrecognizable because they are partially or completely occluded. And, finally, if the objects are defective (Fig. 7.14), the features are even less predictable and hence harder to find.

Since global features are not computable from a partial view of an object, recognition systems for these more complex tasks are forced to work with either local features, such as small holes and corners, or extended features, such as a large segment of an object's boundary. Both types of feature, when found, provide constraints on the position and the orientations of their objects. Extended features are in general computationally more expensive to find, but they provide more information because they tend to be less ambiguous and more precisely located.

Given a description of an object in terms of its features, the time required to match this description with a set of observed features appears to increase exponentially with the number of features. The multiplicity of features precludes the straight-

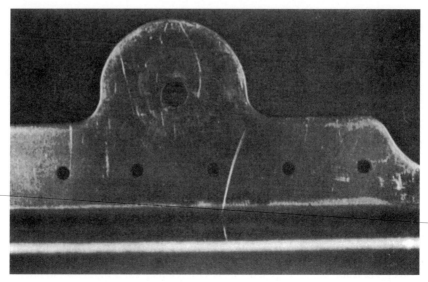

FIGURE 7.11 Portion of an aircraft frame member. (*Courtesy of American SensoR̂, Inc.*)

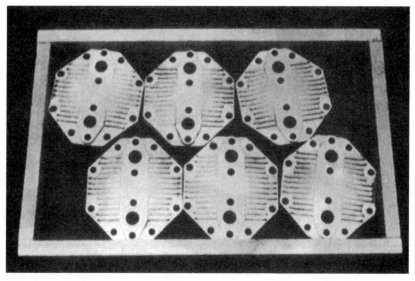

FIGURE 7.12 Objects touching each other. (*Courtesy of American SensoR̂, Inc.*)

forward application of any simple matching technique. Large numbers of features have been identified by locating a few extended features instead of many local ones. Even though it costs more to locate extended features, the reduction in the combinatorial explosion is often worth it. The other approach is to start by locating just one feature and use it to restrict the search area for nearby features. Concentrating

FIGURE 7.13 Objects lying on top of each other. (*Courtesy of American SensoRx, Inc.*)

on one feature may be risky, but the reduction in the total number of features to be considered is often worth it. Another approach is to sidestep the problem by hypothesizing massively parallel computers that can perform matching in linear time. Examples of these approaches include graph matching, relaxation, and histogram analysis. The advantage of these applications is that the decision is based on all the available information at hand.

The basic principle of the local-feature-focus (LFF) method is to find one feature of an image, referred to as the *focus feature,* and use it to predict a few nearby features to look for. After finding some nearby features, the program uses a graph-matching technique to identify the largest cluster of image features matching a cluster of object features. Since the list of possible object features has been reduced to those near the focus feature, the graph is relatively small and can be analyzed efficiently.

The key to the LFF method is an automatic feature-selection procedure that chooses the best focus features and the most useful sets of nearby features. This automatic-programming capability makes possible quick and inexpensive application of the LFF method to new objects. As illustrated in Fig. 7.15, the training pro-

FIGURE 7.14 Trained image.

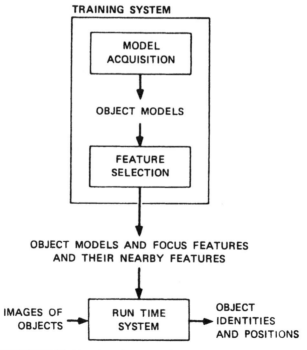

FIGURE 7.15 Run-time phase procedure.

cess, which includes the selection of features, is performed once and the results are used repeatedly.

Run-Time Phase

The run-time phase of the LFF acquires images of partially visible objects and determines their identities, positions, and orientations. This processing occurs in four steps:

1. Reading task information
2. Locating local features
3. Hypothesizing objects
4. Verifying hypotheses

The procedure (Fig. 7.15) is to input the object model together with the list of focus features and their nearby cofeatures. Then, for each image, the system locates all potentially useful local features, forms clusters of them to hypothesize object occurrences, and finally performs template matching to verify these hypotheses.

ULTRASONIC END EFFECTOR

An end effector on a welding robot (Fig. 7.16) contains an ultrasonic sensor for inspection of the weld. An ultrasonic sensor detects such flaws as tungsten inclusions and lack of penetration of weld. The end effector determines the quality of a weld immediately after the weld contact has been made, while the workpiece is still mounted on the weld apparatus; a weld can be reworked in place, if necessary. The delay caused by the paperwork and setup involved in returning the workpiece for rework is thereby avoided.

The ultrasonic end effector can be mounted on any standard gas tungsten arc welding torch. It may also be equipped with a through-the-torch vision system. The size of the ultrasonic end effector is the same as that of a gas cup with cathode.

A set of extension springs stabilizes the sensor and ensures that its elastomeric dry-couplant pad fits squarely in the weldment surface. The sensor can be rotated 360° and locked into alignment with the weld lead. A small force-actuated switch halts downward travel of the robot arm toward the workpiece and sets the force of contact between the sensor and the workpiece.

END EFFECTOR SOUND-VISION RECOGNITION SENSOR

The sound recognition sensor consists of a source that emits sound waves to an object and a sound receiver that receives the reflected sound waves from the same object (Fig. 7.17). The sound recognition sensor array consists of one sound source and one to as many as 16 receivers fitted intricately on an end effector of a robot.

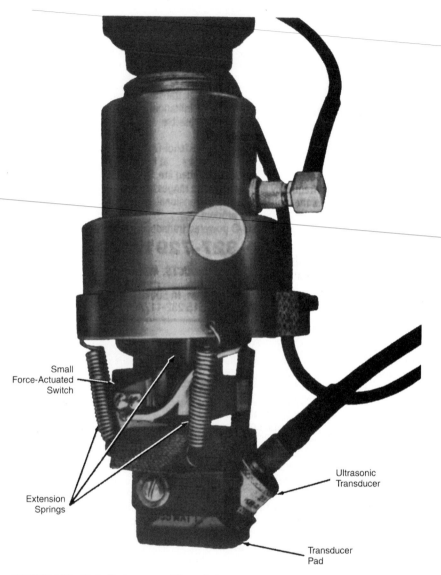

FIGURE 7.16 End effector on a welding robot.

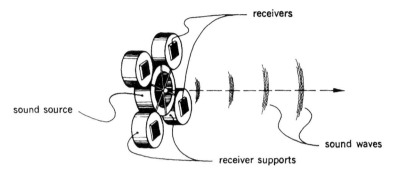

FIGURE 7.17 End effector sound-vision recognition system.

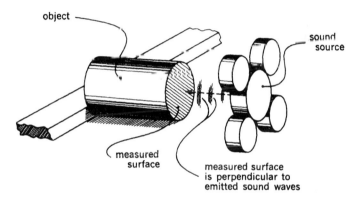

FIGURE 7.18 The measured surface should be perpendicular to the sound waves emitted from the sound source.

The sound-vision recognition sensor array measures reflections from some surface of interest on the object, called the *measured surface,* which is perpendicular to the sound waves emitted from the sound source (Fig. 7.18). There are four conditions governing the performance of sound-vision sensors:

1. Standoff
2. Large surfaces
3. Small surfaces
4. Positioning

Standoff

Standoff is how far the array must be located from the measured surface. The standoff, like other measurements, is based on the wavelength of the sound used. Three different wavelengths are used in the sound-vision sensor recognition system. The array standoff d should be one or two wavelengths λ_s from the measured surface for

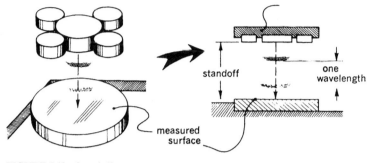

FIGURE 7.19 Standoff.

the highest accuracy. The standoff can be as great as 12 wavelengths, albeit with reduced accuracy (Fig. 7.19).

$$1.5\,\lambda_s \leq d \leq 12\,\lambda_s \tag{7.1}$$

where λ_s is the sound wavelength and d is the standoff distance. The typical standoff distance in terms of frequency and wavelength is described in Table 7.1.

TABLE 7.1 Correlation Functions of Typical Standoff d, Wavelength λ_s, and Frequency f

Frequency, kHz	Wavelength, mm	Standoff distance, mm
20	17	25
40	8	12
80	4	6

Large Surface Measurements

Large surface measurements achieve more accuracy than those made on small surfaces (Fig. 7.20). Whether a surface is large or small, the accuracy depends on the wavelength of the sound waves selected. A "large" surface must be at least one wavelength distance on each side (Table 7.2).

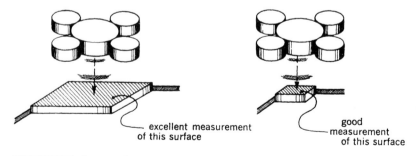

FIGURE 7.20 Large surface measurements.

TABLE 7.2 Correlation of Minimum Size of Large Surface, Frequency, and Wavelength

Frequency f, kHz	20	40	80
Wavelength λ_v, mm	17	8	4
Minimum area of surface, mm^2	275	70	20

The large surface being measured should change its dimension perpendicular to the surface, in the same direction as the sound wave emitted from the sound source (Fig. 7.21).

Sensitivity of Measurements

The sensitivity of the measurements to dimension changes is 5 or 10 times greater when the change is in the same direction as the emitted sound wave (Fig. 7.22).

Small Surfaces

Small surfaces can be measured as long as the robot end effector carrying the sensor array directs the sound wave from the side of the object (Fig. 7.23). Small surfaces are either a small portion of a large object or simply the surface of a small object (Fig. 7.24).

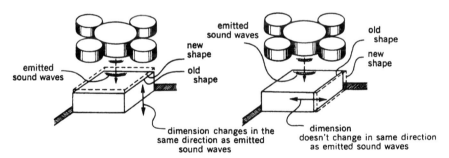

FIGURE 7.21 The large surface being measured should change its dimension perpendicular to the surface.

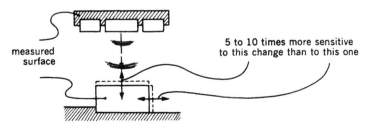

FIGURE 7.22 Sensitivity to dimension changes.

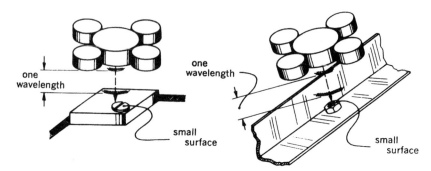

FIGURE 7.23 Small surfaces.

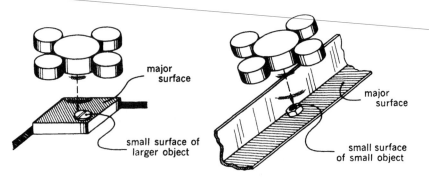

FIGURE 7.24 Small surfaces are either a small portion of a large object or simply the surface of a small object.

For small surfaces, the sound waves diffract or wrap around the surface rather than diverge from it as in a large surface. Similarly, ocean waves diffract around a rock that is smaller than the distance between crests (Fig. 7.25). Because of diffraction, the sound-vision recognition system is sensitive to the volume of the shape change on a small surface (Fig. 7.26).

Volume changes can be positive or negative. Small objects or protrusion from a surface represent positive volume changes, while holes or cavities represent negative volume changes (Fig. 7.27).

Measurement accuracy for small surfaces depends on the change in volume of the surface being measured. Listed in Table 7.3 are the smallest volume change that can be detected and the approximate size of a cube that has that volume for representative acoustic frequencies.

Positioning

The sound-vision sensor array system compares a particular part with a reference or standard part and detects the difference between the two. Objects being inspected—either a part or its standard—must be located relative to the array with at least the

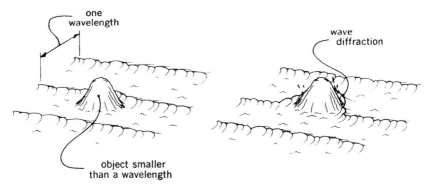

FIGURE 7.25 Ocean waves diffract around a rock that is smaller than the distance between crests.

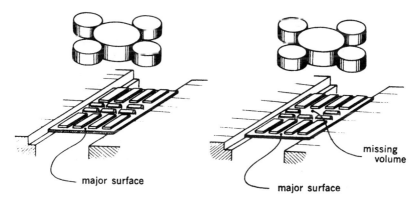

FIGURE 7.26 Sound-vision recognition system.

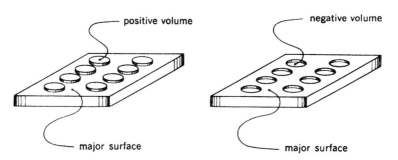

FIGURE 7.27 Volume changes can be positive or negative.

TABLE 7.3 The Least Measurable Volume

Frequency f, kHz	20 kHz	40 kHz	80 kHz
Smallest detectable volume change, μm^3	5×10^{-4}	6×10^{-5}	8×10^{-6}
Smallest detectable cube, mm^3	12×10^2	6×10^2	3×10^2

same accuracy as the expected measurement. Rectangular parts are usually located against stops; rotational parts are usually located in V-blocks on the face of the end effector (Fig. 7.28).

Rotationally unsymmetric parts must be oriented the same way as the standard in order to be compared. The end effector sensor array system can be used to direct a stepper motor to rotate the part until a match between part and standard is found (Fig. 7.29).

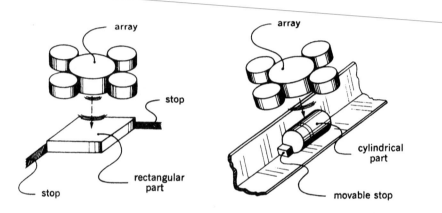

FIGURE 7.28 Diameter and height measurement.

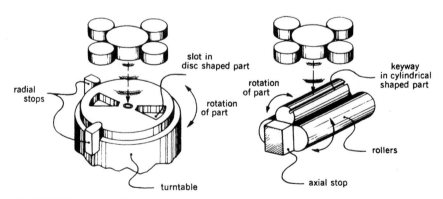

FIGURE 7.29 The end effector sensor can be used to direct a stepper motor to rotate the part until a match between part and standard is found.

END EFFECTOR LINEAR VARIABLE-DISPLACEMENT TRANSFORMER SENSOR

Sensing capability in a robot can have widely ranging degrees of sophistication in addition to a variety of sensing media. For instance, sensing capability can vary from a simple photoelectric cell to a complex, three-dimensional sound-vision system as described in the previous section.

The linear variable-displacement transformer (LVDT) sensor is an electromechanical device that can be attached to a robotic manipulator or can be itself a drive control for a robotic gripper. The LVDT produces an electrical output proportional to the displacement of a separate movable core. It consists of a primary coil and two secondary coils, intricately spaced on a cylindrical form. A free-moving rod-shaped magnetic core inside the coil assembly provides a path for the magnetic flux linking the coils. A cross section of the LVDT sensor and a plot of its operational characteristics are in Figs. 7.30 and 7.31. When the primary coil is energized by an external ac source, voltages are induced in the two secondary coils. These are connected series-opposing so that the two voltages are of opposite polarity. Theretore, the net output of the sensor is the difference between these voltages, which is zero when the core is at the center or null position. When the core is moved from the null position, the induced voltage in the coil toward which the core is moved increases, while the induced voltage in the opposite coil decreases. This action induces a differential voltage output that varies linearly with core position. The phase of this output voltage changes abruptly by 180° as the core is moved from one side of the null to the other.

Extreme Environments

With increasingly sophisticated technology, more and more instrumentation applications have arisen for sensors capable of operating in such hostile environments as extremely cold temperatures, very high temperatures, and/or intense nuclear radiation. The LVDT has been developed with materials that can tolerate extreme environmental requirements. Although the operating environments vary greatly, these LVDT designs use similar materials of construction and share the same physical configurations.

Currently, these LVDT sensors are built entirely with inorganic materials. The coil form is made of dimensionally stable, fired ceramic wound with ceramic-insulated high-conductivity magnet wire specially formulated for the application. Joints between the windings and lead wires are brazed or welded for mechanical reliability and electrical continuity. Ceramic cements and fillers are chosen to optimize heat transfer and bonding between windings, coil form, and housing. The potted assembly is cured at elevated temperatures, fusing the components together into a solidified structure.

Most inorganic insulations tend to be hygroscopic by nature, so the cured coil assembly is encased in an evacuated stainless steel shell that is hermetically sealed by electron beam (EB) welding. This evacuation and sealing process prevents moisture accumulation and subsequent loss of the insulation's dielectric strength. It also seals out surrounding media from the windings, while permitting the core to move freely.

Electrical connections are made to the windings with nickel conductors mutually insulated from each other by a magnesium oxide filler and sheathed in a length of

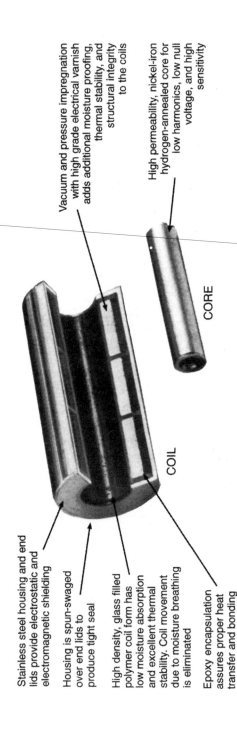

FIGURE 7.30 Cross-section of linear variable displacement transducer sensor.

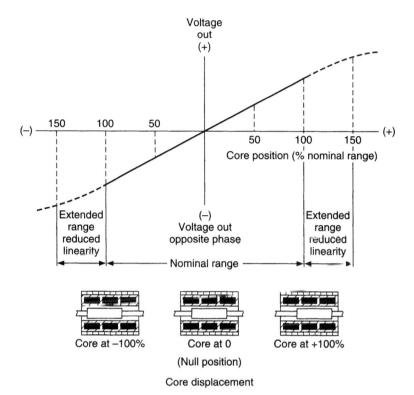

FIGURE 7.31 Plot of LVDT operational characteristics.

stainless-steel tubing. This cable assembly can be terminated by a hermetically sealed header for a connector when the application requires it.

The preceding description gives a brief insight into the material and techniques currently used in constructing the sensor for extremely severe environments. However, the state of the art in materials technology is being continually advanced. As new materials and methods of construction are evaluated, tested, and proved to upgrade performance, they will be incorporated into these sensors.

Cryogenic Manufacturing Applications

An LVDT sensor connected to the gripper of a robot is designed to cover a wide range of cryogenic applications ranging from general scientific research to space vehicle analysis and cryogenic medicine. A significant feature of the LVDT sensor is its ability to withstand repeated temperature cycling from room ambient conditions to the liquefaction temperatures of atmospheric gases such as nitrogen and oxygen. In order to survive such rigorous temperature changes, the sensor is constructed of materials selected for compatible coefficients of expansion while maintaining good electrical and magnetic properties even at −450°F (−270°C). The evacuated and hermetically sealed stainless-steel case prevents damage that could otherwise result from repeated condensation, freezing, and revaporization. Internal magnetic and electrostatic shielding renders the sensor insensitive to external magnetic and electrical influences.

Measurement at High Temperatures in Manufacturing

The LVDT sensor has been developed for measurements involving very high temperatures. It is capable of operating continuously at 1100°F (600°C) and surviving temperatures as high as 1200°F (650°C) for several hours. Typical uses include position feedback from jet engine controls located close to exhaust gases and measurement of roller position and material thickness in hot strip or slabbing mills. In scientific research it can be used to directly measure dimensional changes in heated test specimens without requiring thermal isolation, which could induce measurement errors. The sensor is the culmination of the development of sophisticated construction techniques coupled with careful selection of materials that can survive sustained operation at high temperatures. Because magnetic properties of a metal vanish above its magnetic transformation temperature (Curie point), the core material must be made from one of the few magnetic materials having Curie temperatures above 1100°F (600°C). Another problem is that, at high temperature, the resistance of windings made of common magnet wire materials increases so much that an LVDT sensor using ordinary conductor materials would become virtually useless. Thus, the winding uses a wire of specially formulated high-conductivity alloy. The sensors are made with careful attention to internal mechanical configuration and with materials having compatible coefficients of expansion to minimize null shifts due to unequal expansion or unsymmetrical construction. Hermetic sealing allows the sensor to be subjected to hostile environments such as fluid pressure up to 2500 psi (175 bars) at 650°F (350°C). Units can be factory calibrated in a special autoclave that permits operation at high temperature while they are hydrostatically pressurized.

ROBOT CONTROL THROUGH SENSORS

In order to pick up an object, a robot must be able to sense the strength of the object being gripped so as not to crush the object. Accordingly, the robot gripper is equipped with sensing devices to regulate the amount of pressure applied to the object being retrieved.

Several industrial sensing devices enable the robot to place objects at desired locations or perform various manufacturing processes:

1. *Transducers:*   sensors that convert nonelectrical signals into electrical energy
2. *Contact sensors (limit switches):*   switches designed to be turned ON or OFF by an object exerting pressure on a lever or roller that operates the switch
3. *Noncontact sensors:*   devices that sense through changes in pressure, temperature, or electromagnetic field
4. *Proximity sensors:*   devices that sense the presence of a nearby object by inductance, capacitance, light reflection, or eddy currents
5. *Range sensors:*   devices such as laser-interferometric gages that provide a precise distance measurement

6. *Tactile sensors:* devices that rely on touch to detect the presence of an object; strain gages can be used as tactile sensors
7. *Displacement sensors:* provide the exact location of a gripper or manipulator. Resistive sensors are often used—usually wire-wound resistors with a slider contact; as force is applied to the slider arm, it changes the circuit resistance
8. *Speed sensors:* devices such as tachometers that detect the motor shaft speed
9. *Torque sensors:* measure the turning effort required to rotate a mass through an angle
10. *Vision sensors:* enable a robot to see an object and generate adjustments suitable for object manipulation; include dissectors, flying-spot scanners, vidicons, orthicons, plumbicons, and charge-coupled devices

MULTISENSOR-CONTROLLED ROBOT ASSEMBLY

Most assembly tasks are based on experience and are achieved manually. Only when high volume permits are special-purpose machines used. Products manufactured in low-volume batches or with a short design life can be assembled profitably only by general-purpose flexible assembly systems which are adaptable and programmable. A computer-controlled multisensor assembly station responds to these demands for general-purpose assembly. The multisensor feedback provides the information by which a robot arm can adapt easily to different parts and accommodate relative position errors.

The assembly task may be viewed as an intended sequence of elementary operations able to accept an originally disordered and disorganized set of parts and to increase gradually their order and mating degrees to arrive finally at the organization level required by the definition of the assembly. In these terms it may be considered that assembly tasks perform two main functions: (1) ordering of parts and (2) mating of parts in the final assembly.

The ordering function reduces the uncertainty about the assembly parts by supplying information about their parameters, type, and position/orientation. The performance criterion of this function may be expressed as an entropy. The final goal of ordering is to minimize the relative entropy sum of the components of the assembly. In an assembly station the ordering is performed either in a passive way by sensors or in an active way by mechanical means such as containerization, feeding, fixturing, or gripping.

The part mating function imposes successive modifications of the positions of the parts in such a way as to mate them finally in the required assembly pattern. The modification of the position of the parts is carried out by manipulating them by transport systems and robot arms. The part mating function requires *a priori* information about the parts to be manipulated and the present state of the assembly. During this manipulation, the part entropy may increase because of errors in the robot position or by accidental changes in the manipulated parts.

An example of a multisensor system is a testbed for research on sensor control of robotic assembly and inspection, particularly comparisons of passive entropy reduction to active mechanical means. Programs for the system must be easy to to develop and must perform in real time. Commercially available systems and sensors had to be used as much as possible. The system had to be very flexible in interfacing with different subsystems and their protocols. The flexible assembly system (Fig. 7.32) consists of three main parts:

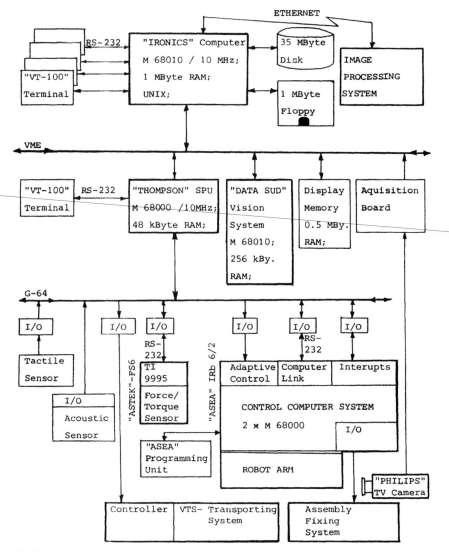

FIGURE 7.32 Structure of a flexible manufacturing system.

1. Robot equipped with a transport system
2. Vision sensors, force-torque sensors, and ultrasound sensors
3. Control system

The system was built with a variable transport system (VTS) (Fig. 7.33). It is a modular system in which product carriers are transported to the system. Two docking stations which can clamp the product carriers are included in the system. Some of the product carriers have translucent windows to allow backlighting of the vision

SENSORS IN FLEXIBLE MANUFACTURING SYSTEMS

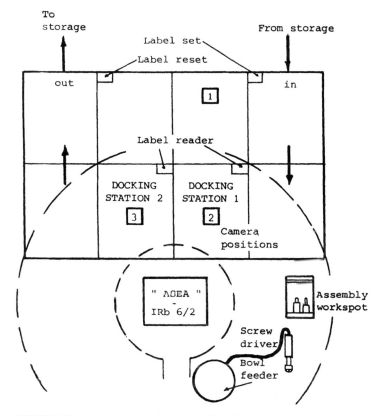

FIGURE 7.33 Layout of robot assembly and variable transport system.

system. The VTS system has its own controller operated by parallel input/output lines.

The robot is equipped with six degrees of freedom and a payload capacity of 4 kg. The absolute accuracy is 0.2 mm. The robot controller consists of two 68000 microprocessors. The standard way of programming the robot is through a teach pendant.

There are three different interfaces between the robot and the control system: (1) computer link at 9600 baud (RS-232), (2) adaptive control, and (3) parallel interrupt lines.

The computer link lets the control computer give commands to the robot such as:

1. Load programs from disk
2. Start programs
3. Set position registers
4. Read out robot positions
5. Take direct control of robot movement
6. Up- and download robot programs

The adaptive control offers search and contour-following modes. Up to three sensor signals can be used to guide the movements of the robot. The direction in which the robot is to move must be given for each signal. The displacement or velocity is proportional to the signal value generated by the displacement sensor. Both digital and analog signals produced by sensors are accepted. Digital signals may consist of 1 to 8 bits.

The interrupt lines are used to start or to stop a robot movement and to synchronize tasks.

Control Computer

The computer used to control the system is microprocessor-based. The main processor unit (MPU) is a VME-bus-compatible single-board computer featuring a 68010 processor, 1-Mbyte dual-ported RAM, a 35-Mbyte Winchester disk, and a 1-Mbyte floppy disk. The system runs under the UNIX V operating system.

Real-time performance is obtained by two special processing units (SPUs). The first SPU realizes the control and interfacing of the robot and the interfacing of the sensors. It links the VME bus to the G64 bus to which the robot, the VTS, the torque-force sensor, and the ultrasound sensor are interfaced. This processor controls the computer link and the adaptive control of the robot. The processor board consists of a 68000 processor and 48-kbit RAM. The second SPU is the vision processing board. This board communicates with the vision boards and the other SPU via the VME bus. It consists of a 68010 processor, 256-kbit RAM, and an NS16081 floating-point processor.

Vision Sensor Modules

The vision module consists of two Primagraphics VME boards. One board contains the frame image of 768×576 pixels and 128 gray values. A frame-transfer solid-state camera is connected to the system.

Force sensing is available and can measure three force and three torque components. Forces up to 200 N are measured with a resolution of 0.1 N. Torques up to 4 N·m are measured with a resolution of 0.002 N·m. The data rate ranges from 1.8 to 240 Hz. The torque-force sensor is controlled by a Texas Instruments 9995 microprocessor. Besides a high-speed RS-232 link 38400 board, an analog output is available.

An ultrasonic range finder sensor has been interfaced to the G64 bus. A tactile array based on the piezoelectric material PVDF has been developed. A stereo vision system is incorporated to the system.

Software Structure

Because an important aim of the testbed is to investigate the use of sensors in an assembly process, it is essential that an optimal environment is present for program development and real-world testing. A user-friendly program environment is provided by the UNIX V operating system. Real-time operation is provided by the special processing units. Programs developed under the UNIX system can be downloaded to the SPUs via the VME bus.

The assembly process consists of a sequence of assembly stages. A succeeding assembly stage is entered when certain entrance conditions are met. This means that the assembly process can be controlled by a state machine. An assembly stage may consist of a number of states. With this approach, a modular and hierarchical control structure is obtained in which certain processes are executed at the lowest level.

The control software for the assembly task consists of three different levels: (1) state machine level, (2) executor level, and (3) driver level.

At the highest level a command interpreter activates certain states of the machine. The state machine activates another (nested) state machine or a process (executor). Executors may run in parallel. For instance, the vision system may analyze an image of a product while the robot is assembling. The synchronization of the processes is obtained from the required entrance conditions for the next state of the state machine. In general, this level requires a moderately fast response and may run under the UNIX system.

At the second level, the executors are activated by the state machine. An executor performs a certain subtask and requires a real-time response. Examples are robot movements and image analysis routines. The executors reside in the SPUs and in the robot controller (robot programs). An executor is the only program that communicates directly with a driver.

The drivers for the different subsystems form the third level and also reside in the SPUs. The drivers take care of the protocol conversion and of the error checking in the communication with the subsystems.

During the assembly task a number of processes (executors) are running, requiring synchronization of data transfer. The data may consist of, for instance, the type, position, and orientation of a detected part by the vision system. The communication between the processes is realized through a shared memory lock in the VME bus.

Vision Sensor Software

The first step consists of a segmentation of gray-value image into a binary image. Because backlighting is used in the system, contrast is excellent and simple thresholding is sufficient for segmentation. During initialization, a threshold is calculated from the image histogram and used until the next initialization takes place. Before the system starts an assembly operation, the vision system must be calibrated to the robot coordinate system. This is done by moving a ring in the vision field. From the known robot coordinates, the ring position, and the computed ring coordinates in the vision system, the necessary transformation is calculated.

A drawback of connectivity analysis is that objects may not touch. The vision package may be extended with graph-matching techniques to allow recognition of touching and overlapping parts.

Robot Programming

The robot is programmed in the teach-in mode with a standard pendant. Different subprograms, which are called from the control computer, are thus realized. The robot has 100 registers in which positions and orientations can be stored by a control computer. In this way position and orientation determined by the vision module can be transferred to the robot controller. The search and contour tracing options of the robot controller are used for adaptive control.

The products arrive in a varying order and in an undefined position. Because the parts (except for bolts) are supplied on a flat surface in their most stable position, i.e., resting on their flat side, they have three degrees of freedom, namely the x and y coordinates and the orientation angle. The different parts are recognized by the vision system on the basis of area, perimeter, and position of the holes in the objects. The position and orientation are found from the center of gravity and the hole positions.

Handling

The assembly starts with the housing, which is to be positioned at a work spot beside the robot where the assembly takes place (Figs. 7.34 and 7.35). The work spot is not suited to a product-carrier feed, since the supply varies, and some carriers would have to be passed over, until the next suitable part arrives. Each of the three variants has its own work spot because each has a different foundation. After it has been picked up, the product-carrier housing has to be rotated 180° around an axis in the x-y plane, so that the flat side is up. Then the diaphragm and the cover are moved and laid down in the correct orientation. For supplying and fastening the bolts, a commercially available automatic screwdriver is used. The sixth axis of the robot adds a considerable compliance in the extreme end of the robot; this is not a disadvantage, since, during placement of the housing, a certain amount of passive compliance is needed for smooth assembly.

Gripper and Gripping Methods

All the products are manipulated by the same gripper, since gripper alternation is not possible. Handling all the products with a multifunctional gripper requires selec-

FIGURE 7.34   Hydraulic lift product family.

FIGURE 7.35 Exploded view of hydraulic lift assembly. (*a*) Bolts, (*b*) cover, (*c*) O ring, (*d*) diaphragm, and (*e*) housing.

tion of a common gripping configuration for the various products. The diaphragm is a particular problem; it does not resemble the other parts in form and composition. The diaphragm has to fit in a groove on the housing and therefore cannot be gripped at its side. Tests showed that the diaphragm could be handled by vacuum from a vacuum line.

Common features of the housing and cover are holes, maximum diameter, and the unprocessed exteriors of the holes. The housing must be positioned on the work spot with high precision in order to obtain a reference for the remaining parts of the assembly. To achieve this precision, the worked surfaces and holes of the housing have to be used, and therefore cannot be used for handling the parts. This leaves the unprocessed exteriors and maximum diameter for handling. The exteriors of the holes were chosen because they allow a smaller gripper and have a larger tangent plane.

As the multifunctional gripper, a pneumatic gripper is used, extended with a suction pad for the diaphragm. The gripper is placed perpendicular to the torque-force (TF) sensor and outside the central axis of the TF sensor. Otherwise, when the housing is rotated 180°, the TF sensor would get in the way as the product is placed on the work spot. The work spots have small dimensional tolerances (about 0.1 mm) relative to the housing. Feedback of information from the TF sensor and the passive compliance of the gripper assist it in placing the housing on the work spot. The gripper can grasp parts in any orientation on the product carrier of the VTS system because the gripper can rotate more than 360° in the x-y plane. The section of the gripper that grasps the scribe on the housing has a friction coating so that a light pneumatic pressure is sufficient to hold the product.

Accuracy

High accuracy in position is not needed to grip the housing or the cover because uncertainty in position is reduced simply by closing the gripper. High accuracy in position for the diaphragm is also unnecessary because the diaphragm can be shifted in the groove by a robot movement. The required accuracy in orientation is about 0.7° for placing the cover. The error in both the robot and the vision system was measured by having the robot present a part to the vision system repeatedly. The maximum error in position was 0.5 mm and in orientation was 0.4° in a field of view of 300 × 400 mm. This amply meets the requirement of automatic assembly.

REFERENCES

1. Ambler, A. P., et al., "A Versatile Computer-Controlled Assembly System," *Proc. of IJCAI-73*, Stanford, California, 298–307, August 1973.
2. Ballard, D. H., "Generalizing the Hough Transform to Detect Arbitrary Shapes," *Pattern Recognition*, **13**(2): 111–112 (1981).
3. Barrow, H. G., and R. J. Popplestone, "Relational Descriptions in Picture Processing," *Machine Intelligence* **6**, Edinburgh University Press, 1971.
4. "Decade of Robotics," 10th anniversary issue of *International Robot*, IFS Publications, 1983.
5. Duba, R. O., and P. E. Hart, "Use of Hough Transform to Detect Lines and Curves in Pictures," *Communications of AQCM*, **15**(1): 11–15, January 1972.
6. DuPont-Gateland, C. "Flexible Manufacturing Systems for Gearboxes," *Proc. 1st Int. Conf. on FMS*, Brighton, U.K., 453–563, October 20–22, 1982.
7. Freeman, H., "Computer Processing of Line-Drawing Images," *Computing Survey*, **6**(1), March 1974.
8. Gilbert, J. L., and V. Y. Paternoster, "Ultrasonic End Effector for Weld Inspection," *NASA Tech Briefs*, May 1992.
9. Ingersol Engineers, *The FMS Report*, IFS Publications, 1982.
10. Karp, R. M., "Reducibility Among Combinatorial Problems," *Complexity of Computer Computations*, Plenum Press, 85–103, 1972.
11. Morgan, T. K., "Planning for the Introduction of FMS," *Proc. 2d Int. Conf. on FMS*, 349–357, 1983.
12. Novak, A., "The Concept of Artificial Intelligence in Unmanned Production—Application of Adaptive Control," *Proc. 2d Int. Conf. on FMS*, 669–680, 1983.
13. Perkins, W. A., "A Model-Based Vision System for Industrial Parts," *IEEE Trans. Computers*, **C-27**, 126–143, February 1978.
14. *Proc. 1st Int. Conf. on Flexible Manufacturing Systems*, Brighton, 20–22 October 1982. IFS Publications and North-Holland Publishing Company, 1982.
15. *Proc. 2d Int. Conf. on Flexible Manufacturing Systems*, London, 26–28 October 1983, IFS Publications and North-Holland Publishing Company, 1983.

16. *Proc. 4th Int. Conf. on Assembly Automation,* Tokyo, 11–13 October 1983.
17. Ranky, P. G., *The Design and Operation of FMS,* IFS Publications and North-Holland Publishing Company, 1983.
18. Roe, J., "Touch Trigger Probes on Machine Tools," *Proc. 2d Int. Conf. on FMS,* 411–424, 1983.
19. Rosenfeld, A., *Picture Processing Computer,* Academic Press, New York, 1969.
20. Suzuki, T., et al., "Present Status of the Japanese National Project: Flexible Manufacturing System Complex," *Proc. 2d Int. Conf. on FMS,* 19–30, 1983.
21. Tsuji, S., and F. Matsumoto, "Detection of Ellipse by a Modified Hough Transformation," *IEEE Trans. Computers,* **C-27**(8): 777–781, August 1978.
22. Tsuji, S., and Nakamura, "Recognition of an Object in Stack of Industrial Parts," *Proc. IJCAI-75,* Tbilisi, Georgia, 811–818, August 1975.
23. Ullman, J. R., *Pattern Recognition Techniques,* Butterworths, London, 1972.
24. Woodwark, J. R., and D. Graham, "Automated Assembly and Inspection of Versatile Fixtures," *Proc. 2d Int. Conf. on FMS,* 425–430, 1983.
25. Zucker, S. W., and R. A. Hummel, "Toward a Low-Level Description of Dot Cluster: Labeling Edge, Interior, and Noise Points," *Computer Graphics and Image Proc.,* **9**(3): 213–233, March 1979.

CHAPTER 8
SPECTR$_x$: AN ONLINE PRODUCTION ANALYTICAL SENSOR DEVELOPED FOR PHARMACEUTICAL, FOOD, PETROLEUM, AGRICULTURE, BEEF, PORK, AND POULTRY INDUSTRIES

PRINCIPLE OF OPERATION

The system is based upon subjecting an organic object with the energy of a near-infrared beam. The amount of energy absorbed by the object provides a unique signature of its contents. The signature may be plotted as a correlation function between the light wavelengths and the amount of energy absorbed (Fig. 8.1).

The system has been developed as a sophisticated sensor for *online* verification and inspection for organic and nonorganic compounds, maintaining an error-free production environment utilizing revolutionary computational techniques such as genetic algorithm and neural network analysis.

INDUSTRIAL APPLICATIONS

The SPECTR$_x$ is an online, nondestructive automated sensor that is rugged, reliable, and simple to use in many different applications throughout pharmaceutical, chemical, food, meat, cosmetic, gas, fertilizer, and petroleum industries. The system provides identification and content analysis of organic-base objects such as tablets, capsules, and candies. It provides instantaneous verification of the presence or absence of foreign substances in any object component.

The system can also be configured to analyze the fat, protein, and water content of beef, pork, poultry, and hamburger.

The industry may use this system to inspect and verify objects contained in plastic cavities prior to the final seal stage using a single probe or multiplexed probes.

FIGURE 8.1 Energy absorbed by an object and reflected through the probe to a detector.

THE PURPOSE OF THE DEVELOPMENT

The SPECTR̥ has been primarily developed to verify, measure, test, monitor, and inspect formulation, chemical composition, potency, contamination, protein, fat, water content, hardness, disintegration, and dissolution of organic base solids, applied in many fields of different industries.

Applicable Industries

Pharmaceutical and chemicals
Nonpharmaceutical chemicals
Prescription drugs
Abused drug analysis
Food additives
Flavor and fragrances
Gas analysis
Coating
Steroids
Minerals and clays
Surface active agents
Meat processing
Pesticides and agriculture
Petroleum
Adhesives and sealant
Monomers analysis
Polymers analysis
Plastic analysis
Polymer processing chemicals
Polyols
Automobile paint chips

SPECTR̥ US/PCT Patent 5,679,954, Dr. Sabrie Soloman. SPECTR̥ is a registered trademark of American SensoR̥, Inc.

Fibers and textile
Dies
Organometallic
Rubber
Water treatment
Enhanced EPA vapor
Crime laboratory analysis
Forensic science analysis

TECHNICAL DESCRIPTION

The SPECTR̲x̲ uses a single-plate, permanently aligned, self-compensating laser beam splitter that provides two input and two output optical ports and yields an inherently symmetric interferogram required for object analysis. See Fig. 8.2.

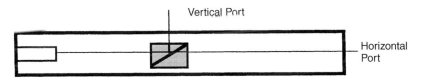

FIGURE 8.2 Beam splitting through prism.

The unique four-port design of the SPECTR̲x̲ offers complete suppression of back-streaming radiation, permitting high photometric accuracy, even for hot or highly reflective samples. The four-port design also permits simultaneous transmission and emission spectroscopy and dual-beam optical null spectroscopy necessary for substance identification.

The SPECTR̲x̲ consists of a modern, completely adjustment-free, carefully sealed spectrometer suitable for harsh environments with high signal-to-noise reproducibility and photometric accuracy performance. The SPECTR̲x̲ may be used in three broad areas:

1. As a series of dedicated analyzers for a variety of monitoring applications in industry:

   - Full **online** 100 percent process monitoring of formulation dissolution/hardness, moisture content, potency, foreign substances, and contamination, to name a few, at a rate of 65 objects per second, using a single probe. A cluster of 4 probes has been developed to provide 260 object analyses per second, using dual port. The system can be configured to use 8 probes yielding an inspection rate of 31,200 objects per second (Fig. 8.3).
   - Semi-continuos **at-line** industrial product verification.
   - Batch quality assurance for composition analysis and quality audit of finished products.

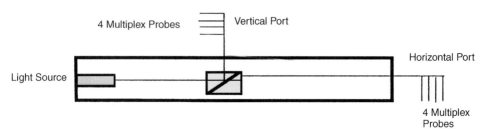

FIGURE 8.3 Multiple-probe configuration.

2. As a series of high-performance bench-top products for the laboratory.
3. As a series of analyzers for environmental monitoring and spectral radiometric remote sensing.

The SPECTR̴ sensor has the capability to achieve from 65 spectra per second per probe. The system shows considerable data stability and reproducibility for high-volume inspection and verification applications. This characteristic will accomplish 100 percent inspection and verification of high-volume production of pharmaceutical products and, in particular, monitor the hardness, dissolution, and potency level of each tablet or capsule throughout production.

These features are equally valuable in other industries such as meat, pork, and poultry processing.

The SPECTR̴ is the only known system to provide **online** instantaneous data of every tablet, capsule, powder, gel, or liquid in the pharmaceutical industry. Also, it is the only known system that can provide instantaneous soil analysis in the agriculture industry.

The Software

The software runs under either Microsoft Windows 95 or Microsoft Windows NT operating systems. The software has been developed for ease of operation. It employs advanced techniques in artificial intelligence, such as neural networks.

Presentation of Data

The simple plug-and-play instructions help the operator to get acquainted with the system quickly. The presentation of the main software—namely interferogram, spectrum, production, database, quantitative spectral analysis, and neural networks—assists the researcher to understand the application and use of the technology.

THEORY OF OPERATION

The SPECTR̴ sensor system consists of the following main components:

- The spectrometer
- The probe and detector

- The computer system
- The software

A schematic diagram of the system is shown in Fig. 8.4.

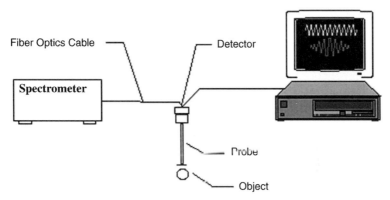

FIGURE 8.4 Schematic diagram of the SPECTR system. *(Courtesy of American Sensors, Inc.)*

The spectrometer sends infrared light (which covers the range of 4000 to 10500 wave numbers) through a fiber optics cable to the probe.
A wave number is defined as

$$10{,}000/\lambda \text{ (wavelength)}$$

Some of the light energy is absorbed by the object, and some of it is reflected back to the probe. Each object absorbs light energy based on its chemical composition and physical characteristics. The spectrum of the reflected light represents a fingerprint of that object. The probe is designed to allow only the diffuse component of the reflected light back into the probe and reflect the specula component. The diffuse component is converted by the detector into an electric signal (interferogram), which is sent to the computer system. The Fourier transformation infrared (FTIR) software resident in the computer performs some transformations (fast Fourier transform or FFT, adopization, and filtering) on that signal and displays the spectrum of the object.

Configuration

The sensor system has evolved into a series of products that are useful for determining the disintegration, hardness, dissolution, and potency values of any coated or uncoated tablet.
The SPECTR is a dual-channel system adaptable to an 8-channel system. Each channel extends 2 meters of specially developed fiber optics cable over the domain of the object to focus a light beam 330 μm in diameter. The fiber optics cable-probe system will segregate diffused reflective components of light energy from specula light energy components.

The SPECTRx is enclosed in a NEMA 4-X cabinet. It consists of:

- The SPECTRx spectrometer
- Fiber optics cables, 2 meters OH free quartz fibers
- Cable interface
- Probe
- InAs detector
- Detector amplifier
- Halogen light source
- Dual-processor computer (Pentium-Pro 300–900 MHz)
- Data acquisition
- 17" monitor (0.27 SVGA monitor)
- 8.07 GB hard drive, 12-speed CD-ROM
- Windows 95 (or Windows NT)
- Advanced quantitative software package
- Deskjet color printer

The Spectrometer

The spectrometer is hermetically sealed but is not vacuum- or pressure-tight. Its cast aluminum housing housed in consists of a lower and upper half separated by a gasket and bolted together. The electrical power, switches, fuses, and data cable are interfaced at a side panel of the casting. See Fig. 8.5.

The analyzer is equipped with internal He-Ne laser reference for digital sampling and mirror velocity control.

Digital sampling is synchronized by means of error-free up/down fringe counting using two references as fringe channels in quadrate.

Permanently mounted/permanently aligned single-plate beam splitters are provided, with splitting and combining surfaces separated vertically, for two input and two output beams of half the diameter of the cube-corner mirrors.

The normal collimated beam diameter is 25 mm. Maximum beam divergence is 90 milliradians full angle when the beam stop is defined at the apex of the cube corner retroflectors. A high infrared quartz halogen source is installed in the normal source position.

Fiber Optics

Optical fibers for the SPECTRx include two fibers of OH-free quartz. Each fiber is 2 m long, with SMA connectors at each end. The core size is 330 µm.

The Detector

An InAs or an MCT detector type is utilized depending on the wavelength range. The detector includes focusing mirrors, specially adapted preamplifiers, and detector housing. All thermoelectrically (TE) cooled detectors require 110 V or 220 V 50/60 Hz (user selectable) for operation.

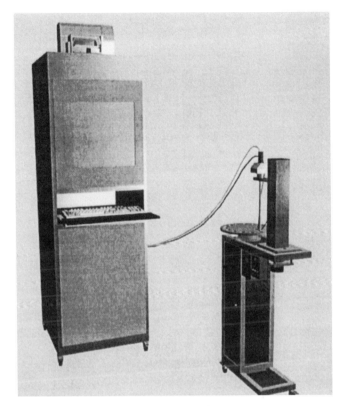

FIGURE 8.5 Courtesy of American SensoR̗.

The system can be supplied with an extended range of 1 mm diameter DTGS detector module, complete with matched preamplifier. The detector is optimized for inspection application of samples varying in length from 4 mm to 15 mm. It has excellent sensitivity and linearity for high-speed data analysis.

SPECTR̗ Software

The system is provided with an advanced FTIR computer program developed by American Senso. **The software runs on a Pentium, Windows 95 or NT-based PC.**

SPECTR̗ is a copyrighted software and contains a fully integrated data collection software permitting operation of most types of interferometers. The software is developed primarily for *online, real-time* production applications. It can also easily handle at-line production analysis. It uses the standard Windows 95 interface and is designed for ease of use. It contains the most advanced spectral analysis and plotting tools, all tailored for real-time production. These tools include:

- Partial least squares quantitative analysis
- Gradient descent nonlinear optimization for quantitative analysis
- Least squares subtraction for pure component presence/amount analysis

- Spectra database
- Database query using SQL language
- Database search
- Database sorting
- Online tablet hardness evaluation using Sabrie's index
- Neural networks for spectral analysis
- Fourier domain smoothing
- Curve fitting
- High-order polynomial baseline correction
- Three-dimensional graphics capabilities
- Rapid multispectral data acquisition
- Area calculations
- Production menu developed for supervisors
- Production menu developed for operator
- Synchronous online inspection-based on a standard
- Upper and lower control limit charts
- Upper and lower working control limit charts
- Audiovisual warning signals
- Automatic optimization of distance between probe and object
- Visual controls for a rotary indexing system

Data Acquisition

The data acquisition system is a single-board electronic module capable of acquiring data at a rate of 2,000,000 sample per second. The sensor is equipped with variable scan-length logic for variable resolution. It is developed with true 12-bit A/D (analog to digital converter) with built-in sample and hold. The data transmission from the electronic module to the computer is unidirectional, whereby the analyzer does not require any commands from the computer. The data applies byte-parallel data transmission; that is, the data is transmitted in byte parallel format. All scanned data is synchronous with all previous scans, which permits direct coadding of spectral data.

Dimensions and Specs

SPECTR requires 508×600 mm² of workbench space, excluding computer and peripherals. The analyzer dimensions are 500 mm wide by 560 mm deep by 190 mm high. It is supplied with either 115 or 230 V AC. The power requirement is approximately 75 watts. The operating temperature ranges from 10 to 40°C. The system will recover from overheat conditions to 50°C and cooling to −15°C.

PLUG-AND-PLAY

Complicated sensor applications may demand simplification of methods of operation. The plug-and-play techniques developed initially by the National Institute of

Standards and Technology may enable unskilled operators to implement sophisticated sensor technology. Each item tailored in the following procedure is connected to an input signal generated by a simple sensor indicating the presence or absence of the item under monitoring. The following procedure is an example of plug-and-play procedure:

1. Ensure that all the power cables are connected and the system is in the off position, input 1.
2. Connect the data acquisition cable to the computer system, input 2.
3. Connect the analyzer cable to the data acquisition board, input 3.
4. Plug the power cord of the indexing table, input 4.
5. Connect the launcher to the analyzer port, input 5.
6. Connect the fiber optics cable to the launcher adapter, input 6.
7. Connect the sensor or the microswitch terminals to the data acquisition board, input 7.
8. Connect the monitor cord to the computer, input 8.
9. Connect the keyboard cord to the computer, input 9.
10. Connect the electronic module to the analyzer.
11. Connect the cables of the electronic module to the detector switch on the analyzer and put it on pause until the system is climatized (red light will appear), input 10.
12. Connect the thermocouple to the data acquisition board, input 11. Eleven simple input sensors will provide an operator a distinctive capability to connect complicated systems without spending intensive time and cost on unnecessary training.

Power-Up

1. Depress analyzer button to run (green light will appear).
2. Switch on electronic module (green light will appear).
3. Switch on computer system (must ensure 300 or 900 MHz).
4. Switch the system synchronous production mode, if desired
5. Ensure all cable connections and inputs.

Start up The SPECTR̽

1. Start from Windows 95 menus or Windows NT.
2. The default values of the system are automatically set.

SPECTR̽ Alignment Procedure

1. Click on **Data Base Form**.
2. Go to the Query line.
3. Type *cat1='Reference'*—cat 1 is an abbreviation of category 1.
4. Click on **First**. This is an abbreviation of first data reference.

5. Find Spectralon Reference.
6. Click on **Set As Current Scan**.
7. Go to *Windows*.
8. Click mouse arrow on *Spectrum*.
9. Click on *Plot*.
10. Place Spectralon under the probe.
11. Press on **Auto Optimum Distance**. The system automatically calculates the optimum distance between the sensor probe and the object.
12. System stops automatically.
13. Click on *Scan-Plot* (use different scan color).
14. Turn the launcher cable left or right until the acquired spectrum is the same as the reference spectralon spectrum. You may also have to adjust the position of the launcher using the X-Y-Z micrometers on the launcher.
15. Tighten the launcher's set screws.
16. The system is calibrated.

SPECTR Reference

1. Place the reference material (spectralon) under the probe.
2. Press on *Auto Optimum Distance*.
3. Click on *Set Reference*.
4. Click on *Scan Reference*.

The reference is divided by each spectrum acquired to correct for day-to-day environment variations.

Optimum Object Distance

1. Put an object under the probe (current production tablet).
2. Choose *Windows* and click on **Spectrum**.
3. Click on *Auto Opt. Distance*, system will be automatically optimize the distance.
4. You may optimize the distance manually through the arrow **Up** or arrow **Down**, until you reach an optimum value.

Setting Standard Scan

1. Go to *Windows*.
2. Place object under probe within 4–6 mm.
3. Press *Auto. Optimum Distance*.
4. System stops automatically.
5. Click on *Set Standard*.
6. Click on *Scan Standard*.

7. System automatically acquires 20 scans and averages them.
8. Standard is set.

Setting Production Run

1. Click on *Set Scan Parameters*.
2. Choose the **number of scans**, from 0 to 1000. The default value is 1.
3. Choose type of smoothing: no smoothing, medium smoothing, or maximum smoothing
4. Choose **Base-Line Correction Type**: Type 1, Type 2, or None. Base Line Correction, if enabled, would make the system less sensitive to distance variation between probe and objects. Type 1 is used for solid hard objects such as tablets. Type 2 is mostly used for soft objects such as meats and gels.
5. **Absorb Start Point** and **End Point** can be set only by management-level authority. Default values are set at 255 to 595 which correspond to a spectral range of 4000 to 10,500 wave numbers.
6. The **gain** can also be set by management level authority. The default value is set at 1.
7. Check **Alternative Sabrie's Index*** box if you want to use this alternative definition, which, for some pharmaceutical tablet applications, gives a better measurement of hardness than the regular definition.

Setting Production Parameters

1. Production tolerance can be set between **0 and 60 percent**. This tolerance is the maximum allowable difference between the Sabrie's index of the object and the Sabrie's index of the standard.
2. Setting the deviation can be done by the operators. The deviation can and should be set between **0.001 to 0.1** depending on the application and the level of noise. This deviation describes the allowable variation between the objects and the standard.

Calibrating Sabrie's Index for Hardness Monitor

1. Place an object of a known hardness under the probe and record **Sabrie's index**.
2. Enter **Sabrie's index** in the first box.
3. Enter its known hardness at its the corresponding box.
4. Place another object of a known hardness under the probe and record also its Sabrie's index.
5. Enter **Sabrie's index** in the second box.
6. Enter its known hardness at its corresponding box.

---

* Discovered by Dr. Sabrie Soloman, Pat. No. US 5,679,954.

7. If the check is marked in the calibration box, the data will be displayed automatically. If the box is not marked, the data will not be displayed at the production screen.

Option for Synchronous Production

1. Synchronous production can be triggered through three types of triggering mechanisms: **analog level** triggering, **analog edge** triggering, and **digital edge** triggering. Choose one of the above triggering mechanisms. The trigger mechanism is dependent on the type of sensor or microswitch.
2. The voltage setting must not exceed 5.0. It can vary between 0 and 5 volts DC.

Applications

1. Go to the menu screen and click on Commands → Set Application Parameters. Enter a check mark in the desired operation:
   - Pharmaceutical tablets
   - Nonpharmaceutical tablets
   - Pharmaceutical liquid
   - Nonpharmaceutical liquid
   - Beef, pork, and poultry
   - Soil fertilizer

Saving Current Scan Plus Settings

1. Go to the menu screen and click on Commands → Save Scans + Settings.
2. The Save Scan screen will appear.
3. Choose a directory and a file name for your scan (SABRIE.scan).
4. Click on Save.

Load Scan Plus Settings

1. Go to the menu screen and click on Commands → Load Scans + Settings.
2. The Load Scan screen will appear.
3. Choose a file name that was previously saved (SABRIE.scan).
4. Click on Open.

Setting User Passwords

1. Go to the menu screen and click on Commands → User and Passwords.
2. Users Passwords will appear.
3. Click on Add New User.

4. Fill in user name.
5. Fill in login name.
6. Mark the authority level of the user: manager, supervisor, or operator.
7. Click on Done.

Changing Internal Parameters

1. The system allows only managers to make any changes in this screen.
2. Go to the menu and click on Commands→ Internal Parameters.
3. Enter the number of the point left of the center burst trigger. The default is set at 1024.
4. Enter the number of the point right of center burst trigger, default is set at 1024.
5. The number of offset points is used as a delay function after the start of the scan signal.
6. The number of sample points is taken from the left and the right of the center burst.
7. The maximum scan rate is 2 million sample points/sec.

The interferogram (Fig. 8.6) of an object represents the raw data coming out of the detector. The interferogram is transformed using the FFT algorithm along with some other transformations into the spectrum. Normally, one needs not to see the interferogram. However, the interferogram is used primarily to check the stability

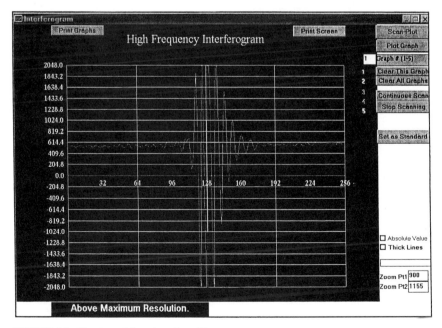

FIGURE 8.6 Courtesy of American SensoRx.

and noise level of the signal as well as for debugging purposes. The object signature that is resident in the interferogram can be manipulated and displayed in a spectrum form. The spectrum has greater advantage to provide meaningful data regarding the characteristics of the object.

The following commands also provide different functions of the sensor.

SPECTR̥ Sensor Commands

The system commands represent different functions:

Scan-Plot. This command will first scan the object under the probe and then plot the interferogram on the screen. The operator can enter the graph number in the text box next to "Graph # (1–5)." See Fig. 8.7.

Plot. This command will plot the interferogram of the last object scanned, which is referred to as "the current scan". This command does not read a new interferogram from the spectrometer; it only displays the last interferogram acquired.

Graph # (1–5). The operator can enter in the adjacent box the number of the graph that is desired to plot or clear.

Clear This Graph. Operator may clear the graph selected in the "Graph # (1–5)" box.

Clear All Graphs. The operator may clear all the graphs.

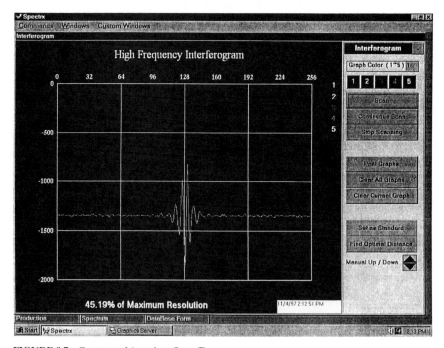

FIGURE 8.7 Courtesy of American SensoR̥.

Continuous Scan. A continuous scanning-and-plotting interferogram of the object under the probe can be achieved.

Stop Scanning. This command may stop the continuous scanning.

Set as Standard. Operator may set the last scan as the standard for production. This function also enables the operator to choose a known good product to be scanned as a gold standard* for objects. The incoming object will always be compared against the gold standard.

Print Screen. The data displayed in the spectrum or the interferogram screen may be sent to the the printer for printing.

Print Graph. The data displayed regarding the spectrum or the interferogram—only without the remainder of the data displayed on the screen—may be sent to the printer for printing.

Text Boxes

Zoom points on the left and the right limit of the spectrum may be selected for plotting. The default internal parameter setting for the SPECTR™ uses 2048 points on the interferogram. The center burst of the interferogram is at point number 1024. Operators may zoom on the center burst, for instance, then select point number 800 and point number 1200 as the left and right limit points.

Data Storage

There are two types of system parameters that control how the data is collected from the interferometer then displayed and stored in the database for analysis. The two types are: scan parameters and internal parameters. All the scan parameters, except the gain, affect only how the spectrum is displayed on the spectrum window. Therefore, the spectrum is always stored in its raw form with no baseline correction or smoothing. The internal parameters, on the other hand, affect how the spectrum is acquired and stored.

SCAN PARAMETERS

The Scan Parameters dialog box is displayed by selecting Commands → Set Scan Parameters from the screen pulldown menu. The box shown in Fig. 8.8 will appear. The following parameters can be set:

- *Number of scans.* Sets the number of scans to be averaged.
- *Smoothing.* There are three levels of smoothing: no smoothing, medium smoothing, and maximum smoothing.

---

* Terminology developed by the FDA.

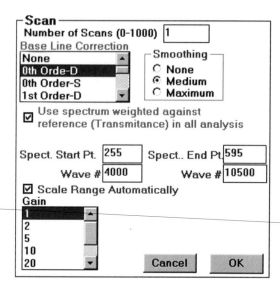

FIGURE 8.8 Scan parameters. *(Courtesy of American SensoRx, Inc.)*

- *Baseline correction.* The following types of baseline corrections can be used: none, nth order-**D**, nth order-**S**, and nth order-**SN**, where $n(=0,1,2,3$ or $4)$ is the degree of the polynomial used as the baseline. In the **S** baseline type, the baseline is subtracted from the spectrum. In the **D** baseline type, the baseline is divided by the spectrum. In the **SN** baseline type, the baseline is subtracted from the spectrum, and the resulting spectrum is normalized.
- *Use spectrum weighted against reference (transmittance) in all analysis.* This box in checked by default. When it is checked, the transmittance (spectrum divided by the spectrum of the reference) is used for the quantitative, neural network, Sabrie's index, and deviation calculations. If this box is not checked, then the unweighted (unreferenced) spectrum is used in the calculations.
- *Spectrum start and end points and corresponding wave numbers (cm^{-1}).* You can set the spectral range for plotting the spectrum, calculating the baseline, and performing Sabrie's index and deviation calculations using this option. The default spectral range is from 4000 cm^{-1} (2.5 µm) to 10,500 cm^{-1} (0.952 µm). The range depends on the spectrometer settings (fiber optics, detector, and infrared source).
- *Scale range automatically.* When this box is checked, the spectrum start and end points are automatically scaled using the following setting: spectrum point 255 is scaled to 4000 cm^{-1}, and spectrum point 595 is scaled to 10,500 cm^{-1}. If this box is not checked, then you can manually calibrate the spectrum start and end points and corresponding wave numbers.
- *Gain.* Sets the gain level for the detector input signal. You should select the gain level such as to obtain a resolution between 75 and 99 percent of the maximum resolution. The gain level depends on the object you are analyzing as well as the particular type and setting of the spectrometer. In general, the gain for pharmaceutical tablets is either 1 or 2, while the gain for meats and soft tissues is either 10 or 20.

INTERNAL PARAMETERS

The internal parameters dialog box is displayed by selecting Commands → Internal Parameters from the pulldown menu. The box shown in Fig. 8.9 will appear. This option is only available for users with manager access privileges. The following parameters that define how the interferogram is acquired from the spectrometer, can be set:

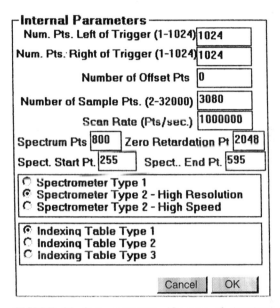

FIGURE 8.9 Scan parameters. *(Courtesy of American SensoR$_x$, Inc.)*

- *Number of points to the left and right of the trigger point.* Number of points on the interferogram to the left and right of the center burst.
- *Number of offset points.* Number offset interferogram points from the center burst.
- *Number of sample points.* Total number of points sampled per interferogram.
- *Scan rate (sample points/sec).* Rate of the scan.
- *Number of spectrum points.* Number of points to store from the spectrum.
- *Zero retardation point on the interferogram.* Location of the zero retardation point on the interferogram.
- *Spectrum start and end points.* Left and right limit points for storing the spectrum to the database.
- *Spectrometer type.* The software supports the following types of spectrometers:
  Type 1—High-speed low resolution
  Type 2—Low-speed high resolution
  Type 2—High-speed high resolution

- *Indexing table type.* The software supports the following types of indexing tables:

  Type 1
  Type 2
  Type 3

THE SPECTRUM

The spectrum (Fig. 8.10) of an object represents a fingerprint of the chemical composition and physical characteristics (such as hardness, dissolution, surface roughness, grain size, and so on) of the object. The spectrum can be plotted in many different ways. The following plot types are available:

- Raw spectrum
- Spectrum
- Log spectrum
- Transmittance
- Absorbance
- Reverse raw spectrum
- First derivative of the spectrum

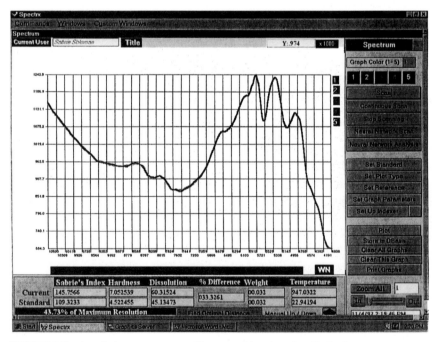

FIGURE 8.10 A typical spectrum screen. *(Courtesy of American SensoRx, Inc.)*

- First derivative of the log spectrum
- First derivative of the transmittance
- First derivative of absorbance

The default plot type is transmittance. In that type, the spectrum is divided by the spectrum of a reference material for the purpose of canceling environment and instrument variations.

Function Commands

Scan-Plot. This command scans the object under the probe and plots the spectrum on the screen. You can enter the graph number in the text box next to Graph # (1–5).

Plot. This command plots the spectrum of the last object scanned, which is referred to as the current scan.

Graph # (1–5). The operator may enter in the adjacent box the number of the graph desired to plot or clear.

Clear This Graph. This clears the graph selected in the Graph # (1–5) box.

Clear All Graphs. This clears all the graphs.

Continuous Scan. This command continuously scans and plots the spectrum of the object under the probe.

Stop Scanning. This stops the continuous scanning.

Store in Dbase. This command stores the current scan in the database and switches to the database screen. The date/time, ambient temperature, and scan title (which you can enter in the title box in the spectrum screen) are automatically entered in the database. The operator should enter the name of the product and the categories (classifications) of the scan in the database screen. The spectrum is always stored in its raw form in the database (i.e., with no baseline correction or smoothing).

Run NN-Analysis. This command runs the neural network analysis on the current scan.

Run NN-Scan. This command also scans the object under the probe, displays its spectrum on the screen, and runs the neural network analysis.

Print Screen. The information displayed on the screen is sent to the printer.

Copy Graph. This function allows the operator to copy the spectrum graph to the Windows clipboard. The operator may insert the graph in another Windows application such as a word processor or a spreadsheet.

Set Standard. Displays the standard dialog box (Fig. 8.11). The standard spectrum is compared with the spectrum of objects in production. The deviation of an object from the standard measures the consistency of the object. There are two buttons in that dialog box:

FIGURE 8.11 Standard dialog box. *(Courtesy of American SensoR_x, Inc.)*

- *Scan standard.* Performs the number of scans specified in the number of scans box, averages them, plots the resulting spectrum, and sets the resulting spectrum as the standard spectrum for production.

- *Set current scan as the standard.* Sets the last scan as the standard for production.

Set Reference. Figure 8.12 displays the reference dialog box. The reference spectrum is divided by the spectrum of an object that yields the transmittance spectrum. The reference object is made out of a standard material. The common reference material in infrared spectroscopy is spectralon, which has a flat IR spectrum. The purpose of using the transmittance spectrum is to cancel the environment and instrument variations. There are three buttons in that dialog box:

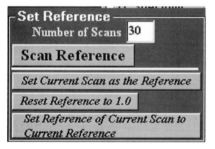

FIGURE 8.12 Reference dialog box. *(Courtesy of American SensoR_x, Inc.)*

- *Scan reference.* Performs the number of scans specified in the number of scans box averages them, plots the resulting spectrum, and sets the resulting spectrum as the reference spectrum.

- *Set current scan as the reference.* Sets the last scan as the reference spectrum.

- *Set reference of current scan as reference.*

Zoom All. This command sets the screen display to the original default view (zoom factor 1).

Zoom In / Zoom Out. This command also increases or decreases the magnification of the magnification of the spectrum. The maximum zoom factor is 5.9 and the minimum zoom factor is 1.

Pan Slider. This command pans the spectrum graph to the left and right.

Write to File. This function enables the operator to write the data of the graphs displayed on the spectrum screen to a file.

Auto. Opt. Distance. The distance between the probe and the object is controlled by a servo motor. This command automatically optimizes the distance between the probe and the object. Keep the mouse on the Stop button to stop the probe in case it comes too close to the object. If the probe hits the object, it may scratch the probe's lens. The control algorithm of the probe has been carefully tuned over a wide variety of objects to prevent the probe from hitting the object. However, to be absolutely safe, you should be ready to press the Stop button.

Manual Up/Down ↑↓. This command manually increases or decreases the distance between the probe and the object. Be extremely careful when lowering the probe in order not to hit the object. If the probe hits the object, the lens of the probe may be scratched, or it may even break.

Speed Slider. This command also controls the speed of the servo motor for the manual control of the distance between the probe and the object.

Option Boxes
  Grid. This check command enables operator to plot the a horizontal and vertical grid on the graph for mathematical comparison.
  Thick Lines. This check command enables operator to plot the spectrum using thick lines for greater display. Displaying data may increase the time for analysis.

Types of Spectra

The raw data plotted between the wave numbers and the energy can be manipulated to obtain meaningful characteristics of objects absorbing or reflecting energy.

1. *Raw spectrum.* The raw spectrum is directly obtained from the interferogram.
2. *Reverse raw spectrum.*
3. *Spectrum.* The reflectance characteristics of the object reflecting energy.
4. *D_Spectrum.* The first derivative of the reflectance spectrum.
5. *Log spectrum.* The log of the reflectance spectrum is calculated using the following formula:

$$L = \log(S)$$

  where S is the reflectance spectrum of the object.
6. *D_Log spectrum.* The first derivative of the log reflectance spectrum.
7. *Transmittance.* The transmittance spectrum is calculated using the following formula:

$$T = S/R$$

  where R is the reference spectrum and T is the transmittance spectrum.
8. *D_Transmittance.* The first derivative of the transmittance spectrum.
9. *Absorbency.* The absorption spectrum is calculated using the following formula:

$$A = -\log(S/R)$$

  where A is the absorption spectrum.
10. *D_Absorbency.* It is the first derivative of the absorbency spectrum.

3D Spectrum Screen

The 3D spectrum screen (Fig. 8.13) contains the same commands as the spectrum screen. The 3D graph can also be animated by using the Animate function.

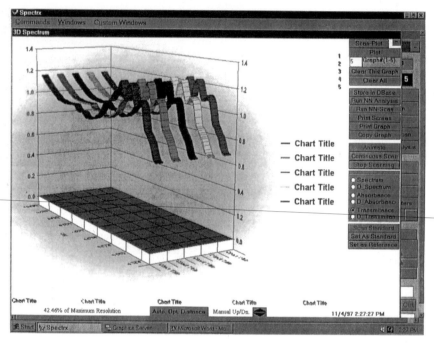

FIGURE 8.13 A typical 3D spectrum screen. *(Courtesy of American SensoRx, Inc.)*

PRODUCTION

The most important function of this sensor is its ability to control and monitor productivity. It is designed with manufacturing agility to enable line operators to program the system without extensive training or previous knowledge of computer language. See Fig. 8.14.

Production Function Commands

Synchronize Function. It runs production in the synchronous production mode. The objects passing under the sensor probe are gated using either a synchronization signal from the light sensor or a signal from the encoder, which is mounted on the axis of the indexing table. The synchronization options can be set from the indexing system dialog box by pressing the indexing system button.

Continuous. This function runs production in the continuous production mode. The spectra are plotted as soon as they are sent by the spectrometer.

Increment. This function runs production for the number of objects specified in the box next to the button and then pauses production.

Pause. Pauses production.

Resume. Resumes production.

SPECTRx: AN ONLINE PRODUCTION ANALYTICAL SENSOR 8.23

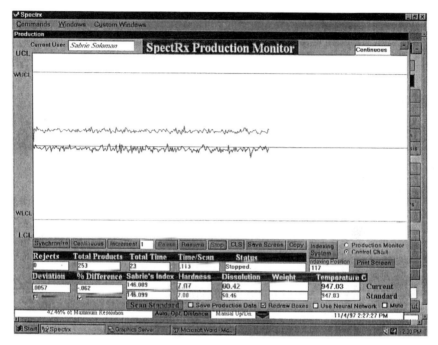

FIGURE 8.14 Production screen. *(Courtesy of American SensoRx, Inc.)*

Stop. Stops production.

CLS. Clears the production screen.

Save Screen. Save the production screen to a file.

Copy. Copies the production screen to the clipboard.

Print Screen. Prints the production screen.

Scan Standard. Performs 20 scans, averages them, and sets the resulting spectrum as the standard production spectrum.

Production Monitor/Control Chart. Toggles between plotting a control chart and a production ball, which gives a quicker indication on the status of production.

Indexing System. The indexing function displays the dialog box for the indexing system, Fig. 8.15. This dialog box is divided into four main areas:

- *Synchronous production control.* Controls the type of gating sensor for production objects. Either a light sensor or an encoder on an indexing machine can be used. If an encoder is used, then the angular step between objects must be specified in the text box next to the encoder. An 8-bit circular encoder is mounted on the shaft of the indexing system with 256 angular divisions per revolution.

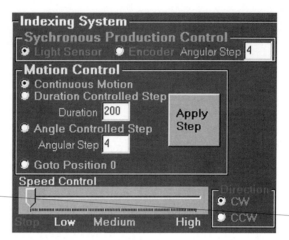

FIGURE 8.15 Indexing system dialog box. *(Courtesy of American SensoRx, Inc.)*

- *Motion control.* Controls the motion of the indexing system. There are four modes of operation of the indexing system:

  Continuous motion. The indexing system rotates continuously.

  Duration controlled step. Supply the control current to the indexing system driver motor for the specified duration in milliseconds.

  Angle controlled step. Steps the indexing system the specified angular step.

  Goto position 0. Moves the indexing system to the 0 position of the encoder.

- *Speed control.* Controls angular speed of the indexing system.
- *Direction.* Controls the direction of the motion of the indexing system: Clockwise or Counter-clock-wise.

Save Production Data. This function enables the operator to save the production data to a file. The operator will be prompted for the file name when starting production by using either the continuous or synchronous modes.

Display Properties. The operator may desire to display several properties simultaneously, such as hardness, disintegration, dissolution, weight, active ingredient, temperature, Sabrie's index, and moisture content. Every time a new spectrum is acquired, all properties will be displayed. Displaying all the data may slow down the computers. This may use up a few valuable microseconds. However, acquiring one spectrum does not inhibit measuring all properties. It is only when requiring to display the data it might take many additional microseconds.

Rejects. This function displays the total number of rejects since the start of production.

Total Time. This function displays the total time in seconds since the start of production.

Time Per Scan. This function displays the approximate time per scan in seconds.

Status. This function displays the status message.

Deviation. This function displays the absolute difference between the spectrum of the standard and current object.

Sabrie's Index. This function displays the value of the Sabrie's index, which is proportional to the hardness, compactness, process of blending, and several other characteristics of pharmaceutical tablets.

Hardness. This function displays the hardness value of pharmaceutical tablets. This value is calibrated using the Sabrie's index from the production parameters dialog box, which can be accessed using Commands → Set Production Parameters.

Dissolution. This function displays the dissolution value of pharmaceutical tablets. This value is calibrated using the Sabrie's index from the Production Parameters dialog box, which can be accessed using Commands → Set Production Parameters.

Percent Difference. This function displays the difference between the Sabrie's index of the standard and the Sabrie's index of the current object.

Temperature. This function displays the ambient temperature for the last scan.

Production Parameters

See Fig. 8.16 for the production parameters dialog box.

- *Production tolerance.* Sets the tolerance in percent for the allowable difference between the Sabrie's index of the standard and the Sabrie's index of the current object. The default value is 5 percent.

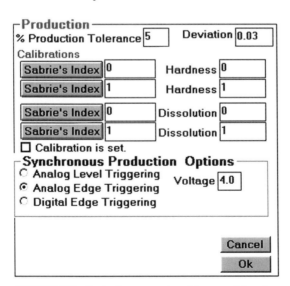

FIGURE 8.16 Production parameters. *(Courtesy of American SensoRx.)*

- *Deviation.* Sets the tolerance on the absolute difference between the spectrum of the standard and the spectrum of the current object. The default value for this tolerance is 0.03. Note that the value for this tolerance depends on the type of baseline correction.
- *Hardness/dissolution calibration.* A linear calibration between the Sabrie's index and the hardness/dissolution of the object can be set using two objects of known hardness/dissolution.
- *Synchronous production options.* Sets the parameters of the gating light sensor.

DATABASE

A typical database screen is shown in Fig. 8.17. The database function consists of the following:

- Navigation between database records.
- Database query using the *query line* and *execute query* functions
- Database sorting
- Commands that are performed on all records satisfying the current query criteria
- Text boxes displaying the content of the fields of the current record
- Table displaying a page of records

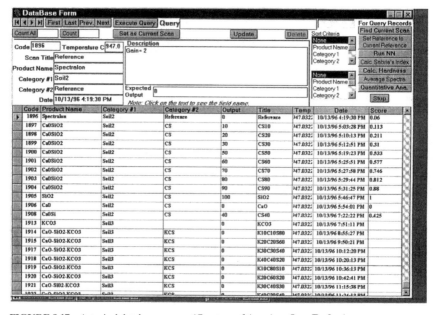

FIGURE 8.17 A typical database screen. *(Courtesy of American SensoRx, Inc.)*

Database Fields

The records stored in the database represent a powerful tool in maintaining the history of single or multiple products, where each product may also contain single or multiple compounds. At any point of time, a product may be recalled from the database for comparison with a similar newly made product. This may verify the compatibility of a newly made product with an older batch. Each record in the spectra database may be stored in the following fields:

- *Code (field name:* "*code*"*).* The code of the record is a unique integer number for each record which uniquely identifies the record from other records. This code is automatically generated for each record as it is added to the database.
- *Product name (field name:* "*name,*" *length = 40 bytes).*
- *Category 1 (field name:* "*cat1,*" *length = 40 bytes).* First classification category of the record. The categories are used to perform queries and identify families of records.
- *Category 2 (field name:* "*cat2,*" *length = 40 bytes).* Second classification category of the record.
- *Output 1 (field name:* "*out,*" *length = 40 bytes).* Second classification category of the record.
- *Title (field name:* "*title,*" *length = 40 bytes).* Title of the scan. The title is an alphanumeric string that should uniquely identify the scan.
- *Temperature (field name:* "*temp*"*).* Ambient temperature when the scan was taken.
- *Date (field name:* "*date*"*).* Date and time when the scan was taken.
- *Score (field name:* "*score*"*).* The score field stores a numeric value that is calculated when one of the following buttons is pressed: Find current scan, "Run NN, Calc. Sabrie's Index, Calc. Hardness, and Quantitative Analysis.
- *Description (field name:* "*desc*"*).* In this field, you can enter a long description or remarks for the database record.
- *Spectrum.* This field stores spectrum. It is displayed when you set the database record as the current scan and then plot the spectrum in the spectrum screen.

Finding the Current Scan. The system can search the database tables and sort the database records according to an ascending order to identify unknown objects by comparing the newly acquired scan with any data stored at an earlier time.

QUANTITATIVE ANALYSIS

In quantitative spectral analysis, we are given a mixture consisting of number of pure components. It is required to find the percent weight of each component in the mixture. The problem can be stated using the following equation:

$$S = w_1 S_1 + w_2 S_2 + w_3 S_3 + \cdots + w_n S_n$$

where S is the spectrum of the mixture, S_i is the spectra of component number i, and w_i is the volume ratio of component i in the mixture. We are given S and S_i, and it is required to find w_i.

A typical quantitative analysis screen is shown in Fig. 8.18. The user enters the database code of the pure components in column B of the table. Then the type of baseline correction and the spectral range over which the quantitative is performed (in cm^{-1}), is selected. There are two problems that can be solved in the quantitative analysis screen:

- *The forward problem.* Find the resulting spectrum of a mixture of pure components given the volume ratios of the pure components.
- *The inverse problem.* Find the volume ratios of pure components in a mixture given the spectrum of the pure components and the spectrum of the mixture.

Adding Spectra

The system can find the resulting spectrum of a mixture of pure components given the volume ratios of the pure components.

Quantitative Analysis Parameters

Baseline Correction. The same type of baseline correction described previously may be used for the quantitative analysis. The final result of the quantitative analysis depends on the type of baseline correction used. In general, it is recommended to use one S type baseline correction along with a consistent distance between the probe and the object. A consistent distance can be obtained by optimizing the dis-

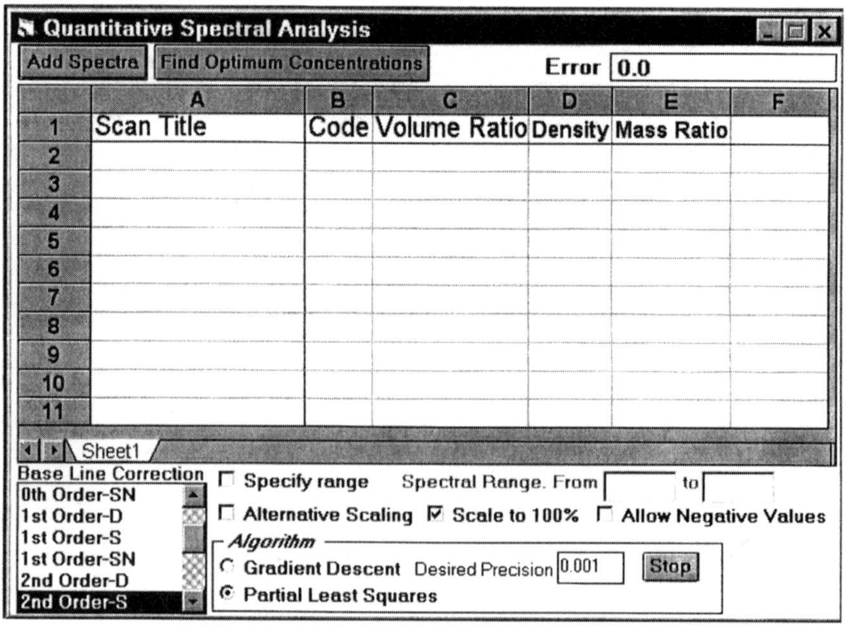

FIGURE 8.18 Typical quantitative analysis screen. *(Courtesy of American SensoRx, Inc.)*

tance between the probe and the object for the pure components as well as the mixture in case the height of the objects varies. The default type of baseline correction is second-order S.

The spectral range over which the quantitative analysis is performed must be specified. In some materials (such as soft materials), a side of the spectrum may be nonlinearly distorted. Thus, limiting the spectral range of the quantitative analysis may result in a more accurate quantitative analysis. If this box is not checked, then the default spectral range defined in the "internal parameters dialog box" is used (default is from 4000 cm^{-1} to 10,500 cm^{-1}).

Use of Quantitative Spectral Analysis (Optimization of Weights)

The objective of the quantitative spectral analysis for a mixture consisting of a number of components is to find the exact amount of each component in the mixture.

Step 1. From the menu, select *Windows → Spectrum*. Scan each component and store the spectrum in the database. Fill the database record with the name of the product and the search categories. You may ignore the rest of the fields. If the spectrum of a component is already stored in the database, skip Step 1 for this component. Write down or remember the database code of each component as you add it to the database. This code is automatically generated. The code field is on the left of the database form. The database code is needed in Step 2.

Step 2. From the menu, select *Windows → Quantitative Spectral Analysis*. In the column of *Dbase Code,* fill the codes of the components of the unknown mixture. As you type the code of a component, its name will appear in the column of *Product Name*. You can find the code of a component by looking in the database table. Select *Windows → Data Base → Data Base Table* and look in the table for the code of the component.

Step 3. From the menu select *Windows → Spectrum*. Scan the unknown mixture by pressing **Scan-Plot**. The spectrum of the mixture is now set as the current scan.

Step 4. From the menu, select *Windows → Quantitative Spectral Analysis*. Press **Find Optimum Weights**. The program will find the percent weight of each substance in the mixture that gives the closest possible fit to the spectrum of the mixture.

If you wish to plot the calculated spectrum to compare with the actual spectrum of the mixture, click on **Add Spectra**. Then, select *Windows → Spectrum* and press **Plot Graph**.

Step 5. If you wish to repeat the analysis for another mixture consisting of the same components, go back to Step 3.

NEURAL NETWORKS

The neural network can be considered as a *transformation function,* as it maps some input into a useful output. In case of spectroscopy, the input is the spectrum of an object, and the output is one or more characteristics of that object that we are interested in. The transformation function of the neural network is obtained by training the network using a large number of objects for which both the spectrum (input) and the characteristics (output) are given. Neural networks consist of processing elements (Pes or neurons) and weighted connections. Figure 8.19 shows a typical neural network. Neural networks are very useful tools in spectroscopy for the following reasons:

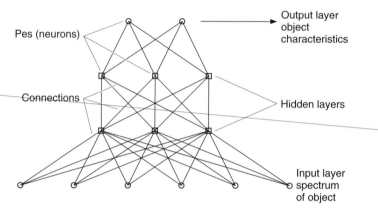

FIGURE 8.19 A typical neural network. *(Courtesy of American SensoRx, Inc.)*

- The number of points in the spectrum is usually large (>300), and, at the same time, the number of characteristics is relatively small (<4).
- The mapping between the spectrum and the desired characteristics is usually not simple and is sometimes even nonlinear.
- Since the present system is designed to be used in real-time productions, decisions must be made in a very short time. Neural networks are among the fastest available spectroscopy analysis tools.
- Although preparing a training set for the network might take a long time and effort, it is usually less effort than to try to derive the mapping, come up with an algorithm for the mapping and write a specialized computer code for the algorithm.

To perform spectral analysis using neural networks, the following steps are required:

1. Create a training set.
2. Generate a neural network.
3. Train the neural network using the training set.
4. Use the network to find the characteristics of a sample.

The training set spectra must be representative of the actual spectra that the neural network will use to analyze and identify. For instance, if the network is going to be used with spectra that consist only of one scan, than the neural network should be trained with spectra consisting of one scan. In order for the network to operate properly, most of the variations in the spectra of the object must be present in the training set. This usually requires at least 6^n spectra in the training set, where n is the number of output characteristics.

Manual Neural Network Generation Screen

The screen in Fig. 8.20 is used to set the parameters and generate a neural network. The following parameters can be set:

- Number of output points
- Spectral range for the neural network analysis
- Network type
- Baseline correction
- Number of hidden layers
- Output names and lower and upper control limits for the output
- Number of neurons in each hidden layer

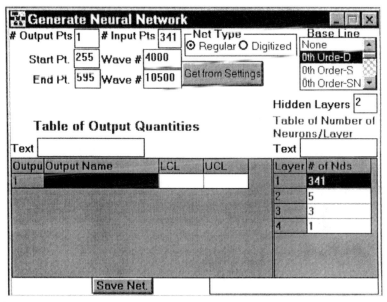

FIGURE 8.20 A typical screen to generate the neural network. *(Courtesy of American SensoRx, Inc.)*

Finish entering the parameters of the neural network. You can generate the neural network by clicking on the **Save Net** button.

Train Manual Neural Screen

The screen in Fig. 8.21 is used to train the neural network. The following commands and options are available:

- **Load net.**   Loads a neural network
- **Save net.**   Saves the neural network.

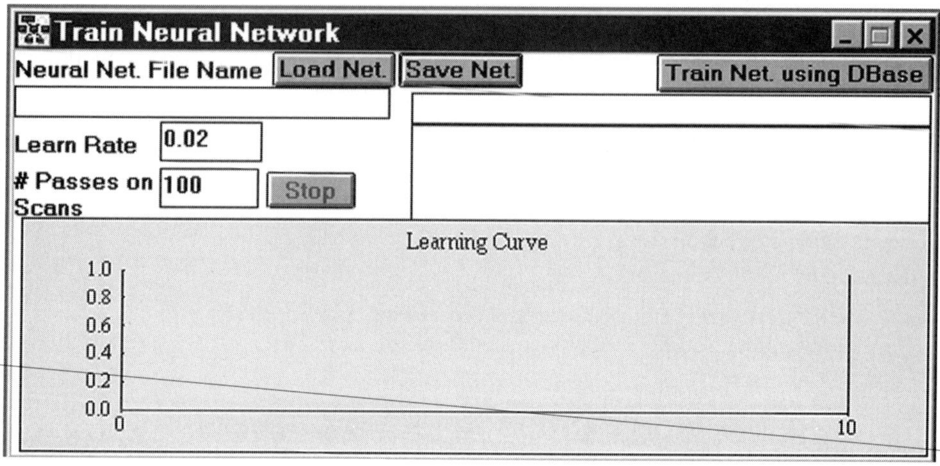

FIGURE 8.21 A typical "train neural network" screen.

- **Train net using database.**
- **Lean rates.** Specify the learning rate (recommended range 0.01 to 0.05).
- **Number of Passes.** Number of training passes on the training set.

Run Network Generation Screen

The screen in Fig. 8.22 is used to display the results and run a neural network analysis. The output of the network is displayed in the table of output quantities in the Value column.

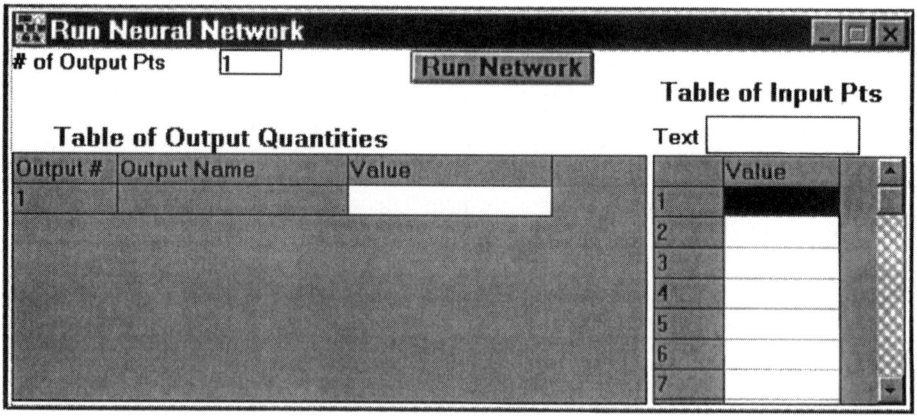

FIGURE 8.22 A "run neural network" screen.

Use of Neural Network Analysis

The objective is to determine some physical characteristics of a substance from its spectrum given the spectra of a set of similar substances and the values of these characteristics for these substances.

Step 1: Create the Training Set

- Select *Windows* → *DataBase* → *DataBase Defaults Form*. Fill the database defaults form with the default values of the database fields (e.g., Scan Title: "CH1," Product Name "Cogentine," Category #1: "Hardness Analysis," Category #2: "Dissolution Analysis," Scan Description: "Analysis of hardness and dissolution of Cogentine tables").
- From the menu, select *Windows* → *Spectrum*. Scan a member of the set by pressing Scan-Plot. Store the spectrum in the database by pressing Store in Dbase. Fill the database record with the scan title, the product name, the search categories, and output. The output field consists of the values of the characteristics separated by spaces. Repeat this step for each member of the set.

Step 2: Generate a Neural Network

- From the menu, select *Windows* → *Neural Network* → *Generate Automatically*. Note that you may want to use the "Manual Neural Network Generation Screen" in order to select baseline correction type and/or spectral range for the neural network.
- You will be prompted to enter the number of characteristics (the default number is 1).
- Next you will enter the name (e.g., hardness, dissolution) and the lower and upper control limits of each characteristic. The control limits of the characteristics are used in the production screen.
- Finally, you will be prompted to enter the file name under which you want to save the neural network (e.g., anyname).

Step 3: Train the Neural Network Using the Training Set

- From the menu, select Windows → DataBase Windows → DataBase Form. In order to specify the members of the training set, you need to put a search criteria that all the members of the set satisfy. For example, if you want to select all the records with Category #1: Hardness Analysis, type in the database criteria line at the button of the database form: "cat1 = 'Hardness Analysis'." In order to verify that you have selected all the members of the training set, press Count and make sure that the total number of members of the set is correct. You can also check each individual record by using the First, Next, Prev., and Last buttons.
- From the menu, select Windows → Neural Network → Train.
- Press **Train Net. Using Dbase**. You will see the learning curve rising as the algorithm iterates over the training set. Once the maximum learning error and the average learning error reach a satisfactory value, press Stop.
- Press **Save Net** to save the trained neural network. You will be prompted to enter the desired file name.

Step 4: Using the Network to Find the Characteristics of a Sample. From the menu, select *Windows* → *Spectrum*. Press **Run NN-Scan**. The values of the characteristics should appear in the *table of output quantities*.

The accuracy of the neural network depends on the accuracy of the training set. Even if one spectrum of the training set is wrong (e.g., the sample was not correctly placed under the probe, the spectrum does not correspond to the specified output, the noise level in the spectrum is too high, or a wrong object is scanned), then the output of the network will be not be correct. Therefore, it is important to check each spectrum visually for accuracy, consistency, and noise.

In order for the neural network to be able to generalize from the mapping that it learns during the training set, the network must be stopped from learning once the average and maximum learning errors reach acceptable values. If the network is allowed to learn until the maximum and average learning errors reach arbitrarily small values, then it will not be able to generalize beyond the training set.

CHAPTER 9
COMMUNICATIONS

INTRODUCTION

The signals supplied by the sensors enable the processing subsystems to construct a true picture of the process at a specific point in time. In this context, reference is often made to a *process diagram*. The process diagram is necessary to logically connect individual signals, to delay them, or to store them. The process data processing can be used in the broadest sense for the processing of a conventional control system, even a simple one. The task of the processing function unit is to connect the data supplied by the sensors as input signals, in accordance with the task description, and pass data as output signals on to the actuators. The signals generated are mainly binary signals.

SINGLE-BOARD COMPUTER

A single-board computer system is illustrated in Fig. 9.1, with input and output units which are suitable for the solution of control problems. The central processing unit is a microprocessor that coordinates the control of all sequences within the system in accordance with the program stored in the program memory.

A read-only memory (ROM) does not lose its contents even if the line voltage is removed. In the system in Fig. 9.1, the ROM contains a program which allows data to be entered via the keyboard and displayed on the screen. This program is a kind of minioperating system; it is called a *monitor program* or simply a *monitor*.

A RAM loses its contents when the voltage is removed. It is provided to receive programs which are newly developed and intended to solve a specific control task.

The input/output units enable the user to enter data, particularly programs, through the keyboard into the RAM and to monitor the inputs.

Binary signals generated by the sensing process can be read in through the input/output units, generating an actuating process as output response. Registers of this type, which are especially suitable for signal input and output, are also known as *ports*.

The system input/output units are connected by a bus system, through which they can communicate with one another. *Bus* is the designation for a system of lines to which several system components are connected in parallel. However, only two connected units are able to communicate with one another at any time (Fig. 9.1).

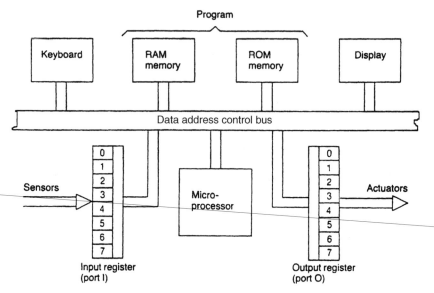

FIGURE 9.1 Single-board computer.

SENSORS FOR INPUT CONTROL

A microcomputer is used for the dual-level input control of a reservoir (Fig. 9.2). The level of the liquid is kept constant within the specified limits. For this purpose, two sensors are placed at the desired maximum and minimum locations:

$$\text{Maximum sensing level} = \text{maximum limit}$$

$$\text{Minimum sensing level} = \text{minimum limit}$$

Maximum and minimum sensors are installed to recognize the upper and lower limits. The motor is switched ON when the minimum level has been reached:

$$\text{Max} = 0$$

$$\text{Min} = 0$$

and OFF when the maximum level has been reached:

$$\text{Max} = 1$$

$$\text{Min} = 1$$

If the level is between the two levels

$$\text{Max} = 0$$

$$\text{Min} = 1$$

the motor maintains current status.

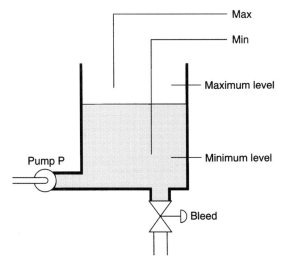

FIGURE 9.2 Dual-level control.

The constant level is disturbed by varying the bleed-off valve. These requirements are shown in a *fact table*, where P represents a current status and PN represents the required pump status (Table 9.1).

The program flowchart in Fig. 9.3 can be used as a programming model. The ports to connect the sensors and actuators are specified in Fig. 9.4.

The programming model is converted into computer program statements. For this, it is necessary for the statement commands to be formulated in machine code. In turn, this makes it necessary to understand the type of microprocessor being used. The well known Z80 type is used in the example in Table 9.2. The program appears in the abbreviated hexadecimal form.

The hexadecimal counting system is based on the number 16. Like the decimal system, the hexadecimal system uses the numerics 0 to 9 and, in addition, the letters A to F, which represent the numbers 10 to 15. Addresses and commands are entered on a keyboard which comprises keys for the hexadecimal alphanumerics 0 to F and various special keys. An address that has been set and the data it contains can be read from a six-digit numerical display.

Once the program has been developed and tested, there is no further need to use keyboard and display. Thus, these components can now be disconnected from the system. They do not need to be reconnected until program editing becomes necessary.

TABLE 9.1 Fact Table

Max	Min	P	Port	PN
0	0	0	I	1
0	0	1	I	1
0	1	0	I	0
0	1	1	I	1
1	1	0	I	0
1	1	1	I	0

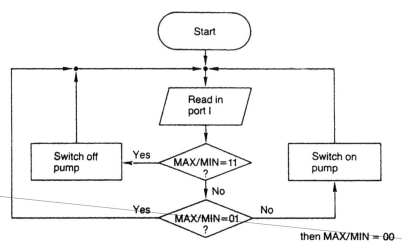

FIGURE 9.3 Program flowchart.

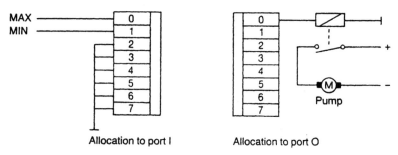

FIGURE 9.4 Ports to connect sensors and actuators.

It always is a good practice to transfer a completed program from a RAM to a memory whose contents are retained even when the supply voltage is switched off. An electrically programmable ROM (EPROM) could be used. Therefore, the RAM is replaced by an EPROM chip on the PC board.

Sometimes it is necessary to develop extensive programs, thus, a comprehensive knowledge and considerable experience are needed. Programs in machine code, as a hexadecimal sequence of alphanumerics, are very complex, difficult to read and document, and hard to test.

MICROCOMPUTER INTERACTIVE DEVELOPMENT SYSTEM

A microcomputer interactive development system consists of the following key elements as illustrated in Fig. 9.5:

TABLE 9.2 Program Statements

Address	Command	Description
0001	DB	Port I (address 10) to the actuator
0002	10	
0003	FE	Compare accu = 0000 0000
0004	00	
0005	CA	If accu = 0000 0000
0006	14	jump to address 14
0007	00	
0008	FE	If accu does not = 0000 0000
0009	01	jump to address 00
000A	C2	If accu does not = 0000 0001
000B	00	jump to address 00
000C	00	
000D	3E	Load 000 000 into accu
000E	00	
000F	D3	Bring accu to port O (address 11)
0010	11	
0011	C3	Jump to address 00
0012	00	
0013	00	
0014	3E	Load 0000 0001 into accu
0015	01	
0016	D3	Bring accu to port O (address 11)
0017	11	
0018	C3	Jump to address 00
0019	00	
001A	00	

1. Software
2. Bulk storage (floppy disk, permanent disk, or other storage)
3. Monitor
4. Keyboard for interaction and data entry
5. Printer

In principle, there are two methods of developing the communication software for an interactive control system:

- The interactive system is located near and directly connected to the installation workstation so that the development may take place directly at the machine under real conditions.
- The interactive system is located away from the installation workstation. In this case, the installation must be totally or partially simulated, otherwise the system must be connected to the installation from time to time.

When the development of the interactive system has been completed, the system will contain a precise image of the hardware and the software, such as programming, memory size, ports, and number of inputs and outputs. The microcomputer system can then be either built as a printed circuit according to customer specifications or assembled from standard modules (Fig. 9.6).

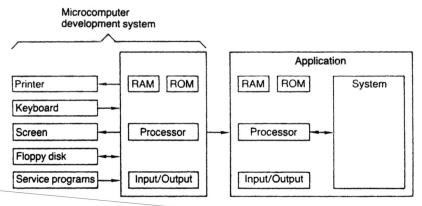

FIGURE 9.5 Microcomputer interactive development system.

PERSONAL COMPUTER AS A SINGLE-BOARD COMPUTER

The personal computer is not a completely new development in the computer evolution. It simply represents a new stage of the development of single-board computers. A single-board computer can play a key role in developing active communication and interfaces for workstations, since it can be effectively integrated with process computer peripherals by the following steps:

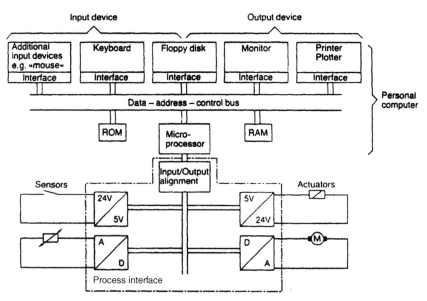

FIGURE 9.6 Microcomputer standard modules.

1. Inclusion of a single-board computer and power supply into a housing
2. Replacement of the simple switch or hexadecimal input by a keyboard similar to that of a typewriter
3. Replacement of the simple LED or seven-segment display by a screen with 24 lines and 40 or 80 characters per line
4. Addition of external mass storage for programs and data (cassette recorder or floppy disk)
5. Connection to a printer or other peripheral device
6. Integration of an interpreter for high-level programming language

A microcomputer system with this configuration is often referred to as a *personal computer*, which is misleading terminology. Although the microprocessor and single-board computer were initially developed for control purposes, the nomenclature directed interest toward purely commercial applications.

The personal computer (or single-board computer) is similar to any peripheral device that requires interfacing to communicate with a central control unit to perform control functions. The personal computer's input and output signals must also be adapted and made available to peripheral devices. Control technology employs a variety of signal forms and levels, e.g., with both analog and binary signals and voltages that may be as high as 24 V. The microcomputer operates with internal voltage levels of 0 and 5 V. Thus, interfacing must provide voltage-level conversion and D/A and A/D conversion. Figure 9.6 is a block diagram of a personal computer and its interfacing to process computer peripherals.

Role of Sensors in Programmable Logic Controllers

The programmable logic controller (PLC) is used in processes where mainly binary signals are to be processed. The PLC first came into the market at the beginning of the 1970s as a replacement for relay circuits. Since then, these systems have been continually developed, supported by the progress in microelectronics. For that reason, the present systems are much more powerful than those which were originally named PLC. Therefore, the terms PLC, process computer, microcomputer, etc., often overlap one another. Technically, these systems are often indistinguishable. In addition, there are often many similarities with regard to their functions. Figure 9.7 shows the functional relations between the components of a PLC.

The input signals supplied by the sensors are passed onto the central control unit via an input module. The signals generated by the central control unit are prepared by the output modules and passed on to the actuators. The program is drawn up by using an external programming device and transferred into the program memory (input of the program varies, however, according to the PLC).

The modules are designed to meet the following requirements on the input side:

1. Protection against loss of the signal as a result of overvoltage or incorrect voltage
2. Elimination of momentary interface impulses
3. Possibility of connecting passive sensors (switches, push buttons) and initiators
4. Recognition of error status

It depends on the manufacturer whether all requirements are met or only some of them. Figure 9.8 shows the basic elements of an input module.

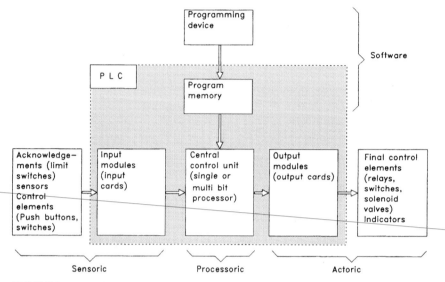

FIGURE 9.7 Relationships between sensor, processor, and actuator activities in the PLC.

The module contains an error voltage recognition facility, which is activated when the input voltage exceeds specified tolerance limits. A trigger circuit with signal delay (added in sequence) ensures that momentary interference peaks are suppressed. Positive and negative supply voltages (+V, −V) are provided at the input terminals for connection of various sensors. Optocouplers separate internal and external circuits so that interference is not able to penetrate the PLC via conductive lines. The input module contains visual indicators (mainly LEDs) which indicate whether a 1 or a 0 is present at the input.

The requirements at the output side are:

1. Amplification of the output signals
2. Protection against short circuiting

Figure 9.9 shows the basic elements of an output module. The optocoupler is used to separate the internal from the external circuit to prevent interference via conduc-

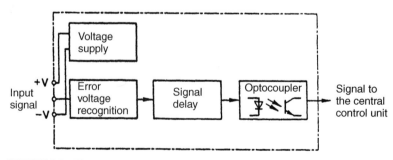

FIGURE 9.8 Elements of an input module.

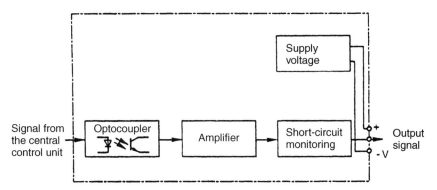

FIGURE 9.9 Elements of an output module.

tive connections. The output signals need to be amplified so that final control elements and actuators which require additional current can be directly connected to the PLC. There are three methods of safeguarding the outputs against short circuiting:

1. Protection by a safety fuse
2. Conditional short-circuit-proof outputs (momentary protection)
3. Sustained short-circuit-proof outputs

Signal status of the outputs can be read from indicators.

Central Control Unit

The first PLCs consisted of a simple single-bit processor. Nowadays, a modern PLC may have at its disposal one or several multibit processors. Thus, in addition to logic comparison operations, a PLC may execute arithmetic, counting, and time functions. By using powerful microprocessors, it is possible to provide PLCs with capabilities which were previously found only in process computers. An example is multitasking, which makes it possible to process a number of unrelated programs simultaneously in the PLC. Of course, the processor itself is still only able to process one command after another. However, because of high processing speeds and complex operating software, it is possible to give the user the illusion of a number of processors working simultaneously.

Almost all PLCs offer important functions like flags, counters, and timers. Flags are single-bit memories in which signal status and program status can be stored temporarily if they are not required until later in the program cycle. Most control tasks require counting and timing functions. Counters and timers help the user in programming such tasks.

With most PLCs it is possible to process simultaneously a number of bits, e.g., 8 bits (rather than just 1 bit such as a switch signal). This is known as *word* processing. Thus, it is possible to connect an A/D converter to the PLC and thus evaluate analog signals supplied by the process. PLCs possess convenient interfaces by means of which they are able to communicate with overriding systems (e.g., control computers) or link up with systems of equal status (e.g., other PLCs).

There are PLC systems available for small, medium, and large control tasks. The required number of inputs and outputs, i.e., the sum of all sensors and actuators, and the necessary program memory capacity generally determine whether a control task is to be designated as large or small. To facilitate matching PLC systems to requirements, they are generally offered as modular systems. The user is then in a position to decide the size of a system to suit the requirements.

Input and output modules are obtainable as compact external modules or as plug-in electronic cards, generally with 8 or 16 inputs/outputs. The number of such modules to be coupled depends on the size of the control task.

A Representative PLC would be composed of plug-in electronic cards in a standard format. The basic housing provides space for system modules and a power supply. Eight card locations might be provided for input/output modules, each containing 16 input/outputs, giving an I/O-connection capacity of 128. If this number were insufficient, a maximum of four housings could be united to form a system, increasing the number of input/outputs to a maximum of 512. With this system, a program memory capacity of 16 kbytes can be installed; i.e., up to 16,000 commands can be programmed. Thus, this controller would be the type used for medium to large control tasks.

Process Computer

The process computer is constructed from components similar to those used in a microcomputer system, with the difference that these components are considerably more powerful because of their processing speed and memory capacity. The supporting operating systems are also more powerful, meaning that many control tasks can be processed in parallel. For this reason the process computer is used in places where it is necessary to solve extensive and complicated automatic control tasks. In such cases, it has the responsibility of controlling a complete process (such as a power station) which is made of a large number of subprocesses. Many smaller control systems which operate peripherally are often subordinated to a process computer. In such cases, the process computer assumes the role of coordinating master computer.

THE NC CONTROLLER

Numerical control systems are the most widely used control in machine tools. A blank or preworked workpiece is machined by NC machine tools in accordance with specifications given in a technical drawing and other processing data. Two-, three-, or multiple-axis machines may be used, depending on the shape of the workpiece and the processing technology. A two-axis machine tool might be an NC lathe, for example. A milling machine is an example of a three-axis machine tool.

In NC machines, sensors and actuators are connected to the processor by the manufacturer. The operator or programmer needs to know the operational principle and the interplay between the controller and machine.

Figure 9.10 illustrates the location of holes to be drilled in a workpiece. The skilled machinist needs to know the material from which the workpiece has been made so that suitable drills and feed values can be selected for machining. A drilling machine is equipped with a work table which is adjustable in two directions. The adjustment is carried out by using a lead screw, and the set value set can be read from a scale. The machine is illustrated in Fig. 9.11.

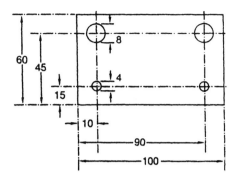

FIGURE 9.10 Location of holes to be drilled in a workpiece.

Manufacturing Procedure and Control

The manufacturing procedure for a machine tool application is as follows:

1. The table is moved up to the reference point. The reference point values are imparted to the computer. The value zero is read from both scales. The drill is located above the reference point.
2. The workpiece is placed on the work table in the precise position required and clamped.
3. The workpiece zero point is determined; a reference edge from which all dimensions are measured is also determined. The drill is positioned at this point. The workpiece zero point can be imagined as the origin of a rectangular system of coordinates: the y axis is in one direction of motion of the table, the x axis is in the other. Similarly, the dimensions in the drawing are referred to this point. Thus, it

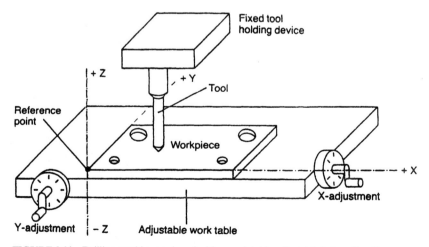

FIGURE 9.11 Drilling machine equipped with a work table adjustable in two directions.

is easy to move the work table to each point in turn where holes need to be drilled. Modifications can be carried out if required.

4. Machining is commenced; the working sequence should be described in detail.

This description can completely replace the technical drawing. Any operator must be able to carry out the working cycle using this exact description.

Machining Program

The machining program is described in terms of sequences:

1. Clamp drill 1
2. Set X to 10 and Y to 15
3. Drill at cutting speed 2
4. Set X to 95
5. Drill at cutting speed 2
6. Unclamp drill 1, clamp drill 2
7. Set Y to 45
8. Drill at cutting speed 1
9. Set X to 10
10. Drill at cutting speed 1
11. Remove workpiece

The positioning data, cutting tool information, and the required cutting speed must be identified to accomplish any machining operation. Usually, the technological data such as speed and feed are included in the tool data sheet. Tool number, speed, feed, and direction of rotation together form a unit. The feed speed has been omitted in the program above; this decision has been left to the operator. If the process is to proceed automatically at a later date, this information will have to be added.

The positioning data is always referenced from the origin of the coordinates. In this case, this is always the point of intersection of the reference lines for the dimensioning of the drawing. The positional information was therefore given in absolute form and can be programmed in absolute dimensions.

The information can also be provided relative to the current position, which is entered as the x value, and a new value added for the next position. This is called *incremental positioning data*. The incremental positioning data may expressed as follows:

1. Set $X + 10, Y + 15$
2. Set $X + 80$
3. Set $Y + 30$
4. Set $X - 80$

Choice of the most suitable method often depends on the dimensioning of an engineering drawing. Sometimes incremental and absolute specifications of positional data are mixed. Then, obviously, it is necessary to describe the type of data in question.

In the example described above, the machining of the workpiece takes place only at four different points; only one machining position is approached. There is no rule as to which path should be taken to reach this position. Thus, no changes have been made at the workpiece when the drilling operation occurs, and the table is moved by a zigzag course to the position for the second drilling operation. This is referred to as a *point-to-point* control.

The situation is different if, as shown in Fig. 9.12, a groove is to be milled in a workpiece. The tool to be used is a milling cutter. The worktable might be the same as that used for the drill. The workpiece is to be contacted on the path from point 1 to point 2, and a zigzag course is, of course, impermissible. However, the problem can be easily solved, since the machining parts are always parallel to one of the two axes. The tool is set to start at position P1. From this point on, only the y value is set to Y2. When the tool arrives there, the x value is set to the new value X2. This is a simple straight-line control.

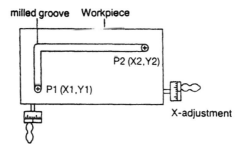

FIGURE 9.12 Groove to be drilled in workpiece.

Figure 9.13 represents a diagonal groove to be milled from point 1 to point 2. A simple solution can be realized using the zigzag course, but it must be calculated with precision, as illustrated in Fig. 9.14.

Without making tedious calculations, it is clear that the machine arrives at point 2 when the tool is moved by an increment of Δx, then by the same amount Δy and then once more by Δx, etc. If the increments Δx and Δy are small, the zigzag line is no longer noticeable and the deviation from the ideal course remains within the tolerance limits. If the required path is not under 45°, the values for Δx and Δy differ from each other, according to the linear gradient. These values must then be calculated. In

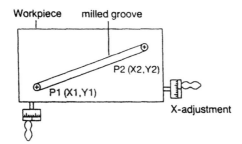

FIGURE 9.13 Diagonal groove to be milled from point 1 to point 2.

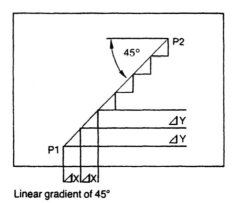

FIGURE 9.14 Zigzag course to mill diagonal groove.

an automatic control system, the computer solves this problem. The straight-line control in this example is referred to as an *extended straight-line control*.

A random course from point 1 to point 2 is illustrated in Fig. 9.15. The increments for the *x* direction and the *y* direction (Δx, Δy) must be formed. However, calculations of this path may be complicated if the curve to be traveled cannot be described by simple geometric equations. Accordingly, a contour control method is used in this manufacturing application. Contour controls can generally travel in arcs, even elliptical arcs. Complicated shapes must be transformed into a straight-line path by coordinating points. A programming workstation is necessary for this application.

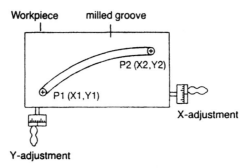

FIGURE 9.15 Random course from point 1 to point 2.

Figure 9.16 illustrates in diagrammatic form the method of automating a drilling machine. The scales are replaced by incremental or absolute displacement measuring systems. Suitable actuators are used in place of the handwheels. Sensors are precisely placed at the extreme ends of the traveled distance. Today, dc motors, which provide infinite and simple adjustments for all axes and for the main drive, are widely used.

There are two different methods of describing the course of a tool or machine slide, incrementally and absolutely. Accordingly, the subcontrol task, movement of a machine slide, is solved differently for each.

The two types of incremental control system are illustrated in Fig. 9.17. In an incremental control system with incremental displacement encoder, the processor counts the pulses supplied by the linear displacement sensor. A pulse corresponds to

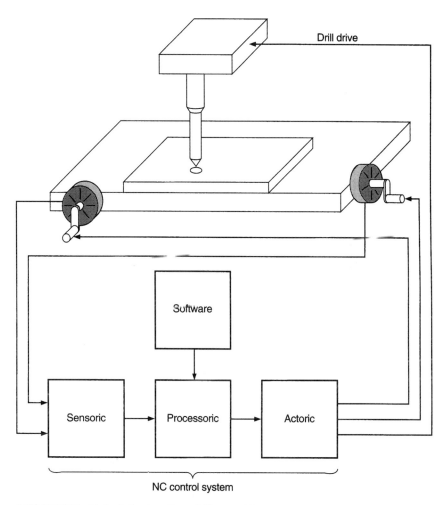

FIGURE 9.16 Method of automating a drilling machine.

the smallest measurable unit of displacement. The distance to be traveled and the direction of travel are passed on to the processor as reference values. From these data the processor forms a control signal for the final control element of the drive motor. If the values do not agree, the motor must continue running. Once they agree, the processor transmits a stop signal to the final control element, indicating that the correct slide position has been reached. When a new reference value is supplied, this process is started up once again. It is important that the slide stop quickly enough. Braking time must be taken into consideration; i.e. before the end position is reached, the speed must be reduced as the remaining gap decreases.

In incremental control with a stepping motor as actuator, there is no acknowledgment of the slide position. In place of the impulse from the linear displacement sensor, a pulse is given straight to the processor. The processor now passes on pulses to the stepping motor until a pulse count adds up to the displacement reference value. As the stepping motor covers an accurately defined angle of rotation for each

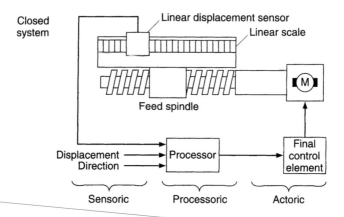

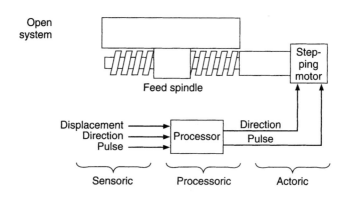

FIGURE 9.17 Two types of incremental control system.

step, each position can be easily approached by the slide. With stepping motors, only very limited speeds are possible, depending on the resolution.

Figure 9.18 illustrates the two types of absolute control system: one with an absolute displacement encoder and the other with an incremental displacement encoder.

Absolute Control

In an absolute control system with an absolute displacement encoder, such as an angular encoder or linear encoder, the actual value of absolute position of the slide is known at any given time. The processor receives this value and a reference value. The processor forms the arithmetic difference between the two values. The greater the difference, the greater the deviation of the slide from its reference position. The difference is available as a digital value at the output of the processor. After D/A

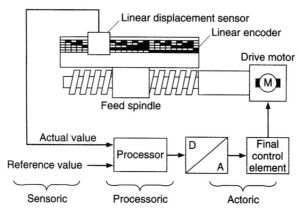
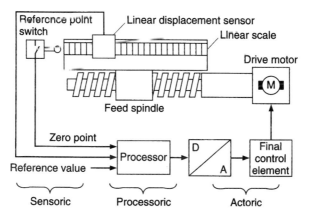

FIGURE 9.18 Two types of absolute control system, with absolute displacement encoder and incremental displacement encoder.

conversion, the difference is presented as an analog value to a final control element such as a transistor that triggers the drive motor. The greater the deviation from the reference value, the greater the motor speed and the adjustment speed of the slide. When there is no further deviation, the motor stops. The motor, of course, has both upper and lower speed limits. At lower speeds the torque may become so low that the tool may falter.

In an absolute control system with an incremental displacement encoder, a sensor acts as a limit switch at the end position of the slide. Once the machine is started, the slide is moved to the end position, where an incremental displacement counter in the processor is set to the reference point value. Each additional movement of the slide is recorded by this counter so that, at any given time, it knows the absolute position of the counter, like the absolute system with a displacement encoder. The absolute control system with incremental displacement encoder is the most widely used today.

NC Software

As with a PLC system, it is necessary to draw up an accurate model for an NC system before a program is written.

The machine must be told which movements it is to execute. For this purpose, displacement data must be prepared which can be taken from a drawing of the workpiece. In addition, the machine needs to know what tools to use and when, the tool feed speed, and the cutting speed. Certain additional functions such as the addition of lubricants must also be programmed.

All these data are clearly of great importance for conventional machining of workpieces. The data are compiled during the work preparation in a factory to create a work plan for a skilled operator. A conventional work plan is an excellent programming model. However, work plans for manual operation cannot as a rule be converted to NC operations.

To ensure that there are no ambiguities in the description of displacement data there is general agreement about the use of cartesian coordinates for NC (Fig. 9.19). Different layers of the coordinate system apply to different processing machines, as shown in Fig. 9.20.

The commands in an NC instruction set are made up of code letters and figures. The program structure is generally such that sentences containing a number of commands can be constructed. Each sentence contains a sentence number. Table 9.3 shows an NC program for an automatic drilling machine.

Displacement conditions are given the code letter G. The displacement condition G00 is programmed in sentence 1, signifying point-to-point control behavior. This condition thus also applies to all other sentences; it does not have to be repeated. The feed is specified with code letter F, followed by a code number or address where the actual feed speed is located. The cutting speed is given with a code letter S and a code number. The code for the tool is T; the following code number 01 in the program indicates that a drill should be taken from magazine 1. The additional function M09 indicates that a lubricant is to be used. M30 defines the program M.

An NC program for a lathe is described in Table 9.4. A part to be turned is illustrated in Fig. 9.20. It is assumed that the workpiece has already been rough-machined. Therefore, it is only necessary to write a finished program.

Path control behavior is programmed in the beginning sentences with displacement condition G01. The lathe is set in absolute values to coordinates X20 and Z100. The displacement condition G91 of the second sentence determines relative addressing; i.e., the x and z values should be added to the previous values. Where the x values are concerned, it should be noted that diameter rather than radius is specified.

The NC programs shown in these examples must, of course, also be put on a suitable data carrier in machine-readable form. The classic data carrier for NC is punched tape. Each character in the program text is converted into a 7-bit combination of 1s and 0s and stamped into the tape as an appropriate combination of holes and spaces (1 = hole, 0 = no hole).

This process of conversion is called *coding*. The totality of all combinations of holes representing a character set is called a *code*. Recommendations have been laid down as to which combination of holes represents a specific character. Figure 9.21 shows the N005 sentence statement in Table 9.3 as a punched tape code to ISO standards. Each 7-bit character is accompanied by a test bit. The test bit is a logic 1 (hole punched) when the number of holes is uneven. Thus, the total number of holes in a character is always even. When the punched tape code is checked to ensure it is even-numbered, errors caused by damage or dirt may be detected. This method of checking for errors is called a parity check.

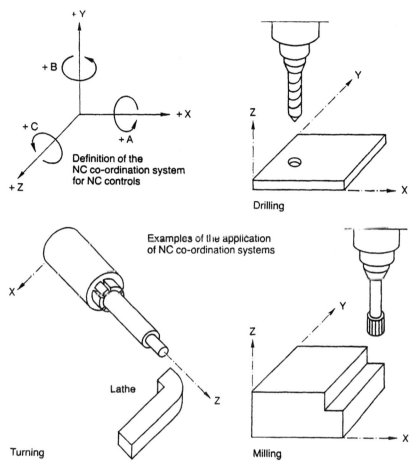

FIGURE 9.19 Cartesian coordinate system for NC and coordinate system applied to different processing machines.

Because punched tape is robust in aggressive industrial environments, it is still preferred by many users. However, modern storage media such as a floppy disks are used more and more. It is also possible to use an economical personal computer as a development and documentation station for NC programs.

Operation of an NC System

Figure 9.22 is a block diagram of a simple NC system. The NC machine is equipped with a punched tape reader into which the program for machining a specific piece is inserted. After starting the program, the processor reads the first sentence. The various displacement and switching data are written into a reference-value memory Next, the commands are processed. The comparator compares the reference values

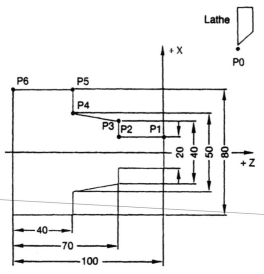

FIGURE 9.20 Parts to be turned.

to the values supplied by the measuring devices (e.g., the current positions of the x- and y-drive units).

If there is a difference between reference values and actual values, the comparator generates suitable control signals for the final control elements of the drives. These are generated only until reference values and actual values agree. The comparator now demands a new sentence from the punched tape reader for processing. This sentence is evaluated in the manner described. Reading and processing are repeated until the final sentence has been processed. As an example of reference value/actual value comparison, Fig. 9.23 shows the adjustment of the x axis of an NC machine.

If the machine slide is moved from its current position representing the actual value to a new position representing the reference value, a continuous comparison

TABLE 9.3 NC Program—Automatic Drilling Machine

Sentence number	Conditions	Displacement data			Feed	Switching data		
		x	y	z		Tool speed	Tool	Addition
N001	G00	X10	Y15	Z2	F1000	S55	T01	M09
N002				Z0				
N003		X909		Z2				
N004				Z0				
N005			Y45	Z2	F500	S30	T02	
N006				Z0				
N007		X10		Z2				
N008				Z0				
N009		X0	Y0					M30

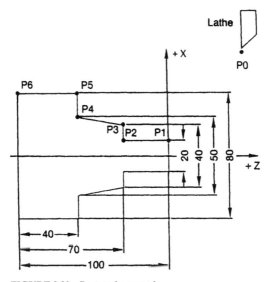

FIGURE 9.20 Parts to be turned.

of the reference value to the actual value is carried out. The comparator determines the difference between the values and supplies a control signal to the final control element of the motor in accordance with the arithmetical sign.

The motor runs in a counterclockwise direction when the actual value is smaller than the reference value. This corresponds to a displacement of the machine slide to the right. The motor runs in a clockwise direction when the actual value is greater than the reference value, which corresponds to a displacement of the machine slide to the left.

The number of rotations of the motor, which is proportional to the displacement of the slide, is counted for the actual-value memory by a measuring device which, in the example, is a light transmitter, a perforated disk on the motor shaft, and a light receiver.

The movement of the slide by the motor drive reduces the difference between the reference value and actual value. When the difference is 0, the comparator sets the control signal to 0, and the motor is stopped.

TABLE 9.4 NC Program—Lathe Machining Operation

Sentence number	Conditions	Displacement data			Switching data			
		x	y	z	Feed	Tool speed	Tool	Addition
N001	G01	X20		Z100	F1500	S60	T03	M09
N002	G91			Z–30				
N003		X + 20						
N004		X + 10		Z–30				
N005		X + 30						
N006				Z–40				M30

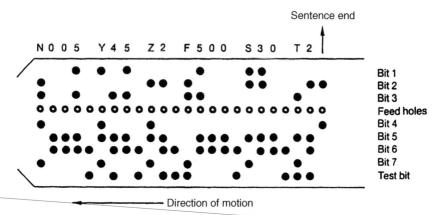

FIGURE 9.21 N005 sentence statement from Table 9.3 (drilling task) as a punched tape code to ISO standards.

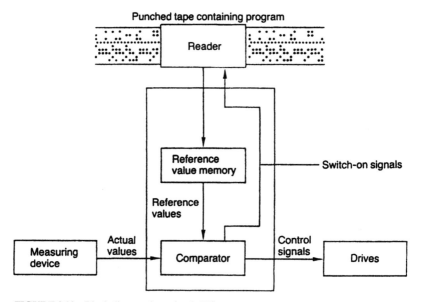

FIGURE 9.22 Block diagram for a simple NC system.

The type of programming in this case is machine-dependent; it corresponds to an assembler language. Application-oriented programming, with which a more general formulation can be written, is also developed for NC systems. An example is the NC programming language Advanced Programming Tool (APT), which can cope with several hundred commands. APT commands are classified into three categories:

1. Geometric commands
2. Motion commands
3. Auxiliary commands

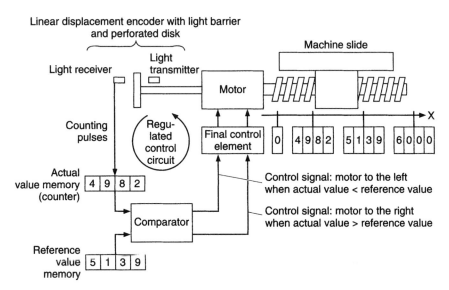

FIGURE 9.23 Adjustment of the x axis of an NC machine.

Geometric Commands. In the NC coordinate system in Fig. 9.24, each point is clearly determined through specification of coordinates x, y, and z. In the APT, each point can be defined symbolically; e.g. the definition for the point $x = 15.7$, $y = 25$, and $z = 13$ is

$$P1 = POINT/15.7,25,13$$

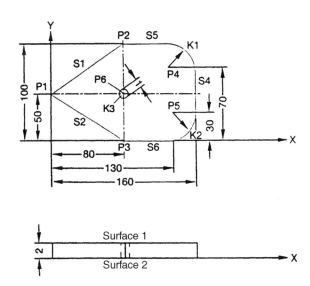

FIGURE 9.24 NC coordinate system for a workpiece.

Thus, the variable to the left of the equals sign can be freely selected. Distances from one point to another are defined as follows:

$$S5 = LINE/P2,P3$$

P2 and P3 must be defined in advance. Points may also be defined as the intersection of two distances, as for example:

$$P1 = POINT/INTOF\ L6,L8$$

A circle can be defined in the following ways:

$$CIRCLE1 = CIRCLE/CENTER,POINT,RADIUS,5.8$$

$$CIRCLE2 = CIRCLE/P4,P5,P6$$

In the first instance, the circle is defined by specifying the center and the radius, and in the second, by specifying three points on the perimeter. A plane surface is defined by three points located on it, e.g.,

$$SURFACE1 = PLANE/P2,P2,P3$$

Another way of specifying a plane is to indicate a point on a surface and a second surface lying parallel to the first one, e.g.,

$$SURFACE2 = PLANE/P1,PARALLEL,SURFACE1$$

The geometric dimensions of a workpiece given in the technical drawing can be fully described using these geometric commands. The workpiece shown in Fig. 9.24 is to be described using the APT geometric commands:

P1 = POINT/0.0, 50.0, 0.0
P2 = POINT/80.0, 100.0, 0.0
P3 = POINT/80.0, 0.0, 0.0
P4 = POINT/130.0, 70.0, 0.0
P5 = POINT/130.0, 70.0, 0.0
P6 = POINT/80.0, 50.0, 0.0
S1 = LINE/P1, P2
S2 = LINE/P1, P3
K1 = CIRCLE/CENTER, P4, RADIUS, 30.0
K2 = CIRCLE/CENTER, P5, RADIUS, 30.0
K3 = CIRCLE/CENTER, P6, RADIUS, 0.5
S5 = LINE/P2, LEFT, TANTO, K1
S6 = LINE/P3, RIGHT, TANTO, K2
S4 = LINE/LEFT, TANTO, K2, RIGHT, TANTO, K1
SURFACE1 = PLANE/0.0, 0.0, −2

The commands for defining paths S4, S5, and S6 form tangents to circles K1 and K2 and are for this reason given the additional functions LEFT, RIGHT, and TANTO. The line S5 starts at P2 and forms a left (LEFT) tangent to (TANTO) circle K1.

Motion Commands. These commands are necessary to describe the path that the tools must travel. A distinction is made between point-to-point (PTP) controls and path controls. The commands for point-to-point controls are as follows:

GOTO   Absolute statement of the point to be approached
GODTLA   Relative statement of the point to the approached

Example: To travel from point 2,5,10 to point 8,10,20, the two commands would have to be used as follows:

Either   GOTO/8,10,20
Or   GODTLA/6,5,10

The drilling operation in this example is programmed as follows:

GOTO/P6
GODTLA/0.0, 0.0, −2.5
GODTLA/0.0, 0.0, 2.5

A path control is programmed with the following commands:

GOFWD   (go forward)
GOLFT   (go left)
GODWN   (go down)
GOBACK (go back)
GORGT   (go right)
GOUP   (go up)

With these commands it is possible, for example, to program the movement of a cutting tool in accordance with the contour in the example.

GOTO/S1,TO,SURFACE1,TO,S2   Go to start point P1
GOFWD/S1,PAST,S5   Go along line S1 to S5
GOFWD/S5,TANTO,K1   Go along line S5 to K1
GOFWD/K1,TANTO,K2   Go from K1 via S4 to K2
GOFWD/K2,PAST,S6   Go from K2 via S4 to S6
GOFWD/S2,PAST,S1   Go along line S2 to S1

Auxiliary Commands. Various switching functions are programmed using commands from this group; for example:

MACHIN   Specifies the tool
FEDRAT   Determines the feed speed
COOLNT   Connects coolant
CUTTER   Diameter of the tool
PARTNO   Part number of the workpiece
FINI   Program end

9.26 CHAPTER NINE

Other application-oriented programming languages for NC programming are EXAPT, ADAPT, and AUTOSPOT. These languages are constructed like APT, and in part build on that language. An NC program written in application-oriented language must, of course, also be translated into machine code and produced in punched tape (Fig. 9.25).

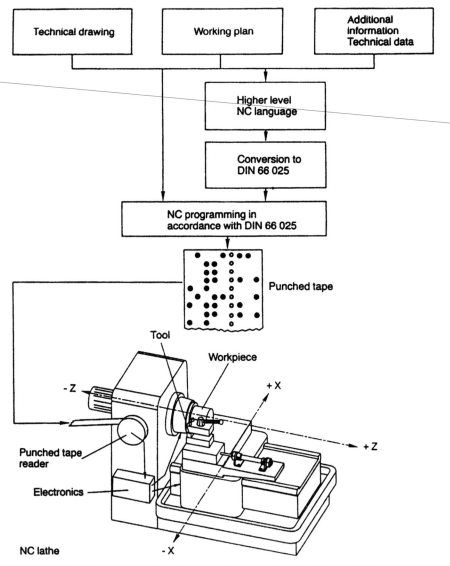

FIGURE 9.25 NC program translated into machine code and produced in punched tape.

Computer Numerical Control System

The program for a specific machining task of an NC machine is located on a punched tape. This punched tape must be read separately for every single workpiece to be machined. Reading of punched tape is a mechanical process that subjects the tape to wear, which can lead to errors.

With the development of computer numerical control, it became possible to overcome these disadvantages and, in addition, to create many advantages. A CNC system contains its own computer system based on modern microcomputer technology. A semiconductor memory is used as program memory. The punched tape is in fact also available with the system. However, it is only required when the program is transferred into the semiconductor memory. In addition to this program entry technique, CNC systems offer the possibility of transferring programs directly from external computers via appropriate interfaces.

Moreover, most CNC control systems are equipped with a keyboard and a screen which make it possible to program directly on the machine. This is referred to as *workshop programming*. Keyboard and screen also offer the possibility of editing and improving programs on site. This is a decisive advantage when compared to NC systems, where modifications and adaptations are much more time-consuming.

With microcomputers it is easy to program the frequently complicated arithmetical operations required for machining complicated curves. Cutting and feed speed can be matched to the best advantage. The computer can check and adapt the values continually and, thus, determine the most favorable setting values.

Integrated test and diagnostic systems, which are also made possible by microcomputers, guarantee a high degree of availability for CNC systems. The control system monitors itself and reports specific errors independently to the operator.

The programming languages correspond in part to those discussed for NC systems but are, in fact, frequently manufacturer-specific. The most important components of a CNC system are illustrated in Fig. 9.26.

9.6 INDUSTRIAL HANDLING

The modern term for handling technology is *industrial handling* or simply *handling*. This concept embraces simple and complex handling and clamping devices.

The best and most versatile handling component is the human hand. Within its physical limitations of weight and size, many possibilities are available: gripping, allocating, arranging, feeding, positioning, clamping, working, drawing, transferring, etc.

Unlike the human hand, a mechanical handling device is able to perform only a very limited number of functions. For this reason, several handling devices have been developed for automatic handling and clamping functions.

It is not necessary to make a full copy of the human hand. A mechanical device is more or less limited to the shape, size, construction material, and characteristics of a specific workpiece to be handled.

Fig. 9.27 shows a device made up of four pneumatic handling units that can insert parts into a lathe for turning.

Cylinder A separates the parts to be turned and assumes the function of allocating parts in the machine-ready position. The part to be turned is brought to the axial center of the machine. At this position the part can be taken over by the gripping cylinder D. Cylinders B and C are used to insert the workpiece into the clamping chuck and remove it again after machining.

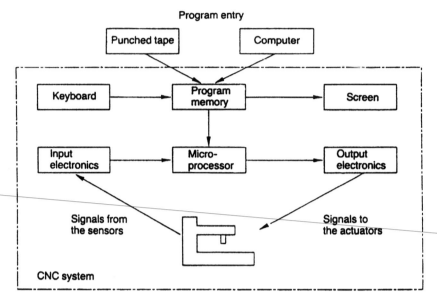

FIGURE 9.26 CNC system components.

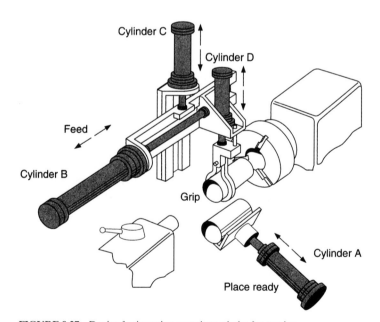

FIGURE 9.27 Device for inserting parts into a lathe for turning.

The fully turned part may be either dropped or picked up by a handling mechanism that is located opposite cylinder A and directly passed on after being oriented.

A PLC is suitable as a control system for the feed unit. A program has to be developed for this purpose.

First, the subtasks (1) "pass on workpiece to the machine tool" and (2) "take workpiece from the machine tool" are to be analyzed. In order to be able to solve these tasks, signal exchange between the PLC and the machine tool (MT) is necessary. Figure 9.28 shows the necessary interface signals.

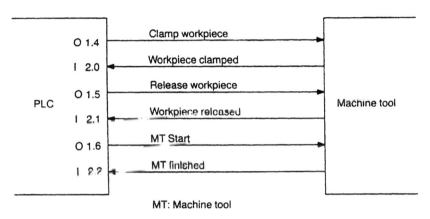

FIGURE 9.28 Interface signals between PLC and the machine tools.

The subtasks are

1. The MT informs the PLC that a workpiece has been completed and can be removed. Signal: MT complete.
2. The gripper travels to the MT and grips the workpiece. The PLC requires the MT to release the workpiece. Signal: release workpiece.
3. The MT releases the workpiece and replies with the signal: workpiece released.
4. The gripper can now advance from the MT with the workpiece.
5. The gripper travels with the workpiece into the MT and requires it to clamp. Signal: clamp workpiece.
6. The MT clamps the workpiece and replies with the signal: workpiece clamped.
7. The gripper opens and retracts from the MT.
8. Once the gripper has left the collision area, the PLC reports that the MT can begin machining. Signal: MT start.

The function chart in Fig. 9.29 illustrates the specified cycle in detail form. Using this chart is necessary to develop a PLC program. Inputs must be assigned to sensors. Outputs from the PLC must also be assigned to actuators. Figure 9.30 shows the appropriate allocation lists. The sequencing program for the feed unit is shown in Fig. 9.31.

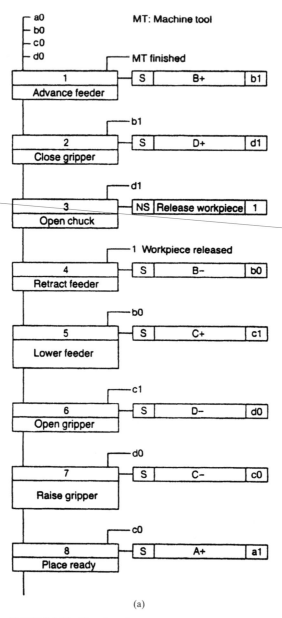

FIGURE 9.29 Function chart.

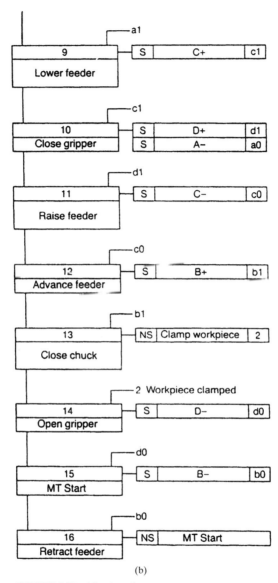

FIGURE 9.29 (*Continued*)

PACKAGING TECHNOLOGY

Arranging and feeding of parts for insertion into boxes, including an intermediate layer, with a feed unit is illustrated in Fig. 9.32. Using this inserting machine, many other functions are carried out either simultaneously with or subsequent to feeding. The various parts which are fed by a conveyer belt are lined up independently and

```
                INSTRUCTION             COMMENT
        ============================================================
        0000 ALLOCATION LISTING MACHINE  V 01
        0001              I 1.0          " LIMIT SWITCH A0
        0002              I 1.1          " LIMIT SWITCH A1
        0003              I 1.2          " B0 HOR. FEED BACK
        0006              I 1.3          " B1 HOR. FEED FORWARD
        0007              I 1.4          " C0 VERT. FEED UP
        0008              I 1.5          " C1 VERT. FEED DOWN
        0009              I 1.6          " D0 GRIPPER OPEN
        0010              I 1.7          " D1 GRIPPER CLOSED
        0011
        0012              I 2.0          " MT WORKPIECE CLAMPED
        0013              I 2.1          " MT WORKPIECE RELEASED
        0014              I 2.2          " MT FINISHED
        0015
        0016              O 1.0          " CYL.A PREPARE
        0017              O 1.1          " CYL.B HOR. FEED
        0018              O 1.2          " CYL.C VERT. FEED
        0019              O 1.3          " CYL.D GRIP
        0020              O 1.4          " MT CLAMP WORKPIECE
        0021              O 1.5          " MT RELEASE WORKPIECE
        0022              O 1.6          " MT START
```

FIGURE 9.30 Allocation list.

then arranged in rows by actuator A, according to a specific pattern. The parts are transferred by the lifting device in accordance with this pattern.

The horizontal movement of the feed unit is executed by actuator B; the vertical movement, by actuators C and D. In the position shown in the diagram, the parts picked up by the lifting device are inserted into the box by extending actuator D. At the same time, actuator C also extends and picks up a square board which is used as an intermediate layer as two layers of parts are being inserted. As soon as actuator C has retracted, holding the intermediate layer, the device is moved by the indexing actuator B. The end position actuator C located above the box and actuator D located above the collection point are used for the arrangement of parts.

When actuator C has extended, the intermediate layer is inserted and the next layer of parts is taken up by actuator D. When actuator B retracts, the position drawn is reached once again and the second layer is inserted into the box. Then, when actuator B extends once again, an intermediate layer is placed on the second layer and the next parts are picked up.

When actuators C and D retract for the second time, box changeover is triggered. Only when a new empty box is brought into position does a new insertion cycle begin.

The feed functions carried out by this inserting machine include arrangement, separation, and insertion cycles.

LINEAR INDEXING FOR MANUFACTURING APPLICATIONS

Linear indexing is best-suited in applications where strip or rod-shaped materials have to be processed in individual stages throughout their entire length. An example of linear indexing is shown in Fig. 9.33.

A machine stamps holes into a strip of sheet metal at equal intervals. After each stamping process, the metal strip is clamped, pushed forward a specified distance, and clamped again. Pushing forward and clamping are carried out by a pneumatic

```
       INSTRUCTION              COMMENT
================================================================
0000 PROGRAM MACHINE   0.0 V 01
----------------------------------------------------------------
0001 STEP 1
0002 IF           I 1.0      " LIMIT SWITCH A0
0003      AND     I 1.2      " B0 HOR. FEED BACK
0004      AND     I 1.4      " C0 VERT. FEED UP
0005      AND     I 1.6      " D0 GRIPPER OPEN
0006      AND     I 2.2      " MT FINISHED
0007 THEN SET     O 1.1      " CYL.B HOR. FEED
0008      RESET   O 1.6      " MT START
----------------------------------------------------------------
0009 STEP 2
0010 IF           I 1.3      " B1 HOR. FEED FORWARD
0011 THEN SET     O 1.3      " CYL.D GRIP
----------------------------------------------------------------
0012 STEP 3
0013 IF           I 1.7      " D1 GRIPPER CLOSED
0014 THEN SET     O 1.5      " MT RELEASE WORKPIECE
----------------------------------------------------------------
0015 STEP 4
0016 IF           I 2.1      " MT WORKPIECE RELEASED
0017 THEN RESET   O 1.1      " CYL.B HOR. FEED
0018      RESET   O 1.5      " MT RELEASE WORKPIECE
----------------------------------------------------------------
0019 STEP 5
0020 IF           I 1.2      " B0 HOR. FEED BACK
0021 THEN SET     O 1.2      " CYL.C VERT. FEED
----------------------------------------------------------------
0022 STEP 6
0023 IF           I 1.5      " C1 VERT. FEED DOWN
0024 THEN RESET   O 1.3      " CYL.D GRIP
----------------------------------------------------------------
0025 STEP 7
0026 IF           I 1.6      " D0 GRIPPER OPEN
0027 THEN RESET   O 1.2      " CYL.C VERT. FEED
----------------------------------------------------------------
0028 STEP 8
0029 IF           I 1.4      " C0 VERT. FEED UP
0030 THEN SET     O 1.0      " CYL.A PREPARE
----------------------------------------------------------------
0031 STEP 9
0032 IF           I 1.1      " LIMIT SWITCH A1
0033 THEN SET     O 1.2      " CYL.C VERT. FEED
----------------------------------------------------------------
0034 STEP 10
0035 IF           I 1.5      " C1 VERT. FEED DOWN
0036 THEN SET     O 1.3      " CYL.D GRIP
0037      RESET   O 1.0      " CYL.A PREPARE
----------------------------------------------------------------
0038 STEP 11
0039 IF           I 1.7      " D1 GRIPPER CLOSED
0040 THEN RESET   O 1.2      " CYL.C VERT. FEED
----------------------------------------------------------------
0041 STEP 12
0042 IF           I 1.4      " C0 VERT. FEED UP
0043 THEN SET     O 1.1      " CYL.B HOR. FEED
----------------------------------------------------------------
0044 STEP 13
0045 IF           I 1.3      " B1 HOR. FEED FORWARD
0046 THEN SET     O 1.4      " MT CLAMP WORKPIECE
----------------------------------------------------------------
0047 STEP 14
0048 IF           I 2.0      " MT WORKPIECE CLAMPED
0049 THEN RESET   O 1.3      " CYL.D GRIP
0050      RESET   O 1.4      " MT CLAMP WORKPIECE
----------------------------------------------------------------
0051 STEP 15
0052 IF           I 1.6      " D0 GRIPPER OPEN
0053 THEN RESET   O 1.1      " CYL.B HOR. FEED
----------------------------------------------------------------
0054 STEP 16
0055 IF           I 1.2      " B0 HOR. FEED BACK
0056 THEN SET     O 1.6      " MT START
0057      JUMP TO S 1
0058      PSE
```

FIGURE 9.31 Sequencing program for the feed unit.

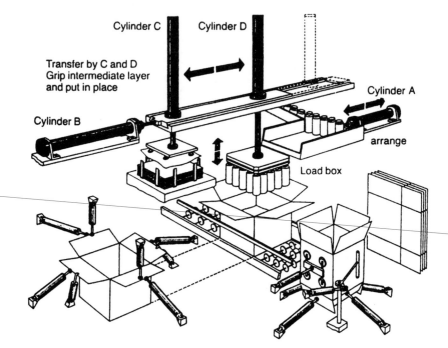

FIGURE 9.32 Insertion machine.

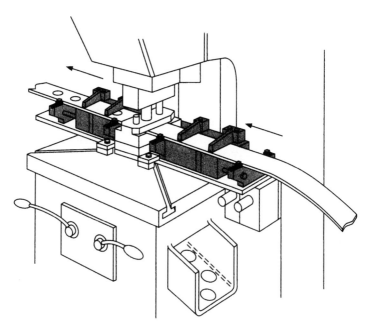

FIGURE 9.33 Linear indexing.

feed unit controlled by various sensors. This feed unit is a compact handling component specifically intended for strip-indexing tasks.

Figure 9.34 illustrates indexing conveyer belts which are also referred to as *linear indexing systems*. Suitable devices for driving such a belt are either motors or pneumatic actuators with a mechanism such as a ratchet to convert an actuator linear stroke into a rotary movement. A pneumatic device is utilized in this case, as an actuator to drive the belt conveyer.

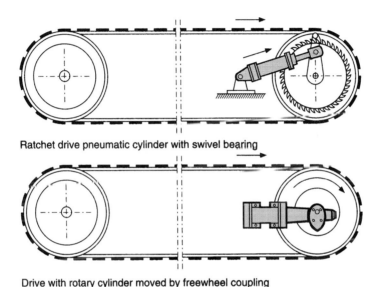

Ratchet drive pneumatic cylinder with swivel bearing

Drive with rotary cylinder moved by freewheel coupling

FIGURE 9.34 Generating indexing motion.

Figure 9.35 illustrates another application of an indexing driving mechanism of a conveyer belt with an electric servomotor as actuator and a sensor as an angular displacement detection.

SYNCHRONOUS INDEXING FOR MANUFACTURING APPLICATIONS

Synchronous indexing systems are used when a number of assembly and fabrication processes are required to be carried out on a product. The product components must be fed once and remain on the synchronous indexing table until all processes are complete.

A vertical synchronous rotary indexing system picks up a minimum of two components. The complete product sequence can be carried out in one position without the need for repeated clamping and unclamping procedures. Rearrangement within the cycle becomes unnecessary. Feeding and withdrawing are limited to the last rotary indexing station; the number of manufacturing functions is thus without significance.

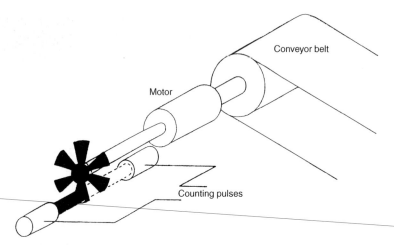

FIGURE 9.35 Conveyer belt with electric servomotor as actuator and sensor for angular displacement detection.

As illustrated in Fig. 9.36, each component is individually processed at six stations:

1. Station 1—insert and remove the workpiece
2. Station 2—drill the front face
3. Station 3—drill the top and base
4. Station 4—drill both sides at a right angle to the top and base
5. Station 5—countersink the front face drill hole
6. Station 6—cut a thread into the drill hole on the front face

A horizontal synchronous rotary indexing table is illustrated in Fig. 9.37. This system consists of 8 stations. Stations 2 to 7 are the machining stations. This is just an example of the many diverse manufacturing processes which can take place in a synchronous fashion. Station 1 is the feed station; station 8 is the eject station.

The synchronous system can be driven by an electric motor with a special driver and cam configuration, by stepping motors, or by pneumatic rotary indexing units.

PARALLEL DATA TRANSMISSION

Figure 9.38 illustrates the principle of parallel-type data transmission. Each data bit has its own transmission line. The data, e.g., alphanumeric characters, are transmitted as parallel bits from a transmitter such as the terminal to a receiver such as the printer.

Some additional control lines are required to ensure synchronization between the transmitter and the receiver. Via these control lines, the printer informs the terminal whether it is ready to take over a character, and the terminal informs the printer that a character is being sent. The distance between the transmitter and the receiver may amount to a maximum of approximately 8 m, as line capacities may lead to coupling and signal deformation. Transmission speed is hardware-dependent

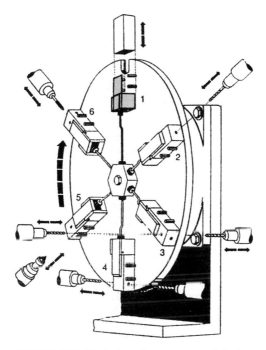

FIGURE 9.36 Vertical synchronous rotary indexing system.

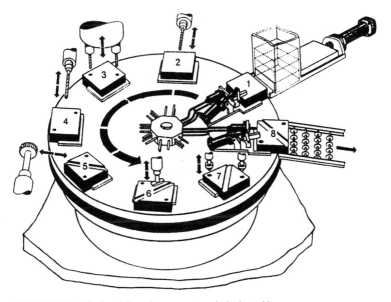

FIGURE 9.37 Horizontal synchronous rotary indexing table.

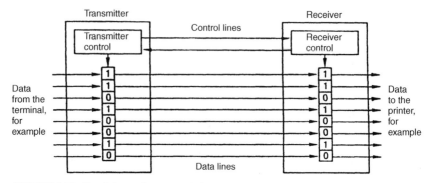

FIGURE 9.38 Parallel-type data transmission.

and may theoretically amount to 1 Mbyte/s, i.e., 1 million characters/s—the line, however, can only be 1 m in length.

If several devices are to communicate with one another according to the principle of parallel transmission, precise coordination of the data flow is necessary (Fig. 9.39).

The devices are divided into four groups:

1. Talkers: able only to transmit, such as sensors
2. Listeners: able only to receive, such as printers
3. Talkers/listeners: can transmit and receive, such as floppy disks
4. Controllers: can coordinate data exchange, such as computers

Each device is allocated a specific address. The controller decides between which two devices data exchange is to be carried out. There are two standards describing the technique of transmitting interface functions: standards IEC-625 and IEEE-488. The number of data and control lines is specified in the standards as eight each. Up

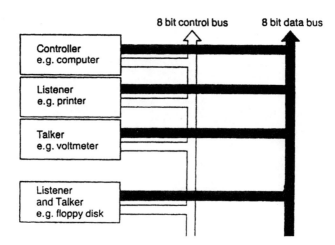

FIGURE 9.39 Several devices communicating with one another using parallel transmission, with precise coordination of data flow.

to 30 devices can be connected to the bus; the complete system of lines may not be longer than 20 m. Maximum transmission speed is between 250 and 500 kbytes/s.

In general, it can be said that relatively fast data transmission, such as several hundred thousand characters per second, can be achieved with parallel interfaces. However, transmission length is limited to several meters.

SERIAL DATA TRANSMISSION

Figure 9.40 illustrates the principle of serial data transmission. Only one line is available for serial data transmission. This means that an 8-bit piece of data, for example, must be sent bit by bit from the send register via the data line to the receiver register.

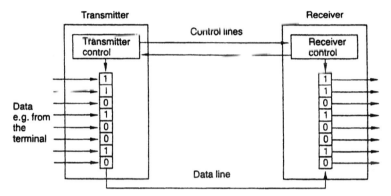

FIGURE 9.40   Principle of serial transmission.

A distinction is made between synchronous and asynchronous data transmission. With asynchronous data transmission, data can be sent at any time; pulse recovery is achieved through the start and stop bits.

In the synchronous process a transmitter and receiver have the same time pulse at their disposal. Control lines are necessary for this purpose.

Functional, electrical, and mechanical characteristics of serial data transmission can also be found in several standards. The most widely used standard versions are the V-24 and RS-232 interfaces (these interfaces are almost identical). With standards, it is guaranteed that different manufacturers' devices are able to exchange data. The V-24 is an asynchronous interface.

Synchronization is achieved by ready signals. The transmitter may, for example, pass data on to the receiver only when the receiver has previously signaled its readiness to receive.

The transmission sequence of the data bits determines the data format (Fig. 9.41). Before transmission of the first data bit, a start bit is sent at the logic zero level. Then 7 or 8 data bits follow, the one with the lowest value coming first. Then a test bit (parity bit) is transmitted, followed by one or two stop bits.

Transmission speeds are freely selectable within a specified framework. Transmitter and receiver must be set to the same speed, or baud rate. Common rates are 75, 110, 135, 150, 300, 600, 1200, 2400, 4800, 9600, and 19,200 bit/s (baud).

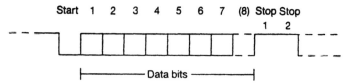

FIGURE 9.41 Transmission sequence of the data bits determines the data format.

The electrical characteristics determine the signal level (Fig. 9.42). If the voltage of a signal on a data line opposite the signal ground is greater than 3 V and is negative, signal status 1 applies, corresponding to the binary character 1. If the voltage of a signal on a data line opposite the signal ground is greater than 3 V and is positive, signal status 0 applies, corresponding to the binary character 0.

Signal status is undefined in the transmission range +3 to −3 V. Connection of the interface line between two devices is effected with a 24-pin D plug. Figure 9.43 shows a V-24 plug with the most important signal lines.

This is one of the oldest serial interfaces. It is used to drive teleprinters [teletypewriters (TTY)]. Transmission takes place along a pair of send and receive lines. Logic

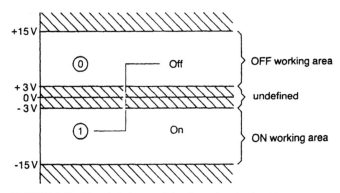

FIGURE 9.42 Electrical characteristics determine the signal level.

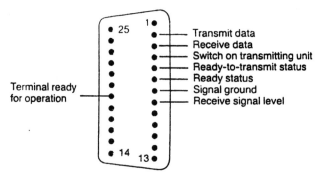

FIGURE 9.43 V-24 plug.

0 is realized via a 20-mA current, and logic 1, via the lack of this current. The functional characteristics of the V-24 interface may also apply to the 20-mA current loop interface.

COLLECTION AND GENERATION OF PROCESS SIGNALS IN DECENTRALIZED MANUFACTURING SYSTEMS

Not long ago it was customary to centrally collect all process signals. The processor and control elements were housed in a central control cabinet (Fig. 9.44). Today, with automation systems, extensive control tasks can be solved effectively if these tasks are distributed among several decentralized self-sufficient systems.

A decentralized structure is illustrated in Fig. 9.45. The sensing, actuating, and processing sections of subsystems are connected to a master computer which evaluates signals applied by each subprocess and sends back control signals.

These subsystems can be realized in different ways; e.g., a PLC or single-board computer could be used as the control system for the subprocesses. A process computer, personal computer, or mainframe can be used as a master computer.

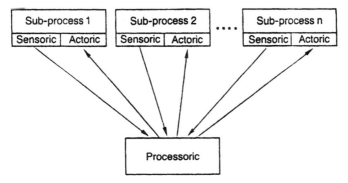

FIGURE 9.44 Centralized control system.

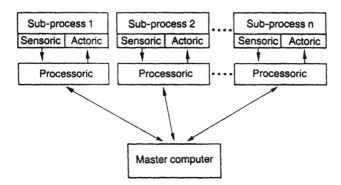

FIGURE 9.45 Decentralized control system.

Figure 9.46 shows a simple control task from chemical process technology as an example of the networking of various systems. In a liquid mixing plant, three different liquids, A, B, and C, are mixed in a specified proportion. Filling is effected by three valves, A, B, and C, and the mixture is extracted via one valve, D. Before the mixture is removed, it is heated to prespecified reference temperature T.

A PLC controls the valves and the mixer. The temperature of the mixer is adjusted to the reference temperature by a two-step controller.

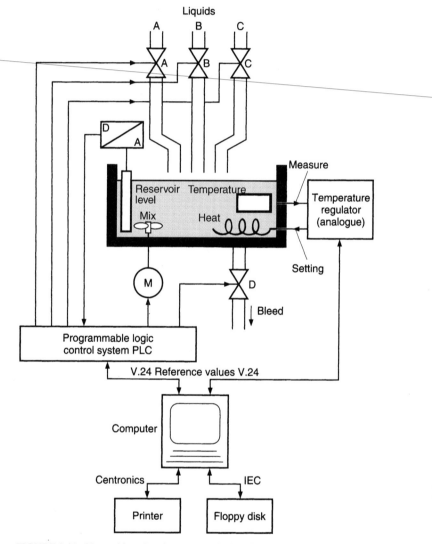

FIGURE 9.46 Networking of various systems in liquid mixing operation.

An overriding computer prepares the reference values for the mixture and the mixing temperature and coordinates the PLC and regulator. In addition, the computer stores operational data and downloads the data to the printer as documentation.

Figure 9.47 illustrates the control process. The reference values 1 and 2, which determine the mixture ratio, are passed on by the computer to the PLC. The PLC controls the remainder of the sequence.

First, valve A is opened and liquid A is poured into the reservoir. The liquid level rises until the level specified by reference value 1 has been reached.

The liquid level in the reservoir is determined by analog means using a capacitive measuring probe. The analog value is transformed into 8-bit digital values. In the PLC, this value is compared with the stored reference value. If reference value and actual value agree, valve A is closed and valve B is opened.

Next, liquid B is poured in until reference value 2 has been reached. Then valve B is closed and valve C opened. Liquid C is poured in until the maximum value (reservoir full) is reached. Reference values 1 and 2 determine the mixture ratio A:B:C.

On reaching the maximum level of the liquid, the mixing motor switches ON and sends on a signal to the computer, which, in turn, switches ON the temperature regulator. The regulator adjusts the temperature to the reference value and informs the computer when the reference value has been reached.

Next, the computer requires the PLC to switch ON the mixing motor and to open the bleed valve D. Once the reservoir has been emptied, the cycle is complete.

A new cycle is now started up with the same or modified reference values. Data communication between the PLC regulator and the computer is carried out serially via V-24 interfaces. Computer and printers communicate via a parallel interface, and computer and floppy disk, via a parallel IEC interface.

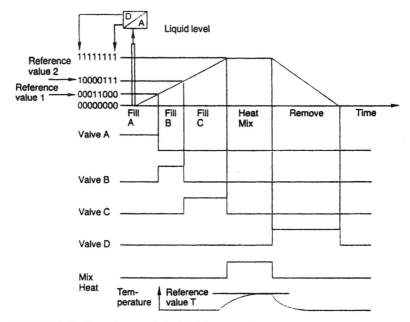

FIGURE 9.47 Control process in liquid mixing operation.

REFERENCES

1. Boyes, G. S., (ed.), *Synchro and Resolver Conversion,* Memory Devices Ltd., Surrey, U.K., 1980.
2. Burr-Brown Corp., "Wiring and Noise Considerations," *The Handbook of Personal Computer Instrumentation,* 4th ed., sec 7, Burr-Brown, Tucson, 1989.
3. Cole, R., *Computer Communications,* Macmillan, London, 1986.
4. Grant, D., "Applications of the AD537 IC Voltage-to-Frequency Converter," application note, Analog Devices Inc., Norwood, Mass., 1980.
5. Hagill, A. L., "Displacement Transducers Based on Reactive Sensors in Transformer Ratio Bridge Circuits," *Journal Phys. E: Instrum.,* **16**: 597–606 (1982).
6. Higham, E. H., *Pneumatic Instrumentation in Instrument Technology: Instrumentation Systems,* Butterworths, London, 1987.
7. Hilburn, J. L., and D. E. Johnson, *Manual of Active Filter Design,* 2d. ed., McGraw-Hill, New York, 1983.
8. Huang, S. M., A. L. Scott, and M. S. Green, "Electronic Transducers for Industrial Measurement of Low Value Capacitances," *Journal of Phys. E: Sci. Instrum.,* **21**: 242–250 (1988).
9. Netzer, Y., "Differential Sensor Uses Two Wires," *Electronic Design News,* March 31, 1982, p. 167.
10. Oliver, B. M., and J. M. Cage, "Electronic Measurements and Instrumentations," McGraw-Hill, New York, 1971.
11. Paiva, M. O., "Applications of the 9400 Voltage to Frequency, Frequency to Voltage Converter," application note 10, Teledyne Semiconductor, Mountain View, Calif., 1989.
12. Tompkins, W. J., and J. G. Webster (eds.), *Interfacing Sensors to IBM PC,* Prentice-Hall, Englewood Cliffs, N.J., 1988.

CHAPTER 10
MICROELECTROMECHANICAL SYSTEMS APPLICATIONS IN ENERGY MANAGEMENT

INTRODUCTION

There are several applications for MEMS (microelectromechanical systems) technology in energy management systems. These applications include low-pressure measurements for detecting excessive pressure drop across filters, noise detection and reduction from motors and dampers, and sensing the presence of chemicals that can be immediately life threatening or eventually cause health problems to occupants. MEMS-based sensing is being investigated for the enabling technologies to allow widespread implementation of all these systems. The increased usage will mean increased energy savings at the national level and improved safety for individuals. This chapter will discuss the systems that are being developed, including the potential applications for MEMS sensors and actuators, and present an available solution for one of the problems using MEMS technology.

TOWARDS IMPROVED EFFICIENCY

Recent studies have shown that motors consume 70 percent of the industrial energy and 50 percent of the electricity generated in the United States. This amounts to a $90 billion electric bill or about 2 percent of the gross national product. It is not surprising that manufacturers are developing more efficient ways to control these motors. Part of the solution is achieved by using variable speed (or frequency) drives (VFD) and zone control directed air flow. VFDs alone have the potential to reduce energy savings energy consumption by almost 40 percent. As shown in Fig. 10.1, a variable speed drive with a pump as the output can typically achieve a savings of 50 percent.[2]

A total view of energy saving technologies is shown in Fig. 10.2.[3] Implementing all these identified improvements could cut the total electricity consumption by 31.3 percent. Buildings in the United States account for 35 to 40 percent of the nation's energy consumption. Building automation will be used to examine MEMS applications for energy management.

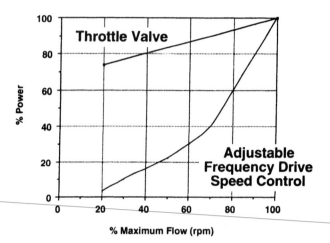

FIGURE 10.1 Efficiency improvement using variable frequency drive control.

THE ROLE OF MEMS IN IMPROVED EFFICIENCY

The energy consumption is directly achieved by using improved efficiency power devices and control techniques such as pulsed width modulation (PWM) that allow VFD to be implemented in building automation. MEMS devices are part of the control solution. The requirement is for medium-accuracy (5 percent or better) products that are low-maintenance and capable of achieving aggressive cost targets.

A number of factors must be considered for comfort and air quality in buildings as shown in Table 10.1. The interaction of these factors provides a comfort level that

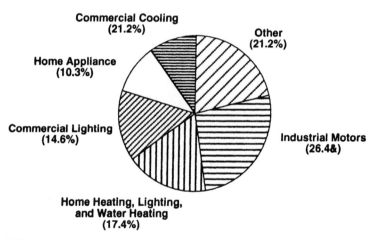

FIGURE 10.2 Energy savings using all available technologies.

TABLE 10.1 Total Environment Quality Factors

Space temperature
Wall, window, and ceiling surface temperatures
Space humidity
Air movement
Outside air ventilation rate (dilution)
Light level

is predictable and therefore controllable. However, individual comfort levels vary considerably, requiring localized control to optimize comfort. Key items for this discussion are humidity, air movement, and outside air ventilation rate.

An example of a system solution in the building automation area is shown in Fig. 10.3. This distributed control environment is interconnected with Neuron ICs that allow a minimal number of wires, infrared, fiberoptics, or even radio frequency transmitters to perform peer-to-peer communications between sensors and actuators via the Lontalk™ protocol. A high-efficiency control strategy would have the VFD controlled motors in the HVAC (heating, ventilating, and air conditioning) system run at their lowest speed and only when required for minimum energy consumption. An alternate strategy could have the motor running continuously at an optimum efficiency point and directing the air flow to specific zones.

A variable air volume (VAV) system controls the volume of air that is heated or cooled, depending on user demand. This reduces the amount of air that must be heated or cooled and consequently reduces energy consumption. Zoned control systems such as the one in Fig. 10.3 combine VFD with a multistage heating or cooling element and the input from sensors such as temperature and air flow.[7]

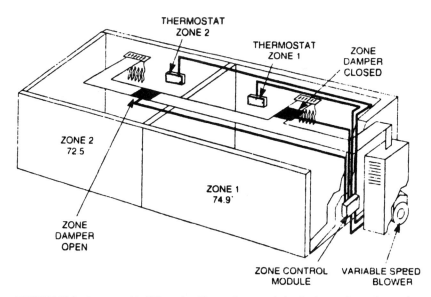

FIGURE 10.3 Automated building using Neuron integrated circuits for sensing and actuating nodes.

Unfortunately, the newest building construction techniques have reduced energy wasting paths through improved sealing so effectively that a problem called sick building syndrome (SBS) has resulted. As a result, the energy management system is also required to exchange air sufficiently frequently to "freshen" the air, even if heating or cooling is not required. Indoor air quality (IAQ) legislation promises to provide additional incentive for improved sensing and control in HVAC systems.

MEMS technology can provide inputs to the system that include a variety of chemical sensors to analyze the air in the building and determine if unhealthy levels of particular gases are present. Excessive levels of CO_2 could initiate air flow only when a threshold level is reached. Since reduced productivity, including absence from work, is attributed to SBS, sensors will be essential to detect unhealthy levels of a number of constituents.

Noise is also a detriment to an employee's ability to concentrate. Excessive noise from fans or dampers can be reduced by lowering the speed of drive motors and by avoiding resonant operating points. Accelerometers or microphones created from micromachined MEMS technology can be used to sense these undesirable noise levels. VFDs provide an unlimited number of operating points and the potential for noise reduction. The accelerometers can also detect bearing failures, prescribe maintenance for off hours, and avoid disruption of normal work activity.

Pressure sensing in the form of barometric, static room pressure, and differential or ΔP for flow measurements is part of the system control. Static room pressure in a sealed office can be used to detect window breakage or unauthorized access during nonworking hours. The most potentially important input that can be realized for even the most cost-sensitive environment is the detection of air flow and excessive pressure drop across a filter. In a highly automated building, the frequency of using the air circulating and filtering system can vary considerably. Routine maintenance to clean filters that may not be operating at an inefficient level can be costly. However, trying to circulate air through clogged, inefficient filters can be even more costly. The signal from a low-pressure sensor can indicate the optimum point to change or clean the filter.

All of the MEMS devices may eventually be integrated into a single device such as the one shown in Fig. 10.4. In addition to the sensors, signal conditioning to amplify the signal, calibrate offset and full scale output, and compensate for temperature effects could be included on this device. Since it most likely will use CMOS technology for the semiconductor process, an onboard analog-to-digital converter (ADC) could also be integrated.

Other system capabilities could include fuzzy logic and neural network interpretation of the input signals. This is especially true if an array of chemical sensors is used to indicate a wide variety of chemical species and overcome many of the problems of available chemical sensing products. Currently, sensors are not this highly integrated and are specified on an as-required basis due to the additional cost that they add to the initial system installation. Furthermore, their inability to function without requiring more maintenance than other parts of the system has delayed their widespread usage.

A LOW-PRESSURE SOLUTION

A major problem with previously available low-level pressure sensors is excessive offset shift. This has required customers to calibrate the output frequently, sometimes even on a daily basis in industrial situations. Obviously, this is not a viable solu-

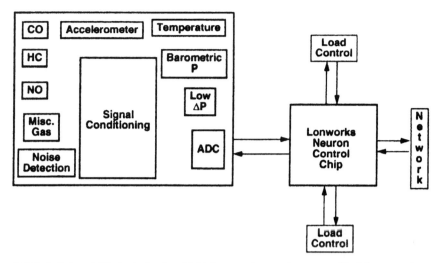

FIGURE 10.4 A MEMS solution for a highly integrated sensor in automated building control.

tion for automation in the home. The short-term goal for industrial users is yearly calibration and ultimately no calibration over the life of the installation.

The largest force acting upon the offset of low-pressure MEMS devices is package-induced stress. Package-induced stress takes many forms: stress transmitted to the sensing element due to mounting of the device; stress induced by the axis of orientation; vibration and shock stress; and thermally induced stress. Two examples of stress seen on typical premolded chip carrier packaged low-pressure, low-level-output sensors are shown. Figure 10.5 shows such a device soldered to a .031-in. FR4 printed wiring board, as is typical of the environment seen in HVAC applications. The board is then torqued and the deflection measured against the zero pressure offset.

If the sensing device is mounted directly to a flexible panel such as air handling ductwork, sheet metal panels, or textured surfaces, stress can be induced on the backside of the device, with the results shown in Fig. 10.6. The measurements are made relative to the cross-section shown.

Externally induced stress can be recalibrated out of the system if it is uniform and repeatable. However, mounting stress is usually highly variable and dependent upon the vibratory frequency and dampening changes within a system. A highly integrated sensor that incorporates on-chip signal conditioning for offset, full-scale output and temperature effects is affected by these stresses more than a basic sensing element. This makes stress isolation critical to elimination of daily recalibration cycles. In order to isolate the highly stress-sensitive element from external forces, a new package standard called Piston-Fit has been developed. When applying the same test methodology to the Piston-Fit package, greater than an order of magnitude improvement in stress isolation is evident, as shown in Fig. 10.7.

An additional evaluation of compressive side loading was tested on the Piston-Fit package, yielding the results seen in Fig. 10.8. The new Piston-Fit package standard also incorporates a radial O-ring seal that further isolates the transmission of stress. The package is easily connected to tubing and hoses by way of a custom port adapter, giving a greater measure of stress isolation. The through-hole-mounted ver-

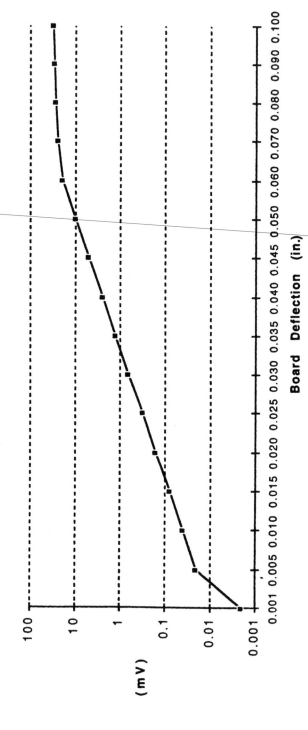

FIGURE 10.5 Change in offset for a sensor mounted on a PC board.

MEMS APPLICATIONS IN ENERGY MANAGEMENT

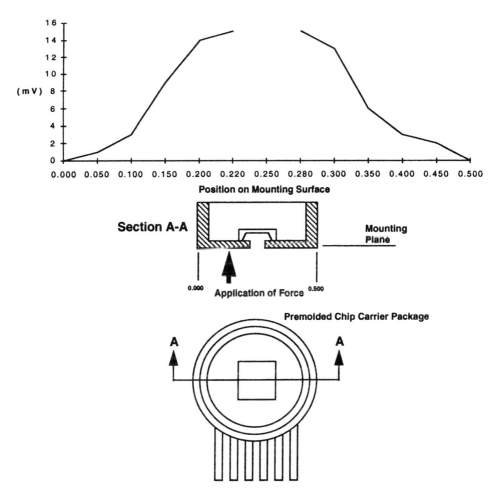

FIGURE 10.6 Change in offset for sensors mounted on flexible surface.

sion of the Piston-Fit package is shown in Fig. 10.9. A surface-mount version for absolute or single sided pressure or vacuum measurement is also possible.

Utilizing a Pro Engineer finite element analysis tool, redundant modeling of the MEMS structure was performed on the silicon die. This analysis led to a design that has proven to be more sensitive than the previous designs and one that overcomes problems common to other available low-pressure units. This MEMS structure is linear over a pressure range of 0 to 2.5 kPa (10-in. water column), with burst pressure in excess of 15 kPa, 40 times the rated pressure.

Initial feedback from targeted customers indicates a differential output of 0.001 in. water can be measured with a usable signal and without problems from noise. Another demonstrated capability of this design is shock resistance. The ruggedness of the package and die allows the sensor to withstand shock. The package has also proven to be insensitive to mounting attitude. This problem is common in MEMS structures that use a boss as a stress concentrator.

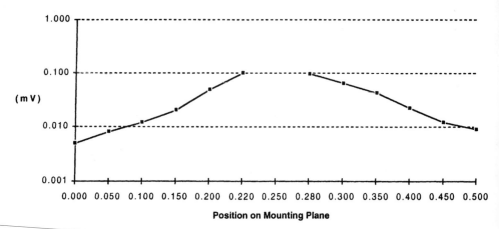

FIGURE 10.7 The Piston-Fit package dramatically improves stress isolation.

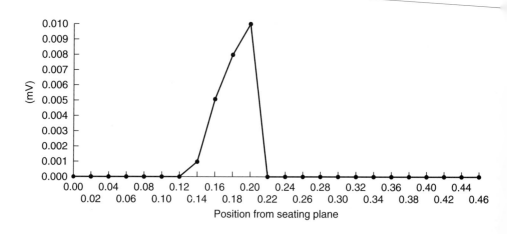

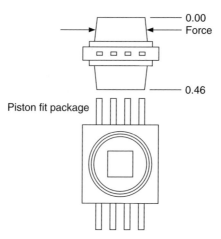

FIGURE 10.8 Results of compressive side loading on the new Piston-Fit package.

FIGURE 10.9 Piston-Fit package on leadframe.

SUMMARY

Energy-efficiency requirements will drive research and development in a number of areas for new MEMS technologies. The most promising areas are pressure, acceleration/vibration, and chemical sensors. Integration of these techniques is forecast but the actual implementation and widespread availability is several years away. Today's technology has allowed low-pressure sensing to add a critical systems element that can be widely implemented due to the cost-effectiveness of the MEMS approach.

LonTalk and LonWorks are trademarks, and Neuron is a registered trademark of Echelon Corporation.

REFERENCES

1. EPRI (Electric Power Research Institute) data.
2. B. Widell, "Saving Energy through Pump Speed Control," *Design News,* 2/20/95, p. 80.
3. EPRI data.

CHAPTER 11
THE MEMS PROGRAM

INTRODUCTION

The programs at Northeastern University in Microelectromechanical Systems (MEMS) Research are described here. In establishing a university program, consideration must be given to the quality of the educational opportunities provided to the graduate student. Students exposed to a broad range of activities are better prepared to deal with the array of problems they may encounter in industry. Northeastern University has traditionally maintained a strong relationship with its industrial partners through its cooperative education. Therefore, the MEMS research program is aimed at practical applications of micromachining and in educating and training young engineers and scientists.

Established in 1991, the program has grown to include a few faculty members, several graduate students, engineers, and a technician. The facilities occupy nearly two thousand square feet of clean room space equipped with modern semiconductor processing equipment. The program has been supported by industrial sponsors including Analog Devices, the Foxboro Company, Kopin Corporation, and Northrop Corporation. Government funding has been provided from ARPA, NASA/JPL, and ONR.

Three main program thrusts are discussed: a bulk micromachined accelerometer, a surface micromachined optical spectrometer, and a surface micromachined Silicon-On-Insulator (SOI) pressure sensor.

SENSOR PROGRAMS

SOI Sensors

Recent advances in silicon-on-insulator (SOI) technology has generated increased interest in its application to high-speed CMOS circuits. The speed advantage is a direct consequence of the insulating layer, which provides isolation from the substrate and thereby decreases capacitance for both devices and electrical interconnections. In addition, total electrical isolation of individual transistors results in complete latch-up immunity leading to smaller device size and higher packing densities.

A new area of interest is the potential application of SOI materials to micromechanical structures. Such structures can be applied to physical, chemical, and optical sensors. The advantages of SOI for these applications are derived from the existence of an insulating layer between the active silicon and the substrate. During the fabrication of mechanical structures, the underlying silicon oxide can become a sacrificial layer that is etched away at the end of the process to *free* the mechanical structure. Structures such as beams, cantilevers, and even diaphragms can be formed in this fashion.

SOI has also demonstrated high temperature capability with CMOS circuits. Its high-temperature performance allows the fabrication of signal conditioning circuits on the same chip and near the sensing element. Sensors are often located in harsh environments where high-temperature operation can be important. The marriage of SOI's mechanical properties with its high-temperature circuit operation would provide performance unobtainable with other technologies.

High-Temperature Sensors

Sensors built on bulk silicon are designed with conventional electronics in mind and are therefore subject to the bulk silicon temperature restrictions (typically <1250° C). This practical limitation reduces the broad application of intelligent sensors using bulk silicon electronics.

In SOI, a single crystal silicon layer is separated from the silicon wafer by a thin oxide. Devices fabricated in SOI using conventional CMOS fabrication techniques have demonstrated substantially better high-temperature characteristics than those of equivalent bulk devices. Large ($k = 50$ J) enhancement mode *p*- and *n*-channel MOSFETs fabricated in zone-melt-recrystallized (ZMR) SOI have demonstrated much lower leakage current for ZMR SOI devices as compared with bulk silicon (Fig. 11.1).

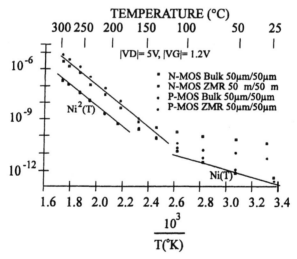

FIGURE 11.1 Leakage current as a function of inverse temperature for bulk and SOI devices. The SOI devices have lower leakage current, thus extending the temperature range for a given circuit design.

Capacitive Pressure Sensor Process

One of the surface micromachined devices we are developing is a capacitive pressure sensor. The device is constructed using ZMR SOI with a 1-micron epitaxial silicon layer and a 1-micron silicon dioxide insulator. The epitaxial silicon layer is used to form the diaphragm of the sensing capacitor and the upper plate of the reference capacitor. The silicon dioxide layer is used as a sacrificial support structure under the diaphragm and as a dielectric for the reference capacitor. The clamp structure used to support the diaphragm can be constructed of polysilicon or silicon nitride. The use of silicon nitride reduces the parasitic capacitances associated with the clamp structure.

This device uses a port etched in the substrate, which allows differential pressure measurements and serves as access to the sacrificial oxide layer under the diaphragm. Twenty-six sensors of varying sizes are included on a die, one half of which are square, and the remainder are circular. The square sensors range in size from 50 microns to 2000 microns on a side, providing a maximum pressure of 265 psi to 1×10^{-4} psi. The circular sensors range in size from 27 microns to 1000 microns in radius, providing a maximum pressure of 250 psi to 1.3×10^{-4} psi. Included with each sensor is a reference capacitor surrounding the device.

The process shown in Fig. 11.2 begins with a 23-mil, 3-in. (100) oriented ZMR SOI wafer with a 1-micron epitaxial silicon film and a 1-micron buried oxide. The wafer is backlapped and polished to 22 mils in preparation for backside processing. The epitaxial silicon is patterned in a RIE. The oxide layer is isotropically etched to remove the exposed oxide down to the substrate and to undercut slightly the silicon structures. This undercut will form the lower portion of the clamp structure, which will support the diaphragm when etched free. A thin silicon nitride layer is deposited to act as a mask for the backside port etch. This layer must protect the substrate and patterned epitaxial silicon layer during the long KOH etch that follows. The backside is patterned in an infrared aligner and then the nitride is etched by RIE. The ports are then etched in KOH to a point. A TMAH etch with a lower degree of anisotropy is used to break through to the oxide under the devices and complete the port openings. The thin nitride is removed in phosphoric acid and a thick (1-micron) silicon nitride layer is deposited to act as the diaphragm clamp structure. The silicon nitride is oxidized, and the front is patterned. This patterning step exposes the silicon diaphragms and reference capacitors. The wafer is etched by RIE to clear the oxide and to begin etching the nitride. After stripping the PR mask, the nitride etch is completed in phosphoric acid. Chromium is deposited and patterned to contact the diaphragm, the reference capacitor, and the substrate. Additionally, a guard ring between the reference and the diaphragm is formed in this metal layer. The wafer surface is coated to protect the nitride clamp during the long HF etch. When the etching is complete, the wafer is rinsed in water, cleaned in sulfuric/peroxide, rinsed in water, and immersed in alcohol. Drying is accomplished at 130° C by evaporating the alcohol.

Material Properties of ZMR SOI

In order to exploit the high-temperature advantages of SOI as a sensor material, its mechanical and electrical properties have to be determined. The following paragraphs provide a brief description of the work performed to date in this area.

Stress Measurement Devices and Results.[2]   A diagnostic mask was prepared for the evaluation of the strain in ZMR films. The majority of the strain diagnostic struc-

Pattern single crystalline silicon layer.

Etch underlying oxide.

Deposit silicon nitride and back etch.

Deposit and pattern clamps.

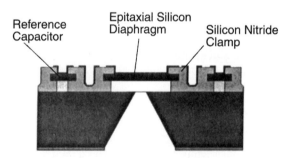

Deposit and pattern metal. Remove sacrificial oxide.

FIGURE 11.2 Five basic process steps used in the fabrication of a SOI capacitive pressure sensor.

tures have been reported by Guckel, et al.[3,4,5] These include doubly supported beams used for the analysis of compressive stresses and ring and beam structures used for tensile stress measurement. In addition, we have included the spiral structures reported by Long-Sheng Fan, et al. for the evaluation of stress nonuniformity with film thickness.

A new test structure was designed at Northeastern for measuring orientation-dependent tensile stress. It consists of two support posts used to anchor long, thin angled beams that connect to two vernier position measuring features, as shown in Fig. 11.3. Three of these structures are needed to cover six orders of magnitude of tensile stress. The operation of the structure is simple. Under conditions of tensile stress, the angled beams, when etched free of the substrate, release their stress by bending. This bending reduces the angle and increases the separation between the two center-position measurement features. The distance these features move can be determined by observing the alignment of the two comb structures. In each comb, subsequent prongs are 2 microns longer than their neighbor to the right. Under unstressed conditions, the rightmost prongs overlap by 2 microns. Increasing the stress will cause the first prong of the lower comb to come into vertical alignment with the second prong on the upper comb. This would indicate a motion of a single micron for each comb. As stress increases further, subsequent prongs come into alignment. Using the vernier arrangement shown, it is a relatively simple matter to measure the relative motion of the combs to within one micron.

In the ZMR SOI, an anisotropy in the strain in the epitaxial layer is found that exhibit no measurable strain in the [100] direction of the scan and a strain of 3.0gx10-4 in the [010] direction.

Piezoresistance of ZMR[7]

SOI wafers with 1-micron silicon epitaxial films on 1 micron of silicon dioxide were obtained on (100) oriented 10-ohm-cm phosphorous doped substrates. The wafers were patterned with a piezoresistor mask set that created a number of test and reference resistors. The resistors were doped at four levels; $5 \times 10^{17}/cm^3$, $2 \times 10^{11}/cm^3$, $7 \times 10^{18}cm^3$, and $1.8 \times 10^{19}/cm^3$, respectively. Each wafer was processed to create five samples, each about 0.5 in. wide and about 1.5 in. long, as shown in Fig. 11.4.

The sample was clamped near its midpoint so that one end of the beam could be deflected. The clamping point was defined by two markers to the left of the piezoresistors. The six piezoresistors to be measured were patterned near the clamp on the side undergoing deflection in pairs of three orientations: longitudinal, diagonal, and transverse. The resistors share a common electrical return. Two additional resistors at the unstressed end of the beam near the contact pads were defined and used as

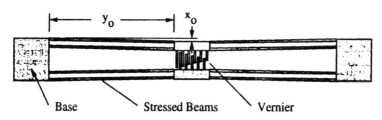

FIGURE 11.3 Top view of tensile test structure.

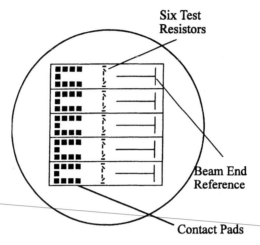

FIGURE 11.4 Layout of wafer with piezoresistance test structures.

controls during the measurement. In addition, a four-point resistivity structure and a metal contact resistance structure were included as a check on our process.

After the wafers were processed and diced, a single sample was loaded into a sample holder for testing. Measurements were made inside an oven whose temperature was adjusted in the range between about 25°C and 150° C. Two terminal resistance measurements were made on all eight resistors at twenty strain levels. In addition, the resistance of the unstrained sample before and after the measurement was recorded.

Measurements of the gage factor (Fig. 11.5) and the temperature coefficients of the gage factor (Fig. 11.6) and the resistance (Fig. 11.7) show that the ZMR material has properties similar to single crystalline silicon as reported by Tufte and Stelzer.[8] The main differences between ZMR and single crystalline silicon are in the temperature coefficients of the resistance and gage factor. Based on these results, it should be possible to fabricate high-quality pressure sensors using SOI wafers based on the ZMR process.

BULK MICROMACHINED ACCELEROMETER

In its simplest form, a conventional accelerometer consists of a proof mass, a spring, and a position detector. Under steady-state conditions, the proof mass that experiences a constant acceleration will move from its rest position to a new position determined by the balance between its mass times the acceleration and the restoring force of the spring. Using a simple mechanical spring, the acceleration will be directly proportional to the distance traversed by the proof mass from its equilibrium position.

In a forced feedback approach, the position of the proof mass is held nearly constant. This is accomplished by feeding back position information to the control electrodes. The resolution of accelerometers is directly proportional to the position detection capability and the effective closed-loop spring constant. Our approach to

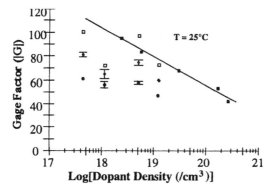

FIGURE 11.5 Gage Factor of ZMR piezoresistors compared to bulk resistors. The open squares represent the data when compensated for transverse stress in the test structure.

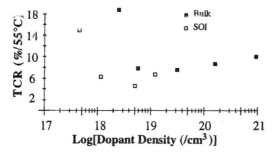

FIGURE 11.6 Temperature coefficient of resistance for ZMR and bulk piezoresistors.

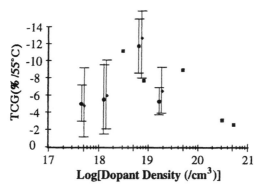

FIGURE 11.7 The temperature coefficient of the gage factor for ZMR piezoresistors (closed boxes with no error bars are bulk data from Tufts, et al).

ultrahigh-resolution devices is to incorporate a weak spring with a sensitive position detector. For position sensing, an electron tunneling tip has been suggested.[11,12] Tunneling tips have been used in sensors with reported resolution below 0.001 Å/√Hz.[13,14,15] Three distinctly different die are fabricated during the process, and these are subsequently assembled using an alloy bonding technique. Electrical contacts are made between layers during the bonding operation. The accelerometer is controlled by electrostatic force plates above and below the proof mass. The lower electrode has a dual role. In operation, it provides a necessary control electrode. When not in operation, it is used to clamp the proof mass and prevent its motion.

A key element of our design is the placement of the tip such that the spacing between the neutral position of the proof mass and the top of the tip is zero. This has two beneficial effects. First, it reduces sensitivity to off-axis acceleration by eliminating torque. In this design, the springs are composed of beam sections. If the proof mass were not centered, the beams would be deflected. Lateral acceleration of the proof mass would then resolve into a normal and axial loads on the springs. The normal component would deflect the beam and proof mass further, thereby giving a false acceleration reading. In a similar manner, thermal effects due to changes in spring stiffness are also significantly reduced.

Figure 11.8 shows a cross section of the final device. This drawing is not to scale, but important features have been indicated. The design is based on four die, bonded together to produce the structure shown in the figure. The top die in the figure is referred to as the tip die and has at its center an electron tunneling tip. The tip is approximately 3.75 µm high. The proof mass is assembled from two identical die. This die is comprised of a border region and the proof mass shown in the center of Fig. 11.9. The mass is 1 cm (10,000 µm) across. The lowest die is referred to as the *force plate die*. The proof mass is located in the exact center of the structure, and the tunneling tip is designed to touch the proof mass. Figure 11.9 shows a top view of just the proof mass and springs, and Figs. 11.10 and 11.11 show top views of the tip and force plate dice.

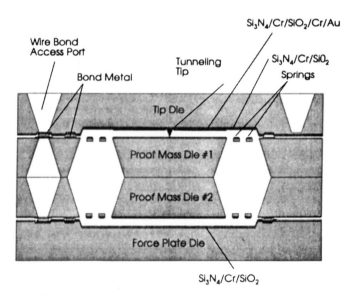

FIGURE 11.8   Cross section of an assembled accelerometer.

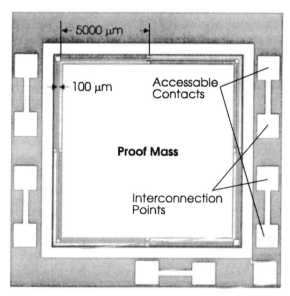

FIGURE 11.9 A top view of the proof mass and spring assembly.

Tip Die

Referring to Fig. 11.10, the tip die is the most complicated structure in the assembly. It not only incorporates the tunneling tip but contains four field plates to control the pitch and roll of the proof mass as well as its position. The tip die must also accommodate the bonding method used to assemble the four separate die that comprise the accelerometer. Two separate metal layers are required to create the tip die. The first layer is isolated from the second by a 0.5-μm oxide layer. The first metal layer is used for interconnecting the tip and quadrant plates to their associated wire bond pads. We are currently using chromium, which survives a subsequent LTO deposition process. This metal layer covers the tip and can be used as the tip metal. Chromium tips have been tested successfully.

The second metal layer is Cr/Au. It forms the quadrant plates and covers the electrical bond pads. It also forms a base layer for the eutectic bond ring that surrounds the center of the die. The center region of the die is recessed with respect to the bond ring and bond pads. This is required to obtain spacing between the field plates and the proof mass. When this die is bonded to the proof mass, the perimeter forms a hermetic seal, and the bond pads are mechanically bonded and electrically connected to their counterparts on the proof mass die. This electrical connection scheme is a key feature of the design. By interconnecting the die in the bonding processes, all electrical connections to the completed accelerometer can be made from its front surface.

To connect the quadrant plates electrically to the bond pads, holes are cut in the oxide layer above the first metal layer in the vicinity of the quadrant plates. The first metal layer is patterned to connect each quadrant plate electrically to its associated bond pad.

Finally, a bond metal layer is deposited and patterned on the bond perimeter. The bond perimeter is also recessed, but not as deeply as the quadrant plates. This recess

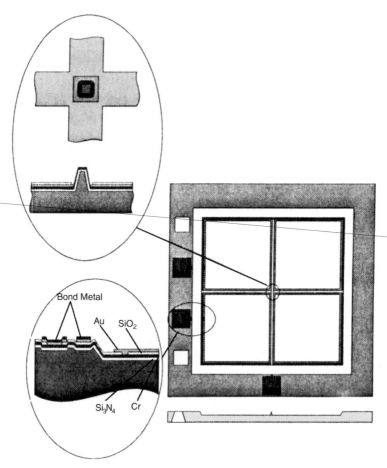

FIGURE 11.10 Various views of the tip die. The tip is located in the very center of the device. A single shielded electrical lead provides continuity between the tip and its associated wire bond pad. The four quadrant plates surrounding the tip can be used to control pitch and roll. Surrounding the four quadrant plates is a eutectic bond ring.

implements the surface-referenced bonding technique we are using. The concept is that the width of the bond metal ring is less than the underlying chrome/gold layer, but its thickness is greater than the depth of the recess. When bonding, the eutectic alloy wets both parts and draws them together through surface tension.

Proof Mass Die

Figure 11.8 shows a cross section of the proof mass, which is assembled from two identical die. As a result of the anisotropic etchant used to machine the proof mass, its sides are tapered at 54.7°. The net weight of the proof mass is 0.18 g. The proof

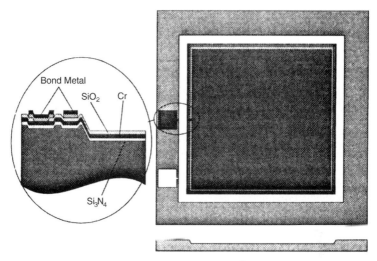

FIGURE 11.11 Various views of the force plate die. The center region contains an oxide-covered force plate. A first metal forms the force plate, lead, and wire bond pad. Surrounding the center depression and force plate is a eutectic bond ring.

mass is held to the surrounding frame by a set of springs referred to as *crab legs*. The target thickness of the crab legs is 25 μm. Figure 11.9 shows a top view of the proof mass and springs.

Each individual crab leg is divided into three spring sections: two short ones and one long one. The length of the short sections is 5000 μm, but the length of the long spring deviates by 200 μm from 10,000 μm due to the requirement to clear the corner ties. The corner ties play no role in the normal motion of the proof mass and do not alter the stiffness of the device in its sensitive axis. These ties are added to increase the stiffness to pitch, roll and yaw.[16]

The proof mass die is maintained at ground potential, and the surface of the proof mass is completely covered by metal. However, at the perimeter of the die, electrical contact between bond pads is routed to facilitate the front-surface contacting scheme described above. Figure 11.9 shows the metal layer layout for the proof mass. Since the anisotropic etching can take place prior to metallization, some freedom exists in the choice of proof mass metal. Our intention is to provide a complimentary layer to the one used on the tip and force plate chips.

Force Plate Die

The force plate die is in many ways similar to the tip die. Again, a two-metal process is used to provide a bond perimeter that is not electrically connected to the force plate. A contact pad to this perimeter metal is provided. The force plate is fabricated in the first metal, which may simply be a single chrome layer similar to that used on the tip wafer. The force plate must be covered with a 1-μm-thick oxide layer, which prevents an electrical contact between the proof mass and the force plate when the proof mass is being electrostatically clamped. Choices for the bond perimeter and contact layer (second metal) are identical to those described for the tip wafer.

Assembly

Figure 11.8 shows the cross section of the assembled accelerometer die with the bond pad regions highlighted. In particular, the nature of the bond pad regions is illuminated. Since the bond pads are meant to contact metal lines on the tip die as well as the force plate die, it is necessary to access bond pads on the force plate and tip dice. To the top right of the figure, an etched groove is shown extending down to the top of the first proof mass die. Contacts to the tip and control plates are made at this level. On the extreme left side of the assembled die, an anisotropically etched hole extends down to the force plate wafer. Bond pads for the force plate and its ground are made through openings of this kind. Figure 11.9 shows the top of the proof mass and the electrical interconnections. Wire bonds are made to one side of each interconnection through the anisotropically etch grooves. The five electrical connections to the tip wafer are made to the opposite end of each interconnection during the eutectic bonding process. Therefore, the pads seen on the tip wafer are connected to the interconnections on the proof mass via the eutectic bond. The opposite side of each interconnection is accessible through an etched groove for wire bonding. The force plate die has only two electrical connections: one to the force plate itself and another to the bond ring. Both are accessible through deep anisotropic grooves.

Results

Figures 11.12 and 11.13 show SEM views of an assembled accelerometer. The corner of the tip die has been broken away to reveal the inner construction of the device. Electrical tests have been conducted on tunneling tips by mounting tips in the drive mechanism for a scanning tunneling microscope. Tips fabricated from chrome and gold have been demonstrated. Additional work is underway to characterize the electrical characteristics of the force plate. It has been determined that the assembled proof mass and force plate may not be parallel over the entire 1-cm area. This requires a higher voltage than originally planned to hold the proof mass during transport. Proof masses have been held against gravity with as little as 2.7 volts.

SURFACE MICROMACHINED MICROSPECTROMETER

Surface micromachining techniques are being used to create a miniature, low-cost replacement for conventional optical spectrometers. Spectrometers are used in scientific instruments for many important measurements including chemical analysis by optical absorption and emission line characterization. A miniaturized instrument or microspectrometer will offer significant advantages over existing instruments, including size reduction, low cost, high reliability and fast data acquisition. Because of these advantages, a broad application of optical measurement techniques may be anticipated in the future.

Previous authors have developed microspectrometers using bulk micromachining techniques.[17] These devices suffer a number of limitations, including size, planarity control, and undesired adsorption in the substrate. To overcome these problems, we are fabricating a surface micromachined device. There are two major advantages of surface micromachining in this application. The first is that there is no fundamental limit to how small the device can be. Devices occupying 30 μm × 30 μm

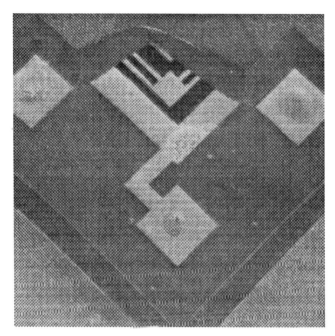

FIGURE 11.12 Top view of completed accelerometer with tip die cut away to reveal the internal structure using a SEM. The bond pad is 1 mm square.

FIGURE 11.13 Close up view showing the 10 g gap between the proof mass and tip quadrant plates.

with active areas as small as 5 μm × 5 μm are feasible. One could easily conceive of multiple spectrometers forming an array or adjacent spectrometers designed for different spectral regions. In fact, small spectrometers are easier to build than larger ones. The second is that our design separates the dielectric mirror and the flexure. This permits the variation of the spacing between mirrors while avoiding deformation of the moving mirror, thus providing better resolution.

Basic Configuration

Figure 11.14 shows a simplified cross section of the proposed visible microspectrometer, and Fig. 11.15 shows the top view. Two elements are required to make a spectrometer: a light detector and a wavelength-selective element. A photodiode is fabricated in a silicon substrate, and a wavelength-selective element is micromachined above the diode. The choice of silicon as the substrate material allows the incorporation of a sense amplifier and drive electronics on the same chip.

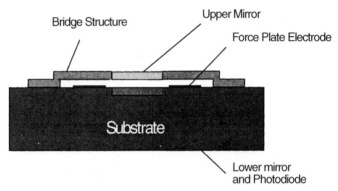

FIGURE 11.14 A cross sectional view of a microspectrometer.

The wavelength-selective element is a Fabry-Perot interference filter with one important difference. The center layer of the interference filter is an air gap created by fabricating a micromechanical bridge above the silicon photodiode. The two mirrors that are components of the interference filter are deposited both in a hole on the bridge and directly on the surface of the photodiode.

The spectrometer described above is a miniature Fabry-Perot scanning interferometer with integral detector. Such spectrometers have been constructed using conventional machining techniques and very high quality optical surfaces. An ordinary optical flat of ¼ wavelength is not sufficient for precise applications. For high-precision measurements, 1/20 to 1/100 wavelength is required. The most significant advantage of the Fabry-Perot interferometer relative to prism and grating spectrometers is that the resolving power can exceed 1 million times (or between 10 and 100 times) that of a prism or grating.

Theory and Considerations

In the schematic diagram shown in Fig. 11.12, an *n*-type emitter layer is diffused into a *p*-type silicon substrate to create a photodiode. The *n*+ layer itself becomes part of

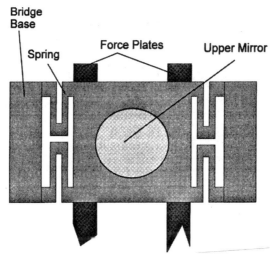

FIGURE 11.15 A schematic top view of the microspectrometer.

the lower interference mirror, which includes a ¼-wave SiO_2 layer and a ¼-wave silicon layer. An air gap width of half the center wavelength must be created. Above this, a second interference mirror consisting of ¼-wave silicon and silicon oxide layers must be created. The choice of silicon and silicon oxide is for convenience and not essential. Other material pairs can be used, where one film has a high index of refraction, such as silicon, and the other is a low-index material such as silicon dioxide.

The number of layers in the mirrors determines the maximum reflectance. The greater the number of layers, the narrower the bandwidth of the interference filter. In our work, we plan to use a seven-layer interference filter.

The optical constants for silicon and silicon dioxide were entered into a computer model, and the transmission of the filter was calculated as a function of the thickness of the air gap. The transmission represents the amount of light that enters the photodiode to be collected and converted to an electrical signal. A center wavelength of 5000 Å was chosen. Figure 11.16 shows the results of these calculations. Note first the curve representing the transmission when the gap is set to 2500 Å. In this case, the curve peaks at exactly 5000 Å or twice the gap spacing as expected. It should be noted that there are second-order responses at large wavelengths and zero-order peaks at short wavelengths. Four other curves show the results if the gap is set to 2000 Å, 3000 Å, 3500 Å, and 4000 Å.

Based on these calculated results, the full width half maximum of the transmitted output is approximately 1/32th of the spacing between the peaks. This suggests a resolution limit for this spectrometer of approximately 160 Å. This result is not as good as that available from conventional spectrometers, which would typically have a resolution exceeding 20 Å.

The resolving power RP of a Fabry-Perot spectrometer can be expressed as

$$RP = N\pi\sqrt{R/(1-R)}$$

where R is the reflectivity of the mirrors and $N = 2nd/l$ with n the index of refraction and d the spacing between the mirrors. The software used in this analysis predicts a

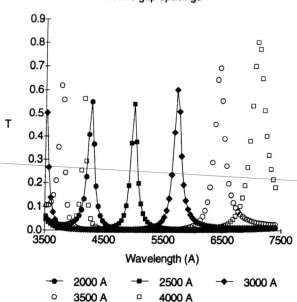

FIGURE 11.16 Calculated optical response of the microspectrometer. The legend indicates the gap spacing for each curve.

reflectivity for a seven-layer mirror centered at 0.5 mm will have a reflectivity of approximately 99 percent. Use of the formula above would result in an estimate of the resolving power RP = 310. By definition, RP = l/dl, and the predicted resolution at 0.5 mm is 160 Å. A typical 30-layer interference mirror would have a reflection exceeding 0.999 and in this application would provide a resolving power in excess of 3000. By increasing d, the resolving power can be increased significantly. Conventional Fabry-Perot spectrometers are capable of RP = 100,000, easily achieving 0.5 Å.

Process Development

The work conducted so far has been aimed at the development of a working microspectrometer. The most important aspect of this work is the establishment of a baseline process for fabricating the device. Developing a process involves, among other things, the verification that the materials and etchants chosen are compatible throughout the process.

In Fig. 11.17, the process for the microspectrometer is outlined. In this process, the first mirror is fabricated on a glass substrate using vacuum evaporation. The silicon and silicon dioxide layers are adjusted to be ¼-wave at the wavelength of interest. An optical thickness monitoring system similar to those used by the optical coatings industry ensures an accurate deposition of the individual layers of the dielectric mirror. A bridge structure is plated above the sacrificial layer. A second

Deposit and pattern base metal layer and lower mirror.

Deposit and pattern sacrificial layer.

Nickel plate bridge structure.

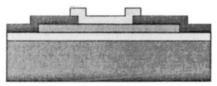

Deposit upper mirror.

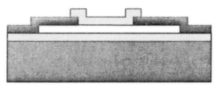

Remove sacrificial layer.

FIGURE 11.17 The basic process flow required to fabricate a microspectrometer.

mirror is plated above the bridge, and finally, the sacrificial layer is removed to release the device.

Figure 11.15 shows a top view of the completed device. At the center of the bridge structure is a transparent circular area. Within this area, the materials that compose the upper mirror are found. Below the bridge, two electrodes are shown. When an electrostatic potential is applied between these electrodes and the bridge, a force is generated between them, which draws the bridge closer to the substrate. In order to reduce the associated stress in the filter area, the bridge is patterned at both edges. The patterned area extends between the center portion of the bridge, which supports the filter and the edge of the bridge, which drops down and contacts the substrate. Most of the bending of the bridge structure will be confined to the patterned area.

Results

Figure 11.18 shows a SEM photograph of nickel bridge structures. These were fabricated using the process described above with copper as the sacrificial layer and electroless plated nickel as the beam material. Figure 11.19 shows a top view of the mirror region of a device. A grid structure is used to improve the flatness of the mirror. Since the mirror is fabricated from multiple thin layers, built-in stresses may cause the mirrors to warp. By reducing the size of the mirror under a critical dimension, mirror bucking can be avoided.

FIGURE 11.18 SEM photograph of nickel-plated bridge structures.

CONCLUSIONS

In this chapter, the research direction of the microsensor program was examined at Northeastern University. The program is aimed at the development of devices with specific applications in mind. Three areas under investigation have been disclosed: SOI high-temperature sensors, a microaccelerometer, and a microspectrometer. SOI high-temperature sensors will find a large market in the automotive and aerospace industries. SOI is an important material because it facilitates the fabrication of high temperature electronics in conjunction with surface micromachined devices leading to smart sensors.

Accelerometers find broad application in the same market areas. The device under development is intended to be used for microgravity measurements by the Jet Propulsion Laboratory. Finally, a microspectrometer will find numerous applications in the process control and chemical industries to aid in the determination of material properties. The miniature device being developed promises a much broader application of optical techniques, because the cost of current sophisticated systems can be significantly reduced.

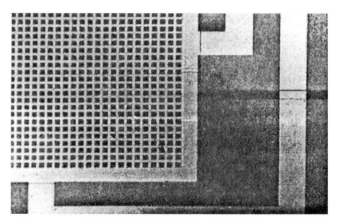

FIGURE 11.19 Top view of a spectrometer, showing the grid structure that supports the upper mirror and a partial view of the springs supporting the device.

REFERENCES

1. D. P. Vu, M. J. Boden, W. R. Henderson, N. K. Cheong, L. Pourcin, L. P. Allen, and P. M. Zavracky, "High Temperature Operation of ISE™ Devices and Circuits," 1989 IEEE SOS/SOI Technology Conference, October 3–5, 1989.
2. P. Zavracky and P. Aquilino, "Strain Analysis of Silicon-on-Insulator Films Produced by Zone Melting Recrystallization," submitted to the *IEEE/ASME MEMS Journal,* 1992.
3. H. Guckel, T. Randazzo, and D. W. Burns, "A Simple Technique for the Determination of Mechanical Strain in Thin Films with Applications to Polysilicon," *J. Appl. Physics,* **57**(5), March 1985, pp. 1671–1675.
4. H. Guckel, D. Burns, C. Rutigliano, E. Lovell, and B. Choi, "Diagnostic Microstructures for the Measurement of Intrinsic Strain in Thin Films," *J. Micromech. Microeng.,* **2**(2), 1992, pp. 86–95.
5. D. W. Burns, *Micromechanical Integrated Sensors and the Planar Processed Pressure Transducer,* Ph.D. dissertation, University of Wisconsin, May 1988.
6. Long-Sheng Fan, R. S. Muller, W. Yun, Roger T. Howe, and J. Huang, "Spiral Microstructures for the Measurement of Average Strain Gradients in Thin Films," *Proc. IEEE Micro Electro Mechanical Systems,* 1990, pp. 177–181.
7. Paul M. Zavracky, Keith Warner, Igor Lassic, and Joshua Green, "Piezoresistivity of Silicon-On-Insulator Films by Zone-Melting-Recrystallization," *J. Micromechanics and Microengineering,* **3**(2), June 1993, pp. 96–101.
8. O. N. Tufte and E. L. Stelzer, "Piezoresistive Properties of Silicon Diffused Layers," *J. Appl. Phys.,* **34**(2), February 1963.
9. P. M. Zavracky, F. Hartley, N. Sherman, T. Hansen, and K. Warner, "A New Force Balanced Accelerometer Using Tunneling Tip Position Sensing," 7th Int. Conf. on Sensors and Actuators, Yokahama, Japan, June 7–10, 1993.
10. A. A. Braski, T. R. Albrecht, and C. F. Quate, "Tunneling Accelerometer," *J. Microscopy,* **152**, 1988, p. 100.
11. S. B. Waltman and W. J. Kaiser, "An Electron Tunneling Sensor," *Sensors and Actuators,* **19**, 1989, pp. 201–210.

12. M. F. Bocko, "The Scanning Tunneling Microscope as a High-Gain, Low Noise Displacement Sensor," *Rev. Sci. Instrum.*, **61**(12), December 1990.
13. T. W. Kenny, W. J. Kaiser, H. K. Rockstad, J. K. Reynolds, J. A. Podosek, and E. C. Vote, "Wide-Bandwidth Electromechanical Actuators for Tunneling Displacement Transducers," *J. MEMS*, **3**(3), September 1994.
14. T. W. Kenny, W. J. Kaiser, S. B. Walkman, and J. K. Reynolds, "A Novel Infrared Detector Based on a Tunneling Displacement Transducer," *Appl. Phys. Lett.*, **59,** 1991.
15. J. J. Yao, S. C. Arney, and N. C. MacDonald, "Fabrication of High Frequency Two-Dimensional Nanoactuators for Scanned Probe Devices," *J. MEMS*, **1**(1), 1992, pp. 14–22.
16. P. M. Zavracky, F. Hartley, and D. Atkins, "New Spring Design and Processes for an Electron Tunneling Tip Accelerometer, 1997.
17. J. H. Jerman, D. J. Clift, and S. R. Mallisnson, "A Miniature Fabry-Perot Interferometer with a Corrugated Diaphragm Support," *Technical Digest IEEE Solid State Sensor and Actuator Workshop*, Hilton Head, June 1990, pp. 140–141.

CHAPTER 12
MEMS IN THE MEDICAL INDUSTRY

INTRODUCTION

Microelectromechanical systems (MEMS) are naturally suited to many applications in the medical industry. Many medical devices have size and weight constraints that MEMS devices are particularly well suited to address. Perhaps more importantly, structures that are impossible to make through any other techniques can be made through MEMS techniques.

Commercial medical applications are also typically sensitive to cost, particularly for retrofit products, or products whose purpose is to displace an older technology in a functionally equivalent way. MEMS devices are well suited to meet these demands through integration of functions on a relatively inexpensive silicon (or other) substrate.

Finally, commercial medical applications are typically high volume. The batch-mode manufacturing techniques that were borrowed from the integrated circuits industry make high-volume manufacturing the nature of MEMS manufacturing. Economies of scale (and in some cases, even economies of scope) naturally drive the application of MEMS devices towards high-volume applications.

The overall sensor market is growing fast, fueled primarily by MEMS (silicon micromachined) devices. Projections for the world sensor market and a breakdown by market are shown in Figs. 12.1 and 12.2.

While automotive and industrial applications dominate today, biomedical use is continuing to grow and remains a significant market for MF-MS devices.

HISTORY

Reusable Blood Pressure Monitoring Transducers

The first use of a MEMS device in the medical industry can be attributed to silicon/quartz micromachined pressure sensors in reusable blood pressure transducers. Before 1980, blood pressure was typically monitored with an external silicon beam or quartz-capacitive type pressure sensor. The pressure was transmitted to the sensing element through a disposable saline-filled tube-and-diaphragm arrangement that was connected to an artery through a needle. The price of these reusable sensors was on the order of $600. The cost of the single use portion (including recalibration and

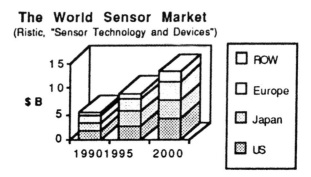

FIGURE 12.1 World sensor market.

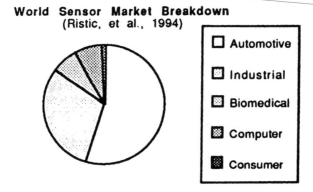

FIGURE 12.2 World sensor market breakdown.

sterilization) was about $501. These sensors were sold at the rate of about 100,000 per year for roughly 7,000,000 procedures per year.

Disposable Blood Pressure Monitoring Transducers

Ongoing developments in the silicon micromachining field made possible, in the early 1980s, the development of a disposable external blood pressure transducer that would duplicate the functions of the reusable sensors without the added cost of recalibration or resterilization. Silicon micromachined piezoresistive pressure transducers were developed to serve this market, and, in 1982, 40,000 units were sold. By 1995, this figure was well over 15,000,000 per year. This period was characterized not only by the rapid adoption of the new technology but also by the evolution of standards for the application.

The Association for the Advancement of Medical Instruments (AAMI) developed a set of standards to ensure compatibility of this function across manufacturers,[2] which has created a commodity-type market for these transducers, the first attempt of its kind for MEMS devices in the medical field.

Disposable pressure sensors have also been developed for other medical applications. These include intrauterine pressure transducers, which monitor the pressure around a baby's head during delivery, and disposable angioplasty devices, which monitor the pressure in a balloon catheter.

Very small devices can be manufactured using silicon micromachined technology. This advantage enabled other developments to take place. Blood pressure must often be measured with a very high and accurate frequency response, as in the case of intracardiac pressure measurements. In these cases, external devices are inadequate because of poor frequency response due to the long (often overly mechanically compliant) signal path between the pressure source (heart, pulse) and the pressure sensor (the external transducer). For such applications, a pressure sensor that is directly mounted in the patient's heart cavity is desired. Silicon micromachining has made it possible to make pressure sensors small enough to accomplish this task. Silicon fusion bonding was utilized to fabricate these sensors. The silicon sensing element is only 0.15 mm × 0.4 mm × 0.9 mm. Three are shown on the head of a pin in Fig. 12.3.

FIGURE 12.3 Three catheter-tip pressure sensors on the head of a pin.

CURRENT USES FOR MEMS DEVICES IN THE MEDICAL INDUSTRY

Although blood pressure monitoring was key in the introduction of MEMS devices to the medical industry, it is by no means the only application. There are several other application areas for which MEMS devices are a natural fit because of cost, size, or performance considerations.

Infusion Pumps

Silicon micromachined devices are being used in large numbers to monitor the pressure in infusion pump lines to check for blockage and air bubbles. A reusable sensor with a physical interface to the drug delivery tube (similar to the first reusable blood pressure monitoring systems) is typical. In this case, the pump is usually a constant-volume-type system (peristaltic pump), and blockage manifests itself as a large pressure increase at the exit of the pump.

Kidney Dialysis

Robust, corrosion-resistant sensors are used in kidney dialysis applications to monitor pressure. Silicon micromachined devices are packaged behind a thin, flexible stainless-steel diaphragm. Pressure is transmitted from the diaphragm to the silicon sensing element with a trapped silicone oil, similar to technologies that have been used for many years in the industrial markets. Micromachined sensors enable the reduction in size and cost of these systems, making portable systems feasible.

Respirators

Very sensitive low-pressure micromachined devices have been developed to monitor the respiratory status of a patient. The sensors are used to measure the differential pressure that is created in the flow paths when air moves in and out of the lungs. This technique can be used both to monitor critical care patients and diagnose respiratory problems.

Other Applications

Other applications that are being studied for MEMS implementation include micromachined accelerometers for heart vibration monitoring, micromachined channels and flow valves for automated drug delivery systems, and micromachined pressure sensors to monitor intracranial pressure and pressure inside the eyeball (tonometry).

FUTURE APPLICATIONS

While the first commercially viable application of MEMS technology to medicine was in the area of pressure monitoring—particularly blood pressure monitoring—many other applications show promise. From interfacing to nerve cells to performing genetic analysis, MEMS devices are likely to begin appearing more frequently in the medical industry.

Neural Interface

Neural interface technologies can be divided into three general categories: penetrating probes, regeneration devices, and cultured neuron devices, as described below.

Penetrating Probes. For over two decades, several research institutions have been developing micromachined probes for recording neural signals in the brain, particularly at the University of Michigan under the guidance of Ken Wise.[3,4] Micron-scale noble-metal electrodes are fabricated on a silicon substrate that is shaped into arrays of needlelike probes on which the electrodes are located. An example of such a device, developed in a collaboration between Stanford University and the California Institute of Technology, is shown in Fig. 12.4. These probes are inserted into brain tissue, and the electrodes can be used to record the electrical firing patterns of neurons. Ultimately, it is hoped that such devices could be permanently implanted in humans to overcome neural deficits in various applications, such as the control of natural or prosthetic limbs in paralyzed patients. In the short-term, it is expected that they will have broad application in research.

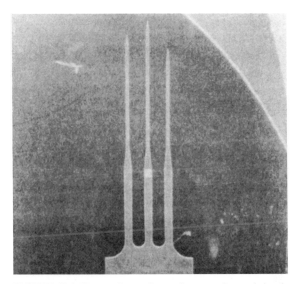

FIGURE 12.4 Penetrating probe used to record neural signals in the brain.

Regeneration Devices. For interfacing to nerves in the limbs (peripheral nerves), an alternative device design is possible that relies on the ability of such nerves to regrow once cut. The approach is to deliberately sever a nerve and allow it to regenerate through a sieve-like silicon substrate on which an array of electrodes is located. The regenerated tissue is then physically locked in place with respect to the electrodes (see Fig. 12.5). Several groups have been investigating this type of device for decades with the hope of using them to interface between an amputee's limb stump and a prosthetic limb.[5,6]

Cultured Cell Devices. By culturing neurons, neuronlike cells, or cardiac cells on arrays of microelectrodes, very useful research can be carried out. The cells can be monitored electrically, and the effects of externally introduced toxins, pharmaceuticals, and so on, can be monitored through the response of an exposed population of cells relative to a nonexposed population. This approach is being used to fabricate sensors for environmental and military toxins as well as toxicity screening tools for pharmaceutical research. Figure 12.6 is an example of such a device.

MEMS devices are critical to the very existence of all of the devices mentioned above. They have all been under development by several groups for decades, and the

FIGURE 12.5 Neuron regenerating device.

FIGURE 12.6 Neuron growth on silicon integrated circuit.

limiting factors in their use have been biocompatibility and packaging issues, sometimes greatly overshadowing the MEMS technologies. For prosthetic applications, there is always the possibility that research into the underlying biological mechanisms of neural repair and regeneration may actually yield biological cures for the disorders that the MEMS devices would address (the *biological end-run phenomenon*). Clearly, from a clinical perspective, fixing biological problems biologically is by far the superior approach, if possible.

Clinical Diagnostics

Several emerging medical MEMS applications are in the area of clinical diagnostics. In certain cases, MEMS devices may be able to increase throughput (of analysis) and decrease cost. For both bedside and clinical laboratory analyses, MEMS offers potential advantages. A new application for MEMS devices is the use of micromachined force sensors to monitor mechanical stimuli when characterizing nerve terminals. The U.S. Department of Veterans Affairs Medical Center, the University of Texas Southwestern Medical Center, and Stanford University have demonstrated effective use of such devices.[7] In order to make precise measurements of mechanical stimuli, it is critical that die tip diameter of the force sensor be well controlled and small (for localized simulation). Silicon micromachining technology is ideally suited to this application. This approach was successfully employed to study corneal nerve action potentials in rabbits. An example of such a device is shown in Fig. 12.7.

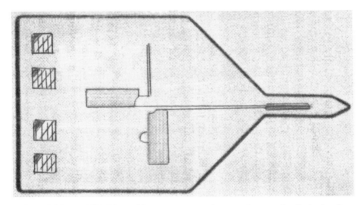

FIGURE 12.7 MEMS-based force sensor used to monitor mechanical stimuli.

Another application of MEMS-based sensors is the measurement of stress and strain in muscle tissue. Figure 12.8 shows micromachined dog bones that are sewn into tissue. These structures have diffused strain gages that output a resistance change with changes in stress.

Chemically sensitive field-effect transistors, despite their availability since the 1970s, have not made great inroads into clinical applications due to difficulties in their reliable and repeatable fabrication. Gas-sensitive devices and other chemical sensors are being successfully developed in many areas and are likely to appear in clinical applications within the next few years. Many development efforts are often focused on chemistry, surface science, and so on, rather than the traditional MEMS areas of engineering. This appears to be a healthy trend in MEMS, since an interdisciplinary approach makes MEMS technology available to those who don't have access to clean rooms, while taking clean-room researchers into projects that would otherwise put them out of their depth.

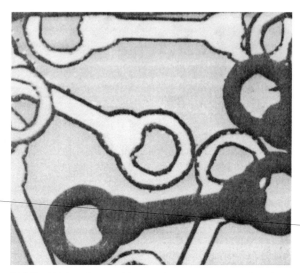

FIGURE 12.8 Micromachined strain gages used to monitor tissue stress.

Another up-and-coming area for applying MEMS in clinical diagnostics is the development of *integrated systems* for chemical analysis. MEMS is increasingly being applied to the miniaturization (often, but not always, with improvements in performance) of traditional analysis devices such as gas chromatography, DNA amplification (through the polymerase chain reaction, or PCR), capillary electrophoresis, and even mass spectroscopy. This approach, if successful, should allow for bedside and field use of sophisticated diagnostic techniques that have been unavailable due to cost and/or physical size considerations.

HURDLES/ENABLERS

Retrofits vs. Enablers

MEMS devices in medical applications can be divided into two general categories: retrofits and enablers. Retrofits are devices that, for cost, size, or performance reasons, can displace an earlier technology. Examples are the disposable blood pressure transducer, which displaced the older technologies largely because of cost considerations; and catheter-tip micromachined sensors, which displaced older technologies through size/performance improvements. Retrofits are typically the fastest to market, and they are the most readily accepted in the medical community because the applications are well understood and tests for safety and efficacy can be designed quickly and thoroughly. A typical cycle-time from design to market introduction is roughly two years for this type of device, which includes roughly one year for development and one year for agency approvals (i.e., FDA).

Enablers, on the other hand, are devices that are not possible except through the advances in MEMS technology. An example is the penetrating neural interface probes that are being developed by several groups. Such devices would be nearly impossible to fabricate with non-MEMS technology. Despite the unique capabilities of MEMS in this area, such devices are still a long way from commercial viability and are subject to biological end-run phenomena, as described above.

Technical Hurdles

Several technical obstacles are present when one is developing MEMS-based systems for use in the medical industry that are not present for other MEMS-based applications. These obstacles fall under the general headings of materials selection and biocompatability.

Materials. While silicon is a well understood, repeatable, and strong material, etched silicon micromachined devices tend to be brittle, and the implanted, diffused, or deposited circuitry tends to be delicate. This is especially true in the physiological environment, which generally contains several biological mechanisms that can lead to MEMS device failure. Such issues include simple mechanical movement, corrosion of metallic and ceramic surfaces, adsorption, and precipitation of protein leading to device "fouling," intrusion of alkali ions leading to the destruction of MOS circuitry (if present), and so on. These issues relate to negative effects of the physiologic L setting on the sensors. The converse, the issue of biocompatibility, is equally important, as it deals with tissue reactions to invasive use of MEMS devices. In order to avoid potentially harmful long- and short-term reactions, materials must be carefully screened and/or modified using surface chemistry. Both issues are being researched with considerable effort, since it is quite clear that they are central to the successful application of MEMS in implantable/invasive devices. Therefore, a premium is put on the development of a packaging technology that simultaneously allows the measurand signal to be transmitted to the sensor while protecting the sensor from unwanted attack. For commercially successful devices like the disposable blood-pressure transducer, this was accomplished with a silicone-based gel whose purpose included:

- Protection of the electrically active circuitry from the saline solution that was used to transmit pressure signals from the artery to the sensor
- Transmission of the pressure signal from the saline solution to the sensor
- Electrical isolation of the sensor from the saline solution for defibrillation and risk current conditions
- Protection of the sensor from light, which could adversely affect performance through electron-hole generation

Such contradictory goals makes the development of suitable packaging materials difficult at best and problematic at worst. This is the primary cause of device development failure. Theoretically useful devices have been designed and fabricated only for their designers to find that suitable protection could not be developed to make these devices clinically or commercially viable.

Biocompatibility. In addition to the performance considerations presented above, several other concerns are introduced when one is developing products for the medical industry. These concerns include toxicity and carcinogenicity. All materials that are in contact with (or could conceivably come in contact with) the human body must be benign. In addition, the materials selected must also be able to withstand sterilization procedures, which range from simple autoclaving to gamma and beta radiation. Typical integrated-circuit processes have difficulties surviving these environments, so special processes must be developed to ensure durability under these conditions. Robustness in only one or two of these sterilization techniques will impact not only cost but also the potential market size for the devices.

Regulatory Hurdles

In addition to the technical hurdles one must overcome, MEMS devices are fraught with the usual regulatory controls that must be addressed. As mentioned previously, a typical design-to-market cycle is about two years for noninvasive retrofit devices, about half of which is the cycle for agency approvals. For more risky or less understood products (i.e., implantables or enablers), the time to demonstrate effectiveness and reliability could be much longer. For breakthrough applications, a more significant hurdle could be determining how to demonstrate this.

Economic Hurdles

Although one may have worked through all the technical and regulatory obstacles, there is no guarantee that the device will be commercially successful. The product must have some demonstrated advantage over other competing technologies that perform similar, if not identical, functions.

For example, there are several competing technologies for measuring pressure. These include bonded-foil strain-gage, metal-plate capacitive, Linear Voltage Displacement Transducer, ion gage, and piezoelectric. Each of these technologies has its own advantages and disadvantages, and the developer of MEMS-based systems must be aware of them. In today's political climate, the cost advantages that could be realized with MEMS-based systems bode well for commercial success. The portability that MEMS devices allow through size and weight reductions is another economic plus.

As MEMS activities continue to evolve toward more application-oriented research (and commercialization) and less toward basic research, practitioners must be constantly aware of what they are trying to achieve and continually assess the appropriateness of MEMS technology.

CONCLUSION/SUMMARY

The use of MEMS-based technologies in medical applications is a natural outcome because of the fit between the advantages that MEMS offers (cost, size, volume, functionality) and the needs of medical practitioners. Several opportunities have already been exploited, particularly for MEMS-based pressure sensors, but the road to technical and commercial success and acceptance is fraught with many obstacles. The authors feel that many, but not all, areas in which MEMS is being applied to medicine, will realize tangible benefits in terms of both improved patient care and economy. But probably the most important issue when developing or contemplating MEMS devices for medical uses is the up-front and thorough analysis of markets, the advantages and disadvantages of the MEMS approach, and the clinical realities of cost and physician acceptance.

REFERENCES

1. Dr. Janusz Bryzek and Dale Gee, "Enabling Technologies for Low-Cost High-Volume Pressure Sensors," *Proceedings of Sensors Expo*, Cleveland, September 20–22, 1994.
2. AAMI, "Blood Pressure Transducers," ANSI/AAMI BP22, 1994.
3. C. Kim, S. J. Tanghe, and K. E. Wise, "Multichannel Neural Probe with On-Chip CMOS Circuitry and High-Current Stimulating Sites," *Proceedings of the 7th International Conference on Sensors and Actuators*, Yokahama, June 1993, p. 454.
4. K. Najafi, "Solid-State Microsensors for Cortical Nerve Recordings," *IEEE Engineering in Medicine and Biology Magazine*, **13**(3), pp. 375–387.
5. G. T. A. Kovacs, C. W. Storment, and J. M. Rosen, "Regeneration Microelectrode Array for Peripheral Nerve Recording and Stimulation," *IEEE Transactions on Biomedical Engineering*, September 1992, **39**(9), pp. 893–902.
6. G. T. A. Kovacs, C. W. Storment, M. HalksMiller, C. R. Belczynski, C. C. Della Santina, E. R. Lewis, and N. I. Maluf, "Silicon-Substrate Microelectrode Arrays for Parallel Recording of Neural Activity in Peripheral and Cranial Nerves," *IEEE Transactions on Biomedical Engineering*, June 1994, **41**(6), pp. 567–577.
7. Bart J. Kane, Christoper W. Storment, Scott W. Crowder, Darrell L. Tanelian, and Gregory T. A. Kovacs, "Force-Sensing Microprobe for Precise Stimulation of Mechanosensitive Tissues," *IEEE Transactions on Biomedical Engineering*, 1995.
8. L. J. Ristic (ed.), *Sensor Technology and Devices*, Artech House, Boston, 1994.
9. J. B. Angell, "Transducers for In-Vivo Measurement of Force, Strain, and Motion," *Physical Sensors for Biomedical Applications*, M. R. Neuman, et al. (eds.), CRC Press, Boca Raton, 1980, pp. 46–53.
10. Dr. Sabrie Soloman, *Sensors and Control Systems in Manufacturing*, McGraw-Hill, New York, 1996.

CHAPTER 13
MEMS: A FUTURE TECHNOLOGY?

INTRODUCTION

In the past twenty years, significant research, development, and money has been invested in microelectromechanical systems (MEMS) or micromachines. The technology has important applications, but we have yet to see major success stories and market breakthroughs. In this chapter, we will explain the obstacles to rapid MEMS growth and suggest ways to circumvent them.

Despite many predictions and hopes to the contrary, silicon micromachining has not revolutionized our lives in the past twenty years. Microelectromechanical system (MEMS) technology emerged as an offshoot of the semiconductor industry, and much of MEMS technology is borrowed from the semiconductor industry. However, MEMS technology has unique manufacturing problems that require solutions that cannot be borrowed from the semiconductor industry. A large demand exists for MEMS technology, but we have failed to manufacture an adequate supply. Fast access to high-quality manufacturing know-how is essential for the MEMS market to reach its potential. The aim of this chapter is to target the obstacles responsible for the slow growth of this high potential technology and to suggest ways of overcoming them.

MEMS: A CURRENT OR FUTURE TECHNOLOGY?

What Are MEMS?

It is important to draw the bounds of MEMS. Defining MEMS simply as microelectromechanical devices or systems ignores too many devices that should fit the MEMS definition. Micromachined integrated-optic sensors, nozzles, fluidic channels, micromachined gas sensors, and other nonmoving devices should fit the MEMS definition. On the other hand, some systems behave like sensors and detectors but really should be called semiconductor devices or integrated circuits. Examples of these include Chem-FETs, charged coupled devices, and flat panel displays. In this chapter, MEMS refers to devices fabricated using bulk, surface, or LIGA microfabrication technology. We use the term *MEMS* interchangeably with the term *micromachines*. *MEMS* should be understood to include the additional subsets.

Current Market

The current market for micromachines is approximately 2 percent of the overall semiconductor market. The revenue from silicon micromachined sensors in 1993 was over $600 million. Automotive pressure sensors account for a large portion of this market. In the United States, the sensor market is growing at a rate between 10 and 20 percent, depending on the focus of analysis: sales or rate of return on investment. At this rate of growth, the market will be somewhere between $1.2 and $2.1 billion by the year 2000.

Most of this revenue is currently generated from silicon pressure and acceleration sensors for automotive, biomedical, and process control applications. The largest quantities demanded have been for low-cost micromachines. Biomedical applications have demanded miniaturization of micromachines. High performance and reliability have proven to be more of a drawback than a selling point of micromachines.

Frost & Sullivan's 1993 report projected that 1994 worldwide revenue from silicon micromachined sensors would be $1.116 billion with a growth of approximately 11 percent over the previous year.[1] The United States captured 54 percent of this revenue or approximately $603 million.

The worldwide breakdown by type of sensors in 1994 was projected to be:

Sensor type	Market share
Accelerometer	49.7
Flow	1.6
Pressure	48.7

The worldwide breakdown by industry type in 1994 was projected to be:

Industry	Revenue
Automotive	43
Medical	23.8
Industrial	1.1
Aerospace and defense	9
Process control	19.4
Energy and environmental	1.4
Education and research	0.6
Others	1.7

Market Projections

There have been a number of efforts to quantify the market potential of micromachines in the past five years. A European survey (July 1993) estimates there are over 300 organizations involved in micromechanics technology.[2] The survey estimates that a rapid proliferation of microsensors will happen by the year 2000, with over 10 billion sensors per year market demand. A market study performed by SPC (July

1994) estimates the micromachines market demand by the year 2000 to be approximately 1 billion units.[3]

The difference between the two studies reflects the difficulty in estimating the market potential. In Transducers '91, IntelliSense prepared a course on the business opportunities in microsensors in which the estimated market for micromachined sensors was over a billion dollars in 1991. Revenues from micromachines today are slightly more than a billion dollars. Five years ago, the annual production of micromachined sensors for engine and transmission controls and diagnostics was more than 12 million units per year in the United States.[4] This number has increased at a rate of less than 20 percent per year since then, far removed from an explosive technology.

Revenues from micromachines will increase as quantities demanded for sensors for the detection of pressure, acceleration, temperature, humidity and angular rate, actuators for control of fluids, positions and light, and microstructures for building various miniature systems meet anticipated market demand. Micromachined devices have diverse applications. We can use them in any system that operates with smarts.

Comparison to Semiconductors

Semiconductor ICs have the advantage of commonality: the technological advances made for high-volume DRAM production spreads quickly to other devices such as microprocessors and other types of memories. No such commonality has benefited micromachines manufacturing yet.

U.S. semiconductor industries invest approximately 12 percent of their $25 billion annual revenue[5] in research and development.[6] Although the level of R&D expenditures is similar in other parts of the world, the focus of the investments is different. For instance, Japan concentrates on the improvements of materials, processes, and automation. The United States spends heavily on product development.

Micromachines research and development expenditure is even higher because the micromachines market is much more fragmented than the semiconductor market. We spend tremendous effort on a variety of devices for different applications. The largest current independent microsensor company has annual revenue of only approximately $20 million. Compared with computer and semiconductor manufacturers like Sun Microsystems, Dell, AST, Intel, and others of similar vintage, investing in micromachines is not very attractive.

WHAT ARE THE OBSTACLES?

With such a large market potential and the advantages of lower cost and widespread knowledge of semiconductor process technology, why have the widely anticipated successes of micromachines not come to fruition? The reason is, we lack manufacturing know-how in making the tremendous variety of miniature components.

Micromachines technology emerged as an offshoot of the semiconductor industry. The machinery used in semiconductor processing technology also proved useful to build various miniature structures, and thus micromachining was born. Rather than using silicon as the substrate, as is the case with integrated circuit devices, micromachines use wafers machined in the form of three-dimensional structures.

These structures are machined often with moving elements and can transfer mechanical motion to electronic response.

The major obstacle to fast growth in micromachines is the lack of manufacturing know-how. The majority of the manufacturing technology used in the development of micromachines is borrowed, and we have adapted the semiconductor technology on a case-by-case basis. Many common micromachining problems were never issues in semiconductor manufacturing. The most important and difficult fabrication problem is the inability to control the process-dependent residual stress in the thin films being deposited with semiconductor machinery. Residual stress was never much of a problem in semiconductor manufacturing, since constituents of integrated circuits are sticking together. However the constituents of a micromachined device often move. When the released constituents of micromachines are under processing-induced stress, they try to relieve the stress by bending back to the unstressed configuration. This is similar to a bent spring which, when released, tries to bend back. Residual stress alone accounts for significant development expenditures in micromachines.

Another obstacle is the lack of prior understanding of how three-dimensional structures will behave when manufactured. We are used to the process of "just do and see what happens." We make a number of the devices and test each. When one method works, it becomes the baseline process. The baseline process is safeguarded with care. No variations from the baseline will be tolerated, and it is often considered important proprietary information. In many cases, the resulting baseline is only one of many ways of accomplishing the same results and often not the best one.

Lack of standards is also an obstacle in the manufacturing of thin films. There are no standards for the measurement, documentation, and reporting of material issues regarding thin films. This has resulted in biased and less than universally applicable data. A measurement of a thin film deposited by a European researcher using machine X may differ from the measurements of a U.S. researcher using machine Y. Machinery dependency and the lack of experimental verification and cross referencing explain why many of the efforts made in the field of micromachines have been less beneficial than they could be.

Proper design and fabrication tools do not exist for micromachines. Today's computer-aided design (CAD) systems are inappropriate because they do not have the necessary micromachine design, material, or fabrication information. As a result, manufacturers are forced to use time-consuming and expensive design-fabrication trial and error to develop micromachines. This enormous expense can be justified today only in applications for industries like the automotive field, which can amortize the costs over large volumes. Lack of micromachining knowledge has resulted in repetitious work. The majority of micromachining processes, especially the good ones, are guarded trade secrets.

CONCLUDING REMARKS

The global race to attain leadership in micromachines has resulted in large government and commercial sponsored research and development investments in many parts of the world. Micromachines will be more attractive business opportunities when we tap into the large market demands. To do this, we must merge high performance and high reliability micromachines with low cost. Better micromachine manufacturing know how and standards must be at investors' and manufacturers' fingertips to accomplish the merger.

REFERENCES

1. *World Emerging Sensor Technologies*, Frost and Sullivan Inc., New York, 1993.
2. *MST News*, VDI/VDE Germany, June 1993.
3. *Microelectromechanical systems (MEMS): An SPC Market Study*, System Planning Corporation, 1994.
4. *High Technology Business*, September/October 1989, p. 29.
5. *WSTS Monterey Forecast*, October 1993.
6. *The Semiconductor Industry Association Annual Report*, 1993.

CHAPTER 14
MEMS ADVANCED RESEARCH AND DEVELOPMENT

INTRODUCTION

During the 1970s and 1980s, fundamental and influential research in micromachining was conducted at Stanford University in fields ranging from pressure sensors to accelerometers to gas chromatography. The world's first silicon accelerometer, demonstrated by Roylance and Angell,[1] was the predecessor of acceleration sensors now of the market. Several of the first pressure sensor chips with onboard circuitry were first demonstrated at Stanford.[2] The first real microelectromechanical system on a chip—a complete gas chromatography including valves, column, flow channels, and detector—was developed by Steve Terry during this time.[3] The world's first totally integrated STM scanning actuator was also demonstrated during this period.[4]

A number of today's leading MEMS companies can trace a portion of their origins to Stanford University graduates, including Microsensor Technology, IC Sensors, NovaSensor, Redwood Microsystems, and SMI.

This tradition of leading-edge MEMS research continues today at Stanford. The micromachined sensor and actuator efforts currently underway are directed by Professors Kovacs, Kenny, Petersen, and Maluf. A guiding philosophical point is to build and grow core competency, not only in micromachining, but also in circuit design, and to be able to combine the best of both to realize true MEM Systems. In addition, it is a central theme to develop transducers with a clear purpose and potential for transfer to industry.

CMOS COMPATIBLE SURFACE MICROMACHINING

The powerful concepts of surface micromachining are being extended to new materials and new applications. The guiding principle of this work is to develop sensor and actuator processes and devices that are fundamentally compatible with standard integrated circuit processing. Typically, a wafer is processed in a standard CMOS foundry process, and micromachined structures are fabricated by low-temperature postprocessing techniques.

This approach is in contrast to those using, for example, high-temperature MEMS materials such as polysilicon, which cannot be added to standard, metallized CMOS wafers as a postprocessing step. If the MEMS processing steps are truly compatible with prefabricated wafers, the optimum underlying circuit process for a given application can be selected independently of the MEMS devices.

Several CMOS compatible micromachining approaches are being investigated. Storment et al.[5] are developing surface micromachined aluminum structures for low-cost optical modulators, actuators, and multiaxis accelerometers. Suh et al.[6] have been investigating the fabrication of thermal-bilayer actuators for the transport, inspection, and other handling of small mechanical parts such as microwave components. While such actuators have been implemented in the past, this is the first effort we are aware of to merge them with circuitry. Another postprocessing strategy is being pursued by Reay et al.,[7] which involves electrochemical etching of standard CMOS devices to yield highly thermally isolated single-crystal silicon islands containing circuitry. These approaches have been applied to the fabrication of thermally stabilized circuits such as a bandgap voltage reference[8] and an RF thermal RMS converter[9] both using a standard foundry CMOS process. In yet another example of CMOS postprocessing, Kane, et al.[10] are developing tactile sensor arrays with on-chip circuitry that are capable of resolving shear and normal forces.

MICROINSTRUMENTATION

One ultimate aspiration of the MEMS community is the development of the fabled microinstrument on a chip. Many research activities at Stanford are focusing on this long-term goal.

Quate et al.[11] recently used MEMS techniques to "write" mechanically submicron lines in photoresist with micromechanical elements. The MEMS lithography instrument on a chip may not be far behind. Kenny et al.[12] are investigating the development of tunneling-based sensors fabricated with MEMS techniques. These extremely sensitive devices are leading the way to high-precision measurement instruments for pressure, acceleration, infrared energy, and other parameters. Reay et al.[13] combined micromachining and advanced analog circuit design to demonstrate a heavy-metal ion measurement instrument that is capable of detecting ppb levels of these contaminants in groundwater, for example. This instrument could ultimately be the size of a pen, and it will replace traditional equipment weighing 60 lbs. and costing nearly $20 thousand. In another system example, Reay et al.[14] successfully demonstrated a new class of micromachined detector for use in capillary electrophoresis analytical instruments.

BIOMEDICAL APPLICATIONS

A traditional unifying theme of the MEMS work at Stanford has always been biomedical applications. For example, the first accelerometers built in the 1970s were intended to be packaged into pills, which, when swallowed, would telemeter out information on the pill's orientation and changes in direction as it traversed the intestinal tract. Biomedically oriented projects continue to be a central theme in MEMS research at Stanford.

There is ongoing research in several types of interfaces between neurons and MEM devices, including several microelectrode array types for probing the nervous system. Neural interface devices that are being developed include penetrating probes for brain tissue (jointly developed with the California Institute of Technology). In addition, micromachined devices have been demonstrated that make use of the regenerative capability of certain nervous tissues (peripheral nerves, found outside the bony casings of the cranium and spinal cord). By fabricating silicon devices perforated by arrays of holes surrounded by micron-scale electrodes, the electrical activity of the regenerated neural tissue can be recorded.[15]

New Process Concepts (DRIE/SFB)

In a search for MEMS fabrication processes that would have wide applicability for problems in biomedicine and microinstrumentation, Stanford University, together with Lucas NovaSensor, has been developing a powerful, new micromachining technique. New developments in reactive ion etching for integrated circuit fabrication now make it possible to etch deep, narrow trenches and holes into silicon with high precision and high aspect ratios. Trenches as deep as 200 μm with features as small as 20 μm have been etched with nearly perfect vertical walls. When this deep reactive ion etching capability (DRIE) is combined with silicon fusion bonding (SFB) processes, extremely versatile sensor and actuator structures can be realized.[16] While this processing technology is only now in its infancy, Stanford and Lucas NovaSensor are aggressively pursuing its development.

STANFORD CIS AND THE NATIONAL NANOFABRICATION USERS NETWORK

MEMS development facilities at Stanford in the Center for Integrated Systems (CIS) are partially funded by the National Science Foundation as a part of the National Nanofabrication User Network (NNUN). As a result of this sponsorship, the MEMS technologies and capabilities developed at Stanford are available to university and industrial partners in a very attractive and easy-to-use format. Any U.S. company can arrange for its own employees to be trained in the use of CIS equipment, then conduct its own public domain development program, using CIS staff as consultants, for a nominal fee. Stanford and the network members, including Cornell, encourage and actively solicit industrial and academic users of this program. Since paperwork and bureaucracy have been reduced to a bare minimum, the NNUN user program has become very popular.

CIS itself is a fully capable integrated circuit and thin-film R&D facility. The 4-in. silicon wafer line supports all the equipment and processes necessary for most MEMS research, including low-stress insulating films and a wide variety of metal films. A quick turnaround analog CMOS process called QUICKMOS has been specifically developed in-house to support the MEMS activities for the integration of circuits and microstructures. Complete thin-film analytical facilities are also available for characterizing circuits and films.

SUMMARY

Stanford remains a key center of development for the increasingly important field of MicroElectroMechanical Systems and is dedicated to enhancing and expanding this strategic technology. In addition, Stanford continues to be a leader in university/industrial partnerships designed to commercialize the emerging field of MEMS.

REFERENCES

1. L. M. Roylance and J. B. Angell, "A Batch Fabricated Silicon Accelerometer," *IEEE Transactions on Electron Devices*, ED-26, 1979, p. 1911.
2. C. S. Sander, J. W. Knutti, and J. D. Meindl, "A Monolithic Capacitive Pressure Sensor with Pulsed Period Output," *IEEE Transactions on Electron Devices*, ED-27, 1980, p. 927.
3. S. C. Terry, J. H. Jerman, and J. B. Angell, "A Gas Chromatograph Air Analyser Fabricated on a Silicon Wafer," *IEEE Transactions on Electron Devices*, ED-26, 1979, p. 1880.
4. S. Akamine, T. R. Albrecht, M. J. Zdeblick, and C. F. Quate, "Microfabricated Scanning Tunneling Microscope," *IEEE Electron Device Letters*, **10**, 1989, p. 490.
5. C. W. Storment, D. A. Borkholder, V. A. Westerlind, J. W. Suh, N. I. Maluf, and G. T. A. Kovacs, "Dry-Released Process for Aluminum Electrostatic Actuators," *Proceedings of the Solid-State Sensor and Actuator Workshop*, Hilton Head Island, North Carolina, June 1994, p. 95.
6. J. W. Suh and G. T. A. Kovacs, "Characterization of Multi-Segment Organic Thermal Actuators," Transducers '95, Stockholm, Sweden, June 25–29, 1995.
7. R. J. Reay, E. H. Klaassen, and G. T. A. Kovacs, "Thermally and Electrically Isolated Single Crystal Silicon Structures in CMOS Technology," *IEEE Electron Device Letters*, **15**(10), October 1994, pp. 399–401.
8. R. J. Reay, E. H. Klaassen, and G. T. A. Kovacs, "A Micromachined Low-Power Temperature-Regulated Bandgap Voltage Reference," *Proceedings of the 1995 International Solid-State Circuits Conference*, San Francisco, February 15–17, 1995, pp. 166–167.
9. E. H. Klaassen, R. J. Reay, and G. T. A. Kovacs, "Diode-Based Thermal RMS Converter with On-Chip Circuitry in CMOS Technology," Transducers '95, Stockholm, Sweden, June 25–29, 1995.
10. B. Kane and G. T. A. Kovacs, "A CMOS Compatible Traction Stress Sensing Element for Use in High Resolution Tactile Imaging," Transducers '95, Stockholm, Sweden, June 25–29, 1995.
11. R. M. Penner, M. J. Heben, N. S. Lewis, and C. F. Quate, "Mechanistic Investigation of Manometer Scale Lithography at Liquid-Covered Graphite Surfaces," *Applied Physics Letters*, **58**(13), pp. 1389–1391.
12. T. W. Kenny, S. B. Waltman, J. K. Reynolds, and W. J. Kaiser, "A Micromachined Silicon Electron Tunneling Sensor," *Proceedings of the IEEE Micro Electro Mechanical Systems Conference*, Napa Valley, California, February 1990, p. 192.
13. R. J. Reay, S. P. Kounaves, and G. T. A. Kovacs, "An Integrated CMOS Potentiostat for Miniaturized Analytical Instrumentation," *IEEE Digest of Technical Papers*, International Solid-State Circuits Conference, San Francisco, February 16–18, 1994, pp. 162–163.
14. R. J. Reay, R. Dadoo, C. W. Storment, R. N. Zaire, and G. T. A. Kovacs, "Microfabricated Electrochemical Detector for Capillary Electrophoresis," *Proceedings of the Solid-State Sensor and Actuator Workshop*, Hilton Head Island, North Carolina, June 1994, p. 61.
15. G. T. A. Kovacs, C. W. Storment, M. HalksMiller, C. R. Belezynski, C. C. Della Santina, E. R. Lewis, and N. I. Maluf, "Silicon-Substrate Microelectrode Arrays for Parallel Recording of Neural Activity in Peripheral and Cranial Nerves," *IEEE Transactions on Biomedical Engineering*, **41**(6), June 1994, pp. 567–577.
16. E. H. Klaassen, K. Petersen, M. Noworolski, J. Logan, N. Maluf, C. Storment, G. Kovacs, and W. McCulley, "Silicon Fusion Bonding and Deep Reactive Ion Etching: A New Technology for Microstructures," Transducers '95, Stockholm, Sweden, June 25–29, 1995.

CHAPTER 15
FUNCTIONAL INTEGRATION OF MICROSYSTEMS IN SILICON

INTRODUCTION

Research on Microsystems at the Delft University of technology is aiming at the fabrication of devices in silicon at the highest level of functional integration (i.e., the ultimate solution would be a device with all the components of the data acquisition unit integrated on chip). However, the desirable level of functional integration is depending on the application. Usually, the economic viability is limiting rather than the technical constraints. The microsystem is not necessarily a micromechanical system, and microinstrumentation systems including thermal, magnetic, and optical sensors are also investigated. The various constraints that limit the level of functional integration are discussed here, along with microsystems presently under investigation.

THE CHALLENGE

The rapid pace of developments in silicon sensor research over the last decade has resulted in a state of the art that is at a sufficiently advanced level to consider the viability of complete microsystems in silicon. The microinstrumentation system contains all the components of the data acquisition system, such as sensor, signal-conditioning circuits, analog-to-digital conversion, and interface circuits to an integrated data processor. Moreover, in principle, actuators could also be incorporated when required in the intended application. The silicon sensor is merely a component in the microinstrumentation system, which is, in turn, a part of a practical measurement and control system. Being able to construct microsystems in silicon is quite essential, as it implies that it would become feasible to replace an entire functional unit in a measurement and control system by plugging in another chip rather than just changing from a sensing element. In the latter case, the silicon sensor would have to compete with industrial standards like the PT-100 temperature sensor, and the main advantage of the silicon sensor—its material compatibility with microelectronic readout circuits—is not really relevant. In case of replacement of an entire functional unit, a significant added value is attributed to the silicon sensor concept, and, as a consequence, practical implementation of silicon-based microinstrumentation systems are more likely to become commercially interesting.

Research on microsystems in silicon can be divided into (1) the on-chip functional integration of the elements of the data-acquisition unit (the smart sensor concept); (2) the multiwafer systems with individually processed components that use wafer-to-wafer bonding or flip-chip bonding as an assembly technology; and (3) microassembly of preprocessed components.

THE APPEAL OF ON-CHIP INTEGRATION

The on-chip microinstrumentation system results from the objective to integrate all the components of the data acquisition unit on the smallest possible material carrier, the silicon chip, in an attempt to have a universally applicable sensor chip available for the measurement of a particular nonelectrical parameter that supplies an electrical output signal in a digital sensor-bus-compatible format and free from nonidealities originating from the basic sensing element such as offset; cross-sensitivities; and a low-level, noise-susceptible analog output signal. The driving force behind this approach is the unprecedented market acceptance of digital and analog integrated circuits in silicon. This is due to the fact that a specification of a digital integrated circuit in terms of the logic levels on the input and output terminals, combined with a functional description of the I/O sequences, including practical timing constraints, along with the range of power supplies, is sufficient for implementation and has strongly contributed to this rapid acceptance.

To a smaller extent, a similar mechanism applies for analog integrated circuits. A specification in terms of transfer function and characteristic input and output parameters, such as offset, impedance, equivalent input noise sources, voltage range, common-mode performance, and frequency range, is sufficient for application without the user having to go into the details of the internal operation.

Obviously, such a user acceptance would be highly desirable in the sensor market as well. One of the major disadvantages of state-of-the-art silicon sensors is that, like nonsilicon sensors, only the basic sensing element is available on the material carrier, and the designer or user of a measurement and control system has to go into the details of the sensor nonideality to add the remaining parts of the data-acquisition system. The electrical output can be a voltage, a change in capacitance, and so on, and it is strongly dependent on the type of sensor. Moreover, only the output signal of the basic sensor element is available, which implies that no compensation of the undesired characteristics mentioned is implemented.

THE TECHNICAL PROBLEMS AND THE ECONOMIC LIMITATIONS

The main technical problem in such an objective is actually to merge the processes that are run to fabricate the various components. Adding one type of device should not interfere with proper operation of the other. This problem was already relevant in the early 80s, when mixed analog/digital designs were to be merged in order to reduce the number of chips in mixed-mode electronic signal processing systems. The merging of fabrication processes for analog and digital circuits has resulted in the BICMOS process. An integrated silicon smart sensor imposes even higher demands on the compatibility of the processes involved. Not only are the analog and digital circuits to be merged, but the sensor also has to be incorporated. The degree to which the fabrication of smart sensors is hindered by the compatibility problem is strongly dependent on the actual nonelectrical signal domain.

Unlike the microelectronic circuits market, the sensor market is a niche market in measurement and control systems implemented in consumer products, professional equipment, and utilities. The measurement and control systems market is, therefore, basically a subcontractor of the final system assembler. This market is highly diverse due to the wide variety of end products and very conservative due to the emphasis on proven reliability often (indirectly) imposed by the final system user. As a consequence, high-volume production is rarely achieved, and the costs to enter the market are very high. The economics of the on-chip microsystem in silicon is determined by costs for testing, dicing, bonding, and packaging rather than wafer yield in mass production.

Dicing and bonding is a general problem in sensors because of the stresses imposed on the fragile micromachined structures. Unlike microelectronic circuits, not only do the electric parameters need to be specified but the nonelectrical parameters also need to be tested extensively in a sensor in order to characterize the sensor performance in terms of the sensitivity to the desired nonelectrical signal and the parasitic nonelectrical signal. Moreover, testing is cumbersome, as it is often only possible on completed sensors, as packaging-induced stress strongly affects the offset specification of a Hall plate. The latter problem strongly adds to the costs, as rejected devices have gone through the complete process.

The sensor packaging is very expensive, not only due to the limited-volume production mentioned already, but also due to the fact that the sensor package should on the one hand (like in microelectronic circuits) provide protection against environmental hazards, whereas, on the other hand (unlike circuits), the sensor by definition requires access to the nonelectrical measurand. These two requirements force a nonstandard package that allows selective access.

All the technological problems mentioned can in principle be solved using available techniques. Therefore, no real technological impediment is imposed to the general introduction of the on-chip functional integration of the data-acquisition unit. However, the market does not always agree with the smart sensor concept. As mentioned, its merit will only prevail in the case of large-series production, and, in practice, single-device postprocessing costs are dominating. The criteria for economic viability are often not met, apart from a few consumer applications. Therefore, the extent of on-chip functional integration should be evaluated at each additional component that is to be integrated in terms of the added value such an integration provides in the intended application. As an example, adding readout circuits to a capacitive sensor usually improves overall performance and makes the device suitable in applications that would otherwise be beyond reach and where the conventional approach (piezoresistive) also has a limited performance. This feature gives such an integration sufficient economic incentive next to the technological challenge. Adding the ADC, however, does not serve any economic purpose.

The research at the Delft University of Technology on on-chip integrated devices is concentrated on capacitive and matrix-type devices. If the operation of the system is based on the coordinated movement of the levers, the object can be moved over the entire extent of the array (Fig. 15.1). The active stroke of the actuators chosen here is upwards, making sure defective or damaged actuators will not inhibit the working of the device by sticking out and obstructing the movement of the object. The levers are implemented as electrostatically actuated surface micromachined structures. The actuators and the array are shown in Figure 15.2.

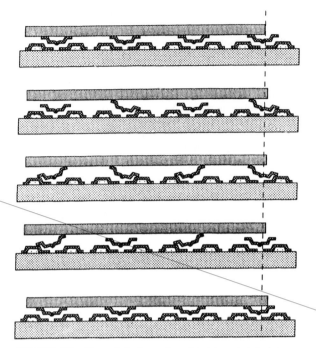

FIGURE 15.1 Upward stack-up of on-chip integrated devices.

FIGURE 15.2 An array of actuators.

This actuator makes it possible to combine differently oriented actuators to provide x-y positioning. By dividing such an array in several sections that can be driven independently, it is possible to provide a reference point to which an object can be transported independent of the original position (Fig. 15.3a), rotated (Fig. 15.3b), and moved to the desired position (Fig. 15.3c). The same technique makes it possible to position several objects relative to each other with great accuracy.

A one-dimensional array has been fabricated using the surface micromachining process. The actuators have an overall size of 32×22 μm including hinges and interconnections, giving a density of 1400 actuators per mm^2. The distance an object is moved in one cycle is 200 nm. Such small step sizes are typical of surface micromachined structures and facilitate accurate positioning.

WAFER BONDING AS A COMPROMISE

Wafer-to-wafer bonding involves the mating of processed wafers either by thermal treatment or by adding suitable intermediate adhesive layers. The main motivation for using wafer-to-wafer bonding is the opportunity to fabricate separately the sensor wafer and the wafer on which the microelectronic circuits are integrated, and thus to postpone any fabrication compatibility infringement, until the very last fabrication step. The sensor wafer can therefore be designed for maximum sensor performance without affecting the performance of active components. Moreover, the separation also allows for the use of different specialized foundries for the processing of the sensor wafer and the readout wafer. Wafer-to-wafer bonding takes place afterwards using relatively simple equipment and can basically be considered as assembly on the wafer level. This does not imply that no compatibility infringement could take place; compatibility problems may arise due to the bonding temperature and the wire bonding. Postponing the fabrication compatibility, therefore, also has important impacts on the economics of smart sensors. Sensor manufacturers often regard the wafer-to-wafer bonding as an intermediate step between fabrication of basic sensing element and integrated silicon smart sensors.

Flip-chip bonding is also employed to bond a completely processed wafer to a functional unit on the wafer level. Solder bumps are formed on both wafers to be bonded. One wafer is flipped over, and the wafers are subsequently brought into contact at an elevated temperature (below 400°C, depending on the solder used). The advantage of this technique is the high flexibility, as not only silicon wafers can be bonded together, but also different materials (e.g., GaAs on Si). Moreover, the

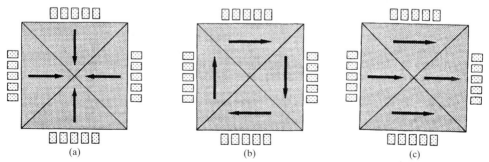

FIGURE 15.3 (*a*) Object transported for x-y positioning. (*b*) Object can be rotated. (*c*) Object is moved to desired location.

cohesive forces between the solder bumps pulls the wafers together after coarse alignment. This is called *self fine alignment*.

The practical application of wafer-to-wafer bonding is presently mainly in passive microstructures due to the fact that reliable low-temperature bonding is still very difficult to achieve. Moreover, the complexity is limited by the achievable wafer-to-wafer alignment. At the Delft University of Technology, research is being performed on gold eutectic bonding at 365°C. The results shown in Fig. 15.4 indicate the feasibility of a hermetic bond. However, the bond quality is strongly depending on bonding temperature, which so far prevents reliable bonding.

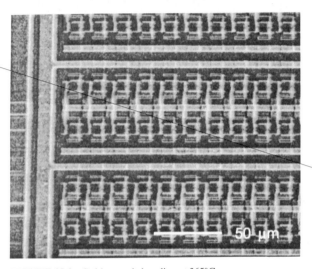

FIGURE 15.4 Gold eutectic bonding at 365°C.

THE MULTICHIP MODULE ON SILICON AS THE OPTIMUM SOLUTION

Generally, an optimum in on-chip functional integration leads to measurement systems composed of on-chip integrated sensor systems supplying a semidigital output signal (such as a frequency or pulse width modulation) and internally compensated for sensitivity for parasitic signals that cannot be accurately reproduced externally (such as chip temperature).

In the approach proposed here, integrated silicon sensors are to be designed with on-chip integrated sensor-related readout circuits, supplying a semidigital output signal such as a frequency that can easily be converted into a digital code using a counter. The on-chip functional integration is therefore pushed towards the level that is economically viable for reasons mentioned already. A multiple-sensor microinstrumentation system is composed of such smart sensors for the various nonelectrical parameters of importance in the intended application. The microinstrumentation system is to be fabricated on a standardized silicon platform. This is a chip-level infrastructure containing a floorplan for individual smart sensor die attachment and an on-chip local sensor bus interface, testing facilities, optional compatible sensors (such as thermal sensors), and a power management unit, as shown in Fig. 15.5.

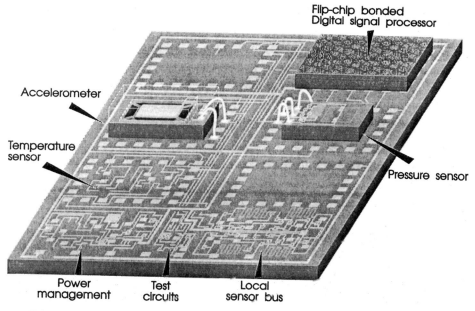

FIGURE 15.5 Power management sensor.

The instrumentation system is controlled by a microcontroller or a digital signal processor (DSP) depending on the amount of data processing required. The criterion for selection is minimum power consumption at sufficient processing power for the intended application. Both the DSP and the μ-controller can be commercially obtained at the unpackaged die level, and they are to be used for signal processing and interfacing a high-level sensor bus in a practical measurement and control system. After separate fabrication of sensor dies and platforms, these are to be bonded along with the DSP using conventional techniques.

CONCLUSIONS

Investigation of the optimum level of functional integration provides the best solution that is tailored to a specific application. Such an approach will maximize market acceptance of the microinstrumentation system.

CHAPTER 16
AUTOMOTIVE APPLICATIONS OF MICROELECTROMECHANICAL SYSTEMS (MEMS)

INTRODUCTION

This chapter will focus on current and future automotive applications of MEMS (microelectromechanical systems) and will present current and future market values to the year 2000 for various MEMS devices. Current product developments in pressure sensors, accelerometers, angular rate sensors, and other MEMS devices will be reviewed. Market and technology trends for MEMS in the automotive sector are addressed.

Microelectromechanical systems (MEMS) have played a major role in automotive engine control in the form of MAP (manifold absolute pressure) sensors since 1979. Today, virtually all automobiles have one of these devices in their electronic engine control system.

Early 1990s vehicles saw the first silicon accelerometer for an airbag crash sensor application. These devices have obtained extensive coverage in the popular and trade press. However, with the exception of these two applications, vehicle design engineers have not opted to include MEMS devices into their systems. The next five to seven years will provide significant opportunities for MEMS devices—whether it be in entirely new applications or in the replacement of traditional technologies. This movement is being fueled by the following factors:

- Longer warranty periods (up to 10 years/100,000 miles) require increased reliability components.
- Continuously developing federal and regional mandated fuel efficiency, emission, and safety standards.
- Higher vehicle performance and comfort.
- Availability of low-cost microcontrollers, memory, and displays.
- Enhanced vehicle diagnostics.

Certainly a significant current and future market for MEMS devices exists in the automotive sector. The worldwide MEMS market in 1995 was $2.7 billion (U.S.) with the automotive sector comprising 41 percent or $1.1 billion (U.S.). Projections for the year 2000 estimate the worldwide MEMS market at $12.2 billion (U.S.), with the automotive sector comprising 24 percent or $3.0 billion (U.S.).[2]

AUTOMOTIVE REQUIREMENTS

A number of stringent performance environmental, reliability, and cost requirements are imposed on automotive components. The harsh underhood automotive environment includes extreme temperature, shock, vibration, humidity, corrosive media, EMC, and a host of other environments. Tests to simulate these environments are outlined in SAE J1211, Recommended Environmental Practices for Electronic Equipment Design.[3]

In addition, automotive components must be able to be produced in extremely large volumes (typically one million or more per year in full production), have operating lifetimes of up to 10 years/100,000 miles, and have very low unit price. Essentially, we can consider automotive components to require the ruggedness of military parts with the price of consumer products. These qualities are inherent in MEMS.

Component cost factors are a significant issue in the selection of criteria of automotive system designers. The total cost of a sensor/actuator frequently includes the MEMS device, signal conditioning electronics (e.g., temperature compensation, filtering, amplification), packaging, connector/cable harness, and testing.

As a result, the cost of the MEMS device itself constitutes a minority of the total delivered component cost. Therefore, a significant challenge to achieve highly efficient design for manufacturability is imposed on all suppliers who wish to be successful in the automotive sector.

UNIQUE MEMS FEATURES

MEMS are well suited for a wide variety of automotive applications. Due to their batch processing manufacturing, large volumes of highly uniform devices can be created at relatively low unit cost. Since MEMS have virtually no moving parts, they are extremely reliable. Silicon has provided itself as a material for sensors in many applications over the last 20 years in applications including military, consumer, and automotive. Silicon's mechanical properties have been well documented.[4]

With the advent of microprocessor compatibility imposed on many automotive sensor/actuator applications, silicon is uniquely qualified to provide high levels of monolithic vertical functional integration using CMOS and classical micromachining processing.

SYSTEM APPLICATIONS

We have reviewed the use of MEMS in the following automotive systems:

- Safety
- Comfort, convenience, security engine/drive train

- Engine/drive train
- Vehicle diagnostics/monitoring

For each system, specific applications are noted (e.g., digital engine control fuel level) and the status of the application is given (i.e., future, limited production, major production). Also noted is the opportunity afforded to a MEMS solution versus that using another technology (e.g., piezoelectric, hall effect). The appropriateness criteria for MEMS selection was based on detailed market research conducted by the author and reported in Ref. 2. The most significant application opportunities will be addressed.

Safety

A summary of MEMS application opportunities in automotive safety systems is given in Fig. 16.1.

Airbag actuation is currently and will continue to be the major application of MEMS. Silicon accelerometers are currently being supplied to U.S. vehicles by I.C. Sensors (Fig. 16.2) and Analog Devices (Fig. 16.3). Sensinor (Norway) is currently providing the majority of silicon accelerometers to the European market, while Nippondenso is a major provider of accelerometers to the Japanese market.

Most recently, a number of manufacturers have investigated the use of compressed gas as a means to replace/supplement the sodium hazide explosive approach to airbag deployment. The use of a pressure sensor to monitor gas cylinder pressure is being actively investigated.

Suspension systems have been configured to provide the driver with optimum vehicle performance in high-speed cornering, rough roads, sudden braking, and acceleration. Numerous systems have been configured using total closed-loop con-

Application	Function	Production status	MEMS opportunity
Vehicle dynamics control	Wheel angle	Limited	Low
	Rotation	Current	Low
	Pressure	Limited	High
	Acceleration	Limited	High
	Valve	Future	Medium
	Angular rate	Limited	High
Air bag Actuation	Acceleration	Current	High
	Pressure	Future	Medium
Seat occupancy	Presence	Limited	Low
	Force displacement	Limited	Low
Object avoidance	Presence/displacement	Limited	Low
Suspension	Displacement	Limited	Low
	Acceleration	Limited	High
	Valve	Future	Medium
	Angular rate	Limited	High
Navigation	Angular rate	Limited	High
	Rotation	Current	Low
Rollover	Accelerometer	Limited	High

FIGURE 16.1 Applications of MEMS safety group.

FIGURE 16.2 Hybrid ASIC and silicon bulk micromachined capacitive accelerometer for ±50G airbag applications. (*Courtesy of I.C. Sensors.*)

trol of the suspension system. The fully active systems are extremely expensive ($5000). They consume significant horsepower to operate the hydraulic pump and add considerable weight to the vehicle. The enhanced performance attained by these systems has been marginal as compared to their cost. As a result, their implementation will be extremely limited. However, numerous suppliers have introduced semi-active systems. Some of these systems use displacement sensors in the shock absorbers and a number of linear accelerometers. This application is ideal for MEMS and companies like Analog Devices, I.C. Sensors, Lucas NovaSensor, and Motorola. These systems are currently offered as an option and cost in the $2000+ range. System manufacturers have a cost target of $900 in the near future.

Comfort, Convenience, and Security

A summary of MEMS application opportunities in automotive comfort, convenience, and security systems is given in Fig. 16.3.

Application	Function	Production status	MEMS opportunity
Seat control	Force	Limited	Low
	Microvalve	Future	Medium
Climate	Mass air flow	Future	Medium
	Temperature	Current	Low
	Humidity	Future	Low
Compressor control	Pressure	Current	High
	Temperature	Future	Low
Security	Proximity	Limited	Low
	Motion	Limited	Low
	Vibration	Limited	Low
	Displacement	Limited	Low

FIGURE 16.3 Applications of the MEMS comfort, convenience, and security group.

AUTOMOTIVE APPLICATIONS OF MICROELECTROMECHANICAL SYSTEMS

The measurement of compressor pressure in the vehicle air conditioning system offers the greatest opportunity for MEMS. Currently, other technologies (e.g., Texas Instrument ceramic capacitive pressure sensor) is being used. Major developments by a number of MEMS companies are actively pursuing this very large opportunity.

Engine/Drive Train

A summary of MEMS application opportunities in automotive engine/drive train systems is given in Fig. 16.4.

Electronic engine control has historically been and is expected to be the major application area of MEMS in automotive applications. Silicon manifold absolute pressure (MAP) sensors are produced by the millions by Ford and Delco. These devices provide an inferred value of air-to-fuel ratio by measuring intake manifold pressure. A great deal of effort has been undertaken to replace these devices with mass airflow (MAF) devices. Currently available on the market are discrete hot-wire anemometer devices (e.g., Hitachi, Bosch). Because of their construction, they tend to be large and expensive. Thick film equivalents of these devices have been recently introduced by Bosch at the 1995 SAE show. A MEMS version of this device is currently under evaluation by a number of research labs. In addition to the MAP/MAF devices, barometric pressure values are needed to provide the engine controller with altitude information to compensate a rich/lean fuel-to-air mixture. MEMS devices are well suited for this application.

Cylinder pressure values are of great importance to optimize engine performance; however, due to the extreme high-temperature levels, piezoelectric and fiberoptic techniques provide a much more pragmatic solution to this application. See Fig. 16.5.

Application	Function	Production status	MEMS opportunity
Digital engine fuel level	Control	Current	Low
Cylinder	Pressure	Future	Low
Manifold	Pressure	Current	High
Barometric	Pressure	Current	High
Engine knock	Vibration	Limited	Low
Air intake	Flow	Limited	Medium
Exhaust	Gas analysis	Future	Low
Crankshaft	Position	Major	Low
Camshaft	Position	Limited	Low
Throttle	Position	Future	Low
EGR	Pressure	Production	High
Fuel pump	Pressure	Future	High
Continuously variable transmission	Temperature	Future	Low
	Pressure	Future	High
	Microvalve	Future	Medium
Fuel	Pressure	Limited	High
Injection	Nozzle	Limited	High
Diesel turbo boost	Pressure	Limited	High

FIGURE 16.4 Applications of MEMS engine/drive train.

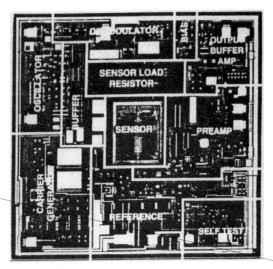

FIGURE 16.5 The ADXL05 ±2G silicon surface micromachined capacitive accelerometer for vehicle dynamic control features a high degree of on-chip functional integration. (*Courtesy of Analog Devices.*)

Silicon pressure sensors are currently being used in master cylinder brake pressure in the 1995 S-class Mercedes. In addition to a piezoelectric angular rate sensor, steering wheel angle and wheel speed sensors are provided for vehicle dynamic control. Currently, most other vehicles only use wheel speed sensors for their vehicle dynamic control (ABS, traction control). Since the unit price of the existing angular rate gyros similar to Bosch's piezoelectric vibrating cylinder[5] or Matsushita's tuning fork[6] are in the $150–$200 range, its implementation is relegated to the top-of-the-line-model vehicles. Currently, much work is being undertaken by vehicle manufacturers and first-tier suppliers (e.g., Bosch, Lucas, Temic, Siemens) to configure systems that are cost-effective. The availability of a low-cost, MEMS-based angular rate sensor similar to that developed by C.S. Draper Lab,[7] shown in Fig. 16.6, or Delco,[8] shown in Fig. 16.7, is expected to propel the adoption of these enhanced systems into less expensive vehicles.

Current navigation system designs use a combination of global positioning satellites (GPS) and CD-ROM maps in addition to wheel rotation sensors and rate gyros. Again, the current cost of these systems are expected to drop due to large-volume enhancement into less expensive vehicles.

Exhaust gas recirculation (EGR) applications exist in Ford and Chrysler systems. Currently, these applications are suitably accommodated by a ceramic capacitive pressure sensor provided by Kavlico. Here again, MEMS devices could provide a lower-cost alternative solution.

Continuously variable transmission (CVT) applications require pressure measurements in hydraulic fluids. MEMS devices that are isolated from the media using various techniques (e.g., isolated diaphragms) could find widespread application. Figure 16.8 shows a Lucas NovaSensor representative solution to this application.

The only known application of a MEMS device in a mechanical structure is in fuel injector nozzles. Here, Ford has micromachined silicon to create highly uniform

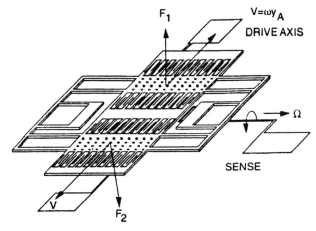

FIGURE 16.6 Schematic drawing of comb-drive silicon micromachined tuning fork gyro. (*Courtesy of Charles Stark Draper Labs.*)

FIGURE 16.7 Photomicrograph of silicon micromachined angular rate sensor. (*Courtesy of General Motors.*)

and circular orifices for fuel injection systems.

Vehicle Diagnostics/Monitoring

A summary of MEMS application opportunities in automotive vehicle diagnostic/monitoring systems is shown in Fig. 16.9.

One of the more interesting applications for MEMS is in tire pressure monitoring. For both safety and optimized fuel performance, proper tire inflation is necessary. A number of systems are currently being offered that provide real-time measure of tire pressure. MEMS devices are ideally suited and are being considered by a number of their manufacturers for this purpose. With the favorable acceptance of run-flat tires, these systems are expected to become very popular by the year 2000. Run-flat tires (e.g., Michelin 60-series) will eliminate the cost and weight of a spare and jack. Engine oil monitoring is a huge opportunity for MEMS. The greatest barrier to the adoption of these systems is price. These pressure sensors must be able to survive the elevated temperature requirements of engine oil and isolate the silicon chip from the media. The price target for this application is in the $5.00–$7.00 range for a fully signal-conditioned, packaged device. Sensor manufacturers currently are not able to meet this price target.

Recent legislation has created a major opportunity for pressure sensors in evaporative fuel systems. In this application, a pressure sensor is used to monitor the pressure level in the fuel tank and ensure that no fuel vapor escapes.

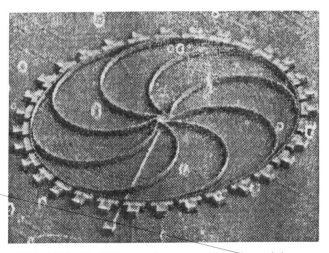

FIGURE 16.8 The NPI series of miniature silicon piezoresistive pressure sensors features isolated diaphragm design for harsh media compatibility. (*Courtesy of Lucas NovaSensor.*)

MARKET FIGURES

Market values in dollars and units for total North American consumption are given in Figs. 16.10 and 16.11, respectively.

According to the Roger Grace Associates Battelle Institute, the total dollar value for automotive MEMS was expected to grow to $912.1 million (U.S.) by the model year 2000, up from $174.4 million (U.S.) in model year 1990 and $423.4 million (U.S.) in model year 1995. This constitutes a compounded annual growth rate (CAGR) of 17.7 percent.

The total number of automotive MEMS was expected to grow to 130.1 million units by the model year 2000 up from 19.5 million units in model year 1990 and 57.3 million units in model year 1995. This constitutes a CAGR of 18.7 percent. Pressure sensor applications far exceed acceleration/angular rate sensor applications. Chemical MEMS and microstructure applications are expected to comprise approximately 5 percent of the total market by model year 2000.

Application	Function	Production status	MEMS opportunity
Tire pressure	Pressure	Limited	High
Engine oil	Pressure	Current	High
	Level	Future	Low
	Contamination	Future	Medium
Brake system	Pressure	Future	High
	Level	Future	Low
Transmission fluid	Pressure	Future	High
	Level	Future	Low
Fuel system	Pressure	Future	High
	Level	Future	Low
	Pressure (evap.)	Future	High

FIGURE 16.9 Applications of MEMS vehicle diagnostics/monitoring.

	1990	1991	1992	1993	1994	1995	1996	1997	1998	1999	2000
Acceleration	20	20	30	35	45	50	55	60	70	80	100
Pressure	0	0	0	0	5	6	7	8	10	12	15
Chem analysis	0	0	0	0	0	2	4	5	5	6	7

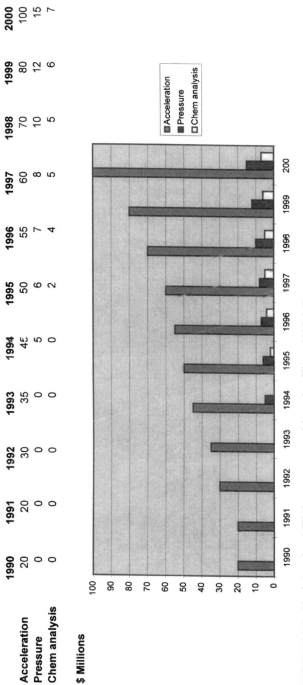

FIGURE 16.10 North American MEMS revenue per model year in millions of U.S. dollars.

	1990	1991	1992	1993	1994	1995	1996	1997	1998	1999	2000
Acceleration	0	0	0	0	5	6	8	14	18	20	22
Pressure	0	0	0	0	1	2	3	5	10	15	20
Chem analysis	200	200	300	350	375	425	500	600	700	750	775

$ Millions

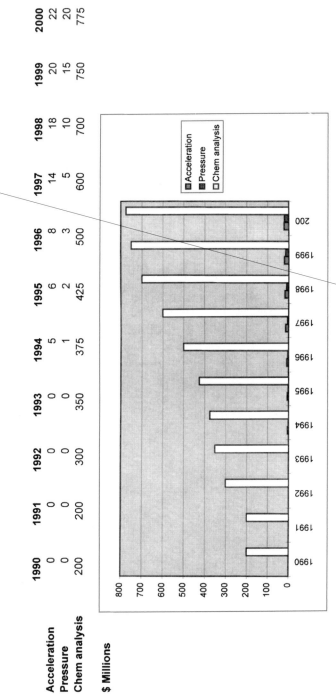

FIGURE 16.11 North American MEMS millions of units.

CONCLUSIONS

The automotive applications of MEMS are expected to constitute a significant segment of the MEMS market by the year 2000. Automobiles provide an excellent application opportunity for MEMS devices due to their low-cost, high-volume, and high reliability. The widespread adoption of the earlier manifold absolute pressure (MAP) sensor and the silicon airbag accelerometer have paved the way for further MEMS applications. The long development cycles of automotive manufacturers (e.g., United States = 5 years, Europe = 4 years, Japan = 3–3½ years) and their extreme cost sensitivity have precluded major adoption of other MEMS-based systems. However, as MEMS technology becomes more mature and its associated costs are reduced, we believe that MEMS devices will proliferate. These applications will be created by a displacement of older technologies by MEMS in addition to the stringent requirements necessitated by new applications.

Vertical functionality, whether monolithic or hybridized, will play a major factor in the proliferation of MEMS. MEMS enjoy the inherent feature of readily enhanced functionality. As the requirement for multiplex bus communication and the sharing of measurement values by different vehicle systems increases, the need for vertical integration will also increase. Implementation of a CAN bus by European auto manufacturers and in Chrysler's C2 D bus in the years 1988 and 1989, respectively, mark the beginning of the communication requirements. It is expected that this communication requirement will migrate from its present electronic control unit (ECU) location to a location within the MEMS device that allows enhanced distributed control.

REFERENCES

1. R. H. Grace, "Semiconductor, Sensors, and Microstructures in Automotive Applications," *Sensors and Actuators: Society of Automotive Engineers International Conference Proceedings,* 1991, pp. 245–260.
2. "Micromechanics: Multiclient Study," Battelle Institute, Frankfurt, Germany, July, 1992.
3. "Recommended Environmental Practices for Electronic Equipment Design SAE J1211," *1994 SAE Handbook,* vol. 2, pp. 23.228–23.244.
4. Kurt Petersen, "Silicon as a Mechanical Material," *Proceedings of the IEEE,* **70**(5), May 1982, pp. 420–457.
5. A. Reppich and R. Willig, "Yaw Rate Sensor for Vehicle Dynamics Control System," *Sensors and Actuators: Society of Automotive Engineers International Conference Proceedings,* 1995, pp. 67–76.
6. I. Toshihiko and J. Terada, *Angular Rate Sensor for Automotive Application, Sensors and Actuators: Society of Automotive Engineers International Conference Proceedings,* 1995, pp. 49–56.
7. N. Weinberg, et al., "Micromachined Comb Drive Tuning Fork Gyroscope for Commercial Applications," *Sensors Expo Proceedings,* September 1994, pp. 187–194.
8. J. Johnson, S. Zarabaldi, and D. Sparks, "Surface Micromachined Angular Rate Sensor," *Sensors and Actuators: Society of Automotive Engineers International Conference Proceedings,* 1995, pp. 77–83.

CHAPTER 17
A BRIEF STUDY OF MAGNETISM AND MAGNETIC SENSORS

INTRODUCTION: QUALITATIVE DESCRIPTION OF MAGNETIC FIELDS

The various units in which magnetic fields are measured are detailed in this section, as well as conversions between those units. The magnetic details of what is happening inside a ferromagnetic material are also detailed.

The general shape of a magnetic field can be illustrated by drawing a set of magnetic flux lines in the vicinity of a magnetic field source. Figure 17.1 shows the external magnetic fields for a variety of permanent magnets. We note several items:

1. These magnetic flux lines are continuous.
2. They form from the north magnetic pole(s) to the south magnetic pole(s) of the magnet.
3. The intensity of the magnetic field is proportional to the density of the flux lines in the drawing.

The magnetic field at each location in space is a vector quantity. At all points, the magnetic field has both a magnitude and a direction. Therefore, for a given coordinate system, the field can be represented by $x, y,$ and z components (see Fig. 17.2). It will be illustrated that this is very important when dealing with different types of magnetometers which are used to measure magnetic fields. Some magnetometers measure the total field intensity (the magnitude of the field vector) while others measure only one of the components of the field. Depending upon the application, each of these types have appropriate purpose.

The fields shown in Fig. 17.1 vary in strength and direction in the space around the magnets. There are some means of creating magnetic fields which are nominally uniform. A uniform field will have the same magnitude and direction over a region in space. This is often very useful in evaluating magnetic field measuring devices.

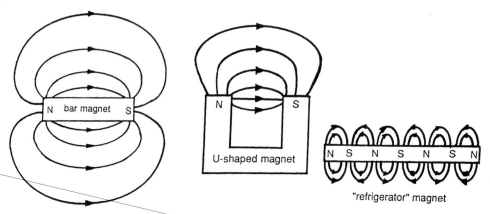

FIGURE 17.1 The magnetic field shapes for several magnet shapes.

THE SI AND GAUSSIAN UNITS

The field of magnetism and magnetic materials has grown up over the past century in labs and manufacturing centers worldwide. The applications of magnetism are so diverse that several different formalisms have been developed to describe magnetism. This means that today there are many different units in which magnetic phenomena are described. Even the defining relations between magnetic quantities differ in the different systems. This makes the field of magnetism most confusing, not only to those new to the field, but also to those who have been working in magnetism, who are asked to explore other applications and/or to communicate with those working in different systems of units. In order to minimize the difficulty in understanding, two approaches will be considered:

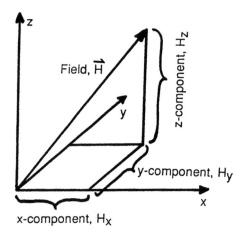

FIGURE 17.2 The x, y, and z components of a field vector H.

1. Detailing of the quantities (and the units) required to describe ferromagnetic materials.
2. Mainly, two systems of units, the *SI* (*System International d'Unites*) system and the *gaussian* system. These are the two most often used systems of units.

When discussing magnetic fields (in air or in non-ferromagnetic materials), two basic quantities are used to describe the field. The first is the *magnetic flux density* (or the *magnetic induction*) which is represented by the letter B. In the gaussian system of units, B is measured in gauss (G). In the SI system, B is measured in tesla M. It should be noted here that one tesla is also equal to one weber/m² (Wb/m²). Note that a weber is a measure of the flux lines previously described. Therefore, a Wb/m² is, as it appears, a measure of density of these flux lines. There are 10,000 gauss in one tesla.

The other quantity which is used to describe magnetic fields is called the *magnetic field strength* which is represented by the symbol H. In the gaussian system, this is measured in units of oersteds (O_e). In the SI system, H is measured in units of amperes/meter (A/m). One oersted is equal to 79.58 amperes/meter. Another unit which is sometimes used for very small magnetic fields is the *gamma*. There are 100,000 gamma in one oersted. Table 17.1 summarizes this information.

One of the convenient relations in the gaussian system is that in air (or where there is no ferromagnetic material) one gauss is equivalent to one oersted. This is why, for many workers in the field, the units of gauss and oersted are often used interchangeably, even though they are really not the same quantity. It should be noted that there is a general movement toward using SI units worldwide for magnetic quantities.

To illustrate the strengths of commonly encountered magnetic fields, one may observe that the earth's field is nominally

$$H = 48 \text{ A/m } (0.6 \text{ O}_e)$$

That gives a magnetic flux density,

$$B = 0.6 \times 10{-4} \text{ tesla } (0.6 \text{ gauss})$$

A large laboratory electromagnet may produce fields on the order of

$$H = 2 \times 10^6 \text{ A/m } (25,000 \text{ oersteds})$$

with

$$B = 2.5 \text{ tesla } (25,000 \text{ gauss})$$

Common bar magnets can provide maximum fields H in the range of 8000 to 12,000 A/m (100 to 1500 O_e) with

$$B = 0.01 \text{ to } 0.15 \text{ tesla } (100 \text{ to } 1500 \text{ G})$$

FIELD SOURCES

One source of magnetic fields, shown in the introduction, are magnets, which are magnetized pieces of a ferromagnetic material. There are north and south magnetic poles located at the surface of a magnet. We now digress briefly into a discussion of what happens inside of these ferromagnetic materials.

A ferromagnetic material (e.g., iron or nickel) is one in which

1. The atoms themselves inherently have a magnetic dipole field (this is due to the collective magnetic field from some of the atoms' electrons which create a magnetic moment).
2. The adjacent dipoles (magnetic moments) from each atom are aligned along the same direction (this is called *ferromagnetic coupling*).

Thus, there are regions of the ferromagnetic material in which the atoms all are magnetically pointing in the same direction. These regions are called *magnetic domains*. If one puts a ferromagnetic material in a strong magnetic field, a majority of the domains can be nominally oriented along the direction of the applied field. This process is called *magnetizing the material*. Figure 17.3 shows a rectangular block of a ferromagnetic material with an indication of what these magnetic domains might look like. Note that the magnetizations of these domains (represented by the arrows) are mostly pointing to the right—this block is magnetized to the right. The magnetized material forms north and south magnetic poles where the magnetization vectors in the material point out through the surface. This also creates the external magnetic field as shown in Fig. 17.1. *Demagnetization* is the reverse process to magnetization. When a material is demagnetized, the magnetic domains are randomly oriented so that there is no net magnetization. This is often done with an AC magnetic field that's amplitude is slowly decreased. Another name for the demagnetization process is *degaussing*.

Note that, for the most part, materials we work with in sensors will fall into one of two (magnetic) categories. They are either ferromagnetic or they are non-ferromagnetic. For our purposes, materials such as air, brass, aluminum, and many stainless steels are non-ferromagnetic. This is in contrast with ferromagnetic materials such as iron, cobalt, nickel, steel alloys, and some stainless steels. Ferromagnetic materials are also known as *ferrous* materials. So called permanent magnets are made of ferromagnetic materials which, once magnetized, are harder to demagnetize.

What does this mean in relation to this tutorial? Non-ferromagnetic materials do not perturb the magnetic field lines (and, of course, these materials are not pulled toward a magnet). The magnetic field goes right on through these materials as if it were a vacuum (or air). On the other hand, when a ferromagnetic material is put into a magnetic field, the magnetic field lines tend to be attracted into the ferrous material. Therefore, the ferrous materials perturb (often strongly) the magnetic field. This is shown in Fig. 17.4.

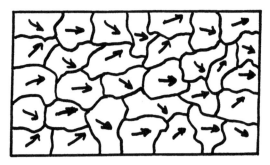

FIGURE 17.3 A block of ferromagnetic material showing the domain structure when the block is magnetized to the right. Magnetic domain walls separate adjacent domains. The average magnetization direction is to the right.

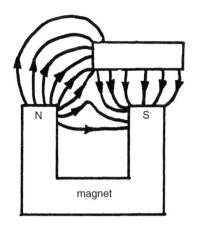

FIGURE 17.4 The magnetic field of a U-shaped magnet being distorted by a piece of ferromagnetic material.

Note that people often put a ferrous block of material across the poles of a magnet when they are not using it. This block is called a *keeper*. This keeper attracts most of the magnetic flux lines. This keeps the magnet from being weakened by nearby ferrous materials.

One application for ferromagnetic materials is to provide magnetic shielding (see Fig. 17.5). Here, we see a ferromagnetic box placed in a magnetic field. The magnetic flux lines in the region around the box will be attracted into the ferromagnetic box. This technique can be used to create regions (e.g., inside the box) where the magnetic field strength is greatly reduced.

The other major source of magnetic fields (other than a magnetized ferromagnetic material) is electric currents. When a DC electric current flows in a wire, a magnetic field is created around that wire (see Fig. 17.6). In this case, the magnetic field wraps around the wire as shown. The intensity of the field is inversely proportional to the distance from the wire. The equation for the magnetic field near a long straight wire is

$$H = \frac{I}{2\pi r}$$

where I is the current in the wire, and r is the distance from the wire to the point of interest. If I is in amperes and r is in meters, the magnetic field is in A/m.

A solenoid coil is formed when wire is wrapped into a long coil whose length is much larger than its diameter. The magnetic field in the center of the solenoid is

$$H = NI/L$$

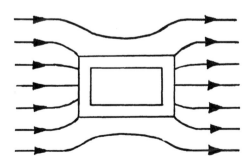

FIGURE 17.5 The shielding effect of a ferromagnetic box in a magnetic field. The magnetic flux lines go into the box and tend to stay within the ferromagnetic material. This leaves the box with very little magnetic field.

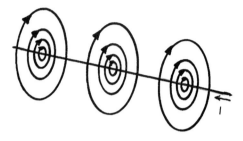

FIGURE 17.6 The magnetic field due to a long straight wire with a current I is circular around the wire. The intensity of the field drops off as $1/r$, where r is the distance from the wire.

where N is the number of turns, I is the current flowing in the wire, and L is the length of the solenoid. This is true for DC currents. Note that if AC currents are used, the inductance created by the coil complicates the situation. The magnetic field created by a current in a solenoid looks just like the magnetic field from a cylindrical bar magnet.

Another coil system of interest is known as a *Helmholtz pair*. Here, two simple, identical, circular coils are arranged such that the spacing between the coils is equal to the radius of the two identical coils. This combination has many applications because the magnetic field in the central region is very uniform. For many applications, such as magnetic sensor testing, it is essential to have a uniform field. The magnetic field in the central region of the Helmholtz pair is

$$H = 0.7155 \, NI/r$$

where N is the number of turns in one of the two coils, I is the current flowing in the coils, and r is the radius of the coils (also the spacing between the coils).

In order to get stronger magnetic fields, it is common to build an electromagnet. The magnetic field for a coil wrapped over a ferromagnetic material is created mostly due to the magnetization in the ferromagnet, but the field can be controlled by varying the current in the coil. Figure 17.7 shows a typical electromagnet where the iron core is formed into a shape such that the two magnetic poles are relatively close together. This not only provides larger magnitude fields, but also can provide a field which is uniform in the central region between the two poles.

The one magnetic field which we all experience is the earth's field. It is believed that the earth's field is formed by electrical currents flowing in the earth itself. This is not fully understood at this time. However, the magnetic field we observe is very similar to the dipole field formed by a bar magnet or a current-carrying coil. Figure 17.8 shows a nominal crosssection of the earth with its magnetic field. Note that the north magnetic pole is in the vicinity of the south geographical pole and the south magnetic pole is in the far northern Canadian island area. This north and south pole terminology developed when people suspended magnets on a string and found that the magnet would rotate so that one end of the magnet would point to the north (the north seeking pole). This end of the magnet was therefore called the north pole of the magnet (and vice versa for the south pole). Since opposite poles attract, the earth's south magnetic pole is the one in the northern hemisphere.

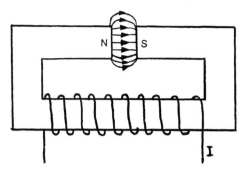

FIGURE 17.7 An electromagnet consisting of a current-carrying coil of wire wound around a ferromagnetic frame (e.g., iron). This design is often used because it creates a nominally uniform magnetic field between the pole pieces.

Note in Fig. 17.8 that near the equator, the field direction is nominally horizontal, pointing to the north. Near the magnetic poles, the field is essentially vertical. Note that in the midlatitudes in the northern hemisphere, the field points downward at an angle into the earth's surface. This angle is called the *magnetic angle of inclination*. Recall that on the surface of the earth, the average strength of the earth's field is about 0.6 O. (48 A/m), but this value does vary. Since we use the earth's magnetic

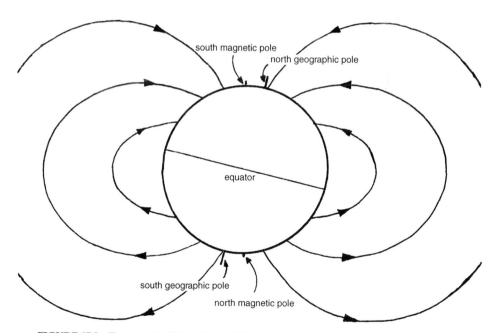

FIGURE 17.8 The magnetic field of the earth looks very much like a dipole field. The north magnetic pole is located in the southern hemisphere and vice versa. Note the magnetic poles and the geographic poles do not coincide.

field to direct our compasses, note that the horizontal component of the field is used for compassing. This is why compasses may provide confusing data near the magnetic poles. Also note that the earth's magnetic field (both magnitude and direction) may be locally perturbed by iron ore deposits, steel structures, and so forth.

AC FIELDS AND DC FIELDS

The preceding discussions assumed that direct current (DC) was flowing in the wires. It should be noted here that magnetic sensors are often used to detect/measure fields which are alternating at some frequency. There are a number of interesting frequency ranges. The first is the *near* DC range of 0.001 to 1 hertz. This is the frequency range of magnetic signals due to ships, submarines, and land vehicles disturbing the earth's magnetic field as they move.

The one to several hundred hertz range is of interest to those looking at the fields created by power lines and electrical machinery. This is also the range of interest for many magnetic proximity sensors. Above several hundred hertz there are many applications of magnetic sensors, some of which operate at frequencies into the megahertz range. These include such applications as geartooth sensors, which measure the rotation of a shaft.

It should be noted that additional questions arise when AC fields are used. In particular, AC magnetic fields generate current flows in nearby conductors. These are called *eddy currents*. In some cases, one does everything possible to minimize these eddy currents because they absorb energy from the system and they provide additional magnetic fields. Remember that all current flows create magnetic fields. In other cases, the presence of the eddy currents is used in the design of some magnetic sensors.

MAGNETOMETERS AND APPLICATIONS

There are many different technologies which are used to measure magnetic fields. We describe several of the more common types and some of their applications here. It should be noted that each of these technologies may have many design variations which are particularly suited to a given application.

A coil of wire becomes a magnetometer when there is relative motion between the coil and the magnetic field being measured. If the field being measured is a DC field, the coil can be rotated to generate an EMF (voltage) in the coil (and a current if the circuit is closed). The amplitude of this signal is proportional to the intensity of the magnetic field present. For AC fields, the coil does not have to be moved at all. The alternating magnetic field will generate an alternating EMF in the coil. Again, the amplitude of the signal is proportional to the magnetic field amplitude. These coils can be easily calibrated and are used for lots of applications. Currently, the majority of antilock braking sensors (ABS) use this approach. A permanent magnet provides the field that is modulated by the gear teeth attached to the axle. This provides an AC electrical signal that's amplitude and frequency is proportional to the wheel speed.

Another application of coils as magnetic sensors is in proximity sensors. Typically, a coil and a ferromagnetic core or frame are used to make a magnetic circuit with a gap. When a ferromagnetic material is placed in or near the gap, the inductance of the coil changes. This is often used, for example, in factory applications where the presence or absence of a ferromagnetic object is to be determined. When an AC signal is provided to the coil, this approach can also be used to detect electrically conducting materials (such as other metals) by looking for the effect of the eddy currents in the object to be detected.

Hall sensors make use of the Hall effect in semiconducting materials. This effect was discovered in 1879 by Edwin Herbert Hall. Today, these sensors have become very common in the marketplace. The most common material for the Hall effect transducer is silicon, although other materials such as indium antimonide, indium arsenide, and gallium arsenide also are used. These transducers are designed and built using integrated circuit technology. In fact, many of them have integrated electronics on the same chip. The most commonly used Hall effect involves passing a current through the semiconductor. When a magnetic field is applied perpendicularly to the surface of the chip, a voltage is developed across the sides of the chip. This Hall voltage is proportional to the applied field. These Hall effect sensors find application in many areas such as automotive sensors, industrial proximity sensors, and keyboards. The span of operation is generally in the milligauss to 2000-gauss range.

Fluxgate magnetometers are used for many applications ranging from laboratory measuring instruments to compassing (both on the earth and in satellites). The fluxgate magnetometer makes use of several coils wound around two ferromagnetic rods (or a single ferromagnetic toroid). One coil set supplies an AC drive field which alternately magnetizes the ferromagnetic rods in opposite directions. A sensing coil which is wrapped around both coils then provides an output signal. When there is no external field (the field being measured), the net signal is zero since the net magnetic flux is zero. However, when there is an external field component along the length of the rods, there is a timing shift when the material becomes magnetized during each half cycle of the AC drive field. This creates a signal in the sensing coil at twice the frequency of the drive frequency, the amplitude of which is proportional to the strength of the field being measured. Note that the fluxgate sensor measures one component of the magnetic (vector) field. By combining three of these devices, all three of the vector components can be measured and the total field magnitude can be calculated.

Magnetic resonance (also known as *nuclear magnetic resonance, NMR*) magnetometers measure the total magnetic field amplitude with no direction information. This technology uses the resonance of the magnetic moment of the nuclei of atoms. Typically, two sets of mutually orthogonal coils are used around a medium containing the nuclei to be resonated. Often water is used as the medium and the resonance is observed on the nuclei of the hydrogen atoms. One of the coil sets is used to provide a radio frequency (rf) signal to excite the nuclei and the other is used to sense when the nuclei are resonating). This type of sensor can be very accurate since the magnetic field value is proportional to the frequency of the signal. Frequencies can be measured very accurately. This technology can be used over a wide range of fields from 10^{-14} to many teslas.

Another group of magnetic sensors make use of the magnetoresistive (MR) effect in which the resistance of a conducting strip is a function of the direction and magnitude of an applied magnetic field. In particular, the applied field rotates the magnetization of the conducting strip. This rotation of the magnetization causes a change in resistance. The most commonly used material for MR sensors is permalloy, an alloy of nickel and iron. Even though the resistance change is only about 2 percent, this technology is finding more and more applications. MR sensors are most commonly implemented using sputtered thin films (thicknesses on the order of 50 nanometers) of permalloy on an insulating substrate. This allows transducer fabrication using integrated circuit technology to be made very small in some cases. MR sensors may be used for compassing, geartooth sensing, proximity sensing, and current sensing. They can be used from the microgauss to thousands of gauss range.

In 1989, a group of French scientists announced the discovery of an effect which they called *giant magnetoresistance ratio (GMR) effect*. At certain temperatures and high magnetic fields they were able to see over 50 percent change in the resistance. Many groups around the world are now studying GMR and trying to see how to make sensors using it. This is still a rapidly changing field; however, it is fair to say that there now appear to be three to five different groups of materials which show GMR. The mechanisms of how they work are not yet fully known. Generally, the deficiencies of most of the GMR devices to date have been thermal effects and the higher magnetic fields required to change the resistance. It is expected that some of these materials will eventually move from the laboratory into manufacturing. Several early devices have been demonstrated and several companies are sampling GMR sensors for evaluation.

One of the most sensitive magnetic sensors is the *superconducting quantum interference device (SQUID)*. These devices use a small conducting loop to gather magnetic flux (much like the coil approach discussed previously). The current from this loop is then detected by a gap-based detector which is cooled to superconducting temperatures (4 to 30 kelvins). This detector can measure the quantified flux units which are very, very small. These sensors are used to detect very small magnetic field changes, such as those from human brain activity or the perturbation of the earth's field by a ship or submarine a long distance away. While these are some of the most sensitive magnetic sensors, they tend to be very expensive due to the superconducting temperatures required.

CONCLUSION

This has been a very brief overview of magnetism, magnetic fields, magnetic sensors, and applications, intended for people new to the field who need to become somewhat literate in the field. Magnetism tends to be a confusing field due to the strange combinations of units which are used to describe it. However, on the positive side there are many, many applications of magnetism in the field of sensors and more are being found every year. As there is greater demand to make better and less expensive magnetic sensors, there will be continued evolution in this most interesting field.

CHAPTER 18
THE FUNDAMENTALS AND VALUE OF INFRARED THERMOMETRY

INTRODUCTION

In the material-processing world, heat and temperature are vital control parameters. Whether it is steel, glass, electronics, cardboard, frozen foods, tires, or paper, at some point in the manufacturing process, heat is applied or removed. Control of this heating process and the material temperature affects product quality, energy consumption, material throughput, operating costs, and profitability. The trade-off for not controlling temperature is often a sacrifice of performance relative to one of the processing factors. Consequently, the control of temperature and these processing factors is critical to optimizing the performance of any material-processing operation.

Considering energy consumption without temperature control, the approach can be to overheat materials in order to insure that the prescribed product properties are achieved. Based on a typical processing heat balance (Fig. 18.1) where processing and equipment factors affect the efficiency of the operation, there is a substantial price to be paid for overheating. As illustrated in the graph, a 5 percent or 100°F increase over the required temperature can represent a 17 percent loss in energy. In a steel mill or glass plant, that represents millions of dollars per year in fuel costs. At lower temperatures, the heat losses are less dramatic, but they are still measurable and significant.

Another approach to operating without temperature control involves processing at reduced throughput rates to insure proper results.

In aluminum extrusion processing, where accurate temperature measurement was not possible in the past, presses where run at very low speeds to insure proper material properties and to minimize the reject rates of materials. Now, with a new infrared technology, noncontact temperatures may be used to maximize throughput rates and eliminate finished material scrap rates. This ability to accurately measure the temperature of the billet entering the press and the extruded product exiting the press has reengineered the process and taken the aluminum extrusion business to a new level of performance using process control and automation.

The benefits for aluminum extruders have been millions of dollars in savings per press by increasing throughput 30 to 50 percent with the elimination of finished product waste and higher press-operating rates. From a capital investment viewpoint, this increased throughput can also postpone capital investments in new presses by allow-

Energy Loss	5	7	10	28	58	110
Furnace Temperature, F	0	500	1000	1500	2000	2500

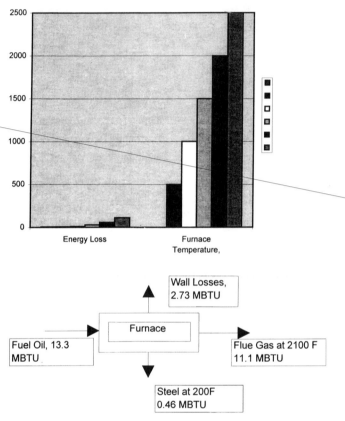

FIGURE 18.1 Typical processing of heat balance.

ing three presses to operate at the capacity of four under the old standards. This is just one example of what people are doing to maintain a competitive advantage in the world markets today using infrared temperature measurement control.

At first glance, some people consider infrared thermometry to be too expensive and complicated to install and maintain. However, this is a misnomer, as these sensors are typically simple to install and operate and are not expensive relative to the benefits of the investment. On average, paybacks have been estimated from two days to two months.

The advantages of infrared thermometers over other temperature-measuring technologies are as follows:

- *Better accuracy* because they measure target temperature (versus its own temperature)
- *Flexible implementation* because noncontact capabilities can be used for measuring moving and intermittent targets, materials in vacuums and electrical fields, as

well as applications involving hostile environments with excessive ambient temperatures and poor operating conditions with smoke, oils, and other obstructions
- *Timely feedback* with fast sensor respond times (10 to 500 ms)

To realize the true potential of the infrared sensor's capabilities, it is best to consider these sensors as a solution to a problem and not just a device to measure temperature.

The following sections describe the fundamentals of infrared thermometry and the different types of sensors and their applications. The objective is to provide the background and the information necessary to properly select and implement the sensor configuration that best meets the requirements of the application.

FUNDAMENTALS OF INFRARED THERMOMETRY

Every object emits radiant energy and the intensity of this radiation is a function of the object's temperature. Noncontact temperature-measuring sensors simply measure the intensity of this radiation. The general relationship of radiant energy (intensity) as a function of temperature and wavelength for a blackbody is illustrated in Fig. 18.2. These blackbody radiation curves are defined by basic laws in physics and are used selectively as the basis for infrared thermometry.

This infrared radiation is similar to visible radiation (0.45 to 0.75 microns) except it has longer wavelengths. It consists of photons that are a form of energy that travel in a straight line at the speed of light (9.83571030×10^8 feet per second) and can be reflected off or transmitted through selective materials. This radiant energy can be seen and felt everyday as the warmth of the sun or the glow of an electric burner or flame. These examples relate to the visible segment of the electromagnetic spectrum where the human eye has sensitivity. The infrared region is an invisible segment of electromagnetic spectrum and represents a substantial form of heat energy.

The infrared region of the electromagnetic spectrum is typically defined in terms of microns and is illustrated in Fig. 18.3 referencing selected infrared filtering used in infrared thermometers. Short-wavelength sensors are typically used for high- and medium-temperature applications as there are high signal levels and technical advantages in this region. For low-temperature applications, the shift is to longer wavelengths and broader band filtering (8 to 14 microns) to maximize the radiant energy measured. In between, a variety of narrowband filters are used to optimize application and sensor-measuring characteristics. For example, the selection of certain filters in atmospheric windows eliminates adverse effects due to intervening moisture and distance sensitivity.

The selection of infrared filtering also determines what type of window materials may be used, if necessary, for a specific application. In the short-wavelength region, a normal glass window (bora silicate) is applicable, while quartz and germanium windows are typically used for mid- and long-wavelength sensors, respectively (Fig. 18.4). These various materials are also used as part of the infrared optics systems as lenses to collect energy from the target and focus it onto the infrared detector.

The adjustable response time of an infrared thermometer will typically cover the range of 100 ms to 10 seconds as required to obtain 99 percent of a reading. For very fast application requirements of 5 to 10 ms, a silicon or germanium detector–based sensor can be used. On average, many instruments use the adjustable response capability to operate in the 1- to 2-second range in order to dampen application noise. However, there are cases involving induction heating and other fast heat sources

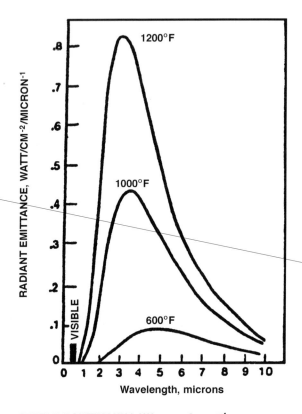

FIGURE 18.2 Blackbody radiation characteristics.

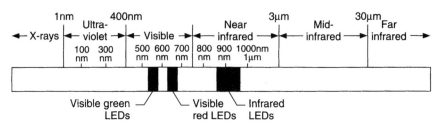

FIGURE 18.3 The light spectrum.

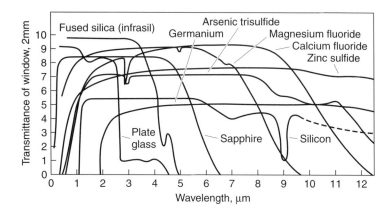

FIGURE 18.4 Behavior of infrared materials.

that require responses in the order of 10 to 50 ms, which is attainable through specialized infrared sensors. Correspondingly, the control system must also be set to operate at the appropriate speed with the sensor in order to implement a complete temperature control system.

Optical resolution or *field of view (FOV)* configuration is another important principle of infrared thermometry. While for a majority of applications it is not a major issue, there are certain applications which require more precise selection of the FOV configuration. A typical sensor will look at an area 1 inch (2.5 cm) in diameter when the sensor is 15 inches (38 cm) away from the target. If an application involves a small target (0.125 inches/0.32 cm) or a very small target (0.030 inches/0.8 mm), then premium optics can be used for precise alignment and accurate temperature measurement. Correspondingly, distant optics, sighting targets 10 to 1000 feet (3 to 300 m) away, also require a special optical configuration.

An alternative optical configuration involves the use of fiber optics. This offers engineering the flexibility to install the sensor electronics remotely from hostile application environments. This can eliminate electrical noise interference and resolve many accessibility and space concerns. The most common application is to locate the electronics in a remote location away from high ambient exposures. Another is to measure temperature where induction, microwave, or RF heaters are used and EMI is excessive for electronic systems. Cable lengths vary from 3 feet (1 m) to 20 feet (6 m) with the occasional applications where 50 feet (15 m) may be required. These fibers are very tough and durable and can handle ambient exposures up to 400°F (200°C) and, for steel mill applications, air-purged heavy duty conduit can operate in temperatures as high as 800°F (425°C). Infrared fiber optics technology continues to improve in transmission and durability, and applications continue to multiply.

The fundamental components of the optoelectronic configuration of the sensor (Fig. 18.5) are divided into a steering design that uses a collecting lens, a selective narrowband infrared filter, a selective detector that converts infrared energy to a form of electrical signal, and electronic circuitry that amplifies, stabilizes, and linearizes the signal to deliver an output that is proportional to the target temperature. An alternative configuration follows the same design chain but utilizes a filter/chopper that sends infrared pulses to the detector where these signals are then conditioned to indicate the target temperature. A major portion of this technology was derived from military and aerospace projects where unique infrared filtering and detectors have been developed as part of defense and space systems.

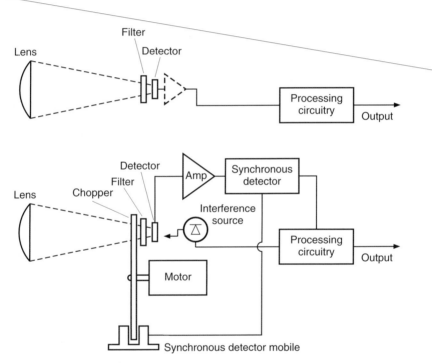

FIGURE 18.5 Fundamental components of optoelectronic configuration of sensor.

The curves in Fig. 18.2 illustrate the theoretical energy distribution for blackbody conditions. However, most materials radiate less energy than a blackbody and this reduction in energy is defined as emissivity. *Emissivity* is a nondimensional factor that is a measure of the ratio of thermal radiation emitted by a graybody (nonblackbody) to that of a blackbody at the same temperature and wavelength. A *graybody* refers to a surface that has the same spectral emissivity at every wavelength, whereas a *nongraybody* is a surface whose emissivity changes with wavelength, such as aluminum.

$$E = L(\text{graybody})/L(\text{blackbody})$$

The law of conservation of energy states that the coefficient of transmission, reflection, and emission (absorption) of radiation must equal 1:

$$t + r + a = 1$$

and the emissivity equals absorbtivity:

$$E = a$$

Therefore,

$$E = 1 - t - r$$

This emissivity coefficient fits into Planck's equation as a variable describing the target's surface characteristics relative to the wavelength. The majority of surfaces measured are opaque, and, consequently, the emissivity coefficient can be simplified to

$$E = 1 - r$$

which implies that a less theoretical definition of emissivity is that it is the opposite of reflectivity. Exceptions are materials such as glass, plastic films, and silicon, but through proper spectral filtering it is possible to measure these materials in their opaque infrared region.

There is typically a lot of confusion regarding emissivity error, but the user need remember only four things:

1. Infrared sensors are inherently color-blind.
2. If the surface is visually reflective (like a mirror or shiny steel), beware—you will measure not only the emitted radiation as desired, but also reflected radiation.
3. If you can see through it, you need to select infrared filtering, for example, glass is opaque at 5 microns.
4. Nine out of ten applications do not require absolute temperature accuracy. Repeatability and drift-free operation yield close temperature control, so fine-tuning the emissivity coefficient is not critical.

If the surface is shiny, there is an emissivity adjustment that can be made either manually or automatically to correct for emissivity error. It is a simple fix for most applications. In the case where emissivity varies (one knows this condition exists when the reflection of the surface is changing) creating processing problems, dual- or multiwavelength radiometry can be used to eliminate the emissivity problem.

THE SELECTION PROCESS

Single wavelength thermometry measures total energy emitted from the target at a prescribed wavelength. These sensors are available as portables, two-wire transmitters, sophisticated online systems and scanning devices. They are available with visual aiming, laser alignment, nonaiming, fiber optics, water cooling, lens air purging, and a variety of bracketry to simplify installation and operation. The online sensors typically have a linear output of 4 to 20 mA that is used to drive remote displays, controllers, data loggers, and/or computers. Applications cover very basic manufacturing operations involving moving webs of paper, plastic, rubber, textiles, as well as a variety of processes where product temperature measurement versus air or heater temperature can improve throughput and deliver a consistent quality product.

The selection of the infrared spectral response and temperature range is determined by the specific application and is very straightforward. Short-wavelength sensors, filtered in the 0.8- and 2.2-micron regions are used for high- and medium-temperature applications, respectively, such as foundries, heat treating, glass melters, steel and semiconductor processing. The 3.43- and 7.94-micron designs are used for measuring various plastic films as most plastic materials have absorption bands (opaque) at these wavelengths. By filtering in this region, the emissivity coefficient is simplified and allows for the use of single-wavelength sensors. Likewise, most glass-type materials become opaque at 4.6 microns and narrowband filtering at 5.1 microns permits accurate measurement of glass surface temperature. On the other hand, to look through a glass window, a sensor filtered in the 1 to 4 micron region would allow easy access via viewing ports into vacuum and pressure vessels. The 5.1-micron filtering is also used for drying and heating applications where quartz infrared lamps are the heat source. The 3.8-micron filter design is insensitive to combustion gases and flames and is used for measuring temperatures inside furnaces, kiln, and combustion chambers where flames are present. For low-temperature (−50°F) applications including frozen foods, ice rinks, race car tires, and printing, the longer wavelength of 814 microns is required due to the lower radiant energy levels.

Dual-wavelength thermometry is used for more difficult and complex applications where absolute accuracy is important, and where the target emissivity is low and/or varying. These sensors also have the unique ability to operate accurately through contaminated intervening media such as dirty windows or with small targets such as a wire which does not fill the sensor's field of view. For example, at high temperatures in steel processing there is accelerated oxidation, causing the surface to have varying emissivity (reflectivity is changing), and there is a lot of airborne dirt, smoke, and moisture between the target and the sensor as well as high ambient temperatures. Using a fiber-optic, dual-wavelength sensor for this application eliminates the effects of varying emissivity, the contaminated atmosphere, and the high ambient operating temperature.

Dual-wavelength (ratio) thermometry involves measuring the spectral energy at two different wavelengths (spectral bands). The target temperature can be accurately measured directly by the instrument regardless of emissivity or intervening media as long as the emissivity is the same at both wavelengths. This is described as a *graybody condition*.

The theory of this design is quite simple and straightforward and is illustrated by the following equations. By using two spectral responses at two adjacent wavelengths and taking the ratio of these signals via Planck's equation, the ratio signal is proportional to temperature and the emissivity coefficients cancel out of the equation.

$$R = L\lambda_1/L\lambda_2 = (\varepsilon_{\lambda 1} \cdot C_1 \cdot \lambda_1^{-5} \cdot e^{-C2/\lambda 1 T})/(\varepsilon_{\lambda 2} \cdot C_1 \cdot \lambda_2^{-5} \cdot e^{-C2/\lambda 2 T})$$

$$R = (\varepsilon_{\lambda 1}/\varepsilon_{\lambda 2}) \cdot [\lambda_1/\lambda_2]^{-5} \cdot e^{[C2/T\,(1/\lambda 1\,-\,1/\lambda 2)]}$$

$$R = [\lambda_1/\lambda_2]^{-5} \cdot e^{[C2/T_r(1/\lambda 1\,-\,1/\lambda 2)]}$$

$$1/T = 1/T_r + \ln\,(\varepsilon_{\lambda 1}/\varepsilon_{\lambda 2})/[C_2 \cdot (1/\lambda_1 - 1/\lambda_2)]$$

If $\varepsilon_{\lambda 1} = \varepsilon_{\lambda 2}$, then $T = T_r$

where: R = Spectral radiance ratio
T_r = Ratio temperature of the surface
ε_1 = Spectral emissivity

The same concept can be viewed in a graphic presentation by taking a segment of the blackbody distribution curve and measuring ratios at various emissivity values as illustrated in Fig. 18.6. Using 0.7 and 0.8 microns as narrowband filters, the ratio factor remains constant at 1.428 for the range of emissivities down to 0.1. This graphically illustrates that with a dual-wavelength sensor, the emissivity is not an issue for graybody targets.

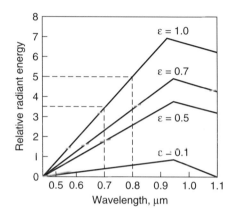

Ratio factor as a function of emissivity

At $\varepsilon = 1$, $\dfrac{5.0}{3.5} = 1.428$

At $\varepsilon = 0.7$, $\dfrac{3.5}{2.45} = 1.428$

At $\varepsilon = 0.5$, $\dfrac{2.50}{1.75} = 1.428$

At $\varepsilon = 0.1$, $\dfrac{0.65}{0.45} = 1.428$

FIGURE 18.6 With a dual-wavelength sensor, emissivity is not an issue for graybody targets.

Similarly, any other changes that are gray in nature will not affect the temperature accuracy determined by the dual-wavelength design. These variations include changes in target size such as a wire or a stream of molten glass where the diameter will vary and/or move. Another example is a case where a target is obscured with smoke or dust or where an intervening window becomes clouded. As long as the obscured medium is not spectrally selective in its attenuation of radiant energy, the analysis remains the same in that the dual-wavelength accuracy is not affected by the application conditions.

Dual-wavelength thermometers have many applications throughout industry and research as simple, unique sensors that can reduce application errors involving graybody conditions. Figure 18.7 illustrates examples of total emissivity for a variety

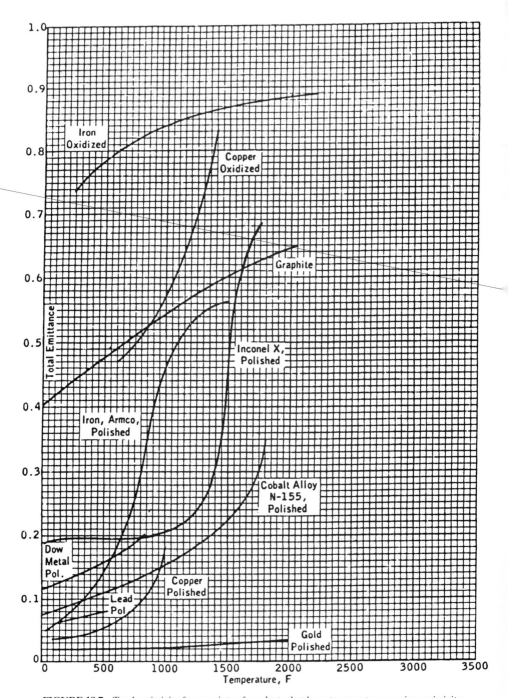

FIGURE 18.7 Total emissivity for a variety of products that have temperature-varying emissivity.

of products that have temperature-related varying emissivity. Most of these are high-temperature applications, but dual-wavelength sensors are functional down to as low as 300°F (150°C).

There are also multiwavelength thermometers available for nongraybody materials where the emissivity varies with wavelength. In these applications, there is a

Application check list:

Williamson has a very broad line of non-contact temperature measuring instruments together with the experience and technology to help you increase yields, lower costs, and improve quality. The more information you provide us about your operation, the more specific we can be about instrumentation to meet your requirements. Your answers to the following questions will help us select the proper instruments and options for you.

Process description: _____

_____

Product to be measured: _____

Reason to measure or control temperature: _____

Functional requirement: ☐ Portability   ☐ Monitoring   ☐ Record   ☐ Control

Temperature range: Minimum ____   Normal ____   Maximum ____

Heating method: _____

Approximate distance from sensor to target: Minimum ____   Maximum ____

Ambient temperature for sensor: _____

Interference between sensor and target: _____

Present temperature measuring method: _____

Present temperature measuring tolerance: _____

Desired temperature measurement tolerance: _____

Estimated cost savings gained by improved temperature measurement: _____

Application sketch

FIGURE 18.8 Application checklist for identifying requirements.

detailed analysis of the product's surface characteristics regarding emissivity versus wavelength versus temperature versus surface chemistry. With this data, algorithms are generated relating spectral radiant energy at various wavelengths to determine temperature.

GETTING STARTED

To begin an evaluation using infrared thermometry, it is very important to establish specific goals for improving the application process and to analyze the value of temperature measurement relative to those objectives. The benefits can be in the form of measurable financial savings based on increased throughput, reduced scrape rates, better and more consistent quality, energy savings, and better utilization of capital equipment. These projects should be viewed for their ability to reengineer the process and assist in product data collection for statistical process control (SPC) and/or certification for ISO 9000.

In terms of selecting a sensor configuration, there are a great number of design features and specifications that can be used to customize the sensors for each application. Consequently, it is important to define a comprehensive set of application requirements in order to develop the best solution. A checklist for identifying application requirements is provided in Fig. 18.8. The general approach is to outline the process in terms of the following:

- Material to be measured
- Typical operating temperature and limits

Design	Single wavelength	Single wavelength	Dual wavelength
Configuration	2-wire	4-wire	4-wire
Temperature range	−50–4500°F	0–4500°F	380–4500°F
	−45–2500°C	0–2500°C	150–2500°C
Accuracy	±0.75% FS	±0.50% FS	±0.75% FS
Repeatability	±0.25% FS	±0.25% FS	±0.25% FS
Response time	50 ms	10 ms	100 ms
Lens optics			
Fiber optics			
Spectral regions			
0.9	•	•	
1.3	•	•	
2.2	•	•	
3.4		•	
3.8	•	•	
4.3	•	•	
5.1	•	•	
7.9	•	•	
8–14	•	•	
8.7&0.8			•
2.0&2.2			•
3.6&4.0			•
Price range	$550–2200	$1400–3850	$2950–4500

FIGURE 18.9 Evaluation criteria for infrared thermometry.

- Type of heat source target size and working distance restraints
- Environmental considerations (i.e., ambient temperature)
- Scope of the system involving data logging, controls, and PLC interfaces

With regard to performance specifications, calibration accuracy is typically in the range of 0.5 to 1.0 percent, while repeatability of temperature measurements of most sensors is in the 0.25 to 0.75 percent range with response times typically less than 1.0 second and adjustable to 5 to 10 seconds (Fig. 18.9). Pricing will vary from approximately $500 to $7000 per sensor, but in the majority of applications should not be viewed as an issue due to short payback periods for properly configured and installed systems.

CHAPTER 19
GMR: THE NEXT GENERATION OF MAGNETIC FIELD SENSORS

INTRODUCTION

Giant magnetoresistance ratio (GMR) materials with magnetoresistance ratios about a factor of 10 higher than anisotropic magnetoresistive ratio (AMR) materials have been reported by many workers since their initial discovery in 1988.[1,2] The initial development of these materials was driven by their potential for thin film head applications, and now initial demonstration of a thin film head using GMR materials has been described in the literature.[3] GMR materials also have potential in many other applications, such as MRAM[4] and the first GMR magnetic field sensor products which were announced in 1994.

The wide variety of possible GMR structures may make it possible for GMR-based sensors to satisfy a broad spectrum of magnetic field–sensing applications. A number of GMR structures have demonstrated high magnetoresistance along with other unusual magnetic and electrical properties, and all of these structures could potentially be used in magnetic field sensors for fields ranging from 1 T (104 gauss) down to below 10 to 9 T. A brief description of GMR material properties and their potential magnetic field–sensing applications are covered in the next two sections. A description of the first commercial GMR magnetic sensor elements and integrated GMR sensors is included in the following two sections to demonstrate how they can be designed and fabricated. The potential for the GMR sensor technology, including sensor-related products, is described in the last section.

GMR MATERIALS

Physics

In general, thin ferromagnetic films have higher resistivity if they are thinner than the mean free path of conduction electrons in the bulk material. This is because electrons are scattered from the surfaces of the films where electrons lose momentum and energy, as well as scattered in the interior (or bulk) of the film; thus, surface scattering plays an important role in resistivity of thin films, and, for sufficiently thin films, surface scattering can be the dominant factor in resistivity. Experiments and theories are relatively well agreed on this point, although some curve fitting for the physical characteristics of the thin film is usually required.[5,6] Figure 19.1 shows a plot of effective resistivity (normalized to bulk resistivity) as a function of film thickness (normalized to mean free path). The mean free path of electrons in many ferromagnetic alloys is on the order of 100 Å, and for films on the order of 50 Å thick, the surface-scattering component is very large. The thicknesses of magnetic films used in GMR structures are often on the order of 50 Å.

Ferromagnetic materials have two types of electrons as carriers. *Spin up* electrons and *spin down* electrons have magnetic moments which are parallel or antiparallel to the direction of magnetization of the material, and, therefore, are opposite to each other. The population of spin up electrons is higher than the population of spin down electrons, and it is more difficult for a spin down electron to act as a carrier in the ferromagnetic material. This characteristic is the fundamental reason for different surface scattering in GMR structures as a function of the magnetic state of the materials.

Consider the simple structure shown in Fig. 19.2, where two magnetic thin films are separated by a conducting nonmagnetic film. Suppose the conducting nonmagnetic film is lattice-matched to the magnetic films and that electrons can pass from a magnetic film into the nonmagnetic film without scattering. Now consider two cases: one where two magnetizations are parallel and the other where they are antiparallel. In the case where the two magnetizations are parallel, the spin up (or majority) electrons should pass relatively easily from one magnetic layer through the interlayer to the other without scattering. In the case where the two magnetizations are antiparallel, the spin up (or majority) electrons in one magnetic layer become spin down (or minority) electrons in the other, and they tend to be scattered when passing from one magnetic layer to the other. Thus, there is more scattering (more resistance) when the magnetizations are antiparallel and less scattering (less resistance) when the magnetizations are parallel. Since scattering has a large effect on thin-film resistance, the difference in resistances for the two magnetic states can be relatively large.

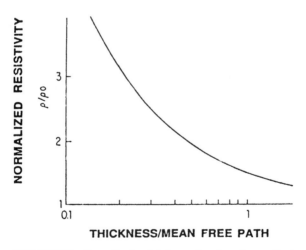

FIGURE 19.1 Normalized resistivity as a function of normalized film thickness.[5]

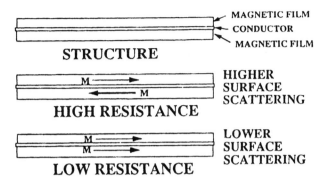

FIGURE 19.2 Scattering as a function of magnetic state in a two-magnetic layer GMR sandwich.

Several different GMR structures have been demonstrated, all of them working on the preceding principle that is, with the magnetizations parallel, the resistance is smaller, and with the magnetizations antiparallel, the resistance is larger. Among these structures are antiferromagnetic-coupled multilayers,[7] spin valves[8] and symmetrical spin valves,[9] unpinned magnetic sandwiches,[10] granular films,[11] and tunneling structures.[12]

When a large enough magnetic field is applied, the parallel-magnetization (low-resistance) state can be reached for any of these GMR structures. The antiparallel state of magnetizations is reached by a variety of mechanisms. Multilayer films utilize antiferromagnetic coupling between nearest neighbor ferromagnetic films, and here the thickness of the nonmagnetic layer is especially critical. Spin valve and tunneling structures make use of high- and low-coercivity layers, and there are several different methods used to get the high-coercivity layers. Unpinned sandwiches use current in a stripe to get the antiparallel-magnetization state. Granular films use a combination of magnetic disorder of isolated magnetic particles along with an antiferromagnetic coupling similar to the multilayer structures.

The magnetic field required to change the magnetic state from the high-resistance state to the low-resistance state is sometimes called the saturation field, and its value depends critically on the coupling mechanisms previously described which are used to attain the antiparallel state. This coupling field is relatively high for multilayer and granular structures and relatively low for spin valve and tunneling structures. The tunneling structure is of special interest for low field applications for reasons to be explained later.

A complete characterization of all of the material attributes of the many different forms of GMR materials has not been published, but some of the characteristics of GMR materials that could be important for magnetic field–sensing applications are discussed here.

GMR and Saturation Field

Sensors utilizing GMR materials have the potential for satisfying a broader range of both static and dynamic magnetic fields than any of the other commercial sensing technologies, at least in part because of the wide variety of GMR structures. Table 19.1 illustrates some of the types of GMR structures with approximate ranges of

TABLE 19.1 Comparison of Magnetoresistance Properties of Various GMR Structures

GHR Structure	GMR (F%)	Saturation field (tesla)
Antiferromagnetic-coupled multilayer	10–70	0.01–0.5
Spin valve	4–14	0.001–0.004
Symmetric spin valve	8–20	0.002–0.008
Granular	8–40	0.02–1.0
Tunneling (projected)*	10–25	10^{-5} 10^{-2}

* The values for the tunneling device are the author's own projections.

room-temperature GMP and saturation magnetic fields for each of the structures. The values in the table are those typically reported in the literature with the exception of the tunneling-device projections, which are the author's own projections.

For tunneling devices, theoretical GMR values have been projected in the 20 percent range,[13] and significant recent progress has been made.[12] In theory, there should be very low coupling fields between the two magnetic layers in a tunneling device, and the GMR values should be relatively independent of thickness. If one of the two magnetic layers is pinned (by coupling to an antiferromagnetic film such as FeMn, for example), and if the other film can be exchange-coupled to an arbitrarily thick soft film, then the device should switch from a low- to a high-resistance state with an applied field equal to the coercivity of the soft film plus a low parallel coupling field. The coupling field may be made arbitrarily small by making the soft film thicker. The coercivity of this soft film could very practically be as low as about 0.1 Oe or 10^{-5} T. The GMR would, in theory, be about the same, regardless of the film thickness. Thus, this tunneling device could be very useful for sensors in very low field ranges.

It should be noted that a similar technique when applied to a spin valve structure or to other horizontal-conduction GMR structures could also result in a low coupling field—but a very thick magnetic layer would result in the majority of the conduction taking place in that layer, and the dominant scattering mechanism would be bulk scattering in that film. Hence, such a structure would have a very low GMR.

It is not necessary that the field to be sensed by a magnetoresistive material be on the same order as the saturation field. For example, AMR sensors which saturate at fields on the order of 10 Oe have been used to sense fields in the 10^{-6}-Oe range.[14,15] Depending on noise, the signals available from much lower fields than the saturation fields can be sensed.

Other factors to consider, besides GMR and saturation field, are also important for sensors. They include hysteresis and linearity, resistivity, thermal coefficient of resistivity, high temperature endurance, and noise.

Hysteresis and Linearity

Most linear sensing applications require that the response be roughly proportional to the stimulus field. Deviations from linearity can be corrected by look-up tables stored in a microprocessor memory, and so long as the sensor output varies monotonically with applied field, the sensor can, in theory, be used for linear applications. If, on the other hand, sensor output depends strongly on magnetic history, then it is very difficult to use the sensor for linear applications. Thus, significant hysteresis of the resistance versus field characteristics for magnetoresistance materials is to be avoided for these applications. Many of the GMR materials described in the literature would be unsuitable for linear sensors, but at least two types of GMR materials have demonstrated suitable properties for linear applications: spin valve (with appropriate easy axis) and antiferromagnetic multilayers.

Figure 19.3 illustrates a spin valve device used in a demonstration read head[3] which shows linear characteristics without hysteresis. The device uses a two–magnetic layer structure with the magnetization of one layer being pinned along one fixed direction, and with the magnetization of the other layer having an easy direction orthogonal to the magnetization of the pinned layer. When a magnetic field is applied, the angle between the two magnetizations changes, and the resistance of the material changes. Symmetrical spin valve structures, where the outside two magnetic layers in a three–magnetic layer structure are pinned, should also be usable for linear sensors using similar construction techniques.

Figure 19.4 illustrates the resistance-field characteristic of an antiferromagnetic multilayer (four magnetic layers and three nonmagnetic interlayers) which demonstrates linear behavior with low hysteresis (except for low fields). This type of material is now being used in commercial magnetic field sensors.[16,17]

Digital applications require the sensor to respond when applied fields reach a fixed threshold value, and the sensor element is connected to circuitry which provides threshold detection. Hysteresis is usually designed into the electronics so that the switch from one state to another is certain (not switched back and forth due to noise near the threshold field). Thus, some hysteresis is quite desirable, and the digital applications are much more tolerant of hysteresis than linear applications.

Repetitive event sensors, such as wheel speed sensors, are even more tolerant of hysteresis. In these applications, the period of time between moments where the magnetic field reaches a threshold can be measured using almost any kind of GMR material provided the sensitivity of the material is compatible with the magnitude of the time-varying field which is to be sensed.

Resistivity

Low power is a requirement for many sensor applications, and this favors using a high value of resistor in the sensor. This high-value resistor can be attained with

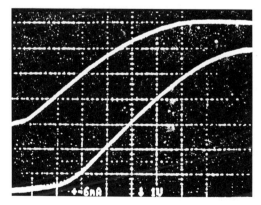

FIGURE 19.3 Spin valve device characteristics. This read-head sensor demonstrates linearity without hysteresis.[3]

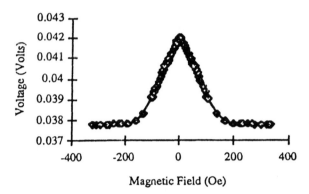

FIGURE 19.4 Antiferromagnetic coupled multilayer characteristics. The voltage output is measured across the resistor using a constant current source.

either a large number of squares or with a high-resistivity material. Using a high resistivity would result in a smaller area device, which could lower cost. Very thick multilayers and granular films would not be favored for these lower-power applications due to their relatively low sheet resistivity (on the order of 1 L2/square) in comparison to spin valve or thinner multilayer materials (10 to 20 ohms per square).

Low power dissipation can also be achieved by pulsing the magnetoresistor at a low duty cycle, say 1 percent. This has the advantage of allowing a lower resistor value for the same or lower power, and since noise increases with about the square root of resistance, it may be possible to improve the sign-to-noise ratio of the sensor.

Temperature Coefficient of Resistivity (TCR)

A low TCR is generally desirable in order to keep the signal relatively constant as temperature varies. Circuitry can be designed to compensate for known TCRs in the sensor. Data on GMR materials have shown about half the TCR of permalloy materials (approximately 3000 ppm/°C for permalloy and 1600 ppm/°C for GMR materials).[18] This is because surface scattering rather than photon scattering is the dominant resistance mechanism in very thin films, and surface scattering is less temperature-sensitive.

The change in resistivity with magnetic field (the ΔR in the $\Delta R/R$ magnetoresistance ratio) changes relatively slowly with temperature.[16] Approximately a 5 percent change in resistance span (ΔR) is typical for a 100°C change in temperature.

High Temperature Endurance

In addition to operating over a wide temperature range, many applications require operation for protracted times at elevated temperatures, and this requirement can be difficult for GMR materials. Automotive applications, for example, often require 150°C operation and higher. Also, sensor fabrication may require the materials to withstand elevated temperatures for a few hours during processing. This requires that the magnetoresistive properties not change at these higher temperatures.

Uncoupled sandwiches with relatively thick interlayers of copper (30 Å) have demonstrated stable properties for temperatures above 250°C.[18] Antiferromagnetic multilayers with 34-Å copper interlayers have shown reasonable stability to 250°C for periods of several hours,[19] but 9.2-Å copper interlayer multilayers were not stable at 150°C for only a few hours. Antiferromagnetic multilayers with 15-Å copper alloy interlayers which are used in commercial magnetic field sensors[17] are stable for 1000 hours at 150°C, but the antiferromagnetic coupling between layers decreases when the material is exposed to more elevated temperatures for hundreds of hours. Little data has been reported relevant to thermal stability of spin valve materials.

Noise

Sensing very low magnetic fields requires both high sensitivity and low noise. Noise in AMR material sensors can easily be a factor of 10 higher than the Johnson noise indicated by resistor values, and, in read heads, great care is taken to pin magnetization at resistor ends to avoid noise associated with wall motion. The noise characteristics of GMR materials have not been reported, and they should be carefully assessed as a prerequisite to using them for low field applications.

Magnetic Field Sensors

GMR materials potentially can be used in a wide variety of sensor applications currently served by a multiplicity of sensor technologies. Magnetic field sensors span a broad range of magnetic fields[20] from 10-10 T and less to more than 1 T. Most of these applications measure magnetic field as a means to sense a second physical factor, such as position, speed, presence or absence of a ferromagnetic body, or a current. Both time-varying fields (up to very high frequencies) and static fields are important for different applications.

These applications, which together are very economically significant, are served by many different kinds of magnetic sensor technologies such as Hall effect sensors, SQUIDS, variable-reluctance sensors, magnetoresistive sensors, and flux-gate sensors. Figure 19.5 shows field ranges for several types of magnetic field sensors.

GMR SENSOR ELEMENT

Sensors using a GMR antiferromagnetic-coupled multilayer have been made for commercial application.[15,18] These sensors use a multilayer whose cross section is shown in Fig. 19.6. There are four magnetic layers, each a composite of several magnetic layers, arranged so that a higher-moment CoFe layer is at the interface to the copper alloy. This arrangement gives relatively good GMR (10–25 percent) and helps prevent Ni and Cu interdiffusion. The thickness of the copper alloy is critical and must be quite uniform and in the 14- to 18-Å thickness range, depending on the desired magnetoresistance properties. All layers were R.F. diode sputtered. The magnetoresistive characteristics of a typical material of this type is shown in Fig. 19.4. The sensor configuration is shown in Fig. 19.7.

The GMR materials are etched into four resistors which are hooked together as a Wheatstone bridge, two of the resistors being shielded from external magnetic fields by a thick, high-permeability shield, and the other two resistors being placed

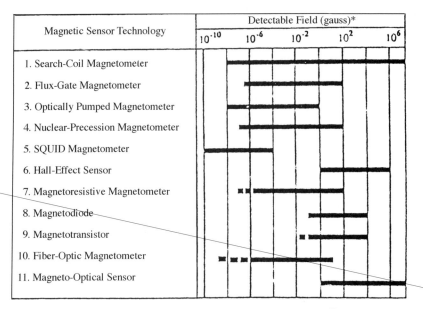

FIGURE 19.5 Magnetic field ranges for various field sensor technologies.[20]

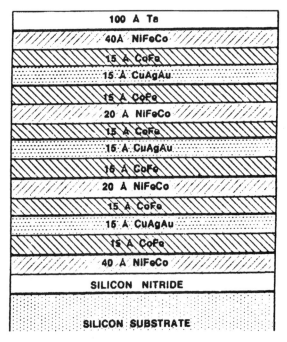

FIGURE 19.6 Cross section of antiferromagnetic-coupled multilayer.

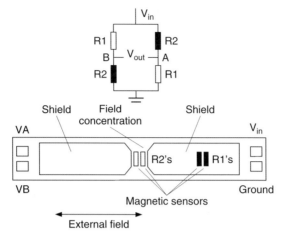

FIGURE 19.7 Sensor configuration of magnetoresistive type.

between two shields so that the external flux is concentrated in the gap between the shields. The external magnetic field is effectively multiplied by a factor of from 2 to 20. Figure 19.8 shows a typical resulting sensor characteristic. A top-view picture of a GMR bridge sensor chip is shown in Fig. 19.9.

Two shields concentrate the flux and magnify the field between the shields. Two sets of two resistors each are visible, one set in the gap between the shields, and the other set is under one of the shields. A specific design of a bridge sensor element for an integrated sensor design is included in the next section.

By tailoring the magnetic material characteristics and the shield and flux concentrator geometries, this type of sensor bridge has been made for both linear and digital sensor functions for magnetic fields in the range of 1 to 1000 Oe. Furthermore, this type of sensor bridge has been integrated with silicon integrated circuits so that the output of the sensor can be customized for particular applications. The most immediate commercial applications for this GMR magnetic field sensor are for DC

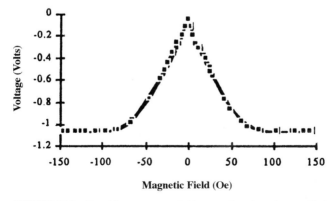

FIGURE 19.8 Signal from a sensor bridge as a function of an applied magnetic field.

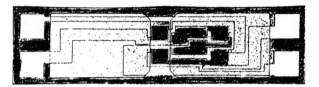

FIGURE 19.9 A GMR bridge sensor chip.

or low-frequency magnetic fields in the 10^{-4} to 10^{-3} T range, where silicon Hall effect sensors and magneto-inductive sensors have difficulty with sensitivity. The GMR sensor also has distinct technical advantages compared with commercial AMR permalloy sensors.

The relatively high interlayer coupling fields of the antiferromagnetic-coupled multilayer make it possible to shrink line widths of the GMR magnetoresistors to 2 micrometers without significant demagnetizing effects, whereas permalloy AMR sensors have line widths restricted to about 15 to 20 micrometers. For a given resistor value, the GMR magnetoresistors are on the order of 1/100 of the area of AMR magnetoresistors. Even after taking into account the area of the shields in the GMR sensor, the total chip area is much smaller for the GMR sensor. As observed earlier, the TCR is also lower for the GMR sensor, which simplifies temperature compensation.

Of special note is the automatic reset characteristic of GMR multilayers. The antiferromagnetic coupling in the GMR multilayer gives a *no field* value of resistance which is the same regardless of magnetic history. AMR permalloy sensors have two possible magnetic states which can give two different magnetoresistance characteristics.[21] AMR permalloy sensors must then either be biased or periodically reset to insure the proper magnetic state. The automatic reset feature of the GMR multilayer sensor makes it easier to apply than an AMR sensor.

INTEGRATED GMR SENSOR

This section describes the first integrated circuit that combines giant magnetoresistive ratio (GMR) materials with integrated circuit electronics.[22] The result is a flexible magnetic field sensor that can be custom-configured for high sensitivity, hysteresis, and output. The sensor element used is of the type described in the previous section. This particular design uses GMR magnetoresistors with about 12 percent GMR at a saturation field of about 200 Oe. The magnetoresistors are connected in a Wheatstone bridge configuration. Design and processing techniques have been developed which allow use of GMR materials on BICMOS integrated circuit underlayers. The completed parts can be configured in many different ways by changing patterning of metallization and GMR layers. They can be designed to perform like a digital or linear Hall effect sensor, but with greatly increased sensitivity and temperature stability.

The IC described here is the first integrated circuit that combines GMR materials with semiconductor devices to make a highly sensitive, temperature-stable magnetic field sensor. The chip uses a BICMOS semiconductor process which is finished through the first layer of metallization at the wafer process in the standard fashion. GMR films are sputter-deposited, patterned into magnetoresistors, and then connected into the transistor array on the chip with metallization layers. The final step is deposition and lithography of a thick magnetic material that concentrates the applied flux around the GMR resistors, allowing for increased sensitivity. This results in an IC that is custom configurable to digitally detect magnetic fields as small as 0.25 Oe or provide linear measurements of applied field in the 10-3-Oe range. The GMR materials made for this IC are composed of thin films of alternating magnetic layers and nonmagnetic conducting interlayers, such as copper. A typical GMR film sandwich is shown in Fig. 19.6.

Integration of a GMR structure such as this onto a silicon wafer with underlying transistors is a challenge because of the topology introduced by the many layers of polysilicon, metal, and oxides over the transistors. This underlying topography is detrimental to the performance of GMR thin films, because the entire GMR structure is only about 200 Å thick, and requires deposition on a very smooth, step-free surface to maintain GMR and prevent breaks in the magnetoresistors. Special design and processing steps have been developed in order to ensure proper integration of a GMR bridge with the transistors on semifinished silicon wafers. The layout of the integrated circuit that forms the foundation of the sensor is designed so that no other structures are found in the same area as the GMR devices. Lower-level oxides that may result in a rough top surface are omitted from this area of the chip.

The layer thickness of the GMR films is controlled very tightly in order to achieve the desired GMR effect; typically, a layer-thickness tolerance of ±2 Å across the entire silicon wafer is maintained. The actual layer structure is variable and is tailored for the application.

For example, the most common layer structure is represented in Fig. 19.6, but a special three-layer structure with lower saturation fields was developed recently for a low-field application. This GMR material provides a 5 percent resistance change at only 10 gauss, making for a very sensitive (2.5 mV/V/gauss) raw bridge structure.

After the GMR material is deposited, it is patterned using standard semiconductor metal processing equipment into serpentine structures that act as resistors. These structures show a sheet resistance of 11-16 ohms/square, depending on the layer structure. The standard layer structure allows construction of a 5K resistor in a 50 μm × 70 μm area on the chip. This small size compares very favorably with a resistor constructed of AMR (permalloy) material; typically, these structures require the serpentine legs of the resistor to be 25 μm wide, thus making the area occupied by a 5K AMR resistor over 100 times larger than the GMR resistor. Since IC area drives cost, a sensor based on GMR materials can be considerably less costly than one based on AMR materials.

A Wheatstone bridge configuration of magnetoresistors comprises a magnetic field sensor whose characteristics can be tailored to an application by varying the GMR and the saturation field of the material. For example, a 5K resistor that drops to a 4.5K resistor with application of the saturation field would have a GMR of (500/4500), or 11.11 percent. Typical saturation field (H_{sat}) values for use in this sensor chip are in the 100–300 or proprietary materials and structures, the GMR response is over 90 percent linear from zero field to the on field value.

The bridge is constructed so that two opposite legs are shielded from the external field by a thick, plated magnetic material and do not change in resistance value. This decreases the maximum bridge output signal by a factor of two, but provides a significant advantage: no magnetic bias condition is required to generate the bridge output. A separate pair of structures, made of the same thick, plated magnetic material as the shield is used for the purpose of flux concentration. The two opposite leg resistors that are exposed to the applied field are placed in a gap between this pair of flux concentrators. The external magnetic field is then magnified in the gap. The multiplication factor is roughly equivalent to the length of one of the flux concentrators divided by the length of the gap. In order to make a GMR sensor with 5× flux concentration, for example, the gap between the flux concentrators might be 100 micrometers, and the length of each flux concentrator would be 500 micrometers. It is important to note that the thick magnetic material that forms the shields and the flux concentrators on these GMR sensors is deposited as the final step in the process and is electrically insulated from the structures underneath it. Therefore, it is not wasted area; any other structure, including transistors, capacitors, resistors, polysilicon or metal lines, and so forth, may be built under the shields and flux concentrators.

For signal conditioning, the bridge output can be interfaced easily into the on-chip amplifier circuitry. For a digital output device, the GMR bridge outputs are connected to a comparator; an on-chip operational amplifier can be used for linear output purposes. For a linear application, the bridge is designed to be balanced with no applied field. Because of the tight tolerances maintained during the deposition of the GMR films, the GMR resistors match very well as deposited; offset voltage of the bridge is maintained within the ± 20-mV range, with a 5-volt supply. For some low-gain applications this is adequate; however, more precise parts require some way to adjust the bridge offset to zero. This is accomplished by designing one of the shielded resistors to be 500 ohms less than the other three bridge resistors, and leaving one leg of this resistor available as an output. A 1K pot is attached to this leg externally and adjusted for zero bridge offset.

For digital applications, the bridge is purposely left unbalanced. The bridge offset is taken into account during the design phase, so that no trimming of the digital parts is required. The output of the bridge goes to a comparator, and the trip point of the comparator is determined by the degree of bridge offset. For example, if a digital output device is desired that turns on at an applied field of 10 Oe, NVE might start with GMR material that exhibits 11.1 percent GMR and saturates at 250 Oe. A flux concentration factor of 5 would be used, so that the GMR resistors in the gap would provide maximum signal at an applied field of 50 Oe. Then, the shielded GMR resistors would be designed to be 4900 ohms and the gap GMR resistors would be designed to be 5000 ohms. This would result in the bridge output transitioning from negative to positive at an applied field of 10 Oe, tripping the comparator and giving a digital output.

Digital parts have been manufactured that trigger at as little as 3 Oe. A part for use as a spin counter in the earth's magnetic field which triggers at 0.25 Oe is currently being designed. Typical digital parts being developed for automotive and industrial applications trigger at 10 to 50 Oe and require a considerable hysteresis to prevent jitter of the output signal due to vibration of the part. There is some hysteresis in the characteristics of the GMR films, but it is small relative to what is required for these applications, so the circuitry on the IC is designed to provide the hysteresis. The amount of hysteresis is adjustable for different applications with a metal mask change, and this illustrates a key point with regard to this IC. It has been designed so that the last few layers can provide complete customization of the part. All the key parameters, such as sensitivity, magnetic range of operation, triggering fields, and hysteresis, are adjustable at the last few steps in the IC processing. The IC is therefore very versatile and can be custom-configured for numerous applications using the same basic chip design techniques.

The output from one of the digital parts is shown in Fig. 19.10 (note the hysteresis). A photomicrograph of this sensor is shown in Fig. 19.11.

POTENTIAL OF GMR SENSOR TECHNOLOGY

Considering the range of available GMR materials, it may be possible for GMR magnetic field sensors to satisfy a broad range of applications which are now served

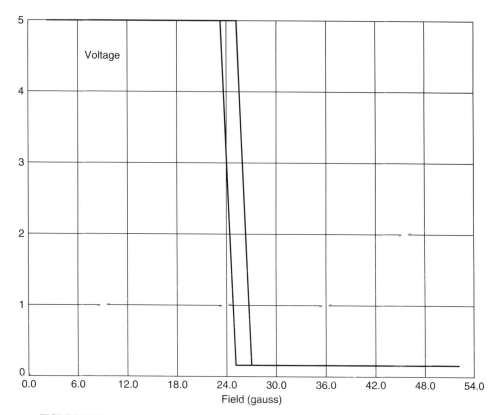

FIGURE 19.10 Digital output from an integrated sensor.

by many different technologies. The first commercial GMR sensors have some distinct advantages when compared to existing magnetic field sensor technologies in the 1 to 100-Oe range. It also appears feasible to make sensors which operate well in both much higher and much lower field ranges. Concepts for examples of these sensors are described briefly. Also, derivatives of these sensor products can be used for functions such as isolators and relay replacements.

High Field Sensors

GMR saturation fields approaching 1 T have been demonstrated, and materials having similar characteristics to materials shown in Fig. 19.6 have been made with saturation fields of well over 0.1 T. Such materials have higher GMR (over 20 percent), and the output from a bridge sensor made of these materials is relatively large, even without signal conditioning. One possible kind of application for this kind of sensor is where error signals due to external fields must be small. Angular and linear position sensors could be one example where external fields must be kept small with respect to the applied field which indicates position in order to maintain accuracy.

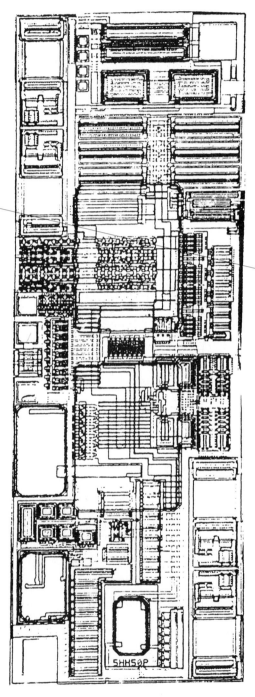

FIGURE 19.11 Photomicrograph of an integrated sensor.

A simple schematic for a linear position sensor is shown in Fig. 19.12, where a magnetic assembly is movable with respect to a stationary sensor. The full strength of the gap field is applied to magnetoresistors in a bridge, and when the gap is centered, the bridge and the output is zeroed. The output then becomes positive or negative when the position of the assembly relative to the sensor is changed, depending on direction. The assembly could use rare earth magnets and be designed to apply a large field (on the order of 1000 Oe). Stray fields of a few Oe should not affect the output significantly.

Low Field Sensors

GMR materials with flux concentration can be used for sensing low magnetic fields. As mentioned in the materials section, tunneling magnetic structures are potentially very sensitive. One proposal for a low field sensor is shown in Fig. 19.13. Here a bridge of tunneling devices or other sensitive GMR sensor elements is placed in the gaps of an electromagnetic circuit so that the elements are biased in opposite directions. An applied field is amplified by the magnetic field concentrator, and the gap fields due to the applied field subtract from the bias field applied to one pair and add to the field in the other gap. The limit to field sensitivity is probably limited by noise, but potentially could reach below 10^{-6} gauss.

Derivative Products

An integrated current sensor can be fabricated by depositing a current strap over a GMR sensor and sensing a magnetic field from the current. With an appropriate linear sensor and conditioning electronics, an analog output proportional to current can be fabricated. Similarly, with a current strap which is electrically isolated front

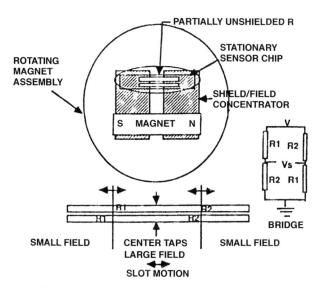

FIGURE 19.12 Example of a high field position sensor.

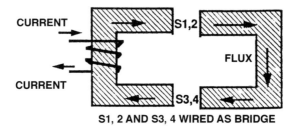

FIGURE 19.13 A proposed low field GMR sensor.

the sensing electronics, the sensor and conditioning electronics could determine a threshold field (and thus threshold current) which could trigger a binary output. This would obviously be a simple isolator which could operate at a high rate (above 10 mHz) and which should be relatively inexpensive because of its monolithic construction. With on-chip power electronics and using similar techniques, solid-state relays and electromechanical relay functions could be replaced. It should be noted that all of these GMR sensors and sensor-derived products use the same fundamental building blocks, that is, GMR materials, integrated circuits, and thick magnetic materials. The same basic competencies are also required, that is, magnetic device and integrated circuit design combined with conductor, dielectric, GMR and thick magnetic material deposition and lithography. The total market to be addressed is very substantial, and therefore the use of GMR materials for magnetic field sensing and derivative products represents a very substantial business opportunity for our industry.

REFERENCES

1. M. Baibich, J. Broto, A. Fert, F. Nguyen Van Dau, F. Petroff, P. Eitenne, G. Creuzet, A. Friederich, and J. Chazelas, "Giant Magnetoresistance of (001) Fe/(001) Cr Magnetic Superlattices," *Physical Review Letters,* November 1992, **61**(21), p. 2472.
2. J. Bamas, A. Fuss, R. Camley, P. Grunberg, and W. Zinn, "Novel Magnetoresistance Effect in Layered Magnetic Structures: Theory and Experiment," *Physical Review B,* November 1990, **42**(13), p. 81.
3. C. Tsang, R. Fontana, T. Lin, D. Heim, V. Speriosu, B. Gurney, and M. Williams, "Design, Fabrication and Testing of Spin-Valve Read Heads for High Density Recording," *IEEE Trans. Mag.,* November 1994, **30**(6).
4. J. Brown, A. Pohm, "1 Mb Memory Chip Using Giant Magnetoresistive Memory Cells," *IEEE Transactions on Components, Packaging, and Manufacturing Technology,* part A, September 1994, **17**(8).
5. L. Maissel and R. Glang, *Handbook of Thin Film Tchnology,* McGraw-Hill, New York, 1970.
6. C. Tellier and A. Tosser, *Size Effects in Thin Films,* Elsevier Scientific Publishing Company, Holland, Amsterdam, 1982.
7. S. Parkin, A. Li, and D. Smith, "Giant Magnetoresistance in Antiferromagnetic Co/Cu Multilayers," *App. Phyg. Lett.,* 1991, **58**(10).
8. B. Dieny, V. Speriosu, S. Metin, S. Parkin, B. Guemey, P. Baumgart, and D. Wilhoit, "Magnetotransport Properties of Magnetically Soft Spin-Valve Structures," *J. Appl. Phys.,* 1991, **69**(8).

9. T. Anthony and J. Brug, "Magetoresistance of Symmetric Spin Valve Structures," *IEEE Trans. Mag.*, November 1994, **30**(6).
10. J. Daughton, P. Bade, M. Jenson, and M. Rahmati, "Giant Magnetoresistance in Narrow Stripes," *IEEE Trans. on Mag.*, September 1992, **28**(5).
11. J. Mitchell and A. Berkowitz, "Dependence of Giant Magnetoresistance on Film Thickness in Heterogeneous Co-Ag Alloys," *J. Appl. Phys.*, May 1994, **75**(12).
12. T. Miyazaki and N. Tezuka, "Spin Tunneling Magnetoresistance Effect," paper at *The 88th Topical Symposium and the 8th Meeting of Specialist Group of Magnetic Multilayers of the Magnetics Society of Japan*, January 27, 1995.
13. J. Slonczewski, "Conductance and Exchange Coupling of Two Ferromagnets Separated by a Tunneling Barrier," *Physical Review B*, April 1989, **39**(10), p. 6995.
14. J. Lenz, G. Rouse, L. Strandjord, B. Pant, A. Metze, H. French, E. Benser, D. Krahn, "A High Sensitivity Magnetoresistive Sensor," paper presented at the *Solid State Sensors Workshop*, June 1992.
15. N. Smith, F. Jeffers, and J. Freeman, "High Sensitivity Magnetoresistive Magnetometer," *J. Appl. Physics*, April 1991, **69**, p. 5082.
16. J. Daughton and Y. Chen, "GMR Materials for Low Field Applications," *IEEE Trans. Mag.*, November 1993, **29**(6), pp. 2705–2710.
17. J. Daughton, J. Brown, R. Beech, A. Pohm, and W. Kude, "Magnetic Field Sensors Using GMR Multilayer," *IEEE Trans. Mag.*, November 1994, **30**(6).
18. J. Daughton, "Weakly Coupled GMR Sandwiches," *IEEE Transactions on Magnetics*, March 1994, **30**(2), p. 3643.68.
19. H. Zhang, R. Cochrane, Y. Huai, M. Mao, X. Bian, and W. Muir, "Effect of Annealing on the Giant Magetoresistance of Sputtered Co/Cu Multilayers," *J. Appl. Phys.*, May 1994, **75**(10).
20. J. Lenz, "A Review of Magnetic Sensors," *Proceedings of the IEEE*, June 1990, **78**(6), pp. 973–989.
21. Philips Technical Publication 268 on Magnetoresistive Sensors.
22. J. Brown, "GMR Materials: Theory and Applications," *Sensors Magazine*, September 1994.

CHAPTER 20
SMART CIVIL STRUCTURES, INTELLIGENT STRUCTURAL SYSTEMS

INTRODUCTION

The area of smart structures involves the use of sensors and actuators to determine the status of an instrumented structure and to take remedial action if necessary. Smart *civil* structures represent an extension of this area into buildings, bridges, dams, and other typically large scale structures. To date, conventional and fiber-optic sensors have been embedded and surface-attached to various structures worldwide with identifiable and tangible results. Specifically, with respect to fiber-optic sensors, various large civil structures have had differing types of fiber-optic sensors installed within and upon these structures leading to measurements not previously available. A review of this smart structures research is presented here.

During the last few years, research institutions throughout the world have reported increasing success in the area of fiber-optic sensor development. Applications of this technology have slowly emerged from those institutions into the more mainstream engineering arena. In general, sensor development, fiber optic and traditional, has been underway for many years. What is unique about this particular application, namely fiber-optic sensing within large civil structures, is that it is trying to marry the advanced technologies associated principally with fiber-optic sensors with the age-old field of construction engineering and building practices. In order to set the stage for this application, a brief review of the original research findings is in order.

During the 1980s, the research area called *smart structures* evolved principally from the aerospace arena. While many particular aspects of aerospace engineering and the sensing within structures was performed during this time frame, efforts seemed to concentrate on using graphite epoxy composite panels and similar materials within an aerospace frame. The carbon fiber panels do not lend themselves to the standard interrogation and sensing methods used in traditional metallic-based structures, such as X rays and other inspection techniques. Therefore, questions of fatigue of such composite epoxy panels, or more importantly, what is the mean time to failure of such panels and how can we investigate their performance, arise. As the composite panel research progressed, so did that of fiber-optic sensors. The application of fiber-optic sensors within the composite panel area principally meant that the optical fibers would be embedded inside, or woven into, the carbon fiber matrix prior to casting. The end result would then be a sensor capable of providing some meaningful measurement being embedded within the panel. Studies investigating such supposedly esoteric topics such as strain field variations across composite, nonuniform composite fiber of panels were carried out to determine if the parameters measured using the optical fiber sensors were useful. Work on that front has continued and continues to this day.

Another application of fiber-optic sensors is within the civil structures arena. The fundamental idea of civil structures and their engineering or building for the past few millennia has basically been "build it and leave it." The structures are designed, typically well beyond their simple tolerance level, fabricated, and then passed on to the owner. It now appears that such practices are changing. Within the United States alone there is a significant amount of civil infrastructure that is in serious need of repair. Highways, bridges, old buildings, and more are constantly being rehabilitated just as new construction continues. The change in basic view, if not practice, is that similar to other products, people who are constructing these large civil structures should also take some particular measures in determining the life cycle, or in the more radical case, warranteeing their products. The notion of life-cycle costing implies that while (with respect to the traditional bidding practices engaged in the construction of buildings) money may be saved in the initial phase through fabricating it with the lowest bid, if it is foreseen that within 10 years major reconstruction of the same facility may arise, it would perhaps be more prudent to spend a bit more on the original construction phase and less in the longer-term costs associated with the maintenance of the building. Through this notion of life-cycle costing of structures, instead of just simple bottom-line initial costing comes one potentially substantial application of fiber-optic sensors: embedded fiber-optic sensors may reveal the building's internal structural integrity. Such sensors may be remotely interrogated thereby allowing a building inspector to check on the health of numerous structures from some remote location. While not totally infallible, such a structural health check would be most useful (and cost-effective) when an area is subjected to some deleterious event (e.g., earthquake, tornado, hurricane). Such lies the promised land for fiber-optic smart civil structures.

SMART STRUCTURE?

This first question may prove to be most important: what does *smart* or *intelligent* mean? This is a general question that has been plaguing the artificial intelligence arena for a very long time (natural versus artificial intelligence). We limit this particular idea with respect to smart civil structures, leading to the following: for the last 15 years there has been an Intelligent Building Institute based in New York City that is specifically looking at instrumentation for residences and major buildings to determine how intelligence may be added to a building. The institute investigates issues that many almost take for granted, such as temperature compensation inside rooms as the number of people come and go, automatic lighting that turns on and off when people enter or leave the buildings, the status of intrusion detectors, and the like. This comprises the *social infrastructure* for the occupant or for the building owners or the people who are actually using or working in this building versus the structure itself.

The question returns, "What is intelligence?" We define *intelligence* to be the fact that the structure itself has some way of measuring or monitoring its internal performance; not necessarily what its occupants are doing or if it should indeed turn on the lights, raise or adjust the temperature inside a room, or balance some particular control feedback system, but rather concerning the building itself and what its structural integrity is.

FIBER-OPTIC SENSING

The idea of using sensors within a structure is not in itself new. What is new is the level of measurement detail presented through the use of fiber-optic sensors. There are two generic flavors of fiber-optic sensors: extrinsic and intrinsic. A brief review follows: an extrinsic fiber-optic sensor uses an input fiber and an output fiber and has some light modulation capability in the internal part. Basically, the fibers merely acts as light pipes transmitting and receiving the light. Some environmental signal encounters the sensitized region along the fiber and modulates the light signal. The modulation may change the light's color, phase, intensity, or polarization state. There are numerous ways in which this modulation may occur. Therefore, known light is injected into the transmit fiber, the light emerges from that fiber segment, encounters an optical modulator of some form, is coupled back into the receive fiber, and photodetected at the output of the fiber. This is an *extrinsic* (or *external*) light-modulating device. This configuration represents an extrinsic fiber optic sensor.

By comparison, there is an *intrinsic* fiber-optic sensor. With an intrinsic fiber sensor, the environmental signal causes some perturbation to occur on the fiber itself, locally changing the boundary conditions. That boundary condition variation changes the fiber's wave-guiding properties and therefore modulates the light. The light is in the fiber. The environmental signal interacts with the fiber, hence changing the light that has propagated through it. One of the problems associated with an intrinsic fiber sensor, more than an extrinsic fiber sensor, is that the environmental signal can influence multiple parameters of the optical field. In certain cases, quite extreme measures are used to localize this parameter. The significant advantage of an intrinsic fiber sensor over an extrinsic one is that the fiber itself is the sensing agent, no additional pieces are necessary. Therefore, there are fewer mechanisms to malfunction.

The generic advantages of fiber apply: lightweight, the size of a strand of hair, nonobtrusive so that they can be painted against the wall and no one even knows that they're there, totally passive, EMI resistant, and so on.

A FEW FIBER OPTICS SMART STRUCTURE RESULTS

The use of embedded fiber-optic sensors to measure physical quantities in concrete was suggested by Mendez (Ref. 1). Mendez indicated that the alkali nature of concrete will damage the silicon in glass, but that damage can be avoided by jacketing the glass fibers with plastic buffers. Some of the early laboratory studies of the use of fiber-optic sensors in reinforced concrete were reported by Huston et al.[2] and Fuhr et al.[3] These studies consisted of embedding optical fibers of various types inside a small $0.1 \times 0.1 \times 1.0$ m^3 reinforced concrete beams. The first test determined whether optical fibers could survive the embedding and curing process. Subsequent tests determined the failure rate of embedded fiber-optic sensors as well as sensitivity measurements for such embedded sensors. The efforts by Huston and Fuhr involved embedded multimode fibers being used to detect vibrations.[4] Additional laboratory studies have been reported by Escobar et al.[5] which used bonded fiber-optic strain gauges on concrete beams and by Kruschwitz et al.[6] which used strain gages in concrete beams.

Microbend technology has been used to measure the tension in a posttensioning strand of a steel reinforcing bar.[7] The fiber is spiral-wrapped around a posttensioning strand. When the strand is placed in tension, the microbend action of the fiber and the strand prevents light from traveling through the fiber. When the tension is relaxed, light can pass through the fiber and the state of relaxation can be detected. These sensors have been installed in a Paris subway and the Marienfeld Pedestrian bridge.

A fiber-optic vibration sensor has also been used on prototype structures in the field. One study reported the use of this sensor that measured the vibrations of a guy wire that was subjected to wind loading.[8] The sensor was also attached to the deck of the I-89 bridge in Winooski, Vermont.[9]

The ability to wrap optical fiber into metal rope also opens the possibility for embedded strain detection along the length of the metal rope. May et al.[10] used Extrinsic Fabry-Fizou Interferometric strain sensors to monitor steel ropes. A similar installation in a linelike structure is reported by Harrison and Funnell who monitored the overheating of power transmission lines with fiber-optic sensors.[11]

Large-scale structures that have actually had fiber-optic sensors embedded into them are still relatively few. Holst and Lessing have installed fiber-optic displacement gauges in dams to measure shifting between segments.[12] Fuhr et al. report the installation of fiber-optic sensors into the Stafford Building at the University of Vermont.[13] Similar installations have been completed at the Winooski One Dam in Winooski, Vermont, in the physics building at the Dublin City University in Ireland,[14] and in a railway overpass bridge in Middlebury, Vermont.

Other field installations use fiber-optic sensors to measure the weight of trucks as they drive over sensitive strips at highway speeds.[15,16] Similar techniques have been applied to intrusion detection systems. Another field stress measurement to be reported is by Caussignac, where the internal stresses in neoprene bridge bearings are measured.[17] Bridge measurements on an overpass in Calgary, Alberta, Canada, have been reported by Measures et al.[18] Wemser et al. are reporting considerable success using OTDR-based fiber-optic sensing of large-diameter pipes at various electricity-generating facilities.[19]

There is a substantial amount of similar work underway throughout the world. In many cases, side-by-side installations of fiber-optic and conventional sensors are involved indicating that the civil engineering world, along with architects, insurers, and others, need more proof as to the usefulness of these typically more-expensive fiber-optic sensors. However, it must be acknowledged that in many instances, fiber-optic sensors are providing meaningful measurements that otherwise would not be made.

SUMMARY

During the past years, research has been underway worldwide investigating development and implementation of smart civil structures. During the course of this effort, numerous fiber-optic and conventional sensor techniques and designs have been developed and tested. Laboratory studies have led to field testing at numerous sites and types of installations ranging from residential homes, conventional multistory buildings, pedestrian footbridges, interstate highway road surfaces and bridges to high-performance structures such as a hydroelectric dam and a nearby railway bridge. The importance of having preliminary efforts involving sensor development and measurements regarding the failure of concrete beams with embedded fibers cannot be under-stated. Many of our findings with respect to actual installation of embedded sensors have been reported, with these efforts continuing.[20] In addition, the actual usage of embedded sensors has led to the addressing of seemingly mundane issues such as who acquires the data and who has to look at the data. Use of the Internet worldwide computer network for such monitoring is now underway.

The smart civil structure area is certainly in its infancy, but given numerous seemingly coalescing factors (e.g., smart structures research, administrative support of rebuilding the U.S. infrastructure, ever increasing awareness of safety issues), it appears that the benefits possible only through the use of embedded sensors will lead to their future use. Undoubtedly, the ultimate extent to which structures are built using intelligent materials will depend on the technology having a positive economic, engineering, and/or safety cost benefit.

REFERENCES

1. A. Mendez, T. F. Morse, and F. Mendez, "Applications of Embedded Optical Fiber Sensors in Reinforced Concrete Buildings and Structures," *Proc. SPIE*, **1170**, *Fiber Optic Smart Structures and Skins II*, 1989.
2. D. R. Huston, P. L. Fuhr, P. J. Kajenski, and D. Snyder, "Concrete Beam Testing with Optical Fiber Sensors," *Proc. of the ASCE Minisymposium on the Nondestructive Testing of Concrete*, F. Ansari, ed., April 1992, San Antonio, Texas.
3. P. L. Fuhr, D. R. Huston, T. Ambrose, and D. Snyder, "Curing and Stress Monitoring of Concrete Beams with Embedded Optical Fiber Sensors," to appear as a tech. note in the *ASCE Structures Jnl.*, October 1993.
4. D. Huston, P. L. Fuhr, J-G. Beliveau, and W. B. Spillman Jr., "Structural Member Vibration Measurements Using a Fiber Optic Sensor," *Journal of Sound and Vibration*, 1991, **150**(2), pp. 1–6.
5. P. Escobar, V. Gusmeroli, and M. Martinelli, "Fiber-Optic Interferometric Sensors for Concrete Structures," *Proc. First European Conf. on Smart Structures and Materials*, Glasgow, 1992, p. 215.
6. B. Kruschwitz, R. O. Claus, K. A. Murphy, R. G. May, and M. F. Gunther, "Optical Fiber Sensors for the Quantitative Measurement of Strain in Concrete Structures," *Proc. First European Conf. on Smart Structures and Materials*, Glasgow, 1992, p. 223.
7. Anon., *The Pedestrian Bridge to Marienfelde Leisure Park*, Strabag Bau-AG, Berlin Branch, Bessemer Str. 42a, 1000 Berlin.
8. P. L. Fuhr, and D. R. Huston, "Guy Wire Vibration Measurements with Fiber Optic Sensors," *Proc. ASCE Structures Congress '92*, J. Morgan, ed., April 1992, San Antonio, Texas, pp. 242–246.
9. D. R. Huston, P. L. Fuhr, and J-G. Beliveau, "Bridge Monitoring with Fiber Optic Sensors," *Proc. Eighth U.S.–Japan Bridge Engineering Symposium*, Chicago, May 1992.
10. R. G. May, R. O. Claus, and K. A. Murphy, "Preliminary Evaluation for Developing Smart Ropes Using Embedded Sensors," *Proc. First European Conf. on Smart Structures and Materials*, Glasgow, 1992, p. 155.
11. B. J. Harrison and I. R. Funnell, "Remote Temperature Measurement for Power Cables," *CIRED Conference*, 1991.
12. A. Holst and R. Lessing, "Fiber-Optic Intensity-Modulated Sensors for Continuous Observation of Concrete and Rock-Fill Dams," *Proc. First European Conf. on Smart Structures and Materials*, Glasgow, 1992, p. 223.

13. P. L. Fuhr, D. R. Huston, P. J. Kajenski, and T. P. Ambrose, "Performance and Health Monitoring of the Stafford Medical Building Using Embedded Sensors," *Smart Materials and Structures,* **1** (1992b), pp. 63–68.
14. B. McCraith, Dublin City University, Ireland, personal communication to P. L. Fuhr, 1993.
15. S. Teral, "Vehicle Weighing in Motion with Fiber Optic Sensors," *Proc. First European Conf. on Smart Structures and Materials,* Glasgow, 1992, p. 139.
16. K. W. Tobin and Jeffrey D. Muhs, "Algorithm for a Novel Fiber-Optic Weigh-in-Motion Sensor System," *SPIE* August 1991, **1589**(12).
17. J. M. Caussignac, A. Chabert, G. Morel, P. Rogez, J. Seantier, "Bearings of a Bridge Fitted with Load Measuring Devices Based on an Optical Fiber Technology," *Proc. First European Conf. on Smart Structures and Materials,* Glasgow, 1992, p. 207.
18. R. Measures et al., "Multichannel Structurally Integrated Fiber Optic Sensing of a Highway Bridge Using Carbon Fiber Prestressing Tendons," *SPIE* February 1995. **2446,** paper no. 4.
19. K. Wanser, as reported in *Laser Focus World,* February 1995, p. 8.
20. P. L. Fuhr, D. R. Huston, P. J. Kajenski, T. P. Ambrose, and W. B. Spillman Jr., "Installation and Preliminary Results from Fibre Optic Sensors Embedded in a Concrete Building," *Proc. of the First European Conf. on Smart Structures,* Glasgow, May 1992.

CHAPTER 21
A NEW APPROACH TO STRUCTURAL HEALTH MONITORING FOR BRIDGES AND BUILDINGS

INTRODUCTION

Structural health monitoring systems have been designed and evaluated for installation in several bridges and commercial buildings. The systems employ solid-state sensor elements which experience a strain-dependent phase transformation from a metastable, nonmagnetic, austenitic phase to the stable, ferromagnetic, martensitic phase. The irreversible phase transformation is useful for indicating the level of peak structural strain experienced in a particular monitored location. Some of the sensor material characteristics and details related to the phase transformation are discussed as applied to structural health monitoring. The design of representative systems for bridges and commercial buildings is included. Important systems features and design capabilities are also discussed.

Infrastructural monitoring can be performed on many different levels with equally many objectives in mind. Depending on the goals of the monitoring system it may be possible to employ fully active, semiactive, or fully passive systems. Large-scale monitoring of existing structures requires that the sensing system employed can be retrofitted at a reasonable cost and with a high degree of reliability. These two requirements are typically difficult to satisfy simultaneously, and so engineers must often choose the most practical method which can provide the necessary information on which to base maintenance or engineering component replacement decisions. The monitoring problem is further complicated by the desire to extend the lives of structures, particularly bridges, beyond their original respective design lifetimes. Economics is, of course, at the heart of life extension, but unless the structural integrity issues can be laid to rest, safety concerns may dictate the choices available resulting in bridge closure or rebuilding. Either of these options affects the local economy and interferes with the normal conduct of business.

The researchers have been engaged in the development of structural-health monitoring systems for bridges and other structures.[1-4] Only a brief summary of the details is presented here. The sensors employed are steel alloys which experience an irreversible, solid-state, strain-dependent phase transition from a nonmagnetic, austenitic parent phase to a ferromagnetic, martensitic product phase in response to the strain experienced. The degrees of ferromagnetic response of the individually positioned sensors can then be correlated with the respective local strains necessary to produce that extent of transformation. The sensors supply an indication of the maximum or peak strain conditions. These data are then coupled with a suitable design model to make decisions regarding structural integrity. The key to the technology is that the sensors can operate entirely passively so that, for a bridge monitoring application, field engineers can simply interrogate the electronics to obtain the ferromagnetic responses of the sensors during routine maintenance operations without the necessity of having a dedicated data acquisition system on the site. These systems are designed to provide peak strain information and are not suitable for obtaining the more detailed information provided by more complex systems, say fiber-optic systems, and so on. In the vast majority of cases related to bridge monitoring the structural peak strain conditions will suffice in determining the engineering options available.

Figure 21.1 shows a generic martensitic transformation curve for TRIP steel. TRIP (transformation-induced plasticity) steels were developed 30 years ago as high-strength, fracture-resistant structural materials. Some of the austenitic stainless steels, particularly AISI 301, and Fe-Mn, Fe-Mn-Cr steels also display the strain-induced martensite transformation that is critical to their application as strain sensors. Both the incubation strain (transformation triggering strain) and transformation rate (sensor sensitivity) can be controlled by thermomechanical treatment or chemistry variation to influence the austenite stability. The sensor materials can be produced to match a wide range of environmental conditions including fully stainless. The long-term durability of these sensing materials is attractive for applications which may extend up to 50 years or more.

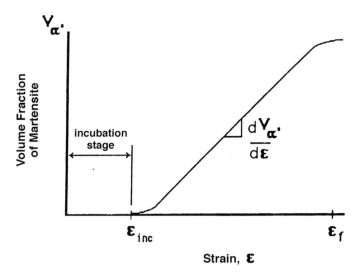

FIGURE 21.1 Volume fraction of martensite versus strain behavior.

Typically, a sensor element of the metastable materials is attached to the structure using custom-designed attachment fixtures. The ferromagnetic response of the sensor element is measured using a Hall effect sensor-permanent magnet combination positioned adjacent to the gage section of the element as shown in Fig. 21.2. The Hall sensor provides an output voltage which is proportional to the magnitude of magnetic flux. A change in the flux due to changes in the ferromagnetic material produced by the strain thereby creates a changing output voltage. The Hall sensors are commercially obtained in extremely small sizes and can be readily positioned next to the sensor element using a number of different approaches.

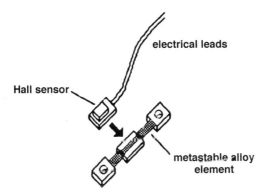

FIGURE 21.2 Schematic of a sheet sensor element with Hall sensor transducer.

The sensor element geometries investigated were stamped from thin (0.13-mm-thick) sheet stock into elements having a reduced gage section in order to control the volume transforming during straining. When small gage lengths are used, it is possible to obtain substantial effect magnifications by using attachment fixture distances significantly longer than the gage lengths. In effect, the deformation experienced by the structure between the attachment fixtures is concentrated into the gage length of the sensor element. Using this approach, extremely low-level peak strains, well within the elastic range, can be easily measured. Figure 21.3 shows the sensor element profiles investigated in this study. Figure 21.4 shows the Hall voltage output for such a case where a strain sensor employing the stamped elements was attached to the grips of the tensile testing machine using a simple tensile gage mounting. The attachment distance becomes the distance between the specimen grips in the strain sensor (38.1 mm) for this testing configuration. Testing was conducted under a range of strain rates and this particular alloy was relatively strain-rate-independent up to at least 1 m/m/s.

Figure 21.5 shows an example of the Hall sensor output data obtained from testing two variations of the Fe-17Cr-9.4Ni sensor materials (labeled B1 and B2) in

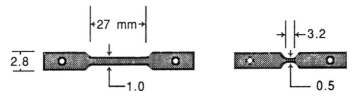

FIGURE 21.3 Sensor element profiles (0.13 nm thick).

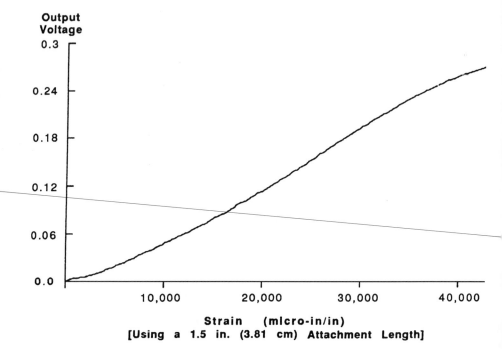

FIGURE 21.4 Hall voltage output.

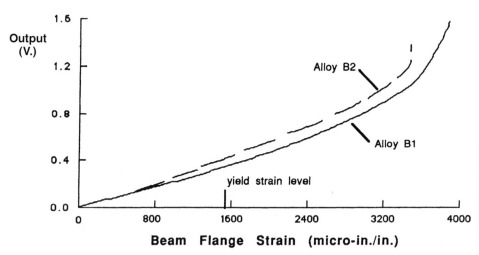

FIGURE 21.5 Sensor transducer output as a function of the peak tensile strain for a steel beam tested in front point bending.

applications involving beam bending. The strain monitor was attached under the bottom flange of a steel box beam using an attachment distance of 152.4 mm and employing sensor elements having gage lengths of 3.18 mm. The beam was loaded at a constant rate up to its strength failure. The upward curve of the output at the far right is due to the inelastic strain during the severe buckling of the beam's compressive flange. In this case, strains of less than 8 percent of the yield value are measurable, a longer gage mount would have provided for an even lower resolution.

Figure 21.6 shows the resolution and maximum measurable strain using this strain magnification approach for the long (26.7-mm) and short (3.18-mm) gage length sensor elements for cases where the strain monitor is attached to the structure and loaded in tension or bending. The sensor elements were fabricated from an early developmental alloy. In normal operation it may be desirable to note the structural peak strain reading, which, of course, would be an elastic strain level, under normal loading conditions prior to the accumulation of any damage to obtain a reference value to include in numerical models. Strains as low as about 1000 microstrains can be measured using the 3.18-mm gage length element with an attachment distance of 38.1 mm, whereas strains well below 100 microstrains can be measured by increasing the attachment distance to about one meter. This approach was verified on a bridge installation where strains as low as 67 microstrains were observed and measured.

The National Science Foundation is participating in a project directed at the monitoring of civil engineering structures for postearthquake damage assessment, particularly bridges and buildings. Currently, it may take days to weeks before a suspect structure can be inspected following an earthquake. Use of that structure in the interim is risky, and there is a need to be able to quickly inform people of its condition and to warn emergency relief support as to problems that need attention. Sim-

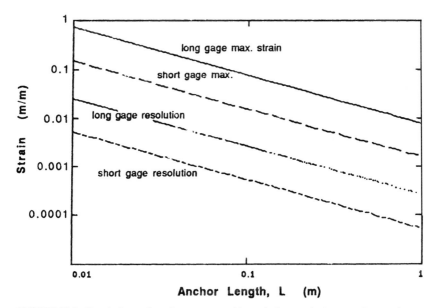

FIGURE 21.6 Resolution and maximum measurable strain for attached sensor elements investigated as a function of the distance between attachment fixtures.

ulation experiments were conducted to determine the strain sensor response characteristics for excitations produced during an earthquake. An axial test machine was programmed in stroke-controlled mode to reproduce the dynamic response at a hypothetical gage site on the new 1-5 Gavin Canyon Bridge (Los Angeles, California). Caltrans uses this earthquake spectrum and others like it as design events. The dynamic response was converted into strain for the purpose of programming the test machine to obtain the earthquake strain-time history. A 3.18-mm gage length sensor element was used with a 152.4-mm attachment distance where the strain monitor was considered to be attached to the vertical side of a column on the bridge at a point near the deck. Figure 21.7 shows the peak strain output from the strain monitor superimposed on the actual strain produced by the earthquake loading spectrum. Note that there are five tensile peaks produced during the earthquake and the peak strain monitor output follows the increasing strain levels and is stable between the individual peaks. An active or semiactive monitoring system would record the entire strain history as shown in the figure, whereas a passive monitoring system would retain only the maximum peak tensile strain during the earthquake, which would correspond to the most damaging strain to the structure. An engineer would be able to use this peak strain value to determine the extent of structural damage.

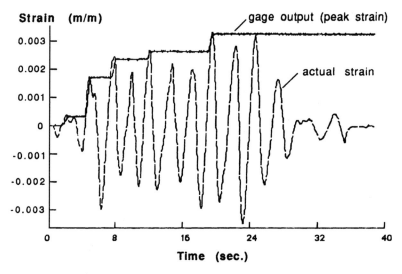

FIGURE 21.7 Earthquake simulation showing the actual dynamic strain with the peak strain history obtained from the response of a column on the 1-5 bridge.

SAVANNAH RIVER BRIDGE PROJECT

The developers installed a prototype monitoring system on the U.S. Interstate 95 Savannah River Bridge near Savannah, Georgia. The Georgia Department of Transportation wished to examine the deflections and stresses created by particularly heavily loaded trucks crossing the bridge. Figure 21.8 shows a schematic of the bridge locations monitored. These were at locations on the long spans and are labeled as sections 1 and 2. The locations were chosen based on three-dimensional

STRUCTURAL HEALTH MONITORING FOR BRIDGES AND BUILDINGS 21.7

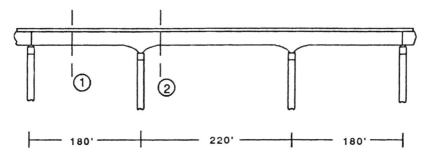

FIGURE 21.8 Elevation of the center spans of the Savannah River Bridge which indicate instrument locations.

finite element modeling of the structural response to the trucks' weights. Two details were monitored by the system: (1) the out-of-plane deflection of the beam web created by the cross-bracing loads, and (2) the longitudinal stress in the beam flange. Three locations were instrumented and connected to signal conditioners and a microprocessor housed on the bridge deck.

Figure 21.9 shows the geometry of the web deflection measurement. The web is approximately two meters (six feet) high and the I-beams were subject to out-of-plane deflections from the nonuniform distribution of the deck loads onto the beams. In a typical failure, the weld on the stiffener plates would fail and cracking in the web occurs. A strain sensor consisting of a tensile element was attached between the stiffener plate and the top of the flange. This monitored the relative deformation causing the weld stress.

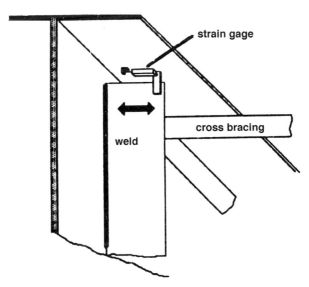

FIGURE 21.9 Cut-away detail of the gaging for the web displacement at the stiffener joint.

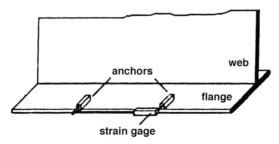

FIGURE 21.10 Configuration for monitoring the beam flange strain.

Figure 21.10 shows the arrangement of the gage for monitoring the strain (and stress) of the lower steel I-beam flange. The strain range of interest here is smaller than that of the alloy's direct transformation range. Therefore, the gage length, or the length over which the transformation occurs, was chosen to be smaller than the anchor point distance. The ratio of these lengths is the magnification factor for the sensor. This method was used to increase the sensitivity by a factor of 63 in the flange monitor.

The system was installed during a period of light traffic, and shortly thereafter it measured the effects of rush hour traffic prior to the trucks crossing the bridge the following day. The strains related to peak afternoon traffic flow were only slightly lower than those measured when the trucks passed across. The peak strains in the beam flange measured for the latter events were 67 to 100 micrometers/m and corresponded to tensile stresses ranging from 14 to 21 MPa, well within the elastic range of the steel beams. The stiffener gage measured a peak out-of-plane deflection of about 0.23 mm (0.009 in) which corresponds to stresses in the beam's web of about 65 MPa. The data were consistent with the stresses predicted by the numerical model.

In this application a microprocessor monitors the gage outputs (peak strains) and stores these values daily into a log file in battery-backed RAM. An RS-232 interface provides for both the routine downloading of the data and control over the logging process. The output from the sensors can be handled in a variety of ways, depending on the purpose of the instrumentation. The sensor output is the raw data and analyses are usually necessary (in a complete monitoring network) to come to any conclusions about the actual structural safety. However, the state of technology of embedded PCs is quite adequate to provide for a complete, on-site decision-making system capable of providing alarms and shut-down control. As an example, a bridge instrumentation system could activate barricades preventing its use if damaged in an earthquake or through foundation failure. The passive operation of the gages also means that on-site instrumentation and power is not necessary, a portable data logger can be carried to the site to interrogate the gages as part of a routine inspection.

ARPA BRIDGE MONITORING PROJECT

A project is currently underway to design and install fully automatic and telemetered strain sensors onto a variety of bridges within Georgia. The ensemble of 10 bridges represent a wide cross section of steel and concrete types of bridge construction. Some of structures are exhibiting distress from previous conditions and these were addressed in the system designs.

An inspection was made by a bridge engineer and the existing and potential problems of each were compiled. The analyses were used to provide information on the locations to monitor, the measurement range, and the number of sensors to be used. Sensors to monitor peak planar strains (stresses), out-of-plane bending, expansion joint movements, crack opening displacements, and deflections induced by weakening were designed. An automatic monitoring system was designed for each bridge which consists of an embedded microprocessor, a twisted-pair wire network connecting the signal conditioners at the gage sites, and a fax/pager/modem interface via cellular phone. Figure 21.11 shows a schematic of a bridge system with its components. A small photocell array provides power for the system if it is not available from the bridge lighting AC line. Two of the bridges are yet to be constructed and provide the opportunity to monitor the condition of the prestressed concrete beams from the fabrication shop, through transport to the site, through construction, and into the operational use of the bridge. The passive behavior of the sensors means that the gages can be attached at the fabrication site and do not need any instrumentation to be present during transit or construction. The final, installation phase of the project was completed in 1997.

EMBEDMENT APPLICATIONS

The passive behavior of the sensor alloys implies that they can be embedded within nonferromagnetic materials to witness the straining going on and can be interrogated via an external magnetometer. This would provide a practical means of inspection for composites, plastics, and concretes. Thin sensor material wires (254-Iim diameter, i.e., 0.01 inch) were embedded in a unidirectionally reinforced carbon fiber-epoxy matrix, multilayer, laminated composite panel for evaluation to detect and monitor tensile straining damage. Two wires were embedded at a distance of about 15 mm parallel to the fiber orientation along the length of the tensile test samples. One specimen was pulled to failure to determine the failure strain which was

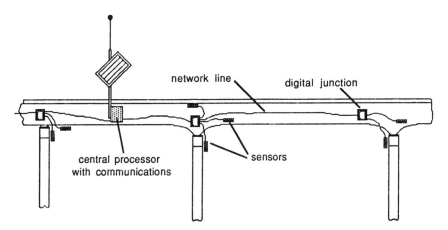

FIGURE 21.11 Schematic of bridge instrumentation network.

about 4 percent in tension. Other identical specimen were tested to a strain of 2 and 3 percent, respectively, at which point no evidence was externally observed to indicate that the composite had been damaged. Compared with control wires, the embedded wires that were tested in tension displayed considerably higher output voltages as shown in Fig. 21.12. The ferromagnetic responses were measured using a SQUID magnetometer with 13 perpendicular passes across the two embedded wires at random locations about 10 mm apart along the length of the embedded wires. Straining was detected at many locations along the wires in the strained sample whereas the readings were very consistent for the unstrained control wires. No external composite damage was detected in the strained panel by visual examination. These data are the first experimental evidence for a feasible, inexpensive technique to monitor composite material damage. Further research is required to perfect both the placement and detection methods although it was possible to detect a change in ferromagnetic response using the Hall sensor chips, but the SQUID magnetometer was much more effective. The 250-gm (0.01-inch) diameter wires are much larger than the anticipated wire size which will later be commercialized for damage-detection applications.

Further testing was conducted using the SQUID magnetometer to measure the

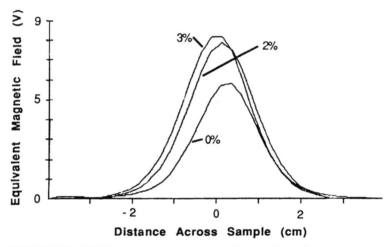

FIGURE 21.12 SQUID magnetometer scan across three tensile specimens.

ferromagnetic responses of 25.4-mm gage length sensor elements after straining 0, 5, 10, 15, and 25 percent in tension. The sensor material investigated did not begin to transform until a strain of about 7 to 8 percent was reached, so the readings for the control (0 percent strain) and 5 percent strain were similar. Once the material began to transform, the SQUID magnetometer easily picked up the increasing levels of martensite as strain increased. No prestraining of the element was performed (as would have been done in an actual installation to increase the sensitivity) since the experiment was also designed to show the effects of tensile plastic deformation without transformation. The minimal differences between readings at 0 and 5 percent indicated that there is little if any effect associated with deformation that does not produce martensite.

REFERENCES

1. Strain Monitor Systems, Inc., "Savannah River Bridge Project," *GADOT SRS No. 218: Final Report to the GA DOT,* SMS Report 25-93, 1993.
2. L. D. Thompson and B. Westermo, "The Development of a New Strain Measurement Methodology for Structural Damage Assessment and Monitoring," *Advances in Instrumentation and Control,* Instrument Society of America, Houston, 1992, **47**(2), pp. 1295–1303.
3. B. Westermo and L. Thompson, "A New Testing and Evaluation Technology for Damage Assessment and Residual Life Estimation in Aircraft Structures," *Proceedings of the 14th Aerospace Testing Seminar,* The Institute of Environmental Sciences and the Aerospace Corporation, Manhattan Beach, California, 1993, pp. 5–10.
4. B. D. Westermo and L. D. Thompson, "Smart Structural Monitoring: A New Technology," *SENSORS,* November 1994, pp. 15–18.
5. L. D. Thompson and B. D. Westermo, "Applications of a New Solid-State Structural Health Monitoring Technology," presented and published in *Proceedings of the Second European Conference on Smart Structures and Materials,* Glasgow, Scotland, October 1994.
6. V. F. Zackay, E. R. Parker, D. Fahr, and R. Busch, "The Enhancement of Ductility in High-Strength Steels," *Trans. ASM,* 1967, **60**, pp. 252–259.

CHAPTER 22
TRUE ONLINE COLOR SENSING AND RECOGNITION

INTRODUCTION

Color is a very critical component of how an object may look or appear to be to the human eye. Most of today's industrial companies rely heavily on high-tech instrumentation to improve the quality of their products and the productivity of their operation.

Electronic sensors with various technologies have been making a lot of progress in today's industrial world. These sensors have made it possible to provide 100 percent product inspection and online sensing and monitoring capabilities.

Initially, companies utilized very expensive inspection equipment in order to improve quality, reduce waste, and increase productivity. With increased competition worldwide, companies began to look for more practical and more economical inspection solutions to replace extremely complex and costly inspection devices.

One of the areas of concern is online color sensing and monitoring applications where true color technologies are required to provide accurate information. Plastic and glass fiber-optic cables in various lengths and design configurations are used in industrial and hazardous areas to provide resistance against environmental adversities including shock and vibration.

SENSING LIGHT AND COLOR

The most widely known detector of color consists of the eye and the brain, including the human nervous system. All technological attempts to sense color are duplicates of the human color detector.

Industrial color sensing and measurement is accomplished using silicon photodiodes and photomultiplier tubes. Their responses vary for different wavelength values (see Fig. 22.1).

DEFINITION OF COLOR

Color is a phenomenon observed by the interaction of light with other objects. Three essential ingredients must be combined to simulate the effect of light (see Fig. 22.2):

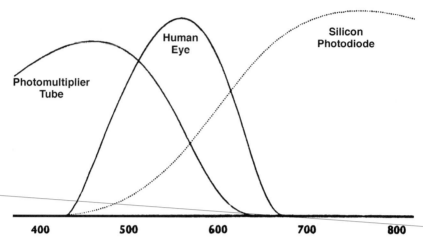

FIGURE 22.1 Responses for the eye, photomultiplier, and silicon photodiodes.

1. Illumination source
2. Object to be illuminated
3. Detector (eye/brain combination)

LIGHT/ENERGY SPECTRUM DISTRIBUTION

Light is a form of radiated energy which is a small segment of a much greater electromagnetic radiation spectrum. Color is a very narrow band of energy known as the *visible light spectrum*.

The human eye is very sensitive to this narrow band of visible radiations (Fig. 22.3). The interaction between the human eye and the visible spectrum results in visual color perception (Fig. 22.4).

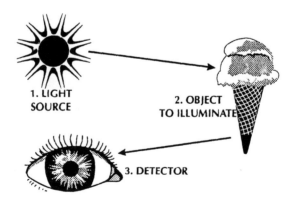

FIGURE 22.2 Light source, illumination, and detection.

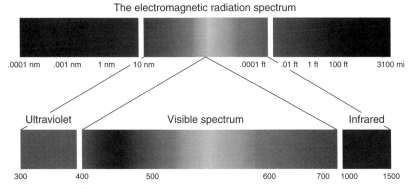

FIGURE 22.3 The electromagnetic radiation spectrum.

LIGHT DISTRIBUTION

The laws of physics state that light distribution varies depending on the object it strikes. In general, when light strikes an object, certain conditions appear depending on the shape, size, texture, opacity, and so forth of that object. These conditions may have an effect on the color and appearance of the object (Figs. 22.5 and 22.6).

Light distribution can be one of the following forms:

1. *Specular reflection:* This is a condition that occurs when a small percentage of the light signal reflects at the first surface (gloss).

Regular Reflection

2. *Diffused reflection:* As the surface becomes rougher, the gloss is reduced and the light will be diffused in all directions.

Diffused Reflection

3. *Diffused transmission:* Light is diffused (scattered) upon exiting the object (haze).

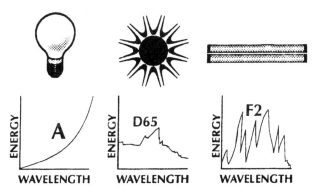

FIGURE 22.4 CIE light sources.

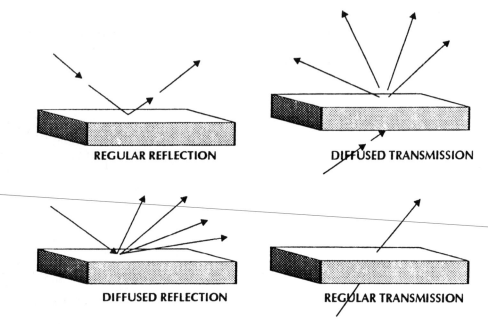

FIGURE 22.5 Regular light reflection.

4. *Regular transmission:* Light passes through the object undistributed, effected only in color.

To have a better understanding of color measurement, certain parameters must be performed under controlled and well-defined conditions. Some of these factors are

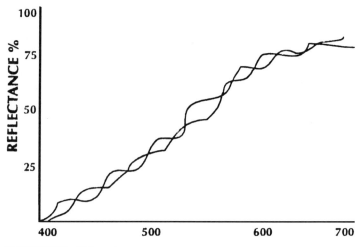

FIGURE 22.6 Metamerism.

1. Level of illumination (Fig. 22.7)
2. Spectral characteristics of the light source
3. The spectral response of the detection element (Fig. 22.8)
4. Background environment
5. Viewing angle

METAMERISM

This phenomenon occurs when two objects with different spectral characteristics match under one type of lighting condition but not under another.

Objects are metameric pairs when they have the same color coordinates, while having different spectral reflectance curves.

BACKGROUND

Even though the human eye is still the most accurate and reliable color detector, it is not a practical solution for today's high-speed and complex manufacturing and production environment. The need for more automation in our factory floors and production lines has prompted many sensing and instrumentation companies to put more emphasis on color and appearance sensing and measurement.

Many types of color instrumentation are available for the purpose of improving product quality and making the process independent of the human factor which is prone to external emotional and physical factors.

Today's color instruments such as spectrophotometers, colormeters, machine vision systems, and other online color-sensing and measuring devices are considered by far the most advanced instruments technology has to offer. This chapter empha-

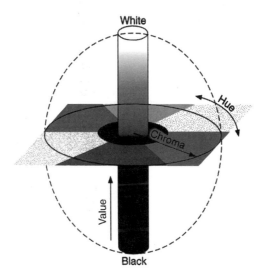

FIGURE 22.7 Hue, value, and chrome.

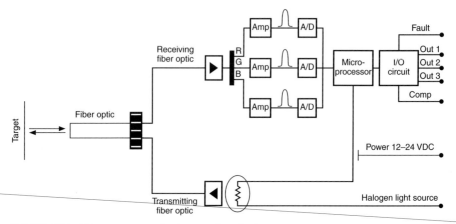

FIGURE 22.8 Discrete outputs of a given number of colors.

sises the use of online true RGB color-monitoring technology, a method that provides 100 percent online color-sensing capabilities at a fraction of the cost of the conventional spectrophotometers, calorimeters, or even color machine vision systems.

SYSTEM DESCRIPTION

Online RGB-based color-sensing devices are high-speed sensors suitable to be used in industrial environments where they can be exposed to the adverse conditions of typical factory production environments. These sensors are microprocessor-based instruments available with either (GO/NO GO) and/or serial and parallel outputs. These outputs provide the interface capability to discrete devices as well as standard computer communication ports. Online color sensors are built specifically for industrial color recognition applications and possess the flexibility required on the production floor.

Most of these sensors are very simple to set up, calibrate, and operate and require no more than a few minutes to put into operation.

ADVANTAGES OF ONLINE COLOR SENSORS

- High-speed operation (1-ms response time)
- Halogen light source for full color spectrum
- Multicolor-sensing capabilities
- Fiber-optic cables with various sizes, lengths, and configurations
- Long range sensing distances
- Teach/calibration mode programming features
- Available with GO/NO GO and computer communication ports
- Can tolerate normal web movement
- Suitable for most industrial manufacturing environments

TRUE ON-LINE COLOR SENSING AND RECOGNITION   22.7

- Easy field installation
- Can be used in hazardous environments

COLOR THEORY

In order to understand how these online color sensors and instrumentations work, we need to touch upon the color theory and its coordinates, hue, value, and chroma. These three components with their alphanumeric values can precisely define a certain color space.

Hue is what the human eye perceives as the color, that is, red, green, yellow, blue, and so on. These are terms used to define the wavelength of certain colors. Hundreds of different variations can be generated by combining different hues together.

Value is defined as the lightness or darkness of a particular color. Light colors have high values and dark colors have low values. For example, by adding black to a certain color in various proportions, the appearance of that color becomes different. In this case, the value has been changed, but the hue and chroma stay constant.

Chroma describes how bright or dull a color is. This number shows the vividness of a particular color. Chroma is the color coordinate that is most affected by the amount of illumination that the color is exposed to. High chroma colors are very bright and vivid.

PRINCIPLE OF OPERATION

Most of today's online color sensors use replaceable halogen light sources to illuminate the target. The light is normally transmitted over a bifurcated fiber-optic bundle which is designed to transmit and receive light to and from the RGB sensing elements. The detector array with the RGB optical filters analyzes the reflected light and produces an analog signal based on the red, green, and blue (RGB) color contents of the reflected beam.

$$R_{component} = (R_{signal})/\text{Total light signal}$$

$$B_{component} = (B_{signal})/\text{Total light signal}$$

$$G_{component} = (G_{signal})/\text{Total light signal}$$

The analog signal of each of the preceding components is then amplified and converted to a digital format through an analog-to-digital converter.

The microprocessor analyzes the digital information provided by the A/D converter and by applying various algorithms, the digital data is then processed to accommodate specific requirements.

Typically these sensors provide multiple discrete outputs for a given number of colors. A simple, easy to follow procedure is all that is required to program the sensor and store the reference color into the EE prom of the sensor.

EXAMPLES OF APPLICATIONS

Obviously the broad needs of modern manufacturing present virtually limitless and challenging situations for such a color sensing system. The versatility of online color sensors is illustrated in the following five examples of unique situations requiring innovative solutions.

1. *Sorting automotive parts with different colors.* A manufacturer of automotive parts needs to differentiate parts whose only visible difference is a slight variation in color. One part is black, the other a dark grey. Because efficiency demands that the parts be sorted at a high rate of speed, the opportunity for mistakes is extremely high. By using advanced color sensing technology, the manufacturer can sort these parts at an enhanced rate of speed, saving time and virtually eliminating errors.

2. *Assembly of medical closures with different color components.* In this situation, containers of liquid medications consist of an aluminum cap with a plastic cover. A complete closure assembly requires the assurance of proper color combination of the two components. Because of the absolute necessity of color accuracy, color sensing technology proves invaluable.

3. *Lumber sorting by color.* Color coding has become a standard method of differentiation in the lumber industry. Not only different types of lumber but the grade, quality, and intended purpose of the lumber is indicated by color. Because the environments in which lumber is sorted can be highly abusive, it is recommended that protective covers be employed to protect the fiber optics used for this process.

4. *Color sensing in the food industry.* Sensing a white target on a white background is challenging using conventional photoelectric sensors. A manufacturer who needs to insure the presence of a white cap on a jar of mayonnaise improves accuracy with a color sensor that employs the RGB color concept. This technology improves the contrast between two slightly different whites.

5. *Ammunition final inspection and sorting process.* An ammunition manufacturer codes the style and caliber of bullets with various colors on the tips. The need to insure that the proper type and caliber of product are correctly packaged necessitates the automation of the product line with color sensing technology. Because of the critical nature of this operation, it is recommended that two color sensing stations be implemented to add an extra safety margin to the operation.

CONCLUSION

Historically, successful online color sensing and measurement solutions have been attributed to the combination of very expensive equipment and visual inspection.

Today's technological innovation has made it possible to use simple easy to operate fiber-optic sensors to precisely sense a specific color, shades of the same color, or combinations of different colors. These advanced optical devices, through the use of fiber-optic technology, can withstand environmental adversities, including shock, vibration, high temperature, and so on.

Proper selection of the sensor and the proper fiber-optic configuration is very critical to the success or failure of the proposed solution. Careful application simulation and bench testing are essential before committing to a specific direction. Environmental factors such as shock, vibration, and ambient conditions should be taken into consideration.

CHAPTER 23
FUNDAMENTALS OF SOLID-STATE PRESENCE-SENSING TECHNOLOGIES

PRESENCE DETECTION

There are two general types of technology for presence detection: (*a*) discrete (on-off), "is it here... or not?" and (*b*) analog (continuous-proportional), "where is it?" Analog gives a continuous response, answering the question "where is it?" They respond best to continuous processes, such as level control, positioning, and the like.

The discrete type answers the question "is it here... or not?" They have contact, and the ON-OFF position corresponds to the discrete states. The discrete type is most applicable to manufacturing, since most sensing applications deal with processes which demand that a target be in a definite position for assembly or at a specific level for filling. In situations such as materials handling or the assembly of products, manufacturing systems need to know when the target is at given location. Period. They don't need to know when the target is almost there or almost gone.

PRESENCE SENSORS

There are two general categories of presence sensors (Fig. 23.1): (1) physical contact (example: limit switch), and (2) noncontact (example: photoelectric sensor).

As an example, you would want a noncontact sensor to detect the presence of freshly painted car doors as a means of counting how many had been sent on to the curing area.

Noncontact Sensors versus Limit Switches

Noncontact Sensors

- Physical contact not required
- No moving parts
- Faster
- No bouncing

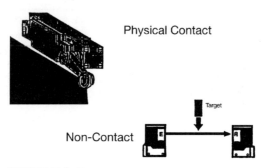

FIGURE 23.1 Presence sensors.

Limit Switches

- More current
- Safer

Comparing noncontact sensors to limit switches shows that with limit switches, the target directly actuates the switch through physical contact. Noncontact sensors are, with few exceptions, solid-state electronic devices which offer distinct advantages:

- Physical contact is not required.
- No moving parts to jam, fail, or be broken by extreme physical contact.
- They are faster.
- There is no bouncing at contact state change.

Limit switches, however, have the following characteristics:

- They carry more current.
- They are safer—when opened, there is an air gap between the contact terminals.

Major noncontact sensing technologies. The major noncontact sensing technologies (Fig. 23.2) are as follows:

Reed relay	Magnetic target
Inductive	Metal target
Capacitive	All materials target
Photoelectric	All materials target

Inductive, capacitive, and photoelectric technologies compose the majority of the industry's applications; however, other technologies are in the field and you should be aware of them. They are

Ultrasonic	All materials target
Magnetic reluctance	Magnetic metals only
Hall effect	Magnetic target

FUNDAMENTALS OF SOLID-STATE PRESENCE-SENSING TECHNOLOGIES 23.3

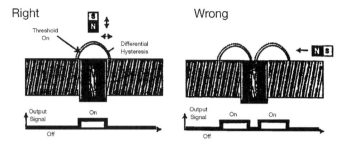

FIGURE 23.2 Reed relay, magnetic targets only. Typical sensing range = 0–1.5 in.

Generally, the more control you have over what passes in front of the sensor, the better. The less control you have over what passes in front of the sensor, the more selective the sensor must be for identifying the target.

Reed relay is among the best in this regard because it senses only magnetic targets—and magnets are rare in the industry (Fig. 23.3). If you differentiate your target by putting a magnet on it and use a reed relay to detect its presence (Fig. 23.4), it won't be fooled by other objects that don't have magnets attached.

Inductive sensors detect metal and would not be fooled by nonmetal objects. This is a highly accurate technology; so, if you are able to control what metal objects pass in front of the sensor, this technology is very good at letting you know precisely where they are.

Photoelectric can be fooled very easily: if you break the beam, something is sensed. That object could be the target, it could be a low flying duck, or nothing at all but a cloud of dust (Fig. 23.5)!

Reed proximity switches are actuated by the presence of a magnetic field caused by a permanent magnet or electromagnet. The typical sensing range is 0 to 1.5 inches.

General Terminologies

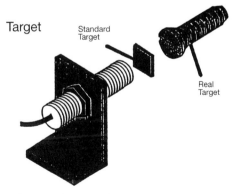

FIGURE 23.3 Magnetic sender.

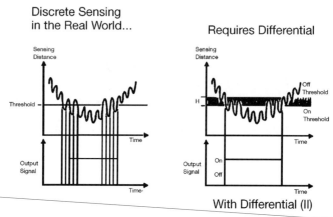

FIGURE 23.4 Regulated magnetic sender.

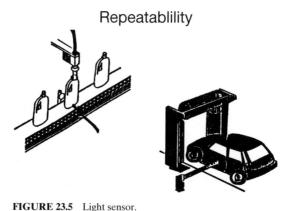

FIGURE 23.5 Light sensor.

The sensor is called a *reed relay* (Fig. 23.6) because it has two little metal foil reeds hermetically sealed in a glass bulb (Fig. 23.7) such that they overlay, but don't touch. When a magnet is brought close to the reed bulb, the foil reeds take on magnetic properties and attract each other. The reeds make contact, closing the circuit.

When the north end of a magnetic field is introduced to the reed head-on, the relay actuates in the area shown by the solid line in Fig. 23.4. This distance will vary according to the strength of the magnet. As the magnet is removed from the area, the reed relay opens at the dotted line. The difference between the point of actuation as the magnetic target comes closer and the point of deactuation as the target moves away is the differential. This is a naturally occurring phenomenon in this technology.

Note: When the magnet approaches the reed in a plane perpendicular to the reed's axis, the area of commutation is altered. Two areas of actuation are created. This is not desired because it gives two signals for one target pass.

Inductive Principles

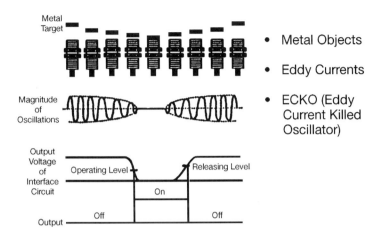

FIGURE 23.6 Inductive sensor may be embedded in objects.

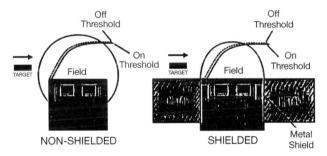

FIGURE 23.7 Shielded and nonshielded inductive sensor. The shielded sensors can even be flush-mounted in metal.

MAGNETIC-ACTUATED SWITCHES APPLICATIONS

As was previously stated, the more selective your sensor can be for the target, the better; and, the more control you have over what passes in front of the sensor, the better. Reed relay is among the best in this regard because it senses only magnetic targets—and magnets are rare in industry. Magnetic actuated switches have the following applications:

- *Security and safety* when nontarget material can accidentally trip other detectors. These switches are ideal for security applications, since magnets are fairly rare in the industrial environment. Because reed relay is noncontact, breakdown from mechanical wear is virtually eliminated.

Examples: Security door interlock, end of travel for elevators, cranes traveling on rails, overhead cranes, secondary switches for safety applications (redundant system for safety; in addition to a switch with positive-opening contact; for very difficult applications).

- *Sensing through metal walls (nonferrous).*
  Example: Piston position in an air or hydraulic cylinder (the piston has a magnetic band).
- *Identification of metal bins, containers.*
  Example: Magnetic-type or small magnets in preassigned slots can uniquely identify a carrier on an automatic assembly line. (Cheaper than photoelectric implementation).
- *Dirty environments* where sand, dirt, dust, oil, or coolants are prevalent.
- High-speed operation.

Features:

- No bouncing effects
- Excellent low-current characteristics
- Low costs

Weaknesses:

- Welds easily
- Sensitivity to welding fields
- Needs a magnet

Magnetic-Actuated Switch Characteristics

Features. Usually these switches have no bouncing effects! They are great for interfacing with PLCs or other solid-state circuits. They have excellent low-current characteristics. They are also a low-cost alternative to proximity sensing in general.

Weaknesses. Welds easily cause failure. As a precaution, use a resistor in series to limit current surges. These switches also exhibit sensitivity to welding fields; do not use them in close proximity. They need a magnet attached to the real target.

General Terminology for Sensing Distance

Nominal sensing distance: The rated operating distance for which the switch is designed. This value should only be taken as a guide, since no manufacturing tolerances or changes in external operating environment are taken into account.

You should be aware that manufacturers arrive at performance standards for their products using standardized criteria so you can compare apples to apples in determining which product is right for your application. These criteria reflect, to a reasonable extent, the real performance which can be expected in an average controlled environment.

Usable sensing distance: This sensing distance is used for specifying a specific sensor for an application. Relative to a standard target, the usable sensing distance takes manufacturing tolerances and external operating conditions into account. This specification gives confidence to the specifier: all devices are guaranteed by the manufacturer to perform within the usable sensing distance.

Effective (or real) sensing distance: This is the actual sensing distance realized by the actual sensor that you take out of the box and install in the actual application. The effective sensing distance will be no less than the usable and is usually closer to the nominal in average circumstances.

COMPONENTS OF A SOLID-STATE SENSOR

What are the components of a solid-state sensor? In generic terms, solid-state sensors will have the following:

- *An interface,* generally an oscillating circuit—the kind which is the main distinction between the technologies. The interface translates the distance into an electrical signal.
- *A trigger* or threshold-level detector. When the signal reaches a predetermined level, it triggers the output circuit.
- *An output circuit* that interfaces with data acquisition systems (PLCs, dedicated controllers) or other control circuits (relays, counters, timers, etc).

General Terminology

Standard target: An object with standardized dimensions or characteristics—common among manufacturers—used in the product laboratory to determine benchmark performance characteristics for a sensor.

Target: The *real* object to be sensed. An example is keyway on a shaft.

Discrete Sensing Requires Differential

The wavy line expresses the real variation of the distance between the target and sensor (Fig. 23.4). The line is wavy because we are dealing with nonuniform movement caused by vibration or jerkiness in the mechanical systems which are found in a real industrial environment. These variations can be a real problem, especially in slow-moving applications.

With only a single threshold, those small changes in distance cause a chattering effect between the ON and OFF states. This is noise that we need to eliminate.

If we introduce a second threshold for OFF above the threshold for ON, the small waveforms between the two are filtered out and the chattering is suppressed. The distance between the ON and OFF thresholds is the differential.

Differential

The differential is also called *hysteresis.* It is the distance between the operating point and the release point as the target is moving away. Differential is necessary in order to make a precise determination of target presence without factors of the environment intervening to create a noisy output signal.

All discrete sensing technologies must have a differential. In some technologies, differential occurs as a by-product of the basic laws of physics; however, in other technologies it must be manufactured through additional circuitry.

Repeatability

Repeatability is the ability of a specific sensor to repeatedly detect a target at a given point. It is related to the precision needed for the application.

For example, are you filling a bottle that requires a positioning accuracy of $\pm\frac{1}{16}$ in at the filling station, otherwise you have product spilling on the floor? Or, are you interested in sensing the presence of a car in an automatic car wash with an accuracy of ± 1 ft?

Rating the Technologies for Repeatability. Rating the technologies for repeatability in descending order gives you the following list. Different technologies offer varying levels of precision.

1. Limit switches
2. Reed relay (magnetic target)
3. Inductive
4. Photoelectric
5. Capacitive

INDUCTIVE PRINCIPLES

An *inductive detector* senses the proximity of a metal object utilizing an oscillator principle. Surface currents—known as *eddy currents*—are induced in the metal target by the electromagnetic oscillating field. The target reacts with the oscillator's field: the closer the target, the stronger the reaction.

When the metal target is outside of the oscillator field, the magnitude of the oscillations is not impeded. When the target is inside the field, it attenuates the magnitude, eventually stopping the oscillator. That's why it's named *ECKO* (*Eddy Current Killed Oscillator*).

Figure 23.4 shows how the proximity of a metal target affects the magnitude of oscillations and the output voltage.

SHIELDED AND NONSHIELDED INDUCTIVE SENSORS

The nonshielded sensor has a field which extends around the front surface to the sides of the unit. When a metal target is introduced to the field, there is a predictable point at which the sensor will sense its presence. The field must be loaded with metal to a threshold point and then the sensor will change states. When the target is closer to the sensor, less metal has to be introduced to the field to make the sensor change

FUNDAMENTALS OF SOLID-STATE PRESENCE-SENSING TECHNOLOGIES 23.9

states. Using a standard target, a very predictable curve can be plotted to define the sensing-distance characteristics.

If a nonshielded inductive sensor is mounted in metal such that the metal is in the field (embedded), this will *preload* the field and the sensing pattern will not be predictable.

The shielded sensor introduces metal into the field within the switch itself. Metal mounting brackets or fixtures will not influence the sensing pattern. The shielded sensor can even be flush-mounted in metal.

Note: A shielded sensor of a given type will have a shorter sensing distance than its nonshielded counterpart.

CAPACITIVE PRINCIPLES

This is an all-material-type of sensor. Capacitive proximity switches contain a high-frequency oscillator having one of its capacitor plates built into the end of the sensor (Fig. 23.8).

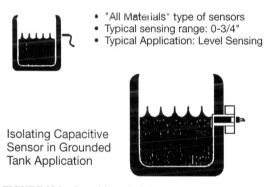

- "All Materials" type of sensors
- Typical sensing range: 0-3/4"
- Typical Application: Level Sensing

Isolating Capacitive Sensor in Grounded Tank Application

FIGURE 23.8 Capacitive principles.

When the target is outside the oscillator field, there is a weak capacitive coupling. When the target is within the oscillating field, a strong capacitive coupling develops.

The sensing zone is influenced by the physical properties of the medium being sensed. All materials in the sensor's proximity are sensed by a change in dielectric characteristics as it relates to air.

Capacitive proximity switches lend themselves ideally to applications requiring the sensing of bulk and liquid level and nonmetallic objects such as glass and plastic. It is the last choice for detecting metal targets because of its nonselectivity (relative to inductive technology).

A typical installation is in the side of a nonmetallic vat; the capacitive sensor can detect the level of the liquid or dry material right through the wall without coming into contact with it.

Capacitive switches have the following weaknesses:

- The actual sensing range of the switch varies widely according to the medium being sensed.
- Very sensitive to environmental factors.
- Not at all selective for its target—easily fooled. Control of what comes close to the sensor is essential!
- If tank is grounded, will need a special well bracket.

GENERAL PHOTOELECTRIC TERMINOLOGY

Emitter: A device that emits a beam of light when an electric current is passed through it. It is usually infrared (invisible).

Receiver: A device that changes its output state when light from the emitter is received.

Light/dark mode: Equivalent to normal open/normal closed functions of a limit switch. Usually, you want the output state to be ON when the target is present because most control systems are set up that way.

Dark mode: Light to receiver is stopped or drops below minimum operating level; the output is ON. Used most often in thru-beam or retroreflex applications, when the target breaks the beam so no light falls on the receiver, output is ON indicating that the target is present. When dark mode is applied to proximity diffuse when the target is Not There, the relay is energized.

Light Mode: Light on the receiver energizes the output. Used most often for proximity diffuse applications, when the beam is reflected from the target and falls on the receiver, the unit signals ON.

Photoelectric Principles

The principles of light emission and reception are used to sense the target (Figs. 23.9 and 23.10).

Types of Photoelectric Sensing. There are three basic configurations for photoelectric sensing:

- *Thru-beam,* where the target passes between an emitting unit and a receiving unit, blocking the beam. These are separate, powered units (Fig. 23.11).
- *Retroreflective,* where the target passes between the sensor and a reflector. The emitter and receiver are in the same housing (Fig. 23.12).
- *Proximity (diffuse),* where the unit senses the light directly from the target. The emitter and receiver are in the same housing, the same as retroreflective; however, the receiver is more sensitive to the weaker light which is diffused by the surface of the target.

The types and usage of photoelectrics are increasing at a very rapid rate because they are extremely versatile. They can be used to detect the presence or absence of

Terminologies

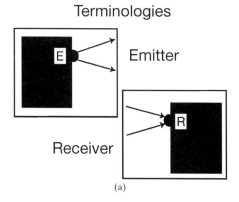

(a)

Types of Photoelectric Sensing

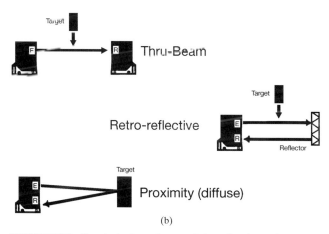

(b)

FIGURE 23.9 Terminologies and types of photoelectric sensing.

nearly any material or object without contacting it physically and can function at a considerable distance from the target.

Photoelectrics are very reliable and have rapid response times because they employ a solid-state design and no-touch actuation. This results in long, low-cost operation.

Thru-Beam Scanning

The transmitter and receiver are in separate enclosures. Of the photoelectric scanning types, thru-beam is the most predictable. Its typical range is 0 to 100 feet. The system is most suitable for the following:

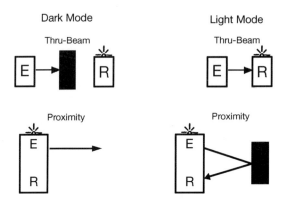

FIGURE 23.10 General terminologies.

Most suitable for:
- Opaque or reflective objects
- Dirty, polluted environment
- Long range
- Accuracy with small objects

Constraints:
- Accurate alignment
- Cannot be used to detect transparent objects
- 2 powered units must be installed

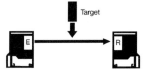

FIGURE 23.11 Thin-beam scanning. Typical range = 0–100 ft.

Most suitable for:
- A power supply is available from only one side
- Rapid, simple installation is required
- A relatively clean environment
- Packaging, conveyor systems, etc.

Not recommended for:
- Shiny or reflective objects
- Dirty or polluted environments
- The detection of very small objects

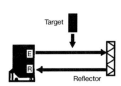

FIGURE 23.12 Retroflective scanning. Typical range = 0–30 ft.

- The detection of opaque or reflective objects.
- Use within a dirty, polluted environment (dust, rain, mist).
- Applications requiring long-range detection.
- Detection where accuracy is important and where small objects are being detected and for counting systems. The target should be larger than the beam and the system must be aligned so that the target completely covers the beam.

It has the following constraints:

- Accurate alignment in setup is essential (see catalog for mounting recommendations).
- Cannot be used to detect transparent objects.
- Two powered units must be installed. This route is more costly and requires more wiring to install.

Retroreflective Scanning

Transmitter and receiver are contained within the same enclosure. The beam is closed by a reflector. A *corner cube reflector* is a unique device that sends light back parallel with the incoming light. This system's typical range is 0 to 30 feet. This system is the most suitable for the following:

- Applications where a power supply is available from only one side of the target.
- Applications where rapid, simple installation is required; the reflector may be mounted at an angle of up to ±15° from the line perpendicular to the light beam.
- A relatively clean environment.
- It is the most common system used on packaging machines, conveyors, and the like.

However, these systems are not recommended for the following situations:

- Where shiny or reflective objects are to be detected.
- For very dirty or polluted environments.
- For the detection of very small objects. Usually, the target should be larger than the reflector; very small reflectors are impractical.

A corner cube reflector does something that a mirror cannot do: it reflects energy back to the source from which it came when the source is not in a perpendicular direction. That's why similar kinds of reflectors are used on cars and road signs to provide a passively lighted warning at night.

Compare the diagrams of how mirrors differ from reflectors (Fig. 23.13).

The mirror reflects light such that the angle of reflection is equal to the angle of incidence. The reflector sends light back parallel to the rays of incidence.

You might wonder if a mirror could be used with an angle of incidence of 90° but you would find this only workable in the laboratory, and only then after much care-

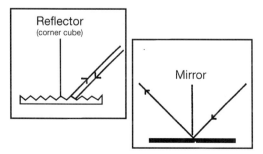

FIGURE 23.13 Terminologies.

ful alignment. The corner cube reflector makes retroreflective photoelectric sensing practical.

Reflectors are available in different sizes and forms to do different jobs. For special situations, there are also reflective tapes.

Retroreflective Polarized Scanning

A polarized photosensor is actually a special retroreflex system which can detect, in addition to normal opaque objects, the shiny objects that fool a normal reflex sensor: mirrors, metal straps, foils, metal boxes, cans, shrink wrap, and mylar tape. This system's typical range is 0 to 24 feet (Fig. 23.14). This type of sensor is used extensively by the canning industry to sense aluminum cans and bottles. Some manufacturers might call it a sensor with an antiglare filter.

Detects shiny objects

Constraints:

- More expensive
- Shorter distance

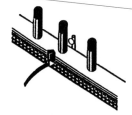

FIGURE 23.14 Retroflective polarized scanning. Typical range = 0–24 ft.

Its constraints are that it is more expensive than standard retroreflective systems and it has a shorter sensing distance.

Proximity (Diffuse) Scanning

Transmitter and receiver are contained within the same enclosure. Detection is based on light from the emitter being diffused from a target back to the receiver. This system's typical range is 0 to 3 feet (Fig. 23.15).

Most suitable for:

- Transparent or translucent objects
- Short range

Constraints:

- Range depends on the color and reflective nature of objects

FIGURE 23.15 Proximity (diffuse) scanning. Typical range = 0–3 ft.

The difference between diffused light and reflected light is that diffused light is scattered light. It's everywhere. We are able to see objects because when light strikes them, the light scatters. Our receivers—our eyes—really don't see the objects themselves, but rather, we see the light diffused by the objects.

Diffused light generates a very small signal, compared with retroreflective sensing. That's why we need very sensitive amplifiers to deal with the weaker signal.

This system is most suitable for the detection of transparent or translucent objects such as empty bottles on a conveyor. It has short-range applications.

The constraints of this system include the following:

- Usable range depends largely on the color and reflective nature of objects to be detected.
- Extremely vulnerable to environmental factors such as dust and humidity.
- An area behind the target should be clear of reflectives surfaces.
- Its use is not recommended for dirty environments or where very small objects need to be detected. Small objects don't bounce enough light to activate the receiver.

Proximity (Diffuse) Background Suppression

Diffuse proximity sensors with background suppression were designed to distinguish a target in front of a background, but a side benefit of the way background suppression works is that the sensor is not fooled by target color patterns and texture. This system's typical range is 0 to 3 feet.

Background suppression utilizes two receivers behind the receiving lens. Just as our eyes provide us with stereo vision by giving our brain two images to compare, the background suppression sensor relies on two receivers that are aimed at a precise point in front of the unit and are adjusted so that they sense the presence of a target *when the output of both receivers is equal* (Fig. 23.16).

- Utilizes 2 receivers behind the receiving lens
- Distinguish a target in front of a background
- Sensor is not fooled by target color patterns or texture

Applications:
- Material handling - conveying systems
- Collision detection for AGVs
- Car/truck wash
- Level sensing

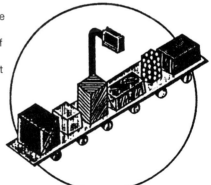

FIGURE 23.16 Proximity (diffuse) background suppression.

Since equality of reflected light amplitude is the only characteristic sensed, it makes no difference whether white, gray, or another color is present or whether it fluctuates in amplitude as it would when encountering a textured surface or lettering or graphics on a container.

Applications for these systems include the following:

- Material handling—conveying systems
- Collision detection for AGVs (automated guided vehicles)
- Car/truck wash
- Level sensing

Color Registration

These systems are widely used in the packaging and material control industry to identify index marks of various colors or the presence of labels on shiny surfaces. These units are characterized by having fast response times and the capability of detecting small marks that pass the unit at extreme speeds (Fig. 23.17).

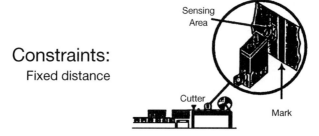

- Identifies index marks of various colors, or the presence of labels on shiny surfaces
- Fast response times

Constraints:
Fixed distance

FIGURE 23.17   Color registration.

A demanding application might be the detection of a 0.03-inch mark that passes the unit at 9 feet per second.

The color registration sensor is a highly specialized diffuse proximity sensor which has the ability to detect fine changes in contrast on a surface. But unlike the standard diffuse proximity sensor, this type of unit uses a powerful lens system and must be positioned at a specific focal distance from the target.

System constraints include the fact that it operates at fixed distances.

FIBER-OPTIC SENSORS

Thru-Beam and Proximity (Diffuse) Scanning with Extension Cords

What do you do when the physical constraints of the application don't allow for installing regular, self-contained sensors? Maybe the target is in a high-temperature or chemically aggressive environment. Perhaps the target is small or very fast moving. Fiber optics, applied to photoelectric scanning, solves these problems (Fig. 23.18).

Fiber Optic
Thru-beam Scanning
Typical range: 0-50"

Proximity (diffuse)
using a bifurcated
fiber optic cable
Typical range: 0-50"

FIGURE 23.18 Fiber optic sensors: scanning with extension cords.

Bending Light around Corners

The principle of conducting light around corners was discovered more than 100 years ago; however, its practical usage is relatively new. It was accelerated by the development of communications, especially the need for noise-free data transmission. Medical imaging systems also played an important role in expanding the technology to the point of making it affordable for industrial and commercial use.

Theory of Operation

A fiber-optic cable is made up of a bundle of hundreds—and sometimes thousands—of glass or plastic fibers which are protected by a flexible armored sheathing. You can think of these as being analogous to conducting electricity using electric cables, only fiber-optic cables conduct light.

Fiber Optics and Sensing

All fiber-optic sensing modes are implemented using one type of amplifier which contains both emitter and receiver in one housing.

Fiber Optic Thru-Beam Scanning

Using two opposed, individual fiber-optic cables, the object to be detected breaks the beam. The target must be at least the same dimension as the effective beam, which in this case is the bundle diameter. Because the beam is very small, the detection can be very precise. A typical application might be edge detection for a web printing press. Needle tips reduce the beam dimension for use with extremely small targets, typical for applications in semiconductors and pharmaceutical industries.

Fiber Optics Applications

Proximity (diffuse) using a bifurcated fiber-optic cable is one application. When you mix the emitter bundle with the receiver bundle at the sensing tip, the result is a bifurcated cable. The setup is rather straightforward: the sensing tip must be placed close enough to the target to directly sense diffused light.

This mode of scanning is the least precise because the properties of the target surface (color, shining, texture, etc.) greatly influence the quantity of light bounced back; however, it is the mode of choice for transparent targets (Fig. 23.19).

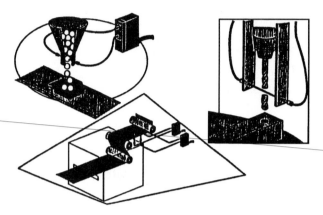

FIGURE 23.19 Fiber optic applications.

The typical range for these systems is 0 to 511 inches.

Typical applications include small parts detection, edge detection, and high-temperature environment (6000°F+). These systems also operate in thru-beam mode and proximity mode.

SOLID-STATE SENSOR TECHNOLOGIES

There are two kinds of output technologies (see Table 23.1). The first is electromechanical. This is a switch which physically opens and closes a circuit (Fig. 23.20). There is no voltage required to power the relay—that's what we mean by *dry contact*. When the circuit is closed, there is no voltage across the switch. Its physical position determines the circuit's status. Electromechanical switches are either *normally open* (*NO*) or *normally closed* (*NC*).

Examples of electromechanical switches are contacts, relays, reed relays, pushbutton operators, and limit switches.

The second technology is solid state; as the name implies, it has no moving parts. Power is required by the solid state at all times. Solid state needs a certain amount of current to energize the circuitry and the output. There is voltage across the switch even when the circuit is closed. When it's open, there is a certain amount of current (leakage current) passing through it.

Examples of solid state include transistors, TRIACs, SCRs.

Electromechanical Contact Advantages

They are universal: the same can be used with either AC or DC. They are excellent for withstanding in the power supply because an air gap is an excellent isolator.

TABLE 23.1 Solid-State Sensor Technologies

	Advantages	Disadvantages	Applications
Magnet operated (reed relay)	Inexpensive Very selective target identification	Magnet required Sensitive to welding fields	Security and safety interlocking Sensing through metal
Hall effect	Complete switching function is in a single integrated circuit Operates up to 150 kHz High temperature (150°C) Good resolution	Magnetic target only Extremely sensitive to industrial environment	Keyboard
Ultrasonic	Senses all materials	Resolution Repeatability Sensitive to background and environment changes	Anticollision on AGV Doors
Inductive	Resistant to harsh environments Easy to install Very predictable	Distance limitation to 60 mm	Presence detection on all kinds of machines and installations Very popular
Capacitive	Senses all materials Detects through walls	Very sensitive to environment changes	Level sensing with liquid and nonmetallic parts
Photoelectric	Senses all materials	Subject to contamination	Parts detection Material handling Packaging Very popular

There is practically no voltage drop when the switch is closed, and there is no leakage current when the switch is open. When the circuit is closed, very little heat builds up in the switch because the voltage drop is negligible. When the circuit is open, no heat is generated by the switch, either because no current flows.

Electromechanical Contact Drawbacks

Wear is always a factor when dealing with devices that have moving parts.

Because moving parts have inertia, the switch is relatively slow. The exception to this is the reed relay which is extremely fast since the reed has virtually no inertia. Reed relays are used for speed detection of gear teeth.

Electronic noise can be generated when the contact opens on inductive loads. RC networks and surge suppressors are recommended for critical applications.

For low-energy applications (5 to 12 VDC, 0 to 15 mA), silver contacts oxidize over time and stop conducting. More expensive gold contacts don't oxidize and can be used for low-energy applications.

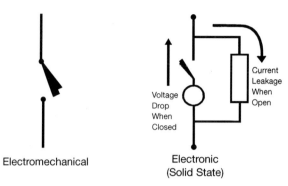

FIGURE 23.20 There are two kinds of output technologies.

Solid-State Advantages

Since there are no moving parts, wear is not a factor. Solid-state switches are extremely fast. Inertia is not a factor. AC switches do not generate noise and transients generated by arcing. They stop conducting when current reaches zero; therefore, there is no energy left to generate arcing. Also, they work extremely well at low energy—especially DC switches.

Solid-State Drawbacks

Solid-state switches must be powered at all times. When they are open, leakage current flows through the load. When they are closed, there is a voltage drop to power the switch which must be considered.

Transients in the power supply can send false signals to the switch or even cause it to fail completely. Clean power is essential.

Solid-state switches, while extremely suited to low-energy applications, handle high-current applications with more difficulty. Because of the voltage drop, they generate heat. At current levels above 0.5 amp, they need bulky heat sinks to dissipate the heat.

TRANSISTOR SWITCHING FOR DC

Transistors are polarity-sensitive. They permit current to pass in only one direction (that's why they can't be used with alternating current). In the environment of DC circuitry, current flows from positive to negative (Fig. 23.21).

There are two kinds of transistors which are used for two basic situations: (1) NPN, which connects the load to the negative; this is also called *Sinking;* and (2) PNP, which connects the positive to the load; this is also called *Sourcing.*

Sourcing and Sinking

Sourcing means that the component is connected to the positive. *Sinking* means that the component is connected to the negative. This is a complementary relationship. If

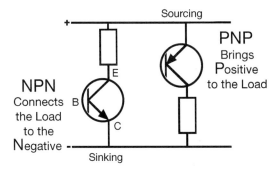

FIGURE 23.21 Transistor switching for DC.

you tried to put together a sourcing load to a sourcing switch, there would be no flow of current because neither is conducting to negative. If you tried to connect two sinking switches, both would be connected to ground and there would be no positive (Fig. 23.22).

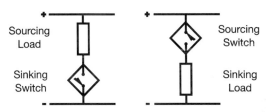

FIGURE 23.22 Sourcing and sinking complementary components.

THREE-WIRE TECHNOLOGY

The elements of three-wire technology are the positive, the negative, and the output. You will usually associate three-wire technology with DC applications; however, if you encounter an AC three-wire, the terms are the *live,* the *neutral,* and the *output* (Fig. 23.23).

As a general rule, three-wire devices have the signal electronics powered separately from the output circuit. The output is usually wired internally to one of the power terminals. The load is wired externally between the output and the other power pole, closing the circuit.

Advantage: Power supply current doesn't pass through the load. The switching characteristics are the best that output switch technology can offer (low voltage drop and very low leakage current).

Disadvantage: Has one more wire than electromechanical switches. (Somewhat higher cost for cable and greater chance for confusion.)

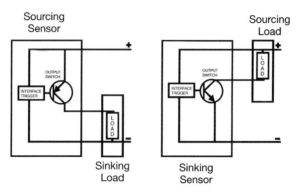

FIGURE 23.23 Three-wire technology.

TWO-WIRE TECHNOLOGY

When a circuit is closed, a small amount of energy is diverted from the main circuit to power the electronics. The electronics requires a voltage across it at all times. The output switch cannot close in parallel because it would short-circuit the electronics. A resistor (R) in series with the output switch represents the power supply circuit. Obviously, the current passing through it creates a voltage drop, which is higher than the three-wire case when we have only the switch in the circuit (Fig. 23.24).

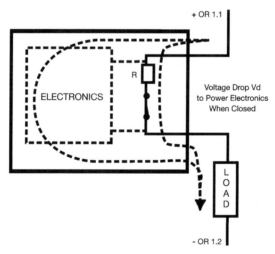

FIGURE 23.24 Two-wire technology.

For most applications, the voltage drop (V_d) is well within the operating range of the load. However, when several sensors are wired in series, this voltage drop can add up to be a consideration. We will illustrate this more fully when we talk about multiple sensor applications.

FUNDAMENTALS OF SOLID-STATE PRESENCE-SENSING TECHNOLOGIES

This also means that an amount of current is moving through the load even when the switch is open. This current is higher than the leakage current of the switch proper because it also has to power the electronics. This is just fine for loads which require a threshold amount of current before they do work. In very low-energy loads, leakage current can become a concern, especially if many sensors are used in parallel. This is illustrated later.

A diode bridge directs current flow through the switch

Universal:
Works as PNP, NPN or AC

FIGURE 23.25 AC/DC two-wire technology.

In AC circuits, the leakage values of current and voltage are not significant for many loads. This explains why two-wire technology is almost exclusively used in AC. In low-energy DC circuits, this might create problems. Three-wire technology is still prevalent in DC, but two-wire DC is starting to gain acceptance, especially when used with PLCs. At low current, the voltage drop is relatively small and the current leakage, using modem technologies, is no longer a problem (Fig. 23.25).

Caution: There are AC two-wire devices with three physical wires. The third wire is the chassis ground and should not be counted since it has no functional role in the circuit.

AC/DC Two-Wire Technology

Diodes only permit current to flow in one direction. When they are put together around a transistor in this configuration, current flows through the switch (i.e., transistor) only in one direction, regardless of the outside voltage polarity.

This diode bridge creates a universal switch: AC and DC can be conducted through it, and it can be used as either a PNP or NPN in a DC circuit.

The advantages of two-wire technology include the following:

I leakage (off) < I threshold (off state)

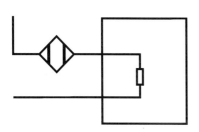

For PLCS:
I threshold = 1.7mA (off state)

FIGURE 23.26 AC direct wiring to PLC.

- More simple to wire
- Less confusion—no polarity considerations

The disadvantages to diode bridges include:

- Higher voltage drop
- Higher cost

Matching Sensors with PLC Input Thresholds

In order for a two-wire AC switch to be directly compatible with a PLC (Fig. 23.26), two conditions must be met:

1. *PLC problem:* Leakage current must be less than 1.7 mA when the switch is off. This represents the standard PLC threshold for OFF. If the leakage current of the sensor is greater than 1.7 mA, the PLC would be on all the time, regardless of the condition of the switch.
2. *Switch problem:* Load current must be greater than the switch minimum when the switch is on. Typical PLC input currents are 12-16 mA. Use the catalog to check your sensor's minimum current requirement for AC loads against the actual load of the PLC.

RADIO FREQUENCY IMMUNITY

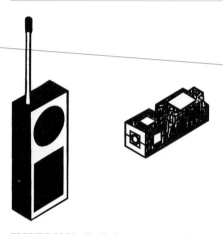

Walkie-talkies and other portable communication devices generate electromagnetic fields that can induce a false output signal from an inductive sensor. The field interferes with the sensor's oscillator coil (Fig. 23.27).

The sensor's wires can also act as an antenna, so when the communication device is transmitting, this modulation of the field causes intermittent sensor trappings.

Electronic filtering achieves immunity against radio transmissions. A walkie-talkie's antenna should be able to touch any part of the switch—except the sensing face, which will detect it as metal, and not change its state.

FIGURE 23.27 Radio frequency immunity.

WELD FIELD IMMUNITY

The problem of welding fields triggering a false sensor signal is similar to RFI. Welding fields are powerful and sensors are often required to operate near a welding process. Weld field immunity (WFI) is a commonly requested protection (Fig. 23.28).

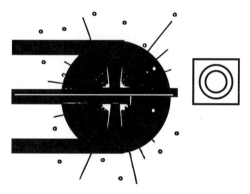

FIGURE 23.28 Weld field immunity.

The weld field must be filtered out both at the oscillator level and the output level because welding is usually achieved by passing a burst of current through the parts on the power line, which also requires protective measures.

Weld-field-immune switches tend to be slower than regular switches because they have a lower operating frequency.

RESPONSE TIME: INERTIA

Response time is defined as being the duration of time required for the interface to trigger an output. When you look under "Specifications" in the Square D catalog, you will find Power-up Delay, On Delay, and Off Delay listed. These are the characteristics of response time.

You can generally think of response time as *inertia* (the nemesis of perpetual motion inventions). There is no such thing as a perfect sensor, one that responds instantly. In fact, such a sensor would be too sensitive to be practical in industrial environments. Because the sensor interface generates a low-energy signal, the device is extremely susceptible to external transients. Filters are imposed between the trigger and the output circuit to suppress transients, however, this also affects overall response time performance.

Terms for response time—*Power-up Delay, On Delay,* and *Off Delay*—are conservation ratings which the manufacturer guarantees as being the maximum performance for that model of device.

Power-up Delay Protection

Power-up delay prevents false signals from triggering an ON condition to the load. Parasitic inductance and capacitance builds up in long cables when switches are open; then they dump a transient spike that can be 2 to 3 times the normal voltage when they close. If a power-up delay was not built in, this transient would pass through the sensor and signal an ON condition to the load—even if there was no target present in the sensing area; this would be a false signal which might cause dire consequences. Because this unsafe condition cannot be tolerated, industry has required that all devices not change state during the power-up phase. The output is intentionally turned off during this period (Fig. 23.29).

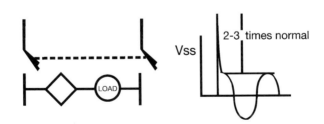

- Parasitic inductance and capacitance generates transients when switch closes

- Transient passes through switch; signals "ON" to load as FALSE SIGNAL

FIGURE 23.29 Power-up delay protection.

Note: Some people call power-up delay *warm-up time*. This term is a holdover from vacuum-tube technology. Solid state requires no warm-up, so *power-up delay* is a more accurate term.

On Delay

Don't confuse on delay with power-up delay. *On delay* is the response time after the unit is powered up. It's the duration of time required for the interface to trigger an output change of state when a target is introduced to the sensing area (Fig. 23.30).

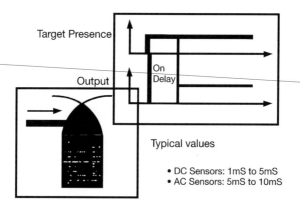

FIGURE 23.30 On delay.

On delay is usually much faster than power-up delay. Typical values for DC sensors are 1 to 5 ms and, for AC, 5 to 10 ms.

Off Delay

Off delay is the amount of time required for the interface to trigger a change of state when a target is removed from the sensing area (Fig. 23.31).

Typical values are 1 to 5 ms for DC output and 8 to 20 ms for AC outputs.

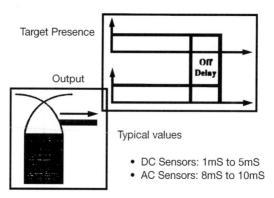

FIGURE 23.31 Off delay.

RESPONSE TIME

We have been discussing the response time of the sensor; however, your application demands that you consider the response time of the entire data acquisition system (Fig. 23.32).

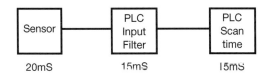

Not just for the sensor; the entire data acquisition system

FIGURE 23.32 Response time—50 ms for the entire data acquisition system.

In this model, consider that there is a 20-ms ON delay at the sensor, a 15-ms delay at the PLC input module, and the signal must be maintained long enough for the PLC to pick up the change of state in its scan cycle. If the signal does not last long enough, the PLC won't see it during the scan (Fig. 23.33).

During a PLC scan, the PLC reads the status of the inputs, solves the logic programmed, and updates the outputs. The duration of the sensor signal must be longer than the PLC scan time, plus the PLC input filter delay.

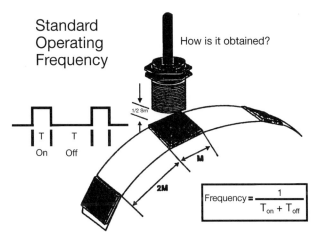

FIGURE 23.33 On-off signal.

The overall system response time is the sum of the ON or OFF delay of the switch proper, the PLC input filter, and the PLC scan time.

STANDARD OPERATING FREQUENCY

Operating frequency is an industry standard, just like a standard target is an industry standard. The method of arriving at this industry standard for inductive sensors is to place standard targets on a nonferrous, rotating disk. The targets are spaced apart a distance of twice their width. Photoelectric sensors use a shutter disk with similar considerations for standard target and space between openings (Fig. 23.34).

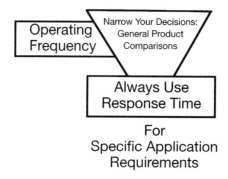

FIGURE 23.34 Operating frequency must be within the sensing frequency.

The sensor is mounted at a distance of one-half of its rated nominal sensing distance from the target at its closest proximity. The disk is rotated and the output of the sensor is displayed on an oscilloscope. The speed of the disk is increased until reaching a frequency where detection stops. The frequency of targets passing the sensor at this point is the maximum operating frequency.

If you look in the catalog, you will find operating frequency under the general description of the sensor. This statistic lets you get an idea about how fast the sensor is in a very general sense. You can compare the products of various manufacturers using this figure, so you can be sure you are comparing apples to apples. You can also use it as a guide for choosing lines in a single manufacturer. (Smaller sensors and DC sensors are faster than larger sensors and AC sensors, respectively.)

Think of standard operating frequency as the EPA sticker on a new car, which gives an estimated fuel economy for city and highway driving. These figures were arrived at on standardized test tracks and have no bearing at all on how you, specifically, will drive the car, how old the car is, and how well you maintained it.

Operating frequency is a useful guide, but do not use it to solve your application problems! Operating frequency relates to a standard target in a standardized laboratory set up.

CHAPTER 24
DESIGN AND APPLICATION OF ROBUST INSTRUMENTATION SENSORS IN EXTREME ENVIRONMENTS

INTRODUCTION

The design and operational techniques required for the exceptionally robust sensor instrumentation operating in the severe environmental conditions at the Mount Washington Observatory can be applied to industrial and sensor product design as well. Here we discuss sensor applications, RFI effects, power-related problems, lightning and static discharge problems, and test and serviceability problems. Detailed electronic and mechanical requirements for robust sensor design including extreme temperatures, humidity, and extreme environmental conditions, with case histories, is discussed. The plans for an expanded industrial and scientific test and development capability are also highlighted.

Meteorology, severe weather testing, and instrumentation development—all in a rigorous climate—are the major areas of interest for several observatories. The Mount Washington Observatory is located on the summit of Mount Washington in New Hampshire at latitude 44° 16′N, longitude 71° 18′W. The geographical summit, at 6288 feet, is the highest point in the northeastern United States. The observatory's observation tower reaches 6309 feet.

The observatory's most important facility is its unique site, which is subject to extremes of meteorological, climatic, biological, and ecological stress not found elsewhere outside the deep polar regions. The summit area is equivalent to alpine and arctic zones and exists as an ecological island, supporting its own biological varieties. Permafrost ground conditions prevail at the summit, there is very rapid frost action at the ground level, and the buildup of rime ice on exposed surfaces is often substantial.

Summit weather is severe. For example, over the 43-year interval from 1935 to 1978, mean annual precipitation at the observatory was 83.55 inches, and mean annual snowfall was 248 inches. A snowfall of 566.4 inches during the 1968–1969 season broke a national snowfall record (Fig. 24.1).

For some part of the day, the summit is in cloud or fog, at least 300 days out of the year. The temperature measured at the summit has never risen over 72°F; the lowest

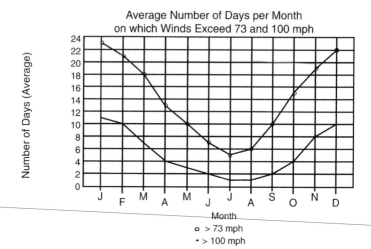

FIGURE 24.1 Average number of days per month on which winds exceeded 73 and 100 mph.

recorded is 47°F, and the yearly average is 27°F. The mean hourly wind speed is 35.1 mph W; "century days," when wind gusts exceed 100 mph, are frequent in winter and early spring. The highest wind velocity ever recorded anywhere in the world was 231 mph, measured at the observatory on April 12, 1934 (Fig. 24.2).

The Mount Washington Observatory is housed in the west end of the Sherman Adams Building, a steel-reinforced concrete structure built at the summit by the State of New Hampshire in 1980. The four-level circular observation tower incorporates a cold lab with a floor area of approximately 75 square feet and a 200-square-foot exposure room with windows specially constructed for accommodating sampling devices and sensors. Instruments can also be exposed on the railing of the tower parapet or the superstructure. In addition, part of the flat roof above the observatory is available as a test platform; a 34 × 18-foot grid of steel girders has been installed there for securing heavy equipment, including instruments and control shelters (Fig. 24.3).

Special facilities within the observatory include a professional darkroom/photographic laboratory; an electronics laboratory; a research library; a workshop; and the weather room, where most of the meteorological instrumentation and radio communications equipment is located. Data processing at the observatory is accomplished by data acquisition software written in the C language. Data is accumulated continuously and is sent by modem and by packet radio to remote sites. A local area network at the summit includes a fiber-optic link to a remote test site on the mountainside below the summit, with an instrumentation shelter for computers and data acquisition equipment, now in use by the FAA for testing airport fog detectors.

The Northern Center offers advanced, state-of-the-art mountaintop facilities for meteorological research, instrumentation studies, and severe weather testing of products accessible year-round. The center is staffed year-round and monitoring of instruments is done 24 hours a day. Continuous weather data since 1932 is available.

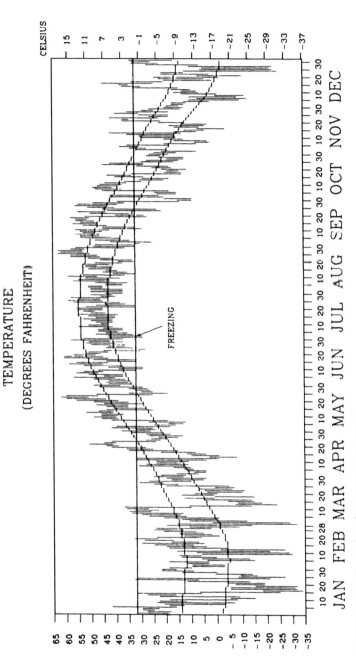

FIGURE 24.2 Fluctuation of temperature per annum.

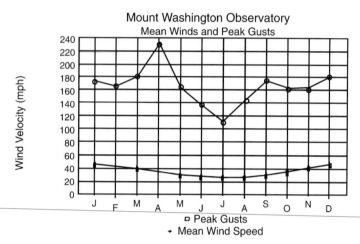

FIGURE 24.3 Mount Washington Observatory—mean winds and peak gusts.

Icing studies, high wind studies, fog studies, and atmospheric chemistry, including ozone studies, as well as product testing and robust instrumentation development and testing are areas of specialization at the Northern Center. Industrial and university collaborative efforts involving comparison studies are an important part of the Northern Center's research and testing program.

The Mount Washington Observatory specializes in the design, testing, and development of robust instrumentation of all kinds. A full complement of meteorological instruments is in place at the observatory, modified for exposure to severe weather. An example is the custom-designed, electrically heated pitot static anemometer and the unique wind direction instrument, designed and tested by the observatory and a group of engineers at University of New Hampshire to monitor winds under icing conditions. In addition, there are multicylinder and impingement-type instruments for measurement of cloud droplet size and liquid water content. Other instrumentation operated at the observatory on recent research programs include neutron and cosmic ray counting equipment, an FAA fog monitor for airports, and ongoing icing studies by the U.S. Army's Cold Regions Research Laboratories.

DESIGN CHALLENGES

Robustly designed instrumentation must meet extreme design constraints in order to survive and to function accurately and reliably in both industrial and field applications. Successful designs in use at Mount Washington Observatory can provide important lessons in dealing with RFI, power-related problems, lightning and static discharge problems, test and serviceability problems, electronic and mechanical design, and extreme environmental conditions. These robust design techniques can be effectively applied to scientific instrumentation, industrial instrumentation, oceanographic instrumentation, airport and highway instrumentation, and other environmentally extreme applications of sensor technology.

EXTREME ENVIRONMENTAL CONDITIONS

Extreme Temperatures

Commercial electronic components operate over a 0 to 70°C range. This is generally a problem in extreme conditions, unless the enclosure is temperature-controlled, so that military grade components may be required. At the observatory, the major problem is the extremely low temperatures, as the highest temperature ever recorded there was 72°F. The normal practice is to heat electronic enclosures, so that commercial parts may be used and so that icing due to condensation is not a problem.

Temperature changes can generate accuracy problems in measurement circuits due to temperature drift of integrated circuits, and component selection and circuit design must consider this variation. An interesting local problem in low-level measurements is thermal gradients from diverse materials found in the signal path which can generate substantial thermal emfs. For example, copper/solder interfaces generate 3 microvolts/°C. Joining wires from two manufacturers can contribute 200 nanovolts/°C. In robust measurement circuits, it is necessary to minimize thermal gradients by thermal shielding, by adding heat sinks, by aligning sensitive circuits along isotherms, and by separating high- and low-level circuits. It is even possible to obtain low-thermal special solder for use at inputs to microvolt-level circuits.

Humidity

The average annual precipitation is 83.55 inches at Mount Washington Observatory, and the mean annual snowfall is 248 inches. High humidity, above 80 percent, is common at the summit, so equipment must be designed to keep out water. Humidity, combined with high winds, can drive water into cables and instruments even 150 feet away and can cause shorts or inaccurate measurements. High humidity can cause corrosion of connectors and other parts of instruments. Because of the high humidity and frequent precipitation, operator safety and convenience must be considered in design and maintenance procedures.

Icing

At the observatory, as well as at sea and in the arctic, rime ice forms on instruments in minutes in the fall, the winter, and the spring. Manual deicing may be required, so that exceptionally robust mechanical design is essential for equipment in northern climates. At the observatory, winter instruments are heated, and energy consumption of circuits is important as a cost consideration. Ice can form on indoor instruments in the tower at night during the spring thaw period when condensation is a problem, so that electronic enclosures and mechanical assemblies need to be watertight and heated to keep the interior humidity low.

High Wind

Equipment in use in high wind conditions, which would include equipment at airports, highway equipment, oceanographic equipment, and many types of construction equipment, may experience great stress and must be very durably built and supported. At the observatory, winds over 100 mph are frequent in winter and spring, so that all outdoor instrumentation experiences great stress, particularly in icing conditions. Many instruments tested for robustness at the Northern Center have not lasted a single day in high wind conditions! High wind makes outdoor maintenance and repair very difficult (particularly with mittens on), so reliability and ease of maintenance is a must.

POWER DISTURBANCES

Power disturbance problems need to be dealt with in robust designs, in order to minimize their effects on electronic equipment. Most of the power disturbance problems are local ones, and design engineers need to provide protection against these effects, and the user needs to be sure that equipment is installed in a way to minimize field problems.

There are many types of power disturbances which can affect equipment. *Voltage variations,* including transients, which last less than one half cycle, swells and sags, which last less than a few seconds, overvoltages and undervoltages, which last more than a few seconds, and interruptions or outages, which are long-term power losses, can all cause problems to electrical equipment. *Frequency variations* are caused by poor generator regulation and can be a concern for international operation or for operation on independent power systems, particularly if the load varies. *Waveform distortion,* which occurs when the power waveform deviates from a sine wave, does not normally disturb electronic circuits but can cause stress on transformers, motors, and electrical distribution systems if the current waveform peaks due to switching power supplies. *Transients* are short disturbances from motors, lightning, relays contacts and switches, or power circuit switching, which last up to a few milliseconds, and which cause spikes or glitches on the voltage waveform. These spikes can be up to several thousand volts and are a serious problem to electronic circuits, as they can cause digital errors or even damage integrated circuits. *Continuous electrical noise* is repetitive disturbances at high frequencies and includes motor brush noise, lighting noise, power supply switching noise, and radio frequency interference.

Robust designs need to consider adding protective circuits to minimize power-disturbance problems. For example, transient protection can be added to limit the transient voltage (such as zener diodes and MOVS) or to provide a crowbar when a threshold is exceeded (such as gas-discharge devices), both line-to-line and line-to-ground. ENU filters will attenuate low levels of continuous RF interference, and a choke in the ground line can reduce common-mode transients such as motor noise or lightning transients. Isolation transformers and optical isolators all help to further protect equipment from power-line disturbances.

At the observatory, power disturbances are a challenge that affects every design. AC power is generated on the mountaintop and is distributed across the summit to the observatory. These long runs across the summit are very susceptible to lightning surges. Other users on the mountaintop heavily load the system during the summer months, so that undervoltages can be a problem. Generators are noisy, and the waveforms are distorted. Power outages are possible and must be considered in all designs. For these reasons, a DC battery system at 12 volts is used for critical instrumentation, and photovoltaic power is sometimes used—especially for remote test sites. For both systems, power consumption is an important design concern, and CMOS logic is preferred, both for low power consumption and for noise immunity.

ELECTROMAGNETIC INTERFERENCE

Electromagnetic interference (EMI) problems are common in equipment designed for use in severe environments and must be dealt with in the design of all robust instruments. In all instruments, there are three sources of interference problems: internal circuits, radiated emissions from external circuits, and conducted noise. The extent of ENU problems depends on frequency, on amplitude, on time, on impedance, and on physical layout of the circuits, since factors not shown on schematics are significant in severe conditions of ENU and determine whether interference will cause problems

Interference problems may be classified into three categories: (1) locally generated problems, (2) problems within a subsystem, and (3) problems caused by the outside world

Subsystem problems are generally caused by neglecting grounding and bypassing and are generally neglected in drawing a schematic. Lines on schematics are assumed to be at the same voltage, but if your sensor circuit is handling microvolts, this assumption is a disastrous one: lines are not at the same potential and they do have resistance, capacitance, and inductance. At high frequencies this can be important. Thus, the art of proper grounding is critical in instrumentation that works.

Grounding is a practical matter in robust instrumentation designs. *Ground* is a term which is more often misused than not. A copper rod driven into the earth is *grounded* (or *earthed,* as the British say), but a connection to *chassis*, a connection to the low side of the power supply, or a connection to the *common* of I/0 signals may not be anywhere near the potential of earth ground! Therefore, we are careful to distinguish between *ground* and *return*. It is essential to keep earth ground, power returns, digital returns, and analog returns quite separate, joining them at a single point with low current interground interconnections. Of course, it is important to avoid *ground loops;* if you control the flow of current carefully, you will control your grounding problems.

Outside world interference occurs when AC signals at high power and high frequency are coupled into analog circuits via stray capacitance and inductance. Keys to minimizing the effects of this type of interference are proper layout and shielding. High- and low-energy circuits should be separated; low-level circuits should be located as close together as possible. Metal enclosures should be used, with careful attention to minimizing dimensions of openings, sealing all seams, and protecting all cable penetrations of the shield. Fiber-optical connections or optoisolators can be effective in avoiding shield penetration problems.

There are over 100 sources of RF interference on the summit of Mount Washington, at many frequencies: a TV transmitter, microwave antenna, FM radio stations, emergency radio, amateur radio, and our own packet radio station. Icing can create some interesting RF patterns in high winds, and RFI frequently causes interference with instruments if they are not properly grounded and shielded. Instruments which succeed in this environment have very robustly designed grounding and shielding systems.

LIGHTNING AND STATIC DISCHARGE

Lightning can be spectacular on the summit of Mount Washington, and several hundred discharges occur there each year. This is typical of mountaintop installations and of operations at sea. Although we are protected from direct strikes by nearby higher structures, transient surges can cause damage to computers, phones, and instruments through grounds and power wires.

Like most mountaintop facilities, we try to avoid operating sensitive instrumentation during storms, but also give attention to making our equipment as robust to transients as possible. Surge protection in the form of clamping or crowbar circuits to protect electronics circuits is helpful in reducing the risk of damage from lightning. We also have found that fiber-optic links and optoisolators provide effective protection for long data cables, particularly when sensors are located outdoors with long data cable runs to the instrumentation.

RELIABILITY AND MAINTENANCE

In any field operation, or in an expensive manufacturing operation, reliability and ease of maintenance are important, as downtime is expensive. A nonfunctional instrument can take weeks to repair in a remote location, spare parts may not be available, and weather may prevent repair during the winter season, so that valuable data opportunities are lost.

Therefore, instruments must be designed for easy troubleshooting and testing. Self-testing features, failure indicator lights, and remote troubleshooting via modem are valuable. Spare parts kits are essential. Complete operator and service documentation are vital. In design, good workmanship standards and correct choice of materials can make the difference between successful robust designs and early failures.

CASE HISTORIES

At Mount Washington Observatory, we have tested many instruments which have operated well in the most extreme conditions and many which have proven insufficiently robust to meet the challenge of extreme conditions. Those which have not worked well often are the most interesting, as there are important lessons in robust design to be learned from them. Some of these include a hot wire anemometer, several propeller-type anemometers, several windmills, and solar panels.

The hot wire anemometer measures wind velocity by measuring the current necessary to maintain a constant temperature in two wires positioned at right angles; this current increases with wind speed. These devices work very well in laboratory conditions, but when very high winds, icing, and fog are added, they have problems. In extreme conditions, rime ice can form at the rate of 8 inches in 30 minutes; under these conditions, the heated wires cannot generate enough heat to keep the ice from forming on the transducer. In high winds, chunks of ice can strike instruments at more than 100 miles per hour, so that the instrument is broken. To protect the delicate sensor, a protective cage was added, and the ice accumulation on the cage became a serious problem. Finally, in fog and rain, the accuracy of the hot wire transducer was compromised.

Another approach to measuring wind velocity is the propeller-type anemometer. In moderate conditions, with warm temperatures and low winds, these work well, but icing conditions caused rapid failure in tests at the observatory, making the instrument unsuitable for use in the winter.

Energy is expensive at the summit of Mount Washington, and occasionally the use of windmills and solar panels has been suggested as an appropriate solution.

Several windmills have been tested at the Northern Center and always with the same result: the unusually high winds and extreme icing conditions in the fall, winter, and spring generally destroy windmills in a matter of days. As for solar panels, winds are a problem in mounting them, but that can be overcome by careful design and installation. The bigger problems are the number of days in the fog and the extreme icing conditions in the winter months.

SUMMARY

The Mount Washington Observatory's Northern Center for Meteorological and Severe Weather Testing in northern New Hampshire is a unique site for field testing of industrial and scientific equipment. The observatory is known internationally as the "Home of the World's Worst Weather." The highest wind ever measured on earth was measured at the observatory: 231 miles per hour. From November to April, the wind equals or exceeds hurricane force (73 miles per hour) on two out of three days, and winds above 100 miles per hour, with temperatures below zero occur frequently on the summit of the highest mountain in the northeast United States. Heavy accumulation of rime ice, up to 6 feet in a 24-hour period, is common in winter, and the center is in the clouds 60 percent of the time.

Meteorological instruments, oceanographic equipment, airport fog and ice detectors, and outdoor products must be very robustly designed to function in these conditions. A robust design for a product or process is insensitive to uncontrollable (noise) factors. Successful designs must meet the following challenges:

- Extreme temperature
- Humidity
- EMI
- Power-related problems
- Lightning
- Static discharge
- Serviceability and maintenance
- Electronic reliability
- Mechanical ruggedness

Techniques of exceptionally robust instrumentation design have been developed at the Northern Center, and these techniques can be applied profitably to sensors for extreme industrial environments.

CHAPTER 25
COLOR MACHINE VISION

WHY COLOR VISION?

Color vision is one of the most powerful and important sensors available to living organisms. Animals, birds, even insects use it to identify food and potential mates. Plants use color vision indirectly; they display their own colors to attract birds and insects which come for food and in return help spread pollen and seeds. Carnivorous plants may even use color in this way to attract their prey.

We humans use color for the basic reasons: rare, medium, or well-done, a poisonous berry or an edible one, ripe bananas or green ones, a red sports car, a golden tan.

Color provides information about virtually everything we see, about chemistry: gold, silver, copper; tarnished or pristine; about physical properties, rough or smooth, wet or dry, metallic or nonmetallic. Distinguishing apples from oranges in Fig. 25.1 is difficult without color. Color may keep us out of trouble by helping us identify conditions as familiar or unfamiliar.

We deliberately color team uniforms, traffic lights, attendee badges, pharmaceuticals, and multipart forms to facilitate classification.

Color-based inspection and sorting are important in many industries. In the food and textile industries, quality is directly related to color. In other industries, parts or entire products may be color-coded to facilitate human inspection. Sometimes an unusual color combination may simply be a strong indicator that something is wrong.

Human inspectors learn quickly and easily from examples of acceptable and unacceptable product and, having once learned, can normally shift from one inspection task to another with little or no effort. To be a useful substitute for a human, automated color-based inspection systems should have the same capabilities. Humans can easily deal with multicolored objects, complex patterns and random orientations under less than ideal lighting conditions. The more of these capabilities possessed by a color machine vision system, the more useful it is.

Until recently machine vision systems with the required capabilities have been unavailable. As a consequence, color machine vision has earned the reputation of being expensive, difficult to program, and unreliable for all but the simplest operations.

Monochrome machine vision systems have, of course, replaced humans in many inspection and sorting applications. They are achieving impressive speed and accuracy in applications such as metrology, shape determination, presence/absence detection, and some greyscale based classification. Unlike humans, machine vision systems don't lose reliability through boredom.

Apples and Oranges

FIGURE 25.1 Color identification—*apples from oranges.* InspectR Vision. (*Courtesy of American SensoR, Inc.*)

Our objective here is to show how inexpensive cameras and frame grabbers combined with new analysis techniques can turn color machine vision into an easy-to-use, cost-effective sensor for a wide variety of inspection and process control applications. We also show how color machine vision systems can be trained in much the same way as human inspectors and, once trained, can perform rapidly and reliably.

To do this, a review of the basic principles of color vision and related sensing techniques must take place. This leads to a brief review of lighting and emerging trends in color CCD camera technology relevant to color machine vision. Next is a discussion of the strengths and weaknesses of some of the traditional approaches to color machine vision and a new approach, minimum description analysis, whose power and generality avoids many of the difficulties of the older methods. Finally, a consideration of some typical applications is discussed, where minimum description is being successfully applied to color machine vision.

PRINCIPLES OF COLOR SENSING AND VISION

In most color vision applications, we are concerned with a surface illuminated by sunlight or artificial light from a fluorescent or incandescent source. Typically, the illumination is white, that is, it contains a broad range of wavelengths which may extend from the infrared through visible red at about 700 manometers, to violet around 400 manometers, and on into the ultraviolet. When the light strikes the surface, some of it is reflected, some of it is absorbed or transmitted into the material. The relative amount of each wavelength reflected is a complicated function of the angle of incidence, the chemistry of the surface, and the polarization of the light. The spectrum of the returned light therefore carries information about the physics and chemistry of the surface.

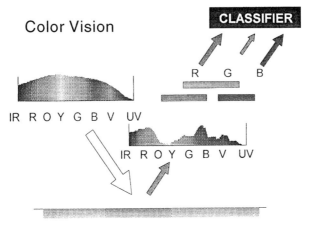

FIGURE 25.2 Color vision.

Information is, of course, contained in regions beyond the visible. Some snakes use thermal infrared sensors, while bees and other insects have vision systems sensitive to the ultraviolet. The techniques discussed here can be applied to images in any spectral region.

A variety of different techniques is available to extract information from the signal returning from the surface. Some of these are listed in Table 25.1. Choice of the sensors and method best suited for extracting information depend on the nature of the analysis to be performed and may involve a trade-off among spectral resolution, spatial resolution, and cost. The characteristics of some of the principal methods are summarized in Table 25.1.

Instruments which measure the spectral properties of the light are known as *spectrometers* or *spectrophotometers*. Analysis of the data obtained with them is generally referred to as *spectrophotometry* when the properties of the light are of primary interest or *spectroscopy* when the properties of the surface are of primary interest.

In most situations of interest to us, the properties of the surface vary in space so the radiation reflected from one region may differ substantially from that of a nearby point. If we were able to gather a complete spectrum from each point in a scene, in principle, we could perform *imaging spectroscopy* to infer the chemical composition at each location in the scene. Imaging spectroscopy has been of particular interest to researchers concerned with remote mapping of the surface of the earth, moon, and other planets. One particularly impressive instrument, AVIRIS

TABLE 25.1 Characteristics of Extraction Techniques

Method	Bands	Locations
Imaging spectroscopy	Many	Many
Color vision	3	Many
Monochrome vision	1	Many
Spectrophotometry	Many	1
Colorimetry	3	1

(Advanced Visible and Infrared Imaging Spectrometer) operates from a high-altitude aircraft. It collects image data from 224 separate spectral bands, each nominally 10 manometers in width. A typical image contains on the order of 200 megabytes of data—so much that it could take several days to transmit through a 9600-baud modem. Unfortunately, the large size of these AVIRIS images makes them difficult to handle, and extracting useful information from single pixels representing complex mixtures has turned out to be somewhat more difficult than originally anticipated.

Some of our recent experiments with AVIRIS images suggest that, for many applications, if bands are properly chosen, use of more than three at any one time offers few advantages.

Compared to imaging spectroscopy, color vision systems may seem primitive. The human color vision system separates the incoming signal using only three sets of detectors with sensitivities peaking in the red, green, and blue respectively. The red and green detectors overlap over much of their range while blue is very insensitive. The red system also has some sensitivity to violet. Because of the relatively broad spectral bands covered by each of the three detector systems, many different signals arriving at the eye can be perceived as the same color. Similarly, two objects which appear to be the same color under one type of illumination may appear quite different under another. In spite of these apparent limitations, the human visual system can differentiate hundreds of thousands of different colors in side-by-side comparisons.

Human color perception is such an effective discriminator that false color images are often used to display relationships between two or three variables over an area. These images are created by assigning each of three primary colors, usually red, green, and blue to one of the variables and making the amount of each color displayed at each location proportional to its corresponding variable at that location. The result is often an image in which subtle relationships between the variables become obvious. Figure 25.3 is a false color image created from Landsat Multispectral Scanner data from a region around Canon City, Colorado. In the presentation used here, a near infrared band (800–1100 nm) is assigned to red in the image, a broad green to red band (500–600 nm) is assigned to green, and a deep red band (600–700 nm) is assigned to blue. The resulting image emphasizes differences between mountain vegetation (primarily coniferous trees), valley vegetation (primarily deciduous trees), and grassland.

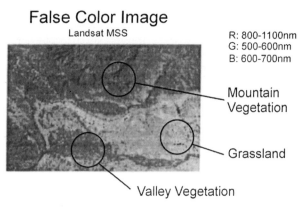

FIGURE 25.3 False color image.

We define a *color vision system* as an imaging system that uses information from three separate and reasonably independent spectral images of the same scene to classify objects within a scene. Although the vast majority of systems use filters which approximate the response of the human eye, this definition is broad enough to include systems which use any three different spectral bands, or even three entirely different sensing techniques.

The science of colorimetry is concerned with color matching. Its sophisticated techniques and instrumentation play an important role in the textile, leather, and paint industries. Colorimetry differs from machine vision in that it is primarily concerned with identifying and/or matching of colors of uniform materials rather than color-based identification in a scene.

Figure 25.4 shows the basic parts of a typical color machine vision system: lighting, camera, and a computer. The computer contains a frame grabber board with dual-ported memory which captures the signal from the computer and makes it available to the classification software. Some sort of digital I/O capability is required to accept a trigger signal to tell the system when to begin inspection and to output the results to control the process.

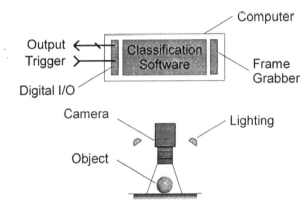

FIGURE 25.4 Color machine vision.

LIGHTING FOR MACHINE VISION

Although our discussion here focuses on cameras and classification methods for color machine vision, a brief mention of lighting is in order. Good results in machine vision are critically dependent on appropriate lighting. Probably more vision system problems can be traced to inadequate lighting than to any other source.

Humans have an amazing ability to correctly infer color even in the presence of strong illumination gradients or under illumination of widely differing colors. Machine vision systems are not so blessed and care must be taken to insure stability in the brightness and spectral content of the illumination. How much care depends on the nature of the classification task. A system to separate red parts from blue parts will probably not require as much attention to illumination as one required to recognize subtle differences in fabric colors.

Most machine vision systems use either tungsten-halogen incandescent lamps or fluorescent lamps. Tungsten-halogen is often used in combination with fiber optics when smaller areas must be illuminated.

Fluorescent lamps are commonly used for larger areas. A variety of effects can cause the light from both types of sources to fluctuate at periods ranging from milliseconds to months or years. Both fluorescent and incandescent lamps are affected by fluctuations in power supply voltage and both exhibit gradual changes in intensity and/or spectrum with age. Fluorescent lamps are also very sensitive to the temperature of the surrounding environment. They are brightest when their surface temperature is about 105°F. Higher or lower temperatures decrease brightness.

For incandescent lamps, combination DC regulated supplies and light sources specially designed for machine vision are available from several manufacturers. To stabilize fluorescent lamps, high-frequency (60 kHz) power supplies are available. Both types of systems may include the provision for real-time feedback to maintain constant brightness in the presence of aging and environmental effects.

Even when a constant brightness is maintained at one spot on the tube, temperature and temperature gradient control may be required to maintain uniform brightness over the entire length.

COLOR CCD CAMERAS

At the present time, color machine vision relies almost exclusively on TV format CCD cameras as a source of images. The design of most of these is based on consumer video cameras which use a single chip to collect all three color signals. This is done by assigning a group of three adjacent pixels on the chip to represent each image location. By overlaying an appropriate pattern of color filters on the chip, it is possible to generate a color signal which, when displayed on a TV monitor, appears to be a reasonable representation of the original scene.

The trade-off of three monochrome pixels for one color pixel obviously results in a loss of spatial resolution in the image. When both high resolution and color are required, three-chip cameras designed primarily for professional TV may be used. These cameras split the incoming light into three separate beams on input and direct them to three separate chips, one for each of the red, green, and blue components of the image. As might be expected, three-chip cameras are substantially more expensive than single-chip cameras. For the vast majority of color classification tasks, single-chip cameras work well.

Some cameras use red, green, and blue filters to generate R, G, and B components of the signal directly. Others use complementary color-based filters. The best images for color machine vision usually come from RGB filter-based cameras which then pass the separate signals directly to the computer.

Consumer cameras and many color machine vision cameras encode the three separate color signals into a single *composite video* signal before output. The NTSC system is used in North America; PAL is used in much of the rest of the world. If this composite signal is passed to the computer's frame grabber, it must be decoded into its color components before being placed in computer memory. This encoding and decoding results in a loss of some color information and potentially a less robust classification. Somewhat higher fidelity color is available from cameras offering S-video signals which are also supported on many newer frame grabbers.

Most cameras generate a TV format *interlaced* signal in which each image frame is captured and transmitted as two successive fields. The first field contains all the odd scan lines in the image, the second field contains all the even scan lines. These TV format cameras have significant limitations for many applications due to the interlace scanning, limited horizontal scanning lines per frame, and relatively low scanning speed. Non-TV-format cameras have now been developed to optimize performance in high-speed machine vision, electronic shutter applications, dynamic motion frame capturing, high-resolution real-time imaging, and direct computer interface applications.

One characteristic of color cameras, which until recently appears to have been neglected, is stability. The sensitivity and spectral response of silicon detectors changes with temperature. This leads to apparent color changes which can be a particularly significant problem for cameras operating in loading docks, recycling facilities, and other open-air environments. Unless the manufacturer has included electronic compensation for these effects it may be necessary to place the camera in a temperature-controlled enclosure. Frame grabber boards may also produce apparent color shifts with temperature.

Line voltage fluctuations can also affect the output of some cameras and for many machine vision applications, a stabilized power source should be used.

TRADITIONAL COLOR-BASED CLASSIFICATION

To distinguish objects of relatively pure color in well-known locations, there are several traditional methods which work quite well. These methods are the basis for several relatively widely used color machine vision systems. We show later that none of them are particularly suitable for classifying complex colored objects such as the apples and oranges of Fig. 25.1.

Traditional classification methods include: table lookup, thresholding, and nearest-neighbor classification. All have been applied successfully to monochrome (gray-scale) machine vision and their extension to color at first glance appears to be straightforward.

Gray-scale classification usually relies on direct table lookup or thresholding. In direct table lookup, each gray level is assigned to a specific class. In thresholding, all pixel intensities falling between two specified values are assumed to belong to a single class. As long as every gray scale has a unique class, these can be very fast and effective techniques whether implemented in general-purpose computers or in specialized intelligent sensors.

The same thresholding and table-lookup approaches used in monochrome classification can be extended to color and again can be very fast and effective sensors for classifying product with simple distinct coloration.

All color classification methods we consider here start with an RGB triplet representing the values of the red, green, and blue signals from a single pixel or the mean of a group of pixels. For ease in visualization, these are often portrayed as a vector in a three-dimensional space. Figure 25.5 shows a color cube representing RGB space. Two vectors from the black origin represent light reflected from two different surface classes as shown symbolically at the bottom of Figure 25.2. The vector pointing diagonally upward to the right represents the purple surface. The vector pointing toward the yellow corner of the cube represents the yellowish surface on either side of it.

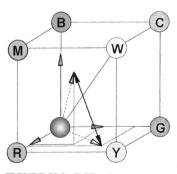

FIGURE 25.5 RGB color space.

As in the monochrome case, the color table–lookup approach uses a table in which each possible color is assigned a class. However, whereas a class table for an 8-bit-per-pixel monochrome image contains 256 entries, a table which uses all the data in an 8-bit-per-pixel per component RGB image will have more than 16 million entries. Even if as few as 5 bits per pixel per component are used in the classification, over 32,000 table entries will be required. Obviously, for most color applications, building color tables can be a substantial task, only justifiable for those applications requiring high speed and for which single value–based classification will work.

The threshold method has more modest memory requirements than table lookup but the performance during actual classification is not quite as fast. Thresholding requires setting upper and lower bounds on each component of the vector in the region of color space occupied by each class; in other words, defining a box around the region.

Thresholding, too, can be time-consuming when many classes are involved and, like table lookup, is restricted to situations where a single color vector is adequate to distinguish the classes.

Successful thresholding in RGB space is made more difficult by the effects of nonuniform lighting. This causes colors associated with a class to be distributed along diagonals in RGB space making it difficult to separate the classes by thresholds normal to the R, G, and B axes.

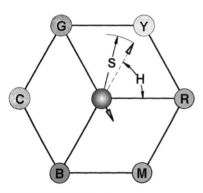

FIGURE 25.6 HSI color space.

To counteract the difficulties from RGB correlation, some systems make use of HSI (hue, saturation, and intensity) space. Definitions of HSI vary slightly among different users but these variations are not significant for this discussion. HSI space (Fig. 25.6) may be regarded as first a rotation of the RGB space so that the intensity (I) axis lies along the black-white diagonals of RGB space, and (2) a transformation to cylindrical coordinates about this axis. It is not hard to see that Fig. 25.6 portrays the cube of Fig. 25.5 as viewed along the diagonal from the black corner. The H and S components are derived from the projections of the RGB vector on the plane normal to the intensity axis. The hue coordinate H is the angle this projection makes with the projection of the R axis. The saturation S varies from 0 at the origin to 1 for fully saturated colors red, yellow, green, cyan, blue, and magenta at the boundaries. Intensity is usually normalized to 1 at full scale when R, G, and B values are all 1.

The major advantage of HSI is that changes in lighting intensity usually result in comparatively small changes in the H coordinate compared to the changes in S and I. Thus, H tends to be a more robust indicator of composition than the other two components and relatively short lookup tables or simple thresholding based on H alone may be adequate to differentiate well-separated classes.

Figure 25.6 also shows the two vectors of Fig. 25.5 projected in the H-S plane. The first appears as a low-saturation vector pointing toward the magenta corner. The second is relatively highly saturated and points toward the yellow corner. Classes this widely separated are good candidates for classification on the basis of hue alone. In such cases, hardware-based HSI conversion, a feature of some high-end image-processing systems, can help greatly in classification. In other situations, when saturations are low (colors lie near the blacks, whites, and grays) a very small change in one of the RGB components can cause a vary large apparent change in hue and thereby cause classification problems.

A third commonly used classification method is nearest-neighbor matching. In this method, a reference RGB (or other) vector is calculated for each possible class. The geometrical distance between the unknown and each of the reference vectors is then calculated and the nearest reference is judged to be the best match. Like the previous methods, nearest-neighbor matching can work well in some situations. It can only be fully justified, however, when the distribution of the colors in each of the classes has an equal, spherically symmetric, and normally distributed variance. Such conditions cannot be counted on for most real classification operations.

APPLES AND ORANGES: A CLASSIFICATION CHALLENGE

Any of the traditional classification methods described previously can work well when applied to simple tasks such as checking in a known location to determine which of several single-color bottle caps are present. Problems appear, however, when color distributions are complex and/or lighting is substantially less than ideal. Figure 25.7 shows the fruit of Fig. 25.7 after random rearrangement. It illustrates features common to many industrial color vision applications which can cause difficulty for traditional color classification methods. These include nonunique colors, nonuniform colors, nonuniformity in lighting, background, overlapping classes, multimodal colors, glints, shadows, and irrelevant regions. These are discussed in more detail following.

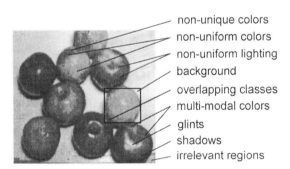

FIGURE 25.7 Apples, oranges, and red herrings.

All the objects in the scene except the rightmost of the oranges and two of the apples have quite nonuniform coloring. The apples show a range from light to deep red as well as various shades of green. The oranges show not only orange but greens similar to those in the apples.

We define *nonunique colors* as colors which appear in more than one class. Neither HSI nor any other color space transformation will make it possible to differentiate in the basis of nonunique colors such as the similar shades of green which appear on two of the apples and three out of four of the oranges.

Some of the apples and the oranges in this scene have multimodal color distributions, distributions which do not cluster in a single region of color space. The apples show red and green with no intermediate colors such as orange or yellow. The oranges show orange and green with no intermediate color such as yellow.

When distributions are multimodal, the mean color can be useless and misleading. The mean color of an apple which is two-thirds saturated red and one-third saturated green would be the same as that of an orange orange. If the amounts of red and green were identical, the mean color of the apple would be banana yellow. In other words, the mean color of an apple or group of apples could be red, orange, yellow, or green (or many shades of brown) depending on the proportions of red and green in the region being classified.

Figure 25.8 shows a mean color computed for each apple and orange in Figure 25.7. Note that the distributions of the mean overlap so that neither hue nor saturation provides a clear differentiation. The intensity values (not shown) also overlap. With overlapping multimodal distributions such as these, none of the traditional methods can be expected to yield reliable results.

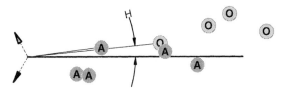

FIGURE 25.8 Apples and oranges in HSI space.

Even if the color of the object itself were uniform, nonuniformity in lighting will cause the range of colors in its image to broaden substantially. Light reflected off nearby objects can also induce apparent color shifts. For example, light reflected from the oranges may be affecting the appearance of the upper-left quadrant of the upper-right apple.

Glints and shadows are extreme examples of nonuniform lighting. The former are reflections of a light source from the surface of the object. Even when care is taken to provide diffuse lighting, glints can be very difficult to eliminate from spherical objects. While glints and shadows provide little information about the intrinsic color of the object, they can provide much information about its shape. They can also have a substantial influence on the intensity and saturation components of the mean.

For efficiency, it is preferable to base classification on rectangular regions of interest such as the one surrounding the rightmost orange. Any region which includes most of the object to be classified will also probably include some of the background. Neighboring objects may also overlap into the classification window as do parts of two apples in this case. A successful classification method cannot be overly influenced by minor amounts of another class.

Finally, we see objects such as the wooden ruler in the image which do not belong to any of the classes for which the system is trained. If these objects are irrelevant to the classification, they should not be allowed to influence it in any significant way.

The preceding effects combine to cause a strong overlap in the colors associated with the various classes. The extent of this overlap is demonstrated clearly in Fig. 25.9 portraying portions of 3D color histograms compiled when training on the apples, oranges, and background. Each histogram contains 512 cells representing eight intensity ranges for each of the red, green, and blue components. The color of

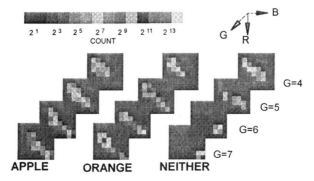

FIGURE 25.9 3D color histogram.

each cell indicates the number of pixels found with the corresponding color combination. The cells for green ranges 4 through 7 are shown, those for ranges 0 through 3 are omitted for clarity.

Attempts to address the effects leading to this overlap individually using traditional table lookup, thresholding, or nearest-neighbor methods can lead rapidly to complex sets of ad hoc rules and exception handlers. What is required is a new approach which will render them, like red herrings, largely irrelevant.

MINIMUM DESCRIPTION: CLASSIFICATION BY DISTRIBUTION MATCHING

It has been described that traditional approaches to color-based classification can encounter difficulties with a task as basic as distinguishing apples from oranges. Yet, in spite of these difficulties, judging from the brain size of some of the birds and insects which depend on color vision for survival, one might infer that color-based classification should be very easy.

One answer to this apparent dilemma comes from taking a fresh look at a old idea, statistical comparison of color distributions. The basic concept is old: comparison of histograms. The comparison functions used are new.

Consider a pair of histograms, one representing the reference distribution R and the other a test distribution T. We define a quantity $T{:}R$ which we refer to as

$$T{:}R = -\Sigma P_{Tk} \log_b (P_{Rk}) \qquad (25.1)$$

where P_{Rk} = the fraction of the test attributes which fall in the kth histogram cell
 P_{Rk} = the fraction of the reference attributes which fall in the kth histogram cell
 b = the base of the logarithms chosen

Shannon[1] showed that $T{-}R$ is the average amount of information per sample necessary to identify a single sample taken from distribution T in terms of a code optimized for identifying samples from distribution R. If $b = 2$, this information will be measured in bits.

The quantities $T{-}T$ and $R{-}R$ are the *entropies* of distributions T and R, respectively. It follows from Eq. 25.1 that

$$T:R - T:T \geq 0 \tag{25.2}$$

with the value of this expression providing a measure of the dissimilarity of distributions T and R. If $T:R - T:T = 0$ then the two distributions must be identical.

The quantity $T:R - T:T$ can be used in classification of a test distribution T by selecting as a best-fitting class, the class j whose reference histogram R results in the smallest value of $T:R_j - T:T$. The term *minimum description* comes from this selection of the class for which information is a minimum.[2] See Eq. (25.2).

$T:R–R:R$ serves as a dimensionless measure of anomalousness. It is straightforward to show that $T–R–R:R \leq 0$ for distribution pairs in which distribution T obviously falls inside distribution R, and that $T–R–R–R > 0$ for pairs in which T falls outside R.

This has led us to propose a definition of the relationship between two distributions T and R such that distribution T *can be regarded as falling inside distribution R if and only if*

$$T:R - R:R \leq 0 \tag{25.3}$$

In all other cases, T is regarded as falling outside R.

When an attribute distribution T falls within distribution R, we generally tend to interpret it as being a member of class R. The value of expression (25.3) can be used to test for the presence of a member of a specified reference class without the necessity of teaching the system to recognize all possible incorrect classes.[3]

Although the roots of the minimum description approach are in information theory, it is also closely related to maximum likelihood estimation.

Figure 25.10 compares the results of a minimum description–based classification of the fruit of Fig. 25.7 with the original image. Class interpretation by the system is indicated by the overlay color. The system was trained by building three reference histograms, one each for apples, oranges, and background using the image in Fig. 25.1 as an example. Training time was of the order of a minute; classification time for all nine objects took less than 0.25 second on a 40-mHz 386-based personal computer.

ORIGINAL INTERPRETATION

FIGURE 25.10 Apple and orange identification.

The system has obviously succeeded in fulfilling the objectives we set out at the beginning. It has learned quickly and easily by example to distinguish apples from oranges. The references have been saved to disk and can be recalled whenever necessary. The system has then made use of its training to correctly classify the fruit in a different image.

A system with such capabilities is a serious candidate to replace human inspectors in a wide variety of industrial inspection tasks.

TYPICAL INDUSTRIAL APPLICATIONS

Minimum-description classification on the basis of color is obviously very general. Vision systems utilizing WAY-2C, our minimum description–based color recognition software package, are successfully operating in a wide variety of installations in the electronics, automotive, wood products, and recycling industries. One of the most recent installations is as part of a high-level nuclear waste disposal monitoring system.

Figure 25.11 shows a practical example of printed circuit board inspection. The objective is to verify that the correct colored wires are soldered in the correct locations even though the order in which the wires emerge from the harness may vary from board to board. The left image shows correctly placed wires: blue, black, red, and gray in order counterclockwise from the lower right. In the image on the right, the system has correctly recognized that the red and black wires are interchanged.

CORRECT   INCORRECT

FIGURE 25.11 Wire placement verification.

Other typical applications, where the objective is to make sure that the right components are present and/or correctly placed, include wire and optical fiber location, sample card and fuse box inspection, and verification that the correct automobile door, instrument panel, or seat is about to be installed.

Many consumer products such as milk, orange juice, and cereal are color-coded for easy recognition by customers. WAY-2C-based systems are in use on loading docks for inventory management and routing of such products.

Appearance matching is important when selecting natural materials such as wood and stone. Attempts to match wood with strongly contrasting (bimodal) grain colors using traditional approaches can produce erroneous results. A minimum description–based system on the other hand is yielding results which are generally in accord with human perception.

Recognition of multicolored fabrics can be difficult with traditional methods but minimum description–based color classification often works well.

The progress of baking, roasting, and similar processes can often be judged from their color distributions. When the color of the product is not intrinsically uniform, minimum description distribution matching can work well while colorimetry-based approaches fail.

Some processes involve application of protective pastes or waxes whose presence or absence is required but whose exact shape and/or position is indeterminate. WAY-2C-based color classification is being used in such processes to verify the presence and location of these materials.

The presence or absence of many coatings can be determined by color and for those which exhibit interference effects, the thickness may also be measured.

Products ranging from food and pharmaceuticals to recycled materials must often be sorted or graded based on color. These are often characterized by irregular shape and broad color distributions, conditions under which WAY-2C is performing well.

Finally, the interaction between illumination and topography can be exploited by setting up lighting systems which generate different color distributions depending on topography. Such systems can be used to measure surface mount lead coplanarity, solder paste thickness, part orientation, and so on.

REFERENCES

1. C. E. Shannon, "The Mathematical Theory of Communication," *Bell System Technical Journal,* 1948, **27** (379) and (623).
2. R. K. McConnell, Jr. and H. H. Blau Jr., "A Powerful Inexpensive Approach to Realtime Color Classification," *Proceedings Soc. Mfg. Engs. Applied Machine Vision Conference '92,* June 1–4, 1992, Atlanta, SME T technical paper MS92-164, Society of Manufacturing Engineers, Dearborn, Michigan, 1992.
3. R. K. McConnell Jr. and H. H. Blau Jr., "Color Anomaly Detection Using Minimum Description Analysis," *Proceedings Soc. Mfg. Engs. Applied Machine Vision Conference '94,* June 6–9, 1994, Minneapolis, Minnesota, SME technical paper MS94-187, Society of Manufacturing Engineers, Dearborn, Michigan, 1994.

CHAPTER 26
MONOLITHIC INTEGRATED PHYSICAL AND CHEMICAL SENSORS IN CMOS TECHNOLOGY

INTRODUCTION

Integration of sensors and electronics on the same substrate becomes more and more important. CMOS circuit design seems to be the most important technique for the monolithic approach. Compared to bipolar circuits, power consumption is much lower; compared to BICMOS, the process is much less complex, therefore having a much lower number of masks. This is very important because in most cases monolithic integration means additional process steps and by this a higher process complexity. This chapter gives a few examples for the monolithic approach in the field of physical and chemical sensors based on CMOS technologies.

Progress in semiconductor technology has been very fast in the last few years. Silicon devices which have a very high functionality and complexity are available. The use of modem CMOS technologies makes the fabrication of microprocessors possible reaching speeds of several 10 million instructions per second. The typical line width in such circuits is 0.5 to 0.8 micrometers. This progress was only possible by a thorough study of the chemical, physical, electronic, and technological aspects of silicon and silicon devices.

Besides thinking of silicon as a material for the fabrication of microelectronic circuits, silicon is also a promising sensor material. Some of its physical properties, such as thermal conductivity, photoconductivity, or piezoresistivity, can be directly exploited to built silicon sensors using silicon devices such as diodes or transistors. The sensing of chemical properties by silicon devices may be achieved by deposition of additional chemically active materials (Table 26.1).

In other applications, the silicon chip can serve as a sophisticated carrier substrate for the deposition of chemically active layers. The full variety of state-of-the-art techniques for fabrication of integrated circuits can be applied to sensor implementation including such techniques as silicon-on-insulator or silicon micromechanics.

The sensing element itself is only part of a whole sensor system including sensor readout electronics as well as interfacing. The use of silicon technologies for sensor fabrication allows, in a variety of applications, the integration of electronic circuits together with the sensing element on one chip. Besides lowering manufacturing

TABLE 26.1 Silicon Devices for Sensing Different Physical and Chemical Quantities

Parameter	Silicon device
Radiation	Photodiode solarcell thermopile
Mechanical quantities	Piezo MOSFET Micromechanical device
Thermal quantities	Thermocouple thermoresistor thermodiode
Magnetic field	Hall device magnetoresistor magnetotransistor
Bio/chemical quantities	IS-FET
	Gas-FET
	Thin film device

costs for high-volume production, the silicon integration of both sensors and electronics offers several significant advantages, such as small size and low weight (e.g., important in medical applications), close physical proximity of sensors and readout (important for low-noise pickup), and the possibility of integrating multisensor arrays in combination with highly complex signal processing. For signal conditioning, CMOS circuit design seems to be the most important technique for the monolithic approach. Compared to bipolar circuits, power consumption is much lower; compared to BICMOS, the process is much less complex and therefore has a much lower number of masks. This is very important because in most cases monolithic integrations mean additional process steps and, by this, a higher process complexity.

This chapter gives a few examples for the monolithic approach in the field of physical and chemical sensors. It concentrates on examples where additional process steps are needed.

PHYSICAL SENSORS

In the field of physical sensors, silicon micromachining is an often-used technique. For pressure and acceleration sensors, mostly bulk micromachining, where silicon is three-dimensionally structured by anisotropic etching, has been used in the past. In the last few years, surface micromachining has gotten more attention. The first products are commercially available.[1]

Surface-Micromachined Capacitive Pressure Sensor

The sensor system presented here has been developed for the measurement of pressure and temperature in very small cavities for a variety of applications in automotive, mechanical engineering, or medical fields. Because of the small dimensions necessary, surface micromachining was used for fabricating pressure sensors. The silicon chip with the dimensions of 0.7 × 7 mm is produced in a 3-gm CMOS process with additional surface micromachining steps which are needed to form membranes for the on-chip capacitive pressure sensor. These additional steps are all slightly modified standard CMOS process steps, that is, steps with altered deposition or etch times. Therefore, only the standard CMOS equipment is required.

The pressure sensor consists of an array of single sensor elements switched in parallel and an equal array of reference elements, which are almost identical to the sensor elements, but having a lower pressure sensitivity.

MONOLITHIC INTEGRATED PHYSICAL AND CHEMICAL SENSORS 26.3

Figure 26.1 shows an electron micrograph of such an array. The array only requires an area of 0.5 × 1 mm. A schematic cross section of such a sensor element and a reference element is shown in Fig. 26.2. The membrane material is polycrystalline silicon. The cavity is obtained by etching away a sacrificial silicon oxide layer that has been deposited and structured before the deposition of polycrystalline silicon. After etching the cavity, under the membrane is vacuum sealed. An n^- doped area in the silicon substrate under the membrane and the membrane form the plates of a pressure-dependent capacitor. The distance between substrate and membrane is only 0.9 jim. The membrane dimensions have been optimized for a pressure range of 1.4, 2, 10, and 20 bars with membrane diameters of 120, 100, 70, and 50 micrometers, respectively. For signal conditioning and amplification of the capacitive pressure signal and the voltage signal of the temperature sensor, a switched capacitor (SC) circuitry is used.

To minimize noise susceptibility of the output signal a conversion to a pulse-width modulated signal is implemented, too. The pulse-width modulated pressure signal, the temperature signal, and some reference signals for compensation purposes are sent via a single connection wire using time multiplexing. The output of the pressure signal is proportional to the quotient of Csens/Cref. By this operation, non-pressure-dependent capacitances as well as temperature dependencies are strongly suppressed.

Figure 26.3 shows a photograph of the integrated sensor chip. The different pressure sensors can be seen in the right part. They were diced and mounted with standard chip technologies in a ceramic housing. No dependence of the output signal on the mounting procedure was found.

In the lowest pressure range, the sensitivity of 425 ms/bar of this sensor is sufficient to resolve a pressure of less than 0.5 mbar (Fig. 26.4). All pressure sensors were

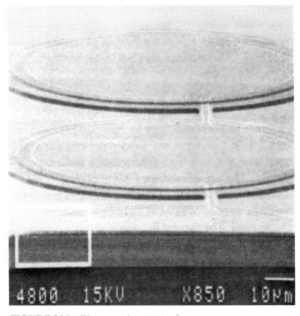

FIGURE 26.1 Electron micrograph of a pressure sensor array.

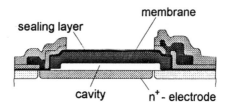

reference element

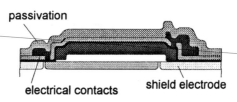

FIGURE 26.2 Schematic cross section of pressure sensor and reference sensor.

found to have a high overpressure resistance at least greater than a factor of 10. The temperature dependence of the offset, that is, the ratio Csens/Cref at a constant pressure, is shown in Fig. 26.4, too. The ratio is corrected by the values of the reference voltages in order to eliminate the temperature dependence of the circuitry. The resulting temperature dependence is lower than 200 ppm/°C full scale.

The pressure elements fabricated with surface micromachining show a variety of advantages compared to other silicon pressure sensors:

FIGURE 26.3 Chip photo of the catheter sensor.

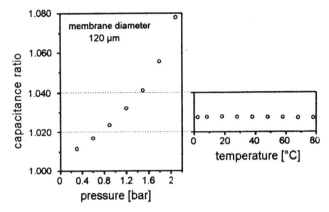

FIGURE 26.4 Output signal of the pressure sensor.

- They can be fabricated with very small dimensions.
- Due to the small area of a single element, an arbitrary chip size can be chosen.
- Due to the low distance membrane-substrate, they show a high overpressure resistance.
- Parasiteric influences are easily compensated by the use of reference elements.
- The pressure sensor is not influenced by the packaging procedure.
- No wafer/wafer or wafer/glass bonding is necessary.

Integrated Flow Sensor

Flow sensors are used in many fields, including industrial process control, biomedical instrumentation, and automotive applications. Most known silicon flow sensors are based on thermal principles. The sensitivity of silicon flow sensors, which use the cooling effect of flowing media along heated parts of the chip, can be improved by an increase of the heat loss into the fluid relative to other heat losses on the chip. This can be done by an improved thermal isolation of the heated area from other parts of the chip. This is why there have been efforts to form silicon membranes, oxide membranes or bridges, and the like with bulk micromaching techniques.[2] A lateral thermal isolation in silicon can also be achieved by trench-etching techniques in the membrane rim.[3] Such a structure combines thermal isolation with the possibility of placing electronic devices directly into the membrane. The presented flow sensor contains such a silicon membrane with trenches in the membrane rim and diodes in the middle. Figure 26.5 shows a sketch of the sensor principle. The flow sensor contains a polysilicon heating element, which heats the membrane.

One diode measures the membrane temperature; a second is located on the chip outside of the membrane region. The fluid flow, in contact with the whole chip, cools the heated membrane. In order to increase the thermal isolation between heated membrane and the other areas of the chip, the membrane has five oxide-filled trenches in the membrane rim with a width of 1.6 gm and a separation distance of 3.2 gm. In the case of perfect thermal isolation between the two diodes, the second diode measures the fluid temperature. The lateral membrane size is 500 × 500 gm and the membrane thickness is 5 gm. This membrane is able to withstand pressures up to 16 bar.

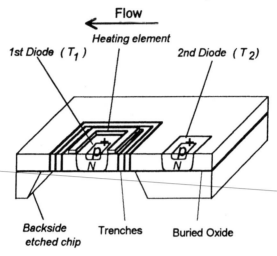

FIGURE 26.5 Measurement principle of the flow sensor.

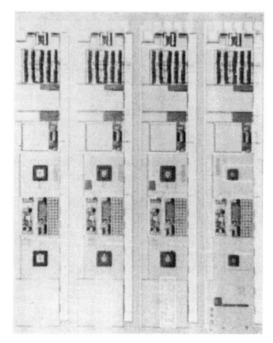

FIGURE 26.6 Chip photo of the flow sensor.

The flow sensor (Fig. 26.6) is fabricated by a slightly supplemented common 2-gm LOCOS-CMOS process. The front side has been passivated by PECVD nitride for the protection of the chip against corrosive attack by liquids and for safety against electrical short circuits. Principally, there are two ways to use a thermal flow sensor: namely at constant temperature or at constant voltage. In the constant voltage mode, the time $t\,90$ between the 10 and 90 percent value of the initial signal has been determined for the sensor in oil to $t\,90 = 8$ ms; for water, we get 50 ms, and in the acid liquid, 280 ms.

The dynamic behaviour of the sensor in the constant-temperature mode was tested, too. The lowest $t\,90$ time in the measurements was 40 ms. This is a factor of 200 better than the value for the constant voltage mode (Fig. 26.7). For water and the acid liquid, the lowest $t\,90$ values of 10 ms were obtained.

CHEMICAL AND BIOCHEMICAL SENSORS

Increasing worldwide importance is seen in the field of sensing chemical parameters. The knowledge of gas concentrations or the concentrations of different substances in liquids is needed in a variety of different fields, for example, engine regulation, medical care, safety, and so forth.

Conductivity Sensor

Invasive measurements in human blood vessels require miniaturized sensors. The sensor presented here was developed for the measurement of the electrolytic conductivity of human blood.[4] The conductivity signal gives information about the amount of hematocrit in the human blood. The chip size is 1×5 mm^2, which makes it possible to embed the chip into a catheter of 2 mm in diameter.

The method for the conductivity measurement uses a 4-electrode structure and a galvanic contact to impress a current into the electrolyte. For this purpose, planar electrodes are placed on the silicon substrate. The electrodes are fabricated in a standard CMOS process with only one additional technological step. Figure 26.8 shows a cross section. The electrode material, which is a thin platinum layer of 100 nm thickness, is deposited by DC sputtering and structured by a lift-off process.

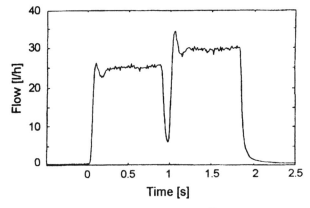

FIGURE 26.7 Flow sensor measurement in oil during a pump cycle of a piston pump. The signal shown is the linearized flow sensor signal.

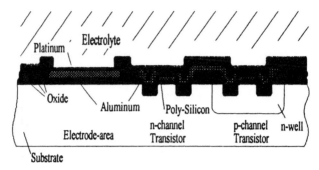

FIGURE 26.8 Cross section of the CMOS process with integrated electrodes.

Four electrodes are placed in a planar arrangement of rectangular shape. Two of the electrodes are used to impress the current, they have a size of 2250 × 500 mm; a second pair of 10 × 250 mm between them is used to measure the voltage drop in the electrolyte (Fig. 26.9). A modified four-electrode method that uses switched-capacitor techniques and offers the possibility of measuring in the audio frequency range has been developed. The voltage drop in the electrolyte is sampled without drawing current from the voltage electrodes. In contrast to traditional methods using sinusoidal signals, a square wave current of 4 kHz, which is generated from the DC current source and integrated analog switches, is injected into the electrolyte at the current electrodes A and D. This leads to a constant voltage drop in the electrolyte for 125 ms in each half-wave.

Measurements in KCl solutions with platinum-type electrodes show a linearity of better than ±1 percent and a reproducibility of ±0.5 percent. The system can be easily adapted to the desired range by changing the measuring current (between 10 and 100 mA) or the amplification (between 6 and 60). The smallest conductivity that could be

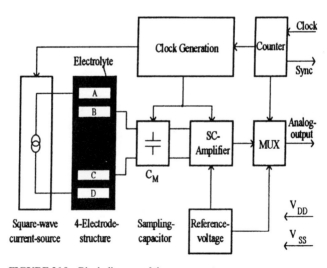

FIGURE 26.9 Block diagram of the sensor system.

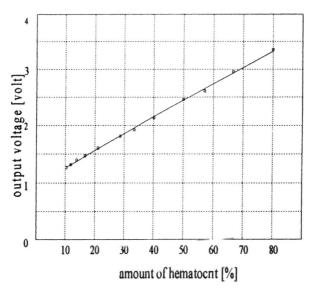

FIGURE 26.10 Measurement of hematocrit in human blood.

measured was 20 ms/cm; the highest was 100 ms/cm. Tests of the sensor system in human blood lead to the results presented in Fig. 26.10. As the hematocrit is approximately nonconducting in this case, there is an increase in conductivity if the amount of hematocrit rises. On the other hand, this allows the determination of the hematocrit. To perform this measurement, the blood sample was thinned with 0.9 percent saline.

Hydrogen-Sensitive MOSFET

Hydrogen gas–sensitive metal oxide semiconductor structures using a palladium gate were first described by Lundstrom in 1975.[5] The general operation principle can be seen in Fig. 26.11. The base device is a MOS transistor. The current between the

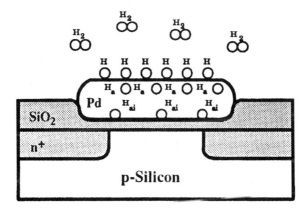

FIGURE 26.11 Hydrogen-sensitive MOSFET.

source/drain contacts is controlled by an appropriate gate voltage. By changing the gate material from polysilicon to palladium, this transistor can be used as a hydrogen sensor. Hydrogen molecules are decomposed at the palladium surface. Hydrogen atoms diffuse through the palladium to the interface, changing there the interface potential. This results in a shift of the drain current to gate voltage characteristics. It is also possible to use MOS capacitances as sensing devices. In this case, the capacitance-voltage characteristics are changed.

These types of gas sensors can very well be combined with standard CMOS electronics. The use of progressive sensor electronics might improve the long-term stability of these devices. Schoneberg et al. developed a switched capacitor circuit for the readout of gas-sensitive MOS capacitances.[6] Due to the differential measurement, cross sensitivities are suppressed. The probability of ionic movement in the oxide or ion accumulation on the surface is very low by avoiding a constant bias potential. Figure 26.12 shows a monolithic integrated hydrogen sensor. The improvement can be seen in Fig. 26.13, which shows a measurement over about one month, using discrete electronics.

Sensor Matrix for Two-Dimensional Measurement of Concentrations

The conventional potentiometric and amperometric measurement methods normally used for chemical and biochemical analysis yield only results about the global concentration of the sample. Observing the spatial distribution of a chemical or biochemical process is not possible using this method. For this purpose, a monolithic chemical sensor array has been developed.[8] It consists of an array of 20 × 20 sensor cells which can be addressed separately. Every sensor cell contains one working electrode, readout electronic, and the sensor cell control that allows the electrode to be used in different modes (Fig. 26.14). For the exact analysis of the electrode signal,

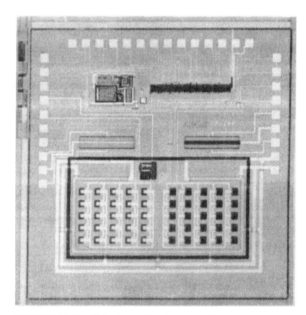

FIGURE 26.12 Chip photo of the integrated H_2 sensor.

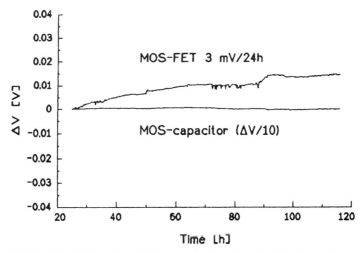

FIGURE 26.13 Comparison of the output signals of a FET and a MOS capacitance.

every electrode is provided with its own readout electronic. Parasitic effects such as leakage currents or wire capacitances are minimized this way.

For addressing the sensor cells, a control unit is integrated into the chip. The output signals of the sensor cells are switched to a single analog output port in a serial way. The sensor chip contains an integrated test mode. The correct function of the control unit of the chip is checked permanently while measuring. Additional test ports allow for applying test signals to the sensor cell readout electronics to guarantee their performances and to calibrate the electronics. Via the test ports, it is also possible to apply a voltage to the sensor electrodes to cause chemical processes, for example, the generation of a membrane by employing electropolymerization.

The 400 working electrodes (50×50 gm^2) are made of 100-nm-thick platinum, while two integrated reference electrodes consist of a silver layer. After dicing the wafer, the individual sensor chips are bonded on a substrate. The chips are encapsulated by a two-component epoxy adhesive to provide electrical isolation during mea-

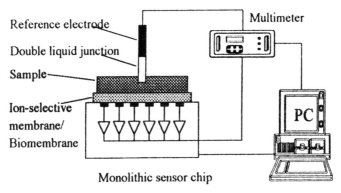

FIGURE 26.14 Measurement scheme.

surements in liquids. For test purposes, the sensor chip was covered with a potassium-sensitive membrane. The membrane was fabricated by dropping a doped PVC cocktail directly on the surface of the sensor array. After a two-day drying procedure, the ammonia-sensitive membrane was ready for use. First, measurements were performed in potentiometric mode. Drops of urea solutions were placed on the sensor chip. Figure 26.15 show the resulting output signals of the sensor array. The results show the spatial distribution of urea on the membrane after dropping samples on it. As a reference, an external Ag/AgCl electrode was dipped into the drop and connected to ground. The drops can be found in the output signal of the chip that is only dependent on the urea concentration of the sample. The chip surface was not affected by the chemicals, so the chip can be reused for further measurements.

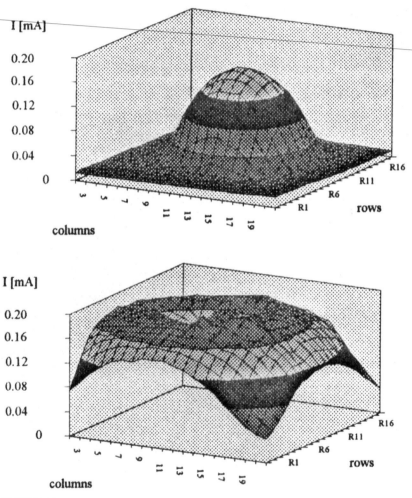

FIGURE 26.15 Imaging of urea. Scans were started 7 and 19 minutes after the contact of the double-liquid junction and the buffer (1.0 M) solution on the chip.

The use of different doped PVC membranes in one measurement allows this chip to become a chemical and biochemical multisensor system. The sensor array will be used in medical applications and pollution control.

SUMMARY

These examples have shown that the monolithic approach using CMOS technologies is possible on a variety of different principles. On the way to market the first products will show up in mass markets such as the automotive field. The application fields will continuously increase if standards can be installed successfully or if production flexibility can improve handling of a lot of different products and processes even for lower volumes.

REFERENCES

1. Analog Devices, product information, 1992.
2. D. Moser, R. Lengenhager, and H. Baltes, "Silicon Gas Flow Sensors Using Industrial CMOS and Bipolar IC Technology," *Sensors and Actuators*, 1991, A27, pp. 577–581.
3. J. Wemo, R. Kersjes, W. Mokwa, and H. Vogt, "Reduction of Heat Loss of Silicon Membranes by the Use of Trench Etching Techniques," *Sensors and Actuators* 1994, A 41–42, pp. 578–581.
4. N. Kordas, Y. Manoli, W. Mokwa, and M. Rospert, "A CMOS-Compatible Conductivity Sensor with Integrated Electrodes," *Digest of Technical Papers Transducers 93*, Yokohama 1993, p. 466, *Sensors and Actuators*, 1994, A43 pp. 31–37.
5. I. Lundstrom, M. S. Shivaraman, L. M. Svensson, and L. Lundkvist, *Appl. Phys. Lett.*, 1975, **26**, p. 55.
6. H. G. Dura, U. Schoneberg, W. Mokwa, B. J. Hosticka, and H. Vogt, "Performance of Hydrogen-Sensitive MOS Capacitances with Integrated On-Chip Signal Conditioning," *Technical Digest Eurosensors* 1991, Rom, **5**, p. 64, *Sensors and Actuators*, 1992, B6, p. 162.
7. U. Schoneberg, H. G. Dura, B. J. Hosticka, and W. Mokwa, "Low Drift Gas Sensor with On-Chip Instrumentation," *Proc. Transducers VI*, San Francisco, 1991, p. 1006.
8. H. Meyer, H. Drewer, J. Krause, K. Cammann, R. Kakerow, Y. Manoli, W. Mokwa, and M. Rospert, "Chemical and Biochemical Sensor Array for Two-Dimensional Imaging of Analyte Distributions." *Sensors and Actuators*, 1994, B1819, pp. 229–234.

CHAPTER 27
A RESEARCH PROTOTYPE OF A NETWORKED SMART SENSOR SYSTEM

INTRODUCTION

The use of distributed systems technology in applications involving measurement and control is a subject of much recent discussion, see Refs 1, 2, 3. As part of this investigation, researchers designed and constructed a networked system of smart sensors and actuators. The goal was to evaluate distributed smart sensor systems with respect to the following:

1. Ease of system configuration
2. Ease of system expansion and modification
3. Ease of data management
4. Ease of application development

BACKGROUND

Traditional test and measurement applications have been based on the use of centralized control and data management. The usual implementation involves a central controller that directs the actions of instruments, sensors, and actuators; polls for any results; and manages the resulting data.

Nowadays, *distributed measurements* usually refers to systems comprising one or more controllers each with one-to-one connections to instruments, sensors, and actuators. Supervisory control and data acquisition (SCADA) systems are a typical example of such a distributed measuring system. Other distributed systems are essentially autonomous measuring devices such as data loggers or electric meters. In this case, the data is brought manually to a central location or it may be polled over a communication link.

In both of these previous cases, the essential determination of the system's behavior resides in the central controller. Communication is one to one between the central controller and each instrument, sensor, or actuator.

In this prototype system, a more general notion of distributed measurements is used. Specifically the nodes, that is, the instruments, actuators, and sensors, determine the overall behavior of the system, rather than depending on a central controller for control. In addition, nodes communicate directly with each other or with groups of other nodes without any restriction to one-to-one communication links. In the rest of this chapter, *distributed measurements* refers to this more general notion. *Distributed measurement systems* as used in this chapter, also includes systems with actuators and more general-purpose nodes such as displays.

It is unlikely that distributed measurement technology will be the best solution for all applications any more than traditional centralized technology is best for all applications. In general, a mixture of technologies will be the optimum for a given application. However, this prototype system and this chapter deal almost exclusively with distributed measurement technology.

OVERVIEW OF DISTRIBUTED MEASUREMENTS

Distributed measurement systems generally require more intelligence to be built into the nodes than a traditional centralized system designed for the same overall task. It is useful to consider three aspects of distributed node intelligence: transducer-related, measurement-related, and system- or application-related intelligence. These aspects might be termed *levels of smartness.*

The following sections discuss some properties of measurements that need to be considered when designing nodes for a distributed system. In general, the more of these properties that are managed by the node, the more useful the node will be as a general component in distributed measurement systems.

For example, consider the transformation of the voltage output of a thermocouple into temperature. If this is not handled in the node containing the thermocouple, then every other node in the system that makes use of the thermocouple output must either make this transformation itself or obtain the result from a node that does make the transformation. Either of these choices provides less system design flexibility than doing the transformation in the node containing the sensor.

Transducer-Related Properties of Distributed Measurement Nodes

Transducer-related properties include the following:

1. *Physical variable:* Temperature, stress, and so forth
2. *Form of transducer input or output:* Voltage, change in resistance, digital signal
3. *Calibration:* Relationship of transducer output to sensible value, for example, converting the value of the voltage from a thermocouple to the measured temperature
4. *Identity:* Transducer serial number, description, and so forth
5. *Limits of use:* Maximum and minimum values, acceptable operating environments, stability of calibrations, repeatability, and so forth

In a distributed system, these items will be handled at the nodes containing the sensor or actuator. Current instruments generally manage these properties, while most available sensors and actuators do not. Many of the smart sensors available today manage only a few of these properties. Examples of such existing smart sensors are as follows:

1. Die Analog Devices ADXL50 accelerometer, which has built-in signal conditioning and self-test
2. The Smartec SMT160-30 temperature sensor with built-in signal conditioning and digital interface
3. The Omega Engineering PX763 pressure sensor with signal conditioning and an interface to the HART communications protocol

Measurement-Related Properties of Distributed Measurement Nodes

Measurement-related properties include the following:

1. *Measurement timing management:* Timed, polled, random and so forth
2. *Local data management:* Store until requested, broadcast upon collection, and so forth
3. *Local computation:* Average, peak value, and so forth
4. *Identity:* Node identification, description, and so forth
5. *Location:* Coordinates or identifier of the measurement point

Instruments manage some but not all of these properties. Management is usually accomplished via external commands or from front panels. Currently available transducers generally do not deal with these properties.

For nodes to be generally useful in distributed systems, these sorts of properties need to be managed within the node.

System- or Application-Related Properties of Distributed Measurement Nodes

System- or application-related properties include the following:

1. Changing measurement properties in response to application-related messages, for example, changing the sample rate in a collection of nodes.
2. Defining the communication patterns among nodes, for example, master/slave, client/server, peer to peer, and so forth.
3. Establishing and modifying communication patterns among nodes, for example, modifying multicast membership.
4. Managing the transport properties of communication among nodes: flow control, reliable delivery, and so forth.
5. Synchronizing the node clocks, if present.
6. Conforming to system data and control models. Few current instruments, sensors, or actuators have the capability to manage these properties. There are a few new LAN-based instruments that have some capabilities in this area.

These properties are often the dominant factors in determining the usefulness of nodes in distributed measurement systems. Without proper support for managing these properties within the nodes, the design and implementation of a distributed system requires impractical levels of low-level programming for each node.

Communication Protocol Issues

Multicast communication capability is desirable for distributed measurements. Here multicast means that M × N communication is supported, that is, multiple sources can communicate with multiple sinks.

In a distributed system, the logic that determines behavior resides in the nodes. In all but the most trivial measurements, this logic requires communication between nodes in a variety of patterns. For example, data collected at a given node may be needed at several other nodes for display, archiving, or perhaps setting an actuator. Likewise, a given node, for example, a display node, may be managing data from a number of sensor nodes. This sort of communication is a good match to a multicast capability. Point-to-point communication is a special case of multicast and will be needed occasionally in distributed measurement systems.

Most existing protocols in use in test and measurement implementations present a point-to-point interface to the designer. However, it is worth noting that at the physical layer most are fundamentally multicast protocols since all nodes share a common multicast medium such as coax or RF. Notable exceptions are RS-232 and the 4-20-ma loop which are dedicated physical links and cannot support a multicast protocol. Point-to-point protocols are not adequate for distributed measurements as envisaged in this chapter.

In the case of the IEEE-488 protocol, information transfer is restricted to a single talker at a time and is not as flexible as a true multicast protocol. The IEEE-488 protocol is not a good candidate protocol for distributed measurement systems.

LAN protocols often have a multicast capability for specialized needs. However, the normal application interface is point to point. The Internet protocol (IP) layer in the protocol stack is usually used for point-to-point links. Most network applications (e.g., FTP, Telnet RPC, NFS) use IP in this manner. There is provision for multicast in the IP layer. This multicast capability is often used for system-level protocols, for example, NTP, and can be used for distributed measurement protocols as well.

One of the challenges in developing distributed measurement systems is finding appropriate multicast implementations.

Data Management Issues

In a distributed system the management of data must be more structured than in traditional systems. In centralized systems, many data management tasks, such as identifying the source and time of a measurement, are based on the properties of point-to-point communication links. In a distributed system using multicast, other techniques must be used for binding the various pieces of information in the system.

An example of such a technique is the data model for physical measurements. It is necessary to include the value or a reference to the value in the data sent from a distributed node for any item normally inferred from the properties of a point-to-point link in a traditional system, such as source node identity. The minimum elements of the relation representing a measurement include attributes for the value, units, time of measurement, the location of measurement, and usually some name. Depending on the type of system being built, additional elements such as accuracy or precision may be included. A more detailed description of the data models used in the prototype system is included in the appendix of this chapter.

The use of data models with the information binding implemented in the nodes provides the best basis for realizing this project's objectives, for example, the "plug-and-play" behavior.

Control Protocol and Real-Time Issues

In a traditional centralized system, behavior is managed by the controller issuing detailed command messages to each of the remote nodes. Such systems, built using existing instruments and sensors, require many of these messages to be concerned with details of the internal operation of the node. These detail messages often dominate network traffic.[4]

In distributed measurement systems, the details of system behavior are determined by the nodes. The control protocol must therefore support each node internally, managing the application details occurring at that node. In addition, the control protocol must support the transmission of synchronization messages between nodes to produce the correct overall system behavior.

Synchronization includes not only the timing of measurements but the overall progress of the application from one sequence of events to another.

In the distributed measurement systems we have built to date, the communication traffic consists mainly of these application synchronization messages and application measurement data.

There are three main real-time considerations in measurement systems:

1. The relative timing of measurements and real world phenomena, that is, synchronization
2. The time needed to transport data
3. The time needed to process data

In traditional systems, synchronization is managed by the central controller, often in combination with hard-wired triggers between nodes. Any notion of real time is imposed by the central controller on the received data.

In distributed measurement systems, each node must explicitly deal with synchronization. Order, but not time specification, can be implemented via messages passed between nodes. If a true time specification is to be imposed, then nodes must have access to a clock. A central time server is one possibility but introduces delays and probably excessive message traffic. For systems with more than a few nodes, a better choice is for each node to have a local clock participating in a synchronization protocol among the nodes.

The real-time requirement for transporting data over the network and processing this data within nodes is an issue of network and node design. In systems with many nodes, network bandwidth may be at a premium for many applications. Consequently, careful consideration of such things as the use of acknowledgments is necessary to avoid saturating the network. Like others,[5] we have found using unacknowledged transport mechanisms with the appropriate end-to-end acknowledgment at the application level, minimizes network traffic and is appropriate for distributed systems.

Simple ordering of events can be accomplished by a message exchange. For example, to ensure that event A1 in node A executes before event B1 in node B (ordering), a message would be sent by node A, after executing A1, to node B. Upon receipt, node B would execute B1. A major deficiency of this scheme is that only the order of the events is specified. The time difference between the two events (e.g., event B1 must follow event A1 by 50 ms) cannot be specified because there is no control over the processing or delivery time of the messages.

To provide this time specification, it is necessary to base the execution of events on time rather than on the receipt of a message. By accessing a sufficiently accurate time reference, each node can be programmed to execute events at the appropriate times. Some event times may be preprogrammed or based on times communicated in messages.

PROTOTYPE SYSTEM

Design Objectives and Specifications

The design objectives define the following specifications of system and node behavior and capability:

1. Connection of a node to the medium is all that is required for the node to become operational in the system (in a sensible way).
2. Deletion of a node for any reason (e.g., disconnection or power-down) does not cause any failure of system behavior other than those effects due to missing the production or consumption of data and messages associated with the node.
3. All necessary protocols for operation of the system are distributed, that is, there is no need for any form of central control during the operational phases of the system.
4. All nodes are capable of accepting an application script that tailors the inherent behavior of the node to meet the requirements of the specific overall system application.
5. Central control is permitted *but is not required* as a means of managing the modification of node behavior for specific applications. However, for this study, the role of the central controller was limited to downloading applications and issuing commands to change the selection or parameters of one of the behavioral models within a node.
6. All nodes implement the same standard data, control, and behavioral models.
7. All nodes contain clocks that are synchronized with the clocks in other nodes of the system.
8. All nodes provide information sufficient to properly interpret communications with the node.

In this report, nodes that meet these design criteria are referred to as *smart* nodes.

General Description of an Application Using the Prototype System

As illustrated in Fig. 27.1, a typical system contains a number of smart sensors and actuators. Laboratory prototype smart sensor nodes have been implemented to measure light intensity, relative humidity, temperature, carbon dioxide concentration, sound level, power line voltage, power line frequency, pH, fluid flow, acceleration, and atmospheric pressure. Laboratory prototype actuator nodes have been implemented to control a relay, control a DC motor, display a voltage, and display a byte of data with a group of LEDs.

In addition to the smart actuators and sensors, a number of laboratory prototype smart system nodes have been built to illustrate more complex functionality, while still being good citizens in the distributed measurement system. The display and archive nodes accept the output of all of the sensors and parse and display the data or save it to a file.

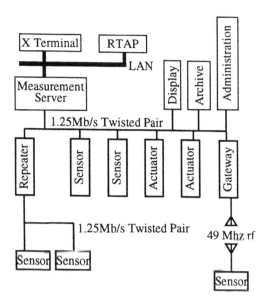

FIGURE 27.1 Prototype system overview.

A variety of communication media are used in the demonstration system. An Ethernet LAN is used in the traditional portion of the system. A 1.25-Mb/s twisted pair and a 49-MHz low-power RF link are used for communication in the distributed portion of the system. A repeater is shown to illustrate the segmentation of the network allowing more nodes than permitted with a single twisted pair.

The administration node downloads applications into smart nodes and sends commands to change the internal model selection and parameters.

The smart sensors, actuators, display, archives, and administration nodes constitute a pure distributed measurement system in the terms of this chapter. The remainder of the components illustrate how such a system might interface to traditional systems.

The measurement server node provides a remote procedure call (RPC)–based interface to the LAN and provides visibility into the distributed measurement system. In this system, the measurement server serves as a gateway by converting LAN packets to and from equivalent packets on the smart sensor network. One of the remaining research issues is the design of measurement servers and the way such servers present distributed measurement systems to the more traditional systems.

The measurement server communicates with RTAP, a Hewlett-Packard SCADA supervisory system. RTAP accepts the data from the smart sensors as delivered by the measurement server node. This data is archived in the RTAP database and is available for display, analysis, trending, and so on. Access to RTAP is from any X terminal on the LAN.

Smart Node Architecture

The architecture for the prototype smart nodes is illustrated in Fig. 27.2.

The communication media access block handles the low-level protocol required to access the physical medium. Included in this block is a media-dependent transceiver. This transceiver is the physical interface to the medium. Ideally, the interface between

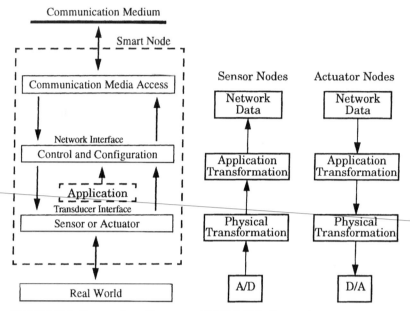

FIGURE 27.2 Smart node architecture.   **FIGURE 27.3** Data path.

this block and the control and configuration block, that is, the network interface, is media- and protocol-independent.

The control and configuration block includes the remainder of the protocol stack, that is, the media- and protocol-independent portion, and the control and computation circuitry needed to implement the functions of the smart node. This control and configuration block is identical in all smart nodes.

The optional application block consists of a ROM containing an application-specific script. Applications may also be loaded over the network. This block is located on a card that plugs into the smart node.

The sensor or actuator block contains the sensor or actuator transducer, any additional circuitry not contained in the control and configuration block, and an electronic data sheet containing specifications and calibration information unique to the specific sensor or actuator.

The two possible data paths within the smart node control and configuration block are illustrated in Fig. 27.3.

The physical transformation converts between the digital representation in the International System of Units, SI, and the raw digitized representation provided by the transducer. This conversion is based on information from the electronic data sheet contained in the transducer. The application transformation is converted between the SI units representation and the application representation which appears at the network. In the absence of any provided application, this will default to an SI unit representation at the network. This application transformation implements the portions of the application specific to the node. The transformation can be quite general. In addition to operations on the data, such as change of units, averaging, filters, and limits, the transformation can generate messages on the network.

In addition to the transformations on the data, the control and configuration block implements the behavior models defined for the smart nodes. These models specify the behavior of the node with respect to such properties as sampling rate and data management.

Operational Aspects of Smart Nodes

The internal operation of the smart nodes may be divided into two phases: start-up and normal operation.

The start-up phase occurs after power-up or reset. During the start-up phase, the following sequence of events takes place within the node:

1. The transducer uploads the information contained in the electronic data sheet.
2. Based on this information, the node configures itself as a sensor or as an actuator. In addition, it configures the physical transformation as well as operating characteristics imposed by the transducer, for example, warm-up time and minimum sampling interval. Thus, it is possible to completely change the nature of a node by substituting a different transducer. For example, a temperature transducer could replace a pressure transducer or perhaps another temperature transducer with lower accuracy. These changes are reflected automatically in the transducer-related node behavior.
3. The contents of the application ROM upload. Based on this information, the node configures the application transformation.
4. The node monitors the network to detect the presence of other nodes. Based on the information received, the node configures the relevant properties, such as data management options.
5. At the end of the warm-up time, the node begins normal operation.

Typical functions occurring during the normal operation mode include:

1. For actuators, the node receives data via a multicast or point-to-point link. This data then moves along the data path illustrated in Fig. 27.3 and normally results in an update of the actuator at the appropriate sampling time.
2. For sensors, the node triggers the sensor at the appropriate sampling time. The data then moves along the data path. Depending on the application transformation, this results in data or messages being placed on the network. The application transformation can also result in more complex behavior such as reporting by exception.
3. For any node, the node monitors the network for messages that indicate a change in the operational behavior, such as multicast membership changes.

INTERFACE DEFINITIONS

The key to successful use of distributed technology lies in the interface and behavioral definitions. All members of a distributed system must adhere to the same set of definitions to permit graceful scaling, modification, and operation of the system.

In smart nodes, the two key interfaces are the transducer interface and the network interface.

Transducer Interface

This interface serves to define all aspects of the smart node that depend only on the transducer. There are four areas of concern: the physical form of the interface, identity specifications, operational specifications, and calibration specifications.

1. *The physical form of the interface* includes definitions for all signals, power levels, connectors, and so forth.
2. *The identity specification* must include a unique identifier that allows the smart node to determine whether the transducer has been changed. The prototype specification includes the manufacturer's identification and an ASCII textual description of the transducer.
3. *The transducer operational specifications of the prototype* include the minimum sampling interval, transducer warm-up time, acquisition time, units, data representation, precision, and accuracy.
4. *The calibration specification* includes the method and parameters to enable the physical transformation to convert between SI units and the transducer signal. Also, the transducer contains the specification of the expiration of the calibration. To allow for recalibration, these items are stored in EEPROM.

This interface is the subject of a current NIST/IEEE standardization effort.

Network Interface

This interface defines all aspects of the smart node that are visible to other nodes on the network. Data, connection, and behavioral models specify the network interface.

The use of common models allows the design of nodes that can operate on the data without further configuration, one of the objectives of this study.

The main feature of the data model is the network-level representation for measurement data. This representation includes value, units, time of measurement, and place of measurement. The data is normally communicated as a multicast message named *data* and is received by all nodes in the multicast group. Receiving nodes filter the incoming data based on their internal application information and on the fields of the data message. For a more detailed discussion see the appendix to this chapter.

EXPERIENCE USING THE PROTOTYPE SYSTEM

The prototypes have been used in a number of applications which are described in the following sections.

Printed Circuit Board Manufacturing

A system much like the one illustrated in Fig. 27.1 was used to monitor conditions in the wet-chemical-processing portion of a printed circuit board manufacturing operation within Hewlett-Packard. The system includes smart nodes to measure the temperature and pH of the solutions in various tanks and the flow rates and line pressure in the automatic spray equipment.

The measurement server transfers measurement results to an RTAP installation for process monitoring.

The interesting feature of this application is that periodic monitoring of the flow rates is inappropriate since the spray process is intermittent. An application transformation was produced which allowed the node to make measurements at the desired rate during spray time, but only place the data on the network when flows above a certain threshold are observed.

Laboratory Ambient Condition Monitoring

A system similar to the printed circuit board system has been installed in several laboratory areas at Hewlett-Packard. This is a straightforward monitoring of atmospheric pressure, ambient temperature, humidity, and concentration of carbon dioxide. RTAP is used for archival recording of the data for later correlation with defect data associated with various operations being conducted in these laboratories.

Miscellaneous Systems

A variety of systems have been built to demonstrate various aspects of distributed measurement such as:

1. Controlling a closed loop system.
2. Adding nodes to running systems.
3. Changing system application behavior based on events observed in one or more nodes and communicated using a multicast operation.
4. Partitioning of systems into multiple multicast groups. All of these systems used the same nodes. Only the application transformations were modified.

Observations

The following observations are based on our experience with the limited number and types of nodes in our current systems.

1. The maximum measured sampling rate possible for the current nodes is 25 samples per second. This figure can be expected to improve with a node processor faster than the Motorola MC 143150 and Toshiba TMPN3150 chips used to implement the nodes.
2. The measured deviation of the sample period for a single node making periodic measurements at a rate below the maximum is on the order of 5 milliseconds.
3. The timing offset accuracy between nodes is on the order of 6 milliseconds based on comparing the time stamps of observations of periodic measurements started at the same time.
4. System application writing and debugging is generally easier than that for corresponding, centrally controlled systems. This appears to be the result of the following:
    - Partitioning the application into node-specific operations which are implemented at the node using synchronization messages between the nodes, using

common data models which promote interchangeability and interoperability by eliminating excessive data manipulation, using multicast communication patterns which provide more flexibility and allow for graceful addition of nodes.
- Modifying the system usually requires modifying the application specification only at the few directly involved nodes.

5. System application writing and debugging would be easier if explicit support for application events had been provided
6. Identifying, partitioning, and modelling transducer, measurement, and system- or application-related issues allows node design to explicitly support these issues, greatly reducing the difficulty in constructing applications.
7. Adopting SI units for all internal operations simplifies the design of the nodes. Application development is also simplified since only those nodes which must present data in human readable form need concern themselves with the variety of unit representations for the same measure. Additional research is needed to design a robust specification technique for compound and dimensionless units.[6]

TOPICS FOR FUTURE RESEARCH

As a result of our investigation, several topics were noted in which additional research is needed to allow more effective design and use of distributed system applications. These include the following:

1. Determining the set of useful models and support structures for implementing the application level of distributed systems. This will result in a more formal definition of an interface between the application and measurement portions of the node architecture.
2. Designing a robust specification technique for compound and dimensionless units and variable naming.
3. Designing an effective method for implementing unique identifications for transducers, nodes, and so forth.
4. Designing the environments for application-development simulation and debugging of distributed measurement systems.
5. Specifying a network interface, as defined in this chapter, that is independent of the actual network transport and physical protocols used.
6. Ensuring low-cost node design.

CONCLUSIONS

This study has shown that the use of distributed system technology is useful and important in measurement monitoring and control systems. With further development, these techniques will enable straightforward construction of applications which are easily configured, expanded, modified, and maintained.

APPENDIX: DETAILED DESCRIPTION OF SYSTEM MODELS

This appendix provides more detail on the top level definitions of the various models and interfaces of the smart nodes. There is considerably more detail to these models than is possible to present here. However, the level of definition given should illustrate how these models support the design objectives of the prototype system.

In the following sections, *italics* indicate that the item is an explanation intended to convey the intent of the underlying detail of the item, that is, it is not a terminal symbol for the definition. A complete definition would extend to the bit structure of the representation.

Network Interface

Network interface:

  network data
  network time
  application script
  node control
  network connectivity

Network data:

  node id
  sequence number
  time stamp
  variable parameters
  value

Node id: *A unique identifier referencing the source of the record.*

Sequence number: *The ordinal number of the set of network packets conveying this item of network data. This may be a compound sequence number if network packet fragmentation is needed.*

Time stamp: *A representation of the local node time. When related to data, it indicates the time at which the data became or is to become valid.*

Variable parameters:

  variable type
  variable name
  variable units
  data model

Variable type: *A type indicating the nature of the data, for example, normal physical data, normal physical data from a node whose calibration is overdue, application level data, and so on.*

Variable name: *A key to either a standard enumerated variable name or an expression defining a user-created name.*

Variable units: *A key to either a standard enumerated SI unit or an expression defining an SI compound or user-created unit.*

Data model: *A type indicating the data type used to represent the data value, such as IEEE-float, 3 IEEE-float vector, and so forth.*

Value: *The value of the data.*

Network time:

node id

time stamp

Application script:

application id
application model
application-variable map
application specification

Application id: *A system-unique identifier referencing this application.*

Application model: *The model used to represent the application. For the prototype: series, categorization, subroutine.*

Application-variable map: *Defines the names, types, units, and relationships between the variables at the input and output of the application transformation defined by the specification.*

Application specification: *As appropriate to the application model, defines either the relevant parameters or an executable subroutine.*

Node control: *A definition of commands which may be used to change node state or to set parameters of various behavioral models, such as reset, stop sampling, set sample rate.*

Network connectivity: *A set of commands and data used to establish or modify the network connectivity.*

Behavioral Models

The selection and definition of the node behavioral models, along with the data models, visible either from the transducer or from the network, determine the interchangeability and interoperability properties of the smart nodes in a distributed system. The following are two of the models used in the prototype to illustrate this point.

Operational State model. The network state modifications possible from the network via node control are as follows (see Fig. 27.4):

reset (or power-up)

d: download a new application

f: calibrate the node

e: accept new calibration

f: use previous calibration

i: reset

All other transitions are governed internally.

Operational state model:

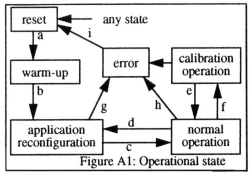

Figure A1: Operational state

The network state modifications possible from the network via "node-control" are:
reset (or power up)
d: download a new application
f: calibrate the node
e: accept new calibration
f: use previous calibration
i: reset

FIGURE 27.4 Operational state.

Sampling state model. The network state modifications possible from the network via node control are as follows (see Fig. 27.5):

a: reset (or power-up)
b: start-polled mode
c: start-periodic mode
d: start measurement
f: start measurement
h: arm measurement
i: start-periodic mode

All other transitions are governed internally.

Transducer Interface

Transducer interface:

   operational parameters
   transducer description
   conversion specification
   calibration specification

Sampling state model:

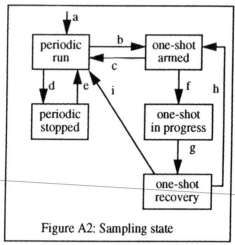

Figure A2: Sampling state

The network state modifications possible from the network via "node-control" are:
  a: reset (or power up)
  b: start-polled-mode
  c: start-periodic-mode
  d: start-measurement
  f: start-measurement
  h: arm-measurement
  i: start-periodic-mode

FIGURE 27.5 Sampling state.

Operational parameters:

  transducer type
  transducer id
  transducer-variable name
  transducer units
  range model
  warm-up time
  acquisition time
  minimum-sampling period
  physics representation
  transducer representation
  precision
  accuracy

Transducer type: *Indicates whether the transducer is a sensor or an actuator and the form of the physical interface. In the prototype, the forms included voltage signal, frequency signal, and a byte stream.*

Transducer id: *A unique identifier for the transducer.*

Transducer-variable name: *A key to either a standard enumerated variable name or an expression defining a user-created name.*

Transducer units: *A key to either a standard enumerated SI unit or an expression defining an SI compound or user-created unit.*

Range-model: *Specifies the representation and values of the range of the physical variable accessible to this transducer.*

Warm-up time: *The time required after power is applied before the transducer is operational.*

Acquisition time: *The time required after the trigger signal applied to the transducer before the data is valid.*

Minimum-sampling period: *The minimum time required between successive samples.*

Physics representation: *The data representation used on the network side of the physical transformation (see Fig. 27.3).*

Transducer representation: *The data representation used on the transducer side of the physical transformation.*

Precision: *The precision of the calibrated data.*

Accuracy: *The accuracy of the calibrated data.*

Transducer description: *An ASCII description of the transducer.*

Conversion specification:

conversion model

default-conversion values

calibrated-conversion values

Conversion model: *Specifies the type of the conversion model used. In the prototype, a series, a categorization, and subroutine models are supported.*

Default-conversion values: *Default value of the conversion values appropriate for the selected model.*

Calibrated-conversion values: *Conversion values appropriate for the selected model.*

Calibration specification:

calibration type

calibration description

calibration script

Calibration type: *Specifies the method used to calibrate the transducer. The prototype supported no calibration possible, self-calibration, and an externally managed N-point calibration.*

Calibration description: *A managed readable set of instructions on the calibration procedure.*

Calibration script: *An executable script defining the externally managed N-point calibration.*

REFERENCES

1. James Pinto, ISA paper no. 94-598, "Networked, Programmable, Intelligent I/O, the 'Truly' Distributed Control Revolution," *Proceedings of the Industrial Computing Conference,* October 23, 1994, **4**(2), pp. 141–147.

2. Gary Tapperson, ISA paper no. 94-569, "Fieldbus: Migrating Control to Field Devices," *Advances in Instrumentation and Control,* October 23, 1994, **49**(3), pp. 1239–1252.
3. James Pinto, ISA paper no. 94-514, "Fieldbus—A Neutral Instrumentation Vendor's Perspective," *Advances in Instrumentation and Control,* October 23, 1994, **49**(2) pp. 669–675.
4. Stan Woods and Keith Moore, "An Analysis of HP-IB Performance in Real Test Systems," Hewlett-Packard Laboratories Technical Report 93-20, August 1992.
5. J. H. Saltzer, D. P. Reed, and D. D. Clark, "End-to-End Arguments in System Design," *ACM Transactions on Computer Systems,* November 1984, **2**(4), pp. 277–288.
6. Stephanie Leichner, William Kent, and Bruce Hamilton, "Dimensional Data," Hewlett Packard Laboratories, private communication, February 1995.

CHAPTER 28
SENSORS AND TRANSMITTERS POWERED BY FIBER OPTICS

INTRODUCTION

A smart fiber-optic interface uses microprocessor control to regulate light energy transfer down optical fibers to power industrial sensors and instrumentation. A power module uses a laser diode to supply light energy to a remote location where it is reconverted to electrical energy to power industry standard sensors and transmitters. Transmitters powered by fiber optics offer the ultimate in isolation, signal integrity, and safety without the need for local power, batteries, and wiring. The advantages of light-powered systems is discussed along with practical applications and system configurations.

The advantages of optical fibers are well known in the communications industry. Fiber-optic systems are also becoming important for process control, often replacing traditional wiring in the plant and in difficult process environments. Information can be transmitted over long distances with unprecedented data security, immunity to EMI and RFI, and reliable lightning protection. Fiber-optic installations provide opportunities for significant cost savings by eliminating the need to install new conduit through new or different areas of the plant.

Smart sensors and microprocessor control located within or near sensors have also resulted in many improvements for the user. New smart sensors and transmitters use microprocessors for reranging, temperature compensation, and greatly improved accuracy. They use communication protocols such as HART that allow both analog and digital signals to be transmitted on a single pair of wires. These are new developments that are finding widespread acceptance around the world. Nearly every process plant now uses some form of smart measurement technology. Fieldbus and other digital standards will almost certainly guarantee the increased use of smart sensor technology.

Fiber-optic power technology makes it both possible and practical to combine the advantages of fiber optics and the latest in sensors and measurement technology. This new technology uses light energy transmitted through optical fiber to be able to eliminate all wiring while at the same time providing electrical power for smart sensors, transmitters, and microprocessor-based systems. The advantages of fiber optics and fiber-optic sensing are retained to provide a unique and more practical solution. Complete electrical isolation is realized without batteries or wires—only nonconductive optical fiber is used. All the functionality of today's best systems is retained with the ability to adapt the latest electronic advances and communication protocols as they become available.

FIBER-OPTIC POWER INTERFACE

Fiber-optic-powered systems (sometimes known as *light-powered, optically powered,* or *power-by-light* systems) have been used to provide electrically isolated power for electronics, sensors, and other remote devices. In the last few years, this new technology has come into its own with the advent of lower-cost, high-output light sources (laser diodes and LEDs), improved optical fiber and connectors, high-efficiency solar cell technology (GaAs), and low-power CMOS electronics used with smart sensors and transmitters.

A typical fiber-optic power interface system (Fig. 28.1) consists of a light source or power module and a remote transmitter module connected together with fiber-optic cable. This kind of system allows for the connection of process transmitters and sensors with optical fibers without special modification. This is done without needing local electrical power, batteries, or wire.

The power module uses an infrared laser diode which supplies light energy through the optical fiber to the transmitter module. The transmitter module accepts the light energy and converts it to electrical power to power conventional transmitters or sensors. The output signal from the transmitter or sensor is transmitted back in optical form through the fiber to the power module. The output is reconverted to a conventional output, often a voltage, two-wire 4- to 20-mA current, or a smart output.

Both the power module and the transmitter module have microprocessors built in (Fig. 28.2).

The light source is controlled with a microprocessor to provide eye safety and maximum laser life by operating at the lowest possible optical output power required by the sensor. For eye safety, continuous checks are made to insure that the remote module is responding with indications that the operating conditions are normal. An inadvertently disconnected fiber will be instantly detected and the light power will be turned off. An auto start-up routine assures full operation once the fibers have been reconnected.

The microprocessor in the transmitter module measures and digitizes the photodiode output voltage, the supply voltage to the transmitter, and the output from the

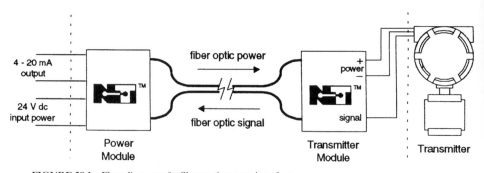

FIGURE 28.1 Flow diagram of a fiber-optic power interface.

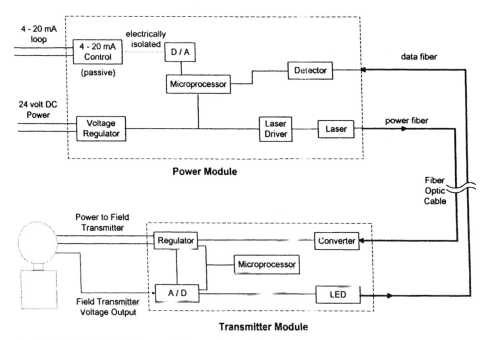

FIGURE 28.2 Block diagram of fiber optic power interface.

transmitter. It also communicates status information on the transmitter module, optimizes the power conversion, and controls the output power to the transmitter. System status is indicted by fault LEDs and out-of-range readings at the power module. Commercially available pressure, level, flow, temperature, and pH transmitters and sensors from leading manufacturers may be connected using interface systems of this type.

ADVANTAGES OF FIBER-OPTIC POWER

In principle, any device now connected with wire could be connected with optical fiber without loss of functionality. Mixed-mode connections (wire and fiber) would not be needed since all the power for active devices can be delivered using light-power technology.

Fiber can be run to nearly any location without regard to electrical isolation, EMI and noise, lightning, transients, or cross talk. Complete electrical isolation is realized without spark and shock hazards or ground loops to be concerned about. Fiber can be used in nearly any existing cable tray for long-distance signal transmission. The elimination of copper and metal conduit would also reduce field failures due to metal corrosion common in many industries. Optical fiber can also reduce weight or size of cable runs.

Perhaps the greatest advantage of fiber-optic power technology is the ability to use today's best electronic devices, sensors, or transmitters securely interfaced and electrically isolated without remote batteries, solar collectors, isolation transformers, pneumatic interfacing, or wiring. This is done using industry standard outputs and protocols with sensors and transmitters requiring local microprocessor compensation or LCD output displays.

Inherent lightning immunity is especially beneficial for long outdoor runs and environmental monitoring. Since optical fiber has relatively low loss, very long runs are possible. Just supply enough light power to overcome the losses (a few db per km depending on the fiber used). Standard fiber-optic cables and components are being field-installed using standard wiring methods. Many companies offer practical fiber and component systems with a growing number of field installations.

PRACTICAL CONSIDERATIONS OF FIBER-OPTIC POWER

Practical electrical power levels using light-powered systems range up to 100 milliwatts today. Much higher levels are possible but are more costly due to the high cost of the laser diode light source. Low-power smart sensors and the newest generation of smart transmitters now use CMOS and ASIC electronics. This improvement allows very reliable operation at practical power levels. Conventional three-wire low-power analog and smart transmitters are easily accommodated today using off-the-shelf fiber power interfaces (Fig. 28.3). Smart 4- to 20-mA transmitters can be interfaced by parking the transmitter current at 4 mA and sending the more accurate digital output over the fiber. The standard protocol is then presented at the power module interface.

The latest laser diodes are lower in cost and can operate at up to +60°C for years of reliable service. The cost and availability of fiber-optic power will continue to improve as large-volume users find new and expanded uses for laser diodes. More and more products will be supplied to the process industry with low-power CMOS, smart, and digital electronics, further reducing the amount of energy needed and thus the number of transmitters and sensors that can be accommodated. As vendors of field-mounted sensors, transmitters, and data acquisition systems integrate power modules directly into their products, further power and cost reductions will be made.

SYSTEM CONFIGURATIONS AND APPLICATIONS

Light-powered systems are being installed in plants for a diverse range of applications. For example, a major chemical company that produces chlorine uses very high and potentially dangerous voltage levels inherent with the process. This application presents a real measurement challenge and a significant safety hazard as well. The installed system used a light-power interface mounted well away from the process. A length of duplex outdoor grade optical fiber was installed between the interface and the smart-level transmitter. Standard fiber-optic connections were made within a weatherproof housing for outdoor use. The return output from the transmitter via the optical fiber is converted to a standard 4-20-mA and HART signal within the interface and run to the control system over the already-in-place twisted pair. Additional smart transmitters are desired for this application including pressure, level, and pH measurement.

An electric power utility installation is used in a high-EMI environment to monitor pressure with an industry-standard three-wire transmitter. This analog transmit-

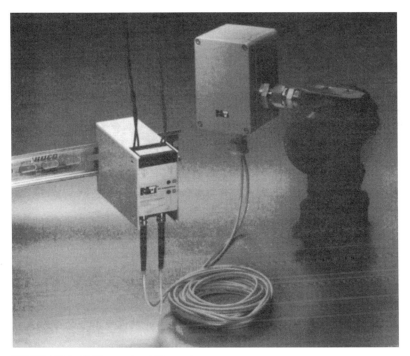

FIGURE 28.3 NT International fiber optic.

ter was supplied from a leading manufacturer without modification. In another application, a temperature transmitter is used to measure temperature in a plant-heating system application using a standard 100-ohm RTD sensor without loss of accuracy. A water and waste application uses an off-the-shelf pH sensor combined with a transmitter providing for local range adjustment. This is an outdoor, high-lightning area where problems in the past have created loss of measurement due to lightning strikes.

An oil refinery is monitoring pressure with a common analog pressure transmitter. This is mounted in a remote location for improved reliability. Another system is being evaluated for application within coal mines. A very long run of optical fiber (several miles) connected to active sensors for gas measurement is now being tested.

CONCLUSIONS

Fiber-optic power is being used today to supply completely isolated electrical energy to conventional electronic devices, sensors, and smart transmitters. These systems can provide unprecedented data security, immunity to EMI and RFI, and reliable lightning protection while eliminating the need for local power, batteries, and wiring. This new interconnect technology will allow for standardized digital communications or operation of sensors, data acquisition systems, or transmitters requiring local compensation or output displays.

Real-world applications of fiber-powered systems will result in increased confidence and industry acceptance of this new technology. Vendors will need to provide hardware that is simple to install and use, cost-effective, with a demonstrated track record of reliable operation. Vendors already producing smart sensors and transmitters will need to help integrate fiber-optic power conversion directly into their products providing more cost-effective systems with environmentally protected optical-fiber connections.

Fiber optics combined with fiber-optic power and the latest in smart sensor and transmitter technology can greatly improve overall system reliability and performance. Light-power technology makes fiber-optic connections more practical and easier to implement since no wiring is needed. Will optical fiber replace copper wiring in the process plant? It is just a matter of time.

REFERENCES

1. A. Ohte, K. Akiyama, I. Ohno, "Optically Powered Transducer with Optical-Fiber Data Link," *SPIE Vol. 478 Fiber Optic and Laser Sensors,* 1984, (11), pp. 33–38.
2. S. Henkel, "Single Optical Fiber Does It All for Smart Transmitters," *Sensors,* January 1992, p. 8.
3. C. Polsonetti, "Smart Transmitters Tear up the Market," *Intech,* July 1993. pp. 42–45.
4. G. R. Cucci, "Light-Powered Systems, a Fiberoptic Interconnect Technology for Process Control," *Advances in Instrumentation and Control,* ISA, 1993, **48**(1) pp. 659–677.
5. J. Bryzek, "A Communications Standard for Low-Cost Sensors," *Sensors,* September 1994, pp. 18–24.
6. P. Cleaveland, "I/O Developers Focus on Communications," *Instrumentation & Control Systems,* December 1994, pp. 25–34.

CHAPTER 29
A PROCESS FOR SELECTING A COMMERCIAL SENSOR ACTUATOR BUS AS AN INDUSTRY INTEROPERABLE STANDARD

INTRODUCTION

An in-depth study has been conducted with the purpose of recommending a sensor actuator bus (SAB) interoperability standard for the semiconductor industry. One major result of this study is a well-defined process for analyzing and selecting an SAB for an industry. Components of this process include a requirements analysis survey, selection criteria derivation, SAB candidate identification, structured candidate presentations/tutorials, follow-up analysis and refinement of selection criteria, comparative weighted analysis of candidate solutions with respect to refined selection criteria, and SAB solution recommendation. The selection criteria derivation process was iterative and flexible; selection criteria were first compiled through analysis of survey responses and updated prior to candidate presentations and again after candidate presentations. Candidate presentation review, follow-up analysis, selection criteria refinement, and comparative weighted analysis of candidate solutions was accomplished by multiple analysts, operating independently. Results from the analysts were reviewed and a consensus on an SAB interoperable standard recommendation was sought. A review of the process has revealed a number of valuable lessons learned that could be applied to future evaluation studies of this type.

The increasing need for interoperability and interchangeability of components in next generation systems has led to a growing emphasis on the development and deployment of standards to specify elements of manufacturing operations. One of these efforts is focused on the realization of a semiconductor manufacturing industry sensor/actuator bus standard. A *sensor/actuator bus* (*SAB*),* or simply *sensor bus*, is defined as "An interface specification that defines a common set of behavior and digital communications protocol to which smart (and dumb) sensors (and actuators) must conform if they are to remain (and function) within a well defined distributed control environment."[1] The SAB standardization effort then addresses the

---

* An in-depth discussion of SAB use and SAB standardization with respect to the advantages afforded over the current environment is presented in Ref. 29.

need for a standard framework to define the intelligent interconnection of sensors, actuators, and their control systems in semiconductor manufacturing tools.

Two organizations within the industry have devoted significant resources to the SAB standardization effort. Semiconductor Equipment and Materials International (SEMI) has been pursuing the development of a standard for one type of sensing system, namely the mass flow controller, with the long-term goal being to expand the effort into other types of sensing systems. Semiconductor Manufacturing Technology (SEMATECH) has undertaken an effort whose primary goal is to provide sensor/actuator bus standard recommendations for semiconductor manufacturing to SEMI. Secondary goals of this effort include education of the industry as to the alternatives, fostering competition between these alternatives, and providing a canonical device description for the industry, that is, providing a taxonomy and process for the definition and characterization of sensors and actuators in SAB systems.[2,3] SEMATECH pursued its goals in coordination with SEMI by developing and implementing a process (or methodology) for selecting an SAB specification best suited as a standard for the industry. The major elements of this process include SAB requirements analysis, requirements derivation, candidate solution evaluation, and final solution recommendation. It is important to note that this process evolved during the SAB specification identification effort and represents a combination of an initial process plan along with modifications resulting from lessons learned; indeed the process itself is considered to be an important result as it may be utilized generically by other industries in search of an SAB standard. It is also important to note that the actual results of the SEMATECH effort, namely the selection of a "best" SAB candidate for the industry, is beyond the scope of this chapter; this information is detailed elsewhere in the literature.[4]

This work contains a detailed description of the SAB selection process and its application to the semiconductor manufacturing industry. Background material on other SAB evaluation efforts and methodologies and on various existing SAB solutions that might be considered during a selection process is presented in the next section. In the section entitled "The Process of Evaluation and Selection," an overview of the entire selection process is given. Key elements of this process, namely a requirements analysis survey, selection criteria, candidate review, and SAB standard recommendation are described both generically and as they relate to the semiconductor manufacturing industry, in the following sections, respectively. This chapter concludes with a summary of the generic process and a discussion of lessons learned with respect to the semiconductor manufacturing SAB evaluation exercise.

BACKGROUND AND RELATED WORK

Sensor/Actuator Bus Evaluation Efforts

Sensor/actuator buses are increasingly utilized in a variety of industries including controls, automotive, building automation, and the like. Thus, issues such as SAB interoperability and standardization are prominent in many arenas. Indeed, there is a growing literature base devoted to the interoperability of SAB systems, notably Refs. 4, 5, 6, and 7. Specifically, there is an emerging literature base devoted to interoperability of SAB systems in the semiconductor manufacturing industry.[2-4,29] However, it is important to note that the degree of SAB technology utilization and standardization for many industries is ahead of the semiconductor manufacturing industry. Representatives in some of these industries have also conducted SAB surveys and evaluations; the requirements and the selection process vary from industry to industry; however, the overall goals are similar. As an example, Wright-Patterson Air Force Base conducted an analysis of the state-of-the-art of SAB hardware and candidate protocol standards in order to enable an Intelligent Distributed Measurement System (IDMS).[5] The Canadian Automated Building Association (CABA) conducted a similar evaluation of an SAB solution for the building automation industry.[6] The evaluation included an identification of guidelines for candidate evaluation, determination of market SAB requirements, development of a communications model, and a candidate protocol analysis and selection process. Recently, an effort in SAB analysis for the control industry has been undertaken by the National Center for Manufacturing Sciences (NCMS, a U.S. consortium). This effort includes candidate analysis and evaluation, but extends the task to support elements of prototype development. Results of this effort may have significant impact on the semiconductor manufacturing SAB selection process because the utilization of an SAB in this industry is closely aligned with the control industry, and SEMATECH and NCMS have had significant interactions on SAB selection. Thus, any NCMS decisions could affect the technology leverage of the candidate solutions being analyzed by SEMATECH.

Sensor/Actuator Bus Candidates

A critical part of the SAB selection process is the identification and investigation of existing protocols that are candidate solutions for the industry. As part of the semiconductor industry SAB selection process, a number of candidate solutions were identified. The protocols vary in many aspects, with each having numerous qualities that make it worthy of consideration. The following is a brief introduction to six candidates that were evaluated as part of the SEMATECH effort. A more in-depth analysis of each candidate and a comparative evaluation is presented in Refs. 4, 7, 9, and 10. Note that this evaluation does not exclude the possible suitability of a candidate not reviewed.

WorldFIP Fieldbus. This protocol specification has been developed (as an open and complete solution successor to FIP) by the WorldFIP organization, representing a large consortia of companies (primarily European) covering a wide variety of industries. The protocol deserves consideration because it is a mature standardization effort, and it is receiving support in the control industry.[11]

CAN. The Control Area Network (CAN) protocol was developed by the Bosch Corporation in the early 1980s for automotive in-vehicle networking. Since that time, CAN has become a standard for high-speed passenger-vehicle applications and is supported by a large consortia of companies from a variety of industries. The protocol deserves consideration because it is a mature standardization effort, and it is receiving support in many industries including the control industry.[12-15] However, CAN does not represent a complete sensor/actuator bus solution as it only specifies a media access method.[12, 16] There are a number of companion standards that have been developed specifying a (more) complete solution utilizing CAN as the media access method. Three such standard specifications were evaluated: J-1939, DeviceNet, and Smart Distributed System (SDS). J-1939 was developed by the Society of Automotive Engineers (SAE) for the automotive industry.[13] DeviceNet was developed by Allen Bradley for the control industry.[14] Similarly, SDS was developed by Honeywell for the control industry.[15]

LONTalk. This protocol was developed by the Echelon Corporation in the late 1980s and is utilized in a number of industries, most notably the building automation industry. The protocol deserves consideration because it is a complete product, is receiving support in the control industry, and has a small installed base in the semiconductor industry.[17]

SERIPLEX. This protocol was developed by Automated Process Control (APC) in the mid-1980s to provide a fast and simple reduced wiring solution for sensing systems. The protocol deserves consideration because it is a complete product (for low-level systems) and has a small installed base in the semiconductor industry.[18]

ISP Fieldbus. Interoperable Systems Project (ISP) SP-50 Fieldbus was developed starting in late 1992 (as an open and complete solution successor to Profibus) by the Interoperable Systems Project Foundation representing a consortium of companies primarily in automation and controls. The protocol deserves consideration because it is a mature standardization effort, and it is receiving support in the control industry.[19]

BITBUS. BITBUS was introduced by Intel in 1984 and is based on its 8044 chip (which contains the entire protocol). In 1990, an extension of the original protocol, BITBUS 11 (specifying a larger message length and more high-level services), was approved as an IEEE standard (1118-1990). The protocol deserves consideration because it is a mature standardized product, is receiving support in the control industry, and has a small installed base in the semiconductor industry.[20]

THE PROCESS OF EVALUATION AND SELECTION

The need for utilization of SAB technology and acceptance of an interoperability standard for sensing systems is clearly indicated.[21] Both SEMI and SEMATECH recognized this industry need and have subsequently pursued the identification and adoption of an interoperability standard for sensors/actuators/controllers for the industry.[2,3] SEMATECH (in cooperation with SEMI) addressed the issue by identifying a task within its Automated Equipment Control (AEC) project; the goal of this task was to recommend an SAB interoperability standard to SEMI. A Project Technical Advisory Board (PTAB) was formed and a kick-off meeting was organized that brought together experts in interoperability standard development, SAB implementation, and SAB use for open discussions.[21] One major conclusion resulting from that meeting was that, in order to effectively pursue an interoperability standard, a *process* had to be derived to govern the interoperability standard recommendation process. This process is depicted in Fig. 29.1. In deriving the process it was first realized that a number of key issues had to be resolved before pursuing standards development. The first issue involved obtaining a consensus on industry requirements for an interoperability standard. This issue was successfully addressed through a sensor/actuator bus requirements analysis survey. Utilizing the results of this survey, an initial set of requirements was derived for an interoperability standard.[1] One of the key requirements derived is that any proposed interoperability standard should be based on an existing productized SAB specification or suite of specifications (i.e., for many reasons including overhead of R&D and loss of cost and technology leverage, the standard should not be developed from scratch by the semiconductor industry). Thus, the indicated course of action was to analyze existing SAB solutions with respect to the remaining requirements derived from the survey and arrive at the best solution.

SEMATECH pursued this course of action (analysis and evaluation) by organizing a set of SAB presentations. Each presentation was devoted to a single SAB solution candidate. The one-day presentation included a general description of the SAB protocol, a detailed description of the protocol with respect to the set of require-

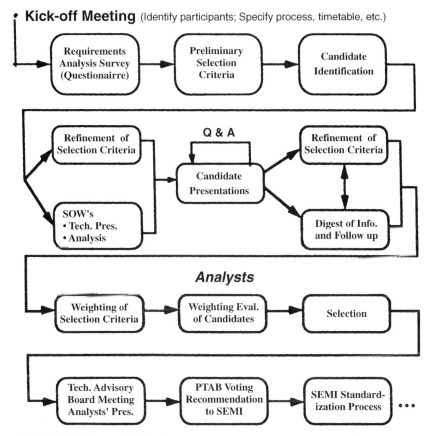

FIGURE 29.1 Schematic of selection process.

ments, and a discussion and demonstration of the implementation of the protocol in various systems utilized in the semiconductor manufacturing industry. As a preparation aid, presenters were given a presentation outline so that they would necessarily gear their presentations toward issues identified in the requirements specification. Representatives of the following candidate solutions were invited: WorldFIP Fieldbus, CAN-J19392,3, LONTalk, SERIPLEX, ISP Fieldbus, and BITBUS. An introduction to each candidate is provided in the previous section; a summary of information derived from the candidate presentations and subsequent follow-up is provided in Refs. 4, 7, 9, 10. Additionally, two analysts were commissioned to aid in the selection process.[22] Specifically, their duties included moderating the presentations, analyzing the information presented with respect to the requirements specification derived, conducting necessary follow-up, deriving a (quantitative) weighted selection criteria, and providing a comprehensive weighted comparative evaluation of the candidate solutions to the PTAB. A third analyst was commissioned to investigate SAB cost-of-ownership issues including development and implementation costs associated with the candidate solutions.[23]

Following the candidate presentations, the two analysts used the presentation material along with the survey results and initial requirements specification to derive a more quantitative weighted selection criteria. Each candidate was then mapped into this set of criteria. Follow-up conversations were conducted with candidate representatives as necessary. A report was compiled by each analyst for submission to the PTAB that contained the following elements:[22]

- A brief discussion of the benefits of an interoperability standard (i.e., justification)
- A brief review of the evaluation process
- A description of each of the candidate solutions
- A description of the weighted selection criteria derived, including justification for weights

Initially, CAN was perceived as a candidate solution. However, after discussions with CAN representatives, it became clear that CAN was technically a specification for just the media access (layer 2 of the ISO-OSI seven-layer model)[1,6] portion of a sensor bus solution. Complete candidate solutions were subsequently identified that utilized CAN. The standard solution that was presented was J1939.

Two additional CAN solutions, DeviceNet CAN and Smart Distributed Systems (SDS) CAN, were identified during the initial round of (six) presentations and subsequent discussions. A presentation was given at a later date on each of these two candidates.

- A quantitative mapping of the candidate solutions into the selection criteria
- A comparative evaluation of the candidate solution
- A proposed interoperability standard solution

Note that under guidelines imposed by the project leader, the two analysts did not share information on derivation of selection criteria or candidate evaluation while compiling their reports. This step was taken to ensure the validity of each report as an independent evaluation.

As the final step in the process, a meeting of the PTAB was held in which the analysts submitted their reports and gave in-depth presentations of their results. The PTAB then attempted to reach a consensus on an interoperability specification to submit to SEMI as a recommended standard. Also, a few follow-up items (e.g., product searches and select data verification) as well as future directions (e.g., test-bed development and a canonical device description for sensors and actuators) were identified. The entire process (from kick-off meeting to PTAB consensus meeting) spanned approximately one year; the analysts performed their duties during the final three months of the process.

Although the survey process was very iterative and involved, it contained four basic critical elements, namely conducting sensor/actuator bus surveys, derivation and description of selection criteria, presentation and review of candidate solutions, revision of selection criteria, and selection of a best solution through comparative candidate evaluation. These key elements are described in detail in the following sections.

SENSOR/ACTUATOR BUS SURVEY

The initial step of the process following the kick-off meeting was the development and conducting of a sensor/actuator bus requirements analysis survey. The objective of the survey was stated as follows:

To gather sensor/actuator communications requirements from the semiconductor manufacturing community including: (SEMATECH) member companies, tool and equipment suppliers, sensor and actuator suppliers, control system suppliers and others, and present these findings in a document format that will enable independent third parties to perform the following activities:

- Analyze information gathered during the requirements analysis and propose a digital communication standard for smart sensors and actuators for the international semiconductor manufacturing industry; and develop generic and specific sensor and actuator application models for adoption by the international semiconductor manufacturing industry.

Survey questions were iteratively derived from the PTAB members and their interactions with sensor/actuator systems suppliers and users. The survey was conducted by Harbor Research Corporation; a total of 83 contacts were interviewed. Contacts included sensor/actuator suppliers, tool suppliers, subsystem suppliers, and university researchers.

The 25-page survey document contained questions on a variety of topics including company and interviewee profile, interest in SABs, clarification of definitions, expected future trends, performance requirements, architecture, reliability, cost-of-ownership needs and expectations, familiarity with and use of various existing candidate solutions, and so forth. Topics were addressed both separately and in a comparative mode, where appropriate, to derive a prioritization of issues.

A detailed discussion of the results of this survey are beyond the scope of this chapter.[1] However, the following is a sample of survey results:

- 80 percent of sensor/actuator response time requirements were between 1 and 100 ms.
- A 25 percent increase in cost per device is acceptable.
- Tool control systems typically can have the following ranges of control points: 30–650 digital, 20–300 analog, 5–100 presence, and 0–75 intelligent.

SELECTION CRITERIA

An analysis of survey results along with further investigation of the PTAB into potential solutions revealed that there were a number of existing SAB solutions that mapped well to the industry requirements identified. Representatives of each of these candidate solutions were invited to give a one-day technical presentation to the PTAB members.

Utilizing the information given by the presenters, the results of the survey, and follow-up investigation, each of two analysts commissioned to aid in the selection process (see the section entitled "The Process of Evaluation and Selection") derived weighted (i.e., prioritized) selection criteria that could be used to evaluate the suitability of a candidate interoperability standard specification to the semiconductor manufacturing industry. The criteria and associated weights derived by each analyst are summarized in Table 29.1. Note that, of course, the criteria overlap to a large extent. In the following paragraphs, the union of the selection criteria derived by the analysts is described in more detail in general order of decreasing importance to the industry.

Performance/Speed/Determinism. Results of the SAB analysis survey indicate that respondents view *I/O response time* as one of the most important communica-

TABLE 29.1 Weighted Selection Criteria Derived by Each Analyst

Analyst no. 1	
Selection criteria	Weight assignment
Performance (Asynchronous, synchronous and loaded network)	10
Leverage (cost and technology)	8
Product availability	7
Cost (development and implementation)	7
Peer-to-peer communication capability	7
Memory requirements	5
User-friendliness	5
Ownership	5
Installed base in semiconductor manufacturing	4
Board size	3
Message-passing capability	2

Analyst no. 2	
Selection criteria	Weight assignment
Interoperability	10
Availability	9
Outlook	9
Reliability	9
Speed	8
Flexibility	8
Cost per node	7
Proven	7
Determinism	6
Size per node	5
Tools	4
Remaining work	4
Cost to develop	2

tion performance requirements.[1] If an SAB does not meet response time requirements, it cannot be utilized, irregardless of its other qualifications. Although response times down to 10 microseconds were indicated as necessary by a few respondents, most indicated requirements in the 1 to 100-ms range.

There are many types of response times that are important, the most important of which is the ability to quickly report an alarm or operate in a time-critical environment (such as drive by wire). This is referred to as *asynchronous response time*. An important issue with this type of response time is *determinism* of the protocol, or the ability to support predictable and stable transmission of high-speed control parameters. This capability is critical to the design of stable closed loop for current control problems and especially for next generation control systems. A second important type of response time identified is the ability to poll a point quickly and deterministically. This is referred to as *deterministic polling response time*. A final important form of response time is the capacity of a network and its performance under heavily loaded situations. This includes the ability to minimize latency of low-priority messaging under these heavy loads. This is referred to as *loaded network performance*.

Interoperability. Without a solid foundation to provide for the *interoperability**
and *interchangeability*† of sensors, actuators, and controllers, the specification will be
of little use to the industry as a whole. (Note that both interoperability and interchangeability are critical in next generation smart-sensing systems.) Issues to be considered include definitions of data structures and behavior rules, completeness of the
protocol and conformance to the OSI model, support from a variety of companies
and interests, and a capability to provide specification conformance certification.

Cost and technology leverage. SABs are being utilized in a number of industries
from housing to automotive to controls. Interoperability standardization efforts are
also being pursued in these areas; indeed many of these standardization efforts are
ahead of the semiconductor industry effort. It is important to note that many of
these industries have a much greater SAB technology market potential than the
semiconductor manufacturing industry. In fact, the semiconductor manufacturing
market is probably not large enough to act as a technology and cost-driving force for
SABs. It is important then to realize that any choice of an interoperability should
leverage as much as possible off of other industries. Further, any choice should not
only consider current leverage, but future long-term leverage.

There are two major types of leverage that must be considered, namely cost and
technology. *Cost leverage* is basically a function of current and projected future component demand. It is important to note that demand among the components
required for vertical integration of a technology (i.e., transceiver, protocol chip,
application development software, diagnostic tools, etc.) could vary greatly depending on an application. As an example, an SAB utilized in a production vehicle results
in high chip and transceiver demand, but low application development software
demand. Thus, demand at all levels of the technology must be considered.

The second type of leverage, *technology leverage,* results from high-volume applications requiring more performance from an SAB, thus providing motivation for technology improvements in that bus. It includes leveraging of technology that utilizes the
bus, such as smart sensors, control systems, and so forth. It is important to note that
high demand and cost leverage do not necessarily imply technology leverage in all
areas. As an example, the building automation industry is unlikely to be a force driving the speed performance of a protocol. Thus the types of technology leverage
expected of an SAB must be compared to the specific requirements of this industry.

Product availability. Before an SAB standard is to be accepted and implemented
across an industry, it should be determined that there is a sufficient infrastructure (or
clear potential for an infrastructure) to support the product implementation and
maintenance. This support should span all levels of the protocol (from physical media
to application development environment) and be available from multiple sources at
all levels so that there is no reliance of the standard at any level on one company. Thus,
issues to be considered in assessing availability include status of the standard, number
of sources for components, software libraries and/or firmware, and so forth.

Development and implementation cost. One of the major benefits anticipated in
SAB utilization is the reduction in cost of ownership of systems in which they are
utilized.[24] However, it is important that the cost of development and implementation
of an SAB not outweigh this cost-of-ownership improvement. There are a number of

---

* *Interoperability* is the ability of multiple systems of devices to exchange information and mutually use the information that has been exchanged between systems and devices.[1]

† *Interchangeability* is defined as the capability to replace component parts or devices of different manufacturers and different product families in a computing or control system while maintaining the full range (or partial range) of optimal system functionality.[1]

costs that must be considered. The most important, of course, is the typical SAB node cost, including the cost of a dumb slave node. The second is the development platform costs. Finally, miscellaneous costs such as training, conformance testing (if available), yearly maintenance, and user membership fees (if necessary for implementation development) should be considered. Note that all cost factors must be considered in the context of current and future leverage off of volume production, development, and implementation.

Outlook. An important consideration in any SAB selection is the future prospect of the candidate. The evaluation of candidates using this criterion is somewhat subjective and is based on future expectations of the quality of the candidate with respect to the other selection criteria.

Reliability. Reliability is a key issue that directly affects the safety, time, and cost of manufacturing. Issues to be considered when assessing reliability include fault tolerance and recovery, noise immunity, redundancy, error detection, data flow control, and message acknowledgment.

Peer-to-peer capability. This capability is needed to provide for the *flexibility* in communications that generally cannot be provided by traditional master/slave systems, but that is required for distributed intelligence, distributed programming, and distributed operation in an SAB system. A distributed environment is ideal for a number of reasons such as: it allows the realization of more complex and robust controllers; it reduces host controller software overhead, thereby providing faster, lower-cost, and more versatile solutions; and it reduces the problem of single point of failure (host). Smart sensors, which will be increasingly utilized in future (distributed) applications, generally require the capability to asynchronously generate events to the control system in which they operate. Further, many sensors/actuators that are defined in the SAB survey had a level of intelligence that is suited to peer-to-peer communications. Thus, because the SAB interoperability standard is a long-term solution that should not restrict the utilization of current or future sensor technology, the underlying SAB protocol should be able to support peer-to-peer communications.

Memory requirements. Many of the devices expected to be communicating via SABs in the near future will have limited memory space, due in large part to physical size limitations placed on many transducers by an operating environment and cost limitations of many simple transducers (e.g., proximity switch). Many of these devices will operate in a slave mode and thus will require only a partial implementation of most protocols. It is important that the candidates have a capability to meet these requirements while still providing the services specified by the other criteria.

User-friendliness/tools. The ease of use of a product impacts greatly on its degree of acceptance. It also results in reduction of development and implementation costs, as well as time to market. Thus, the current and potential user-friendliness of a solution must be considered. Issues that impact user-friendliness include object-oriented (or function-block) specification at the application layer, availability of graphical user interfaces and point-and-click application developers, node simulators, other application development tools, training, product support, and alignment of product development with the control industry.

Ownership. As previously stated, the ultimate objective of the entire SAB analysis effort is to specify (to SEMI) an SAB standard for the industry. Thus, one issue of concern that must be addressed is the ownership and openness of the specification. Open, nonproprietary specification ballots can only be rejected with technically persuasive arguments. However, in general, proprietary solutions can only be balloted in a standard if they are clearly the best solution. Although this is admittedly a gray area, it nevertheless provides grounds for possible rejection of a ballot and thus should be considered in the evaluation process.

A second and equally important issue of concern related to ownership is the potential lack of protection against product pricing and loss of supply. Clearly, if a single-source proprietary solution is chosen and adhered to as an SAB standard, there would be no competing product available to guarantee (long-term) competitive prices as well as adequate product supply and support. Thus, any proprietary SAB solution should be seriously considered for standardization only if there are clear guarantees against unequatable pricing and inadequate supply. These guarantees could include a well-defined path toward openness (specification and manufacturing) at all levels of a specification.

Installed base in semiconductor manufacturing and a proven record. Many standards in existence today (Ethernet, VHS, etc.) have arisen out of de facto standards. Industry investment in an SAB solution (current implementation and planning) must of course be considered when choosing a standard. Currently, no candidate solution has achieved critical mass in the semiconductor industry so that it may be considered a de facto standard; however, there is a degree of industry investment in a few of the candidates.

Board size. Many of the SAB nodes will be small devices that have limited space available for bus communication hardware (e.g., proximity switches). Therefore, compact SAB node implementations must be achievable.

Message-passing capability. All of the candidate protocols have a capability to send and receive a message/variable over the network that has a length less than or equal to the protocol packet length. However, not all of the protocols have a capability to communicate longer messages as this requires breaking up or "packetizing" the message at the sending end and reconstructing the packets (in order) back into a message at the receiving end. Although SABs are not expected to be utilized for large amounts of message passing, *flexibility* should exist to support applications where this is needed, such as in a start-up dialog.

Issues that must be considered in evaluating the candidates with respect to this criterion include: (1) whether or not the service is provided by the protocol, and (2) the degree of blocking required (a higher degree of blocking results in lower performance).

Remaining Work. The amount of work remaining to complete an interoperability standard is important because the extra time and effort required would hinder the standardization process, possibly to the extent that the narrow window of opportunity of industry adoption of a standard specification would be lost to the de facto standardization process. It is also important to note that specifications resulting from any additional work may be unique to the semiconductor manufacturing industry and thus would have limited technology and cost leverage.

CANDIDATE PRESENTATION AND REVIEW

As stated in "The Process of Evaluation and Selection" section, representatives of each of these candidate solutions were invited to give a one-day technical presentation to the PTAB members. A critical element of this portion of the process is the statement of work (SOW) supplied to each candidate presentation team prior to the presentation.[25] This statement of work served to guide both the presenters and reviewers by identifying the playing field and audience for the presenters and providing a common forum for candidate analysis for the analysts. The following were key elements of the statement of work given to the presentation teams:

- A requirement that the presenter read the related interoperability paper and SAB requirements survey so as to better understand SAB requirements for the specific industry.
- A requirement that the presenter review the proposed list of selection criteria. Although this list is preliminary, it represented the views of the selection committee that would be brought to the presentations.
- A suggested presentation general outline. The outline is presented as a set of general category titles, each followed by a list of appropriate questions and directives. Category titles are "General Issues," "Business Issues," "Memory Requirement Issues," "Cost Issues," "Development and Product Deployment Issues," "Support Issues," "Protocol Layer Summary," and "Present Software/Hardware Examples." Typical questions in these categories include: "What mechanisms are available to guarantee a reliable, timely response?" and "Who provides conformance testing?"
- A requirement that the presenter address audience questions.
- A qualification plan that gives SEMATECH ownership and distribution rights of all presentation material, as well as videotapes of presentations.

Each presentation team described the protocol of its candidate in detail and identified the advantages and disadvantages of the candidate especially with respect to the (preliminary) selection criteria identified. (Most of the disadvantages were derived from thorough questioning of the presenters by the audience.) A brief summary of each of the candidates with appropriate references is presented in the section entitled "Background and Previous Work."

Following the candidate presentations, the two analysts, operating independently, analyzed the information presented along with documentation on each candidate, as well as information forwarded in previous SAB selection efforts. During the analysis, there was frequent contact with representatives of each candidate solution. The output of each candidate was an in-depth analysis document and an accompanying presentation that included the following key elements:[9, 10, 22]

- An introductory section that identifies the benefits for having an interoperability SAB standard, motivates the need for the technical SAB presentation, and summarizes the technologies that will be analyzed.
- A technology comparison table that maps the qualities of each SAB analyzed (columns) to the set of comparison criteria derived for the evaluation (rows).
- A section containing a detailed discussion of each SAB protocol analyzed. This information maps to the columns of the technology comparison table.
- A section containing a detailed description of the evaluation criteria used (see the section entitled "Selection Criteria"). This information maps to the rows of the technology comparison table. Note that the description of the evaluation criteria includes a weighting, depicting the relative importance of the criteria to the industry.
- A conclusion section that contains a numerical evaluation of each candidate with respect to the weighted criteria, a final numerical ranking that identifies the candidate(s) best suited as an SAB standard for the industry, and a general discussion of factors that had significant impact on the ranking.

A review of the conclusions of the analysts indicates that the results were very consistent, despite the fact that the conclusions were arrived at independently. Indeed the ranking of the top four (out of six) candidates were consistent not only in order, but in relative total weights. Additional conclusions forwarded by each analyst included the following points: (1) each candidate has advantages that make it well suited for some aspects of SAB operation and disadvantages that render it less optimally suited for other aspects; an important result of the review process is the identification of the trade-offs of each candidate and a determination which candidate's qualities best map to the requirements of a semiconductor manufacturing SAB; (2) the information contained in the output of the analysts is very temporal, thus, the reader is encouraged to contact candidate representatives to obtain more detailed and current information.

SAB INTEROPERABILITY STANDARD SELECTION

The final and perhaps most difficult step in the evaluation process is the comparative evaluation of the candidates as a candidate for an SAB interoperability standard. The conclusions of each of the two analysts were related to the PTAB via the summary report and accompanying presentation.[9, 10, 26] Both analysts definitively concluded the same best SAB solution among the candidates. After in-depth discussions, the majority of the PTAB members concurred with the recommendations of the analysts. The PTAB (in the role of representative of SEMATECH) subsequently forwarded these recommendations to SEMI. The impact on SEMI was significant; indeed the subsequent revision of SAB standard ballot documents complies with the PTAB recommendations.[27]

CONCLUSIONS

Lessons Learned

A review of the evaluation and selection process and supporting methodology reveals that there are many noteworthy elements that greatly facilitated and enhanced the quality of the evaluation process. For example, the elements in the process leading to the initial and iteratively revised set of (weighted) selection criteria are of paramount importance. One of the most importance of these elements is the initial survey and the subsequent digest of information. The survey provided the selection committee with a view of SAB requirements as seen by a complete cross section of parties involved in SAB development, (potential) implementation, integration, and use. The relative flexibility of the SAB selection criteria and interoperability requirements also proved advantageous; the review process was a learning process, thus the continuous updating of the requirements set allowed the additional knowledge to lead to a more complete and cohesive final set. Another important element identified in the process is the decision to utilize multiple analysts and the instruction that these analysts conduct their evaluations independently. This was important because a consensus of results between the two analysts helped to verify the quality of the evaluation process as well as the accuracy of the analysts, both to the PTAB as well as to representatives of the candidates. A third important element is the presentation outline used to help guide the candidate presentations toward better addressing issues related to the consideration of each candidate as an SAB interoperability standard. Similarly, guidelines for analyst report preparation were also crucial. A final important element is the timely obtainment of permission from each candidate to publish and distribute results. The avenues for dissemination of information of a review process should be thought out and guaranteed (both technically and legally) well ahead of the actual analysis.

This same retrospective view also reveals a number of issues that, if dealt with more completely and timely, would have helped to further facilitate the evaluation process. For instance, an issue that should have been emphasized was instructing the candidate representatives to comment exclusively on their candidate; too often negative comments that proved untrue were made by a candidate representative of a competitor. Removal of candidate representatives from the PTAB decision-making process was also under-emphasized. Although this was eventually accomplished, it was not clear until the final stages of the decision-making process that this course of action was appropriate. Also, it should be realized that if there is an installed base of any candidate within members of the decision-making body, then a consensus of that body toward any other candidate as a solution is probably impossible. Thus, although installed base is an important consideration, the decision-making body should rely on majority decision or similar (as opposed to unanimous consensus) when an installed base of at least one candidate exists within the industry.

Summary

An in-depth study was conducted by SEMATECH with the purpose of recommending a sensor actuator bus (SAB) interoperability standard solution for the semiconductor industry. One important result of this study is the derivation of a generic process for selecting an SAB standard for an industry. Such a process has been described in this chapter. The major components of the process include a requirements analysis survey, selection criteria derivation, SAB candidate solutions identification, structured candidate presentations/tutorials, follow-up analysis and refinement of selection criteria, comparative weighted analysis of candidate solutions with respect to refined selection criteria, and SAB solution recommendation.

The entire analysis effort yields many benefits to the semiconductor industry and to manufacturing in general. In the semiconductor industry, it provides a qualified SAB interoperability standard recommendation; relates the operation benefits and disadvantages of each of the candidates as an SAB for the industry; helps to educate the industry as to the advantages of an SAB; relates to the developer and user the expected present and future requirements and directions of an SAB; and further opens the standardization process up to the industry, thus providing a potential for development of a complete, agreeable, trusted, and utilized semiconductor manufacturing SAB interoperability standard. As for manufacturing in general, this effort yields a generic template that can be used in the comparative analysis of SAB alternatives, thus shortening the path toward developing a sensor/actuator bus standard for an industry.

APPENDIX: LISTING OF ACRONYMS

APC	Automated Process Control, Inc.
CSMA	Carrier sense multiple access
CAN	Control area network
GEM	Generic equipment model
HSMS	High-speed message service

ISO	International Standards Organization
ISP	Interoperable systems project
MTBF	Mean time between failures
MTTR	Mean time to repair
NCMS	National Center for Manufacturing Sciences
OSI	Open Systems Interconnection (also ISO-OSI)
SAB	Sensor actuator bus
SDS	Smart distributed system
SECS	SEMI Equipment Communication Standard
SEMI	Semiconductor Equipment and Materials International
SEMATECH	Semiconductor Manufacturing Technology
SAE	Society of Automotive Engineers

REFERENCES

1. "Expanded Survey Report, Sensor/Actuator Interoperable Standard Project," Document no. 4012196A XFR, SEMATECH, February 1994, pp. 204.

2. A. Stock and D. Judd, "Interoperability Standard for Smart Sensors and Actuators Used in Semiconductor Manufacturing Equipment," Document 93091800A-ER, SEMATECH, February 1994; also see *Proceedings of Sensor Expo West*, February 1994, pp. 298–311.

3. A. Stock and D. Judd, "Interoperability Standard for Smart Sensors and Actuators Used in Semiconductor Manufacturing Equipment," *Sensors Expo '93*, Philadelphia, October 1993.

4. J. Moyne, N. Najafi, D. Judd, and A. Stock, "Analysis of Sensor/Actuator Bus Interoperability Standard Alternatives for Semiconductor Manufacturing," *Sensors Expo '94*, Cleveland, September 1994.

5. K. Wehmeyer and J. Wulf, "Appendix A: Intelligent Distributed Measurement System Technology Forecast and Analysis," technical report WL-TR-92-8006, Wright Laboratory, Wright-Patterson Air Force Base, Ohio 45433-6533, April 1992, pp. 27.

6. K. Wacks, "The Challenge of Specifying a Protocol for Building Automation," *Canadian Automated Building Association Newsletter*, 1994, 6(1), pp. 2–6.

7. J. R. Moyne, N. Najafi, D. Judd, A. Stock, "Analysis and Selection of a Sensor/Actuator Bus Interoperability Standard for Semiconductor Manufacturing," to be published, *Special Issue on Intelligent Instrumentation of the International Journal of Measurement and Control*, 1995.

8. K. D. Wise and N. Najafi, "The Coming Opportunities in Microsensor Systems," (invited paper), *Technical Digest, IEEE TRANSDUCERS '91*, June 1991.

9. D. Judd, "Analysis of Sensor/Actuator Bus Communication Standard Alternatives," SEMATECH internal document, March 1994.

10. J. Moyne, "Analysis of Sensor Bus Technology for the Semiconductor Industry: Final Report," SEMATECH internal document (to be released), February 1994.

11. WorldFIP North America Inc., P.O. Box 13867, Research Triangle Park, NC 27709.

12. R. Bosh GmbH, "Control Area Network Specification: Version 2," *Postfach 50 D7000*, Stuttgartl, Germany, 1991.

13. "Recommended Practice for Serial Control and Communication: Vehicle Networks, Class C; J1939 Committee Draft," Society of Automotive Engineers, August 1993.

14. Allen-Bradley: Global Technical Services, 6680 Beta Drive, Mayfield Village, OH 44143.
15. Honeywell MICRO SWITCH Division, 11 West Spring Street, Freeport, IL 61032.
16. ISO/7498-1984, "OSI Basic Reference Model," American National Standards Institute, 1984.
17. Echelon, 4015 Miranda Avenue, Palo Alto, CA 94304.
18. Automated Process Control, Inc., 106 Business Park Drive, Jackson, MS 39213.
19. ISP Foundation, 9390 Research Boulevard, Suite 100, Austin, TX 78759.
20. Industrial Service Technology, 3286 Kentland Court S.E., Grand Rapids, MI 49548.
21. "Sensor Bus (J94A) PTAB Meeting Minutes," Document no. 93081753A-MIN, SEMATECH, February 1993.
22. "Statement of Work for Analysis of Sensor Bus Technical Presentations," SEMATECH internal document, December 1993, contact Dr. Allen Stock, AMD, Austin, Texas.
23. "Statement of Work for Cost of Ownership Analysis of Sensor Bus Alternatives," SEMATECH internal document, November 1993, contact Dr. Allen Stock, AMD, Austin, Texas.
24. R. Gyurcsik, "Report on Enhancements to SEMATECH Cost-of-Ownership (COO) Model for Sensor Bus Entries: Vers. 2.0," Document no. 94062421 B-ENG, SEMATECH, December 1994.
25. "Statement of Work for Technical Presentation of (Candidate) Communication Standard," SEMATECH internal document, November 1993, contact Dr. Allen Stock, AMD, Austin, Texas.
26. "Sensor Bus (J94A) PTAB Meeting Minutes," Document no. 94032273A-MIN, SEMATECH, March 1994, pp. 5.
27. "Draft Document 2250: Interoperability Guideline for Sensor/Actuator Network," "Draft Document 22 Communication Standard for Sensor/Actuator Network," "Draft Document 2252: Common Device Application Model for the Sensor/Actuator Network," and "Draft Document 2253: Mass Flow Device Application Model for the Sensor/Actuator Network," SEMI Subsystems Committee—Sensor/Actuator Network Task Force, Semiconductor Equipment and Materials International, December 1993, 1994.
28. "Validation Plan for the SEMATECH Sensor/Actuator Bus Interoperable Standard Project," Document no. 94102564A-TR, SEMATECH, March 1995, pp. 21.
29. J. Moyne, N. Najafi, D. Judd, and A. Stock "A Generic Methodology Used in the Selection of a Sensor/Actuator Bus Standard for Semiconductor Manufacturing," (submitted to) *IEEE Transactions on Semiconductor Manufacturing,* (submitted April 1995).

CHAPTER 30
A PORTABLE OBJECT-ORIENTED ENVIRONMENT MODEL (POEM) FOR SMART SENSORS

INTRODUCTION

This chapter discusses the use of object-oriented (OO) technology in smart sensors applications that integrate networking and signal processing with an adaptable system control infrastructure. We propose a Portable OO Environment Model (POEM) and underlying architecture that supports system integrator needs. OO techniques are suitable for complex applications where software objects may be used to encapsulate hardware devices and abstract over the details of low-level interfaces. Another emerging technology is script-based *intelligent agents*. The use of scripts allows for a dynamic reconfiguration of the system to adapt to changes imposed by environment. Real-time synchronization required by smart sensors applications is supported by *reactive objects*. In POEM, we integrate these three technologies to create a powerful system integration environment.

Transducers that convert physical variables into electrical signals are the key components of sensors. Transducers are analog devices that are often hard to use by an average computer user. Smart sensors integrate a number of traditional modules such as network, microprocessor, and analog transducer into an easy-to-use digital element. By adding a microprocessor, adaptive control software, and standardizing communication protocols, sensors can talk to each other, recognize system configuration, and become ready for plug-and-play operations. This emerging smart sensor standard will simplify system integration and reduce system cost by assuring interoperability and interchangeability of components.[1] At the same time, the designers must develop new applications, system design techniques, and development tools to take full advantage of the increased processing capabilities offered by smart sensors; see Table 30.1.

The success of a smart sensor depends on the widespread acceptance of communication standards. Compliance to a standardized networking architecture will change the ways the sensor-based systems are designed, maintained, and used. Complex computer systems such as networks impose sets of system concepts, operational models, and introduce patterns of system use that are new to the designers of sensor-based systems. They must recognize that smart sensors will no longer serve as read-only devices

TABLE 30.1 Key Attributes of Smart Sensors

Function	Smart sensors offer data acquisition with local signal processing *capabilities*.
Heterogeneity	Smart sensor systems can collaborate with other components by dynamically adapting to a variety of signaling and interfacing *requirements*.
Distribution and scale	Smart sensors will dynamically react to changes in the network topology and to changes in system requirements. Incremental additions to the *system structure* and functions will be *easily manageable*.
Performance	Smart sensor will offer predictable bandwidth. As new features and sensors are added to the network, smart sensors will be able to be reconfigured to adapt to the changes with the predictable impact on the *system response*.
Reliability and availability	Smart sensor will offer a reliable and self-healing operation. Software will support high availability of measurement data through indirect *estimation of missing data using other measurements*.

or sources of data stream, instead they will provide services to other networked elements such as controllers, control panels, and other sensors. As a networked server, they will provide the user with a number of services such as histogram collection, gathering of statistics, and alarm functions. These services can be dynamically modified and combined with services that are offered by other components of the network to create a flexible application. Furthermore, at the system software level, standards are needed to unify access to those services.

Smart Sensor System Integration Issues

The adaptable nature of smart sensors will allow more services to the systems. A new, information-based model of control systems must be created to reflect changes of roles in a system—a smart sensor becomes a server and a controller acts as a client. The development of smart sensor system technology will follow the tracks of development techniques used in telecommunication systems that usually quickly outgrow initial requirements in terms of size and complexity. As systems become bigger and more flexible, more and more attention and money will be spent on basic system integration tasks; see Table 30.2.

An Outline of the Approach

Control systems are the main users of sensors and they are often designed for their algorithmic correctness. Traditionally, algorithms are executed at the central location that has access to all sensors. As systems grow in size and processing complexity, their designers must increasingly rely on heuristics and adaptive techniques. Real-time systems built using such techniques are hard to develop and they are the subject of active research.[2,3] Introduction of smart sensors allows us to mix off-the-shelf software to handle communication and control algorithms. Both these tasks are time-critical. In the design of POEM, we provide special support in this area.

TABLE 30.2 System Integration Tasks Associated with Smart Sensors

Specification	Each new service is based on predefined system assumptions. Adding new services may affect assumptions made by existing services. We must be able to *assert* completeness of the specifications, *detect* potential conflict, and *resolve* it. *Example:* Event (measurement) information is simultaneously handled differently by two or more active services (with *a potential for different expected outcomes*).
Design	As we develop complex networks, the growing number of services will interact with each other. *Example:* Performance characteristics of the growing networks will be monitored to detect failures; this will add traffic, which further reduces the bandwidth available for the regular *system functions*.
Testing	Interoperability and exchangeability of systems based on smart sensors require well-designed testing procedures. *Example:* A service A is available only at part of the network, but the new service B assumes that A is available all over the network.
Runtime	Effective runtime environment is critical for the user. Heterogeneity will inadvertently introduce critical interactions that must be resolved in real time in a fashion that is transparent to the user. *Example:* Predictable behavior is required but hard to achieve in an environment that uses a mixture of proprietary services (with hidden features) with services for which *the algorithms* are known.

In POEM, we integrate OO technology, intelligent agents, and reactive objects into a powerful system integration environment (Fig. 30.1). Reuse of code and expendability are widely recognized characteristics of OO systems. Objects may be used to encapsulate the great variety of hardware devices used in such applications and to abstract over the details of low-level interfaces. This is especially important in the area of smart sensors that merge digital networking with signal processing and advanced system control schemes. All of these schemes may be encapsulated within objects making them easier to reuse across applications.

We extend the OO approach to include real-time synchronization. Real-time synchronization required by smart sensor applications is supported by reactive objects. Real-time synchronization is an essential aspect of smart sensors which sample and process information from an analog sensor and process and transfer data to the rest of the system over a communications network. In this chapter, we propose a set of programming abstractions and an approach to address real-time synchronization requirements in an object-oriented framework. In our approach, *active objects* encapsulate signal processing activities. Real-time synchronization is maintained by *reactive objects* that control the execution of control and signal processing objects within a smart sensor. A key advantage of our approach is that it allows the separation of synchronization from the behavior of objects. Both objects and synchronization specifications may be reused in different contexts. In addition, the approach enables the specification of real-time synchronization in a high-level notation that has proven well suited to this task.

Scripts as a Mechanism for Adaptation. A smart sensor provides adaptable services to the rest of the system and therefore must be capable of local processing. Interactive control is needed to offer the flexibility necessary to adapt to the changes in the environment and to the changes in offered services. In particular, the user may modify the signal processing requirements and add or delete functions. To support such flexibility, a smart sensor has the processing capabilities of a general-purpose

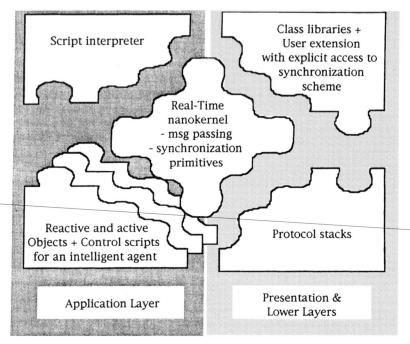

FIGURE 30.1 Object-oriented technology, smart agents, and reactive objects are the basis for POEM.

computer, but, in spite of a great degree of interactivity, it lacks many typical services, such as terminal, file system, or I/O devices]. To support such services, a small operating system (OS) kernel with a limited number of features is enough; we plan to develop such a kernel, referred to as a *nanokernel*.

New ideas such as intelligent agents, which are able to independently perform tasks, to negotiate services with other agents, to spawn agents, and to deliver results to the originator, need flexible, yet powerful implementation mechanisms. To access the device, the users must interpret a state of the sensor or use a remote agent to handle local control. In the limited processing environment offered by a smart sensor, we provide the needed flexibility by offering scripting capabilities. We adapt Tcl, a popular scripting environment to automate the control of smart sensor resources.[4]

Tcl is a glue language that uses a script interpreter to link existing services with new ones, or it may be used to define new services from scratch. If needed, a script can be transformed into its functionally equivalent and faster C code. Tcl is well described and has a number of implementations running from DOS to UNIX, and we port it to smart sensors. The following example illustrates an ability to dynamically reconfigure a system using a script.

> *Example:* A smart sensor containing a thermometer sends messages to a water pump that circulates cooling water in an electronic module. Due to the pump failure, the temperature of the module rises and the excess heat may affect other modules nearby.

The built-in script reroutes the data to a newly installed power switch that controls the power. In this case, the thermometer sends a message when the temperature exceeds a predefined threshold over 5 minutes. The script in a switch cuts off the power supply only if more than three modules are turned on during the last hour.

The use of Tcl allows the user to remotely access the sensor using a powerful GUI interface. (A full-blown Tcl includes a complete graphic environment.) Tcl has mechanisms to spawn processes (threads) and it is capable of supporting distributed applications. In distributed applications developed in Tcl, users can download scripts on the fly. Once loaded, the script can communicate with other scripts and this can be used to coordinate group computation. To further facilitate the system to the needs of the sensor system users, we plan to provide CAN drivers for Tcl and make it truly network-independent (currently it supports TCP/IP).

Software Engineering Issues. Smart sensor systems have to cope with a variety of hardware devices, a large number of communication protocols, and complex algorithms such as, compensations, parameter estimation, distributed system debugging, and system monitoring schemes. Objects can encapsulate the details of particular devices and data representations so that they may be manipulated through simpler higher level interfaces. In addition, encapsulation and polymorphism support the incorporation of new devices and data representations in a way that is transparent to preexisting applications. This allows preexisting applications to benefit from improvements in devices and data processing schemes without modification.

In our approach, Fig. 30.2, a smart sensor application is structured as a set of active objects whose execution is coordinated by a set of reactive objects. *Active objects* encapsulate signal and control processing activities.[5] They provide support for concurrent execution within objects and for the synchronized use of the shared resources they encapsulate. This is accomplished in a way that promotes object-oriented software reuse and is achieved by three novel features: abstract states, state predicates, and state notification. The first feature, *abstract states*, is used to constrain the execution of the methods of an active object so that invocations are only accepted when the object is in an appropriate abstract state. The second feature, *state predicates*, provides a mapping from abstract states to concrete implementations of the state in a way that supports a flexible interpretation of abstract states. Finally, *state notification* supports the coordination of multiple objects based on abstract states by allowing objects to synchronize their execution with abstract state changes in other objects. The separation of behavior and synchronization supported by these features promotes reuse by allowing the interaction between components in a complex system to be designed, in principle, independently of the behavioral components of the system.

Reactive objects support the real-time requirements of smart sensor applications by encapsulating application-defined real-time synchronization constraints over the execution of active objects.[6] Reactive objects react in real time to the reception of some combination of input events (called signals) by emitting a corresponding set of output events (signals). The real-time execution of reactive objects is ensured by a combination of the use of special languages for their implementation. The proposed model can be used to specify real-time support infrastructure at the operating system level.[7,8] The languages used in reactive objects are so called *synchronous languages*.[9] The defining property of these languages is that, although they contain richly expressive high-level constructs, they can nevertheless be compiled into simple finite state machines with a bounded execution time between the occurrence of any particular set of input signals and their corresponding outputs. This property

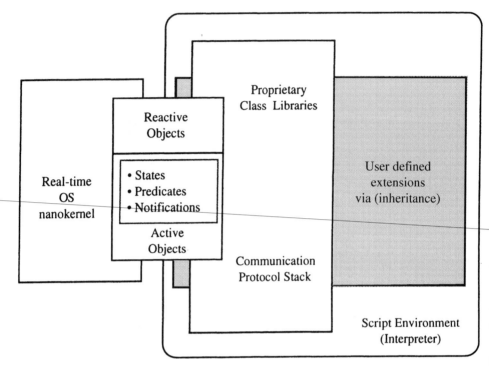

FIGURE 30.2 Synchronization patterns are external to class libraries and this allows us to use inheritance to modify processing and, at the same time, retain a black box model of classes.

leads to a number of attractive formal characteristics. In particular, the designer of a reactive object can assume that the execution of his/her code takes *zero time* as long as the infrastructure can ensure that the object can complete its reaction to one set of incoming events before another set arrives.[9]

In the next section, an outline of a typical example illustrates the interaction between active and reactive objects. Following this, in the section entitled "The Object Model in Detail," a smart sensor is presented and a programming model is detailed. The fourth section presents a C++ class framework that provides low-level programming support for our model. We then return to the example and show how it can be realized in terms of our C++ classes ("The Example Revisited"). Finally, we discuss related work in the sixth section and present our conclusions in the last section.

AN ILLUSTRATIVE EXAMPLE OF OO TECHNOLOGY FOR SMART SENSORS

OO technologies address the requirements of smart sensors. This chapter focuses on the particular problem of real-time synchronization which is required to maintain the temporal integrity of sampling continuous physical parameters such as acceleration, pressure, or temperature, and also to maintain intersensor synchronization in executing control schemes.

Example of Programming Model

A sensor consists of a transducer which continuously measures and electronics which periodically sample the transducer's readings and then converts them into a digital format for further local processing and transmission to other nodes (Fig. 30.3). Active object (*a*) samples and converts physical parameter readings into a digital sample placed at the shared memory buffer. The samples are then processed by active object (*b*), packetized, and transmitted by an active object (*d*). The time taken by object (*b*) to process a data sample may vary. However, samples have to be transmitted by object (*d*) at a fixed rate.

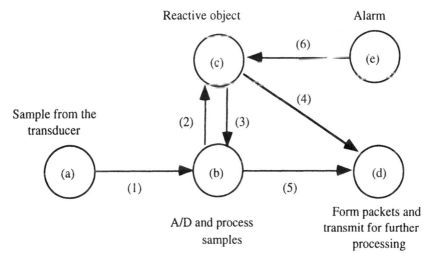

FIGURE 30.3 Smart sensor operations.

The reactive object (*c*) coordinates the operation of the other components to maintain the fixed sampling rate. An alarm active object (*e*) is used to inform the reactive object when it is time to transmit the next packet. Arrows (1) and (5) represent the video data flow between the active objects. Arrows (2), (3), (4), and (6) represent signals entering and leaving the reactive object. The realization of the signal-based communication between the active objects and the reactive object uses the abstract state and state notification features to minimize the burden of programming the interactions. Signals (2) and (3) allow the reactive object to monitor the processing. Signal (6) is sent periodically by the object alarm (*e*) to indicate that it is time to display the current sample. At this time, the reactive object sends a signal (4) to the transmits object (*d*). The actual signal sent depends on whether or not the current processing and compensation operations have completed and determines the action of the transmitter (*d*). This issue is further examined in a later section where we reexamine the realization of the example in detail.

THE OBJECT MODEL IN DETAIL

Active objects are based on a concurrent object-oriented model designed to address some generally accepted difficulties in combining concurrency and object orientation.[5,10-12] An important problem addressed by the design of active objects arises in connection with object reuse. In concurrent object-oriented systems, it is necessary to correctly synchronize the execution of methods to protect the internal object state and coordinate the execution of methods. The obvious solution of embedding such concurrency control within object classes leads to difficulties in reuse, for example, through inheritance, as synchronization assumptions are not visible to the user of the object. The solution proposed in this paper is to provide an object model with explicit synchronization constraints exposed to the programmer. The same constraints are also used internally allowing full access to the synchronization policy for objects.

Active Objects

Active objects are multithreaded objects (resembling server processes) whose threads are synchronized through the concept of *abstract states*. Constraints defined over these abstract states provide the expressive power needed for managing concurrency and provide the explicit synchronization of methods required for reuse. An *object manager*, inherited by all active objects, controls the execution of requests. This manager ensures that concurrent requests are executed in a way consistent with the state of the object. A second concept, *state notification*, allows other objects to monitor and control state changes and to synchronize their own execution with activities encapsulated in independently developed objects.

The key features of abstract states, state predicates, and state notification are described in more detail in the following sections.

Abstract States and State Predicates. Abstract states allow the expression of conditions on the state of active objects independently of the realization of this state. Synchronization constraints on the invocation of methods may then be expressed abstractly and be reused to constrain the execution of those methods for alternative implementations or subclasses. In the following example of a bounded buffer (shown in pseudo code based on C++), the abstract states Empty and Full are used to specify the states where a bounded buffer is empty or full, irrespective of its implementation (e.g., as a linked list, array, etc.). In the example, the method execution constraints are whennot: Full and whennot: Empty.

```
Class AbstractBoundedBuffer: public Activeobject
       estates:       {Empty, Full};
                      constraints:
                                Insert                 whennot: Full;
                                Remove                 whennot: Empty;
                      public:
                                void       Insert(eltype) = 0;
                                eltype     Remove(void) = 0;
```

Abstract states are also used to express condition synchronization in the execution of methods. This makes it possible to employ class inheritance and to define subclasses with additional methods and state variables; abstract states provide the necessary indirection for synchronizing methods that have been developed independently. In our model, active objects can benefit from this aspect of abstract states by using methods (inherited from the object manager class) to synchronize the execution of their own methods using abstract states. For instance, a thread executing a method of a bounded buffer object may suspend itself, as the following shows, until the buffer is empty by calling the method suspendUntil (defined in the object manager class), supplying as an argument an object representing the state Empty.

```
this->suspendUntil( State( Empty) );
```

State predicate objects are used to anchor abstract states in concrete implementations. For example, the state predicate associated with the abstract state Full could be implemented as an object with a method returning a boolean value depending on whether or not the underlying data structure was fully occupied. New kinds of state predicate can be defined when existing state predicate classes are inadequate. An example of defining state predicates for a new class of active object is discussed in the section entitled "Time-Based Activities," which follows.

State Notification. State notification allows active objects to monitor and synchronize with state changes, expressed in terms of abstract states, occurring in other objects. Together with abstract states, state notification may be used to describe abstractly, as a sequence of abstract state changes, practically any activity encapsulated within an active object.

State notification is supported by a protocol inherited by every active object. This protocol is implemented in such a way that different *notification policies* may be defined to take advantage of the features of particular state predicates and also to support different implementations. Notification may be *asynchronous* or *synchronous*. In the asynchronous case, when the object changes state, its object manager services all notification requests and then proceeds with the normal execution of the object's methods. The synchronous variant allows the object that requested the notification and the notifying objects to be synchronized in a way similar to "rendezvous." This variant guarantees, by postponing the execution of the monitored object methods, that the notified object will get the chance to invoke methods of the notifying object while it is still at the requested state. In the rest of the chapter, we use the term *notification* to refer to the asynchronous variant.

An example of using (asynchronous) state notification follows. In this example, an object of type NotifyEvent is created to represent the event that the active object sensor, reaches the abstract state sensor.state(Paused). This is associated with a state predicate encapsulated in the sensor class and representing the state at which the sensor operation is paused. The wait method of the NotifyEvent object may then be called to suspend the thread until the sensor object has been paused.

```
NotifyEvent sensIstop(this, sensor, sensor.state(Paused)); sensIstop.waito;
```

Figure 30.4 shows a conceptual view of the architecture used to support state notification. The figure shows two active objects A and B. B's object manager has made a state notification request to A's object manager. Following this, A's object manager has created a local notification server object that represents the notification request. The object manager maintains a list of notification server objects (notification requests); each time the object state changes, it goes through the list and activates each notifier server. The notifier server invokes the associated state predicates) and informs the notify client if appropriate. When the notify client is informed, it requests

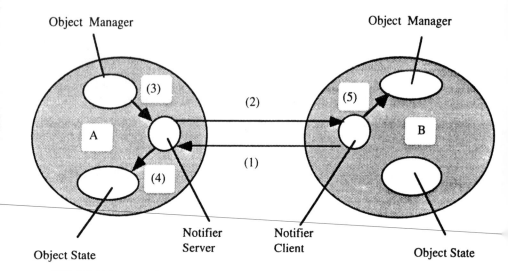

FIGURE 30.4 An architecture for state notification.

its object manager to take the appropriate action; for instance, schedule a suspended method for execution (as previously shown).

Note that this is a general architecture and can thus be instantiated with different interfaces and/or implementations. For example, support for real-time notification can be provided by specifying a time-out at the notifier client and/or by requesting a bound on the notification delay.

Reactive Objects

Reactive objects are real-time controllers that accept events from an outside environment and instantaneously respond by generating new events. They can be programmed in any concurrent programming language as long as the semantics of the reactive constructs satisfy the following properties:

- Events internal to the reactive object are visible to all internal activities.
- Events can be propagated to all internal activities in zero time.
- Statement execution within the object takes zero time.

These semantics, known as the *synchrony hypothesis*,[9] are feasible in practice as *zero time* effectively means "less time than the interarrival gap between events from the surrounding environment" and reduces to ensuring that the implementation of a reaction is atomic with respect to the reactive object's environment. One class of language capable of meeting the synchrony hypothesis is the class of synchronous languages. These languages compile into table-driven finite state automata and uphold the synchrony hypothesis as long as the transitions are faster than external event occurrences. It is possible to verify statistically this condition as there are a fixed and known number of machine instructions executed for each transition. Note that any language capable of being reduced to such an automaton could also be used in the programming of reactive objects.

In our work, we have adopted the synchronous language Esterel[9] to develop reactive objects. Some particular benefits from the use of Esterel for programming reactive objects are

- Esterel provides high-level constructs allowing a concise description of the complex relationships between events governing the behavior of a reactive system.
- The language has clearly defined semantics with respect to temporal execution and allows the behavior of reactive systems to be specified formally.
- Programs in Esterel are deterministic. They can be translated into a deterministic finite automata that can be used for an efficient implementation. The automaton may be also used to formally analyze behavioral properties and derive execution bounds.

Esterel can be characterized as an imperative parallel synchronous language. An Esterel program consists of a set of parallel processes that execute synchronously and communicate with each other and with the environment by sending and receiving signals. The signals present at each instant are broadcast instantaneously to all processes. Signals may carry a value, in which case they are called *valued signals*, or be used just for synchronization, in which case they are called *pure signals*.

PROGRAMMING SUPPORT

In this section, we present a class framework designed to support the programming model previously described.[13,14] The framework has been designed to take advantage of the real-time guarantees provided by a particular infrastructure[7] and to incorporate an implementation of reactive objects in Esterel[15] also designed for this infrastructure.

To accommodate the real-time requirements of smart sensors, POEM provides programming support at a fairly low level of abstraction. At this level, the programmer is aware of, and specifies explicitly, the resource requirements of active objects, reactive objects, and their bindings. Although we briefly discuss some real-time and resource allocation aspects, detailed discussion of these aspects is outside the scope of this chapter.

Active Object Classes for Supporting Smart Sensors

We now provide an indication of how the active object model can be used to define object classes for smart sensor applications. We also briefly indicate how real-time support may be provided and how specialized state notification policies, important for smart sensor applications, can be incorporated in our framework.

Smart Sensor Objects. In the proposed object-oriented framework, *smart sensor objects* are active objects capable of producing and/or consuming data and supporting an interface with the methods start, stop, pause, and resume (used to start, pause, and resume the signal processing activity they encapsulate). We assume that a sensor may consume data for compensating its own measurements or for estimating readings from missing or malfunctioning sensors.

To support the development of such objects, we define an abstract class *Activity* which can be inherited to provide an interface to control the activity of smart sensor objects. In addition, this class defines the abstract states *idle, active,* and *paused* which allow independent objects to monitor and be synchronized with the data processing activity through the state notification mechanism. The class maintains a simple state predicate that stores a value representing the activity states and also defines methods start, stop, pause, and resume. Application-specific actions to be executed when these methods are called can be specified in the corresponding methods redefined in subclasses.

Time-Based Activities. We adapt time-based activities as particular kinds of Activity classes that progress in a logical time system by executing a sequence of discrete steps.[16] Such an abstraction of activities reflects activities in communication protocol stacks or in sensor systems and their algorithmic support environments. For example, a step could correspond to a time step needed to sample the data from a transducer and convert it into a digital format. The following briefly discusses the class TBActivity that is designed to support such activities. A subclass of this class that supports real time is used in the example discussed in the next section.

The following functionality is provided by the class TBActivity to its subclasses:

- It inherits the interface and abstract states from the Activity class. Specifically, it supports basic control functions such as start, pause, resume, and stop. It uses the abstract states defined for state notification on the progress of activity.

- It annotates points on the activity's time line using abstract states. It uses a state predicate, TBPredicate, that can be combined with state notification to allow other objects to be notified when the activity's execution has reached a certain point.

- It supports a number of methods to its subclasses to query and modify the current state of the activity. These methods can be used in the method step to further control the progress of the activity.

- It provides the specification of the duration of the activity and the user can use the activity's virtual step method to maintain the progress of the activity along its time line. The programmer can further specialize the methods by including code of the actions that the activity performs at each step. If the activity is paused or stopped, the class method automatically stops executing the step method.

- It suspends the thread if the activity is not in the Active state. Furthermore, it uses endhandler that suspends the activity and may activate user-defined termination or halt procedures. For example, a termination condition may be associated with crossing a threshold or a combination of conditions determined by local measurements or measurements from associated smart sensors. This feature is especially useful to aid value-driven system debugging (an equivalent of a break point in a traditional debugger).

The object manager uses TBPredicate to support synchronization and state notification. The method state maps integers representing points on the activity's time line into abstract states so that they may be used by the object manager. The methods offered in TBActivity are quite flexible, for example, the user can change the nextstate function to produce some acceleration effects that are useful in, for example, the oversampling process used in some filters. The main instance variables and methods needed to understand the overall design of these classes is described in Ref. 16.

Real-Time Support. The class RtActivity provides support for the real-time execution of activities. Real-time requirements are specified in terms of the instance variables stepTime, period, and jitter. These specify the time it takes for a typical step, how often a step should run, and the acceptable jitter. An exception method is called when the real-time requirements have been violated. This method can be redefined by subclasses that may want to take some specific action in this case. An exception object, passed as an argument to the exception method, provides more information about what has happened (i.e., the execution of step took too long or the infrastructure failed to provide the required support). An example of the use of this class is given in the section entitled "Active Objects."

Reactive Object Classes

Creating Reactive Objects. The creation of a reactive object is supported by the class ReactiveObject shown in Fig. 30.5a. This class provides some protected configuration methods to tailor the execution of the reactive object and also some methods supporting interaction with active objects.

```
class ReactiveObject: ActiveObject
{ protected:
        // execution parameters
        int ClockTickSignal(inputSignal&);
        int SetCollectIntervall(int);
        int SetIntClockPeriod(int);
// access to signals
public:
        ReactiveObject(int reactiveActorId);
        InputSignal& getInputSignal(SIGNALS);
        OutputSignal& getOutputSignal(SIGNALS);
        int attachInputSignal(SIGNALS InputSignal&);
        int attachOutputSignal (SIGNALS outputSignal&); }
                            (a)

class Signal: ActiveObject
{
        friend class ReactiveObject;

        states: { Disconnected, Connected, On, Off};
        protected:
        virtual connect(ConnectionInfo) = 0;

        class InputSignal: Signal
        {
            constraints:
                set when: Connected;
            public:
                virtual void set() = 0; };
            class OutputSignal: Signal
            public:
                virtual int IsItset() = 0;
        }
```

FIGURE 30.5 Classes for reactive objects and signals.

The configuration methods specify policies for the collection of signals for a given reaction of the reactive object and also place constraints on the exact timing of reactions. These policies are crucial in realizing the semantics of reactive objects in interfacing to the external environment. Further details can be found in Ref. 15.

The methods supporting the interaction between reactive and active objects are related in the following discussion.

Interfacing Reactive Objects to Active Objects. The mapping between events in active objects to signals in reactive objects is supported by particular active objects of abstract class Signal (with associated subclasses InputSignal and OutputSignal) as shown in Fig. 30.5*b*. Signals support the abstract states Off, On, Connected, and Disconnected. They are created by invoking the getInputsignal and getOutputSignal methods on the ReactiveObject class.

Abstract states Off and On correspond to the signal being absent or present (to simplify the presentation, we do not consider valued signals in this paper), respectively. Input signal objects generate their associated reactive object signal when their method set is called. They are at the state On from the time their set method is called until the signal has been collected by the Esterel automaton behind the reactive object. They then return to the state Off. Similarly, output signal objects are updated at each reaction to reflect whether or not the associated output signal was emitted. The normal state notification mechanism can be used to inform interested parties when an output signal object moves into the On state.

The abstract states Connected and Disconnected are necessary to reflect the underlying distribution. Reactive objects are effectively proxies for potentially remote processes implementing Esterel automata. In such an environment, it is necessary to create a *binding* between the proxy and the potentially remote execution machine. The connected state then corresponds to the existence of such a binding. Note that a default binding is created implicitly by the methods getInputsignal or getOutputSignal.

Rather than using the default Signal classes returned by getInputsignal and getOutputSignal, it is also possible to define and use subclasses of InputSignal and OutputSignal.* This facility can be used, for example, to specify alternative (application specific) strategies for queuing signals as discussed in Ref. 15. If specialized Signal classes are required, the methods attachInputsignal and attachOutputSignal are used to create the necessary underlying binding.

THE EXAMPLE REVISITED

We now revisit the example of the previous section and show how it can be realized in terms of the facilities described in this chapter. We first describe, in the section entitled "Active Objects," the active object components and then describe the reactive object in the following section. The main program, presented in Fig. 30.6*f,* simply creates the component objects, associates the signal objects with the reactive object, and starts the active objects. Note that the example has been refined from the version in the previous section by adding new objects as described in the following (error checking and program termination issues are ignored).

---

* Note, however, that the definition of new classes of input and output signals requires knowledge of a specific signal protocol. Although we are not discussing implementation of a finite state automaton used to implement protocol stacks explicitly in this chapter, a protocol stack can be implemented using the same synchronization mechanisms as TBA.

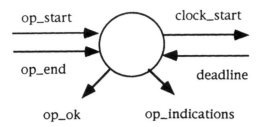

FIGURE 30.6 Active objects OpMonitor.

Active Objects

Initially, samples converted into digital format are placed in a shared memory buffer by the TransducerReader (object *a* in Fig. 30.1). The TransducerReader and the SignalProcessor are synchronized in a *producer-consumer* pattern. We assume that TransducerReader keeps the bounded buffer filled with A/D converted samples and do not discuss this issue any further.

The *signal processing* is an active object of class SignalProcessor (shown in Fig. 30.6c). It processes samples into a buffer supplied to it through its method setbuffer. The SignalProcessor may be in one of the abstract states Idle, Ready, Active, and Finished; the invocation of its methods is constrained by these abstract states according to the constraints in its class definition (Fig. 30.6a). The overall behavior of the SignalProcessor is depicted in the state diagram in Fig. 30.6b. The SignalProcessor is initially at the abstract state Idle. It becomes Ready after its method setbuffer is called. Its method start may then be called to start the signal-processing process and move into the Active state, where it remains until the end of the signal-processing process. At this point, the SignalProcessor moves spontaneously (indicated by the label T associated to this transition) into the Finished state, from where its method getbuffer may be called to obtain the decompressed buffer. The SignalProcessor then moves to the Idle state from where it may restart the same cycle. Sampling rate and processing algorithm used by SignalProcessor will affect the state diagram (external data may be required from the network.)

The *transmit* object has a method transmit which is called to packet and send the data over the network. It has states Active and Idle which are used to serialize calls to send packets.

The *synchTransmit* object has been added to the original configuration to maintain the required synchronization between the SignalProcessor and the transmit object. It has an internal thread (Fig. 30.6c) that starts executing at object creation time and repeatedly waits, using state notification, for the occurrence of either one of the signals indication or ok (i.e., until the associated InputSignal goes to the On state). The method waitAny is inherited from the ActiveObject class and is used to wait for the occurrence of the first event that occurs among a set of notification events. Depending on the signal received, it either transmits the last successfully processed sample, saved in an internal buffer, or gets a newly processed A/D converted sample from the SignalProcessor.

The *alarm* object, used to generate the signal deadline as required by the reactive object, is implemented as a periodic real-time activity (see Fig. 30.7c). An active object of class SignalEachState (see Fig. 30.7) is used to monitor periodic state changes and generate the signal deadline. This class provides a general way to create on-the-fly objects that monitor other objects and then send an input signal to a reactive object. An instance of this class receives at its creation a reference to an active object, an abstract state reference, and an InputSignal object reference. It then uses state notification to suspend itself until the object reaches the desired state. When resumed, its thread calls the InputSignal object's set method to generate the associated signal. SignalEachState objects are also used to interface the SignalProcessor to the reactive object.

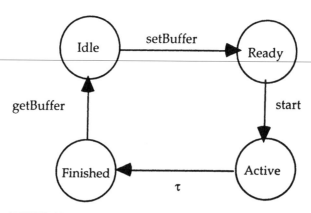

FIGURE 30.7 Reactive objects OpMonitor.

The Reactive Object

The reactive object with its associated signals is illustrated in Fig. 30.7a. This object has the general purpose of monitoring the execution of some operation and reporting at some precise point in time, indicated by an input signal, deadline, if the operation has completed successfully. The input signals op-start and op_end are used to indicate to the reactive object when the operation is initiated and when its execution terminates, respectively. The signals clock-start and deadline are used to initialize an external clock and for signaling the deadline. Finally, the emission of signals op_ok and op_indication indicates whether or not the operation terminated before the deadline.

The corresponding Esterel module OpMonitor is shown in Fig. 30.7b. This module waits for the occurrence of the signal op_end and records its occurrence using the variable ok. When the signal deadline is received, the module responds by emitting the signals op_ok or op_indication depending on whether or not it has already received the signal op_end. The do_upto statement allows a signal (in this case op_end) to be awaited until the occurrence of some other signal (deadline, in our case). It should be noted that this is different from a time-out; the execution of the do_upto statement does not terminate before the occurrence of deadline, irrespective of whether or not op_end occurred.

Following the do_upto statement, the presence of the signal op-end is checked once more. The reason for this is that op_end may have occurred at the same instant as deadline. In this case, the do-upto statement is aborted and it is not possible to detect the occurrence of op_end.

```
class SignalProcessor: public Activeobject

    states:      ( Idle, Active, Ready, Finished);
    constraints:
```

The signal clock-start is emitted by the program when it receives the signal opstart. It can be used to start an external clock that may do some time keeping related to the emission of deadline.

RELATED WORK

Kafura and Lee (Ref. 17) have presented an approach that uses abstract states to express synchronization constraints on the invocation of object methods. Their proposal, however, does not address the synchronization of multiple objects. Moreover, their interpretation of abstract states is more restrictive than ours in that objects have to be in exactly one abstract state at a time. Our more flexible approach in defining abstract states avoids a number of problems related to the use of class inheritance which are known as *inheritance anomalies*.[18] The coordination of multiple active objects developed independently has been addressed by other research; Synchronizers (Ref. 19) and Abstract Communication Types (ACTS, Ref. 20) are two relevant proposals. An essential difference of these proposals from our work is that they both use method invocations, rather than abstract states, to monitor and control the execution of active objects. We have several reasons to believe that abstract states are more appropriate to support reuse in the specification of multi-object coordination. For instance, in different realizations of object behavior, different method invocation sequences may lead to the same abstract state. These issues are further discussed in Ref. 21.

Real-time synchronization problems similar to smart sensors occur in the area of multimedia. The active object model proposed for sensors has been used in our previous work for synchronization in multimedia applications (Ref. 16).

Object-oriented frameworks for the development of globally distributed applications are the subject of intensive research and development work (Ref. 22). From the smart sensor perspective, the most important development and standardization work was led by X/Open and Object Management Group (OMG). This effort attempts to provide a standardized distributed-object technology. This goes beyond networking and plug-and-play function postulated by Smart Sensor group and provides a standard way to define the interfaces to components (objects) in a distributed system. POEM offers *hooks* to integrate smart sensors with OO technology for large distributed systems proposed by OMG. The approach discussed in this chapter provides a migration path for smart sensors to become first-class objects in the rapidly emerging globally distributed OO environment.

Synchronous programming has been an area of intense activity in the real-time community. A number of researchers have investigated the use of the synchronous language Esterel in an asynchronous environment. For example, work has been carried out in the fields of robotics and automatic control systems (Ref. 23). In contrast, the work presented in this chapter focuses on providing a general way to support the implementation of synchronous programs in a distributed environment.

CONCLUSIONS

We have presented a programming model that addresses two essential issues in the development of real-time applications for smart sensors. First, it promotes software reuse by being based on an object-oriented approach. Second, it allows the specification of real-time synchronization in a high-level synchronous language that is especially suited for this purpose. A set of programming abstractions supporting this model has been proposed. These abstractions are based on a general concurrent object-oriented programming model and have been incorporated in an object-oriented framework. The main goals in the design of this framework are to support the design of higher-level programming abstractions and to make effective use of the real-time support provided by a particular platform without being totally committed to it; and to allow the incorporation of further techniques for expressing and supporting the real-time requirements of smart sensors.

The OO architecture of POEM has been designed to support protocol stacks needed by different smart sensors. In POEM, protocol stacks and time-critical operations are uniformly treated by *using abstract states, state predicates, and state notification*. These mechanisms externalize all synchronization structures and, combined with inheritance mechanisms, allow the modification of proprietary solutions at the networking level or proprietary application object class libraries to be provided by third-party software developers.

To provide a flexible system, we propose using script language. We employ a Tcl scripting environment that contains a built-in X-based graphic script system, Tk.[4] This system integration environment model is especially suited for *visual programming* through graphic-based development tools which are in high demand by today's system users. The OO nature of POEM will allow it to be emulated on standard workstations with Graphic User Interface (GUI) where the user can take full advantage of the OO approach and its visual support.

Key elements of POEM have been prototyped and will be refined to match the needs of smart sensors. We plan to implement a selected protocol, for example, CAN. Consistent use of OO technology in POEM assures a high degree of *code reusability* and simplifies its internal structure. Specialized classes in POEM play the role of executable *object specification* as defined in Ref. 1. Through the use of classes, software that is resident in a sensor can be compiled to configure its functions. The software will provide a uniform and scalable execution environment for a number of microprocessors used in smart sensors. POEM brings state-of-the-art software techniques to the development of smart sensor applications in accordance with emerging smart sensor standards.

The C++ framework presented in this paper has been partly implemented on top of Chorus and SunOS 4. The implementation of the classes that take advantage of the real-time infrastructure is in progress. We intend to further refine the framework by using it to develop more sample applications.

REFERENCES

1. IEEE/NIST Working Group on Transducer Interface Standards, Standardization Working Group Charter, draft rev., February 6, 1995.
2. *Proceedings of the IEEE, Special Issues on Real-Time Systems,* January 1994, **82**(1).

3. J. Maitan, A. Mrozek, R. Winiarczyk, and L. Plonka, "Overview of the Emerging Control and Communication Algorithms Suitable for Embedding into Smart Sensors," *Proceedings of the SensorsExpo*, Cleveland, Ohio, September 20–22, 1994, pp. 485–501.
4. J. K. Ousterhout, "Tcl and the Tk Toolkit," Addison-Wesley, New York, 1995.
5. P. Wegner, "Concepts and Paradigms of Object-Oriented Programming," *ACM, OOPS MESSENGER*, August 1990, **1**(1).
6. D. Harel and A. Pnueli, "On the Development of Reactive Systems," *Logics and Models of Concurrent Systems*, ed. K. Apt, NATO ASI series, Springer, 1985.
7. G. Coulson, G. S. Blair, P. Robin, and D. Shepherd, "Extending the Chorus MicroKernel to Support Continuous Media Applications," *Proc. of the 4th International Workshop on Network and Operating Systems Support for Digital Audio and Video*, November 1993, Lancaster, United Kingdom.
8. C. Nicolaou, "An Architecture for Real-Time Communication Systems" *IEEE, Journal on Selected Areas in Communications*, April 1990, **8**(3), pp. 391–400.
9. G. Berry and G. Gonthier, "The ESTEREL Synchronous Programming Language: Design, Semantics and Implementation," *INRIA*, 1988, p. 842.
10. G. Agha and C. J. Callsen, "ActorSpace: An Open Distributed Programming Paradigm," *4th ACM Conference on Principles and Practice of Parallel Programming, ACM SIGPLAN Notices*, 1993, **28**(7), pp. 23–32.
11. B. Meyer, "Systematic Concurrent Object-Oriented Programming," *ACM Communications*, April 1993, **36**(9).
12. M. Papathomas, "Language Design Rational and Semantic Framework for Concurrent Object-Oriented Programming," Ph.D. dissertation, Universite de Geneve, 1992.
13. L. P. Deutsch, "Design Reuse and Frameworks in the Smalltalk-80 System," *Software Reusability*, eds. T. J. Biggerstaff and A. J. Perlis, ACM Press, 1989, **2**, pp. 57–71.
14. R. J. Wirfs-Brock and R. E. Johnson, "Surveying Current Research in Object-Oriented Design," *Communications of the ACM*, September 1990, **33**(9), pp. 104–123.
15. G. S. Blair et al., "Supporting Real-time Multimedia Behavior in Open Distributed Systems: An Approach Based on Synchronous Languages," *Proc. of ACM, Multimedia '94*, 1994.
16. M. Papathomas, G. S. Blair, G. Coulson, and P. Robin, "Addressing the Real-Time Synchronization Requirements of Multimedia in an Object-Orinted Framework," *Proc. of SPIE, Multimedia Computing and Networking 1995*, San Jose, February 1995, **2417**.
17. D. G. Kafura and K. H. Lee, "Inheritance in Actor-Based Concurrent Object-Oriented Languages," *Proceedings of the Third European Conference on Object-Oriented Programming Languages*, ed. S. Cook, British Computer Society Workshop Series, Cambridge University Press, 1989.
18. S. Matsuoka, K. Taura, and A. Yonezawa, "Highly Efficient and Encapsulated Reuse of Synchronization Code in Concurrent Object-Oriented Languages," *Proc. OOPSLA'93, ACM, SIGPLAN Notices*, October, 1993. **28**(10), pp. 109–129.
19. S. Frolund and G. Agha, "A Language Framework for Multi-Object Coordination," *Proc. ECOOP'93, LNCS*, July 1993, **707**, pp. 346–360.
20. M. Aksit et al., "Abstracting Object Interactions Using Composition Filters," *Object-Based Distributed Programming*, LNCS 791, Springer-Verlag, 1993.
21. M. Papathomas, G. S. Blair, and G. Coulson, "A Model for Coordinating Object Execution Based on State Abstractions and Synchronous Languages," *Proc. ECOOP '94, Workshop on Coordination of Parallelism and Distribution*, Bologna, Italy, July 1994.
22. "Special Report on Client-Server Computing," *Byte*, March 1995, pp. 108–151.
23. E. Coste-Maniere, "Utilisation d'Esterel dans un Contexte Asynchrone: Une Application Robotique," research report 1139, *INRIA*, December 1989 (in French).

CHAPTER 31
NEW GENERATION OF HIGH-TEMPERATURE FIBER-OPTIC PRESSURE SENSORS

INTRODUCTION

A new generation of fiber-optic pressure sensors is reported that is suitable for dynamic pressure measurements at temperatures up to 500°C. A durable metal diaphragm design ensures fatigue-resistant long-term sensor operation. Refined sensor housing design and assembly techniques result in improved sensor performance at reduced cost. An intelligent autoreferencing technique replaces more complicated and expensive compensation techniques needed to combat high temperature and other environmental effects. Performance data collected in customer tests indicate accurate and drift-free sensor performance under high-pressure and high-temperature conditions. The FiberPSI pressure sensor system, commercially viable at this time, meets or exceeds the performance of research-grade pressure sensors at a fraction of the cost.

Numerous applications exist for pressure measurement in high-temperature environments. These applications include automotive engine cylinders, large-bore natural gas–burning engines, aircraft turbines, or steamline pressure in turbine generators. While an intermittent or short-term measurements are important in testing and research applications, the largest demand exists for sensors for open- and closed-loop control systems. Existing or proposed control applications include power balancing in natural gas–burning large-bore engines,[1] lean-burn combustion in passenger cars,[2] or stall control in aircraft engines. Other continuous-use applications of high-temperature pressure sensors include engine health and parametric emission monitoring.[3]

Conventional electronic pressure transducers, such as piezoresistive or piezoelectric, are not suited for continuous high-temperature pressure measurements. Fundamentally, strain or capacitance gages used in pressure transducers exhibit large, unrepeatable, and unpredictable changes in gage output at temperatures typically greater than 125°C or 250°C, respectively. These changes are caused by such effects as alloy segregation, phase changes, selective oxidation, and diffusion, and ultimately lead to premature failure of the gage or lead wires. While water cooling has been used to increase temperature ranges of piezo sensors, uncooled sensors are preferred for industrial applications.

Fiber-optic sensors are potentially very well suited for high-temperature applications. Due to the resistance of fused silica to extreme temperatures, fiber-optic sensors can potentially operate at temperatures up to 800°C. Electrical passiveness makes them immune to EMI and ground-loop problems. Due to very low transmission losses from optical fibers, sensor interface devices can be located away from high-temperature areas, as far as hundreds of meters.

Several fiber-optic sensors have been described in the literature which are potentially applicable to high-temperature pressure measurements.[4, 5] However, the majority of these sensors have been designed for specialized military applications and their prices are too high for large-scale commercial applications.

Among fiber-optic sensors suitable for low-cost commercial applications, the simple intensity-modulated sensor we reported in the past[6] offers the most promise. It utilizes an optical fiber in front of a flexing diaphragm for optical reflection measurement of pressure-induced deflections. By employing this sensing principle coupled with a hermetically sealed sensor structure to eliminate diaphragm oxidation under high temperatures, we recently demonstrated that such a sensor can operate under prolonged exposure to high temperatures, and that the sensor can readily detect misfire or knocking in an automotive engine.[6]

In this chapter, we describe a new generation of high-temperature fiber-optic pressure sensors intended for long-term operation and dynamic pressure measurements. The sensor combines the benefits of high-temperature capability and small size of our original design with the low cost of the second-generation optoelectronic interface unit, reduced fiber-link environmental susceptibility, and excellent durability under continuous high-pressure and high-temperature conditions.

SENSOR SYSTEM DESCRIPTIONS

The fiber-optic pressure sensor reported here is primarily intended for applications in industrial control and monitoring systems. In these types of applications, very high accuracy is typically not required; instead, critical sensor specifications include repeatability, long-term durability, and low cost. Due to our particular focus on combustion cylinder applications, the present system has been designed for dynamic pressure measurements. While the present fiber-optic approach is suitable for static pressure detection, dynamic pressures require considerably less-expensive sensor and related optoelectronic designs. Basic specifications for the present sensor system are summarized in Table 31.1.

TABLE 31.1 Sensor System Specifications

Pressure range	0 to 1000 psi
Over-pressure range	3000 psi
Operating temperature	–40 to 500°C
Frequency response	0.1 to 15 Hz
Linearity and hysteresis	±1%
Temperature coefficient of sensitivity	+0.015%/°C
Sensor housing size	5 mm diameter
Output	0–5V or 4–20 mA

Higher or lower maximum pressure ranges are possible with the present sensor design; maximum pressure range can be simply adjusted by changing the diaphragm's

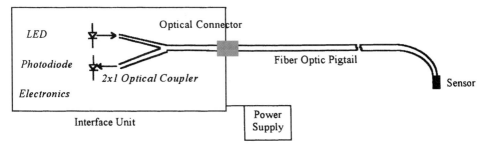

FIGURE 31.1 Sensor system block diagram.

thickness. The optoelectronic interface unit can accommodate a family of sensors of different pressure ranges and accuracy.

A schematic diagram of the present fiber-optic sensor system is shown in Fig. 31.1. It consists of an optoelectronic interface unit, a fiber-optic sensor with a ruggedized pigtail, and an optional fiber-optic patch cable. Compared to our previously reported design,[6] the present interface unit is considerably simpler, smaller, and of lower cost. As shown in Fig. 31.1, it consists of an optoelectronic transceiver, external power supply, and a minimum amount of analog circuitry. The transceiver contains a 2 × 1 fiber-optic coupler, a near-infrared LED, and a PIN photodiode. Note that no feedback detector nor second LED is used in the present design. As described later, our proprietary autoreferencing technique requires only a single optical source and a single detector, thus simplifying the design and reducing cost considerably.

SENSOR HEAD DESIGN

In the sensor head portion depicted in Fig. 31.2, three elements are shown: (1) diaphragm, (2) sensor housing, and (3) ferrule. The sensing diaphragm is the most critical element of the sensor. It must maintain excellent mechanical properties at extreme operating conditions and function repeatably over as many as tens of thousands of hours. Its reflectivity must also remain nearly unchanged over the sensor lifetime. Compared to the previously used flat disk construction,[6] the present sensor uses a more durable and rugged hat-shape diaphragm with varying thickness across its diameter. A high-strength alloy (Inconel) has been used as a diaphragm material. This proprietary diaphragm design has been selected so it can withstand a large number of deflections without yielding or mechanical creep. Other benefits of the present construction include excellent linearity of the pressure response and reduced sensitivity to direct flame effects. In addition, the diaphragm is welded to the sensor housing away from the combustion area, thus improving long-term weld stability and strength. Using finite element analysis, we have established that, under optimal conditions, maximum stresses are reduced almost an order of magnitude in the hat-shaped diaphragm versus the disk-shaped design.

A metalized fused-silica optical fiber is bonded inside a metal ferrule to form a hermetic joint between the metal and the fiber surfaces. The ferrule extends beyond the surface of the sensor housing and typically is kept about 25 microns away from an undetected diaphragm. Such position provides adequate optical signal level and excel-

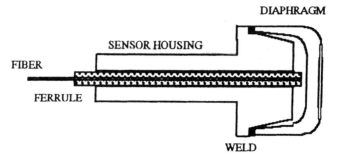

FIGURE 31.2 Sensor head construction.

lent linearity. In addition, the ferrule provides a mechanical stop to the diaphragm movement under overpressure conditions. This proprietary technique enables the sensor, designed for 1000 psi nominal pressure, to withstand pressures as high as 5000 psi.

A graded index, 0.28-NA, 62.5-micron core diameter fiber was used in sensor construction. With a typical full-scale diaphragm deflection around 15 microns, the resulting optical modulation is very linear and equals approximately 20 percent. Figure 31.3 shows a theoretical relation between diaphragm deflection and optical modulation for the fiber used in the sensor.

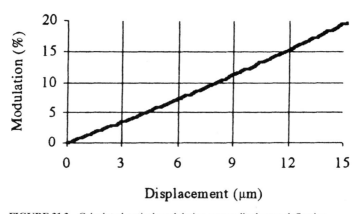

FIGURE 31.3 Calculated optical modulation versus diaphragm deflection.

AUTOREFERENCING TECHNIQUE

While fiber-optic sensors can operate up to very high temperatures, they are nevertheless subject to errors resulting from thermal effects on the sensor package and the fiber itself. In the present design, thermal effects are primarily concentrated in two sensor elements: (1) diaphragm and (2) fiber-ferrule subassembly. The diaphragm errors result primarily from the fundamental temperature dependence of its material Poisson's number and Young's modulus.[7] The fiber error is primarily due to the microbending effect resulting from a large thermal mismatch between a fused silica optical fiber, a metal ferrule, and a bonding material. Under the condition of varying stress on fiber, due to temperature changes, both transmitted intensity and the modal distribution of guided modes may change. These effects will affect the sensor's offset and sensitivity.

Other sources of error in intensity-modulated sensors, including the one described here, are fiber-bending losses and optical-connector mechanical and/or thermal instability. If compensation for light intensity fluctuations is not implemented, errors in the sensor offset and gain will result.

Many compensation techniques have been reported in the past that reduce susceptibility of intensity-modulated sensors to environmental and handling errors. In the past, we reported dual-wavelength compensation techniques for both microbend[8] and diaphragm-type sensors.[6] Others have reported numerous alternative techniques including multifiber,[9] time of flight,[10] and multiwavelength approaches.[4] Unfortunately, the preceding approaches are typically quite complex and unacceptably expensive for large-scale commercial applications.

In this chapter, we report a novel referencing technique which is applicable to dynamic pressure measurements and provides effective compensation for intensity fluctuations, reduces thermal errors, and allows for calibrated sensor operation. In a smart way, the sensor's input parameter, LED light intensity, is adjusted in response to environmental signal changes. A functional block diagram of the electronic circuitry is shown (Fig. 31.4).

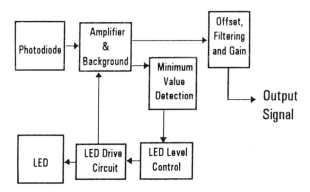

FIGURE 31.4 Autoreferencing functional block diagram.

At initial power-on, the LED launches light into the fiber coupler. The returned intensity is measured at the photodiode. Any offsets (optical and electrical) are eliminated by the background offset circuit. The remaining signal is used for two functions, control and dynamic output signal. The dynamic output portion of the circuit performs three functions, offset elimination, gain, and filtering. The control portion of the interface unit measures and holds the minimum value at a fixed reference value. This is accomplished by slowly adjusting the current to the LED (bandwidth DC to 0.1 Hz).

Besides the offset control, the present technique compensates for sensitivity errors that otherwise may result from undesirable fiber transmission changes. Since light intensity is adjusted such that the resulting reference detected voltage is always maintained at a constant level, the sensitivity of the sensor, which is proportional to that reference voltage, remains unchanged. In a similar way, sensor calibration is also maintained even if the fiber-optic link is changed. For example, fiber-optic connectors may exhibit intensity transmission changes as much as 10 percent with multiple reconnections. Even with such changes, the output will remain calibrated, without any user adjustments.

The present technique offers another important benefit not available to alternative approaches. Through continuous monitoring of the LED current, the status of the sensor may be assessed. Under nominal sensor operating conditions, the LED current is typically at 50 percent of its maximum rating.

When the sensor and fiber link are subjected to environmental and handling effects, the LED current varies. In the event, for example, of a thermally induced sensor failure, optical transmission would degrade slowly before complete failure. By monitoring the LED current level or its rate of change, one can identify potential sensor failure before it occurs. This ability is particularly important in control applications where sensor failure may cause malfunction or even failure of the controlled device.

SENSOR CALIBRATION AND LABORATORY TESTS

Before field installation, the sensor system was extensively tested under laboratory conditions. The basic tests included linearity and hysteresis comparisons against commercially available piezoelectric and piezoresistive sensors, temperature soaking and cycling, and sensitivity-dependence on temperature. Figure 31.5 demonstrates a comparison of the typical room-temperature dynamic pressure responses, over 0 to 500 psi pressure range, of OPTRAND's OPS 300 sensor and a reference piezoelectric transducer (PZT) (Kistler Model 6121).

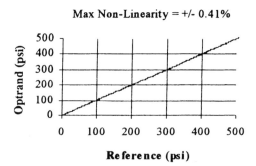

FIGURE 31.5 Dynamic pressure comparison between OPS 300 and a piezoelectric transducer.

Typical fiber-optic sensor linearity was of the order of +1 to 0.5 percent. Fiber-optic sensors showed negligible hysteresis under room-temperature conditions. Signal-to-noise ratio (SNR) of the OPS 300 is comparable to the reference transducer; approximately 2000. When the fiber-optic sensor was connected to different interface units, a typical full-scale error was of the order of ±1 percent.

Pressure detection sensitivity dependence on the sensor housing temperature was investigated under laboratory conditions. An unheated PZT (Kistler Model 6051) was used as a reference. The fiber-optic and the reference sensors were installed in a common pressure chamber. Sensor sensitivity was established by heating the fiber-optic sensor housing over a 200°C temperature range and monitoring peak pressure changes with a data acquisition system. A typical value of temperature coefficient of sensitivity was around +0.015 percent/°C. Note that this coefficient is positive in contrast to negative values for both piezoelectric and piezoresistive sensors. The primary reason for fiber-optic sensors' positive coefficient is the increase in diaphragm deflection at elevated temperatures.[7]

While diaphragms in piezoelectric and piezoresistive sensors show similar behavior, the gages they use degrade much faster with temperature than the rate of increase in diaphragm's deflection. In contrast, changes in optical fiber transmission have relatively little effect on sensor sensitivity provided that minimum light intensity is maintained at a constant level.

ENGINE TEST RESULTS

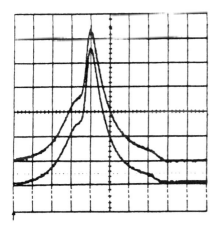

FIGURE 31.6 Pressure traces from a large-bore engine for OPS 300 and a reference piezoelectric transducer.

Two types of engine tests were performed. The first one was on large-bore engines used in natural gas pipeline compressor stations. In one test, OPTRAND's OPS 300 fiber-optic pressure sensor was installed directly into a cylinder head with a reference PZT (Kistler Model 6121) located in a different port on the same engine. Figure 31.6 demonstrates the results in a comparison chart. Pressure traces are vertically offset for clarity.

The temperature of the fiber-optic sensor housing was measured to be approximately 2500°C, whereas the reference transducer was at approximately 150°C. After a small gain adjustment, the sensors were indistinguishable. It is to be noted that the fiber-optic sensor was connected to the interface unit through an 80-ft fiber-optic patch cable while a much shorter coaxial cable was used with the Kistler sensor.

Six-month sensor endurance tests have yielded impressive data on a natural gas compressor station engine. After 12 months of continuous 24-hour-a-day operation, the sensor has not shown any signs of deterioration.

The second type of test was performed on passenger car engines. In various trials, OPS 300 was threaded into a commercially available modified spark plug with an off-axis pressure port. Prior to the tests, sensor housing temperature was measured. Depending on the engine operating conditions, maximum temperatures did not exceed 300°C. Figure 31.7 demonstrates sensor performance obtained on a four-cylinder engine under idling and unloaded engine operation. During the course of engine tests, we observed repeated PZT output drifts. In comparison, the FiberPSI system showed no drift.

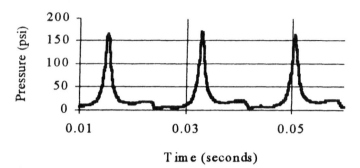

FIGURE 31.7 Real-time pressure waveform collected in a four-cylinder passenger car engine.

CONCLUSIONS

We have presented a second-generation high-temperature fiber-optic sensor system employing a durable hat-shaped diaphragm and a low-cost optoelectronic interface unit. A smart autoreferencing technique compensates for sensor and fiber-link environmental and temperature effects and guarantees drift-free operation. We have shown that, at room temperature, the present sensor demonstrates a typical ±0.5 percent linearity, negligible hysteresis, and overall accuracy of 1 percent of full scale. Typical SNR for the sensor is approximately 2000. Large-bore natural gas-burning engine test results, obtained at sensor housing temperatures around 250°C, have demonstrated excellent sensor performance against a considerably more expensive piezoelectric transducer. Over a one-month period of continuous 24-hour-a-day stationary engine operation, the sensor has not shown any signs of deterioration.

REFERENCES

1. "Cylinder Balancing System for Gas Engines," *Diesel & Gas Turbine Worldwide*, March 1995.
2. N. Sugitani and K. Tsukada, "A Combustion Pressure Sensor for TOYOTA Lean Burn Engine Control System," SAE paper no. 930882, presented at the *1993 SAE International Congress and Exposition*, Detroit, Michigan, March, 1993.
3. G. M. Beshouri, "On the Development of Modern Analysis Techniques for Single Cylinder Testing of Large-Bore Engines," *ASME Trans.*, July 1991, **113**.
4. J. W. Berthold and S. E. Reed, "Flight Test Results from FOCSI Fiber Optic Total Pressure Transducer," *Proc. SPIE*, 1993, **2072**.
5. M. Lequime and C. Lecot, "Fiber Optic Pressure and Temperature Sensor for Down-Hole Applications," *Proc. SPIE*, 1991, **1511**.
6. G. He, A. Patania, M. Kluzner, D. Vokovich, V. Astrakhan, T. Wall, and M. Wlodarczyk, "Low-Cost Spark Plug-Integrated Fiber Optic Sensor for Combustion Pressure Monitoring," SAE paper no. 930853, presented at *SAE International Congress and Exposition*, Detroit, Michigan, March 1993.
7. G. He and M. T. Wlodarczyk, "High Temperature Performance Analysis of Automotive Combustion Pressure Sensor," *Proc. SPIE*, 1992, **1799**.

8. M. T. Wlodarczyk, "Wavelength Referencing in Single Mode Microbend Sensors," *Opt. Lett.*, 1987, **12;** pp. 741–743.
9. M. Trimmel, "Fiber Optic Pressure Sensor for Medical Applications," *Proc. SPIE*, 1993, **2070.**
10. J. W. Berthold, W. L. Ghering, and D. Varshneya, "Design and Characterization of a High-Temperature, Fiber-Optic Pressure Transducer," *J. Lightwave Tech.*, 1993, **LT-5**(7).

CHAPTER 32
PEER-TO-PEER INTELLIGENT TRANSDUCER NETWORKING

INTRODUCTION

Distributed control systems offer several benefits, including scalability, flexibility, higher performance, and robustness. Implementing intelligent transducers that comprise distributed systems has become very easy. Completeness of a protocol, availability of integrated low-cost ICs, an open framework for 100 percent interoperability, complete and seamless device and system development tools, and tailored field commissioning and diagnostic tools are key factors for easy implementation. This chapter discusses the benefits of distributed control systems and explains these key factors. Over the past four decades, transducers have evolved mechanically as well as electronically toward compact, intelligent devices. During the 1960s and 1970s, they evolved from pneumatic and other mechanical devices toward mechanical miniaturization. Migration to electronic devices was the next phase, followed by inclusion of point-to-point communication (e.g., 4-20-mA loops, RS-232 links). Toward the late 1980s and into the early 1990s, electronic communication evolved first to the use of multiplexed buses (RS422, RS485), then to buses with communication protocols (e.g., HART, Opto-22, Interbus, Seriplex, and others), and further to next-generation protocols such as SDSTM, DeviceNetTm, and NWORKSO technology.

In most cases, these protocols are run on standard processors such as MCUs. The spare cycles of these MCUs are applied toward transducer intelligence, evolving them to smart sensors.

Further IC integration has allowed more processor power to evolve the transducers to intelligent devices to allow highly distributed control. It has also allowed integration of more sophisticated, peer-to-peer communication protocols suited for distributed control. LONWORKS D technology is an example of a means for implementing distributed control systems.

Following are several useful definitions for this area.

Transducer: Includes sensors, actuators, instrumentation, and so forth. The term *sensor* is often used for *transducer.*

Smart transducer: Device-integrating transducer element and basic electronics for linearization, calibration, and, in many cases, communication. May include self-documentation and diagnostics.

Intelligent transducer: Includes greater intelligence than smart transducer. More importantly, includes all or part of a control loop, plus operating kernel to manage control loop tasks and communicate with the network.

Distributed control system: Systems where control intelligence is colocated with transducers, that is, distributed across the system. This term applies across industries, including factory automation, process controls, building automation, transportation, homes, and medical instruments. It differs from *distributed control systems* (*DCSs*), a term specific to the process controls industry, where it refers to systems with multiple central controllers distributed across the plant and connected together via a higher-level network, with each controller controlling remote and local, dumb and smart instrumentation and I/O.

Peer-to-peer communication: Direct communication among transducers without mediation by a master (in a master/slave system) or time slicing (in a token-passing scheme).

WHY PEER-TO-PEER TRANSDUCERS ARE BETTER

Form Follows Function

Distributed control is the most logical architecture for control systems. Peer-to-peer communication is the most logical communication scheme for distributed control systems.

The first conclusion can be drawn by examining the anatomy of a control loop (Fig. 32.1). Control loops originate at sensors and terminate at actuators. The control function may be packaged in a separate unit and consolidated for several loops (centralized), or it may be colocated (distributed) with the transducers.

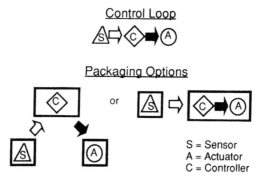

FIGURE 32.1 Control loop functions and packaging options.

The former choice arose during the late 1970s and early 1980s as an artifact of IC technology (see Table 32.1). Due to the high cost of ICs relative to transducers, it was more cost-effective to centralize control intelligence and amortize it across multiple loops, even with the expense of a dedicated controller housing, power supply, extra to-and-from wiring, and other disadvantages.

TABLE 32.1 Chronological Evolution of Transducers

Decade	Key Trend
1960s	*Development of pneumatic* and other mechanical transducers.
1970s	Miniaturization of mechanical transducers. Replacement of mechanical transducers with electromechanical and electronic devices. Use of transducers and instruments with addressable I/O. Inclusion of point-to-point digital and 4-20-mA loop-based electronic communication *capability in* transducers.
1980s	Use of digital buses such as RS-232 and, later, RS-422 and RS-485 for communicating between transducers and central controllers. Emergence of buses with lower-level protocols, such as CAN, Bitbus. Emergence of buses with higher-level protocols, such as, Interbus-S, Opto-22, Seriplex, Hart, ASI, and others.
1990s	Emergence of other buses with higher-level protocols optimized for discrete transducers, such as DeviceNet, SDS. Emergence of other buses with higher-level protocols optimized for process controls, such as, FIP, ProfiBus. Emergence of peer-to-peer distributed control networks such as LONWORKS technology; optimized for discrete as well as analog transducers. *Progress in* microelectromechanical structures (MEMS).
2000 to 2010	Implementation of >50 percent of all transducers as intelligent sensors with ability to operate *in peer-to-peer* distributed *systems.*

Over time, as the cost of silicon has dropped, it has become viable to package ICs with control intelligence in the transducers. Over this same time period, the cost of packaging has remained more or less constant, so that in effect, packaging as a percentage of total system cost has become more significant. As a result, it has also become desirable to eliminate the cost of the extra chassis, power supply, connectors, and wiring associated with the separate controller housing.

Of course, devolving control to the transducers also removes the central controller, a potential system bottleneck and point of failure, or relegates its role to less time-critical, supervisory tasks. Plus, as is discussed in the following sections, distributed systems offer better scalability, performance, and flexibility.

Clearly, all of these attributes make distributed control a desired architecture for control systems. And, as the IC industry races ahead at a rate even faster than that predicted by Moore's Law ("IC complexity doubles every 18 months") to provide chips with 100 million transistors by the turn of the century, is it inconceivable that in due course, the power of the Pentium will reside at every transducer at the cost of today's 8051 MCU? Distributed control is already available at costs economical enough to make it broadly compelling. The march towards higher levels of IC integration will add to the universality of its appeal.

The second conclusion, that peer-to-peer communication is the best scheme for distributed control, can be seen by looking at the communication needs of distributed systems. If intelligent devices communicate directly with one another, eliminating extra passes through a central controller, loops close with minimum elapsed time, minimum bandwidth use, and minimum risk of error or other failure. This is discussed further in the following sections.

Easier to Build Small or Large Systems . . . and Expand Them

Centralized systems become difficult to implement as the number of I/O points increases. The reasons include wiring congestion from point-to-point links to the central controller, the maximum I/O address space, and the scan rate (or other equivalent performance parameter) of the central processor. Although central controller manufacturers are migrating to multiplexed buses to mitigate problems of wiring congestion, cost, and maintenance, it is worth noting that with such buses, physical congestion may be replaced with bus saturation. The bus carrying traffic from all the sensors to a controller and back to actuators is susceptible to congestion, especially during critical conditions (e.g., multiple simultaneous alarms) since the total capacity needs of the devices to communicate to and from the controller can exceed its bandwidth. Compared to peer-to-peer communications, for equivalent control functionality, master/slave systems require at least twice the 305 bandwidth for unacknowledged traffic, and four times the bandwidth for acknowledged traffic. Errors and retransmissions further increase this multiplier. Loops closed within an intelligent device (e.g., a compound device that may contain a sensor and transducer) also exacerbate the difference.

One attempt to overcome the limits of a central controller is to connect multiple centralized systems together. For example, in process controls, this architecture is reflected in DCSS. In factory automation, it is reflected in the ControlNet/DeviceNet hierarchy recently announced by Allen-Bradley. It was also the canonical architecture of the erstwhile MAP (Manufacturing Automation Protocol).

In systems containing multiple central controllers, controllers executing the centralized control strategy typically also act as communication controllers for the smart transducers. Plus, they perform the role of surrogate application-layer gateways or bridges to other controllers. Given that they are designed more for control than transparent communications, the data available from one subsystem across the gateway to another subsystem may be limited and inflexible. This prevents the vision of total system control from being realized, since in these architectures, it is extremely hard if not impossible to view any sensor or control any actuator from any controller.

Scaling a centrally controlled system is also difficult. Consider a conveyor system consisting of tens or even hundreds of proximity sensors, limit sensors, motors, and so on, running at a nominal speed of, say, 0.5 ft/s. A web of wiring radiates from the controller box to the transducers. Now expand the picture to a complete assembly line over 100,000 square feet of manufacturing space in a small car plant. Or, consider running the small system at twice the speed, 1 ft/s, or both.

Scaling the size of the system requires addition of more controllers. Likewise, speeding it up requires replacing the controller and communications link with faster ones (assuming the transducers can communicate faster) or splitting up the task among multiple controllers and buses to maintain the needed scan rates. neither type of task is easy with centralized systems.

Now consider a distributed system. Wiring in distributed system increases at a far slower rate even as system size increases, since a new device can be attached with a single connection to the nearest point in the system or subsystem. Thus, the complexity disadvantage of centralized systems grows with size.

Also, multiplexing of traffic does not pose a problem with distributed systems, since traffic per device is lower. Small, highly compact, and low-cost routers that fit in enclosures as small as a 3-in-wide DIN box can also be used easily to filter traffic and contain most of it locally, where it generally pertains. The principle of locality applies as much to networking as to cache memory design (successive memory accesses are more likely within the cached block).

Peer-to-peer systems also allow scaling of the control intelligence easily. The intelligence for additional loops gets added simply as you go.

Better Loop Performance or Lower Cost or Both

In a single scan, a central controller such as a PLC has to read every input and write every output, all in a finite amount of time. Many, if not most, of the evaluations and actuations in a scan cycle are mutually unrelated. Yet their *individual* performance is limited by the scan time for *all* the I/O and by the performance of the *shared* controller.

Let us return to the example of a conveyor system. Using a PLC, multiple proximity sensors and a global ON/OFF switch report their state to the PLC, which then communicates with multiple motors and an alarm. This potentially creates a large number of evaluations and actuations per scan for the PLC. Increasing the speed of the conveyor from 0.5 to 1 ft/s may require a new PLC or may simply not be possible.

In a peer-to-peer system, the data from the proximity sensors is fed directly to the motors for the succeeding and preceding stages *simultaneously*. Both motors can act on the same data in their own respective ways that differ by context, such as the preceding by stopping and the succeeding by starting. Other proximity sensors in the line similarly communicate with neighboring motors. Occasionally, a motor or a proximity sensor may also send a message to a central alarm or a common message to other motors to stop and avoid a jam. Each such message is sent to all the pertinent devices in a given loop, directly and simultaneously, and only when needed.

As mentioned earlier, *peer-to-peer* communication clearly uses much less bandwidth. Plus, the control loops also execute faster. With no other unrelated intervening evaluations or commands distracting each loop controller, each loop closes faster. The sum total of faster communications and faster loop closure is higher conveyor speed or, viewed another way, the needed speed at much lower cost.

Better Flexibility from à la Carte Computing to à la Carte Controls

In a centralized system, adding more loops or devices to improve a manufacturing line is not a simple matter of connecting them into the system. It is limited by the capacity of the controller. Adding a sensor in one subsystem to the list of devices monitored in another subsystem or even at a supervisory station a level higher in the system hierarchy is even more difficult. As discussed earlier, communication through controllers which also act as surrogate gateways among subsystems is subject to the limitations of the controller.

Yet, the preceding represent common needs. In a processing plant, adding a few sensors in critical locations to better monitor the thickness of a coating may improve its durability and product quality or significantly improve manufacturing yield. In a factory, monitoring the feeder lines to all the conveyor subsystems via a common monitor station may involve monitoring individual devices across several subsystems. A printing press might require new feeders and sorters for extended use with new publications. In a building, tenant churning may require remapping climate control, lighting, and security zones. Evolving traffic in a growing neighborhood might require upgrading traffic lights at an intersection and timing them to a different set of relationships to other intersections.

These needs, and the impeding inflexibility of central control systems, are reminiscent of the problems with mainframe and minicomputer systems of a decade ago.

Those of us who remember being told by an MIS department to wait six months before a change in the monthly budget report could be implemented at a transfer cost of several thousand dollars welcomed the flexibility of desktop computing and the convenience of à la carte spreadsheet design at our own desks within minutes and can readily draw an analogy to the migration from centrally controlled systems to distributed ones.

As with PCs, control loops in distributed control systems can be added or changed autonomously. The programs in individual nodes and loops do not interfere with others.

"But We Don't Think We Can Completely Replace Our Controller"

This plaintive comment is heard often from systems suppliers, integrators, and end users. Yet distributed control does not imply complete *exclusion* of central controllers. Rather, it offers a *mixed* world, with distributed control used for the overall fabric as well as specific loops where it fits best, coexisting with central controllers managing other loops, often in the same system.

Central controllers with concentrated computing power may be required for computation-intensive loops. For example, a supervisory loop coordinating many subloops in a plant may process complex control strategies and handle a large amount of data. A PC, or a workstation, may be necessary for such a task. A very fast, fixed scan may be best served by a PLC connected via point-to-point links to the pertinent I/O (given that direct wires communicate faster than multiplexed buses) and linked into a larger distributed system. Or, two unrelated transducers (e.g., an imager involving a sensor array and an instrument involving floating point calculations in a complex PID loop) may each require substantial math performance from matrix transformations and floating-point operations and may be run on a high-performance microcontroller. Likewise, a person-machine interface (PMI) for the distributed control network may be hosted on a Windows PC.

Thus, distributed control adds a key architectural option to the existing framework of centralized solutions. More fundamentally, it offers a new open, multiplatform, multisourced framework into which various tailored solutions, including intelligent transducers, PLCs, microPLCs, DCSs, PCs, workstations, and other off-the-shelf and custom platforms can be easily added. For example, LONWORKS interfaces to parallel buses such as ISA, VME systems, PC/104, STD32, and Centronics parallel port already exist. Gateways to serial links such as EIA-232 and buses such as Opto-22, Modbus, Ethernet, and several others have also been implemented. These allow the integration of mix-and-match systems with a fabric of peer-to-peer controls and intelligent transducers.

The Bottom Line: A Matter of Pure Economics

As observed in the foregoing sections, distributed control saves money in several ways. Elimination of a central chassis, replacement of expensive controllers with lower-cost PCs, easier and lower-cost multiplex wiring, equivalent performance at lower cost, and longer system life cycles due to easy upgrades all save money.

Suppliers using distributed, peer-to-peer technology have been able to realize considerable cost savings over centralized systems. For example, a supplier of control systems for prisons has experienced cost savings of 40 percent. Imagine such cost savings for a large manufacturing corporation with a $100 million annual installation and replacement budget for control systems. The cost savings resulting from distributed control contribute directly to the bottom line and boost shareholder value.

IMPLEMENTING INTELLIGENT TRANSDUCERS

Whether migrating from smart transducers or leapfrogging from dumb devices, transducer manufacturers can make an easy transition to intelligent devices.

Implementing the Hardware: It's All in the IC, Network Transceiver, and I/O Objects

Imagine a networkable microPLC or an embedded computer implemented with an 8-bit microcontroller running a real-time kernel, integrated with powerful networking hardware and software as well as transducer interface logic and drivers, integrated into a single compact IC, and available at a total cost comparable to today's standard 8-bit microcontrollers, such as $3 to $4 in large quantities (100,000 units). Such a device is all that is needed to graduate a dumb transducer directly to an intelligent device.

The Neuron Chip, which is at the heart of every LONWORKS node, is an example of such a device. It has already been available for three years now.

Ongoing advances in MEMS (microelectromechanical systems) that allow integration of sensor and actuator microstructures with solid-state electronics, as well as the inexorable march of ICs toward higher levels of integration and lower cost per transistor, will further simplify hardware implementation over time. It is not inconceivable that in several years, the power of today's PC will reside in a single gang junction box or a small 12-mm proximity switch.

But What About the Software Development and System Integration?

For many transducer manufacturers, software has historically been someone else's problem. It is done in the controller and is therefore supplied partly by the controller supplier, the systems integrator or consultant, and the user (e.g., usage parameters). In companies that make controllers as well as transducers, the same cultural barrier must be crossed, since for a transducer/instrumentation division, software has always belonged to the platform or processor division.

In the absence of suitable aids, the transducer engineers would have reason to worry as network and control software invade the transducer package, since they must replicate a mini-PLC software environment in the transducer. Of course, as always, this software must be written quickly and should be bug-free. Even if it is available off the shelf, there is the task of developing the code to make the sensor smart (i.e., add calibration, linearization, and diagnostic software) or intelligent (add control strategy). Will this software operate bug-free? Will it work seamlessly with the operating and communications firmware?

There are also many system-level questions. How will the system be partitioned? Who will do systems integration? How will compatibility with code in complementary devices manufactured by other suppliers and even competitors be achieved for plug-and-play operation? Who will provide compatible user interfaces that will allow everyday users to use the system?

Interoperability

Before discussing the solutions, it is worth spending a little time on the subject of interoperability. For many, when they see the term *interoperability,* they think of *interchangeability.* The two are different. *Interchangeable* refers to devices that are exactly the same in all respects and can be exchanged with one another. *Interoperable* means devices can share data with one another and can be integrated into a system without the development of custom tools or gateways. Thus, interoperability addresses the issues of system integration and maintenance.

In most control systems, roughly 75 percent of the system's cost relates to system integration. Only 25 percent of the cost comes from the hardware.

Integration, installation, and commissioning cost factors include wiring, physical and logical installation of transducers, writing or debugging software to make devices work together, and addition of other system software such as PMIs and the like. Affecting hardware cost clearly provides severely diminished returns compared to these costs. Yet, transducer manufacturers may sometimes place a greater emphasis on hardware costs. This results from habits both users as well as transducer suppliers formed over the years in purchasing stand-alone devices, where the transducer is used by an operator as a part of a manual control loop.

The transducer manufacturer can affect the 75 percent factor in several ways. For example, the network medium supported by the transducer can impact physical installation. A free topology transceiver that eliminates the wiring restrictions of a traditional multiplexed bus, used with other transducers supporting such a choice, allows easier and lower-cost wiring.

However, as mentioned earlier, the interoperability quotient impacts the 75 percent factor in perhaps the biggest way. Designing for plug-and-play interoperability offers a transducer manufacturer a significant competitive advantage.

Parameters comprising the interoperability quotient include such items as use of standard interoperable transceivers, 100 percent compatible protocols, standard application-level interfaces, electronic self-documentation embedded in devices and standard formats for their extraction and use by standard installation and diagnostic tools, and support for a standard network database architecture.

Let us review each of these items briefly.

The framework of interoperability in a distributed control system must be built inside out, with the protocol as the innermost layer. The higher tasks in a device will work properly only if the underlying layers are interoperable.

Viewed in the framework of the seven-layer OSI reference model for networking protocols, standard transceivers known to meet physical layer (Layer 1) specifications of a standard are clearly the first rung in protocol compliance.

The implementation of the higher layers is usually done in firmware running on a processor host, with hardware assistance from a dedicated communications/control IC. Certainly, writing a compliant implementation of the protocol stack is an alternative that would be unproductive at best and a competitive disadvantage at worst, even for transducer suppliers with large economies of scale. Embarking on such a task is like building highways to get into the car business. Or, to use a more direct analogy, it is like having to write the microcode in a microprocessor to conform to a standard machine language, plus the compiler and assembler to conform to a mnemonic or HLL standard, before proceeding to program it in a higher-level language. Even if the time were available, such efforts are subject to the vagaries of incomplete documentation and individual interpretation, which cause incompatibilities. It is small surprise that microcodable processors went out of style nearly a decade ago.

In using an off-the-shelf protocol implementation, it is further important to ensure that it is either "inherently golden" or verified for 100 percent compliance to a gold standard. The latter is a very difficult if not impossible task for all but the simplest protocols. Yet 99 percent interoperability does not work, for the same reason that you and I do not buy 99 percent compatible PCs today in order to avoid the frustrations of mysterious and unpredictable incompatibilities and failure modes even with mainstream applications.

In practical terms, this means that some mechanism needs to be in place to certify that all implementations and interpretations of layers 2 through 6 of a protocol specification behave 100 percent identically in all ways under all circumstances. For example, in LONWORKS technology, this behavior is guaranteed by incorporating layers 2 through 6 into a silicon device, the Neuron Chip, much the way that microprocessors guarantee compatibility by implementing basic functions in microcode.

At layer 7, semantic conventions need to be defined to allow devices to exchange data with one another and to make reasonable interpretations of those data. Most modern systems accomplish this using some sort of data-driven device object model rather than older style command languages. Within objects, the use of a data- rather than command-driven paradigm facilitates interoperability.

Messages in traditional systems consist of commands issued by a controller and status returned by a device. The responsibility of constructing and interpreting the full repertoire of commands understood by each device in the system is that of the controller, resulting in cumbersome and extensive code for processing them. Addition of new devices requires the addition of new commands to the controller's repertoire. While such a paradigm is necessary for controller-centric systems, it is cumbersome and wasteful for peer-to-peer systems. Peer-to-peer systems communicate better with data messages.

Data messaging delivers three main benefits. First, it simplifies interpretation tables. Second, it allows easier device interoperability. Each destination can operate on this data depending on its context. A display can display it, an alarm can sound an alarm if a setpoint is exceeded, and a cooling fan an activate if its setpoint is exceeded. Third, it reduces network traffic since multiple devices can operate in their own ways on the same data, allowing simpler, multicast messaging.

Data sharing also delivers a fourth benefit for the transducer application programmer. It represents the same paradigm as data sharing using variables among programs located on the same node. Thus, from the programmer's point of view, he or she can use both variables the same way, with the exception of a simple modifier that distinguishes network data from local data.

Data messaging is also consistent with object-oriented programming methods. A change in the instrument or its linearization table or internal operational repertoire only affects its internal code. It does not require any modification at the device interface or at another device. This allows easier porting of a transducer model objects across devices and suppliers.

A canonical or cardinal repertoire of data messages with fixed semantic interpretation (e.g., temperature in Celsius) ensures correct use and common interpretation among multiple device types and suppliers, and thus facilitates interoperability further. For example, in LONWORKS technology, devices communicate using LONMARK objects. LONMARK objects contain control members (for example, the output reading of a sensor) and configuration members (for example, high and low limits and linearization tables). Data members use Standard Network Variable Types (SNVTs) to express their contents in standard engineering units. To further simplify development of devices and integration of devices into systems, functional profiles group sets of LONMARK objects into higher-level functionality. The contents of LONMARK objects, SNVTs, and functional profiles are administered by the LONMARK Interoperability Association, an independent group of OEMs, system integrators, and end users.

Perhaps the *least* important component in the interoperability framework is the standardization of user interfaces. While ensuring consistently and clearly defined data structures and data types within such a framework is important, the specification of a universal one-size-fits-all user interface is almost antithetical to the customization inherently desirable for UIs. In this regard, the focus of the IEC 1131-3 standard on consistent data types, and its recommendation of five choices for the user interface, delivers the right emphasis.

Developing Software for an Individual Transducer

The task of developing application code is simplified with the use of high-level languages such as Neuron C, which is based on ANSI C, with built-in extensions for control.

Modern code-generation methods and visual aids simplify this process even further. For example, the NodeBuilderTm development system for LONWORKS technology uses modern techniques such as wizards and code generation to simplify the creation of devices and to facilitate interoperability. By answering a set of questions and selecting choices from a series of menus, the transducer manufacturer can quickly create a complete program template compliant with a function profile. Such a template, using pretested code generators, automatically complies with the standard definitions of the LONMARK objects invoked. Within the template, the transducer supplier can focus on application-specific code such as diagnostics, linearization, and so on, that give a competitive advantage while offering a standard application-level interface. Interoperability at the physical layer is supported by selection of one of the standard LONMARK transceivers. Of course, the intermediate layers are inherently interoperable in the LonTalk protocol.

Using state-of-the-art tools such as the NodeBuilder tool, functional transducer software can be developed and debugged within hours.

System analysis tools can be added to the NodeBuilder tool for a complete toolbox. If device manufacturers want to examine network traffic, log packets from a device and so on, they can use the integrated browser or obtain a stand-alone protocol analyzer for more detailed yet symbolic high-level packet analysis.

Toolboxes for Verifying Multidevice Operation

System and subsystem suppliers and system integrators can similarly assemble a toolkit or use an integrated toolbox. LONWORKS technology offers both options. The LonBuilder@ Development System offers an integrated environment for code and system development, debug, and installation. It also supports simultaneous emulation of multiple nodes.

Systems Integration and Maintenance

Three types of installation-specific information is necessary to turn a set of intelligent transducers into a functioning control system: (1) control strategies, setpoints, or other aspects of the control loops that are not generic to a transducer, but rather unique to a control system, (2) logical connections among devices based on how the control loops straddle the intelligent transducers, and (3) transducer calibration, PID tuning, and so forth, as a part of the commissioning process.

A basic set of core services and system database architectures constitute the skeleton around which a rich set of industry- and application-specific tools can be deployed for the preceding tasks. In the emerging open framework of control systems, this area is expected to experience an onslaught of innovative third-party software with tailored UIs, based on advanced graphics.

The range of such software includes products which conform to industry-standard user interfaces, such as IEC 1131 for machine control and factory automation, to yet-to-be-invented paradigms down the Holy Grail of highly user-friendly UIs for home automation. Of course, with minor modifications, these UIs and paradigms can be applied to centralized or distributed systems. Thus, methods familiar to users of centralized systems such as ladder logic are being carried forward and make the transition to distributed systems transparent to end users.

Interapplication and internode communication and embedding methods such as DDE (Dynamic Data Exchange), netDDE, and increasingly in the future, OLE (Object Linking and Embedding) carry the richness and flexibility of the preceding tools even further, especially in distributed environments. Object-oriented frameworks such as Microsoft's COM (Common Object Model), OMG's CORBA (Common Object Request Broker Architecture), and others will help further with reusable code and interoperable objects. Visual programming tools have already entered the world of programmers. Packaged suitably, they will also provide toolboxes for the end user to extend their user interfaces. In the near future, home control systems may allow users to create a graphical representation of their security systems. Once done, they can integrate them into their graphical navigators and invoke them seamlessly to arm the systems while sitting at their computers in the late evening exchanging e-mail with their stockbrokers over the information superhighway. Or, while at work, users will be able to change the temperature in their offices by merely invoking the microclimate agent and giving it colloquial instructions in the middle of their spreadsheet analyses. Or, while away from home on a business trip, it will allow users to visually rescript the lived-in-profile of their homes, including reprogramming the timing of their entertainment systems, lights, and other items of use, easily.

As building blocks, these frameworks and tools will facilitate distributed control systems. As components of open architectures, they will in turn flourish with open, interoperable, distributed systems.

The skeleton of services provided by the distributed control protocol will manage the databases, carrying the configuration and programming data seamlessly to their target transducers, and reflecting it in system databases. In large multivendor systems where multiple operators may be trying to access or modify these databases simultaneously, suitable services will again support the demands of these multiple operators simultaneously, thus offering total building control or total plant control.

INTELLIGENT TRANSDUCERS AND SELF-DOCUMENTATION

Intelligent transducers can carry information in the form of a built-in data sheet or an integration instruction manual. Provided in a standardized fashion, this can be used by smart third-party installation tools and cuts systems-integration cost.

For example, a sophisticated yet efficient and flexible self-documentation scheme is defined in LONWORKS technology. The transducer manufacturer can include *network-readable* self-documentation in the form of ASCII strings at three levels—

transducer description, description of each of the objects that comprise its network interface, and description of each of the variables in each object. Other self-documentation can also be included in the program ID. Additional documentation can also be included in the field to describe location, or system information, in the form of the location ID.

During commissioning, the peer-to-peer system allows installation tools to be attached anywhere in the system and communicate with devices under installation.

PROBLEMS AND DIAGNOSIS

Centralized systems with dumb transducers allow limited diagnosis. At a central controller, there may be no distinction between the symptoms for transducer failure, line failure, or failure of the hardware or software related to the affected sensor at the controller. On the other hand, with intelligent transducers, failures can be pinpointed very precisely. Intelligent devices also provide much more data to diagnostic tools and, in fact, can even debug themselves.

Communication errors can also be isolated from device errors. Diagnostic services relating to communication errors may be built into the protocol to allow diagnostic tools to use them in a standard manner and to eliminate the need for the sensor developer to do extra work. This is the case with the LonTalk protocol, where local databases at each node record CRC errors, time-outs, resets, and so on.

Application-specific diagnostics can be added by the transducer manufacturer without restriction and can be tailored to the failure modes. For interoperable use by diagnostic tools, such services can be defined in function profiles of specific devices such as servo motor drives or pH meters. Diagnostic tool suppliers can use these standard services to offer a range of capabilities from simple shutdowns to neural net-based diagnoses.

SUMMARY

Distributed control systems with peer-to-peer communications among intelligent transducers and controllers offer many benefits.

With integrated low cost ICs, frameworks geared for plug-and-play interoperability and complete development, commissioning, and maintenance toolkits, the process of developing and deploying intelligent transducers has already become easy. As a result, end users are beginning to experience the tangible benefits of distributed control systems, such as cost savings of 40 percent.

In parallel, the cost of integrated computing and communication for intelligent transducers is dropping and will continue to drop.

Driven by demand-side economic benefits and supply-side technology advances, the march toward open, distributed control networks is reminiscent of the computing revolution away from mainframes and toward personal computer-based distributed systems that started 15 years ago and is now nearly complete.

Although 95 percent of transducers today are still dumb, transducer suppliers are using the opportunity to leapfrog their products directly to the intelligent, distributed world.

CHAPTER 33
PRINCIPLES AND APPLICATIONS OF ACOUSTIC SENSORS USED FOR GAS TEMPERATURE AND FLOW MEASUREMENT

INTRODUCTION

Since the early 1970s, sonic (acoustic) instrumentation systems have found a growing number of applications for measuring distance, temperature, and velocity in industrial processes. Of particular interest for this chapter is the field of acoustic pyrometry. The development of acoustic pyrometry into a commercially available technology has been underway in the United States and England since about 1984. Today, these systems are used in a wide range of applications. Acoustic pyrometry provides a practical technology for the online continuous temperature measurement of gas temperature and velocity in hostile furnace and stack environments. The technique provides average line-of-sight measurements between an acoustic transmitter and receiver. Multipath, side-to-side, front-to-back, and diagonal measurements within a furnace volume provides information on plane wide average temperatures or, using deconvolution methods, produces isothermal maps at any given plane of a furnace volume. The theory and application of acoustic pyrometry is discussed in some detail with many practical applications presented. The reader will gain an understanding of this new sensor technology and the benefits that it can bring to many industrial processes.

HISTORICAL REVIEW OF TEMPERATURE AND FLOW MEASUREMENTS [1, 2, 3, 4, 5]

Temperature is one of the basic variables in science and engineering. Throughout recorded history, humans have been aware of degrees of heat with what has appeared to be an infinite variation between hot and cold. The blazing sun, the frigid waters, the cool wind, and other such descriptions are used as a qualitative measure of temperature. During the early days of the Renaissance, scholars slowly accepted the need to observe the natural world to practice science. Some sources credit the discovery of the first instrument for measuring degrees of hot and cold to Galileo, the Italian professor of mathematics, who is best known for his invention of the telescope in 1608. One of his pupils, Vincenzo Viviani, in the *Life of Galileo* (1718), wrote: "that about the time Galileo took the chair of mathematics in Padua at the end of 1592, he invented the thermometer, a glass containing air and water." In addition, Francesco Sagredo of Venice wrote to Galileo on May 9, 1613, stating, "The instrument for measuring heat, which you invented, I have made in several convenient styles so that differences in temperature between one place and another can be determined." Instruments of the type used by Galileo were influenced by barometric pressure and now are called *barothermoscopes*.

The word *thermometer* first appeared in the literature in 1624, in a book by J. Leurechon, *La Rereation Mathematique*. The author described a thermometer as "an instrument of glass which has a little bulb above and a long neck below, or better a very slender tube, and it ends in a vase full of water.... Those who wish to determine changes by numbers and degrees draw a line all along the tube and divide it into eight degrees, according to the philosophers."

The first temperature-sensing instrument independent of pressure, which we would now recognize as a true thermometer, was invented by Ferdinand II, Grand Duke of Tuscany, in about 1654.[3] This sealed thermometer had a column of liquid that expanded as it was heated and indicated temperature by the movement of the leading edge of the liquid as it traversed a graduated scale (Fig. 33.1). The liquid used was alcohol, and the instrument consisted of a glass bulb to which a fine tube was attached. A glassblower named Iacobo Marioni made a number of thermometers of this type for the Academia del Cimento in Florence during the period 1657–1667. The design of the instruments evolved by a process of trial and error. The glassblower would fill up the tube to a certain point with alcohol, heat it to expel the air, and then seal up the end. The use of standard fixed points—that is, the freezing and boiling points of water, for example—was not generally accepted. An example of the complete arbitrariness of the times is the description of the construction of a thermometer scale by Magalotti (1667) who, as a member of the Academia del Cimento, wrote, "The next thing is to divide the neck of the instruments or tube into ten equal parts with compasses marking each of them with a knob of white enamel, and you may mark the intermediate divisions with green glass, or black enamel: These lesser divisions are best made by the eye, which practice will render easy."

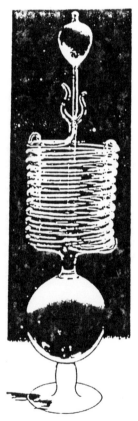

FIGURE 33.1 A Florentine thermometer, 17th century (300° Spiral Thermometer) (from [3]).

It was not until 1694 that Carlo Renaldini, who held the same chair of mathematics at Padua as did Galileo, suggested using the melting point of ice and boiling point of water as two fixed points on a thermometer. As with many good ideas, Renaldini's contribution to thermometry was unappreciated and subsequently forgotten. Newton, in 1701, independently defined a temperature scale based on two fixed points so that at such landmarks no ambiguity in the scales of various thermometers was apparent. For one fixed point, he chose the melting point of ice and labeled this as zero on his scale. As the second point, he chose the armpit temperature of a healthy Englishman and labeled this as 12. Based on his scale, Newton set the boiling point of water as 34.[2]

In 1706, Daniel Gabriel Fahrenheit, an instrument maker in Amsterdam, began making thermometers. All of Fahrenheit's instruments had one thing in common: "The degrees of their scales agree with one another, and their variations are within fixed limits." Fahrenheit found that, on his scale, the melting and boiling points of water at normal atmospheric pressure were approximately 32 and 212, respectively. These were soon adopted as the most reliable landmarks on the Fahrenheit scale. In 1742, Anders Celsius, professor of astronomy at the University of Uppsala, proposed a scale with 100 divisions between the melting and boiling points of water. The following year, Christin of Lyons independently suggested the familiar centigrade scale (now called the *Celsius scale*).

It was gradually realized that, no matter how many points were chosen, no one could define an acceptable temperature scale. A standard interpolation instrument and a standard interpolation procedure were essential if ambiguity at other than the fixed points was to be avoided. By the end of the eighteenth century, there were as many instruments for interpolating between the fixed points, making use of as many thermometric substances, as there were glass-blowing scientists; and, in general, no two empirically labeled thermometers in thermal equilibrium with each other gave the same indication of temperature.[1,21]

Finally, in the first half of the nineteenth century, a thermometer was developed based on the work of Boyle, Gay-Lussac, Clapeyron, and Regnault, which was founded on the expansion of air. These air thermometers were soon recognized as the instrument least liable to uncertain variations and became generally accepted as a standard for comparing all types of thermometers.

In 1848, William Thomson (Lord Kelvin) recognized that Sadi Carnot's analysis of 1824 of a reversible heat engine operating between isotherms and two adiabats provided a basis for defining an absolute thermometric scale. Thomson succeeded in defining completely the absolute thermodynamic temperature scale and related it to the ordinary Celsius by

$$K = °C + 273$$

Thomson, however, knew that his thermodynamic scale would be of little importance in practical thermometry unless some means were found to realize these absolute temperatures experimentally. Since it is known that no real gas follows the ideal gas law, the gas thermometer as suggested by Kelvin failed to provide a unique experimental determination of the thermodynamic temperature. Nevertheless, in 1887, it was agreed internationally that the constant-volume hydrogen gas thermometer should define the absolute temperature scale.[21] This was done even though such a scale (uncorrected) did not agree exactly with Thomson's thermodynamic scale or even with similar scales defined by other gas thermometers. In this sense, the hydrogen gas thermometer was certainly not universal. The hydrogen gas thermometer and the standard scale, however, did find use up until 1927. The International Practical Temperature Scale of 1927 (with several minor revisions in 1948, 1954, and 1960) is

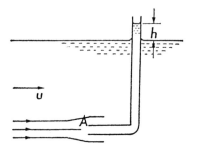

FIGURE 33.2 A Simple Pitot tube, (from [4]).

based upon six fixed and reproducible equilibrium states to which are assigned numerical values corresponding to the respective Celsius temperatures. The six primary fixed points are as defined by the Advisory Committee on Thermometry of the General Conference on Weights and Measures National Bureau of Standards (1961).

The measurement of fluid flow has always been a mark of civilization. For example, the Egyptians used indications of their Nile meters at selected sites along the river to estimate water availability to predict flow. One of the basic premises of fluid mechanics is that mass is conserved. This principle of continuity can be credited to Leonardo da Vinci at about 1502. Isaac Newton's second law of motion was first proposed in 1697. This law, along with the choice (arbitrary) of a standard of mass, fixes the force (hence of weight) and, on a given volume basis, sets the relation between specific weight and density. Christian Huygens and Gottfried Wilhelm Leibniz at about 1700 first discussed the conservation of energy during an elastic impact. It remained for Leonhard Euler, friend and contemporary of D. Bernoulli, to generalize the continuity principle on its current form.[3] Henri Pitot developed a method in 1732 for measuring velocities in the River Seine.[4] Figure 33.2 shows the device he used. A right-angled glass tube, large enough for capillary effects to be negligible, has one end (A) facing the flow. When equilibrium is attained, the fluid at A is stationary, and the pressure in the tube exceeds that of the surrounding stream by pu, where ρ = *the fluid density* and u = *the fluid velocity*.

The liquid is forced up the vertical part of the tube to a height h above the surrounding free surface, where

$$h = \frac{\Delta P}{w} = \frac{1}{2}\frac{\rho u^2}{w} = \frac{u^2}{2g}$$

$$\frac{\Delta p}{w} = \frac{1}{2}\frac{\rho u^2}{w} = u^2/2g$$

and where Δp = *differential pressure*
w = *specific weight of the fluid* = ρg
g = *weight per unit mass*

Such a tube is termed a Pitot tube and provides one of the most accurate means for measuring the velocity of a fluid.[4] Jean Charles Borda in 1766 introduced the concept of elementary stream tubes and showed that not only did the contraction of a jet influence the flow rate through a small opening, but also that a loss of energy accompanied such a flow.[2] In the 1840s, the principle of conservation of energy was extended to include the energies of heat and work by such men as Julius Robert Mayer and James Prescott Joule. In this same period, Julius Weisbach did important work concerning flow losses, contraction coefficients, and approach factors. He also greatly popularized the use of the Bernoulli equation in flow analyses. In 1895, Clemens Herschel invented the Venturi meter (Fig. 24.3) for measuring the discharge of a liquid.[2] In the twentieth century, Buckingham and Bean (NBS) and Beilter (OSU) advanced the art and science as flow measurement by their careful analyses

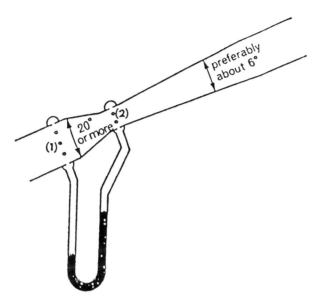

FIGURE 33.3 A Venturi-Meter, (from [4]).

and experiments, primarily on the determination of discharge coefficients of nozzles and orifices.[2]

There is an almost endless variety of instrument types used in the measurement of flow rates: variable-area or float types, electromagnetic types, force or fixed vane types, turbine or rotating vane types, thermal types, and now acoustic or sonic types. Whatever instrument is used, the inherent difficulties of accurate flow measurement make the process a complex one, and it is necessary to take into consideration the characteristics of the fluid being measured, the fluid surroundings, and the characteristics of the barriers, pipes, or channels used to control the flow.

For purposes of this chapter, the main emphasis will be the measurement of volumetric gas flow in the stack and/or large ducts associated with power plants.

It is beyond our present scope to discuss all of the many techniques and technologies devised to measure temperature and flow, because, in most cases, the selected method is specific to the application at hand. The excellent works by Benedict[1,2] however, will provide an in-depth discussion of the many classic sensors and methods available. The main intent of this chapter is to introduce the use of sound-wave propagation for the measurement of temperature and flow in gases. The main contribution that acoustics (acoustic pyrometry) offers is its ability to measure (non-contact) average temperatures over selected paths through a gas and provide useful isothermal temperature profiles for various industrial processes. Acoustic pyrometry is a new and exciting field of study that offers both qualitative and quantitative measurements of temperature. The availability of other than point measurements of temperature in near-real-time presents a new form of temperature data for many industrial processes.

The three main contributions that acoustics offers for the measurement of flow in large stacks and ducts is the fact that it can be used to measure average flow, it is not susceptible to pluggage, and it can meet the stringent requirements of the Clean Air Act for accuracy, reliability, and self calibration.

HIGH-TEMPERATURE GAS MEASUREMENTS

Gas temperature is one of the most important parameters to measure in a number of industrial processes. For example, the progressive design and operation of modern utility boilers, chemical recovery, and refuse boilers depends to an increasing extent on the critical evaluation of gas temperature conditions in the furnace and superheater sections of the equipment. Successful performance of such boilers must take into account the limitations imposed by metal temperatures of boiler tubes and the proper and complete combustion of the various fuels involved. The direct and accurate measurement of the gas temperature can provide valuable information to optimize the design and the operation of boilers and furnaces.

In all cases of contact gas temperature measurement, the temperature sensor approaches a temperature in equilibrium with the conditions of its environment. Although the sensor receives heat primarily by convection transfer from the hot gas in which it is immersed, it is also subject to heat exchange by radiation to and from the surrounding surfaces and by conduction through the sensor itself.

Thermocouples

An acceptable method for measuring high-temperature gases such as those found in boilers is through the use of *high-velocity* and *multiple-shield high-velocity thermocouples* (HVT and T). These probes were developed to correct for radiation effects and have been considered the most widely accepted method for measuring high-temperature gases. A typical high-velocity thermocouple probe is shown in Fig. 33.4. This portable assembly is used primarily for making test traverses in furnaces by insertion through inspection doors or other available openings.[6]

The thermocouple is supported by a water-cooled holder, consisting of concentric tubes of suitable length to span the traverse. The measuring junction is surrounded by a tubular porcelain radiation shield through which gas flow is induced at high velocity by an aspirator attached to the external connections. The gas aspiration rate over the thermocouple can be checked by an orifice incorporated with the aspirator and connected to the probe by a length of hose. The gas mass velocity over the thermocouple junction should be not less than 15,000 lbm/ft^2 h (see Fig. 33.4). Heat transfer to the junction and shield by convection is simultaneous, so that both approach the temperature of the gas stream. Radiation transfer at the junction is diminished by the shield. However, since the shield is exposed externally to the radiation effect of the surroundings, it may gain or lose heat, and its temperature may be somewhat different from that of the junction.

With the increasing size of steam generators, handling HVT probes of suitable length has become more difficult, but up until the invention of acoustic pyrometry, no acceptable substitute method of measuring high-temperature gases has been available. To minimize the manual effort involved in HVT gas temperature measurement, the probe itself must be lightened by using lighter gage stainless steel tubing and fittings, sheathed-type thermocouples, and a track-type support. Temperatures are recorded on a strip chart, reducing labor and providing a permanent record.

The use of HVT probes provides an intrusive method for measuring high-temperature gases. However, the procedure is very expensive and difficult to apply because of the equipment and labor involved, especially in the larger furnaces. Another limitation is the relatively long time required to gather temperature information, since a desired resolution may take many traverses. HVTs cannot be used for real-time control.

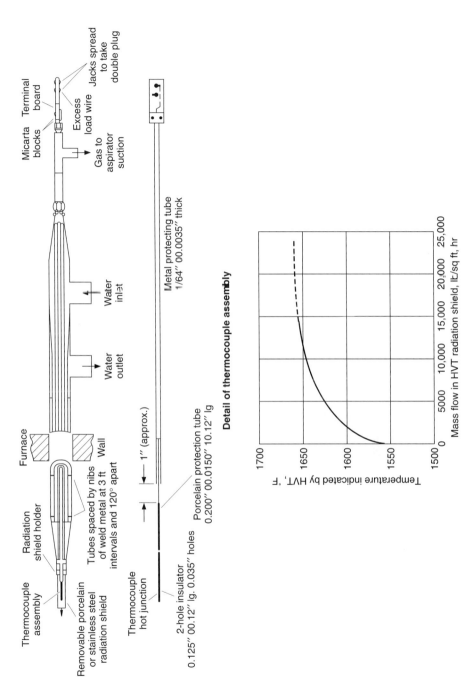

FIGURE 33.4 Gas temperature measurements using high-velocity thermocouple probes.

Optical Pyrometers and Radiation Pyrometers

There has been a continuing interest in developing nonintrusive techniques to obtain high-temperature gas measurements in near-real-time. Most nonintrusive methods to date are based in principle on optical or radiation pyrometers. All solid bodies emit thermal radiation. The amount is very small at low temperatures and large at high temperatures. The quantity of radiation may be calculated by the Stefan Boltzmann formula:[8]

$$\frac{q}{s} = \alpha \varepsilon T^4$$

where q = radiant thermal energy
 s = surface area, ft^2
 α = Stefan-Boltzman constant, $1.71 \times 109, h, T^4$
 ε = emissivity of the surface, a dimensionless number between 0 and 1
 T = absolute temperature, $R = F + 460$

By sighting an optical pyrometer on a hot object (Fig. 33.5), the object's brightness can be compared visually with the brightness of a calibrated source of radiation within the instrument, usually an electrically heated tungsten filament. A red filter may be used to restrict the comparison to a particular wavelength. This instrument is designed for measuring the temperature of surfaces with an emissivity of 1.0, which is equivalent to a blackbody (Fig. 33.6). By definition, a blackbody absorbs all radiation incident upon it, reflecting and transmitting none. When accurately calibrated, the pyrometer will give excellent results above 1500°F, provided its use is restricted to the application for which it is designed. Measurement of the temperature of the interior of a uniformly heated enclosure, like a muffle furnace, is such an application.

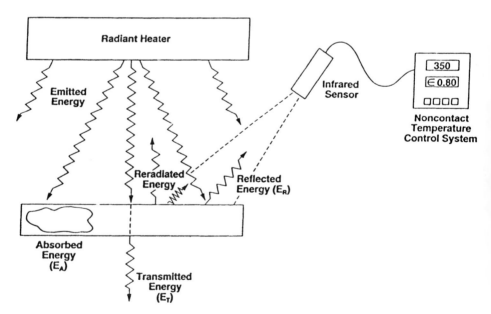

FIGURE 33.5 When an object intercepts energy, it absorbes, reflects, or reradiates that energy. How much energy the source emits depends upon temperature and emissivity, (from [8]).

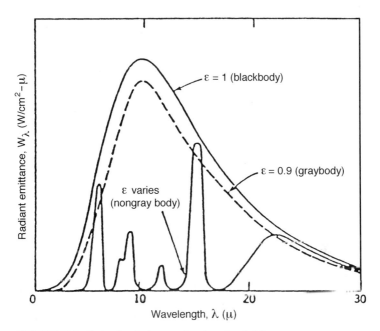

FIGURE 33.6 Emitted radiation as a function of emissivity.

When used to measure the temperature of a hot object in the open, the optical pyrometer will always read low, the error being small (20°F) for high-emissivity bodies such as steel ingots, and considerable (200–300°F) for unoxidized liquid steel or iron surfaces. The optical pyrometer has a wide field of application for temperature measurements in heating furnaces and around steel mills and iron foundries. It is of no value for the commercial measurement of gas temperature, because clean gasses do not radiate in the visible range.[7] In one type of radiation pyrometer, a radiation from the hot body, regardless of wavelength, is absorbed by the instrument. The heat absorption is measured by the temperature rise of a delicate thermocouple within the instrument, calibrated to indicate the temperature of the hot surface at which the pyrometer is sighted, on the assumption that the surface emissivity equals 1.0. The hot surface must fill the entire field of view of the instrument.[8]

The radiation pyrometer has been developed into a laboratory research instrument of extreme sensitivity and high precision over a wide range of temperature. The usual industrial instrument gives good results above 1000°F when used to measure temperatures of high-emissivity bodies such as the interiors of uniformly heated enclosures. Since operation is independent of human judgment, radiation pyrometers may be used as remotely operated indicators or recorders or in automatic control systems. Errors in measuring the temperatures of hot bodies with emissivities of less than 1.0, especially if they are in the open, are extremely large.[9,10]

Radiation pyrometers sensitive to selective wavelengths have been developed and give good results, measuring temperatures of bodies or flames utilizing the infrared band, for example. Infrared radiation is produced by all matter at temperatures above absolute zero, and a detector sensitive to infrared, such as lead sulfide, may be used to sense the radiation. Using a system of lenses, a lead sulfide cell, an amplifier, and an indicator, temperature measurements may be made of radiating bodies.

The radiation characteristics of hot gases, however, are not like those of hot solids, in that gases do not emit a continuous spectrum of radiation. Instead, the thermal emissivity of a gas exhibits a rapid variation with optical wavelength—that is, gases radiate or emit strongly only at certain characteristic wavelengths that correspond to absorption lines (see Figs. 33.6 and 33.7). It is clear that principles of radiation pyrometry entirely different from those just described are called for in the measurement of gas temperatures, and the appropriate literature should be consulted for further information.[11,12,13]

The accuracy of a temperature determination by the single-color optical pyrometer just discussed is based on blackbody furnace sightings or on known emissivities. A two-color radiation pyrometer, on the other hand, is used in an attempt to avoid the need for emissivity corrections. The principle of operation is that energy radiated at one color increases with temperature at a different rate from that at another color. The ratio of radiances at two different effective wavelengths is used to deduce the temperature. The two-color temperature should equal the actual temperature whenever the emissivity at the two wavelengths is the same. Unfortunately, this is seldom true.[2] Dual-wavelength (ratio) thermometry involves measuring the spectral energy at two different wavelengths (spectral bands).[12,13] The target temperature can be read directly from the instrument if the emissivity has the same value at both wavelengths. This type of instrument can also indicate the correct temperature of a target when the field of view is partially occluded by relatively cold materials such as dust, wire screens, and gray translucent windows in the sight path.

The theory of this design is quite simple and straightforward, and it is illustrated by the following equations, where we take Planck's equation for one wavelength and proportion it to the energy at a second wavelength.

$$R = L_{\lambda 1}/L_{\lambda 2} = [\varepsilon_{\lambda 1} C_1 \lambda_1^{-5} e_2^{-c/\lambda_1} T]/[\varepsilon_{\lambda 2} C_1 \lambda_2^{-5} e_2^{-c/\lambda_2} T]$$

$$R = (\varepsilon_{\lambda 1}/\varepsilon_{\lambda 2}) \cdot [\lambda_1/\lambda_1]^{-5} \cdot e^{[-(C2/T)(1/\lambda_1 - 1/\lambda_1)]}$$

$$R = [\lambda_1/\lambda_1]^{-5} \cdot e^{[-(C2/T)(1/\lambda_1 - 1/\lambda_1)]}$$

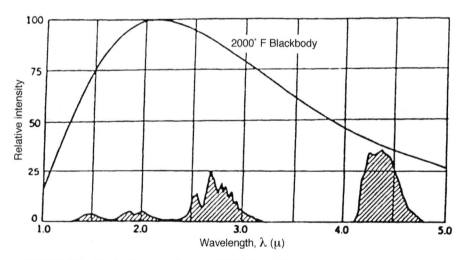

FIGURE 33.7 Combustion gas radiation at 2000°F (from [2]).

$$1/T = 1/T_r + [\ln(\varepsilon_{\lambda 1}/\varepsilon_{\lambda 2})]/(C_2)(1/\lambda 1 - 1/\lambda 2)$$

If $T_r = \varepsilon_{\lambda 2}$, then $T = T_r$

where R = spectral radiance ratio
T_r = ratio temperature of the surface
ε_λ = spectral emissivity

In this process, if the emissivity at both wavelengths is equal (graybody condition), the emissivity factor cancels out of the equation, and we find the ratio is directly proportional to temperature. However, there are always limits that must be recognized. The dual-wavelength does not perform on nongraybodies like gases; it also has difficulty looking through nongray windows or heated Pyrex; and it tends to measure background temperatures where the background is hotter than the target.[8,13]

An adaptation of an optical pyrometry has been developed for monitoring high-temperature gas streams in pulverized coal boilers. This instrument deduces the gas temperature by sensing the radiation emitted by the ash particles entrained within the gas stream.[14] Unlike infrared pyrometers, this instrument senses visible wavelengths that distinguish the radiation emitted by the hot ash from that emitted by the relatively cool boiler walls. The system has been installed at several pulverized coal fired utilities and appears to be working. The major drawbacks appear to be that this system cannot measure clean gases such as those found in gas fired or oil-fired boilers. Also, it appears that the measurement time of 90 seconds may present some real-time control problem. It is also noted that the temperature of the ash in the gas stream must be assumed to be the same as the average gas temperature within the field of view of the instrument. One other limitation is the operating temperature range. The minimum temperature that it can sense is 1850°F, which limits its usefulness for lower-temperature applications and boiler startup. The maximum temperature for operation is 2900°F.

The main point to be made here is that, in their current configuration, optical and radiation pyrometers are not suitable for directly determining gas temperatures.

In the next section, we will discuss a relatively new field of study called *acoustic pyrometry*, wherein sound waves can be used to provide temperature measurements in hot gasses.

ACOUSTIC PYROMETERS[15-37]

Background Information

It is well known that the speed of sound is a strong function of the temperature of the medium through which the sound wave travels. Acoustic pyrometry is not a new concept, as the earliest suggestion of its use may be found in a paper by Mayer (1873).[15] In ranging systems, this variation in sound speed is treated as an error that requires an appropriate correction; however, in acoustic pyrometry, the changes in sound speed provide the desired measurement.

The measurement of the flight time of sound for distance measurements in meteorological, hydrological, and industrial applications has been known for some time. In most of these cases, however, air was the gas through which the acoustic pulse passed, and distance (depth) was the desired result of the measurement. The determination of temperature, on the other hand, requires the measurement of the flight time of an acoustic pulse over a known distance. The result of this measurement, therefore, is the average temperature of the entire acoustic path. The principles involved in acoustic pyrometry are quite straightforward from a theoretical viewpoint. The speed of sound in a gas is related to gas temperature by the equation

$$c = (\gamma RT/M)^{1/2}$$

where γ = ratio of specific heats
 R = universal gas constant, 8.314 J/mole-K
 T = temperature, K
 M = molecular weight, kg/mole

In theory, all that is required for the measurement (see Fig. 33.8) is to install a sound source (transmitter on one side of the boiler and a receiver or microphone on the opposite side). A sound pulse is emitted from transmitter and received by the receiver. Since the distance is known and fixed, we can then simply compute the average temperature of the path traversed by the acoustic pulse. The practical application of this method, however, offers many challenges. For example, it has been experimentally determined that the practical frequency range for an acoustic

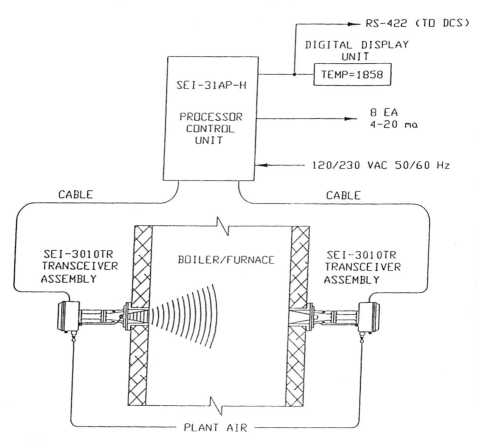

FIGURE 33.8 A single path acoustic pyrometer using air bursts as the source of sound.

pyrometer (in large-scale power plants) is between 500 Hz and 2000 Hz. The temperature of the gases involved can range up to 3000°F, resulting in acoustic velocities in excess of 879 and wavelengths on the order of 1 m. This means that the sound pulse flight time must be resolved to a fraction of one wavelength in order to obtain a practical temperature resolution and system accuracy. Problems are encountered when the acoustic path is disturbed by severe thermal and velocity gradients as well as by cavities between tube banks located at various points throughout the boiler.

The speed of sound is determined by measuring the flight time of an acoustic wave and dividing it by the distance traveled. Once the speed of sound is known, the temperature can be computed, as shown in following equations:

$$c = \frac{d}{\tau}$$

where c = speed of sound
  d = distance over which sound wave travels, m
  τ = flight time of acoustic wave m

Combining the last two equations gives an expression relating gas temperature to distance and flight time:

$$T_c = (d/B_1\tau)^2 - 273.16$$

where T_c = gas temperature, °C
  B_1 = acoustic constant = $(\gamma R/M)^{1/2}$ (SI)
  d = distance, m
  τ = flight time

For example, an acoustic pyrometer is to be operated across a furnace that is 18.29 m wide. The furnace is a coal-fired utility boiler where the flue gas is known to be:

Flue gas composition—dry volume (%)

Nitrogen (N_2)	82.0
Oxygen (O_2)	6.0
Carbon dioxide (CO_2)	12.0
Total	100.0

Percent moisture content by weight (kg moisture/kg dry gas)

Now say you want to calculate the path average temperature of the flue gas when the flight time of the acoustic wave is measured to be $r = 25.94 \times 10^{-3}$ seconds. Using the relationships described in Refs. 7 and 8, it can be shown that, for this flue gas composition of $y = 1.28$, $M = 29.24$, and $R = 8.314$, we can calculate a value for B:

$$B = \sqrt{(\gamma R/M)} = \sqrt{[(1.28 \cdot 8 \cdot 314)/(29.24 \cdot 10^{-3})]} = 19.08$$

Then, using an equation, we can calculate the average path temperature Tc as

$$T_c = \left(\frac{18.29}{(19.08) \cdot (25.94 \cdot 10^{-3})}\right)^2 - 273.16 = 1092.5°C$$

Acoustic pyrometry provides a practical approach for the online continuous temperature measurement of gas temperature and velocity in hostile furnace and stack environments. The technique provides average line-of-sight measurements between an acoustic transmitter and receiver. Multipath, side-to-side, front-to-back, and diagonal measurements within a furnace volume provide information on planewide average temperature. Using deconvolution methods can produce isothermal maps at any given plane of the furnace volume. See Fig. 33.9.

Acoustic pyrometry is a useful tool for the measurement of gas temperatures in the furnace and superheater regions in boilers. Results have shown that acoustic pyrometry is easy to apply, accurate, and noninvasive. Gas temperature measurements are provided and displayed on a continuous real-time basis for single path measurements in various furnace regions and for multiple path arrays in the furnace exit gas plane. The determination and presentation of isothermal maps of gas temperature provide a new tool for operation and maintenance diagnostics, not achievable by any other currently known method. Acoustic pyrometers are now being used to help identify and correct burner problems, slagging problems, and furnace overheating before these conditions can adversely affect operations. Acoustic pyrometers have also begun to replace thermal probes for startup and are used for the routine control of sootblowing. Alternate applications for acoustic pyrometers are the control of sorbent injection systems and heat rate determination.

Applications

***Thermal Probe Replacement.*[24]** Acoustic pyrometers can be used to measure furnace exit gas temperature (FEGT) during boiler startup, thus replacing conventional retractable thermal probes. Because of its wide dynamic range, the acoustic system can continue to provide valuable temperature information regarding FEGT after the boiler is placed online and over the full range of load conditions. Acoustic pyrometers may be used in several locations throughout a boiler. The drawbacks (short life, high maintenance) of conventional mechanical retractable thermal probes are widely known.

***Burner Control and Optimization.*[28, 29]** Acoustic thermal mapping is being used to provide operators with information for fuel and air balancing, detecting malfunctioning pulverizers, and optimizing tangential burners. One immediate benefit is in keeping the fireball off the walls in order to prevent wall tube fireside corrosion. The real-time visualization of temperature distribution in the control room allows meaningful operational changes to be made with an immediate indication of the response. An example of this use of acoustic pyrometer thermal mapping is shown in Fig. 33.9. The burners can be adjusted during various load conditions in order to achieve optimal performance and minimize wall tube fireball impingement. There is a savings in plant availability as well as plant life extension benefits due to less wall tube fireside erosion. For example, it has been recently reported that exposing tubes to 150°F excess temperature will shorten their design service from 117 years to only five months.[28] Fuel savings and related heat-rate savings have not as yet been determined. This particular system is now fully operational and has been online for several years.

A several month investigation in Germany on a 600-mw brown coal-fired boiler was recently conducted.[29] The effects of burner control, recirculated flue gas, staged air, and coal quality effects on temperature were studied. The acoustic pyrometer generated temperature distributions that could be correlated with boiler fouling behavior and emission values. NO, CO, and O_2 concentrations and the arithmetic mean value of the 24-path acoustic pyrometer-generated temperatures were shown plotted as a function of overfire air valve position. All parameters reached a mini-

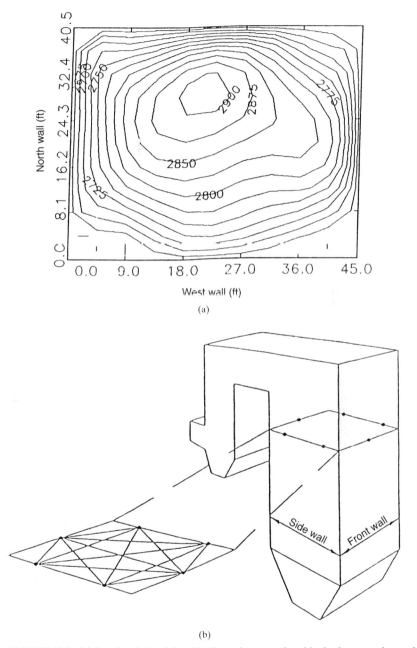

FIGURE 33.9 (*a*) Sample of a "real time" isothermal map produced in the furnace volume of the boiler described in (*b*); (*b*) Multiple path planar array using acoustic transceivers installed on a large utility boiler.

mum at a valve position of 70 percent, which was determined to be the usual desired operating status.

Slagging Measurements and Control[19–23, 27, 29]

Temperature measurements taken between superheater tube banks can provide valuable real-time indications of tube-bank performance and, hence, indicate the onset of tube fouling. This is possible because a baseline tube-bank in/out temperature differential trend is indicative of tube-bank fouling, since heat is no longer being transferred at the same rate from the gas side to the steam side of the tubes. Continuous measurement of FEGT in utility boilers provides a valuable operating tool with regard to furnace wall slagging and eventual secondary superheater fouling. Under certain operating conditions, a rise in FEGT indicates a reduction in furnace wall heat absorption due to slagging.

There are many benefits to be derived from slag monitoring, especially when a system can be operated in real-time, be noninvasive, and provide the data necessary to effect closed-loop control of the slagging problem. Acoustic pyrometry meets the requirements to solve these problems. The direct benefits from this are:

Improved boiler performance and plant thermal efficiency

Automatic closed-loop control of soot blowers, providing soot blowing only on an as-needed basis

Better control of slagging and/or fouling for different fuels, reducing the number of tube failures by properly controlling the duration and frequency of sootblower operations

Emission Reduction Using Sorbent Injection[25, 30, 31]

A major environmental problem confronting industry and the utilities today, is acid rain and the emissions of SO_2 and NO_x, particularly from coal-fired plants. The projected impact to the U.S. utility industry (and consumers) of applying conventional SO_2 control equipment, such as wet scrubbers, is measured in billions of dollars. With the recent need to reduce NO_x emissions, the problem has now been compounded dramatically.

With the current motivation to reduce simultaneously both SO_2 and NO_x, a single cost-effective sorbent is an attractive option for retrofit applications. A process has been reported wherein a lime-urea hydrate is injected into the combustion gases in the upper furnace region of a boiler.[30] This process has been shown to reduce simultaneously both SO_2 and NO_x emissions. This process does, however, require that the lime-urea hydrate be introduced within a critical temperature range ($850° \leq T \leq 1200°C$) in order to be effective. For example, at an injection temperature of 950°C, SO_2 and NO_x removals of 50 percent are achievable, which makes this process comparable to low-NO_x burners (when combined with selective catalytic reduction methods). However, since both SO_2 and NO_x are reduced, the new process may be much more cost-effective.

There are several other injection systems currently under study.[31] The selective noncatalytic reduction (SNCR) of NO_x using urea has proven to be an effective method in controlling NO_x from various stationary combustion sources. The SNCR of NO_x with urea is very temperature-dependent. At temperatures lower than the optimum reduction temperature, NO_x reduction is low, and reaction intermediates such as ammonia, CO, and H_2O result as byproducts. As temperature approaches the optimum, NO_x reduction increases, and by-product (slip) emissions decrease. At even higher temperatures, NO reduction and by-product emissions decrease with increasing temperature.

Refuse Fired Boilers[25, 32–35]

There are a number of serious problems that occur during the operation of WTE boilers that relate directly to the gas temperature. These problems range from serious waterwall and superheater tube failures[32–34] to emission control systems.[35] It appears that the primary interest at present in the United States is with the waterwall and superheater tube failure problems, while in Europe it is with tube wastage as well as the emissions generated by the combustion process. Acoustic pyrometry will prove useful in the future in WTE boilers by providing temperature distribution data to a control system that will minimize hotspots and thus reduce tube wastage. The automatic control of NO_x emissions via ammonia injection has already been extensively field tested in Germany.[35]

The injection process is a selective noncatalytic reduction (SNCR) process that involves the injection of ammonia (NH_3) into the high temperature flue gas in the 900°–1000°C temperature range. It is well known that gas temperature is the most critical factor in this process. There is a typical balance or trade off that must be considered for ammonia (NH_3), urea, or another nitrogen bearing compounds used for NO_x reduction. The requirement is to balance the ammonia slip against the NO_x removal efficiency. Lowering reagent costs is one obvious benefit. Also, the trade-off between emitting a pollutant (NO_x) or the hazardous gas (NH_3 slip) must be considered.

There have been a number of acoustic pyrometers applied to WTE boiler. Uses have ranged from single-path measurements for compliance reasons to full mapping or zonal systems for use with ammonia injection. One of the most dramatic applications to date has been recently reported.[25, 35]

It has been determined that the acoustic pyrometer data can be used to control directly the feed rate and grate with the following results:

- Stable heat output
- Reduction in the range in temperature variation from approximately 300°C to 100°C
- Reduction of O_2 variations
- Reduction of CO peaks

The dampening of the temperature fluctuations has also led to stabilizing the flue gas conditions prior to entering the SNCR unit. The furnace is then able to operate within the narrow window of opportunity for the NH_3 spraying process, thus providing a maximum of NO_x removal with the least amount of NH_3 slip.

Online Measurement of Boiler Performance and Unit Heat Rate

The input/output method, used widely for determining heat rate, requires measurement of coal flow rate and coal heating value. Although users have determined accurate average values for these coal parameters over periods of time, they have found it difficult to measure instantaneous coal flow rates and heating values. This method does not provide data that is sufficiently accurate or timely for online use.

Existing methods of measuring coal-fired unit heat-rate and boiler efficiency do not use instantaneous data and do not provide the accuracy needed for online economic dispatch of units. They also do not provide the accuracy needed for optimizing fireside operating parameters.

A new technique determines unit heat rate and boiler efficiency online using measurements of output and losses. The method is much less sensitive to measurement errors than existing methods.[36]

THE MEASUREMENT OF GAS FLOW IN LARGE DUCTS AND STACKS

The simplest method for determining the rate of flow of a fluid is to catch it in a suitable container and weigh it over a given interval of time. Such a procedure avoids, for example, consideration of how specific weight varies with temperature; it obviates the need for point-velocity measurements and an averaging technique; and it makes superfluous a measurement of the area involved in the flow.

It is not possible, however, to determine the flow of a gas using this method, since many gases cannot be condensed into a liquid under ordinary conditions. Also, the gas, if not condensed, would simply escape easily from any open container of the type needed to catch the flow. Also, in the case of a large flow such as found in a power plant stack, one would have to trap the entire flow, which, of course, is not possible.

This fundamental problem has led to an almost endless variety of instrument types to measure gas flow rates. However, recent experience has shown that three basic techniques can be applied with varying degrees of success to large ducts and stacks. These are ultrasonic transducers, thermal sensors, and differential pressure sensors. The problem with the application of these existing methods can be summed up in one statement: that is, if one uses one of these methods, a measurement is obtained; however, if two or more of these methods are applied, then one obtains an argument.

It is instructive to return to fundamentals in order to get a proper understanding of the basic measurement problem and what is needed to solve it.[37,38]

It is possible to define an average volumetric flow rate as

$$U_m = \text{total discharge/area} = \int u \, dA/A$$

For example, in a circular stack where the velocity profile $u(r)$ is given as a function of the radius, the average volumetric flow rate is calculated using

$$U_m = \text{total discharge/area} = \int u \, dA/A$$

where: $A = \pi R^2$
$d_A = 2 \pi r \, dr$

$$U_m = \frac{\int ur \, 2\pi r \, dr}{\pi R^2} = \frac{2}{R^2 \int_0^R r u r \, dr}$$

The currently accepted reference method for measuring volumetric flow in stacks involves the use of pressure differential sensors or pitot tubes to obtain the velocity at points of equal area of the cross sectional area of the stack. For example, Fig. 33.10 shows a case where the circular stack of cross-sectional area A has been

Traverse point	Distance, % of diameter
1	4.4
2	14.6
3	29.6
4	70.4
5	85.4
6	95.6

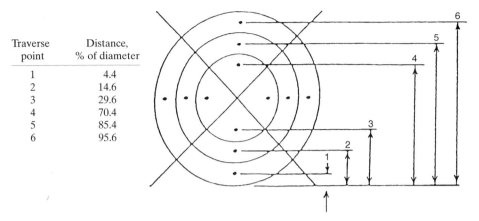

FIGURE 33.10 Example showing Circular Stack Cross-Section divided into 12 equal areas with location of traverse points indicated.

divided into twelve equal areas. An estimate of the average volumetric flow velocity is determined by using the following relationship:

$$U_m = \frac{\Sigma U_n A_n}{A} = \frac{A_i \Sigma U_n}{N A_i} = \frac{1}{N \Sigma U_n}$$

where: A_i = one segment of the equal area segments
N = the number of equal area segments
U_n = velocity measured at each point of equal area segment A_n

This relationship shows that one can estimate the average volumetric flow velocity by taking velocity measurements at each point of equal area and then taking the arithmetic mean of these measurements. It is clearly seen that a different result would be obtained if one were simply to take velocity measurements of equidistant points across the measurement plane and then take the arithmetic mean of these measurements. What would result in this case would be the *path averaged velocity*, which would be in error from the required measurement.

INSTRUMENTS USED TO MEASURE GAS FLOW IN DUCTS AND STACKS[37-42]

There appears to be at the present time three methods for measuring volumetric flow rate in stacks under serious consideration. These are instrument systems based on thermal dispersion, differential pressure, or ultrasonics. In order to understand better the benefits and limitations imposed by each approach, a brief overview is in order. It must be pointed out that, at the time of the writing of this chapter, there has been a lot of confusion and finger pointing going on among the principle manufacturers of the instruments based on each of these different methods. Each manufacturer claims to have solved the problem; however, to date few results are being made public on the tests that are being conducted by many utility companies. With this in mind, the following overview is presented.

Thermal Dispersion

This is the generic terminology for sensors that use *hot wire* or *hot film* cooling for operation. Fluids passing the heated sensor cool it at a rate proportional to the fluid velocity. The sensor is heated to a constant temperature approximately 75–100°F above the temperature of the gas flow being measured. The electrical power at any given stack temperature required to maintain a constant temperature in the velocity sensor is proportional to the convective cooling effects of the measured gas flow, and hence velocity is inferred.

A typical probe and installation is shown in Fig. 33.11. These type of sensors tend to get coated with materials that are present in the gas stream and have no inherent ability to rid themselves of buildup. This can greatly affect the heat transfer and hence the velocity measurement. Moisture, which is also usually present in the gas stream, is a problem in that moist fluids remove sensor heat at a rate dependent more on the percent of moisture than on the rate of flow. Other gas characteristics that vary with process conditions can also affect the thermal capacity measurement. These instruments comprise a bulky, cumbersome cross-stack array and appear to have a proven record only in clean environment applications.[39]

Differential Pressure Sensors

This is the generic terminology for sensors that are placed in a flow stream to extract flow-rate data from flowing fluids as a differential relationship between the upstream and downstream pressures. The differential pressure methods (i.e., pitot tubes and annubar) also require bulky, cumbersome arrays of primary measuring elements, subject to plugging and additional compensation.

A typical installation for a pitot tube assembly is shown in Fig. 33.12 and a typical installation for the annubar system is shown in Fig. 33.13.

Ultrasonic

This is the generic name for a class of sensors that infers gas flow using sonic pulses.

A recent paper[36] describes an ultrasonic method that is being used for measuring the velocity of a stack. This system is shown in Fig. 33.14. This ultrasonic anemometer relies on two transceivers, one on side of the stack, with one transceiver located downstream from the other by a *distance* of approximately one stack diameter. When a tone burst is transmitted from one of the transceivers, it is carried by the gas medium to the other transceiver. It can be easily shown[26] that, by transmitting and receiving the sound wave in opposite directions, the path average velocity of the gaseous medium can be determined from:

$$U_p = d/2 \cos\theta \left[\frac{(\tau_2 - \tau_1)}{\tau_1 \tau_2} \right] \text{m/sec}$$

where U_p = path average velocity of the gas (m/sec)
 d = distance between the transceivers (m)
 θ = angle in degrees (shown in Fig. 33.14)
 τ_1 = flight time of the sound with the gas flow (sec)
 τ_2 = flight time of the sound against the gas flow (sec)

ACOUSTIC SENSORS USED FOR GAS TEMPERATURE AND FLOW MEASUREMENT

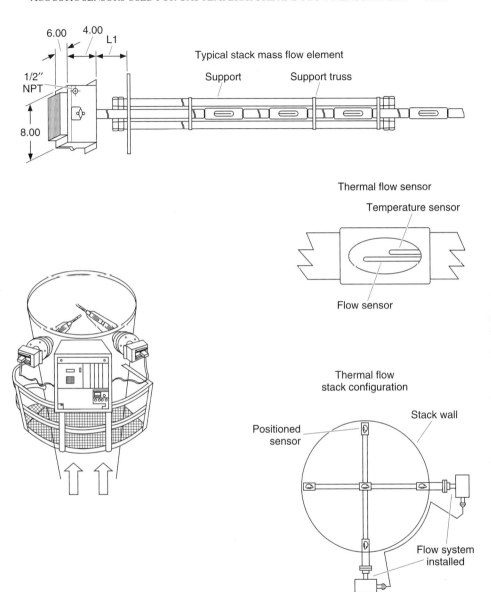

FIGURE 33.11 A typical thermal dispersion sensing system.

There are a number of errors that result from using this approach. For example, the path average velocity measured by a sonic instrument, for example, may be calculated using Ref. 26:

$$U_p = 1/R \int_0^R u(r)\, dr$$

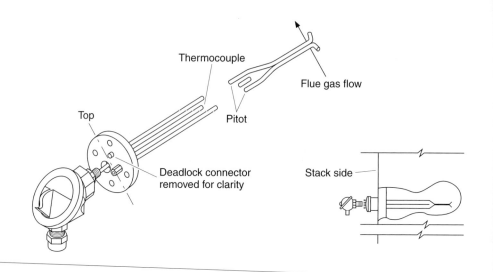
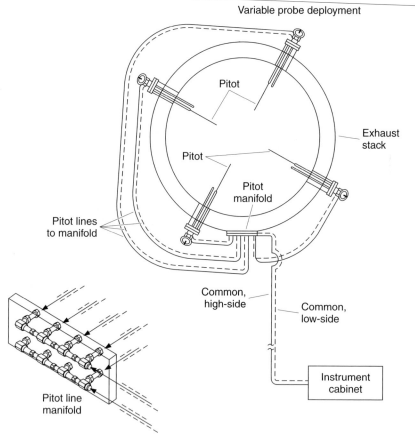

FIGURE 33.12 Typical Pitot tube installation.

ACOUSTIC SENSORS USED FOR GAS TEMPERATURE AND FLOW MEASUREMENT   33.23

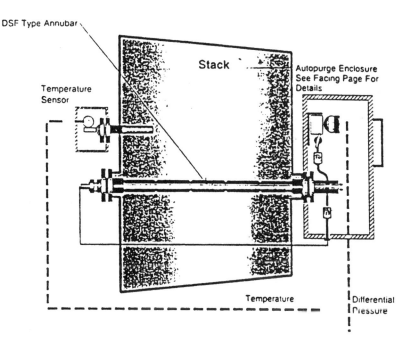

The DIAMOND II SENSOR is a multiple tube, diamond shaped, rigid structure with dual averaging chambers. These averaging chambers are of maximum size for easy purging to preclude any liklihood of plugging. Meets or exceeds the EPA's draft specification requirement for relative accuracy.

FIGURE 33.13   Typical Annubar installation.

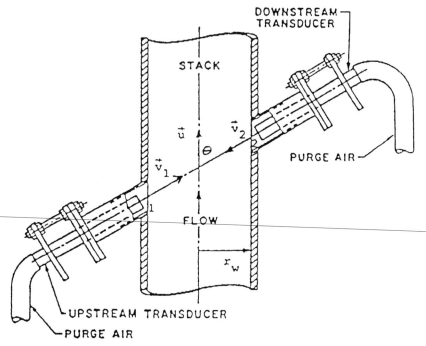

FIGURE 33.14 Ultrasonic Anemometer discussed in [18].

which differs from U_m by the factor shown below:

$$U_p/U_m = \frac{1/R \int_0^R u(r)\,dr}{2/R \int_0^R u(r)\,dr}$$

$$= R/2 \left[\frac{\int_0^R u(r)\,dr}{\int_0^R r\,u(r)\,dr}\right]$$

Note that the only case where the sonic measured flow equals the true flow is for the very special case where velocity profile is constant; that is

$$= R/2 \left[\frac{\int_0^R \int U\,dr}{\int_0^R rU\,dr}\right] = 1$$

This is a most important result, since an ultrasonic system measures only the path average velocity and not the desired volumetric average flow velocity unless the velocity profile is perfectly uniform over the entire cross-section of the circular stack. This, of course, is never true due to the effects of viscosity at the stack walls. For example, consider an ideal, fully developed turbulent flow in a circular stack that varies approximately in accordance with Prandtl's one-seventh power law.

$$u = U_{max} (y/R)^{1/7}$$

where u = velocity of the fluid at a distance $y(<R)$ from the stack wall
U_{max} = maximum velocity of the fluid that occurs at the center of the stack

The path average velocity is written as

$$U_p = \left(\frac{1}{R}\right) \int_0^R U_{max} \left(\frac{y}{R}\right)^{1/7} dr = \frac{7}{8} U_{max}$$

while the volumetric average velocity U_m is given by

$$U_m = \int u dA \,/\, \pi R^2 = (49 \, \pi \, R^2 \, U_{max})/(60 \, \pi \, R^2) = 49/60 \, U_{max}$$

The ratio of U_p to U_m is then given as

$$\frac{U_p}{U_m} = \frac{(7/8 \, U_{max})}{(49/60 \, U_{max})} = 1.07$$

which indicates an error of 7 percent.

It can be seen that an ultrasonic sensor only measures the path average velocity and not the volumetric average flow velocity, thus requiring some method to correct for this inherent error.

Other concerns that have been expressed by users of the ultrasonic approach relates to the need for having the transducers face each other at an angle to the stack wall. This must be done in order for the ultrasonic unit to take advantage of the beam-forming properties of the system, which improves the signal-to-noise ratio between the transceivers. The main drawback to this configuration is that the lower transceiver is exposed to moisture and/or contaminants that can run down the stack wall and into the transceiver unit. A purge air system (shown in Fig. 33.14) is used to minimize this problem. The system uses a fan or other source of air to blow out the contaminants. How well this will work and/or how one corrects for the effects of this extra flow of air is not yet reported.

One of the major limitations of the ultrasonic approach is the requirement that the instrument be located far enough from the duct inlet so that the velocity profile may be assumed to be symmetric and well-developed. This is required so that the axial component of velocity does not change with axial position. In practice, this usually requires that the sensors be located at a distance of at least 20 to 25 stack diameters from the flow inlet.[4] As a consequence, the accuracy of the ultrasonic instrument will suffer if it is utilized too close to the stack inlet, where the profile can be highly irregular and subject to changes with variations in plant load. This is also a problem where multiple furnaces are attached to a single stack, and the ducts involved are not long enough to assure plugged-flow conditions.

For rectangular ducts, it is necessary to use multiple sensors, since the velocity profiles in such flow passages have complex, two-dimensional shapes.

It is with these limitations in mind that the system described in the following paragraphs was developed. It will be shown that it is possible to use acoustic (sonic) techniques in conjunction with pitot tube and thermocouple technology to obtain a system that can accurately measure gas mass flow rates under varying conditions of velocity profile and also (given the proper data inputs) estimate the moisture content of the gas in real time.

A Practical Method for Obtaining Accurate and Reliable Measurements of Volumetric Flow Rates in Large Ducts and Stacks

The novel patented volumetric flow rate measurement system described in this paper known as the STACKWATCH® system uses sound waves much like an ultrasonic system but with some very important differences. The sound waves used in this system are in the mid audio range (less than 10 kHz) and hence are well below the high-frequency ultrasonic band. This provides a number of distinct advantages over the use of ultrasound.

There is much less attenuation of the transmitted signal in the audio range than there is at the ultrasonic frequencies.

Transducers based on ultrasound must be pointed at one another in order to function. This is because of the beam effect caused by the higher frequencies (shorter wavelengths). The lower frequency STACKWATCH® system, on the other hand, does not require that its transducers be pointed, since there is no beam formed at the lower frequencies used.

The system described herein uses as its sound source a simple blast of compressed air and not a complicated ultrasonic sound source. The low-frequency sound wave is generated by momentary operation of a standard air valve. The air blast is also used to clear away any debris and/or other contaminants that may accumulate on the transducers.

The STACKWATCH® system uses a combination of well-proven techniques (RTD thermal sensors and pitot tubes) in conjunction with acoustics in order to achieve a combined system that can obtain accurate and reliable volumetric flow-rate measurements. This system represents a major breakthrough in large duct/stack volumetric flow-rate measurement technology.

The system is shown mounted in a utility power plant stack in Fig. 33.15. A cross-section of the stack is shown. Details of the acoustic probes are shown in Fig. 33.16. The system operates using SEI's standard acoustic pyrometer (BOILERWATCH® 31AP-H) techniques. A blast of air is used as the sound source, which is picked up at both the transmitter and receiver probes. Special impulse response signal-processing methods are then used to measure the flight time of the random noise pattern generated by the air blast. RTD thermal sensors are also imbedded in the end of each probe. These thermal sensors are used to improve the reliability of the flight time measurements and also to provide the gas temperature measurement. Note that, by using these thermal sensors, reading the gas temperature measurement is not a function of the gas composition, which, as noted before, is a major source of error for the ultrasonic device.

Because the STACKWATCH® system is intrusive, the probes can also be used to support a pitot tube. The air blast that creates the sound source is also used to keep the pitot tube clean and free of fouling.

Notice that, because each probe is the same, there are four known velocity points. That is, the velocity is known to be zero at each wall, and the two other points are measured using standard pitot tubes. See Fig. 33.17. The STACKWATCH® system, in addition, measures acoustically the path average velocity as described earlier. Using the known velocity points and the acoustic average measurement, the system then determines an approximation for the actual flow distribution across the stack. Using this information, the system then continually upgrades the flow distribution and calculates the actual volumetric flow in real time.

ACOUSTIC SENSORS USED FOR GAS TEMPERATURE AND FLOW MEASUREMENT 33.27

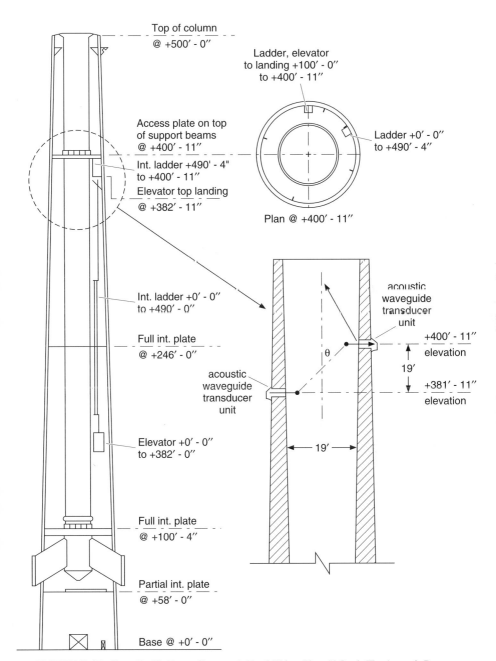

FIGURE 33.15 Sierra Pacific Power Company's North Valmy Plant #1 Stack. The Acoustic Pyrometer is located at the 382' and 400' levels.

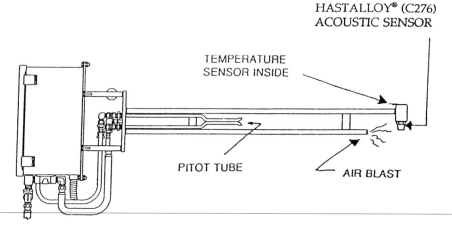

FIGURE 33.16 SEI Acoustic Probe.

This adaptive system provides a direct measurement for volumetric flow not available before. The system is inherently more accurate and reliable than any other technique devised thus far.[42]

It is also worthy of note that the STACKWATCH® system is the only truly self-contained volumetric flow-monitoring system on the market today. It measures all of the necessary parameters to calculate the flow in SCFH (wet); that is, T_s, P_s, M_{wg}, and U_m. This is truly a benefit, since the system errors are all combined in one single instrument package.

The STACKWATCH® system has been installed on several types of large ducts and stacks. Ratas have been conducted, with the STACKWATCH® system being compared to the reference method. All testing that has been completed to date indicates that, when the STACKWATCH® system is properly installed, it will operate well within the 7.5 percent relative accuracy flow-rate standards set by the Clean Air Act for the year 2000 and beyond.

CONCLUSIONS

Many installations and extensive field tests in utility and industrial boilers have shown that acoustic pyrometry is a practical approach for the online continuous measurement of gas temperature and velocity in hostile environments. This technique gives accurate, and timely measurements of average gas temperatures and velocities in a line-of-sight volume between an acoustic transmitter and receiver. Multiple path arrays provide side-to-side, front-to-back, and diagonal measurements as well as temperature distribution profiles in a plane.

Acoustic pyrometers are being used to measure furnace exit gas temperatures (FEGT) during boiler startup as a cost-effective replacement for conventional retractable thermal probes. Because of its wide dynamic range, the acoustic system can continue to provide valuable temperature information regarding FEGT after the boiler is placed online. Acoustic pyrometers may be used in several locations throughout a boiler, making them more applicable than optical pyrometers, which

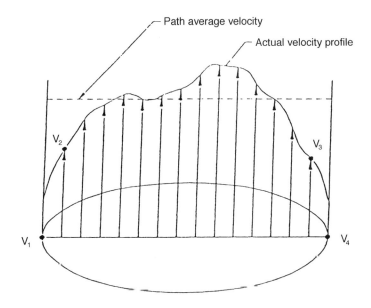

$V_1 = V_4 = 0$ at walls
V_2 = velocity at point, measured by pitot ube
V_3 = velocity at diametric point, measured by pitot tube

FIGURE 33.17 Actual velocity profile, path average velocity and four (4) points of known velocity.

require a minimum gas temperature of 1600°F. Temperature measurements taken between superheater banks can provide valuable real-time measurements of tube-bank performance and, hence, indicate the onset of tube fouling. This is possible because a differential trend in baseline tube-bank in/out temperature is indicative of tube-bank fouling, since heat is no longer being transferred at the same rate from the gas side to the steam side of the tubes. Continuous measurement of FEGT in utility boilers provides a valuable operating tool with regard to furnace-wall slagging and eventual secondary superheater slagging. Under certain operating conditions, a rise in FEGT indicates a reduction in furnace-wall heat absorption due to slagging.

For municipal refuse-boiler applications, continuous measurement of FEGT provides the operator valuable information needed to assure clean operation with reduced emissions. Certain toxic compounds can be thermally decomposed given sufficient time and temperature. An acoustic pyrometer provides an online method for monitoring combustion-zone temperature so corrective action can be taken.

Real-time direct measurement of furnace exit and convection-pass gas temperatures in process-recovery boilers provides operators the needed information so that operational adjustments can be made in a timely manner to avoid boiler operational upsets.

Acoustic pyrometry affords the boiler operator and the boiler design engineer a new source of information about the heat transfer driving potential of the combustion gases in the furnace enclosure and superheater tube banks. The real time and continuous measurement of gas temperature in areas hostile or unavailable to other sensor techniques, provides valuable data to affect control adjustments impacting combustion efficiency, emissions, the need for sootblowing, and temperature excesses that result in shortened tube life.

One of the most critical challenges for continuous emission monitoring systems (CEMS) has been the continuous real-time measurement of flue gas volumetric flow rate and moisture content. Flow-rate monitoring in CEMS still remains one of the least understood and hence the most controversial aspect of the new Clean Air Act regulations. This is primarily due to the fact that the utility industry still has only very limited experience in measuring volumetric flow rate in large ducts and stacks. The two major issues that must be understood when selecting a flow monitor for CEMS applications are compliance and allowance. It is essential that one clearly understands the differences between these two important aspects of the regulations. This chapter has also presented a practical method for the online measurement of volumetric flow rate in large ducts and stacks for CEMS application. The system uses proven methods based on thermal sensors and pitot tubes in combination with a proven acoustic pyrometer. This unique combination of field-proven technologies provides an online estimate for the actual flow-rate distribution in the stack. The final output of the system is a true measurement of the volumetric flow rate rather than the path average value. The equipment is simple in structure and has a minimum of its electronics on the stack. It requires a small amount of shop air to operate and is self-cleaning. The system is also economically expanded to multiple paths, should the complexity of the flow problem require additional data. A single-path SEI system can provide accuracy and resolution that could only be matched by using four conventional single-path systems. It is felt that the accurate measurement of volumetric flow rate will require a combination of techniques rather than one single technology.

Acoustic (Sonic) technology has advanced significantly within the past several years, with commercial systems now providing solutions to long-standing problems as well as finding their way into new and innovative practical applications. It is expected that this trend will continue.

REFERENCES

1. R. P. Benedict, "Review of Practical Thermometry," ASME paper 57-A-203, August 1957.
2. R. P. Benedict, *Fundamentals of Temperature, Pressure, and Flow Measurements*, 3d ed., John Wiley and Sons, New York, 1984.
3. N. Hawkes, "Early Scientific Instruments," Abbeville Press, New York, 1981, p. 157.
4. B. S. Massey, "Mechanics of Fluids", Van Nostrand, New Jersey, pp. 88–90.
5. S. Carnot, *Reflections on the Motive Power of Fire,* 1824, R. H. Thruston, trans., Dover Books, New York, 1960.
6. F. G. Ely, "Measurement of Gas Temperature in Boiler Furnaces," Babcock and Wilcox Co. Bul 3-555, 1969, pp. 1–15.
7. *Steam: Its Generation and Use,* 39th ed., Babcock and Wilcox Co.
8. R. Bushing, "Understanding and Applying IR Temperature Sensors", *Sensors,* October 1994, pp. 32–37.
9. E. K. Matthews, "Infrared Temperature Measurement with Emissivity Errors," *Proceedings Session 51 ISA 90 Conference,* New Orleans, October 1990.
10. H. J. Kostkowski, "The Accuracy and Precision of Measuring Temperature Above 1000°K," *Proc. Int. Symp. on High Temp. Tech.,* McGraw-Hill, 1959, p. 33.

11. R. H. Tourin and W. S. Tandler, "Monochromatic Radiation Pyrometry for Gas Temperature Measurement," paper presented at 11th Annual Symposium ISA, April 1959.
12. V. Lappe, "Applying Two-Color Infrared Pyrometry," *Sensors*, August 1994, pp. 12–15.
13. W. R. Barron, "The Principles of Infrared Thermometry," *Sensors*, December 1992, pp. 10–19.
14. H. R. Carter, J. T. Huston, M. R. Frish, P.G. Tremblay, A. A. Boni, J. R. Morency, M. E. Morgan, B. Dyas, and R. F. Afonson, "An Optical Monitor for Furnace and Boiler Gas Temperature Measurement and Control," presented at EPRI Workshop on Optical Sensing in Utility Applications, Philadelphia, August 1992.
15. A. M. Mayer, "On Acoustic Pyrometer," *Philosophical Magazine*, 45, 1873, pp. 1822.
16. S. G. Green and A. U. Woodham, "Rapid Furnace Temperature Distribution Measurement by Sonic Pyrometry," Central Electricity Generating Board, Marchwood Engineering Laboratories, Marchwood, Southampton, England, 1983.
17. J. A. Kleppe, S. P. Nuspl, E. P. Szmania, and P. R. Norton, "Applications of Digital Signal Processing to Sonic Measurements of Temperature in Large Power Plants," *Proc. Lasted Conf. on High Tech. in the Power Industry*, Bozeman, August 1986, pp. 50–54.
18. J. A. Kleppe, *Engineering Applications of Acoustics*, Artech House Press, Boston, 1989.
19. J. A. Kleppe, "The Measurement of Combustion Gas Temperature Using Acoustic Pyrometry," *Proceedings Instrument Society of America ISA 90*, New Orleans, October 1990, pp. 253–267.
20. G. E. Mizzell and J. A. Kleppe, "Applications of Acoustic Pyrometry in Boiler Testing," *Proceedings EPRI 1991 Heat Rate Conf.*, Scottsdale, May 1991.
21. J. A. Kleppe, "Acoustic Pyrometry Ups Boiler Efficiencies," *Instruments & Control System*, June, pp. 35–40, 1991.
22. L. G. Yori, F. H. Dams, and J. A. Kleppe, "Acoustic Pyrometers—Picture Windows to Boiler Performance," *Power*, August 1991, pp. 65–67.
23. L. G. Yori, "Computer Helps System Keep Close Tabs on Boiler Temperature," *Instruments & Control Systems*, November 1992, pp. 61–62.
24. D. D. Hilleman, R. J. Marcin, and J. A. Kleppe, "Application of Acoustic Pyrometry as a Replacement for Thermal Probes in Large Gas and Oil Fired Utility Boilers," *Proceedings Power-Gen 93*, November 1993, pp. 83–112.
25. J. A. Kleppe, "The Reduction of NOx and NH3 Slip in Waste-to-Energy Boilers Using Acoustic Pyrometry," *Proceedings Power-Gen 93*, November 1993, pp. 421–438.
26. J. A. Kleppe, "Sonic Instrumentation for Use in CEMS and Utility Boiler Applications," *Proceedings 1994 EPRI Heat Rate Improvement Conference*, Baltimore, May 1994.
27. J. A. Kleppe and L. G. Yori, "Acoustic Pyrometry—A New Tool for the Operation and Maintenance Diagnostics of Fossil-Fueled Utility Boilers," 1990.
28. Palermo, "Operating Temperatures Have a Strong Effect on Pressure Port Life," *Power Engineering*, March 1989, pp. 30–32.
29. W. Derechs, F. I-leb, K. Menzel, and E. Reinarz, "Acoustic Pyrometry: A Correlation Between Temperature Distributions and the Operating Conditions in a Brown Coal Fired Boiler," presented at the 2d International Conference on Combustion Technologies for a Clean Environment, Portugal, July 1993.
30. M. E. Teaque, R. E. Thompson, and L. J. Muzio, "Simultaneous NO_x/SO_2 Removal with Dry Sorbent Injection," presented at 1st Combined FGO and Dry SO_2 Control Symp., St. Louis, October 1988.
31. W. H. Sun, P. G. Carmignani, J. E. Hofmann, D. A. Prodan, D. E. Shore, L. J. Muzio, G. C. Quartucy, R. C. O'Sullivan, J. W. Stallings, and R. D. Testz, "Control of By-Product Emissions through Additives for Selective Non-Catalytic Reduction of NO_x with Urea," *Proceedings Power-Gen 93*, vol. 4, November 1993.

32. G. Sorrell and C. M. Schillmoller, "Alloys for High-Temperature Service in Municipal Waste Incinerators," A 5M First International Conf. on Heat Resistant Materials, Lake Geneva, Wisconsin, September 1991.
33. T. P. Steinbeck, "Managing Tube Metal Wastage in RDG-Fired Boilers at Elk River Station," presented at ASME National Waste Processing Conference, 1992.
34. S. Collins, "Slay the WTE Plant Dragon: Boiler Tube Wastage," *Power,* October 1992, pp. 42–46.
35. M. Deuster, G. Hentschel, and H. Voje, "Measurement of High Gas Temperatures with the Help of Sound Pyrometry—Measurement Principles, Installation Possibilities, and Operation Experience with Waste Incineration Plant," VDI Berkhte NM 1090, 1993.
36. E. K. Levy, N. Sarunac, J. Fernandes, S. Williams, and R. Leyse, "On-Line Performance Monitoring Using the Output/Loss Measurement System," *Proceedings 1992 EPRI Heat Rate Improvement Conf.,* Birmingham, November 1992.
37. J. A. Kleppe, "An Instrument That Measures Flow Rate in Large Ducts and Stacks," *Proceedings 1992 EPRI Heat Rate Improvement Conf.,* Birmingham, November 1992, pp. 20-1 to 20-20.
38. L. C. Lynnworth, "Ultrasonic Measurements for Process Control—Theory, Techniques, Applications, Academic Press, Inc., Boston, 1989.
39. C. E. Retis and R. E. Henry, "Continuous Emission Monitoring: A New Emphasis and Importance," presented at Sargent and Lundy Engineering Conference, Spring 1991, pp. 1–6.
40. R. L. Gielow, J. G. Konings, M. R. McNamee, and E. M. Petrill, "A Numerical Flow Modeling Technique for Positioning Flow Rate Monitoring Equipment in Breeching Ducts and Stacks," *Proceedings Power-Gen 93,* 1993, pp. 131–140.
41. E. Levy, N. Sarunac, M. D'Agostini, R. Curran, D. Cramer, S. Williams, and R. Leyse, "The Application of Ultrasonic Anemometry for Measuring Power Plant Stack Gas Flow Rate," presented at EPRI CEM Workshop, Atlanta, October 1991.
42. McRanie and S. S. Baker, "EPRI Flow Monitor Database," presented at CEMS Users Group Meeting, Baltimore, April 1993, pp. 1–8.

CHAPTER 34
PORTABLE PC-BASED DATA ACQUISITION: AN OVERVIEW

INTRODUCTION

Over the last several years, PC-based hardware and software tools have revolutionized the data acquisition and instrumentation market. PC-based data acquisition (DAQ) systems, able to take advantage of standard, inexpensive PC hardware, powerful software tools, and flexible instrumentation hardware, are giving users the ability to customize and build their own virtual instruments that are optimized for their particular application.

Traditionally, the cutting edge of PC technology was universally packaged in the standard desktop PC, which included several expansion bus slots—XT/ISA/EISA for IBM and compatible PCs, NuBus for Macintosh computers, or Micro Channel for IBM PS/2 computers, for example. Data acquisition instrumentation was easily integrated into the desktop PC with data acquisition (DAQ) boards that plugged directly into these bus slots. Accordingly, you can now choose from a great variety of plug-in multifunction I/O boards with a wide range of capabilities. Equipped with today's innovative software tools, users can easily combine the high performance of plug-in DAQ boards with the power and flexibility of the PC to build virtual instruments.

Virtual instrumentation has fundamentally changed the way scientists and engineers solve instrumentation tasks.

A recent, major development in virtual instrumentation has been the explosion of the compact, portable, and increasingly powerful notebook PC. Market researchers forecast that half of the PCs shipped in the next 12 months will be notebook PCs. With the computing power on par with high-end desktop systems, notebook PCs appear to be a natural platform for a whole new generation of portable DAQ systems.

PORTABLE APPLICATIONS

The need for portable, mobile instrumentation systems cuts across a very wide range of industries and applications.

- *Automotive/aerospace.* A portable PC-based DAQ system forms the basis of a powerful in-vehicle or in-flight data acquisition and monitoring system.
- *Manufacturing/process monitoring.* A portable DAQ system can be used as a very flexible diagnostic tool that the operator carries to the plant site and connects to the machine or process in question. The portable system is also useful in environments where standard AC power is not available or where space is too limited for a full-sized PC-based DAQ system.
- *Service/maintenance.* Many organizations are realizing the benefits of portable DAQ systems over traditional handheld instruments used by service personnel. Portable DAQ systems are carried by personnel to monitor tanks in a tank farm periodically, check wells in an oil field, or diagnose an automobile brought in for service.

The simple desire to make a system smaller and lighter, whether or not the system is intended to be portable, has also motivated the conversion of many traditional instrumentation and desktop PC systems to portable PC-based DAQ systems based on a notebook PC.

A LACK OF SLOTS

While notebook PCs open up a great variety of new applications to PC-based DAQ, notebook PCs universally lack the one feature that has been integral to the PC-based DAQ market: bus expansion slots for plug-in DAQ boards.

Initially, this void was filled with portable PCs that had the ability to accept one or two standard ISA bus plug-in boards. Often called *luggables* because of their hefty size and weight, these PCs were still relatively bulky and not really convenient for field use. Another option is an expansion chassis that adds bus slots to a notebook PC. These expansion chassis often dock directly underneath the notebook PC and accept standard ISA boards. However, these chassis only work with certain notebook PCs and tend to be relatively expensive. In addition, this approach is often not optimal for portable applications that must be battery-powered. The typical plug-in DAQ board can consume several watts of power, which can quickly drain a notebook PC battery.

ALTERNATIVES TO PLUG-IN DAQ

While the greatest number of choices and widest range of DAQ functionality can be found in plug-in DAQ boards, alternatives that are better optimized for portable applications are becoming more and more common. The traditional alternative to plug-in boards has been an external DAQ device cabled to the serial port. This solution, although convenient, cannot deliver the level of performance that is typical of plug-in DAQ boards. Fortunately, new standard PC technologies have emerged that are filling the void left by the disappearing ISA slots. Specifically, faster, more fully functional parallel ports and PCMCIA slots now provide high-performance expansion options for portable data acquisition applications. The main advantages and disadvantages of each of the three main approaches to portable DAQ applications—serial port DAQ devices, parallel port DAQ devices, and PCMCIA DAQ cards—are summarized in Table 34.1 and discussed in the following sections. See also Fig. 34.1.

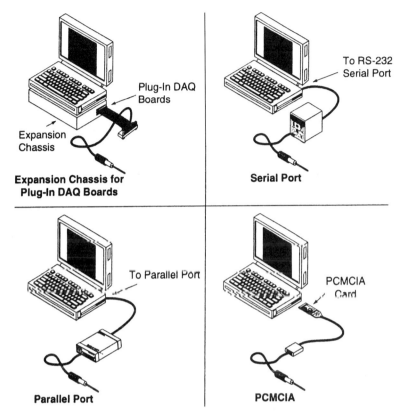

FIGURE 34.1 Alternatives for portable PC-based data acquisition.

SERIAL-PORT DAQ DEVICES

A traditional interface for external instruments and DAQ equipment is the RS-232C serial port, found on just about every PC. External instruments with an RS-232C serial port interface have been available for years. However, most of the existing serial port DAQ devices were designed for general-purpose PC interfacing and not for portable applications. Therefore, serial-port DAQ devices tend to be bulky and consume large amounts of power.

Most PCs include one or more RS-232C ports that can transfer data at rates at up to 19,200 baud (bits per second). At eight to ten bits per byte (allowing for one or two bits for communications and error checking), this is equivalent to a transfer rate of about 2 kilobytes per second. If the serial-port DAQ device is communicating binary data (12 or 16 bits per data point), each data point will require at least two bytes. However, many serial port devices do not transfer data this fast because they communicate over the RS-232C link using a high-level ASCII protocol, which requires the transmission of data *character by character*. Therefore, the data point value of "−9.309 V" requires 8 characters, plus any required delimiter characters. For this example, the practical data rate, ignoring other communications overhead, will be approximately 200 readings per second.

TABLE 34.1 Comparison of DAQ Systems Based on the Serial Port, Parallel Port, and PCMCIA

	Serial port devices	Parallel port devices	PCMCIA cards
Maximum raw data transfer rate*	Typically <2 kB/s	Standard (SPP): 150 kB/s EPP: 2 MB/s	5 MB/s
Form factor	Device external to PC; size and form factor vary	Device external to PC; size and form factor vary	PCMCIA card; <6 sq. in. of circuit space
Power scheme	Externally powered	Externally powered	Draws power from slot
Main advantages	Standard on PC and Macintosh computers	• Standard on PCs (EPP use is increasing) • High data transfer rate	• Most compact packaging • High data transfer rate • Additional power source not required • Hot insertion • Becoming standard on PC and Macintosh computers
Main disadvantages	• Slow data transfer rate • Most devices not designed for portability	• Lack of multiple parallel ports (unless device includes second pass through port)	• Functionality limited by card size

* The sustainable data rate will typically be much less than the maximum raw data rate and will depend heavily on *the device* (speed, onboard FIFO size, bytes per data point), PC, and software.

The relatively slow transfer rate of the serial port makes it necessary for many of these serial devices to include expensive onboard memory to store data until it can be later downloaded to the PC. Therefore, many serial port DAQ devices are designed as *data loggers*—standalone DAQ devices that can log large amounts of data to onboard memory.

The nature of the serial communications protocol lends itself well to long-distance communications, as opposed to local, high-speed data acquisition applications. Accordingly, different versions of the serial port implementation, such as RS-422 and RS-485, add capabilities for long-distance communication and attaching multiple devices to a single port. To connect a PC to an RS-422 or RS-485 port requires either an RS-232C adapter device or plug-in adapter card.

PARALLEL-PORT DAQ DEVICES

Like the serial port, the parallel port is standard on just about all desktop and notebook PCs. The parallel port, however, is designed for higher speed data transfer, making it very attractive for portable DAQ applications.

The original, standard parallel port (SPP) implemented on XT and ISA PCs, often referred to as a Centronics port, is an 8-bit port designed to communicate data from the PC to a printer. Designed for unidirectional data transfer, the SPP port can be adapted for both input and output by using four control lines of the parallel port as input lines. Taking software overhead into account, the SPP port is capable of a maximum data rate of 150 kilobytes per second in unidirectional mode. At two bytes per data point, this translates to a maximum data rate of 75 kS/s, and typical transfer rates in bidirectional mode will be closer to about 20 kS/s.

With the goal of modifying the parallel port to make it better suited for high-speed, bidirectional communications with a peripheral device, the IEEE Standard 1284-1994 recently defined the Standard Signaling Method for a Bidirectional Parallel Peripheral Interface for Personal Computers. This IEEE 1284 standard included the definition of the bidirectional Enhanced Capabilities Port (ECP) and the Enhanced Parallel Port (EPP), which add high-speed handshaking. Previously, all handshaking with Centronics ports was performed under program control by the CPU. These enhance speed transfer rates up to 2 MB/s, which is comparable to plug-in DAQ transfer rates. The 1284 standard also defines new, more compact connectors and cables.

Many of the newer notebook PCs now implement EPP ports, and the 80386SL microprocessor chip set includes EPP capability. The 1284 standard also includes a negotiation scheme that allows the PC to determine the type of parallel port capability of the external device and then use the proper mode—SPP, PS/2 bidirectional, ECP, or EPP. Therefore, DAQ devices designed with EPP capabilities can also work when connected to SPP ports.

Taking advantage of the speed of the parallel port, a number of external DAQ devices are available that can acquire data directly to the PC at relatively high data rates. The use of a high-speed communications interface removes the need for large amounts of memory in the DAQ device, and these parallel-port DAQ devices tend to operate with performance comparable to plug-in DAQ boards. However, the parallel port (SPP or EPP) does not include direct memory access (DMA) capabilities. Therefore, the sustainable transfer rates will depend heavily on the efficiency of the PC operating system and software, and the size of the onboard FIFO buffer in the DAQ device. An onboard FIFO buffer is a data buffer in the DAQ device that buffers acquired data before it is transferred to the PC, preventing loss of data during periodic delays in the servicing of the parallel port by the PC's CPU.

One potential drawback of the parallel port is the general lack of more than one port on a single notebook. Therefore, some parallel-port DAQ devices include a second pass-through port, which is a second parallel-port connector on the device with the ability to pass communications on to a second parallel-port device such as a printer (see Fig. 34.2). Therefore, with DAQ driver software designed to use the pass-through port, you can use both a parallel port DAQ device and a printer from a single parallel port.

PCMCIA DAQ CARDS

The PCMCIA (PC Memory Card International Association) "standard" was originally established in 1989 by an association of computer and peripheral manufacturers for building removable PC memory cards. In 1991, the group added I/O and interrupt capabilities, opening PCMCIA up to all types of I/O peripherals. Today, you can find PCMCIA cards for just about any type of application, including fax/modems, pagers, LAN adapters, cellular telephony, and even data acquisition and control. Although a relatively recent standard, most new, high-end notebook PCs today have PCMCIA slots. The trend is catching on for desktop PCs, also. According to estimates from BIS Strategic Decisions, 70 percent of all desktop PCs will include PCMCIA slots by 1998.

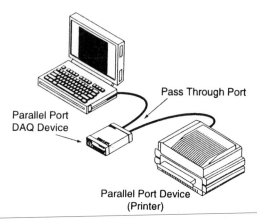

FIGURE 34.2 Parallel-port devices with a second pass-through port allow daisy-chaining to a second parallel-port device such as a printer.

The PCMCIA standard defines credit card-sized devices that come in three thicknesses: 3.3 mm (Type 1), 5.0 mm (Type 11), and 10.5 mm (Type III). All cards include a 68-pin connector, which interfaces to the address, data, control, and power lines of the PCMCIA bus. Most notebook PCs include a Type III slot, which by definition includes two connectors and can therefore accommodate either two Type II cards or one Type III card. PCMCIA also defines two layers of system software—socket services and card services—to manage the system hardware. For example, this software allows the very convenient feature of *hot insertion,* or the capability to insert, use, and remove PCMCIA cards without having to reboot the PC.

Data transfer from a PCMCIA card can reach burst rates up to 5 MB/s. Like with the parallel port, however, data transfers to and from the PCMCIA slot must be handled through processor interrupts without the benefits of DMA. Therefore, sustainable data acquisition rates can be limited by the interrupt latencies of the PC, which can be especially problematic under Windows. Therefore, high-speed acquisition requires the use of onboard FIFO buffers and efficient interrupt management.

Although probably the most attractive feature for DAQ applications, the ultimate compactness of PCMCIA is also the root of one of the big disadvantages. The limited area on a PCMCIA card (less than 6 sq. in.) constrains the amount of DAQ functionality that can be implemented on a card. While board layout and component technology continue to ease this constraint, an external DAQ device will always have access to more area on the circuit board and potentially to more DAQ functionality. See Fig. 34.3.

Cardbus, the successor of PCMCIA, will add many advanced capabilities that will be very important for portable DAQ applications. Cardbus, among other things, adds DMA capabilities and expands the bus to 32 bits.

POWER CONSIDERATIONS

Many portable instrumentation systems will be used where power outlets are not available, such as in-vehicle testing. These situations require systems that can run off

FIGURE 34.3 Example of a PCMCIA DAQ board.

batteries or other DC voltage source and therefore use this power efficiently. Therefore, low power consumption is a requirement for portable DAQ hardware. Plug-in DAQ boards and PCMCIA cards will draw the needed power from the notebook PC bus or slot. The average notebook PC consumes only about 8 W in normal operation. A rule of thumb is that the DAQ board or card should draw one-tenth or less this power so that it does not drastically affect battery life.

As mentioned earlier, plug-in DAQ boards that are not designed specifically for portable applications can draw anywhere from 5 to 10 W or more. PCMCIA cards designed for low-power consumption, on the other hand, are available with power requirements of 0.5 W or less, constituting a negligible drain on a normal notebook PC battery.

External DAQ devices that connect to the serial port or parallel port typically do not draw power from the PC. External parallel port DAQ devices designed for portability are typically designed with a DC power input, and include an optional AC adapter. Many of these DAQ devices are available with a rechargeable battery. The operating life of the battery will depend on the power of the battery, the DAQ device, and the nature of the DAQ device activity. Continuously scanning analog input channels will consume much more power than occasional monitoring of a digital input line.

CHAPTER 35
UNDERSTANDING AND APPLYING INTRINSIC SAFETY

INTRODUCTION

Intrinsic safety is a commonly used, but often misunderstood term in the instrumentation and sensor industry. Intrinsic safety prevents instruments and other low-voltage circuits in hazardous areas from releasing sufficient energy to ignite volatile gases. Intrinsic safety was invented around 1915 in Great Britain as a result of a series of mine explosions. A bell signaling circuit in the mine shaft created sparks which ignited the volatile gases in the mine. Intrinsically safe barriers were invented to prevent excess energy from reaching the hazardous area while still allowing the low voltage to operate properly.

Although intrinsic safety has been used in North America for many years, it has only been part of the National Electric Code (NEC) since 1990. Since that time there have been more and more intrinsically safe products introduced into the market making the proper selection of the barrier and instrument seem very difficult. The purpose of this chapter is to explain intrinsic safety technology so that sensors in hazardous areas operate safely and properly.

WHERE CAN INTRINSIC SAFETY BE USED?

First, one must define the hazardous area where the instruments or sensors are located. In North America these areas are defined by classes, divisions, and groups.

CLASS I, II, III: The class defines the type of materials that are in the hazardous area:

Class I   Flammable gases and vapors
Class II   Combustible dusts
Class III   Fibers and flyings

DIVISION 1 or 2: Hazardous areas are further broken down into two divisions:

Division 1   Normally hazardous
Division 2   Not normally hazardous

GROUPS A to G: The group designates the type of vapor or dust in the area:

Group A Acetylene
Group B Hydrogen
Group C Ethylene
Group D Propane
Group E Metal dust
Group F Coal dust
Group G Grain dust

The class and group ratings are relatively simple to define. However, the division rating is more subjective and will influence the equipment requirements needed to protect the sensors in the hazardous area. If the area is classified as division 1, more stringent safety means of protection are usually required than for division 2 locations.

METHODS TO PREVENT EXPLOSIONS

There needs to be three elements for an explosion to occur: oxygen, fuel, and ignition (Fig. 35.1). This is known as the *explosion triangle*. If one of these three elements is missing, the explosion cannot occur.

In North America, there are three primary means to prevent explosions: purging, explosion-proof enclosures, and intrinsic safety.

Purging removes the fuel from the area by forcing air or inert nitrogen through an interlocked enclosure. *Explosion-proof enclosures* contain the explosion. The resulting hot gases cool as they escape through specially designed threaded or machined flat joints. These cooled gases are not hot enough to ignite the gases in the hazardous area. Intrinsic safety is preferred by many in the sensor industry because it allows the user to maintain, calibrate, and work on the circuits while they are live without any danger of the circuit creating a spark large enough to ignite the gases or dusts in the hazardous area. This can reduce the amount of downtime in the process industry.

Intrinsic safety removes the ignition source of excessive heating or sparking by keeping the energy levels of voltage and current below the ignition points of the material. (See Fig. 35.1 of ignition curve.) As an example, a combination of 30 volts and 150 mA would fall on the ignition curve for hydrogen. Under the right conditions, this could create a spark large enough to ignite the gases. For intrinsically safe circuits, the maximum voltage and current produced under a fault condition on the safe side must always be below the curve. In this case, a combination of 24 volts and 100 mA would be considered safe. Intrinsically safe applications always stay below these curves where the operating level of energy for sensors is about 1 watt or less. There are also capacitance and inductance curves which must be examined in intrinsically safe circuits.

LIMITING THE ENERGY TO THE HAZARDOUS AREA

The intrinsically safe barrier is an energy-limiting device. Under normal conditions, the barrier is passive and allows the field device to function properly. Under fault conditions the barrier will limit excess voltage and current from reaching the field

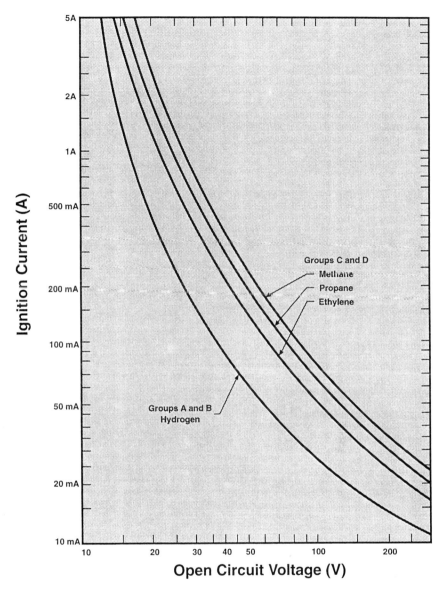

FIGURE 35.1 The explosion triangle.

device and igniting the gases in the hazardous area (Fig. 35.1). There are three components to a barrier that limit current and voltage: a resistor, at least two zener diodes, and a fuse. (See Fig. 35.2 of the barrier.) The resistor limits the current to a specific known value, known as *short circuit current*, I_{sc}. The zener diode limits the voltage to a value referred to as the *open circuit voltage*, V_{oc}. The fuse will blow when the diode conducts. This interrupts the circuit, which prevents the diode from burning and allow-

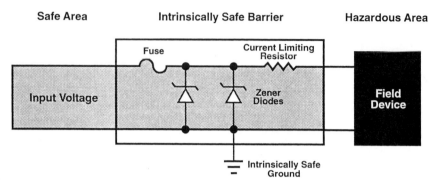

FIGURE 35.2 Barrier circuits.

ing excess voltage to reach the hazardous area. There always are at least two zener diodes in parallel in each intrinsically safe barrier. If one diode should fail, the other will operate providing complete protection.

A simple analogy is a restriction in a water pipe with an over-pressure shut-off and relief valve. The restriction prevents too much water from flowing through, just as the resistor in the barrier limits current. If too much pressure builds up behind the restriction, the over-pressure shut-off valve turns off all the flow in the pipe. This is similar to the zener diode and fuse with excess voltage. If the input voltage exceeds the allowable limit, the diode shorts the input voltage to ground and the fuse blows, shutting off electrical power to the hazardous area.

WHICH SENSORS AND INSTRUMENTS CAN BE MADE INTRINSICALLY SAFE?

When designing an intrinsically safe circuit, begin the analysis with the sensor or instrument which is referred to as the *intrinsically safe apparatus* or *field device*. This will determine the type of barrier that can be used so that the circuit functions properly under normal conditions but is still safe under fault conditions. The types of field devices that are most commonly used are shown in Table 35.1.

TABLE 35.1 Most Common Types of Field Devices

Intrinsically apparatus	Intrinsically safe applications (%)
Switching	32.0
Two-wire transmitters	22.0
Thermocouples and RTDs	13.0
Load cells	8.5
Solenoid valves	4.5
Potentiometers	2.5
LEDs	2.0
I/P transducers	2.0
Other devices	13.5
Total field devices	100.0 percent

More than 85 percent of all intrinsically safe circuits involve commonly known instruments and sensors. These devices, also referred to as *intrinsically safe apparatuses*, must be further classified as either simple or complex devices. *Simple apparatus* is defined in paragraph 3.12 of the ANSI/ISA-RP 12.6-1987 as any device which will neither generate nor store more than 1.2 volts, 0.1 amps, 25 mW, or 20 microJ. Examples are simple contacts, thermocouples, RTDS, LEDs, noninductive potentiometers, and resistors. These simple devices do not need to be approved as intrinsically safe. If they are connected to an approved intrinsically safe apparatus (barrier), the circuit is considered intrinsically safe.

A *nonsimple* or *complex device* can create or store levels of energy that exceed those previously listed. Typical examples are transmitters, transducers, solenoid valves, and relays. When those devices are approved as intrinsically safe, under the entity concept, they have the following entity parameters: V_{max} (maximum voltage allowed); I_{max} (maximum current allowed); C_i (internal capacitance); L_i (internal inductance). The V_{max} and I_{max} values are straightforward.

Under a fault condition, excess voltage or current could be transferred to the intrinsically safe apparatus (sensor). If the voltage or current exceeds the apparatus' V_{max} or I_{max}, the device can heat up or spark-ignite the gases in the hazardous area. The C_i and L_i values describe the sensor's ability to store energy the form of internal capacitance and internal inductance. A comparison of the entity values of intrinsically safe apparatus and associated apparatus is as follows:

Associated Apparatus (barrier)	Apparatus (sensor)
Open-circuit voltage	$V_O \leq V_{max}$
Short-circuit current	$I_{sc} \leq I_{max}$
Allowed capacitance	$C_a \geq C_i$
Allowed inductance	$L_a \geq L_i$

MAKE SURE THE CIRCUIT WORKS

It is very important to make sure that the circuit functions properly under normal conditions. With a current-limiting resistor in the barrier, a voltage drop will be between the input and output of the barrier. A voltage drop must be accounted for in the circuit design. The purpose of this chapter is to show how easy it is to make a circuit intrinsically safe.

Temperature Sensor: Thermocouples and RTDs

First, test your understanding of intrinsic safety examining a thermocouple circuit. Consider the ignition curves to demonstrate a point about thermocouples. A thermocouple is classified as a simple device. It will not create or store enough energy to ignite a mixture of volatile gases. However, if a thermocouple is installed in a hazardous area, it will not be considered intrinsically safe without a barrier between it and the recorder in the safe side. Why?

The reason is that a fault of 110 volts could be introduced into the recorder which could reach the hazardous area and ignite the gases. A barrier must be installed to limit the energy which could be created by a fault on the safe side.

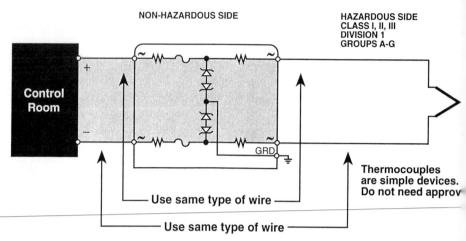

FIGURE 35.3 Typical values—thermocouple circuits.

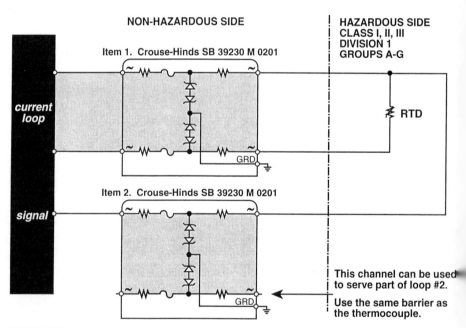

FIGURE 35.4 Intrinsically safe three-wire RTD circuit.

BARRIER TYPES

The most common barriers are classified as DC barriers which are rated for a DC voltage with a positive or negative potential and AC barriers which are rated for an AC voltage input. Since a thermocouple has both a positive and negative leg, one positive and one negative DC barrier or one AC barrier could be used. *To avoid polarity problems, select an AC barrier.*

RATED VOLTAGE

A thermocouple will generate a millivolt signal. The rated voltage (V_n) of a barrier must be equal to or larger than the voltage supplied to it. *Choose an AC barrier with a rated voltage* V_n *of 1 volt or higher.*

INTERNAL RESISTANCE

Since the millivolt signal has a very small current and is going to a high-impedance voltmeter, the resistance of the barrier will not influence circuit function. A simple rule of thumb is that when a signal is going to a high-impedance voltmeter, an internal resistance of less than 1000 ohms will not affect the mV signal. It is usually good practice to select a barrier with the lowest internal resistance in case the circuit is modified later.

For RTDs, the same analysis will draw similar conclusions. Selecting the correct barrier to make all thermocouples and RTDs intrinsically safe is not difficult. Use a double-channel AC barrier with a rated voltage of at least 1 volt with the lowest internal resistance. (See Figs. 35.3 and 35.4.) All thermocouples and RTDs are classified as simple devices and do not need approval. If they are connected to an approved intrinsically safe barrier, the circuits are intrinsically safe.

CHAPTER 36
APPLICATION OF ACOUSTIC, STRAIN, AND OPTICAL SENSORS TO NDE OF STEEL HIGHWAY BRIDGES

INTRODUCTION

Current bridge condition determination is based almost entirely on the use of visual inspection. This approach to bridge inspection provides data that is subjective and not traceable. Nondestructive evaluation (NDE) is a tool that in actuality is little used on bridges, but could eliminate much of the subjectivity of the input data for the bridge condition determination. A critical task for NDE on these structures is to detect and locate flaws that are growing and may eventually lead to serious impairment of the structure's ability to perform its designed function. This problem area is the focus for a bridge NDE program currently being conducted by Northwestern's Infrastructure Technology Institute (ITI). Under this program, all elements of the bridge inspection problem are being investigated by an interdisciplinary group consisting of members of Northwestern's faculty and BIRL staff. One of the major research areas of the program is the application of acoustic emission (AE) monitoring to steel bridges. AE monitoring is being combined with strain and optical sensors to develop a practical bridge inspection tool. This paper will presents the latest results of field tests conducted recently on bridges in California and Wisconsin.

Steel bridges may develop cracks in structural members resulting from a variety of causes. The cracks may have been produced during fabrication or grow from fabrication flaws, or they may be the result of fatigue damage. Not all cracks grow to failure. Most NDE methods currently in use can detect, locate, and to some degree size a crack, but they cannot determine if the crack is growing. Acoustic emission monitoring has the capability of detecting crack growth in real time. In fact, AE only responds to active flaws. This unique feature of AE makes it a prime candidate for crack characterization in highway bridges.

The acoustic emission monitoring discussed in this chapter was done using a computer-based monitor with six input channels. This device is a hardened field portable unit. The key feature of this AE monitoring system is the powerful pattern-recognition system that is applied in real time to the AE signals. This pattern-recognition algorithm was originally developed for in-process weld monitoring. It is based on empirical results that key on signal characteristics that allow crack-related information to be separated from a noisy background. The algorithm tests the rate of occurrence of the AE bursts, and when a group of bursts is received that exceeds the preprogrammed rate limit (typically 3 Hz), the algorithm evaluates the locational spread of the group of signals. If the high-rate group all came from a tightly clustered location (typically less than a 1-inch spread), the algorithm counts this group as one indication. The algorithm has been successfully applied to in-process weld monitoring on virtually every type of weld process and material that is commonly encountered in heavy fabrication. Since 1982, the same approach has been successfully applied to the in-service monitoring of steel highway bridges. This approach is the only known way that AE can be successfully applied to bridge details that are adjacent to or part of bolted splices. The fundamental problem with AE monitoring of these details is the noise produced by the bolts. The bolt-fretting imitates AE very well, and if the area to be monitored is not locationally isolatable from the bolts, the noise-rejection algorithm must be used to eliminate the irrelevant bolt noise.

Strain-gage monitoring in conjunction with the AE information provides additional useful information on cracks. The strain gage data can indicate the magnitude of the live strains in the vicinity of the flaw being characterized. In the following sections, we will discuss the application of this AE monitoring system to four different steel-bridge NDE problems.

WIDOT Structure B-5-158, Green Bay, Wisconsin

Wisconsin Department of Transportation Bridge B-5-158 is located in the city of Green Bay in Brown County, Wisconsin. The structure carries east and westbound I-43 traffic over the Fox River at the southern end of Green Bay. The total length of the structure is 7982 feet, including a 450-foot-long tied arch. The bridge was constructed in 1980.

In-depth inspection of the bridge by WIDOT personnel detected visual cracks inside the tie girders in the tied arch. Figure 36.1 shows a close-up of this detail.

The cracks were located in welds at the ends of 1-by-6-inch bars that join the bars to the hanger diaphragms at two sites. The bars, which serve as horizontal stiffeners,

FIGURE 36.1 Close-up of cracked detail.

are welded to the inside of the tie girder at the point of attachment of the floor beams. The welds that join the 1-by-6-inch bars to the tie girder web and the hanger diaphragm were fabricated using shielded metal arc welding (SMAW). They have rough unfinished reinforcements, which makes ultrasonic inspection very difficult to perform. The welds were supposed to have been full-penetration, and both visual as well as ultrasonic inspection indicate that this is not true.

WIDOT expressed an interest in gaining a better understanding of the nature of the visible cracks as well as additional information on the condition of the stiffener to web welds. The following discussions between BIRL and WIDOT regarding this project was initiated. The test program utilized a combination of acoustic emission and strain-gage monitoring to provide the needed information on live load and crack activity. The tests were performed by BIRL with assistance from the Kentucky Transportation Center (KTC).

On May 3, 1993, BIRL commenced testing. Test sites included hangers 4 and 6 on both the north and south tie girders. A total of six sites were monitored (the west side of hanger 4 on the north girder, the east side of hanger 6 on the south girder, and the east and west sides of hanger 6 on the north and hanger 4 on the south girder). These sites included all of the known cracks, sites that were adjacent to known cracks but with no known cracks, and sites that had no known cracks present either in or adjacent to the test site. The acoustic emission setup monitored both the stiffener-to-diaphragm and stiffener-to-web welds at each test site. Two strain gages were monitored at each of the test sites. Testing continued through May 13, 1993. Traffic loading during the tests included many large and obviously heavy loads. A wide range of environmental conditions were encountered, including high gusty winds and temperatures ranging from 37°F to 80°F. Test results are summarized in Table 36.1. These results show no detectable crack activity at any of the test sites and very low live strains. These test results imply that some other mechanism besides fatigue is responsible for the visible cracks.

TABLE 36.1 I-43 (B-5-15 9)

Site	Condition	Strain gage	AE
NG4W	No visible cracks	30 to 50 µ in./in.	No flaw indicated
NG4E	Visible cracks	30 to 50 µ in./in.	No flaw indicated
NG6W	No visible cracks	30 to 50 µ in./in.	No flaw indicated
SG4E	Visible cracks	30 to 50 µ in./in.	No flaw indicated
SG4W	No visible cracks	30 to 50 pin./in.	No flaw indicated
SG6E	No visible cracks	30 to 50 µ in./in.	No flaw indicated

Caltrans Structure B-28-153, Benicia Martinez, California

California Department of Transportation structure B-28-153 carries Interstate Highway 680 traffic across the Sacramento river at the east end of Carquinez strait thirty miles northeast of San Francisco, California. This 1.2-mile-long, high-level structure

FIGURE 36.2 The I-680 Bridge.

consists of ten steel deck-truss spans ranging in length from 330 to 528 feet. The bridge was designed by CALTRANS in the late 1950s and has been in continuous service since it was opened in 1963. An overall view of the structure is shown in Fig. 36.2.

The design uses built-up steel H-sections for the truss members and bolted connections. The steel H-sections were fabricated using T-I and ASTM A242 plates. Stay plates were welded to the flanges of the H-sections at the joints of the truss members. Cracks have been detected in these welds. Subsequent retrofit was performed on these details that included removal of the ends of the cracked plates by coring and adding bolted doubler plates. A typical crack site is shown in Fig. 36.3

AE has been applied with strain-gage monitoring to two of these crack sites. The first site was located in span 7 of the west truss at location L11, bottom plate. This crack was approximately ½ inch long. The second test site was located in span 5 of

FIGURE 36.3 Typical crack site.

the west truss at location U14, top. This crack was approximately ¾ inch long. The test results are summarized in Table 36.2 below.

TABLE 36.2 I-680 (B-28-253)

Test site	AE results	Strain gage
L11	No flaw activity	80 tin./in. max.
U14	2 flaw indications	360 tin./in. max.

The U14 flaw indications were at the end of the weld and near its midpoint. The AE indications coupled with relatively high live stresses indicates that this site (U14) has an active fatigue crack and should be closely watched.

Caltrans Structure B-22-26 R/L (Bryte Bend), Sacramento, California

The Bryte Bend bridge carries I-80 traffic over the Sacramento river near Sacramento, California. The bridge consists of two 4050-foot trapezoidal steel boxes 36 feet wide. Its approaches are 146.5-foot simple spans 8.5 feet deep with main spans of 370 feet and 281.5 feet in length at a depth of 15.5 feet. Flanges on the sloped side and vertical center web support the composite concrete deck. A view of the internal construction of the trapezoidal box is shown in Fig. 36.4. In-depth inspection by Caltrans personnel led to the discovery of cracks in the web of the trapezoidal box at the lower attachment point for the stiffener cross frames. BFRL engineers applied acoustic emission and strain gage monitoring to three crack sites to determine the nature of the cracks. Since this bridge is an all-welded structure, we were able to apply the acoustic emission using a simple guard channel approach (no extraneous noise sources). Strain gages were mounted on the web of the girder near the crack site. Two gages were mounted at each test site, and data was recorded in the rainflow

FIGURE 36.4 Internal construction of Trapezoidal Box.

mode. In the relatively short period of time taken for these tests, significant live strains were recorded with ranges of 200 microstrain and higher. The conclusion reached from the combined AE and strain gage tests were that the cracks were growing under fatigue loading. Discussions with Caltrans engineers subsequently led us to apply strain gages over longer time periods to obtain more statistically significant live strain histograms. This additional work was performed in June of 1994. An analysis of the strain-gage data further confirmed the fatigue findings. A summary of these tests is shown below in Table 36.3. Channel 1 was mounted on the horizontal stiffener transverse to the bridge axis. Channels 3 and 4 were mounted on the vertical web, with 3 as horizontal and 4, vertical. The counts to date are based on the life of the bridge, assuming uniform traffic volume.

TABLE 36.3 June 1994 Strain Gage Results
(range > 100 microstrain)

	Ch. 1	Ch. 2	Ch. 4
Total counts	4831	577	1389
Counts/hour	39.72	4.74	11.42
Counts/day	953.27	113.86	274.08
Total counts to date	8,002,687	955,817	2,300,918

WIDOT Structure B-47-40, Prescott, Wisconsin

Wisconsin DOT structure B-47-40 carries east and westbound traffic on U.S. Highway 10 over the St. Croix River in the town of Prescott in Pierce County, Wisconsin. The bridge consists of five spans and has an overall length of 682.7 feet. The center span (span 3) is a two-leaf, rolling bascule lift bridge with an overall length of 205.5 feet. The bridge deck is 66 feet in width and has four lanes of vehicular traffic and a pedestrian sidewalk.

The three-piece segmental castings that the bridge rolls on are attached to the bascule girders with high-strength friction bolts. This is a relatively recent design modification (8 to 10 years ago) and replaces the traditional turned bolts or rivets that were commonly used in this application. This design change is basically a cost-cutting measure. Wisconsin DOT bridge inspectors have observed cases of bolt failure and casting slippage on similarly constructed bridges. These occurrences can lead to dangerous operating conditions or structural failure.

Prior to this test, concerns for safe bridge operation by Wisconsin DOT bridge inspection personnel led them to call upon BIRL engineers to apply advanced NDE technology to the Tayco Street lift bridge (WI-DOT structure B70-97-93) in Menasha, Wisconsin, in September of 1993. This bridge utilized the same design modification as the Prescott bridge. Bridge operation and inspection personnel had observed loud audible impact noises during bridge operation. BIRL engineers applied acoustic emission monitoring techniques to determine that the source of the impact noises was the high-strength bolts. Strain gages were applied to the casting/girder interface, and large permanent displacements of the casting with respect to the girder were observed.

The loud impact noises that were observed on the Tayco Street bridge were not observed during operation of the Prescott bridge.

However, continuing concerns on the part of WI-DOT inspection personnel over the performance of the friction bolts led to an agreement with BIRL to perform the

APPLICATION OF ACOUSTIC, STRAIN, AND OPTICAL SENSORS TO NDE **36.7**

type of testing on this structure that was previously applied to the Tayco Street bridge.

Tests utilizing acoustic emission and strain gage monitoring were performed. Additional experiments were performed using a displacement sensor based on laser triangulation.

The tests were completed on November 8 and 9, 1994.

ACOUSTIC EMISSION TESTING

AE testing was performed on each of the four casting assemblies using a field-portable AE monitoring system. The test procedure was developed during a series of experiments performed on a similar bridge in Menasha, Wisconsin, during September of 1993. AE sensors were attached to each of the three casting segments. An additional sensor was attached in the vicinity of the pinion gear on the upper part of the bascule girder. The sensors were 175-kHz resonant piezoelectric devices. Silicone grease was used for acoustic complant, and magnetic holddowns were used to clamp the sensors to the structure. Unity gain-line-driving preamplifiers were used in each signal line to eliminate cable-loading effects on the sensors. AE system gain was set at 40 db on each channel and was checked using an electronic pulser and AE transducer as a simulated AE source. AE data were recorded on disk files using a portable PC attached to the AE monitor via an RS232C serial port. The recorded data was analyzed posttest. The casting mounted sensors were used to detect any impact or fretting events related to the casting attachments. The pinion mounted sensor was utilized to intercept drive gear and deck-related AE signals and to act as a guard for the casting-mounted sensors. The AE parameters of interest for these tests are hits and average relative energy. A hit is defined as the receipt by one sensor of an AE burst. The sensor that first receives the burst is the one most closely located to the AE source. The arrangement of sensors for these tests ensures that any signal that hits sensor 1 first (sensor 1 is mounted on casting segment 1) has to originate in that casting segment. The same holds true for the remaining two casting-mounted sensors. Bursts that arrive at the sensor mounted near the pinion gear first, originate either in the drive gear or the upper bridge structure.

Examination of the three casting-mounted sensor's hits and energy shows that the Prescott bridge had both lower hit counts and much lower energy. The difference in AE test data between the two bridges is even more apparent if we look at the product of event counts and their average energy, which is a relative measure of the total detected AE activity resulting from a complete bridge operating cycle. Table 36.4 shows this result for the two bridges. Clearly, the Prescott bridge has less overall AE activity.

TABLE 36.4 Summary of AE Activity

AE sensor	Tayco Street	Prescott NE	Prescott NW	Prescott SE	Prescott SW
1	0	2325	646	53	417
2	111,544	178	544	504	0
3	0	4734	0	0	4730
4	1776	0	136	8640	5150

STRAIN GAGE TESTING

A Measurements Group quarter-bridge-type CEA-06-W250A-350 weldable foil strain gage was mounted diagonally across the center casting to bascule girder mating surface on both the north and south corners at the east end of the bridge. This application is a nonstandard approach to using strain gages. The purpose is to attempt to observe displacements between the casting segment and the girder flange. The observations are more qualitative than quantitative because the gage is not subjected to a uniform strain field across the gage width. This mounting approach was developed during experiments performed on the Tayco Street Bridge in Menasha, Wisconsin.

Strain gage data was recorded using a Somat model S2000 field computer in the time history mode. The S2000's programmable digital low-pass filter was set at a 15-Hz cutoff to eliminate electrical noise. The strain gage signals were sampled at a 100-Hz rate with 8-bit resolution.

The strain gage data taken on the Prescott bridge showed three significant departures from similar tests on the Tayco Street bridge. The peak strain on Prescott was 50 to 75 microinches per inch while Tayco Street itself was 600 to 1200 microinches per inch. Secondly, the Prescott strain gage data showed even symmetry and good repeatability on the strain wave forms recorded during raising and lowering, while poor odd symmetry and no repeatability was observed in the Tayco Street data. The Prescott strain gages returned to their original zero within the quantization error of

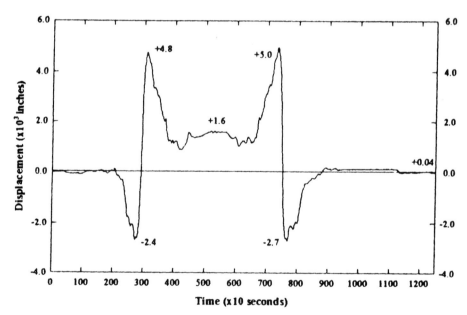

FIGURE 36.5 Typical laser gage plot.

the monitoring system (approximately 10 microinches per inch) while the Tayco Street strain data was offset by as much as 150 microinches per inch following a complete bridge cycle.

LASER DISPLACEMENT GAGE TESTING

The laser displacement gage (Aeromat LM300) had source and target mounted on opposite sides of the casting to girder interface and aligned parallel to this interface. It allowed us to observe the elastic deformation of the casting under dynamic loading conditions during bridge open and closure and was easily capable of detecting slippage of the casting. The shape and symmetry of both the laser gage and the strain gage data agreed well and the laser gage showed no appreciable slippage between the girder flange and the casting following a complete bridge operating cycle. A typical laser gage plot is shown in Fig. 36.5. The laser gage typically showed some offset when the bridge was at the maximum opening point (midrange). The gage was mounted on the center casting in all tests. We believe the offset observed at the bridge up position is the result of load sharing between the castings that is fostered by the wedges applied between castings. These test results indicate that under current operating conditions the segmental casting attachments on the Prescott bridge (B47-40) exhibit no abnormal behavior as evidenced by AE, strain gage, and laser displacement testing. However, future retesting would be prudent based on past experience with this design.

SUMMARY AND CONCLUSIONS

The four tests discussed in this paper are examples of the useful information that application of AE, strain, and optical sensors can provide to the bridge owner. The I-43 Green Bay bridge had both visible as well as suspected flaws. AE and strain gage tests clearly showed that the visible cracks were not of fatigue origin. The lack of crack-related AE coupled with low live stresses indicate that the cracks were most likely an example of early failure of a weld flaw. AE further confirmed that there were no active cracks in the stiffener to web welds. The tests performed on the Benicia Martinez on the other hand clearly confirmed that one of the visible cracks is being driven by fatigue. Similarly, the AE and strain measurements on the Bryte Bend bridge clearly showed that the cracks are active fatigue cracks being driven by significant live strains.

The Menasha and Prescott lift bridges are examples of an application of this sensor technology that is quite different from the usual crack characterization. In this example, these sensors were used to diagnose a problem in a new structure that was simply a case of poor mechanical design. In one case, the AE was able to clearly pinpoint the source of the loud impact noises while strain-gage monitoring confirmed that the castings were moving tangentially with respect to the bascule flange. This diagnosis was made early enough to allow corrective action to be taken before bolt failure and potential jamming of a casting which would render the bridge inoperable. In the other case, the sensor technology showed that, at least for the present, the attachments are performing properly.

CHAPTER 37
LONG-TERM MONITORING OF BRIDGE PIER INTEGRITY WITH TIME DOMAIN REFLECTOMETRY CABLES

INTRODUCTION

Time domain reflectometry (TDR) techniques under development can be employed to detect hidden cracking of bridge piers induced by seismic excitation decades after installation. This paper reviews the installation of TDR cables in laboratory-constructed, reinforced concrete columns and as part of a retrofit reinforcement program for existing bridge piers. Specific types of metallic, coaxial cables are selected to measure accurately cable deformation from shearing and extension within columns. Even though TDR cable monitoring requires a metallic cable to detect distortion caused by shear cracking, its robust construction and ease of attachment allow it to withstand the rigors of the construction environment. Shear distortion of the column and embedded cables during an earthquake can be measured by postseismic pulsing of the TDR cables, which reveals the extent of concrete fracture that is either internal or obscured by the exterior retrofit materials.

Recent earthquakes in Northridge, California and Kobe, Japan have demonstrated that bridge supports designed to resist earthquake shaking can be severely damaged yet remain standing. It is therefore important to be able to evaluate easily and rapidly the integrity of standing bridge piers following an earthquake. While physical damage can be obvious, often it is internal or obscured by exterior reinforcement.

Time domain reflectometry (TDR) technology may offer a simple and rapid measurement technique for identifying internal cracking of bridge piers with embedded coaxial cables and portable, exterior electronics. These robust cables can remain embedded within the structure for decades and be probed at any time to detect shearing. This technology has been successfully deployed in a number of measurement situations, which include monitoring of rock mass deformation and ground subsidence, soil moisture measurements, and groundwater level monitoring, described in many papers in the Proceedings of the Symposium and Workshop on Time Domain Reflectometry in Environmental, Infrastructure, and Mining Applications.[7] A synthesis of these applications reveals that metallic cables have been found to withstand the rigors of emplacement and attachment under conditions of field construction. This robustness and ease of instrument attachment appear to be potentially advantageous compared to glass fibers.

This work describes installation techniques to be employed for TDR cable installation during new construction and retrofitting of existing bridge piers. Techniques for installation during new construction are being developed through installation of 0.5 in. (1.27 cm) diameter, coaxial cables in reinforced concrete columns constructed in the laboratory for subsequent shear testing. Techniques for installation during retrofitting are being developed through installation of coaxial cables in conjunction with wrapping of steel cables around existing columns. This field demonstration involves in-service columns that will not be deformed.

BACKGROUND

Bridge Column Failure

Field observations of seismic failure of bridge columns in Northridge, California, involve shearing of the upper roadway connection and the base connection. Field shearing of two-column bridge piers in East St. Louis, Illinois, near the New Madrid fault revealed that tensile deformation occurs at the interface of the column and its base, and that vertical slippage or shearing occurs between spliced steel reinforcement.[4] Figure 37.1 shows the geometry of the lower connection of the laboratory models of the East St. Louis bridge piers. The field bridge columns were sheared after demolition of the upper roadway; laboratory columns also remain free of upper connections.

Time Domain Reflectometry (TDR)

TDR is an electromagnetic technique originally developed to detect faults along power transmission lines. To locate faults, a cable pulser is connected at one end of a cable, sends a step voltage pulse along the line, and records the travel time of any voltage reflections resulting from cable discontinuities. Knowing the pulse propagation velocity, which varies from cable to cable, the time records are converted to distance measurements. Accurate location of cable faults can be determined with this technique.

O'Connor and Dowding[6] noted that the amplitude of reflected voltage, called *reflection coefficient,* at a shearing fault can be directly related to the magnitude of cable shear deformation. These initial findings have been extended by others, such as Aimone-Martin et al.,[1] to include extension and combined deformation modes. To measure cable defoliation accurately, the correlation between reflection coefficient and deformation magnitude must be known for the selected cables prior to field installation.

For example, shear, tensile, and combined shear and tensile deformation of rock salt has been studied in the laboratory.[1] Experiments consisted of a coaxial cable grouted into a centrally drilled hole in a rock salt core that extends from the core to a cable pulser. Cables were pulsed as individual cores were subjected to shear and tensile loadings to establish the reflection coefficient/cable deformation relationships for each mode.

A TDR measurement system provides several advantages for long-term monitoring of bridge piers. Coaxial cables are commercially available and inexpensive (approximately $1.50/ft. or $5.00/m), and installation of these flexible cables can be accomplished without special equipment. Properly protected cables can be left in

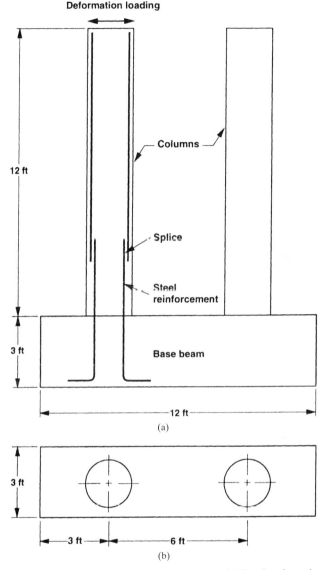

FIGURE 37.1 Laboratory model of two-column bridge pier, shown in (a) front view and (b) plan view.

place for a number of years or even decades. Although long-term performance has not been proven yet because of obvious time constraints, metallic cables have already been in place underground and below the water table for a number of years.[2] Experiments are underway to define more precisely the conditions under which cables might be susceptible to corrosion.

Furthermore, it is only necessary to monitor cables after an earthquake, although they can be evaluated at any time. All electronics are exterior and can be attached intermittently (portable system) or housed on the bridge and accessed through remote telecommunications (fixed system). Most importantly, shearing is easily detected by its unique voltage reflection and provides a simple integrity check.

TDR CABLE INSTALLATION IN NEW COLUMN CONSTRUCTION

Project Description

As part of research on the behavior of bridge columns subjected to cyclic lateral loading, six half-scale columns, two per base beam, are being constructed in Newmark Laboratory at the University of Illinois.[4] Each column has a unique reinforcement layout, but all exterior dimensions, which are shown in Fig. 37.1, are the same. Columns are 12 ft. (3.66 m) high and 2 ft. (0.61 m) in diameter, with 6 ft. (1.83 m) center-to-center spacing on the beam. Base beams are uniformly 12 ft. (3.66 m) long by 3 ft. (0.91 m) high by 3 ft. (0.91 m) wide.

Dowel reinforcement consists of #11 (1.38 in. or 3.50 cm diameter) steel bars that extend up from the beam and into the bottom 3 ft. (0.91 m) of the column. Column reinforcement also utilizes #11 steel, but these bars are spliced with steel wire ties to the dowel bars, where they overlap in the bottom fourth of the column. Column splicing is a common field construction practice, and this zone is often the weakest section of a laterally loaded column.

Columns will be displaced laterally at the top at a controlled rate. Displacements will be applied cyclically, as illustrated by the double arrow in Fig. 37.1, and increased over time to failure. Given this loading scheme, it is expected that both shear and tensile deformation will occur in the bottom fourth of the column, particularly where the dowel and column reinforcement are spliced.

Cable Selection

Coaxial cables must be appropriately selected for TDR monitoring based on the expected deformation modes and engineering properties of the deforming medium. Two types of 0.5-in. (1.27-cm) diameter coaxial cables were selected for installation in both the laboratory columns and existing bridge piers. These cables were chosen based on their selective sensitivity to shear and tensile deformation.[1]

The cable employed in shear-sensitive regions consists of a smooth wall aluminum shield with a solid copper-clad aluminum inner conductor separated by a polyethylene foam dielectric. The cable employed in tensile or extension regions has the same inner conductor and dielectric material, but it is surrounded by a corrugated copper shield. Both cables are commercially available without a protective jacket.[3] These cables are robust yet flexible for easy installation. For simplicity, the two cables described here will be referred to as *shear* and *extension* cables throughout this chapter.

Cable Installation

Shear and extension cables are installed with two distinct orientations within the columns. Two extension-sensitive cables are placed vertically inside the columns, and up to three cables will be installed transversely through the column diameters. Both shear and extension cables will be placed in the transverse orientation to determine the cross sensitivity. Figure 37.2 shows a sketch of final cable layout in a single column. Dashed lines indicate that cables are encased in the concrete; solid lines represent the exposed lengths of cable.

Cables were cut cleanly with a hacksaw to desired lengths of 6.5 ft. (1.98 m) for vertical placement and 3 ft. (0.91 m) for transverse placement. The end of each cable exposes a cross section of the conductors and dielectric, producing an open circuit reflection when pulsed. An open circuit reflection serves as a marker for the cable terminus. The other end will be connected to a lead transmission cable extending from the cable pulser.

One end of each vertically oriented extension cable is encased in the column, as shown in Fig. 37.2. As a precaution, the metallic conductors at this end were sealed to prevent infiltration of water and to mitigate cable degradation. A plastic tip, 0.5 in. (1.27 cm) in diameter and 0.75 in. (1.90 cm) in length, was placed over the end and taped to the cable.

All cables were spray-painted before installation to deter chemical interaction. It has been shown that aluminum corrodes in a high pH (approximately 12) environment of curing concrete

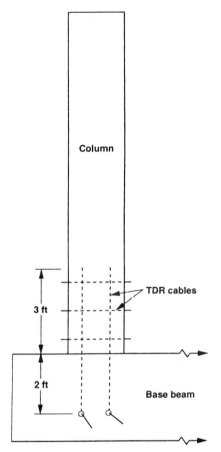

FIGURE 37.2 Orientation of TDR cables in a column.

before the water becomes immobilized.[5] The corrosivity of copper in concrete was unknown at the time, so copper cables were painted as a precautionary measure. Even for the aluminum, the interaction declines greatly when water becomes fixed after curing. Therefore, a thin coat of paint suffices to insulate the cables until the concrete has set. However, a more durable coating may be needed for field installation if water is expected to penetrate the bridge piers during its decades of service.

An example of vertical cable placement before concrete casting is provided by the lengthwise photograph of a reinforced beam in Fig. 37.3. Two extension cables are attached to the dowel reinforcement by thin, steel wires. For long-term field conditions, plastic ties should be employed to avoid establishing an iron-aluminum or iron-copper corrosion cell. The exposed cable lengths for the distant column can be seen in the bottom right corner of the photograph. Each cable was bent easily by hand through a 90° arc with a 1.3-ft. (0.40-in.) radius of curvature inside the beam to allow a perpendicular exit from the beam and formwork. A continuous, non-kinked curve does not produce a voltage reflection and so will not interfere with TDR testing.[8]

FIGURE 37.3 TDR extension cables attached vertically to dowel reinforcement for two columns.

A cross-sectional plan view of the column in the spliced zone, shown in Fig. 37.4, reveals reinforcement and TDR cable details. All details in this figure are to scale. The dashed line represents the encased portion of one transverse cable; the solid lines are the transverse cable ends protruding from the column. Recall that one or two more shear or extension cables are aligned in the same direction at other vertical distances along the column. Each transverse cable passes through the column center, near but not parallel to the direction of deformation loading to weaken the reinforcing splice zone, where slip is expected.

TDR CABLE INSTALLATION IN EXISTING COLUMNS

Project Description

The Illinois Department of Transportation (IDOT) is currently reinforcing bridge piers along Interstate Highway 55 in East St. Louis, Illinois. Under this program, hundreds of bridge columns supporting the highway and its exchanges will be retrofitted with circumferential reinforcement as a preventive measure to mitigate potential distress should the New Madrid fault move and produce a rare midcontinent earthquake.

Retrofitting will provide additional confinement at the bottom and top sections of bridge piers, where excessive bending moments can result from earthquake-

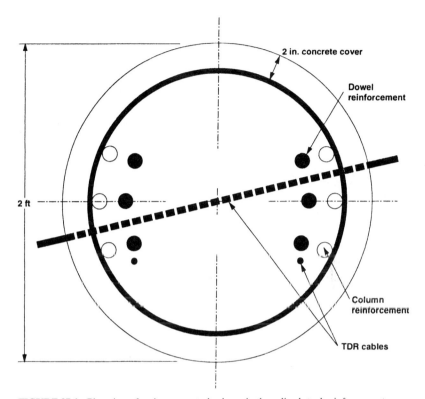

FIGURE 37.4 Plan view of an instrumented column in the spliced steel reinforcement zone.

induced ground shaking. IDOT is retrofitting columns by wrapping steel reinforcement cables around the column at regular vertical spacing, as shown in Fig. 37.5. Nine steel cables are wrapped around the base of the column at approximately 1-ft. (0.30-m) spacing. Formwork is then placed around the reinforced area to secure the concrete, which is poured to encapsulate the reinforcement.

Other reinforcement techniques involve either a fiberglass composite, shown in Fig. 37.6, or steel jacket that is wrapped around the column. The composite jacket shown here is prefabricated and placed around a polyurethane coating on the column surface to bond the jacket to the column tightly. Alternatively, steel jackets are secured to the column by grouting the annular space between the jacket and column.

Proposed Cable Installation

TDR cables will be installed as part of the IDOT retrofit reinforcement program described above to determine the challenges of field installation. Shear and extension cables will be emplaced vertically on at least one column during the retrofitting process. It is expected that these cables can be attached to the steel reinforcing cables before placement of the encapsulating concrete.

FIGURE 37.5 Steel reinforcing cables wrapped around an existing bridge column in East St. Louis, Illinois. (*Courtesy of David Prine.*)

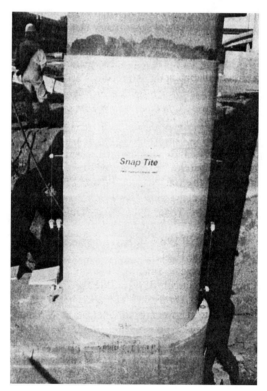

FIGURE 37.6 Fiberglass composite jacket wrapped around an existing bridge column in East St. Louis, Illinois. (*Courtesy of David Prine.*)

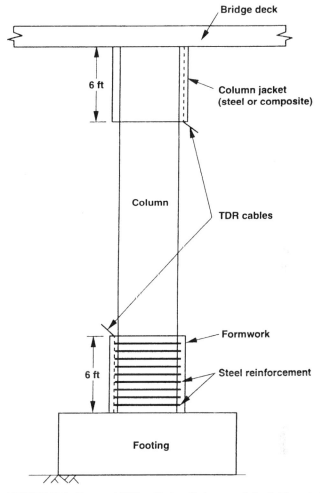

FIGURE 37.7 Proposed TDR cable installation on existing bridge column.

A diagram of proposed cable installation within both forms of retrofit reinforcement (cables and jacket) is shown in Fig. 37.7. This sketch does not suggest that steel cables and jackets are necessarily placed together on individual columns; it is only used to show the cable placement within each type of reinforcement. Cables will be approximately 6.5 ft. (1.98 m) long and attached at the top and bottom sections of the column. Dashed lines represent the embedded cable length, and the solid lines represent the exposed sections needed for electronic attachment.

TDR cables must be prepared for long-term performance in potentially harsh environments. Thus, they will be protected with a durable but thin surface coating, and the embedded ends of the cable will be sealed with plastic tips, as previously described. The monitoring ends should be housed in a plastic box attached to the pier. During installation, plastic ties will be used to attach the TDR cables to steel reinforcing cables.

SUMMARY

This chapter described the emplacement techniques necessary to install metallic, coaxial TDR cables to assess bridge pier integrity long after installation. This approach may provide simple and rapid measurements of hidden, internal column fracture resulting from seismic activity. Installation of aluminum-conducting and copper-conducting coaxial cables during reinforced-concrete-column construction and retrofitting of existing bridge piers has been described. Long-term performance of metallic cables was considered throughout the installation process.

Newly constructed columns will be tested to failure by laterally displacing the free ends at the top. Although data are not yet available, it is expected that results will confirm that shearing and extension of the column occurs at the footing connection. Instrumenting the column with transverse cables at multiple vertical levels should provide data on the development and progression of shear cracking within the spliced zone. Future data analysis will help refine the original cable placement scheme to monitor more accurately internal column deformation.

Existing field bridge columns outfitted with experimental TDR cables will not be deformed since they remain in service. However, TDR cable installation during retrofitting will allow identification of problems and refinement of installation procedures. In the future, exposed cable lengths can be visually inspected and cable degradation can be checked to monitor the long-term serviceability of the embedded cables. These checks will indicate the degree of success that the protective measures provided for long-term performance.

This project has been supported by the Infrastructure Technology Institute (ITI) at Northwestern University. David Prine of the Basic Industrial Research Laboratory (BIRL) at Northwestern University provided photographs and project details at East St. Louis, Illinois. Finally, the assistance and cooperation of Dr. William Gamble and graduate student Yongqian Lin of the University of Illinois, and Neil Nauthaus of the Illinois Department of Transportation (IDOT) are gratefully acknowledged.

REFERENCES

1. C. T. Aimone-Martin, K. I. Oravecz, and T. K. Nytra, "TDR Calibration for Quantifying Rock Mass Deformation at the WIPP Site, Carlsbad, New Mexico," *Proc. of Symposium and Workshop on Time Domain Reflectometry in Environmental, Infrastructure, and Mining Applications,* SP 19-94, U.S. Bureau of Mines, September 1994, pp. 507–517.
2. T. R. C. Aston, M. C. Betournay, J. D. Hill, and F. Charette, "Application of TDR for Monitoring the Long-Term Behavior of Canadian Abandoned Metal Mines," *Proc. of Symposium and Workshop on Time Domain Reflectometry in Environmental, Infrastructure, and Mining Applications,* SP 19-94, U.S. Bureau of Mines, pp. 518–527, September 1994.
3. *Cablewave Systems,* Catalog 720, North Haven, Connecticut, 1989.
4. W. Gamble, Department of Civil Engineering, University of Illinois at Urbana Champaign, personal communication, December 1994.
5. M. H. Kim, "Quantification of Rock Mass Movement with Grouted Coaxial Cables," M.S. Thesis, Northwestern University, Evanston, Illinois, 1989.

6. K. M. O'Connor, and C. H. Dowding, "Application of Time Domain Reflectometry to Mining," *Proc. of 25th Symposium on Rock Mechanics,* Northwestern University, Evanston, Illinois, 1984, pp. 737–746.
7. K. M. O'Connor, C. H. Dowding, and C. C. Jones, *Proc. of Symposium and Workshop on Time Domain Reflectometry in Environmental, Infrastructure, and Mining Applications,* SP 19-94, U.S. Bureau of Mines, September 1994.
8. C. E. Pierce, C. Bilaine, F. C. Huang, and C. H. Dowding, "Effects of Multiple Crimps and Cable Length on Reflection Signatures from Long Cables," *Proc. of Symposium and Workshop on Time Domain Reflectometry in Environmental, Infrastructure, and Mining Applications,* SP 19-94, U.S. Bureau of Mines, September 1994, pp. 540–554.

CHAPTER 38
NONDESTRUCTIVE EVALUATION (NDE) SENSOR RESEARCH, FEDERAL HIGHWAY ADMINISTRATION (FHWA)

NDE FOR BRIDGE MANAGEMENT

This chapter summarizes FHWA's research and development program in nondestructive evaluation for bridges. This program is driven by the need to develop new and better ways of collecting condition information about the nation's bridges. Such information is essential to support bridge management.

Objectives

FHWA's NDE research and development program has two main objectives. First, to develop new tools and techniques to solve specific problems. Some examples are locating, quantifying, and assessing fatigue cracks on steel bridges; quickly, efficiently, and quantitatively assessing the condition of reinforced concrete bridge decks even though they are covered with asphalt; and dealing with the 100,000 bridges where we do not know how deep the foundation piles extend or in fact, in some cases, if there are piles or not. Any type of scour or seismic assessment is meaningless without that information. The second main objective is to develop technologies for the quantitative condition assessment of bridges in support of bridge management. Today, the data that feeds all bridge management systems is based upon visual inspection and subjective condition assessment. We are developing technologies for quick, efficient, and quantitative measurement of global bridge parameters such as stiffness and load-carrying capacity.

Background

The NDE program is targeted at the most important and pressing needs. The specific projects underway and those that we will undertake in the future are designed to

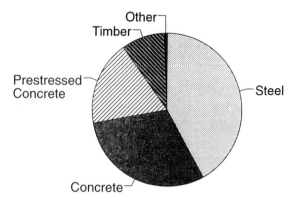

FIGURE 38.1 Number of bridges by material (excludes culverts and tunnels).

solve specific problems. An overview of the bridge inspection and management problem is provided to explain the priorities we have established for the NDE program.

There are about 570,000 highway bridges in the National Bridge Inventory. If we exclude culverts and tunnels for this summary, there are about 470,000 bridges in the inventory. The proportions by superstructure type are shown in Fig. 38.1.

There are more steel bridges than any other type, followed by concrete, prestressed concrete, and timber. There are a few other types of bridges, such as masonry, iron, and aluminum. This is interesting but does not help focus on the problems.

Figure 38.2 shows when the bridges in the inventory were built, by type and age. We can see the two bridge-building booms in the post-Depression era and the interstate years. We also see that the majority of bridges built prior to 1970 were steel bridges and that the proportion of prestressed concrete bridges has been increasing

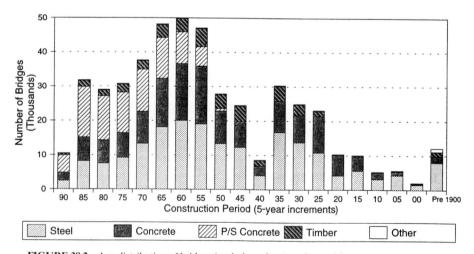

FIGURE 38.2 Age distribution of bridges (excludes culverts and tunnels).

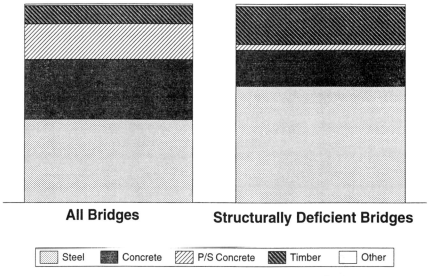

FIGURE 38.3 Proportion of bridges by type (excludes culverts and tunnels).

steadily. There has also been a small but steady use of timber bridges over the years. This is also interesting but again does really focus on the problems.

Figure 38.3 begins to show where the most critical problems are. The proportion by type of all bridges is shown on the left, and the proportion of structurally deficient bridges is shown on the right. A structurally deficient bridge is one where deterioration has progressed to the point where the load-carrying capacity of the bridge has been significantly reduced and represents the most serious types of deterioration. There are about 105,000 structurally deficient bridges in the inventory. As can be seen, while steel bridges represent about 40 percent of the overall bridge population, they represent about 60 percent of the structurally deficient bridges. Reinforced concrete bridges are underrepresented, and a relatively small proportion of the pre-stressed concrete bridges are structurally deficient. Timber bridges, while representing only 9 percent of the overall population, represent 20 percent of the structurally deficient bridges. About half of the nation's timber bridges are classified as structurally deficient.

Taking this analysis one step further and looking at why the bridges of different types are classified as structurally deficient leads us to Fig. 38.4. The most frequent reason for steel and timber bridges being classified as structurally deficient is a low structural adequacy rating. This means the bridge has a very low load rating. It is also worth noting that more structurally deficient steel bridges have bad substructures than they do bad superstructures, and just as many have bad decks.

The NDE program, as its first priority, is developing technologies for steel bridge inspection and bridge deck inspection.

Figure 38.5 shows a bar chart of the nation's bridges by age. Large culverts and tunnels are included. The chart has the same general shape as Fig. 38.2. The black line shows the percentage of each age group that is classified as structurally deficient or functionally obsolete. There is a steady increase with age, with 80 percent of those bridge built between 1905 and 1910 classified as deficient. About 1 percent, or about 5000 bridges, become deficient each year. Today, about 200,000 bridges are classified

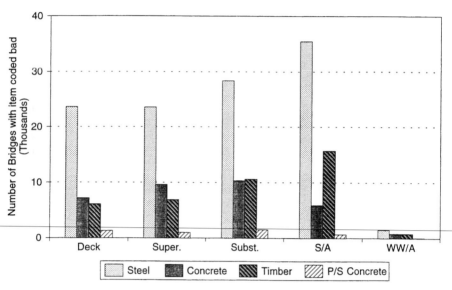

FIGURE 38.4 Structurally deficient bridges: comparison of ratings for key items.

as deficient. This number has been reduced somewhat over the last few years, but only after federal bridge funding was increased to over $3 billion per year in 1985. We are currently building or rehabilitating about 10,000 bridges per year. To deal with the backlog of 200,000 deficient bridges, the 5000 bridges that become deficient each year and to ensure that the $3 billion spent on bridges each year is spent in an

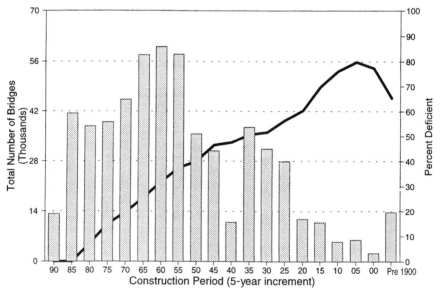

FIGURE 38.5 Age distribution of bridges.

optimal manner, FHWA is mandating the implementation of bridge management systems.

The essence of a bridge management system is shown in Fig. 38.6. Data is collected from a number of sources but primarily from periodic bridge inspections; it is transferred to a large database; a sophisticated analysis of the data is performed, which generates prioritized lists of candidate projects, optimizes bridge replacement and maintenance strategies for various available funding and resource scenarios, predicts the deterioration of bridges with time, allows managers to evaluate different management options and in general provides powerful decision support tools to help formulate the best program for management of a bridge population. The systems that have been implemented to data, have been a tremendous benefit to decision makers.

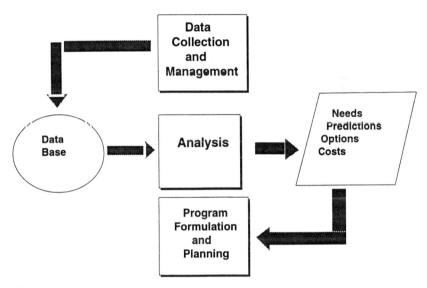

FIGURE 38.6 Bridge management system.

However, bridge management systems are still driven by the data that is collected about the condition of the bridges. No matter how sophisticated and elaborate the analysis and no matter how elegant the algorithms employed, in the final analysis the recommended decisions cannot be any better than the data upon which they are based. Today, this data is based almost entirely upon a visual bridge inspection with condition evaluation based upon visible indications of deterioration and distress. Deterioration that does not manifest some visible symptom is not detected or quantified.

Since bridge management systems are driven by *quantitative* data about the condition of bridges, NDE technology can and should be used to collect effective and efficient quantitative data about bridge condition. This is especially true for certain types of hidden deterioration such as corrosion of reinforcement in concrete or cumulative fatigue loading in steel bridges.

The development of NDE technology to evaluate quantitatively the condition of a bridge is a significant part of FHWA's NDE program.

Current NDE Research Program

A few of the current research and development projects we are funding are listed below.

- Dual-band infrared imaging for bridge decks
- Ground-penetrating radar imaging system for bridge deck inspection and fatigue crack detection system
- Global bridge evaluation using laser scanning
- Wireless bridge monitoring system

The first two projects are developing new bridge deck inspection capabilities. We are transferring technologies developed for mine detection and using it to inspect bridge decks. By scanning a bridge deck using two different infrared wavelengths at the same time, and using advanced image analysis techniques, significant improvements over traditional infrared scanning have been demonstrated. A prototype system is to be delivered in late 1995.

The second project represents a breakthrough in bridge deck inspection technology. Taking advantage of very recent developments in radar miniaturization at Lawrence Livermore National Laboratory and using phased array and synthetic aperture techniques, we are developing a ground penetrating radar imaging system. The system will be able to provide three-dimensional images of the interior of the full lane of asphalt covered bridge deck with resolution of individual rebar. The concept has been proven in the laboratory and a prototype for bridge deck inspection will be delivered by the summer of 1996. The same system, when driven over a bridge deck at traffic speeds will be able to quickly quantify delaminations and other distress a full lane width at a time.

FHWA is also developing a new fatigue crack detection system that combines ultrasonic and magnetic inspection capabilities into a single instrument. This system consists of a backpack computer and heads-up display and features one-hand operation (essential for use on a bridge). This system will greatly improve our capabilities to detect and quantify fatigue cracks in steel bridges, even though they may be covered with paint.

The next two projects are new initiatives and are the result of a broad area announcement in 1994 for global bridge monitoring and evaluation. These projects will provide new capabilities for the collection of quantitative information on bridge condition.

The first is an adaptation of a system developed for NASA. It is a portable laser scanning system that will quickly measure the deflected shape of a bridge with submillimeter accuracy. It will also measure the vibration of the bridge and has potential to facilitate the application of modal analysis as a bridge inspection tool. The system was delivered to FHWA in the summer of 1996.

The last project is developing a wireless bridge monitoring system. It will consist of a number of sensor/transponder modules that will communicate via microwaves to a local controller. The system development emphasizes the use of off-the-shelf components developed for the cellular telephone and automotive applications and consequently minimizes technical risk. There will be modules measuring strain, rotation, deformation, and vibration. It is the goal to develop technology that will make it possible to instrument a bridge at a dozen locations for a cost of less than $5000. The implications for bridge management are obvious.

Future NDE Research Program

Some projects we will be initiating, subject to available funding, are listed below:

- Improved acoustic emission monitoring for bridges
- Telerobotic system for visual inspection of steel bridges
- Fatigue crack detection using remote sensing
- Broad area announcement for nondestructive bridge evaluation technology

We are teaming with our Office of Technology Applications to develop an improved acoustic emission system specifically packaged for application on bridges. The system will take existing AE technology and build a small, rugged, battery-powered, multichannel acoustic emission monitoring system that can be installed on a bridge and left unattended for up to one month. AE has been proven to be useful in evaluation and assessment of fatigue cracks, but the current systems are too bulky and require line power. There is a strong need for such a system.

A small telerobotic system has been developed. The telerobotic system has been developed for visual inspection of steel bridges. There are many locations and details on steel bridges that should be inspected but are not because they are just too difficult to get to, or the traffic disruption and expense of an under-bridge inspection unit is too great. Similar systems, using magnetic wheels and video imaging capabilities have been used in the power industry for some time. We propose to adapt this technology for use on highway bridges.

The Small Business Innovative Research program is established to develop a noncontact system for the detection of fatigue cracks. There are a number of technologies that can feasibly be used to detect such cracks using remote sensing. There are large area laser holographic systems commercially available that are used on aircraft skin inspections. Active thermography (infrared imaging of dynamic heat flow) is currently used to detect defects in the aerospace industry. Another potential technology is thermoelastic heating. Recent advances in infrared scanning technology provide the capability of measuring temperature differences to $0.04°C$. The localized heating at the ends of fatigue cracks undergoing cyclic loading could be detected using such scanners.

Applications that will request proposals for innovative, but implementable solutions to some of the other problems we face. Examples include the unknown foundation problem, inspection of cable stay anchorages and evaluation of embedded prestressing strands and tendons.

Conclusion

The Federal Highway Administration strongly believes that advanced technologies can help manage the nation's highway bridges. FHWA is focusing on specific problems such as fatigue and corrosion of steel bridges and more rapid and quantitative bridge deck inspection capabilities. We are also sponsoring research and development to technologies to provide more quantitative data on bridge condition to improve the data used as input to bridge management systems. FHWA has several innovative projects underway will continue to sponsor the development of solutions for the management of the nation's aging highway bridges into the next century.

CHAPTER 39
SENSORS AND INSTRUMENTATION FOR THE DETECTION AND MEASUREMENT OF HUMIDITY

INTRODUCTION

In this chapter, a general overview will be given on humidity measurements, sensors, and instrumentation. We will describe the many different methods, limitations, and features of various sensors, typical applications, calibration, maintenance, and operational problems.

DEFINITION OF HUMIDITY

What Is Humidity?

The term *humidity* refers to water vapor, which is a gas. It is water in gaseous form. This chapter does not cover *moisture,* which relates to water in liquid form that may be present in solid materials or other liquids.

Humidity is present everywhere in the atmosphere. Even in bone-dry areas and in boil-off from liquefied gases, there are traces of water vapor that in some applications could cause problems. Measurement of humidity is more difficult than the measurement of most other units such as flow, temperature, level, and pressure. One reason for this is the extremely broad dynamic range, starting from say 10 parts per billion, representing a partial vapor pressure of 10^{-9} in Hg to steam at 100°C, representing a dynamic range of more than 10^9. Another reason is that measurements may have to be made in widely varying atmospheres, for example from temperatures of –60°C to 1000°C, in the presence of a wide range of atmospheres that could be corrosive or noncorrosive, and in the presence of a variety of contaminants, particulate or chemical.

The average person, when faced with a humidity measurement problem, is confronted with a variety of measurement techniques and instrument types. The problem is to decide which sensor types are best suited for a particular situation, and of these, which is most useful for the application at hand.

The measurement of humidity is not a trivial task, nor is it intuitively understood. It is the intent of this chapter to give the user of humidity instrumentation the knowledge to make an informed decision as to which measurement technique is the best choice.

What Is Its Importance?

Humidity measurements have become increasingly important, especially in the industrialized world, since it has been recognized that humidity has a significant effect on quality of life, quality of products, safety, cost, and health. This has brought about a significant increase of humidity measurement applications and, concurrent with this, an increase in research and development activities to improve measurement techniques, accuracy, and reliability of instrumentation.

Despite the recent search in R&D activity to improve humidity sensors, the present state of the art is such that humidity measurements require more care, more maintenance, and more calibration than other analytical measurements. Furthermore, there is no sensor available that can even come close to measuring the full range of water vapor levels. For this reason, many different methods and sensors have been developed through the years, each having certain advantages and limitations and being suitable for some but not all applications. For someone not skilled in the field of humidity, it is often very difficult to make an intelligent choice of sensor, and when this is not done, disappointing results will occur.

There are no humidity manufacturers who manufacture all possible types of sensors and instruments. Some offer only one type to meet requirements of a certain market segment, while others offer a broad range to cover a wider range of applications. Manufacturers offering the broader ranges are generally in a better position to assist a user in selecting the right instrument and sensor for an application. However, for those involved in important humidity measurements or in new applications requiring such measurements, it is important to be able to obtain an adequate knowledge of this subject.

SENSOR TYPES

Relative Humidity

Percent relative humidity is the best known and perhaps the most widely used method for expressing the water vapor content of the air. Percent relative humidity is defined as the ratio of the prevailing water vapor pressure, to the water vapor pressure if the air were saturated, multiplied by 100:

$$\% \text{ RH} = \frac{e_a}{e_s} \cdot 100\%$$

where e_a and e_s are the actual and saturation vapor pressures, respectively. A measurement of RH without a corresponding measurement of dry bulb temperature is not of particular value, since the water vapor content cannot be determined from percent RH alone.

It has become evident in recent years that the influence of humidity, or the effect of humidity control, is of paramount importance in many types of industries, such as in the manufacture of moisture-sensitive products, storage areas, computer rooms, energy management, hospitals, museums, libraries, and so on.

It is not easy to measure humidity exactly. Many methods have been used to measure relative humidity: the method using human hair or nylon film, the wet and dry bulb method, the semiconductor sensor method, and so on. All of these methods have their advantages and limitations, and none of them can be labeled perfect. In addition, there are very few highly accurate calibration devices for humidity generation and measurement, making it relatively difficult to confirm the accuracy of a humidity measurement.

Bulk Polymer-Humidity Sensor

The most advanced humidity sensors on the market today are the so-called *bulk polymer sensors*, consisting of a miniature electrode base-plate coated with a humidity-sensitive macro polymer. An electrical grid structure is vapor deposited upon the element, and an electrical measurement is made, which is a function of the relative humidity.

The two most popular polymer sensors are the bulk resistance and the capacitive sensor, depending on whether the sensor is designed for change of resistance measurements or capacitance measurements. The polymers used and the manufacturing techniques for resistive and capacitive sensors are different, and the sensors have somewhat different characteristics, with advantages and limitations for both.

Specifically, the resistive sensors have a slower response, though generally they're fast enough for most applications. Resistive polymer sensors on the market today are generally suitable for a narrower ambient temperature range, typically from 0° to 80°C, whereas capacitive sensors are presently available for temperature ranges from –40°C to over 150°C. Most capacitive sensors have a very fast response but tend to be more expensive than the resistive types. The use of a resistive or capacitive sensor is often dictated by the application.

Resistive Polymer Sensor

In a typical resistive sensor, the sensitive material is prepared by polymerizing a solution of quaternary ammonium bases and the reaction of this functional base with a polymer resin. This produces a three-dimensional thermosetting resin characterized by its excellent stability in extreme conditions.

Humidity is measured by the change in resistance between anode and cathode. When the quaternary ammonium salt contained in the electrode shows ion conductivity caused by the presence of humidity, mobile ion changes in are created. The more the humidity in the atmosphere increases, the greater the ionization becomes, and, as a result, the concentration of mobile ions increases. Conversely, when humidity decreases, ionization is reduced, and the concentration of mobile ions decreases. In this way, electrical resistance of the humidity-sensitive film responds to changes in humidity absorption and desorpsion. The movement of such ions is measured using the variation in impedance caused by the cell as typically measured in a Wheatstone bridge. Contrary to the Pope cell (to be discussed later), which is a surface-resistance element, the bulk polymer sensor is a bulk effect sensor, which means the bulk resistance of the polymers is related to the relative humidity.

Features of the Resistive Polymer Sensor

- 100 percent solid-state construction
- Low power input
- Compact size
- Low cost
- Bulk polymer minimizes accuracy degradation resulting from contamination
- Broad humidity range (15 to 99 percent)
- Temperature range from −10° to 60°C (typical)
- Fast response
- Excellent reproducibility (better than 0.5 percent RH)
- High accuracy (= ±2 percent RH, ±1 percent RH in narrow range)

Like all relative humidity sensors, when exposed to contaminants and/or extreme environmental conditions, accuracy degradation could result. Polymer-type sensors have proven to be more resistant to such errors or drift than the older Pope cells and Dunmore cells.

Resistance to Contaminants. Because of the good chemical stability of polymer sensors, the sensors exhibit excellent resistance to solvents and are therefore often the preferred sensor for industrial applications, except those having the highest concentrations of corrosive chemicals. Many common substances such as petrochemicals have little effect.

Since different polymers are used, sensitivity to contaminants for resistive and capacitive sensors are different: some contaminants cause greater adverse effect on resistive sensors, and others affect capacitive sensors more. Hence, the choice of sensor is often also dictated by the presence of certain contaminants.

Pope Cell. An older type of resistive RH sensor is the so called Pope cell, named after its inventor, Dr. Pope. The sensor is comprised of an insulating ceramic substrate on which a grid of interdigitated electrodes is deposited. These electrodes are overlayed with a humidity-sensitive salt imbedded in a polymer resin. The resin is then covered by a protective coating that is permeable to water vapor. Water vapor is able to proceed throughout this surface layer and enter the polymer region.

This infiltration of water vapor allows the imbedded salt to ionize and become very mobile within the polymer resin. An AC excitation is provided to the electrodes, and the impedance (AC resistance) of the sensor is measured.

Sulphonated polystyrene sensors have a large signal change with humidity. A typical sensor will change its resistance four orders of magnitude with relative humidity. Signal changes of 1 kiloohm at 90 percent and 10 megaohms at 10 percent RH are representative of this type of sensor. This large signal change enhances the RH resolution. Resistive RH sensors are generally suited to the higher RH ranges due to the extremely high impedance below 10 percent and are most often used at RH levels above 15 percent.

The Pope cell consists of a conductive grid over an insulative substrate. This substrate consists of polystyrene that has been treated with sulfuric acid. This acid treatment causes ulphonation of the polystyrene molecules. The sulfate radical SO_4 becomes very mobile in the presence of hydrogen ions (from the water vapor) and readily detaches to take on the H^+ ions. This alters the impedance of the sensor as a function of humidity.

A disadvantage is that, due to the extremely active surface of the sensor, the sensing surface can be easily washed off or contaminated to cause errors in readings.

The sensor offers the advantage of covering the entire range of RH measurements, although it has a very nonlinear curve that is difficult to linearize electronically with good accuracy. It is sensitive to contaminant errors, and the older sensors exhibit a considerable amount of hysteresis. The more advanced resistive bulk polymer sensors offer better performance with virtually no hysteresis and far better resistivity to contaminants. Pope cells are therefore finding less and less applications.

Dunmore Cell. The Dunmore cell, named after its inventor Dr. Dunmore, is also an older type of resistive sensor that is being replaced more and more by today's polymer sensor.

The Dunmore element is comprised of a bifilar wound inert wire grid on an insulative substrate. This is then treated with a known concentration of lithium chloride in solution. Due to its hygroscopic nature, the lithium chloride will take on water vapor from the surrounding atmosphere. This modifies the surface resistivity (impedance) of the sensor.

The Dunmore element is a good RH sensor, but it suffers one major drawback. It is only active over a small RH range due to the doping concentration of the salt. This is usually remedied by placing multiple elements into an array and designing crossover circuitry into it for range transition switching.

Like the Pope cell, the Dunmore array can be readily contaminated. However, if proper application techniques are employed, this type will give good service.

Capacitive Polymer Sensor

The use of high-temperature thermosetting polymers have resulted in the development of a new generation of capacitive RH sensors that can perform continuous measurements at temperatures up to 185°C with continuous exposure to 2100°C permissible. Because thin-film sensor layer processing is performed at over 400°C, the maximum temperature is determined by the materials used in the sensor packaging.

The basic construction of a typical circuit for measuring capacitance and thus RH using a capacitive sensor is shown in Fig. 39.1.

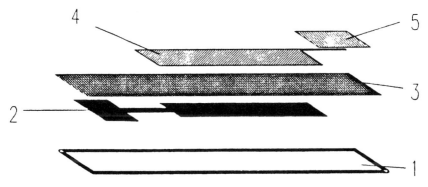

FIGURE 39.1 Operating principal of capacitive polymer sensor.

1. The base is a substrate, typically glass. Its main function is to support the other layers of the sensor.
2. One of the electrodes is made of conductive and corrosion-resistant material.
3. A thin polymer layer is the heart of the sensor. The amount of sorbed water in the film varies as a function of the surrounding relative humidity. The thickness of this film is typically 1 to 10 micrometers.
4. The upper electrode also plays a role in determining the performance and characteristics of the sensor. For fast response, it must have good permeability for water. It must also be electrically conductive and have good corrosion resistance.
5. A contact pad for the upper electrode is also required. Since there are many constraints on the design of the upper electrode, a separate metallization for making reliable contacts is often required.

The capacitance of the sensor is determined by the overlapping area of the upper and lower electrodes according to:

$$C = \frac{\varepsilon \cdot \varepsilon_o \cdot A}{L}$$

where: ε = Relative permittivity of the polymer film (typically in the range 2–6)
ε_o = Permittivity of vacuum ($8.85 \cdot 10^{-12} \cdot Fm^{-1}$)
A = overlapping area of the upper and lower electrodes
L = thickness of the polymer film

Area A is constant and well-defined. The thickness of the polymer film can be considered constant. Therefore, only the relative permittivity of the film changes as water is sorbed into it.

Since the water molecule is highly polar, even small amounts of water can cause a significant change in the sensor capacitance. The relative permittivity of water is 80 compared to 2 to 6 for the polymer material. This makes capacitive measurement a good choice for humidity measurement, and it makes the sensor less prone to interference from other atmospheric gases.

Temperature Dependence. All RH sensors are temperature-sensitive and, when calibrated at one temperature, exhibit errors at other temperatures. One advantage of the polymer sensors is that they generally exhibit far less temperature dependency (temperature coefficient) than the older RH sensors such as the Pope cell and Dunmore cell. Hence, if the operating temperature is not too far different from the calibration temperature, the error is small and can often be ignored. If the sensor is to be used in extreme temperature environments, or if optimum accuracy is desired, electronic temperature compensation must be incorporated. Temperature compensation can, as a rule, be easily accomplished in temperature ranges of about 50°C, but it is more difficult and less accurate over broader temperature ranges. Hence, when used in a very broad temperature range, polymer tensors lose some of their accuracy. Nevertheless, modern polymer sensors are available with accuracies of ±1 percent RH in narrow ranges and, at calibration temperature, to ±3 percent RH when newly calibrated over a broad temperature and humidity range. When operated over a period of time, one should expect some accuracy deterioration, and, when contaminated, significant errors can occur, requiring recalibration or sensor replacement.

Displacement Sensor

Perhaps the oldest type of RH sensor still in common use is the displacement sensor. These devices use a strain gage or other mechanism to measure expansion or contraction of a material in proportion to changes in relative humidity. The three main materials in use are hair, nylon, and cellulose. The advantages of this type of sensor are that it is inexpensive to manufacture and highly immune to contamination. Disadvantages include a tendency to drift over time and to take a set if remaining at a given humidity for a protracted period, thereby losing sensitivity. These properties are due to the mechanical nature of the sensor.

Below are several considerations to keep in mind when selecting an RH instrument or transmitter:

1. All instruments employ a sensor as well as varying amounts and types of supporting circuitry. The sensor itself is transparent as long as the instrument performs as required.

2. All of the sensor types discussed above are empirical (secondary measurement device); that is, they measure a change in some material as a function of relative humidity. They do not measure any fundamental property of water vapor. All empirical sensors are subject to drift and loss of calibration.

3. Relative humidity is an expression of a ratio of vapor pressure. Thus, difficult RH applications can sometimes be tackled with instruments that measure fundamental properties of water vapor. The results can be converted to RH either by formula or by use of appropriate charts and tables. One should seek the technology that makes the most sense for the application.

Aluminum Oxide

Aluminum oxide humidity instruments are available in a variety of types, starting with low-cost single-point systems, including portable battery-operated models to multipoint microprocessor-based systems with the capability to compute and display humidity information in different parameters such as dew point, parts-per-million, and percent relative humidity.

A typical aluminum oxide sensor is a capacitor, formed by depositing a layer of porous aluminum oxide on a conductive substrate and then coating the oxide with a thin film of gold. The conductive base and the gold layer become the capacitors electrodes. Water vapor penetrates the gold layer and is absorbed by the porous oxide layer. The number of water molecules absorbed determines the electrical impedance of the capacitor which is, in turn, proportional to water vapor pressure.

Applications. The aluminum oxide sensor is also used for moisture measurements in liquids (hydrocarbons), and, because of its low power usage, it is suitable for intrinsically safe and explosion-proof installations. Aluminum oxide sensors are frequently used in petrochemical applications where low dew points are to be monitored in line and where the reduced accuracies and other limitations are acceptable.

The advantages of the aluminum oxide sensor are as follows:

1. The sensor can be small and is suitable for in-situ use.
2. It can be used very economically in multiple sensor arrangements.

3. The sensor is suitable for very low dew point levels without the need for cooling. Typical dew points down to −100°F (−73°C) can be easily measured.
4. The aluminum oxide sensor can operate over a wide span.

The limitations of the sensor are as follows:

1. The aluminum oxide sensor is a secondary measurement device and must be periodically calibrated to accommodate aging effects, hysteresis, and contamination.
2. Sensors are nonuniform, calling for separate calibration curves for each sensor, and are nonlinear.

Many improvements have been made over the last several years by various manufacturers of aluminum oxide sensors, but it is important to remember that, even though this sensor offers many advantages, it is a lower accuracy device than any of the fundamental measurement types. It is a secondary measurement sensor and therefore can provide reliable data only if kept in calibration and if incompatible contaminants are avoided.

Electrolytic Hygrometer

The electrolytic hygrometer is desirable for dry gas measurements because it is one of the few methods that gives reliable performance, for long periods of time, in the low PPM region. Electrolytic hygrometers do require a clean input gas. They should not be used with gases that react with a hygroscopic material.

An electrolytic hygrometer electrolyzes water vapor into its components, hydrogen and oxygen. The amount of electrical current required to dissociate water vapor at a particular temperature and flow rate is proportional to the number of water molecules present in the sample.

The cell has a bifilar winding of inert electrodes on a fluorinated hydrocarbon capillary. Direct current applied to the electrodes dissociates the water, absorbed by the P2O5, into hydrogen and oxygen. Two electrons are required to electrolyze each water molecule, and the current in the cell represents the number of molecules dissociated. A further calculation, based on flow rate, temperature, and current, yields the parts-per-million concentration by volume of water vapor.

In order to obtain accurate data, the flow rate of the sample gas through the cell must be known and constant. Since the cell responds to the total amount of water that enters it, the PPMV calculation depends upon knowledge of the volume of dry air. An error in the flow rate causes a direct error in the measurement.

A typical sampling system, to ensure constant flow, maintains constant pressure within the cell. Sample gas enters the inlet, passes through a stainless steel filter and enters a stainless steel manifold block. (it is most important that all components prior to the sensor be made of an inert material, such as stainless steel, to minimize absorption of water). After passing through the sensor, the sample gas pressure is controlled by a differential pressure regulator that compares the pressure of the gas leaving the sensor to the gas venting to the atmosphere through a preset valve and flow meter. Thus, constant flow is maintained in spite of nominal pressure fluctuations at the inlet port.

Typical minimum sample pressure, which is required for this type of control, is 10 psig (1.7 kg/cm^2) to 100 psig (8 kg/cm^2). Sensors are presently also available for higher pressures. A filter is normally located upstream to prevent contaminants from reaching the cell.

A typical electrolytic hygrometer can cover a span from 0 to 1000 ppmv with an accuracy of ±5 percent of the reading. This is more than adequate for most industrial applications. Special sensors have been built during the last few years that can measure in the PPB (parts per billion) range. The sensor is suitable for most industrial applications, but only with low levels of contaminants.

Applications. These are mostly used for PPM measurements in dry cylinder gases and for semiconductor manufacturing. They are also used for natural gas measurements, though this function requires much maintenance.

Gases to Avoid. The electrolytic hygrometer cannot be used for humidity measurements in gases that are corrosive (chlorine, etc.) or that readily combine with P_2O_5 to form water (alcohols). Also, gases such as certain acids, amines, and ammonia react with P_2O_5 and must be avoided. Unsaturated hydrocarbons (alkynes, alkadienes, and alkenes higher than propylene) polymerize to a liquid or solid phase, clogging the cell.

When the electrolytic hygrometer is applied to clean gases in the low PPM region, the instrument offers long and reliable service. However, if misapplied, the electrolytic cell is easily ruined. Although procedures are furnished by the manufacturers of these cells for unclogging the capillary and recoating with phosphorouspentoxide, this is not a trivial procedure. Often, a failed electrolytic cell must be replaced. On the other hand, since the electrolytic cell is essentially a primary device, no calibration is required when replacing the unit.

Chilled Mirror Hygrometer

The chilled mirror sometimes called the condensation-type hygrometer, is the most accurate, reliable, and fundamental hygrometer and is therefore widely used as a calibration standard. First commercial applications of the hygrometer started in the early 1960s after it became practical to use thermoelectric coolers and electrooptical detection methods. Prior to 1960, the use of chilled mirror hygrometers was primarily confined to laboratories because of the difficulties in cooling the mirror with cryogenic liquids, controlling its temperature, and detecting the onset of dew, usually through visual means.

The largest source of error in a condensation hygrometer stems from the difficulty of measuring the condensate surface temperature with accuracy. Typical industrial versions of the instrument can be made with accuracies of up to ±0.1°C (±0.20°F) over a wide temperature span. Chilled mirror hygrometers can be made very compact using solid-state optics and thermoelectric cooling and with extensive capabilities using microprocessors.

In its most fundamental form, dew point is detected by cooling the condensation surface (mirror) until water begins to condense, and by detecting the condensed fine water droplets optically with a light-emitting (LED) detection system. The signal is fed into an electronic feedback control system so as to control the mirror temperature so that a certain thickness of dew is maintained.

Since the first introduction of the chilled mirror hygrometer, this type of instrumentation, despite its higher cost, has been widely accepted for precision laboratory calibrations and standards and also in many industrial applications where high accuracy and traceability are required. This has resulted in many improvements, new developments, and several patents.

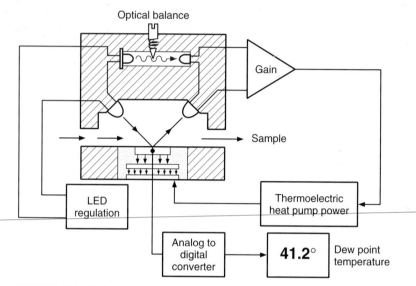

FIGURE 39.2 Optical dew point sensor.

Operation of the basic optical dew point hygrometer is shown in Fig. 39.2. The surface temperature of a small gold or rhodium plated mirror is controlled by a thermoelectric cooler (heat pump). The mirror is illuminated by a high intensity light emitting diode (LED). The quantity of reflected light from the mirror surface is detected by a phototransistor or optical detector.

A separate LED and photo transistor pair are used to compensate for thermally caused errors in the optical components. The phototransistors are arranged in an electrical bridge circuit with adjustable balance that controls the current to the thermoelectric mirror cooler and, therefore, the mirror temperature.

Reflectance is high when the mirror surface temperature is above the dew point and maximum light is received by the optical detector. However, if the thermoelectric cooler reduces the mirror temperature below the dew point (or frost point, if below 0°C), moisture condenses on the surface, scattering the light and thereby reducing the amount of light received by the sensor.

The system is designed so that the bridge is balanced and stable only when a thin layer of dew or frost is maintained on the mirror surface. Under these equilibrium conditions, the surface temperature is precisely at the dew point of the gas passing over the mirror. A precision NIST-traceable platinum or other thermometer is imbedded within the mirror surface to measure the surface temperature. The dew point temperature is displayed on the front panel.

When the mirror is clean, a perfect layer of condensation can be maintained, and high accuracy and repeatability results.

Sensitivity to Contaminants. The main difficulty with chilled mirror hygrometers is that, like any other humidity instrumentation, the sensor is sensitive to contaminants. Even in clean applications, ultimately some contaminants will appear on the mirror surface, thereby influencing the optical detector and servo-balancing functions of the instrument. A common practice to minimize this problem is to open the servo feedback loop, causing the mirror surface to heat to a dry state, and then readjusting the balance or reference of the optical circuit to compensate for the reduced reflectance of the mirror.

In earlier models, this procedure was performed manually. Subsequent improvements have been the so-called *automatic balance* and *continuous balance* by EG&G and *PACER* by General Eastern. In all of these, the mirror is continuously at the dew point temperature and has at all times, except during the short balancing cycle, a dew or frost layer on it. A new type of hygrometer was developed and patented by Protimeter plc in the U.K., which cycles the mirror temperature and only briefly keeps the mirror at the dew point and a dew layer on it—just the opposite of the conventional continuous chilled mirror hygrometer. The advantage is that a dry mirror does not hold on to contaminants as readily as a wet mirror. On the other hand, some of the fundamental measurement features are sacrificed, which makes this type of instrument, even though very high accuracies can be obtained, less suitable for use as a *fundamental* calibration standard.

Automatic Balance. The automatic balance method of standardizing a chilled mirror hygrometer was first developed and patented by EG&G. This method, with some variations, was subsequently adopted by most other manufacturers and became a widely accepted method of checking and standardizing the mirror of a condensation hygrometer.

The method consists of circuitry to interrupt the control circuit automatically at a prescribed interval that usually can be selected by the user, for instance once every 12, 24, or 48 hours, depending on the expected contamination level. The mirror is then heated to a temperature above ambient, typically 60° to 80°C, causing any dew to be removed. Reflectance from the dry but contaminated mirror is then compared with a standard, based on a dry but clean mirror, and a correction is made to standardize the mirror without cleaning it. The mirror is then cooled down to the dew point again and goes into control like it was before the automatic balance cycle. This operation typically takes 3 to 8 minutes, depending on how low a dew or frost point is being monitored. A longer time is required for low frost points because it takes longer for a mirror to go into the control mode when the mirror has to be cooled far below ambient temperature.

The automatic balance cycle does not remove or disturb any contamination on the mirror. It only allows the instrument to be restandardized based on a contaminated mirror, thereby extending the time period between required mirror cleanings. As will be shown later, contaminants can also contribute to errors in measurement. The automatic balance reduces but does not eliminate such errors.

PACER Circuit. As shown above, the automatic balance circuit extends the mirror-cleaning cycle and reduces (but does not eliminate) contaminant-induced errors. The PACER technique, an acronym for Programmable Automatic Contaminant Error Reduction, permits accurate long-term operation of a thermoelectrically cooled, optically detected, dew point hygrometer without need for shutdown for cleaning the mirror. This is an important technological advance in a fundamental measurement instrument.

After operating the sensor for some time, water-soluble salts and other contaminants, if present in the sample gas, can accumulate on the mirror surface. They create two error-causing effects, illustrated in Fig. 39.3.

1. The operating dew density decreases as contaminants like the dew layer decrease the light received by the detector. The system eventually becomes inoperative

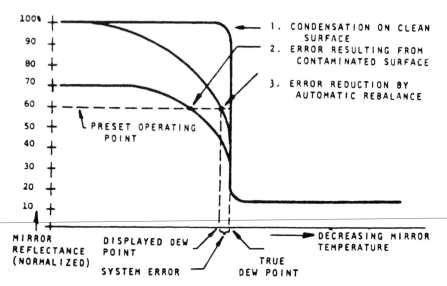

FIGURE 39.3 Errors caused by mirror contamination.

because it no longer controls on dew. At this point, the system must be shut down and the mirror cleaned to attain proper operation.

2. Soluble contaminants, after dissolving in the operating dew layer, modify the vapor pressure; the control loop compensates for this by elevating the mirror temperature. Since the thermometer measures the actual mirror surface temperature, a measurement error results. The temperature gradually increases five or six degrees Celsius (typically undetected), then finally takes a dramatic jump, at which point it is observed and the mirrors cleaned.

Systems outfitted with automatic balance correct the first type of error-causing effect only—that is, reduced mirror reflectance due to contaminants. This has been accomplished by periodic automatic or manual interruption of the temperature control loop, and heating the mirror to evaporate the dew. The reflectance of the dry mirror, although reduced by the contaminant deposits, is then compensated by rebalancing.

This is effective in maintaining control because it reestablishes the operating dew layer. However, it does not address the measurement error problem that occurs long before the system goes out of control.

PACER Technique. Using the PACER technique, the time interval between mirror cleanings can usually be extended significantly. The PACER circuit, unlike conventional automatic balance circuits, first causes the mirror to cool below the prevailing dew point, which results in an excess of water deposited on the mirror surface. In essence, it allows a puddle to form on the mirror. This excess of solvent (water) encourages soluble materials to go into solution.

The PACER circuit then causes the mirror to heat quickly, evaporating the water. The large puddle breaks up into small puddles that shrink in size, holding higher and higher concentrations of solute in solution, leaving bare, uncontaminated surfaces behind.

THE DETECTION AND MEASUREMENT OF HUMIDITY   39.13

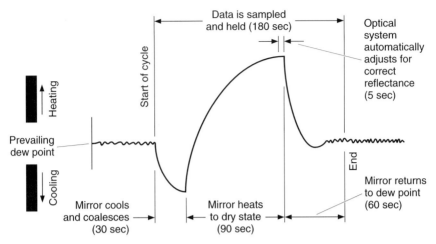

FIGURE 39.4 Typical PACER cycle.

Eventually, each puddle becomes saturated, and the solute begins to precipitate out in polycrystalline clusters. The amount of the salt has been redistributed or clustered in such a fashion that most of the mirror surface is clean; the resulting reduction in reflectance is only 15 or 20 percent, even in severely contaminated situations.

The PACER circuit automatically adjusts the control loop offset to operate at a level of condensate that corresponds to a preset reduction of reflectance below that of a clean mirror. The temperature control loop, calling for this reflectance level, causes the mirror to be maintained at true dew point because it is at that temperature only so that such a reduction can be realized. A typical PACER cycle is shown in Fig. 39.4.

Figure 39.5 shows a contaminated mirror surface before and after the PACER cycle. In Fig. 39.5, a mirror which, for instance, has been allowed to become contaminated for a period of one week in a heavily laden lithium-chloride air stream is shown after the drying process of a typical balancing procedure [Fig. 39.4(*a*)]. The salt is uniformly distributed over the entire mirror surface. The reflectance of this surface is only 40 percent of that of a clean mirror in the dry state.

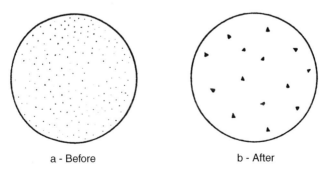

FIGURE 39.5   Mirror condition before and after the PACER cycle.

In Fig. 39.5(*b*), the mirror has first been cooled for thirty seconds, allowing the salt to dissolve into the puddle, and then heated to the dry state. The salt is now grouped into isolated colonies, and the reflectance of the surface is 85 percent of that of a clean mirror. The result is a hygrometer with improved long-term accuracy in the presence of high levels of soluble contaminants.

Figure 39.6 shows graphically the elimination of dew point error. Note that the operating point after the PACER cycle and rebalancing has returned to the true dew point.

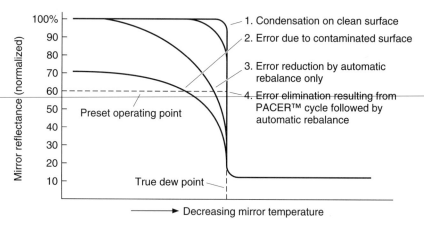

FIGURE 39.6 Error elimination with the PACER cycle.

Typically, a PACER cycle can be programmed automatically every 2, 6, 12, or 24 hours; cooling and heating times are also programmable. During the short interval of the PACER cycle, the output data and display are held at the last prevailing dew point.

Continuous Balance—Dual-Mirror Twin-Beam Sensor

Dew Point Hygrometer with Continuous Balance Control. As indicated above, optical condensation-type hygrometers must be periodically readjusted or balanced versus a reference to compensate for contaminants that always deposit on the reflectance mirror of the sensor. In conventional systems, this is done manually or automatically with time intervals of a few times a day to once every several days depending on contamination levels. Using the PACER circuit described above, a cool or coalescence cycle is added to redistribute soluble contaminants on the mirror and thereby prevent measurement errors.

In a dual-mirror sensor, one mirror is always kept at a temperature above the dew point (dry mirror) while the other is controlled at the dew point (wet mirror). The sample gas is passed over both mirrors. Two light sources are used—one for each mirror. See Fig. 39.7.

Reflectance from the dry mirror is used as a reference for controlling the wet mirror, and it is therefore no longer necessary to interrupt the control circuit periodically for a balance cycle (typically 2 to 4 minutes), and the method offers the advantage that loss of data during this 2-to-4-minute period every 12 or 24 hours is eliminated. However, limitations of this method are:

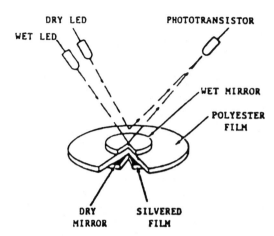

FIGURE 39.7 Twin-beam dual-mirror sensor.

1. It is evident that when a dry and a wet mirror are exposed to the same contaminated gas stream, the wet mirror will collect contaminants at faster rate than the dry mirror, so that effective balance control is only possible for a limited period of time. Some improvements may be possible by keeping the temperatures of both mirrors in close proximity, with the dry mirror temperature at just above the dew point. However, the problem of contaminant imbalance cannot be eliminated, which is a disadvantage of this method.
2. The continuous balance method does not address the soluble contaminant error problem as described above. Contaminants remain on the mirror and are undisturbed by the continuous balance operation. Long-range measurement errors are therefore larger, and time intervals between mirror cleanings are shorter than for instruments outfitted with the PACER circuitry.

Cycling Chilled Mirror Dew Point Hygrometer (CCM)

This section describes a significant advancement in the measurement of dew point by means of chilled mirror sensors. Specifically, a patented cycling chilled mirror (CCM) control system and integral filter are used, offering the user a sensor requiring virtually no mirror or filter maintenance. Also, for the first time, a new sensing technique makes it possible to always measure dew point below 0°C, solving the well known problem of dew versus frost point uncertainties that could cause errors of 1° to 3°C in conventional hygrometers. Signal processing inside the sensor allows longer and lower cost cables to be used when the sensor is operated remotely. For high-temperature applications, a sensor has been developed utilizing fiber optic bundles that keep the temperature-sensitive electrooptical components away from the high-temperature environment.

Almost all of the chilled mirror hygrometers that have been on the market are of the *continuous condensation* type. As we have seen, the condensate surface (reflective mirror) is maintained electronically in vapor pressure equilibrium with the surrounding gas, while surface condensation is detected by optical techniques. The surface temperature is then the dew point temperature by definition. Typical industrial versions of the instrument are accurate to ±0.2°C over wide temperature and dew point spans. Such instruments provide fundamental measurements that can be made traceable to NIST (National Institute for Standards Testing).

The CCM hygrometer uses a cycling chilled mirror method. The mirror temperature is lowered at a precisely controlled rate until the formation of dew is detected. Before the dew sample is able to form a continuous layer on the mirror, the mirror is heated and the dew on the mirror surface is evaporated. Hence, the mirror is almost always (95 percent of the time) in the dry state and contains a dew layer for only a short time (5 percent of the time), when a dew point measurement is made. Typically the measurement cycle is once every 20 seconds. Faster rates are possible. Another important benefit is that the dew layer is maintained for such a short time that it will never convert into frost, even when below 0°C. As shown above, this eliminates the problem with conventional optical hygrometers, that one cannot be certain whether the instrument reads dew point or frost point when below 0°C. The reason is that with supercooling, a dew layer often remains for a long time, or even continuously, depending on flow rate, mirror contaminants, and other factors.

Mirror Cycling. Figure 39.8 shows how the mirror is cycled in four specific steps.

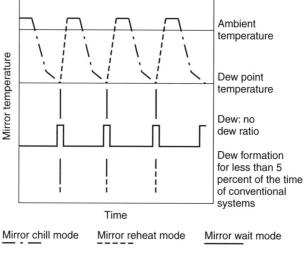

FIGURE 39.8 Dew point determining cycle.

Step one consists of the rapid cooling of the mirror from a level above ambient temperature to approximately 1.5°C above the last known dew point. This reduces the total cycle time, especially when the difference between the dew point temperature and ambient temperature is large. During step two, starting at about 1.5°C above the dew point, the cooling rate is slowed to approach and cross the dew point as slowly as possible, allowing dew to form in a uniform and repeatable manner, so that the correct formation of dew is detected.

Step three starts when the dew detection is completed. The current to the cooling device is reversed, which causes the mirror to rise rapidly in temperature to a few degrees above ambient temperature.

The fourth and final stage of the cycle represents a short period of time between successive cooling stages that allows the mirror to stabilize. This period is usually a few seconds and could be varied from seconds to minutes, if desired.

It is evident from Fig. 39.8 that dew is present on the mirror surface for only a very short time, causing contaminant buildup on the mirror to be kept at an absolute minimum. Also much less power is used, and therefore the dissipated energy is very low. This means that any heating of the sensor or surrounding air is kept to a minimum. This is especially important when the sensor is operated in a small, confined area, such as when used for water activity measurements in small containers.

The optical system is arranged in such a way that the absolute light level is unimportant. Only changes in light level, as to whether the mirror is dry or contains dew are important. Although the mirror is allowed to cool below the dew point, the correct dew point is stored in memory and will be used if the downward optical trend persists long enough to be real dew formation.

All measured signals from the sensor are converted into current levels and can be connected to the instrument. This arrangement of current drive makes it possible to operate the sensor over a long distance from the mainframe (up to 1000 ft.) while using low-cost nonshielded cables.

CCM Sensor. Top and bottom views of an industrial CCM sensor are shown in Fig. 39.9. A pressure-tight, O-ring-sealed pressure cap covers and protects the mirror and filter compartment. The cylindrical filter surrounds the mirror and can easily be replaced if needed. The optical bridge assembly is mounted in a separate, removable unit that easily plugs into the sensor mirror compartment.

Sensors are generally available in anodized aluminum or in stainless steel for corrosive applications. Most chilled-mirror sensors, conventional as well as CCM, are designed for pressures up to 300 PSI. The sensor shown is rated for 180 PSI.

Three electronic circuit boards are mounted on the bottom of the sensor. One is the optics control board, the second is a temperature board, and the third is a pressure compensation board. The sensor can be installed for online measurements or for use with a sampling system. Because of temperature limitations of the optical components and the thermoelectric cooler, optical condensation sensors are generally not suitable for operation at ambients above 80°C to 95°C.

Dew Point/Frost Point Conversion. As pointed out above, misinterpretation of dew versus frost point could result in errors of up to 3.7°C at $-40°C$, as is shown in Table 39.1.

In conventional continuous chilled mirror hygrometers, when measuring below 0°C, initially a dew layer is established on the mirror, and the instrument reads dew point. It is generally assumed, though this is not always the case, that the dew layer converts to frost. However, this could take a few minutes, several minutes, or even several hours, and it is unpredictable. In some cases, the dew layer continues to exist indefinitely, even when the equilibrium temperature is well below 0°C. This is called supercooled dew, a condition that could exist because of certain types of mirror contamination, a high flow rate, or a number of other reasons. In continuous hygrometers, it is usually *assumed* that, when the mirror temperature is below 0°C, the measurement represents frost point, but this is by no means certain, and even if this were so, it could take a long time before the real frost point is measured. In the interim, the measurements could be in error, unknowingly, to the user.

When such errors cannot be tolerated, it is customary to use a sensor microscope mounted on top of the sensor to make it possible to observe the mirror and deter-

FIGURE 39.9 CCM Sensor showing top view with mirror and integral filter (left) and bottom view showing integral signal processing electronics.

THE DETECTION AND MEASUREMENT OF HUMIDITY

TABLE 39.1 Dew/Frost Point Conversion

Frost point	Dew point	Deviation
0°C	0°C	0°C
−5°C	−5.6°C	0.6°C
−10°C	−11.6°C	1.2°C
−15°C	−16.7°C	1.7°C
−20°C	−22.2°C	2.2°C
−25°C	−27.7°C	2.7°C
−30°C	−33.1°C	3.1°C
−35°C	−38.4°C	3.4°C
−40°C	−43.7°C	3.7°C

mine whether dew or frost is measured. This is, of course, impractical and by no means an automatic industrial procedure. The CCM hygrometer, therefore, offers a significant advantage when measuring below 0°C because one *knows* that the measurement is dew point. If the user wants to read the frost point, it is usually a simple matter to program the onboard microprocessor to display frost point.

Maintenance Requirements. The combined use of the CCM technology and the unique circular filter (shown in Fig. 39.9) in the industrial sensor, which is not consumed by the sample flow, has resulted in an instrument requiring considerably less maintenance (i.e., less mirror cleaning). In cases where typical conventional hygrometers require mirror cleaning at intervals of a few days or weeks, the CCM hygrometer has in some cases been used without mirror cleaning for periods of up to or more than one year, while maintaining good accuracy.

A typical CCM hygrometer sensor is shown schematically in Fig. 39.10.

Surrounding the mirror is the cylindrical, 40-micron filter. This filter, unlike the inline filters used in conventional hygrometer systems, does not require 100 percent

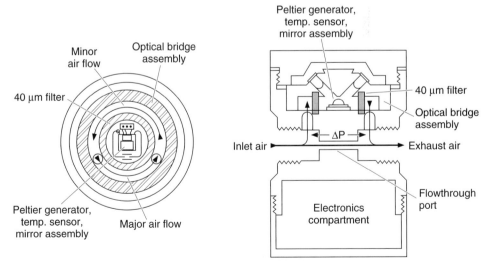

FIGURE 39.10 Typical cycling hygrometer sensor with integrated flow-around filter.

of the total sample gas to pass through its element. Instead, sample gas circulates around the outside of the element, and it is measured by means of convection across the filter element. Because most particulates circulate freely around the filter and exit the measurement chamber, the filter is not consumed by the sample flow and therefore does not contaminate as rapidly. Furthermore, the usual slow-down of the sensor response time as a result of filter contamination is greatly reduced, causing this kind of sensor to have a much faster response than conventional systems. This technique often works out satisfactorily in cases where a conventional system is ineffective.

Benefits of the CCM Hygrometer. The advantages of CCM technology may be summarized as follows.

- A unique patented optical slope detection system which virtually eliminates the need for regular re-calibration of optics. The mirror contains dew for less than 5% of the time and is therefore almost always dry. Hence particulate contaminants do not attach to the mirror and less mirror cleaning is required.
- A unique integral cylindrical 40 micron filter that is not consumed by the sample air as in conventional systems but that causes most of the sample air to bypass the mirror and filter and thereby further reduces mirror contaminants. Mirror cleaning cycles go as long as one year in some cases.
- A sensor/filter design that does not significantly increase the response time. In conventional chilled mirror hygrometer systems, the filter is mounted in the sample line. It must therefore pass all of the sample gas, and all contaminants present in the sample flow will be trapped in the filter, thereby slowing the sensor's response time and requiring considerable filter maintenance.
- A chilled mirror sensor design (shown in Fig. 39.9) that includes the signal processing electronics, eliminating the need for heavy and costly shielded cables and allowing the use of low-cost standard nonshielded cables and at lengths of 1000 ft. or more as compared to the 300-ft. cable-length limit in most conventional systems.
- Guaranteed dew point measurement at all times, even below 0°C. A common drawback of the chilled mirror hygrometer is that at dew/frost points below 0°C, it is unclear whether supercooled dew or frost is measured. The new slope detection system appears to have solved this problem.
- Integral pressure sensors provide dew point compensation from vacuum to 150 PSIG.

In general, the new advances discussed in this chapter eliminate almost all of the traditional operational problems and make this type of hygrometer a user-friendly instrument.

The CCM hygrometer is believed to be suitable for most industrial applications requiring automatic and relatively maintenance-free operation while providing high accuracy with excellent repeatability. Such applications include heat treating, automobile emissions testing, dryers, humidity chambers, environmental areas, plant air systems, process control, and semiconductor manufacturing, where low maintenance and ease of operation are essential.

Chilled Mirror Dew Point Transmitters

The cost of chilled mirror hygrometers is prohibitive for many applications such as air lines, air conditioning, and room dew point monitoring. In such applications, RH transmitters using polymer-resistive or capacitive sensors are presently very popular. Though these sensors have attained a high degree of performance and accuracy, they are secondary measurement devices requiring periodic calibration against a certified calibration standard that is time-consuming and costly. Furthermore, such sensors cannot be relied upon if measurements have to be traceable and certified. In such instances, the fundamental chilled mirror sensor is the ideal choice.

To accommodate such applications, where only a current or voltage signal is needed and no readout, several manufacturers are offering chilled mirror type transmitters. Though still more expensive than the polymer RH transmitters, they are considerably less expensive than the conventional chilled mirror hygrometers and offer fundamental and traceable measurements.

The first company to offer such a product was General Eastern Instruments. Its Dew-10 is still the lowest cost chilled mirror transmitter on the market, but it lacks some features of the later product entries, and the Dew-10 appears sensitive to contaminants, flow rates, and other environmental and operating conditions.

The EG&G Dew Track offers more features and has shown better performance in a number of applications, but it is very expensive compared to the standard RH transmitters and is therefore not widely used.

The most recent introduction has been by Protimeter, who offers a transmitter called the Dew Tector with CCM control and the additional new improvements of a scratch-resistant sapphire mirror and an integral automatic wiper system to clean the mirror at predetermined intervals. The combination of CCM, sapphire mirror, and wiper system has reportedly resulted in a virtually maintenance-free and very reliable chilled mirror transmitter, selling at a cost only slightly higher than the General Eastern Dew-10 and lower than the EG&G DewTrack.

The Dew Tector is quite new, and limited information on field performance is available. However, the technology suggest a major improvement in chilled mirror technology, especially with regard to low maintenance, ease of operation, and reliability.

SUMMARY OF BALANCING METHODS

Manual Balance

Advantages.   It is simple and reliable. It is the least expensive method.

Limitations.   Since an operator is periodically required, it is not desirable for automatic industrial applications. This method of balancing has been almost completely abandoned in favor of one of the other automatic systems.

The hygrometer does not measure dew point during the balance cycle.

Automatic Balance Control (ABC)

Advantages.   No operator required until the system goes "out of control." In modern systems, a warning signal is given in advance to indicate that mirror cleaning is required.

Limitations.   Although the time interval between mirror cleanings is extended by using the ABC method, the contaminants on the mirror are not removed. Hence, measurement errors resulting from contaminants (Raoult effect) are not eliminated.

The hygrometer does not measure dew point during the ABC cycle, which is performed at predescribed intervals regardless of important measurements are being made when the cycle is initiated.

PACER Cycle

Advantages. PACER is essentially an automatic balance cycle but with the addition of a cool cycle to redistribute contaminants on the mirror and thereby reduce measurement errors.

In many cases, the time interval between mirror cleanings is longer than with the ABC circuit and measurement errors caused by "soluble" contaminants are in some cases significantly smaller.

Limitations. The system must be operated to control on a relatively thick dew layer, which in some cases causes flooding of the mirror and a crash of the control system.

The system has proven to be unreliable in outdoor, meteorological applications, mainly due to the flooding problems, especially during fog conditions.

Continuous Balance

Advantages. The hygrometer makes measurements continuously. The interruption of 2–8 min. during balancing is eliminated.

For low frost point measurements, this is an advantage since it takes a long time to heat the mirror, balance it, and then cool it down again to the very low frost point. When using "continuous balance," the mirror always stays at low temperature.

Limitations. Like automatic balance, and unlike PACER, contaminants on the mirror are not disturbed and hence contaminant (Raoult effect) errors are not reduced or eliminated.

The time interval between mirror cleanings is generally not reduced significantly since a dry and a wet mirror contaminate at different levels and at different speeds.

This technique has not been widely used. It is used primarily for low frost point measurements where it has the advantage of faster response.

Cycled Mirror (CCM) Technique

Advantages. The mirror is dry 95 percent of the time. Therefore, it contaminates at a much slower rate. The time interval between mirror cleanings is much shorter than for any of the other balancing methods.

Limitations. The cycling parameters must be carefully set, and the instrument must be calibrated against a NIST Traceable Hygrometer to ensure accurate measurements.

The method is basically fundamental. However, by its nature, the CCM hygrometer is of far greater importance in industrial applications than for use as a laboratory calibration standard, where the continuous chilled mirror method is still the preferred choice in most cases.

CCM with Sapphire Mirror and Wiper

Advantages. This has the same advantages as in the CCM technique, plus the advantage of automatic mirror cleaning and a virtually indestructible mirror (sapphire).

OTHER TYPES OF DEW POINT HYGROMETERS

Dew Cup

A dew cup is the simplest instrument for the measurement of dew points. The gas sample is drawn from the furnace or generator across the outside of a polished cup made of chromium-plated copper. The cup is enclosed in a glass container so that the moisture can be seen condensing on the cup surface when the dew point is reached. Dew is observed by the operator, and measurement differences can occur due to operator interpretation.

The cup surface is cooled progressively by dropping small pieces of dry ice in acetone (or methanol) until the dew point is reached, as indicated by condensation on the cup surface at a temperature indicated by a thermometer in the acetone. The dew cup is most accurate for dew points above the freezing point of water. At dew points lower than 32°F, there is the possibility of supercooling, with a resulting low dew point reading. The use of the dew cup requires a considerable amount of skill and consistency on the part of the operator, and it is not recommended for close control. Incorrect dew point readings may result if the atmosphere is sooty, if there are leaks in the dew cup or sampling lines, or if the change of atmosphere is too fast. If the temperature is lowered too rapidly, or if the lighting conditions are poor in the area where the dew point is being observed, errors could result. A dew cup measurement is a one-time measurement, which may be repeated one or more times a day. It is not suitable for continuous monitoring.

Fog Chamber

The fog chamber is another type of manual instrument for measuring dew point. It is used throughout industry because of its portability, simplicity, and ability to cover a broad range of dew points without the need for cooling. The fog chamber operates on the principle that a rapidly expanding gas cooled adiabatically will produce a fog only when specific requirements of pressure drop, ambient temperature, and moisture content in the gas sample are satisfied. The atmospheric sample to be tested is drawn into the apparatus and held under pressure in an observation or fog chamber by a small hand pump. A pressure ratio gage indicates the relationship between the pressure of the furnace atmospheric sample and the ambient atmospheric pressure. The temperature is indicated by a high-grade mercury thermometer that extends into the observation chamber. The atmosphere sample is held in the observation chamber for several seconds to stabilize the temperature, after which the quick opening valve is depressed, releasing the pressure and creating adiabatic cooling, which causes a visible condensation or fog to be suspended in the chamber. The fog is easily observed with the lens system that provides a beam of light in the fog chamber when the quick opening valve is depressed. The procedure is repeated to find the end point, the point at which the fog disappears. The dew point is then determined by referring to a chart based on the initial temperature reading of the thermometer and the pressure ratio gage reading at the point where the fog disappeared. Though simple in construction and easy to operate, the method has the same drawback as the dew cup in that it is subject to human interpretation and the skill of the operator. Measurement errors of 5°C are not uncommon. It is also a one-time measurement.

Piezoelectric Hygrometer

The piezoelectric sensor measures moisture by monitoring the vibrational frequency change of a hygroscopically sensitive quartz crystal that is exposed alternatively to wet and dry gas. Sample gas is divided into two streams, sample and reference, which are alternately passed across the measuring crystal. The reference gas is passed through a molecular sieve dryer, which removes virtually all its moisture.

As sample gas is passed over the measuring crystal, moisture is sorbed by the crystal's hygroscopic coating, thereby causing a vibrational frequency change. At the end of the 30-second period, the microprocessor reads and stores the frequency difference (ΔF_1) between the crystal and a sealed local oscillator crystal, then switches the gas flow to expose the crystal to dried reference gas for the next 30 seconds to dry the crystal. At the end of this 30-second period, the microprocessor again reads and stores the vibrational frequency difference (ΔF_2) between the two crystals. The two different frequency values, ΔF_1 and ΔF_2, are then subtracted from each other to produce ΔF_3, which is proportional to the amount of moisture sorbed from the sample gas during that cycle. The moisture level is related to ΔF_3 through a polynomial expression with constants determined during the factory calibration of the cell. The microprocessor uses this expression to calculate the moisture level for each measured value of ΔF_3 and multiplies the results by a field-adjustable sensitivity factor to obtain the final output signal.

Both crystals are temperature-controlled at 60°C for optimum repeatability and accuracy. All electronics are solid state for highest reliability.

The instrument offers excellent accuracy and is especially useful for accurate low-level (PPM and PPB) measurements. However, such a system tends to be quite expensive.

Wet Bulb/Dry Bulb Psychrometer

Psychrometry has long been a popular method for monitoring humidity, primarily due to its simplicity and inherent low cost. A typical industrial psychrometer consists of a pair of matched electrical thermometers, one of which is maintained in a wetted condition.

Theory of Operation. Water evaporation cools the wetted thermometer, resulting in a measurable difference between it and the ambient, or dry bulb, measurement. When the wet bulb reaches its maximum temperature depression, the humidity is determined by comparing the wet bulb/dry bulb temperatures on a psychrometric chart. In a properly designed psychrometer, both sensors are aspirated in an airstream at a rate of 600 feet per minute for proper cooling of the wet bulb, and both are thermally shielded to minimize errors from radiation.

Applications. A properly designed and utilized psychrometer such as the Assman laboratory type is capable of providing accurate data. However, very few industrial psychrometers meet these criteria and are therefore limited to applications where low cost and moderate accuracy are the underlying requirements.

The psychrometer does have certain inherent advantages:

- The psychrometer attains its highest accuracy near 100 percent RH. Since the dry bulb and wet bulb sensors can be connected differentially, this allows the wet bulb depression (which approaches zero as the relative humidity approaches 100 percent) to be measured with a minimum of error.
- Although large errors can occur if the wet bulb becomes contaminated or improperly fitted, the simplicity of the device affords easy repair at minimum cost.
- A properly designed psychrometer can be used at ambient temperatures above 212°F (100°C), and the wet bulb measurement is usable up to a few degrees below the boiling point.

Major Shortcomings of the Psychrometer

- As relative humidity drops to values below about 20 percent RH, the problem of cooling the wet bulb to its full depression becomes difficult. The result is seriously impaired accuracy below 20 percent RH, and few psychrometers work well below 10 percent RH.
- Wet bulb measurements at temperatures below 32°F are difficult to obtain with any high degree of confidence. Automatic water feeds are not feasible due to freeze-up.
- Because a wet bulb psychrometer is a source of moisture, it can only be used in environments where added water vapor from the psychrometer exhaust is not a significant component of the total volume. Generally speaking, psychrometers cannot be used in small, closed volumes.
- Most physical sources of error, such as dirt, oil or contamination on the wick, insufficient water flow, and so on, tend to increase the apparent wet bulb temperature. This results in the measured relative humidity being higher than the actual RH.

Saturated Salt (Lithium Chloride)

The saturated salt (lithium chloride) dew point sensor was one of the most widely used dew point sensors. It was a popular choice in air-conditioning systems due to its inherent simplicity, ruggedness, low cost, and ability to be reactivated.

Theory of Operation. In this type of sensor, a bobbin covered with an absorbent fabric and a bifilar winding of inert electrodes is coated with a dilute solution of lithium chloride. An alternating current is passed through the winding and hence through the salt solution, causing resistive heating. As the bobbin heats, water evaporates into the surrounding atmosphere from the diluted lithium chloride solution, the rate of evaporation being controlled by the water vapor pressure in the surrounding air. As the bobbin begins to dry out due to evaporation, the resistance of the salt solution increases, causing less current to flow through the winding, and this allows the bobbin to cool. In this manner, the bobbin alternately heats and cools until an equilibrium is reached where the bobbin neither takes on nor gives off any water. This equilibrium temperature of the bobbin is a function of the prevailing water vapor or dew point of the surrounding air. A simple offset calculation is needed to arrive at the correct dew point.

Properly used, a saturated salt sensor is accurate to ±2°F (±1°C) between dew point temperatures of 10°F (−12°C) and 100°F (38°C). Outside these limits, small errors may occur due to the multiple hydration characteristics of lithium chloride.

If a saturated salt sensor does become contaminated, it can be washed with an ordinary sudsy ammonia solution, rinsed, and recharged with lithium chloride. It is seldom necessary to discard a saturated salt sensor if proper maintenance procedures are observed.

Limitations of saturated salt sensors include a relatively slow response time and a lower limit to the measurement range, imposed by the nature of lithium chloride. Also, once the cell is coated with lithium chloride, electric power must be applied continuously, or it must be kept in an environment with less than 11 percent relative humidity. Left unpowered in a humid environment, a lithium chloride cell can absorb enough water to dissolve the salt that will drip off the sensor.

The sensor cannot be used to measure dew points when the vapor pressure of water is below the saturation vapor pressure of lithium chloride, which occurs at about 11 percent RH. In certain cases, ambient temperatures can be reduced, increasing the RH to above 11 percent; but the extra effort needed to cool the gas usually warrants selection of a different type of sensor. Fortunately, a large number of scientific and industrial measurements fall above this limitation and are readily handled by the sensor.

CALIBRATION

National Calibration Laboratories

The standards for humidity in the United States are held at NIST (National Institute for Standards Testing), formerly the NBS (National Bureau of Standards) in Gaithersburg, Maryland. There are several similar Institutions overseas, such as the NPL (National Physics Laboratory) in the United Kingdom, PTB in Germany, CETIAT in France, and the National Research Laboratory of Metrology in Japan.

The primary responsibility of NIST is to provide the central basis for the National Measurement System, to coordinate that system nationally and internationally, and to furnish the essential services leading to accurate and uniform measurements throughout the United States. This section summarizes the activities of NIST and the humidity measurement systems.

The hierarchy of humidity standards is shown in Fig. 39.11. There are two routes by which the units of humidity are propagated. Both originate with the national standards for the base units. One route leads directly to the primary standard for humidity and then through precision generators, secondary standards, and fixed points, to working instruments and controls. The second route leads to the national standard for derived units, such as pressure, which, in turn, is used for measuring water vapor pressure over fixed points and for characterizing the pure water substances. With the enhancement factor or the correction for the real gas behavior of water vapor and gas mixture, water is used as the reference material in precision generators to produce known (predictable) levels of humidity for calibrating secondary and working standards.

The NBS Standard Hygrometer

The NBS standard hygrometer is based on the well-known gravimetric method of water vapor measurement. In the hygrometer, the mass of water vapor is mixed with a volume of gas that is absorbed by a chemical desiccant and weighed. The volume

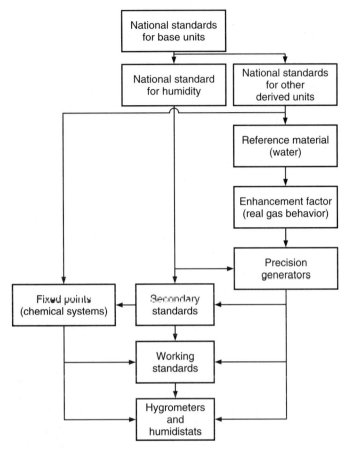

FIGURE 39.11 Hierarchy of humidity standards.

of dry gas is measured directly. Since mass and volume are fundamental quantities, this method yields an absolute measure of humidity. The unit of measurement is mixing ratio; that is, mass water/mass dry gas.

The graphimetric method is the most accurate system available, but it is very cumbersome and time-consuming to use. Hence, most calibrations are performed against other humidity systems.

Precision Humidity Generators

There are several practical methods for producing atmospheres of known humidity with sufficient precision and accuracy that the use of an auxiliary hygrometer for direct measurement is not required. Equipment incorporating these methods may be called precision humidity generators. Precision generators based on three different principles were developed at NBS (now NIST). The methods used are called the two-flow, two-temperature, and two-pressure methods.

Two-Flow Method

In this method there are two streams of gas. One stream is dry gas and the other is saturated with respect to water or ice. The two streams are combined in the test chamber, The humidity in the test chamber can be calculated from the known flow rates of the two streams of gas.

Two-Temperature Method

The principle of the two-temperature method is to saturate a stream of air with water vapor at a given temperature and then to raise the temperature of the air to a specified higher value. Figure 39.12 is a simplified schematic diagram of a two-temperature recalculating humidity generator.

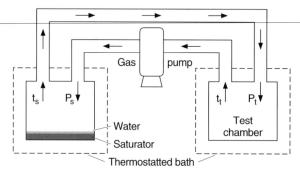

FIGURE 39.12 Simplified schematic diagram showing the two-temperature recirculating method.

Two-Pressure Method

In this type of generator, a stream of air at an elevated pressure is saturated at a fixed temperature and then isothermally expanded into a test chamber at a lower pressure, usually atmospheric pressure. Measurements of the temperature and pressure within the saturator and the test chamber or other test space establish the humidity of the test air.

Figure 39.13 is a simplified flow diagram that illustrates the principle of operation. The air enters at A, through a pressure regulator B, saturator C, expansion valve E, and test chamber F. A two-pressure generator was designed and built to encompass the relative humidity range of 20 to 98 percent and ambient temperature of $-40°C$ to $+40°C$.

After careful evaluation of the performance and operational characteristics of the two-flow, two-temperature, and two-pressure humidity generators, it was determined that the two-pressure method was the most suitable generator for providing an improved humidity calibration and testing facility at NIST.

Secondary Standards

Hygrometers suitable for use as secondary standards are instruments that have high precision and long-term stable response and whose output is predictable by means

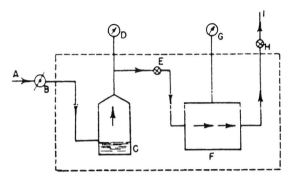

FIGURE 39.13 Basic Schematic of two-pressure humidity generator.

of a theoretical or empirical formula. Some of the devices suitable for use as a secondary standards are the condensation dew point hygrometers, electrolytic hygrometers, pneumatic bridge hygrometers, adiabatic saturation psychrometers, and aspirated psychrometers. The most commonly used secondary standards at NIST and industrial laboratories are the continuous chilled mirror hygrometers.

APPLICATIONS

There are numerous applications for measuring humidity, and through the last three decades there has been an increasing need for such measurements. There are many applications where humidity measurements were ignored in the past and where such measurements are now considered very important and in fact often required by regulatory agencies (EPA, Food & Drug Administration, Nuclear Agencies, and so on).

When confronted with such requirements, the most important decision is the selection of the best measurement method. There are many ways humidity can be measured, as we have seen, and there are many excellent humidity instruments on the market. However no one instrument is good for all applications and if an otherwise excellent instrument is applied in the wrong application, unsatisfactory results will be obtained.

It is highly advisable to review this matter carefully and discuss it with the manufacturer. Manufacturers of a large variety sensors are likely to give a more objective appraisal than companies making only one type of instrument who may be inclined to push their own type, even if it is not the best solution.

It is impossible to review all conceivable applications, so I will list some of the more common applications and the types of sensors used.

Automobile Emissions

The EPA requires accurate and verifiable measurements that, as a rule, require the use of NIST traceable chilled mirror hygrometers. The measurements could be made with lower cost instruments, but EPA generally insists on traceability of measurements, which can only be provided by the primary chilled mirror hygrometer.

Computer Rooms

Accurate and reliable RH measurements are usually called for. Most applications can be met with accurate bulk polymer-type RH transmitters. In some cases, NIST traceable measurements are needed, and we must then resort to chilled-mirror-type transmitters. If it is necessary to control on RH rather than dew point, a conversion from dew point to RH can be made with a built-in or remote microprocessor.

Nuclear Power Stations

Because of federal regulations, a chilled mirror hygrometer is almost always called for, even though the application usually does not require chilled mirror accuracy.

Petrochemical Gases

In most cases, measurements are required at many points. Measurements are for trace moisture in the low PPM range and do not need to be extremely accurate. In such applications, aluminum oxide sensors are often the choice.

Natural Gas

Humidity measurements in natural gas are usually very difficult to make with a high degree of reliability and usually require much maintenance due to contamination. Both aluminum oxide and P_2O_5 sensors have been used with only modest performance results. The need exists for a more suitable sensor for this application.

Semiconductor Manufacturing

Manufacturing of semiconductors requires a very dry environment for assurance of quality and long life expectancy. Required measurements are often in the PPM or PPB range. Phosphorous pentoxide sensors have often been used successfully in such applications. Aluminum oxide sensors have also been used for semiconductor applications, but they require calibration, which is very difficult at these low levels. Where a reliable calibration system is needed, the piezoelectric hygrometer has been successfully used, despite its high expense. The three- or four-stage chilled mirror hygrometer has also often been used for calibration purposes, but the measurements are difficult and time-consuming to make.

Pharmaceutical Applications

Because of their reliability, accuracy, and NIST traceability, the chilled mirror dew point hygrometers have usually been the sensor of choice.

Energy Management

This ordinarily requires RH measurements at many points. The bulk polymer sensor is ideally suited for this; both resistive and capacitive types are frequently used.

Heat Treat Applications

The best method, which is used in applications where accuracy, continuous monitoring, and NIST traceability are required, is the chilled mirror. For more routine applications, the fog chamber is very popular. The fog chamber only provides a one-time measurement, the accuracy of which is somewhat dependent on the operator's skill, but the cost is considerably less than a chilled mirror hygrometer. In the past, lithium chloride sensors have also been used with limited success, but lately there are very few lithium chloride sensors in use by heat treaters.

Meteorological Applications

For precision measurements, the chilled mirror hygrometer is always the right choice. For most applications, this is too expensive. In the past, the lithium chloride sensor was widely used by the weather bureau, but this type is now for the most part being replaced by the bulk polymer sensor, especially the capacitive types that operate over a broader temperature range than the resistive.

Laboratory Calibration Standards

The chilled mirror hygrometer is almost exclusive used for this application, because of its accuracy and NIST traceability.

Engine Testing

Because of the importance of accuracy, usually the chilled mirror is chosen.

Museums

The more reliable and accurate bulk polymer transmitters (resistive as well as capacitive types) are usually the choice and have been successfully used. A good calibration systems is needed to check the sensors and recalibrate them periodically as required. Portable calibration units for this purpose are presently on the market.

CHAPTER 40
THERMAL IMAGING FOR PROCESS AND QUALITY CONTROL

INTRODUCTION

The work presented in this chapter represents several important factors that need to be evaluated when designing a thermal imaging system. Topics include infrared camera selection, processor selection, and system design techniques.

As manufacturers strive for zero defects in production, more and more sophisticated techniques need to be employed during inspection and testing. Thermal imaging provides additional information to the quality control process.

There are a number of ways a temperature-related failure can occur: thermal runaway, gate dielectric junction fatigue, electromigration diffusion, electrical parameter shift, and package-related failure due to internal faults or assembly faults. When these faults cause complete device failure, the defects can be detected by traditional methods such a functional test. However, when these faults are not immediately catastrophic, functional tests will not reveal the defect. Thermal imaging hotspot analysis can detect latent thermal defects and aid in the quest for zero defects through a process which correlates thermal variations to long term failures.

The following sections will provide information on camera and processor selection as well as system design techniques.

CAMERAS

There are three important aspects to evaluate in an infrared camera:

- Spectral response
- Detector type
- Cooler type

The infrared spectrum covers approximately 1 μm through 0.1 cm. Most infrared cameras operate in either the midband spectrum (3 μm–6 μm) or in the far-band spectrum (6 μm–15 μm). Detector materials and optical filters determine the operational bands of the cameras. Detector material and fabrication techniques have traditionally caused a large price differential between the midband and far-band cameras. Traditionally, midband cameras have been less expensive than their far-band counterparts.

Midband cameras service applications that present mid- to high-temperature scenes (50°C–1000°C). Far-band cameras are better suited for lower temperature ranges (0°C–50°C).

In general, there are two types of infrared cameras:

- Scanning
- Focal plane array (staring)

Scanning cameras produce an image by optically scanning the field of view with a small number of detectors (usually less than 50). The scanning is typically carried out by sweeping the field in a raster scan fashion.

Focal plane arrays (FPA) produce an image by exposing a 2D array of infrared detectors to a scene. FPAs typically contain more than 5000 detectors and are organized in square arrays (i.e., $128 \times 128, 256 \times 256$).

Until recently, most commercial cameras were of the scanning type due to difficulties in fabricating large arrays of working detectors. Faults in both detector fabrication and readout electronics made yields on FPAs extremely low and thus the cost quite high.

The fabrication of infrared detectors for both scanning cameras and staring cameras introduces another problem for users. The detectors are not uniform in either offset or response. This nonuniformity is generally corrected in the digital electronics downstream from the detector. Correcting the nonuniformity of the detectors generally requires a two-point correction.

By exposing the array to two different uniform energy sources, say a hot frame and a cold frame, the gain and offset for each pixel (detector element) can be calculated. The gain and offset are stored for each pixel. During operation, the offset and gain are applied on a pixel-by-pixel basis to provide a uniform response to the current scene.

Again, scanners have an advantage here for two reasons. First, a scanning system has fewer detectors, so the number of gains and offsets that need to be calculated and stored and subsequently applied is smaller than the FPAS. Second, it is easier to incorporate internal calibration sources into a scanner. Typically, this is done at the start and end of each raster scan. With this design, the detectors are being corrected continuously. For a FPA, it is more difficult to provide a source, since the array is larger and there are no moving optics. The FPA tends to require external sources.

FPA correction may represent a problem in industrial service, since the correction is typically manual. This means that a correction procedure must be included as part of the maintenance of the system. Typically, these calibrations must be made every few hours (or less).

Most infrared cameras use cooled detectors. There are three common types of cooling:

- Liquid cooling (nitrogen or argon)
- Stirling cooling (cryoengine)
- Thermoelectric cooling

Liquid cooling involves filling the camera with liquefied gases. Nitrogen and argon are the typical choices. The detector is enclosed in a vacuum-sealed dewar which is formed into a reservoir for the cooling liquid. Liquid-cooled cameras tend to be the least expensive since they require no support hardware. However, the user must continuously fill the camera with the cooling liquid. Generally, these cameras are unsuited for use outside a laboratory environment.

Stirling coolers are used for a closed cycle mechanical cooling system based on a compressor. There coolers are suited for operation in almost any environment. But since they are a mechanical device, they must be serviced every 5000–10,000 hours of operation.

Thermoelectric (TE) coolers have no moving parts and require no maintenance. Until recently, TE coolers could not cool the large thermal mass of a focal plane array, so this was not an option for cooling FPAS.

Based on these three parameters, an infrared camera can be selected for a particular application.

1. Midband infrared cameras are suitable for most applications, unless near-ambient temperature measurements are required.
2. Scanning cameras are suitable for low-cost applications, while FPAs should be used for the highest image quality or for high-speed acquisition.
3. For most industrial applications, a stirling or thermoelectric cooler is required.

Not all cameras provide a temperature readout. This feature requires that the cameras have an internal temperature source to calibrate the detector output. If the application requires absolute temperature readings rather than relative readings, this is an important feature.

Almost all infrared cameras have an analog video output (RS-170). Some of the newer models also have a digital output that is compatible with the RS-422 standard. This is an important feature in order to provide high-fidelity images and to facilitate all digital processing.

PROCESSORS

The selection of the processor on which to run a thermal imaging application must be based on the desired throughput of the system.

It must be noted that the limiting time frame for execution may not be driven by the processing but rather by the thermal aspects of the unit under test. In other words, the unit may require a long time constant to reach thermal equilibrium. In other cases, the system may require that every frame be processed.

There are two major categories of processing hardware: computers and dedicated hardware. The computer class includes general purpose computers such as the Intel x86, RISC processors such as the SPARC, and digital signal processors (DSP) such as the C40, C80, 56K, SHARC, and i860. The general distinction of this group is the ability to program the processing elements extensively.

The dedicated processor class includes application-specific integrated circuits (ASICs) and custom hardware.

ASICs generally take a specific process or algorithm and optimize the function in hardware. The resulting device is typically much faster than generic processors, but it is limited to only one function. The only programmability of such a device is parameter adjustment; thus, there is a tradeoff between flexibility and speed (and perhaps cost). Examples of common ASICs for image processing include: convolution chips, histogramming chips, ranking chips, and feature extraction chips. Dedicated processors have the advantage of speed but give up flexibility.

Custom hardware operates on the same principles as ASICs but may be implemented in programmable logic devices or discreet components. One simple example is a two-input lookup table (LUT). This small piece of hardware can outperform many commercial processors for functions such as complex math. The LUT allows the results to be precomputed once, and the input data sets form a look-up of the correct answer.

Processor selection must begin with a throughput requirement analysis of the imaging application. The total processing time allotment is often determined by the action of the UUT or the speed of the production line.

The next determination of the system is how many operations need to be completed to obtain the desired results. System analysis and algorithm selection determine this number. Often functions similar in terms of processing result, require different numbers of operations.

The combination of processing time and number of operations yields the required operations per second. This number can be used to select the appropriate type of processor.

If the required number of operations per second is very high, the design often has to be targeted to dedicated hardware. When the number is low, the task is generally targeted to a general-purpose processor. The exception to this is extremely high volumes where the cost of dedicated or custom hardware can be minimized via economies of scale.

The processing functions can be divided between dedicated and general-purpose processors in order to take advantage of both methods. The division is often labeled preprocessing and postprocessing. Preprocessing involves high data rates and repetitive operations and is assigned to dedicated processors. Postprocessing takes the results from the preprocessor and performs higher level operations on smaller sets of data.

SYSTEM DEVELOPMENT

There are several techniques that will improve the quality of a thermal imaging defect detection system.

One of the most common and effective enhancement methods is frame integration. This involves averaging several sequential frames and works because the valid image content sums at a higher rate than random noise. This technique has become so popular that many cameras offer frame integration as a built-in feature. One of the major problems with frame integration is the reduction in the apparent response time of the imaging system. This is because a new event in the scene must persist long enough to make it through the frame-averaging process.

Another effective technique for improving image quality is a subtraction process. Subtracting a cold or ambient frame from current frames will remove residual offset terms from the resulting images. This technique is very effective in time sequence applications where an image can be taken at time 0. Each subsequent frame has the time-0 frame subtracted from it. In addition to removing ambient temperature offsets, this method can also account for varying initial thermal conditions. A time-0 subtraction creates a delta temperature image.

An effective method of detecting defects in a unit under test (UUT) is to compare the current thermal readings to a master template or model. In order to compare the UUT to a template or model, the UUT must be aligned with the template or the design database (or both). Aligning the UUT to the design database allows the defect locations to be translated into actual component designators and part numbers. Furthermore, database cross referencing allows globally detected objects to be postprocessed with individual component thresholds.

Alignment can be accomplished by locating device fiducials that can be detected in the infrared image. This may involve illuminating or backlighting several fiducials on the UUT. These points are then located in the image, and their relative locations are used to compute scale factors, rotation, and pan/scroll offsets, which will align the image data to the design database.

This alignment information can also be used to align the current image to a template image. The template image is generally composed of an average of several defect-free images. This template provides the basis for comparison to the UUT. The template and UUT images must be aligned with each other, or the comparison will generate potential defects in the misaligned areas. A simple subtraction of the template frame from the UUT frame will generate a delta frame, which is a map of thermal deviation from an accepted standard.

This deviation map is evaluated with region-based thresholds to determine possible defects. Defects include components running both hotter and colder than the accepted thresholds. The deviations are stored in a defect database for trend analysis.

If absolute temperature measurements are to be used and compared to accepted template values as a function of operating time, care must be taken to stabilize the starting conditions. Because of varying thermal time constants of the components on the UUT, the starting temperature of the UUT may significantly alter the recorder temperatures. For example, circuit boards coming straight from a solder process during assembly may generate significantly different results than circuit boards taken from storage. This situation can be minimized or avoided by either working with delta temperatures or by providing a thermal soak of the UUTs prior to inspection. The soak process involves storage of the UUTs in a thermally controlled environment to provide a common starting point for the test process.

CONCLUSIONS

There are many factors to consider when developing a thermal imaging system for quality control. Camera selection and processor design are the key to an effective system. However, other considerations, such as operating conditions, maintenance cycles, and system flexibility must be evaluated before committing to a system design.

CHAPTER 41
THE DETECTION OF ppb LEVELS OF HYDRAZINE USING FLUORESCENCE AND CHEMILUMINESCENCE TECHNIQUES

INTRODUCTION

Two methods for continuously monitoring low levels of hydrazine in air (detection limits less than 5 ppb) are discussed. The first method is based upon a chemical derivatization scheme in which hydrazine is chemically reacted with 2,3-naphthalene dicarboxaldehyde to form a highly fluorescent derivative, 2,3-diazaanthracene. This system operates by concentrating the hydrazine vapor into a small volume of aqueous reagent and monitoring the green fluorescence arising from the hydrazine derivative via a spectrofluorometer. The second detector utilizes a chemiluminescence reaction between luminol and a radical peroxo species arising from the catalytic oxidation of hydrazine by colloidal platinum. The major components of this system are simply a photomultiplier tube and a peristaltic pump for transporting the liquid reagent. A linear dynamic range for hydrazine has been obtained for both of these techniques from 10 to 2000 ppb in air with a response that is fully reversible and achieves plateau response in less than 2 minutes.

Hydrazine is a widely used propellant and a common precursor in the synthesis of a number of polymers, plasticizers, and pesticides. Due to the toxic nature of hydrazine, there is a growing need in both industrial and government laboratories for the development of a highly sensitive, online monitor for low levels of hydrazine in air. This is exacerbated by the American Conference of Governmental Hygienists' (ACGIH) proposal to lower the threshold limit value (TLV) for hydrazine from 100 ppb to 10 ppb in air.[1] While other analytical techniques (e.g., coulometry, flow injection analysis[1]) have shown success for the detection of hydrazine, none combines the necessary qualities of an inexpensive, online, sensitive, and selective monitor for hydrazine that is necessary for protecting the safety of workers in hazardous workplace environments.

The similarity in properties between hydrazine (N_2H_4) and NH_3 makes it reasonable to presume that many of the reagents designed for the fluorescent derivatization of amines, amino acids, and peptides might also be applicable to the trace analysis of hydrazine. Among the different derivatizing agents developed for use in fluorescent amino acid analysis, particular emphasis has been placed in recent years upon the use of the dicarboxaldehydes, ophthalaldehyde (OPA) and naphthalene-2,3dicarboxaldehyde (NDA). OPA and NDA react with peptides and primary amino acids in the presence of nucleophiles, such as 2-mercaptoethanol and the cyanide ion, respectively, to form highly fluorescent isoindole rings.[4,5,6] It has been recently demonstrated that NDA is also an effective derivatizing reagent for the fluorescent detection of extremely low levels of hydrazine in solution.[7,8]

Chemiluminescence techniques have been successfully applied to the detection of numerous analytes.[9] These methods typically rely upon the oxidation of a chemically reactive species like luminol or lucigenin, and the subsequent emission of a photon from an electronic, excited-state intermediate. Chemiluminescence sensors are free of Raman and Rayleigh scattering, interferences typically associated with fluorescence techniques, thus permitting the operation of a photomultiplier tube at maximum sensitivity. One of the most widely investigated chemiluminescence reagents is 3-aminophthalhydrazide or luminol. Luminol has been effectively used for the detection of many different metals and oxidants due to the catalytic behavior these metals have on the chemiluminescent oxidation of luminol. Examples include the detection of $Cr(III)$,[10] $Co(II)$,[11] $Fe(II)$,[12] H_2O_2,[13] NO_2,[14] and ClO^-.[15]

There have been no reports of chemiluminescence methods for the detection of hydrazine vapor, although Faizullah and Townshend report a flow injection analysis scheme based upon hypochlorite for the chemiluminescence detection of hydrazine in solution.[16] Pilipenko and Terletskaya demonstrated that the detection of trace amounts of platinum (0.02 µg/ml) is possible in a basic solution of lucigenin and hydrazine.[17] They proposed that hydrazine is catalytically oxidized at the surface of colloidal platinum to form a radical, peroxo intermediate. This reactive intermediate subsequently oxidizes lucigenin, allowing the detection of platinum. We demonstrate here the sensitive detection of hydrazine using a basic solution of luminol and colloidal platinum.

Chemiluminescence and fluorescence techniques present the possibility for the development of inexpensive, online monitors of hydrazine that are rapid, sensitive, and reversible over a wide dynamic range.

THE EXPERIMENT

Apparatus

All static, fluorescence measurements were made on an SLM-8000 double-beam scanning spectrofluorometer using a Rhodamine B reference cell to correct for instrument response and fluctuations in the lamp intensity (450-W Xenon arc lamp). Initial characterization of the chemiluminescence system was performed using the same spectrofluorometer, operating in the chemiluminescence mode of measurement. The injection port consisted of a lid with a septum permitting injection of different levels of hydrazine to a 4-cm^3 cuvette, while maintaining complete darkness within the sample compartment. A magnetic stirring bar allowed continuous mixing of the reagents. An EMI (PT9635QA) UV/VIS-selected photomultiplier tube was used for these experiments at a potential of 900–1200 V.

The flowthrough fluorescence results were obtained using a GTI/Spectrovision FD-300 dual monochromator fluorescence detector. This system employs a 10-W Xenon flash lamp (operated at 30 Hz) and a 100-µl flowthrough cell. The peristaltic pump used in these studies was a Masterflex 7521-50, which is capable of pumping 10 different channels simultaneously. The vacuum pump was an SKC Airchek Sampler, and the organic removal columns were supplied by Barnstead/Thermolyne, D8904. Pharmed tubing was used for the peristaltic pump, while all other tubing was teflon-coated.

The chemiluminescence vapor detection system consisted of an Anspec three-channel peristaltic pump, a vacuum pump (MSA Flow-Lite Turbo) for controlling the air flow from 0–5 L/min, and a photomultiplier tube (PMT). A 1000-V power supply was used to drive the PMT, and a Keithley 485 picoammeter was used to monitor the anodic current.

National Instruments' LABView for Windows software was used to control the data acquisition from the analog output of the picoammeter to a National Instruments' analog/digital board (Lab-Pc+). Pharmed tubing was used within the peristaltic pump, while all other tubing within the system was teflon PFTE. Tubing distances used were 60 cm for the gas/liquid sampling line leading to the impinger and 40 cm for the tubing leading to the PMT.

The hydrazine vapor generation system, which has been described elsewhere, was used to generate hydrazine concentrations from 0–2000 ppb in air.[18] The air stream consisted of house air scrubbed to remove water, CO_2, NO_2, and organic impurities. All flow rates were controlled using Matheson 8200 mass flow controllers. Humidity was controlled through the use of bubblers and line mixers, with the humidity being quantitated with a Hygrodynamics hygrometer. The hydrazine concentration was determined using the coulometric method described earlier. The flow rate to the MSA vacuum pump was determined using an American Meter Model 802.

Chemicals and Stock Solutions

All chemicals were used as received from the suppliers. 2,3-naphthalene dicarboxaldehyde (99%) was obtained from Molecular Probes, Inc., and 3-aminophthalhydrazide (luminol) (97%) was purchased from Aldrich. Stock solutions of NDA (10^{-2} M in anhydrous ethanol) and luminol (5×10^{-3} M in 0.1 MKOH) were prepared. A colloidal platinum stock solution was prepared by chemically reducing a 10^{-3}-M solution of hydrogen hexachloroplatinate (IV) with ascorbic acid at a pH of 3.5, with subsequent dilution to give a 10^{-4}-M solution of platinum. Hydrazine standard solutions were prepared on a daily basis by diluting the anhydrous hydrazine in purified water (Millipore) to give the following concentrations: 1 /µg/l, 100 µg/l, 100/µg/l, 1 mg/l, 10 mg/l, 100 mg/l, and 1000 mg/l. All pH studies were accomplished using 0.1 M buffer solutions of either boric acid (pH 6–11) or sodium dihydrogen phosphate (pH 3–5), wherein each buffer solution was adjusted to the appropriate pH with either NaOH or HCl. Those solutions with a pH above 11 were made using either NaOH or KOH, while those with a pH below 3 were prepared using either HNO_3 or H_2SO_4 acid.

A number of interference gases were tested using the hydrazine monitoring system. Gas mixtures were obtained from Matheson and Scott Specialty Gases and diluted with air to give varying concentration levels. The standards employed include: 99 ppm NO_2 in N_2; 110 ppm NO in N_2; 99.8 ppm SO_2 in N_2; 10% HCl in Ne; 7.2% H_2 in N_2; 450 ppm NH_3, in N_2; 99.99% CO_2; Freon 12; multicomponent mixture 1 [51.2 ppm Freon 12, 50.7 ppm Freon 11, 49.3 ppm Freon 114, 20.4 ppm CH_4, and 9.96 ppm hexane in He]; and mixture 2 [1.004% CO, 1.098% CO_2, 0.9640% CH_4, 0.9996% ethane, 1.014% ethylene, and 1.014% acetylene in He].

Procedure

For those experiments utilizing the SLM-8000 spectrofluorometer, 2 ml of the appropriate buffer solution were pipetted into a quartz cuvette, along with 10–100 µl (Eppendorf microburette) of the NDA stock solution, or 10–100 µl of both the platinum and luminol stock solutions. All examinations were done at room temperature. Subsequent spiking of these solutions with varying levels of hydrazine was accomplished by injecting 1–100 µl of solution from the appropriate dilution standard via the injection port of the sample compartment. The final concentration of the hydrazines in solution ranged from 50 ng/l to 50 mg/l. Unless mentioned otherwise, the fluorescence measurements were made at pH 2.5, while the chemiluminescence measurements were done at pH 13. The following excitation and emission wavelengths were used for the detection of the fluorescent derivative formed: λ_{ex} = 403 nm, λ_{em} = 500 nm.

Chemiluminescence Mode of Detection. The addition of colloidal platinum to a basic solution of luminol permits the indirect quantitation of hydrazine via a two-step process.

1. $2O_2(g) + N_2H_4(aq) \rightleftharpoons 2H_2O_2(aq) + N_2(g)$
2. $luminol(aq) + 2H_2O_2(aq) + OH^- \rightleftharpoons$ 3-aminophthalate(aq) + $3H_2O + N_2 + h\nu$
3. $N_2H_4(aq) + 2H_2O_2(aq) \rightleftharpoons N_2(g) + 4H_2O$ (side reaction)

Both reactions 1 and 2 are catalyzed by colloidal platinum.[19,17] Reaction 1 is the first step in the oxidation of hydrazine at the platinum surface. Pilipenko et al. have indicated that, although hydrogen peroxide is formed in the net reaction of hydrazine with oxygen, it is likely an intermediate, radical peroxo species (and not H_2O_2) that is actually responsible for the oxidation and subsequent chemiluminescence of lucigenin.[17] The presence of a radical species was inferred from the immediate inhibition of chemiluminescence following the addition of methyl methacrylate, a free radical acceptor. An important side reaction is the reaction of hydrogen peroxide with a second molecule of hydrazine, which completes the oxidation of hydrazine to nitrogen and water (reaction 3).

Initial evaluation of the analytical potential of this system was accomplished by observing the complete, chemiluminescent response obtained following the introduction of 49 µg/l of hydrazine to a basic solution (pH 13, NaOH) of platinum (0.25–2.9 µM) and luminol (7.4 × 10⁻⁵ M). The concentration of colloidal platinum plays an important role in determining the shape and size of the response curve (see Fig. 41.1). Note the difference in response obtained for solutions containing 0.25 µM and 1.0 µM colloidal platinum. For the lowest concentration of platinum examined, 0.25 µM, the time necessary to reach a maximum in signal intensity is longer, the peak height is substantially lower, and the chemiluminescence persists for up to an hour following the injection of hydrazine. As the concentration of colloidal platinum increases (up to 1.0 µM), the decay of the chemiluminescent signal becomes much faster, the maximum in signal intensity is attained in seconds, and the intensity of the chemiluminescent signal is more than four times that obtained for the 0.25 µM platinum solution.

It is evident from Fig. 41.1 that, as the concentration of platinum is increased, the total integrated chemiluminescence signal decreases dramatically. This result is attributed to the catalytic effect platinum has upon the oxidation of hydrazine to water and nitrogen (reactions 1 and 3), a side reaction decreasing the total luminol oxidation possible. A maximum in the instantaneous, chemiluminescent intensity is

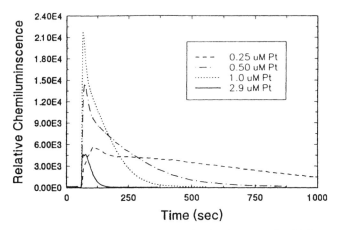

FIGURE 41.1 Time dependence of the chemiluminescence signal following the introduction of 49 µg/l hydrazine to a pH 13, 7.4×10^{-5} M luminol solution containing 0.25, 0.50, 1.0, and 2.9 µM colloidal platinum.

observed for 1.0 µM platinum. The signal intensity decreases for either higher or lower concentrations of platinum. The optimal concentration of colloidal platinum requires a compromise between having sufficient metal to catalyze the oxidation of luminol (reaction steps 1 and 2) promptly and having an excess of platinum that catalyzes the oxidation of hydrazine to water and nitrogen (reaction 3). As we shall see in the discussions pertaining to the continuous flow system, the concentration of platinum, the liquid flow rate and the distance from the vapor/liquid interface to the photomultiplier tube are all important parameters influencing the overall sensitivity of the system.

The pH and luminol concentrations are important parameters that determine the magnitude of the chemiluminescence response. A basic solution is required to initiate the reaction sequence. Typically, luminol systems used in the detection of metals or hydrogen peroxide operate optimally at a pH of 10–11.[9] It has been suggested that this is necessary in order to form a reactive diazine by abstraction of the second amino proton from luminol.[20] The optimal pH in the detection of hydrazine is pH 13, with the chemiluminescence signal intensity decreasing on either side of this pH. Several factors in the reaction sequence are known to be pH-dependent.

The autoxidation of hydrazine is maximized at a pH of 12 to 12.5,[19] while the production and lifetime of the intermediate product (peroxide) is highest at a pH of 13 to 13.5.[17] As shown later, operating at a higher pH (greater than 12.5) probably gives rise to some selectivity, by lowering the sensitivity of luminol to interferents such as NO_2 and O_3. Increasing the concentration of luminol results in an increase in the integrated chemiluminescence obtained for a given concentration of hydrazine. The increase in response with luminol concentration is nonlinear, however, due to self-absorbance of the chemiluminescence by luminol, or to luminol acting as a bidentate chelate, complexing with platinum, and perhaps rendering it inactive. The latter effect has been seen for the cobalt-catalyzed reaction with luminol.[21] The possibility remains, though, that a high concentration of luminol may be used in the hydrazine vapor detection system to improve the lifetime characteristics of the system.

The development of a hydrazine vapor monitor employing the Pt/luminol system requires first a means for concentrating hydrazine vapor into solution. A simple method for accomplishing this task is diagrammed in Fig. 41.2. A peristaltic pump delivers the luminol/platinum/KOH reagent through a short length of teflon tubing which concurrently transports sampled vapor. It is this gas/liquid interface which initiates the oxidation of hydrazine at the Pt surface, and ultimately, the oxidation of luminol. The liquid is efficiently separated in a modified, teflon impinger and pumped through a short section of glass tubing coiled directly in front of the PMT. The rapid decay of the chemiluminescent signal back to the background level following the addition of hydrazine to the line allows the reagent to be recirculated. The lifetime of the reagent is ultimately dependent upon the concentration of luminol in solution and the lifetime of the catalyst. Because of problems associated with the slow oxidation of luminol by $O_2(g)$ and/or the gradual change in pH from absorbed $CO_2(g)$ in the airstream, it may be necessary to change the solution daily to prevent shifts in the sensitivity.

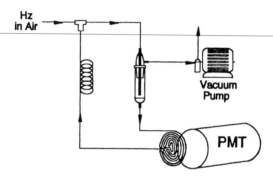

FIGURE 41.2 Diagram of the continuous-flow hydrazine monitoring system, consisting of a vacuum pump for sampling the atmospheric air, a peristaltic pump for transporting the luminol/platinum reagent, and a photomultiplier tube for detecting the chemiluminescence.

As demonstrated earlier, the concentration of active platinum in solution determines the magnitude of the chemiluminescence signal seen at the photomultiplier tube. The optimum concentration of platinum used for the continuous flow system is closely linked to the liquid flow rate and the length of tubing from the sampling point to the photomultiplier tube. Shown in Fig. 41.3 is a three-dimensional plot of the chemiluminescence response with respect to the liquid flow rate and the concentration of platinum while maintaining a constant air pumping rate and luminol, KOH, and hydrazine concentrations. For the tubing distance used in our experimental setup, low concentrations of platinum (10^{-7} M) resulted in a weak chemiluminescent signal that decreased with increasing liquid flow rate. An examination of Fig. 41.1 indicates that, at these concentrations of platinum, while the signal has a long lifetime, its maximum peak intensity is not very high. The slower the flow rate, then, the longer the liquid remains in view of the PMT, and the larger the chemiluminescent signal will be. In addition, at a constant air flow rate, decreasing the liquid flow rate concentrates more hydrazine vapor into solution, thereby giving a larger signal at the PMT.

As the concentration of platinum in solution is increased (5×10^{-7} M), the chemiluminescent signal becomes larger and the maximum in peak intensity occurs at higher flow rates. Referring again to Fig. 41.1, at these platinum concentrations, the

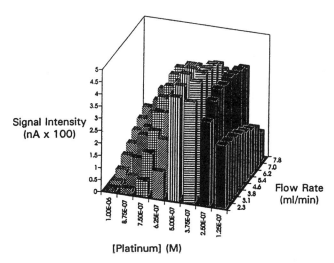

FIGURE 41.3 Plot of the chemiluminescent signal obtained from a 4.9 × 10^{-1}-M luminol solution at pH 13 as a function of the concentration of colloidal platinum and the liquid flow rate.

instantaneous chemiluminescent signal obtained is higher, therefore, giving a larger signal at the PMT. This is providing, of course, that the liquid flow rate is rapid enough to transport the solution to the PMT before the signal decays. If the flow rate is too high, though, the signal will decrease for two reasons: (1) less hydrazine vapor is concentrated into solution; and (2) the solution remains in view of the PMT for a shorter period of time. Ultimately, as the platinum concentration is increased to even higher concentrations (10^{-6} M), the signal intensity drops substantially for low flow rates. At these concentration levels, the chemiluminescence signal decays extremely rapidly—long before the solution reaches the PMT. In order to use these higher concentrations of platinum effectively, it is necessary to use extremely rapid flow rates so that the signal is sampled before the chemiluminescence decays away.

Just as the liquid flow rate controls the concentration factor for bringing hydrazine vapor into solution, so too does the air flow rate. As the air flow increases at a constant liquid flow rate, the amount of hydrazine concentrated into solution increases. There are several problems associated with operating at high flow rates. First, there is difficulty in preventing air bubbles from entering the tubing leading to the PMT, a result that leads to excessive noise. In addition, at the higher air flow rates, our vacuum pump rate was very uneven and the level of liquid in the impinger was difficult to maintain constant, both of which are factors leading to inconsistent results. Finally, the efficiency for concentrating hydrazine vapor into solution is expected to decrease at very high flow rates; the hydrazine has less time to interact with the solution within the liquid/gas sampling line. An optimum pumping rate can be determined for a particular platinum concentration that maximizes the trapping efficiency without sacrificing the stability of the measurement.

Our studies have established the following optimum experimental parameters for the detection of hydrazine: 10^{-6} M platinum, 10^{-4} M luminol, 0.1 M KOH, 7 ml/min liquid flow rate, 2 l/min air flow rate. In order to test the analytical response of this system, the chemiluminescent response was monitored for successively higher concentrations of hydrazine in air. Shown in Fig. 41.4 is an example of the

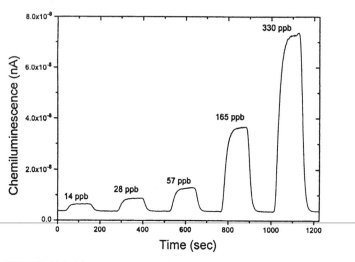

FIGURE 41.4 Time responses obtained using the hydrazine-vapor detection system for 14, 28, 57, 165, and 330 ppb hydrazine in air.

time responses obtained following the introduction and cessation of 14, 28, 57, 165 and 330 ppb hydrazine in air. A linear, dynamic range was obtained from 10–2000 ppb hydrazine ($r = 0.99$). The time necessary to reach 90 percent of a plateau response is 1.3 minutes, and the return to background requires 1.2 minutes. The detection limit was calculated to be 1.4 ppb, based upon a signal-to-noise ratio of 3:1.

Interference effects were studied by sampling various gases at concentration levels at least twice that of their regulated exposure levels. A solution of luminol/platinum was prepared at pH = 12.6 and exposed to the gases listed previously. The majority of gases examined, with the exception of NH_3, NO_2, SO_2, and HCl, either gave no signal or actually produced a slight decrease in the baseline signal. This latter effect arises from a decrease in the concentration of oxygen. Oxygen results in a small baseline chemiluminescence signal due to the direct oxidation of luminol. Positive interferences were seen for NO_2, SO_2 and HCl, but the sensitivities were much lower than for hydrazine. Concentrations of 9 ppm NO_2, 124 ppm SO_2, and 210 ppm HCl gave signals comparable to 10 ppb hydrazine in air. Considering that the regulated exposure levels for NO_2, SO_2, and HCl are 1 ppm (STEL), 2 ppm (TLV), and 5 ppm (TLV), respectively, these interferents are not expected to be a problem for hydrazine vapor detection. The presence of NH_3 can give a false positive signal if the sampling line is not constructed correctly. Hydrazine will stick to any portions of tubing on the end of the sampling tee that are not being continually washed by the reagent. Ammonia, a strong Lewis base, competes very well with hydrazine for adsorption to the sides of the tubing leading to the impinger. Any dislodged hydrazine will react to give a measurable chemiluminescent response. The signal is spurious, returning to the baseline again after all of the hydrazine coating the tubing has been eliminated.

Reports in the literature have demonstrated that the chemiluminescence of luminol systems can be enhanced by the introduction of a high concentration of halide ions.[22] We corroborated this effect in the detection of hydrazine via the addition of 0.56 M KBr to a 5×10^{-5}-M luminol and a 1×10^{-6}-M platinum in pH 12.6 solution. These experiments established that the sensitivity to hydrazine can be more than doubled via the introduction of a large concentration of the bromide salt. It is thought that the halide facilitates electron transfer from luminol to the active peroxide.[22]

Fluorescence Derivatization. Shown in Fig. 41.5 is a derivatization scheme for the reaction of hydrazine, N_2H_4, with 2,3-naphthalene dicarboxaldehyde (NDA). The initial reaction between Hz and NDA involves the nucleophilic attack of one of the lone pair of electrons on the Hz toward an aldehyde group on the NDA. The alcoholic intermediate (II) formed is believed to be unstable, quickly losing a molecule of water to form the hydrazone shown as molecule III. The proximity of the remaining lone pair of electrons on the hydrazone to the remaining aldehyde group enables a similar set of reaction steps to complete the cyclicization of the NDA reagent to form 2,3-diazaanthracene V.

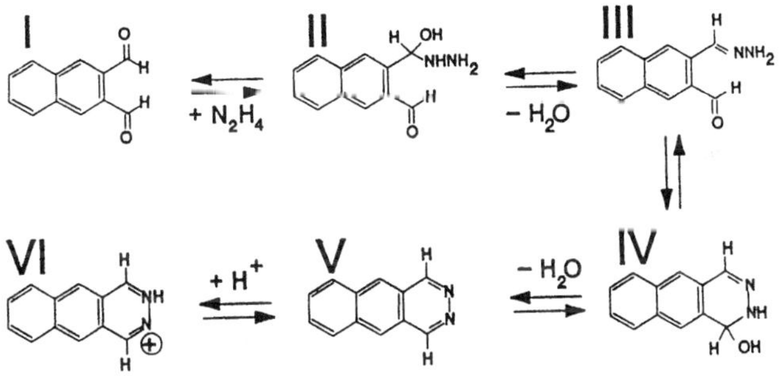

FIGURE 41.5 Proposed derivatization scheme for the reaction between hydrazine and 2,3-naphthalene dicarboxaldehyde.

The incorporation of hydrazine into an aromatic framework such as benzene, naphthalene, or anthracene as an additional, fused, six-membered heterocycle, results in the formation of a highly efficient fluorophore whose λ_{em} is red-shifted away from the fluorescent spectrum of the original compound. Figure 41.6 contains a set of emission spectra collected for NDA (λ_{ex} = 403 nm) following the sequential addition of 0, 4, 25, and 105 µg/l of hydrazine to the solution. Prior to the addition of any hydrazine, the fluorescence emission at this excitation wavelength is minimal and nearly within the background. The resultant derivative formed, on the other hand, is characterized by a broad fluorescence emission centered at 500 nm, which increases in proportion to the concentration of hydrazine in solution. The benefit of using fluorescence detection over absorbance techniques is that there is a distinct advantage to measuring a small, fluorescent signal set upon a dark background, as opposed to measuring an equally small difference in signal set upon a large signal background (i.e., absorbance).

The reaction rate measured for the formation of the fluorescent NDA/hydrazine derivative reaches a maximum at a pH of 2.5, as well as at concentrations of H^+ above 1.9 M. It should be pointed out, though, that as the concentration of H^+ goes above 0.7 M, the overall sensitivity toward hydrazine decreases due to a progressive decrease in the fluorescence intensity seen for a given concentration of hydrazine.

Because of the success in determining trace levels of hydrazine in solution using the NDA reagent, further efforts were made to extend this technology to the detection of hydrazine in air. We have been working in cooperation with NASA/Kennedy

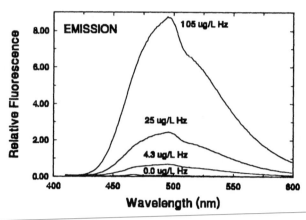

FIGURE 41.6 Emission spectra obtained for NDA following the addition of 0, 4.3, 25, and 105 µg/l of hydrazine.

Space Center on the development of a highly sensitive detector for hydrazine intended for use in a new satellite payload spin test facility (PSTF). The specifications for this sensor mandate that it be continuously operable for 90 days (maintenance free), with a detection limit of 10 ppb for hydrazine and a response time less than fifteen minutes. Shown in Fig. 41.7 is a schematic for the continuous flow, fluorimeter system we propose to use for trace-level hydrazine monitoring. This system is comprised of several different components. A peristaltic pump with multichannel pumping capability is used for the transport of the liquid reagent (1.0 ml/min) at various points throughout the system. A concentrated, stock solution of the NDA reagent (4 l at 5.4×10^{-3} M) is prepared in a 50/50 water/ethanol solution and is slowly diluted into acidic water (pH = 2.5) in order to provide for a 1×10^{-4}-M active solution over the 90-day inspection period. The peristaltic pump delivers the diluted NDA reagent through a short length of teflon tubing that is concurrently transporting vapor from the outside room due to the placement of a vacuum pump upstream.

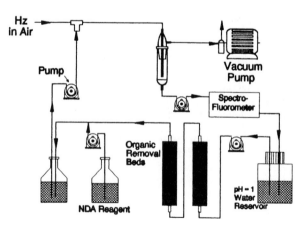

FIGURE 41.7 Schematic of the hydrazine-vapor detection system.

It is this gas/liquid interface that enables the derivatization reaction between NDA and hydrazine to take place. The liquid is then collected in a modified glass bubbler, whereupon the NDA/hydrazine derivative is pumped through the spectrofluorometer and detected. In order to prevent the production of large quantities of waste, the liquid reagent is recycled through a set of organic removal columns (charcoal) and reused once again. Initial characterization of these columns has shown that the fluorescent NDA/hydrazine derivative (VI) is quantitatively removed from the liquid stream, while the starting reagent, NDA (1), is reduced in concentration by a factor of 50 (from 1×10^{-4} M to 2.3×10^{-6} M). Because the starting reagent is nonfluorescent at the wavelengths being probed, this should not present a problem over the course of the 90-day measuring period.

The quantitative response of this system is demonstrated in Fig. 41.8, which is the response obtained following the introduction of 15 and 150 ppb hydrazine in air. In this experiment, the active reagent was a 1×10^{-4}-M aqueous solution of NDA adjusted to a pH of 2.5 in nitric acid, with a liquid flow rate of 1.0 ml/min. For each different concentration of hydrazine, a cycle of two minutes on and two minutes off was followed. The response tested from 10 to 1000 ppb is very linear in nature ($r > 0.999$), bearing a detection limit for hydrazine in air ($S/N = 3$) of 2 ppb. The time response for this sensor is less than two minutes.

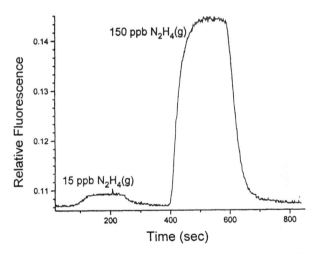

FIGURE 41.8 Quantitative response curve to the introduction of 15 and 150 ppb hydrazine in air.

We examined several likely interferents, including NH_3, Freon 113, isopropyl alcohol, ethanol, and methyl ethyl ketone. Each of these different interferents was evaluated at a concentration of twice the threshold limit value (TLV). In all cases, there was no measurable or detectable fluorescent response at 500 nm. It is particularly noteworthy that NH_3 is not an interferent. Apparently, at the pH being used (pH 2–3) and without the presence of added nucleophiles such as the cyanide ion or thiol, ammonia does not form a fluorescent derivative with NDA. Because of the nature of the reaction between NDA and hydrazine and the fact that we can monitor a particular fluorescent wavelength, this reaction appears to be highly selective towards hydrazine.

CONCLUSION

The fluorescent tagging of hydrazine with NDA has been shown to be a highly sensitive and selective method for the detection of these structurally similar hydrazines. The superior sensitivity of the NDA reagent toward hydrazine (detection limit = 50 ng/l) has enabled the use of simpler, less expensive instrumentation for our monitoring applications. We have demonstrated the operation of a fluorescence monitoring system for hydrazine that is capable of determining the vapor levels of hydrazine within a given working environment below the design goal of 10 ppb hydrazine in air.

Chemiluminescence, utilizing the colloidal platinum/luminol system, is an alternative method for determining trace levels of hydrazine vapor in air in real-time. The real-time detection limit is 1 ppb hydrazine in air; further improvements in the sensitivity can be accomplished by adding KBr salts, increasing the rate of air flow, or increasing the number of coils in front of the photomultiplier surface. A linear dynamic range was obtained from 10-2000 ppb hydrazine, with response times for the detection and return to baseline of less than two minutes. The response time of the system can be improved substantially by increasing the liquid flow rate, although the platinum concentration must also be adjusted to ensure a comparable sensitivity. Selectivity is the biggest potential disadvantage of this system. Our tests indicate that of the gases examined, only NO_2 and NH_3 gave a positive response, the ammonia response being spurious and the nitrogen dioxide sensitivity being low enough that, at the TLV level or below, it is undetectable.

This research was supported by NASA/Kennedy Space Center (DL-ESS-24, CC-82360A).

REFERENCES

1. American Conference of Governmental Industrial Hygienists (ACGIH) report on proposed changes to TLV and BEI lists, 21, May 1989.
2. E. C. Olsen, *Anal. Chem.,* **32,** 1960, p. 1545.
3. W. D. Basson and J. F. Van Staden, *Analyst,* **103,** 1978, p. 998.
4. M. Roth, *Anal. Chem.,* **43**(7), 1971, p. 880.
5. S. M. Lunte and O. S. Wong, *LC-GC,* **7**(11), 1989, p. 908.
6. P. J. M. Kwakman, H. Koelewijn, I. Kool, U. A. Th. Brinkman, and G. J. De Jong, *J. Chromatography,* **511,** 1990, p. 155.
7. G. E. Collins and S. L. Rose-Pehrsson, *Anal. Chem. Acta,* **284,** 1993, p. 207.
8. G. E. Collins and S. L. Rose-Pehrsson, *Analyst,* October, 1997.
9. K. Robards and P. J. Worsfold, *Anal. Chem. Acta,* **266,** 1992, p. 147.
10. R. Escobar, Q. Lin, A, Guiraum, and F. F. de la Rosa, *Analyst,* **118,** 1993, p. 643.
11. J. L. Burguera, A. Townshend, and S. Greenfield, *Anal. Chem. Acta,* **114,** 1980, p. 219.
12. W. R. Seitz and D. M. Hercules, *Anal. Chem.,* **44,** 1972, p. 2143.
13. F. Shaw, *Analyst,* **105,** 1980, p. 11.
14. P. Mikuska and Z. Vecera, *Anal. Chem.,* **64,** 1992, p. 2187.
15. D. Gonzalez-Robledo, M. Silva, and D. Perez Bendito, *Anal. Chem. Acta,* **228,** 1990, p. 123.
16. A. T. Faizullah and A. Townshend, *Anal. Proc.,* **22,** 1985, p. 15.
17. A. T. Pilipenko and A. V. Terletskaya, *J. Anal. Chem. USSR,* **28,** 1973, p. 1004.

18. S. L. Rose and J. R. Holtzclaw, NRL Report 8848, NTIS ADBO91299.
19. E. W. Schmidt, *Hydrazine and its Derivatives: Preparations, Properties, Applications,* Wiley, New York, 1984.
20. T. P. Vorob'eva, Yu N. Koziov, Yu V. Koltypin, A. P. Purmal´, and B. A. Rusin, *Izv. Akad. Nauk SSSR, Ser. Khim.,* **9,** 1978, p. 1996.
21. T. G. Burdo and W. R. Seitz, *Anal. Chem.,* **47,** 1975, p. 1639.
22. D. E. Bause and H. H. Patterson, *Anal. Chem.,* **51,** 1979, p. 2288.

CHAPTER 42
SENSITIVE AND SELECTIVE TOXIC GAS DETECTION ACHIEVED WITH A METAL-DOPED PHTHALOCYANINE SEMICONDUCTOR AND THE INTERDIGITATED GATE ELECTRODE FIELD-EFFECT TRANSISTOR (IGEFET)

INTRODUCTION

A novel gas-sensitive microsensor, whose design is based upon the interdigitated gate electrode field-effect transistor (IGEFET), was realized by integrating it with a selectively deposited, chemically active, electron-beam-evaporated copper phthalocyanine (CuPc) thin film. When isothermally operated at 150°C, the microsensor can selectively and reversibly detect parts-per-billion (ppb) concentration levels of two environmentally sensitive pollutants—nitrogen dioxide (NO_2) and diisopropyl methylphosphonate (DIMP). Although the CuPc thin film chemically and electrically interacts with NO_2 and DIMP, just as it will likely interact with other electrically active gases or combinations thereof, the selectivity feature of the microsensor was established by operating it with 5 V peak amplitude, 2 μs duration, and 1000 Hz repetition frequency pulse, and then analyzing its time and frequency domain responses. As a direct consequence of this analysis, the envelopes associated with the normalized difference Fourier transform magnitude frequency spectra reveal features that unambiguously distinguish the NO_2 and DIMP challenge gas responses. Furthermore, the area beneath each response envelope may correspondingly be interpreted as a metric for the microsensor's sensitivity to a specific challenge gas concentration. Scanning electron microscopy (SEM) was used to characterize the CuPc thin film's morphology. Additionally, infrared (IR) spectroscopy was employed to verify the alpha and beta phases of the sublimed CuPc thin films and to study the NO_2 and DIMP-CuPc interactions.

In the course of daily activity, the living and working environment exposes mankind to a variety of toxic gases and vapors. Two critical groups of environmentally sensitive contaminants include the oxides of nitrogen and the organophosphorus pesticides and their structurally affiliated compounds. Today, it is recognized that even trace amounts of these pollutants can pose an adverse effect on many ecological systems. This acknowledgment has motivated the development of sensitive and selective personal monitoring sensors that are capable of detecting subtoxic concentration levels of these contaminants.

The oxides of nitrogen, particularly nitrogen dioxide (NO_2), are emitted into the atmosphere from a variety of industrial stacks and the exhaust of automobiles. Because of its detrimental environmental impact, nitrogen dioxide was identified as one of the model compounds used to evaluate the performance of a novel gas microsensor, whose design is based upon the interdigitated gate electrode field-effect transistor (IGEFET).

On the other hand, the organophosphorus pesticides and their structurally affiliated family of compounds are intentionally synthesized and valued for their deleterious biochemical effect and persistence. Additionally, the distribution of these compounds is essential for their efficacy. Fortunately, a significant portion of the environmentally sensitive organophosphorus contaminants contain either the phosphoryl or thiophosphoryl group, and the thiophosphoryl pesticides readily oxidize to produce phosphoryl compounds. Consequently, diisopropyl methylphosphonate (DIMP), a phosphoryl-containing compound that has also been utilized in prior organophosphorus compound sensor research,[1-7] was selected as the second model compound evaluated in this investigation.

Recently, coated bulk-wave piezoelectric quartz crystal microbalances (QCMS)[1-5,8,9] and surface acoustic wave (SAW) devices[10-14] have been investigated as candidate sensor technologies for detecting NO_2 and the environmentally sensitive organophosphorus compounds. However, performance limitations associated with both technologies generally include the lack of selectivity, sensitivity to moisture, and response reproducibility. These results are not totally unexpected, since the physiochemical sorption of any environmental gas or vapor will increase the mass of the sensor's coating and generate a response. Hence, selectivity remains an elusive performance feature of these mass-sensitive detector technologies.

In order to achieve selectivity with a sensor that is exposed to an ensemble of gases, a more robust, selective, measurable, and quantifiable interaction between a challenge gas of interest and the sensor must occur, and this event must be decipherable from the sensor's response. This chapter describes the IGEFET gas microsensor concept and the progress achieved toward satisfying this goal.

In contrast to the generally less specific physiochemical sorption process, the IGEFET gas microsensor research is focused on investigating the electronic properties (specifically, conductivity and permittivity) of an organic semiconductor that are modified when the semiconductor is exposed to a dehumidified ambient air atmosphere (typically less than 2 percent RH) containing small concentrations of NO_2 or DIMP. To accomplish this objective, electron-beam-evaporated thin films of copper phthalocyanine (CuPc) have been sublimed onto the interdigitated gate electrode (IGE) structure of the IGEFET to realize a gas microsensor that can selectively detect parts-per-billion (ppb) concentration levels of NO_2 and DIMP.

The challenge gas chemical-to-electrical transduced response is generated when the IGEFET microsensor is configured to detect variations of the in situ metal-oxide-semiconductor field-effect transistor's drain-to-source current as it responds to changes induced in the electronic structure of the gas-sensitive thin film deposited on the interdigitated gate electrode (IGE). As illustrated in Fig. 42.1, the electronic changes induced in the gas-sensitive CuPc thin film may occur as the result of an electronic dipole (polarization) or charge-transfer interaction with either an adsorbed electron acceptor or donor gas. The electronic changes are correspondingly manifested in the gas-sensitive thin film's dielectric relaxation function, which, fortuitously, can be measured as a change in the film's electrical impedance.

The concept of utilizing a planar interdigitated electrode chemiresistor to monitor electrical impedance changes caused by a bulk chemical reaction (most notably the cure of epoxies) has been reported in the literature.[15–43] However, the overwhelming majority of these investigations have focused on measuring the irreversible direct current electrical conductivity changes manifested in chemically active films. A significantly smaller number of studies have examined the film's alternating current impedance behavior, but generally only with respect to a small

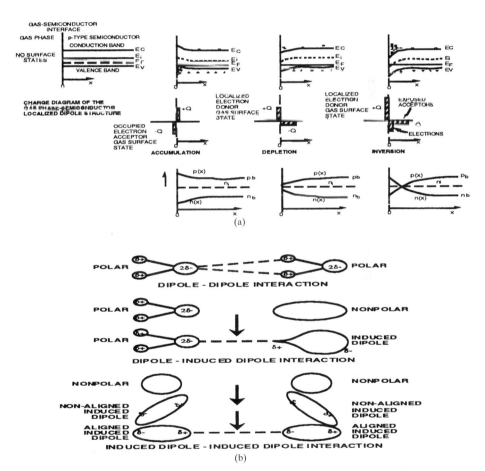

FIGURE 42.1 Depiction of the feasible electronic dipole interactions that may occur between an adsorbed gas molecule and the surface of a chemically active thin film: (*a*) energy band diagram illustrating the electronic changes induced by the adsorption of an electron acceptor or donor gas on the surface of a *p*-type semiconductor such as copper phthalocyanine (CuPc); (*b*) physical model of the interaction.

subset of discrete frequencies. Hence, since the interaction between a gaseous adsorbent and the thin film has been shown to produce selectively a distinct change in the film's electrical resistance and reactance at specific frequencies for different challenge gases,[44] the IGEFET microsensor concept was motivated, and its response is processed to produce a detailed signature or fingerprint of the corresponding chemical interaction in both the time and frequency domains.

The physical phenomenon that motivates this unique perspective is the chemically active thin film's dielectric relaxation response. For the IGEFET device, calculating the chemically induced normalized difference Fourier transform magnitude spectrum has been experimentally demonstrated to be unique for the two model gases used in this investigation.[44] The unique ensemble of the closely spaced frequency domain data comprises the microsensor's response signature and notably contributes to its selectivity.

By contrast, the QCM, SAW, and planar interdigitated electrode chemiresistors have, most frequently, been operated in an attempt to discern sensor selectivity by simply comparing changes that are associated with a single measurand (for example, a resonant frequency shift, a change in the direct current electrical resistance, or an impedance change at a specific frequency); certainly, a near impossible endeavor. That is, the sensor's performance often reveals that, for a given concentration of a specific challenge gas, there likely exists a different gas (or combination of gases) at some concentration that evokes an identical sensor response; hence, the apparent lack of selectivity. To further support this observation, a specific example illustrating this dilemma, which also happens to occur when the IGE's direct current electrical resistance change, is reported in an attempt to discern the microsensor's selectivity for exposures to two different concentrations of NO_2 and DIMP. It is discussed in the section on sensor performance. To resolve the selectivity dilemma, an analysis of the IGEFET microsensor's time and frequency domain responses for a voltage-pulse mode of excitation is demonstrated as a favorable solution.

SENSOR CONCEPT

As depicted in Fig. 42.2(*a*) the IGEFET microsensor consists of a planar interdigitated gate electrode (IGE) structure, which forms the gate contact of a conventional metal-oxide-semiconductor field-effect transistor (MOSFET). The IGE structure is composed of a driven electrode component that envelopes the entire microsensor, and thus functions as a guard ring to minimize stray surface leakage currents. The corresponding floating electrode component is used to establish the IGE's high-input electrical impedance contact via the MOSFET's gate oxide. Prior to the deposition of the chemically active thin film, the electrical isolation between the driven- and floating-electrodes is typically on the order of 100 Ω; this high degree of electrical isolation is accomplished by supporting the electrodes on a 1-μm-thick silicon dioxide layer whose resistivity is on the order of 10^{14} $\Omega\cdot$cm.

As illustrated in Figs. 42.2(*b*) and (*c*), the IGEFET's gas sensitivity is realized by depositing a chemically active thin film on the surface of, and between, the IGE components. When the chemically active thin film is exposed to a challenge gas of interest, its electrical impedance (via C_e and R_e) is perturbed. These impedance perturbations can readily be measured and quantified via the microsensor's characteristic response time constant ($t = R_e \times C_e$) when the IGEFET's driven-electrode is excited with a voltage pulse. Because the microsensor's floating electrode component is electrically connected to the MOSFET's gate oxide, charge may be trans-

SENSITIVE AND SELECTIVE TOXIC GAS DETECTION

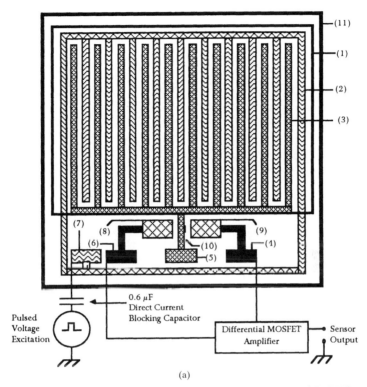

(a)

FIGURE 42.2 (*a*) Integrated electronic gate field effect transistor (IGEFET) physical structure and electrical connections. **1:** Copper phthalocyanine (CuPc) thin film. **2:** Driven electrode conductor. **3:** Floating electrode conductor. **4:** MOSFET drain contact. **5:** MOSFET gated floating electrode contact. **6:** MOSFET source contact. **7:** Driven electrode contact. **8:** MOSFET source region. **9:** MOSFET drain region. **10:** MOSFET drain to source channel. **11:** Host silicon substrates.

ferred through or polarized in the chemically active thin film, and this behavior is manifested as a temporally dependent potential applied to the high-input impedance gate contact of the MOSFET.[15] By incorporating the microsensor in an *in situ* MOSFET differential amplifier configuration [Fig. 42.2(*d*)], the small, temporally dependent gas-exposure-induced potential variation that appears at the IGEFET's gate oxide contact can be amplified and ported for subsequent processing. Thus, a critical feature of the IGEFET microsensor is the electrical isolation afforded by the dielectric-supported interdigitated electrode structure and the gate oxide combination. This feature facilitates the IGEFET functioning as an *in situ* observation window that is capable of monitoring the electrical behavior of the chemically active thin film while correspondingly minimizing its influence on the changes induced by the challenge gas. Finally, because the microsensor's response is amplified *in situ* [Figs. 42.2(*a*) and (*d*)] rather than remotely, the performance limitations associated with electrical noise and the parasitic impedance effects of long lead lengths are minimized.

The IGEFET microsensor's selectivity feature is realized by computing a normalized difference Fourier transform magnitude spectrum. The rationale for this

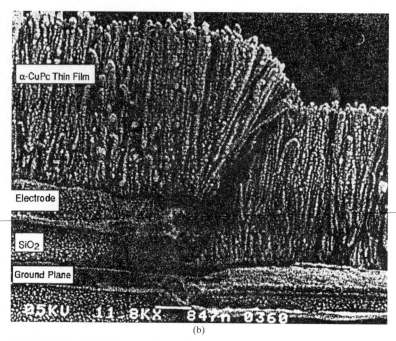

FIGURE 42.2 (*b*) SEM photomicrograph of the cross sectional view of a sublimed α-phase CuPc thin film deposited on an IGE structure.

computation was motivated, as illustrated in Fig. 42.3, by the highly detailed frequency domain spectral characteristics of a time domain voltage pulse, and after it was experimentally observed that the long-term pulse excitation characteristics of an unexposed or purged CuPc-coated IGEFET microsensor were reproducible and pseudo-time-invariant.[44] The normalized difference Fourier transform magnitude spectra are computed by Fourier-transforming the microsensor's time domain voltage response due to a challenge gas exposure and then normalizing it with a factor that is also used to normalize the corresponding Fourier transform of the microsensor's unexposed or purged time domain voltage response. Finally, the overall microsensor's response is determined by subtracting the two normalized spectra on a point-by-point basis (at common frequencies), and the magnitude of each difference is used to generate an envelope whose signature reveals the detector's sensitivity and selectivity.

In practice, the area beneath a normalized difference Fourier transform magnitude spectrum can be interpreted as a metric that corresponds to the sensitivity of the microsensor to a specific challenge gas concentration. Additionally, the large number of components that define the detailed shape of the spectral response envelope yields valuable guidance for unambiguously specifying the particular challenge gas that has triggered the microsensor's response. That is, although different concentrations of NO_2 and DIMP may manifest an identical change in the direct current resistance of a CuPc-coated IGE structure, the shape of the response envelope defined by the normalized difference Fourier transform magnitude spectrum clearly manifests the unique dielectric relaxation behavior caused by the two challenge

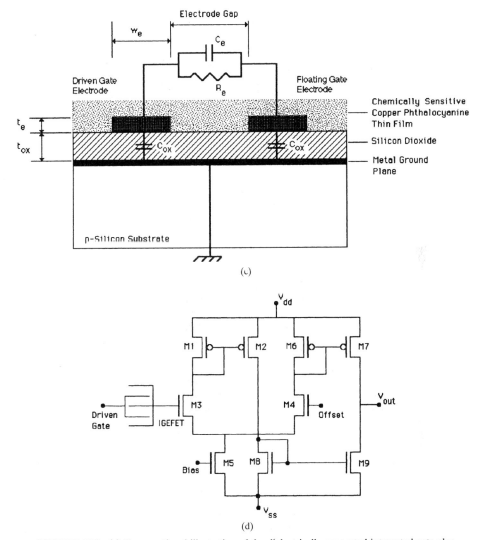

FIGURE 42.2 (c) Cross sectional illustration of the dielectrically supported integrated gate electrode (IGE) structure showing the physical and electrical relationship of nearest-neighbor-driven and floating gate electrodes (C_e is the interelectrode capacitance, R_e is the interelectrode resistance, C_{ox} is the electrode to ground plane coupling capacitance attributable to the silicon dioxide dielectric support, t_e is the aluminum electrode's thickness, w_e is the electrode's width, and t_{ox} is the thickness of the silicon dioxide dielectric supports). (d) MOSFET differential amplifier designed with independent bias and temperature compensation offset features.

gases. Furthermore, the derivative of the normalized difference Fourier transform magnitude spectrum with respect to frequency also reveals a unique and useful signature for each challenge gas.

The selection of the IGEFET microsensor's chemically active thin film is critical for establishing the class of compounds to which the detector will manifest a favor-

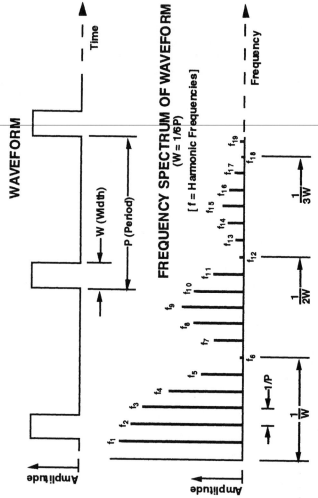

FIGURE 42.3 Frequency-domain spectral features that are influenced by the characteristics of a corresponding time-domain voltage-pulse waveform.

able degree of sensitivity and selectivity. In this IGEFET microsensor investigation, the well-known CuPc organic p-type semiconductor was utilized because it is a member of the metal-substituted phthalocyanine family (which also includes Ag, Al, Au, Be, Co, Cr, Fe, Ga, Ge, Hf, Ir, Li, Mg, Mn, Na, Nd, Ni, Os, Pb, Pt, Rh, Ru, Sc, Sn, Ta, Ti, V, Y, Zn, and Zr), whose electrical conductivity has been experimentally reported to increase upon exposure to electron acceptor challenge gases (for example, BCl_3, BF_3, Cl_2, F_2, and NO_2).[34,35,38,39,41–43,45–47] Additionally, CuPc is an exceptionally stable compound, as evidenced by its low proton affinity and resistance to dissolution by concentrated mineral acids, not to mention the fact that it can be sublimed at temperatures as high as 580°C without decomposing. As depicted by the two concentric circles superimposed upon the metal-substituted phthalocyanine chemical structure illustrated in Fig. 42.4, the predominant electron donor regions formed by the planar π-electron bond system that symmetrically surrounds the central copper atom has been postulated to be responsible for the experimentally observed gas-induced electrical conductivity changes.[41,43,45,48] Since NO_2 is rapidly (on the order of a few seconds) and reversibly adsorbed on heated CuPc thin films (typically at temperatures spanning 100–200°C), the interaction site has been postulated to occur at the film's intercrystallite interfaces. It is not a true bulk diffusion mechanism.[41,42] As depicted in Fig. 42.1(b), the strength of the chemical bond between the CuPc thin film and the electrically active adsorbed gas is likely to be a strong dipole interaction or weak coordination bond; that is, a bond whose energy is stronger than a purely physical adsorptive interaction (less than 40 kJ/mole), but weaker than a true covalent bond (on the order of 300 kJ/mole). Thus, it is anticipated that the selectivity of the interaction between a CuPc thin film and a particular electrically active adsorbed challenge gas will be strongly influenced by the electronic and steric features of the interaction. In this IGEFET microsensor investigation, both NO_2 and DIMP were reversibly adsorbed, and NO_2, being the more robust electron-acceptor challenge gas, induced a correspondingly stronger electrical interaction relative to identical exposure concentrations of DIMP. This observation was further supported by the infrared (IR) spectroscopy (Perkin Elmer, model 683, Norwalk, Connecticut) studies of NO_2 and DIMP exposed CuPc thin films. For example, Figs. 42.5(a) and (b) illustrate the reversible interaction that occurs between NO_2 and an alpha-phase CuPc thin film. A much less significant NO_2 interaction is observed with the corresponding beta-phase CuPc thin film and with comparable exposures of the DIMP challenge gas.

SENSOR FABRICATION

The IGEFET microsensor design was implemented utilizing the scalable 2-μm p-well, double-metal, double-polysilicon, complementary metal oxide semiconductor (CMOS) integrated circuit technology. The p-well feature was incorporated to enhance electrical isolation between the IGE structure and the MOSFET, and n-channel enhancement-mode devices were utilized. Aluminum was used as the second-level metal for the IGE array and the microsensor's bonding pads. Except for the IGE structure and the peripheral IC bond pads, the surface of the IGEFET was passivated with a 3-μm-thick silicon dioxide layer to minimize undesirable environmental interactions. The overall dimensions of the IGEFET microsensor integrated circuit are 4466 × 6755 μm. The IGE structure (Fig. 42.1) consists of 34 floating-electrode fingers and 35 driven-electrode fingers. Each finger is 10 μm wide, and the interelectrode finger spacing is 10 μm. The chemically active area of the IGE measures 1370 × 1370 μm.

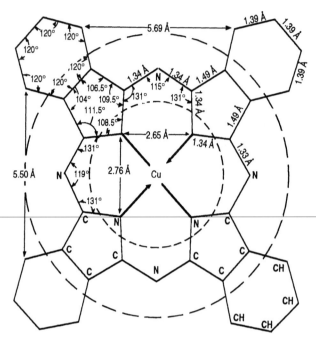

FIGURE 42.4 Chemical structure of the metal-substituted phthalocyanine compound. The two concentric circles depict the relative locations of the strong donor formed by the planar π-electron surplus. Common candidates for the control metal (M) atom include: Ag, Ab, Au, Be, Co, Cr, Cu, Fe, Ga, Hf, Ir, Li, Mg, Mn, Na, Ni, Os, Pb, Pt, Rh, Ru, Sc, Ta, Ti, V, Y, Zn, and Zr.

In order to couple the chemically active thin film with the IGEFET, a supply of high-purity CuPc bulk material (Fluke Chemical Corp., Ronkonkoma, New York) was first prepared by the entrainer sublimation process (Fig. 42.6), and then it was deposited at a rate of 10 Å/s on the dielectric-supported IGE structure using a high-vacuum (10^{-6} Torr), cryogenically pumped (helium), electron-beam thermal evaporation process (Denton Vacuum, Inc., model DV-602, Cherry Hill, New Jersey). An etched metal mask was used to confine the CuPc thin film deposition within the boundaries of the interdigitated electrode structure. As measured with an *in situ* QCM and later verified with an independent ellipsometric measurement,[44] nominal CuPc film thicknesses on the order of 1000 Å could be reproducibly deposited (±5 percent thickness variation). To facilitate handling the IGEFET microsensor, it was mounted in a standard 64-pin dual-in-line ceramic IC package (Kyocera Corp., part number KD-83578, Edina, Minnesota).

SENSOR OPERATION

The electronic instrumentation configuration for operating the CuPc-coated IGE-FET microsensor is illustrated in Fig. 42.7. A 5-V peak-amplitude, 2-μs duration,

SENSITIVE AND SELECTIVE TOXIC GAS DETECTION

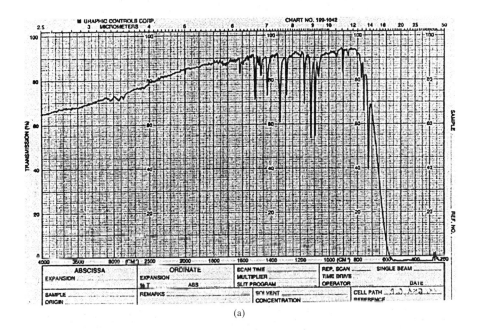

(a)

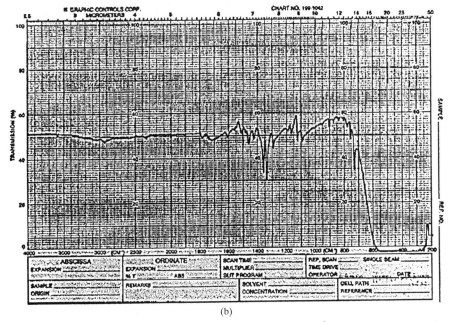

(b)

FIGURE 42.5 (*a*) Infrared (IR) spectra illustrating the α-phase of 6.5-kÅ-thick CuPc film sublimed on a sodium chloride substrate before exposure to NO_2. (*b*) The effect of a 15-min., 200-ppb N_2O exposure.

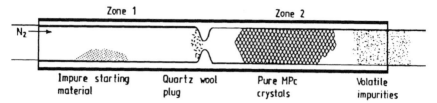

FIGURE 42.6 Entrainer sublimation furnace configuration utilized to purify the copper phthalocyanine CuPc bulk material that was subsequently deposited on the IGE structure of the IGEFET gas microsensor.

1000 Hz repetition-frequency pulse excitation signal (Wave-Tek Corp., model 148, San Diego, California) was applied to the IGEFET's driven electrode. The time-domain pulse excitation characteristics were experimentally determined to produce a corresponding Fourier transform magnitude spectrum possessing frequency components with essentially equal mV-level magnitudes (to within 2 percent) spanning the frequency range corresponding to the 3-dB cutoff frequency (15 kHz) of the

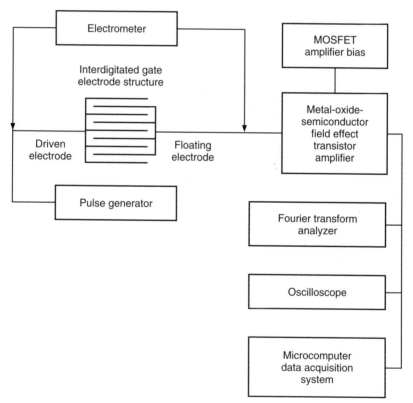

FIGURE 42.7 Instrumentation arrangement for measuring the electrical performance of the IGEFET microsensor.

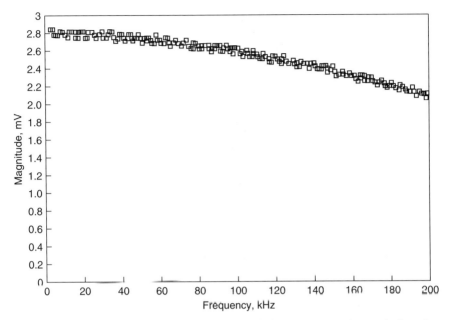

FIGURE 42.8 Fourier transform magnitude plot (partial) of the IGEFET's voltage excitation pulse (5 V) peak amplitude, 2 microseconds duration, 100 Hz repetition frequency pulse.

microsensor's *in situ* MOSFET amplifier (Fig. 42.8). This operating condition implies that the IGEFET microsensor is being equivalently excited with an array of nearly constant amplitude harmonic excitation signals whose frequencies are established by the Fourier transform mathematical operator. The microsensor's response correspondingly reveals how the pulse excitation energy is stored, dissipated, and propagated as a consequence of the gas-induced electrical impedance changes that are collectively manifested via the level of attenuation that is associated with each harmonic component.

In practice, the microsensor's excitation and response time-domain signals were captured (Fig. 42.7) on a dual-trace oscilloscope (Tektronix Corp., model 475, Beaverton, Oregon). To complement the time domain measurements, a dual-channel Fourier transform analyzer (Bruel and Kjaer Instruments, Inc., model 2032, Marlborough, Massachusetts) was used to generate the corresponding microsensor excitation and response frequency domain spectra (0–25 kHz bandwidth). For completeness, the direct current resistance of the CuPc-coated IGE structure was independently monitored with an electrometer (Keithley Instruments, Inc., model 617, Cleveland, Ohio) to determine the microsensor's equilibrium response during the cyclical challenge gas exposure and purge processes. Data collection was automated using a microcomputer (Zenith Data Systems Corp., model Z-248, St. Joseph, Minnesota) equipped with an IEEE-488 interface plug-in card (Capital Equipment Corp., model 0100060300, Burlington, Massachusetts).

As shown in Fig. 42.9, the dynamic gas generation and delivery system incorporating calibrated permeation tubes (GC Industries, models 23-7502 and 23-7392, Freemont, California) was configured to create a broad-challenge gas concentration range spanning 20–400 parts per billion (ppb) for NO_2 and 40–4000 ppb for DIMP.

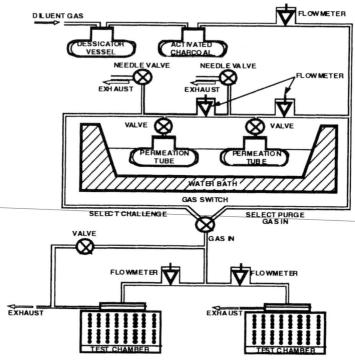

FIGURE 42.9 Challenge and purge gas generation and delivery system. The performance of two microsensors can be simultaneously analyzed relative to a single gas component or a binary mixture.

A complete description of the thermostated microsensor test cell design is reported in a related publication.[6]

SENSOR PERFORMANCE

The electrical transfer function characteristics of the *in situ* MOSFET amplifier portion of the IGEFET microsensor [Fig. 42.1(*d*)] were established with a semiconductor parameter analyzer (Hewlett-Packard Corp., model 4145, Palo Alto, California). The low-frequency gain was typically 12 ± 1 dB, and the 3-dB cutoff frequency was approximately 15 kHz. To illustrate the selectivity feature attributed to the IGEFET microsensor when it is excited with a voltage pulse and its response is examined in the time and frequency domains, the corresponding direct current conductivity measurements establish a useful reference. To implement the direct current conductivity measurements, the microsensor was thermostated at 150 ± 0.5°C, and the film's exposure and purge dynamics were measured. Figure 42.10 illustrates the CuPc thin film's conventional direct current electrical resistance versus time plots for cyclical challenge gas exposures and purges, and Fig. 42.11 depicts the film's equilibrated resistance values after exposure to an ensemble of NO_2 and DIMP concentration levels. From the gradients of the least-squares fitted exposure polynomials, an 800

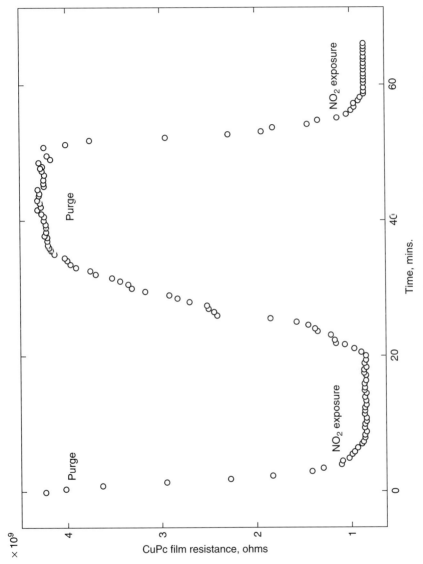

FIGURE 42.10 Isothermal (ISO ± 0.5°C) direct current resistance of the interdigitated gate electrode (IGE) structure coated with a 1000° thick film of CuPc upon exposure to alternating purges and challenges of 40 ppb of NO_2 mixed with filtered laboratory air.

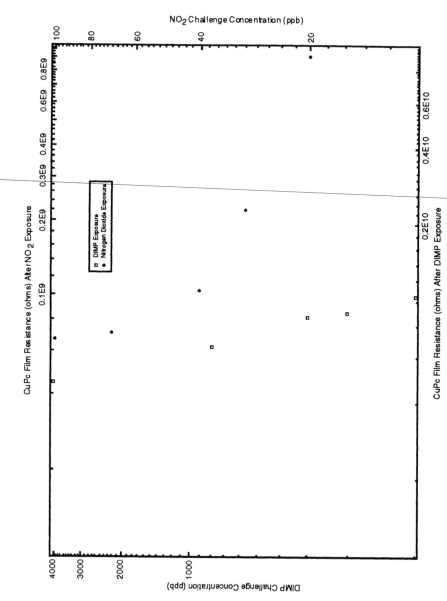

FIGURE 42.11 Isothermal (ISO ± 0.5°C) equilibrated electrical resistance of a 1000-Å-thick CuPc coated IGE structure as a function of the challenge gas concentration.

ppb DIMP challenge concentration produces the equivalent resistance change predicted by the 30 ppb NO_2 challenge; hence, this simple example illustrates the selectivity dilemma that typically plagues the single measurand class of sensors.

On the other hand, the corresponding isothermal (150 ± 0.5°C) IGEFET microsensor time domain voltage-pulse excitation and response waveforms measured with respect to cyclical exposures and purges with NO_2 and DIMP challenges are illustrated in Fig. 42.12, and this information is further supplemented in a related publication.[44] The preexposure and purged responses attest to the IGEFET microsensor's reversibility [Fig. 42.12(b)]. More important, however, is the observation that the shape of the time domain responses corresponding to the 100 ppb NO_2 and DIMP challenges are distinctly different [Figs. 42.12(c) and (d)]. That is, by considering the exposed CuPc film to be a lossy dielectric (the loss component tends to increase upon exposure to increasing concentrations of an electron acceptor gas), the corresponding influence on the film's equivalent electrical RC time constant ($r = ReCe$), as reflected in the microsensor's time domain response, can be appreciated. In particular, a specific concentration of a challenge gas can be interpreted to be the cause for modifying the film's electrical impedance (electrical conductivity and permittivity). Correspondingly, this change results in the manifestation of a specific ensemble of RC time constants where members simultaneously interact to produce the characteristic exponential rise and decay features in the IGEFET's time domain response.

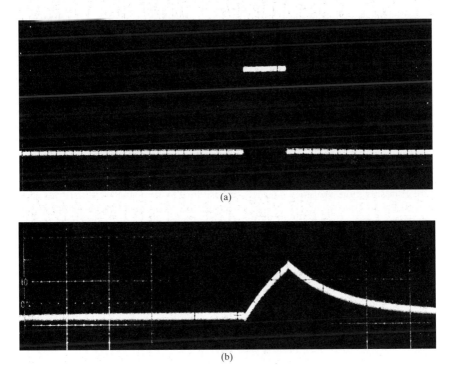

FIGURE 42.12 (a) Time domain response of the IGFET microprocessor when excited with the pulsed-voltage waveform due to 100 ppb NO_2 and DIMP challenge gas concentration (CuPc 1000 Å) thick film, ISO ± 0.5°C operating temperature and 2 percent r.h. filtered ambient air carrier. (b) Filtered ambient air response that is reproducible after purging from either the NO_2 or DIMP challenges.

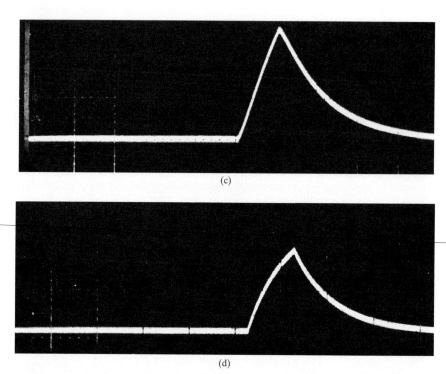

FIGURE 42.12 (c) NO$_2$ exposure. (d) The horizontal scale in each frame is 2 μsec. per major division, and the major division vertical scales are 2.5 V for a, 0.5 V for b, and 0.5 V for c.

Rather than implementing the complex analysis required to deconvolve a set of RC time constants from the microsensor's time domain response,[49] the alternative of directly generating the corresponding equivalent frequency domain response (via the Fourier transform mathematical operator) was implemented. In order to be able to compare the frequency domain responses with respect to a specific gas and exposure concentration, the microsensor's reversibility feature [Fig. 42.12(b)] was capitalized upon. That is, the Fourier transform of the microsensor's purged response was correspondingly measured and normalized with the same factor used to normalize the microsensor's frequency domain challenge-gas-induced response. The corresponding normalized difference Fourier transform magnitude spectra associated with a set of NO$_2$ and DIMP time domain responses are shown in Fig. 42.13.

The distinct merit of the voltage-pulse excitation and Fourier transform signal processing scheme is that it produces a detailed spectral response envelope which unambiguously ascribes a selectivity feature to the IGEFET microsensor. Since the spectral envelopes for the two challenge gases only intersect at a very limited number of points, this feature clearly emphasizes the distinction (independence) between the two responses. In particular, the first peak in the NO$_2$ response occurs at approximately 1000 Hz, while that for DIMP occurs at 3400 Hz. Additionally, the rise and decay rates associated with the first peaks are more pronounced in the NO$_2$ spectra. Both challenge gas spectra also contain a secondary peak; the NO$_2$ secondary peak occurs at approximately 8200 Hz, while that for DIMP occurs in the

SENSITIVE AND SELECTIVE TOXIC GAS DETECTION   42.19

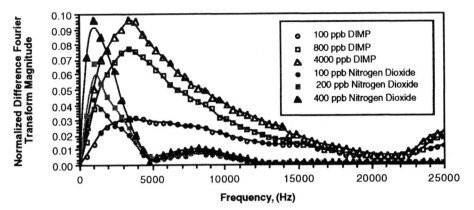

FIGURE 42.13 Normalized difference Fourier transform magnitude spectra for the DIMP and NO$_2$ challenges. (Each data point represents the arithmetic average of six independent measurements.)

vicinity of 25 kHz. These low-frequency (acoustic) resonant peaks suggest that a long-range dielectric polarization interaction is being facilitated by the adsorbed challenge gases. Qualitatively, the electronic and steric features of the challenge gases, along with the interstitial crystallite adsorption model discussed earlier, are postulated to account for the observed interfacial dielectric relaxation behavior.

Additionally, because the normalized difference Fourier transform magnitude spectra peak widths and gradients are different (Fig. 42.13), an alternative distinct response signature of each challenge gas can be generated by computing the corresponding spectral derivatives with respect frequency. For example, Figs. 42.14(a) and (b) depict the corresponding first derivative plots of the 4000 ppb DIMP and 400 ppb NO$_2$ challenge gas exposure responses, respectively. Each challenge-gas-induced response can readily be identified by the characteristic ensemble of fre-

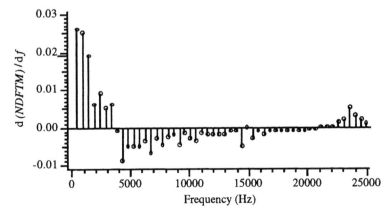

FIGURE 42.14 (a) Normalized difference Fourier transform magnitude (NDFTM) spectra derivatives (with respect to frequency) for the 4000-ppb NO$_2$ challenge gas concentration.

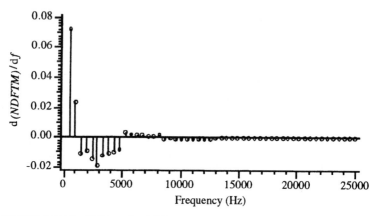

FIGURE 42.14 (b) Normalized difference for the 400-ppb NO_2 challenge gas concentration.

quencies where the first derivative changes sign; for example, DIMP has linearly extrapolated zero-valued derivatives at 3600 Hz, 14,880 Hz, and 20,880 Hz, while NO_2 has its zero values at 1200 Hz, 5040 Hz, 8400 Hz, and 13,200 Hz.

Finally, considering the 25-kHz spectral bandwidth limitation of the measurement instrumentation (Fig. 42.7), the area beneath each response in Fig. 42.13 can be used as a metric to report the microsensor's sensitivity to a particular challenge gas concentration.

CONCLUSION

When compared to the conventional direct current or single-frequency alternating current modes of operating chemically active, thin film, IGE detection devices, the voltage pulse mode of operating the IGEFET microsensor and the associated Fourier transform signal processing technique yields an ensemble of frequency components that defines a unique challenge gas response envelope for small concentrations of NO_2 or DIMP mixed with a common dehumidified ambient air source (diluent). The ambiguity associated with differentiating specific challenge gas responses is minimized with the voltage pulse excitation mode because the microsensor's complete frequency domain response envelopes are considered rather than simply comparing the changes associated with a single measurand (for example, a change in the direct current electrical conductivity). Although the normalized difference Fourier transform magnitude and first derivative spectral features reported for NO_2 and DIMP unambiguously differentiate the challenge gas responses, it is acknowledged that an interferant likely exists that will yield a spectral envelope that would be statistically unresolvable from those associated with either of the two model challenge gases. If the ambiguity between microsensor responses for several challenge gases manifests itself as a serious performance limitation, it should be possible to minimize its influence by utilizing a larger two-dimensional array of IGEFETs arranged on a common substrate where each microsensor element is coated with a different chemically active thin film. Figure 42.15 illustrates the most recent version of the IGEFET microsensor that incorpo-

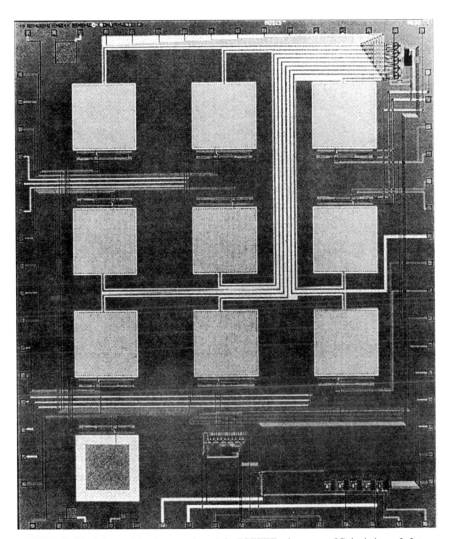

FIGURE 42.15 Advanced implementation of the IGEFET microsensor IC depicting a 3×3 two-dimensional array of identical sensors whose independent set of responses can be electrically multiplexed in the *in situ* circuitry. (The overall dimensions of the IC are 4466×6755 micrometers. Each IGE structure consists of 34 floating-electrode fingers and 35 driven-electrode fingers. Each finger is 10 micrometers wide, and the interelectrode finger spacing is 10 micrometers. The chemically active area of the IGE measures 1370×1370 micrometers. Each IGE structure may be coated with a different metal-substituted phthalocyanine thin film or an identical candidate if response redundancy and reproducibility are of interest.)

rates an electrically multiplexed array of identical IGEFET devices that may be coated with different metal-substituted phthalocyanine thin films. By utilizing the same Fourier transform signal processing technique and implementing a pattern recognition algorithm, it is envisaged that this evolutionary IGEFET microsensor concept will emerge as an electronic nose technology for detecting and indentifying the constituents of a multicomponent gas mixture. Finally, considering a broad spectrum of practical applications, such as is the situation for many industries, laboratories, and manufacturing facilities who must comply with OSHA, NIOSH, and EPA employee exposure and environmental pollution standards, this technology possesses the adaptability, sensitivity, and selectively to detect a broad spectrum of toxic gases—a potential limited only by the successful deposition of stable, chemically active thin films.

REFERENCES

1. E. P. Scheide and G. G. Guilbault, "Piezoelectric Detectors for Organophosphorus Compounds and Pesticides," *Anal. Chem.* **44**, 1972, pp. 1764–1768.
2. W. M. Shackelford and G. G. Guilbault, "A Piezoelectric Detector for Organophosphorus Pesticides in Air," *Anal. Chem. Acta* **73**, 1973, pp. 383–389.
3. Y. Tomita and G. G. Guilbault, "Coating for a Piezoelectric Crystal Sensitive to Organophosphorus Pesticides, *Anal. Chem.* **52**, 1980 pp. 1484–1489.
4. G. G. Guilbault, Y. Tomita, and E. S. Kolesar, Jr., "A Coated Piezoelectric Crystal to Detect Organophosphorus Compounds and Pesticides," *Sens. Actuators,* **2**, 1981, pp. 43–57.
5. G. G. Guilbault, J. Affolter, Y. Tomita, and E. S. Kolesar, Jr., "Piezoelectric Crystal Coating for Detection of Organophosphorus Compounds," *Anal Chem.* **53**, 1981, pp. 2057–2060.
6. E. S. Kolesar, Jr. and R. M. Walser, "Organophosphorus Compound Detection with a Supported Copper + Cuprous Oxide Island Film: (1) Gas-Sensitive Film Physical Characteristics and Direct Current Studies, *Anal. Chem.,* **60**, 1988, pp. 1731–1736.
7. E. S. Kolesar, Jr. and R. M. Walser, "Organophosphorus Compound Detection with a Supported Copper + Cuprous Oxide Island Film: (2) Alternating Current Studies and Sensor Performance," *Anal. Chem.* **60**, 1988, pp. 1737–1743.
8. K. H. Karmarkar and G. G. Guilbault, The detection of ammonia and nitrogen dioxide at the parts per billion level with coated piezoelectric crystal detectors, *Anal. Chem. Acta,* **75**, 1975, pp. 111–117.
9. L. M. Webber, J. Hlavay, and G. G. Guilbault, "Piezoelectric Detectors for Specific Detection of Environmental Pollutants," *Mikrochimica Acta,* **1**, 1978, pp. 351–358.
10. M. S. Nieuwenhuizen, A. W. Barendsz, E. Nieuwkoop, M. J. Vellekoop, and A. Venema, "Transduction Mechanisms in SAW Gas Sensors," *Electron. Lett.,* **22**, 1986, pp. 184–185.
11. M. S. Nieuwenhuizen and A. W. Barendsz, "Processes Involved at the Chemical Interface of a SAW Chemosensor," *Sens. Actuators,* **11**, 1987, pp. 45–62.
12. M. S. Nieuwenhuizen, A. Nederlof, and A. W. Barendsz, "Metallophthalocyanines as Chemical Interfaces on a Surface Acoustic Wave Gas Sensor for Nitrogen Dioxide," *Anal. Chem.,* **60**, 1988, pp. 230–235.
13. A. Venema, E. Nieuwkoop, M. J. Vellekoop, W. J. Ghijsen, A. W. Barendsz, and M. S. Nieuwenhuizen, "NO_2 Gas-Concentration Measurement with a SAW-Chemosensor," *IEEE Trans. Ultrasonics, Ferroelectrics, and Frequency Control,* UFFC-34, 1987, pp. 148–155.
14. A. W. Barendsz, J. C. Vis, M. S. Nieuwenhuizen, E. Nieuwkoop, M. J. Vellekoop, M. J. Ghijsen, and A. Venema, "A SAW Sensor for NO_2 Gas Concentration Measurements," *Proc. IEEE Ultrasonic Symposium,* San Francisco, 1985, p. 586.
15. S. D. Senturia, C. M. Sechen, and J. A. Wishneusky, "The Charge-Flow Transistor: A New MOS Device," *Appl. Phys. Lett.,* **30**, 1977, pp. 106–108.
16. S. L. Garverick and S. D. Senturia, "An MOS Chip for Surface Impedance Measurement and Moisture Monitoring," *Tech. Digest, IEEE Int. Electron Devices Meet.,* Washington, DC, December 8–10, 1980, pp. 685–688.

17. N. F. Sheppard, S. L. Garverick, D. R. Day, and S. D. Senturia, "Microdielectrometry: A New Method for In-Situ Cure Monitoring," *Proc. 26th Natl. SAMPE Symp.*, Los Angeles, April 28–30, 1981, pp. 65–76.

18. S. D. Senturia, N. F. Sheppard, S. Y. Poh, and H. R. Appelman, "The Feasibility of Electrical Monitoring of Resin Cure with the Charge Flow Transistor," *Polym. Eng. Sci.*, **21**, 1981, 113–118.

19. S. L. Garverick and S. D. Senturia, "An MOS Device for AC Measurement of Surface Impedance with Application to Moisture Monitoring," *IEEE Trans. Electr. Dev.*, ED-29, 1982, pp. 90–94.

20. S. D. Senturia, N. F. Sheppard, H. L. Lee, and D. R. Day, "*In-situ* Measurement of the Properties of Curing Systems with Microdielectrometry," *J. Adhesion*, **15**, 1982, pp. 69–90.

21. N. F. Sheppard, D. R. Day, H. L. Lee, and S. D. Senturia, "Microdielectrometry," *Sens. Actuators*, **2**, 1982, pp. 263–274.

22. S. D. Senturia, "The Role of the MOS Structure in Integrated Sensors," *Sens. Actuators*, **4**, 1983, pp. 507–526.

23. S. D. Senturia, N. F. Sheppard, H. L. Lee, and S. B. Marshall, "Cure Monitoring and Control with Combined Dielectric/Temperature Probes," *SAMPE J.*, **19**, 1983, pp. 22–26.

24. M. C. W. Coln and S. D. Senturia, "The Application of Linear System Theory to Parametric Microsensors," *Digest of Papers for the 1985 International Conference on Solid-State Sensors and Actuators (Transducers '85)*, Philadelphia, June 11–14, 1985, pp. 118–121.

25. S. D. Senturia and N. F. Sheppard, "Dielectric Analysis of Thermoset Cure," in *Epoxy Resins and Composites IV*, K. Dusek (ed.), Springer-Verlag, New York, 1986, pp. 1–48.

26. D. R. Day, "Cure Control: Strategies for Use of Dielectric Sensors," *Proc. 31st Natl. SAMPE Symp.*, Las Vegas, April 7–10, 1986, pp. 1095–1103.

27. S. D. Senturia, S. L. Garverick and K. Togashi, "Monolithic Integrated Circuit Implementations of the Charge-Flow Transistor Oscillator Moisture Sensor," *Sens. and Actuators*, **2**, 1982, pp. 59–71.

28. S. D. Senturia and M. T. Fertsch, "A Charge-Flow Transistor Oscillator Circuit," *IEEE J. Sol. St. Circuits*, SC-14, 1979, pp. 753–757.

29. S. D. Senturia, J. Rubinstein, S. J. Azoury, and D. Adler, "Determination of the Field Effect in Low-Conductivity Materials with the Charge-Flow Transistor," *J. Appl Phys.*, **52**, 1981, pp. 3663–3670.

30. S. D. Senturia, M. G. Huberman, and R. Van der Kloot, "Moisture Sensing with the Charge-Flow Transistor," *Proc. ARPA/NBS Workshop on Moisture Measurement in Integrated Circuit Packages*, Gaithersburg, Maryland, March, 1978, pp. 108–114.

31. N. F. Sheppard and S. D. Senturia, "Molecular Contributions to the Dielectric Permittivity of Unreacted Epoxy Amine Mixtures," *Technical Program Summary of the Adhesion Society Annual Meeting*, Savannah, Georgia, 1985, p. 22a.

32. N. F. Sheppard and S. D. Senturia, "Chemical Interpretation of the Relaxed Permittivity during Epoxy Resin Cure," *SPE Tech. Papers*, **31**, 1985, p. 321.

33. D. R. Day, "Effects of Stoichiometric Mixing Ratio on Epoxy Cure—Dielectric Analysis," *SPE Tech. Papers*, **31**, 1985, p. 327.

34. F. W. Kutzler, W. R. Barger, A. W. Snow and H. Wohltjen, "An Investigation of Conductivity in Metal-Substituted Phthalocyanine Langmuir-Blodgett Films," *Thin Solid Films* **155**, 1987, pp. 1–16.

35. S. Baker, G. G. Roberts, and M. C. Petty, "Phthalocyanine Langmuir-Blodgett Film Gas Detector," *IEE Proc. Part I*, 130, 1983, pp. 260–263.

36. H. Wohltjen, W. R. Barger, A. W. Snow, and N. L. Jarvis, "A Vapor-Sensitive Chemoresistor Fabricated with Planar Microelectrodes and a Langmuir-Blodgett Organic Semiconductor Film," *IEEE Trans. Electr. Dev.*, ED-32, 1985, pp. 1170–1174.

37. R. H. Tregold, M. C. J. Young, B. Hodge, and A. Hoorfar, "Gas Sensors Made from Langmuir-Blodgett Films of Porphyrins," *IEE Proc. Part I,* 132, 1985, pp. 151–156.
38. T. Jones and B. Bott, "Gas-Induced Electrical conductivity in metal phthalocyanines," *Sens. and Actuators,* **9,** 1986, 27–37.
39. B. Bott and T. A. Jones, "A Highly Sensitive NO_2 Sensor Based on Electrical Conductivity Changes in Phthalocyanine Films," *Sens. and Actuators,* **5,** 1984, pp. 43–53.
40. J. P. Blanc, G. Blasquez, J. P. Germain, A. Larbi, C. Maleysson and H. Robert, "Behavior of Electroactive Polymers in Gaseous Atmospheres," *Sens. and Actuators,* **14,** 1988, pp. 143–148.
41. R. L. van Ewyck, A. V. Chadwick, and J. D. Wright, "Electron Donor-Acceptor Interactions and Surface Semiconductivity in Molecular Crystals as a Function of Ambient Gas," *J. Chem. Soc. Faraday Trans. I,* **76,** 1980, pp. 2194–2205.
42. P. B. M. Archer, A. V. Chadwick, J. J. Miasik, M. Tamizi and J. D. Wright, "Kinetic Factors in the Response of Organometallic Semiconductor Gas Sensors," *Sens. and Actuators,* **16,** 1989, pp. 379–392.
43. J. D. Wright, "Gas Adsorption on Phthalocyanines and Its Effects on Electrical Properties," *Prog. Surf. Sc.,* **31,** 1989, pp. 1–60.
44. E. S. Kolesar, Jr. and J. M. Wiseman, "Interdigitated Gate Electrode Field Effect Transistor for the Selective Detection of Nitrogen Dioxide and Diisopropyl Methylphosphonate, *Anal. Chem.,* **61,** 1989, pp. 2355–2361.
45. J. Simon and J. J. Andre, *Molecular Semiconductors,* Springer-Verlag, New York, 1985.
46. F. H. Moser and A. L. Thomas (eds.), *The Phthalocyanines* vols. 1 and 2, CRC Press, Inc., Boca Raton, Florida, 1983.
47. C. C. Leznoff and A. B. P. Lever (eds.), *Phthalocyanines—Properties and Applications,* VCH Publishers, Inc., New York, 1989.
48. P. M. Burr, P. D. Jeffery, J. D. Benjamin and M. J. Uren, "A Gas-Sensitive Field-Effect Transistor Utilizing a Thin Film of Lead Phthalocyanine as the Gate Material," *Thin Solid Films,* **151,** 1987, LI 11–LI 13.
49. F. I. Mopsik, "Precision Time-Domain Dielectric Spectrometer," *Rev. Sci. Instr.,* **55,** 1984, pp. 79–87.

CHAPTER 43
MOLECULAR RELAXATION RATE SPECTROMETER DETECTION THEORY

The Molecular Relaxation Rate Spectrometer is a Chemical Sensor that has a detection sensitivity at the parts per trillion (ppt) concentration level in ultrapure air. This has already been shown in laboratory demonstrations at the University of Maryland over a decade ago.[1,2] Ambient molecules present in ordinary room air establish a clutter background barrier to detection at the parts per billion (ppb) concentration level. Recent developments have restored the ppt detection capability for preselected molecular species in ordinary air.[3] Here I discuss basic phase fluctuation optical heterodyne (PFLOH) detection theory, establish its detection sensitivity at the ppt level, illustrate a method for processing the PFLOH signal, and show the obscuring effects of an interfering ambient molecular species present in ordinary air.

Molecules may absorb energy from the emissions of a laser that is tuned to a molecular absorption resonance. This absorbed electromagnetic energy is stored in the form of molecular vibrations and/or rotations. The exchange of this molecular vibrational-rotational energy to molecular translational kinetic energy is mediated by collision. The collision may be with a surface interface, such as may exist between the fluid medium containing the absorbing molecules and a distinguishable other medium that might be liquid or solid. The collision may also be between the molecule that absorbed the optical energy and other molecules in the fluid medium. The rate of molecular energy exchange (relaxation rate) is a unique property of the molecular species in a specific fluid mixture at a particular temperature and pressure.

An alteration of the local refractive index occurs subsequent to molecular absorption of modulated optical energy and the resulting molecular energy exchange mediated by molecular collisions. This alteration of refractive index is detectable using an interferometer means and can be used to sense the presence of particular molecular species.

Consider the example of an interferometer with a path length of 10 cm, illuminated by a 2-mm diameter single-frequency CW laser beam. Suppose that a molecule has an absorption feature at 3×10^{13} Hz but does not absorb at the frequency of the CW laser beam. Suppose this molecule is situated within the interferometer volume illuminated by the single frequency CW laser and absorbs (from another laser source) power modulated electromagnetic energy having a frequency of 3×10^{13} Hz. For a single absorption event, the subject molecule absorbs 1.986×10^{-20} J of energy.

Suppose further that the fluid medium is air and has an index of refraction at the CW laser frequency of 1.0003 and a heat capacity of 33 J per mole of medium per kelvin, (which are approximately the values for air at standard temperature and pressure). Then the single molecular absorption event would eventually result in increasing the temperature of the 10-cm interferometer beam volume ($.1 \times .1 \times 10\pi$ cm^3) by 4.3×10^{-17} kelvins. The average time to convert the absorbed energy of the molecule to heat in the medium would typically be about a microsecond. The average time for the heat to be transmitted from the volume of space being illuminated by one of the single optical mode laser beams in the interferometer would typically be about 200 msec. The transmission of energy from the absorbing volume may be caused by any of a number of processes. For example, the energy may be radiated away, conducted away, or convected away from the absorbing fluid medium volume. Thus, if the exciting laser source emission is output power modulated in a way that is periodically varying with a period that is longer than the average time to convert the absorbed energy of the molecules to heat in the medium, the thermal modulation of the medium will resemble the power modulation of the exciting laser.

The thermal variations of the medium occur proximate the molecular absorption location, take place at a time t subsequent to the time of absorption, and are attended by refractive index variations of the medium proximate the molecular absorption location r. These refractive index modulations, $\Delta n_v(r,t)$, depend on the medium refractive index at the laser optical frequency v, n_v; the temperature T; and the induced variation in the local temperature modulation $\Delta T(r,t)$ as represented by

$$\Delta n_v(r,t) = -(n_v - 1) \times \left(\frac{\Delta T(r,t)}{T}\right) \tag{43.1}$$

Under ideal conditions, the signal-to-noise ratio of a coherent detection process depends on the quantum efficiency of the photodetector η, the single optical mode laser photon energy h_v, the detection bandwidth of the signal processing electronics Δf (Hz), and the laser signal power associated with the optical phase fluctuation sidebands of the refractive index variations of the sample space being probed P_s. The expression for signal to noise is represented by

$$\frac{S}{N} = \frac{\eta \times P_s}{h_v \times \Delta f} \tag{43.2}$$

The laser signal power associated with the optical phase fluctuation sidebands of the refractive index variations of the sample space being probed P_s depends on the length of the laser path through the medium sample l, the equilibrium refractive index at the laser frequency n_v, the single mode laser optical frequency v, the speed of light c, and the laser power P. The expression representing the laser signal power is given as

$$P_s = \frac{P}{2}\left(\frac{2\pi l_v \Delta n_v}{c}\right)^2 \tag{43.3}$$

For a signal-to-noise ratio of unit y, the amplitude of the minimum detectable refractive index fluctuation Δn_v (min.) is approximately given as

$$\Delta n_v \text{ (min.)} = c/(2\pi v\, l) \sqrt{[2h\, v\, \Delta f(Hz)]/\eta\, P}$$
$$= (1.74\ 10^{-17}/l) \sqrt{[(\Delta f(Hz)]/v\eta\, P)} \tag{43.4}$$

For a system having a 10-cm length ($l = 10$ cm), a 100-mW single optical mode laser at an optical frequency of 3×10^{14}, and a square law photodetector quantum efficiency of 80 percent, the minimum detectable refractive index variation is $3.54 \times 10^{-15} \sqrt{\Delta f}$ (Hz).

This means that a sinusoidal refractive index variation Δn_v of 3.54×10^{-15} may be detected with a signal processing detection bandwidth of 1 Hz. At 300 K, this refractive index variation corresponds to a temperature difference between the interferometer paths of 3.5×10^9 K. This is the amount of temperature variation that would occur in a 10-cm standard temperature and pressure air sample being probed by a 2-mm diameter laser beam (a volume of 0.314 cm^3) containing a one-part-per-trillion concentration of absorbing molecular species.

Interferometers always have two output ports. Complementary portions of the optical power applied to the input port of the interferometer appear at either of the two output ports. The output optical power of an interferometer output port is related to the phase difference between the two laser beams being compared by a classical interferometer response function. The two output ports have phase responses 180° out of phase with each other. The response functions of these two ports and the power output in response to the same phase modulation is shown in Fig. 43.1.

Optical heterodyne interferometry is one of the most sensitive detection methods known. But typical Mach Zehnder, Michelson, or Fabry Perot designs are susceptible to drift and external vibrations. These cause variations in the arm lengths of the interferometer and introduce noise in the output signals. The keys to practical use of interferometry are phase stabilization and noise suppression.

Under normal operating conditions, the largest noise contribution in a heterodyne signal is due to mechanical noise sources. Such sources have signal content from the degrees of freedom of the mechanical structure. Drift may be due to movement of the optical components in an unconstrained direction. Vibrational noise may be due to excitation of the various modes of the solid structures comprising the interferometer. These modes may be torsional, shear, or longitudinal.

The design shown in Fig. 43.2 provides excellent mechanical noise immunity. Due to the topology of the interferometer shown, excursions of the longitudinal and shear sort are common to both the interferometer arms. Thus there is almost no coupling of longitudinal or shear mechanical noise to the interferometer optical output signal. The torsional motions of the optical components relative to one another are the only excursions to which the interferometer shown is susceptible. The degree of coupling of the torsional modes to the optical output of the interferometer depends on the amplitude of excursion and the separation distance between the two parallel interferometer laser beams (6 and 7).

In Fig. 43.2, a tunable single mode laser (1) produces a CW laser beam (2) that illuminates a specially coated optical component (5). The optical component produces two output beams of nominally equal power (6 and 7) and redirects them toward a mirror/Brewster angle window (8) that redirects the two beams through a device (9) toward another mirror/Brewster angle window (10) that redirects the two beams to another specially coated optical component (11). The specially coated optical component (11) produces three output beams (12, 14, and 16). Output beam 12 is the surface reflection of beam 7. Output beams 14 and 16 are produced from the interference of beams 6 and 7. The CW laser power is proportional to the power of beam 12, which is monitored by a photodetector (13). The heterodyne signals (14 and 16) detected by photodetectors (15 and 17) depend on the power of laser 1 and the relative phase difference between particular optical path lengths.

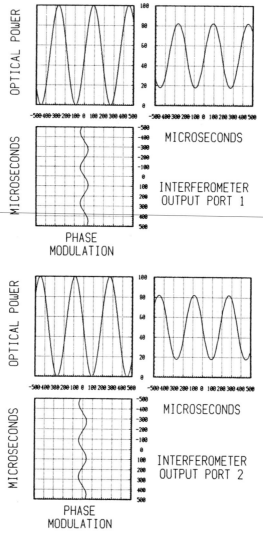

FIGURE 43.1 Plots of the interferometer output port's response to an oscillating phase fluctuation.

A power-modulated exciting laser (3) is tuned so its optical frequency coincides with an absorption feature of the molecule whose detection is desired. The optical output from this laser (4) is passed through a beam shutter (18) to a mirror/Brewster angle window (8) that passes the laser signal (4) along a path that is coaxial with one or the other of the laser beams (6 or 7). The exciting laser beam (4) passes through the common interferometer path to illuminate another mirror/Brewster angle window (10) that passes the laser beam (4) to a beam dump (19), which absorbs and dissipates the laser energy.

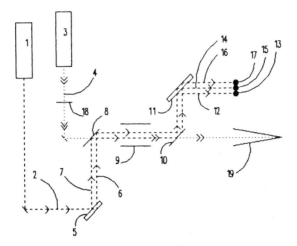

FIGURE 43.2 Optical system design molecular relaxation rate spectrometer.

PFLOH Demonstrations have been performed where excitation was accomplished by driving infrared active[1] and Raman active[2] transitions of molecules in the gas. Either method of excitation is suitable for use in this instrument. However, resonant Raman excitation has significant practical advantages over infrared excitation that are peculiar to this instrument.

The interferometer is phase-stabilized by proportionally controlling the continuous single mode laser optical frequency v to maintain the long time constant interferometer output at a set point in the linear portion of the interferometer response function.

The phase difference $\Delta\Phi(v_{\text{laser}}, t)$ between optical paths of the interferometer OPL (v_{laser}, t) and resonant Raman excitation is accomplished by mixing the signal of an idler laser of optical frequency v, with the signal of a drive laser of optical frequency v_2 in the fluid medium. Molecules having a Raman active mode coinciding with $|v_1 - v_2|$ absorb energy from the optical field of the two lasers.

OPL_2 (v_{laser}, t) may drift and jitter with time t and is given as

$$\Delta\Phi\left(v_{\text{laser}}, t\right) = \Phi_1\left(v_{\text{laser}}, t\right) - \Phi_2\left(v_{\text{laser}}, t\right) \tag{43.5}$$

This may be rewritten as

$$\Delta\Phi\left(v_{\text{laser}}, t\right) = \Delta OPL(v_{\text{laser}}, t) \cdot 2\pi/\lambda_L(t) \tag{43.6}$$

Thus, in the case where the interferometer paths are not of equal length (i.e., $\Delta OPL(v_{\text{laser}}, t) \neq 0$), the long time constant phase of the interferometer output ports may be maintained at a preselected set point in the linear response region using laser optical frequency control. Adjustment of the laser optical frequency of the single-mode laser by an appropriate amount maintains the interferometer output power at the desired steady-state value. This can be accomplished automatically.

The interferometer phase difference changes as the single optical mode laser vacuum wavelength changes, as given by

$$\delta\Delta\Phi = -\Delta OPL(v_{\text{laser}}, t) \cdot \left(\frac{2\pi}{\lambda_L^2}\right) \delta\lambda_L(t) \tag{43.7}$$

An interferometer with a ΔOPL of 1 mm illuminated by a laser may be tuned through one cycle of interferometer output power by altering the laser wavelength from 830 nm to 830.69 nm.

An interferometer with a ΔOPL of 5.74 mm illuminated by a laser may be tuned through a half cycle of interferometer output power by altering the laser wavelength from 829.99 nm to 830.05 nm, as shown in Fig. 43.3.

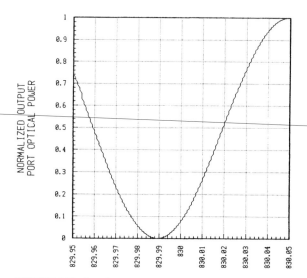

FIGURE 43.3 Interferometer port output power as a function of laser wavelength, in nm.

The interferometer response signal will be a superposition of phase shifted sinusoidal frequency components of the fundamental ($n = 1$) and harmonic ($n > 1$) frequencies of the modulation frequency v_p:

$$A_o(t) = \sum_j \sum_n A_{o,j} (2\pi n v_p) \cdot \sin [2\pi n v_p (t - \tau_j)] \tag{43.8}$$

where $\tau_j = 1/k_j$ is the time constant for relaxation of the jth species.

At modulation frequencies higher than the thermal diffusion rate of heat from the volume of medium being probed by the CW laser (1 in Fig. 43.2), the amplitude of the modulation frequency component due to molecular species j, $A_j (2\pi v_p)$, of the interferometer response signal will vary with the power modulation frequency v_p according to

$$A_j(2\pi v_p) = \sum_j \alpha_j \times N_j \times (1 - \alpha_j e^{-k_j/v_p})(1 - \alpha_j e^{-2k_j/v_p})(1 - \alpha_j e^{-3k_j/v_p})(1 - \alpha_j e^{-4k_j/v_p}) \tag{43.9}$$

where α_j is the fraction of j molecules excited in time τ_j, k_j is the relaxation rate of the jth molecular species in the fluid medium, and N_j is the amount of the jth molecular species in the fluid mixture.

The interferometer signal component at the power modulation frequency of the exciting laser beam (4 in Fig. 43.2) will be given by

$$A_o(v_p, t) = \sum_j A_{o,j}(2\pi 1 v_p) \cdot \sin[2\pi v_p(t - \tau_j)] \qquad (43.10)$$

The interferometer output signals are converted to electrical signals by the photodiodes (15 and 17 in Fig. 43.2) and subsequent circuits that have amplitude and phase proportional to that of the interferometer output signal.

A heterodyne technique can be used to detect modulation induced by the power-modulated exciting laser in the electronic signals from the photodiodes. This heterodyne technique may be implemented as a synchronous detection circuit comprised of a four-quadrant multiplier and a low-pass filter circuit. If, for example, an electrical sinusoidal signal $B \cdot \sin(2\pi v_p t + \phi)$ and the modulated infrared laser optical signal (4 in Fig. 43.2) derive from the same reference frequency source, then both will have the same frequency v_p and phase ϕ as the reference source.

The A_E $(2\pi v_p, t)$ electrical signal may then be detected by taking the product of $B \cdot \sin(2\pi v_p t + \phi)$ and the total electrical signal representation of the interferometer output signal. This detection may be performed by applying the signals to the appropriate inputs of a four quadrant multiplier circuit. The output of the four-quadrant multiplier may then be passed through a low pass filter to produce the amplitude of $A_E(v_p)$ shown in Eq. (43.14).

As an example of this signal detection method the interferometer response signal $A_o(t)$ may be converted to an equivalent electrical signal $A_E(t)$ that also depends on the optical power modulation frequency v_p, the molecular relaxation time constants τ_j, and those amplitude, $A_{E,j}$ $(2\pi v_p)$—components unique to particular molecules comprising the fluid medium. The total signal $A_E(t)$, may be represented as

$$A_E(t) = \sum_j \sum_n A_{E,j}(2\pi n v_p) \cdot \sin(2\pi n v_p(t - \tau_j)) \qquad (43.11)$$

The product may be obtained by feeding signal $A_E(t)$ to one input of a multiplying electrical circuit and feeding the signal $B \cdot \sin(2\pi v_p, t)$, to a second input of the same multiplying electrical circuit. The output of the multiplying electrical circuit depends on, and is proportional to, the product of these signals and may be represented as:

$$A_E(v_p, t) = \sum_j \sum_n B \cdot A_{E,j}(2\pi n v_p) \cdot \sin(2\pi n v_p(t - \tau_j)) \cdot \sin(2\pi v_p t) \qquad (43.12)$$

This representation may be rewritten as:

$$A_E(v_p, t) = \sum_j B \cdot A_{E,j}(2\pi n v_p) \frac{\cos(2\pi v_p \tau_j)}{2} - \sum_j B \cdot A_{E,j}(2\pi n v_p) \cdot \frac{\cos(4\pi v_p t)}{2}$$

$$\times \cos(2\pi v_p \tau_j) + \sum_j \sum_{n \neq 1} B \cdot A_{E,j}(2\pi n v_p) \sin(2\pi n v_p t) \cdot \sin(2\pi n v_p t) \cos(2\pi v_p \tau_j)$$

$$- \sum_j \sum_n B \cdot A_{E,j}(2\pi n v_p) \cdot \sin(2\pi n v_p, \tau_j) \cdot \cos(2\pi k v_p, t) \qquad (43.13)$$

This signal may then be passed through a low-pass filter to produce the DC component of the signal. cases (such as those illustrated in Figs. 43.5 and 43.6) where the electrical filter passes electrical signals of one oscillation frequency than v_p, the electrical signal that is output by the low-pass filter may be represented as:

$$A_E(V_p) = \sum_j \tfrac{1}{2} B, A_{E,j}(2\pi v_p) \cdot \cos(2\pi v_p \tau_j) \qquad (43.14)$$

A computer program shown in Fig. 43.4, was written to generate $A_E(v_p)$ for two representative cases where only one species is present in the air. The plots of these two cases appear in Figs. 43.5 and 43.6. The combined $A_E(v_p)$, shown in Fig. 43.7, is representative of the case where both are present in the air. Note that each point on the curve associated with expression 14 represents a DC output signal corresponding to a particular optical power modulation frequency.

Figure 43.5 shows such a plot for the case where a single target species is excited by the power-modulated laser signal. This plot is highly characteristic of the particu-

```
OPEN "d:pfloh5.dat" FOR OUTPUT AS #1
pi = 4 * ATN(1)
INPUT "Number of Molecules: nn"; nn
INPUT "Fractional Excitation per Pulse: a"; a
INPUT "Relaxation Rate: k"; k
FOR m = 1000 TO 1000000 STEP 1000
nu = m * LOG(m)
Ao = 1
FOR n =        1 TO 10
Bo = Ao * (1 - a * EXP(-n * k / nu))
IF ABS(AO - Bo) <.0001 THEN GOTO 10
Ao = Bo
NEXT n
10: Ao=Bo* A * nn/2 * COS(2 * PI * NU / K)
PRINT #1, nu, Ao
NEXT m
```

FIGURE 43.4 Program used to model $A_E(v_p)$.

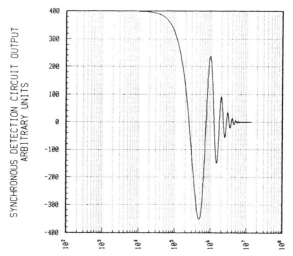

FIGURE 43.5 Synchronous detection circuit output vs. laser modulation frequency: $\tau = 1$ μsec, $\alpha = 80$ percent excited in time = τ, $N = 1000$ absorbing molecules.

MOLECULAR RELAXATION RATE SPECTROMETER DETECTION THEORY

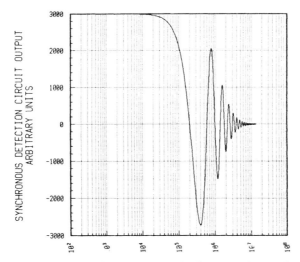

FIGURE 43.6 Synchronous detection circuit output vs. laser modulation frequency; $\tau = 1.25$ μsec, $\alpha = 60$ percent excited in Time $= \tau$, $N = 10,000$ absorbing molecules.

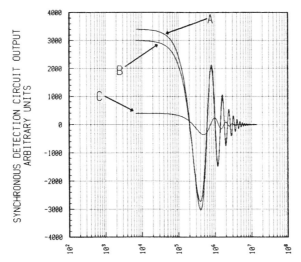

FIGURE 43.7 Synchronous detection circuit output vs. laser modulation frequency for the sum A of the cases B and C.

lar target species. Figure 43.7 shows a similar plot for the case in which a second species shown in Fig. 43.6 (representative of a chemical interferent) is also present and excited by the power-modulated laser signal. Here, the curve that is produced is A. The curve characteristic of the target species C is obscured by the interferent species signal contribution B. However, signal processing techniques drawn from radar and image processing technology[4,5] can be applied to detect the presence of the target species characteristic signal.[3] An optimal clutter suppression filter can be used for detecting the characteristic signature of a specific target molecular species, thus recovering sensor PPT detection performance even in the presence of PPB molecular background clutter.

REFERENCES

1. C. C. Davis and S. J. Petuchowski, "Phase Fluctuation Optical Heterodyne Spectroscopy of Gases," *Applied Optics,* **20**(14), July 1981.
2. J. T. Siewick, "The Development of a New Technique for the Study of V → T Relaxation of Infrared Inactive Vibrations and its Application to Molecular Hydrogen," University Microfilms International.
3. J. T. Siewick, "Substance Analyzer Device," Patent Application Serial Number 07/882, 134, May 13, 1992.
4. Irving S. Reed, "Detection of Moving Optical Sources Against a Moving Background," Hughes Aircraft Co. Interdepartmental Correspondence, June 1978.
5. C. F. Ferrara, R. W. Fries, and H. H. Mansour, "Three Dimensional Matched Filtering on IRSTS Scenes," *Proceedings IRIS Targets, Backgrounds and Discrimination* vol. 1, 1989.

CHAPTER 44
CURRENT STATE OF THE ART IN HYDRAZINE SENSING

INTRODUCTION

Hydrazine is an extremely useful chemical for industrial and propellant applications, and, at the same time, is extremely hazardous. Current requirements for hydrazine detection are at three concentration levels: explosive levels (percent), toxic levels, and threshold limit value (TLV) levels. Various applications require fixed-point, portable, and personal detectors. Although most of the research and instrumental progress in the last decade has been devoted to detecting hydrazine at the TLV levels, a new field-deployable FTIR system has been developed for toxic levels. Electrochemical sensors have historically been used for both toxic and TLV levels. The last five years have seen significant improvements that allow electrochemical sensors to meet the proposed TLV limits of 10 ppb. Progress has been made in colorimetric methods used for fixed-point monitoring at the TLV levels. An important addition to low-cost detection has been the development of colorimetric, real-time dosimetry systems for personal monitoring. Ion mobility spectrometry has been successful for hydrazine detection in space at very low levels. Research is currently underway on two different systems to develop a fixed-point monitor for the proposed TLV using both electrochemical analysis and fluorescence techniques. Other sensors systems are also being developed and investigated to meet the demands for small, inexpensive, reliable sensors at low concentrations levels.

The three hydrazines currently used by the Department of Defense as hypergolic fuels are hydrazine, monomethylhydrazine (MMH), and unsymmetrical dimethylhydrazine (UDMH). Hydrazines are used as propellants in space-launch vehicles, satellites, and aircraft emergency power units. Because of this widespread use, concern has developed over the toxicological properties of the compounds. Hydrazine is considered toxic at 5 parts per million (ppm). The American Conference of Governmental Industrial Hygienists (ACGIH) has categorized the hydrazines as suspected human carcinogens and has recommended threshold limit values (TLV) for hydrazine, MMH, and UDMH of 100, 200 and 500 parts per billion (ppb), respectively.[1] Potentially, this level could be lowered to 10 ppb for all three hydrazines.[2] To minimize the risk of exposure, monitoring of employees who come into contact with hydrazines and of the associated work environments is conducted to insure that the presence of hydrazines remains below hazardous levels. The Department of Defense and NASA require air monitoring for hydrazines in areas where they are handled and/or stored.

HYDRAZINE DETECTION INFRARED SPECTROMETER

A Midac Fourier transform infrared spectrometer was modified by NASA/Kennedy Space Center (KSC) to meet detection requirements for hydrazine at leak detection levels of 1–20 ppm on the space shuttle launch pad. The instrument provides real-time leak detection and area monitoring capable of measuring MMH, hydrazine, and nitrogen dioxide simultaneously. The instrument uses optical absorbance in a long-path gas cell with multiple component data analysis. Detection limits have been demonstrated at 1 ppm with an operating range of up to 100 ppm for all three components. The instrument is capable of rejecting interferences from several compounds. Interfering vapors known to be in the sampling environment have been tested, and the instrument has been calibrated to reject ammonia, isopropyl alcohol, methanol, freon 113, and water up to 2 percent by volume.[3]

ELECTROCHEMICAL SENSORS

Interscan Corporation has developed a new electrochemical instrument capable of detecting hydrazine, MMH, and UDMH at the proposed threshold limit values of 10 ppb. The instrument is designed as a portable survey monitor for intermittent use. The instrument uses an amperometric gas sensor consisting of two electrodes in contact with a solid matrix saturated with cesium hydroxide for the electrolyte. Four production prototype Interscan 4000DX instruments have been evaluated at KSC. The standard test conditions were 10 ppb hydrazine, MMH, or UDMH at 45 percent relative humidity and 25°C. The instruments are capable of reliably detecting a 10-ppb hydrazine, MMH, or UDMH vapor. The precision of the instruments are within +/– 10 percent maximum deviation. Zero air drift is within +/– 2 ppb for 4 hours. Span drift was evaluated at 10 ppb for 4 hours. All of the instruments exceeded the desirable criteria of +/–2 ppb for hydrazine and MMH, and the instruments drifted from 3 to 6 ppb in 4 hours. When testing the linearity of the instruments, MMH produced a linear response over a range of 0 to 1000 ppb with deviations up to +/– 15 percent. The linear range meeting this criteria for hydrazine was 0 to 200 ppb. A range of 0 to 700 ppb was observed for UDMH. The average response times for 10 ppb MMH, hydrazine, and UDMH were 5.5, 5.2, and 3.2 minutes, respectively. The time required to recover from a vapor exposure for MMH, hydrazine, and UDMH were 2.5, 3.8, and 4.8 minutes. The electrochemical sensors have a limited lifetime and must be replaced after about 1000 hours of operation.[4]

COLORIMETRIC DETECTORS

Two companies—MDA Scientific, Inc. and GMD Systems, Inc.—make instruments that use colorimetric indication for hydrazine detection. The instruments have paper tape impregnated with phosphomolybdic acid. Upon exposure to hydrazine, MMH, or UDMH, the indicator on the paper turns blue, and the intensity of the color change is proportional to the concentration. These instruments have excellent selectivity. Continuous and portable instruments are commercially available. Previous instrument designs had a limited linear range, but manufacturers now provide two ranges. One fixed-point monitor, MDA 7100, extends the linear range automatically by controlling the sample volume. The sample volume is controlled by monitoring the rate of color change, and lower concentrations are given more sampling time than higher concentrations. For example, if the sample changes quickly, the sample time is reduced before it goes off-scale, a fresh piece of tape is advanced into the sample position, and a new sample begins. Tape is also conserved by reusing portions where no color change was indicated. In addition, these instruments can be calibrated using a calibration card. Gas vapor calibration is recommended, but if the sample volume is verified, a good calibration can be achieved without the use of a calibration laboratory. The accuracy is dependent on the ability to control precisely the sample volume. Leaks in the sample inlet could result in artificially low responses. Long sample lines are not recommended to minimize leaks. The biggest limitation of these instruments is the detection limit. The fixed-point detectors have lower detection limits of 10 ppb, which is the proposed TLV level. Sensitivity can be improved by increasing the sample volume, but there are practical limitations on sample time and pump requirements. Some sample restrictions are necessary to maintain the paper tape integrity. The paper tape provides approximately two weeks of continuous sampling and has a limited shelf life. In addition, the relative response for each of the three hydrazines of interest is different for the MDA instruments, requiring calibration for the hydrazine of interest and inaccurate readings if the other vapors are present.[5]

COLORIMETRIC DOSIMETRY

The extreme reactivity of the hydrazines is responsible for a variety of technical problems encountered in performing ambient air monitoring. One approach that utilizes this reactivity is derivatization of the hydrazine to a species that is easier to analyze. One method is based on the condensation of a hydrazine and an aldehyde, resulting in a product known as a hydrazone. In the case of unsubstituted hydrazine, two moles of aldehyde can react with one mole of hydrazine to form an azine. The mechanism involves the nucleophilic addition of the nitrogen base, followed by the elimination of water. This reaction is acid catalyzed by protonation of the carbonyl. A well-known ASTM method uses para-N,N-dimethylaminobenzaldehyde (PDAB) in a condensation reaction.[6] Condensation reactions have been investigated for detection applications. Two chemistries are particularly attractive: 2,4-dinitrobenzaldehyde and vanillin. Both aldehydes react with hydrazines through condensation reactions to form yellow-colored products that absorb in the visible region. While vanillin reacts rapidly with MMH and hydrazine, it does not react sufficiently with UDMH for use as a real-time dosimeter. Acidification is required in the 2,4-dinitrobenzaldehyde reaction with UDMH and in the vanillin reaction with hydrazine to form the yellow-colored products.

A real-time dosimeter using vanillin to detect hydrazine and MMH was developed and patented at the Naval Research Laboratory (NRL).[7] A two-site badge incorporating both PDAB and vanillin was developed for NASA applications and is commercially available from GMD System, Inc. The badge has been fully characterized through extensive laboratory evaluation and field testing.[8] The intensity of the color correlates to the dose with a detection limit of 7 ppb-hours. The vanillin-based dosimeter possesses the necessary sensitivity for ppb detection, is unaffected by the relative humidity of the sampling environment and sunlight, and the only known interferent is cigarette smoke. An active sampling card called a *Sure-Spot* and a card holder allowing one to pump air through the card were also developed by KSC and GMD Systems, Inc. using vanillin chemistry. The card is sampled at 1–2 l/min using a personal sample pump to provide accurate screening of an area. Actively sampled hydrazine vapors has a detection limit of < 30 ppb-l.[9]

Because vanillin does not detect UDMH, a similar dosimeter for reliable, real-time detection of UDMH was also developed at NRL using 2,4-dinitrobenzaldehyde. With the joint efforts of NRL, Geo-Centers, Inc., and GMD Systems, Inc., the dosimeter was developed and evaluated at the current and proposed TLV levels of hydrazine, MMH, and UDMH in air.[10] The dosimeter consists of a replaceable dosimeter card and a reusable, polypropylene badge housing that contains a clear, UV absorbent shield.[11] The replaceable dosimeter card has two reaction sites, each containing one of two chemistries on a paper substrate. One chemistry uses 2,4-dinitrobenzaldehyde and the other, vanillin. When 2,4-dinitrobenzaldehyde is coated on porous filter paper and exposed passively to UDMH, the detection limit is < 20 ppb-hrs. Actively drawing UDMH vapors through the paper gives a detection limit of < 50 ppb-l. The second reaction utilized in the dosimeter system involves vanillin for the detection of hydrazine and MMH. Neither chemistry is hindered by a sensitivity to relative humidity changes. Badges remain sensitive to hydrazines vapors for at least one year when stored in a freezer. Interferences with the chemistries are limited to tobacco smoke and sunlight for the vanillin and 2,4-dinitrobenzaldehyde chemistries, respectively. The incorporation a of polyester sheet impregnated with UV inhibitors controls the sunlight effect. A field test of the passive system indicated that it can be used for two consecutive days in the sunlight. For indoor applications, the passive badges can be used for at least one week. There were no false positives or negatives for any sites tested.

ION MOBILITY SPECTROMETRY

Hydrazine and MMH are used as fuels in satellites. Astronauts inside the space shuttle can be potentially exposed to hydrazines from the off-gassing of space suits if astronauts returning from repair missions outside the shuttle are exposed and hydrazine is condensed on the suit. To prevent the hazard, the airlock must be monitored before the astronauts enter the shuttle cabin. The monitoring equipment must be sensitive and rugged for use in the airlock, and ideally it will operate at a reduced pressure. Current procedures require an exposed astronaut to leave the airlock and reenter after allowing time for the hydrazine to dissipate. Air can be preserved if the detection of the hazard can be made at a reduced pressure. A Graseby Ionics ion mobility spectrometer (IMS), adapted to monitor hydrazine, and MMH was tested. Using 5-nonanone as the dopant, hydrazine and MMH can be detected. Individually, they are easily observed, but identification is more complex if both hydrazines are present. Concentrations as low as 9 ppb have been detected, and, with sample inlet modifications, lower limits may be possible. Calibration was found to be stable for at least one month. At low concentration, response times were very slow, but 55-ppb MMH can be detected in one minute. At high concentrations, recovery times were observed as great as 35 minutes. Modifications of the sample inlet could improve these times. The instrument can be used in reduced pressures, down to 5 psia, but at low pressures one cannot discriminate between the analytes.[12]

KSC contracted Graseby Ionics to determine the feasibility of IMS as a portable instrument for detecting hydrazine and MMH at 10 ppb and to build prototype instruments to demonstrate the performance. Studies were conducted to design these instruments with an improved sample inlet. It was necessary to detect the hydrazines individually in the presence of ammonia. Laboratory testing at KSC determined that the instruments did not perform sufficiently to proceed in further development.

HYDRAZINE AREA MONITOR

The hydrazine area monitor, a fixed-point continuous monitor for low levels of hydrazine using the MDA/POLYMETRON hydrazine analyzer, is currently being developed by KSC. The system is designed to monitor hydrazine continuously, 24 hours/day, for up to 90 days under microprocessor control with self calibration. The hydrazine area monitor is a complete sampling system that consists of the MDA/POLYMETRON analyzer, a microprocessor, a dynamic acid/base reagent generation and vapor sampling system with alarms to monitor its functions. The analyzer is a three-electrode amperometric electrochemical liquid analyzer that oxidizes hydrazine on the surface of a platinum electrode. The current flow produced is proportional to the concentration of hydrazine in the sample. A method was developed that simultaneously pulls hydrazine vapors and a dilute sulfuric acid solution down a sample tube from a remote location to the analyzer. The hydrazine-laden dilute acid stream is separated from the air and adjusted to a pH greater than 10.2 prior to analysis. To achieve a 90-day maintenance period, reagents are automatically regenerated as needed by dilution of stock solutions. The system has a dual range, 1–1000 ppb and 1–10 ppm. Remote sampling has been successfully tested up to 40 ft.; tests to 50–60 ft. are ongoing. Fluids in the system are recycled to minimize waste, and the liquid samples are passed through Barnstead filters prior to reuse. The system does not respond to TLV levels of isopropyl alcohol, ethanol, freon 113, and ammonia. The instrument is linear over the ranges given with a detection limit of 1 ppb. Excellent repeatability has been demonstrated over 90 days. The response and recovery times to 90 percent of the output signal at 50 ppb is less than 10 min. and above 1 ppm is 2 min. or less.[13]

FLUORESCENCE DETECTION

A fluorescence detection system is currently being developed at the Naval Research Laboratory under KSC sponsorship to provide continuous fixed-point detection of hydrazine at 10 ppb. Chemical derivatization schemes based upon the use of 2,3-naphthalene dicarboxaldehyde (NDA) and 2,3-anthracene dicarboxaldehyde (ADA) have been devised for the sensitive and selective determination of hydrazine and MMH levels at the parts-per-trillion (ppt) levels. The incorporation of hydrazine into an aromatic framework, such as naphthalene or anthracene, as an additional, fused six-member heterocycle, results in the formation of a highly efficient fluorophore whose emission wavelength is red-shifted away from the fluorescent spectrum of the original compound. Prior to the addition of any hydrazine, the fluorescence emission for both of the reagents, NDA or ADA, is minimal. The resultant derivative formed on the other hand is intensely fluorescent. After much study, NDA was selected for development of a real-time detection system. Like the system above, the detection system should operate 24 hours/day for 90 days without maintenance. The reagent is pumped to the end of the sample line so that hydrazine vapor is pulled down the sample line with the liquid reagent. The air is separated from the liquid sample containing the hydrazine derivative. The fluorescence is measured in real-time as the reagent is drawn through a GTI/Spectrovision FD-300 dual monochromator fluorescence detector. A linear range has been demonstrated for 0–1000 ppb with a detection limit of 2 ppb in two minutes. If the flow rate is decreased, part-per-trillion detection is possible within 10 minutes. The current design has a high flow configuration with response and recovery times of less than 2 minutes to 90 percent of the final value. This detection system recycles the fluids to minimize waste. Fluorescent species are removed by passing the liquid through Barnstead filters. New reagent is added prior to sampling.[14]

CONDUCTIVE POLYMER HYDRAZINE SENSOR

Through the U.S. Air Force Small Business Innovative Research program, a prototype personal dosimeter for monitoring worker exposure to hydrazine and UDMH during field operations was developed. Spectral Sciences, Inc. (SSI), under contract from Spire Corporation, developed a conductive polymer thin film sensing element for ppb level detection. The prototype is called *personal alarm thin-film conducting-polymer hydrazine exposure sensor* (PATCHES) and consists of a gold interdigitated electrode array on a quartz substrate coated with poly(3-hexylthiophene) and doped to the conductive state using nitrosonium hexafluorophosphate. Exposure to hydrazine or UDMH causes an irreversible increase in the film resistance (decreased conductivity), which depends on the exposure dose. The observed high sensitivity is a direct result of the strong electron-donating ability of the hydrazines, which act as an antidopant. The extreme sensitivity, potential low cost, and small size of these devices have made them very attractive sensors for a variety of applications. Detection limits of 0.7 ppb-hr for UDMH and 1.6 ppb-hr for hydrazine and a dynamic range of approximately 250 ppb-hr with +/−25 percent variation in sensors are reported by the manufacturer.[15] Relative humidity has a big effect on the sensor until they are appropriately aged. Further work is necessary to improve response characteristics. Tests at KSC indicate slow response times and behavior that is inconsistent with a dose response.[16]

CONCLUSIONS

Many detection requirements are necessary for the hydrazines to protect personnel and property because the hydrazines are explosive, toxic, and carcinogenic. A variety of techniques are available, each having their strengths and weaknesses. The best methods depend on the user application. High sensitivity and selectivity are difficult to achieve in a single instrument. Research continues to explore new techniques that will provide sensitivity, selectivity, and small size at a low cost. At low concentrations, hydrazine is difficult to transport. Therefore, it would be desirable to have sensors to detect the hydrazine at the remote site rather than relying on sample lines to transport the hydrazine to a central detector. Portable and fixed-point monitors that require less maintenance are also attractive.

REFERENCES

1. American Conference of Governmental Industrial Hygienists, "7LVs—Threshold Limit Values for Chemical Substances and Physical Agents in the Workroom Environment with Intended Changes for 1981," Cincinnati, 1981.
2. American Conference of Governmental Industrial Hygienists, "1992–1993 Notice of Intended Changes," Cincinnati, 1989.
3. C. B. Mattson and R. C. Young, "Evaluation of FTIR Methods for the Second Generation Hypergolic Fuel Vapor Detection System (HVDS II)," *Proceeding of the JANNAF Safety and Environmental Protection Subcommittee Meeting,* CPIA Publication 569, July 1991, p. 191.
4. D. Lueck and D. Curran, "Development of a Portable Vapor Detector for NH, MMH, and UDMH at 10 ppb Concentration," *Research and Technology 1993 Annual Report,* NASA Technical Memorandum 109200, p. 110.
5. J. R. Wyatt, S. L. Rose, and C. M. Hawkins, "An Analytical Comparison of the GMD Autostep and the MDA 7100 Hydrazine Detectors," *NRL Ltr Rpt* 6110-134:SLR.1, April 1986.
6. National Institute for Occupational Safety and Health, "Method #149" in *NIOSH Manual of Analytical Methods* vol. III, 2d ed. DHHEW/NIOSH Pub. No. 77157c. Cincinnati, 1977.
7. P. A. Taffe and S. L. Rose, Hydrazine detection. U.S. Patent #4,900,681, February 1990.
8. K. P. Crossman, P. T. Carver, J. R. Wyatt, S. L. Rose-Pehrsson, and A. Thurow, "Laboratory Evaluation of a Colorimetric Hydrazine Dosimeter," NRL Memorandum Report 6668, 1990.
9. D. M. Blaies, N. A. Leavitt, and R. C. Young, "Development of an Interim Active Vanillin Sampler Used to Detect 10 Parts per Billion of Hydrazine and Monomethylhydrazine in Air," *Proceeding of the JANNAF Safety and Environmental Protection Subcommittee Meeting,* CPIA Publication 569, July 1991, p. 203.
10. K. P. Brenner and S. L. Rose-Pehrsson, "Development of a Dosimeter System for Unsymmetrical Dimethylhydrazine, Monomethylhydrazine, and Hydrazine," *NRL Memorandum Report,* July 1994.
11. Courtaulds Performance Films Industrial Products, P.O. Box 5068, Martinsville, Virginia 24115, (703) 629-1711. Purchased through Read Plastics, Inc., 12331 Wilkins Ave., Rockville, Maryland 20852, (301) 8817900.
12. John H. Cross, Thomas F. Limero, Steve W. Beck, John T. James, Nathalie, B. Martin, Harry T. Johnson, Rebecca C. Young, and Carl B. Mattson, "Monitoring Hydrazines in the Space Shuttle Airlock with a Proof-of-Concept Ion Mobility Spectrometer," *Proceeding of the JANNAF Safety and Environmental Protection Subcommittee Meeting,* CPIA Publication 588, August 1992, p. 501.
13. B. Meneghelli and D. Lueck, "Evaluation of the MDA/Polymetron Analyzer as a Monitor for Hydrazine Vapor Samples," *Proceeding of the JANNAF Safety and Environmental Protection Subcommittee Meeting,* CPIA Publication 600, August 1993, p. 1.
14. G. E. Collins and S. L. Rose-Pehrsson, "The Fluorescent Detection of Hydrazine, Monomethylhydrazine, and 1,1-Dimethylhydrazine by Derivatization with Aromatic Dicarboxaldehydes," *Proceeding of the JANNAF Safety and Environmental Protection Subcommittee Meeting,* CPIA Publication 600, August 1993, p. 11.
15. Spire Corporation, "A Final Report for Personal Alarm Thin-Film Conducting-Polymer Hydrazine Exposure Sensor," FR-60193, June 1993.
16. D. J. Curran, M. R. Zakin, and D. Lueck, "Evaluation of a Conductive Polymer-Based Sensor for Detection of Hydrazines at the Parts-per-Billion Level," *Proceeding of the JANNAF Safety and Environmental Protection Subcommittee Meeting,* CPIA Publication, August 1994.

CHAPTER 45
MICROFABRICATED SENSORS: TAKING BLOOD TESTING OUT OF THE LABORATORY

Microfabricated sensors have proven beneficial in the development of portable nonlaboratory diagnostic devices. While the clinical laboratory provides blood testing and other services within a hospital, costs are high. The lab is typically 15 percent of a hospitals total budget and turnaround time for "stat" testing is too long. Portable nonlaboratory testing responds to the costs and time concerns of major hospitals. Biosensor chips—capable of measuring critical "stat" blood tests such as sodium, potassium, chloride, ionized calcium, hematocrit, pH, glucose, and others—are the development that makes this portable diagnostic technology possible.

This work will discuss sensor design considerations and performance requirements for nonlaboratory blood testing systems, including the issues of simplicity, reliability of use by nonlaboratory workers, speed of results, ability to be easily worked into health-care provider's workload, and the automation of quality control and data management. The impact technology has on these issues will also be discussed.

BIOSENSOR FOR AUTOMATED IMMUNOANALYSIS

Biosensors integrate a sensing element (optic or electronic device) with a biological coating to provide novel analytical instruments. Biosensors with antibody coatings called *immunosensors* can provide a rapid, quantitative immunoanalysis. The many applications of immunosensors include general clinical diagnostics (especially point-of-care medical testing), health and safety compliance, environmental monitoring, food contamination detection, illicit substance interdiction, and biochemical defense. A specific immunosensor, the biorefractometer, performs a one-step, label-free (homogeneous) assay to detect various materials including toxins and viruses. The device measures refractive index changes at the surface of a planar waveguide using interferometry.

A prototype instrument was assembled and used to measure various analytes of interest. Staphylococcus enterotoxin B, ricin toxin, human immunoglobulin-G, and adenovirus were used as model analytes for these studies. Because binding of the biological agent antigen occurred on the surface of the antibody-coated waveguide, there was a time-dependent phase shift of the helium-neon laser light beam. The rate of change in this shift was measured over a 10-min. period. A detection limit of 50–100 ng/ml was obtained for SEB, ricin, and hIgG, while the adenovirus could be detected at 250–2500 viral particles/ml. These encouraging results indicate that the biorefractometer could provide the sensing platform for a variety of practical measurement applications.

CHAPTER 46
CLOSED-LOOP CONTROL OF FLOW RATE FOR DRY BULK SOLIDS

INTRODUCTION

Closed-loop control yields considerably better performance indices for faster response to changes and closer tracking of the desired operating variable than open-loop control. But closing the loop also demands a higher price in terms of complexity and attention to detail. If the controller gains are not well chosen, the resulting system may require a longer time to settle into steady-state behavior after changes in the desired value of the output operating variable. The system may even exhibit decidedly oscillatory behavior that actually degrades its performance and renders it unacceptable for its intended application.

Flow rate control of dry bulk solids is accomplished in many different ways, some with the use of closed-loop control and some by using open-loop control. This chapter focuses on the use of closed-loop control to achieve an actual flow rate that closely matches the desired flow rate at any point in time. Two machines that serve as the structure for implementing closed-loop flow control of a dry powdered or granulated solid are weigh belt feeders and loss-in-weight feeders. Volumetric feeders are also frequently used to achieve flow rate control, but these machines use open-loop control and will not be discussed.

STRUCTURE AND NATURE OF CLOSED-LOOP CONTROL

Figure 46.1 shows the basic elements of a general closed-loop control system. The output variable is measured by a suitable transducer, converted to a proportional electrical signal, and fed back for comparison with the input command signal that represents the desired value of the output variable. The error signal is the difference between the input command signal and the measured value of the output variable. It is fed to an amplifier where an adjustable gain is applied and the resulting gained-up signal drives the actuator that modifies the output variable. If disturbances upset the output variable, the feedback transducer senses the altered output variable, and the system makes the appropriate and necessary corrections. Thus, the output variable

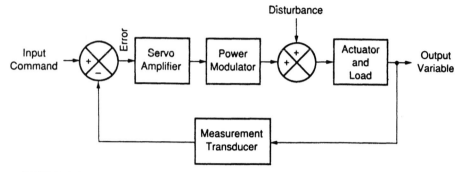

FIGURE 46.1 Feedback control system block diagram structure with disturbance input.

follows the input command in spite of disturbances that are always present in real systems. But because of the loop structure, excessively high gain applied to the error signal may cause oscillatory unstable behavior. (See Ref. 1, Chap. 23, for discussion and description of this unstable performance.)

Most closed-loop control systems have provisions for maintaining a tradeoff between a sufficiently high gain to maintain good following accuracy between the input command and the output variable without the undesirable oscillatory unstable behavior. A three-term controller is inserted between the error signal and the control actuator as shown in Fig. 46.2. This type of controller or control algorithm is also called *PID* for proportional, integral, and derivative control action. Operation on the error signal is exactly as suggested by the names of the three terms *proportional gain*, *integration* of the error signal over time, and *differentiation* of the error signal with respect to time. Each of these three processes has its own gain so that the rela-

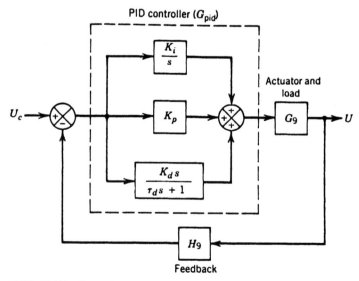

FIGURE 46.2 Generalized PID controller.

tive amount of each type of control action can be operator-selected. (See Ref. 1, Chap. 26 for a discussion of this type of controller.)

The next two sections show how the closed-loop control structure and its associated three-term controller merge with a loss-in-weight or a weigh belt feeder to yield flow rate control. References 2 and 3 describe the fundamentals of weigh belt and loss-in-weight feeders. Reference 5, Chap. 8 also describes these two types of dry solids delivery machines.

WEIGH BELT FEEDER AND ITS FLOW RATE CONTROL LOOP

Figure 46.3 shows the components of a weigh belt feeder and its flow rate control loop. The weigh bridge is essentially a pair of load cells that carry one or more rollers

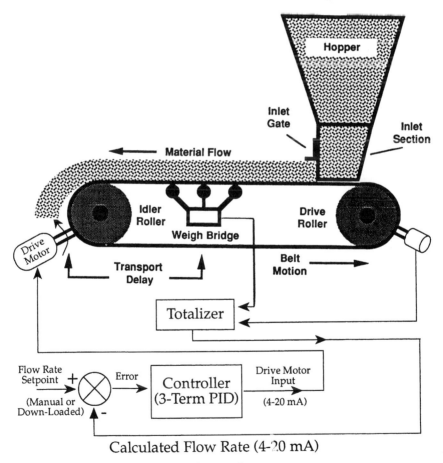

FIGURE 46.3 Closed-loop weigh belt feeder control.

or a slider pan to support a short section of belt as R transports material from the hopper to the discharge point. The weight signal generated by the load cells is fed to the totalizer for calculation of the live load per unit length of belt. It is called a totalizer because it also computes a running total of the weight of material that has passed over the weigh bridge since some starting time. Flow rate requires another measured variable—the motion of the belt itself. This is typically accomplished by a shaft encoder coupled to one of the roller shafts; a nondriving shaft if possible. The internal clock of the microprocessor totalizer and a gating circuit adds yet another measurement to the determination of flow rate: the time between pulses of the shaft encoder. From these three separate measurements, the flow rate of the material conveyed by the belt is calculated by the totalizer. Reference 2 calls the weigh belt feeder an "inferential measurement machine" because flow rate is inferred from several other measurements.

Control of the flow rate in a weigh belt feeder is accomplished by comparing the calculated flow rate with the flow rate set point and generating an error signal that is processed by the controller, typically a three-term controller as shown in Fig. 46.2. The flow rate set point may be a constant value, a preprogrammed function of time, or a signal downloaded from another manufacturing or processing portion of the plant. For example, if the flowing material in Fig. 46.3 is one of several ingredients in a proportioning system, this flow rate may need to be metered in proportion to the other ingredient flow rates. In this case, the flow rate set point would be downloaded from another machine in the proportioning network. See Refs. 4 and 5.

Modulation of the flow rate is accomplished by speeding up or slowing down the belt drive motor while keeping the belt material loading as uniform as possible. Note that there is a transport delay time between the belt load sensed at the weigh bridge and the discharge point. Therefore, a well-designed controller would build in a similar time delay between the time it makes a decision to change the belt speed and the time that it actually carries out the speed change.

LOSS-IN-WEIGHT FEEDER AND ITS FLOW RATE CONTROL LOOP

Figure 46.4 shows the components of a loss-in-weight feeder and its flow rate control loop. A loss-in-weight feeder is so named because the measured weight in the hopper continuously decreases as the feeder discharges. Measurement of the live material is implemented by mounting the entire machine on load cells (called a scale in Fig. 46.4). Just as in weigh belt feeders, flow rate is also a derived or calculated variable. But in this machine, the flow rate calculation is considerably less complicated because flow rate is simply the change in weight of the hopper divided by the increment of time over which the loss in weight is measured. Increases or decreases in flow rate are achieved by increasing or decreasing the rotational speed of the feeder screw that delivers material. Except for the method of calculating flow rate, the structure of the control loop is substantially the same as for a weigh belt feeder. A three-term controller operates on the error signal and delivers a 4–20-mA signal to the drive motor. One important difference in the control strategy of the two machines is that the loss-in-weight feeder must switch to a constant screw feed speed when the hopper runs empty and during refilling of the hopper. During the fill time, the scale obviously cannot generate a signal indicative of the outflow from the feed screw.

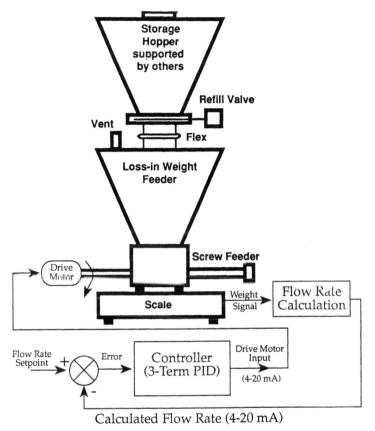

FIGURE 46.4 Closed-loop loss-in-weight feeder control.

CLOSURE

The application of closed-loop control to weigh belt feeders and to loss-in-weight feeders is both simple and complicated. Both machines must rely on a computed variable as the feedback signal. On the other hand, the control requirements for either machine are generally not particularly difficult to meet.

REFERENCES

1. C. L. Nachtigal (ed.), *Instrumentation and Control: Fundamentals and Applications,* John Wiley & Sons, New York, 1990.
2. David Wilson, "Weigh Belt Feeders: Principles, Design Strategies, and Applications," 1996.
3. David Wilson, "Loss-in-Weight Feeders: Principles, Design Strategies, and Applications," 1996.
4. C. L. Nachtigal, "Multiingredient Proportioning For Dry Solids," 1996.
5. Hendrik Colijn, *Weighing and Proportioning of Bulk Solids* 2d ed., Trans Tech Publications, Federal Republic of Germany, 1983.

CHAPTER 47
WEIGH BELT FEEDERS AND SCALES: THE GRAVIMETRIC WEIGH BELT FEEDER

OVERVIEW

Introduction

This chapter will describe the principles of operation for the standard weigh belt feeder as applied to the process market.

Why Feeders?

What Are They?

Why are feeders needed? Feeders control the flow of powders, granules, and other types of bulk solids into processes that require either a specified, constant flow or precise metering and control for blending and preparing formulations of a final product.

Feeders are either volumetric or gravimetric bulk solids metering devices. Volumetric feeders control only volumetric flow of a product. Gravimetric feeders control the mass flow rate—the rate of material addition to the process—of a product by knowing something about its bulk density.

Gravimetric feeders come in two basic types with certain derivatives. Gravimetric weigh belt feeders and gravimetric loss-in-weight feeders are the most common in processes today. These feeders use force or mass sensors to obtain information about the bulk solid being metered and use a sophisticated control system to calculate accurately mass flow and control the metering device to the exact set point. The control system must be responsive to changes in flow characteristics of the bulk solid and to changes in operating conditions. Feeder selection is one of the process requirements to determine the best volumetric feeder that will meter the bulk solid while considering such issues as installed cost, maintained cost, process flexibility, and so on.

Definitions

Variable	Description	Units
a	Cross sectional area of material pile on belt, vibratory tray or disk feeder	dM^2
k	Factor that converts belt weight into belt loading	m or ft.
MF	Mass flow rate	kg/min.
v	Belt or material velocity	m/min.
∂	Product density	kg/dM^3 or kg/li
Δ	Belt loading—the weight per length of belt of product on the belt	kg/m or lb./ft.
A/D conversion	The process of electronically converting an analog signal to a digital one that can be used directly by a computer	
belt tracker	The mechanical servo that maintains belt position on the machine	
conveyor assembly	The mechanical assembly that houses the components of the weighing system	
drive pulley	The lagged pulley that drives the belt	
gravimetric	The use of weight measurement in the operation of a belt feeder	
idler pulley	The driven pulley propelled by the weigh belt	
inlet section	The portion of the belt feeder that converts vertical flow to horizontal controlled cross section flow onto the belt	
run charts	Graphs that show trends of data over time	
SFT®	Smart Force Transducer	
side skirts	Plates or blades that retain material on the weigh belt	
transport delay	The delayed use for the new belt loading value to update belt speed until that slice of material is at the feeder discharge.	
WBF	Weigh belt feeder—gravimetric	
weigh bridge	The mechanism that contains the weight sensor and converts the belt loading into a weight measurement	

THE BASICS

This section presents some basic concepts of gravimetric weigh belt feeders and introduces the idea of the technology triangle for feeding.

Technology Triangle

One must observe the ideas presented in the concept of the feeding technology triangle to design and build a well functioning feeder. It is shown in Fig. 47.1.

The linkages on this graphic are important if one wishes to build an accurate and dependable gravimetric feeder.

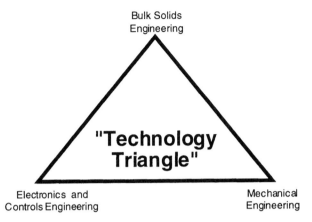

FIGURE 47.1 Technology triangle.

Since bulk solids engineering tends to be looked upon as an art by those not skilled in the field, and since feeder selection has to be made based upon how well that device can meter the bulk solid in question, this topic should be addressed first. One must understand the impact of how a particular bulk solid may affect feeder performance. For example, if the material does not flow out of the bin above the feeder, the feeder will not serve its purpose, or, more commonly, if the flow is erratic, feeder performance may be degraded from a weigh belt feeder because of insufficient material supply at the feeder inlet. The mechanical design of the feeder and supporting bins and hoppers must be of prime concern. With any feeder, knowing the flow properties of the bulk solid plays a critical role in the success of the installation.

The electronics and controls engineering portion is where many engineers focus their time, since many gravimetric feeders require sophisticated controls and may be linked into process computers for supervision. Electronics design is where the bulk of the engineering time is spent as designers continue to optimize equipment performance by the implementation of improved algorithms and interfaces.

The mechanical engineering aspects of feeder design include the gear boxes, metering devices, and, in the case of gravimetric feeders, the weighing system. The metering device must handle the bulk solid reliably and predictably. The weighing system also plays a key role. In industrial weighing systems, certain attributes must be available to the designer if a high-performance gravimetric feeder is required.

Here is a sample of some of the criteria for a feeder weighing system.

- Linear output
- Stable zero point
- No hysteresis or creep
- Adequate resolution
- Compact dimensions
- Small measuring deflection
- Suitable for installation in explosion risk zones
- Output that is compatible with the control system

One can be judged by the critical parameters for each type of feeder system. Finally, the quality of the feeder installation can affect feeder performance.

Basics of Feeding

Feeding is defined as the addition of substances to production processes according to density, flow cross section, and time. The general case is reviewed:

$$\text{mass flow} = a \cdot v \cdot \partial = \text{throughput per unit time}$$

where a = material cross section
 v = material flow velocity
 ∂ = Density

The task of the designer and constructor is to meet the requirements defined in the above equation. It is important to realize that a constant mass flow has to be formed on the basis of variable factors. Certainly the issues of accuracy—repeatability, linearity, and stability—must be addressed in the final analysis. One needs to deal with the three variables of the equation, presented here in a slightly revised order.

Product Density. The density can vary over a wide range, since raw material characteristics may fluctuate and preliminary mixtures can differ in composition. Atmospheric humidity and compression also have an effect on density. And certainly the issue of how density is measured can result in calculations that are in error of reality. Is it measured loose or packed? If packed, to how much pressure and consolidation time is the sample subjected?

Material Cross Section. Material cross section is also a parameter that cannot be taken as constant. Deposited material on the metering or measurement device, material bridging, and excessive rotational speeds can easily affect the material cross section. The angle of repose will have an influence, particularly when the material is unconstrained.

Product Flow Velocity. In order to ensure constant mass flow, one of the two factors a or v must be continuously adjusted. The easiest parameter to adjust is the velocity v; in other words, the mass flow is controlled by the speed adjustment. Many early feeding devices varied a because low-cost, high-performance motion control systems were not available. Pneumatic or other mechanical servo systems controlled the height of a gate (for example, to set the value of a), while the motion control system maintained a fixed speed. Current designs establish a belt velocity to set mass flow rate based upon a fixed opening onto the weigh belt.

Controlling Mass Flow

In the following example, the weigh cell measures the weight of material passing over it. The control system based upon that information controls the speed of the belt conveyor to maintain a given set point of mass flow. The speed sensor merely provides closed-loop feedback for the motion control system (see Fig. 47.2).

This model can be transposed for use with vibratory tray feeders, screw feeders, rotary valve feeders, and others. One merely restates the formula for the particular

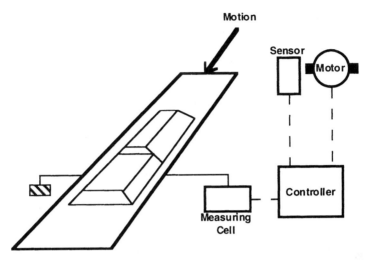

FIGURE 47.2 Vibrating transport.

device. In most cases, velocity becomes RPM of a feed screw, the a value becomes dM^3/rev.

PRINCIPLES OF WEIGH BELT FEEDER OPERATION

Basic Function of the Weigh Belt Feeder

Weigh-belt feeders are relatively simple and reliable gravimetric feeders that provide high feeding precision (+0.50–1.0 percent) as well as efficient process monitoring particularly for high flow rates. They are used in both continuous and batch operations. Figure 47.3 shows a weigh belt feeder in operation.

The gravimetric weigh belt feeder is designed to fit under a bin or silo and extract material on a continuous basis. Flow is controlled by the feeder so that the downstream process has a regulated input of bulk solid. Using a speed-controlled weigh belt feeder, the product is delivered in a continuous flow onto a conveyor belt, through an inlet slide gate or automatic prefeeder (see Fig. 47.4).

A load sensor continuously measures the weight of the product over a defined length of belt. Belt loading is defined as the weight of material on a unit length of belt (kg/meter); belt speed is the linear speed of the belt (meter/min.). These values are converted into material velocity and belt loading as described below.

$$MF \text{ (kg/min.)} = \text{material speed (BS)} \cdot \text{belt loading (BL)}$$

where BS = Velocity of material in meters per minute = v
BL = Belt loading in kg/m $10 \cdot \partial \cdot a = \Delta$

If MF = SP (set point), with substitution then BS = SP/BL if BS is the controlled variable (see Fig. 47.5).

The weigh belt feeder is an inferential measuring and controlling device. That means that the measurement of belt loading and material velocity are not made

FIGURE 47.3 Gravimetric weigh belt under silo.

directly but are inferred from other measurements. It is evident from the equation that one needs to measure material velocity and material loading. This is where problems can arise, since it is customary to measure material weight and belt velocity or motor speed as an indication of belt velocity.

The transport delay function holds off the specific use of the measured value of belt loading until that slice of material is nearly ready to fall off the weigh belt. This delay ensures that the belt speed change correctly reflects the value of belt loading and thus that of the mass flow that is immediately discharging into the process.

Mechanical Design Strategies for Weigh Belt Feeders

Mechanical Systems. The mechanics of the weigh belt feeder must focus on converting the measurements made of the process into those of belt loading and material velocity. The mechanical systems, including the weighing system, must be of such design, that they must maximize the potential accurately to achieve the conversion to these values. From an earlier discussion, one may remember that the feeder sensors are belt velocity and material weight. These must be converted to material velocity and belt loading. To ensure the best possible conversion, the belt must be precisely positioned, velocity-controlled, tensioned, and cleaned of waste material.

Inlet Section. The inlet section is the device that shapes the material cross section onto the weigh belt from the storage bin above. The material flow changes direction from vertical to horizontal, and, in some cases, the material velocity increases. The material is sheared through the inlet gate, and the pile shape is defined. An inlet sec-

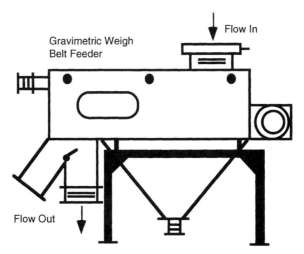

FIGURE 47.4 Load sensor for continuous measurement of weight on a belt.

tion has a fixed width and an adjustable height. Different configurations of inlet gates may be used for materials of differing flow characteristics.

The inlet section is used to establish a desired value of belt loading. Lowering the gate height will lower the belt loading and increase the belt speed. Raising the gate height will increase the material opening, increasing belt loading.

An example of material flow from an inlet section and the discharge pattern onto the weigh belt is shown in Fig. 47.6.

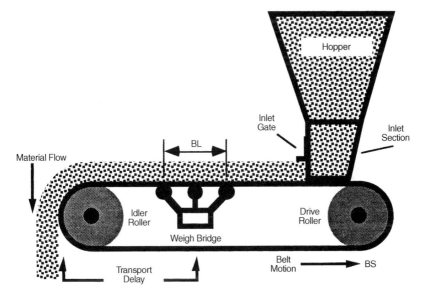

FIGURE 47.5 Weigh bridge.

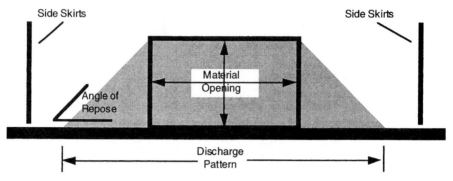

FIGURE 47.6 Discharge pattern.

The bulk material, the inlet section configuration, and the belt velocity will affect the pattern of material flow onto the belt. You want a pattern as shown above with no voids in the material stream.

Gear Reduction Assembly. The gear reduction is used to reduce the maximum motor speed to the maximum required belt speed. The gear reduction often consists of an inline gear reducer as well as a chain drive with pulley reduction (Fig. 47.7).

The total reduction can be calculated as follows:

$$\text{reduction} = \text{inline reduction} \cdot \left(\frac{\text{pulley teeth}}{\text{reducer teeth}} \right)$$

The overall drive system looks like Fig. 47.7. Therefore

$$\text{belt velocity} = (\text{motor RPM/reduction}) \cdot (\text{drive pulley circumference})$$

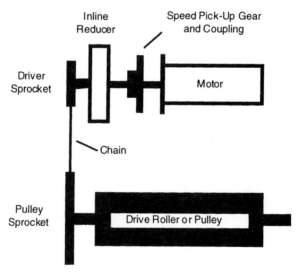

FIGURE 47.7 Driving system.

Belt velocity is normally taken from an encoder on the idler roller or by an independent encoder that uses a wheel to rotate against the belt surface. Others use motor speed as the indication of belt velocity and hope that no belt slip occurs. In the above example, the speed pickup provides velocity feedback for the motor control and can also provide data for belt velocity.

Weigh Belt. The weigh belt is a critical part of the feeder and is often overlooked in the design and later in the operational phase of the feeder. The weigh belt characteristics are somewhat environmentally dependent.

Each weigh belt will have a different tare value, and therefore the belt feeder must be tared after a belt has been replaced. Belts also stretch during warmup, so it is recommended that a belt feeder be run for several minutes with a new belt before taring (Fig. 47.8).

Some weigh belts are precision weigh belts. To replace them with ordinary belts compromises the weighing accuracy of the system. Careful attention to achieving constant weight profiles and proper tracking are key concerns to design engineers (Fig. 47.9).

Remember that the weight measurement taken by the feeder includes the belt weight in that measurement. If a large variation of weight caused by the belt were to occur, significant fluctuation of mass flow may occur. Some vendors have gone so far as to design a system that memorizes weight values of the belt in 60–200 increments that are later subtracted from the overall weight measurement during normal operation to improve performance.

Conveyor Assembly. The conveyor assembly is the critical portion of the weigh belt feeder. This unit contains the weigh bridge, the belt cleaning systems, the belt tracker, and the tensioner (see Fig. 47.10).

1. Drive roller
2. Idler roller
3. Weigh belt
4. Conveyor frame
5. Inlet stationary platform
6. Weigh bridge assembly
7. Belt scraper
8. Tracker assembly
9. Cleanout screw assembly
10. Counterweight for automatic belt tension

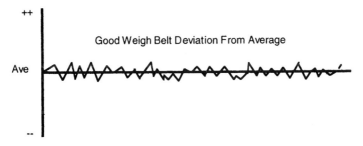

FIGURE 47.8 Belt deviation—good weigh.

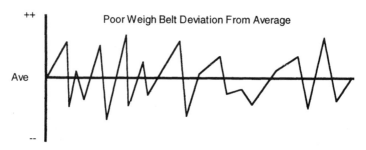

FIGURE 47.9 Belt deviation—poor weigh.

Weigh Bridge Assembly (6). A typical weigh bridge utilizes a three-bar system with two bars fixed, one at the inlet side of the conveyor and the other at the exit. The middle bar or roller for large feeders is moveable in the vertical plane but fixed in the other planes by flexures. As the material on the belt moves across the plane of the weigh bridge, the middle or load bar senses the weight and provides an output proportional to the weight per unit length of material on the belt.

In fact, the amount of weight it measures is equal to the quantity of material from either fixed bar to the center bar if the geometry is symmetrical. While the sensor output is weight, the control system has to convert that value to kg/M or lb./ft., values called *belt loading.* The conversion factor k is the distance from the middle bar to the end or fixed bar or roller.

$$BL = \frac{kg}{k}$$

The weigh bridge is a precision device and should be handled as such. Do not lay heavy items on it, walk on it, or connect welding wires to it in any way. Shock can damage the unit.

It has been mentioned that either bars or rollers can be used on the weigh bridge assembly. It merely depends upon the application. With high belt speeds of more than 30 M/min. and belt loads exceeding 50 kg/M, vendors use rollers that are precision manufactured to reduce the frictional or wear component. For lower capacities, hardened bars are adequate (see Fig. 47.11).

Tracking Assembly (8). The tracking unit is designed as a mechanical servo that monitors the position of the weigh belt and provides a corrective action when the belt moves too far in either lateral direction.

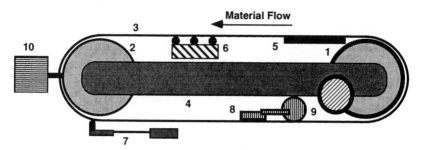

FIGURE 47.10 Conveyor assembly.

THE GRAVIMETRIC WEIGH BELT FEEDER

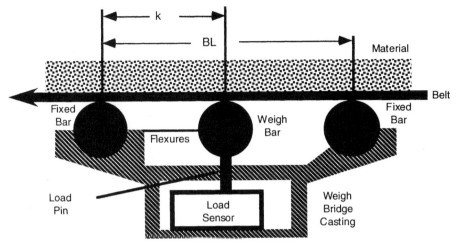

FIGURE 47.11 Precision weigh bridge.

1. Pivot
2. Rollers
3. Tracker frame
4. Inside belt scraper
5. Belt guide fingers

For the tracker to function properly, it must be free to rotate in its bearing in the vertical support pivot (1), and the individual wheels (2) that contact the belt must be free to rotate. Also, the belt fingers (5) that are secured around the edge of the belt must be in place.

As the belt moves out, the unit pivots in bearing (1), the arm with the fingers (5) moves out also, the outside rollers increasing tension on the belt edge on the same side the belt is moving to, relieving tension on the opposite side of the belt. This force causes the belt to move into place. The scraper (4) cleans the inside of the belt (see Fig. 47.12).

Belt Tensioning System (10). Belt tension is also critical when a weighing system must weigh through the belt. In other words, when the weight sensor is under the belt surface, it is important to control the catenary effect precisely of the belt so that the zero value is stable. The parallelogram system used on some conveyor systems guarantees that the belt tension will be stable over the normal length variations and stretching of the belt. The tension system also assures proper traction of the belt with the polyurethane lagged drive roller to minimize belt slip another cause of error. A counterweight provides the belt tension. This weight can be locked in the up position during belt replacement (Fig. 47.13).

1. Counterweight
2. Idler roller
3. Bearing housing
4. Lower bearing pivot-fixed

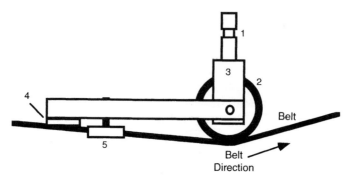

FIGURE 47.12 Belt scraper—cleaning belt.

5. Cross link
6. Arm weldment
7. Lower pivot—fixed
8. Counterweight arm

The belt provides the restoring force working against the counterweight unit (1) that pivots the roller at point (4). The bearing support (3) rotates to the left around pivot point (4) as the belt stretches. The counterweight arm (8) pivots on bolt (7) with arm (6) that is welded to the counterweight arm (8) causes the rod (5) to push against the bearing support (3), moving the pulley (2) to the left tightening the belt (see Fig. 47.13).

Belt Cleaning System (7,9). Belt cleaning is necessary to minimize tare changes to the belt and to ensure minimum belt slip and tracking errors.

1. Scraper
2. Trough weldment
3. Plastic screw—powered by gear attached to drive roller

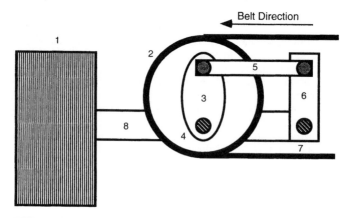

FIGURE 47.13 Counterweight.

4. Inlet stationary platform
5. Rubber-coated drive roller
6. Inlet section with material

Four scrapers condition the belt. The leading edge of the inlet stationary platform (4) scrapes the inside of the belt and drops the material into the cleanout channel (2) for removal by the drive-pulley-driven cleanout auger (3). Debris is removed by the auger (see Fig. 47.14).

Additionally, a tracker assembly scraper cleans the inside of the belt. At the discharge end of the feeder, a plastic counterweighted removable scraper cleans the outside of the belt, and the scraped material falls to the discharge. A clean feeder is truly necessary for accurate weighing and control.

Side Skirts for Material Control. It is also appropriate to mention another adjustment that is critical for proper weighing operation. Setting the side skirts is critical for success. They must contain the material but not restrict the weighing system. Do not set them to pinch the weigh bar or cut the belt. Incline the skirts in the direction of product flow. This assures the operator that material will not become pinched and cause a false weight reading. For powders, a slight incline is all that is required. Containing the product on the belt is an important part of housekeeping. Also reducing belt speed to control dust generation will also aid the housekeeping effort (Fig. 47.15).

Sensors and Controls

The Basic Control Mechanism. Figure 47.16 outlines a typical cascade control loop used with weigh belt feeder control. The outer loop computes mass flow from

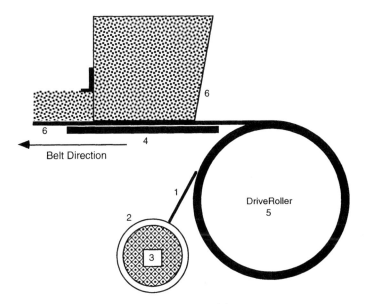

FIGURE 47.14 Belt cleaning for accurate weigh.

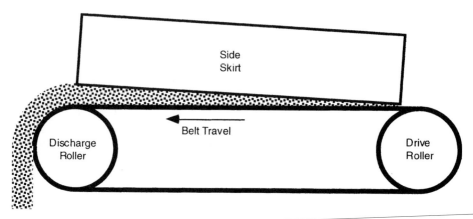

FIGURE 47.15 Side skirt—material control.

the variables of weight and belt velocity (or motor speed). This control system output provides the operating set point for the motion control system. The process variable for the motion system is motor velocity and is measured directly at the motor shaft. Motor velocity is controlled to maintain the desired set point of mass flow. As with any gravimetric feeder, any variation in the density of the material is reflected as a change in belt loading, which is compensated for by adjusting the belt speed (Fig. 47.16).

Sensors. Since feeder controllers are generally driven by microcomputer, sensors are preferred if they have a digital output compatible with the control input. This eliminates the need for A/D conversion.

Two primary sensors are used in a weigh belt feeder. One is the load sensor, sometimes located under the belt in a weigh bridge assembly. The weigh bridge supports a volume of bulk solid, and the load sensor measures the supported weight. Critical parameters for the load sensor include:

- Low deflection
- Stable zero point
- No hysteresis or creep
- Resolution of at least 10,000 parts
- Linear over its operating range—0.05 percent of reading or better
- Rugged design
- Ability to work in Class 1 and Class 2 hazardous areas

The weight sensor may be used individually in some types of weigh bridges or in pairs that eliminate the need for a complex flexure system when wide weigh belts are employed.

Many device types are used. Most designers use a strain gage device that normally requires signal amplification and analog-to-digital value conversion to be used by a digital controller. K-Tron uses a vibrating wire sensor called an SFT, a digital force measurement device, that is described at the end of this chapter.

The other sensor required to compute mass flow is the material velocity sensor. Since we cannot accurately measure material velocity, belt velocity is the next best

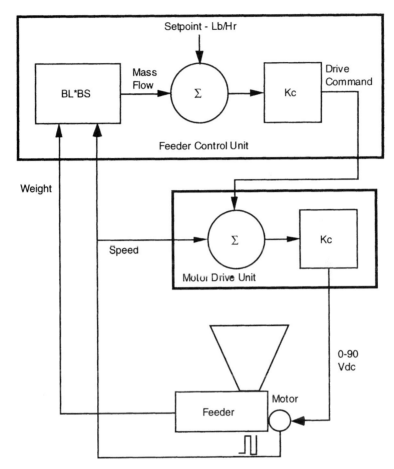

FIGURE 47.16 Weigh belt control mechanism.

bet. However, assumptions are made. What are they? It should be noted here that some vendors also use motor speed to indicate belt speed. Without care, problems can arise. Figure 47.17 shows a belt velocity sensor.

Most velocity sensors are zero-speed digital sensors, either magnetic or optical, that supply a pulse stream to the controller, the rate of which is a function of belt velocity obtained by converting rotation of the sensor wheel against the belt or idler roller. The same type of sensor can be applied to measure motor velocity if required. Some may still use analog tachometers, but these are less desirable.

Since the motor that drives the belt must have velocity feedback, a speed pickup is used. A similar one can be used as a belt velocity sensor.

The typical motor speed pickup may generate a square wave with a frequency proportional to the motor speed. The proportionality constant GT represents the number of teeth on the speed pick up gear. The speed pick up is used to calculate the speed, in RPM, of the feeder drive motor.

Motor speed is calculated as

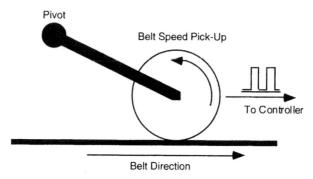

FIGURE 47.17 Velocity control sensor.

$$\text{motor (RPM)} = \frac{\text{speed pickup frequency} \cdot 60}{GT}$$

The feedback signal is returned to the motor drive unit and matched with the correction signal. Any deviation is immediately corrected (Fig. 47.18).

Velocity control must be stable, with a fixed correction signal and quick response to a changing correction signal. If the output of the feeder controller is stable—K-Tron calls it drive command or the correction signal, as shown in Fig. 38.18—motor speed must be stable. Variation of motor speed, when the correction signal is fixed, results in variation to flow that the weigh belt feeder control system cannot easily correct.

SFT—The Smart Force Transducer. The sensors that are used in the flexure, knife-edge, and suspension scales can be of many types. However, K-Tron uses the SFT, the smart force transducer.

The SFT is a vibrating wire force transducer of high accuracy. The tremendous advantage is that this device is a fully calibrated unit with the weight processing microelectronics located internally. The output is a computer data stream that is compatible with feeder controller electronics. Ranges are from 6 kg to 1000 kg. See Fig. 47.19.

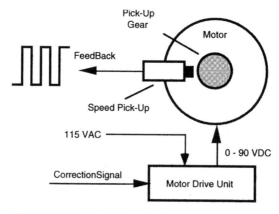

FIGURE 47.18 Feedback mechanism.

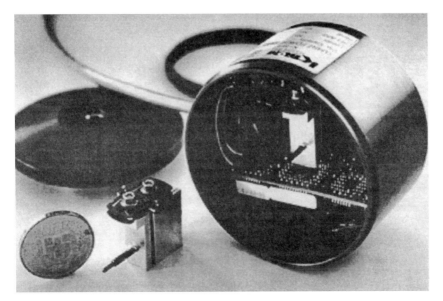

FIGURE 47.19 Force transducer

Employed in all K-Tron continuous gravimetric feeders and blenders, the SFT Smart Force Transducer provides accurate, stable, and reliable digital weighing under a broad range of operating conditions. Three SFTs may be used per feeder to provide the weight information to the controller. The controller sums the results of one or two SFTs to produce the total weight information used in the control of the feeding device. If a stabilized weigh bridge is employed, the SFT can be used singularly.

Features. The following are some features of the SFT.

- Direct digital weighing
- Onboard microprocessor
- Excellent repeatability and stability
- High resolution
- Ultra low deflection
- High tolerance to vibration and electrical noise
- Electronic linearization
- Electronic temperature compensation
- Stiff, rugged construction
- No dampers required
- Serial channel interface (RS-422/485)

Principle of Operation. The SFT exploits the dependency of a vibrating wire's resonant frequency on its tension to measure applied loads. Through mechanical means, the applied load is transmitted to the wire, causing a change in its resonant frequency from which the value of the applied load is computed by the onboard microprocessor. A fully calibrated (linearized, spanned, and temperature-compensated) signal is transmitted via RS-422/485 serial communication to the K-Tron feeder controller. Up to twelve SFTs can be handled on the same serial bus, transferring weight data with a 2400 to 38,400 baud range, but for weigh belt feeders, normally only one, maybe two, are used.

Employing the principle of induction, the wire is electronically excited to vibrate at its resonant frequency. A magnetic field, provided through the use of permanent magnets, is created across the wire's plane of vibration. (The wire possesses a rectangular cross section to constrain vibration to a single plane.) The wire is electrically conductive and is connected at its ends to an exciter circuit. Through special feedback techniques, the exciter circuit simultaneously senses the wire's frequency and produces an alternating current in the wire to maintain resonant vibration, regardless of its frequency value. Because a high-vibration frequency is desirable to avoid signal contamination, the wire is prestressed and forced to vibrate at its second harmonic (two antinodes) at all times.

The formula for string vibration is an approximation of the following:

$$\text{Frequency (Hz)} = 0.5 \times [T/M \times L]^{1/2} \text{ where}$$

where T = tension in the string
M = mass of the string
L = length of the string
T = mass of applied weight · (gravity) or where T = weight

The wire's resonant frequency (Fig. 47.20) is not a linear function of applied tension but, in practice, is almost parabolic. Its representation is shown as F_w. However, over the range of loads, the curve is straight to within 2–3 percent. To calculate the applied load from the sensed frequency, microprocessor-based regression techniques are used. Typically, the range of frequencies from no load to full load is from 11 kHz to 18 kHz. While the range is fixed, the zero point is not but is taken into account in the conversion. However, rather than computing the weight value from F_w directly, a high frequency is counted for a number of integral periods of F_w. This generates a representation of F_w in a high-precision digital form that can be counted in a short period. This allows the calculation of 1-ppm weight values with certainty. A four-step regression is calculated in order to straighten the curve. To assure the utmost precision, linearization coefficients for each SFT are individually established at the factory and reside in an E²PROM that is mounted internally to the SFT.

Of the possible environmental sources of error, only temperature variations require compensation. The effects of ambient magnetic field, changes in atmospheric pressure, relative humidity, and so on, have been found to be negligible. To adjust for changes in ambient temperature, a thermal sensor is installed within the SFT. The thermal response characteristics of each are determined through testing at manufacture, and appropriate coefficients are registered in the E²PROM for continuous electronic temperature compensation.

The temperature detector located inside the SFT provides a frequency output of about 6 kHz for ambient temperature, and its output is based upon a formula calculated in kelvins. This internal frequency is shown as F_t and is used by the microcom-

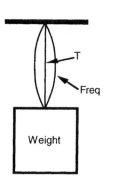

FIGURE 47.20 Vibrator effects.

puter to correct the weight value for temperature effects. A graphic shows the internal electronic and physical components (Fig. 47.21).

Startup Operation. When the SFT is powered up, it must be able to communicate once the serial channel with the other SFTs that are connected. Very much like a telephone party line, the controller that, as the buss-master, controls the communication activity on the weight channel must also distinguish between SFTs. It does this by assignment of an address that acts like a unique telephone number in the party line analogy. Once the assignment has been made, the data (address) is stored in the onboard E^{2}PROM along with the linearization data that was computed in the factory calibration procedure. The linearization activity is not alterable by the user (Fig. 47.22).

The communication hardware protocol is RS-485 and is a four-wire transmission scheme with differential receive and transmit lines connected as shown to the controller weight channel. One merely connects all SFT receive signals in parallel, sign to sign, and the same is done for the transmit signals. The result is connected directly to the controller. The communication protocol is 9-bit serial binary and is decoded by the controller as absolute weight from each SFT. The ninth bit is used to identify the following character as an address. This type of identification is particularly efficient. No addresses may be the same (Fig. 47.23).

Electrical and Physical Specifications. A list of specifications follows:

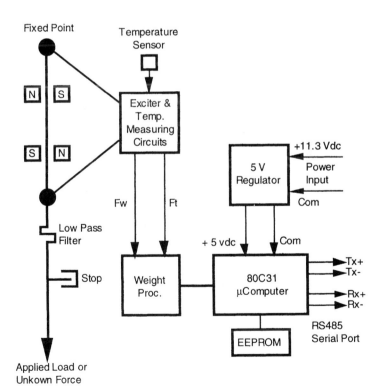

FIGURE 47.21 Temperature and weigh sensor.

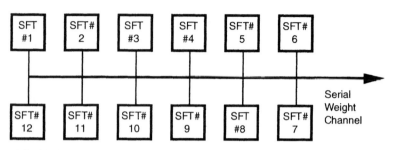

FIGURE 47.22 Special weigh channel.

Resolution	1,000,000 parts
Combined Error	≤±0.03%
Repeatability	≤0.001% (std. deviation of 30 measurements, 2 sec)
Creep	0.02% of full scale (20 min at 72°F [22°C])
Hystersis	0.02% max
Measurement interval	150 msec to 4500 msec
SFT ranges	6 kg to 1000 kg
Operating temperature range	14°F to 140°F (−10°C to +60°C)
Storage temperature range	−13°F to 176°F (−25°C to +80°C)
Compensated temperature sensitivity	Span 0.002%/kelvin Zero 0.003%/kelvin
Shock tolerance	75G in all directions (without recalibration); 100G in all directions (with recalibration)
Signal transmission type and distance	differential; 1,640 ft. (500 m) maximum
Warmup time	30 minutes
Output	RS422/485 full duplex
Power consumption	0.5 W @ 9 vdc
Voltage requirements	9 vdc ± 1 vdc
Enclosure rating	NEMA 4, IP65

APPLICATIONS OF WEIGH BELT FEEDERS

Introduction

This section will present some typical applications of weigh belt feeders to the food, plastics, and chemical processing markets.

Food Processing. One client uses weigh belt feeders to deliver flows of up to 25,000 kg/hr. of both sugar and flour into a cake mix process. While these powders

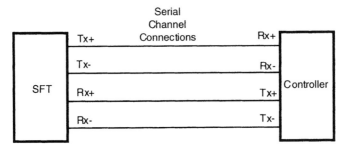

FIGURE 47.23 Serial channel connectors.

and fine granular materials might best be handled by loss-in-weight equipment, the cost of that equipment might be 2–3 times the cost of an equivalent weigh belt feeder. These folks are used to dealing with weigh belt feeders, and they clean their equipment on a periodic basis to minimize feeding errors.

Plastics Processing. Many plastics producers use weigh belt feeders to meter granular materials like plastic pellets into an extruder to make a final compound. The accuracy of the weigh belt feeder is high, and the product is clean, making the belt feeder a good choice for this application. Resin addition may approach 5000 kg/hr. Accuracy of flow is better than 1 percent. A typical application uses a weigh belt feeder to meter 500–1000 kg/hr of PVC pellets into an extruder for cable jacket extrusion.

Chemical Processing. Chemical processors use weigh belt feeders when high flow rates are encountered. In the production of gypsum wallboard, weigh belt feeders meter high rates of calcined gypsum powder exceeding 25,000 kg/hr. that is added to the mixer for the wallboard slurry. Handling high flow rates with low headroom is a benefit this customer needed.

Weigh belt feeders, because of their method of computing mass flow, are inherently unpredictable. One way to minimize problems is to provide periodic cleaning and retare. I recommend that users maintain run charts of weigh belt calibration values—tare and span—to determine the minimum time between cleanings and calibration. Additionally, the use of these charts can help predict mechanical problems in the feeder weighing system. With good preventative maintenance programs, gravimetric weigh belt feeders can maintain accuracies of better than 1 percent and deal with flow rates from 22 kg/hr. to 1200 ton/hr. with a turndown of 20:1.

A prefeeder is preferable to an inlet gate when discharging materials such as powders that flood or bridge from the inlet section onto the weigh belt. The use of single or twin screws, a rotary valve, or a vibratory tray gives the prefeeder extra metering control.

The prefeeder discharges volumetrically. Material extracted from the feed hopper is deposited in a smooth and even ribbon onto the belt just prior to the weighing section of the feeder. Gravimetric control comes into play once the material is on the weigh belt.

There must be a direct correlation between the rate at which the prefeeder discharges and the speed of the weigh belt. By slaving the prefeeder to the desired flow rate via the feeder controller, the volumetric discharge of the prefeeder is increased or decreased in proportion to changing belt speed. Since the belt speed is continuously adjusted during gravimetric feeding, this ensures the prefeeder does not over- or underfeed materials.

Weigh Belt Feeder Calibration Issues

The Calibration Process. The first issue of weigh belt feeder setup is the tare calibration. This procedure establishes the zero point for the operation of the feeder. If the weight of the belt was exactly constant, the material weight seen by the load sensor after calibration would only be that of the material. Once the tare is completed, material is added to the feeder, catch samples are produced and checked against set point, and, if any deviation of the average of the samples is found, a span change is made.

How often is the calibration process run? It depends upon a number of factors. Feeder cleanliness, required accuracy, types of materials being fed, and so forth are key issues. The use of run charts can help determine calibration frequency.

Run Charts. Run charts aid in the feeder calibration or other instrument calibration, and, over a number of calibration periods, it can give you much information about the equipment. First, the data will give capability. Capability is how much variance is expected as normal in the instrument. Also, changes either in the positive or negative directions can give evidence of a problem that can be addressed—not covered up—with a new calibration. Trending the data can also provide you with recommended intervals for calibration. When the trends of zero and span begin to shift outside the upper and lower control limits, it is time to calibrate. To calibrate before that is likely a waste of money and time. Figure 47.24 shows what to expect for calibration.

Zero Run Chart. Every time a belt tare is done, the weight value of the tare should be recorded on this chart with the date of check and the feeder ID number. The actual value should be recorded above the data point.

Over time, the chart will reveal the natural variation of the process and the inherent stability of the feeder.

Span Run Chart. Every a span check is done, the span value of the feeder should be recorded on this chart with the date of check and the feeder ID number. The actual value should also be recorded above the data point.

Over time, the chart will reveal the natural variation of the process and the inherent stability of the feeder. In Fig. 47.25, the span is quite stable over time. The dates of calibration are not filled in, but the time interval between span checks was 30 days.

Run Chart for Span—Belt Scales Overview. There is a special adaptation of weigh belt feeders called *belt scales*. These devices use many of the components of the weigh belt feeder but use no motor control. They are a sensor of mass flow rate rather than a controller of mass flow. We treat these more fully in the next section.

Basics of Belt Scales

Belt Scale Basics. Conveyor belt scales fit into all types of bulk material handling facilities. They are a valuable tool for inventory and process control activities.

Their benefits are:

- Require little space
- Don't interrupt material flow
- Are inexpensive
- Generally rugged in construction

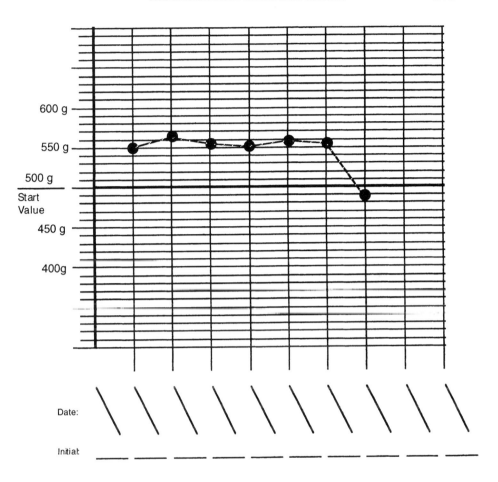

FIGURE 47.24 Run chart for zero.

The advantage of belt scales is to be able to weigh high tonnage of materials without material flow stoppage. However, the challenge of producing a high-quality belt scale is not trivial. For example, weighing 5000 kg/min. on a belt scale may mean that only 50 kg is on the scale mechanism. For 0.25 percent accuracy, the sensitivity or accuracy of the weigh system must be better than 0.1 kg. The qualities of a good belt scale are as follows:

- High weight sensitivity
- High weight resolution
- Fast response time

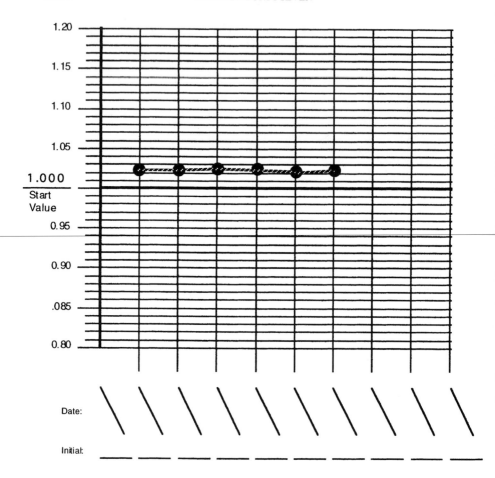

FIGURE 47.25 Run chart for span.

- Low hysteresis
- Good linearity of the weighing mechanism
- High stability
- Stable zero point
- Low deflection of the weigh system
- Good alignment
- Rugged design
- Easy servicing

While the belt scale must meet this difficult group of requirements, it is a valuable tool for process and inventory measurement of high tonnage of bulk solids.

Belt Scale Concepts. Belt scales consist of one or more idlers mounted to a weigh bridge where the weight sensing element may be electronic, mechanical, or hydraulically/pneumatically operated. This signal is multiplied with a signal that represents belt speed to give mass flow as:

$$\text{mass flow} = \text{belt loading} \times \text{belt speed}$$

This is as we have seen with our weigh belt feeder.

Belt loading is converted from the weigh idler signal by the geometry of the weigh bridge. The belt speed measurement is generated by a mechanical system in contact with the belt surface. The product of the two signals is integrated over time to produce the totalized value of weight delivered over the scale.

Colijn's Cardinal Rule for Conveyor Weighing. "For accurate conveyor weighing, conveyors should be as short as practical and loaded uniformly to near capacity."

Basics of Scale Suspensions. The scale suspension is the most critical design element in the belt scale system. This system must resolve only forces normal to the load and not any forces lateral to the load. That means that force vectors created by the moving belt must not be recognized by the weigh system.

The design criteria of the belt scale requires:

- Rigid, minimal deflection of the weigh unit
- High torsional stability
- Immunity to lateral forces
- Immunity to effects of eccentric belt loading
- Simple alignment procedure
- Minimal tare load on the sensor
- Maximum belt load portion on the sensor for higher accuracy
- Minimal horizontal surface for dirt collection and zero shifts
- Easy installation
- Frictionless pivots for all weigh parts
- Ability to accept high overloads without a calibration shift

Differing Belt Scale Weigh Bridges. An example of a type of weigh bridge system is shown by the single idler system in Fig. 47.26. The load cell is connected by a lever to the idler, which is in contact with the belt. The measured weight is converted to belt loading by the relationship to distance of the idler to the fixed idlers. The quality of the pivot is critical for continuous high-performance weighing. Both flexures and knife-edges are used. Flexure pivots are probably better, since they are not affected as much by dirt. This approach to weigh bridge design is inexpensive and works well with lower belt speeds and lower capacities (Fig. 47.26).

The next style is the approach-retreat system with multiple idlers on the weigh section. This design is used for higher loadings and for higher belt speeds. This design is well suited for long spans and multiple idler setup. As with the single-idler, pivot-arm system, the fulcrum is not inline with the belt, and thus some moment will be coupled vertically and cause weighing errors (Fig. 47.27).

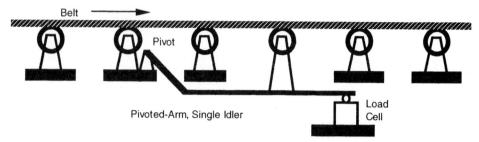

FIGURE 47.26 Simple idler system.

The fully suspended system uses multiple load cells or a combination of load cells and suspension to produce a scale system that is not affected by horizontal moments that couple forces in the vertical direction.

A closeup look at the single-idler, pivot arm system gives an idea of how to compensate for loadings of tare by the use of a counterweight. Additionally, electronic tare can handle the tare variations, thus leaving the bulk of load cell capacity for weighing material. Also notice that the pivot is close to the plane of the weigh belt. This reduces the vertical moment into the load cell that be caused by horizontal forces around the pivot point.

Another type of system uses a torque tube arrangement to transmit and reduce the forces applied to the idler roller weigh system to the measurement transducer. In the following case, the weigh bridge assembly causes a twisting movement to be applied to the supported tubes, and the summed forces are coupled to the lever system, and the force reduction unit is supported by knife edges or other frictionless pivots (Fig. 47.28).

Sensors and Controls for Belt Scales

Load Sensors for Belt Scales. The same weighing devices that work in smaller feeder systems can be made to work in a belt scale. The weigh system must be rugged and stable. The unit must be fitted into the weigh bridge with overload protection that works. Strain gages are generally the devices of choice for many belt scale applications since they are electronic, compact, and stable with low deflection, and they can have ranges to many hundreds of pounds or kilograms per device. As long as overload protection is provided, they are a good choice. Also, vibrating wire technology and force/distance devices using LVDT sensors are also useful. Additionally, the SFT is now also being used for high-accuracy applications.

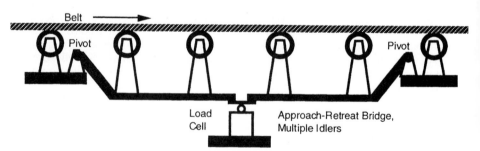

FIGURE 47.27 Multiple idler system.

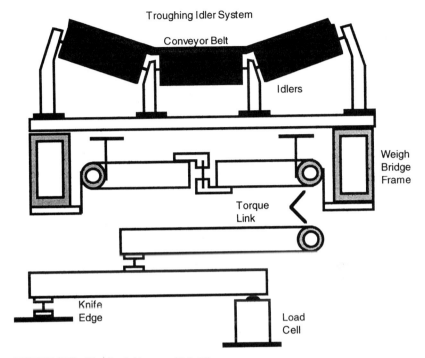

FIGURE 47.28 Floating bridge—multiple idlers.

Belt Speed Measurement. A belt speed sensor can take many forms. In older designs, AC or DC tach generators gave an output voltage that was directly proportional to belt speed. Newer styles incorporate digital approaches to provide higher accuracy and a design that works directly with a digital integrator. At least two digital types exist. First is the ferrous tooth wheel used in conjunction with a coil around a magnet. This device provides pulse output as long as the wheel has sufficient speed to provide adequate flux change. If a sixty-tooth gear is use and the gear is turning at 500 rpm, the output frequency would be 500 Hz. This frequency output is linear with gear speed. The output voltage, however, varies with the rate of gear tooth passage. As gear tooth speed increases, so does the voltage. While this type of transducer cannot resolve zero rpm because of the nonexistent flux, for normal belt scale applications, this is not a problem because of the nearly constant speed of the conveyor belt (Fig. 47.29).

Other types of velocity feedback devices exist. One style uses magnetoresistors in a Wheatstone bridge. The excited bridge drives a comparator that switches state when a gear tooth is sensed. This device gives an output that is either 5 vdc or 0 vdc based upon the absence or presence of the ferrous metal gear tooth. Both devices are hermetically sealed in a metal or plastic container with provisions for locating and locking its position in relation to the pickup gear. Typically, the inductance type pickup is used with gears of fewer than 100 teeth. The magnetoresistor pickup can handle gears of 60 pitch, and we find that 240 teeth per revolution are normal.

Another type of pickup used in this application is the optical encoder. The optical encoder works by the shining of a lamp or solid-state LED through a grating on

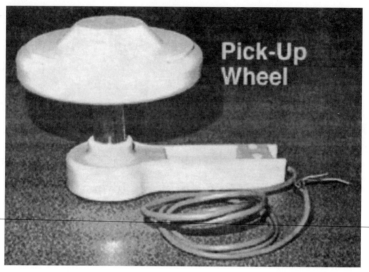

FIGURE 47.29 Belt speed measurement.

a glass or plastic wheel or by shining the light through a metal wheel with small slits. As the light passes the slits or grates, it is detected by a photocell or phototransistor. Its output is amplified and shaped to create a stable square wave output signal generally at a 5- or 12-vdc level. High resolution is possible. More than 2000 pulses per revolution are normal.

The challenge of using any speed-measuring device is to contact the belt surface properly so that linear belt speed may be accurately translated into a frequency from the pickup unit. Protection of the device from damage caused by falling debris is critical. The pickup wheel or linear revolving mechanism must be kept clean and in contact with the belt at all times.

The Belt Scale Integrator. The integrator is the device that intercepts the load and belt velocity signals and converts the information to mass flow and integrates it into material delivered.

While in older systems, the integrator might have been mechanical, this discussion will focus on electrical devices. Some of the criteria for an integrator are:

- Negative integration function
- High resolution for calibration
- Noninteracting span and zero settings
- Automatic tare
- Optional low belt load deadband—adjustable
- Simple controls and display
- Integration into computer data collection systems
- Belt speed compensation
- High stability and accuracy over wide temperate and humidity extremes
- High MTBF
- Low MTTR

While analog systems are still used, most systems are now digital, at least in the computing portion. When analog load cells are used as measuring devices in the weigh bridge, the amplification and the conversion to a digital value is critical. A couple of methods exist. If a microcomputer is used, the analog signal of belt loading can be converted by a 12- or 16-bit A/D converter and read by the computer. The belt speed signal in frequency form can be read in to a counter and valued on a periodic basis. If a microcomputer is not used for computation, the analog signal from the weigh bridge can be conditioned and fed to a high-performance voltage-to-frequency (V/F) converter that has high resolution and linearity. The product of the two frequency streams can be multiplied in a couple of ways. One is to count the frequency of the V/F converter into a counter that provides the loaded values for a rate multiplier circuit. The belt speed is then provided as the ripple frequency to the rate multipliers. The output frequency is a product of the two values. By scaling the frequency through counters, the value can be integrated. Figure 47.30 shows an integrator done with digital components. More likely, the design would be implemented by microcomputer.

Since, in some cases, ease of data exchange to a process computer is required, the integrator using the microcomputer is a better choice. The choice is probably better when the load cell system is digital.

Belt Effects. The conveyor belt has a great influence on the performance of the belt scale. The factors of stiffness and the action of belt tension extend to create weighing errors if the weigh idlers are out of alignment. Belt effects can be minimized if deflections of the weighing system can be reduced to near zero. Problems come about when calibrating the unit with no material on the belt and then loading it. The additional deflection and the stress influence of the belt from the cantilever effects result in zero shifts as the belt loading increases from zero. There are a number of formulas that have been developed to give a measure of error that can be expected when misalignments are present. For example, in a single idler system, the error is valued at:

$$E(\%) = \left(24 \times \frac{KT}{QL^2}\right) D \times 100$$

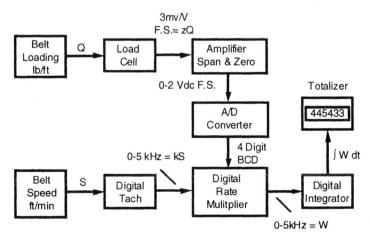

FIGURE 47.30 Load and speed integration.

where E = error in percent of belt loading
 K = belt stiffness factor
 T = tension in the belt in lb at the scale
 Q = belt loading in lb/ft. of length
 L = spacing between idlers in in.
 D = displacement from the belt in in.

The belt stiffness factor runs from a value of 1 when the belt has little stiffness to a number that exceeds 50 or more for a stiff belt. There are additional complexities that will not be developed here with regards to the complexities of belt stiffness and elasticity and the issue of troughing.

Design Criteria.

1. Conveyor should be horizontal and the length should be less than 200 feet. At least 10 complete belt circuits should be made in each weighment to minimize the inaccuracies due to a nonuniform belt.
2. Conveyor belts should be designed to use at least 80 percent of the volumetric capacity. Maximum troughing angle should be 35° and the belt material should be as uniform and as flexible as possible for the intended service.
3. Conveyor stringers should be strong enough to prevent any deflection of the weigh system from exceeding 0.02 in. Deflection between adjacent idlers should be less than 0.010 in. Structural alignment should be maintained to 0.02 in.
4. Conveyor should have some type of automatic belt takeup to maintain a relatively constant belt tension and to help keep the belt tracking properly.
5. Belt loading should run between 50–90 percent, and the mechanical design should provide adequate and constant feed.
6. The belt, mechanics, and electronics should be protected from the weather. A 100°F change in temperature can change the stiffness of a conveyor belt by 50 percent.
7. There should be no power takeoffs from the conveyor.
8. All idlers in the weighing region and the lead on and off idlers should be carefully aligned and sealed in place. Idler concentricity should be better than 0.010 in., and troughing profile should match within 0.015 in.
9. The scale bridge location should be more than 20 ft. but less than 50 ft. from the loading point. Skirting and training idlers should not be located closer than 20 ft. to the weigh bridge.
10. The belt scale should use multiple weigh idlers to provide a long weigh span for higher accuracy at high belt loadings. The weigh bridge under load should deflect less than 0.030 in. The scale system should provide for off-center loadings and side loadings such that either create no error. Belt effects should produce no more than 0.1 percent of the total error.
11. The electronic integration system should operate on both the weigh and speed measurement. The integration should have a high-resolution function that allows negative as well as positive counting. The readout should have an accuracy of better than 0.1 percent of reading.
12. The belt speed measurement should be made with a wheel contacting the belt, either on the measurement or return side.
13. Calibration chains should be selected to permit calibration at about 75 percent maximum belt loading.

14. The layout of the weighing system should be such that a quantity of material may be drawn off into a rail car or separate weigh hopper to be checked independently.

MULTIINGREDIENT PROPORTIONING FOR DRY BULK SOLIDS

Figures 47.31–47.34 have to do with proportioning two or more dry bulk solid ingredients continuously on a belt delivery system. It includes both weigh belt feeders and conventional three-roll troughing-idler conveyor belts. Weigh belt feeders are often inserted into a process stream in order to enable the measurement of flow rate and totalized weight. When appropriately equipped with additional components, weigh belt feeders can also control the flow rate of the product or ingredient stream. Conveyor belts on the other hand are specifically installed to play the roll of material deliverers and are rarely used as flow rate control devices. Therein lies some of the purpose of this chapter: to share with a larger audience the character and quality of the results obtained in using an existing conveyor belt to proportion multiple ingre-

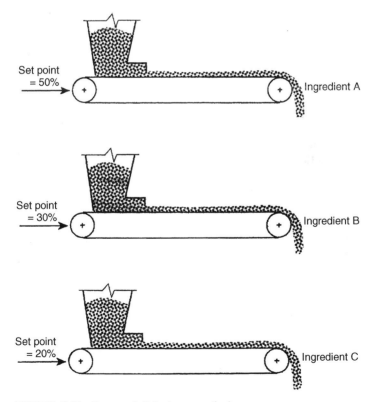

FIGURE 47.31 Conveyor belt feeder proportioning.

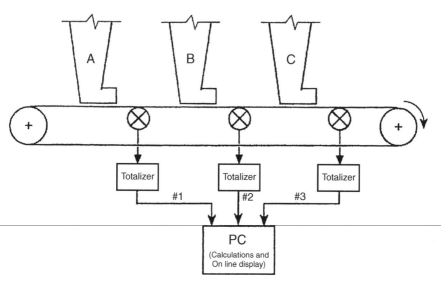

FIGURE 47.32 Conveyor belt proportioning.

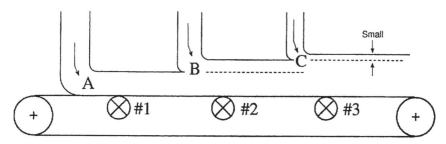

FIGURE 47.33 Graphic illustration.

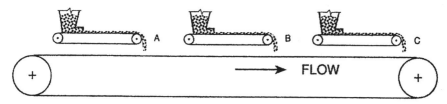

FIGURE 47.34 Main conveyor belt with feeder belts.

dients. Additional purposes are to present briefly different options for proportioning dry solids and to show the general layout of belt measurement systems for a new plant.

Proportioning is the process of delivering two or more ingredients in the correct ratio by weight. It does not include blending, which means to carry out a mixing operation to make the entire formula reasonably consistent in its material makeup.

REFERENCES

1. Hendrik Colijn, *Weighing and Proportioning of Bulk Solids,* Trans Tech Publications, Federal Republic of Germany, 1983.
2. C. L. Nachtigal, W. H. Mann, and R. H. Inwards, "Improving a Sand Blending Process by Continuously Displaying Ingredient Weight and Flow Ratios," *The International Journal of Storing and Handling Bulk Materials,* **8**(2), April 1988.

CHAPTER 48
LOW-COST INFRARED SPIN GYRO FOR CAR NAVIGATION AND DISPLAY CURSOR CONTROL APPLICATIONS

INTRODUCTION

Although gyroscope technology has existed for many years, these sensors are typically found in military and aviation applications due to their high cost and large size. Gyration, Inc. has developed a new spin gyroscope called the GyroEngine. This gyroscope offers advantages of low-cost, low-power consumption and small size. The extensive use of injection-molded polycarbonate components avoided the high cost associated with machined parts. Polycarbonate has the added benefit of being an optically clear material. This allowed for the optical encoding system that provides a digital output for convenient interface to common microcontrollers.

With the availability of a low-cost motion sensor new markets are emerging. Gyration is actively involved in remote cursor control and car navigation applications. The GyroEngine is an attractive solution for these and other motion sensing applications.

THEORY OF OPERATION

The GyroEngine operates on the principle of a conventional two-degree-of-freedom spin gyroscope, shown schematically in Fig. 48.1. A rotating flywheel provides the angular momentum to define a reference axis. The flywheel is mounted in the inner gimbal, which in turn is mounted in the outer gimbal. This assembly is then enclosed in a frame. A pendulous mass placed on the inner gimbal orients the spin axis with respect to the local gravity vector. The GyroEngine is available in two configurations, vertical and directional. The spin axis of the vertical GyroEngine is nominally vertical while the spin axis of the directional GyroEngine is horizontal. Changes in the positions of the gimbals are reported by an optical sensing system which uses Gyration's patented optical encoder. The output from the GyroEngine is digital phase quadrature for each gimbal. This allows the developer to interface directly to an inexpensive microcontroller and avoids the added cost and complexity associated with analog-to-digital conversion. The vertical GyroEngine provides relative pitch and roll angles. The directional GyroEngine, shown in Fig. 48.2, provides relative yaw and the inner gimbal relative to outer gimbal angle. The resolution of the output is 0.20°.

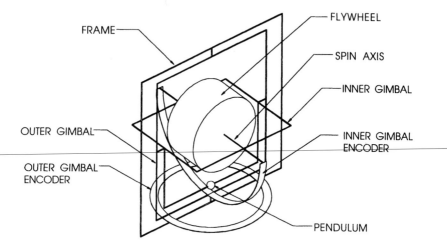

FIGURE 48.1 GyroEngine schematic.

CURSOR CONTROL APPLICATIONS

With the emergence of multimedia and interactive television markets, there is a need for intuitive cursor control devices. These devices should be user-friendly and provide the user with as much freedom of mobility as possible. Gyration has developed a free-space pointer called the GyroPoint that utilizes two GyroEngines to manipulate the cursor. The device is illustrated in Fig. 48.3. A vertical GyroEngine is

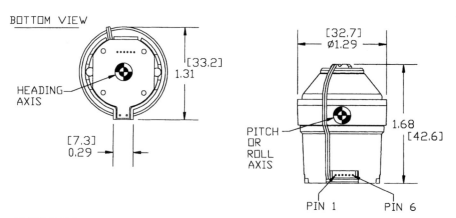

FIGURE 48.2 Directional GyroEngine.

used to measure the relative pitch angle of the user's hand and wrist. A directional GyroEngine is used to measure the relative yaw. The output from the GyroEngines is monitored by a common 8-bit microcontroller, which in turn communicates to a host computer via serial protocol. The pitch and yaw motions of the user are translated into left/right and up/down movement of the cursor. A block diagram of the system configuration is shown in Fig. 48.4.

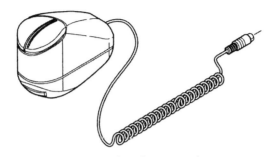

FIGURE 48.3 GyroPointer free-space pointer.

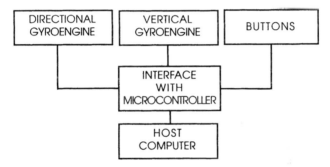

FIGURE 48.4 GyroPoint block diagram.

By using inertial sensors, the user is not limited to a flat surface, as is the case with a conventional mouse. There is also a distinct advantage over feedback or infrared-based devices in that the user is not required to have line of sight to a receiving station or point directly at the screen. In many cases, the user will have his back to the screen and face an audience. This situation is no problem when using a gyroscope-based pointer such as GyroPoint.

CAR NAVIGATION APPLICATIONS

Long used for marine and aircraft navigation, gyroscopes are now being utilized in car navigation systems for the consumer and fleet management applications. The gyroscope allows for dead-reckoning in situations where GPS is either inaccurate or unavailable. The Directional GyroEngine is an excellent sensor for these applications. Since navigation requires more accuracy than a pointer application, the developer should be aware of compensation techniques that will improve performance.

THE EFFECT OF THE PENDULUM ON PERFORMANCE

To achieve optimal performance, the spin axis of the GyroEngine should be nominally level. The GyroEngine utilizes a pendulous mass on the inner gimbal to ensure a level position during startup. However, during normal operation, the spin axis may be displaced from level due to friction acting on the outer gimbal axis. The pendulum will then act to torque the spin axis back to level. But, due to the nature of gyroscopes, this torque will induce a precession of the outer gimbal. This relationship is described by the following equation:

$$T = \Omega \times H \quad (48.1)$$

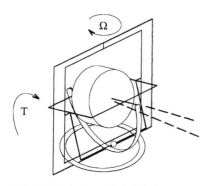

FIGURE 48.5 Inner gimbal displacement.

where the vector T, the torque due to the pendulum, is the cross-product of the vector Ω, precession rate, and H, angular momentum. This principle is illustrated in Fig. 48.5. In scalar form, this equation is expressed as

$$T = \Omega \times H \times \cos(\text{IG Angle}) \quad (48.2)$$

where IG Angle (inner gimbal angle) is the angle of the spin axis with respect to level. The precession will usually act in the same direction as the turn, and a smaller displacement angle will be reported due to the differential motion between the outer gimbal and frame. This effect is referred to as *undershoot*.

SOFTWARE COMPENSATION

Gyration has applied this knowledge of the GyroEngine's behavior to develop a software algorithm to improve upon the raw output. By reading the inner gimbal encoder, the drift caused by the pendulum can be calculated. Taking Eq. (48.2), we can express the relationship as

$$W \times d \times \sin(\text{IG Angle}) = H \times \Omega \times \cos(\text{IG Angle}) \quad (48.3)$$

where W is the weight of the pendulum and d is the distance between the inner gimbal axis and the pendulum. For the GyroEngine, W, d and H are all parameters that are held constant by process control. We can now combine these constant parameters and define a drift compensation term C as

$$C = K \times \tan(\text{IG Angle}) \, (\text{deg/sec}) \quad (48.4)$$

where K is a single constant accounting for pendulosity and angular momentum and IG Angle is the accumulated inner gimbal reading since one second after startup. The compensated heading is then calculated by

$$\text{Current\_Heading} = \text{Previous\_Heading} + \text{Heading\_Change} + C\,\Delta t$$

where Heading_Change is the change in yaw angle reported by the GyroEngine and Δt is the sampling period. For this algorithm, it would be ideal to know the angle of the spin axis with respect to level. The GyroEngine only provides a relative reference, which is measured on the inner gimbal. In practice, using this relative information provides good results due to the fact that the majority of roads are level or average to level over the long term. In other words, the inner gimbal serves as an adequate level sensor for these purposes.

In the case in which the GyroEngine starts up on an incline, the spin axis will kick to a neutral position: that is, spin axis, outer gimbal axis, and inner gimbal axis are all perpendicular to one another. This occurs because the spin axis has a natural tendency to seek a torque-free position. Therefore, the spin axis is starting off in a position displaced from level. When the GyroEngine reaches a level position, the pendulum will act to bring the spin axis to level, and the algorithm will compensate for the effect this has on the output.

NAVIGATION SYSTEM CONFIGURATION

In order to navigate by dead reckoning, both heading angle and odometer information are required. To do software compensation, we need to collect the inner gimbal angle data as well. GYRATION has developed a prototype NavBox. The NavBox is used in conjunction with the GyroEngine, odometer sensor, and a laptop computer for navigation. A block diagram of the system is shown in Fig. 48.6. This configuration does not include an expensive GPS receiver commonly found in other navigation systems.

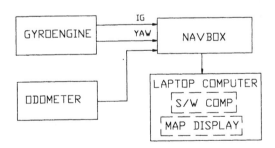

FIGURE 48.6 System block diagram.

ROAD TEST RESULTS

Gyration has conducted extensive road testing with the Directional GyroEngine using the prototype NavBox and software compensation. Testing was conducted in different environments in order to cover a variety of typical road conditions including residential areas, urban commercial districts, and highways.

Figure 48.7 shows the dead-reckoned ground path of the vehicle compared with the actual road. With compensation, the percent error over distance travelled was 4.5 percent for a 10-km course. Typical values for percent error range from 2–5 percent. Typical drift for the gyroscope sensor range from 1–5°/min. with compensation. This performance combined with other compensation techniques such as map-matching can produce a robust car navigation system without the need for a GPS receiver.

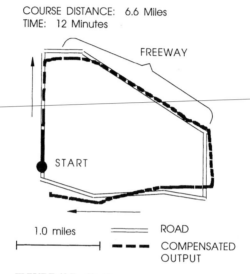

FIGURE 48.7 City/Freeway course.

CONCLUSION

Injection-molding and optical encoding technology has been combined to produce a new, commercially available gyroscope called the GyroEngine. The basic concept is based on conventional spin gyro technology; however, the innovative design has provided advantages in terms of cost, size, and performance. The gyroscope that is approximately the size of a roll of 35-mm film has drift characteristics on the order of 5°/min. Software compensation techniques have been used to reduce the drift to 1–3°/min. typically. The GyroEngine should play a major role in the expansion of consumer markets such as interactive television and car navigation.

CHAPTER 49
QUARTZ ROTATION RATE SENSOR: THEORY OF OPERATION, CONSTRUCTION, AND APPLICATIONS

THEORY OF OPERATION

The use of a vibrating element to measure rotational velocity by employing the Coriolis principle is a concept that has been around for more than fifty years. In fact, the idea developed long ago out of the observation that a certain species of fly employs a pair of vibrating antenna to stabilize its flight. This sensing technique has been successfully employed in a practical embodiment: the quartz rate sensor or QRS.

To understand how the QRS works requires familiarity with the Coriolis principle. Simply stated, this means that a linear motion within a rotating framework will have some component of velocity which is perpendicular to that linear motion. The handiest example of the Coriolis effect is that exhibited by wind patterns on earth (refer to 49.1). Convection cells in the atmosphere set up a wind flow from the poles toward the equator (with a north-south orientation). However, as the earth rotates, these linear flows develop a sideways (orthogonal) component of motion due to the earth's rotation.

This bends the wind from a north-south direction to an east-west direction. It is the Coriolis effect that creates the east-west trade winds and is responsible for the spirals of clouds observed in satellite photos.

Now let's apply this principle to our rotation sensor. Looking at Fig. 49.2, you can see that the QRS is essentially divided into two sections: drive and pickup. The drive portion looks and acts exactly like a simple tuning fork. Since the drive tines are constructed of crystalline quartz, it is possible to ring this tuning fork electrically just like a quartz watch crystal. Each of the tuning fork tines has a mass and an instantaneous radial velocity which changes sinusoidally as the tine moves back and forth. As long as the base of the tuning fork is stationary, the momentum of the two tines exactly cancels and there is no energy transfer from the tines to the base. In fact, it only takes about 6 µW of power to keep the fork ringing.

As soon as the tuning fork is rotated around its axis of symmetry, however, the Coriolis principle exerts a profound influence on the behavior of this mechanism. By convention (the right-hand rule), the rotational vector ω_i is described by an arrow that is aligned with the axis of rotation.

FIGURE 49.1 Earth's rotation causes the wind patterns to spiral. This is called the Coriolis Force.

The instantaneous radial velocity of each of the tines will, through the Coriolis effect, generate a vector cross product with this rotation vector. The net effect is that a force will be generated by each tine that is perpendicular to the instantaneous radial velocity of each of the tines and is described by the following equation:

$$F = 2m\omega_j \times V_r$$

where m = tine mass
ω_j = rotation rate
V_r = radial velocity

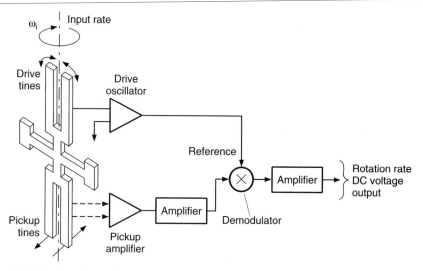

FIGURE 49.2 Tuning fork diagram.

Note that this force is directly proportional to the rotation rate, and, since the radial velocity of the tines is sinusoidal, the force on each tine is also sinusoidal. Since the radial velocities of the two tines are equal and opposite, the Coriolis forces are equal and opposite, producing an oscillating torque at the base of the drive tine fork that is proportional to the input angular rate.

The pickup portion of the QRS now comes into play. The sinusoidal torque variation causes the pickup tines to start moving tangentially to the rotation and at the same frequency as the drive vibration. Since the forces causing the pickup tines to move are directly proportional to the rotation rate d, there is no rotation; the pickup tines will not move. The QRS can therefore truly detect a zero rotation input.

Once the pickup tines are in motion, it is a simple matter to amplify the pickup signal and then demodulate it using the drive frequency as a reference. One additional stage of amplification allows for some signal shaping and produces a DC signal output that is directly proportional to the input angular rate. All of the electronics are fairly simple and can be contained within the same package as the sensing element.

CONSTRUCTION

The ORS is constructed from a wafer of single-crystal synthetically grown quartz. This choice of material is very significant for a variety of reasons. The most obvious is that the material is piezoelectric. Quartz, however, has the advantage of being the material with the most stable piezoelectric properties over temperature and time. No other material can match quartz's stability, making it the most suitable choice for precision instrumentation.

A second significant property of quartz is that it exhibits a high modulus of elasticity and therefore can be made to ring very precisely with a high Q (quality factor). It is this property that also makes R such a desirable material for the companies that produce frequency standards and watch crystals. This combination of stable piezoelectric properties and high Q means that quartz embodies the two most significant characteristics that are needed to produce a high-quality sensor.

A third and more subtle feature of quartz is that R can be chemically etched using conventional wet-etch production techniques. In fact, the manufacturing process is very similar to that used in the semiconductor industry for producing silicon chips. Figure 49.3 shows the steps involved in manufacturing a QRS. First, a flat, polished wafer of quartz has a metal layer deposited on R. The metal serves as a photo mask and is chosen so as to be resistant to the etch bath chemicals that will be used to etch the quartz. Next, a light-sensitive emulsion is placed on the metal layer, and R is exposed using a film representation of the finished part to create a pattern on the metal layer. The unexposed portions of the emulsion can now be washed away, revealing the metal layer below. With the metal exposed, R can now be etched away to uncover the quartz below. The quartz material can now be removed through a wet chemical etch thereby creating the desired mechanical configuration. After cleaning off the remaining emulsion and metal layers, the final step is to plate the quartz on gold electrodes, which will serve to drive the drive tines and be used to pick up charge off of the pickup tines. The whole sensing element is then sealed into a her-

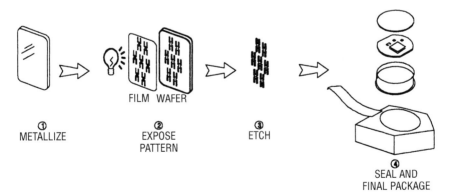

FIGURE 49.3 Construction of QRS.

metic package, connected to its electronics, and packaged for easy handling and electrical connection.

APPLICATIONS

Until now, the most common rotation sensor available that relied on principles of inertial mechanics was a spring-restrained spinning-wheel gyroscope. These tend to be large and heavy, and they consume lots of power. They also tend to wear out after only a few thousand hours of operation and so cannot be used continuously for long periods of time. Their use has been restricted to highly specialized applications such as military aircraft and missiles where the short mission times and availability of maintenance personal made their use practical. By contrast, with a MTBF in excess of 100,000 hours, the QRS technology is finding its way into military flight control applications, and with the low cost of ownership, it has become available to industrial commercial customers as well.

Applications for the QRS fall into two broad categories: open-loop or instrumentation applications and closed-loop or control applications (Fig. 49.4). Each of these applications categories has two subsets, discussed below. A description of each of these classes of use will be presented with a specific focus on implementing a navigation and a stabilization application.

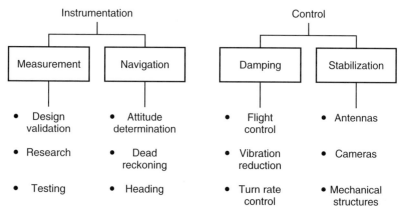

FIGURE 49.4   Application categories for rate sensors.

Instrumentation

This class of application involves either instrumenting a structure for purposes of determining its rates of rotational motion (measurement) or processing of that information in real time to generate information about orientation (navigation). Typical situations in which QRS technology has been used in measurement of rotational velocity include instrumentation of vehicles for crash studies, determination of dynamics of specific platforms (i.e., boats, trains, robots, even human beings), and environmental measurements such as earthquakes and wave motions.

Measurement. One of the key elements in design of measurement systems is to determine the peak rotational velocities involved to ensure that an instrument with the proper range is used. If too low a range is chosen, then the output will be clipped, and valuable information will be lost. A fairly straightforward way to determine this range requirement is to establish two parameters: the frequency of movement of the structure to be instrumented and the peak angular displacement of that movement. As an example, let's assume that we want to determine the dynamics of a vehicle body roll while taking a turn. The body roll motion can be described by the equation

$$\theta = A \times \sin(2\pi \times F_n \times t) \text{ in degrees}$$

where A = amplitude of movement
 F_n = frequency of movement

The parameter of interest for measuring angular velocity is the change in angular position with time or $(d\theta/dt)$. By taking the differential of the above equation, the result is:

$$(d\theta/dt) = A \times 2\pi \times F_n \times \cos(2\pi \times F_n \times t)$$

In our example, let's assume that the natural frequency of the vehicle suspension system is 6 Hz, and the peak body roll is 10°. By substituting these into the above equation:

$$(d\theta dt) = 10 \times 2\pi \times 6 \times \cos(2\pi \times 6 \times t)$$

$$= 377 \times \cos(37.7 \times t)°/\text{sec}$$

Since the cosine term has a maximum value of 1, the peak rotational velocity is 377°/sec. So even a seemingly benign environment—a 10° roll at 6 Hz—generates fairly high peak velocities.

Navigation. Navigation applications are becoming increasingly interesting for the QRS, especially in light of the availability of GPS receivers at a reasonable cost. In principle, by reading the output from the rotation sensor (rotational velocity) and integrating this output over time, one can determine the angular displacement of the sensor. In Fig. 49.5, a block diagram is shown in which a QRS is used for sensing vehicle yaw as part of a navigation package.

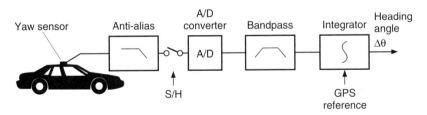

FIGURE 49.5 Sensing vehicle yaw.

Antialiasing Filter. Since a computer interface requires use of an analog-to-digital converter (ADC), the output from the QRS becomes part of a sampled-data stream. In order to prevent aliasing of the output, a filter needs to be used with the corner frequency usually set at ¼ to ½ of the sampling frequency.

ADC. The analog-to-digital conversion should come after the antialiasing, since this puts the converter close to the QRS and reduces the overall noise of the system, yielding the most stable results. A 12-bit converter is adequate for most uses. The sample frequency should be appropriate for the system, but typical values range from 100 to 1000 Hz.

Bandpass Filter. This is tailored to the specific application. As an example, when used as part of a head-mounted display for a virtual reality application, very small, high-frequency head movements do not need to be tracked, since they may be part of the normal jostling associated with interactive game playing. Only larger, definite swings of the head need to be tacked. Similarly, low-frequency variations in the output from the QRS, which are usually associated with changes in environmental temperatures or warmup, are not meaningful tracking information and need to be rejected. These two scenarios determine the lower and upper ranges of the bandpass filter. A reasonable start point would be to choose upper and lower corner frequencies of 0.1 Hz and 10 Hz.

Integrator. This is where the angular velocity information is turned into angular position. Since the initial conditions are indeterminate at startup, it is recommended to include a reset capability. This allows you to initialize the integrator to zero or some known position at startup. Usually during the initialization, the portion of the body that is to be measured needs to be held very steady so that the initial conditions represent as close to a true zero input state as possible. Any residual error at start up will cause the apparent output from the integrator to drift. One method to reduce the startup error is to average the input to the integrator for a few seconds during the initialization sequence and then subtract this average value to establish the zero point.

As a practical matter, it is virtually impossible to measure the pure rotational velocity without introducing or reading some error at the same time. This accumulation of errors means that, over time, the true angular position and the calculated angular position will diverge. The sensor output is not drifting, but the apparent calculated angle is. The rate of this divergence is determined by a variety of factors including: how well the initial conditions are established, the accuracy of the alignment of the sensor to the true axis of rotation, the quantization errors of the signal (if it has been digitized), and the stability of the environment in which the measurement is being done.

For most practical applications, therefore, the QRS is used only for short-term navigation. In order to prevent these incremental errors from growing too large, it is the practice to update, or correct, the calculated angle periodically through the use of a fixed, external reference as shown in the block diagram. The reference used will depend on the situation, but some examples include: a GPS signal, a corner cube with optical line-of-sight, or an encoded magnetic signal. In fact, the combination of dead reckoning between fixed reference updates is a nearly ideal means to navigate through a variety of environments. This method has been employed to maneuver autonomous delivery robots through hospital corridors, automated forklift navigation in warehouses, and emergency vehicles deployed in urban environments.

Control

To employ the QRS in control applications requires an understanding of how it works as part of a system. Typically, a system model is employed that takes into account the magnitude and phase relationships of the sensor response. A typical plot of this relationship is shown in Fig. 49.6. As can be seen from the figure, the QRS acts very much like a typical second-order system with a comer frequency of about 75 Hz. An alternative method of describing this behavior is through the use of a Laplace representation. The equivalent s-plane representation is shown in the figure as well.

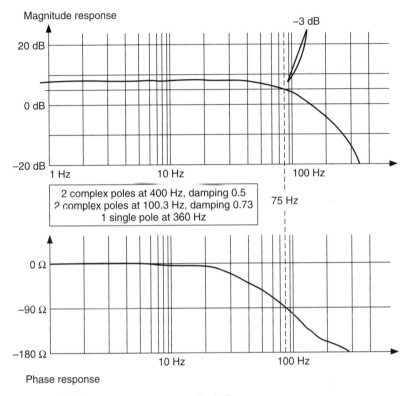

FIGURE 49.6 Frequency response curves (typical).

Damping. Being able to measure rotational velocity accurately opens up new possibilities for control of structures. One of the most useful types of control applications is to damp out the resonant behavior of mechanical systems. Very few mechanical systems involve pure linear motion, as most practical machines have parts that rotate or pivot. In the same vein, most vehicles (airplanes, trucks, ships, and automobiles) are controlled by means of roll, pitch, and/or yaw controls. By motoring and controlling these motions, it is possible to provide active roll damping on ships, remove Dutch roll from aircraft flight, reduce body roll on a car as it takes a turn, or damp out end-effector shake in an industrial robot.

Stabilization. There is a special instance of closed-loop control in which the item being controlled is intended to remain stationary, even while the platform to which it is attached moves. This class of application is referred to as *stabilization*. For these applications, it is important that the QRS be tightly coupled mechanically to the platform and that the item to be controlled, usually a camera or antenna on a multi-axis gimbal, be tightly coupled to the structure with no mechanical resonances in the bandwidth of the servo mechanism. The system designer needs to take into account the transfer function of the system servo loop and ensure enough phase margin to prevent oscillation. For these applications, it is often desirable to be able to move the camera or antenna independently, so a provision must be made to provide a commendable DC offset in the control loop so that an operator can rotate and point the camera in the gimbal. This method has been employed successfully to stabilize antennas aboard ships and land vehicles as well as cameras aboard helicopters and survey airplanes.

An example of such an application is shown in block diagram form in Fig. 49.7. Here, the QRS is used as part of a servo control loop to provide an absolute pointing angle in attitude as well as image stability for a mobile telescope. For simplicity, it is assumed that the telescope is mounted on a platform that can rotate only in attitude, and the control mechanism is therefore an attitude control system only. The principle described can be applied to the other axes of rotation.

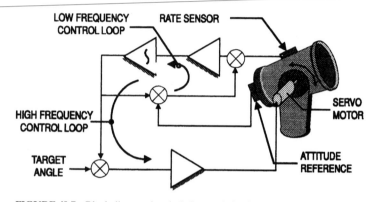

FIGURE 49.7 Block diagram for gimbal control circuit.

Refer first to the high-frequency control loop portion of the diagram. Assume that this circuit is designed to operate at 10 Hz, which is a typical value for a servo control. Let's further assume that the telescope has a rotational inertia $J = 12$ slug-ft².

Since $\omega_n2 = K_s/J$, then $K_s = (10 \times 2 \times \pi) \times 12 = 47,300$ ft.-lb/rad. K_s represents the servo stiffness.

This implies that an external applied torque of 10 ft.-lbs will allow a movement of only $10/47,300 = 0.0002$ radians or 0.7 arc-minutes.

Now let's look at the low-frequency control loop portion of the diagram. This will act as a vertical reference unit and make sure that the absolute pointing angle of the telescope matches with the commanded (or target) angle. In order to do this, a stable, long-term attitude reference needs to be provided. For most common systems, gravity does the job quite nicely. A simple tilt sensor is always referenced to local gravity and, over a fairly narrow range, will behave quite linearly. To avoid coupling in any high-frequency movements that are by definition not gravity-related, this reference is part of a control loop that has a time constant, typically, of 100 seconds or so. This allows the attitude reference to follow closely the typical platform motions you might find on most common mobile platforms: ships, airplanes, and automobiles. Generally, this control loop will incorporate a proportional and differential control element that is not shown as part of the block diagram.

In summary, a new type of sensor is now available that can add significantly to the capabilities of engineers and designers alike. Based on inertial-sensing principles, the quartz rate sensor provides a simple, reliable measurement of rotational velocity that can be used to instrument structures in new ways and gain a more in-depth insight into designs, to aid in short-term navigation of autonomous mobile platforms, and to allow for improved methods of stabilizing structures.

CHAPTER 50
FIBER OPTIC RATE GYRO FOR LAND NAVIGATION AND PLATFORM STABILIZATION

INTRODUCTION

Fiber optic gyroscopes are now beginning to replace mechanical gyroscopes in new design. The technology is based on the Sagnac effect and draws many of its components from concepts developed for telecommunications applications. The device described in this paper is engineered for characteristics desirable for angular rate sensors, with performance balanced against cost considerations.

Measurement of angular rotation is essential in inertial navigation and stabilization systems. Traditionally, the angular momentum of a spinning rotor was used to determine angular rate or displacement. As the technology matured, the cost and reliability continued to be a concern. Several new classes of nongyroscopic angular sensors have been developed that show promise of replacing mechanical gyros in new design and perhaps some existing equipment as well. Some of these depend on Coriolis force effects on tuning forks, or hemispherical quartz sensors, but Sagnacring interferometers can be built with no moving parts.

Known for 80 years, the Sagnac effect was first put to practical use in the ring laser gyroscope (RLG), which is now a mature technology incorporated in aircraft inertial navigation systems. The differential phase shift between two counterpropagating laser beams in an evacuated cavity measures angular rotation. More recently, it has been shown that the same effect can be employed in a fiber coil, eliminating the high-voltage and vacuum technology and making low-cost, accurate, inertial rotation sensors practical.

Although the individual components of the optical circuit—the directional couplers, polarizer, fiber coil, laser, and detector—bear a strong resemblance to those used in telecommunications applications, there exist differences in detail requirements. This has impeded somewhat the development of the fiber optic gyro (FOG) as a sensor. The technology now exists to produce both inertial navigation grade and rate FOGs. After describing the two basic configurations of the FOG, the application of the open-loop FOG to inexpensive rate gyros is described.

GYRO DESIGN

Figure 50.1 illustrates the two most often used configurations of the FOG: closed loop and open loop. Common to each type of gyro are the fiber coil, two directional couplers, a polarizer, the optical source, and the photodetector. The configurations differ in the method of applying a phase modulation bias and in the electronic signal processing.

Light from the laser propagates through the first directional coupler, polarizer, and second directional coupler to enter the coil. The directional coupler attached to the coil divides the light equally, and two light streams travel around the coil in opposite directions, recombining when they return to the directional coupler. The light then returns through the polarizer and first directional coupler to be received

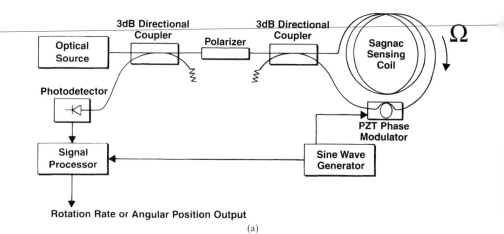

(a)

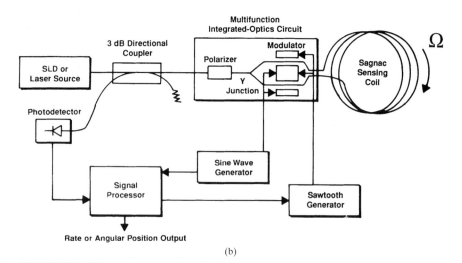

(b)

FIGURE 50.1 Fiber optic gyro configuration: (*a*) Open-Loop; (*b*) Closed Loop.

by the photodetector. Although it may seem that a simpler arrangement may be possible, this is the minimum configuration necessary to ensure stable operation.[2]

At the directional coupler attached to the coil, the two waves returning from their transit through the coil comprise an optical interferometer whose output is a raised cosine function of the phase shift (Fig. 50.2). There is no difference in the phase of light traversing the closed path if there is no rotation, but a phase difference of

$$\Delta S = (2\pi\, LD\Omega)/(\lambda c) \qquad (50.1)$$

where ΔS = Sagnac phase shift
L = coil length
D = coil diameter
c = speed of light in free space
λ = optical wavelength
Ω = rotation rate

is experienced when the path rotates. This effect can be shown to be independent of the shape of the path, and of the propagation medium.[3] For low rotation rates, the change of interferometer output with Sagnac phase shift is small, operating at the maximum of the cosine function, and it is necessary to apply a dynamic bias to the light path in order to optimize the sensitivity. The method of applying this bias and the electronic signal processing technique are the essential differences between the two basic methods. Our discussion will focus on the open-loop method.

A piezoelectric transducer, wrapped with multiple turns of the fiber is incorporated in one end of the coil. The transducer is modulated with a sinusoidal voltage, impressing an optical differential phase shift between the two counter-propagating light beams at the modulation frequency. The interferometer output now exhibits the periodic behavior of Fig. 50.2(a). Since the modulation is symmetrical, only even-order harmonics of the modulating frequency are present; the ratio of the harmonics depends on the optical modulation amplitude. When the coil is rotated, the Sagnac phase shift unbalances the modulation, as shown in Fig. 50.2(b), and the fundamental and odd harmonics will also be present. A simple FOG may only measure the output at the modulation frequency

$$S(f_m,\Delta S) = k\, J_1(\emptyset_b)\, \sin(\Delta S) \qquad (50.2)$$

where $S(f_m,\Delta S)$ is the gyro output power at the modulation frequency, k is proportional to the optical power, $J_1(\emptyset_b)$ is a Bessel function of the first kind of order 1, whose argument is the biasing phase modulation amplitude, and ΔS is the Sagnac phase shift given by Ref. 1.

Open-loop configurations are often criticized for the sinusoidal nature of the input-output characteristic. However, the sinusoidal characteristic can be electronically linearized. Information contained in the higher order frequency terms can be used to obtain a linear output,[4] even when the Sagnac phase shift is large enough to translate the interferometer output through the null output state.

The enclosed area of the coil determines the constant of proportionality between the rotation rate and Sagnac phase shift at any operating optical wavelength. When the effects of noise are considered, the minimum sensitivity is proportional to the coil area and optical source power. This is a design tradeoff, since the cost of fiber and the laser source are often significant contributors to the product cost. Often, the maximum physical size is a design constraint, and the length of the coil must be increased in order to compensate for the smaller diameter. Increased attenuation in the coil then may result in a requirement for a higher-power laser source.

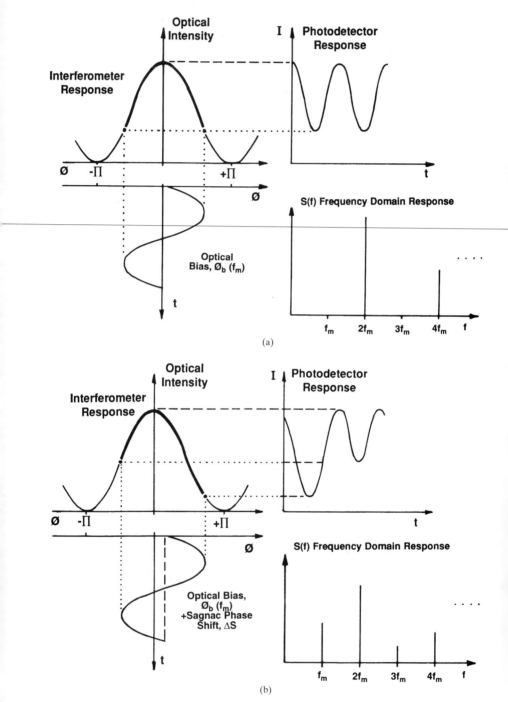

FIGURE 50.2 Sagnac interferometer response for open-loop configuration: (*a*) Stationary; (*b*) Rotating.

PERFORMANCE

We have produced a FOG whose characteristics are suitable for use in stabilization of platforms such as antenna pedestals and optical mounts and which may also be used as a heading angle change sensor for land navigation systems. In the latter application, it is an input to the dead-reckoning position computation, often used to complement global positioning system (GPS) position fixes, or with map-matching. The dead-reckoning computation fills in holes in GPS coverage that often occur in urban areas and heavily forested areas.

The gyro (Fig. 50.3) operates at a wavelength of 820 nanometers, with a 150-meter coil of elliptical-core polarization-maintaining fiber. Several different types of semiconductor lasers have been found to be satisfactory light sources. For this application, a rather broad source spectral width, perhaps 20 nanometers, is desirable to suppress parasitic reflections at splices. In both this respect and the use of polarization-maintaining fiber, the FOG departs from telecommunications technology.

We have chosen a type of open-loop signal processing that has been termed *pseudo-closed-loop*, since it generates a form of replica gyro signal and uses a phase detector to derive an error signal in a feedback loop. All of the operations are performed without altering the modulation imposed on the light, a function necessary in true closed-loop designs. Performance of the open-loop signal processing is limited by drifts with temperature and the accuracy of initial adjustment and does not seem capable of reaching performance levels attained by closed-loop designs. The cost is significantly less, however.

The overall performance of the gyro at rest and room temperature is summarized (Fig. 50.4) by a signal-processing technique known as the Allan variance.[5] It evaluates the noise (termed *angle random WAIW*, ARW), the short-term bias instability, and the long-term bias drift (*rate ramp*).

FIGURE 50.3 Fiber optic rate gyro.

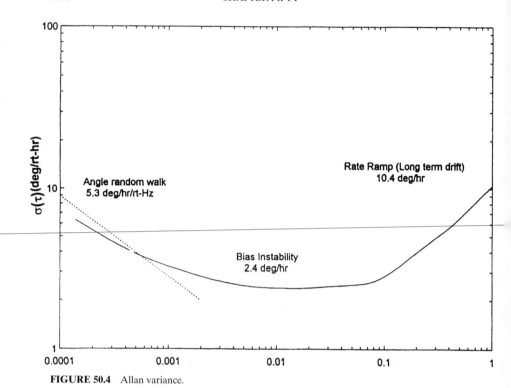

FIGURE 50.4 Allan variance.

It may be easier to visualize the gyro performance by measuring the response to small changes in rate input (Fig. 50.5). The gyro is placed on a rate table, and the input rotation rate is changed in discrete steps. With an integration time of 1 second, the minimum change of 0.015°/sec (54°/hr) is easily discerned, as would be expected from the ARW data. This gyro is actually capable of clearly measuring the earth's rotation rate of 0.0042°/sec (15°/hr).

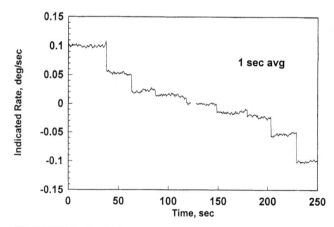

FIGURE 50.5 Sensitivity.

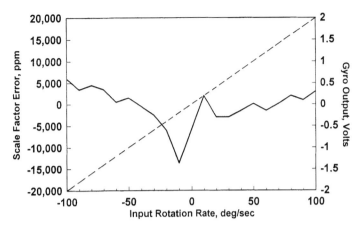

FIGURE 50.6 Scale factor linearity.

If the gyro is to be used in a navigation system rather than as a stabilization sensor, the linearity of the input-output characteristic, often expressed as V/(deg/sec), is important. Figure 50.6 illustrates the performance of this gyro, whose maximum input rate of 100 deg/sec corresponds to the maximum value expected from a sports car. The typical scale factor linearity of 5,000 parts per million is satisfactory for land navigation, particularly when used in conjunction with the other sensors previously described.

CONCLUSION

Fiber optic gyroscopes are an alternative to spinning wheel gyroscopes in existing applications, and they are expected to be both more reliable and less costly as production increases. Use of FOGs in new designs and in new applications such as land navigation are likely. The lack of vibration sensitivity and solid state design are significant advantages.

REFERENCES

1. V. Vali and R. W. Shorthill, "Fiber Ring Interferometer," *Appl. Optics* **15**, pp. 1099–1100.
2. R. Ulrich, "Fiber-Optic Rotation Sensing with Low Drift," *Optics Letters*, **5**, 1980, pp. 173–175.
3. H. Lefeve, *The Fiber-Optic Gyroscope*, Artech House, Boston, 1993.
4. A. B. Tveten, A. D. Kersey, E. C. McGarry, and A. Dandridge, "Electronic Interferometric Sensor Simulator/Demodulator," *Optical Fiber Sensors 1988 Technical Digest Series* vol. 2, part 1, pp. 277–290.
5. G. W. Erickson, "An Overview of Dynamic and Stochastic Modeling of Gyros," *Proceedings of the 1993 National Technical Meeting of the ION*, January 1993.

CHAPTER 51
A MICROMACHINED COMB DRIVE TUNING FORK GYROSCOPE FOR COMMERCIAL APPLICATIONS

INTRODUCTION

For automotive applications, Draper Laboratory and Rockwell are developing small, low-cost, single-crystal tuning fork gyroscopes using microfabrication technology. These tuning-fork gyros are extremely rugged, inherently balanced, and easy to fabricate. When packaged with application-specific integrated circuits (ASICs), the gyroscope plus electronics will fit in a 3-cm-per-side flat pack. Excitation will be 0 and 5 VDC.

The gyroscopes are fabricated from single-crystal silicon anodically bonded to a glass substrate (dissolved wafer process). For a 1-mm silicon gyroscope, projected performance is 10 to 100°/hr. for bias stability and resolution in a 60-Hz bandwidth. Resolution of 470°/hr. in a 60-Hz bandwidth (0.6°/hr. angle random walk) and bias stability of 55°/hr. overnight have been demonstrated with self-drive oscillation. In special tests, resolution better than 150°/hr. has been recorded. Without compensation, the bias changes 3600°/hr., as instrument and electronics temperatures vary from −40 to +85°C. These data were taken with the Engineering Development Model 2 (EDM2) electronics.

In this chapter, the following topics are covered: the principle of operation, fabrication of silicon gyroscopes, and projected and measured performance. Associated electronics and control issues will also be addressed.

The Charles Stark Draper Laboratory, Inc., has invented and developed inertial guidance systems for earth and space applications for over 50 years. Draper's silicon gimbaled gyroscope[1] was the first demonstration of rate sensing with a silicon micromachined unit.

Micromachining utilizes process technology developed by the integrated circuit industry to fabricate tiny sensors and actuators (with active areas typically less than 1 mm^2) on silicon chips. In addition to shrinking the sensor size by several orders of magnitude, integrated electronics can be added to the same silicon chip, creating an entire system on a chip. These instruments will result in not only the redesign of conventional military products but also new commercial applications that could not exist without small, inexpensive, angular rate sensors and computing.

Although most mechanical gyroscopes employ a rotating wheel as the inertial element, micromachined gas or ball bearings are not yet available. Sliding contacts cause frictional forces that are large in comparison to inertial forces; therefore, micromechanical gyroscopes vibrate with no sliding contacts. Macroscopic tuning fork gyroscopes have been used for many years.[2]

Silicon tuning fork rate gyroscopes fabricated on glass substrates are described herein. To demonstrate the concept, initial devices were fabricated on silicon wafers using electroformed nickel for the active vibrating elements. The production units are fabricated using single-crystal silicon on glass technology. The Draper gyros are the first to employ comb drive structures, which Tang and Howe[3] at the Berkeley Sensor and Actuator Center applied to polysilicon microresonators.

Previous work on micromachined gyroscopes includes both silicon devices[1,4,10] and quartz devices.[5,11] This paper will present the theory, fabrication, and initial test results of the comb drive tuning fork gyro.

THEORY OF OPERATION

The gyro's operation is shown in the photograph of Fig. 51.1 and the schematic of Fig. 51.2. The combs are excited so that electrostatic forces are generated that only weakly depend on the lateral position of the masses. The resulting large amplitude vibrations increase gyro sensitivity and reduce errors from external forces such as Brownian motion.

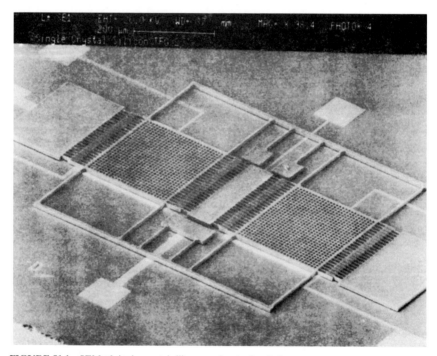

FIGURE 51.1 SEM of single-crystal silicon on glass tuning fork gyroscope.

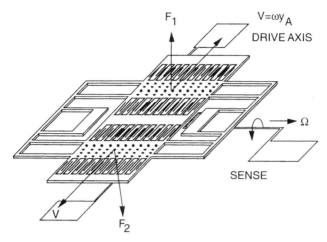

FIGURE 51.2 Schematic drawing of the comb drive tuning fork gyro, showing input Ω, Coriolis forces labeled F_1 and F_2, and horizontal drive velocities $V = \omega y_A$.

The in-plane vibration results in a drive velocity $V = \omega y_A$, where ω is the drive angular frequency and y_A is the peak amplitude of vibration. Angular rate Ω in the plane of the substrate lifts one mass up and the other down through Coriolis forces (labeled F_1 and F_2 in Fig. 51.2); thus, the sense and input axes are collinear (Fig. 51.2). Capacitors below the proof masses are used for gyro sense and perhaps force rebalance, although operation is now open loop. To optimize resolution and bandwidth, the separation between eigenfrequencies of the resonant drive and of the sense axis modes is designed at 5 percent of the drive eigenfrequency. This tuning fork gyroscope is mounted to the substrate through a stiff mechanical structure so that vibration errors are small and the device is rugged. The tuning fork mode, where the tines move in opposite directions, further rejects linear acceleration.

The masses are driven with forces 180° out of phase. The flexures are designed and sized to ensure that the tuning fork antiparallel mode is excited and that the translation modes are attenuated. The flexure geometry renders the eigenfrequency of the tuning fork mode (masses move in opposite directions) higher than the eigenfrequency of the mode where the masses move parallel to one another. The frequency separation allows the tuning fork mode to be easily excited, even with mismatched masses or springs. Its higher gains enable the tuning fork mode to be excited by a simple, self-excited oscillator loop.

With gyroscopes whose active elements are roughly 1 mm, projected performance is 10 to 100°/hr. for bias stability and for resolution in a 60-Hz bandwidth. The wide bandwidth resolution is limited by the instrument gain combined with input voltage noise in the preamplifier, which senses proof mass displacements along the sense axis (normal to the substrate).

FABRICATION

Draper now builds tuning fork gyros by a dissolved wafer process that results in crystal silicon structures anodically bonded to glass substrates. Compared to conducting substrates, the silicon on glass technology has low stray capacitance (less than 5 fF with 1-pF sense capacitors) limited by the symmetry of the bond wires, a major advantage for AC devices. Although on-chip electronics are not yet possible with P^{++} silicon on glass, it is ideal for hybrid or flip chip bonding technology.

The single-crystal tuning fork is fabricated using a dissolved wafer process using reactive ion etching (RIE) and boron diffusion to define the final structure. The fabrication sequence involves both silicon and glass processing, as shown in Fig. 51.3. Processing starts with a P-type (100) silicon wafer of moderate doping (>1 ohm-cm). A recess is etched into the silicon using KOH. This recess defines the gap spacing of the conducting plates. A high-temperature (1150°C) boron diffusion, which defines the thickness (5 to 10 μM) of the structure, follows. The features of the structure are defined by using RIE. By etching past the P^{++} etch stop, the structures are released. A CF_3Br chemistry was used for the etching in a parallel plate reactor, which gives straight side walls and high aspect ratios.

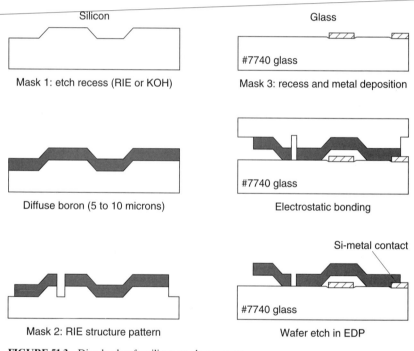

FIGURE 51.3 Dissolved wafer silicon on glass process.

The glass processing involves recessing the glass 1600 Å, then depositing and lifting off a multimetal system (400 Å Ti/700 Å Pt/1000 Å Au) on a #7740 Corning glass wafer. This results in a planar structure with the metal protruding only 500 Å above the surface of the glass. The metal forms the sense and drive plates of the capacitor and the output leads from the transducer. The silicon and glass are then electrostatically bonded together.[7,8] The bonding takes place at 375°C with a potential of 1000 V applied between the glass and silicon. In order to make electrical contact to the silicon, a metal lead on the glass is allowed to overlap the silicon rim over a small area.

During the electrostatic bonding process, the silicon and glass are drawn tightly together, a process that ensures a low-resistance silicon to gold contact (40 ohms for a 40 μm × 20 μm area). The final step is a selective etch in ethylene diamine pyrocatechol, which dissolves the silicon and stops on the heavily doped (P++) diffused layers. The overall fabrication sequence requires only single-sided processing with two masking steps on silicon and one on the glass. The process is high yield and is compatible with batch processing.

ELECTRONICS

A block diagram of the tuning fork gyroscope and electronics is shown in Fig. 51.4. The electronics consist of an oscillator loop to drive the proof mass and detection for the sense axis motion.

The drive axis operates in a self-resonant oscillator loop to allow the resonant frequency to shift with temperature, age, stress, or pressure. This ensures that the gyro is on resonance despite environmental disturbances and eliminates external frequency sources. Nonlinearity in the double-ended tuning fork springs causes the resonant frequency to increase with vibration amplitude. The self-oscillation circuit provides lock in and the correct phase to drive at large amplitudes.

The critical task of the electronics is the detection of the proof mass position. Circuitry for detecting the TFG proof mass motion is shown in Fig. 51.5. The sense capacitor plates are excited with constant bias voltages or carrier excitation applied to the opposing plates. A differential motion of the masses results in current flow into the operational amplifier. Critical requirements for this preamplifier are low noise, high impedance, and high gain bandwidth product. Detecting drive axis position for the oscillator loop is similar.

A constant angular rate results in the proof mass moving in the sense direction at the drive frequency. The input rate is modulated by the proof mass velocity. Assuming DC bias voltage, the sense axis position signal is demodulated by the drive veloc-

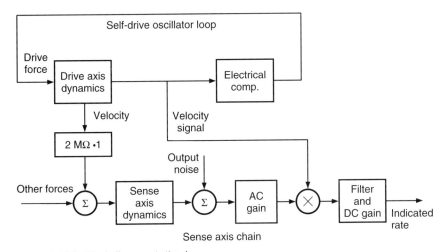

FIGURE 51.4 Block diagram of vibrating gyroscopes.

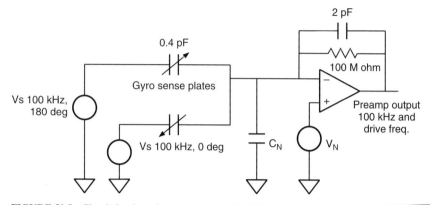

FIGURE 51.5 Circuit for detecting sense axis motion. Node capacitance C_N and input voltage noise V_N are employed for modeling performance. The excitation can be a carrier as shown or a DC bias.

ity. (If a carrier is employed, an additional demodulation step is required.) To a first approximation, demodulation is multiplication as shown in Fig. 51.4. Along with the desired information in phase with the proof mass velocity, signals generally exist in quadrature—that is, 90° out of phase with the rate signal.

For most practical gyros, the baseband transfer function is a second-order system whose resonant frequency occurs at $|\omega_s - \omega_d|$, where ω_s and ω_d are the eigenfrequencies of the sense and drive modes.

TEST RESULTS

Through July 1994, 90 gyros of different configurations had been tested on three generations of electronics. These data were taken with the Engineering Development Model 2 (EDM2) electronics.

Earlier testing with probe stations and strobed microscopes verified that the correct modes were being excited. Good agreement has been achieved between calculated (finite elements) and observed mode frequencies. Amplitude of motion in air was observed to be on the order of 10 μm zero to peak, using a DC bias voltage of 30 V and an AC drive voltage of 30 V peak on the comb drive electrodes. For gyro operation in vacuum, the drive amplitude is set to 8 μm. The reported results were obtained with units evacuated to 200 mTorr.

In-plane motion is lightly damped by air, while out-of-plane motions are strongly damped due to squeeze film effects. Perforations in the masses (Fig. 51.1) minimize air trapping; however, the output mode is still heavily damped in air. For out-of-plane modes, Q rises rapidly as pressure is reduced. In contrast, the in-plane drive direction shows a smaller increase as the air is removed. At pressures of 100 mTorr, the drive axis Q is 40,000 and the sense is 5000.

Scale factor, output versus angular rate input at the preamplifier output (Fig. 51.5), for a single-crystal tuning fork gyro type TFG 11 (Fig. 51.1) is 1.25 mV/rad/s. After demodulation and gain, the gyro scale factor is 0.4 V/rad/s, as seen in Fig. 51.6. This gyro was driven at a 10 μm-amplitude. The quadratic term at input rates of

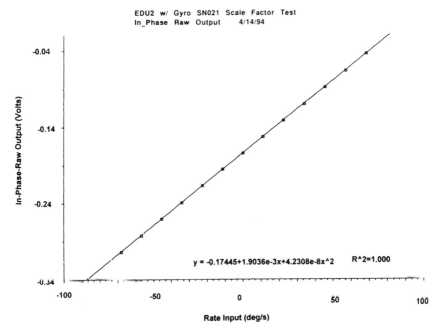

FIGURE 51.6 Output versus input angular rate.

90°/sec. is 0.2 percent of the linear term; that is, nonlinearity is less than 0.2 percent. For a week of testing across temperature and shut down, the scale factor repeated to better than 0.1 percent (1000 ppm).

Shown in Fig. 51.7, the minimum resolvable rate (noise-equivalent rate) is 470°/hr. (0.13°/sec.) in a 60-Hz bandwidth with the oscillator loop in high yield units. This is equivalent to 0.02°/sec. in a 1-Hz bandwidth or angle random walk of 0.72°/$\sqrt{hr}$. In special units, resolution of 150°/hr. was measured with open-loop drive. Resolution is limited by the gyro gain combined with preamplifier input noise. Larger gyros and improved preamplifiers will bring the noise equivalent rate down under 100°/hr. in a 60-Hz bandwidth within the next year.

Gyro output, essentially bias, and temperature are plotted versus time in Fig. 51.8 for gyro SN 041. In 10°C steps, the temperature was cycled from −40 to +70°C. At the temperature extremes, the raw bias deviates from that at room temperature by 1.5°/sec., acceptable for many automotive applications. At 9 and 25 hours, the 70°C does not repeat, so that the residual after fitting bias with linear and quadratic temperature is better than 0.1°/sec. The oven temperature shown in Fig. 51.8 may differ from that of the gyroscope or its electronics. Because performance is acceptable for automotive applications, thermal testing has focused on additional units. In the coming months, bias stability should be better than 100 to 200°/hr. over −40 to +85°C with compensation.

In Fig. 51.8, the data was passed through a 0.1-Hz low-pass filter. Through precision micromachining and attention to electronic details, the bias magnitude is roughly 2°/sec. With adjustment for drive amplitude, the bias uncertainty, "the thickness of the line," corresponds to the 60-Hz resolution (Fig. 51.7). In the latter portion of Fig. 51.8, some trending in bias is evident. For 5-h samples at room temperature,

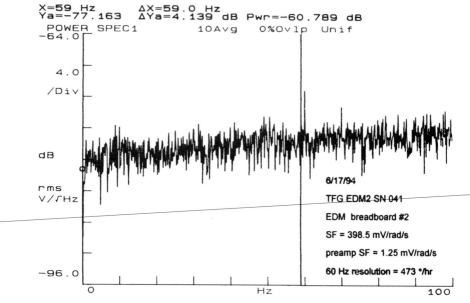

FIGURE 51.7 Tuning fork gyroscope rate equivalent noise.

the rms uncertainty is 28°/hr. with 0.1-Hz bandwidth without removing the trend. For 16-hour samples, the uncertainty is 55°/hr. with 0.1-Hz bandwidth without removing the trend.

Acceleration sensitivity was measured by tumbling low-gain tuning fork gyroscopes with respect to gravity. In the four units tested, the acceleration sensitivity could not be measured because it was lower than the test's threshold limit; that is, acceleration sensitivity is less than 200°/hr./g. Based on construction tolerances and measured eigenfrequencies, the calculated acceleration sensitivity is 25°/hr./g.

CONCLUSIONS

A novel micromachined comb drive tuning fork gyroscope has been fabricated using the single-crystal silicon on glass, dissolved wafer process. Draper Laboratory and Rockwell have allied to produce the comb drive tuning fork gyro for automotive applications. Excellent resolution, bias stability, and scale factor have been demonstrated. Current effort is focusing on thermal modeling, improved preamplifiers, and scaling to production quantities.

Possible benefits of the tuning fork approach include: (1) simpler fabrication than other configurations, (2) better performance because of the relatively large drive amplitude, and (3) static balance and stiffer springs that reduce sticking and vibration or acceleration sensitivity. The low stray capacitance of the glass substrate makes the dissolved wafer gyros easy to mate with hybrid electronics.

Silicon gyroscopes whose active elements are roughly 0.7 mm have demonstrated 55°/hr. bias stability and 470°/hr. resolution in a 60-Hz bandwidth. Scale factor sta-

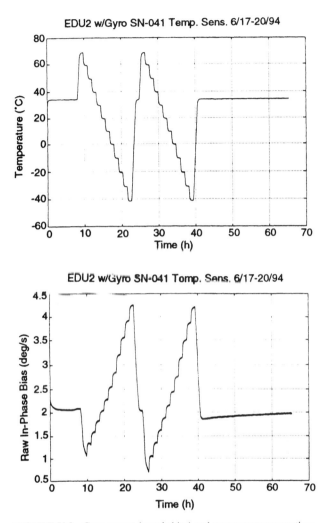

FIGURE 51.8 Gyro output (mostly bias) and temperature versus time.

bility is less than 1000 ppm. Within the coming year, performance should be less than 50°/hr. over temperature. These inexpensive instruments satisfy many current and future applications, such as tactical missile, automotive guidance and control, and camcorder stabilization.

REFERENCES

1. P. Greiff, B. Boxenhorn, T. King, and L. Niles, "Silicon Monolithic Gyroscope," *Transducers '91, Digest of Technical Papers,* 1991, pp. 966–969.

2. G. Newton, "Comparison of Vibratory and Rotating Wheel Gyroscopic Rate Indicators," *Proceedings AIEE*, July 1960.
3. W. C. Tang, M. G. Lim, and R. T. Howe, "Electrostatically Balanced Comb Drive For Controlled Levitation," *IEEE Solid State Sensor and Actuator Workshop*, Hilton Head Island, South Carolina, June 4–7, 1990, pp. 23–27.
4. B. Boxenhorn, "A Vibratory Micromechanical Gyroscope," *AIAA Guidance and Control Conference*, p. 1033.
5. J. Soderkvist, *A Mathematical Analysis of Flexural Vibrations of Piezoelectric Beams with Applications to Angular Rate Sensors*, Ph.D. Thesis, Uppsala University, Sweden, 1990.
6. W. C. Tang, T. H. Nguyen, and R. T. Howe, "Laterally Driven Polysilicon Resonant Microstructures," *Sensors and Actuators*, **20**, 1989, pp. 25–32.
7. L. Spangler and I. C. D. Wise, "A New Siliconon-Glass Process for Integrated Sensors," *IEEE Sensor and Actuator Workshop Tech. Digest*, June 1988, pp. 140–142.
8. Y. Gianchandani and K. Najafi, "A Bulk Silicon Dissolved Wafer Process for Microelectromechanical Systems," *Technical Digest, International Devices Meeting (IEDM)*, December 1991, pp. 757–760.
9. M. Weinberg, J. Bernstein, S. Cho, A. T. King, A. Kourepenis, and P. Maciel, "A Micromachined Comb-Drive Tuning Fork Rate Gyroscope," *Proceedings of the 49th Annual Meeting, "Future Global Navigation and Guidance,"* The Institute of Navigation, June 21–23, 1993.
10. M. Putty and K. Najafi, "A Micromachined Vibrating Ring Gyroscope," *Technical Digest, Solid-State Sensor and Actuator Workshop*, Hilton Head, South Carolina, June 13–16, 1994.
11. F. Aronowitz and S. Hammons, "Micromachined Quartz Sensors for Tactical Missions," *Proceedings of the 49th Annual Meeting, "Future Global Navigation and Guidance,"* The Institute of Navigation, June 21–23, 1993.

CHAPTER 52
AUTOMOTIVE APPLICATIONS OF LOW-G ACCELEROMETERS AND ANGULAR RATE SENSORS

INTRODUCTION

Automotive volumes have provided the impetus for several sensor manufacturers to develop new sensor products and extend the range of existing products. Accelerometers for crash sensing have received significant attention to meet legislated requirements and customer demand for increased safety. However, additional accelerometer applications exist and are being developed for lower operating ranges and different performance criteria. Additionally, the combination of low-G accelerometers and angular rate sensors under development will provide the capability for low-cost, commercial-grade inertial measurement units. These units can be used in automotive navigation and integrated vehicle dynamic applications. The large production volumes associated with the small, low-cost gyroscopes used to sense angular rate in automotive applications also offer the potential for applying these inertial devices to sense roll/pitch/yaw in new commercial applications that have not previously been considered. This paper will explore the expanded automotive applications for accelerometers, the impact of the development of low-cost angular rate sensors, and the performance requirements associated with these applications.

The high volume applications in the automotive industry is one of the main driving forces behind the current development of solid-state mechanical sensors. The main thrust in technology to meet the requirements of automotive accelerometer applications in the semiconductor industry has been twofold: using bulk micromachining and surface micromachining technologies. Other technologies such as piezoelectric devices are available, but in the applications under consideration do not have the acceptance of micromachined devices. Described in this chapter is the surface micromachining approach to cost-effective accelerometers and angular rate sensors using capacitive sensing techniques. Capacitive sensing and control is the only method that can make use of electrostatic force feedback to minimize the variations in the mechanical properties of the sensor. Surface micromachining was selected, as it is the most cost-effective approach to meet the requirements, to process integration with MCU-compatible technology for the development of a smart sensor and self-test features, and to meet the rugged small size requirements and reliability needed in such devices. A cost-effective 50-G automotive crash sensor has been developed and shown to meet the specification. This technology is now being expanded to low-G devices to meet other automotive requirements, such as ride control, ABS, traction, cornering, and vibration-detection applications. Novel structures are also being developed for angular rate sensors. These sensors will initially be used in low-cost inertial navigation systems, but their use could be extended to include active suspension, ABS, crash recorders, and headlight-leveling system applications.

SENSING TECHNOLOGIES

There are two basic method of sensing used with micromachined sensors. These are capacitive and piezoresistive. Capacitive sensing, however, is the only method that can be used in force feedback systems for closed-loop operation. In this kind of system, external forces are balanced by electrostatic forces. Capacitive sensing can also be applied to many other different types of sensors (e.g., pressure, humidity, motion, and vibration). Differential capacitive sensing also has the following advantages:

- Temperature insensitive
- Very high sensitivity
- Small size
- Wide operating temperature range
- Can have a DC response
- Can use closed-loop techniques to minimize the variations in mechanical properties of the sensors
- Self-test features available

To obtain the mechanical movement required in acceleration and angular rate sensors using silicon, there is a choice of micromachining the seismic mass from either bulk silicon or using surface micromachining techniques. Table 52.1 gives a comparison of the characteristics of the two types of micromachined structures.

Because of smaller size, lower cost, and the possibility of expansion to other types of devices, surface micromachining was chosen. This technology also has the poten-

TABLE 52.1 Characteristics of Bulk and Surface Micromachining

Bulk	Surface
Large	Small
Crystal oriented etching	Isotropic etching
Etch stops	Selectivity
Integration	Easier integration
High cost	Lower cost
No annealing	Annealing required
Capacitive or piezoresistive sensing	Capacitive sensing

tial of easier integration with CMOS processing as used in MCU's.[1,2] This will make the evolution of the smart sensor a reality.

ACCELERATION SENSOR

Figure 52.1 shows the basic structure of a surface micromachined accelerometer. It consists of three layers of polysilicon. The second layer contains the seismic mass, which is suspended by two cantilever arms and is free to move between the first and third layers. Movement is sensed by measuring the differential capacitance between the second layer and the first and third layers of polysilicon. This type of three-layer polysilicon structure is fabricated by depositing and shaping alternate layers of polysilicon and glass. The glass is then etched away to leave a free-standing polysilicon mass as shown. This structure has a number of inherent advantages in addition to those described above. The third layer of polysilicon constrains the active second layer polysilicon, or seismic mass. This structure has good mechanical stability against cross axis forces, high shock tolerance, and low cross-axis sensitivity. The sensor can be capped with a known internal pressure during processing in a clean room. This will eliminate particulates from later processing steps to ensure high reliability and longevity. The internal pressure provides squeeze film damping and is used to control bandwidth[3] and critically damp any resonant frequencies of the structure.

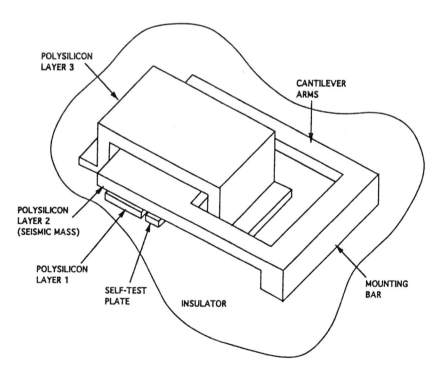

FIGURE 52.1 Three-layer micromachined accelerometer.

ANGULAR RATE GYROSCOPE

Micromachining technology has made possible very small cost effective devices to sense angular rotation. The implementation of one such gyroscopic device for measuring roll/pitch/yaw is shown in Fig. 52.2. This device is fabricated on a silicon substrate using the same surface micromachining techniques as used for the accelerometer. In this case, three layers of polysilicon are also used with the first and third layers being fixed and the second layer free to vibrate about its center. The center is held in position by four spring arms attached to four mounting posts as shown. This device can sense rotation about two axes; that is, the x- and y-axes; and sense acceleration in the direction of the z-axis. The center layer of polysilicon is driven into oscillation about the z-axis by the electrostatic forces that are produced by voltages applied between the fixed comb fingers and the comb fingers of the second polysilicon. Capacitor plates as shown are formed between the first and third polysilicon on the x- and y-axes and the second layer of polysilicon. Differential capacitive sensing techniques are used to sense any displacement of the vibrating disc caused by angular rotation. For example, if angular rotation takes place about the x-axis, Coriolis forces produce a deflection of the disc about the y-axis. This deflection can then be detected by the capacitor plates on the x-axis. The sensing of the three functions is achieved by using a common sensing circuit that alternatively senses the x-rotation, y-rotation, and acceleration. The gyroscope is designed to have a resolution of <1° per second for angular rate measurements and an acceleration resolution of <20 mg.

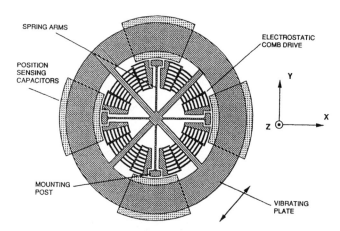

FIGURE 52.2 Three-layer micromachined gyroscope structure.

CIRCUIT TECHNOLOGY

Automotive control systems contain MCUs for signal processing. By choosing CMOS signal conditioning circuits, future integration with the MCU becomes a reality, which can also give a reduced package count.

Other advantages that are obtained from the use of the CMOS processing are as follows:

- Low current consumption
- Switched capacitor filters for noise reduction and bandwidth control
- Switched capacitor techniques for sensing
- EPROM for trimming
- Analog or digital control, with either output formats
- Minimized package count
- Lower system cost and better reliability due to the elimination of external components
- Customization of MCU (CSIC device) for specific customer requirements

Figure 52.3 shows a block diagram of the system. The accelerometer is self-contained. The control die has an onboard voltage reference, an oscillator, driver circuits, a detection circuit, a sample and hold, a switched capacitor low-pass filter used to set the bandwidth and reduce noise, an output buffer, and the trim network necessary for adjusting the offset and sensitivity. Self-test features are also incorporated in the system for diagnostics. The device assembled on a 16-pin dual in line plastic package lead frame is shown in Fig. 52.4. There are a number of approaches to signal conditioning that can be used in this and other applications.[4] These range from simple open loop differential capacitive sensing systems through bridge schemes to complex closed loop servo systems. These systems can use analog or digital control loops and have analog or digital outputs. The control circuit chosen for the low-G accelerometers and angular rate sensors were closed-loop systems. These systems have the advantage of minimizing the effects of variations in the mechanical parameters of the sensor. The feedback is normally achieved by using electrostatic forces to counterbalance the forces on the seismic mass produced by acceleration. This maintains the mass in its at-rest position.

Figure 52.5 shows the block diagram of an analog closed-loop system.[5] In this system, the top and bottom plates are held at a high and low DC reference voltages,

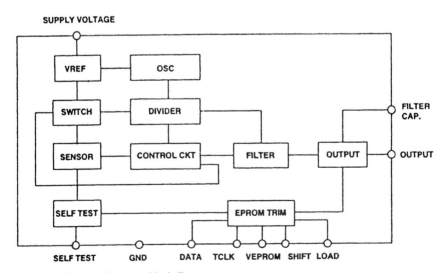

FIGURE 52.3 Accelerometer block diagram.

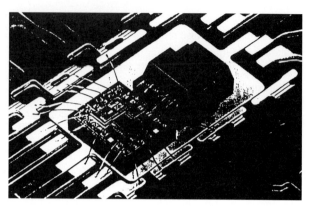

FIGURE 52.4 Packaged accelerometer.

respectively. An AC voltage is also superimposed on the top and bottom plates. The AC voltage applied to the two plates is in antiphase and is used to sense the position of the middle plate. The signal from the middle plate is amplified and fed through a sample and hold circuit and a filter. The signal is then fed back on to the center plate as a DC voltage. This voltage produces electrostatic forces between the middle, top, and bottom plates to counterbalance any deflection produced by an accelerating force. It can be shown that, in such a system, there is a linear relationship between acceleration and output voltage.

In a digital control system, a digital output can be obtained directly. Such a system using a delta modulator system is shown in Fig. 52.6(a).[6] In this case, the voltage on the center polysilicon is set at one of the plate voltages VF, while the other plate

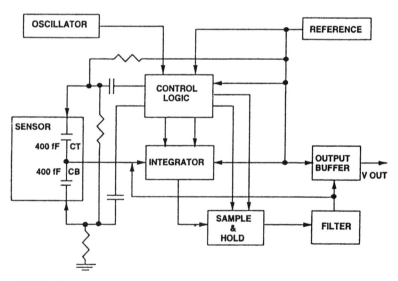

FIGURE 52.5 Analog closed-loop system.

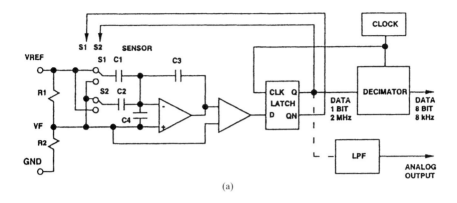

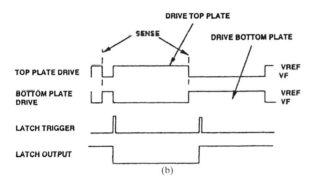

FIGURE 52.6 Digital closed-loop system.

is held at a reference voltage V_{ref}. The signal from the middle plate is used to set a latch, which feeds the V_{ref} voltage back to either the top or bottom plates.

This voltage will produce a counter balancing electrostatic force to offset the accelerating force. This system is periodically sensed and the latch set. Acceleration is thus converted into the time domain; i.e., an output period proportional to acceleration is obtained. This can be seen in Fig. 52.6(*b*), which shows the waveforms associated with the delta modulator. The output from the latch can either be fed through a low-pass filter to convert it to an analog signal or fed out directly as a serial digital signal. The digital signal has the distinct advantage of maintaining signal integrity in remote sensing systems and high noise environments.

LOW-G ACCELEROMETER APPLICATIONS

Low-G linear accelerometers are being developed for ride control, ABS, traction, and inertial navigation applications. Solid-state acceleration sensors have special mounting requirements that are different from normal integrated circuits. These are to ensure that acceleration forces are transmitted to the sensor package. In acceleration applications, the devices are required to operate over the temperature range –40 to 85°C (125°C under the hood), and to withstand >2000 G shock. The low-G accelerometers used on the axles of the four wheels in ride control systems to detect the load changes on the wheels have the following specification: ±2 G full scale; accuracy, ±5 percent over temperature; bandwidth, DC to 50 Hz; and cross-axis sensitivity, <3 percent. The acceleration information and data from the wheel speed sensors is used to provide the information necessary for the MCU to operate the dynamic ride control valves. Acceleration sensors can also be used to minimize the roll angle during cornering. The roll angle of a typical vehicle fitted with fully active suspension can be reduced from 6° to less than 5° with a lateral acceleration of 8 m/sec.² A combination of sensors is used for active suspension. These are accelerometers, speed sensors, chassis to ground sensing, and level sensing in the suspension system.

In ABS and traction control systems, the sensors measure deceleration when braking and acceleration when the throttle is opened. If skidding occurs during braking, the brake pressure is reduced and adjusted for maximum deceleration, or if wheel spin occurs during acceleration, the throttle is adjusted for maximum traction. A typical specification for the accelerometer required in this application is ±1 G full-scale output; accuracy, ±5 percent over temperature; bandwidth, 0.5 to 50 Hz; and cross-axis sensitivity, <3 percent. Other low-G applications exist for accelerometers in the automotive arena such as headlight leveling and crash recorders.

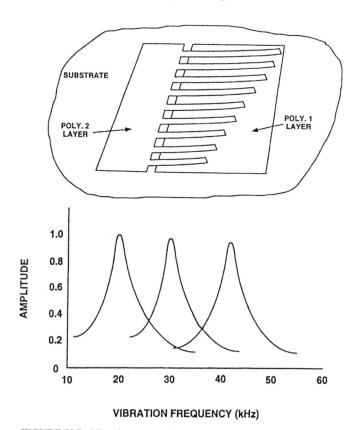

FIGURE 52.7 Vibration sensor output.

A type of acceleration sensor is the vibration sensor. This device contains a number of fingers of varying length that will vibrate at their resonant frequency when that frequency is encountered. The resonance is capacitivly coupled to the sensing circuit, and the outputs shown in Fig. 52.7 are obtained. Vibration sensors can be used as an engine antiknock sensor, in adjustable engine mounts for vibration reduction, or for vibration monitoring in maintenance applications.

ANGULAR RATE GYROSCOPE APPLICATIONS

Long-range inertial navigation systems and inertial measurement units (IMU) rely to a large extent on high accuracy accelerometers and gyroscopes. These units are used to provide location data for intelligent vehicle highway systems.[7,8] A typical accelerometer and gyroscope specification for this application is given in Table 52.2.

TABLE 52.2 IMU Gyroscope Specification

	Gyroscope	Accelerometer
Input range	±100°/sec.	±20 m/sec.^{2} (2 G)
Sensitivity	20 mV/°/sec.	100 mV/m/sec.^{2}
Output	0.5–4.5 V	0.5–4.5 V
Offset	2.5 V	2.5 V
Offset drift	<360°/hr.	±0.5 m/sec.^{2}
Operating temperature	−40 to +85°C	−40 to +85°C

A centrally located IMU can be expanded to cover other applications such as suspension, ABS, traction, and working with crash avoidance sensors. This may be the way for cost-effective system design in the future.

CONCLUSION

Surface micromachining technology has been developed and shown to be cost-effective for use in low-G acceleration devices and angular rate sensors. These devices are being designed into automotive systems. The systems applications include ride control, ABS, traction, and inertial navigation. The angular rate sensor possibly has a greater application potential than the accelerometer. Its usage can be expanded from IMUs to cover most of the applications using accelerometers. It is possible in the long term that a single, cost-effective, centrally located IMU can be used to provide the sensing and control signals for many of the above applications now using accelerometers.

REFERENCES

1. Lj. Ristic, et al., "Surface Micromachined Polysilicon Accelerometer," *IEEE Solid State Sensor and Actuator Workshop,* Hilton Head, South Carolina, 1992, pp. 118–121.

2. H. Steele and L. May, "Attributes of Surface Micromachining as shown by a Monolithic Accelerometer," *Sensors Expo West Proceedings 1992,* C53A-1, Anaheim, California, 1992.
3. S. Terry, "A Miniature Silicon Accelerometer with Built-In Damping," *Tech. Dig., IEEE Solid State Sensor and Actuator Workshop,* Hilton Head, South Carolina, 1988, p. 114.
4. R. Payne and K. Dinsmore, "Surface Micromachined Accelerometer: A Technology Update," *SAE,* Detroit, Michigan, 1991, p. 127.
5. F. Rudolf et al., "Precision Accelerometers with μg Resolution," *Sensors and Actuators* **A21, A23**, 1990, p. 297.
6. Weijie Yun, R. T. Howe, and P. R. Gray, "Surface Micromachined, Digitally Force-Balanced Accelerometer with CMOS Detection Circuitry," *Solid State Sensor and Actuator Workshop,* Hilton Head, South Carolina, 1992, pp. 126–131.
7. J. Hellaker, "Prometheus-Strategy," *Proceedings of the International Congress Transportation Electronics,* SAE 901139, 1990, pp. 195–200.
8. H. Okamoto and M. Hase, "The Progress of AMTICS—Advanced Mobile Traffic Information and Communication System," *Proceedings of the International Congress Transportation Electronics,* SAE 901142, 1990, pp. 217–224.

CHAPTER 53
MICROFABRICATED SOLID-STATE SECONDARY BATTERIES FOR MICROSENSORS

INTRODUCTION

A thin-film solid-state Li/TiS$_2$ microbattery has been developed at Eveready Battery Company (EBC). It is fabricated using sputtering for deposition of the metal contacts, TiS$_2$ cathode, and oxide-sulfide glassy electrolyte. High-vacuum evaporation is used to deposit a LiI layer and Li anode. EBC microbatteries range from 8 to 12 μm in thickness and have a capacity between 35 and 100 μA-hr./cm^2, depending on the amount of anode and cathode deposited. The microbattery OCV is approximately 2.5 V. EBC microbatteries routinely go more than 1000 cycles between 1.4 and 2.8 V, with greater than 90 percent cathode utilization, at current densities as high as 300 μÅ/cm^2. Several EBC microbatteries have gone over 10,000 cycles at 100 μÅ/cm^2 with greater than 90 percent cathode utilization on each cycle. The microbatteries will cycle at temperatures as low as −10°C at a current density of 100 μÅ/cm^2, are capable of supplying pulse currents of several mÅ/cm^2, and have excellent long-term stability, both on shelf and while cycling. Sets of five microbatteries have been assembled in either a series or parallel configuration and cycled more than 1000 times. EBC microbatteries as large as 10 cm^2 have been fabricated and cycled over 1000 times at close to 100 percent cathode efficiency.

The miniaturization of electronic devices has, in many cases, resulted in extremely low current and power requirements. This has made possible the use of thin-film rechargeable microbatteries as power sources for these devices. Some of the advantages offered by thin-film microbatteries for these applications include: (1) they are manufactured by the same techniques as currently used in the microelectronics industry; (2) the extreme thinness of the electrolyte layer allows the use of relatively poor ionic conductors such as glassy, solid-state electrolytes; (3) the sequential vacuum deposition processes provide cleaner and more intimate interfaces between layers; and (4) the microbattery can be constructed in almost any two-dimensional shape. The fabrication of a thin-film rechargeable microbattery has been investigated for many years.[1-5] However, no one has yet reported the development of a microbattery with properties suitable for use as a long-life, rechargeable power supply. We feel that the performance of the EBC microbattery has accomplished this goal. The outstanding secondary performance with thousands of deep discharges, the capability of current density as high as several mÅ/cm², and the long-term stability both on shelf and during cycling make the EBC microbattery superior to any other microbattery reported to date.

EXPERIMENTAL

The EBC microbattery is constructed by use of appropriate masks in the following sequence:

1. DC magnetron sputtered chromium contacts
2. An RF magnetron sputtered TiS_2 cathode
3. An RF magnetron sputtered oxide-sulfide glass electrolyte
4. A vacuum-evaporated film of LiI
5. A vacuum-evaporated lithium anode

A recently developed protective coating used to isolate the EBC microbattery from the environment is a proprietary polymer covered by a metallic layer.[6] The polymer coat is produced by vapor-phase deposition with polymerization of the conformal coating occurring at RT. The metallic coating is deposited by sputtering or vacuum evaporation. These layers may be repeated if required. The microbattery configuration is shown in Fig. 53.1. The masks are simple line-of-sight masks machined from aluminum or Macor ceramic. The substrate for the microbattery is usually a glass microscope slide, but any reasonably smooth surface can be used. For example, alumina, mylar, and paper have also been used as the substrate. The use of mylar and paper have shown that the EBC microbattery has good flexibility in spite of its metallic and glass components.

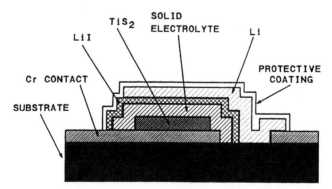

FIGURE 53.1 Cross section of the EBC microbattery.

The sputtered layers are deposited using a Materials Research Corporation 903 Sputtering System. The sputtering gas is ultra high purity argon from Linde. The chromium target was purchased from MRC and is used with an MRC 900 cathode. The TiS_2 target and the solid electrolyte target are made at EBC Technology. The electrolyte target composition is $6LiI-4Li_3PO_4-P_2S_5$. The targets are constructed from the powdered material that has been pressed into tiles of 1 in². These tiles are mounted on special Vac-Tec cathodes that require only half the normal target area to be covered with target material (40 in^2 vs. 75 in^2). All target preparation at EBC is done in a stainless steel Vacuum/Atmospheres Dri-Lab. An MO-40 Dri-Train keeps the argon atmosphere of the Dri-Lab below 1 ppm H$_2$O.

The vacuum evaporation system was built at EBC Technology and installed on a second stainless steel Vacuum/Atmospheres Dri-Lab such that the bell jar is inside the Dri-Lab. All other components of the system (turbomolecular pump, LN$_2$ trap, power supply, etc.) are outside the Dri-Lab. An MO-40 Dri-Train keeps the argon atmosphere of the Dri-Lab below 1 ppm H$_2$O and 10 ppm O$_2$.

A custom-built stainless steel glove box was installed over the load lock area of the MRC 903. Circulation through a canister of molecular sieves keeps the argon atmosphere of the glove box below 5 ppm H$_2$O. This glove box is attached to the antechamber of the Dri-Lab containing the vacuum evaporation system. It is therefore possible to prepare substrates, change masks, sputter components, and vapor-deposit components without removing the microbattery from a dry, inert atmosphere. The LiI (Aldrich), Li$_3$PO$_4$ (Pfaltz and Bauer), and P$_2$S$_5$ (Alfa) are dried before use. The TiS$_2$ (Degussa) and Li (Foote Mineral) are used as received.

The microbattery discharge and cycling data are acquired using a Hewlett-Packard 3852A Data Acquisition Unit controlled by a Hewlett-Packard 319M Computer. The constant current sources for the test system were made at EBC Technology. Because of the higher current values needed for the pulse data, a Keithley 224 Programmable Current Source is used for these measurements. Complex impedance measurements are done using a Solatron 1170 Frequency Response Analyzer and a Solatron 1186 Electrochemical Interface controlled by a Hewlett-Packard 9845B Computer.

RESULTS AND DISCUSSION

Contacts

Sputtered films of aluminum were originally used for the contacts of the EBC microbattery. However, the sputtered aluminum films were found to be too reactive to function as reliable contacts. They reacted with the sulfides contained in the other battery components and also formed Al$_2$O$_3$ crystals on their surface with the small amount of residual H$_2$O and O$_2$ in the Dri-Lab. These reactions increased the resistance of the Al contacts and also created a rough surface for deposition of the subsequent films of the microbattery.

Because of the potential problems that aluminum presented as a contact material for the microbattery, a chromium target was substituted. Contacts made from sputtered chromium do not have the problems associated with the aluminum contacts, but care must still be exercised with the chromium contacts due to what appears to be a decreased adherence to the glass substrate and an increased film brittleness. However, if the substrate cleanliness and the sputtering conditions are properly controlled, sputtered chromium makes an adequate contact for the EBC microbattery.

Cathode

The intercalation compound TiS$_2$ was chosen as the microbattery cathode due to EBC's experience with it as a secondary cathode. Analysis of sputtered TiS$_2$ show the films have good stoichiometry (TiS$_{2.09}$), with only a slight excess of sulfur.

The sputtered TiS_2 cathode is the limiting electrode of the microbattery. The theoretical capacity is used to calculate the microbattery's discharge efficiency. Because the volume of the TiS_2 film was used to determine the cathode capacity, the density of sputtered TiS_2 film was gravimetrically determined. The density was found to be 1.47 g/cm^3, about 45 percent of the theoretical density of single crystal TiS_2. SEM photomicrographs of the sputtered TiS_2 films show that they are similar in appearance to the CVD TiS_2 film reported by Kanehori et al.[2] The CVD TiS_2 had a reported density of 65 percent of theoretical. The major difference between the CVD TiS_2 films and the sputtered TiS_2 films is the small crystallite size of the sputtered films (0.3–0.4 μm vs. 10–12 μm). Because of this small crystallite size, the sputtered films appeared amorphous to X-ray diffraction, and therefore the orientation of the sputtered material could not be determined. To obtain the least hindered Li insertion, one would prefer the TiS_2 to be oriented such that the c-axis was parallel with the substrate surface. However, as will be shown later in this paper, the sputtered thin films of TiS_2 will cycle at relatively high current densities. This indicates that the orientation of the TiS_2 crystallites is not a problem with the sputtered films, probably because of the high surface area of the small crystallites.

Electrolyte

The first electrolyte to be tried had the composition $5LiI\text{-}4Li_2S\text{-}1.95P_2S_5\text{-}0.05P_2O_5$. This electrolyte was chosen because it had been used successfully at Eveready to make thick-film solid state lithium coin cells. Thin films sputtered from this target had a conductivity of 2×10^{-5} S/cm. This is about an order of magnitude less then the 5×10^{-4} S/cm conductivity of bulk glass of the same composition.[7] Although microbatteries were constructed with this electrolyte, the high sulfide content of this target presented a maintenance problem with the sputtering system. Therefore, the use of this target was discontinued.

A second target was constructed with a reduced sulfide level and had the composition $4LiI\text{-}Li_2S\text{-}2Li_3PO_4\text{-}P_2S_5$. A sputtered film of this electrolyte had a conductivity of 2×10^{-5} S/cm, equivalent to that of the higher sulfide-containing electrolyte. However, many of the microbatteries constructed using this electrolyte were shorted. SEM photomicrographs showed that the sputtered films of this electrolyte had a tendency to crack due to internal stress, and this was resulting in shorted cells.

The next target that was constructed had the composition $4LiI\text{-}Li_2S\text{-}3Li_2O\text{-}B_2O_3\text{-}P_2S_5$. Thin films of this electrolyte did not show any cracks in SEM photomicrographs; however, the conductivity of the sputtered film was reduced by an order of magnitude to 2×10^{-6} S/cm. Because of the thinness of the electrolyte layer in a microbattery (1–3 μm), a solid electrolyte with a conductivity of 10^{-6} S/cm would be acceptable. For example, a battery of 1 cm^2 area using this electrolyte would only add 50 Ω to the cell IR for every μm of electrolyte thickness.

However, it was still felt that the sulfide content of this electrolyte was causing equipment maintenance problems. Therefore, another electrolyte sputtering target was constructed that further reduced the sulfide level. It had the composition $6LiI\text{-}4Li_3PO_4\text{-}P_2S_5$. SEM photomicrographs of sputtered films of this electrolyte did not show any cracking, and the conductivity of these films was again found to be about 2×10^{-5} S/cm. Impedance spectroscopy was used to determine the reactivity of the sputtered TiS_2 film with this sputtered electrolyte. These two layers did not seem to be reacting to any significant degree. It was therefore decided that this electrolyte would be used in the fabrication of the EBC microbattery.

LiI Layer

LiI is a Li ion conductor and could be used as the solid electrolyte. However, it has a relatively low conductivity of 1×10^{-7} S/cm.[8] Also, LiI forms electronically conductive "color centers" if the stoichiometry of the material is modified.[9] It is very possible that this could occur during deposition of the Li layer; therefore, the LiI is considered an ionically conductive protective layer over the oxide-sulfide electrolyte. This layer was necessary to prevent a passivating film, probably Li_2S, from forming when the Li was vapor-deposited directly onto the oxide-sulfide electrolyte. Complex impedance measurements on EBC microbatteries show the total ionic conductivity (glassy electrolyte + LiI) to be 2×10^{-6} S/cm. This conductivity value correctly falls between the 10^{-7} S/cm conductivity of the LiI layer and the 10^{-5} S/cm conductivity of the oxide/sulfide electrolyte layer.

Anode

Li films are vapor-deposited rather then sputtered due to the low melting point of the Li metal. Vapor deposition of the Li film is relatively fast with Li being deposited on the substrate at a rate of 0.1 μm/s. The bottom of the vapor-deposited Li film, as viewed through a glass microscope slide substrate, has a mirrored appearance. However, despite the facts that the water and O_2 content of the Dri-Lab are kept very low by the Dri-Train, and that Li foil is placed throughout the Dri-Lab to scavenge reactive molecules, the top surface of the vapor-deposited Li film quickly turns a dull gray. This rapid filming over of the exposed Li surface is the reason that the Li is the last layer of the microbattery to be deposited.

Microbattery

Typically, EBC microbatteries have a total thickness around 10 μm and an OCV between 2.4 and 2.5 V. The primary discharge of the EBC microbattery shows a classically shaped curve and close to 100 percent cathode utilization to a 1.8-V cutoff, based on the amount of TiS_2 present undergoing a 1-electron change at current densities from 10–135 μÅ/cm^2. This is illustrated in Fig. 53.2. For these microbatteries, this calculates to an energy density of 140 W h/l and a power density up to 270 W/l. The energy density can be increased substantially by increasing the cathode thickness. For example, increasing the cathode thickness from 2 μm to 4 μm would result in an energy density of 230 W h/l.

The EBC microbattery is capable of producing current densities as large as several mA/cm^2. Figure 53.3 shows the closed circuit voltage of a microbattery at the end of a two-second current pulse up to a current density of 4 mÅ/cm^2. The 2-mÅ/cm^2 pulse would be equivalent to a power density of greater than 4000 W/l. Shorting the microbattery through a 0.1-Ω resistor produces a current density of approximately 12 mA/cm^2. It would be possible to discharge the EBC microbattery at these rates, but the discharge would only last 1–2 minutes due to the μÅ-h capacity of the microbattery.

The EBC microbattery shows outstanding secondary performance. Microbatteries achieve well over 1000 cycles with depths of discharge greater than 70 percent of one electron at current densities of up to 300 μÅ/cm^2 to the 1.8-V cutoff. A typical result is shown in Fig. 53.4. As can be seen, even after 6000 cycles at more than a 70

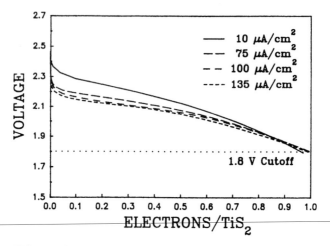

FIGURE 53.2 Primary discharge of the EBC microbattery at various current densities.

percent depth of discharge per cycle and a current density of 100 µA/cm^2, there has been little fade in the cell capacity and very little change in the curve shape. It was observed with the EBC microbattery that discharges subsequent to the initial discharge were usually lower in depth of discharge. In intercalation systems, it is commonly found that the first charge does not restore the same capacity to the cathode as was removed on the first discharge. This is illustrated in Fig. 53.5. It appears that some of the Li$^+$ becomes trapped in the cathode and is not removed during the charge. For the EBC microbattery, after the first cycle the discharge efficiency is reduced to about 70–80 percent of its initial value to the 1.8-V cutoff. Though the cathode efficiency is reduced, cycles after the initial one are very reproducible and

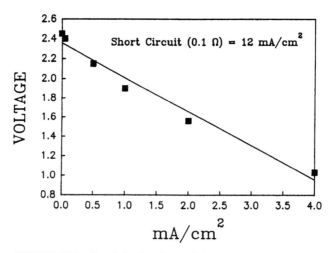

FIGURE 53.3 Closed-circuit voltage of the EBC microbattery after 2-sec. pulses of various current densities.

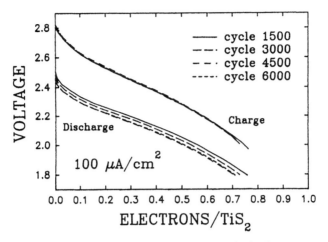

FIGURE 53.4 Secondary performance of the EBC microbattery to a 1.8-V cutoff.

give essentially 100 percent cycling efficiency, meaning that all the capacity put in during the charge is removed on the subsequent discharge. No additional Li⁺ becomes trapped after the first cycle.

It has been found that, by changing the lower cutoff voltage from 1.8 V to 1.4 V, the discharge capacity can be brought up to approximately 100 percent cathode utilization with no detrimental effect on the microbattery's secondary performance. An example of this is shown in Fig. 53.6. This shows an EBC microbattery that was running at a current density of 40 µA/cm^2 and about 80 percent cathode utilization to the 1.8-V cutoff for the first 3000 cycles. The lower cutoff voltage was then changed to 1.4 V, and the next 3500 cycles gave approximately 100 percent cathode utiliza-

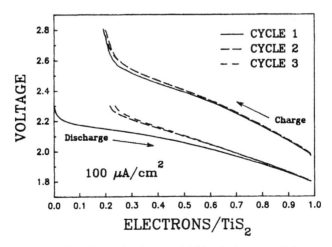

FIGURE 53.5 Comparison between initial and subsequent discharges of an EBC microbattery.

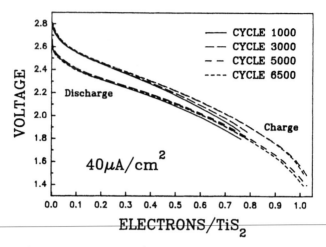

FIGURE 53.6 Secondary performance of an EBC microbattery after changing the lower cutoff limit from 1.8 to 1.4 V.

tion. Figure 53.7 shows an EBC microbattery that has gone 10,000 cycles at 100 µÅ/cm^2 and is still running at about 90 percent cathode utilization to a 1.4-V cutoff. These results show that the reduced cutoff voltage has no damaging effect on the secondary performance of the microbattery. Even at the high drain rates of 200 and 300 µÅ/cm2, microbatteries have been cycled thousands of times at better than 90 percent cathode utilization between 1.4 and 2.8 V.

In other rechargeable Li systems, the charge current must be significantly lower than the discharge current to achieve reasonable cycle life. As shown by the cycling data presented above, the EBC microbattery cycles well when the charge rate is equivalent to the discharge rate. It has also been found that the EBC microbattery

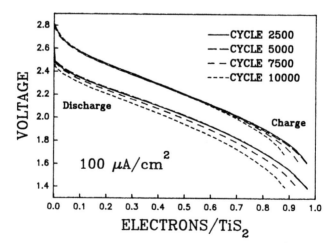

FIGURE 53.7 Secondary performance of the EBC microbattery to a 1.4-V cutoff.

cycles well when the charge rate is significantly higher than the discharge rate. An example of this is shown in Figs. 53.8(a) and (b). This EBC microbattery has gone 500 cycles at better than 90 percent cathode utilization at a discharge rate of 0.15C (.5 µÅ/cm^2) and a charge rate of 5C (192 µÅ/cm^2). In another example, an EBC microbattery has been cycled for over 21 months and more than 500 times at a discharge rate of 0.035C (1 µA/cm^2) and a charge rate of 3.5C (100 µÅ/cm^2). This type of result indicates that an EBC microbattery could be designed to give days or even months of power at a low drain rate and then be fully recharged in a matter of minutes. This cycle could then be repeated hundreds or even thousands of times.

The EBC microbattery can be cycled at the relatively high current density of 100 µÅ/cm^2 in the temperature range of –10°C to 90°C. The microbattery efficiency at various temperatures is shown in Fig. 53.9. This microbattery was originally taken from ambient to lower temperatures. The cycle time is significantly shortened as the temperature is lowered (45 percent of the RT capacity at –10°C). As would be expected for a solid-state battery, this reduced cycle time appears to be primarily the result of increased cell IR. At –20°C, the microbattery would no longer cycle at this current density. The microbattery was brought back up to RT, where it showed no

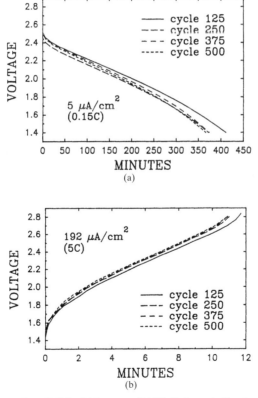

FIGURE 53.8 (a) Low-rate (0.15C) discharge half-cycle of the EBC microbattery; (b) high-rate (5C) charge half-cycle of the EBC microbattery.

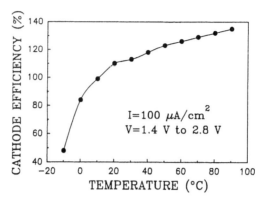

FIGURE 53.9 Discharge efficiency of the EBC microbattery at various temperatures.

detrimental effect in its secondary performance from the low temperature excursion. The microbattery was then taken to higher temperatures and successfully cycled up to 90°C. Temperatures higher than 90°C were not investigated due to the temporary packaging available for the microbattery at the time, but there was no indication that the microbattery would not function up to the melting point of Li.

Possible applications for the EBC microbattery require a higher voltage and/or more capacity than a single microbattery can supply. Therefore, groups of five EBC microbatteries have been tested in both series and parallel configurations. (Five microbatteries were used because this is the amount fabricated per microscope slide.) The series configuration gives an OCV around 12.5 V and was cycled between 14 and 7 V. Five microbatteries in series were cycled for over 10,000 times at 100 μA/cm². The cathode efficiency of this set of microbatteries was only about 70 percent initially and faded to less than 50 percent after 10,000 cycles. The low efficiency is probably the result of one cell in the group being only marginally good and adding a disproportionate amount to the series IR. Another group of five microbatteries that was recently put on test in a series configuration has gone over 250 cycles at close to 100 percent efficiency. Five microbatteries in parallel have been cycling between 1.4 and 2.8 V for over 12 months and have gone more than 1000 cycles at close to 100 percent cathode efficiency. The total current through the parallel microbatteries is 100 μÅ which equates to a current density of 20 μÅ/cm² due to the larger cathode area involved with the parallel configuration.

The standard size of the EBC microbatteries presently being tested is 1 cm². To demonstrate that it would be possible to scale up the size of the microbattery, a set of masks were made that increased the area of the microbattery to 10 cm². More than 1500 cycles between 1.4 and 2.8 V have been obtained with better than 90 percent cathode efficiency from a 10 cm² microbattery. This is shown in Fig. 53.10. The current density at which this microbattery is cycled is 100 μA/cm². Because of the larger area of this microbattery, the total current is 1 mÅ. This indicates that the scale-up in the area of the EBC microbattery is not be a problem. We are now developing a set of masks to demonstrate that any two-dimensional shape can potentially be used for an EBC microbattery.

After RT storage for an average of 3.2 years, EBC microbatteries still show 93 percent voltage maintenance and pulse performance similar to when they were fresh. This small voltage loss indicates that the electronic conductivity of the glassy

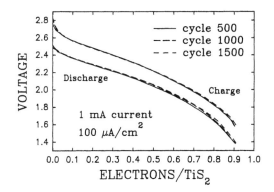

FIGURE 53.10 Secondary performance of a large-area (10 cm^2) EBC microbattery.

electrolyte/LiI combination must be very low. Assuming all the loss to be from self-discharge through these layers, the volume resistivity is estimated to be 10^{13}–10^{14} Ω-cm. At this rate of loss, the microbatteries should be discharged to 1.8 V after 11.5 years on shelf. The components of the EBC microbattery also appear to be very stable during cycling. More than 21,000 cycles have been obtained from an EBC microbattery during 29 months of cycling between 1.4 and 2.8 V at 100 µA/cm^2 with about a 20 percent loss in cathode efficiency. Another EBC microbattery has been continuously cycled at 40 µA/cm^2 for more than 39 months and 12,000 cycles with about a 15 percent loss in efficiency during this time.

Due to the recent development of the protective coating, none of the testing reported to date has been done on a microbattery that includes this feature. Further performance testing, shelf-life testing, and thermal cycling must now be done on EBC microbatteries that include the protective coating.

SUMMARY

The EBC microbattery has been shown to be a thin-film solid-state system that demonstrates high cathode utilization, the capability of producing current densities in the mA/cm^2 range, outstanding secondary performance, and excellent long-term stability. In the near future, it will be possible to incorporate the EBC microbattery with many types of microdevices (microsensors, CMOS-SRAM, and so on) during their manufacture to provide a rechargeable, long-term power supply for the device.

REFERENCES

1. K. Kanehori, K. Matsumoto, K. Miyauchi, and T. Kudo, *Solid State Ionics*, **9–10**, 1983, p. 1445.
2. G. Meunier, R. Dormoy, and A. Levasseur, *Materials Science and Engineering*, **B3**, 1989, p. 19.
3. R. Creus, J. Sarradin, R. Astier, A. Pradel, and M. Ribes, *Materials Science and Engineering*, **B3**, 1989, 109.
4. E. K. Shokoohi and J-M Tarascon, *U.S. Patent 5,110,696*, 1992.

5. J. B. Bates, N. J. Dudney, G. R. Gruzalski, R. A. Zuhr, A. Choudhury, and C. F. Luck, *Journal of Power Sources,* **43–44**, 1993, p. 103.
6. J. B. Bates, Oak Ridge National Laboratory, personal communication.
7. J. R. Akridge, S. D. Jones, and H. Vourlis in G. Nazri, R. A. Huggins, and D. E. Shriver (eds.), *Solid State Ionics, Mat. Res. Soc. Proc.,* **135**, Boston, Massachusetts, 1988, pp. 571–584.
8. A. M. P. Jelfs, *Ph.D. thesis,* Leicester Polytechnic, U.K., 1987.
9. C. C. Liang, J. Epstein, and G. H Boyle, *J. Electrochem. Soc.,* **116**, 1969, p. 1452.

CHAPTER 54
HIGH-TEMPERATURE CERAMIC SENSORS

INTRODUCTION

Ceramic materials with interesting electrical characteristics can be synthesized for a wide range of electronic applications, both as passive and active components.[1-3] Their electrical conduction behavior, in particular, has been extensively used for sensor applications. In principle, by careful selection of base materials, additives, and processing conditions, electrical properties of ceramics can be engineered to meet specific applications. Ceramic gas sensors, for example, can be designed to be sensitive to a specific gas, while ceramic thermistors can be made sensitive only to temperature changes. Availability of reliable and rugged sensors that can operate in industrial environments for extended periods and cycles is crucial to automation, application of artificial intelligence, and enhanced productivity with emission control.

Recent consideration of pollution control and fuel efficiency for a variety of combustion processes and increasing concern over safety in industrial activities involving inflammable and poisonous gases have necessitated the development of sensors for reliable detection of these gases.[4] Carbon monoxide is one of the important reducing gases to be detected in combustion of hydrocarbon fuel and petroleum, and also in the first stages of fire. Besides CO, there are several other pollutants such as H_2, NO_x, SO_x, CO_2, volatile organic compounds (VOCS), and so on.[5-9] Besides gas sensors, availability of stable and reliable thermistors capable of continuously monitoring temperatures beyond the capability of the present-day thermocouples is crucial to the success of advanced manufacturing technology.

While the low-temperature sensors have been commercially successful, less success has been achieved by their high-temperature counterparts. This is due mainly to the problems associated with sensitivity, stability, and reproducibility at high temperatures. Reproducibility depends, to a large extent, upon the processing conditions. Device instability can be attributed to the change of material properties and/or contact degradation with time or thermal cycling. While the problem of sensitivity, in principle, can be solved with suitable selections of base materials and dopants, the problem of reproducibility and stability can only be resolved with a sufficient knowledge of the underlying physical processes. Although the literature is abundant with a variety of sensor development efforts, it is important to note that most of these developments are based on rather empirical and/or trial and error methods. Consequently, a fundamental understanding of the sensing mechanisms, in most cases, is lacking, and this makes the optimization of the sensing behavior a challenging task, if not impossible.

This chapter summarizes the results of some of the recent work on ceramic-based CO sensors[10-17] and thermistors.[18,19] Several candidate materials were studied with respect to sensitivity, recovery, reproducibility, response time, and stability. The effect of dopants and second-phase addition was also investigated. Prototype sensors were developed based on promising candidates and tested for stability and reproducibility. Sensing and degradation mechanisms were also investigated.

CERAMIC GAS SENSORS

There are two major problems with the currently available sensor technology. Sensors capable of *in situ* monitoring of gas composition in high-temperature environments are practically nonexistent. The other limitation is that the majority of the available sensors show various degrees of interference from other gases and hence are not selective. This work has led to the development of new ceramic oxides as reliable and rugged CO and/or H_2 gas sensors at high temperatures.

Several candidate materials were investigated with respect to their sensitivity to different gases, recovery, reversibility, and response time. Based on such screening tests, two promising candidates were identified: molybdenum trioxide (MoO_3)- and titania (TiO_2)-based systems. The sensors were made in thick-film configurations using pastes prepared by sensor materials in 1-heptanol solvent. Sensor films were printed on alumina substrate using a 325-mesh screen and subsequently heated in air to burn off the solvent. Gold wires were attached to the film with the help of gold paint for electrical measurements.

It was demonstrated that, while MoO_3 can be used as an on/off-type detector, the two-phase mixture (1:2 molar ratio) of MoO_3 and ZrO_2 can be used as an online monitor for up to 3 vol.% of CO.[10,13] The anatase form of pure TiO_2 was found to sense both CO and H_2.[11] Significant change in the sensing characteristic of the base material (TiO_2) was observed when admixed with an insulating second phase, such as α-alumina (Al_2O_3) or yttria (Y_2O_3). In the case of TiO_2-10 wt.% Al_2O_3 (TA), the sensor response was found to be exclusively dependent on hydrogen alone; the presence of CO or CO_2 did not affect the sensitivity. On the other hand, the sensor based on TiO_2-10 wt.% Y203 (TY) showed increased sensitivity to CO and decreased interference due to H_2. Addition of elemental iron or palladium in small concentration to the two-phase mixture of titania and yttria further improved the sensitivity and selectivity to CO; up to 400 ppm, it becomes exclusively a CO sensor. Neither TA nor TY showed any interference from NO_x. Measurements were made on various samples prepared in an identical manner, and also measurements were repeated several times on a single sample. It was demonstrated that the sensor material showed very good reproducibility, reversibility, and recovery.[11,13]

Two prototype sensors were designed, one for application as an alarm (on/off) and the other for *in situ* (online) detection in high-temperature environments. These prototypes were tested in environments containing CO gas in concentrations ranging from several ppm to about 5 percent and showed good reproducibility, thermal stability and sufficiently long shelf-life. Figures 54.1(*a*) and (*b*) are the schematic diagrams of the on/off and online sensor assembly, respectively. The tooth-comb configuration of interdigitated electrodes, on which the sensor film was screen-printed, is shown in Fig. 54.1(*c*). Figures 54.2 and 54.3 depict the typical CO concentration dependence of the resistance of such prototype sensors. In the present form, these devices are bulky but still intricate in design and likely to draw large amounts of power to operate. However, with microfabrication of the components and screen-printed, miniaturized heaters and thermometers, the overall size could be greatly reduced. This would eventually lead to their commercial viability and device operation at much lower power consumption.

With collaboration from engineers at Orton Ceramic Foundation (a small business in Westerville, Ohio) work is already underway to package the sensor into a device. Figure 54.4 shows the block diagram of the prototype CO sensor with electronic interfacing. The CO sensor is connected to the gain resistor input of the amplifier, which has a reference voltage applied to its input. The output of the amplifier is thus solely a known function of the sensor resistance. A K-type thermocouple is used to monitor the temperature in the vicinity of the sensor. These two signals are converted to digital numbers and sampled by the computer once every second. The computer calculates the resistance of the sensor and the temperature and applies the generic parametric equations to determine the concentration of carbon monoxide in the ambient.

Extensive field testing of the above CO sensors was carried out in the Center for Automotive Research (CAR) at the Ohio State University. This facility has a V6 test

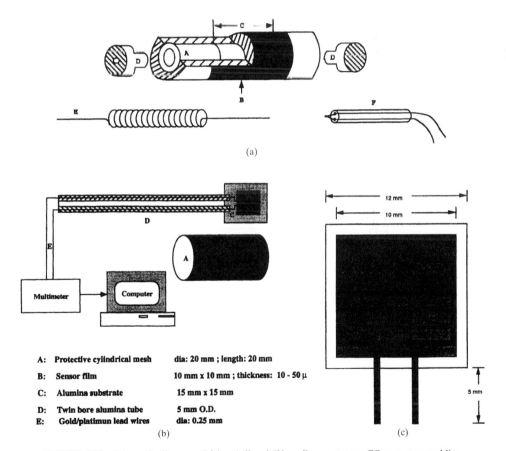

FIGURE 54.1 Schematic diagram of (*a*) on/off and (*b*) on-line prototype CO sensor assemblies. Comb-tooth configuration of the miniaturized sensor design with interdigitated gold electrodes, employed in on-line CO meters, is shown in (*c*). (*Courtesy ESTC, Cleveland, OH.*)

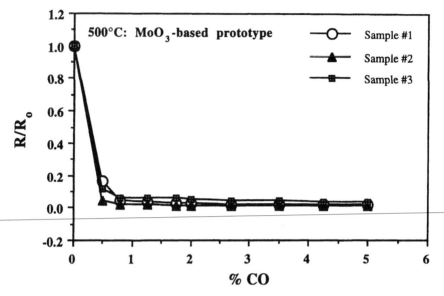

FIGURE 54.2 Response of MoO_3-based prototype (ON/OFF) sensors.

engine provided by Ford Motor Company, where the fuel combustion was programmed to be under near stoichiometric conditions ($\lambda = A/F \sim 14.7–15.0$). The facility also has an FT-IR analytical instrument (Nicolet, Rega 7000), which is capable of analyzing and quantifying about 20 possible components of the gas mixture online while the engine is running. Two locations were chosen for the testing of our sensors: in the vicinity of the zirconia sensor in the combustion chamber and near the exhaust pipe.

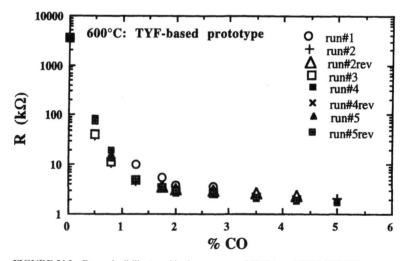

FIGURE 54.3 Reproducibility trend in the response of TYF-based ON-LINE CO sensors.

HIGH-TEMPERATURE CERAMIC SENSORS

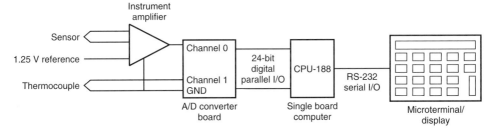

FIGURE 54.4 Block diagram of prototype CO sensor with electronic interfacing.

For protection against high-velocity gases exerting pressures of up to ~20–30 kpsi, the sensor assembly was inserted in a one-end-closed, round alumina tube (50 mm long, 14.1 mm ID, 15.1 mm OD), with a small slit (~2 mm) at the center of the closed end. See Fig. 54.5. During the four-minute cycle of the engine operation, the gas temperature varied very swiftly and widely: from room temperature (when started cold; the initial temperature was higher in subsequent cycling) up to as high as ~700°C. A type-K (chromel-alumel) thermocouple was attached to the sensor to follow the variation of temperature as the combustion proceeded. This assembly was positioned inside the combustion zone with the direction of flow at a right angle to the slit in the tube. This design has been successfully tested for 25 cycles in the engine and seems to have stood up to the severity of the engine atmosphere so far. The sensor has responded very quickly and reversibly, both under revving and idling conditions. The signals obtained from the sensors do not seem to be affected by the generation of steam, carbon dioxide, and other components during the combustion. Some of the typical preliminary responses of the sensor as a function of time, gas temperature, and concentration, are shown in Figs. 54.6 and 54.7.

Along with the development of the sensor material and the demonstration of the device, attempts were made to gain a fundamental understanding of the sensing behavior, which is a gray area in the field of chemical sensors. Simple surface-reaction-based models have been proposed to explain the sensing mechanisms.[10,11] An attempt was also made to elucidate the microstructure-property correlation by means of immittance spectroscopy (IS). The application of IS has, so far, shown that the individual contribution of grain interiors, grain boundaries, and electrode/specimen interfaces to the total resistance could be resolved via a complex plane analysis.[14,15] Further analyses based on these measurements are anticipated to clarify the role of additional phases to the sensitivity of the sensors.

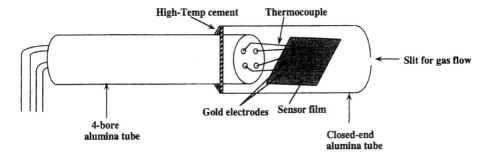

FIGURE 54.5 CO sensor assembly for field testing in Ford V6 engine at OSU.

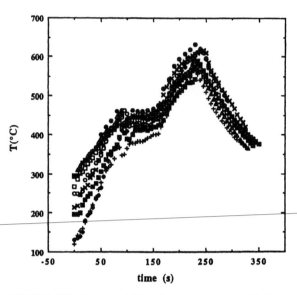

FIGURE 54.6 Rapid variation of gas temperature with time and progress of combustion.

CERAMIC THERMISTORS

Several oxide systems were evaluated as potential candidates for therinistors.[19] The resistivity (ρ) followed the familiar Arrhenius-type behavior for all except for the silica-containing mixtures.

$$\rho = A \exp\left(\frac{Q}{kT}\right) \quad (54.1)$$

where Q is the activation energy for electrical conduction, k is the Boltzmann constant, T is the absolute temperature, and A is the preexponential factor. Normalized thermal sensitivities—namely temperature coefficient of resistance (TCR)—can be calculated using the definition

$$\alpha = \left(\frac{1}{R}\right)\left|\frac{\partial R}{\partial T}\right| \quad (54.2a)$$

or

$$\alpha = \left(\frac{1}{\rho}\right)\left|\frac{\partial \rho}{\partial T}\right| \quad (54.2b)$$

From Eqs. (54.1) and (54.2b), one obtains

$$\alpha = \frac{Q}{k}\left(\frac{1}{T^2}\right) \quad (54.3)$$

The value of normalized sensitivity can be misleading, because it only reveals a relative change in a variable, either resistance for thermistors or EMF for thermo-

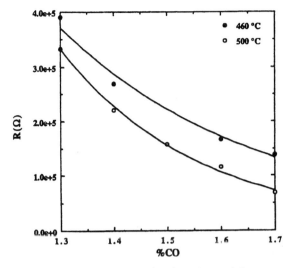

FIGURE 54.7 CO concentration dependence of the sensor film resistance as a fuel combustion progressed in the car engine.

couples, as a function of temperature. For example, it is recommended that the type-K thermocouple with a sensitivity of 0.0388 mV/K (shown in Table 54.1) be used to measure temperatures around 1000°C, instead of a type-S with a sensitivity of 0.0115 mV/K; even though the normalized sensitivity of type-K, 0.094 percent, is less than that of type-S, 0.12 percent. Sensitivity of a thermistor s can be defined as the change in resistivity with temperature and can be calculated mathematically as follows:

$$s = \left|\frac{\partial \rho}{\partial T}\right| = \frac{AQ \exp(Q/kT)}{kT^2} \qquad (54.4)$$

The activation energy Q, preexponential constant A, sensitivity s, and normalized sensitivity a at 1000°C for the various systems studied are summarized in Table 54.1.

Two promising candidates, Y_2O_3 and $CaZrO_3$, were selected for further studies such as the aging behavior and effect of dopants. Minimization of the aging behav-

TABLE 54.1 Summary of the Oxide Systems Studied in this Research

Composition	$Q(eV)$	A	s (Ω-cm/K)	$\alpha\%$
Y_2O_3	2.3	7.4×10^{-4}	16,000	1.7
$CaZrO_3$	2.2	8.6×10^{-4}	7400	1.6
Al_2O_3	1.6	2.5×10^{-1}	6500	1.2
Dy_2O_3	1.5	1.3×10^{-1}	3000	1.1
MgO	1.4	6.6×10^{-1}	2300	1.0
47% TiO_2, 53% SiO_2	0.7 T<826°C 0.9 T>826°C	984 T<826°C 58.2 T>826°C	1130	0.7
70% Al_2O_3, 30% SiO_2	1.0 T<840°C 1.3 T>840°C	110 T<840°C 3.7 T>840°C	1000	0.9
68% ZrO_2, 32% SiO_2	1.3 T<600°C 0.7 T>600°C	0.04 T<600°C 42.9 T<600°C	103	0.9

ior was obtained by stabilizing the microstructure through an optimized processing route. Dopants were found not only to affect the sensitivity but the stability as well. For example, it was established that a small amount of ZrO_2 addition can make Y_2O_3 stable in environments containing humidity.

The response time of a sensor is crucial for many applications. For transient conditions, the governing equation for the thermal equilibrium of a sensor in relation to the surrounding environment can be represented as:

$$\frac{CdT}{dt} + k_t (T - T_a) = P \qquad (54.5)$$

where C is the heat capacity of the sensor, k_t is the thermal conductance or dissipation constant (determined by both the configuration of the sensor and its operation environment), T_a is the ambient temperature, and P is the thermal power consumed by the sensor, which numerically equals to the product of current and voltage. The resistance of a thermistor at a given temperature depends not only on the thermal equilibrium but also on the electrical equilibrium. For example, electrical equilibrium was not reached after 30 minutes for a thermally equilibrated calcium zirconate-based thermistor under a DC potential of 5 V. Under AC conditions, however, electrical equilibrium was achieved almost instantly when a potential of an appropriate amplitude was applied.

Frequency of the AC power source must be determined in order to measure the temperature accurately. It was found that, at low frequencies, the resistance fluctuated significantly, as shown in Fig. 54.8, giving uncertainties in the measurements. At higher frequencies, the fluctuation diminished; however, the sensitivity decreased due to the drop in the resistance value. Therefore, an optimum frequency range must be determined in order to achieve the highest possible sensitivity. Also, stabilization of the microstructure plays a critical role on the aging behavior of thermistors as demonstrated by Fig. 54.9.

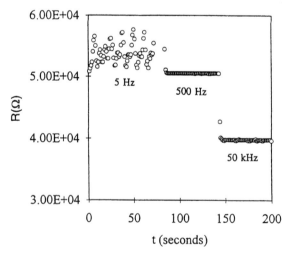

FIGURE 54.8 Resistance of a calcium zirconate sample at 1000°C at frequencies indicated.

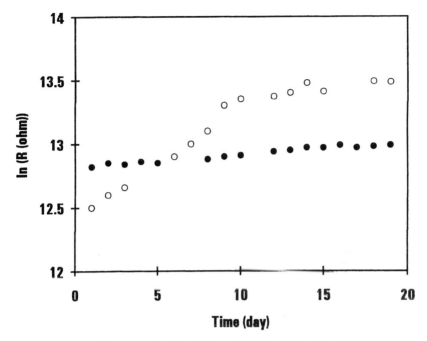

FIGURE 54.9 Resistance as a function of time at 1100°C for yttria sintered at 1550°C for 24 hours (open circles) and 168 hours (filled circles).

REFERENCES

1. R. C. Buchanan (ed.), *Ceramic Materials for Electronics: Processing, Properties, and Applications,* Marcel Dekker, Inc., New York and Basel, 1986.
2. B. C. H. Steele (ed.), *Electronic Ceramics,* Elsevier, New York, 1991.
3. A. J. Moulson and J. M. Herbert (ed.), *Electroceramics: Materials, Properties, Applications,* Chapman & Hall, New York, 1992.
4. A. M. Azad, S. A. Akbar, S. G. Nffiaisalkar, L. D. Birkefeld, and K. S. Goto, *J. Electrochem. Soc.,* **139,** 1992, p. 3690.
5. S. H. Schneider, *Sci. Amer.,* **256,** 1987, p. 78.
6. C. W. Seigmund and D. W. Turner, *Trans. ASME,* **96A,** 1974, p. 1.
7. R. A. Dalla Betta, D. R. Sheridan, and D. L. Reed, "Development of Reliable SO_2 and NO_x Sensors and an Integrated In-situ Analyzer for Flue Gas Monitoring," Final Technical Report #DE-AC0386ER80421, DOE SBIR Phase II Program, 1991.
8. G. Adachi and N. Imanaka in *Chemical Sensor Technology,* vol. 3, Kodansha Ltd., Tokyo, 1991, p. 131.
9. O. W. Bynum, D. R. Sheridan, and J. A. White, GRI Report 92/0373, 1992.
10. A. M. Azad, S. G. Mhaisalkar, L. D. Birkefeld, and S. A. Akbar, *J. Electrochem. Soc,* **139,** 1992, p. 2913.
11. L. D. Birkefeld, A. M. Azad, and S. A. Akbar, *J. Am. Ceram. Soc.,* **75,** 1992, p. 2964.

12. A. M. Azad, L. B. Younkman, and S. A. Akbar in *Proceedings of 1993 USEPAIA & WMA International Symposium on Field Screening Methods for Hazardous Wastes and Toxic Chemicals,* Las Vegas, Nevada, 1993, pp. 94–104.
13. A. M. Azad, L. B. Younkman, S. A. Akbar, and R. J. Schorr, *Solid State Gas Sensors for Combustion Control,* Final Project Report #EMTEC/CT33/TR-94-33, 1994.
14. A. M. Azad, L. B. Younkman, S. A. Akbar, and M. A. Alim, *J. Am. Ceram. Soc.,* **77**, 1994, p. 481.
15. A. M. Azad, S. A. Akbar, L. B. Younkman, and M. A. Alim, *J. Am. Ceram. Soc.,* October 1994.
16. A. M. Azad, S. A. Akbar, and L. B. Younkman, "Ceramic Sensors for Carbon Monoxide and Hydrogen," *Interface,* December 1994.
17. S. A. Akbar, A. M. Azad, and L. B. Younkman, "A Solid-State Gas Sensor for Carbon Monoxide and Hydrogen," patented February 1996.
18. C. C. Wang, W. H. Chen, S. A. Akbar, and V. D. Patton, "A Review on Electrical Properties of High-Temperature Materials: Oxides, Borides, Carbides, and Nitrides," *J. Mater. Sci.,* **12**, 1995.
19. S. A. Akbar, W. Chen, V. D. Patton, and C. C. Wang, "High-Temperature Thermistor Device and Method," patent application filed June 1994.

CHAPTER 55
MICROFABRICATED AND MICROMACHINED CHEMICAL AND GAS SENSOR DEVELOPMENTS

INTRODUCTION

The increasing needs of environmental, automobile, and process-control-related monitoring and detection of various chemicals and gases add new impetus to the research and development of chemical and gas sensors. Chemical and gas sensors, in principle, can provide continuous or periodic real-time, online point measurements. Furthermore, it is desirable that the sensor not require additional chemical reagents or sample preconditioning. Characteristically, the sensor should have good selectivity, sensitivity, and fast response time. For practical purposes, the sensor needs to be produced at modest cost. Silicon-based microfabrication processes have demonstrated the capability of manufacturing geometrically well-defined, highly reproducible microstructures at modest cost in integrated circuit (IC) production. It is feasible to employ microfabrication techniques to produce chemical and gas sensors to meet the desirable sensor properties described. This approach has been used by researchers in the development of various chemical and gas sensors.[1-3]

In recent years, the advancement of micromachining technology has added new dimensions to silicon-based microfabrication processes for the development of chemical and gas sensors. Micromachining processes, particularly anisotropic and plasma etching, and the sacrificial layer method make possible the construction of three-dimensional devices. Although micromachining techniques have been used for the development of physical sensors, their applications to chemical or gas sensors have not yet been realized.

In this study, silicon-based microfabrication and micromachining processes are used to develop chemical and gas sensors. Chemical and gas sensors operated at different principles can be produced by these fabrication techniques. In our laboratory, chemical and gas sensors produced by microfabrication and micromachining techniques include tin oxide or metal oxide-based sensors, Schottky diode-type devices, solid electrolyte electrochemical sensors, and calorimetric sensors. These sensors, with proper modifications and utilization of chosen catalysis and operating temperatures, can be applied in many practical situations. In this chapter, examples of these chemical and gas sensors are discussed.

TIN-OXIDE-BASED SENSORS

Tin-oxide-based sensors are currently widely used for gas sensing. The seminal work on tin oxide gas sensors by Taguchi[4] and Seiyama et al.[5] demonstrated their effectiveness in gas sensing. Sintered tin oxide is used, which is a rather labor-intensive manufacturing process. It is known that, in many instances, the selectivity of tin oxide sensors for detecting gas can be enhanced using various catalysts such as platinum, palladium, and metal oxides, such as CuO. It is also known that sensor performance can be enhanced by controlling the operating temperature of the sensor.[6,7] In some cases, it is necessary to decompose thermally the detecting gas prior to measurement.[8] Therefore, it is desirable to be able to control and measure the operating temperature of a tin-oxide-based sensor. Thus, the incorporation of heating and temperature sensing elements on a tin oxide sensor is desirable. A micromachined tin oxide gas sensor integrated with heating and sensing elements is fabricated in our laboratory and is shown in Fig. 55.1. Details of the fabrication are given elsewhere[1] and will not be repeated here.

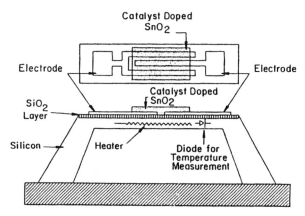

FIGURE 55.1 Structure of a micronmachined tin oxide sensor with on-chip heating and temperature sensing elements.

The selectivity of this sensor can be improved by adding the proper catalyst. For instance, the introduction of CuO as a catalyst to the tin oxide film provides good selectivity to H_2S. In addition to CuO, other catalysts can be applied to the tin-oxide-based sensor for the detection of other gases. Platinum black and palladium are commonly used for hydrocarbon detection. The incorporation of a heating element permits operation of the sensor at a desirable elevated temperature. The bulk of the silicon substrate where the sensor is constructed is selectively removed by anisotropic etching. This minimizes thermal mass loss and allows accurate temperature control and low heating energy requirement for the sensor. The unique features of the sensor can only be realized by using microfabrication and micromachining techniques.

SCHOTTKY-DIODE-TYPE SENSORS

When gas or chemical vapor molecules diffuse toward the interface between the metal and the insulating layers of a diode, the height of the Schottky barrier of the

CHEMICAL AND GAS SENSOR DEVELOPMENTS

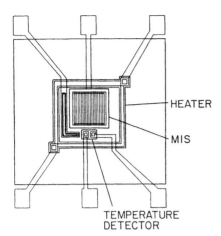

FIGURE 55.2 Structure of a micromachined Pd-Ag metal insulator semiconductor diode hydrogen gas sensor.

diode is diminished. This leads to a change in either the forward voltage or the reverse current of the diode, forming the mechanism of the gas or chemical vapor diode. Lundstrom et al.[9] described a palladium gate Schottky-diode-type sensor for hydrogen detection. In our laboratory, silicon-based microfabrication and micromachining processing are employed to produce a silver-palladium Schottky-diode-type hydrogen gas sensor.

Figure 55.2 shows the schematic structure of the device, and a more detailed description of the sensor can be found elsewhere.[1] This sensor also incorporates temperature-sensing and heating elements. In order to ensure minimum energy consumption and thermal mass loss, the backside of the silicon substrate is selectively removed. Consequently, the sensor is fabricated on a 6–10-μm thick silicon-based membrane structure. As anticipated, the sensitivity and response time of the sensor improve markedly when operated at elevated temperatures. This illustrates the importance of using micromachining processing in producing the sensor, and the advantage of incorporating temperature-sensing and heating elements into the sensor are evident.

SOLID ELECTROLYTE ELECTROCHEMICAL SENSORS

Many chemical species can be detected using an electrochemical sensor. Electrochemical sensors, based on the mode of operation, can be classified as conductivity, potentiometric, and voltammetric (including temperometric) devices. For the detection of chemical vapors or gases, the operation of the sensor may have to be at an elevated temperature. Consequently, solid-state electrolytes are required. A well-known example of the solid electrolyte electrochemical sensor for gas sensing is the ZrO_2-based high-temperature oxygen sensor. This sensor can be operated either in potentiometric or amperometric mode,[10–11] and a sintered ZrO_2 disc is used in production of these oxygen sensors. This has the disadvantage of labor-intensive production costs. Also, because the sensor is heated and operated at 650°C in order to ensure the ionic conductivity of ZrO_2, the amount of energy required to heat the sensor is relatively large.

To overcome these shortcomings, we have developed a ZrO_2-based amperometric oxygen sensor using microfabrication and micromachining processing. Figure 55.3 shows the basic structure of the device. The electrode elements are thin platinum films, as are the temperature-sensing and heating elements. The formation of the ZrO_2 layer is done by ion beam coating using a yttria-stabilized ZrO_2 target. Thermal isolation as well as minimization of energy consumption of the sensor are accomplished using selective chemical etching, both from the front and the backsides of the silicon substrate. The advantages of such a device in terms of size, energy consumption, and potential fabrication cost are overwhelming compared to a conventional oxygen sensor. The use of microfabrication and micromachining processing in producing solid electrolyte electrochemical sensors for other chemical and gas detection are scientifically and commercially viable.

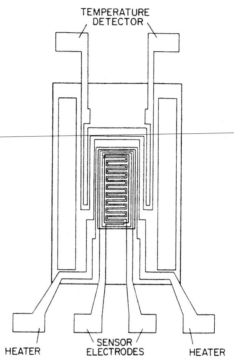

FIGURE 55.3 Structure of a micromachined high-temperature solid electrolyte oxygen sensor.

CALORIMETRIC SENSORS

The oxidation of a combustible gas produces heat (in calories), which can be measured and used to quantify the gas presented. This concept has been employed over the years in the form of a so-called *pellistor* device. In general, two platinum resistance thermometers in coil form are used. One of the thermometers is coated with the proper catalyst for the oxidation of the detecting gas. The other thermometer does not have any catalyst and will not promote the oxidation reaction. Thus, the temperature difference between these two thermometers indicates the heat produced by the oxidation reaction, and, consequently, the amount of the gas. This is the general principle of a calorimetric sensor. Normally, the thermometers of a calorimetric sensor are heated to an elevated temperature, enhancing the oxidation reaction.

Conventional methods in producing the temperature-sensing elements of the calorimetric sensor are relatively large in size and labor-intensive. In our laboratory, microfabrication and micromachining processes are used to produce calorimetric sensor prototypes. Figure 55.4 shows the schematic structure of a calorimetric sensor. Thin-film platinum resistance thermometers are used as the sensing elements, which also can be heated by an applied voltage. The thermometers have a suspension bridge structure created by anisotropic etching. This results in a low-energy-consumption, low-thermal-loss device. One of the thermometers is coated with a catalyst suitable for the desired oxidation of the detected gas. The temperature difference between the thermometers is then used to quantify the detected gas.

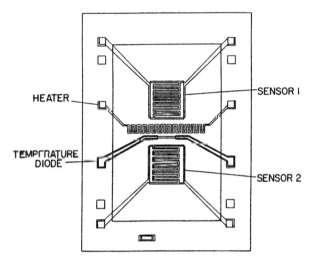

FIGURE 55.4 Structure of a microfabricated calorimetric sensor.

In summary, silicon-based microfabrication and micromachining processes are very promising in the manufacture of chemical and gas sensors. Low-thermal-mass-loss, low-energy-consumption devices can be produced at modest cost. The high degree of reproducibility and relatively small size of the sensor will enhance both sensor performance and the potentials for practical applications.

REFERENCES

1. Q. H. Wu, K. M. Lee, and C. C. Liu, *Sensors and Actuators, B*, 13–14 (1993) 1–6.
2. O. J. Prohaska, *Transducers '85*, 1985, p. 402.
3. S. C. Chang and D. B. Lecks, in D. Schuetzle and R. Hammerle (eds.), *Fundamentals and Applications of Chemical Sensors*, ACS Symposium Series 309, ACS, Washington, D.C., 1986, pp. 58–70.
4. N. Taguchi, Japan Patent No. 45-38200, 1962.
5. T. Seiyama, A. Kato, K. Fuji, and M. Nagatani, *Anal. Chem*, **34**, 1962, pp. 1502–1503.
6. T. Seiyama and N. Yamazoe, in D. Schuetzle and R. Hammerle (eds.), *Fundamentals and Aplications of Chemical Sensors*, ACS Symposium Series 309, ACS, Washington, D.C., 1986, pp. 39–45.

7. M. Egashira and Y. Shiazu in N. Yamazoe (ed.), *Chemical Sensor Technology* vol. 3, Kodansha, Tokyo, Japan, 1991, pp. 1–17.
8. J. Mzsei, *Sensors and Actuators B,* **2**, 1990, 199–203.
9. I. Lundstro, M. S. Shivaraman, and C. M. Svensson, *J. Appl. Phys.,* **46**, 1975, pp. 3876–3381.
10. S. Kimura, S. Ishitani, and H. Takao, in D. Schuetzle and R. Hammerle (eds.), *Fundamentals and Aplications of Chemical Sensors,* ACS Symposium Series 309, ACS, Washington, D.C., 1986, pp. 101–120.
11. E. M. Logothetis and R. E. Hetrick, in D. Schuetzle and R. Hammerle (eds.), *Fundamentals and Aplications of Chemical Sensors,* ACS Symposium Series 309, ACS, Washington, D.C., 1986, pp. 136–154.

CHAPTER 56
ELECTRO-FORMED THIN-FILM SILICA DEVICE AS OXYGEN SENSOR

INTRODUCTION

Silica thin films are prepared between metal electrodes from a resinous H-silsesquioxane precursor by spin-coating and subsequent low-temperature pyrolysis. After electro-forming, the device exhibits a nonlinear current-voltage characteristic that gives rise to a high-resistance off-state and a low-resistance on-state. The device can be switched between these two states with suitable voltages applied to the electrodes. The switching parameters, in particular the threshold voltage for the off to on transition, depend on the oxygen content of the ambient. Concentrations of oxygen in nitrogen down to 50 ppb (5×10^{-8}) could be detected. Details of device preparation and the measurement procedure will be presented.

The electronic conduction through dielectric materials is characterized by a variety of nonlinear current-voltage relationships which have been investigated and described[1] extensively, also with respect to the important phenomenon of dielectric breakdown in insulators.[2] There were also some reports about investigations of metal-insulator-metal (MIM) devices[3] that exhibit the effect of a voltage-controlled negative differential resistance (NDR). Such devices are additionally characterized by two stable regimes of conduction between which the device can be reversibly switched. This effect has to be induced in the MIM device by applying a forming voltage to the electrodes,[4] a process called electro-forming. While several mechanisms have been proposed to explain the observations, the particular mechanism underlying these effects is not yet fully understood.[5] It was observed, however, that performing the electrical measurements under exclusion of air[6] improves the response of the device to the electro-forming process.

For a study of these phenomena, a chemically and structurally well characterized dielectric material was chosen for preparing the MIM devices in order to identify the electronic processes in the material. It was indeed found that oxygen inhibits the effect of electro-forming and switching. It was possible to quantify the phenomenon and utilize this effect in an application as a gas sensor.

DEVICE PREPARATION

MIM devices were fabricated in a four-step process consisting of (1) deposition of metal-back electrodes on a glass substrate by vacuum evaporation, (2) partial masking of electrodes and then the spin-coating from solution of a thin film of a silica ceramic precursor onto the substrate, (3) conversion of the precursor to silica by pyrolysis at $\approx 400°C$ in air, and (4) evaporation of the top electrodes. The substrate used was Corning® 7059 glass, and for the electrodes, gold or nickel thin films were evaporated.

PRECURSOR CHEMISTRY

The precursor to the silica ceramic is a resinous hydrogen-silsesquioxane polymer[7] that has the general chemical formula of $(HSiO_{3/2})_n$. It can be prepared by hydrolysis from trichlorosilane. This polymer, also called H-resin for short, forms cagelike structures of various sizes and conformation and is soluble in a range of hydrocarbon solvents. As a solid, it is readily converted to silica by pyrolysis; that is, by a low-temperature heat treatment between 150°C and 450°C, usually in air. The reaction may proceed by the following steps (other reaction schemes are also possible):

$$2\ HSiO_{3/2} + O_2 \xrightarrow{\Delta} 2\ HOSiO_{3/2} \qquad (56.1)$$

This first reaction product, silanol, either reacts with itself

$$2\ HOSiO_{3/2} \xrightarrow{\Delta} 2\ SiO_2 + H_2O \qquad (56.2a)$$

or with another silsesquioxane unit

$$HOSiO_{3/2} + HSiO_{3/2} \xrightarrow{\Delta} 2\ SiO_2 + H_2 \qquad (56.2b)$$

to form near-stoichiometric porous silica material with a density of typically 1.7 g/cm³ depending on the temperature, the ambient, and the duration of the conversion process. In comparison, fused quartz—that is, dense amorphous silica—has a density of 2.2 g/cm³.[8] Conversion conditions for the silica used in the present sensor device were 425°C for 2 hr. in air.

DEVICE STRUCTURE

A schematic diagram of the device is shown in Fig. 56.1 in cross section and layout. Table 56.1 lists the physical properties and features of a typical device. The thickness of the metal electrodes range from 80 nm to 150 nm, depending on the chosen electrode metal. Their width is ≈ 3.5 mm. They form a device area of 0.15 cm²; narrower electrodes are also possible if the device area needs to be smaller.

The thickness of the silica ceramic thin film ranges from 100 nm to over 1 µm, but the typical insulator thickness in the devices reported here is between 400 nm and 500 nm. This parameter is controlled by the concentration of H-resin in the solution (≈ 20 percent) and by the spin-on conditions. The pyrolysis (heat treatment) does not change the film thickness substantially.

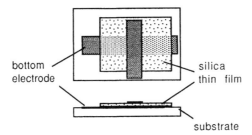

FIGURE 56.1 Schematic diagram of device configuration. The substrate is a 0.5-mm Corning 7059 glass slide (25 mm × 38 mm), bottom and top electrodes are Au or Ni, forming a device with an area of 0.15 cm^2 or smaller, the insulator is silica derived from H-silsesquioxane by low-temperature pyrolysis.

Electrical Measurements

After preparing a device, it is necessary to electroform it[4] by applying a sufficiently large voltage pulse to the electrodes of the device; a voltage of 15 V to 25 V, applied for <1 sec., is generally adequate. The procedure is performed in an inert atmosphere, that is, one free of oxygen (or other reactive gases), while the device is mounted on a sample holder in a shielded enclosure in which an inert atmosphere can be maintained by purging the device ambient with a purified inert gas (nitrogen or argon). The electrical characteristics of the device as described below can be observed only after it is electro-formed in this way. All the data were collected with the device kept at room temperature.

Electrical connection to the electrodes was made with spring-loaded copper contacts; the potential drop across the device was measured independently from the voltage supplied to the current-carrying parts of the device electrodes (four-wire method). The electrical data were collected by automated device measurement equipment consisting of the HP 9920S (HP 9000 series 200) computer with an HP 3497A data acquisition and control system as well as other peripherals which were all controlled through the HPIB (IEEE 488) bus. Full voltage sweeps through both quadrants (first and third) between −10 V and +10 V were programmed in voltage increments from 30 mV to 1 V with typically 3 to 7 data points acquired per second. In addition, voltage pulses of selectable plateau, polarity, and duration could also be applied to the device.

TABLE 56.1 Physical Device Characteristics

Substrate	Corning 7059 glass
Electrodes Au, Ni	Thickness = 100 nm
	Area = 0.15 cm^2
Dielectric	SiO$_2$
Precursor	H-silsesquioxane
Thickness	420 nm
Conversion	Pyrolysis in air at 425°C, 2 hr.

Device Characteristics

Before a description of the device performing as a gas sensor can be given, it is necessary to establish the background of the electrical characteristics of an electro-formed silica thin-film device in an inert gas. This will serve to introduce the device features on which the operation of the gas sensor is based.

Inert Ambient. The electro-formed device typically exhibits a high conductance at low voltages (i.e., below ≈3 V). The current depends superlinearly on the applied voltage in this range. Upon increasing the voltage, however, it goes through a maximum at $V = V_{mx}$, beyond which the current decreases as the voltage is increased further. This part of the current-voltage (jV) characteristic is known as negative differential resistance (NDR) and occurs as a typical behavior of electro-formed thin-film insulator devices. Eventually the current reaches a minimum at $V = V_{mn}$, and from there it increases again; at and beyond the minimum, the device has a high resistance (low conductance). The upper trace in Fig. 56.2 shows a representative device jV characteristic that demonstrates this feature.

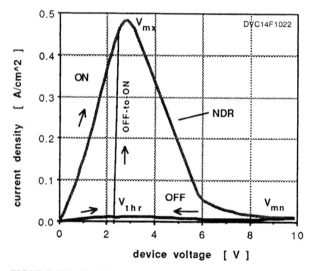

FIGURE 56.2 Typical jV curve of a thin-film H-resin silica device showing its three major features: (1) on trace (top curve) with negative differential resistance part (NDR) leading to current minimum at high voltage; (2) off trace obtained by fast scan back from high to low voltage; (3) off to on transition at the threshold voltage V_{thr} where the device switches itself back on.

Negative Differential Resistance (NDR). The NDR connects the high conductance or on-state of the device with its insulating or off-state. Its importance lies in the fact that it provides reversible access to these two device states. While the current along the NDR regime is stable at all voltages (for sufficiently small-voltage source impedance), it is found that it can be traced backwards only for sufficiently slow rates of decrease in the magnitude of the applied voltage. In this case, the device current goes through the maximum again, and at low voltages, $V < V_{mx}$, the device is still on.

Switching the Device Off. For a rapid voltage decrease from an applied voltage beyond the NDR regime, that is, from $V \approx V_{mn}$, the device current follows a different jV characteristic—see the lower trace in Fig. 56.2. Thus, when reaching the low voltage range, the device is then in its off state.

Threshold Voltage. The device will remain in its off state as long as the applied voltage stays below a certain value. When this value, referred to as the threshold voltage V_{thr}, is exceeded, the device undergoes a switching process in which the current rises by orders of magnitude over a small voltage interval and returns to the on-state device characteristic. This is indicated by the near-vertical off-to-on transition in Fig. 56.2 where the threshold voltage V_{thr} is located. Typical values for the threshold voltage are from ≈ 2 V to ≈ 6 V, depending on device characteristics and mode of operation. To illustrate the difference between the on and off state current for this particular device, the low-voltage portion of the data shown in Fig. 56.2 are replotted with a logarithmic current scale in Fig. 56.3. For this device, a comparatively small factor of ~40 for the on-to-off current ratio was observed; values up to 2000 have been found.

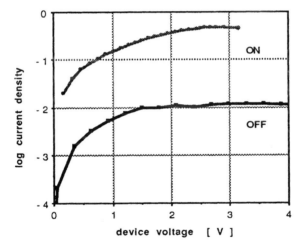

FIGURE 56.3 A plot of the on- (top trace) and the off-current (bottom trace) semilogarithmically against device voltage, showing a factor of ≈ 40 difference between the two device states in the voltage region where switching from off to on occurs at the threshold voltage. (Data from Fig. 56.2.)

Oxygen Sensitivity. For the switching processes to be observed as described above, an inert device ambient is required; in the experiments, this condition was maintained by purging the sample enclosure (volume 5.75 l) with purified nitrogen (flow rate 2 l/min). In the presence of oxygen, however, the device characteristics change in two parameters: (1) the threshold voltage V_{thr} moves to higher values as the oxygen content in the device ambient increases, (2) the on-current becomes increasingly diminished.

While the effect that oxygen has on the on-current depends on the applied voltage as a parameter and also on time, and thus is not readily amenable for quantitative measurement purposes, the threshold voltage of the device varies in a simple way with the O_2 concentration in the ambient.

Figure 56.4 shows this dependence for an H-resin silica device with Ni electrodes; the insulator thickness is 415 nm. The oxygen concentration (as a volume fraction) is plotted logarithmically (base 10) on the abscissa. The data point at the lower left of the diagram was obtained for an O_2 impurity of 5×10^{-8} (50 ppb). This is the residual O_2 concentration in the N_2 gas used to purge out the sample enclosure. The straight lines are drawn in to aid the eye. The change in slope at 0.3 percent is not yet fully understood; it coincides with the onset of the device NDR.

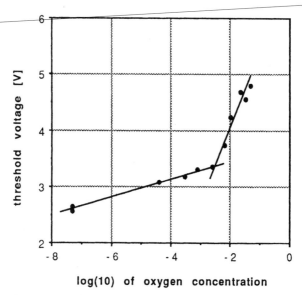

FIGURE 56.4 The threshold voltage V_{thr} of the device (at which it switches from off to on) depends on the amount of oxygen in the gaseous ambient of the device. The experimental data are plotted versus the logarithm of the oxygen concentration given as a volume fraction. The low value, 5×10^{-8} (50 ppb), corresponds to the residual O_2 impurity in the purge gas. The two lines connecting the data points are drawn in to aid the eye.

SENSOR OPERATION

For each data point, the device was first switched off by applying a voltage pulse of $(-)9.9$ V to the device electrodes. Then the voltage was slowly raised from zero ($V > 0$) while monitoring the current. The rapid rise of the current at the threshold voltage was evaluated from a plot of the jV curve to obtain the value of V_{thr}. The applied voltage was then again reduced to zero, and the oxygen concentration in the ambient

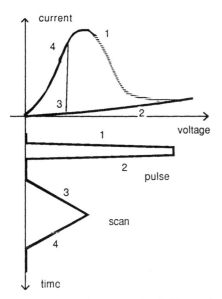

FIGURE 56.5 Sequence of switching and scanning the device in oxygen-sensing mode. Top: device jV curve. Bottom: voltage applied to device as function of time. The parts of the diagrams labeled with the same numbers correspond to each other. 1,2: pulse to switch device off (by trailing edge of pulse); 3: scan (along lower trace) to threshold voltage where device switches on; 4: return to zero along upper trace of jV curve.

changed to a new value. This procedure was repeated for each data point. The data shown in Fig. 56.4 were recorded for decreasing O_2 concentration starting from air that was purged out by the flow of N_2.

As follows from the preceding description of the device characteristics, a voltage pulse of this magnitude reaches into the current minimum from where the voltage is rapidly removed. Thus, the device is switched off prior to scanning its off-jV characteristic (lower trace in Fig. 56.2) until the threshold voltage is reached where the device switches on again. The sequence through these steps, and the parts of the device characteristic involved, is schematically illustrated in Fig. 56.5. The voltage is along the horizontal axis; time proceeds downwards while the device current is plotted upwards. Corresponding parts are labeled by the same numbers.

The control electronics to be used with this type of gas sensor has to provide the sequence and adjustments of the switching parameters of pulse voltage and duration, the slew rate (of the order of 100 V/s), as well as the value of the voltage increments ΔV and timing between steps for the voltage applied to the device. The evaluation electronics includes an A/D converter with a dynamic range between 2×10^3 and $\approx 10^4$ but only relatively modest sensitivity requirements, since currents for typical devices are in the >10 µA range. For the evaluation of the current increase, a suitable rate detector is needed to locate the voltage at which $\Delta j/\Delta V$ for the transition between the two device states is a maximum.

DISCUSSION

The electrical nonlinearity and bistable conduction state that led to the switching effect utilized in the gas sensor described has also been observed in other insulating thin films.[3,9] While the details of the electro-forming of thin dielectric films and the conduction mechanisms in the on state are not fully understood, the reproducibility of the effects, namely switching between the two stable states, and the sensitivity of the on condition to the ambient gas, has been sufficiently demonstrated for these silica devices to make applications feasible. In particular, the sensitivity of the threshold voltage to reactive gases was also found for ammonia. Selectivity and specificity, as well as the effect of the dielectric's porosity on the electric properties, are currently being investigated. Results so far indicate that the metal of the top electrode and its polarity have no effect on the device performance.

The development of a new preparation method for the fabrication of electroformed devices based on dielectric material that is derived from a polymeric precursor offers several advantages over other approaches: (1) the low-temperature conversion process (H-silsesquioxane to silica) is compatible with established IC manufacturing technology, (2) the sensor can be integrated with the control and data acquisition electronics on a common chip that contains also the evaluation logic for the threshold voltage, (3) the concept of a polymeric precursor can be extended to prepare dielectrics with sensitivities to other gases or fluids.

SUMMARY

Thin-film MIN4 devices were prepared with a nondense silica dielectric derived from a resinous H-silsesquioxane precursor. The device can be electro-formed at low voltage in an inert atmosphere and exhibits strong nonlinearity of the current-voltage curve including a region of negative differential (NDR) resistance. It was found that the switching parameters of the device, especially its threshold voltage at which the off to on transition occurs, are changed by the presence of reactive gases in the ambient such as oxygen or ammonia. The dependence of the threshold voltage on oxygen concentration was measured down to 50 ppb (in nitrogen), which demonstrates a high sensitivity; the concentration range extents to an ambient of air. A procedure is described by which these device properties can be utilized in an oxygen sensor. Its advantages include compatibility with IC manufacturing processes so that the device can be integrated on electronic chips.

REFERENCES

1. M. A. Lambert and P. Mark, *Current Injection in Solids*, Academic Press, New York, 1970; I. Bunget and M. Popescu, *"Physics of Solid Dielectrics,"* Elsevier, Amsterdam 1984.
2. J. J. O'Dwyer, *The Theory of Electrical Conduction and Breakdown in Solid Dielectrics*, Clarendon Press, Oxford 1973.
3. J. G. Simmons, "Electronic Conduction through Thin Insulafing Films," in L. I. Maissell R. Glang (ed.), *Handbook of Thin Film Technology* McGraw-Hill, New York, 1970.
4. T. W. Hickmott, *J. Appl. Phys.*, **33**, 1962, p. 2669; *J. Appi. Phys.*, **35**, 1964, p. 2118.
5. G. Dearnaly, D. V. Morgan, and A. M. Stoneham, *J. Noncryst. Sol.*, **4**, 1970, p. 593; J. Beynon and J. Li, *J. Mat. Sci. Lett.*, **9**, 1990, p. 1243.
6. C. A. Hogarth and M. Ilyas, *Thin Solid Films*, **103**, 1983, p. 267.
7. C. L. Frye and W. T. Collins, *J. Am. Chem. Soc.*, **92**(19), 1970, p. 2286.
8. R. C. Weast (ed.), *CRC Handbook of Chemistry and Physics*, 67th ed., p. F-56, CRC Press, Boca Raton, Florida, 1986.
9. G. Dearnaley, A. M. Stoneham, and D. V. Morgan, *Rep. Prog. Phys.*, **33**, 1970, p. 1129.

CHAPTER 57
USING LEG-MOUNTED BOLT-ON STRAIN SENSORS TO TURN A TANK INTO A LOAD CELL

INTRODUCTION

The weight of material in a large vessel can be monitored by sensing the strain in its support structure; for example, its vertical legs. As material is drawn from or added to the vessel, the resulting stress changes in the support members cause small but measurable strains. It is not reasonable to expect these stresses to be of the same order of magnitude as those achieved in metal foil strain gage load cells. At rated load conditions, load cell stresses are typically as high as 30,000 psi. Stresses of this magnitude in the supports of vessels would endanger their structural integrity. A strain-gage module called a *Microcell sensor* has to overcome the problem of accurately sensing and transducing these relatively low-level stress changes (typically of the order of 3,000 psi change from empty vessel to full-scale). Although this sensor is also based on strain gage technology, it overcomes the low stress problem in two ways: (1) using semiconductor strain gages with much higher sensitivity to strain and (2) a frame design that results in no-moving-parts amplification of vessel stress.

Creating an onsite load cell using the vessel's support structure introduces some application characteristics that must be overcome in order to achieve acceptable weighing performance. The two most prominent characteristics are: (1) thermally induced stresses internal to the vessel legs themselves and (2) unequal expansion of X-braces relative to that of the legs. Both problems have been addressed and can be overcome by some rather simple but technically sound solutions. The leg error stresses are greatly reduced by proper placement of Microcell sensors at the nodes of the induced stresses and by wrapping the vessel's legs in the neighborhood of the sensor mounting locations. The brace expansion problem is greatly reduced by wrapping the X-braces, thereby slowing down the heating effect of the solar radiation.

Instrumentation for measuring the quantity of solid material in bulk storage vessels generally falls into one of two categories: either level or weight sensing. If the vessel contents have variable density, level sensing is more effective in preventing overflow. The most common level-sensing devices include ultrasonic, microwave, capacitive, mechanical obstruction devices such as paddle wheels and tuning forks, and cable/bobbin systems. However, users most often prefer weight sensing because the quantity of raw material used in a manufacturing process depends on weight rather than on volume. Weight sensing is more difficult and/or more expensive, judging by the much smaller selection in this category—load cells installed under the vessel legs and strain sensors installed on the support structure. Both methods are predominantly strain-gage-based. The structure-mounted strain-sensing method relies on detecting load-induced strain in the support members. Electrical signals resulting from conversion of strain-gage-resistance changes are combined, conditioned, and then displayed as weight information. This technique can only be applied on vessels supported by metallic structures such as vertical or horizontal I-beams or H-beams, vertical pipe legs, or sheet-metal skirts, commonly referred to as *skirted silos*. Load cells installed under the legs of the vessel also rely on load-induced strain; the strain to be sensed is in the load cell structure rather than the vessel support structure. This difference is discussed further elsewhere in this chapter.

BOLT-ON WEIGHT SENSING

A measurement system able to determine the weight of a vessel's contents simply by attaching strain sensors to an existing structure is extremely attractive in any industry dealing with large volumes of raw materials such as coal, minerals, plastics, chemicals, or food materials. Bolting a number of strain sensors to the support structure is markedly easier, safer, and more economical than lifting the silo and setting it on load cells. This approach also features some significant advantages compared with level sensing systems. The sensors can be installed outside the vessel and often at ground level with no need of climbing to the top of the vessel to install a probe or some type of mechanism. Since the sensors do not contact the contents of the vessel, there is no danger of contamination or corrosion. In a properly installed system, the sensor life is practically unlimited and service-free. The system has no moving parts. But this bolt-on weighing technique, when applied to outdoor vessels, has some fundamentally different application characteristics than a load cell array mounted underneath the vessel legs. These issues, once they are understood, can normally be dealt with in order to achieve acceptable weighing performance, particularly in tackling inventory measurement problems.

The Bolt-On Microcell® Sensor

A Microcell sensor, described in Ref. 1, is a two-active-arm Wheatstone bridge solid-state strain-gage module. It amplifies the strain occurring across its bolt-down mounting holes by the unique no-moving-parts mechanical design of its base frame. If the vessel structure to which it is bolted is steel, the sensor has a sensitivity to stress of approximately 70 mV/1000 psi at 12 Vdc excitation. Although the Microcell sensor has two strain gages, it is a single-axis strain sensor operating in push-pull, thereby achieving good linearity and temperature compensation.

In most vessel weighing applications, more than one bolt-on strain sensor is used. For best results, two sensors are installed on each leg and all legs are instrumented. Two sensors per leg on opposite sides of the H-beam is optimal from the point of view of compensating for any slight bowing of the vessel support column as the load in the vessel changes. Also, the two sensors work together to compensate for nonuniform stresses in the legs created by solar radiation heating.[2] Instrumenting all legs is important from the standpoint that bulk-solid material-fill variations will shift the center of gravity laterally. If only a single leg is equipped with Microcell sensors, a lateral shift is equivalent to a weight change. A Wheatstone bridge used in all Microcell sensors are connected in parallel to form the active push-pull half of the bridge circuit. Normally a strain-gage bridge circuit operates with the highest sensitivity possible by using active strain sensing elements in all four arms. But because a Microcell sensor has a rather large sensitivity, two active arms are sufficient. A typical vessel support structure is designed for approximately 3000 psi of stress change from empty to full. This yields full-scale output = (70 mV/1000 psi) × 3000 psi = 210 mV output. All sensors are connected in parallel; therefore, the entire vessel is sensitive to changes in weight. The output voltage is made up of two parts: the live-load voltage and the offset voltage. The latter is controlled during installation and can be tailored for particular needs. For example, if the sensors receive impact loading from live material falling into the vessel, the measurement system can be set up with additional overload protection by pretensioning the sensors as they are bolted onto the vessel support structure as depicted by the line of negative offset voltage.

Two-Axis Strain Sensors

A dual-axis sensor can be formed by bolting two Microcell sensors in a T configuration, called a rosette array. For example, this can be installed on the flanges of an H-beam leg. The purpose of creating a two-axis sensor array is to compensate for very rapidly changing temperature in the structure to which the sensors are bolted. Because heat conduction takes time, the Microcell sensor frame does not remain in perfect temperature equilibrium with its environment. This mismatch in the expansion or contraction between the metal of the leg and the similar metal of the sensor frame causes an error output voltage during periods of temperature difference. When two such sensors are bolted at right angles to each other, one sensor (the vertical one) has normal-polarity excitation voltage, and the horizontal sensor receives reversed-polarity excitation. In the face of an unsteady temperature fluctuation, both sensors are simultaneously pushed into compression or pulled into tension. Since the two are connected in parallel, the positive-going voltage of one single-axis sensor is cancelled by the negative-going voltage of the other single-axis sensor. The sensitivity to changes in weight is lowered when using a two-axis array. This is true because Poisson's ratio dictates that the horizontal expansion or contraction is 30 percent of the vertical strain and of opposite sign. The two sensors connected in parallel produce [(1.0 + 0.3)/2] × single-axis sensitivity = 0.65 × 70 mV/1000 psi = 45.5 mV/1000 psi.

BOLT-ON WEIGHT SENSORS VERSUS LOAD CELLS

When viewed from a broad perspective, an array of strain sensors mounted on multiple support columns of a storage vessel represents all the elements found in a conventional strain-gage load cell:

- A metal structure designed to carry the entire load of the vessel and its contents
- One or more strain gages to sense the strain in the support structure when the quantity of live load changes
- Opposite-going points of strain in the structure
- Strain gages connected in a Wheatstone bridge configuration

But there are also numerous differences between a load cell and Microcell sensors installed on an existing support structure. An obvious difference—relative size of the support structure—is an important one. Because a vessel's structural size is so much larger than that of a load cell, it is far more vulnerable to changes in its environment, and it has so many more entry points for spurious effects, a few of which are mentioned here. A well-designed load cell structure is nearly immune to stress changes due to uneven heating or cooling; a vessel structure, on the other hand, has several significant mechanisms whereby the stresses change due to solar radiation.[2-4] A well-designed load cell structure has no bolted joints that can cause hysteresis and nonlinearity; a vessel structure has many such bolted or riveted joints that can and do give rise to hysteresis and nonlinearity. The designer of a load cell structure pays a great deal of attention to minimizing the stress response to off-axis loading; a vessel structure designer most assuredly gives no thought at all to the measurement problem caused by stress responses due to side loading when the vessel undergoes thermal expansion. The shape of a load cell is carefully planned in order to optimize the stress response to changes in vertical loads impinging upon it; the shapes of the structural members supporting a vessel were never intended to produce a straight-line relationship between applied vessel weight and structural member stress response.

In spite of these significant differences, a bolt-on Microcell weight measuring system has proven to be a very satisfactory alternative to installing load cells under each support point. Also, some vessels simply do not have a structure with a few discrete load points that is amenable to load cell installation. One example is the skirted silo, where the contact between the vessel's skirt structure and its foundation is a line contact. Another example that is more amenable to bolt-on strain-sensor weighing is the Horton sphere. Its leg structure is typically eight to twelve large-diameter pipes. Often its contents are hazardous. Installing a load cell under each leg is both costly and not very desirable. Invading the shell of the sphere in order to install a level probe is not always possible. So the solution of measuring the strain in the support columns with bolt-on sensors is not only economically attractive but may also be the only alternative.

In concluding this section on comparing a bolt-on strain sensor to a load cell solution, some remarks about the relative accuracy of the two weighing solutions are in order. A single load cell, when tested in an environment where no side effects detract from its best possible accuracy, can deliver linearity performance in the range of errors no greater than 0.05 to 0.1 percent of its rated load. When installed under a real outdoor vessel with a constant temperature environment and with typical inlet and outlet ducts attached to the silo, the error range for the system may typically be as low as 0.1 to 0.3 percent of rated load. A Microcell sensor can easily deliver nonlinearity errors no greater than 0.1 percent of its typical full-scale range of 3000 psi of stress change. (A Microcell sensor's rated stress measurement range is −8500 psi to +8500 psi.) When the appropriate number of sensors are installed on an outdoor vessel with a solid, well-built support structure, the weight measurement system can deliver nonlinearity errors in a range as low as 0.3 to 0.5 percent of the full scale.

The largest sources of error in outdoor vessels with bolt-on weighing systems are due to solar radiation causing nonuniform expansion of portions of the support structure. This in turn creates stresses internal to the structural members unrelated to the stresses caused by changes in the live load. Bolt-on sensor users have mistakenly associated daytime drift errors with sensor sensitivity to temperature changes. No doubt because of this mistaken perception, other product manufacturers in the vessel-weighing industry have claimed that their two-axis sensors are immune to environmental temperature changes. (For the reasons described in the last section.) But these claims are flawed because temperature changes also cause stress changes, and a two-axis strain sensor installed on a vessel structural support member is just as vulnerable to a temperature-induced stress as is a single-axis sensor. The next section describes these temperature-induced stresses and what can be done about them.

VESSEL LEG AND BRACE TEMPERATURE-INDUCED STRESSES AND THE CURE

H-Beam Leg Effects

When the sun shines directly on the web surface of H-beam vessel legs, it becomes markedly warmer than the flanges because only the leading edges of the flanges are exposed to the sun's radiation. As a result of this uneven heating and consequent differential temperature within the leg structure, the web expands but is unable to expand freely because of its juncture with the flanges. Therefore, the web experiences an internal compressive stress; a very bad location for mounting sensors designed to sense only the strains caused by material weight changes. When the sun shines on the outside surface of a flange, it, too, wants to expand but is not able to do so freely because of its integral joint with the web. In this case, the web creates a compressive internal stress in the flange that is receiving direct radiation. This problem is described in greater detail in Refs. 2 and 4. Two solutions for minimizing this error effect resulted from this prior research work: (1) installation of two Microcell sensors on the intersection of the web centerplane with the middle of the flanges, where the stress tension on the flange essentially cancels with the stress compression on the web; (2) installation of a sun barrier around a section of the vertical support member where the Microcell sensors are installed (leg-wrapped region). Since indoor vessels are not generally exposed to this nonuniform heating, they do not require the leg-wrap treatment.

X-Brace Effects

Almost all outdoor leg-supported vessels have some type of X-bracing between the legs. These braces are always made of a thinner metal than the legs. As a result, solar radiation raises the brace temperature faster than the leg temperature during the daylight hours.[3] The differential expansion of the X-braces compared to the legs causes an increase in the compressive longitudinal stresses in the braces and a decrease in the compressive stresses in the legs. Consequently, strain sensors bolted only on the legs interpret the force shift from leg to brace as a decrease in the material weight of the vessel. This causes a negative error in the weight reading during the morning hours when the differential heating effect is strongest. The cure for this error source is both simple and complex, depending on the X-brace-to-leg cross sectional area ratio. If the ratio is approximately 0.25 or less, a simple vinyl wrap around the X-braces is used (brace-wrapped region).

This wrap slows down the brace heating effect so that the legs and braces expand nearly at the same rate, thereby eliminating or greatly reducing the loading shift from leg to brace during the thermally active period of the day.

If the X-brace-to-leg cross sectional area ratio is greater than 0.25, Microcell sensors are also installed in the longitudinal direction on the X-braces. The X-brace sensor array forms an output signal, and the leg sensor array forms a second output signal. These two signals are mixed in the correct proportion so that the positive-going portion of the brace signal compensates the negative-going portion of the leg signal. When the X-brace-to-leg cross sectional area ratio is greater than 0.40, both strain sensors and brace wrap are installed on the X-braces.

LOAD CELLS USING MICROCELL STRAIN SENSORS

A load cell style that was specially designed for very large weighing applications and is well adapted to using Microcell sensors for measuring its internal strain is described in the next two sections. Some advantages of this load cell over other load cells with similar load capacities are that it is field-reparable, it requires no additional hardware such as stay rods or check rods, and it meets UBC requirements.

Physical Description of a Load Stand® Transducer

A Load Stand transducer is one of six load cells in a large vessel-weighing application. Its principal features are a cylindrical steel column welded to a pair of heavy-duty top and bottom flanges. The vertical column is designed so that a selected load rating generates a desired rated stress in that member. Stress created in the column causes a strain (contraction under increased loading, expansion under load reduction) that is sensed by the Microcell strain sensors bolted directly to the column. These sensors are behind the environmental covers vertically mounted on the central column. The strain-responsive frame of the Microcell sensor is illustrated in Ref. 1 along with its corresponding half-bridge electrical equivalent circuit. As the sensor frame contracts due to increased stress in the column of the Load Stand transducer, the cross beam arches downward. In the case of a totally relaxed load cell, the beam is straight. This type of sensor can also be pulled into tension, causing the cross beam to arch upward. But this cannot occur in weight transducers under a vessel even under high wind velocities because the vessel's dead load keeps the strain sensors in compression.

Electrical Characterization

The compressive loading condition increases the lower strain gage resistance and decreases the upper strain gage resistance by an equal amount. Therefore, the output voltage increases under compressive loading. Completing the bridge with two fixed resistors and connecting all the individual half-bridge sensors in parallel results in an equivalent two-active-arm bridge circuit. A beneficial effect arising from connecting the half-bridge Microcell sensors in parallel is their averaging effect, an important consideration because of the inevitable variations in column stress from sensor location to sensor location. For example, noncentral loading on the top flange causes uneven column stress. In this case, the top flange applies a variable torque along the contact circle between the column and the top flange, thereby creating unequal stresses and strains at the various sensor-mounting locations.

A four-active-arm full-bridge sensor array would cause twice the change in voltage generated by the parallel-sensor two-active-arm array. Generally strain-gage sensors benefit by the largest possible change in bridge response voltage. However, this type of Load Stand transducer offers an adequately large sensitivity, even with the present two-active-arm half-bridge sensor array, 400-mV output at rated load as compared with 24 to 36-mV at rated load for four-active-arm metal foil strain gage load cells. This high sensitivity is partly brought about by the unique no-moving-parts amplification of column strain by the Microcell sensor frame design. Additionally, semiconductor strain gages are used instead of metal foil gages, contributing significantly to the transducer's high sensitivity.

CALIBRATION WITHOUT MOVING PREMEASURED LIVE MATERIAL

Calibrating the weight sensors/transducers installed on or under very large vessels is typically very difficult because of the need for premeasured live material up to 25 percent of the capacity live-load weight of the vessel. Even small vessels are inconvenient to calibrate with live material because of the disruption of production and/or the problem of introducing the material into the middle of the process chain of events. References 5 and 6 describe a means of calibrating such difficult vessels without using premeasured live material. For vessels with bolt-on strain sensors, the procedure consists of calculating the calibration parameter values based on the sensitivity of the vessel structure and the installed strain sensor array. For vessels with load cells installed under the support structure, only the sensitivity of the array of load cells is required. The calculated scale factor for bolt-on strain sensors is:

$$\text{vessel weight sensitivity} = \frac{\text{Microcell sensitivity to stress}}{\text{equivalent support structure}}$$

$$\text{area scale factor} = \frac{1}{\text{vessel weight sensitivity}}$$

For load cells:

$$\text{load cell array sensitivity} = \frac{\text{load cell rated output}}{N \times \text{load cell rated input}}$$

where N = number of live + dummy load cells.

$$\text{scale factor} = \frac{1}{\text{load cell array sensitivity}}$$

These values are entered directly into a microprocessor-based weight signal processor's firmware via the keyboard. Actual operating data can then be used to tune the scale factor parameter values.

SUMMARY AND CONCLUSIONS

Measuring the live material in an outdoor bulk storage vessel by installing bolt-on strain sensors is not simply restricted to the sensor's ability to measure strain accurately in the face of varying live-material weight. The thermal dynamics of the vessel's support structure must also be reckoned with. A rather simple solution of wrapping the section of each leg where the strain sensors are installed is quite effective in greatly diminishing if not eliminating fluctuations caused by solar radiation impinging on the legs. An equally simple solution was devised for the differential temperature changes between legs and X-braces. If not corrected, this causes a shifting of some of the vessel loading from the legs to the X-brace support structure. A vinyl wrap around the X-braces equalizes the temperature changes of the two support members and thereby eliminates the differential expansion between leg and brace. For extreme cases of large X-brace-to-leg cross sectional areas, additional strain sensors are installed on the X-braces. A simple tuning algorithm is used to compensate the leg array signal error trend during periods of solar radiation.

REFERENCES

1. Microcell, Strain Gage Sensor Data Sheet KM 02-O1.MCL.02-01, Kistler-Morse Corporation, Redmond, Washington.
2. W. R. Kistler and C. L. Nachtigal, "Errors Encountered when Using Leg-Mounted Strain Sensors to Measure Vessel Contents," *Proceedings of the 14th Annual Powder & Bulk Solids Conference,* Rosemont, Illinois, May 1989.
3. C. L. Nachtigal and W. P. Kistler, "Bolt-On Weighing: Compensating the Errors Caused by Dissimilar Support Structures (DSS)," *Proceedings of the 15th Annual Powder & Bulk Solids Conference,* Rosemont, Illinois, June 1990.
4. C. L. Nachtigal and W. P. Kistler, "Bolt-On Strain Sensor Technology: A Look at Applications and Vessel Structure Compensation Techniques," *Weighing & Measurement Magazine,* August 1993.
5. C. L. Nachtigal, "Calibrating Weight Measuring Systems without Using Premeasured Live Material," *The International Journal of Storing, Handling & Processing Powder,* **4**(4), November 1992.
6. C. L. Nachtigal (ed.), *Instrumentation and Control: Fundamentals and Applications,* John Wiley & Sons, Inc., New York, 1990.

CHAPTER 58
FIVE NEW TECHNOLOGIES FOR WEIGHING INSTRUMENTATION

INTRODUCTION

Process weighing instrumentation systems are now available with several new technologies, that, in combination, address many of the emerging requirements for processors in the global marketplace. Implementation of total quality programs has placed a greater demand for more precise process information and control. Open system communication of process data throughout the enterprise is needed for better control of inventory and improved customer response. And finally, continuous price pressure places demands for higher productivity and lower cost to install and maintain equipment.

The new technologies to be discussed all individually offer benefits, but only in combination do they synergize to raise the current state of the art in weigh system technology.

The combination of the new technologies to be discussed includes:

Sigma Delta A/D conversion. Provides increased measurement precision and speed.

Dynamic digital filtering. Statistically analyzes high-speed/resolution process data and uses the results to optimize filter configuration.

Multichannel Synchronous A/D control. Allows the use of individual transducer data in a multi-cell system.

Expert system diagnostics. Concept uses relative measurements between several transducers to identify system and sensor faults.

Digital communication network. A simple two-wire local area network (LAN) communicating via a gateway to a host computer, using industry standard MODBUS or other protocol.

SIGMA DELTA A/D CONVERSION

Load cell systems produce an electrical output, proportional to load, in the range of 0 to 30 mv. Precise digitization of these low-level analog voltage signals, at a sample rate fast enough to control set points accurately, and so forth, requires well-designed analog-to-digital conversion techniques. Several A/D conversion methods have been successfully used in weighing instrumentation, with the most common probably being the dual-slope integrating A/D.

Dual-slope integrating converters used for electronic weighing applications have typical performance in the range of 16-bit resolution at sample rates of 200 msec. In many applications requiring precision set point control, a relatively slow, two-speed cutoff was required.

At least one integrated circuit manufacturer has introduced single-chip Sigma Delta A/D converter products that offer up to 20-bit resolution at sample rates as fast as 50 msec. The latest models also combine some level of signal conditioning that reduces the size and cost of circuit design while also minimizing noise problems. By using these new technologies, some electronic weighing instrumentation manufacturers now claim resolutions as high as one million counts.

The achievement of higher resolution at a faster rate is a benefit to production capacity. It is now possible to increase material fill and feed rates four times faster with greater set point precision, resulting in not only increased production capacity but also improved batch quality and consistency.

A secondary but important benefit of an increased resolution capability is the opportunity it creates to perform true statistical evaluation of process noise, with large quantities of data, and to adjust filter parameters for optimum stability and response (Fig. 58.1).

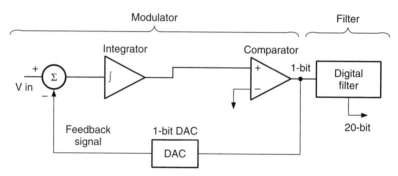

The modulator's output provides small packages of information (1-bit) at a high rate and in a form that the digital filter can process to extract high resolution (20-bits) at a lower rate.

FIGURE 58.1 Sigma Delta analog-to-digital converter.

DYNAMIC DIGITAL FILTERING

Random process noise can be difficult to filter while maintaining adequate response to actual weight changes. In order to make accurate set point cutoffs without the use of two-speed valve/motor control, high-speed stable weight measurement is required. Therefore, an effective filter will provide stable weight information when weight is not changing, and it will also provide immediate response when weight does change.

Traditionally, weigh systems have used either fixed-frequency hardware-type filters, digital averaging, or a combination of the two to stabilize noisy processes. In most cases, the magnitude of filtering had a corresponding decrease in set point response, which in turn produced inconsistent batch results.

A partially effective solution to this problem has been to use software algorithms to establish filter cutoff limits that would in effect decrease or eliminate digital averaging whenever weight changed beyond a predetermined magnitude. While substantially better than hardware of simple digital averaging filters, they were not always easily optimized for the application.

The availability of very-high-resolution weight data makes it possible to use the large quantities of data produced to perform reliable statistical characterization of the process noise. The results of this characterization is then used to tune more finely the two-dimensional filter algorithms, resulting in a more optimized relationship between stability and response (Fig. 58.2).

Statistical analysis of weight data that includes random process noise will reveal the mean, or average value, as well as the standard deviation or distribution of the data. That information is used to adjust digital filter averaging for data stability and to establish the magnitude of filter cutoff bands for maximum response to actual weight changes. The greater the quantity of data derived from higher resolution at faster sample rates, the better the reliability of the calculated values. The development of a well performing digital filter will contribute significantly to process/batch production and quality by increasing feed/fill rates with better set point control. The benefit of these new filters is the ability to maximize process speed without diminishing product quality.

MULTICHANNEL SYNCHRONOUS A/D CONTROL

The efficient packaging and impressive cost versus performance ratio of the new A/D converters now make it more economically feasible to digitize individually each transducer in a multicell system instead of using a summed analog average. This has resulted in the development of new transmitter instrumentation products that provide all of the performance benefits of digital transducers while maintaining compatibility with the large installed base of analog load cells. This allows users to upgrade existing weigh systems to the latest technology benefits without replacing the existing analog load cell sensors (Fig. 58.3).

In a dynamic process application, it is possible for movement of material to cause accumulated measurement errors in sequentially measured systems. This same problem can occur on outdoor vessels that are subjected to wind loads. A key requirement for using multiple converters within a single weigh system, when process dynamics are involved, is the ability to measure simultaneously the signal from each transducer. A new method uses a single microprocessor to synchronize the operation of multiple A/D converters, thereby making the simultaneous measurement of several transducers within a single weigh system possible.

EXPERT SYSTEM DIAGNOSTICS

Load cell transducers generally have a reputation for outstanding reliability. However, when one transducer in a multiple load cell system fails, it can be time-consuming to identify. In addition, complex systems with connected piping or other mechanical influences can make it difficult to identify the source of system perfor-

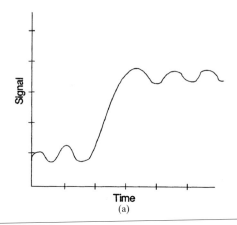
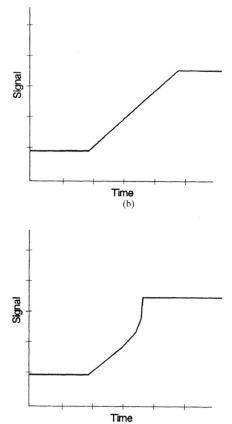

FIGURE 58.2 Dynamic digital filtering: (*a*) raw signal with step change; (*b*) digitally averaged signal; (*c*) digitally averaged signal with cutoff filter bands.

FIVE NEW TECHNOLOGIES FOR WEIGHING INSTRUMENTATION 58.5

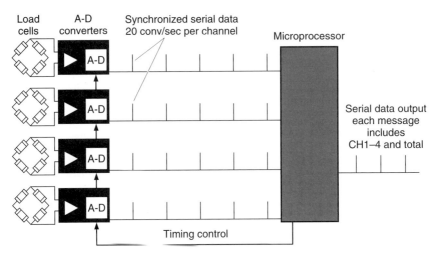

FIGURE 58.3 The microprocessor initiates a timing control pulse that triggers A/D conversions for each load cell simultaneously. The microprocessor reads the results of each conversion, sums them together, and outputs the results of each channel along with the total.

mance problems. With the introduction of transmitters able to digitize each transducer individually, it is now possible to perform several levels of diagnostics. An *expert system* compares the information from one transducer against others within the same system or against known limits of acceptable performance and makes an inferred decision regarding performance. This application of the expert System concept results in true online diagnostics of both individual transducers as well as the entire weigh system inclusive of mechanical interaction (Fig. 58.4).

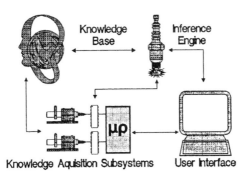

FIGURE 58.4 Expert system model.

The types of diagnostics tests currently performed include:

1. *Load shift.* Compares the ratio of force measured at each point on the weigh system and identifies changes beyond preselected limits. Possible causes are bridged material within the vessel, excessive support deflection, mechanical shunting from connected pipes, or a defective load cell.

2. *Zero shift.* Compares the zero load output of each load cell against the original reference measurement and identifies each transducer that has changed beyond a preset limit. Possible causes are a severely unbalanced system, causing one or more cells to be overloaded; heel buildup within the vessel; or a defective load cell.
3. *Drift.* Measures the signal change of each transducer during periods of no scale activity and identifies load cells that drift beyond preset limits. Possible causes are load cell moisture contamination, structural problems, heat-related structural changes, and the influences of EMI/RFI.
4. *Noise.* Performs statistical analysis of data from each transducer and determines the standard deviation. The standard deviation of each transducer is compared against the others and identified if it exceeds preset limits. Possible causes are resonating supports, loose cable connections, or load cell moisture contamination.
5. *Overload.* The peak value measured on each individual transducer is updated continuously. Measurements that exceed the original capacity setting are identified. Possible causes are bridged material within the vessel, excessive support deflection, mechanical shunting from connected pipes, or a defective load cell.

When problems are identified and communicated to either an operator or host computer, it is possible for either one to command the system to operate in a degraded mode. In a degraded mode, one or more faulted data sources are eliminated from the system weight total, resulting in the ability to continue production while the problem is being solved.

DIGITAL COMMUNICATION NETWORKS

Communication of weight information to process control computers has often been accomplished using the tried and true 4–20-ma analog signal. While this benefits from its simplicity, reliability, and low cost, it is not consistent with high resolution or online diagnostics capabilities. A digital serial connection, on the other hand, has no resolution limitation, is multivariable and therefore capable of sending diagnostics information, and is two-way, or duplex, which makes it possible to download new filter setups, calibration, and so on. The primary reasons for resistance to using digital serial communication links have been cost and risk. Both these concerns are being addressed with both hardware and software solutions.

In terms of hardware, a simple local area network communicating via a gateway device to the host computer reduces costs associated with individual serial ports for each scale system and lowers the cost of plant wiring. It also increases system performance by reducing network operating overhead in the host computer, freeing processing time for more important tasks (Fig. 58.5).

Model

The perception of risk and expense associated with the software portion of the interface is reduced with the availability of industry standard protocols like MODBUS RTU. MODBUS was originally developed for communication between PLCs but has gained widespread acceptance among users of DCS and PC-based equipment as

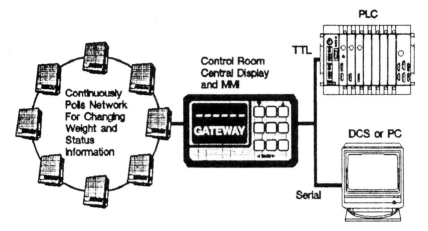

FIGURE 58.5 Responses to requests from the host with the latest information in standard MODBUS or simple binary TTL formats.

well. Turnkey MODBUS drivers are currently available from virtually all the major DCS, PLC, and PC control software vendors.

In addition, *open systems* has become a buzzword for some of the most progressive PLC and DCS suppliers, and there are many programs in place to allow weigh system manufacturers to license previously proprietary interface technologies. Concurrent with this trend is the international consolidation of various fieldbus development groups to establish a single worldwide digital communication and control standard.

SYNERGY

All of the technologies discussed combine together to meet the changing needs of process weighing instrumentation users. Sigma Delta A/D conversion and multichannel synchronous A/D control combine to improve system speeds and accuracy and make it possible to implement expert system diagnostics. Dynamic digital filtering takes advantage of the A/D technology to increase production capacity with better filter performance. Finally, the digital network capability makes it possible to improve performance while minimizing installed cost, maintenance tasks, and risk (Fig. 58.6).

All of these technologies are currently available in various instrumentation products.

REFERENCES

1. Bill Schweber, "The Pros and Cons of Sigma-Delta Converters," *Machine Design*, July 25, 1991.

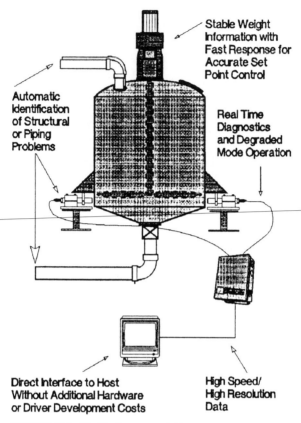

FIGURE 58.6 The ideal process weighing system.

2. *DXp-40 Expert Weight Transmitter,* BLH publication number PD-672, May 1994.
3. Ron Burke, "New Technologies for Process Weighing" BLH Weighing and Measurement, December 1993.
4. R. Holmberg and F. Ordway, System for Synchronous Measurement in a Digital Computer Network, patent number 4982185, January 1, 1991.
5. *Easy Digital Interface—MODBUS,* BLH publication number TD-075, November, 1992. (MODBUS is a trademark of Modico).
6. Global Fieldbus and Technology Forum, Boston, Massachusetts, June 21, 1994.

CHAPTER 59
MULTIELEMENT MICROELECTRODE ARRAY SENSORS AND COMPACT INSTRUMENTATION DEVELOPMENT AT LAWRENCE LIVERMORE NATIONAL LABORATORY

INTRODUCTION

The increasing emphasis on environmental issues, waste reduction, and improved efficiency for industrial processes has spurred the development of new chemical sensors for field or in-plant use. Specifically, sensors are needed to gage the effectiveness of remediation efforts for sites that have become contaminated, to effect waste minimization, and to detect the presence of toxic, hazardous, or otherwise regulated chemicals in waste effluents, drinking water, and other environmental systems. In this regard, electrochemical sensors are particularly useful for the measurement of inorganics in aqueous systems. Electrochemical sensors have the attractive features of high sensitivity, low cost, small size, and versatility of use, and they are capable of stand-alone operation. This chapter reviews our work on the development of microelectrode array sensors and the user-friendly, compact instrumentation we have developed for environmental and process control applications.

RESULTS AND DISCUSSION

Microelectrodes have characteristic dimensions in the micrometer range.[1] A typical example is platinum or gold disc, 10 µm in diameter, inlaid in an insulating plane.

Electrodes of these dimensions have some very unique properties that make them useful for sensors. Most importantly, they have high signal-to-noise ratios (especially when used in arrays); mass transport to microelectrodes is enhanced, resulting in rapid establishment of equilibrium; they are relatively unaffected by stirring effects; they can be used with small solution volumes (μl to ml); and they minimize electrochemical cell resistances, allowing them to be used in unconventional, high-resistance environments.[1–3]

We have developed unique multielement, microelectrode arrays for electrochemical sensors.[4,5] The array sensors were fabricated on a single insulated silicon substrate (standard 2-in. wafer) using photolithography. While a variety of methods have been used to fabricate microelectrodes, microfabrication using photolithography and thin-film deposition techniques offers the most convenient and reproducible means for fabricating arrays of microelectrodes. A typical process flowchart for the fabrication of a microelectrode array is shown in Fig. 59.1. By repetition of the sequence shown in this figure, a variety of sensor materials can be deposited. The sensors use a variety of electrode materials that are deposited sequentially as the scheme in Fig. 59.1 is repeated. Various modifications to this fabrication sequence have been used.

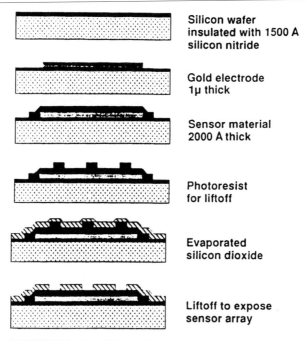

FIGURE 59.1 Photolithographic fabrication sequence for multielement microelectrode arrays.

Several different microfabricated microelectrode arrays are shown in Fig. 59.2.* The choice of size, number of elements, and geometries is virtually unlimited. In part, the sensor layout will depend upon application. Microelectrodes as small as 5

---

* Note that the sensor in the upper right hand corner is a chemiresistor, which consists of a set of interdigitated gold "finger" electrodes. This sensor functions somewhat differently from the others shown in the figure. The pads on the right and left hand sides are for electrical contact.

µm in diameter are easily fabricated using photolithography. Even smaller elements are possible using advanced techniques. For the arrays shown in Fig. 59.2, the number and size of the elements ranges from four 50-µm microelectrode disks per array, using up to four different sensor materials (sensor at the top center), to up to thirteen different materials and approximately one million microelectrodes per array (array shown at bottom right, where each square is a subarray containing approximately 76,000 individual disk microelectrodes, each 5 µm in diameter, separated by 14.5 µm center to center). In addition to the sensing electrodes, the arrays also contain a counterelectrode and a solid metal "pseudo" reference electrode. It has been an ultimate goal to develop reproducible mass fabrication procedures for the arrays so that they could be produced inexpensively and therefore could be disposed of following use.

In the earliest sensor design (sensor in lower left in Fig. 59.2), five different electrode materials were used: platinum, gold, iridium, vanadium, and carbon (graphitic). Note that only the contact pads at top are visible for this sensor; the actual microelectrode sensors are tick marks coming off the lead lines from the contact pads. These prototype sensors were used for qualitative analysis of both aqueous and nonaqueous solutions. It was shown that an improvement in information content for electrochemical (voltammetric) measurements resulted when a matrix of different electrode materials was used, relative to the case where all measurements were made using a single electrode material.[4] For example, electrochemical data was obtained for a group of four explosive and four nonexplosive (but structurally similar) compounds using the microsensor array. The information content of these measurements was then evaluated using the probabilistic model developed by Shannon.[6] It was found that the information content was improved using the multi-

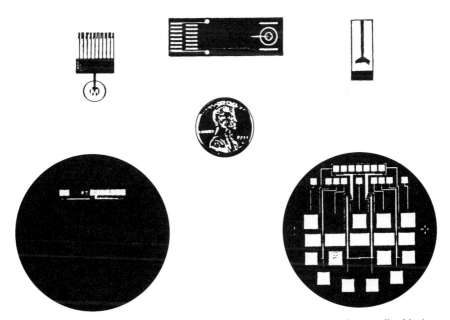

FIGURE 59.2 Representative microelectrode arrays fabricated using procedures outlined in (or similar to those in) Fig. 59.1.

element approach, with the average information content increasing by *at least* 25 percent relative to the case where a single platinum microelectrode array was used, with the same number of measurements being made in each case. Different systems and experimental protocols yielded even larger enhancements of information content. While such a result may be intuitively obvious, it was demonstrated that a quantitative figure of merit could be assigned. Application of more powerful pattern recognition methods would be very useful in conjunction with the microsensor arrays. Among other applications, such a sensor strategy may be valuable in industrial process control.

Depending upon application, without the use of pattern recognition, the arrays with bare metals may not provide much intrinsic chemical selectivity (i.e., would respond to just one or a select group of molecules or ions). Selectivity is an important consideration for field environmental sensors. In order to impart greater selectivity, various chemical or physical modifications to the individual microelectrode surfaces may be done. Each element can be partially tuned to respond to a given ion or molecule in the analyzed environment, resulting in a multielement array capable of multiple species detection. Techniques such as polymer modification, ion implantation, deposition of thin films of mercury, oxide film formation, and so on, can be used for selectivity enhancement. For instance, to form a chloride ion-selective sensor, a thin silver chloride film may be formed on a silver microelectrode substrate. Either electrochemical anodization in chloride ion media or ion implantation of chloride ions can be used to accomplish the surface modification. We have previously used the latter method of surface modification (although not on an actual array but on an individual electrode) and obtained good results.[7] Rapid response is obtained for these implanted electrodes because of the relative thinness of the implanted (membrane) layer, which can be on the order of 100 mm thick. Similarly, an iridium microelectrode can be electrochemically oxidized to form an iridium oxide film, which can function as a pH electrode.[8,9] Further, an iridium or platinum surface can be plated with a film of mercury to form a surface at which anodic stripping voltammetry (ASV) can be performed.[10,11] Such an electrode can be used to analyze for Cu^{+2}, Pb^{+2}, Cd^{+2}, Zn^{+2}, and other amalgam-forming ions. An example of the use of a mercury thin-film Pt array element for ASV is shown in Fig. 59.3, where a multiple-component solution containing Pb^{+2}, Cd^{+2}, and Cu^{+2}, all at 1 ppm concentration, was analyzed. In this figure, the potentials at which peaks occur identify the metal ion and the current peak heights are used for concentration measurement. The data were obtained using the custom experimental control/data-acquisition system described below. Detection limits using ASV are often below 1 ppb. Polymer modification can be used to form ion-selective electrodes for a variety of anions and cations.[12-14] In this way, using a variety of methods to modify the individual elements, one can tailor the array to form a multiple-species sensor. Such a sensor has enormous application for field environmental analysis. It would also be possible to use such a sensor to monitor corrosion phenomena, since corrosion is an electrochemical process tied to environmental chemical factors (e.g., pH and high concentrations of chloride ion).

For use in conjunction with the arrays, we have developed a compact electronics package for experimental control, data acquisition, and processing. At the heart of this instrument is a multichannel potentiostat, which can sequentially access and gather information from all the sensor elements on the array. This system is capable of operating in the field for on-command measurement of the analytes of interest. The system is very compact, weighing less than 0.45 kg. It comes equipped with an optional battery pack for stand-alone operation. The system is capable of monitoring up to eight elements of a sensor electrode array (more are possible). In addition

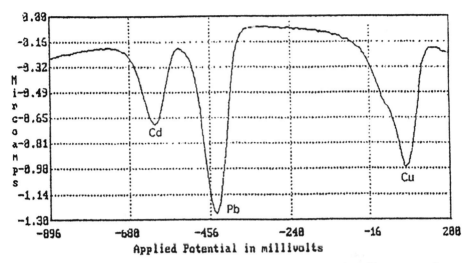

FIGURE 59.3 Osteryoung square-wave anodic stripping voltammogram of a multicomponent solution of Cd^{2+}, Pb^{2+}, and Cu^{+2}, each at a concentration of 1 ppm in pH 4.0 acetate buffer. Mercury thin film platinum microelectrode (diameter = 50 m) element on the array was used as working electrode.

to stand-alone operation, the system can be interfaced with a laptop computer, or it can communicate with a remotely placed host computer over an RS232 serial line, or it can use a UHF link. In fact, we have even embedded it in a laptop computer for a completely self-contained system. A distributed monitoring system, with units placed at various sites throughout a manufacturing plant or environmental system, is one potential application for this system. In all, the hand-carryable system will provide a complete, extremely versatile, user-friendly analytical device suitable for process control, effluent monitoring, and on-site environmental analysis. Specific applications include monitoring electroplating baths (process control) and associated waste streams, for instance, for printed wiring board manufacture; contamination from mining operations; inlet and outlet from water treatment facilities for ions such as Cu^{+2} and Pb^{+2}; providing a real-time gage of the progress of cleanup of sites contaminated with metals; and many others.

CONCLUSIONS

New sensors and analyzers are needed for environmental work and for industrial process control. In this regard, electrochemical devices have several attractive features, particularly for the analysis of inorganics (although electroactive organics can also be targeted). Our work has focused on the development of a prototype system that includes a hand-held experimental control and data acquisition/analysis system interfaced to a multielement microelectrode array. This system can potentially be used in a variety of applications. In summary, we have shown that:

1. State-of-the-art microfabrication (employing photolithography) and thin-film deposition methods are available for fabricating multielement microelectrode arrays. Almost unlimited design options are possible to suit a particular application.

2. Both microelectrodes and multielement microelectrode arrays have advantages for sensors.
3. The arrays can be used in both the (chemically) nonselective as well as selective modes.
4. Methods are available for fabricating and surface-modifying multielement microelectrode arrays to make them more selective.
5. A compact experimental control was built with data acquisition for use in system analysis with arrays. Taken as a whole, the sensor system is very versatile.

REFERENCES

1. M. Fleischmann, S. Pons, D. R. Rolison, and P. P. Schmidt, *Ultramicroelectrodes,* Datatech Science, Morganton, North Carolina, 1987.
2. S. Pons and M. Fleischmann, *Anal. Chem.,* **59**, 1987, p. 1391A.
3. P. M. Wightman, in *Electroanalytical Chemistry* vol. 15, Marcel Dekker, New York, 1989.
4. P, S. Glass, S. P. Perone, and D. R. Ciarlo, *Anal. Chem.,* **6.2**, 1990, p. 1914.
5. R. S. Glass, S. P. Perone, D. R. Ciarlo, and J. F. Kimmons, U.S. Patent Nos 5120421 and 5296125.
6. C. E. Shannon, *Bell System Technical Journal,* **27**, 1949, p. 379, 623.
7. R. S. Glass, R. B. Musket, and K. C. Hong, *Anal. Chem.* **6.3**(19), 1991, p. 2203.
8. A. Fog and R. P. Buck, *Sensors and Actuators* **5**, 1984, 137.
9. P. H. Huang and K. G. Kreider, NBS report NBSIR 88-3790, 1988.
10. J. Wang, *Stripping Analysis,* VCH Publishers, Deerfield Beach, Florida, 1985.
11. P. R. De Vitre, M.-L. Tercier, M. Tsacopoulos, and J. Buffle, *Anal. Chem. Acta.,* **24.9**, 1991, p. 419.
12. L. D. Whitely and C. R. Martin, *Mol. Cryst. Liq. Cryst.,* **16.0**, 1988, p. 359.
13. J. Ye and R. P. Baldwin, *Anal Chem.* 60(18), 1982 (1988).
14. S. Daunert, S. Wallace, A. Florido, and L. G. Bachas, *Anal. Chem.,* **6.3**(17), 1991, p. 1676.

CHAPTER 60
ENABLING TECHNOLOGIES FOR LOW-COST, HIGH-VOLUME PRESSURE SENSORS

INTRODUCTION

Within the past decade, advances in silicon sensor technology have opened up many new markets. Shipments of silicon pressure sensors grew from about 3 million units in 1983 to about 45 million units expected in 1994, with the forecasted growth to over 125 million units/year within the next decade. Almost all the growth will occur in the area of low-cost pressure sensors and transducers. Several market segments demand this type of product, including automotive, medical, and consumer.

This chapter will overview new developments of the low-cost pressure sensor technology. These developments include the new generation of disposable blood pressure transducers, silicon-fusion-bonded (SFB) piezoresistive sensor die technology (addressing consumer and automotive markets), two sensor packaging technologies utilizing SFB sensor die (plastic leadframe and small IsoSensor), and a smart, DSP-based pressure sensor technology addressing industrial and automotive applications.

MEDICAL DISPOSABLE PRESSURE SENSORS

Market Overview

The medical market is the second largest (in terms of units shipped) and one of the faster growing segments of the silicon sensor market. The focus of the industry has been on lowering the cost of medical care. This trend accelerated growth of the silicon sensor applications.

The first large volume application in the medical market was invasive blood pressure measurement. Prior to using silicon micromachined pressure sensor chips, blood pressure was sensed with expensive ($600), unreliable (easily damaged when dropped on the floor), primarily silicon-beam-based pressure transducers. These devices required sterilization and calibration before each use, with an average cost in 1982 for these procedures of about $50.

The technical and commercial success of the automotive MAP sensor, combined with the trend toward safer, disposable medical products, inspired sensor manufacturers to develop a disposable blood pressure transducer (DPT).[1,2] This device was used as a part of the patient's IV system and measured heart blood pressure transmitted by a column of a saline solution. The sensor is required to provide at least 8 kV electrical isolation between the fluid port and electrical output and deliver a 1 percent measurement accuracy over an operating life of about 72 hours. The objective for a disposable sensor was to deliver the measurement function at lower cost than the sterilization and recalibration cost. As a by product, disposability eliminated a possibility of disease transfer from a previous patient.

The first disposable blood pressure sensors were introduced in the United States in 1982, and approximately 40,000 units sold in that year. In 1993, the market grew to about 17 million units. During this time frame, the unit price paid by the hospital dropped from about $50 to $8, and the price of the basic fully packaged, compensated, and calibrated sensor module dropped from about $15 to below $2.

Other emerging medical applications utilizing technology developed for disposable blood pressure sensors include the intrauterine sensor (IUP), which monitors pressure during child delivery, and the angioplasty sensor, monitoring pressure during the catheterization procedure. These sensors ship in a cumulative volume of about one million units. It is expected that several new applications will emerge in coming years.

The market for disposable medical pressure sensors grows by approximately 12 percent per year.

Disposable Sensor Technology Overview

Two key technologies enabled Lucas/NovaSensor development of the new generation of disposable blood pressure sensors. These were temperature compensated sensor die and snapstrate-based packaging (supporting batch-mode high-volume manufacturing).

Sensor Die. Performance specifications of the blood pressure transducer are outlined by the Association for Advancements in Medical Instrumentation (AAMI).[3] One of the recommended parameters is temperature error of zero lower than 0.3 mmHg/°C. Previous generations of Lucas/NovaSensor's die required temperature testing to define the magnitude of temperature error and then a follow-up laser trimming to provide temperature compensation. To reduce the cost, a target for the new design included elimination of the production temperature test.

Design of a TC Zero below the required AAMI limit is not a major problem by itself. Superimposing this requirement on the overpressure requirement of at least 6500 mmHg for a 100-mmHg nominal measuring range, however, created a challenge. To survive the overpressure, the diaphragm's pressure sensitivity had to be reduced. This increases the temperature coefficient of zero. Another restriction was the emerging requirements for disposable sensors to be compatible with gamma radiation sterilization. To meet this requirement, a radiation-hardened sensor die process was selected. A drawback of this process is that it results in less sensitive silicon die.

The die was extensively modeled to deliver a high die attach stress reduction (one of the TC zero contributing factors), high conversion efficiency (controlling signal to temperature error ratio), and maximum burst pressure by controlling processing aspect ratios. This new silicon piezoresistive die was designed as a fully active Wheatstone bridge. Die size was set at $2.05 \times 2.05 \times 0.35$ mm, yielding about 1500 sensors per

4-inch wafer, thus supporting low-cost and high-volume manufacturability. One production wafer cassette typically contains 25 wafers, with about 37,500 sensors.

The new sensor die delivered outstanding performance with a very tight distribution of zero and TC zero (see Fig. 60.1), which made feasible the elimination of production temperature testing. Sensors manufactured with the radiation hard process have also demonstrated device stability after exposure to the radiation levels in excess of 7.5 Mrad.

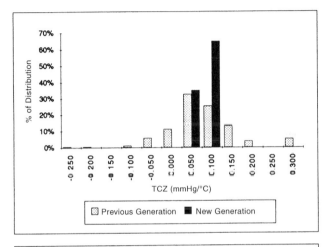

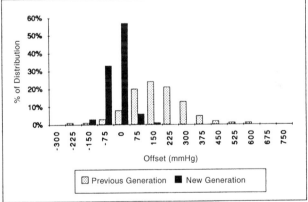

FIGURE 60.1 Tight distribution of zero and TC zero for silicon sensor die eliminated the need for production-temperature testing of disposable blood pressure sensors.

Snapstrate Packaging. The major advantage of wafer processing is batch-mode manufacturing, in which a large number of devices are fabricated simultaneously. This process delivers not only an outstanding price/performance advantage over other technologies in which each unit has to be manufactured individually, but also a very high volume capability. To duplicate these advantages for the packaged sensors, a snapstrate technology was developed for the new generation of disposable sensors.

A *snapstrate* is defined as a ceramic plate with laser prescribed lines. The plate has multiple identical cells, or substrates, each representing a package for a pressure sensor. Each substrate also has thick resistors to enable laser-trimmed calibration. Throughout the entire manufacturing process, a group of snapstrates is handled in a cassette, equivalent to the wafer-handling cassette. As a final operation, ceramic plate is snapped, or broken along prescribed lines, yielding singulated devices.

The individual sensor design includes a direct sensor die attached to a ceramic substrate. To reduce the cell size, trimmable thick-film resistors are located on the opposite side of the substrate and are connected to the sensor via the through-hole interconnects. The dielectric isolation that is necessary to support 8 kV minimum breakdown voltage is achieved through a silicone-gel-filled plastic cup. The cup also serves as the pressure connection port for the sensor. Figure 60.2(*a*) shows a cross-section of the new disposable sensor, and Fig. 60.2(*b*) shows a completed sensor.

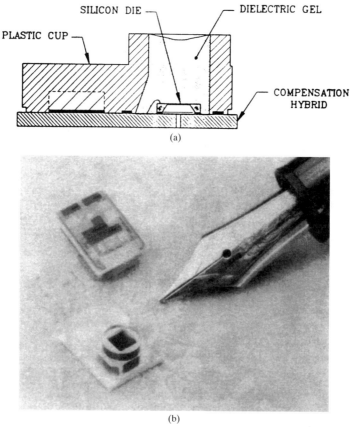

FIGURE 60.2 (*a*) Cross-section view of the new generation disposable medical sensor; (*b*) singulated disposable blood pressure sensor.

To maximize the batch mode advantages, a 4.5 × 4.5-in. ceramic snapstrate size was selected. This size was limited by the worst-case operating area of the available high-speed automated assembly equipment. The small sensor size enabled packaging of 120 sensors on a single plate (Fig. 60.3). Automated manufacturing equipment was designed to handle the snapstrate through all operations in a cassette containing typically 12 (maximum 30) snapstrates, thus creating a typical production batch of 1440 sensors (Fig. 60.4).

FIGURE 60.3 The ceramic snapstrate contains 120 disposable blood pressure sensors. One side of the snapstrate has a plastic cup attached with a gel-filled sensor cavity to provide dielectric isolation. The other side of the snapstrate has thick film resistors, providing calibration of sensor output. Visible snapstrate bar coding is used to facilitate test data transfer between the pressure test system and laser trimmer.

Manufacturing operations performed on the snapstrate form include sensor die attach (Fig. 60.4), gold wire bond, plastic pressure port attach, dielectric isolation gel fill, pressure testing, laser trimming, and final testing. The last operation is singulation, which yields individual sensors that are ready to be packaged in a plastic housing.

MINIATURE PRESSURE SENSORS

The consumer market currently represents a small fraction of the sensor market, but it is expanding rapidly. Successful examples of consumer applications include scuba diving computers, which employ two pressure sensors (to measure and display water pressure and tank's air pressure), digital tire pressure gages, barometric and diving watches, bicycle computers with elevation measurement capability, barometric stations, fitness equipment (based on hydraulically controlled resistance), washers with water level control, and vacuum cleaners that feature an automatic switch of suction pressure between carpet and hardwood floor.

The consumer segment of the market is characterized by a very short implementation cycle, often measured in months from product concept to volume production.

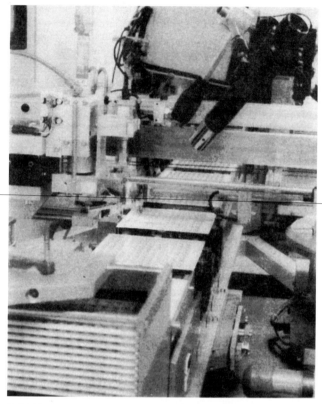

FIGURE 60.4 Automated cassette-to-cassette handling is shown for die attach operation. A snapstrate is automatically removed from the feeding cassette, the sensor die is attached to all locations on the ceramic plate with the help of pattern recognition system, and the snapstrate is inserted into a receiving cassette. Operator involvement is restricted to the insertion of a cassette and its removal after processing 1440 sensors.

It is quite feasible that consumer application may start dominating the growth of the sensor market within the next decade. One of the key aspects of this market is a drive for low cost, as the market is very price-elastic. Lower prices open potential for new applications, which increase the total available market size.

Sensor Die

The cost of the pressure sensor die is restricted by its available size, as is the case for all other semiconductor devices. The smaller the die size, the more sensors can be placed on a silicon wafer. Since the number of wafer processing steps is independent of the number of devices located on a wafer, smaller die size translates to lower die cost.

Lucas NovaSensor developed a silicon-fusion-bonding (SFB) technology[4,5] that allows for much smaller silicon die than previously possible through conventional

Conventional Pressure Sensor

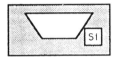

Equivalent Silicon-Fusion Bonded Sensor

FIGURE 60.5 Silicon-fusion-bonded die enables reversing of the etched slope, yielding reduced size and improved mounting package stress reduction.

silicon sensor manufacturing techniques (Fig. 60.5). Smaller size results from the capability of etching the constraint plate prior to bonding a diaphragm wafer, thus reversing the slope of the anisotropic cavity etch. This type of etch also creates a stronger mounting base for the diaphragm, opening the possibility for a so-called hard die attach.

With classical sensor design the sensor die must be mounted with a soft RTV to eliminate errors resulting from the mounting stress. A reinforced silicon constraint reduces the mounting stress transfer, so even a hard epoxy-mounted die will not exhibit large errors. This is a manufacturing advantage as epoxies can be cured quickly, and most RTVs require a 24 cure. These attributes are ideal for servicing the consumer market

Figure 60.6 shows a silicon-fusion-bonded die with respect to previous generations of silicon sensing die.

FIGURE 60.6 Comparison of several generations of micromachined silicon pressure-sensor die. The smallest die are silicon-fusion-bonded.

Another benefit of silicon-fusion-bonded die is the minimization of thermal stresses from die attach. A thick layer of silicon effectively insulates the sensing diaphragm from unwanted thermal coefficient of expansion mismatch. This allows the user to design with simpler, less costly signal conditioning circuitry. It can also lead to a completely uncompensated or statistically compensated sensor that is suitable for many applications. An added feature is Lucas/NovaSensor's SenStable™ technology for short- and long-term device stability, which significantly improves performance at a minimum of added cost.

Leadframe Packaging

Packaging for the consumer market must be high-volume, low-cost, *and* easily and inexpensively integrated into further automated assembly processes. One of the silicon sensor community's major benefits is the support from mainstream IC industry. For consumer sensors, the concept of the Mini-DIP was adopted. This type of package is compatible with existing automated handling equipment and enables the advantages of very-low-cost manufacturing to support the consumer market.

The SFB pressure sensor die was used in this type of a package. To accommodate customer demand for mounting flexibility, package pins can be formed in different ways, supporting both through-hole and surface mounting (Fig. 60.7).

FIGURE 60.7 Lucas/NovaSensor family of NPP-201 lead-frame-based pressure sensors.

IsoSensor

Very small sensor die bring a major advantage to the design of all media isolated IsoSensors. In this type of a design, input pressure is isolated from direct contact with the silicon sensor via a metal diaphragm. This diaphragm transmits a pressure on a sensor die with the help of a silicone oil. The larger the sensor die size, the larger the oil volume and the greater the effect of temperature-induced volume expansion. This, in turn, mandates a larger diaphragm and package size for a given accuracy level. Large package size for high operating pressures generates large forces within the sensor package, including welds, which reduces the number of operating full pressure cycles (due to fatigue failure).

Small SFB sensor die size eliminates all these problems and enables development of a nugget, which is an ultraminiature IsoSensor package that can easily fit into a variety of pressure port packages (Fig. 60.8). This package is one of the world's smallest media-isolated pressure sensors. First implementations targeted higher pressure ranges, 1000 to 5000 psi, supporting applications such as scuba diving computers, automotive oil pressure sensing, and hydraulic controls.

FIGURE 60.8 Ultraminiature IsoSensor (NPI-21A) using silicon-fusion-bonding technology.

SMART SENSOR TECHNOLOGY

Traditional temperature compensation and calibration for sensors is performed in the analog domain. Individual transducer data are stored in an analog memory, such as potentiometers or laser trimmed resistors. It was always difficult to achieve a batch compensation of nonlinear silicon-sensor temperature errors with such a technology. This created a limit on the available performance and the higher performance transducer cost.

Evolution of the technology enabled the development of a single-chip compensation/calibration algorithmic processor that utilizes an E^2PROM to store individual sensor calibration compensation data. The sensor signal is amplified, converted to the digital domain, digitally compensated and calibrated, and then converted back to the analog domain. The nucleus of this technology, the Analog Engine ASIC, was developed by in 1992. Using a novel design approach, this single chip ASIC implementation integrated all (with the exception of nonvolatile memory) sensor signal processing functions in a 3μ CMOS technology.[6,7] No external microprocessor was used. The sensor DSP algorithm was hardwired onto the ASIC chip.

The revolutionary aspect of this design was the integration of the manufacturing test system interface (in the form of a local bus) on the ASIC chip. This interface enabled batch-mode manufacturing for signal-conditioned pressure transducers. Several printed boards are tested simultaneously. The test system interface manages a bidirectional information flow between the supervisory test system and each of the transducers to control operational parameters of key functional ASIC blocks. The system outputs all necessary data from a specific transducer address to the test system for compensation and calibration data acquisition and for the final test, as well as data from the test system, to a specific transducer for downloading E^2PROM calibration and compensation coefficients. All the system communication and addressing are performed via a proprietary test system bus.

The important byproduct of the use of mixed analog-digital ASIC technology is the capability of transducer operation at a single 5-V supply voltage. Furthermore, the output voltage range 0.5 V to 4.5 V at 5 V supply voltage maximizes the utilization of the input range for most low-cost microprocessor-compatible A/D converters.

Multi-up handling technology was incorporated in the ASIC to reduce manufacturing cost via batch processing. The multi-up manufacturing mode was enabled by the use of the proprietary test bus. The interface to this bus was integrated into each ASIC. This bus allows connection of multiple transducers and multiplexing analog and digital signals from and to transducers under the control of a test computer. Transducers are assembled on a snapstrate equivalent *snap-boards:* printed circuit boards that carry 25 transducers each (see Fig. 60.9). Operations such as die attach, wire bonding, temperature testing, calibration and final testing can be performed with a single load operation on large number of transducers with automated test equipment. After calculation of calibration/compensation coefficients, the coefficients are loaded on the transducers memory via the same bus. Singulation delivers a fully calibrated and compensated transducer subassembly.

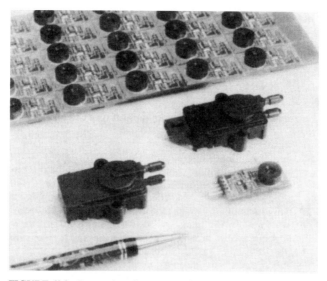

FIGURE 60.9 Integration of test system bus-enabled batch-mode manufacturing of smart pressure transducers (Lucas/NovaSensor NPS Series).

SENSOR COMMUNICATION

Networking is gradually entering the fields of sensors, controllers, and actuators. This networking evolved from, but is different than, computer system networking. Computer networks use large data packets and relatively infrequent transmission rates, with high data rates to support the transmission of large data files. Control networks, however, must shuttle countless small but frequent data packets among a relatively large set of nodes. Rapid growth of control-related networking rapidly changes the architecture of sensor-based measurement, control, and display systems. Traditional configuration of sensor-based systems is a point-to-point hardwired architecture, where each system change requires corresponding hardware rewiring changes.

Smart transducers with digital communication capabilities are currently available primarily at the high end of the market, priced around $1000 and more. Most available low-cost transducers (priced in $10–$200 range) still provide an analog output.

Major control communication developments dedicated to support smart sensor and actuator applications and are possibly related to the low-cost applications include the following:

- *InterOperable Systems Project (ISP)*. Merging process control ISA SP-50 Fieldbus physical layer with the German Profibus and French FIP data link and application layers.
- *WorldFIP*. Global open fieldbus based on IEC/ISA standard. WorldFIP recently merged with ISP.
- *CAN (Controller Area Network)*. A message-based networking system developed for the automotive industry.
- *LonWorks™*. Networking system introduced by Echelon Corporation in December 1990.
- *J1850*. Automotive networking standard approved by SAE in February 1994.
- *DeviceNet*. Open communication network based on CAN, introduced by Rockwell Automation/Allen Bradley Co., Inc. in March 1994.
- *SDS*. Communication network based on CAN and introduced by Honeywell in March 1994.

There are many benefits to sensor network implementation, including reduced system development, significantly reduced cost of wiring, and increased flexibility of reconfiguration. IEEE TC-9 and NIST initiated an effort aimed at establishing a hardware-independent networking standard. It may be expected that, in the future, the majority of currently analog output transducer applications will be converted to use a network output, bypassing the generation of transducers with a simple digital output.

CONCLUSIONS

The fastest-growing segments of pressure sensing are in high-volume/low-cost applications. The ability to serve these markets must start at the silicon level. Silicon-Fusion bonding technology and advanced packaging technologies enable a very low-cost, high-performance sensor that is ideal for these markets. It is not only important for the cost of the chip to be much lower but also the cost of subsequent packaging and signal conditioning.

Historically, the lessons that were learned in silicon manufacturing from the integrated circuits industry have been successfully transferred to silicon manufacturing in the sensor industry. It is now very important that the lessons of high-volume back-end IC manufacturing also be transferred to the sensor industry. Batch-type ceramic processing or leadframe-based processing are natural and appropriate extensions of these technologies for sensor manufacturing and must be incorporated for the sensor industry to address these high-growth markets successfully.

REFERENCES

1. Janusz Bryzek, Joseph R. Mallon, Jr., and Kurt Petersen, "Silicon Sensors and Microstructures in the Health Care Industry," *Proceedings of Sensors Expo,* Cleveland, Ohio, 1989.
2. Janusz Bryzek, "New Generation of Disposable Blood Pressure Sensors," *Proceedings of Sensors Expo,* Detroit, Michigan, 1987.
3. AAMI, "Standard for Interchangeability and Performance of Resistive Bridge Type Blood Pressure Transducers," AAMI Standard, Arlington, Virginia, 1986.
4. Kurt Petersen, Phillip Barth, John Poydock, Joseph R. Mallon, Jr., and Janusz Bryzek, "Silicon Fusion Bonding for Pressure Sensors," *Technical Digest Solid-State Sensor and Actuator Workshop,* Hilton Head Island, South Carolina, June 9, 1988.
5. Janusz Bryzek, et al., *Silicon Sensors and Microstructures,* Lucas/NovaSensor, 1991.
6. Janusz Bryzek, "Evolution of Smart Transducers Design," *Proceedings of Sensors Expo West,* San Jose, California, March 2–4, 1993.
7. Ali Rastegar, "Electronics for Smart Sensors," *Proceedings of Sensors Expo West,* San Jose, California, March 2–4, 1993.

CHAPTER 61
A TWO-CHIP APPROACH TO SMART SENSING

BACKGROUND

Great advances have taken place over the past few years in silicon micromachining technology that have enabled new classes of sensors to be produced. In addition, sensors can now be produced at costs low enough for new applications to emerge. Many of the new products that have been reported have incorporated new silicon micromachining processes and new sensing mechanisms. However, most sensor suppliers have overlooked the interface between these new microstructures and the customer's application. Specifically, developments in signal conditioning electronics and the packaging have not kept pace with the sensing technology.

The primary reason for this growing gap between sensing technology and sensor products has been the lack of ability or willingness for sensor makers to design their product to be more easily used by customers. Sensor users have been asking for products tailored to their needs, products in packages that are convenient for them to assemble into their systems, and products that are designed to solve reliably a measurement problem and not involve experimental technology.

In response to these market demands, new sensor products have been developed that employ a two-chip architecture that is more flexible and reliable than single chip approaches, smaller and much lower cost than discrete solutions, and higher performing than either approach. The issues involved with the decisions to select this two chip approach are discussed in this chapter, along with two product examples.

APPROACHES TO SOLVING PROBLEMS

Discrete Component Approach

Traditionally, sensing elements have been used along with discrete amplifier circuits to provide an interface between the sensor output and the input to the customer's system. Typically, sensor outputs of 0–100 mV or +/–50 mV need to be amplified and scaled to provide 0–5 V, 4–20 mA, or other analog input to the customer's electronics. This burden of designing and implementing these interface circuits has historically been largely with the sensor user. While these circuits have performed well, they have drained resources from customers that could have been used to solve problems more relevant to the customer's application or system. Further, there is a tremendous cost associated with hundreds of sensor users duplicating the same task of designing, implementing, and manufacturing very similar discrete solutions.

The historical advantages of discrete solutions are that they can be developed and implemented in relatively short time. Further, for small and medium volume applications, the development cost of a customized solution cannot be justified.

Another issue that arises when sensor users implement discrete solutions is that the division of responsibility is not always clear during a performance issue or sensor failure. Since the performance of the sensing function depends on the performance of both the sensing element and the associated circuitry, there is often an added element of difficulty (a.k.a. finger pointing) in identifying the source of a problem and in implementing corrective action.

Single-Chip Approach

In the single-chip approach, the sensing element is fabricated on the same silicon chip as its signal processing electronics. It requires that the sensor fabrication be completely compatible with the integrated circuit (usually CMOS) fabrication that dominates the processing of the chip.

The high-performance IC circuitry that is utilized to provide signal conditioning and amplification requires the use of complex CMOS or possibly BICMOS IC processes. These processes cannot accommodate significant variations in their manufacturing flow that are required for the optimized fabrication of the tile sensor element. In the single-chip approach, the combination of the sensor and IC elements together results in major compromises in the design and fabrication processes of both the sensor structure and electronics. The dimensions of the sensor elements are reduced due to the available IC process steps. The addition of deposited layers and their etching puts the IC components through process steps that are usually not included in the manufacturing flow.

Separate optimization of the sensor and IC processes for high yield is also not possible with the single-chip approach. The yields of the complex IC process and the sophisticated mechanical structure formation are multiplied together when the two are integrated on the same chip. For example, a 70 percent yielding IC process and a 70 percent yielding mechanical structure process result in a 49 percent cumulative yield. In the two-chip approach, both designs and processes can be optimized for high sensor and electronics yield, and only known good chips are put together. Here, a 90 percent yielding IC and a 90 percent yielding sensor can be placed together in the same package with no additional yield fallout.

As a result of the compound yield loss and the performance compromises related to the single-chip fabrication approach, the theoretical advantages of lower cost for full integration are not achieved. Further, the sensor may end up with worse performance than a two-chip sensor or a discrete approach.

Two-Chip Approach

In a two-chip approach, today's well-developed ASIC (application-specific IC) technology is used to create a custom IC chip to perform the signal conditioning along side a separate sensor chip. Both the ASIC and the sensor chips are packaged together in the sensor package.

TABLE 61.1 Summary of Benefits and Limitations of Signal Conditioning Approaches

Approach	Flexibility	Development time	Customer burden	Cost	Performance	Size
Discrete	High	Low	High	High	Good	Large
Single-chip	Low	Long	Low	Medium	Poor	Good
Two-chip	Good	Moderate	Low	Low	Good	Good

The advantages of this two-chip approach are that the size and cost of the circuitry are much smaller than the discrete approach. By having the sensor supplier combine the volumes of a number of small, medium, and large customers, the lower cost of an ASIC-based solution can be realized by sensor users who could not afford to develop and manufacture the ASIC themselves. Further, the reduction in size from a discrete solution using 10 through-hole or even surface-mount components to a single-chip-level component is substantial (Table 61.1).

Another advantage of ASIC-based solutions is that the ASIC can be designed to match the output characteristics of a specific sensor, improving the performance compared to discrete solutions built with general purpose analog circuit components.

PRODUCT EXAMPLES

Two examples of sensors using a two-chip architecture are the IC Sensors Model 3255 accelerometer and the Sensors Model 1230 pressure sensor. Each of these products was developed to reduce the size and cost of the complete product and relieve the burden of signal conditioning from the sensor user.

Accelerometer

The sensor uses an ASIC that includes all of the amplification and signal conditioning functions that previously had either been provided by IC Sensors using discrete circuit components or were the responsibility of the customers. The ASIC amplifies the +/−50mV sensor output to a 0.5–4.5V calibrated output range while providing temperature compensation and self-diagnostic functions. The ASIC is specifically designed to work with this accelerometer, so its performance is optimized.

The reduction in component count to provide the conditioned sensor output allows innovative package designs to be implemented. For example, a unique surface mount package has been developed that allows this accelerometer to be mounted either on its side or base. The side mounting allows the sensing axis to be parallel with the plane of the PC board, which is the measurement direction on most automotive air bag systems. Alternatively, the sensor can be mounted more conventionally for side impact or ride-control applications or for general instrumentation applications. These packaging options were not viable until the sensor package size could be reduced.

The primary applications of the Model 3255 accelerometer are in automotive air bag systems. Previous generations of IC Sensors accelerometers used discrete electronics to amplify and calibrate the sensor output. While these approaches performed well from a technical standpoint, the resulting products are larger and more costly than those produced with a two-chip approach.

Pressure Sensor

The Model 1230 pressure sensor is similar to the IC Sensors Model 1210 dual inline package. It maintains the same footprint (0.6 × 0.6 inches) yet adds amplification and full calibration. By eliminating the thick film resistors on the Model 1210 and using the electronic trimming on the ASIC, a substantial size reduction is realized. Adding the amplification functions inside the sensor package greatly reduces the space required for the pressure-sensing function and reduces the total cost of use for the customer by reducing component count and providing a more integrated solution.

Previous versions of the pressure sensor did not include the signal conditioning circuitry. This part of the sensing function was the responsibility of the customer. The pressure sensor included thick film resistors for calibration and temperature compensation when used with the customer's circuit. As with the accelerometer, this approach meets the technical performance requirements of the applications but requires extra space and higher cost than the two-chip approach.

CONCLUSIONS

To take full advantage of the size, cost, and performance benefits of silicon micromachining technology, sensor products need also to include well-designed signal-conditioning electronics and application-driven packages. Multiple approaches exist for implementing the signal conditioning electronics. Historically, discrete solutions have been implemented. Recently, products based on single-chip architectures have been publicized. However, the approach giving the customer the most flexibility, lowest cost, smallest package size, and best performance is one in which an optimized sensor and an optimized ASIC are combined in a single package.

Products are now coming on the market that utilize this approach. Sensor users will be able to concentrate more on their own systems and sensor applications instead of providing the interface to the sensing element. Companies using these two-chip products will have more time and resources to spend on their own products.

CHAPTER 62
SPECIFYING AND SELECTING SEMICONDUCTOR PRESSURE TRANSDUCERS

GENERAL FACTORS

Semiconductors are one of several sensing technologies used for pressure transducers. Although the internal designs will vary substantially among manufacturers, the specifications provide a common ground for comparing these designs. While this chapter focuses on specifying and selecting transducers using semiconductor technology, many of the points described apply to other technologies (Fig. 62.1).

Because pressure transducers find their way into a multitude of applications, the transducer manufacturers have developed many variations of mechanical packages, specifications, performances, and prices. Many will be suitable for a given application. Often the difficulty in selecting a transducer is not finding one, but finding one that can do the job and fits the budget. For the designers of OEM equipment, you have additional considerations of supplier reliability in addition to the reliability of the product. Can the supplier deliver the quantity needed when it's needed with no incoming rejects?

As part of the design team, your first task is to identify the preferred and the minimum performance for the proposed system. The reason for considering both possibilities is that your final selection will likely be a compromise of performance and cost. By knowing the parameters that can be compromised, you will have greater possibility of cost saving on the transducer. Take care not to overspecify. It's easy to try to improve performance by seeking superior specifications, but those improved specs can cost substantially more.

Details

A transducer's performance is described on a manufacturer's data sheet. The names of the specifications will vary slightly among the suppliers, but the units of measure and grouping of values will guide you. Generally, you will be able to separate the specs into three broad areas: physical or mechanical, electrical, and performance. These broad groups are not as important as the detailed specifications they contain.

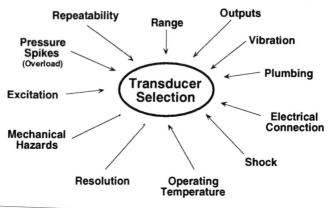

FIGURE 62.1 Selection factors.

Physical / Mechanical

Range, proof pressure,* and burst are the three pressure ratings for a transducer. The range is the minimum to maximum pressure the transducer is intended to measure. The transducer is calibrated for these pressures. The calibration process sets the output at zero (minimum pressure) and the output at full-scale (maximum) pressure. A key part of the range, especially for pressures less than 100 psi, is the reference given with the units of measure. These are vented gage (PSIG), sealed gage (PSIS), absolute (PSIA), and differential (PSID). The measured pressure is compared to one of these references (Fig. 62.2).

---

* Proof pressure is also referred to as overload or over range.

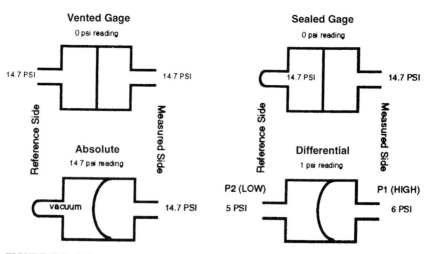

FIGURE 62.2 Reference pressures.

Proof pressure is pressure above the range that can be applied to the transducer without damaging it. You will have an output, but it will have an unspecified error (Fig. 62.3).

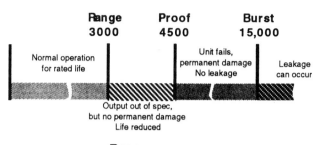

FIGURE 62.3 Pressure ranges.

Applying pressures between full-scale and proof can shorten the life of the unit. Pressures above proof and below burst will permanently damage the transducer. This damage can occur with as little as one over pressure cycle of only a few milliseconds. These pressure spikes are often found in liquid media systems that have fast-acting valves or gear pumps. Burst pressure is the maximum pressure that can be applied without causing an internal leak of the transducer; i.e., rupturing the diaphragm. Proof pressure and burst are often expressed as multiples; e.g., 2×, 10×, of the full scale.

The physical specifications also include items like temperature range, shock, vibration, response time and media compatibility. You will see two or sometimes three temperature ranges listed: compensated, operating and storage. The most important of these is the compensated range. The transducer is guaranteed to meet its performance specifications within this range. Operating temperature is the temperature range over which the unit will function with power and pressure applied, but not guaranteed to perform within published specifications. Temperatures in this range will not permanently damage the transducer, and the readings, while out of spec, will often be repeatable and hence usable. The storage temperature is that range where no damage will occur without power and pressure applied. Permanent damage can occur if the temperature goes outside the operating range while power or pressure is applied, or goes outside the storage range. These temperatures are for the temperature of the transducer, not the temperature of the pressure media (Fig. 62.4).

Values for vibration and shock should be considered figures of merit. Generally, more is better, but you pay for that. Besides the value, be sure to compare the vibration test methods. There are test differences such that a lower value from a method that dwelled at a resonance point may indicate better performance than a higher value for a simple sweep through a frequency range. Vibration is a complex area, and things like mounting method and orientation can impact the transducer's ability to survive a given application.

Response time is the time it takes the output of the transducer to reach a specified percentage of the final output from a step change of the applied pressure. The mechanical design of the pressure port, internal cavity, and the electrical circuits with the transducer may control this value. Response time is often important in testing applications.

Also in mechanical (and it's important these are not overlooked) are the case dimensions, pressure port thread, and electrical connection or cable length.

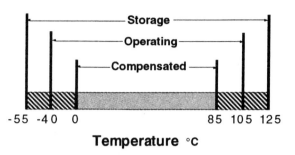

FIGURE 62.4 Temperature ranges.

Electrical

The group of electrical specifications includes excitation, full-scale output, zero output, and input and output impedance, among others. Full-scale output, or the type of output—amplified or unamplified—is usually the second most important transducer specification, as it must be compatible with the system's signal conditioning or analog-to-digital conversion circuitry. The millivolt output is often used where cost is critical and the system's signal conditioning is positioned close to the transducer. Voltage or milliamp output devices are more suitable for longer cable runs or high noise environments. Note that it is often easier to specify the transducer and then match the signal conditioner and power supply. Specifying the transducer last will limit your selections and may force you into a more expensive device. The other electrical specifications are normally less critical, but they will affect the electrical interface.

Performance

The performance specifications are where confusion among manufacturers' specifications is most likely to occur. There are two terms that are noteworthy. These are *accuracy* and *typical*.

We all use accuracy. The transducer manufacturers use it, as do the specifying engineers. But, undefined accuracy is at best a figure of merit because it does not define the source and type of error you can get. The transducer manufacturer must define the errors that comprise accuracy. Most manufacturers include linearity,[*] hysteresis, and repeatability at room temperature. The other word to be alert for is *typical*. Again, the manufacturer should define it. "Typically," it should mean that greater than 75 or 80 percent of a lot of transducers will meet this performance.

The performance specifications can be grouped into pressure and temperature. We have mentioned accuracy being made up of linearity, hysteresis, and repeatability. Linearity is the closeness of the output of the transducer at various pressures to a specified straight line. The problem is that the specified straight line can be one of five. Most manufactures will use independent linearity, also called a best-fit line. This gives the lowest value. The best-fit line is midway between the two parallel straight lines closest together and enclosing all output versus measured values on the calibration curve[†,3] See Fig. 62.5.

---

[*] While linearity is the term often used, the value given in most data sheets actually states the amount in nonlinearity of the device. A transducer with 1 percent linearity is actually 99 percent leanear and 1 percent nonlinear.

[†] A calibration curve is a graph of the output of the transducer versus known pressures applied to it.

Hysteresis is the maximum difference in output at any pressure within the specified range when that value is approached first with increasing and then decreasing pressure (Fig. 62.6).

Repeatability is the ability of a transducer to reproduce output readings when the same pressure is applied to it consecutively under the same conditions and in the same direction (Fig. 62.7).

For systems with cycling pressure, this is frequently the most important specification. Fortunately, most pressure transducers have very good repeatability. If one of these three is critical in your application—repeatability for example—ask the prospective vendor for the internal or manufacturing limit, as there is likely one.

Resolution has been an unimportant spec for many years because the analog compensation and amplifier circuits used for the last 25 to 30 years provided infinite resolution. In the past, a system's resolution was dictated by noise threshold or analog to digital conversion that was downstream of the transducer. There are new transducer designs that use digital compensation. These designs have an analog-to-digital converter, an ASIC, and a digital-to-analog converter to provide the traditional analog output. These designs use values loaded into the ASIC to compensate

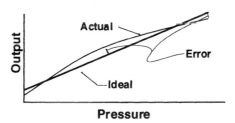

FIGURE 62.5 Independent linearity.

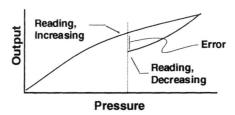

FIGURE 62.6 Hysteresis.

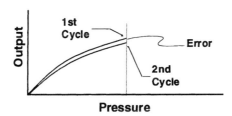

FIGURE 62.7 Repeatability.

the unit rather than trim resistors on the analog designs. What this means for you is that there will be a resolution value for the digital devices that will be expressed as a percentage of full scale.

Temperature is usually the largest cause of measurement error. There are two temperature specifications that are critical for wide temperature swings. These are zero temperature coefficient or thermal zero shift and full-scale temperature coefficient or thermal sensitivity shift. Thermal zero shift is the zero shift due to changes of the ambient temperature from room temperature to the specified limits of the operating temperature range. Zero temperature coefficient generates an offset (Fig. 62.8).

For cyclic operations, an auto zero circuit or auto zero algorithm can eliminate this error. The thermal sensitivity shift is the change in sensitivity due to changes of the ambient temperature from room temperature to the specified limits of the operating temperature range. This causes a gain error; the higher the pressure, the greater the error (Fig. 62.9).

For both of these specifications, notice that the error from minimum to maximum of the compensated temperature range could be twice what is listed since the test starts at room temperature. And when comparing the temperature errors and ranges among manufacturers, consider the error and the compensated range. A large error over a broader range may perform better than a device with a smaller error over a narrower compensated range.

All of these errors plus the tolerance of zero and full-scale combine to produce a total error band (TEB). TEB describes the absolute boundaries of the transducer's performance relative to an ideal output. The limits of performance vary with the number of test variables (Fig. 62.10).

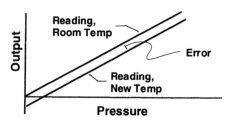

FIGURE 62.8 Temperature effect on zero output.

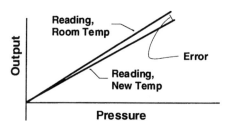

FIGURE 62.9 Temperature effect on full-scale output.

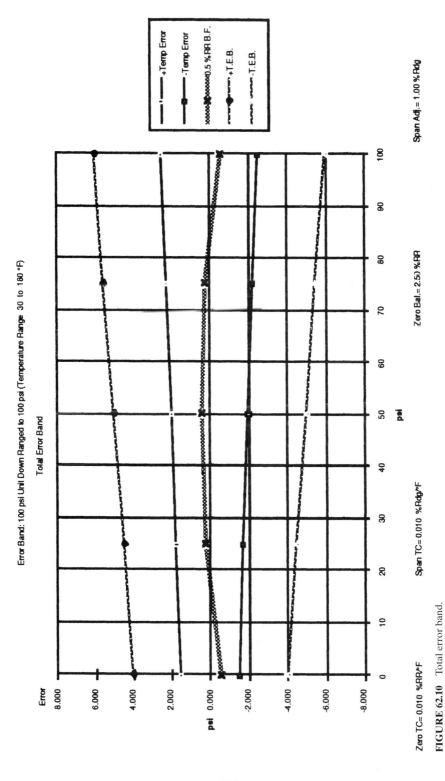

FIGURE 62.10 Total error band.

At room temperature and not including zero error, the transducer's output should follow the middle curve above with the "accuracy" spec. This test at any temperature or combination of temperatures should give outputs that fall within the inner curves in Fig. 62.10. For this transducer, when you include the tolerance on zero balance and any temperature within the compensated range, the output must fall between the outer curves.

CHAPTER 63
INTRODUCTION TO SILICON SENSOR TERMINOLOGY

INTRODUCTION

The growth of the sensor industry occurred simultaneously in several industries (such as aerospace, process control, medical, automotive, and industrial controls), utilizing a broad range of technologies (e.g., metal strain gage, quartz, vibrating cylinder, metal capacitive, silicon piezoresistive). Initially, there was no common standard defining the terminology and units of measure, and thus each industry created its own. As a result, many terms used to define sensors and their performance have different meanings depending on the manufacturer or industry.

A major effort to standardize terminology and units was undertaken by the Instrument Society of America (ISA). The ISA published a series of standards that have been adopted by some industries. A new effort is underway by IEEE. Selecting a transducer from a data sheet or catalog, however, can still be a major project, especially for new users of sensors.

This note defines the major parameters used to specify pressure and acceleration sensors following the ISA recommendation whenever possible.

The most relevant standards to the discussed subjects are:

ISA-S37.1	Electrical Transducer Nomenclature and Terminology
ISA-S37.3	Specifications and Tests for Strain Gage Pressure Transducers
ISA-S37.5	Specifications and Tests for Strain Gage Linear Acceleration Transducers
ISA-S37.8	Specifications and Tests for Strain Gage Force Transducers

GENERAL DEFINITIONS

Accuracy. The ratio of the error to the full scale output expressed in percent. The error is defined as the algebraic difference between the indicated value and the true value of the measurand. Typically, accuracy is defined as a combination of linearity, hysteresis, and repeatability, and it is used to describe static error band.

Anisotropic Etching. Etching that is direction-dependent. This chemical etching technique is used to form structures in silicon that are bounded by slow-etching <111> planes.

Anodic Bonding. The bonding of silicon to glass by heating the two in contact and applying high voltage.

Best (Fit) Straight Line. A line midway between the two parallel straight lines closest together and enclosing all output versus measurand values on a calibration curve.

Boron. Dopant used to form *P*-type resistors in piezoresistive sensors.

Burst Pressure. The pressure that may be applied to the sensing element or the case of the transducer without rupture of either the sensing element or transducer case. The sensing element does not have to operate after application of the burst pressure.

Center of Seismic Mass. The point within the acceleration transducer where acceleration forces are considered to be summed.

Compensation. Procedure of providing a supplemental device, circuit, or special material to counteract known sources of error (e.g., temperature).

Cross Axis Sensitivity. A characteristic of accelerometers used to describe the sensitivity of the device to accelerations perpendicular to the primary sensing direction.

CVD. Chemical vapor deposition; a technique for depositing thin films of materials involving the reaction of gaseous mixtures at high temperature.

Dead Volume. The total volume of the pressure port cavity of a transducer with a room barometric pressure applied.

Diffusion. Process of thermally induced distribution of impurity atoms throughout the silicon crystal lattice structure, thereby changing the electrical properties of silicon.

Die Down. Process of attaching a silicon chip to the carrier (e.g., header, ceramic substrate).

Dopant. A trace element found in silicon that alters the conductivity (e.g., boron and phosphorus).

EDP. Ethylenediamine pyrocatechol; a chemical solution used as an anisotropic etchant for silicon. It has an extremely low etch rate for silicon dioxide.

Error Band. The band of maximum deviations of output values from a specified reference line or curve. Error band should be specified as applicable over at least two calibration cycles to include the effect of repeatability.

Excitation. The external electrical voltage/current applied to a transducer for its operation.

Frequency (Natural). The frequency of free (not forced) oscillations of the sensing element of a fully assembled transducer. It is also defined as the frequency of a sinusoidal applied measurand at which the transducer output lags the measurand by 90° and is applicable at room temperature unless otherwise specified.

Frequency, Resonant. The measurand frequency at which a transducer responds with maximum output amplitude. When major amplitude peaks occur at more than one frequency, the lowest of these frequencies is the resonant frequency. A peak is considered major when it has an amplitude of at least 1.3 times the amplitude of the frequency to which specified frequency response is referred.

Frequency Response. The change with frequency of the output/measurand amplitude ratio (and of the phase difference between output and measurand) for a sinusoidally varying measurand applied to a transducer within a stated range of measurand frequencies.

Input Impedance. The impedance (presented to the excitation source) measured across the excitation terminals of a transducer. Unless otherwise specified, input impedance is measured at room conditions, with no measurand applied and the output terminals open-circuited.

For silicon bridge sensors without compensation, input resistance equals output resistance. For shear gage sensors and compensated and normalized sensors, output resistance is often lower than input resistance.

Ion Implantation. Process of placing dopants (impurities) ions in semiconductor layers at various precisely controlled depths and with accurate control of dopant ion concentration, much better than available from a diffusion.

Insulation Resistance. The resistance measured between specified insulated portions of a transducer when a specified DC voltage is applied at room conditions, unless otherwise stated.

Isotropic Etching. Etching that proceeds at the same rate in all exposed directions.

KOH. Potassium hydroxide; a common anisotropic silicon etchant.

Laser Trimming. Process of changing a geometry of thin film or thick film resistors and metals by an evaporation of a trace within a given structure, using a focused laser beam, to change its resistance.

Lattice. The regular array of atoms found in single-crystal substances such as silicon.

LPCVD. Low-pressure chemical vapor deposition; a technique for depositing thin films such as silicon nitride by the reaction of source gases at low pressures and high temperatures.

Mask. Photographically or electronically created image of one layer of the integrated circuit. Several layer of masks, each used to perform a different operation (such as oxidation, diffusion, and etching) are needed to form a functional device.

Maximum Pressure (Proof Pressure). The maximum pressure that may be applied to the sensing element of a transducer without changing the transducer performance beyond specified tolerance.

Measurand. A physical quantity, property, or condition that is measured (e.g., pressure, acceleration).

Micromachining. A process of three-dimensional batch processing of silicon wafers for fabricating mechanical structures. Micromachining represents an evolution of integrated circuit processing.

Micromachined Sensor. A silicon sensor fabricated with the use of micromachining.

Modeling. Process of the mathematical representation of certain device behavior. The most frequently used modeling for sensors and microstructures includes integrated circuit process modeling (e.g., diffusion), mechanical performance modeling (such as deflection, stress, resonant frequency, heat distribution, flow), circuit modeling and functional performance modeling (e.g., for temperature compensation).

Output Impedance. Differential impedance across the output terminals of a transducer presented by the transducer to the associated external circuitry.

Pascal (Pa). Newtons per square meter, a unit of pressure in the SI system. Often used in multiples of 1000. Pa = 1 kpa, or multiples of 1,000,000 Pa = 1 Mpa.

Piezoresistance. The effect in semiconductor materials producing a resistivity change with applied stress. The equivalent resistance change is about two orders of magnitude larger than in metals.

PSI. Pounds per square inch, a common unit of pressure.

Pressure: Absolute. Absolute pressure is differential pressure measured relative to an absolute zero pressure (perfect vacuum). The output of the absolute sensor will change as a result of barometric pressure change and thus can be used as a barometer.

Pressure: Differential. Differential pressure is a difference in pressure between two points of measurement. Differential pressure is often used to measure flow via a pressure drop measurement across the orifice inserted into a flow path.

Pressure: Gage. The gage pressure is a differential pressure measured relative to an ambient pressure. The output of a gage pressure sensor is not sensitive to a changing barometric pressure.

Range. The measurand values over which the transducer is intended to measure, specified by their upper and lower limits.

SEM. Scanning electron micrograph; a high magnification picture obtained using an electron beam as an imaging source.

Sensor Die (Chip). A silicon sensor die or chip is a micromachined silicon sensing element. This element requires microelectronic packaging before it becomes useful for most applications.

Sensor. A sensor is a first-level packaged sensor die with or without compensation and normalization.

SFB. Silicon fusion bonding; the term conceived by Dr. Kurt Petersen to describe a process of hermetically bonding two silicon wafers together under high temperature.

Silicon Sensor. Sensor built on silicon wafers utilizing both electrical and mechanical properties of silicon.

Smart Sensor or Transducer. A smart sensor or transducer is a device that employs digital signal processing to provide improved sensor performance and/or additional features such as communication.

Sputtering. Process of vacuum deposition of ionized materials that creates thin films (thickness on the order of hundreds of angstroms).

SRP. Spreading resistance profile; a plot of electrical resistivity versus depth obtained by lapping of samples and electrical probing.

Static Pressure. The average pressure on the inputs of a differential-pressure transducer (sometimes referred to as reference, common mode, or working pressure).

Terminal Line. A theoretical slope for which the theoretical end points are 0 and 100 percent of both measurand and output.

Transducer. A transducer is a fully packaged, signal-conditioned, compensated, calibrated, and interchangeable device.

Transmitter. Transmitter is a transducer with a current loop output, typically 4–20 mA.

Transverse Acceleration. An acceleration perpendicular to the sensitive axis of the transducer.

Vee Groove. The V-shaped structure formed in a (100) silicon wafer when a rectangular opening is exposed to anisotropic etchants. The sidewalls slope inward to form a self-terminating channel in the silicon.

Vibration Error. The maximum change in output, at any measurand value within the specified range, when vibration levels of specified amplitude and range of frequencies are applied to the transducer along specified axes.

Wire Bonding. Process of electrical connection between a silicon chip and a substrate. It is based on attaching a thin gold or aluminum wire (diameter in the range of .00111), using either a thermosonic or thermocompression approach.

PERFORMANCE-RELATED DEFINITIONS

Compensated Temperature Range. The temperature range over which the transducer specifications are specified.

Damping. The energy dissipating characteristics that, together with natural frequency, determines the limit of frequency response and the response-time characteristics of a transducer. In response to a step change of measurand, an underdamped system overshoots and/or oscillates about its final steady value before coming to rest at that value. An overdamped system comes to rest without overshoot. A critically damped system is at the point of change between the underdamped and overdamped conditions.

Damping Ratio. The ratio of the actual damping to the damping required for critical damping.

Full Scale (FS). The upper limit of sensor output over the range (Fig. 63.1).

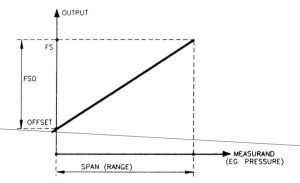

FIGURE 63.1 Definition of basic sensor transfer function terms.

Full-Scale Output (FSO). The algebraic difference between the end points of the output (Fig. 63.1).

Hysteresis. The maximum difference in output at any measurand value within the specified range when the value is approached first with an increasing and then decreasing measurand (Fig. 63.2). Hysteresis is expressed in percent of full-scale output (FSO) during one calibration cycle.

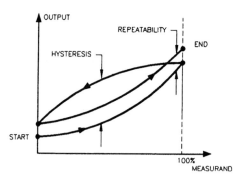

FIGURE 63.2 Definition of hysteresis and repeatability.

Life Cycling. The specified minimum number of full range excursions or specified partial range excursions over which the transducer will operate as specified without changing its performance beyond specified tolerances.

Linearity (Often Called Nonlinearity). The closeness of a calibration curve to a specified straight line. Linearity is measured as the maximum deviation of any calibration point on a specified, straight line during any one calibration cycle and is typically expressed as a percent of FSO.

There are four basic nonlinearity definitions used in the industry (Fig. 63.3):

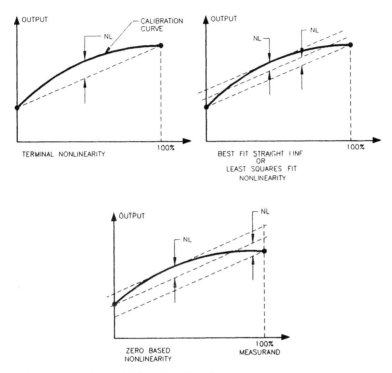

FIGURE 63.3 Three definitions of nonlinearity.

Independent or Best-Fit Straight Line (BFSL) Linearity. Measured in reference to a best-fit straight line fitted through a calibration curve. It represents the symmetrically distributed measurement error in reference to a straight line.

Least Squares Linearity. Measured in reference to a least square fit line through a calibration curve. For the majority of silicon sensors, for all practical purposes, it is equivalent to a BFSL nonlinearity.

Zero-Based (ZB) Linearity. Measured in reference to a straight line anchored at zero and positioned in such a way that maximum positive and negative distances are equal. It minimizes the error with no measurand applied, which is often the easiest point to monitor by the user.

Terminal-Based (TH) or End Point Linearity. Measured in reference to the end point line. It represents the easiest (from nonlinearity options) calibration procedure based on a calibration of the transducer at the end points.

For most silicon sensors, pressure linearity (or nonlinearity) can be approximated by a second-order curve, and thus the following approximate relationship exists between different nonlinearities:

$$NLBFSL = .5\ NLTB$$

$$NLZB = .7\ NLTB$$

The above relationship will not apply to a third-order (S-shaped) calibration curve.

Offset (Null or Zero Output). The output of a transducer, under room conditions unless otherwise specified, with normal excitation and zero measurand applied (Fig. 63.1). Often expressed in mV/V, normalized to the excitation voltage.

Operating Temperature Range. The temperature range over which the transducer will function.

Range (Span). The measurand value over which transducer is intended to measure, specified by upper and lower limits (Fig. 63.1).

Repeatability. The ability of a transducer to reproduce output readings at room temperature, unless otherwise specified, when the same measurand is applied to it consecutively, under the same conditions and in the same direction (Fig. 63.2). Repeatability is expressed as the maximum difference between output readings—it is expressed as percent of full-scale output (FSO).

Resonance. Amplified vibrations of transducer components within narrow frequency bands, observable in the output as vibration is applied along specified transducer axis.

Sensitivity. The ratio of the change in transducer output to a change in the value of the measurand (e.g., 100 mV/15 psi), often normalized to unity excitation and measurand (e.g., 20 mV/V/psi)

Stability. The ability of a transducer to retain its performance characteristics for a certain period of time. Unless otherwise stated, stability is the ability of a transducer to reproduce output reading obtained during the original calibration, at constant room conditions, for a specified period of time; it is then typically expressed as percent of full-scale output (FSO).

TCR. Temperature coefficient of resistance: this parameter characterizes the resistance change of any resistor as a function of temperature, usually in %/°C.

TCS. Temperature coefficient of sensitivity, usually expressed in % per °C.

Temperature Compensation. Procedure of reducing temperature errors of sensors and transducers. Most often, two parameters are compensated: offset and sensitivity. For silicon sensors, compensation often takes advantage of the temperature dependence of a sensor bridge. The simplest procedure employs a two-temperature compensation, bringing the value of a given parameter to the same level at two temperatures (Fig. 63.4).

Temperature Nonlinearity. This term quantifies the nonlinearity of certain parameters over the compensated temperature range (Fig. 63.4). Most often, parameters characterized for temperature nonlinearity are zero output, sensitivity, bridge resistance, and FSO. It is expressed in percent of FSO. Nonlinear temperature coefficients

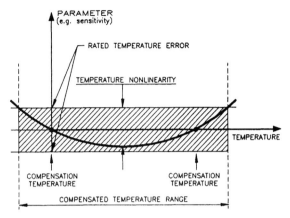

FIGURE 63.4 Temperature nonlinearity.

produce a residual temperature error after a simple temperature compensation that limits the overall accuracy. Compensation of temperature nonlinearity requires a more sophisticated compensation technique.

Thermal Hysteresis. Nonrepeatability of parameters at reference temperature before and after the temperature cycles. Due to different sources of thermal hysteresis, three groups of this parameter can be distinguished:

Cold-Cycle Thermal Hysteresis. Defined as a change of a given parameter after the negative temperature exposure in comparison to the reference temperature.

Hot-Cycle Thermal Hysteresis. Defined as a change of a given parameter after the positive temperature exposure in comparison to the reference temperature.

Full-Cycle Thermal Hysteresis. Defined as a change of a given parameter after both cold and hot temperature exposures in comparison to the reference temperature.

Thermal Sensitivity Shift (Temperature Error). The sensitivity shift due to a change in ambient temperature from reference temperature to the specified limits of the operating temperature range.

Thermal Zero Shift (Temperature Error). The zero shift due to changes in ambient temperature from reference temperature to the specified limits of the operating temperature range.

Transverse Sensitivity. The sensitivity of a transducer to transverse acceleration or other transverse measurand. It is measured as a maximum transverse sensitivity when a specified value of measurand is applied along the transverse plane in any direction, and is usually expressed in percent of the sensitivity of the transducer in the sensitive axis.

Warm-up Time. Time in which the device stabilizes its performance to a predefined accuracy after application of rated excitation, due to temperature change resulting from applied power. For silicon sensors, this time is typically on the order of several seconds. There are several other effects such as space charge distribution, also related to the time of application of excitation, that take much longer (e.g., days) to stabilize.

CHAPTER 64
SILICON SENSORS AND MICROSTRUCTURES: INTEGRATING AN INTERDISCIPLINARY BODY OF MATERIAL ON SILICON SENSORS

INTRODUCTION

This chapter represents an effort to integrate an interdisciplinary body of materials to generate a composite of the silicon sensor and its microstructure technology. The book correlates the importance of the mechanical microstructure of the silicon materials and its mechanical properties.

The first systematic characterization of mechanical applications of silicon was published in the 1982 *IEEE Transactions on Electron Devices* by Dr. Kurt Petersen. The work published in this chapter provides an update on the progress of the technology and introduces an overview of markets, sensor basics, signal conditioning, and packaging.

Several commercial endeavors have taken advantage of the mechanical properties of the silicon composite and created profitable ventures. This, in turn, has lead to the advancement of the state-of-the-art in the field of silicon sensors and microstructures.

The technology that creates a foundation for silicon sensors and microstructures is called *micromachining* (µ-machining), which is widely believed to have the potential to revolutionize the field of sensors and create a large new industry for silicon microstructures. Micromachining is the basis of the current generation of silicon sensors and is beginning to revolutionize the world of mechanical components, including both passive mechanical elements (such as nozzles and optical masks) and actuators (such as pressure valves, servo-actuators, and motors). Much has been written about the technology. The excitement with this is visible not only in both the scientific and business communities but also more recently in the mass media. It is obvious that much more work remains to be done to exploit the technology fully.

It is believed that a major hurdle in the growth of this is the awareness of technologists, particularly the OEM designers, regarding the full possibilities of the technology, both in the future and, perhaps more importantly, in the present.

This chapter will help to accelerate the growth of the micromachining applications. In the final analysis, the success of silicon technology will depend on the efforts of universities and companies to bridge the gap between research, industrial environments, and the customers and their technologists who have the vision to employ the latest technology in their next generation of products.

MARKETS AND APPLICATIONS

Introduction

Integrated circuits have changed every aspect of our life. However, while the microelectronic revolution has spread, a more subtle technology—silicon micromachining—has gradually transformed the way today's engineers think about physical sensing devices and miniature mechanical structures. The field of mechanical structures in silicon—specifically sensors—has received much attention in recent years, not only from technologists developing new generations of devices, but also from professional business and financial communities as well as TV and daily press.

Characteristics of Sensors and Transducers

Sensors and transducers convert physical variables (such as pressure) into an electrical signal. They are often the critical components that determine the feasibility of new products. As such, they have a significant leverage factor, enabling production of sensor based systems exceeding sensor cost by several orders of magnitude. For example, all modern cars rely on a fuel delivery systems based on the manifold absolute pressure or flow sensors sold in volume under $15.

Almost all new sensor applications are silicon-based. The silicon technology is also displacing older technology in traditional applications. Only silicon can meet the aggressive cost or cost-performance requirements of most new applications. The declining cost of electronic components, especially microcontrollers, combined with the low cost of silicon sensors, makes new sensor-based applications feasible. From the strategic point of view, the stability and performance of a sensor supplier became of the utmost importance to high-volume, highly sensor leveraged users.

The high cost of production equipment, R&D, and engineering, together with the availability of customers with large requirements, is causing sensor manufacturers to focus on winning the market share to drive up volume. As in the integrated circuit industry, higher volumes allow for increased yields, lower costs per sensor, and greater visibility in the industry. Winning a few large contracts makes a company more cost-competitive in other applications.

Originally an expensive and exotic technology, silicon-based pressure transducers were initially used only in specialized aerospace, testing, and instrumentation functions. Within the past ten years, however, advances in silicon micromachining manufacturing technology have opened up many new uses for the technology. The need to provide microprocessor-controlled systems with information about mechanical variables such as pressure and acceleration has resulted in the dramatic growth of silicon sensor technology in recent years. The microprocessor is deaf, dumb, and blind without suitable sensors providing input to a surrounding world of physical variables, such as pressure, acceleration, level, flow, humidity, and temperature. Widespread application of microprocessors has created the need for low-cost, high-performance, high-volume sensors.

The sensor market has exhibited extreme price elasticity. For example, it had been known for many years that Manifold Absolute Pressure (MAP) sensors could make a dramatic improvement in gas consumption and performance in automobiles. Spurred by the government gas mileage standards for new cars, the number of MAP sensors shipped went from under 10,000 units in 1976 at prices about $40 to 40,000,000 units in 1996 with prices under $15. Another example is the disposable blood pressure transducer. Introduced in 1982 at $40 as a replacement for a reusable blood pressure transducer, 40,000 such units were shipped in that year. In 1989, the shipments grew to 8,000,000 units with prices about $10.

Classification of Silicon Sensor

Classification of silicon sensor users reveals dynamic changes in the sensor industry. There are four classes of users of silicon sensors. They can be characterized as follows:

- *Existing transducer companies* using older, more expensive semiobsolete technologies, purchasing chip or low-level sensor as the heart of transducer lines.
- *Captive users* of pressure sensors looking for either a second source or next generation of sensor technology.
- *Existing users* of sensors seeking a second source or the next generation of devices replacing their existing technology.
- *New users* of sensors with clearly defined products with known market potential.

While the first three categories of users move through implementation cycles of new sensor-based products relatively efficiently, the new users of sensors underestimate the effort necessary to bring this type of products to the market. It is clear, however, to all these groups that micromachining already has become the technological choice for today's silicon pressure sensors and accelerometers.

GENERIC SENSOR CLASSIFICATION

Electrical sensors can be characterized in several different ways. One of the more interesting approaches is summarized below.

There are six basic signal domains with the most important physical parameters that are used to characterize sensors. These six signal domains of sensors can be also characterized as follows:

- *Radiant signal domain* is related to electromagnetic radio waves, microwaves, infrared, visible light, ultraviolet, X-rays, and gamma rays. Measured variables include light intensity, wavelength, polarization, phase, reflectance and transmittance.
- *Mechanical signal domain* includes two forms of energy—gravitation and mechanical. Gravitational energy concerns the gravitational attraction between a mass and the earth. Mechanical energy pertains to mechanical forces—displacement flow, and so on. Measured variables include force, pressure, torque, vacuum, flow, volume, thickness, mass, level, position, displacement, velocity, acceleration, tilt, roughness, acoustic wavelength, and amplitude.

- *Thermal signal domain* relates the kinetic energy of atoms and molecules. Measured variables include temperature, heat, specific heat, entropy, and heat flow.
- *Electrical signal domain* deals with electrostatic fields, currents, and voltages. Measured variables include voltage, current, charge, resistance, inductance, capacitance, dielectric constant, electric polarization, frequency, and pulse duration.
- *Magnetic signal domain* relates with magnetic fields, currents, and voltages. Measured variables include field intensity, flux density, moment, magnetization, and permeability.
- *Chemical signal domain* includes both molecular and atomic energy. Molecular energy is the bond energy that holds atoms together in a molecule. Atomic energy is the binding energy that is related to the forces between nucleus and electrons. Measured variables include composition, concentration, reaction rate, toxicity, oxidation-reduction potential, pH, and so on.

A sensor may utilize a direct conversion of specific energy into electric signal or may use an indirect way of conversion, utilizing a different transitional form of energy. The first case can be illustrated by a piezoresistive effect, which is used in creating pressure transducers, where pressure directly changes resistance which in turn represents the value of pressure.

The second case can be illustrated by a thermally based flow sensor, which first converts mechanical into a temperature difference, which in turn is converted into an electrical signal.

Physical and chemical effects in silicon available for construction of sensors are briefly characterized in Table 64.1. These effects can be grouped into two major categories: self-generating and modulating. The generating effects create a direct electric energy as a result of exposure to measured variable. The modulating effect changes one of the electrical parameters; for example, resistance of the strain gage under stress; without generating electricity. To measure the modulating effect, an external energy has to be supplied (e.g., power supply voltage).

TABLE 64.1 Physical and Chemical Effects in Silicon Useful for Sensing

Signal domain	Self-generating effect	Modulating effect
Radiant	Photovoltaic effect	Photoconductivity Photoelectric effect
Mechanical	Acoustoelectric effect	Piezoresistivity Lateral photovoltaic eff Lateral photoelectric eff
Thermal	Seebeck effect Nernst effect	Thermoresistive
Magnetic		Hall effect Magnetoresistance Suhl effect
Chemical	Galvano effect	Electrolytic conduction

Different physical and chemical effects can be utilized in sensor design. A brief characterization of these effects that are or could be used in the direct conversion of nonelectrical signal into an electrical signal in silicon sensors is discussed below.

Radiant Signal Domain

When a material is exposed to electromagnetic radiation, several effects can be observed; most effects are dependent on light intensity and wavelength.

Photovoltaic Effect. This effect stands for the process by which a voltage is generated by incident radiant energy at the junction of two dissimilar materials such as a Schottky contact or a *p-n* junction.

Dember Effect. When a semiconductor is locally illuminated, owing to the differing diffusion coefficient electrons and holes, a voltage is generated between the illuminated and the nonilluminated part.

Photomagnetoelectric Effect or Photo-Hall Effect. This effect describes the generation of an electric field perpendicular to both an applied magnetic field and an incident flux of radiant energy.

Photoconductivity. When a semiconductor or insulator is illuminated by radiant energy, the electrical conductivity increases.

Photoelectric Effect. When a semiconductor junction is illuminated by radiant energy, electrons and holes are generated in the junction, decreasing its resistance.

Photodielectric Effect. Some materials show a change in the dielectric constant and loss properties when illuminated by radiant energy.

Photocapacitive Effect. When a metal-insulator-semiconductor capacitor is illuminated, its capacitance changes.

Mechanical Signal Domain

When a solid-state material is subject to isotropic or anisotropic strain, several effects can occur that are based on the change of the mobility (resistivity), band-gap (junction voltage), or the polarization of the material.

Deflection Effect. Mechanical energy causes silicon to bend, creating a stress in a mechanical plate or beam or changing distance to a another plate, affecting capacitance and changing resonant frequency.

Piezoelectric Effect. This mechanical force, acting on a crystal that lacks a center of symmetry, causes surface charges to appear.

Acoustoelectric Effect. An electric current can be generated in a crystal by a traveling longitudinal acoustic wave interacting with the conduction electronics.

Triboelectrification. Rubbing suitable materials together produces positive or negative charges on the surfaces.

Piezoresistivity. A mechanical force acting on a semiconductor causes an appreciable change in the resistivity.

Piezojunction Effect. The characteristics of a *p-n* junction in a semiconductor are changed when the junction is subject to a mechanical stress.

Lateral Photovoltaic Effect. When a junction of two dissimilar materials is locally illuminated by radiant energy, a lateral voltage arises along the junction.

Lateral Photoelectric Effect. The current division across two contacts at one side of a reverse-biased semiconductor junction is dependent on the location at which radiant energy is incident on the junction.

Faraday-Henry Induction. When a closed electric circuit is moved in a magnetic field, a current is induced proportional to the embraced flux change per unit time.

Thermal Signal Domain

Most materials change their electrical properties when the temperature is changed. Many effects describe changes that depend on the absolute temperature, whereas another group of effects are only observed when a temperature gradient is present.

Seebeck or Thermoelectric Effect. In an electrical circuit composed of two dissimilar metals or semiconductors, voltage is generated when the junctions are kept at different temperatures.

Pyroelectric Effect. Polar crystals show a change of the polarization when the temperature is changed.

Nernst Effect. When a temperature gradient is established in a conductor with a magnetic field normal to the gradient, an electric field arises, mutually perpendicular to the gradient and the magnetic field.

Thermodielectric Effect. The dielectric constant of a ferroelectric material is a function of the temperature.

Thermoresistivity. The resistance of metals increases while that of semiconductors and insulators decreases with increasing temperature.

Thermojunction Effect. The forward voltage across a p-n junction at constant current decreases for increasing temperature.

Superconductivity. Below a specific transition temperature, some materials have no resistivity.

Thermoelastic Effect. When a metal is subjected to mechanical strain, an electric potential occurs between two regions of the metal, which is a function of the temperature difference between these regions.

Magnetic Signal Domain

A moving charge carrier experiences a Lorentz force in a magnetic field. Many of these so-called galvanomagnetic effects in solid-state materials are based on this Lorentz force.

Hafl Effect. When a current in a conductor flows in the x-direction and a magnetic field is applied in the z-direction, an electric potential will arise in the y-direction.

Magnetoresistivity. The electrical resistivity of materials changes when a magnetic field is applied.

Suhl Effect. The conductivity of a surface region in a semiconductor is changed when a magnetic field is applied normal to the current direction.

Magnetoelectric Effect. In composite materials consisting of a piezoelectric and a magnetostrictive material, a magnetic field can cause electrical polarization of the material.

Faraday-Henry Induction. A varying magnetic field induces an electric current in a closed electric circuit.

Superconductivity. Materials change from the superconductive to the normal state at a specific magnetic field.

Photomagnetoelectric Effect. This effect describes the generation of an electric field perpendicular to both an applied magnetic field and an incident flux of radiant energy.

Nernst Effect. When a temperature gradient is established in a conductor with a magnetic field normal to the gradient, an electric field arises, mutually perpendicular to the gradient and the magnetic field.

Chemical Signal Domain

The definition of this signal domain is still not very clear, so that an enumeration of effects belonging to this domain is not easy. Chemical signals nearly always have something to do with the concentrations of one material in another, often in different aggregation states. Listed below are a few effects that occur in a solid and that can be used to monitor chemical signals outside the solid. When a gas, liquid, or other solid is brought into contact with the solid, a slight change in the electrical, optical, thermal or magnetic properties of the solid might occur.

Volta Effect. When two dissimilar metals are brought into contact with each other, a contact potential is established.

Galvanoelectric Effect. When two plates, each made from a different conducting material, are immersed in an electrolyte, a voltage between the plates, which is material-dependent, can be measured.

Chemodielectric Effect. This effect describes the change in the dielectric constant and the loss properties of a material when surrounded by a gas or immersed in an electrolyte.

Chemofield Effect. The conductivity of a surface region in a semiconductor is changed when immersed in an electrolyte or subjected to a gas.

From all these effects, the most often used are piezoresistive effect and capacitive (deflection), both creating foundation for almost all pressure and acceleration sensors.

Evolution and Growth of Silicon Sensor Technology

The technology of processing mechanical structures in silicon is called *silicon micromachining* (μmachining). It is more precisely defined as a three-dimensional sculpting of silicon, using standard or modified semiconductor batch processing technology.

The roots of this technology date back to Bell Laboratories. The same team of people working on the basics of semiconductor technology discovered a piezoresistive effect in silicon. The first paper was published in 1954, outlining the basic piezoresistive properties of silicon and germanium. This effect is responsible for a large resistance change in semiconductor material subjected to stress, larger by approximately two orders of magnitude than the corresponding resistance change in metals used previously for strain gage applications.

The evolution of silicon as a sensing material for measuring pressure occurred in six phases.

At the beginning, the first silicon sensors were developed by cutting a silicon bar out of the wafer to form resistive strain gages. For pressure measurement, these sensing elements were adhesively bonded (by hand) onto a metal diaphragm that was in direct contact with the media to be measured (Fig. 64.1, Phase I). The finite stresses and subsequent deflection of the diaphragm were transmitted to the bonded strain gages. The change in resistance of a particular strain gage is linearly proportional to the applied force. This first design suffered not only from low yields but also high temperature errors and poor stability as a direct effect of thermal mismatch of the silicon-glue-metal interface. The first industrial applications of silicon-based pressure sensors using this technology were introduced in 1958 by three companies: Kulite, Honeywell, and MicroSystems.

A quantum jump in technology came from Kulite when the strain gages were directly diffused into the silicon diaphragm (Phase II). It was no longer necessary to attach brittle gages by hand to the diaphragm, and one of the key factors of time-dependent drift (glue) was eliminated. Strain gages were buried in the atomic structure of the silicon diaphragm. The entire silicon diaphragm now had to be bonded to a metal constraint that provided package stress isolation. The performance of the sensor was improved. The size of the sensor was still large. However, it was limited by the strain gage size and the positioning accuracy of the silicon disk mounted on the metal constraint. At this point, silicon still did not produce performance comparable with other technologies such as thin-film strain gage or force balance.

To achieve better performance and lower cost, new packaging techniques had to be developed, and the size of the sensor chip had to be reduced. The solution was delivered with the introduction by Honeywell of a cavity (cup) milled directly into the silicon to form a thin diaphragm (Phase III). A combination of mechanical milling supplemented by a final chemical etching process provided this function. Additional performance improvement came from the introduction of a silicon constraint bonded to a silicon cup through a gold-silicon eutectic bond. Sensors based on this technology were successfully implemented in aerospace and process control applications. The major drawback of this process was the cost. Cavity formation was a mechanical process, milling one cavity at a time on a wafer. Bonding a constraint also was done in one-at-a-time mode. Another problem was in the dimensional tolerance of the location of the diffused strain gage with respect to the stress/strain region of the diaphragm due to rounded diaphragm corners through isotropic etching. Positioning accuracy of the mechanical milling process was also a concern.

Consecutive improvement was introduced again by Kulite (Phase IV). The diaphragm was formed entirely through a selective anisotropic chemical etching process. This etching was performed on an entire wafer of sensors simultaneously.

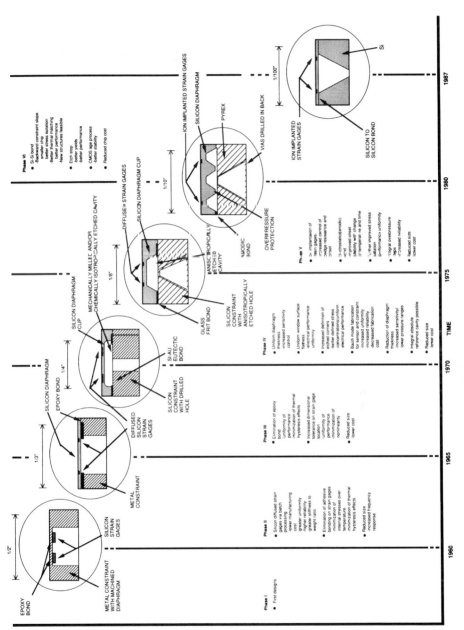

FIGURE 64.1 Evolution of silicon pressure sensor technology.

Cavity size was defined by a masking process identical to other IC masking steps. The thickness of the diaphragm was defined by the etch time. This process offered the advantage of precisely aligned crystallographic planes in silicon to provide an order of magnitude improvement in strain gage positioning, allowing reduction of the sensor die size.

Simultaneous improvement was introduced to the constraint technology. Glass first was introduced to bond a silicon constraint wafer to the diaphragm wafer. For the first time in history, the entire sensor manufacturing process was performed in a batch mode on a wafer, yielding multiple devices per wafer. Pressure sensor technology entered new era: lower cost manufacturing.

Further improvements (Phase V) were brought by the ion implantation of strain gages. This increased the control of electrical parameters (offset, bridge resistance, and so on). A new technology of electrostatic bonding of silicon wafers to special glass (matching silicon's thermal expansion) simplified the constraint manufacturing process and improved sensor stability and temperature errors. More complex micromachining yielded on-chip overrange protection. Costs continued to decrease.

The latest advances in sensor technology (Phase VI) brought a new revolutionary achievement introduced by NovaSensor: silicon-to-silicon wafer lamination, a process allowing the creation of a molecular bond between two or more silicon wafers. The technique opened new horizons for sensor development, allowing micromachining of structures not feasible before, reducing their size, improving performance, and making higher levels of mechanical integration on silicon chips for sensors and microstructures possible.

New dimensional control tools such as programmable etch stop techniques further improved performance. The implementation of CMOS research results in the area of space charge control improved the stability of sensors.

The chronological summary of milestones in the development of silicon pressure sensor technology is summarized in Table 64.2.

Development of silicon sensor technology is closely related to the growth of the market for silicon sensors and microstructures, which can be divided into four phases (Fig. 64.2).

During the discovery period, only very crude devices were made, using silicon as the mechanical material. These types of transducers offered higher sensitivity than metal strain-gage-based transducers, although they suffered from very high instability and temperature errors combined with high cost. The technology, however, was already sufficiently exciting to support a major effort within Honeywell and give birth to companies like Kulite Semiconductor and MicroSystems.

During the commercialization and market development phases, there was a major effort by several industrial companies to bring this technology to practical applications. Major technological progress was made at Kulite, Honeywell, and Fairchild.

During the next phase, cost reduction and application expansion, when the practical applications were already feasible, the prime area of application was aerospace and industrial controls. The early pioneers of this technology were companies such as IC Transducers, National Semiconductor, Honeywell Microswitch, and Kulite Semiconductor. Their efforts resulted in the creation of the OEM market for solid-state sensors.

During the consecutive phase of growth, micromachining and rapid market growth, the technology gradually started to respond to widespread application of microprocessors and their dramatically decreasing cost. Users trying to upgrade the sophistication level of microprocessors were looking for component-type sensors comparable in both packaging and price with microprocessors. The only technology capable of supplying the required price/performance ratio was machining of silicon.

TABLE 64.2 Chronology of Silicon Sensor Technology

1950–1960: Discovery Period	
1954	Smith discovers piezoresistivity at Bell Labs
1958	Discrete silicon strain gages commercially available

1960–1970: Commercialization and Market Development	
1961	Integration of strain gages into silicon diaphragm (Kulite)
1966	Mechanically milled cavity introduced (Honeywell)

1970–1980: Cost Reduction and Application Expansion	
1970	Isotropic micromachined window introduced (Kulite)
1970	First piezoresistive acceleration sensors developed (Kulite)
1971	Sensors employing anodic bonding introduced (Kulite)
1974	First high-volume hybrid sensor (National Semiconductor)
1976	Anisotropic micromachined windows introduced (Kulite)
1976	First under-hood automotive silicon sensor (Honeywell)
1977	First silicon capacitive pressure sensor demonstrated (Stanford University)
1978	Bossed silicon diaphragms introduced (Endevco)
1979	Passive on-chip temperature compensation (Kulite)
1979	Ion implantation of strain gages (Honeywell)

1980–1990: Micromachining and Rapid Market Growth	
1982	First disposable medical transducer (Foxboro/ICT, Honeywell)
1982	Active on-chip signal conditioning (Honeywell)
1985	Polysilicon diaphragm sensor by additive process (Wisconsin University)
1987	Si-Si and Si-SiO$_2$ lamination implemented into sensor designs (NovaSensor)
1988	SFB based acceleration sensor with 1000:1 overload protection (NovaSensor)
1990	Introduction of SenStable: new superstable piezoresistive process/design (NovaSensor)

While existing industrial companies were focusing efforts primarily on existing applications, there was a great deal of effort at the university and research level in the development of new sensor technology. While industrial companies were focusing their effort primarily on the utilization of technology developed for diodes and transistors in the mid 1960s, academic-level sensor technology was focusing on the utilization of the newest tools brought by the VLSI age. As a result, selected companies started to focus their attention on the incorporation of recent advances of micromachining and VLSI processing into sensors. The objective was to create the next generation of devices that would be able to create not only a better price/performance ratio, but also open up new applications not previously possible with the old technology. The world's first research-oriented company in this field, Transensory Devices, was formed in Silicon Valley in 1982.

The genealogy of solid-state sensors in the U.S.A. is shown in Fig. 64.2. As can be seen, it is still a small community, and there is a strong personal interaction between different firms. It is also visible that solid-state sensor technology is being developed within a limited number of companies.

The development of silicon micromachining is complex, since it involves interaction between the mechanical and electrical properties of silicon. It can be compared to the development of high-performance analog IC technology in a certain way: there are a limited number of gurus capable of contributing to the development of this technology, and their availability is a limiting factor on the market.

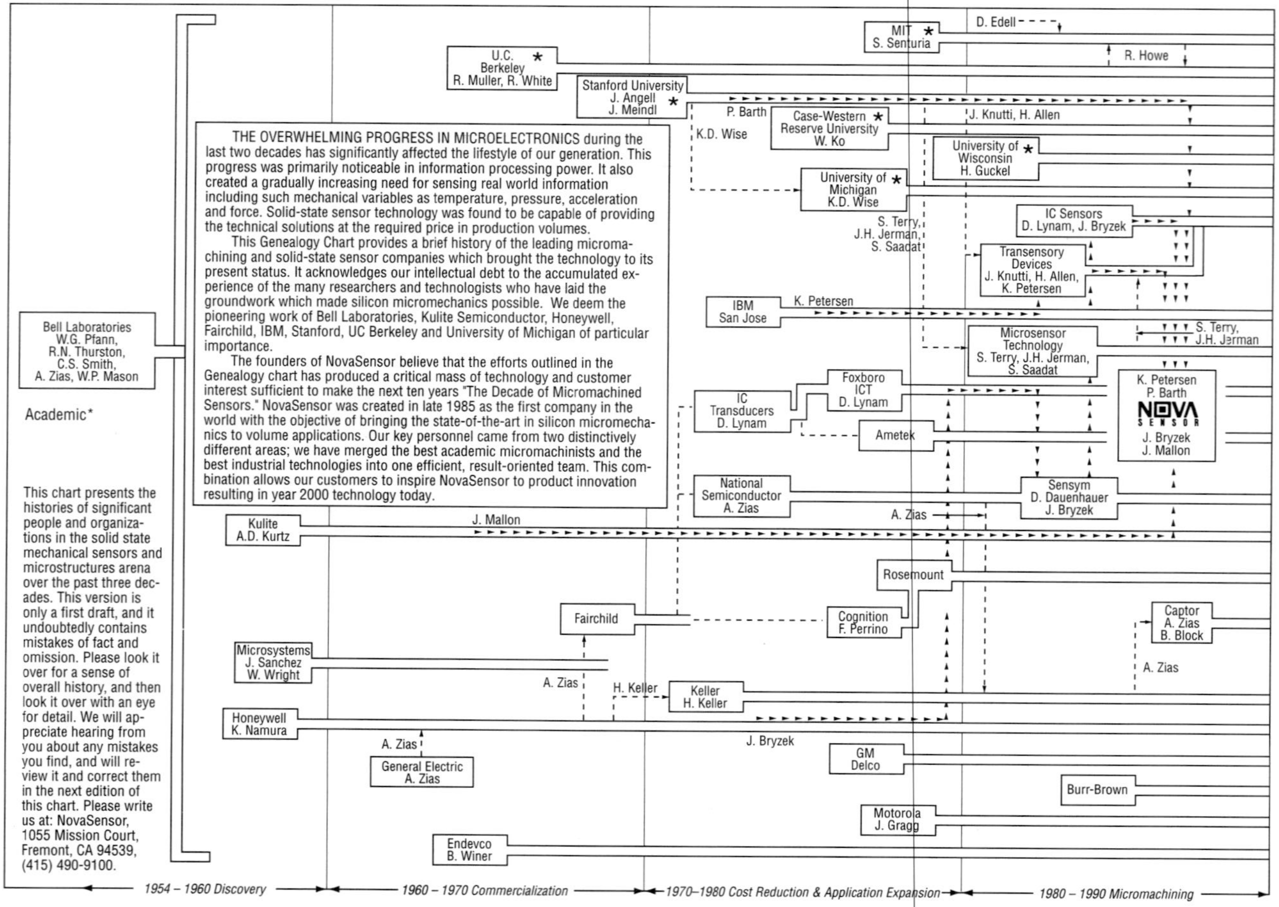

FIGURE 64.2 A genealogy of solid state sensors in the United States.

The trend visible from Fig. 64.2 indicates the increasing activity of the research community and also increased activity of the startups, most of them in Silicon Valley. As in the early days of the development of IC technology, Silicon Valley is already housing most of the young and fast growing companies that develop silicon sensors and microstructures.

SILICON MICROMECHANICS: ADVANTAGES AND OBSTACLES

Silicon is widely used in manufacturing of purely electronic integrated circuits where only its electrical properties are exploited. Micromechanics utilizes both electrical and mechanical properties of silicon and thus creates a new generation of electromechanical silicon chips. While it should be expected that this technology can benefit from the older developments, it should also be expected that some problems with the mechanical aspect of the applications need to be solved by this new industry.

There are two major factors making silicon micromachining technology so attractive. The first one is that silicon is almost a perfect mechanical material, especially for sensors. It's stronger than steel, it does not show mechanical hysteresis, and it is highly sensitive to stress (Table 64.3).

More detailed analyses of mechanical properties of silicon are included in Chap. 65, "Mechanical Performance of Silicon." In addition to excellent mechanical performance, silicon brings significant advantages from the established mainstream

TABLE 64.3 Silicon as a Mechanical Material for Sensors

High sensitivity to physical variables
Essentially perfect elasticity, no mechanical hysteresis
High modulus of elasticty—same as steel
High tensile yield strengh—stronger than steel
Low density—same as aluminum
Hardness—same as quartz

electronic industry, specifically semiconductor manufacturing processes. Ultrapure materials, advanced processes, and high-volume manufacturing and packaging technologies, in addition to educated silicon-processing technologists, can be borrowed for free for sensors from the semiconductor industry (Table 64.4). The fantastic leverage silicon sensor technology enjoys becomes clear when one compares total R&D effort in sensor and IC technologies. With the exception of solid-state sensor technology, the total sensor R&D budget may be on the order of $500 million per year worldwide. A single integrated circuit technology is spending on the order of $10 billion yearly on R&D.

Silicon-based sensors and microstructures are benefiting from vast resources developed not for sensors but for mainstream electronics: microprocessors, memories, and advanced linear microcircuits. The advantage of using silicon sensor technology is not only in an increased price/performance ratio but also in an incredible volume manufacturing capability that IC manufacturing brings.

Silicon wafers are used in the manufacturing process of silicon sensors and microstructures. For example, a four-inch wafer used by NovaSensor (Fig. 64.3) han-

TABLE 64.4 Leverage of Mainstream Electronics on Sensor Technology

Mature mainstream electronics technology
Supported by $10 billion yearly R&D effort
Batch processing—high volume manufacturing capability at low cost
Excellent dimensional reproducibility
Miniature/precise geometries
Commercialy available in extremely pure, defect-free form (no fatigue effect)
Advanced debugged manufacturing processes
High volume sophisticated equipment at relatively low cost
Utilization of developed packaging technology
Educated technologists

dles up to 16,000 sensors. Wafers are normally processed in cassettes, each of them containing approximately 25 wafers (Fig. 64.4). Thus, in one cassette, 400,000 mechanical sensors can be moving through the wafer production line, fabricating, say, 400,000 diaphragms simultaneously. This compares with one-at-a-time manufacturing for most of other technologies. The basic semiconductor processing equipment with cassette-to-cassette handling can easily support production exceeding the total size of the present sensor market.

FIGURE 64.3 A four-inch wafer using SFB technology contains up to 16,000 sensors.

Educated Technologists

Utilizing materials processes, equipment developed for mainstream electronic industry brings down significantly the development time and cost for mechanical devices as compared to original effort in IC industry. Sensor companies are exploring new application-specific micromachining techniques and technologies, including those built on the existing high-level infrastructure. While mainstream electronics brings

FIGURE 64.4 Wafers are processed in cassettes of 25, creating the bases for batch processing.

substantial advantage to the sensor industry, several aspects of sensor-related technology are unique to the sensor industry and create a substantial technological barrier to entry. These technologies required in a debudded manufacturing condition include the following:

- Deep silicon etch
- Chip stress isolation
- Dimensional control of silicon structures
- Stability of silicon resistors
- Wafer lamination
- High-volume low-cost pressure/acceleration/, temperature testing packaging

Deep Silicon Etch

The necessity of creating 3-dimensional structures on silicon wafers requires implementation of deep etching precisely positioned in reference to both surfaces of the wafer (Fig. 64.5). These etching techniques are not needed in mainstream IC industry. The most widely used deep silicon etching is the anisotropic wet etching of wafers. Protection of the active circuit side of the wafer from both etching and contamination is a major challenge.

Chip Stress Isolation

Piezoresistive sensing elements on each chip transform the mechanical stress on the chip's surface into an electrical output signal. Packages can transfer unwanted

FIGURE 64.5 Silicon micromechanics uses a deep silicon etching not used in the IC industry. Both sides of the pressure sensor wafer shown here include deep cavities, forming diaphragms.

mechanical stresses to the sensing elements. These undesirable stresses cause errors in the sensor signal and must be eliminated by proper design of both the chip and the package.

The optimum rejection technique should be incorporated directly in the chip design. Several techniques have been developed to meet this requirement. The most common is anodic bonding of the silicon chip to a thick glass constraint wafer (up to 0.25″ for high performance applications), which acts as a stress isolator (Fig. 64.6). This technique works well; however, some stress problems persist because the glass has different mechanical properties than the silicon chip itself.

Silicon fusion bonding (SFB), a new technology developed at NovaSensor, allows a direct bonding of two silicon wafers and thus makes it possible to replace the glass as a stress isolator. Due to a better thermal match and greater stiffness, SFB sensors offer a better stress isolation than equivalent glass-bonded structures.

FIGURE 64.6 A Variety of silicon sensing elements manufactured at NovaSensor shows both silicon-glass and silicon-silicon bonded structures.

Dimensional Control of Silicon Structures

Batch manufacturing of mechanical components on silicon wafers has required the development of a manufacturing technology that would provide a high degree of dimension control for these mechanical structures. One of the simplest devices is a membrane manufactured on silicon wafers in a batch mode. Such thin diaphragms are the fundamental mechanical elements for silicon piezoresistive pressure sensors.

The first micromachining technique implemented in this application was time-controlled anisotropic etching. Etching was stopped after a predefined amount of time. Parallelism of the wafer was extremely important, since a thickness variation in the wafer was directly translated into a variation in the thickness of the silicon-sensing diaphragm. Since the sensitivity of the silicon sensor changes as a function of the square of the diaphragm thickness, a huge sensitivity variation on each wafer occurred with this etching technology.

A new approach uses a computer controlled etch-stop that allows the designer to program a precise diaphragm thickness into the wafer through the use of a well-controlled layer grown on the silicon wafer. The other dimensional problem in sensor industry is the accuracy of front-to-back wafer alignment. Special double-sided mask aligners are developed for this application (Fig. 64.7).

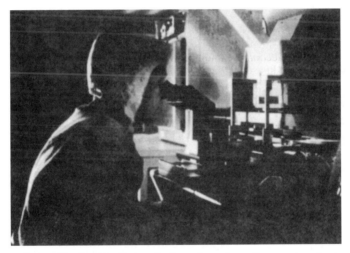

FIGURE 64.7 Double-sided mask aligner shown here allows a unique front-to-back wafer pattern alignment critical to sensor industry.

Stability of Silicon Resistors

In mainstream IC electronics, resistors are usually targeted to a nominal 50 percent accuracy. The IC industry has not developed sufficient know-how related to a high stability of implanted or diffused silicon resistors, especially over long periods of time under hostile environmental conditions.

The pressure sensor industry requires high stability silicon resistors. Typically, a 1 to 10 percent implanted resistor (strain gage) resistance change is obtained for a full-scale applied pressure. If a 0.1 percent pressure measurement stability is required, then a 10 to 100 ppm stability of the resistors is necessary, even when the resistor may be in direct contact with a contaminating gas or liquid media. Although such performance seems very challenging, even higher levels of stability have been achieved by selected suppliers.

Wafer Lamination

Micromachining creates three-dimensional structures and needs new techniques not used in the IC industry to make these structures possible. One of the techniques is wafer lamination technology. Classical technique relied on a single or multiple silicon-glass anodic bonding (Fig. 64.6). Besides anodic bonding, NovaSensor has pioneered the use of silicon fusion bonding for sensors and microstructures. In fusion bonding, two specially treated silicon wafers are pressed together and heated to over 1000 C. The wafers will form a perfect atomic bond. The perfection of this process has provided an unprecedented degree of design flexibility for sensors and microstructures. Many chips have been built with this process that would not have been possible otherwise. Some of them are discussed in a later chapter.

High-Volume, Low-Cost Pressure/Acceleration/Temperature Testing

Sophisticated test systems that have been developed for integrated circuits and transistors are not suitable for high-volume, low-cost testing of sensors. Sensors are sensitive to both the measured variable (such as pressure) and temperature. Sensor test systems must simultaneously precisely control and deliver both variables (such as pressure and temperature) to the device under test. The precision application of pressure is very demanding since even the smallest pressure leaks degrade the accuracy of testing. Also, pressure has to be delivered over usually a broad temperature range which creates additional problems with reliable pressure seals. Testing of this type creates a major source of production yield reduction and thus has tremendous effect on a sensor cost.

Packaging

Since silicon sensors interface with a variety of environments from which standard ICs are protected, such as salt water, acids, and bases, a variety of new techniques had to be developed in order to make sensor applications feasible. Typical, less demanding applications such as the pressure measurement of clean, dry air use a protection layer of silicon gel, which isolates the sensor from the input media. For more demanding applications, however, many media isolated-packages have been developed using a stainless steel diaphragm for protection.

Development of efficient packaging requires significant resources that critically affect the final cost of a sensor.

SENSOR MARKET DEFINITION

A number of sensor market surveys have been performed over the last 10 years. Various analysts use terms defining sensor market segments in quite different ways. To provide a consistent frame of reference, we will define the terms used in this market.

Sensor Die. A sensor die is a micromachined silicon chip. It is typically sold for $.50–$2 as a commodity product, but as high as perhaps $50 for a high-performance die in smaller quantity.

Sensor. A sensor is defined as a first level packaged sensor die with or without a basic compensation and normalization. Typical selling price may range from $2.50 in a million-unit yearly quantity to over $100 for more complex sensors in smaller unit quantities.

Transducer. A transducer is a fully packaged, compensated, calibrated, and signal-conditioned sensor suitable for use by an end user. In typical OEM volumes of several thousand, the selling price ranges from $10–$125.

Smart Transducer. A smart transducer is a packaged sensor employing digital signal processing to provide performance enhancements to the user. Such a device employs a microprocessor to provide digital temperature compensation and digital calibration plus some other functions, such as remote programmability and networking. Typical cost might be $15–$150 in OEM quantities.

Transmitter. Transmitter is a transducer with a current (e.g., 4–20 mÅ) output. It allows transmission of information over a long distance without loss of accuracy. Transmitters for the process control market are priced at $400–$2000.

Sensor-Based Subsystem. A sensor based subsystem is an upgraded smart transducer to perform an application-specific function. Typical costs are $20–$500 in OEM quantities.

Pressure Switch. A pressure switch is a device providing an on/off signal as a function of pressure. The device may be entirely mechanical or electronic. A mechanical switch may use a silicon chip, but the majority of applications are switched by nonsilicon devices. Electronic pressure-switch technology employs a pressure sensor and signal electronics. Prices vary from $2 for automotive switches to $150 for industrial-media-compatible switches.

Microstructure. A microstructure is a silicon chip used for its mechanical properties (e.g., ink jet nozzle). Selling price for OEM volume may range from $1 to $100 in OEM quantities.

Microactuator. A microactuator is a silicon mechanical component designed to perform an actuation function (like a pressure microvalve). Selling price in OEM quantities may range from $5–$200.

THE WORLD'S MARKET SIZE AND GROWTH

There are both traditional markets for silicon sensors (some developing for decades) and new emerging markets for silicon sensors and microstructures that are already visible in several market segments (Table 64.5).

The recent increased interest in sensors is stimulated by the fact that we live in an analog world. Real-time control applications of microprocessors are limited by the availability of cost-effective digital-compatible sensors. Many manufacturers of instruments such as watches, home blood pressure monitors, automobiles, and pro-

TABLE 64.5 Traditional and Emerging Markets

Traditional markets	Emerging markets
Automotive	HVAC
Process controls	Automotive (new applications)
Medical	Consumer
Industrial controls	Test
Aerospace/military	Communication
	Factory automation
	Computer peripherals

cess control instrumentation are pushing the sensor manufacturers to develop the sensor that would make new final applications feasible. The only technology compatible and capable of meeting the challenge is silicon-based sensor technology.

When one compares the ratio of prices (Fig. 64.8) between CPU's and sensors for the last few decades, it becomes clear that several years ago the prices of microprocessors or microcontrollers available in high volume fell below the cost of sensors in large volume. This is very significant because presently many new applications are driven by microprocessor users who are trying to find cost-effective solutions for tying microprocessors to the real (analog) world of pressure, acceleration, flow, velocity level, and so forth.

When one compares the percentage of the total market by silicon sensors, it becomes clear that most market growth comes from the silicon segment of the market.

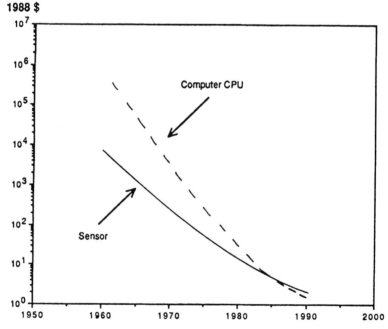

FIGURE 64.8 Changes in sensor and CPU prices.

Year	Applications served by silicon pressure sensors
1980	16%
1985	40%
1990	60% (estimated)

This short history illustrates a visible conversion of the market from all other sensing technologies to a silicon sensing technology. This conversion is similar to that which happened three decades ago in the market, when metal strain gage transducers started to penetrate new applications and eventually took over the majority of the market.

Sensors are a highly diverse and fragmented market. The two most important competitive factors market include sensor technology level and manufacturing technology. Advanced sensor technology opens markets by offering improved price/performance levels. Advanced sensor manufacturing technology helps customers by being able to deliver product in the requested volume and on time. Both factors are still problems in the sensor industry, since manufacturing engineering did not receive proper attention in the past, lagging behind the mainstream semiconductor industry.

The total worldwide market for sensors and transducers exceeds five billion dollars. The largest market segment (40 percent) is for pressure measurement. Acceleration and vibration account for 13 percent (Figure 64.9). About 50 percent of the total market is in the United States; a significant fraction of this segment ($900M) is already served.

The largest segments of the sensor market are aerospace/military, automotive, medical, process control, and industrial control. During the last ten years, a number

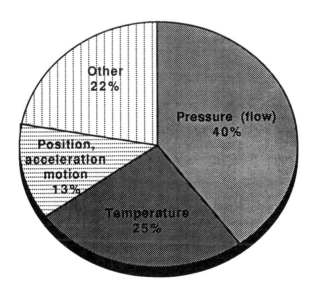

1985 Total market value = $2.5 Billion (U.S.)

FIGURE 64.9 Breakdown of sensor market.

of market research studies by leading marketing companies have attempted to characterize the sensor and transducer. One of the most comprehensive reports was developed by Market Intelligence Research Company (MIRC), estimating the sensor/transducer and transmitter market at $6 billion, with a split approximately 50/33/17 percent, respectively, among the United States, Europe, and Japan. The market covers measuring devices for a broad range of measured variables, technologies, applications, and industries.

Neglecting temperature, the largest silicon sensor market segments are pressure and acceleration. The silicon sensor market for these two variables is characterized as shown in Table 64.6 (pressure) and Table 64.7 (acceleration).

The silicon pressure sensor market is expected to grow 20 percent/year, with the fastest growth of 75 percent/year in the consumer segment of the market. The silicon acceleration market is expected to grow faster than pressure, on average 27 percent/year, with the automotive segment growing 30 percent/year. It is indicated that years following 1992 will see a step increase in usage of automotive acceleration sensors due to volume implementation of smart suspension and smart airbags.

Silicon micromechanics market development is following the pattern established previously by silicon piezoresistive pressure sensors in the early 1970s. The first phase includes the learning period, with slow growth, geared primarily to support the development of relatively specialized and expensive new applications. In the second phase, products emerge to the market, and the final user is educated regarding the advantages of the new products and techniques of applying them. The third phase is a volume market emergence.

The total market development for silicon pressure sensors has occurred over 30 years. It is expected, however, that the growth of the silicon micromechanics markets for other sensors as well as microstructures may be much faster, primarily because the techniques of manufacturing, packaging, signal conditioning, and environmental

TABLE 64.6 Silicon Pressure Sensor Market (Source: Market Intelligence Research Company [MIRC], 1992)

	$000s					
	1988	1989	1990	1991	1992	AAGR*
Automotive	$228	$274	$ 328	$394	$473	20%
Process Control	$288	$337	$ 394	$461	$540	17%
Aerospace/mil	$140	$161	$ 185	$213	$245	15%
Medical	$ 90	$108	$ 130	$156	$187	20%
Consumer	$ 14	$ 32	$ 57	$ 89	$130	75%
Total	$760	$912	$1094	$1313	$1575	20%

* Average annual growth rate (by silicon-based devices).

TABLE 64.7 Silicon Acceleration Sensor Market (Source: MIRC, 1989)

	$000s					
	1988	1989	1990	1991	1992	AAGR
Automotive	$10	$12	$16	$20	$28	30%
Total	$30	$36	$47	$61	$79	27%

restrictions are known in more detail than in the case of pressure sensors two decades ago. Additionally, this market is being pulled presently by developments in electronics, where low-cost distributed intelligence is calling for low-cost interfaces to the real world full of mechanical variables. Sensors and microstructures create the foundation for this interface.

A common theme is resulting from the market studies. It can be summarized as follows:

1. The market is growing very vigorously, from 10–300 percent/year or more for the various sectors.
2. Silicon sensor technology is universally perceived as the key for advancing sensors, with the expected growth much faster than the total market.
3. The best growth and opportunities are for low-cost, high-volume sensors for use with microprocessors.
4. Pressure sensors represent the largest established market for sensors.
5. Acceleration sensors represent the largest potential for significant growth from emerging markets.
6. While the market for sensors is clearly visible and characterized, the market for silicon micromechanics is barely noticeable. The estimated 1989 size is about $10 million, about three orders of magnitude smaller than the sensor market. It is predicted, however, that this market segment will experience the fastest growth rate.

CHARACTERIZATION OF EMERGING MARKETS

During the last two decades, the cost of computers has decreased by seven orders of magnitude, bringing the prices of microprocessors in large volumes to under $1. Simultaneously, the prices of sensors decreased approximately only two orders of magnitude. (Fig. 64.8). The low cost of processing power created the demand for a low-cost sensor technology. Many new applications would not be feasible if the cost of the transducers had remained at the previous level. This brief overview characterizes the most important applications in selected sensor market segments.

Automotive Market

Sensor applications in the automotive segment of the market are without a doubt the largest current and near-future applications for silicon sensors.

The market for silicon sensors for the automobile industry is presently dominated by captive and semicaptive suppliers. Delco, which supplies GM with a variety of electronics, produces 6–8 million silicon pressure sensors per year. Ford, Chrysler, and several European auto makers are sourcing their pressure sensors from Motorola for a total of another 7 million. Nippon-Denso supplies about 4 million pressure sensors to Toyota. Auto makers demand high-volume, low-cost, reliable sensor products. Typical pressure sensor ASP runs $7–$18.

The automotive design-in cycle is typically 3–5 years. Several major design changes are the norm, requiring a large amount of dedicated engineering and design resource. The transfer and ramp-up to full-scale production is also time-consuming and costly. Most costs other than dedicated tooling and equipment (usually the responsibility of the customer) is born of the sensor maker. The total cost for development of a major design can run from $500,000 to $5 million.

Recent new regulations for exhaust emissions in Europe, enforced in June 1989, have spurred great demand for engine sensors. It is expected that eastern European countries and gradually other parts of the world will follow this trend, further increasing the market.

Acceleration sensor applications will begin to increase dramatically after 1993, when the current designs for near-term applications such as air crash bags and smart suspensions make their way into production. The most attractive new application is the smart suspension system. This application will employ from 210 acceleration sensors per car to monitor the vibration of each wheel and the car body in order to define automatically the required response of the suspension system. The first car of this type was developed by Lotus. The first cars using simplified smart suspension were introduced in 1989 model year (Ford, Citroen). The first fully active suspension in the United States was introduced in 1993. It is expected that, within the next ten years, a substantial percentage of all cars manufactured and sold in the U.S. market may feature this type of system. In the year 2000, close to 100 million sensors could possibly be used worldwide for this application in passenger cars, trucks, and trains. The major automotive applications are listed in Table 64.8.

TABLE 64.8 Silicon Sensors—Markets and Applications

Manifold absolute pressure (MAP)
Barometric pressure
Turbo boost pressure
Oil pressure
Continuously variable transmission pressure
Tire pressure
Antilock braking pressure and acceleration
Smart suspension acceleration
Engine knock acceleration
Smart airbag acceleration
Smart engine mount
Air-conditioning compressor pressure fuel level pressure
Mass air metering flow
Diesel engine pressure
Fuel pump pressure
Auto security systems
Diagnostic systems in service stations

Medical Market

The medical market is one of the fastest growing segments of the silicon sensor market. The focus of the industry has been on lowering the cost of components and increasing the sophistication level of the equipment. Both trends accelerated growth of silicon sensor applications.

Disposable blood pressure sensors were introduced in 1982. In the first year of sales, approximately 40,000 units were sold. In 1989, over 8 million units were delivered to the market.

There are a number of other new medical applications that are emerging and are expected to increase the market size significantly. One of them is the intrauterine pressure sensor (IUP) that monitors the pressure around a baby's head during birth, verifying the doctor's decisions. Another one is angioplasty, where a sensor monitors the pressure during catheterization procedure. Infusion pump pressure sensors to control the flow of IV fluid create the requirement for another type of sensor. All of these applications need disposable sensors, and all have a multimillion-unit-a-year production potential. A list of more interesting applications is given in Table 64.9.

TABLE 64.9 Biomedical Applications of Silicon Sensors

Catheter tip pressure sensors
Disposable blood pressure sensors
Hospital sphygmomanometers, manometers
Respirators
Lung capacity meters
Kidney dialysis equipment
Infusion pumps (implantable and external)
Barometric correction of medical instrumentation
Reusable blood pressure transducers
Angioplasty
Patient movement monitoring
Hydraulics for patient bed
Intrauterine transducers
Intracranial transducers
Transducer manufacturers
Tonometry
Drug manufacturing (bacteria growth)

Process/Industrial Controls Markets

The process control and industrial market encompasses a highly diverse range of participant-control applications. Silicon sensors, along with a variety of other technologies, have long been used to measure and control processes within the industrial sector. Silicon pressure sensors are commonly employed in transducer transmitters. Companies differ in their focus and type of end product—from producing component die to systems.

Elevator Vibration Monitoring

The principal variables measured in control systems are pressure, flow, temperature, and level. Silicon micromachined sensors can be used for all of these variables to improve cost effectiveness and reaction speed of control systems. Manufacturers in the process control arena have large capital investments in current technology and are generally slow in incorporating new technologies. Much faster implementation pace is visible, however, in smaller scale instrumentation arenas, such as factory automation.

Pressure sensors will continue to be the dominant application in the industrial sector. Micromechanical switches will find their way into control applications, replacing relays. Accelerometers will find use in monitoring machinery for vibration and shock. The list of more important applications in this field is listed in Table 64.10.

Consumer Market

The consumer market, while presently only about $30M, is expanding rapidly. An excellent successful example of a consumer application is the new generation of

TABLE 64.10 Process/Industrial Applications of Silicon Sensors

Closed-loop hydraulics
Paint sprayers
Agricultural sprayers
Pressure calibrators
Pressure meters
Digital pressure indicators
Compressors
Refrigeration equipment
HVAC
Water level measurement
P/I converters
I/P converters
Pressure switches
Transducer manufacturers
Food processing equipment
Level controllers
Strain gages
Telephone cable leak detection

scuba-diving computers used to measure and display diver depth, diving time, need for decompression, oxygen usage, and remaining air supply. Microprocessors are used to control the display and process information from two pressure sensors; one measuring water pressure and temperature, the other measuring air tank pressure. This information is compared against the decompression tables stored in the computer memory.

The list of new emerging consumer applications is staggering, as can be seen in Table 64.11. The consumer segment of the market is characterized by a very short implementation cycle, measured in months from concept to production, which helps to deliver a 75 percent/year growth rate.

HVAC Market

The HVAC market is potentially the second largest market for silicon sensors. A low-pressure differential pressure or flow sensor will be required for each air conditioning vent to control a solenoid actuated damper for smart distributed heating and cooling. Such systems are expected to reduce energy costs by 30–50 percent. Total potential market size is 100 million units/year in the year 2000.

With new developments of very-low-pressure sensors based in silicon, it looks like they offer the technical advantage over the silicon flow sensors.

Aerospace

The aerospace/military market is one of the first and largest markets for sensors. Entry in this market segment requires strong engineering support to move small quantities of products through extensive testing and qualification. Prices for transducers are high, $500–$5000.

TABLE 64.11 Consumer Applications of Silicon Sensors

Automotive Aftermarket	Recreational
Tire pressure gages	Fitness equipment
Tire pressure controllers	Scuba-diving computers
Remote readout of tire pressure	Scuba-diving watches
Lateral G-force indicators	Hikers watches
Aftermarket MAP sensors	Bicycle computers
Car security	Watch barometer stations
Altimeter/barometer	Tennis rackets
Navigation systems	Golf clubs
Home	Hang Gliders
Smart home controls	Boat speedometers
Smart air conditioning systems	Sail tension controllers
Remote monitoring of natural gas	Personal
Home security	Wristwatch instrumentation (depth,
Natural gas leaks	altitude, barometric pressure, weather
Compressor pressure in refrigeration	station)
Systems	Scales
Swimming pool safety equipment	Weather stations
Swimming pool maintenance equipment	Clock barometers
Earthquake monitoring	PC-interfaceable weather stations
Appliances	General Aviation
Microwave and conventional ovens	Altitude meters
Washing machines	Autopilot instrumentation
Dryers	Transponders
Refrigerator hydraulics	Microphones with noise cancellation
Air conditioners	headphones
Vacuum cleaners	Musical
Personal Medical Instrumentation	Electrical pickups (guitars, wind
Home sphygmomanometers	instruments)
Lung capacity monitoring	Breath controllers for MIDI
PC-based home diagnostic systems	
Audio Video Computers	
Camera image stabilizers	
Microphones	
Smart mouse	
Speakers feedback	

    Typical applications include cockpit instrumentation for oil, fuel, transmission, and hydraulic pressures; air speed measurement; and test instrumentation. These applications include supplying die for wind-tunnel testing instrumentation.

    New components, such as fiber optic sensors, closed-loop acceleration sensors, fuses, and microstructures, are just emerging and will create new applications in the aerospace market. A number of typical applications are listed in Table 64.12.

Micromachining Market

The market for micromechanical devices, while small, is already visible. It exists primarily in the form of development contracts for feasibility studies and prototypes, however several micromechanical devices are already used on a production scale. These devices include RMS converters for digital voltmeters, 35-mm refrigerators, and many others.

TABLE 64.12 Aerospace/Military Applications

Ejection seats
Inertial reference systems
Altimeters
Modules to aerospace transducer manufacturers
Sonobuoys
Radiosondes
Diagnostic systems in military vehicles
Tire pressure
Altimeters and barometers for weather stations (gun aiming systems)
Disposable weapons (steerable ejectors, low-cost missiles)
Hydrophones
Cockpit instrumentation
Military scuba diving

A partial list of micromechanical devices fabricated using micromachining is listed in Table 64.13. Several of these devices (such as fuel injectors and ink jet nozzles) have a very large volume production potential within the next five years, while others (such as motors) may not find volume markets during the next decade or two.

TABLE 64.13 Typical Applications for Micromechanical Structures in Silicon

Cryogenic microconnectors	Microprobes
Fiber optic couplers	Micropumps
Film stress measurement	Microswitches
Fluidic components	Microvacuum tubes
IC coolers	Microvalves
Inkjet nozzles	Nerve regenerators
Laser beam deflectors	Photolithography masks
Laser resonators	Pressure switches
Light modulators	Pressure regulators
Membranes	Programmable RMS converters
Microaligners	
Microbalances	Thermal print heads
Microfuses	Thermopile
Microgears	Torsional mirrors
Micromolds	Vibrating microstructures
Micromotors	Wind tunnel skin friction
Gages	
Micropositioners	

There are only two industrial companies actively participating in this market segment so far: NovaS and IC Sensors. However, many universities are very active in this field, and selected R&D groups of companies are visibly increasing the effort in this field. We estimate that, by the year 2000, this market segment will exceed $200 million.

In summary, the emerging applications create incredible market potential. Table 64.14 summarizes expected quantities and average selling prices for many of the new applications that were visible on the market as of the beginning of 1990. The silicon market size increase based on these projections is a staggering $3.6 million, sufficient to create at least a one-billion-dollar company. Some of these applications may not deliver the expected potential in the year 2000. There will be, however, without doubt, a myriad of new applications that will be visible in coming years.

TECHNOLOGY TREND

The silicon sensor market is gradually moving in the direction analogous to other electronic components: higher integration level. In the case of silicon sensors, there are two options for a possible on-chip integration:

TABLE 64.14 Growth of Emerging Markets for Silicon Sensors and Microstructures

Market Segment	Application	1988 Units	2000 Units	ASP	2000 Size
Automotive	Smart suspension: 4 to 6 acceleration sensors/car	2K	100M	$10	$1,000M
	Air conditioning: 2 pressure sensors/car	1K	20M	$10	$200M
	Fuel Level: 1 pressure sensor/car	1K	4M	$10	$40M
	ABS brakes: 1 pressure sensor wheel/car	1K	10M	$10	$100M
	ABS brakes: an acceleration sensor/car	0K	10M	$5	$50M
	Diesel engine control: 1 pressure sensor	1K	1M	$10	$10M
	Mass air flow: 1 flow sensor	1M	10M	$10	$100M
	Microstructures	1K	50M	$0.40	$20M
	Tire pressure switch: 4 switches/car	1K	50M	$2.50	$125M
	Air bag actuation	1K	10M	$8	$50M
HVAC	Air conditioning controllers: 1 flow or low differential pressure sensor/vent	20K	100M	$10	$1,000M
Consumer	Scuba diving computers: 1 to 2 pressure sensors	50M	2M	$5	$10M
	Barometers: 1 pressure sensor	1K	2M	$5	$10M
	Inclinometer/G Indicators	1K	1M	$10	$10M
	Watches: 1 pressure sensor	5K	10M	$3	$30M
Appliance	Water level: 1 pressure sensor	1K	10M	$2	$20M
	Humidity sensor	1K	10M	$2	$20M
	Vacuum cleaners: 1 pressure sensor	10K	3M	$2	$6M
Industrial Controls	Microactuators	1K	5M	$30	$150M
Test	Microrelays	1K	1M	$10	$10M
Aerospace	Fiber optic sensors	20K	$500	$10M	
	Smart and low-cost weapons	1K	200K	$50	$10M
	Closed loop (Servo) acceleration sensors	20K	200K	$200	$40M
Communication	RMS converters	20K	200K	$20	$4M
	IR power meters	20K	200K	$50	$10M
Factory Automation	Tactile sensors	1K	10M	$5	$50M
	Pressure sensors	1K	10M	$5	50M
	Vibration sensors	10K	10M	$25	$250M
Computer Peripherals	Tape drive control	20K	1M	$10	$10M
	Joysticks control	10M	$5	$50M	
	Disk drive control	0M	50M	$5	$250M
	TOTAL				$3,665M

- Electronic integration
- Mechanical integration

The summary of technology trends is given in Table 64.15.

Electronic On-Chip Integration

In the area of electronic integration, there are several possible levels of vertical integration enhancing performance of sensors. All of them rely on developed technology form on monolithic and hybrid integrated circuits. The basic sensing element includes only four resistors. These resistors require additional signal conditioning in order to provide useful signal for a final user application. Traditionally, all signal conditioning was performed outside the sensor chip using discrete components.

The first level of signal conditioning that can be integrated on a chip is balancing of the offset. Sensor offset variation maybe as high as 10mV/V, and final applications may require a much tighter distribution (e.g., 0.1 mV/V). To perform this function, either ion implanted or thin-film resistors are interconnected into the bridge, which adjusts the offset to zero through either laser trimming or electrical link blowing.

Higher levels of on-chip integration may include temperature compensation of both offset and sensitivity. This is done in a way similar to offset balancing but includes a more complex network of resistors, usually interacting with each other during the trimming procedure. The highest level of passive (no active components) on chip circuit integration also includes signal normalization. As all semiconductor devices, silicon sensors deliver sensitivity variation on the wafer and between different wafers. To provide interchangeability of sensors, sensitivity variations have to be reduced. This can be achieved through on-chip trimmed resistors.

The second large category of signal conditioning that is possible to integrate on the chip includes active circuitry for signal amplification. Typical passive (low-level) sensors deliver signals on the order of 100 mV. Integration of amplifiers promises a high-level output signal on the order of 5 V.

An additional aspect of on-chip integration may include incorporation of temperature-sensing resistors that can be used for either temperature measurement or for external temperature compensation (i.e., using a microprocessor).

Different levels of on-chip signal conditioning offer both advantages and disadvantages. The higher the electronic integration content of the chip, the higher is the engineering and tooling cost required to provide efficient manufacturing flow of the device. However, a higher integration level offers a lower price in high volume.

TABLE 64.15 Technology Trends

Vertical Integration (On-Chip)	Smart Sensors	Secondary Standard Performance Sensors Application-Specific Sensors
• Offset balancing (on-chip) • Temperature compensation (on-chip) • Signal normalization (on-chip) • Signal amplification (on-chip) • Temperature sensing (on-chip)	• Digital compensation/ normalization • Communication • Diagnostics • Application specific functions	• Packaging • Sensors • Signal conditioning

The crossover point above which higher level of integration offers lower cost sensors is a function of the existing technology infrastructure; for some manufacturers, implementation of on-chip integration may be prohibitively expensive. For others, hybrid compensation is not feasible. From a performance point of view, there is a tradeoff between required performance and integration levels. High integration level usually lowers overall performance of the chip as a result of the diversity of temperature coefficients of different materials, including metals that are incorporated on the stress-sensing chip. For a very-high-performance (.01 percent) application, full on-chip signal conditioning is not feasible.

From the economic point of view, every level of integration on the chip increases the number of masks and wafer processing steps and directly increases the cost of the chip. An additional indirect increase of the cost results from the effect of lowered overall yield. The economic feasibility of the integration can be estimated by comparing the cost of competing technology. Competing technology in this case is a hybrid thick-film technology. Designs using thick-film resistors on ceramic substrates are usually less expensive in smaller volumes. Fully on-chip calibrated and temperature-compensated sensors become less expensive in quantities exceeding a million units per year. However, there may be other factors, such as reliability or size or existing in-house technology that can bias the decision toward on-chip integration.

The integration of active electronics may add as many as 12 masking steps to the sensor process, increasing direct manufacturing cost and decreasing overall yield. It also creates a new difficulty level in final testing and trimming. When all the components (including strain gages, normalization, and compensation resistors) and active electronics are prewired together, efficient compensation and calibration creates a major technical problem, since all adjustments are interacting with each other.

From the cost point of view, the attractiveness of integration is not clear. When volumes exceed several million units a year, it is usually less expensive to buy standard IC chips (debugged and manufactured in much larger volume) and use a hybrid solution than to integrate everything onto one single chip. As was mentioned previously, other factors can decide in favor of active electronics integration. To date, there is no single sensor in volume production with on-chip active circuitry, to the best of our knowledge.

Mechanical On-Chip Integration

The area of mechanical integration on silicon provides a potential for substantial benefits to the users. The existing mechanical designs are manufactured using traditional one-at-a-time technology. The potential of scaling down mechanical components on a single chip offers access to a high-volume, low-cost, batch-mode manufacturing technology despite increased complexity of wafer processing.

An example of on-chip mechanical integration (or enhanced micromachining) content is a new generation of acceleration sensors. NovaSensor's design for a 2 G full scale range incorporates on-chip shock protection in excess of 1000 G in each axis. This is achieved through creative micromachined silicon structures integrated on the chip. Moreover, air damping is also incorporated on the same chip. Both these features are batch manufactured on the wafer level, thus attaining high volumes needed for new markets and lowering the cost by more than an order of magnitude, when compared to older technologies. On-chip mechanical integration is expected to be the fastest growing segment of silicon micromachining.

Application-Specific Sensors (ASS)

The diversity of applications spread over a range of different industries creates situations similar to that of applications specific integrated circuits (ASIC). Technology progress makes feasible efficient customization even in a relatively small quantity (10,000 units/year). Customization or semicustomization includes packaging, sensing chips, and signal conditioning.

Smart Sensors

Another visible technology trend is the incorporation of microprocessor power for performing different functions at the sensor level. The low cost of digital components combined with the necessity for temperature compensation for pressure sensors helped to create the first high-volume smart-sensor application. These first developments were incorporated in process control transmitters. The microprocessor, in addition to temperature compensation and normalization of the sensor signal, also performed other functions such as remote communication, remote deranging, and diagnostics of the transducer. The incorporation of the microprocessor allowed the improvement of performance of silicon sensors. New digital temperature compensation techniques challenged the accuracy of the secondary standards such as dead weight testers. For the first time in the history of the transducers, a total error band over an extended temperature range (−550° to 960°C) and the full pressure range was reduced below 0.01 percent.[2]

MARKET TRENDS

It becomes clear that the sensor is assuming the function of an electromechanical component instead of an instrument. Emergence of the sensor as a commodity component affects dramatically its positioning on the market. Demand for an enhanced price/performance ratio encourages the start-up of new companies utilizing new technology not available within existing sensor companies. It also gives incentive for design engineers to include sensors in applications that were not feasible, even several years ago (for example, scuba-diving watches). New markets are being created as the result of this enhanced price/performance ratio.

As prices go down in the microsensor market, new markets and applications are expected to develop even more rapidly then now. Potential high-volume applications that await new sensor technology include: consumer electronics, medical diagnosis in the home, automotive applications, consumer appliance controls, and regulation of heating and cooling systems. Also paving the way for this expected growth is the increase of sensor integration level, which allows more functions per sensor die and consequently a better price/performance ratio for the end user.

This growth is not accidental. While the last ten years can be defined as the decade of microprocessors, there are visible indications that the next ten years may be named the decade of the sensors. Several signs point to this strong market potential.

- Introduction of complex micromachined features (e.g., overload protection and air damping for acceleration sensors), significantly improving price/performance ratio, reliability, performance, size, and weight

- Increase of the sensors' level of integration (e.g., thin-film on-chip laser calibration)
- Growing number of sensor startup companies
- R&D efforts by most leading world universities and major corporations
- Entry into the market by major Japanese manufacturers (Hitachi, Nippon Denso, NEC, Toshiba Fujikura)
- Development of new sensors based on state-of-the-art processing technology

As a result of the leverage given silicon sensors by mainstream electronics, there is a clearly visible convergence of the market from traditional to solid-state sensors and a visible proliferation of new sensor companies, especially in the area of Silicon Valley, California. Dramatically increased research activity is visible at leading universities and research laboratories of large corporations. Silicon sensors are being incorporated into consumer products, promising new high-volume levels. There is a visible integration of microprocessors into sensor-based systems and an increased international participation of suppliers. It is also significant that there is a move of academicians into the commercial market with the objective of creating the next generation of sensors and sensor-based systems, utilizing the state-of-the-art in silicon processing technology (Table 64.16).

TABLE 64.16 Market Trends of New Sensor Companies

Emergence of the sensor as a commodity component
Enhanced price/performance ratio
Creation of new markets via lower unit costs
Integration into consumer products
Conversion from traditional to solid-state sensors
Proliferation-increased research activity at leading universities and labs
Improved accuracies
Integration into microprocessor-based systems
Increased international participation of suppliers
Academicians entering commercial market

Taking all of the above into account, it becomes clear that sensors are entering into a new era of development, an era marked by the rapid growth of new applications and of the overall sensor market. The decade of sensors is ahead.

REFERENCES

1. C. S. Smith, "Piezoresistive Effect in Germanium and Silicon," *Physical Review,* **94**(1), April 1954.
2. P. DuPuis, "A Novel Primary Data Quality Pressure Transducer," *Proc. Air Data Systems Conference,* 1985.
3. S. Middelhoek, et al., "Silicon Sensors," Academic Press, 1989.
4. Venture Development Corporation, "Pressure Transducer and Transmitter Industry Strategic Analysis," 1986.
5. McIntosh International Limited, "Solid State Sensors and Micromachined Structures," 1985.

6. Batelle Institute D. V. Frankfurt, "Sensors: Miniaturization and Integration Techniques and Markets, Cost Effectiveness and Trends," 1987.
7. Frost and Sullivan, "Biosensor Market," A1788, 1987.
8. Frost and Sullivan, "Transducer Market," A1795, 1987.
9. Frost and Sullivan, "Semiconductor Sensor Market," A1668, 1986.
10. Frost and Sullivan, "The Fiberoptic Sensors and Associated Communication Markets in U.S. vs. Control," A1599, 1986.
11. Frost and Sullivan, "Transducers," E908, 1987.
12. Frost and Sullivan, "Semiconductor Sensors Market," E877, 1987.
13. Frost and Sullivan, "Industrial Sensors, E878," 1987.
14. Frost and Sullivan, "Biosensors Market E868," 1986.
15. Frost and Sullivan, "Position Sensors Market," E818, 1986.
16. Frost and Sullivan, "Fiberoptics Sensors," E799, 1986.
17. MIRC, "Solid State Sensors Markets," 1987.
18. MIRC, *Sensors Market and Technological Newsletter.*
19. MIRC, "Biomedical Electrode and Sensor Market Analysis," 1985.
20. MIRC, "Fiberoptic Sensors in Medical Markets," 1985.
21. MIRC, "Fiberoptic Sensors Markets," 1984–1994.
22. Futuretech, "Special Report on Smart Sensors," 1987.
23. Technical Insights, Inc., "Industrial Sensors," 1987.
24. James Winters Associates, "The U.S. Transducer Market," Report no. 2233, 1987.
25. Venture Development Corporation, "The Automotive Sensor Market: A Migration from Conventional to Solid-State," 1988.
26. MIRC, "The European Sensor Market," 1989.

CHAPTER 65
UNDERSTANDING SILICON PROCESSING AND MICROMACHINING

WHAT IS SILICON?

Silicon is an element with a colossal number of uses. It is the second most abundant element (after oxygen) in the earth's crust, making up 25.7 percent of it by weight. It is present in a wide variety of materials, including sand, clay, glass, and bone. Pure silicon, however, is not found in nature but must be refined from its oxides. The most useful form of silicon for sensors and electronics is single-crystal silicon. Silicon in single-crystal form is a unique material, both for its electrical[1,2] and mechanical[3] properties.

Single-crystal silicon is composed of a very regular array (lattice) of atoms. The atoms are arranged in a diamond structure with cubic symmetry as shown in Fig. 65.1. In this crystal structure, each atom is bonded to four of its nearest neighbors. The various directions in the lattice are denoted by three indices. The (100) direction, for example, is along the edge of the cube, as shown in Fig. 65.2. The electrical, chemical, and mechanical properties of silicon all depend on the orientation. For example, certain planes of the lattice will dissolve much more rapidly than others when exposed to chemical etchants.

Single-crystal silicon is grown from a very pure melt. A tiny single crystal seed is dipped into the melt and slowly withdrawn under conditions of tight temperature and motion control. By controlling the rate of withdrawal, a sausage-shaped *boule* of silicon is produced with a fixed diameter, typically 100, 150, or 200 mm. The orientation of the crystal with respect to the axis of the boule is determined by the orientation of the seed. The electrical properties of the crystal are controlled by small amounts of impurities introduced into the melt. It is quite impressive that a boule of silicon weighing perhaps 100 pounds can be supported by the tiny seed only millimeters across. These conditions correspond to over 100,000 psi of tensile stress.

After crystal growth, the boule is sliced into wafers about 0.5 mm thick. These wafers are then chem/mechanically ground and polished to a very high surface quality in preparation for use in producing integrated circuits or silicon sensors. Although for integrated circuits, one side is typically polished, for sensors, it is more common to begin with wafers that are polished to a mirror surface on both sides. The surface orientation of the silicon wafers is determined by the boule. The most common types of

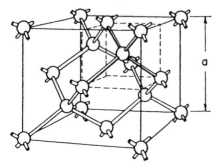

FIGURE 65.1 Arrangement of atoms in a single-crystal silicon lattice.

wafers produced have surfaces oriented along the (100) or (111) direction. Silicon sensors almost always occupy (100) wafers because of their unique autisotropic etching characteristics.

Electrical Properties

Because it is a semiconductor, the electrical resistivity of silicon can be varied many orders of magnitude by the incorporation of minute quantities of impurities. In order to create resistors, diodes,

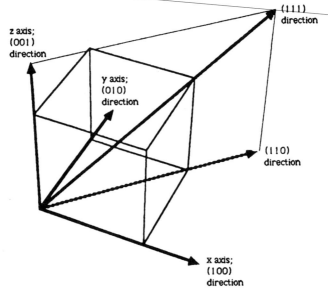

FIGURE 65.2 Silicon wafers can be oriented with respect to the single crystal in different directions.

and transistors in silicon, it is necessary to add these impurities to selected regions of the silicon wafer. These impurities are called dopants. For silicon, there are two types of dopants: *p*-type and *n*-type. The dopant atoms are substituted for the silicon in the crystal lattice and thus try to form bonds with four other silicon atoms. However, *p*-type dopants have one electron too few, while *n*-type dopants have one electron too many to make a perfect crystal fit. These extra and missing electrons (electrons and holes) act as charge carriers and thus increase the conductivity of the various doped silicon regions. Boron is an example of a *p*-type impurity, while phosphorus and arsenic are examples of *n*-type impurities. If a boundary exists between a *p*-type region and an *n*-type region, a *p-n* junction or *diode* is formed. Such a junction will allow an electric current to pass in one direction only. This useful property is the basis for many integrated circuit and sensor technologies.

Silicon also exhibits an important electromechanical effect: piezoresistivity. When the silicon crystal is strained, its resistivity changes by a reproducible amount. This effect is the basis for piezoresistive devices such as pressure sensors and accelerometers. In these devices, a force such as acceleration or pressure from a gas or liquid deflects a thin piece of silicon. The strain induced causes a change in electrical resistance, which can be sensed by external electronics.

The electrical properties of silicon are very temperature-dependent. As a result, most devices will operate accurately only over a limited temperature range. Typically, silicon sensors require temperature compensation. By incorporating appropriate temperature-independent circuitry with the sensor, the temperature range over which sensors will operate accurately can be extended substantially.

Mechanical Performance of Silicon

Electronics has become a near-magical technology in recent years. Almost any imaginable visual or audio sensation can be produced by modern electronic techniques. This amazing capability has been made possible by the nearly perfect electronic properties of single-crystal silicon. Enormously expensive research efforts have been invested by the electronics industry during this time to understand, control, and commercialize the electronic performance of silicon. Unwittingly, these research efforts have simultaneously optimized and perfected the mechanical properties of single-crystal silicon.

Although it is well-known that silicon is an ideal electronic material, it is not generally recognized that silicon is also a remarkably high-performance mechanical material. The strength-to-weight ratio of silicon is about 5 times higher than stainless steel.[4] The hardness of silicon is the same as quartz. Its thermal conductivity is near that of aluminum. See Table 65.1 for a comparison of the mechanical properties of various important mechanical materials.[5] Silicon is also readily machineable with a

TABLE 65.1 Mechanical Properties of Silicon

	Yield strength (10^{10} dyne/cm^2)	Knoop hardness (kg/mm^2)	Young's modulus (10^{12} dyne/cm^2)	Density (gy/cm^3)	Thermal conductivity (W/cm°C)	Thermal expansion (10^{-6}/°C)
Diamond	53	7000	10.35	3.5	20	1.0
Sic*	21	2480	7.0	3.2	3.5	3.3
TiC	20	2470	4.97	4.9	3.3	6.4
Al$_2$O$_3$*	15.4	2100	5.3	4.0	0.5	5.4
SI$_3$N$_4$*	14	3486	3.85	3.1	0.19	0.9
Iron*	12.6	400	1.96	7.8	0.803	12
SiO$_2$ (fibers)	8.4	820	0.73	2.5	0.014	0.55
Silicon	7.0	850	1.9	2.3	1.57	2.33
Steel (max. strength)	4.2	1500	2.1	7.9	0.97	12
Tungsten	4.0	485	4.1	19.3	1.78	4.5
Stainless steel	2.1	660	2.0	7.9	0.329	17.3
M$_o$	2.1	275	3.43	10.3	1.38	5.0
Aluminum	0.17	130	0.70	2.7	2.36	25

* Single crystal.

variety of chemical and electrochemical etchants as well as chem/mechanical machining, as described below. And, it is significant that silicon is the basis of a materials system that includes such compounds as silicon dioxide, silicon nitride, and silicon carbide. Since these materials are compatible with and readily produced from silicon, they provide a powerful degree of flexibility for mechanical devices and structures fabricated on silicon. They produce excellent passivation coatings. Silicon nitride, in particular, is a commonly used, extremely tough, hard, and corrosion-resistant material that can greatly enhance the mechanical performance of a single-crystal silicon micromechanical part.

When all these mechanical attributes are combined with the electronic performance of silicon and its sensitivity to light, magnetic fields, and stress, it becomes obvious that silicon provides a unique fundamental base technology for all types of micromechanical sensors and actuators.

Because silicon is a brittle material, some additional comments are necessary related to its strength and hardness. When a steel mechanical component is stressed beyond its breaking strength (about 350,000 psi), it fails by plastically deforming. When silicon is stressed beyond its breaking strength (about 10^6 psi), it fails by fracturing. These differing failure modes influence the design and/or use of a component. Another crucial aspect that modifies the effective strength of silicon mechanical components is surface finish. Silicon surface preparation technology is capable of creating near-perfect surfaces on the basis of mechanical and electronic characteristics. When microstructures are etched from the wafer, however, sharp corners are usually formed. Because such abrupt features tend to concentrate stresses, they will prevent the component from achieving its full potential mechanical strength. This factor is a critical consideration in the design and fabrication of components that will be subjected to high stress levels. Wherever possible, abrupt corners should be avoided, and smoothing techniques such as wet isotropic etching, plasma etching, or thermal oxidation should be incorporated into the process.

Finally, this section concludes with the description of a device that uniquely illustrates the impressive mechanical capabilities of silicon. Figure 65.3 shows a silicon wafer, mounted on a spindle and rotating at 360 ORPM.[6] The wafer is coated with silicon nitride only 49 nm thick. In cross-section, this thin-film sandwich and an MNOS device is known to have charge-storage behavior. Researchers have contacted such wafers with a tungsten carbide probe, similar to a phonograph needle, and electroni-

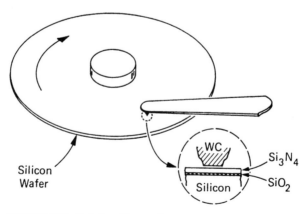

FIGURE 65.3 Rotating silicon wafer.

cally stored charge and readback information. Virtually no mechanical deterioration of this incredibly thin nitride layer was observed despite continuous, extended abrasion by the tungsten carbide probe over the same data track. This demonstration is one of the more aggressive applications of silicon as a high-performance mechanical material.

The excellent mechanical behavior of silicon in a diverse range of micro electromechanical systems assures its continued expansion into advanced, demanding uses as a microstructural material.

BASIC SENSOR MATERIALS AND PROCESSING TECHNIQUES

The production of silicon sensors and microstructures requires a number of materials and processes. Many of these materials and processes are common to both sensor and integrated circuit manufacturing. This is an advantage to sensor manufacturers, since many more dollars are spent on silicon processing research for integrated circuits than are spent on sensor research. Many of the developments from the integrated circuits industry are directly applicable to silicon sensor technology.

Photolithography

Photolithography is a means by which only selective areas of the wafer are processed. For example, any etchable film (such as metal, insulator, or silicon) that might be deposited on a wafer can be patterned using the processor photolithography. This process is illustrated in Fig. 65.4. Using photolithography, features as small as 1 μm can be defined. In addition, layer upon layer of such features can be successively aligned to one another with the same high accuracy.

Photolithography begins when a light-sensitive film called *photoresist* (or simply *resist*) is spun onto the wafer. After a heat cycle, the wafer is brought into close proximity with a glass mask plate that has chrome deposited in the desired pattern. The wafer is then precisely positioned so that any pattern on the wafer is aligned with the additional pattern on the mask. (Sometimes it is necessary to align two masks simultaneously so that a pattern on one side of the wafer is aligned to a pattern on the other side.) The mask/wafer combination is then locked in place and exposed to ultraviolet light. The light passes through the openings in the mask and produces a chemical change in the photoresist. The resist is then developed in a chemical bath. This removes resist in either the exposed areas (positive resist) or the unexposed areas (negative resist). In this way, the pattern is transferred from the glass mask to the resist.

Once the resist pattern is formed and hardened by a heat treatment, the wafer is ready for further processing. Usually, the wafer will be put through an etching step that will selectively remove material from the wafer. After processing, the photoresist is stripped, either using concentrated sulfuric acid or an oxygen plasma.

Diffusion and Ion Implantation

The two major methods of introducing dopants into the silicon crystal structure are diffusion and ion implantation. Diffusion is done by depositing a large concentration

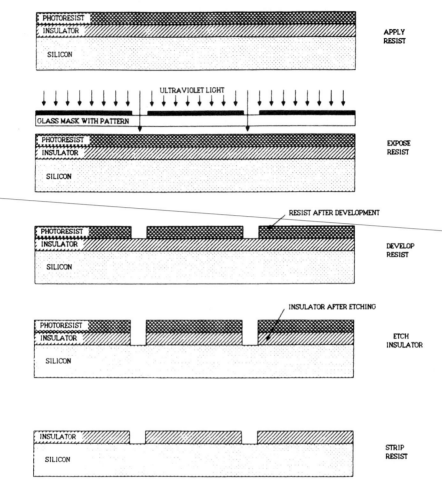

FIGURE 65.4 Silicon photolithography process.

of the dopant on the surface of the silicon. This is done either in a high-temperature furnace or in the form of a liquid that is spun onto the wafer, then dried on the wafer surface. When the wafer is then brought to high temperatures (usually above 800°C), the dopant will diffuse into the silicon due to the concentration gradient from the surface into the silicon. The depth and final surface concentration of the dopant profile are controlled by the time and temperature of the diffusion.

Ion implantation is a process in which the dopant atoms are ionized (1 or more electrons removed), then accelerated through an electromagnetic field and physically blasted into the silicon. The acceleration energy and ion mass determine how far the dopant ion will penetrate into the silicon. With this method, the wafer needs to be processed in a high temperature anneal for the dopant atoms to become incorporated into the crystal structure and become electrically activated. This high-temperature operation may also be used to diffuse the dopant from the initial implanted depth deeper into the wafer.

Although ion implantation is usually more expensive than diffusion, it has the important advantage that the amount of dopant can be precisely monitored and controlled. As a result, resistors made by ion implantation, for example, have very tight tolerances over resistance values and their temperature coefficients. Equivalent tolerances are usually impossible to obtain with diffusion techniques.

Metal Layers and Interconnects

Electrical components fabricated on silicon need to be connected in order to form a circuit. These interconnection lines should be as low a resistance as possible and should also allow wire bonding to various packages. Even the most conducting silicon is orders of magnitude too resistive for these interconnect lines, so that a different material is required. For silicon devices, aluminum is by far the most widely used interconnecting material. Aluminum is usually deposited onto silicon wafers by direct evaporation or by sputtering. In an evaporation process, a crucible of aluminum is heated above the boiling point of aluminum.

The crucible and the wafers are in a very low pressure chamber. The aluminum boils off and coats the wafers as well as the entire chamber. In a sputtering process, high-energy ions, usually argon, are accelerated into an aluminum target. The ions are energetic enough to cause aluminum atoms to be sputtered off the target. This is also done in a low-pressure chamber, but with argon present. The wafers are situated so that aluminum that is sputtered off the target will coat the wafers.

Other metals are used in semiconductor and sensor processing when unique requirements are needed. For example, platinum, palladium, titanium, tungsten, and gold are used because they form unique alloys when in direct contact with silicon. These interconnect metals also withstand higher temperatures than aluminum.

Resistors can also be constructed from metal alloys or metal components. These materials include nickel-chrome, silicon-chrome, and tantalum nitride. Such resistive metal layers can be integrated with diffused or ion-implanted silicon resistors to form temperature-compensated circuits. An advanced NovaSensor chip which incorporates this principle will be described in Chap. 67.

Insulators

Insulators are also very important in the processing of devices. If the aluminum were placed directly on the silicon surface, all the devices would be electrically shorted together. In order to prevent this, an electrically insulating material is needed to separate the metal from the silicon. In addition, this insulating material must be easily patterned to allow windows or contacts for the metal to reach the silicon where desired. Insulators are also used as masking layers. Masking layers are necessary so that the silicon can be selectively doped or etched into the desired electrical and mechanical configuration. Fortunately, silicon forms two compounds that ideally fit these requirements. These compounds—silicon dioxide and silicon nitride—are excellent electrical insulators and are easily patterned and etched.

Silicon dioxide (SiO^2) is usually formed directly from the silicon itself. By heating the wafer to temperatures over 800°C in an oxidizing gas (e.g., oxygen or steam), a thin layer of silicon is converted to SiO^2. This reaction can be controlled to yield extremely uniform films of SiO^2. SiO^2 can also be deposited on the wafers, usually by reacting SiH^4 with O^2. In addition to being an excellent insulator, SiO^2 acts as a mask to the diffusion of dopants used to change the silicon conductivity and to the etchants used to shape the mechanical structure of the silicon device.

Another common insulator is silicon nitride, Si^3N^4. This insulator is deposited on the wafers rather than being grown. This is usually done by reacting SiH^2Cl^2 with NH^3 above the wafers at a temperature near 850°C. Silicon nitride has the unique property of being almost impervious to most impurities and to all of the silicon etches used to produce sensors and microstructures.

These two materials have similar insulating properties but different mechanical properties. For example, as-grown SiO^2 films are generally in compression while deposited nitride films are in tension. Si^3N^4 is very impervious to most silicon anisotropic etchants, whereas SiO^2 is not. Si^3N^4 will also effectively block the oxidation of underlying silicon at high temperatures, a process known as LOCOS (local oxidation of silicon).

Additive Processes

To form mechanical structures, silicon can be added to the starting wafer using several techniques. A process very common in the integrated circuits industry is *epitaxy*. During this process, silicon is deposited at high temperature (typically 1100–1500°C) from a gaseous source such as $SiCl^4$. The atoms deposit on the bare silicon surface in the same regular array, so the deposited layer is also single-crystal silicon. The conductivity of the epitaxy can be controlled by the introduction of dopants into the gas. The layer can also be of opposite doping type as compared to the substrate, leading to a *p-n* junction. In addition, buried layers of various doping concentrations can be formed on the substrate prior to epitaxy. In this way, buried electrodes of high conductivity can be formed. One disadvantage of epitaxy is that it can only be deposited easily on a bare silicon surface. If an insulator is present, the silicon will deposit in a polycrystalline form except in very well controlled circumstances. Generally, epitaxy will not work very well if large topologies exist on the wafer. Epitaxy layer thicknesses are generally limited to less than 100 microns, with 20 microns or so being typical.

Thin layers of silicon can also be added by depositing polycrystalline silicon (polysilicon). This type of silicon is deposited by chemical vapor in a process very similar to that described for silicon nitride above. Both doped and undoped polysilicon can be deposited. The electrical properties of polysilicon are substantially different from single-crystal silicon, however. Its mechanical properties are generally very similar to single-crystal silicon, but there may be subtle differences in long-term stability. This is an area of active investigation.[7] Unlike epitaxy, polysilicon is easily deposited on insulators. Once deposited, it is easily patterned using plasma etching. Polysilicon thicknesses are usually limited to a few microns.

When thicker layers are needed or when large topologies exist on the substrate wafer, the process of silicon fusion bonding (SFB)[8] can be used. In this process, two wafers are bonded together to form a single piece of silicon. One or both of the wafers may be partially processed before the bonding. This allows one to produce wafers with unique features such as buried channel sand cavities. This technique has the important advantage that only silicon is involved in the bond, and thus the two parts are exactly matched in their mechanical, chemical, electrical and thermal properties. For example, the electrical resistivity of a diffusion across the bonded interface is shown in Fig. 65.5. Notice that the diffusion profile is disturbed very little by the presence of a bond interface.

Another useful material for sensors is Pyrex glass. To reduce packaging stresses, silicon sensors are often bonded to a glass substrate using a process called anodic bonding.[9] In this process, silicon and glass are brought together and heated on a hot

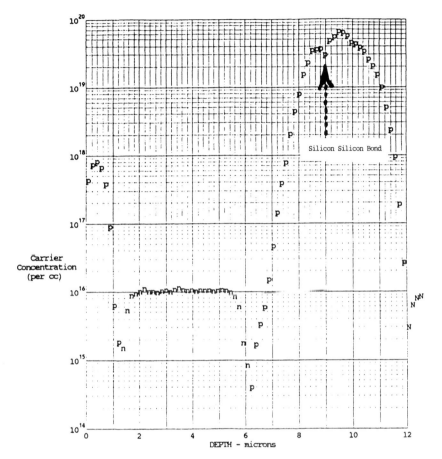

FIGURE 65.5 Resistivity profile through a silicon fusion bonded wafer assembly.

plate to temperatures in the range of 300–500°C. A high voltage (typically 500 V) is then applied between the silicon and the glass. Sodium ions in the glass are repelled from the interface, causing an electrostatic attraction between silicon and glass. A nonreversible, very strong bond occurs in a few minutes. One must be aware, however, that the thermal expansion mismatch between the silicon and glass can cause changes in the temperature coefficients of the device.

Subtractive Processes

The removal of materials is also an essential process for sensor fabrication. Many thin films, including aluminum, oxide, nitride, and polysilicon, are patterned using photo lithography and etching. Most etching of these materials has been done with the use of acids or other chemicals, so-called *wet etching*. As device geometries and cleanliness requirements increase, however, the technology has moved to plasma etching, so-called *dry etching*. In this technique, the wafers are placed into a process chamber that is first evacuated and then backfilled to a low pressure (perhaps 100 mT) with a process gas mixture. Typical process gases contain chlorine or flourine and oxygen. The wafers are located between two plates of a high-frequency plasma generator. When the plasma is energized, the process gases are ionized and react with the material to be etched. With careful control of gas mixture and pressure, plasma power, electrode configuration, and wafer positioning, very fine geometries are possible. Plasma etching is an important and versatile tool in the micromachinist's toolbox. A typical example of a plasma-etched pattern is shown in Fig. 65.6. Under certain conditions, anisotropic (vertical walled) etching is also possible, although the etch rates are typically limited to 1–2 microns per minute. A similar process using oxygen can be used for stripping photoresist.

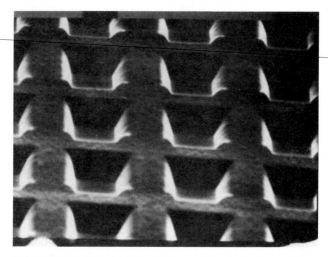

FIGURE 65.6 A micromachined pattern produced with plasma etching.

A critical process for the production of microstructures and sensors is the anisotropic chemical etching of silicon.[3] Certain chemical etchants such as hydrazine, EDP, and KOH attack the (100) and (110) planes of silicon much faster than the (111) planes. In a wafer with a (100) surface, this fact is used to produce a number of accurately defined shapes in silicon. Typically, the wafer is first oxidized, and then this oxide is patterned using photolithography. When such a wafer is immersed in an anisotropic etchant, silicon is removed only in the areas when no oxide is present. The etch proceeds downward in the (100) direction very rapidly, but when (111) planes are encountered, the etching effectively stops. In a (100) wafer, the (111) planes are oriented at an angle of 54.7° with respect to the surface. Thus, V-shaped grooves and cavities can be precisely formed in the wafer. Several of the etch profiles that are possible with different etching techniques and wafer orientations are shown in Fig. 65.7.

BASIC PRESSURE SENSOR PROCESS

A very simple piezoresistive pressure sensor is illustrated in Fig. 65.8. The device consists of a silicon "chip" into which a cavity has been anisotropically etched from

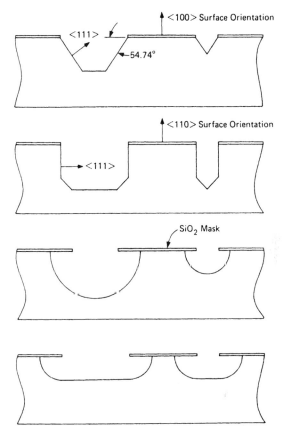

FIGURE 65.7 Cross sections of various etch profiles possible with silicon micromachining.

the backside. The bottom of the cavity forms a thin silicon diaphragm. When a pressure difference exists across the diaphragm, it deflects. This deflection causes strain in the diaphragm with maximum strain occurring near the edges. Small resistors are fabricated in the top surface of the diaphragm, near its edge. The values of the resistors change as a function of the existing strain. Depending on how the resistors are oriented with respect to the diaphragm edges, the change in resistance can be either positive or negative. These resistance changes with pressure can be sensed with external electronic circuitry. Typically, a Wheatstone bridge configuration is used as the sensing circuit. This results in an output that is near zero at zero pressure and changes linearly with pressure. It is instructive to consider a simplified process for forming such a sensor. The sequence of events is illustrated in Fig. 65.9. The process starts with an n-type (100) wafer that is polished on both sides. As a first step, a dielectric layer such as silicon dioxide or silicon nitride is grown or deposited on both surfaces. A photolithography step is next used to define oxide cuts in the top surface. The conductive piezoresistors that act as the strain-sensitive electrical elements of the device are then formed by introducing boron into the silicon through the oxide cuts. Either ion implantation or diffusion techniques are used to introduce

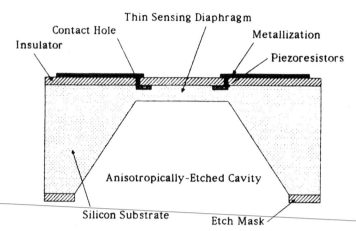

FIGURE 65.8 Simple silicon micromachined piezoresistive pressure sensor chip.

the boron. During this doping, some additional insulating oxide is grown over the openings.

The piezoresistors now must be connected to the outside world. Another photolithography step is used to open small contact windows in the oxide at the ends of the resistors. Then metal is deposited onto the front surface of the wafer. Photolithography is again used to define the metal into paths leading from the resistors to larger pads. Later, when the chip is packaged, small gold wires are bonded to these pads which allow connection to the pins of the final packaged device.

The next step is to create the thin silicon diaphragm that transmits strain to the resistors when pressure is applied. Typical wafer thicknesses of 500 μm or so are too thick and so would not deflect substantially for most pressure ranges of interest. To form the diaphragm, a photolithography step using front-to-back alignment is necessary. A window in the insulator on the back of the wafer is opened. The wafer is then immersed in an anisotropic etchant, and a cavity is formed. The process is designed so that the edges of the cavity on the backside of the wafer are precisely aligned to the strain-sensitive resistors on the front side. Special photolithographic equipment is necessary to perform this alignment step on a production basis. Careful control of the diaphragm thickness is also necessary if the sensitivity of the device is to be predictable.

Last-generation technologies control the diaphragm thickness by simply timing the etch. For very thin diaphragms produced in large volume, however, timed etching is not sufficiently reproducible. In this case, etch-stop techniques must be employed.

Several methods can be used to stop the silicon etching at a particular thickness. Ethylenediamine-pyrocatechol (EDP) for example, will stop etching silicon completely if the boron concentration in the silicon is above a critical level. Thus, if one produces a very heavily doped p-type layer near the top surface, the etchant will stop there, as the cavity is etched from below.

Another technique is the electrochemical etch stop. In this technique, a p-n junction is used as the stopping interface. If a voltage is present on the junction during etching, the etching will stop when the p-n junction is exposed. This active electrochemical etch-stop technique works well for hydrazine etching.

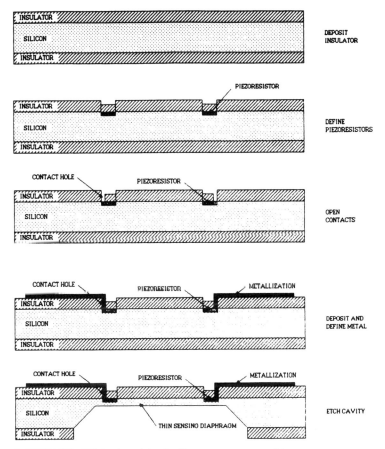

FIGURE 65.9 Fabrication of simple silicon pressure sensor chip.

Active and passive (no applied voltage) electrochemical etchants also work for isotropic etchants such as mixtures of nitric and hydroflouric acids. Etch-stop processes provide unparalleled control over diaphragm thickness (and therefore sensitivity) for pressure sensors. Such control makes it economical to manufacture chips with tighter tolerances on sensitivity than ever before possible. A graph of sensitivity distribution over several hundred chips from a typical lot of wafers is shown in Fig. 65.10.

After the silicon chip is fabricated, it must be carefully packaged. Care must be taken to protect and isolate the chip from any stresses that are induced from sources other than the pressure to be measured. Thermal expansion effects, for example, can be very detrimental, causing large drifts in the signal with temperature.

As described above, the silicon chip is often mounted to a glass support using anodic bonding. The resulting hermetic seal is very strong and irreversible. A typical, simple package using the anodic bond process is illustrated in Fig. 65.11. Silicon fusion bonding may also be used to bond the device to a silicon substrate early in the process.

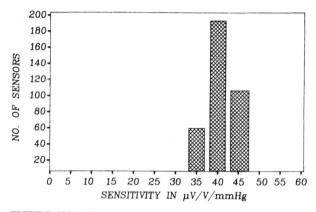

FIGURE 65.10 Graph showing sensitivity distribution over several hundred chips from a typical lot of wafers.

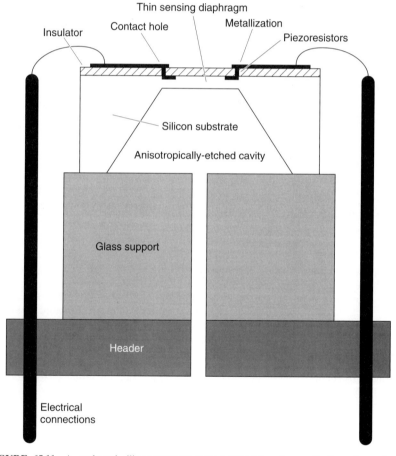

FIGURE 65.11 A packaged silicon pressure sensor illustrating the application of anodically bonded glass constraints.

CONCLUSIONS

The fabrication techniques described above can be used to form a number of devices in addition to piezoresistive pressure sensors. For example, a structure that consists of two conductive plates that are in close proximity to one another, but electrically insulating, can be used as a capacitive pressure sensor. Enclosed channels can be fabricated together with minute resistance heaters to produce a gas mass flow meter. Silicon accelerometers are becoming an increasingly important product in the marketplace, especially for automotive applications. By forming a singly or doubly supported beam with sensing elements at its edges, a device that detects acceleration can be fabricated. Other devices such as IR detectors, microwave detectors and vacuum sensors are also possible using silicon micromachining. Other examples of micromachined sensors, actuators, and microstructures are described in later chapters.

REFERENCES

1. Tsai Beadle and Plummer, *Quick Reference Manual for Silicon Integrated Circuit Technology,* John Wiley & Sons, 1984.
2. Sze, *Physics of Semiconductor Devices,* John Wiley & Sons, 1969.
3. K. Petersen, "Silicon as a Mechanical Material," *Proceedings of the IEEE,* **70**(5), May 1982, pp. 420–457.
4. G. L. Pearson, W. T. Reed, Jr., and W. L. Feldman, "Deformation and Fracture of Small Silicon Crystals," *Acta Metallurgica,* **5**, 1957, p. 181.
5. H. F. Wolf, *Silicon Semiconductor Data,* Pergamon Press, New York, 1969.
6. S. Iwamura, Y. Nishida, and K. Hashimoto, "Rotating MNOS Disk Memory Device," *IEEE Trans. Electron Devices* vol. ED-28, 1981, p. 854.
7. H. Guckel, D. W. Burns, H. A. C. Tilmans, D. W. DeRoo, and C. R. Rutigliano, "Mechanical Properties of Fine Grained Polysilicon—The Repeatability Issue," *Technical Digest of the IEEE Solid State Sensors and Actuators Workshop,* Hilton Head Island, South Carolina, June 6–9, 1988, p. 96.
8. K. Petersen, P. Barth, J. Poydock, J. Brown, J. Mallon, Jr., and J. Bryzek, "Silicon Fusion Bonding for Pressure Sensors," *Technical Digest of the IEEE Solid State Sensors and Actuators Workshop,* Hilton Head Island, South Carolina, June 6–9, 1988, p. 144.
9. K. Albaugh, "Mechanisms of Anodic Bonding of Silicon to Pyrex Glass," *Technical Digest of the IEEE Solid State Sensors and Actuators Workshop,* Hilton Head Island, South Carolina, June 6–9, 1988, p. 109.

CHAPTER 66
UNIVERSAL SENSORS TECHNOLOGY: BASIC CHARACTERISTICS OF SILICON PRESSURE SENSORS

SILICON PIEZORESISTIVE PRESSURE SENSORS

Silicon pressure sensors are by a wide margin the most successful commercial implementation of solid-state mechanical sensors or silicon micromachined structures. In fact, silicon micromachining initially evolved largely as a tool for the creation of such sensors.

The development of these devices had its origins in the 1950s with the discovery of the relatively large piezoresistive effect in silicon.

This section discusses the fundamental properties of silicon sensors, particularly piezoresistive devices, including basic operating mechanisms, performance, geometry, the effect of dopant concentration, and the dynamic performance of typical structures. Another important technology, silicon capacitive-based sensing, is also discussed.

Piezoresistance in Silicon

The relatively large piezoresistive effect in silicon was discovered in 1954 by C. S. Smith,[1] a professor at Case Western Reserve University, while visiting at Bell Laboratories in Murray Hill, New Jersey. Smith's field of interest was crystal symmetry. Piezoresistance is one of the few electrically anisotropic properties of materials. Smith did little additional work in piezoresistance after his initial activity in the early 1950s. An interesting aside to this discovery was the fact that, years later, a silicon piezoresistive catheter-tipped pressure sensor was inserted into Smith's body to diagnose a heart condition. Thus, the invention came full cycle from inception to practical use in the health care of its discoverer.

Piezoresistance is defined as the change in resistance of a solid when subjected to stress. The fractional change of resistance in silicon resistors is proportional to the increase of the strain to which it is subjected. For small change, this relation can be expressed by the equation:

$$\frac{\Delta R}{R} = K \cdot \Delta \cdot \varepsilon$$

where R = resistance
K = proportionality constant (gage factor)
e = strain

For conventional metal strain gages the gage factor is about 2; however, for silicon resistors, a gage factor of 200 is achievable.

The overall resistance change for a silicon piezoresistor can be written as:

$$\frac{\Delta R}{R} = \frac{\Delta \rho}{\rho} + \frac{\Delta l}{l} - \frac{\Delta \tau}{\tau} - \frac{\Delta \omega}{\omega}$$

where $\rho, l, \tau,$ and ω are, respectively, resistivity, length, thickness, and width of silicon sensor.

The effect of change in resistivity has generally the most predominant effect; thus, for a resistor, Ohm's law is written as:

$$\{E\} = [r]\{j\}$$

where E is the electric field vector, r is the resistivity tensor, and j is the current density vector.

The relationship between the tensor of resistivity change and stress can be expressed as:

$$[\Delta r] = [\pi][s]$$

where $[\Delta r]$ is the matrix of resistivity change, $[\pi]$ is the piezoresistive coefficient matrix, and $[s]$ is the stress matrix. For the cubic crystal structure of silicon, the piezoresistivity coefficient matrix is symmetric in the following form:

$$[\pi] = \begin{vmatrix} \pi_{11} & \pi_{12} & \pi_{12} & 0 & 0 & 0 \\ \pi_{12} & \pi_{11} & \pi_{12} & 0 & 0 & 0 \\ \pi_{12} & \pi_{12} & \pi_{11} & 0 & 0 & 0 \\ 0 & 0 & 0 & \pi_{44} & 0 & 0 \\ 0 & 0 & 0 & 0 & \pi_{44} & 0 \\ 0 & 0 & 0 & 0 & 0 & \pi_{44} \end{vmatrix}$$

As shown above, the change in resistance is approximately equal to the change in resistivity, which in turn is a predominant function of the plane stresses. In general, it can be written as:

$$\frac{\Delta R}{R} = \pi_l \sigma_l + \pi_\tau \sigma_\tau$$

where l and τ indices refer indicating longitudinal and tangential directions with respect to resistor orientation. The values of π_l and π_τ are a function of the crystal logic orientation.

The simplest piezoresistive device is a semiconductor strain gage, as shown in Fig. 66.1. Such a device is a two-terminal sensor. Strain is applied parallel to the axis of the device; a strain gage is relatively insensitive to cross-axis strain. Silicon piezoresistive devices are in fact stress sensors, since strain is related to stress by Hook's law:

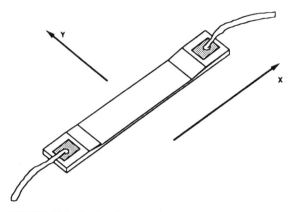

FIGURE 66.1 Semiconductor strain gage.

$$s = e \cdot E$$

where E is Young's modulus and e is the strain.

The discovery of piezoresistance was immediately recognized to be of fundamental importance. Conventional strain gages had achieved success in the late 1940s, both as a tool for stress and strain analysis of mechanical structures and also as the sensing element for a new generation of mechanical sensors. Semiconductor strain gages offered a sensitivity 50–100 times higher than conventional devices. Moreover, semiconductor sensors could be manufactured using the tools that were rapidly evolving for the fabrication of transistors.

Thus, substantial effort was applied to the rapid development of these devices by the inventors and developers of the transistor.

Piezoresistive devices are typically fabricated from a single crystal silicon of high purity. Silicon exhibits diamond cubic symmetry. To model the piezoresistive effect, it is necessary to specify the crystallographic directions in the x- and y-directions and the plane of the sensor.

Silicon semiconductor piezoresistive sensors in general are sensitive to stress in three principal axes plus shear stress in three planes. Voltage and current must be specified in three principal directions to characterize fully these devices. For a simple strain gage geometry subjected to planar stress, attention can be limited to the voltage and current in the x-direction and stress in the x-direction. For a strain gage integrated into a planar force collector, however, the principle stresses are in the x- and the y-directions.

A number of orientations are useful for sensors. Some typical orientations are given below:

$$(100)\text{Plane} \quad \langle X \rangle = \langle 111 \rangle \quad \langle Y \rangle = \langle 211 \rangle$$

This orientation is commonly used for semiconductor strain gages. The $\langle 111 \rangle$ direction is the axis of densest packing; the maximum piezoresistive effect is observed along this axis.

For the purpose of this chapter, p-doped silicon only will be discussed, since the largest and most linear piezoresistive effect is observed in p-type silicon. For p-type silicon, the largest observed piezoresistive coefficient is π_{44}, the shear piezoresistive coefficient. The relative sensitivity is normalized to this value.

For this orientation, the on-axis sensitivity in the x-direction is $+.66\pi_{44}$. The sensitivity in the y-direction is $-.33\pi_{44}$. In the absence of cross-axis stress of a sign opposite to the on-axis stress, this is the most sensitive orientation.

$$(110)\text{Plane} \quad \langle X \rangle = \langle 110 \rangle \quad \langle Y \rangle = \langle 100 \rangle$$

It can be appreciated that cross-axis sensitivity is undesirable for certain geometrical situations. Consider, for instance, that stresses are equal and orthogonal in the center of a square or round silicon diaphragm. Net resistance change for a piezoresistive sensor is the sum of the response to the on-axis and cross-axis response to stress. Since the sign of the cross-axis sensitivity is negative and positive for on-axis sensitivity, the net change in resistance is reduced. For such a situation, it is desirable to maximize the on-axis sensitivity while minimizing the cross-axis effect.

For this orientation, the on-axis sensitivity is $+0.5\pi_{44}$, while the cross-axis sensitivity is close to zero. This orientation provides the maximum on-axis sensitivity under the constraint of zero cross-axis sensitivity.

$$(100)\text{Plane} \quad \langle X \rangle = \langle 110 \rangle \quad \langle Y \rangle = \langle 110 \rangle$$

This orientation is perhaps the most widely employed orientation for silicon-integrated sensors. It was employed by investigators at Stanford and later at Delco.[2] (100) material is widely available and commonly used for conventional integrated circuits.

Moreover, for a silicon diaphragm, the normal maximum stress is at the edge of the diaphragm, while the stress is relatively small parallel to the edge of the diaphragm. Thus, a piezoresistive sensor geometry that has a large cross-axis sensitivity is desirable.

$$\langle Y \rangle = \langle 211 \rangle$$

This orientation provides a maximum cross-axis sensitivity of $-.5\pi_{44}$. The on-axis sensitivity is equal and opposite and is $+.5\pi_{44}$. Thus, a radial and a transverse gage combination allows equal and opposite resistance changes when placed at the high stress edge of a diaphragm.

$$(100)\text{Plane} \quad \langle X \rangle = \langle 100 \rangle \quad \langle Y \rangle = \langle 100 \rangle$$

In this orientation, a relatively small piezoresistive effect of approximately $+.02\pi_{44}$ is observed. This orientation is useful for fabricating a temperature sensor.

Such a temperature sensor, if fabricated simultaneously with the piezoresistive elements, will have a temperature coefficient of resistance (TCR) equal to that of the silicon piezoresistors. This type of device is useful as a temperature-compensating element or temperature input to a microprocessor.

Another geometry useful for the fabrication of silicon-integrated piezoresistive sensors is shown in Fig. 66.2. Typically, silicon piezoresistive devices are formed from four piezoresistors integrated into a Wheatstone bridge. On the other hand, it is possible to fabricate a four-terminal shear sensor of the geometry shown. The piezoresistive sensitivity of such a sensor is given by

$$V_o = i \cdot R \cdot \pi_{44} \cdot \sigma_{xy}$$

Where V_o is the output voltage, i is the excitation current, and σ_{xy} is the shear stress in the plane of the sensor along the gage axes. This geometry was developed in the early 1960s at Bell Labs and has recently been exploited commercially by Motorola.

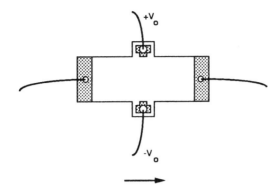

FIGURE 66.2 Shear piezoresistive sensor.

The shear geometry has both advantages and some disadvantages. This approach effectively integrates a Wheatstone bridge at a single point. On the other hand, the geometry is identical in performance to a square-patch piezoresistive geometry, which produces a similar output signal while dissipating less current.

Typical Integrated Sensor Geometries

The various orientations and gage geometries discussed above are effectively a set of building blocks that can be used in the construction of many useful devices, particularly integrated silicon diaphragm pressure sensors.

Early devices used epoxy to attach bar-type silicon strain gages to a metal diaphragm. While simple in implementation, such devices were poor in performance due primarily to the difficulty of transmitting stresses from the diaphragm surface to the relatively difficult geometry of such a strain gage. The great success of silicon piezoresistive devices has been based on the integration of the strain sensors into a silicon diaphragm. A typical configuration is shown in Fig. 66.3.

This device, while simple in concept, has revolutionized the field of pressure measurement, created the basis of a new industry, and firmly established the direction of the development of modern sensors.

The four piezoresistive elements are fabricated, using the techniques of integrated circuit manufacture, to form a Wheatstone bridge. Previously these elements were formed by solid-state diffusion, but the industry has largely converted to ion implantation. The piezoresistive elements are typically passivated by a dielectric layer and isolated from the insulating mechanical diaphragm by the presence of a *p-n* junction.

The diaphragm is formed by milling out a cavity to form a relatively thin deflecting membrane. The technique of fabrication is typically wet chemical etching, or *micromachining*. Though originally simple in concept, this machining operation has developed into a diverse fabrication technology, providing the ability to machine complex, three-dimensional structures while allowing the precise control of diaphragm thickness at the micron level through the use of electrochemical etch stop techniques.

Typically, aluminum metallization is used for contact areas, but refractory or noble metals can be employed for extended temperature requirements or for additional corrosion resistance.

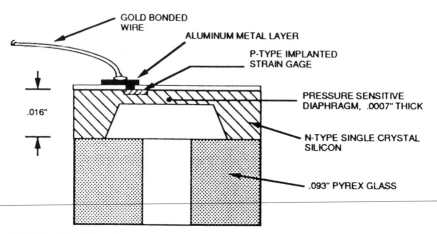

FIGURE 66.3 Silicon integrated piezoresistive sensor.

In addition to the piezoresistive Wheatstone bridge, other devices are readily integrated and are becoming increasingly common, including temperature sensors for input to microprocessors, static pressure sensors for use at high common-mode pressures, and active electronics for signal conditioning.

Figure 66.4 shows various basic sensing geometries possible for silicon piezoresistive pressure sensors. Often, a manufacturer chooses a particular geometry for all applications. NovaSensor has taken the approach of using the most appropriate geometry for a given application to maximize performance for that application.

There are two fundamental pressure scales: absolute and gage (Fig. 66.5). The absolute pressure scale is based on zero pressure defined at a perfect vacuum. The ambient pressure value using this scale is about 100 kPaA (A stands for absolute) and changes as a result of barometric pressure changes. The gage pressure scale is based on zero pressure existing at ambient atmospheric pressure. The value of perfect vacuum is about −100 kPa and changes with barometric pressure.

From an application point of view, different pressure scales result in five possible measurement configurations for pressure sensors:

- Absolute pressure
- Sealed gage pressure
- Gage pressure
- Vacuum
- Differential pressure

The major difference between these five configurations is the construction of a pressure reference port. A pressure-sensitive diaphragm responds to a differential pressure across it. If we denote a pressure on the top of the diaphragm (circuit side) by P_1 and a pressure on the bottom of the diaphragm (cavity side) by P_2, all these sensor configurations will be characterized by the same input pressure P_1 (Fig. 66.4). Configuration of the other pressure port P_2 will differ significantly.

Absolute Sensor. The absolute pressure sensor has a zero deflection of the diaphragm at P_1 equal to an absolute vacuum. To achieve this, the pressure port P_2

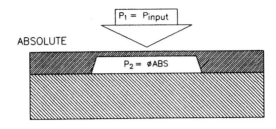

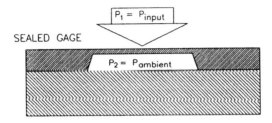

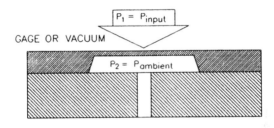

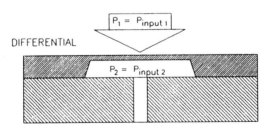

FIGURE 66.4 Basic configurations of pressure sensors include two designs with a sealed reference cavity (pressure port P_2) for absolute and sealed gage sensors and three designs with open reference cavity for gauge, vacuum, and differential sensors.

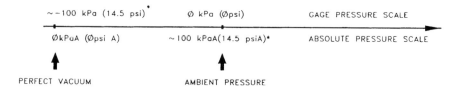

* EXACT VALUE DEPENDS ON THE ACTUAL VALUE OF BAROMETRIC PRESSURE.

FIGURE 66.5 A comparison of gage and absolute pressure scales reveals the key difference in anchoring the zero value: zero pressure (absolute) is at perfect vacuum, while zero pressure (gage) is at ambient pressure.

should be hermetically sealed at a zero absolute pressure. The sensor output at ambient pressure will vary proportionally to barometric pressure changes. This configuration is most often used for measurements of barometric pressure.

Sealed Gage. The sealed gage sensor configuration is the same as the absolute pressure sensor. The reference cavity P_2, however, is sealed at ambient pressure instead of vacuum. This configuration is frequently used for higher pressure-gage ranges, where the variation of barometric pressure creates the negligible measurement error.

Gage Sensor. To create the gage pressure sensor, the diaphragm should have a zero deflection at ambient pressure; thus, an ambient atmospheric pressure is supplied to the pressure port P_2. This is achieved by creating a hole in the constraint wafer. The sensor output is proportional only to the difference between the input and ambient pressure and is not sensitive to barometric pressure variations. It is possible to reverse the order of pressure port configurations and connect the input pressure to the cavity side (P_2) of the sensor. The circuit side of the sensor (P_1) is then exposed to an ambient pressure. Such a configuration allows a better media compatibility for port P_2, since no sensor electrical circuit is exposed in the cavity.

Vacuum Sensor. Vacuum sensor is designed as a gage pressure sensor, but the output terminals are reversed as compared to a gage configuration to provide an increasing output with the increasing vacuum (decreasing pressure). At ambient pressure, a vacuum sensor indicates a zero output with a full-scale output produced at a full vacuum. Output is not dependent on a barometric pressure.

Differential Pressure Sensor. Differential pressure sensor configuration is the same as the gage sensor; however, the reference cavity is exposed to another input pressure P_2. Zero deflection of a diaphragm is achieved at equal input pressures $P_1 = P_2$. This condition may occur at a significantly higher pressure then the ambient pressure. In such a case, the sensor output is proportional not only to an applied differential pressure ($P_1 - P_2$) but to a certain degree to a static pressure ($P_1 + P_2$)/2 as well. This last effect is called a static pressure sensitivity.

Figure 66.6(*a*) shows a rectangular diaphragm with four gages employing transverse piezoresistance. Two gages are placed parallel to and near the edge of the diaphragm; two coparallel gages are placed at the center.

A rectangular diaphragm subjected to an applied pressure produces a relatively large stress parallel to the short axis of the diaphragm. If the aspect ratio of the diaphragm is greater than 3 to 1, stress along the major axis is limited to Poisson's

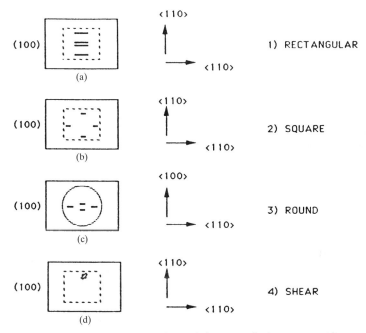

FIGURE 66.6 Typical integrated piezoresistive sensor diaphragm geometries.

ratio or approximately +.2 times the stress along the minor axis. Such a device can achieve acceptable output from piezoresistive sensors that have equal and opposite longitudinal and transverse piezoresistive coefficients.

Using the technique of computer modeling, it is possible to position the gages to produce equal and opposite output, achieving good linearity. Such a device can typically produce .05 percent nonlinearity over pressure ranges from 15 to 500 psi.

The square geometry of Fig. 66.6(b), on the other hand, has the advantage of placing all strain gages at the periphery of the diaphragm, where the stress is maximum. The geometry was employed in early devices at Stanford University.[2] Cross-axis stress parallel to the diaphragm edge is approximately Poisson's ratio. A fully active Wheatstone bridge can be formed by a combination of two longitudinally and two transversely active strain gages. Once again, gages can be positioned for equal and opposite resistance change, producing outstanding nonlinearity.

Such a device is readily useful over ranges of 15 to 500 psi, producing linearity similar to Fig. 66.6(a). This device exhibits a high degree of thermal symmetry and is useful to relatively higher excitation voltages without undesirable self-heating effects. The full utilization of the available stress allows for a higher output. This configuration is probably the most widely exploited geometry.

Figure 66.6(c) uses a round diaphragm and employs gage geometry with zero transverse sensitivity. While some manufacturers still use this geometry, it has become largely obsolete with the impact of VLSI processing and the move toward conventional (100) materials and the use of micromachining. Of course, the gage pattern of Fig. 66.6(b) is equally appropriate to the round diaphragm configuration.

Figure 66.6(d) shows a square diaphragm with a shear-type pressure sensor. The maximum shear stress for a diaphragm is at 45° to one of the edges of the diaphragm. Such devices are manufactured by Motorola Semiconductor.[3] While this geometry is quite useful and has been commercially successful, it does have several negative features inherent in the design, including a lack of thermal symmetry and difficulty in making it insensitive to stresses transmitted to the base of the chip from the package.

Low-Pressure Sensors

The geometries discussed above are useful for sensors in the range of 5–500 psi. For pressures below 5 psi, however, nonlinearity of the silicon diaphragm limits performance. A review of the various approaches to the problem is given in Ref. 3.

Diaphragms are highly linear when the deflection under pressure is a small fraction of the diaphragm thickness. For silicon piezoresistive sensors, the pressure range is adjusted by thinning the diaphragm to provide the desired or design stress. Thinner diaphragms are employed for lower pressures. The maximum stress at the edge of a diaphragm is given by

$$\sigma = P \cdot K_1 \cdot r^2/t^2$$

where P is the pressure, K is a constant, r is the diagram radius, and t is the diaphragm thickness. The deflection is given by

$$d = P \cdot K_2 \cdot r^3/t^3$$

It can be seen that, if the design full-scale operating stress is fixed at a given level, the thickness must be decreased to maintain that stress as the pressure range is lowered. However, the deflection is inversely proportional to the cube of the thickness, while the stress is inversely proportional to the square of the thickness. Thus, the relative deflection increases for lower pressure ranges. Eventually, the deflection becomes excessive, and a highly nonlinear membrane stress is superimposed on the linear bending stress. This effect, along with the difficulty of fabricating a thin, thermally stable silicon diaphragm, has in the past limited the use of silicon sensors for low pressure ranges.

The solution is to employ the three-dimensional technique of micromachining to stiffen sections of the diaphragm. This problem was recognized for conventional strain sensors by investigators such as Stedman[5] at Statham and Pien[6] at Consolidated Controls. Modern low-pressure silicon sensors adapt these techniques to silicon geometries. Several typical geometries are shown in Fig. 66.7.

Figure 66.7(a) is a double-bossed configuration, first proposed in metal for conventional sensors by Pien and employed for silicon sensors by Wilner at Endevco.[7] Such a configuration employs relatively thick bossed areas or stiffeners to limit the diaphragm deflection while transversely active piezoresistors are placed in the narrow slots at the center and edge of the sensor.

This technique is sometimes incorrectly thought of as stress concentration. The purpose of the bosses serves primarily to limit the deflection of the diaphragm. This is apparent from the fact that the bosses actually decrease the stress for membrane of a given thickness rather than increase it. Such bosses should be thought of as deflection-limiting structures rather than stress-concentrating structures.

The configurations shown in Figs. 66.7(b) and (c) employ central bosses to limit and linearize deflection of the diaphragm. In the configuration of Fig. 66.7(b), a central square boss is employed. The use of a central boss was suggested originally by Stedman of Statham and a square micromachined boss for use in silicon was proposed by Mallon, Kurtz, and Nunn[8] at Kulite. Round central bosses are used by

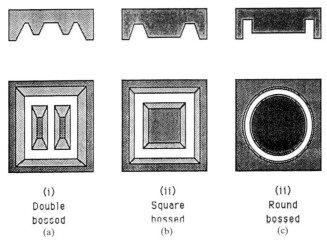

FIGURE 66.7 Typical low pressure geometries for silicon piezoresistive sensors.

Hitachi and Siemens. The use of such techniques results in substantial improvement in linearity of piezoresistive sensors at low pressures.

Various gage geometries can be employed with the central bossed configuration, including longitudinally or transversely active gages. The pattern of Fig. 66.6(*b*), for instance, is very appropriate.

Using the stiffened or bossed geometries shown above, pressure sensors with high linearity can be fabricated for use at pressures of 2^{11} H_2O and below. Typically, a 1-psi sensor produces a full-scale output of 100 MV with a linearity of better than 0.1 percent.

The issue of linearity is important, but many other factors are equally important in fabricating a low-pressure sensor, including the thermal stability of the diaphragm, insensitivity from packaged induced stresses, and so on.

High-Pressure Sensors

At pressures above 500 psi, another factor affecting nonlinearity begins to become important. Such high-pressure sensors employ diaphragms with thicknesses comparable to their diameters. For instance, a 10,000-psi sensor may have an aspect ratio (diameter:thickness) of only 2:1. For such a geometry, linear bending stresses no longer predominate and shear stresses become important. The deflection and resulting stress of such a sensor is highly nonlinear.

To achieve acceptable performance from such devices, it is necessary to model the geometries properly. NovaSensor uses finite element analysis (FEM) for this purpose. Since analytic solutions for the complex geometries of micromachined devices are difficult, NovaSensor has employed this technique of analysis to model such structures precisely.

Using a maximized gage geometry and precise positioning of the strain gages, it is possible to achieve sensitivities of 200 mv full-scale with linearities of better than .1 percent at full-scale pressures as high as 10,000 psi.

Effect of Dopant Concentration on Piezoresistive Sensors

The basic properties of piezoresistive devices are controlled largely by the doping level or concentration of added or extrinsic charge carriers.

A pure semiconductor is an insulator at room temperature. The substrate material employed for diaphragms of piezoresistive devices is typically 1–10 ohm-cm n-type.

Piezoresistive elements are formed by adding a donor dopant, such as boron, to the silicon lattice. Modern VLSI fabrication techniques employ ion implantation to control the doping level of active devices. Ions of boron are accelerated in a field and penetrate the surface of the silicon wafer. A subsequent anneal at high temperature (typically 1100°C) is used to control the final depth of penetration and resultant concentration of the dopant atoms.

The operating characteristics of piezoresistive sensors are highly dependent upon the relative concentration of these extrinsic atoms. Central to the understanding of the effect of the various dopant concentrations is the need for temperature compensation of the devices.

When silicon piezoresistive sensors were first introduced, it was apparent that their temperature sensitivity was extreme. Considerable skepticism was raised concerning the viability of these sensors, because of these large temperature coefficients.

The skeptics failed to realize, however, the tremendous advantage of precise control of temperature coefficients of the individual elements that was achievable using the semiconductor processing techniques. When piezoresistive elements are integrated into a silicon diaphragm and formed into a Wheatstone bridge, temperature becomes a common mode input. Process control is such that the rejection ratio is high.

Typically, silicon sensors have a temperature effect that, over 55°C (a typical operating span), is five times larger than the desired piezoresistive effect. Uncompensated TC-zero of modern sensors are typically 1 percent of full-scale over this temperature range. Thus, precise matching of temperature coefficients and use of devices in the Wheatstone bridge largely eliminates temperature sensitivity of these devices.

Presented in Fig. 66.8 for reference is a typical compensation circuit for constant voltage or for constant current that is generic of compensation schemes typically employed for these devices.

In general, the requirements of such a circuit are determined by the following parameters:

V_o	Offset at room temperature with no applied pressure expressed as a ratio of applied excitation in units of mV/V.
FSO	Pressure sensitivity at full scale applied pressure expressed in units of mV/V.
TCR	Temperature coefficient of resistance expressed as a fractional or percentage change of resistance versus temperature in units of %/°C.
TCV_o	Temperature coefficient of offset expressed as a fractional change relative to full-scale output in units of %FS/°C.
TCS	Temperature coefficient of sensitivity expressed as fractional or percentage change of sensitivity versus temperature in units of %/°C.

For convenience, temperature coefficients are often expressed as %/100°F or %/55°C.

The technique of temperature compensation and normalization must achieve room temperature normalization of V_o and V_{Fso} and reduction to near zero of the

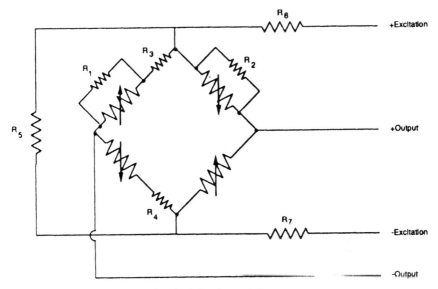

FIGURE 66.8 Basic compensation circuit for piezoresistive sensor.

temperature coefficients TCV_o and TCS. Thus, to effect compensation, four independent variables are required.

The compensation circuit of Fig. 66.8 uses four independent resistors or resistor pairs to achieve this end. It should be noted that the series sensitivity-setting resistors and the zero balance resistors are often split for convenience and common-mode rejection, but conceptually they can be replaced by a single resistor.

Typical uncompensated values for a silicon piezoresistive sensor are given below.

V_o	3 mV/V
V_{fso}	20 mV/V
TCR	+0.26%/°C
TCS	−0.21%/°C
TCV_o	±0.02%/°C

The circuit of Fig. 66.8 achieves compensation of these parameters by the use of passive resistors to shunt differentially or load the effect of the temperature coefficients and voltage applied to the various arms of the Wheatstone bridge.

Adjustment of zero balance of the sensor is achieved by placing resistors R3 and R4 in series with the bridge arms. These external resistors have zero temperature coefficients of resistance compared to the relatively high temperature coefficient of bridge. These change the temperature coefficient of the bridge arm into which they are inserted, thus inducing an additional component of TCV_o.

A series and shunt resistor in a given arm will affect the offset in opposite directions while affecting TCV_o in the same direction. In general, a unique combination of a series and shunt resistor in an appropriate arm exists. This will compensate TCV_o and set V_o to zero at room temperature. For convenience in achieving the relatively low values required for the series resistor, two resistors are often employed.

The normalization of sensitivity and compensation of TCS for piezoresistive sensors is readily illustrated by a constant voltage compensation circuit. In such a circuit, a series resistor is used to set the sensitivity and to provide near-constant current to the Wheatstone bridge. Since the temperature coefficient of sensitivity is negative, an excitation voltage which increases with temperature will compensate for the negative decrease in sensitivity. The positive coefficient of resistance causes the bridge input impedance to increase with temperature. If excited by constant or near constant current, the bridge voltage increases proportionally to temperature. By proper choice of the series resistor, and equally importantly, the TCR and TCS of the bridge, the effect of this circuit induced voltage increase can be made to be equal and opposite the sensitivity decrease. Proper adjustment of TCR/TCS ratio can also provide compensation for constant current.

On the other hand, since two constraining conditions must be met: TCS of zero and normalization of V_{Fso}, the circuit of Fig. 66.8 incorporates for constant voltage compensation an additional shunt resistor R5. The circuit thus provides two independent variables that can be used to reduce the sensitivity coefficient to near zero to adjust for normal manufacturing tolerances. A similar equivalent circuit can be employed for constant current excitation.

The use of such a passive circuit compensation technique provides a flexible method of compensating the performance of piezoresistive sensors. It should be noted that the temperature compensating resistors are passive and serve only to adjust differentially the temperature coefficient of the silicon piezoresistive elements. Only the sensor responds to temperature.

This is contrasted to the techniques of using active elements employed with the undesirable effects of making the devices sensitive to temperature gradients and slowing their thermal response. It is highly desirable to provide all or most of the required compensation with temperature-insensitive elements as shown.

Piezoresistive Temperature Coefficients versus Doping Level

It can be appreciated from the above brief discussion of thermal compensation that it is important to understand and control the temperature coefficients of the piezoresistive devices. These coefficients are a function of the sensor dopant level. The only published parametric study over broad temperature ranges and doping concentration levels of these parameters was given by Tufte and Selzer of Honeywell in the early 1960s. Subsequent work by many investigators has generally confirmed the trend of their results.

The temperature coefficients are a function of the concentration of extrinsic charge carriers that are intentionally added to the silicon lattice. Silicon piezoresistive devices typically have a dopant concentration that is high near the surface, decreasing to a lower value at the junction. Their performance is approximately a function of surface dopant concentration or, more precisely, of average dopant concentration.

Figure 66.9 plots the average concentration versus the piezoresistive coefficient π_{44}, the TCR, and the TCS. The following trends can be noted as the concentration is increased.

Parameter	Trend
π_{44}	↓
TCR	↓ then ↑
TCR	↓

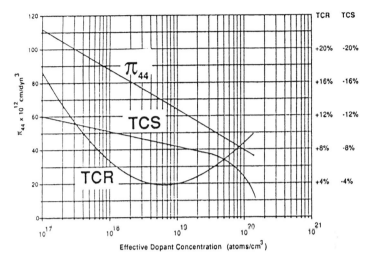

FIGURE 66.9 Basic piezoresistive properties (π_{44}, TCR & TCS) as a function of dopant concentration.

Several regions of this curve are of interest to the device designer. For instance, a concentration of approximately 2×10^{18} atoms/cm^3 produces a TCR of approximately +14%/55°C and a TCS of approximately −12%/55°C. This concentration is useful for constant current excitation, since TCR and TCS are approximately equal. To the left of that point, the TCR is larger than the TCS. To achieve constant voltage compensation in the circuit of Fig. 66.8, it is necessary to have a TCR > TCS. This region is useful for constant voltage compensation (Fig. 66.9).

On the other hand, these conditions can also be met at an average concentration in the area of 1×10^{20} atoms/cm^3. Thus, both constant current and constant voltage compensation can be achieved at appropriate doping levels.

It should be noted, however, that, at higher doping levels, the piezoresistive coefficients are reduced by a factor of 2 or more; thus, sensitivity is lower. However, the temperature coefficients are much more linear over a broader temperature range. Thus, compensation using simple circuits with limited ability to reduce second-order effects can be used to achieve a more precise compensation over extended temperatures. (See Fig. 66.10.)

Once the surface concentration and the basic output level are chosen, the device designer can use many techniques of temperature compensation to obtain a high-performance sensor. Passive techniques such as those shown in Fig. 66.8 are slowly being replaced by microprossessor compensation. In either case, a knowledge of the device parameters is essential. Figure 66.12 (in a subsequent section of this chapter) shows a three-dimensional representation of the performance surface of a representative device.

It should also be noted that, at the high concentration levels, the semiconductor is essentially degenerate. At these dopant concentrations, the device is much less sensitive to external effects such as ionization by light or radiation. It is essentially equivalent to a wire strain gage with comparable stability. Moreover, the device is less subject to undesirable effects of surface charges and voltages that can affect the *p-n* junction isolation. Very precise high-accuracy sensors can be made using these doping levels. The significant advantages of degenerate level piezoresistors was recognized early by Kurtz.[12]

It can be seen from the above discussion that a designer of a piezoresistive sensor can employ a wide range of variables, including dopant concentration, sensor geometry, and three-dimensional structuring to achieve the required sensing goal. As needed, additional flexibility can be achieved by the use of n-type silicon, germanium, or II-VI and III-V semiconductors. The full range of capability of this class of devices has, in fact, just begun to be explored.

Piezoresistive Sensors for Extended Temperature Operation

By far the largest requirements for the application of sensor is near room temperature, typically from 0–70°C or –40°C–125°C.

Many industrial and military applications require the use of sensors over an extended temperature range. Fortunately, the piezoresistive properties of silicon are quite useful over a very broad temperature range.

Figure 66.10 is a graph of the piezoresistive and temperature coefficients of p-type silicon for an extended temperature range for both high and low concentrations. It can be seen that, while the lower concentration provides a more sensitive device, the smaller temperature and relatively more linear nature of the temperature coefficients make high concentration desirable for certain applications.

Silicon piezoresistors are useful from liquid oxygen temperatures to an upper range that has not been determined but is well in excess of 500°F, perhaps as high as 800°F.

The use of piezoresistive sensors at very low temperatures is essentially an engineering and packaging problem. Relatively few commercially interesting applications of this type exist. For those applications, semiconductor sensors have been successful. In fact, their basic properties are often improved at low temperatures.

Because of the nonlinearity of temperature coefficients over such an extended temperature range, compensation of these devices is more difficult than for room temperature applications. Often nonlinear elements are employed, although this is undesirable because of the sensitivity to temperature gradients. A more effective means of compensation is the use of a microprocessor to store the temperature/performance-related parameters for an individual sensor and to compensate for temperature effects by an appropriate algorithm.

Piezoresistive Sensors for Use at High Temperatures

The use of conventional piezoresistive sensors at higher temperatures, however, requires a solution of a more basic problem. Typically, gage to substrate isolation is accomplished by means of a p-n junction. This isolation is effective in reducing the flow of majority carriers generated in the depletion region but not minority carriers across the p-n junction. At temperatures above 150°C, the effect of thermal energy in the silicon lattice is to generate hole-electron pairs. The presence of a substantial number of minority carriers causes reversible degradation of the p-n junction. In addition, the substrate conductivity increases dramatically.

At temperatures above 150°C, the piezoresistive bridge is essentially shorted out, and the device becomes relatively useless as a sensor. The precise point at which a device is no longer serviceable depends on the level on the dopant and substrate concentrations. Higher concentration piezoresistive devices are useful over a more extended temperature range.

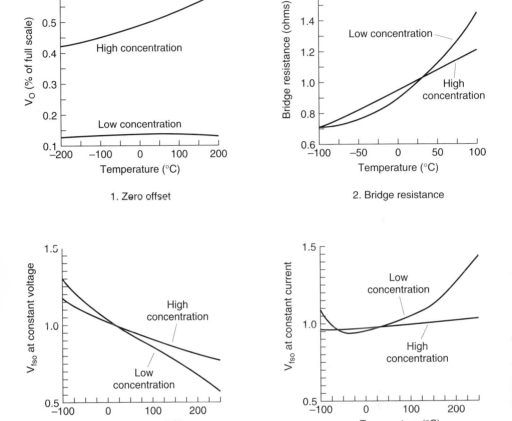

FIGURE 66.10 Temperature performance of piezoresistive sensors over an extended temperature range.

The solution lies in eliminating the *p-n* junction isolation and replacing it with a dielectric layer effective at higher temperatures. Mallon and Germanton[13] reported on the use dielectric isolation and two other technologies in 1970. Since then, piezoresistive high-temperature devices have evolved through a number of generations, with many investigators using a variety of techniques to produce high-temperature sensors.

The popularity and use of these techniques parallels the development of isolation technology in integrated circuits today, referred to as SOI (silicon on insulator) technology. As usual, piezoresistance has drawn directly from this IC technology with many such techniques including silicon on sapphire, grapho-epitaxy, dielectric isolation, SIMOS, and lateral regrowth employed. The basic resulting structure is shown generically in Fig. 66.11.

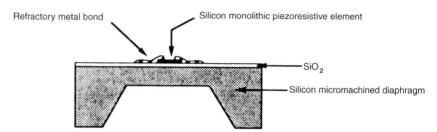

FIGURE 66.11 High-temperature piezoresistive sensor.

The modern technique of micromachining has resulted in the ability to be able to refine the simple structure above into a complex, small, high-performance device. The device is based on the use of silicon wafer lamination techniques to achieve a unique, state-of-the-art, high-performance, high-temperature pressure sensor.

SOI high-temperature piezoresistive sensors are useful for a wide range of applications, including particularly low-cost, high-temperature automotive sensors. Exploitation of this technology can be expected over the next several years.

Total Error Band Concept

The compensated and uncompensated performance of a silicon sensor can be represented by a curved surface formed along three axes: output error, temperature, and pressure.

Silicon sensor output is affected by both pressure and temperature. The error is related to initial offset (V_o) and pressure sensitivity (full-scale output). From the application point of view, a total error band that defines maximum error over an entire operating temperature and pressure range sometimes makes more sense. Such an approach is especially useful in the implementation of a microprocessor-based temperature compensation, where in essence a surface that defines the output as a function of both variables has to be converted ideally into a plane with no temperature error that responds to pressure only.

A graphic representation of a typical sensor output is shown in Fig. 66.12. The sensor output is a function of the excitation mode and temperature-compensating resistors. The three cases are shown for the same sensor.

In the constant voltage mode with the sensor output at a reference temperature of 20°C, as shown in Fig. 66.12(a), the error is nonlinearity. This error is barely visible in the drawing due to a low 1 percent value for NovaSensor's sensors. Temperature causes both zero and sensitivity to deviate from initial values. Zero change will be random; and for the example shown, the T_cV_o is approximately −1 percent at −40°C and +2 percent at 120°C.

In the constant current mode, the sensor's TCR approximatly compensates for TC_{od} sensitivity, since bridge voltage changes with temperature. This change compensates FSO change but introduces a new offset component (T_cV_o). Zero now shifts [Fig. 66.12(b)] from a reference 20°C value by −3 percent FSO at −40°C and +5 percent FSO at 120°C.

Introducing passive temperature compensation in the constant current mode discussed earlier reduces the deviations from 20°C to about +1 percent FSO at −40°C for both zero at full-scale pressure reading and to about +2 percent FSO at 120°C for both errors. See Fig. 66.12(c).

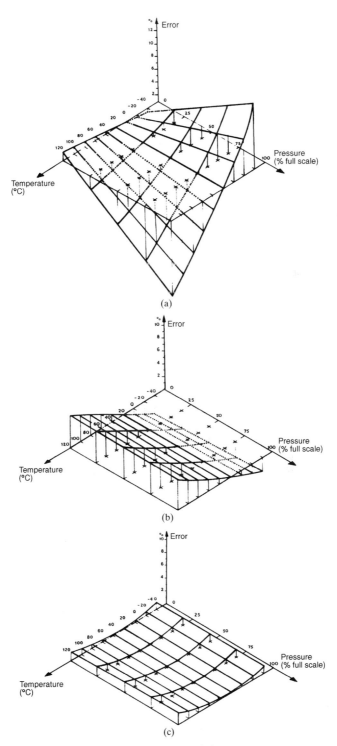

FIGURE 66.12 Surfaces of error piezoresistive sensor.

The summary of these data is as follows:

Configuration	Zero deviation		FS deviation		Total error ban
	−40°C	120°C	−40°C	120°C	
Constant voltage	−1%	2%	+11%	−19%	30%
Constant current	−3%	5%	−2%	4.5%	6.5%
Compensated	+1%	2%	+1%	2%	2.0%

Note that for a digital error compensation, a temperature sensing should be performed directly sensing chip to avoid temperature gradients between pressure and temperature sensors. The shape and magnitude of the surface defining sensors output will be a function of both sensor process and design.

The basic resulting structure is shown generically in Fig. 66.12.

The modern technique of micromachining has resulted in the ability to refine the simple structure above into a complex, small, high-performance device. The device is based on the use of silicon lamination techniques to achieve a unique state-of-the-art, high-performance, high-temperature pressure sensor.

SOI high-temperature piezoresistive sensors are useful for a wide range of applications, including particularly low-cost, high-temperature automotive sensors. Exploitation of this technology can be expected for the next several years.

Overpressure Capabilities of Piezoresistive Sensors

Silicon is an exceptionally strong material. In fact, many of the desirable characteristics of silicon have been governed by its excellent mechanical properties. When subjected to over pressure, silicon typically fails in tension, in a brittle fashion at approximately 150,000 psi. Well-prepared samples have exhibited tensile failures of 300,000 psi or higher. Shown is a photograph (Fig. 66.13) taken by researchers at Upsala University Sweden to show the incredible strength of silicon.

Silicon piezoresistive pressure sensors are typically rated conservatively at three times their full pressure limit for low to mid pressure ranges. In fact, they are often much stronger.

The overpressure capability of low-pressure-range sensors benefits from the fact that the diaphragm behaves in a highly nonlinear fashion when stressed excessively and becomes substantially stiffer, thus limiting the stress generated at high overpressure.

Figure 66.14 plots overpressure as a function of full-scale pressure range for some typical samples of style sensors. It can be seen that the overrange factor decreases with increasing pressure, but in all cases fail is observed at many times full scale. (It should be noted that the upper useful pressure is a function of pack constraints, and the NPH sensor should not be used above their rated pressures without consulting the factory.)

It can be seen that silicon piezoresistive sensors exhibit very desirable overpressure characteristics. This excellent overpressure capability has been useful in many applications of these devices, including particularly disposable blood pressure transducers. This application often requires a 5-psi transducer to be subjected to pressures 150 psi or higher.

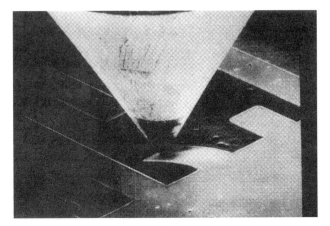

FIGURE 66.13 The strength of silicon.

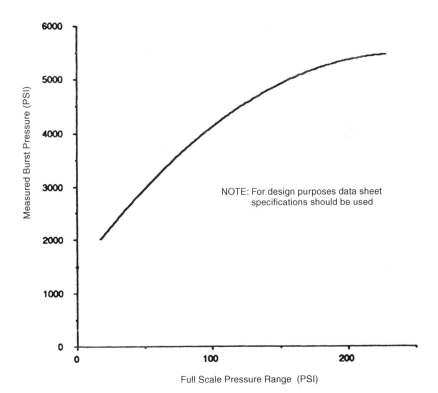

FIGURE 66.14 Overpressure performance of silicon piezoresistive sensors.

Linearity Performance

The very high sensitivity of piezoresistive devices in general allows their use at relatively low design stresses. Typical design stresses are 7500 psi or lower.

As explained previously, pressure nonlinearity is a strong function of the pressure range and the mechanical properties of the diaphragm. Additionally, the nonlinearity of the piezoresistive effect itself must be taken into account. Finally, the linearity of the circuit configuration in which the device is employed is important.

Overall, silicon piezoresistive devices are outstanding in their linearity performance. Figure 66.15 shows a plot of 5, 15, and 100 psi sensors calibrated at up to six times the normal full-scale range.

It can be seen that these devices are quite useful at several times full-scale pressure range and may be employed in that fashion under certain conditions.

If linearity is compensated using microprocessor techniques, the overall accuracy of the transducer is much improved, because fixed errors such as zero drift, TCV_o, and noise are reduced as a fraction of full-scale output. This approach can be the basis of the development of a highly accurate sensor.

High-Stability Piezoresistive Sensors

The development of piezoresistive devices at NovaSensor and elsewhere has proceeded through many generations from early devices to modern high-performance sensors.

The temperature and stability errors once thought intractable for high-performance devices have been largely overcome through the introduction of sophisticated VLSI processing and modern micromachining techniques. These techniques precisely control the mechanical properties of the device, allowing the elimination of undesirable, destabilizing package stresses.

In addition, the modern techniques of process control introduced through VLSI have allowed the creation of highly stable piezoresistive elements. A number of sophisticated techniques have been employed to control undesirable charge buildup in layers adjacent to the semiconductor.

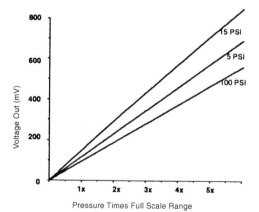

Best Fit Straight Line Non-linearity			
Overrange Factor	Pressure Range		
	5PSI	15PSI	100PSI
1x	.047	.030	.022
2x	.038	.066	.014
3x	.117	.120	.021
4x	.288	.174	.022
5x	.482	.256	.019

FIGURE 66.15 Overrange linearity of piezoresistive sensor.

Figure 66.16 plots short-term (24-h) stability of an extremely high stability sensor. This performance is typical of sensors using high-stability processing. It can be seen that this overall stability over 1000 minutes is approximately 1 mV/V or better. In order to achieve this level of performance, care must be taken to use very good instrumentation and shielding of the device from sources of external noise. Another effect that can be very important is thermal voltage generated by temperature gradients at junctions.

Investigators at NovaSensor and elsewhere have improved the performance of piezoresistive devices to the point where they compete with the best of alternate technologies, including secondary standard quartz instruments. For instance, Honeywell announced several years ago an expensive air data-class sensor that achieves a total error band over an extended temperature range and a long-term stability of less than .01 percent FSO. This illustrates the very high accuracies possible with piezoresistive devices.

In 1989, NovaSensor introduced a new sensor/process design called SenStable. It yields outstanding stability from the moment it is activated.

Overall performance is, of course, a price/performance tradeoff. Semiconductor sensors can offer very acceptable performance at low cost or state-of-the-art performance at higher cost.

Another aspect of stability is warm up time. Shown in Fig. 66.17 are two plots of warm-up for a piezoresistive device employing a high-stability and extremely high

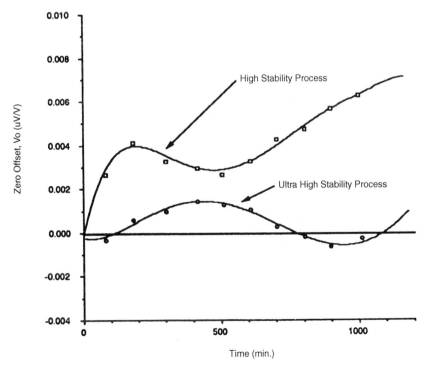

FIGURE 66.16 Short-term stability of V for piezoresistive sensor.

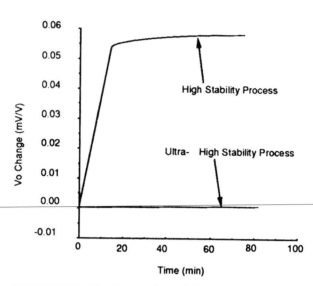

FIGURE 66.17 Warming up a piezoresistive sensor.

stability process, respectively; it can be seen that the best techniques of processing result in the potential for very short warm-up times for these devices.

The performance of silicon sensors has also been improved in terms of their stability after thermal cycling and multiyear stability; extremely inexpensive production devices offer overall stabilities of a 0.1–0.25 percent over several years and many thousands of temperature cycles and millions of pressure cycles.

Voltage Sensitivity of Semiconductor Devices

It is highly desirable that pressure sensors be completely ratiometric. Since silicon piezoresistive pressure sensors are simple first-order passive resistors, a high degree of ratiometricity is obtained.

Since power is dissipated in the piezoresistive elements, self-heating causes a rise in the sensor temperature above the surrounding ambient temperature. This temperature rise results in the chip operating at a temperature somewhat above room temperature. The temperature rise is proportional to the square of the applied excitation voltage.

Figure 66.18 plots the resistance of a 350-Ω silicon sensor as a function of applied excitation voltage. It can be seen that the resistance change for the 350-Ω sensor exhibits a noticeable nonlinearity. Since the bridge temperature coefficient resistance is known, the thermal impedance of the device may be calculated. It is found to be approximately 300°C/watt.

The best and most stable performance of piezoresistive sensors is obtained at operating temperatures of less than 20°F above room temperature. On the other hand, the devices will operate quite acceptably at considerably higher temperatures.

Another factor to be considered is the possible effect of voltage-induced change in V_o at increasing excitation voltage. This factor is largely a function of the resistors and thermal symmetry of the piezoresistive pattern. Figure 66.19 is a plot of V ver-

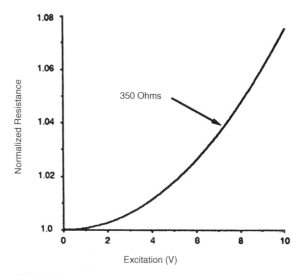

FIGURE 66.18 Normalized resistance vs. excitation.

sus excitation voltage for two patterns: one of high symmetry and one of much lower thermal symmetry (see Fig. 66.6). It can be seen that the effect of thermal symmetry is to provide a smaller and more linear change in V_o as a function of excitation.

Dynamic Performance of Piezoresistive Sensors

Silicon was chosen as a sensor material strictly because of its electrical properties. It is clear from the above discussions that nature has been unusually kind in providing the transducer designer with an excellent mechanical material. The dynamic performance of semiconductor sensors benefits from these excellent mechanical properties.

Silicon is relatively stiff, with the elastic modulus of steel (28–106 psi). On the other hand, it has the density of aluminum.

The resonant frequency of structure is given by

$$f = \frac{1}{2\pi}\sqrt{\frac{K}{M}}$$

where K is the stiffness and M is the mass.

The stiffness of a diaphragm sensor is a direct function of r^4/t^3, while the stress at the diaphragm edge is a function of r^2/t^2.

The mass of the diaphragm is a function of the density (silicon has the density of aluminum and the stiffness of steel) and the volume, which is proportional to $r^2 t$.

It can be seen that if r is decreased and t is adjusted to maintain a fixed design stress level, the difference in the exponent of t results in a higher resonant frequency for a smaller diameter diaphragm. Silicon piezoresistive devices, of course, employ very small diaphragms. A typical diaphragm diameter is .060 in., with diaphragms as small as .010 in. in volume for commercial production.

Thus, the frequency response of these devices is outstanding. In fact, the dynamic performance of such devices has been extremely useful for wind tunnel and flight

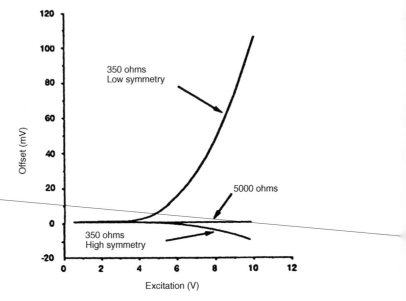

FIGURE 66.19 Offset change versus excitation.

test measurement of frequencies in the 20- to 100-kHz range and has resulted in the rapid development of modern military aircraft.

A plot of diaphragm resonant frequency versus pressure range for NovaSensor's NPH series sensors is seen in Fig. 66.20.

It can be seen from the graph that the resonant frequency of the diaphragm-sensing element is very high in all cases and increases with increasing pressure range.

The useful response time of a typical low-cost device is largely a function of its pneumatic interconnection. This is somewhat variable, depending on a particular customer's packaging configuration. The effective response is limited by the resonance of the pneumatic tubing; the response for average geometries in most applications is faster than 1 msec.

A calculation for a typical geometrical situation of 6 in. of 0.187-in. diameter tubing connected to an NPH-8 yields a tube resonance of 1400 kHz. Such a sensor can be expected to show a response time of less than 1 msec. Measurement of fluctuating pressure over 100 Hz should be possible without excessive error.

For special situations requiring a very high response, appropriate custom packaging is available.

SILICON CAPACITIVE PRESSURE SENSOR

Silicon piezoresistive pressure sensors have developed extensively through many generations since 1954. Capacitive pressure sensors, on the other hand, are relatively newer. Highly stable quartz-based pressure sensing elements were employed at several aerospace companies, including Bendix and Sperry. In the 60s and later, they were used to make highly stable air-data sensors. Silicon capacitive sensors were

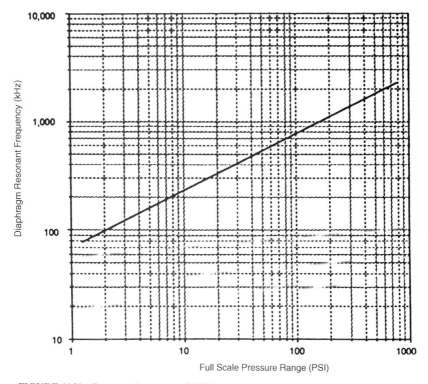

FIGURE 66.20 Resonant frequency of NPH pressure sensors.

developed at many institutions, especially at Stanford University, Case Western University, and Ford Motor Company. This device is represented generically in Fig. 66.21.

It can be seen that the structure is rather simple and includes no junction-isolated elements in the simplest implementation.

The capacitive approach has many potential advantages; however, it is far behind silicon piezoresistive devices in its commercial exploitation. This comparatively slow development of a viable technology is a phenomenon seen elsewhere in mainstream electronics. For instance, the development of gallium arsenide and indium antimonide and other compound semiconductors has proceeded slowly, even where

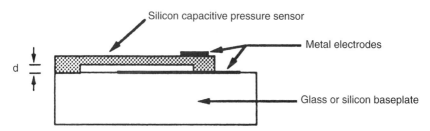

FIGURE 66.21 Silicon capacitive pressure sensor.

such devices had fundamental advantages in specific applications, because the knowledge base and leverage of silicon processing gives it an overwhelming advantage at any given point in time for a given application.

The comparison between piezoresistance and silicon capacitance does not give a clear fundamental advantage to either at the present time, but each approach does offer a number of substantial advantages.

Principle of Operation

The silicon capacitive device shown in Fig. 66.21 operates by the flexing of a micromachined silicon diaphragm under pressure. Flexing serves to vary the capacitance of the sensor as follows:

$$C = K_2$$

where C is the area of the capacitor plates and K is the separation between the plates.

It can be seen that the capacitance is inversely proportional to the separation between the plates. Since, to a first order, the deflection is linear with pressure, the capacitance varies inversely with applied pressure. Additional, although smaller, nonlinearity is caused by second order effects, including curvature effects and midplane stretching of deflecting diaphragm.

The first-order response of the device is highly nonlinear. On the other hand, the device is readily employed in circuit configurations, which eliminates this $1/P$ dependence.

Relative Advantages of Capacitance versus Piezoresistance

Silicon capacitive and piezoresistive sensors share to a large extent the same fabrication technology. Because of the long prior history of piezoresistive sensors, capacitive sensors are often contrasted with piezoresistive devices in order to compare relative performance advantages that may be exploited. The performance characteristics contrasted below are relevant to the application suitability of these devices.

Stability and Temperature Sensitivity. A piezoresistive device is a stress sensor, sensitive to stress in the plane of the diaphragm. A capacitive sensor is a deflection sensor. Such a sensor should have a reduced base sensitivity to stress induced by packaging. Commercial devices exhibit comparable performance.

To the first order, capacitive sensors are relatively insensitive to temperature. The major effect is a relatively small change in dimension with temperature and the change in the elastic modulus of silicon. Typically, temperature effects on sensitivity are on the order of <1%/100°F compared to −12%/100°F for silicon devices. Potentially, capacitive devices could be made to be more stable than piezoresistive devices, other factors being equal.

The current performance of commercially available devices however is comparable for TCV_o. Current data indicates no temperature or stability advantage of small micromachined capacitive versus piezoresistive sensors.

On the other hand, larger capacitive sensors manufactured by conventional technology have, for a number of years, achieved the required stability for air data applications, while diffused sensors have only recently begun to meet these requirements. This indicates, perhaps, a higher degree of difficulty in achieving comparable performance. On the other hand, the performance of the best commercially available air data sensors using both technologies is very comparable.

Manufacturability. Silicon piezoresistors have a definite edge in manufacturability and are much more readily manufactured than capacitive sensors for a number of reasons, especially the difficulty in providing electrical feed thrust for capacitive sensors from the inner capacitive chamber.

Also, since the capacitance is proportional to area, it is very difficult to design a relatively small capacitive sensor; thus, silicon piezoresistive devices can be made substantially smaller than capacitive sensors with comparable performance, achieving a tremendous advantage based on the leverage of batch processing by using larger numbers of die per wafer.

Media Compatibility. Capacitive pressure sensors for use in moist air are not suitable for use in a gage configuration, due to the change of the dielectric constant of air. Media compatibility can be obtained using the relatively expensive technique of pressure coupling through hydraulic oil. However, in such a case, difficulty is encountered with the change of dielectric constant of the oil with temperature. Furthermore, it is difficult to oil-fill extremely small silicon diaphragm structures.

Signal-to-Noise Ratio. The question of signal-to-noise ratio is a complex one, because to date no analysis has been performed based on fundamental energy considerations of silicon versus capacitive sensors. The noise level for both devices is a function of fabrication technology, particularly for silicon piezoresistive devices, and a function of circuit configuration for silicon capacitive devices, both of which are subject to improvement.

Also, in comparing the signal-to-noise ratio it is necessary to make certain assumptions regarding acceptable linearity and over pressure performance. With capacitive sensors there is a tradeoff between initial capacitance, full-scale capacitive change and over pressure. Thus, in determining signal-to-noise ratio, it is necessary to make a large number of arguable assumptions. Recent work, however, by researchers at Stanford[15] does indicate that silicon piezoresistive devices may have some advantage in signal-to-noise ratio.

Summary. It can be seen from the above that capacitive-based sensing has a number of potential advantages that may be exploited and some disadvantages. To date, the most successful commercial use of capacitive sensors has been the manufacture of manifold absolute pressure sensors produced annually in multimillions by Ford and Motorola. In addition, silicon micromachined capacitive pressure sensors have been successfully employed for process control. It seems likely that the future would see an increase in the use of capacitive micromachined sensors, with silicon piezoresistive sensors continuing to dominate the market.

SILICON ACCELEROMETERS

Both piezoresistive and capacitive sensing principals can be employed in a wide variety of devices. Any physical or mechanical parameter that can be converted into a deflection or a force can, in principle, be measured by such devices. An example is acceleration, which can be readily converted. Generically, the structure of such an acceleration sensor is shown in Fig. 66.22.

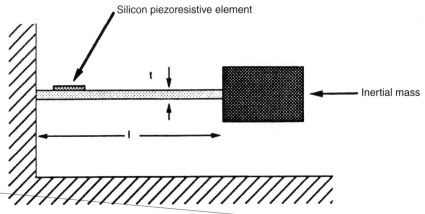

FIGURE 66.22 Basic silicon acceleration sensor.

This structure is classic, with a seismic mass bending a cantilever and producing fiber stresses near its fixed end. The challenge in getting maximum performance out of an accelerometer is to maximize the product of the two parameters of primary interest in an accelerometer: sensitivity and resonant frequency.

The sensitivity is a function of the stress of a given acceleration level, and it is given by

$$s = k \cdot g \cdot l/t^2$$

where g is the applied acceleration, t is the thickness, and l is the length.

On the other hand, frequency is given by

$$f_n = \sqrt{(E/M) \cdot (l/t^3)}$$

The silicon technologist can exploit these equations to produce a very superior device. It can be seen from the above that a material of high stiffness, such as silicon, is very desirable.

In order to maximize the performance of the device, the designer seeks to maximize the stress in the gage while minimizing the deflection, thus maximizing the product of the resonant frequency and the sensitivity. Since the stress is measured only at the support end of the beam in the area of the gages, the techniques of micromachining may be employed to stiffen the rest of the cantilever while leaving the cross section small in the gage area. Thus, the deflection can be reduced without decreasing the sensitivity. It can be seen that, as in the case of the pressure sensor, a small structure is highly desirable, other factors being equal. This is because the sensitivity is a function of l/t^2, while the resonant frequency is a function of the square root of l/t^3. Thus, if the design stress is fixed, and t is reduced to keep the stress constant as l is decreased, the product of the sensitivity and the resonant frequency is increased. Small silicon accelerometers thus enjoy a fundamental advantage in figure of merit.

Moreover, the high sensitivity exhibited by semiconductor piezoresistive sensors allows the design stress levels to be kept low while maintaining an excellent signal-to-noise ratio. Low stress levels mean low deflection and thus high resonant frequency.

Silicon has a high elastic modulus and a low density. Thus, the noninertial mass part of the structure does not substantially decrease the resonant frequency.

Silicon is an essentially perfectly elastic material. Thus, accelerometers of very high Qs can be manufactured. Qs for silicon microstructures have been observed as high as 200,000.

Given the above advantages, one might question why silicon microaccelerometers have not taken over the market since their early introduction in the late 1960s by Kulite, Endevco, and others. A number of reasons exist for this slow development, including the relatively small market size for accelerometers in the past as compared to pressure sensors. Naturally, micromachining technologists have concentrated on the more lucrative pressure-sensing market. It should be noted, however, that silicon microaccelerometers have, even so, until recently been taking market share consistently from the traditional quartz devices, with silicon piezoresistive accelerometers accounting today for perhaps 30 percent of the market, with a growing market share.

Two fundamental factors, however, have limited the development of these devices. The first factor is related to the relatively small deflection of such tiny structures, which makes it difficult to provide effective overrange stops. A second factor, also associated with the small deflection, is that to achieve maximum performance it is highly desirable to damp accelerometers. Damping is very difficult to achieve with small deflection. These difficulties have been overcome with the development of the new generation of micromachined silicon accelerometers just coming to market. A subsequent section will discuss an advanced, revolutionary silicon microaccelerometer.

SUMMARY

This section has sought to point out some of the fundamental considerations involved in the use and design of silicon sensors. The development of such devices follows a path that is governed by the interaction of fundamental performance factors, the direction of university research, the availability of trained technologists and manufacturing processes, and the commercial requirements for such devices. There has been a continuing interest in the development and exploitation of these devices over a number of decades. This interest is accelerating because of the fundamental requirement to provide sensing of mechanical and environmental parameters to the low-cost CPUs of modern electronic systems.

It can be expected that this trend will continue. The full capability of silicon mechanical sensors has, to date, not pushed the basic physical limits of the technology.

REFERENCES

1. C. S. Smith, "Piezoresistance Effect in Germanium and Silicon," *Physical Review,* **94**, November 1954, pp. 42–49.
2. T. A. Nunn, "Silicon Absolute Pressure Transducer for Biomedical Applications," Stanford Technical Report 461 0-1.
3. J. E. Gragg, U.S. Patent 4,317,126, April 1980.
4. J. R. Mallon, Jr., F. Pourahmad, K. Petersen, T. Vermeulen, and J. Bryzek, "Low Pressure Sensors Employing Etch Stopping," *Sensors and Actuators,* A21–23, 1990, pp. 89–95.
5. C. K. Stedman, U.S. Patent 3,341,794, September 1967.

6. H. S. Pien, U.S. Patent 3,341,794.
7. L. B. Wilner, "A Diffused Silicon Pressure Transducer with Stress Concentration at Transverse Gauges," presented at 23rd International Instrumentation Symposium, Las Vegas, Nevada, May 1977.
8. A. D. Kurtz, J. R. Mallon, Jr., T. A. Nunn, U.S. Patent 4,236,137, March 1979.
9. M. Nishihara, H. Yamada, and I. Matsuoka, "Recent Semiconductor Pressure Sensors," *Hitachi Review,* June 30, 1985.
10. J. Binder, K. Becker, and G. Ehrler, "N Silicon Pressure Sensors for the Range2 k Pato4OMPa," *Siemens Components,* **xx**(2), 1985, pp. 64–67.
11. O. N. Tufte, and E. L. Stelzer, "Piezoresistive Properties of Silicon Diffused Layers," *J.A.P.,* **34**(2), pp. 313, February 1963.
12. A. D. Kurtz, "Adjusting Crystal Characteristics in Semiconductor and Conventional Strain Gauges," Academic Press, New York, pp. 259–272.
13. J. R. Mallon, Jr., and D. C. Germanton, "Advances in High Temperature Ultra-Miniature Solid State Transducers," presented at the ISA Silicon Jubilee Conference, Philadelphia, PA, October 1970.
14. K. E. Petersen, J. Brown, P. W. Barth, J. R. Mallon, Jr., and J. Bryzek, "Multrastable High Temperature Sensors Using Silicon Fusion Bonding," *Sensors and Actuators,* A21–23, 1990, pp. 96–101.
15. R. R. Spencer, B. M. Fleischer, P. W. Barth, and J. B. Angell, "A Theoretical Study of Transducer Noise Piezoresistive an Capacitive Silicon Pressure Sensors," *IEEE Transactions on Electron Devices,* **35**(8), August 1988.
16. P. W. Barth, F. Pourahmadi, R. Mayer, J. Poydock, and K. E. Petersen, "A Monolithic Silicon Accelerometer with Integral Air Sampling and Overrange Protection."

CHAPTER 67
ADVANCED SENSOR DESIGNS

INTRODUCTION

Silicon pressure sensors similar to those described thus far have been in production for nearly two decades. With the help of the latest techniques in silicon micromachining as well as those of established integrated circuit manufacturing methods, remarkable improvements in performance, sophistication, functionality, size, and cost can be delivered. This chapter describes in detail several different chips manufactured with advanced, well-established techniques. These chips represent the highest level of sensor performance and technology presently available anywhere in the world.

The first chip described demonstrates the high degree of integration which is possible with state-of-the-art sensors. This device is a fully temperature-compensated and calibrated pressure transducer containing on-chip low-TCR resistors. The resistors are laser-trimmed on each device to provide chip-to-chip interchangeability. Such complex, highly integrated silicon transducers typify the future of silicon sensors. The primary driving force behind this trend is to simplify implementation by the customer and facilitate sensor incorporation into products.

Very small pressure sensors for consumer applications are described as a simple example of how this process allows a further degree of miniaturization. A low-pressure chip made with this process is one of the world's smallest commercially available pressure transducers, used as a medical, catheter-tip sensor. A high-pressure version of this process exhibits exceptionally low-temperature coefficients. High overpressure tolerance has also been demonstrated in a device fabricated using SFB. A capacitive pressure sensor is also described. The techniques used to fabricate all of these sensors provide operational improvements not otherwise possible.

A combination of unique processes were integrated to accelerate the production of silicon sensors. High sensitivity, integral overrange protection, independent control of damping, and manufacturing compatible with pressure sensors are a few of the advantages of this device. Advanced micromachining is also used to produce transducers for use at very low pressures (<1 psi).

FULLY ON-CHIP COMPENSATED, CALIBRATED PRESSURE SENSORS

Introduction

Disposable blood pressure sensors were introduced to the market in 1982. The only technology capable of the required price/performance ratio for this application ($40/1% accuracy) was based on silicon sensor chips. The first of these products used hybrid technology, in which sensing was performed by a silicon monolithic chip with calibration and compensation performed by a thick-film laser-trimmed circuit. The market for this type of product increased from approximately 40,000 units delivered in 1982 to about 6 million units in 1988. This dramatic volume increase, combined with a simultaneous sensor price drop to below $3 in large volume, stimulated a transition to new technology with a higher level of vertical integration on the chip, including laser-trimmable thin-film resistors.[1]

The challenge to chip manufacturers has been to achieve designs with planar or near-planar circuit topology, rapid trim capability, good stability after trimming, high yield at the die level as well as the package level, and low cost. The chip introduced here meets those goals and is the smallest and most advanced to date for this market (Fig. 67.1).

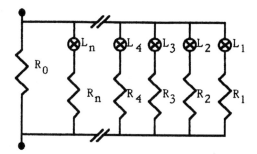

FIGURE 67.1 Illustration of quasi-analog digital trimming showing an array of parallel resistors and laser trimming links.

Integration of additional circuitry on a chip necessitated efforts geared toward:

- Developing and combining processes for piezoresistors and temperature-stable laser-trimmable thin-film resistor networks without degradation of the characteristics of either
- Optimizing the chip design and fabrication process to improve parameter distributions (die to die, wafer to wafer, and lot to lot), thus simplifying the trimming network
- Minimizing chip size to minimize fabrication cost
- Designing the logistics of pressure/temperature testing and laser trimming for high throughput and low cost
- Designing capabilities to trim chips in either wafer or package form

ADVANCED SENSOR DESIGNS

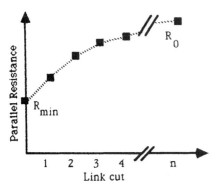

FIGURE 67.2 Monotonic change in resistance obtained by trimming the array in Fig. 67.1.

The packaged, trimmed product must deliver a set of well-controlled parameters. The initial guidelines were provided by AAMI (Association for the Advancements of Medical Instrumentation). Marketing issues have pushed the performance even further, necessitating trimming for initial offset, offset temperature coefficient, sensitivity and its temperature coefficient, output symmetry and impedance, input impedance, and shunt calibration. Pressure nonlinearity, offset, and sensitivity shift with excitation voltage (over a 1–12-V range), and overpressure capability had to be provided by design (Fig. 67.2).

Sensor Design

A 900-μm square diaphragm was selected with a fully active Wheatstone bridge configuration of strain gages. This small diaphragm size requires a thickness of 12 μm to deliver the necessary pressure sensitivity. To provide effective stress transfer, a supershallow (1 μm) piezoresistor process was developed using ion implantation.

The doping level for the resistors was chosen with the aid of modelling to optimize temperature coefficients and pressure sensitivity. A controllable etch stop technique was then selected to obtain a tight sensitivity distribution and predictable, repeatable overpressure performance across the wafer. The chip size of approximately 2.6 mm square was selected as a compromise between real-estate cost and isolation from package stress.

A new thin-film resistor process with low temperature coefficient was developed based on sputtered, passivated Si-Cr. A TCR of 20 ppm/°C was achieved, exceeding the original target of 100 ppm/°C.

The resistor network implementation is based on quasianalog digital trimming techniques. These techniques allow rapid sequential cutting of links while network resistance is monitored by a comparator until a threshold resistance value is reached.

The fastest available laser trimmers were developed for trimming and repairs of semiconductor memory chips through link blowing technology. In that application, metal links are evaporated by a pulse of laser energy, which changes a short circuit to an open circuit. The fastest lasers can achieve link-blowing speeds as short as 10 ms. These link-blowing techniques are readily adaptable to sensor technology.

The concept of the resistor design is shown in Figs. 67.1 and 67.2. Resistor R is shunted by N parallel resistors. Initial resistance R_{min} is equal to the parallel combination of all resistors. The laser begins blowing links at link L_1 introducing step changes in the value of effective resistance R, as shown in Fig. 67.2. When all links are open, the maximum value of the network is equal to R. Throughout the trimming range R_{min} to R_0, the resistance increases monotonically with each consecutive link blown. This type of trimming process is equivalent to analog laser trimming in that it allows monotonic trimming to a required value for parameters such as offset, sensitivity, and so on.

To use quasianalog digital trimming over a broad range of resistance change without an excessive number of links, several parallel resistor networks are placed in series, each of them delivering a different trimming range. The layout of this complex chip is shown in Fig. 67.3.

FIGURE 67.3 Layout of a compensated, calibrated pressure chip.

Sensor Characteristics

The combination of laser-trimmed calibration and advanced piezoresistor fabrication techniques has resulted in a chip with outstanding performance characteristics, including:

Full-scale pressure	300 mmHg
Trimmed sensitivity	5 tL V/V/mmHg + 1%
Nonlinearity	<0.2% (terminal-based)
Offset shift	<1 mmHg for excitation voltage shift of 1–12 V
Backside burst pressure	>10,000 mmHg
Stability	<0.1 mmHg/8 hours typical

Packaging

Because laser trimming of the chip can be accomplished after packaging, the effects of package stresses can be compensated for. The sensor has been successfully mounted, tested, and trimmed on a variety of substrate materials including alumina, kovar, and various thermoplastics. Performance on all of the above materials has been well within AAMI guidelines. Mounting on almost any substrate material appears practical as long as the material has reasonable dimensional stability in the temperature range expected during packaging and use. The stable performance and good untrimmed temperature characteristics of the sensor allow efficient high-speed testing in the package before and after laser trimming.

Conclusion

The design of the chip and the on-board thin-film resistor network are the result of extensive thermal, mechanical, and electrical modelling (both analytical and finite element) to assure an optimum price/performance ratio. In addition, exceptional backside over pressure capability has been achieved without sacrificing intrinsic sensitivity or linearity.

The technology used for this sensor design is suitable for a wide variety of additional applications in automotive, HVAC, consumer, and industrial markets. The device creates a new level of price/performance for blood pressure sensing and has the potential for similar advantages in other applications.

PRESSURE SENSORS USING SI/SI BONDING

Introduction

The versatility of micromachining technology is often severely constrained because of the structural limitations imposed by (1) the traditional backside etching process and (2) the fact that few practical methods have yet been demonstrated for true silicon/silicon hermetic bonding. The techniques that have been employed for sensors (such as thermomigration of aluminum, eutectic bonding, anodic bonding to thin films of pyrex, intermediate glass frits and "glues") suffer from thermal expansion mismatches, fatigue of the bonding layer, complex and difficult assembly methods, unreliable bonds, and/or expensive processes. In addition, none of these processes provide the performance and versatility required for advanced micromechanical applications. For example, the (backside) cavity side walls slope outward from the diaphragm in anisotropically etched pressure sensor chips, forcing the overall chip size to be substantially larger than the active sensing diaphragm. Another problem area is that narrow, hermetic gaps between two single crystal wafers are difficult to fabricate with precision because of the intermediate layers required. Finally, the most exciting class of recently developed micromechanical structures are actually thin-film moveable structures that are incompatible with current silicon/silicon bonding techniques. Workers have addressed these problem by fabricating beams, diaphragms, and (recently) more complex structures such as mechanical springs and levers from polycrystalline silicon using sacrificial-layer etching.

The silicon/silicon bonding technique described here has been previously discussed in several papers.[2] SFB is becoming a very powerful process for creating a wide range of all single-crystal mechanical structures, which can replace polysilicon in many of the currently used sacrificial-etching methods. Several sensor designs are presented below, illustrating the practical implications of this technology.

1 × 1-mm Pressure Sensor for High-Volume Applications

A simple and inexpensive pressure sensor device that measures only 1 mm square has been fabricated using SFB. Standard pressure sensors have diaphragm cavities with side walls sloping outward, leading to large size. When SFB is used, a shallow cavity can be etched in a substrate wafer, and then a second wafer can be bonded to the first. The second wafer is thinned to become the sensor diaphragm. The comparison is illustrated in Fig. 67.4. The diaphragm for this chip measures 400 microns square. A 50-psi version with a 16-micron diaphragm yields an output of 20 m V/V full-scale, with linearity better than 0.1 percent. This chip has found applications in low-cost consumer markets such as tire-pressure sensing. With over 6000 die on a 100-mm wafer, such a process has enormous cost advantages. Other pressure ranges are being developed for a multitude of applications where small size and low cost are important. An electron micrograph view of the chip is shown in Fig. 67.5.

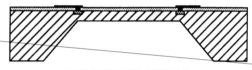

CONVENTIONAL SENSOR

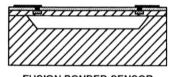

FUSION BONDED SENSOR

FIGURE 67.4 Comparison of conventional and fusion-bonded sensors showing the size advantage that fusion-bonded sensors enjoy.

Catheter Tip Pressure Sensors

One important application of silicon/silicon-bonded sensors is ultra miniature, catheter-tip transducer chips for *in vivo* pressure measurements. A typical chip using standard processing for this vital product is shown in design 1 in Fig. 67.6. This device has dimensions of 2400 μm by 1200 μm by 175 μm thick. It has a 600-μm square active diaphragm 8.0 μm thick. Diaphragm thickness is controlled to within 0.5 μm by an electrochemical etch-stop process. The silicon chip is anodically bonded to a glass constraint wafer that is only 125 μm thick. Much of the processing of this product requires lithography, etching, bonding, and general handling of these extremely thin wafers. A piezoresistive half-bridge is located on two edges of the diaphragm for a pressure sensitivity of 18 μV/V/mmHg. Linearity of these chips is about 1.2 percent of the full-scale output.

A new generation of silicon fusion-bonded (SFB) chips makes it possible to fabricate much smaller chips with equivalent or better overall performance. Schematic drawings of these chips are shown in Fig. 67.6, and the fabrication procedure is outlined in Fig. 67.7. The bottom, constraint substrate is first anisotropically etched with a square hole that has the desired dimensions of the diaphragm. In the most recent generation of chips, the bottom wafer has a (standard) thickness of 525 μm, and the diaphragm is 250 μm square, so the anisotropically etched hole vees out at a depth of about 175 μm.

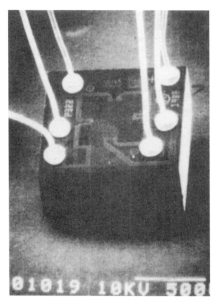

FIGURE 67.5 SEM micrograph of a 1-mm × 1-mm pressure sensor die suspended by gold wire bonds.

At the same time that the pattern for these holes is defined, an alignment pattern is produced on the backside of the wafer in a double-sided aligner. Next, the etched constraint wafer is SFB bonded to a top wafer consisting of a p-type substrate with an n-type EPI layer. The thickness of the EPI layer corresponds to the required thickness of the final diaphragm for the sensor. After the substrate from the second wafer is removed by an electrochemical etching process, resistors are ion-implanted, contact vias are etched, and metal is deposited and etched. These patterns are aligned to the buried cavity by using a double-sided mask aligner, referenced to the marks previously patterned on the backside of the wafer at the same time as the cavity. All these operations have a high yield because they are performed on wafers that are standard thickness. In the final step, the constraint wafer is ground and polished back to the desired thickness of the device, about 140

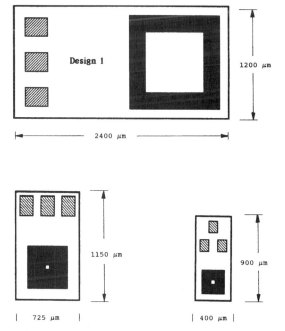

FIGURE 67.6 Three generations of catheter-tip transducers. Design 1 is a "conventional" chip, while the other two use the SFB process.

μm. The bottom tip of the anisotropically etched cavity is truncated during this polishing operation, thereby exposing the backside of the diaphragm for a gage pressure measurement. In an optional configuration, the initial cavity need not be etched to completion. Figure 67.8 shows an SEM of a device with a sealed reference cavity only 30 μm deep. This design can be directly compared to the polysilicon sacrificial-layer pressure sensor techniques currently under investigation.

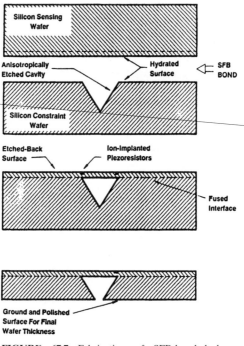

FIGURE 67.7 Fabrication of SFB-bonded low-pressure sensors for catheter-tip applications.

Despite the fact that all dimensions of this chip are about half that of the conventional chip described above, the pressure sensitivity of the SFB chip is identical to the larger device, and its linearity is actually improved, 0.5 percent of full scale. In addition, the implementation of SFB technology increases wafer yield because the wafers are not as thin (fragile) during most wafer processing steps.

A comparison of conventional and SFB technology (as applied to ultra-miniature catheter-tip transducers) is shown in Fig. 67.9. For the same diaphragm dimensions and the same overall thickness of the chip, the SFB device is almost 50 percent smaller. In this special application, chip size is critical. SFB fabrication techniques make it possible to realize extremely small chip dimensions that permit this catheter-tip pressure sensor to be used with less risk to the patient in many medical procedures.

High-Pressure Sensors

High-pressure absolute sensors have been constructed in a modified SFB process to create much thicker diaphragms and to provide a larger mass of silicon in the con-

FIGURE 67.8 Typical electron micrograph of a single-crystal silicon diaphragm bonded to a single-crysal silicon substrate.

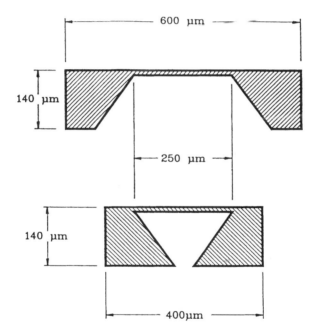

FIGURE 67.9 Comparison of conventional and SFB process. The SFB process results in a chip that is at least 50 percent smaller than the conventional chip.

straint wafer, thereby optimizing material compatibility and minimizing thermal mismatch problems.

First, the oxidized constraint wafer is aligned and exposed on both sides. The bottom surface has a marker pattern that will be used as an alignment reference later in the process in a manner similar to the low-pressure sensor. The top surface has a round cavity pattern that will correspond to the diameter of the sensing diaphragm.

After the oxide is wet-etched, the silicon is plasma-etched in CF_4 or anisotropically etched in KOH or EDP on both sides of the wafer simultaneously to a depth of about 10 μm. Next, all photoresist and oxide layers are stripped from the constraint wafer, and the top surface of this wafer (the surface with the round depression) is bonded to another *n*-type wafer at 1100°C. This operation is illustrated in the top part of Fig. 67.10. After bonding, the top wafer is mechanically ground and polished back to a thickness corresponding to the desired pressure range. For a 900-μm *dia*phragm with a thickness of 200 μ*m*, this chip will result in a pressure sensor with a full-scale output of 130 mV at 4000 psi. Linearity is exceptional at about 0.2 percent of full scale.

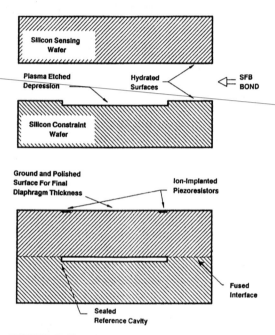

FIGURE 67.10 Fabrication of SFB-bonded high-pressure sensor suitable for pressures from 1,000 to 10,000 psi.

Finally, an insulator is deposited, piezoresistors are implanted and annealed, contact holes are opened, and metal is deposited and etched. This completes the process for the pressure sensor. No backside etching and no anodic bonding are required. In addition, 100 percent of the bulk of the chip is single crystal silicon. Creating a monolithic single-crystal silicon sensor chip provides important performance advantages in temperature coefficient of offset. The SFB chip, with a total silicon thickness of over 600 μm, exhibits a very low temperature coefficient of bridge offset of about 0.3 percent/100°C of full-scale output. A similar chip made with conventional backside etching and anodic glass/silicon bonding for the constraint, exhibits a typical TC of zero higher by a factor of 3.

Low-Pressure Sensor with High Overpressure Protection

Some applications require a sensor that can tolerate extremely high levels of overpressure on the order of 100 times full-scale or more. Conventional pressure sensors will not survive such extremes. By utilizing SFB, a pressure sensor has been fabricated that survives over 500 times full-scale pressure without failure.[3]

The pressure sensor is fabricated in a way similar to that described for the high-pressure sensors. The diaphragm is thinner, larger, and circular. The cavity that is etched into the substrate wafer is very shallow and well controlled. This shallow cavity acts to stop diaphragm motion when the device is subjected to high overpressures. At low pressures, the response is just like other fusion-bonded pressure sensors. A typical low-pressure output characteristic is shown in Fig. 67.11. At about 10 psi, the diaphragm contacts the stop and output saturates. The high-pressure characteristic is shown in Fig. 67.12. Output is still evident at 5000 psi. The diaphragm survives multiple exposures to such pressures.

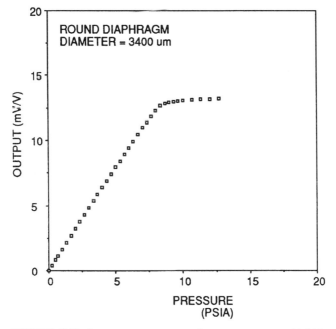

FIGURE 67.11 Low-pressure response of a pressure sensor with high overpressure protection.

High-Temperature Pressure Sensors

The chip operates over an extremely large temperature range (–40°C–250°C) and has been produced in low (0–15 psig), mid (0–300 psia), and high (0–5000 psia) pressure ranges. Compatible with piezoresistive silicon pressure sensor technology, the high-temperature device is constructed from single-crystal silicon pressure-sensing elements isolated from a single-crystal silicon substrate by a dielectric layer. This structure combines the performance of dielectrically isolated single-crystal silicon technology with the cost-effective manufacturing of anisotropically etched silicon membranes.

The high-temperature sensor has evolved from commercialized and widely accepted silicon pressure sensor technology. The chip size is small, only 2.5 mm by 2.5

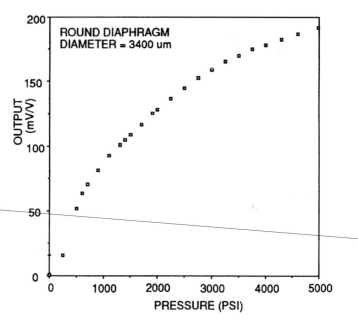

FIGURE 67.12 High-pressure response of a pressure sensor with high overpressure protection.

mm. The sensing elements are in a standard open-bridge configuration that allows for temperature and output compensation. They are located on a thin flexible silicon membrane, balanced in a square layout for enhanced offset and linearity performance.

In order to accommodate the extremely large temperature range, a process has been developed to isolate the silicon sensing element from the underlying substrate (see Fig. 67.13).

This SOI (silicon on insulator) process has been made possible by silicon fusion bonding, whereby a thermally oxidized silicon wafer is bonded to a single-crystal silicon wafer without any intermediate adhesive, as described in the preceding sections. After controlled thinning of one wafer, the resistor sensing elements are then formed on the thinned silicon layer, and the flexible membrane (which varies in thickness depending on the pressure range) is formed on the opposing side.

The sensor has excellent repeatability and long-term stability. Repeatability/hysteresis is less than 0.05 percent FSO over the whole range. Long-term stability is less than 0.1 percent over 6 months. Typical over pressure tolerance exceeds 300 percent of full scale. Uncompensated nonlinearity is less than 0.1 percent of FSO.

Capacitive Absolute Pressure Sensors

Some applications require a sensor whose capacitance, rather than resistance, varies with pressure. Silicon fusion bonding has also been used to fabricate such a device. In this case, an oxide interlayer is grown on one wafer before bonding. The thickness of this oxide serves to define the gap of the capacitor. High-temperature metals or the silicon itself can be used as the plates of the capacitor. A schematic view of such a sensor is shown in Fig. 67.14.

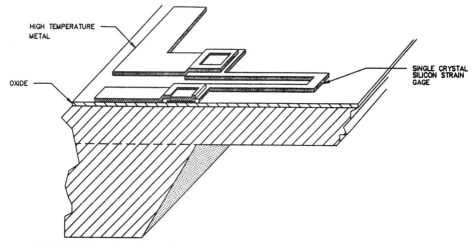

FIGURE 67.13 Schematic drawing of a dielectrically isolated structure used in a high-temperature pressure sensor.

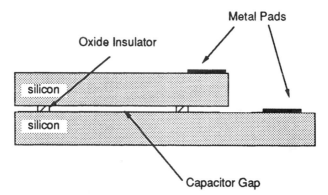

FIGURE 67.14 Schematic illustration of a fusion-bonded capacitive pressure sensor.

Silicon Accelerometers

Several market forces have pushed the development of low-cost batch-fabricated accelerometers over the past few years. Primary among these are automotive needs for crash sensors (for air bag deployment) and ride motion sensors (for active suspension components). Additional markets include military components (e.g., smart weapons) and aviation (e.g., rate-of-climb indicators). Accelerometer development efforts have pushed for medium performance (less than required for inertial navigation or gravimeters), reproducible characteristics, and low cost.

The primary obstacle to the development of such sensors has been the fragility of the fabricated devices: adequate sensitivity for low accelerations (0.5–1 G) has historically resulted in easy breakage of the sensor during and after fabrication. This breakage problem drives yield down and price up. The problem has been avoided in the accelerometer reported here by the introduction of mechanical overrange stops, batch-fabricated at the wafer level, into the silicon chip structure.

Piezoresistive accelerometers generally place strain-sensing resistors at points of maximum stress on a bending beam attached to a *seismic mass,* or a mass that moves in relation to a supporting substrate under the influence of acceleration. While many beam arrangements are possible, most designs have used either a cantilever,[5] in which the seismic mass is suspended along one edge, or a doubly fixed beam (DFB), in which the mass is suspended from points along two opposing edges.[6]

For a given seismic mass, and for a given width, length and thickness of the suspending beam, a cantilever displays at least eight times the sensitivity (beam stress per G) as that of a design with a doubly fixed beam. This advantage offers the potential for low chip cost, because a cantilever design of a given sensitivity can be fabricated in a smaller area.

However, basic cantilever designs are also less resistant to shock than DFB designs. Under overload accelerations, a cantilever design continues to bend in the same way that it bends under normal accelerations. In this bending mode, stress is concentrated along one surface of the cantilever beam, and eventually the beam breaks (at a surface tensile stress of approximately 150,000 psi). In contrast, a doubly clamped design under overload can go into linear tension as well as bending stress, distributing the overload through the thickness of the beam, and so can support a greater overload than a cantilever design of equal sensitivity.

Cantilever designs thus require some additional overrange protection features if they are to compete with DFB designs and retain advantages in terms of small size. While such protection has been added during packaging in some previous designs, no previous accelerometer has been able to incorporate overrange stops directly into the sensor at the wafer level before fabrication of the cantilever beams. The chip reported here contains such protection as illustrated in Fig. 67.15.

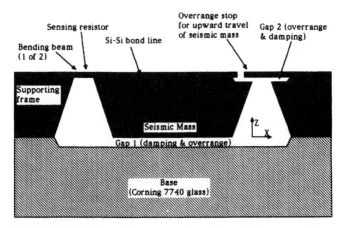

FIGURE 67.15 Schematic cross section of a dual-beam cantilever accelerometer.

The novel overrange protection in this chip is provided by a series of interdigitated tabs that extend over underlying shelves. A tab on the frame next to the seismic mass and extending over a shelf on the seismic mass prevents the mass from moving upward past a preset limit. Similarly, a tab on the seismic mass extending over a shelf on the supporting frame prevents the mass from moving downward past a preset limit. The tabs themselves are thin enough to be somewhat flexible so that they don't shatter when they hit the stopping shelf.

Overrange protection exists in all three axes. In the z-axis (perpendicular to the chip surface), protection is provided by the interlocking tabs described above.

In the x-axis (in the plane of the chip and parallel to the length of the beams), protection is provided by the strength of the beams themselves in linear tension and compression. The cross-sectional area of such beams is greater than it would be for a DFB design of equal sensitivity, so that overrange protection in the x-direction is greater.

In the y-axis (in the plane of the chip and perpendicular to the length of the beams), protection is provided by the great width of the cantilever beams in relation to their thickness, plus the wide separation of the beams, both of which provide great strength and stiffness in this direction (again, greater than for a DFB design).

All accelerometer designs force compromises among sensitivity, resonant frequency, and damping of undesirable resonances. Increased sensitivity generally results in lowered resonant frequency, and the lowest resonance limits the available measurement bandwidth. The sensitivity-bandwidth tradeoff is basic to any design (including the present design) in which the sensing element is placed directly on the bending beam and can be avoided to some extent only by structures that remove the sensing element from the plane of the bending beam. The sensor reported here uses a fully active Wheatstone Bridge to achieve the highest resonant frequency and highest sensitivity simultaneously possible.

Undamped accelerometers suffer serious measurement problems from oscillations occurring at resonance. (Undamped buildup of these oscillations can also lead to destruction of the device if no overrange protection is present.) To avoid measurement errors, the oscillation must be damped to some extent; typically, critical damping is preferred, which avoids oscillation while preserving signal amplitude to a frequency near the cutoff (−3 dB) frequency, and results in a well characterized phase lag at frequencies near cutoff.

Both liquid damping and gas damping are possible. Liquid damping suffers from several problems: reduction of sensitivity due to density of the liquid (which tends to buoy the seismic mass), viscosity variation with temperature, and costs associated with both the liquid material and the liquid filling procedure. Gas damping, in contrast, has minimal problems related to density and viscosity, and if the gas is air at atmospheric pressure, it costs nothing. Air damping is thus the method of choice if it can be efficiently designed into the sensor structure.

The present sensor design achieves air damping by using controlled gaps in the chip structure (Fig. 67.15). Beneath the seismic mass, gap 1 between glass and silicon constitutes one air-damping area. Gap 2 between the overrange protection elements constitutes another such area. Gaps 1 and 2 can be set independently, as can the areas bounded by those gaps. The result is great freedom in designing the air-damping characteristics for the chip. For low-G-range devices, the gaps can be fairly wide and can occupy minimal areas, while for high-G-range chips the gaps can be narrower and can occupy larger areas. Of course, there are also engineering tradeoffs involved between the overrange protection and the damping provided by the gap regions.

The design process for the cantilever accelerometer was set up to ensure maximum success of the first fabricated silicon and minimum effort devoted to redesign. Initial analytical models were developed to rough out chip dimensions for the desired sensitivity ranges. Next, finite element models were developed and compared to the analytical models for simple cases. Then the finite element models

(FEM) were used to predict quantities such as torsional resonances and cross-axis sensitivity, which are difficult or impossible to model analytically. Finally, the FEM results were used as inputs to the mask design and the fabrication process design.

The ANSYS program was used for the FEM effort. Running on a MicroVAXII computer, this program predicted stress patterns on the cantilever beams (aiding in resistor placement), sensitivity (stress versus acceleration), first resonant frequency, higher-order resonances (including off-axis resonances and torsional resonances), and overrange protection. Typical output plots from the analysis are shown in Chap. 69. Figure 69.9 displays the fundamental bending mode of the chip at a resonant frequency of 839 Hz. Figure 69.11 displays the second resonance of the structure, a torsional mode at 67,345 Hz. The modelling makes it obvious that only the fundamental mode falls anywhere near the measurement bandwidth; the higher-order modes present no perceptible problems.

The fabrication process is designed for high yield and volume production at low cost. It incorporates standard techniques of silicon fusion bonding, plasma etching, and programmed etch-stopping. Batch processing techniques are used throughout, up to the point where devices are placed into individual packages.

The use of silicon fusion bonding permits great freedom in the fabrication process. For example, the cross section of Fig. 67.15 illustrates a cavity (gap 2) beneath the overrange stop tab; this cavity can be created by etching before two silicon wafers are laminated together.

The two-layer silicon structure shown in Fig. 67.15 avoids placing any capping layer over the circuit side of the silicon chip as has been used in some previous designs. Because the top side of the chip is completely exposed, there are no constraints on the placement of metal lines, trim networks, and bonding pads on the chip surface. This freedom in the placement of interconnections permits ease in creating design variations for specific applications and also permits laser trimming of the chip as part of the packaging and calibration process after fabrication is complete.

The finished chip is illustrated in the scanning electron micrograph of Fig. 67.16. The central mass is surrounded by the supporting frame, and the two cantilever beams are evident near the far edge of the chip. The chip also contains a laser-trimmable resistor network for setting the offset voltage, plus extra resistors for temperature compensation of the output characteristic.

Chip size (3.4 mm^2) and operational characteristics correspond closely to those of silicon piezoresistive pressure sensors, permitting packaging techniques, excitation voltages, calibration procedures, and temperature compensation methods to be compatible with those used for high-volume pressure sensor production. Devices have been packaged in T05 and T08 cans and on hybrid circuit boards.

Typical characteristics for one chip configuration are as follows:

Bridge resistance	2000 ohm
Sensitivity	10 mV/G at 5V excitation
−3 dB frequency	500 Hz
Nonlinearity	<1 percent at 2 Gs
Cross-axis sensitivity	<1 percent (any axis)
Overrange protection	>1000 G (any axis)

The cantilever accelerometer introduced here avoids many of the cost and performance limitations of prior designs. The basic cantilever structure provides high sensitivity for a given chip size. Overrange protection in all three axes is available as a basic property of the chip structure. Air damping is controllable by the use of adjustable gaps.

FIGURE 67.16 Scanning electron micrograph of a completed silicon accelerometer chip.

These advantages translate to small size, low cost, and good performance. The low cost constitutes an applications breakthrough that will permit penetration into new areas including automotive air bag actuation, automotive active suspensions, motion detection in industrial process controls, military and aerospace components, and consumer and medical applications. The small size ensures that the device will be useful in many applications where space constraints are important.

Conclusion

The successful development of these products shows that SFB bonding can be applied on a commercial basis. Not only are the processes employed here simpler than those currently used to build conventional pressure transducers for the same applications, but the yields are higher, the costs are lower, and the chip performance is improved compared to conventional technologies.

Beyond the devices shown here, however, the potential of SFB bonding in other micromechanical structures is enormous. This process totally eliminates all the disadvantages of previous silicon/silicon-bonding methods.

- It is hermetic.
- It does not require any intermediate bonding layers.
- It accurately preserves any pattern previously etched in either or both bonded wafers.
- It has a high yield strength, as much as double that of anodic bonds.
- It can be used to create vacuum reference cavities.
- It can be used in place of sacrificial layer technology in many applications.
- It eliminates thermal expansion and Young's modulus mismatches in bonded wafer structures.

The bond itself is a true single-crystal/single-crystal interface. We have demonstrated, for example, that pn junctions with low leakage and sharp breakdown voltages can be formed by bonding a wafer with a p-type diffusion to an n-type wafer.

During the next few years, silicon fusion bonding will revolutionize the field of silicon microstructures and will have a vast impact on high-performance silicon microsensors.

VERY LOW PRESSURE SENSOR

An example of silicon micromachining at its best is NovaSensor's low-pressure sensor chip.[7] It is difficult to obtain high performance from conventional pressure sensor designs, as the diaphragms become very thin. As the rated pressure of a conventional sensor chip is decreased to 1 psi full-scale, the diaphragm must be made extremely thin (on the order of 10 µm) in order to get a full-scale output of 50 or 100 mV. Such thin diaphragms are extremely flexible. Maximum deflections of 10 µm or more occur at full-scale pressures of 1 psi. When diaphragm deflections become comparable to diaphragm thickness, however, membrane self-stiffening effects become important, and the sensor output becomes nonlinear. The combination of low pressure and high output are not compatible with high linearity in traditional silicon pressure sensor designs.

This dilemma has been resolved by developing advanced micromachining techniques to form two thick, stress-concentrating elements on the backside of an extremely thin, rectangular diaphragm. These elements are called *bosses*. They are situated in such a way as to form three flexible thin regions, two of which are in compression and one in tension as pressure is applied to the top surface of the wafer. A view of a chip with two such bosses is shown in Fig. 67.17. The four resistors are placed in these three regions to produce the typical full-bridge output (two resistors in the tensile region and one resistor in each of the compressive regions).

FIGURE 67.17 Scanning electron micrograph of a dual-bossed micromachined structure for stress concentration on the diaphragm for a very low pressure sensor.

Although this design minimizes stiffening effects and results in a very linear response, fabricating the chip accurately and consistently is a real challenge in micromachining manufacturing technology. The thin areas must have a tightly controlled thickness, typically 7.0 + 0.5 µm; this requirement is only possible by using etch-stop techniques. The width of the thin areas must be controlled to 20 ± 5 µm, which requires the thickness of the wafer to be precise within 5 microns. The alignment accuracy between the front and backside patterns must be within 2 microns. A particularly important and difficult parameter that must also be minimized and precisely reproduced is the built-in stress on the diaphragm. Indeterminant thin-film stresses cause large and unpredictable offset voltages. The causes of these stresses can be elusive without a fundamental understanding of solid state materials and processes.

CONCLUSIONS

A wide variety of advanced sensor designs have been described in this chapter. In addition to these devices, however, an entire class of microstructures can also be fabricated from silicon. A sampling of these are described in the next chapter.

REFERENCES

1. J. Bryzek, R. Mayerand, and P. Barth, "Disposable Blood Pressure Sensors with Digital On-Chip Laser Trimming," *Technical Digest of the IEEE Solid State Sensors and Actuators Workshop,* Hilton Head Island, South Carolina, June 6–9, 1988, p. 121.
2. K. Petersen, P. Barth, J. Poydock, J. Brown, J. Mallon, Jr., and J. Bryzek, "Silicon Fusion Bonding for Pressure Sensors," *Technical Digest of the IEEE Solid State Sensors and Actuators Workshop,* Hilton Head Island, South Carolina, June 6–9, 1988, p. 144.
3. L. Christel, K. Petersen, P. Barth, J. Mallon, Jr., and J. Bryzek, "Single-Crystal Silicon Pressure Sensors with 500× Overpressure Protection," *Sensors and Actuators A21*(1–3), 1990, p. 84.
4. K. Petersen, J. Brown, P. Barth, J. Mallon, Jr., and J. Bryzek, "Ultra-Stable High Temperature Pressure Sensors Using Silicon Fusion Bonding," *Sensors and Actuators,* A21(1–3), 1990, p. 96.
5. P. Barth, F. Pourahmadi, R. Mayer, J. Poydock, and K. Petersen, "A Monolithic Silicon Accelerometer with Integral Air Damping and Overrange Protection," *Technical Digest of the IEEE Solid State Sensors and Actuators Workshop,* Hilton Head Island, South Carolina, June 6–9, 1988, p. 35.
6. S. Terry, "A Miniature Silicon Accelerometer with Built-In Damping," *Technical Digest of the IEEE Solid State Sensors and Actuators Workshop,* Hilton Head Island, South Carolina, June 6–9, 1988, p. 114.
7. J. R. Mallon, Jr., F. Pourahmadi, K. Petersen, P. Barth, T. Vermeulen, and J. Bryzek, "Low Pressure Sensors Using Bossed Diaphragms and Precision Etch-Stopping," *Sensors and Actuators,* A21(1–3), 1990, p. 89.

CHAPTER 68
SILICON MICROSTRUCTURES

INTRODUCTION

In a very real sense, the technology behind silicon sensing chips is derived from integrated circuit fabrication methods. At the same time, the widespread success and commercialization of high-volume silicon sensors since 1980 has inspired a new technology based on silicon sensor fabrication methods. While silicon sensors detect and measure external environmental parameters, new classes of silicon microstructures and microactuators interact with and physically modify their environment.

The driving force behind sensor development efforts has been the proliferation of the microprocessor. We have entered an age when microprocessors are inexpensive, and engineers have installed them into every conceivable instrument, appliance, and piece of equipment from automobiles to typewriters to coffee makers to bathroom scales. Unfortunately, however, the microprocessor is deaf, dumb, and blind. It relies completely on one form of external data input to perform its job. Currently, the bulk of this data input is provided by the common, bulky, slow keyboard. Microsensors represent the cheapest, quickest, and most efficient way for microprocessors to obtain information about their environment.

Obtaining information, however, is only half the story. Next, the microprocessor is expected to act on its environment; the complete system requires actuators. Some actuators are simply other electronic devices such as displays, speakers, or lights. Advanced, high-technology systems, however, increasingly demand more sophisticated devices such as automated microplumbing, arrays of miniature electromechanical switches, hydraulic and pneumatic microvalves, miniature machines for the manipulation of optical fibers, light deflectors and scanners, micropositioning devices, and other actuation mechanisms. Traditional mechanical fabrication techniques such as machining, molding, and so on are not precise enough, are not capable of producing such small structures, or are too expensive for the intended applications. It is often impossible to employ many aspects of traditional mechanical engineering disciplines in the design and fabrication of even the simplest miniature high-precision mechanical components required in many new electromechanical systems. For example, arrays of nozzles used in inkjet printers are extremely small (typically 50 microns), and their dimensions must be controlled to unparalleled precision (better than 1 micron).

Fortunately, the same processes and techniques that are now used routinely to manufacture silicon sensors can be exploited to fabricate a broad range of silicon actuators and micromechanical components. The list of potential devices and structures that have been proposed in the literature is extensive. A brief version is given below. New markets are being opened, and new structures are constantly being developed as the technology evolves.

In this section, we will explore several representative microactuator and micromechanical components. The wide variety of devices discussed below includes examples of many microactuators, microstructures, and micromechanical applications that have been demonstrated with silicon micromachining technology. (See Refs. 1, 2, and 3 for technical reviews on this field.) They are the precursors of future generations of advanced silicon microinstruments that are destined to alter the principles and the scope of mechanical engineering as we now understand it.

MICROPLUMBING

Synoptic View

Most laboratory, medical, and industrial instrumentation as well as test and manufacturing equipment make use of hydraulic and pneumatic components for measurement, mechanical drive, and general transport of gases and fluids. Micromachining technology is an ideal tool for constructing miniature, high-precision parts and assemblies for these plumbing applications. Components fabricated with silicon processing technology bring advantages in miniaturization, performance, reproducibility, cost, and ease of integration with electronic components. An application market particularly suited to this technology but not yet commercialized is disposable micro-fluid components for medical testing and diagnostics. For blood testing, such components can be precise, easily manufactured, cost-effective, and biologically inert.

Nozzles

In order to create precise fluid flow, very high accuracy orifices and tubes must be reproducibly fabricated. Fluid flow rate through a tube, for instance, depends on the diameter of the tube raised to the third power. This relation demands that the tube and orifice dimensions be very tightly controlled. The simplest example of these structures is the silicon nozzle. Large arrays of nozzles (up to 5000 or more per wafer) can be made with better than 1 micron precision. Even more importantly, this precision can be maintained from wafer to wafer because the inherent accuracy of the fabrication process depends only on the crystal planes in the silicon and on standard photolithographic techniques based on VLSI processing techniques and manufacturing equipment. Several configurations of nozzles are possible. Square nozzle shapes[4] are easily made from the silicon crystal itself, as shown in Fig. 68.1. Other arbitrary shapes (such as round) are also possible[5] by using the silicon as a substrate to support thin membranes (such as oxide or thin silicon layers) with arbitrarily shaped holes etched through them.

Channels and Flow Restrictors

Fabrication of high-precision microminiature tubes (and integrating such tubes with nozzles) is another straightforward application of micromachining technology. In this case, the characteristics of anisotropic etching are employed to etch precise grooves in silicon. Again, the dimensions of the groove are controlled by the crystal

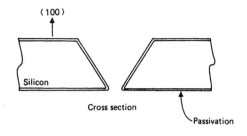

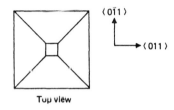

FIGURE 68.1 High precision, anisotropically etched nozzle in silicon.

planes of silicon and by the principles of VLSI photolithography. After the grooves are etched, a Pyrex glass wafer (or another silicon wafer) is hermetically bonded onto the etched wafer to form an enclosed channel. Since the tubes can be configured into spirals or interleaved patterns, up to one meter of tubing length (75 microns in diameter) can be fabricated on a 1-cm^2 chip.

An important practical application of micromachined miniature tubes is flow restrictors. Implantable drug infusion pumps are usually designed with long, small diameter tubes serving as large fluid flow resistors (flow restrictors) to limit the rate of drug delivery. These systems are used for insulin infusion (for diabetics),[6] hypertension treatment, and cancer chemotherapy. Micromachining technology provides the precision, reliability, and reproducibility to assure that such critical medical functions are consistently achieved.

Another important application of micromachined miniature plumbing is the IC-cooling technique originally developed at Stanford.[7] A key limitation in the implementation of high-speed, high-density integrated circuits is cooling capability. Integrated circuits must be maintained below about 100°C in order to operate reliably Fig. 68.2. Unfortunately, however, high-speed operation demands high energy dissipation. The faster the chip, the hotter it gets. A promising method for operating ultra-high-speed circuits includes grooves etched on the backside of integrated circuit chips. After bonding the chips to a fluid manifold plate, fluid can be pumped through the assembly, resulting in highly efficient cooling. See Fig. 68.2. More than an order of magnitude increase in power dissipation (from 25 watt/cm to 400 watt/cm) was demonstrated while maintaining an acceptable temperature rise in the chip. Modifications and refinements to this basic technique can be applied to many types of IC chips.

Two other remarkable structures that make use of silicon microplumbing are the gas-chromatograph-on-a-chip[8] and the fluidic-amplifier-on-a-chip, both also origi-

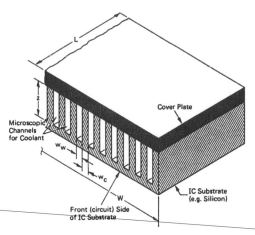

FIGURE 68.2 Micromachined integrated circuit-cooling structure first demonstrated at Stanford.

nated at Stanford.[9] In the gas chromatograph, combinations of gas flow channels, etched nozzles, valve seats, and thermal detectors were integrated on a single silicon wafer. Figure 68.3 presents a photo of such a wafer, courtesy of MicroSensor Technology. This chromatograph increases the speed of gas analysis by at least an order of magnitude because the gas volumes are so small and accurately defined. Such microinstruments are currently employed for online BTU analysis during gas custody transfers. Shown in Fig. 68.4 is a plasma-etched silicon fluidic amplifier chip.

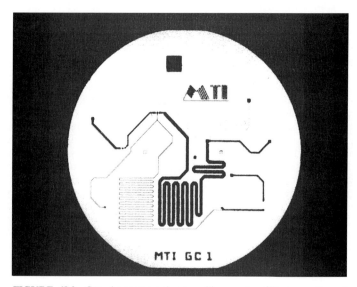

FIGURE 68.3 Gas chromatograph on a silicon wafer. (*Photo courtesy of Microsensor Technology.*)[8]

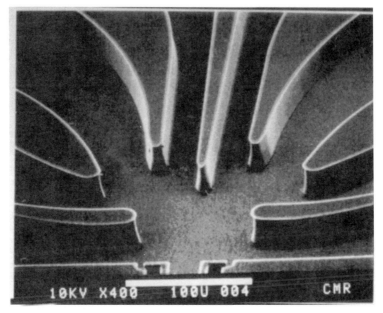

FIGURE 68.4 Plasma-etched fluidic amplifier chip fabricated at Stanford University.[9]

Power gains as high as 3.5 were demonstrated with this micromachined device, which is the world's smallest fluid amplifier.

Valves

Besides simple, passive fluidic components such as holes and grooves, extensive development work at many laboratories is underway for active components including valves. Several types of valves have been demonstrated by a number of micromachining research groups around the world. Case Western Reserve University and the University of Upsalla in Sweden have both tested micromachined fluid flow diodes[10,11] in which a silicon cantilever beam or diaphragm, suspended in a fluid flow channel, is positioned over an orifice etched in silicon. When fluid flows through the orifice toward the beam, the beam is pushed away from the orifice, and the flow is unrestricted. When the fluid flows in the opposite direction, the beam is pushed against the orifice, thereby restricting the flow. Figure 68.5 shows a schematic of the device made at Case Western.

Another type of micromechanically actuated valve was built at Stanford University.[12] In this device, a flexible diaphragm similar to a pressure sensor diaphragm is spaced a small distance (typically 20 μcrons) above a flow channel. The cavity behind the diaphragm is filled with a low-vapor-pressure liquid such as alcohol. Inside this sealed cavity is a miniature, thin-film heating resistor Fig. 68.6. When a high current is passed through the resistor, sufficient amounts of liquid will vaporize to create a high pressure within the sealed cavity. The high pressure will cause the silicon diaphragm to flex upward and seal down over the flow channel. Gas flow rates of 100 cc/sec have been realized with on/off ratios higher than 10,000.

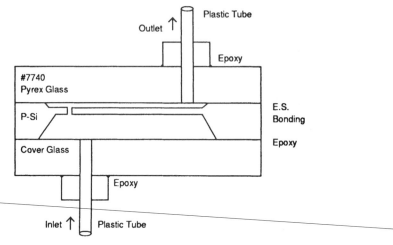

FIGURE 68.5 Fluid "diode" fabricated in silicon at Case Western Reserve University.

Other similar valves have also been fabricated that employ electromagnetic actuation. Early experiments performed at Stanford at Microsensor Technology, and more recently at Tohuko University in Japan,[13] use silicon diaphragms or more sophisticated micromachined flexures attached to conventional, external electromagnetic solenoids or piezoelectric stacks. The high forces generated by the solenoid causes the flexure to open up or clamp down over holes etched in silicon. The Stanford device incorporated a series of etching steps to create sealing rings and very low dead volume chambers.[8]

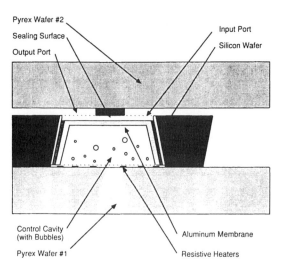

FIGURE 68.6 Silicon microvalve fabricated at Stanford University by Mark Zdeblick.

Although silicon micromachined valves and flexures have not yet been commercialized, recent advances in etching techniques, new structures, and strength-enhancing processes have been impressive. New applications will benefit greatly from these technical innovations during the coming years.

THERMALLY ISOLATED SILICON MICROSTRUCTURES

Introduction

Temperature, gas flow, electrical power, and thermal radiation are often measured by devices that maintain a temperature difference between a heated element and the surrounding environment. Such sensing structures are perfectly suited for implementation in silicon because silicon itself provides high thermal conductivity, while the common insulators deposited on silicon provide low thermal conductivity.

In addition, silicon can be micromachined in precise configurations to limit and control thermal flow. As a result, it is possible to place thermal sensing elements in a controlled thermal environment in which the paths for thermal conduction, convection, and radiation are repeatable from device to device.

In combination with the extremely small size of silicon components, these micro-engineered thermal parts exhibit high sensitivity and extremely fast thermal response times. In addition, such thermally isolated structures can be suitable as radiation emitters and indeed have been used experimentally to create miniature incandescent filaments.[14]

Typically a thermally isolated sensor will place the sensing element at the center of a thin area (a bridge, membrane, mesh, or cantilever) supported by a thick peripheral silicon substrate; a typical doubly supported bridge structure is illustrated in Fig. 68.7 courtesy of Innovus. The thin area provides maximum thermal isolation from the substrate and ensures that heat transfer to or from the sensing element is primarily by convection or radiation rather than by conduction to the substrate. It also ensures that parasitic heat-flow paths are dominated by the high thermal resistance of the thin area. On the thin area, an element such as the resistor shown in Fig. 68.7 can be heated to some point above room temperature and its response to changes in the ambient gas (due to convection or conduction) can be measured. In these types of microstructures, the thin area is most often composed of an insulator such as silicon nitride for maximum thermal and electrical isolation. These thermally isolated features have also been fabricated from thin silicon or polycrystalline silicon layers that provide less thermal isolation but may simplify fabrication.

Sensors for Thermal Conductivity and Flow

Structures for monitoring gas flows have been developed at Stanford,[15] Innovus,[16] Honeywell,[17] the University of California at Berkeley,[18] Tohoku University,[19] and Chalmers University in Sweden,[20] among others.

The Stanford structure, developed for monitoring the thermal conductivity of a passing gas stream in a miniature gas chromatography system, consists of a thin pyrex mesh in which a nickel film resistor is embedded. A shallow cavity beneath the mesh permits the passing gas flow to penetrate the mesh so that little heat is conducted directly to the silicon substrate beneath. The resistor is held at a constant

FIGURE 68.7 Thermally isolated mass flow sensor structure. (*Photo courtesy of Innovus.*)

temperature, and the power required to maintain that temperature is monitored as a measure of the thermal conductivity of the gas flowing at a constant velocity past the sensor. With a thermal response time on the order of 200 microseconds, the sensor provides extremely rapid measurement of changes in the gas stream.

The Innovus structure[16] uses a single metal film resistor embedded in an insulator bridge that contains a mesh of holes. Its action is similar to that of the Stanford sensor, but it is used for measuring mass flow of a gas of known composition rather than for monitoring thermal conductivity of that gas. A schematic layout of the Innovus chip is shown in Fig. 68.8. Also illustrated in the figure is the integral flow channels micromachined on the chip.

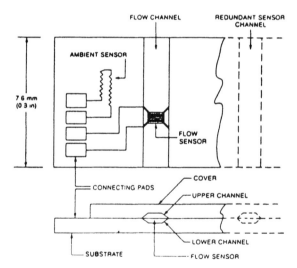

FIGURE 68.8 Schematic layout of Innovus mass flow sensor chip. (*Drawing courtesy of Innovus.*)

The Honeywell structure[17] uses two closely spaced bridges of silicon nitride in which permalloy thin-film resistors are embedded. A resistor on each bridge acts as a heater, while a separate resistor on each bridge monitors the temperature of the gas stream flowing past the sensor. The device provides a rapid directional measurement of the gas flow past the sensor and a flow channel of known resistance can provide a measure of differential pressure between two separated volumes.

At UC Berkeley, polycrystalline silicon has been used as the material for a bridge structure and has also been used as the material for a resistor element in this bridge. The response time of this sensor is slower than for the insulator-bridge sensors discussed above, probably because substantial heat conduction occurs through the polycrystalline silicon to the supporting substrate.

Radiation Detectors

Researchers at the University of Michigan have developed a thermopile radiation detector in silicon.[21] The device consists of a thin silicon membrane on which the hot junctions of an array of thermocouples are placed and a supporting silicon frame on which the cold junctions of the thermocouple array are placed. The junctions on the membrane are covered by a black polymer absorber, and under incident infrared radiation, they heat slightly above their surroundings. The device provides a means of sensing IR using an inexpensive silicon device despite silicon's large band gap (1.1 eV), which normally makes silicon transparent to IR. See Fig. 68.9, courtesy of the University of Michigan.

Vacuum Sensors

The thermal conductivity of a gas provides a sensitive measure of the gas pressure in the low pressure ranges (approximately 0.1 kPa and less). Thermally isolated silicon resistors or thermopiles can provide a measurement of thermal conductivity in these low pressure ranges and in this way act as vacuum gages in many applications. Figure 68.10 schematically illustrates how a suspended micromechanical beam can be configured as a vacuum sensor. Such sensors have been developed at Delft University in the Netherlands.[22]

Heated Filaments

Thermally isolated structures also have the potential to act as incandescent filaments. Researchers at the University of Wisconsin have demonstrated polycrystalline silicon bridges, coated with silicon nitride, that glow when resistively heated.[14] With a melting temperature of 1412°C, silicon can emit a bright orange blackbody light without melting. These devices have potential applications as optical isolator components that can send a signal to a nearby photodiode. Other structures, such as isolated refractory metal film resistors on silicon nitride membranes, have the potential to operate at even higher temperatures. This concept has recently been extended at UC Berkeley[23] to include an integral, transparent cap over the filament, sealing the microfilament in an inert microcavity for protection and improved light dispersion.

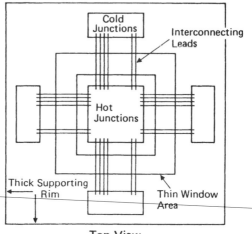

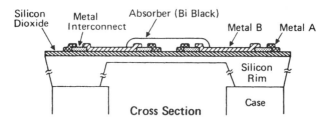

FIGURE 68.9 Infrared detector on silicon using thin-film thermocouples on a thermally isolated silicon microstructure. (*Drawing courtesy of Ken Wise, University of Michigan.*)

Gas Sensors Using Suspended Structures

In many solid-state gas sensors, the gas is absorbed into a thin layer of material coating the sensing element. After exposure to high gas concentrations for long time periods, this layer will become saturated with the gas molecules. In order to purge and recover the sensitivity of the sensor, it is common to heat the element and thermally desorb the gas. For this reason, it is important to fabricate such gas sensors on thermally isolated structures. In other types of gas sensors, such as oxygen sensors and combustible gas sensors, the sensing material is actually heated several hundred degrees Celsius during the operation of the device. Thermally isolated structures are also important in these applications.

A group at MIT has solved this problem by fabricating a single-crystal silicon plate suspended over a hole etched through a silicon wafer.[24] The silicon plate contains a polysilicon heating resistor and a tin-oxide sensing element. Although the realization of this device required many complex process development issues, the final structure successfully demonstrated a temperature rise of 300°C in less than 2 msec.

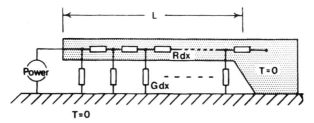

FIGURE 68.10 Micromechanical vacuum sensor based on thermally isolated silicon structures. This structure was demonstrated at Delft University, the Netherlands.

Microwave Power Sensing

The measurement of microwave power produced by a signal generator is made possible by thermally isolated structures such as one developed by Hewlett-Packard, shown in Fig. 68.11. This device suspends a silicon mass on thin silicon arms connected to a surrounding silicon frame.[25] On the mass are a resistor and a temperature-sensing diode. In an enclosed package, heat loss from the mass is primarily by conduction along the silicon arms to the frame and from there to the header on which the frame sits. The arms constitute the highest thermal impedance in the heat flow path. In operation, the microwave signal is fed to the resistor, which dissipates the microwave power, heating the suspended mass. The temperature rise in the mass is linearly proportional to the input power and is measured by the temperature-sensing diode.

ELECTRICAL SWITCHES

Electrostatically deflectable thin membranes fabricated with silicon micromachining techniques are not yet commercially available but hold great potential. Among the more intriguing applications are electrical switches of very small size. Switches of this nature can neatly fill the gap that exists between conventional silicon transistors and electromagnetic macromechanical relays.

Quite simple in principle, the device is a cantilever beam approximately 0.5 microns thick and 100–350 microns long. These metal-coated insulating membranes are attached to the silicon at a single end and suspended over a pit in the silicon substrate[26] as shown in Fig. 68.12. The depth of this shallow pit is important, so that one can obtain large electrostatic forces using relatively small voltages during the deflection of the device.

Fabrication of such devices is made possible through the use of a unique anisotropic etchant, EDP. Besides its anisotropic etching properties (discussed in previous chapters), the etch rate of EDP is also dependent on the doping level in the silicon. Intentional introduction of a high boron concentration (a common silicon impurity) will affect etch rate, in some instances stopping it almost completely. These highly doped silicon regions, silicon dioxide, and silicon nitride all have extremely low etch rates in EDP, as do metals like chromium and gold. With this combination of characteristics, the following process sequence can be used to fabricate the devices.

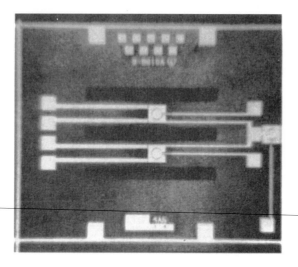

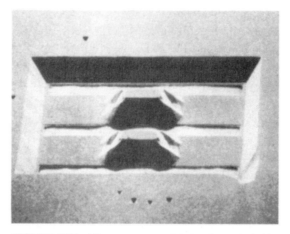

FIGURE 68.11 Microwave power detector demonstrated by Hewlett-Packard. (*Photo courtesy of Peter O'Neil.*)

1. Implant a highly boron-doped etch-stop layer on the wafer and then grow a lightly doped epitaxial layer of silicon the thickness of the desired pit depth.
2. Grow or deposit a layer of insulator desired for the beam (e.g., silicon dioxide, silicon nitride). Pattern and etch the insulator into the shape of a beam.
3. Deposit metal layer, typically very thin layers of chrome and gold. Pattern and etch the metal.
4. Spin-coat another layer of photoresist. Pattern windows in the photoresist and deposit another thin layer of metal as an electroplating base.

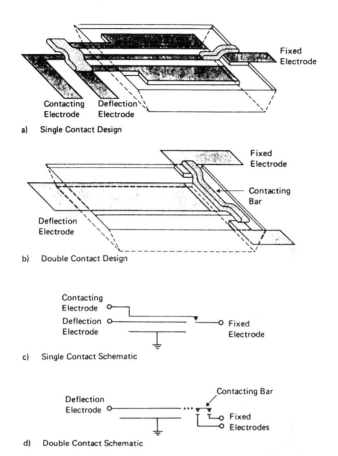

FIGURE 68.12 Miniature electromechanical switch fabricated on silicon. This switch is activated by electrostatic forces.

5. Spin coat and pattern another layer of resist. Holes in the resist define regions to be gold-plated. Plate up thick metal regions for switch contacts and high current lines.
6. Strip away photoresist. Etch the metal lightly to remove the plating base, leaving thick plated regions intact.
7. Anisotropically etch out silicon from underneath the insulating cantilever beam using EDP.

This process sequence is summarized in Fig. 68.13.

Since the metal lines are electrically isolated from the substrate, the substrate can be used as the ground plane. Voltages applied to thin metal on the cantilever will then deflect the beam electrostatically. When the beam is deflected, the overhanging gold layer suspended at the end of the beam make electrical contact with the opposite gold contact on the substrate, thus closing the relay. Switches of this type can typically be closed at 40–100 V, with 10-microsecond response times, 100

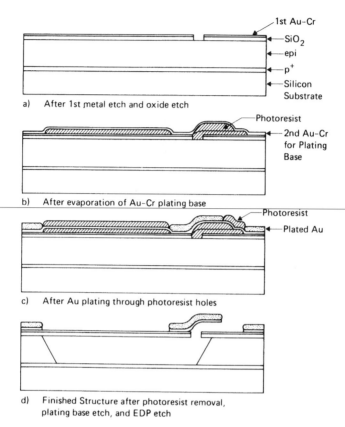

FIGURE 68.13 Fabrication sequence for the micromechanical electrical switch.

mÅ current capability, an on-state resistance of about 1 ohm, and a very high off-state impedance (many megaohms). The world's smallest such electromechanical switch is shown in Fig. 68.14.

Although very small, these devices are quite robust. Similar metallized amorphous SiO_2 membranes in preliminary life tests have been observed to survive 100 billion deflections without noticeable change in behavior. Amorphous elements are not likely to fail due to fatigue like their polycrystalline metal counterparts. Devices of this sort can also survive accelerations of hundreds of GIs without failing.

It should be noted that these devices are subject to significant contact wear if not designed properly and can be quite temperature sensitive, since the metal and insulating layers have differing thermal expansion coefficients. These devices cannot carry the large currents of reed relay switches, nor do they have the speed of transistors; but their tiny dimensions, and the fact that they are at least an order of magnitude faster than reed relays, make them perfect for many applications where neither alternative will work. In addition, the fact that they can be fabricated in interconnected arrays on a single silicon chip will be crucial in applications such as communications, data transmission, and testing.

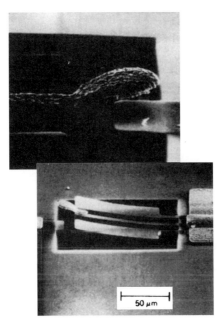

FIGURE 68.14 Scanning electron micrograph of a micromechanical electrical switch first demonstrated at IBM. This switch is 125 μm long.

LIGHT MODULATORS AND DEFLECTORS

There are other unique applications of electrostatically deflectable SiO_2 membranes. One of the most interesting of these is an optical element similar to a light valve. Using the same basic fabrication techniques described above for the switches, many electrostatically deflectable cantilevers can be fabricated side by side, as shown in Fig. 68.15. A voltage applied to the beam metallization with respect to the underlying substrate will electrostatically deflect the beams. One-dimensional[27] or two-dimensional[28] arrays of such deflectable elements can be used to modulate light and laser beams, forming high contrast displays using Schlieren-type optical systems. In earlier work at Westinghouse, arrays of clover-leaf-shaped membranes supported by silicon pedestals were deflected in vacuum systems by electron beam raster scans.

Another unique optical element that has been used in imaging applications is the silicon torsional mirror.[29] This is not a thin membrane structure in the sense of the cantilevers discussed earlier, but it similarly uses electrostatic forces to deflect a 100-micron-thick mirror for optical beam scanning. The mirror comprises a square of silicon suspended at either end by long narrow beams, all of single-crystal silicon. A typical mirror element with integral torsion beams is shown in Fig. 68.16. Electrodes on either side of the line of symmetry can rock the silicon mirror element from side to side along a fulcrumlike ridge running beneath the square, thereby vibrating the structure in a torsional mode.

The torsional mirror is fabricated using standard silicon anisotropic etching techniques. In this case, the mirror and beams are fabricated by etching completely

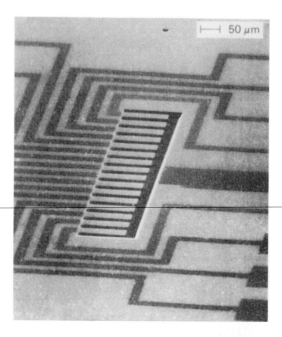

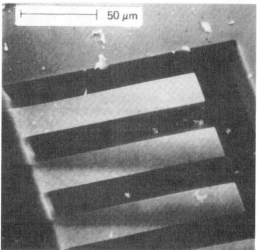

FIGURE 68.15 Array of electrostatically deflectable light-modulator elements fabricated on silicon. Each element is 100 μm long, 25 μm wide, and 0.5 μm thick.

through a thin silicon wafer. The electrodes and the supporting fulcrum ridge are fabricated on a glass plate bonded to the silicon chip.

As an optical material, silicon possesses intrinsic advantages over glass or quartz mirrors in moving applications because of silicon's high modulus-to-weight ratio. This number is typically three times larger for silicon than for quartz, meaning that distor-

SILICON MICROSTRUCTURES

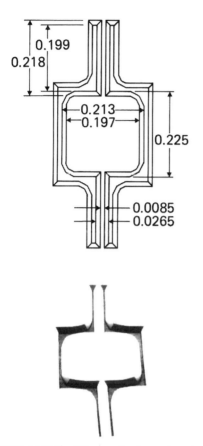

FIGURE 68.16 Silicon torsional scanning mirror. This device is micromachined from single-crystal silicon. Light images are scanned by resonating the structure using electrostatic forces.

tions are three times smaller. The silicon structures are also highly reliable, since single crystal silicon has virtually no mechanical imperfections or hysteresis.

MICROMOTORS

Perhaps the most amazing silicon micromachined creations to date are the efforts by several research groups to construct tiny solid-state motors. The first successful micromotor, which is the smallest man-made rotating machine ever built, was operated by the UC Berkeley Sensor and Actuator Center during August of 1988.[30] This motor, and other motor designs now being built,[31,32] are only about 3 microns thick and 75 microns in diameter. Figure 68.17 shows a typical design. The fabrication techniques required to make a successful micromotor are the most challenging in the field of micromechanics. One major impediment to engineering progress in this field is that familiar mechanical concepts become ill-defined and uncertain as mechanical design and fabrication principles are reduced to such minute dimensions. Material strength, static and viscous friction, levitation, and air resistance are all concepts that are clearly defined macroscopically but that present an altogether different face in this microminiaturized scale. Continued research on such microactuators will lead to greater understanding of these concepts on a micro scale.

All micromotors demonstrated to date operate by electrostatic driving principles instead of the familiar electromagnetic forces of the macroscopic world. There is good reason for this. As electric forces are scaled down from the millimeter range to the micrometer range, large force levels can be maintained even if the driving voltage decreases. On the other hand, as electromagnetic forces are scaled down this far, current density must be increased by a factor of a million to maintain equivalent forces. This is presently impossible. For this reason, electrostatic driving force is the mechanism of choice for micromotors.

Many different motor designs that employ electrostatics exist. The very first electric motors known were designed by Benjamin Franklin and Andrew Gordon in the 1750s; these were electrostatic. The designs most commonly employed today by micromachinists are variations on the motor designed by Karl Zipernowsky in 1889. Tangential force on a rotor is generated by the phasing of attractive and repulsive potentials between the rotor and the stators.

FIGURE 68.17 Scanning electron micrograph of an electrostatic micromotor fabricated at the University of California at Berkeley. (*Courtesy of R. S. Muller.*)

Fabrication of these motors requires high-quality dielectric and conductive films. These are used not only in the structure of the motor itself but also as molds in the construction of the motor using a technique known as sacrificial micromachining. In this technique, alternate layers of polysilicon and silicon dioxide are deposited and patterned into different shapes. The silicon dioxide is etched away at the end of the process, leaving only the polysilicon behind, the polysilicon microstructure having been molded by its complementary silicon dioxide form. With such a lost-wax process, self-restraining joints, flanges, pins, gears, springs, rotors, and the like have been fabricated by several groups, including UC Berkeley, AT&T Bell Laboratories, and MIT.

The micromachining community has not yet demonstrated micromotors in a practical, commercially viable application. However, the potential implications of these developments transcend immediate commercial requirements. The ability to fabricate devices as sophisticated as motors that are nearly the size of biological cells is a revolutionary accomplishment with vast, future significance.

RESONANT STRUCTURES FOR MEASUREMENT AND ACTUATION

The intrinsically strong mechanical qualities of silicon make it ideal for high-performance resonating structures. Large Q factors have been observed for both single-crystal silicon and polysilicon vibrating mechanisms. Such devices will have a pivotal impact on future sensors and actuators. Two important features of resonant devices are:

1. The resonant frequency depends only on the mechanical characteristics and stability of the materials and not its electrical performance. Mechanical parameters are inherently more stable, for example, than the values of diffused piezoresistors.
2. The output of resonant devices is fundamentally digital. Digital signals are intrinsically more compatible with modern signal processing electronics and more accurate than conventional analog voltage or current measurements.

The accuracy of common quartz frequency sources is largely dependent on the quality factor (Q) of the device. Q depends both on the details of the structure, the surface characteristics of the material, and the strength and physical stability of the material. Intrinsic Q values for silicon are expected to be at least as good as quartz. indeed, Q's near-50,000 have been demonstrated in single-crystal silicon bulk (millimeter-size) structures,[33] which are similar to those realized in quartz resonators.[34] Even polysilicon has been shown to be capable of Q factors over 100,000.[35]

Over the past few years, an impressive variety of resonant devices have been built and tested; resonant chemical sensors,[36] resonant pressure sensors,[37,38] resonant acceleration sensors,[39] arrays of resonant vibration sensors,[40] and resonant lateral actuators.[41] Resonant sensor and actuator concepts based on these principles will become increasingly important as the development of silicon microstructures continues.

APPLICATIONS IN MICROBIOLOGY

By its very nature, silicon micromachining is ideally suited to applications in microbiology. Many types of silicon microstructures can be fabricated on the same scale as living cells. These structures have been used to manipulate and measure individual cells and biological molecules. Early efforts in this area include the development of the neural probe array fabricated at the University of Michigan[42] and the etched structures for nerve regeneration studied at MIT.[43]

The future possibilities in this area are diverse and exciting. They have even inspired science fiction-type proposals such as intelligent, self-guided, micromachines that would be injected into the bloodstream to seek out and destroy foreign organisms. Although these fanciful ideas are currently impossible and impractical, numerous realistic micromachined devices are currently in development that will radically transform our perceptions of and our technical capabilities in the fields of medicine and microbiology.

Cell Manipulation

At least three micromechanical techniques have been shown to be capable of manipulating individual cells. Tweezers are a simple, obvious device that might be scalable for manipulating microscopic objects. Indeed, microtweezers have been built and tested on a silicon substrate at Cornell University.[44] These devices are made from deposited thin films of tungsten and are remotely opened and closed by applied electrostatic voltages. The prongs of the tweezers are typically 3 microns in cross section and 200 microns long.

A second technique for accurately manipulating large numbers of cells was developed by Hitachi[45] for use in a cell-fusing apparatus. In this system, a silicon wafer containing an array of hundreds of minute etched orifices is lowered into a solution of A-type cells. When suction is applied to the back of the wafer, individual cells will become trapped at the orifice locations (these orifices are smaller than the cells). The array of cells can now be transferred into the small depressions in a fusing plate. The fusing plate consists of a similar array of etched depressions, at the bottom of which are two electrodes. Next, another suction picks up and transfers an array of B-type cells into the depressions in the fusing plate. At this point, each depression contains one A cell and one B cell. When a voltage is applied to the two electrodes, A will fuse with B. Fusing efficiencies as high as 60 percent have been demonstrated with this silicon micromechanical system, compared to 2 percent with conventional techniques in use today.

Finally, a fascinating cell manipulation concept has been demonstrated at Seikei University.[46] In this scheme, cells are steered and maneuvered with the use of electrostatic fields. A periodic structure fabricated on silicon consists of binding sites where the cells can reside, as well as electrode patterns. By applying proper voltage sequences to the electrodes, cells can be made to move from site to site in the form of a cell shift register. High-frequency, high-voltage signals, together with properly contoured electrode structures, are employed to attract the cells in an aqueous solution. With such a basic cell handling method in place, we can envision microminiature cell assembly lines in which various cell types are transported to processing areas; the individual cells are subjected to various operations or measurements, one at a time; and finally automatically sorted into proper reservoirs. Such micromachined, automated cell assembly lines will revolutionize the field of microbiology.

Atomic Force Microscopes

One of the most amazing technical accomplishments of the past 20 years has been the invention and demonstration of the scanning tunnelling microscope (STM)[47] and its derivatives, the various versions of the scanning atomic force microscopes (AFM). These devices are based on the principle that atomically sharp points approaching a conducting surface will draw an electric current at a distance of a few angstroms. In fact, the value of the current is a precise measurement of the separation distance. Resolution of such systems can be better than 0.1 angstrom, or smaller than an atomic diameter. If the sharp point is precisely raster-scanned over a surface, an atomic image of the surface can be electronically constructed. Modern STM instruments are capable of resolving and even identifying individual atoms on well-prepared surfaces. In addition, recent versions of these instruments can even operate in an aqueous solution.

Precise, microminiature actuators with integral atomically sharp points have been fabricated on silicon with micromachining techniques for STM applications. IBM and Stanford University are pioneers in this field. IBM has concentrated on the reproducible fabrication of atomically sharp probe tips,[48] while Stanford has built entire scanning systems, including flexures, probe tips,[49] and three-dimensional, deflectable piezoelectric raster-scanning beams.[50] IBM has even demonstrated the manipulation of individual atoms using microprobes for information storage.

The significance of micromachined scanning atomic microscopes is enormous. Many future thin-film measurement instruments will be based on scanning mechanical microscopy systems—which are, in turn, based on micromachined mechanical components. Many microbiology instruments, designed to decipher complex molecular structures and molecular reactions, will also be based on scanning mechanical microscopy systems. In the future, information storage systems (with over 10^{12} bits/cm^3) may even be possible with micromachined STM instruments.

CONCLUSION

Silicon microactuator technology is only in its infancy. Although few silicon actuator devices are commercially available as yet, the potential for making high volumes of ultrasmall, ultraprecise mechanical devices at low cost is enormous. And the implications for advanced products of all types are remarkable, both technically and commercially. It is an exciting time to be involved in silicon micromechanics. Together, business, university, government, and customers are inventing new industries. The way the world thinks about mechanical devices will never be the same.

REFERENCES

1. *Proceedings of the IEEE Micro Electro Mechanical Systems,* Salt Lake City, Utah, February 1989.
2. *Proceedings of the IEEE Micro Electro Mechanical Systems,* Napa Valley, California, February 1990.
3. "Small Machines, Large Opportunities: A Report on the Emerging Field of Microdynamics," Report on the NSF Workshop on Microelectromechanical Systems Research, 1988.
4. E. Bassous, H.H. Taub, and L. Kuhn, "Inkjet Printing Nozzle Arrays Etched in Silicon," *Appl. Phys. Lett.,* **31**. 1977, p. 135.
5. E. Bassous, "Fabrication of Novel Three-Dimensional Microstructures by Anisotropic Etching of (100) and (110) Silicon," *IEEE Trans. Electron Devices,* vol. ED-25, 1978, p. 1185.
6. M.G. Guvenc, "A v-Groove Silicon Planar Capillary for Insulin Microinfusers," *IEEE Frontiers in Engineering in Health Care,* 1982, p. 300.
7. D.B. Tuckerman and R.F.W. Pease, "High-Performance Heat Sinking for VLSI," *IEEE Trans. Electron Devices,* EDL-2, 1981, p. 126.
8. S.C. Terry, *A Gas Chromatography System Fabricated on a Silicon Wafer Using Integrated Circuit Technology,* Ph.D. dissertation, Stanford University, 1975.
9. M.i. Zdeblick, P.W. Barth, and J. B. Angell, "Microminiaturefluidic Amplifier," in *Proceedings of the IEEE Solid-State Sensors Workshop,* Hilton Head, South Carolina, June 1986.
10. J. Tiren, L. Tenerz, and B. Hok, "A Batch-Fabricated Nonreverse Valve with Cantilever Beam Manufactured by Micromachining of Silicon," *Sensors and Actuators,* **18**(3), June 1989.
11. S. Park, W.H. Ko, and J.M. Prahl, "A Constant Flow-Rate Microvalve Actuator Based on Silicon and Micromachining Technology," in *Proceedings of the IEEE Solid-State Sensors Workshop,* Hilton Head, South Carolina, June 1988.
12. M.i. Zdeblick and J.B. Angell, "A Microminiature Electric-to-Fluidic Valve," in *Proceedings of Int. Conf. on Solid-State Sensors and Actuators,* Tokyo, 1987, p. 827.
13. S. Nakagawa, S. Shoji, and M. Esashi, "A Microchemical Analysing System Integrated on a Silicon Wafer," *Proceedings of IEEE Micro Electro Mechanical Systems,* Napa Valley, California, June 1990, p. 89.
14. H. Guckel and D.W. Burns, "Integrated Transducers Based on Blackbody Radiation from Heated Polysilicon Filaments," *Proceedings of Int. Conf. on Solid-State Sensors and Actuators,* Philadelphia, Pennsylvania, 1985, p. 364.
15. S.C. Terry, J.H. Jerman, and J.B. Angell, "A Gas Chromatograph Air Analyzer Fabricated on a Silicon Wafer," *IEEE Trans. Electron Devices,* ED-26, 1979, p. 1880.
16. K.E. Petersen, J. Brown, and W. Renken, "High-Precision, High-Performance, Mass Flow Sensor with Integrated Laminar Flow Microchannels," *Proceedings of Int. Conf. on Solid-State Sensors and Actuators,* Philadelphia, 1985, p. 361.
17. R.G. Johnson, R.H. Higashi, P.J. Bohrer, and R.W. Gehman, "Design and Packaging of a Highly Sensitive Microtransducer for Air Flow and Differential Pressure Sensing Applications," *Proceedings of Int. Conf. on Solid-State Sensors and Actuators,* Philadelphia, 1985, p. 358.
18. Y. Tai, R.S. Muller, and R.T. Howe, "Polysilicon Bridges for Anemometer Applications," *Proceedings of Int. Conf. on Solid-State Sensors and Actuators,* Philadelphia, 1985, p. 354.

19. M. Esashi, S. Eoh, T. Matsuo, and S. Choi, "The Fabrication of Integrated Mass Flow Controllers," *Proceedings of the Int. Conf. on Solid-State Sensors and Actuators,* Tokyo, 1987, p. 830.
20. G.N. Stemme, "A Monolithic Gas Flow Sensor with Polyimide as Thermal Insulator," *Proceedings of IEEE Solid-State Sensors Workshop,* Hilton Head, South Carolina, June 1986.
21. G.R. Lahiji and K.D. Wise, "A Monolithic Thermopile Detector Fabricated Using Integrated Circuit Technology," *Proceedings of Int. Electron Devices Meeting,* Washington, D.C., 1980, p. 676.
22. A.W. Van Herwaarden, "Integrated Vacuum Sensor," in *Proceedings of Int. Conf. on Solid-State Sensors and Actuators,* Philadelphia, 1985, p. 367.
23. C.H. Mastrangelo and R.S. Muller, "Vacuum-Sealed Silicon Micromachined Incandescent Light Source," *Proceedings of IEEE Int. Electron Device Meeting,* Washington, D.C., December 1989.
24. M.A. Huff, S.D. Senturia, and R.T. Howe, "A Thermally Isolated Microstructure Suitable for Gas Sensing Applications," *Proceedings of IEEE Solid-State Sensors Workshop,* Hilton Head, South Carolina, June 1986.
25. P. O'Neill, "A Monolithic Thermal Converter," *Hewlett-Packard Journal,* May 1980, p. 12.
26. K.E. Petersen, "Dynamic Micromechanics on Silicon: Techniques and Devices," *IEEE Trans. Electron Devices,* ED-25, 1978, p. 1241.
27. K.E. Petersen, "Micromechanical Light Modulator Array Fabricated on Silicon," *Appl. Phys. Lett.,* **31**, 1977, p. 521.
28. R.N. Thomas, J. Guldberg, H.C. Nathanson, and P.R. Malmberg, "The Mirror Matrixtube: A Novel Light Valve for Projection Displays," *IEEE Trans. Electron Devices,* ED-25, 1975, p. 765.
29. K.E. Petersen, "Silicon Torsional Scanning Mirror," *IBM J. Research and Development,* **24**, 1980, p. 631.
30. L.S. Fan, Y.C. Tai, and R.S. Muller, "IC-Processed Electrostatic Micromotors," *Proceedings of IEEE Int. Electron Devices Meeting,* San Francisco, California, December 1988.
31. M. Mehregany, S.F. Bart, L.S. Tavrow, J.H. Lang, S.D. Senturia, and M.F. Schlecht, "A Study of Three Microfabricated Variable-Capacitance Motors," *Proceedings of the 5th International Conference on Solid-State Sensors and Actuators,* Montreux, Switzerland, June 1989, p. 106.
32. L.S. Tavrow, S.F. Bart, M.F. Schlecht, and J.H. Lang, "A LOCOS Process for an Electrostatic Microfabricated Motor," *Proceedings of the 5th International Conference on Solid-State Sensors and Actuators,* Montreux, Switzerland, June 1989, p. 256.
33. R.A. Buser and N.F. de Rooij, "Tuning Forks in Silicon," *Proceedings of IEEE Micro Electro Mechanical Systems,* Salt ake City, Utah, February 1989, p. 94.
34. L.D. Clayton, E.P. Eernisse, R.W. Ward, and R.B. Wiggins, "Miniature Crystalline Quartz Electromechanical Structures," *Proceedings of IEEE Micro Electro Mechanical Systems,* Salt Lake City, Utah, February 1989, p. 121.
35. H. Guckel, J.J. Sniegowski, T.R. Christenson, and F. Raissi, "The Application of Fine Grained, Tensilepolysilicon to Mechanically Resonant Transducers," *Proceedings of the 5th International Conference on Solid-State Sensors and Actuators,* Montreux, Switzerland, June 1989, p. 143.
36. R.T. Howe and R.S. Muller, "Integrated Resonant-Microbridge Vapor Sensor," *Proceedings of IEEE Int. Electron Devices Meeting,* San Francisco, December 1984, p. 213.
37. T.S.J. Lammerink and W. Wlodarski, "Integrated, Thermally Excited Resonant Diaphragm Pressure Sensor," *Proceedings of Int. Conf. on Solid-State Sensors and Actuators,* Philadelphia, 1985, p. 97.
38. K. Ikeda, H. Kuwayama, T. Kobayashi, T. Watanabe, T. Nishikawa, and T. Yoshida, "Silicon Pressure Sensors with Resonant Strain Gauges Built into Diaphragm," *Proceedings of the 7th JIEE Sensor Symposium,* Shigaku Kaikan, Tokyo, 1988, p. 55.

39. S.C. Chang, M.W. Putty, D.B. Hicks, C.H. Li, and R.T. Howe, "Resonant-Bridge Two-Axis Microaccelerometer," *Proceedings of the 5th International Conference on Solid-State Sensors and Actuators,* Montreux, Switzerland, June 1989, p. 142.

40. W. Benecke, L. Csepregi, A. Heuberger, K. Kuhl, and H. Seidel, "A Frequency-Selective, Piezoresistive Silicon Vibration Sensor," *Proceedings of Int. Conf. on Solid-State Sensors and Actuators,* Philadelphia, 1985, p. 105.

41. W.C. Tang, T.H. Nguyen, and R.T. Howe, "Laterally Driven Polysilicon Resonant Microstructures," in *Proceedings of IEEE Micro Electro Mechanical Systems,* Salt Lake City, Utah, February 1989, p. 121.

42. K. Najafi, K.D. Wise, and T. Mochizuki, "A High-Yield IC-Compatible Multichannel Recording Array," *IEEE Trans. Electron Devices,* ED-32, 1985, p. 1206.

43. D.J. Edell, "Micromachined Silicon Neural Information Transducers," presented at the Annual Meeting of the Biomedical Engineering Society, Anaheim, California, April 1985.

44. L.Y. Chen, Z.L. Zhang, J.J. Yao, D.C. Thomas, and N.C. MacDonald, "Selective Chemical Vapor Deposition of Tungsten for Microdynamic Structures," *Proceedings of IEEE Micro Electro Mechanical Systems,* Salt Lake City, Utah, February 1989, p. 121.

45. K. Sato, Y. Kawamura, S. Tanaka, K. Uchida, and H. Kohida, "Individual and Mass Operation of Biological Cells Using Micromechanical Silicon Devices," *Proceedings of the 5th International Conference on Solid-State Sensors and Actuators,* Montreux, Switzerland, June 1989, p. 268.

46. M. Washizu, "Electrostatic Manipulation of Biological Objects in Microfabricated Structures," presented at the Third Toyota Conference on Integrated Micro Motion Systems, Nissin, Aichi, Japan, October 1989.

47. G. Binning, H. Rohrer, C. Gerber, and E. Weibel, "7×7 Reconstruction of Si(111) resolved in Real Space," *Phys. Rev. Lett.,* **50**, 1983, p. 120.

48. H.W. Fink, "Monoatomic tips for STM," *IBM Journal of Res. and Development,* **30**, 1986, p. 460.

49. T.R. Albrecht, S.A. Akamine, T.E. Carver, and C.F. Quate, "Fabrication of Microcantilever Stylus for the Atomic Force Microscope," *Journal Vacuum Science and Tech.,* **11**, 1995.

50. S. Akamine, T.R. Albrecht, M.J. Zdeblick, and C.F. Quate, "Microfabricated Scanning Tunneling Microscope," *IEEE Electron Device Letters,* **10**(11), November 1989, p. 490.

CHAPTER 69
COMPUTER DESIGN TOOLS

INTRODUCTION

The design process for a micromachined device is based on a host of scientific and engineering principles combined with the designer's engineering experience. Typically, the initial ideas for a device are subjected to many rounds of modification and improvement before the design is set in concrete and implemented as hardware. For successful competition in the marketplace, it is vital to use tools that shorten the design cycle and ensure successful operation of the hardware on the first attempt.

Computer-aided design (CAD) is an all-embracing term for the many computerized design tools used in this design process. CAD includes several engineering disciplines that facilitate and automate the design process, including drafting, solid and finite element modeling, tolerance studies, and documentation.

The development of the CAD systems is directly linked to progress in microelectronics and computers, which traces back to the discovery of the transistor at Bell Labs in 1947. In the mid 1950s the CRT console/computer connection was demonstrated at MIT. The concept and viability of interactive computer graphics was demonstrated by Sutherland of MIT in 1962. By the mid 1960s, corporations like GM, McDonnel Douglas, and Lockheed were actively involved in computer graphics research based on mainframe computers.

The ongoing development of efficient, inexpensive software and hardware has dramatically increased the use of CAD in engineering design in recent years. Figure 69.1 shows the decline of computer usage cost over the past several years.

Computers running CAD software cover a broad range, from low-cost affordable personal computers (PCs) to powerful supercomputers that routinely perform computation-intensive tasks.

Standalone PC-based CAD systems display great versatility and enjoy tremendous popularity. Compared to the truly high-powered computers, PCs are slower and have smaller memories; however, their lower price/performance ratio makes them especially attractive for small, growing firms. They provide an economical way to get into CAD process with the advantages of troublefree installation, program variety, and ease of operation.

Available software can enable PCs to emulate graphics terminals, provide connections to larger computers, or link computers in a common communication environment called Local Area Networks (LANs). The major CAD software packages for PC-based systems are AutoCAD, VersaCAD, CADKEY, and CADVANCE. The capabilities include drafting, dimensioning, wireframe, and solid modeling.

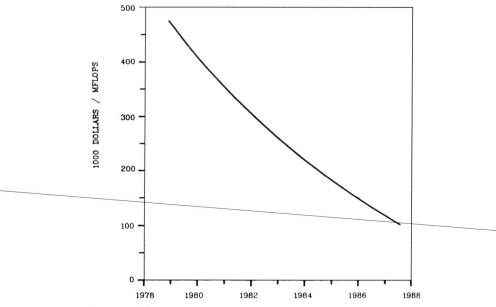

FIGURE 69.1 Decline in computer usage cost, 1978–1988.

COMPUTER MODELING

Designing an engineering structure or device requires evaluation of the projected design performance. This requires not only design experience but also close familiarity with the engineering principles involved. CAD techniques help the designer to record and codify this experience in the shortest possible time and to perform analyses of proposed structures. Computer modeling for engineering design can be divided into geometrical modeling and numerical modeling. Geometrical modeling includes graphical modeling, drafting, **3D** solid modeling, and wireframe modeling. The basis for geometrical modeling is that the information related to the design can be better understood when presented visually. Prior to incorporating of computers into the design process, the design engineer had to work with physical models and sheets of drawings. By using geometrical modeling techniques, one can interactively access the computer database and the particular design file to perform design modifications in a relatively short time.

In solid modeling, the graphics workstation represents the complete 3D shapes of device parts and assemblies. The software in solid modeling ensures that a solid model will duplicate its physical model so closely that the critical aspects of the design will be preserved, and the 2D drawing ambiguities and interferences will not be represented. Design functions such as interference studies and kinematic analysis are therefore typically included in solid modeling packages.

In numerical modeling, the designer tries to simulate the physical process and therefore predict the behavior of the device under predefined and realistic loading conditions in order to achieve the expected design performance. Use of numerical modeling in the design process offers a powerful tool to optimize the design in a minimum number of iterations.

Numerical modeling can further be divided into analytical and computational (finite element and finite difference) modeling.

Analytical Modeling

Analytical modeling requires understanding of the geometry, operating conditions, physical properties, and the functions to be performed by the device that is to be modeled. Very often, for a complex geometry or a simple geometry with complicated boundary conditions, a satisfactory analytical model is difficult to obtain. In general, the designer breaks down a complex structure to smaller, more primitive structures for which classical theories can be used. The speed of the design process and the functionality of the design depend to a great extent on the personal judgment of the designer, which originates from his analytical knowledge and design experience in that particular field.

The analytical information available in today's literature is the result of investigations of many researchers and theoreticians over many years. The theory of bending and the moment distribution for beams and plates were initially presented by Coulomb in 1836 and Navier in 1820, respectively. These theories are still extensively used in design analysis.

The analysis of a pressure sensor diaphragm illustrates the principles of analytical modeling. The mechanical structure of a typical micromachined silicon pressure sensor is shown in Fig. 69.2. The operation of this device is based on the deflection of a square diaphragm under applied pressure from either top or bottom. Deflection of

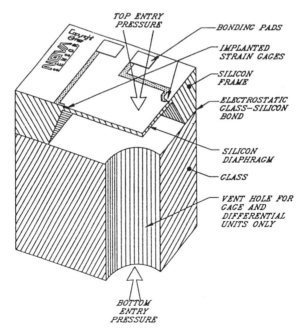

FIGURE 69.2 Mechanical structure of typical micromachined silicon pressure sensor.

the diaphragm generates bending moments that in turn generate stresses in the plane of the diaphragm. By placing stress-sensitive piezoresistive elements in an optimal position on the diaphragm, one can design a pressure sensor with relatively high sensitivity, good linearity, and good overpressure capabilities. Therefore, the first step is to analyze the device under given pressure.

The pressure sensor in Fig. 69.2 is a relatively complex structure for which an analytical closed-form solution is difficult to derive. The natural tendency is to break down the structure into smaller structures that can be more easily modeled. The diaphragm part of the device can be isolated from the rest, and the results of classical plate theory can be applied. The governing equation for displacement of a thin plate under lateral applied pressure in Cartesian coordinates is

$$\frac{\delta^4 W}{\delta x^4} + \frac{2\delta^4 W}{x^2 \delta y^2} + \frac{\delta^4 W}{\delta y^4} = \frac{q}{D}$$

where W, D, and q are plate deflection, rigidity, and lateral loading. Subject to the correct boundary conditions, the solution of the equation yields deformation information necessary to calculate the stress field throughout the plate. (The thin-plate equation is derived under the following assumptions: thin plates with small deflection, no deformation in the plate's midplane (neutral axis), and negligible second-order effects due to shear and transverse normal stresses.)

The next step is to choose an appropriate boundary condition for the diaphragm. The existing classical solutions for thin plates are for clamped and simply supported boundary conditions. For a rigidly clamped boundary condition, the plate's edge can neither translate nor rotate, while for a simply supported case, the rotation is allowed. In reality, the boundary condition, in our case, is neither rigidly clamped nor simply supported but is elastically clamped. In other words, the translation and rotation at the edge depend on the deformation of the support structure surrounding the diaphragm. An approximation to the real condition is to assume a clamped edge. This can be justified due to the relative high rigidity of support structure around the diaphragm.

In certain cases when lateral stress and/or strain effects can be neglected, it is logical to approximate a full 3D problem as a 2D model and seek solutions for plane stress, plane strain, or an axisymmetric equivalent of the original 3D problem.

There are certain limitations to full reliance on analytical modeling. Though simple and quick, the analytical model is only an approximation of the real problem in the majority of the applied design cases. Its results, however, can be used for design initiation and cross-checking with the more powerful computational techniques such as finite difference and finite element modeling.

Finite Element Modeling (FEM)

History. In 1678, when Robert Hooke presented his theory of elastic body deformation, he probably couldn't predict that his simple stress/strain relationship would be used as the foundation for a powerful analysis technique called finite element modeling.

In analyzing a complex structure, it is quite natural for the analyst to decompose it into an ensemble of discrete, structurally known entities. In early attempts to capitalize on this concept, the elastic body was reduced to an assembly of small beams that are well known engineering structures. The overall solution was the collection of solutions for each individual beam. In 1941, Hrenikoff introduced a so-called framework method in which elastic problems could be numerically solved. This and the studies of McHenry (1943), Courant (1943), Newmark (1949), and Page and Synge (1947) are the pioneering work in formulating a technique that was later known as the finite element method. Finite element analysis achieved its commercial acceptance in the aerospace industry. During the 1960s, the aerospace industry in general accepted the technique as a standard analysis method, and it has developed significantly ever since.

Theory and Application.   The finite element method is a computer-based numerical technique that, when applied to solid mechanical problems, evaluates the stress/strain relationship for almost any complex structure of engineering importance. In FEM, the structure is subdivided into many smaller, well defined components called *elements,* which are interconnected by *nodes.* The elements can be 3D (rectangle, tetrahedron), 2D (rectangle, triangle) or 1D (line segment). One can visualize a meshed finite element model as a system of solid springs, as shown in Fig. 69.3, and the nodes as numerical molecules where the primary output information (i.e., nodal displacements in most FEM software) are stored. When a load is applied to the structure, all the springs (elements) deform until the system will reach an equilibrium. The general form of Hooke's law is:

$$\{F\} = [K]\{U\}$$

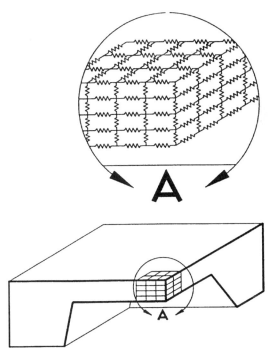

FIGURE 69.3   Meshed finite element model.

where $\{F\}$ is the nodal force vector composed of both external loads and reaction forces applied on the structure. $[K]$ is the global stiffness matrix formed from the contribution of the stiffness of every single element in the model, and $\{U\}$ is the nodal displacement vector. The equation above presents a set of algebraic equations for the solution of nodal displacements in various directions. After applying the boundary conditions, standard numerical techniques are used to solve these equations. With the nodal displacements on hand, the strains are derived from displacements using:

$$\{e\} = [B]\{U\}$$

where $\{e\}$ is the strain vector and $[B]$ is the strain-displacement matrix. In order to derive elemental strains from nodal displacements, an interpolation function throughout the elements is used, usually referred to as the *shape function*. The stress is then obtained from strain:

$$\{s\} = [D]\{e\}$$

where $[D]$ is the constitutive law matrix containing material properties.

It is apparent from the above discussion that the analysis of a complex structure can be reduced to searching for solutions to a large number of algebraic equations that should be solved in a reasonable time on a computer. In general, as an integral part of FEM procedure, the emphasis is to minimize the computer storage space and processing time. Various numerical techniques are used to minimize computer storage of matrix elements and increase the solution efficiency.

Solution of an engineering problem by FEM requires a step-by-step process. These general steps are

1. Formulation of governing equation and boundary conditions
2. Selection of elements, their shape function, and material properties
3. Meshing (i.e., subdivision of the structure into elements)
4. Assembly of global equations
5. Solution
6. Results validation

In today's FEM packages such as **ANSYS** from Swanson Analysis Systems, there are only three major steps for the user:

1. Preprocessing phase: analysis type, element type, structure and meshing, boundary conditions, loading conditions, etc.
2. Solution phase (requires no user interaction)
3. Postprocessing phase: extraction of output data, displacements, stress, temperature, voltage, etc.

Although FEM was initially developed for structural problems, it has recently been generalized to cover design problems in a wide variety of engineering fields. FEM software packages such as ANSYS have a large element library for application, not only to solid mechanics but also to heat transfer, fluid mechanics, electrostatics, magnetics, and other fields.

The following table shows analogous FEM terms of the general forms of Hooke's law for various engineering fields.

Engineering Field	{F} Vector	[K] Matrix	{U} Vector
Statics	Force	Stiffness	Displacement
Heat transfer	Heat flow rate	Heat conductivity	Temperature gradient
Electrostatics	Current density	Electrical conductivity	Electric field density
Fluid mechanics	—	—	—
Irrotational flow	Flow rate	—	Potential gradient

Some of the available commercial FEM software packages are ANSYS, NASTRAN, Algor, PATR and COSMOS/M.

Accuracy. Finite element modeling is a powerful technique for solving design problems. However, it is import to remember that FEM is a numerical approximation rather then an exact solution.

In general, the accuracy of FEM models depends on the complexity of the structure and the amount detail imposed on the model. To model a physical assembly accurately, one must recognize and understand the modeled geometry, loading, boundary conditions and material properties. Other important factors are the selection of appropriate element types, their meshing configuration, their density and the total number of elements (or nodes). In general, the finer the mesh density in the high stress regions, the more accurate the results will be in that region. An alternative is to use higher-order elements with midside nodes that can calculate variable strain fields inside the elements. Unfortunately, in both cases, the computer processing time and storage needed to accommodate the model data increases. The plot in Fig. 69.4 shows the increase of computer processing time versus the number of elements in a particular model (static analysis).

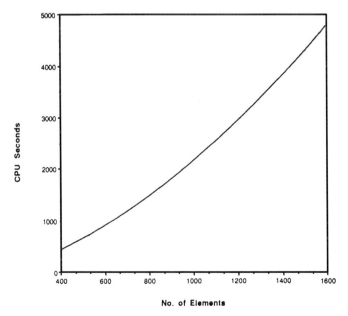

FIGURE 69.4 Representation of number of elements and CPU sensor.

Submodeling. The objective if submodeling is to achieve solutions in the small regions of a FEM model. To accomplish that, the segment containing the region of interest is numerically cut out of the model based on which a second model is built (Fig. 69.5) for which boundary conditions are obtained from the solution of the original model. This channeling down process can be repeated many times with as many models to achieve the desired refinement in the region of interest.

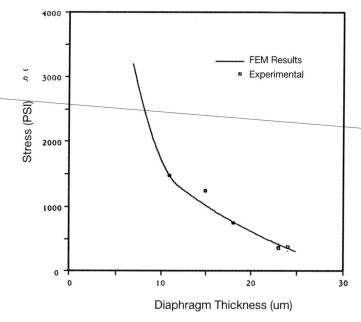

FIGURE 69.5 Representation of diagram thickness (μm) and stress (psi).

Modeling Temperature Effects. Modeling temperature effects is an important step in the design process of a silicon pressure sensor and microstructure. A silicon microstructure typically goes through several process steps in which different films of silicon oxide, nitride, and metal is deposited on. Depending on the sensitivity of the device and/or the geometry of the structure, the mechanical and thermal properties of these films could have a pronounced effect on the certain characteristics of the device. Modeling in these cases can help optimize not only the chip design, but also the process associated with that design. Figure 69.6 shows the modeling concept for multilayer, multitemperature modeling concept, which can be modeled in several steps using FEM analysis.

Finite Difference Modeling

Finite difference modeling (FDM) is the classical numerical approach to complex mathematical problems. In this approach, the physical region is descretized with a regular mesh network and the governing equation approximated. The approxima-

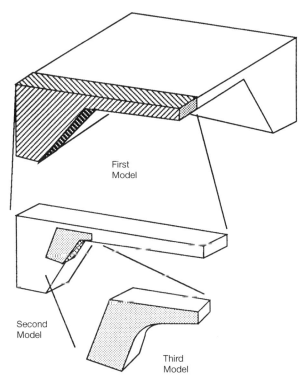

FIGURE 69.6 Multilayer, multitemperature modeling concept.

tion is based on the derivatives in the governing equation and its boundary condition using finite spatial and temporal steps. The resulting algebraic equation is iterated and solved in the meshed region.

The finite difference method is well suited for cases with regular boundaries, since boundary irregularities usually necessitate complicated mathematical conditions for each special case, thus limiting their general use in solid mechanics. FDM modeling is, however, very popular in the areas such as fluid mechanics and heat transfer. Some of the available commercial packages are FIDAP, FLUENT, and PHOENICS.

PROCESS MODELING

A wide variety of process steps are used to fabricate devices on single crystal silicon wafers. These steps include oxidation, ion implantation, dopant (impurity) deposition, diffusion of impurities, silicon nitride deposition, and polycrystaline silicon deposition, to name just a few. Understanding how the different process steps affect previous and subsequent steps and how the combination of all of them affect the final device characteristics is of prime importance to the designer when planning a fabrication sequence. There are processing models available that simulate not only the types of process steps mentioned above but also the photolithography processes that are used to define patterns in the wafer (Fig. 69.7).

SUPREM (Stanford University Process Engineering Model) is a well known processing model for silicon. Both one-dimensional (calculating oxide thickness and impurity concentrations as a function of depth only) and two-dimensional (calculating impurity profiles laterally as well as vertically) versions are available. The available process steps in SUPREM include not only oxidation, diffusion, and ion implant processes, but also thin-film processes like silicon nitride deposition or polysilicon deposition. SUPREM output includes oxide thickness and impurity concentration data based on well known process models (e.g., the Deal-Grove oxidation model). The user can define special models for use when the classic formulas are not appropriate. SUPREM also calculates certain device parameters for a simulated process (i.e., oxide capacitance, junction capacitance, MOS threshold voltage, and sheet resistance).

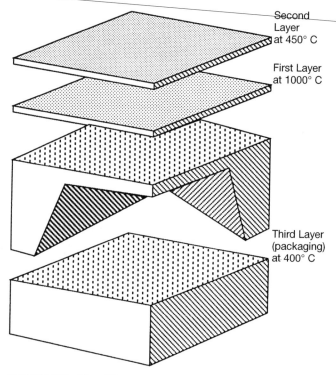

FIGURE 69.7 3D multilayer patterns.

COMPUTER-AIDED LAYOUT OF SENSORS AND MICROSTRUCTURES

The patterns and dopant profiles designed and modeled using the computer techniques discussed above must somehow be transferred from concepts to silicon hardware by pattern-creating techniques suitable for batch fabrication. Photolithography is the main pattern-creating technique for silicon sensors and microstructures. Photolithography for micromachined structures, like that for integrated circuits, uses high-quality photographic halftone plates (*photomasks*, or simply *masks*) on which the design of a single process level of a single silicon chip is replicated hundreds or thousands of times. Computer-aided layout (CAL) provides great precision in the design process for these masks and allows rapid design revisions.

Since silicon microstructures are three-dimensional (3D) patterns while photomasks are two-dimensional (2D) patterns, the designer's first step in creating a new design is to envision a 3D shape and abstract somehow that shape into a 2D pattern that will, when combined with available micromachining tools, create the desired 3D structure.

This envisioning process is similar to that followed by the traditional mechanical engineer, who envisions a three-dimensional part that may be formed by a combination of casting, turning, drilling, and milling. in the same way, a silicon 3D structure may be formed by a combination of epitaxy, oxidation, photolithography, etching, and bonding on of another flat layer. The designer must allow for chemical etching effects such as erosion of overlying masking layers, undercutting of masking layers (which leaves overhanging shelves of masking material), rounding off of corners by the etchants (which is desirable in some situations and undesirable in others), and sloping of etched sidewalls to create the final 3D shape. This portion of the design process requires a strong intuitive understanding of the micromachining tools available.

Once a 3D shape compatible with a 2D mask has been roughed out (either mentally or with the aid of pencil and paper; as yet there are no CAD tools for this portion of the design process), 2D photomasks can be created using layout tools that were first developed for the microcircuit industry. A host of layout systems are available on various hardware platforms, including DEC MicroVAX, SUN, IBM PC and PS/2, Apple MacIntosh, and Silicon Graphics IRIS workstations. These systems allow color drawings of photomasks to be created and laid atop one another on the computer screen in order to check for the compatibility of various process steps, then transferred to magnetic tape, which can be transported to a mask-making shop.

There is an unmet need for a combined 2D and 3D capability in this mask design process. The available mask design programs, including offerings from Schlumberger, Calma, DKL technology, CAECO, and others, are strictly 2D systems. Various 3D drafting programs exist but are limited in their ability to serve as 2D mask design systems. AutoCAD running on various workstations has some possibilities for combining 2D and 3D capabilities and, in the future, may improve in this regard.

Once a set of masks has been designed and sent to the mask-making shop, the designer must track this design through the mask-making process to make sure that patterns are replicated correctly and that they result in high-quality photographic plates. These plates are typically made of 5-in.^{2} soda-lime glass coated with a thin opaque chromium layer in which the mask is patterned. Once the masks for a given process are complete, the actual fabrication of the silicon device can begin.

ELECTRICAL MODELING FOR SILICON SENSORS

Many design tools are available today to help the sensor engineer with circuit design, from simple passive temperature compensation networks to full custom signal conditioning ICs.

Digital Circuit Simulators for Analog Circuits

One of the most generally used tools is the SPICE circuit simulator. This digital simulator for analog circuits is available in numerous forms and various levels of sophistication. Versions are available to run one very conceivable computer from the IBM PC to a supercomputer like the CRAY. While small circuits can be simulated on a PC, medium to large circuits require the power of an engineering workstation to achieve reasonable convergence and simulation times.

SPICE provides several types of circuit analysis, including AC response, DC response, and transient response. The SPICE program uses numerical iterative solutions of a NET of components described to SPICE in a SPICE deck. This term is a holdover from the days of punched card data entry in the form of a deck of cards. Today, the deck is replaced by a simple ASCII text file. Each card in the deck describes to the SPICE program the characteristics and nodal connections of one basic circuit element or subcircuit such as resistors, capacitors, diodes, or transistors.

SPICE uses a method of calculation called *relaxation* to force the sum of currents at each node in the circuit to zero on an instantaneous time interval basis. During each time interval, the program solves the equations describing the current and voltage in each component in the net iteratively until all of the nodes in the circuit satisfy the conditions for equilibrium within an adjustable tolerance. Additional enhancements are usually added to adjust dynamically the size of the time interval to suit the rate of change of the node voltages and currents during the simulation. These changes are used to minimize simulation time.

Stored within SPICE are predefined models for most of the common electronic components. SPICE models include numerous electrical and thermal parameters for each device. This allows full modeling of circuits over all possible voltage, current, and temperature extremes. Devices that are not predefined can be modeled by the user using fundamental elements such as resistors, capacitors, inductors, and dependent sources. If these elements are not sufficient, more sophisticated models utilizing polynomial equations are available.

Various enhancements have been added to SPICE by third parties licensed to sell the SPICE product. Typical of these enhanced versions of SPICE is the Analog Workbench by Analog Design tools. has added, in addition to improved convergence techniques, a graphic circuit entry format (sche entry) graphic screen and hardcopy parametric output (performance data), onscreen icons for typical work instruments, and aids for checking design integrity.

The design integrity of modules for worst-case and Monte Carlo analysis and parametric variation an as well as checks for excess component stress enhances the design engineer's probability of a success when first circuits are fabricated.

With very few exceptions, digital circuit simulators for analog circuits can provide excellent results for almost any circuit. In some cases the simulator is the only method of design verification due to the inability to build suitable *breadboards,* or prototypes of circuits that will be integrated on a single chip in the final product.

Digital Simulation and Design Aids for Digital Circuits

- Faster computers
- Smarter computers
- Application-specific integrated circuits (ASIC)
- Microcontrollers

- PALS
- EPLDS
- Memories

What do all of these have in common? All have become mainstays of the high-volume electronics industry, and all would be virtually impossible to design with were it not for simulations and design aids.

Every day, design of digital ASICs has become so complicated due to the ever increasing circuit density that the chance of a working design on the first try with manual design and layout techniques is almost negligible.

Digital simulators provide the ability to build a complex circuit in software and verify its performance before fabrication. As with the analog simulators, all levels of performance are available from PC-based to dedicated hardware simulators that run in near real time.

Digital simulation packages check the correctness of logic elements, analyze timing errors, accept simulated digital input signals, and can accept information directly from a schematic capture program that can also provide the required information to a computer aided layout program. This single database approach ensures that the simulation program and the layout program are each working with exactly the same circuit.

A trend toward mixed circuit ASICs (combined analog and digital circuits) will place additional demands on the analog and digital circuit simulators, with all-in-one simulators coming to market within the next year or two.

Software Development Aids

Sensors, like most other products, benefit from the computer revolution. Just as in improved test capability, computers—or, more specifically, microcontrollers—will revolutionize the sensor business by providing improved performance at lower cost.

Embedded microcontollers that provide signal conditioning and control algorithms have traditionally been programmed in assembly language. Assembly language is the lowest level of readable English programming language. Development in assembly language is slow and difficult to simulate and debug.

Since about 1986, a programming aid new to the microcontroller but available for three decades on large computers, allows a program to be written in a high level language such as Basic, Fortran, or C. This tool is the compiler, a translator from the high-level language to the assembly language of a specific computer. High-level languages provide a degree of portability to the software developed so a change from one processor to another does not involve a total rewrite of already developed software.

Sophisticated debugging tools are available for these high-level languages. They are called source-level debuggers. These debuggers can interrupt program execution on error or user-specified conditions and point directly to the source code that cased the condition. Development cycles are shortened through this computer design aid.

A new branch of the CAE field is computer-aided software engineering (CASE). This program development aid provides assistance to the engineer in the form of a structured program-development environment. Programming is done top-down, in the strictest sense of the word, starting with descriptions of the program, modules, subroutines, and finally down to actual code generation and debugging. CASE also allows development of large programs by multiple programmers while assuring integrity of the overall program by taking care of the housekeeping tasks associated with the interactions of the various program modules.

CHAPTER 70
SIGNAL CONDITIONING FOR SENSORS

INTRODUCTION

Silicon sensors, as well as other silicon devices, exhibit high temperature sensitivities and variation of performance from unit to unit. Some of these variations are introduced as a result of intrinsic properties of silicon, and others reflect the nature of semiconductor processing and fabrication. Signal conditioning compensates for these variations and typically provides the following functions:

- Initial offset balancing
- Offset temperature compensation
- Sensitivity normalization
- Sensitivity temperature coefficient compensation

For more dedicated and sophisticated applications that require either a higher accuracy or an additional functional signal processing, other types of signal conditioning may be required. More frequently used functions include the following:

- Pressure nonlinearity correction
- Temperature nonlinearity correction for offset and sensitivity
- Static pressure effect compensation (process control applications)
- Shunt calibration, input resistance, symmetry correction (medical applications)
- Temperature output processing
- Temperature compensation feedback to the next level electronics

Taking into account all the preceding factors, the design of signal conditioning for silicon piezoresistive sensors becomes a relatively complex task that requires an indepth understanding of both the capabilities and limitations of silicon sensor technology.

Efficient compensation/calibration procedures are essential to low-cost applications. Silicon-based sensors produce output signals on the order of 1 to 10 percent of a supply voltage. This is typically equivalent to a strain gage resistance change due to temperature variations by 2 to 40°C. While the silicon sensor output level is high, as compared to some of the technologies such as metal film strain gages (with typical output of 0.2 percent of a supply voltage), it is most often not sufficient for a direct interfacing to systems and instruments. Signal-to-temperature error for silicon sensors is less favorable than for other technologies. This requires correspondingly a more sophisticated signal conditioning.

The effect of the temperature on all four resistors is almost identical. The first level temperature compensation is thus achieved by selecting the operating configuration of a sensor as a fully active Wheatstone bridge. However, most of the applications require better accuracy and thus additional signal conditioning has to be employed.

Implementation of the signal conditioning traditionally relied on analog circuitry. Recently, a new class of digital signal processing started to emerge initially in a high end (e.g., process control transmitters) and more recently low end (e.g., bicycle altitude computer) of the market.

It should be pointed out here that most of sensor errors can be compensated for using more or less sophisticated signal conditioning, and thus even temperature errors are not a factor limiting a potential for measurements accuracy. The only errors that cannot be compensated for, and thus limit the potentially achievable accuracy, are time-related stability and thermal hysteresis of a sensor. Digital signal processing does approach this limit in selected aerospace applications. While discussions in this section focus primarily on pressure sensor applications, the basic ideas are valid for all piezoresistive sensors.

CHARACTERISTICS OF PRESSURE SENSORS

In order to provide a measurement of an input variable, such as pressure or acceleration, the ideal sensor should exhibit a transfer characteristic dependent only on a measured variable. In practical implementations, silicon sensors, in addition to a broad variation of characteristics between units within production lots, also exhibit sensitivity to several environmental variables and do not deliver the required transfer function as manufactured. The origin of the measurement errors related to the preceding factors is either mechanical or electrical.

Examples of mechanically originated error sources related to pressure sensors include nonlinearities of diaphragms, package stress transfer, difference of temperature coefficient of expansion of materials used in sensor die construction (e.g., silicon and glass), and the like.

Examples of electrically originated error sources include band gap voltage and holes mobility change with temperature, junction leakage currents, dielectric mobile charge effect, and so on.

From the application point of view, the errors can be categorized into five functional groups:

- Calibration
- Static
- Dynamic
- Temperature
- Stability

Calibration errors define an initial offset and sensitivity variation between different sensors within production lots.

Offset voltage (or zero pressure output voltage) represents the bridge output voltage without any input pressure. It is one of the more unpredictable sensor parameters. Offset voltage shows values different from zero as a result of an electrical mismatch of the sensing resistors and from mechanically induced stresses in the sensor. Typical full-scale output for silicon sensors ranges between 1 to 20 mV/V, thus an offset of 1 percent FSO requires an effective bridge resistor matching of 0.001 to 0.02 percent. Even with today's state-of-the-art IC processing technology, this level of matching tolerance cannot be achieved directly and requires some form of auxiliary adjustment network either on or off the sensor chip.

Pressure sensitivity represents a normalized sensor response to applied pressure. Normalization is in respect to a unity bridge voltage and unity pressure and it is often expressed as mV of FSO per one volt of bridge voltage and one psi (mV/V/psi). For a given sensor, the pressure sensitivity is independent of the type of excitation (voltage or current) or pressure range.

Pressure sensitivity is directly proportional to a gage factor that defines a relationship between a relative gage resistance change and the applied stress. It is also proportional to an active diaphragm area and inversely proportional to a squared diaphragm thickness.

Full-scale output (FSO) voltage is defined as a product of pressure sensitivity S and bridge voltage V_b. FSO variations from die to die and wafer to wafer result from the nature of semiconductor processes used to fabricate silicon sensors. Parameters that affect the sensitivity include primarily diaphragm thickness, strain gage placement, doping levels, and bridge resistance if the sensor is excited in a constant current mode.

Static errors relate to a measurement accuracy of a constant (static) pressure. Errors in this category include pressure nonlinearity, hysteresis, and repeatability.

Deviation of sensor's transfer function from a straight line is defined as *pressure nonlinearity*. This nonlinearity is generally affected by sensor design parameters such as gage geometry and doping levels, diaphragm design, operating stress levels, operating power level, and electrical and mechanical gage symmetry. Pressure nonlinearity for silicon sensors is stable with time and therefore can be compensated by the use of an active signal conditioning circuitry. Nonlinearity exhibits certain levels of temperature dependency.

The maximum difference between sensor output at the same pressure for increasing and decreasing pressure is called *pressure hysteresis*. The maximum difference of the output for the same direction of pressure is called *repeatability*.

Dynamic errors characterize dynamic measurement accuracy and include such error categories as frequency response, rise time, phase shift, overshoot, settling time, and so on.

Temperature errors describe variation of all sensor characteristics with temperature. Most frequently used ones include offset, sensitivity, pressure nonlinearity temperature variations, and bidirectional temperature repeatability (often called thermal hysteresis).

Stability errors define a sensor's ability to maintain the same transfer function over a certain time period. This time frame can be defined as eight hours for disposable blood pressure sensors or ten years for some military applications. Traditionally, the most troublesome parameter for silicon sensors was stability of the initial bridge offset. Long-term instability was the least understood parameter of silicon sensors that created major application delays in several markets, including process control. Recent developments, such as NovaSensor's SenStable process, brought major progress in this area.

CONSTANT CURRENT VERSUS CONSTANT VOLTAGE EXCITATION

Primarily the intrinsic properties of silicon strain gages affect the temperature dependance of the bridge resistance and pressure sensitivity.

In a *constant voltage* excitation mode, the FSO is directly proportional to pressure sensitivity. Since pressure sensivity has negative temperature coefficient, FSO will change in the same mode.

$$FSO = SV_b$$

The temperature coefficient (called TCS) of pressure sensitivity S is dominated as previously mentioned by piezoresistive properties of silicon. Within limits, the value of TCS can be adjusted over -0.05 to -0.25 percent °C by changes in the doping level and process conditions used to create the strain gages. This adjustability is utilized to provide a simple compensation option in a constant current mode.

Compensation of this negative change can be achieved with temperature-sensitive external or on-chip components, such as thermistor or diode.

With a *constant current* supplied to a bridge, its voltage is proportional to the bridge resistance R_b, and thus FSO can be expressed as:

$$FSO = SR_b I$$

Bridge resistance R_b changes with temperature and is characterized by its temperature coefficient TCR that can be adjusted within limits over 0.05 to 0.35 percent/°C by changes in the semiconductor processes used in the sensor fabrication. There is a strong correlation between TCR and TCS, so changing one of these parameters automatically readjusts the other one. Bridge resistance (and thus bridge voltage) increases with temperature and partially compensates a decrease of pressure sensitivity, providing foundation for a simple temperature compensation technique.

While both the constant current and constant voltage excitation configuration provide viable options, their differences can be summarized as follows:

- Constant current mode provides a possibility for a simpler temperature compensation.
- Constant voltage mode provides a potential for better accuracy, since bridge resistance variation does not affect the output.

ANALOG ELECTRICAL MODELS OF PIEZORESISTIVE PRESSURE SENSORS

There are three basic techniques for providing compensation for sensor errors:

- Calculated guess (sometimes from a look-up table) of a needed solution, repeated until the desired compensation is achieved. This technique requires long test cycles and is suitable only for small-volume production.
- Active trimming, utilizing often repetitive iterations to trim interactive components (e.g., offset trim changes sensitivity).
- Passive trimming based on a modeled test data. This approach offers not only a quick throughput, but allows a software solution of simultaneous equations to calculate exact values of interacting components.

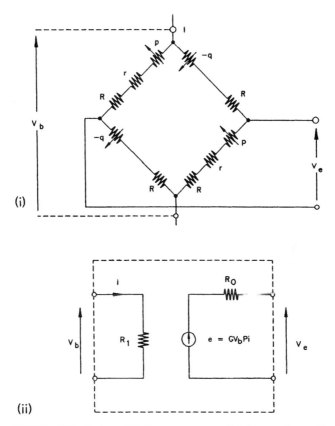

FIGURE 70.1 Basic model of pressure sensor: (i) bridge configure; (ii) equivalent electrical circuit.

Modeling sensor performance to compensate the errors seems to be the most efficient high-volume production approach.

The modeling technique introduced by Bryzek in Ref. 1 converts the primary sensor parameters into values of equivalent resistors which enable the modeling of a sensor's performance in essentially every possible electrical configuration. The three sensor parameters used in modeling include the following:

- Offset voltage
- Normalized pressure sensitivity
- Bridge resistance, created typically by the four arm resistors, two of which increase and the other two decrease with applied pressure

Each of these parameters is affected by several variables, out of which the most important are the following:

- Differential pressure across the diaphragm
- Operating temperature

- Excitation voltage or current level
- Static pressure

In order to build a suitable model, a number of measurements of sensor performance must be performed. The minimum number of measurements is set by the required accuracy level. More accurate models require more measurements to be made.

If certain functions can be defined as results of known correlations with some other variables (e.g., temperature coefficient of pressure sensitivity and bridge resistance as a function of doping concentration), then the required number of measurements can possibly be reduced.

The limit of a possible modeling accuracy is set by the long-term stability and thermal hysteresis of the sensor.

Basic Model

The simplest model of a piezoresistive pressure sensor has to include three components providing a degree of freedom to represent each: offset voltage, pressure sensitivity, and bridge resistance. The model shown in Fig. 70.1 is a little more complex, including four elements. These four elements have to be calculated for each test temperature and pressure. Each of them can be made sensitive to other variables such as static pressure and excitation current level, if needed, and number of test points may be increased accordingly. Most often, however, these values are calculated only at two test temperatures.

- Resistance R represents bridge resistance of a perfectly balanced bridge with no pressure applied.
- Resistance r represents bridge unbalance (offset) at zero pressure applied.
- Resistance q represents pressure sensitivity of the strain gages subjected to compression.
- Resistance p represents pressure sensitivity of the strain gages subjected to tension.

Differentiation between p and q sensitivities is justified by possibly different stress levels in the areas of tension and compression on the diaphragm. For example, in the rectangular diaphragm, the maximum stress at the edge can be more than two times higher (and of different polarity) than the stress in the center. If p and q are not equal, bridge resistance will change with an applied pressure.

The model introduced here works best if the following assumptions are valid:

- Bridge resistance is symmetrically distributed between left and right half-bridges.
- Pressure sensitivity of strain gages increasing and decreasing with pressure is respectively the same (symmetrical) in both half-bridges.

This model is sufficiently accurate for the basic temperature compensation applications. It should be noted, however, that the accuracy of offset compensation will degrade if a bridge symmetry is poor.

All four elements R, r, p, q of the model can be calculated based on the results of simple measurements of the bridge output performed with a constant current excitation 1. Denoting V_0 and V_1 the output voltage and V_0 and E, the bridge voltage, both respectively at zero input pressure and full-scale input pressure, the values of model's components can be found as follows:

$$R = (E_0 - V_0)$$
$$r = 2V_0/I$$
$$q = (V_1 - V_0 + E_0)/I$$
$$p = (V_1 - V_0 - E_1 + E_0)/I$$

This procedure should be repeated for each test temperature. If a more complex model is used, then it should be repeated also for each static pressure, test current, and each additional pressure needed to model pressure nonlinearity. Each of the model's four elements can then be defined as a respective function of these variables. Any curve-fitting technique can be then used to make an interpolation for the other points.

For basic temperature compensation applications, it is usually sufficient to know values of the elements R, r, p, q only for two temperatures in order to calculate the required compensating components. This requires measurement of bridge output and bridge voltage with and without applied pressure at two temperatures, for a total of eight measurements.

Measurement of bridge arm resistance can be performed either directly or indirectly in a closed bridge configuration. The selection of a closed bridge test configuration offers several major advantages:

- The model will represent the distribution of voltage potentials and leakage currents across the diaphragm very close to a final operating distribution. Since typically all strain gages are isolated from the silicon substrate through a p-n junction, this configuration includes all parasitic effects that would not be included by a direct resistance measurement. This issue is specially important for a new generation of sensors with very shallow (submicron) junction geometries.

- Excitation current can be turned on for a long enough time before the measurements to allow a proper warm-up of the sensor, without affecting the testing cycle time.

- Smaller number of data acquisition channels is required (two instead of four for each sensor) resulting in major savings on cabling and input modules for production test systems that support simultaneously large numbers (several hundred) of sensors.

- Better accuracy of measurements is achieved. Typical resistance change with a full-scale pressure change is typically in the range of 2 percent of the bridge's arm resistance. Requirement for a pressure sensitivity modeling accuracy of 0.1 percent creates a demand for a bridge arm resistance measurement accuracy of 10 ppm. Typical sensor has approximately 2500 ppm/°C temperature coefficient. To obtain a direct resistance measurement accuracy of 10 ppm, the sensor's temperature should be kept constant to about 0.001°C during measurements, which is not a simple task.

- Voltage measurement mode in a closed bridge configuration offers a distinctive advantage over a resistance measurement mode: ability to measure directly a differential output voltage with a simultaneous rejection of a common mode signal. For the same modeling accuracy as previously mentioned, only 0.05 percent voltage measurement accuracy is required as long as a voltmeter has high enough common mode rejection.

- Closed bridge configuration rejects a majority of other common mode errors as well, such as first-order temperature sensitivity and RFI interference, simplifying requirements for stabilization of temperature and shielding.

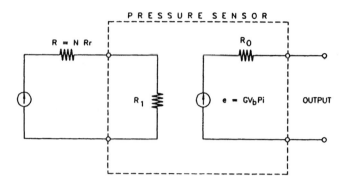

(i)

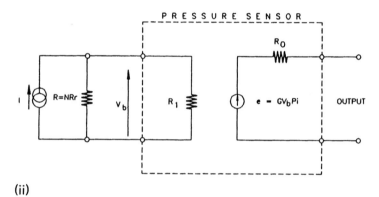

(ii)

FIGURE 70.2 FSO span calibration resistor for constant voltage (i) and constant voltage (ii) excitation.

Using the calculated model's components we can easily create the equivalent electrical circuit useful in modeling applications as shown in Fig. 70.2. The circuit is analogous to an instrumentation amplifier model and it consists of three elements: input resistance R_1, output resistance R_0, and pressure-controlled voltage source e. The values can be easily calculated as follows:

$$e = S\,V_b\,P_1 \text{ for voltage supply}$$

$$\text{or } e = S\,R_i\,I\,P_1 \text{ for current supply}$$

$$R_i = R + (r + p - q)/2$$

$$R_0 = R + (r + p - q)/2$$

where S is a normalized pressure sensitivity and P_1 is input pressure. Please note that input and output resistance are the same; this is a result of the insufficient degrees of freedom built into the model.

To demonstrate the effectiveness of the introduced model, we evaluate the effect of a single resistor FSO calibration on temperature coefficient of FSO. There are two basic supply configurations for the sensor: constant current and constant voltage. From an electrical point of view, both circuits require the same value of a resistor R to provide the temperature compensation of FSO. With a constant voltage excitation source [Fig. 62.2(a)], the FSO calibrating resistor R should be connected in series with the sensor bridge and with constant current excitation [Fig. 70.2(b)], it should be connected in parallel to the bridge.

Insertion of the calibration resistor R introduces a loss of output FSO for given supply voltage or current. There are two different expressions for output FSO as a function of excitation mode. For a voltage mode, the bridge voltage is defined as follows:

$$V_b = E\, R_i/(R + R_i)$$

and thus FSO is expressed as follows:

$$\text{FSO} = S\, E\, R_i\, P_i/(R + R_i)$$

For the current mode, the equations are modified as follows:

$$V_b = I\, R_i\, R/(R_i + R)$$

and thus FSO:

$$\text{FSO} = S\, I\, R_i\, R\, P_i/(R_i + R)$$

Denoting by R input resistance of the bridge at reference temperature and expressing calibrating resistor R as a multiple N of this reference bridge resistance:

$$R = N\, R_r$$

we can now plot the FSO normalized to the original FSO without the normalizing resistor R at the same excitation level, as shown in Fig. 70.3. For small value N, a constant voltage (CV) mode provides a smaller loss of the output signal. For high values N, a constant current (CC) mode provides lower loss of an FSO.

To perform temperature errors evaluation, we have to include temperature coefficients of bridge resistance and pressure sensitivity. Two popular diffusion processes used in manufacturing of pressure sensors yield the following:

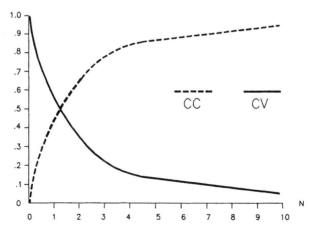

FIGURE 70.3 Loss of FSO as a function of calibrating resistor R expressed in multiples of bridge resistance.

	Process 1	Process 2
TCR:	0.25%/°C	0.14%/°C
TCS:	−0.21%/°C	−0.11%/°C

Bridge resistance R and pressure sensitivity S can now be related to the temperature deviation dT from a reference temperature condition, respectively R_r and S_o:

$$R = R_r(1 + dT\,\text{TCR})$$

$$S = S_o(1 + dT\,\text{TCS})$$

FSO temperature error can then be plotted as shown in Fig. 70.4. As can be seen, for a certain value of R, a zero temperature coefficient of FSO is obtained, providing temperature compensation. To show the effect of possible process variations on FSO, the error for several other possible temperature coefficients of bridge resistance is shown. For higher values of N, a sensitivity of the FSO temperature error to process variations such as TCR is smaller. Also, for process 2, a lower value of N is required to achieve the zero temperature coefficient of FSO, Fig. 70.5.

As shown in Fig. 70.2, the implementation of this simple analog model made the temperature compensation of FSO easy.

High-Performance Analog Model

The first model assumed a symmetry within a bridge. This allowed a reduction of the number of measurements necessary to create the model. The major drawback of that model is a relatively low accuracy of offset and input and output resistance, decreasing the effective sensor accuracy in applications that require absolute sensor calibration.

Major improvements of modeling accuracy can be achieved when a model is upgraded to four degrees of freedom, one for each of the four bridge resistors R1,

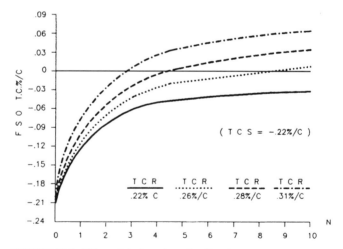

FIGURE 70.4 Effect of single resistor R calibration on span temperature coefficient. The resistor value is expressed in multiples N of input resistance at reference temperature. This is a high TCS process.

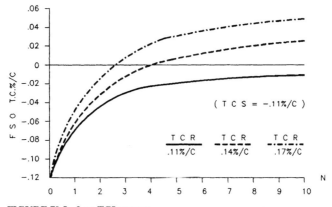

FIGURE 70.5 Low TCS process.

R2, R3, R4 (Fig. 70.6), each changing independently with differential pressure, static pressure, temperature, and excitation power level. The equivalent electrical circuit is the same as shown for a simpler model (Fig. 70.2); however, the expressions for differential input and output resistance are different:

$$R_i = (R1 + R4)(R2 + R3) / (R1 + R2 + R3 + R4)$$

$$R_o = (R1 + R2)(R3 + R4) / (R1 + R2 + R3 + R4)$$

In order to calculate all four bridge resistors, we need two times as many measurements as before: at least four measurements for each pressure, temperature, and supply voltage or current value. To compare with a previous example, two-pressure and two-temperature compensation would require at least 16 measurements, four at each pressure-temperature test point.

The new technique of closed sensor bridge testing and modeling was developed by Bryzek.[1] This unique method allows an efficient testing of sensors in a closed bridge configuration to derive values of all four of the model's resistors with a full accuracy of the test system used. This test system uses two current sources I1 and I2 (Fig. 70.6) for sensor excitation. Two measurement cycles are employed.

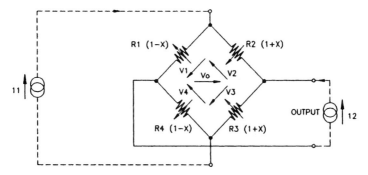

FIGURE 70.6 High-performance model pressure sensor. Dashed line indicates interconnection of current during testing.

During the first measurement cycle, only the main current source I1 is used (I2 = 0). Five measurements are made: offset voltage V_o and four bridge arm voltages V_1, V_2, V_3, and V_4. During the second measurement cycle, the additional crosscurrent I2 is injected to the bridge output and two additional voltage measurements across resistors R1 and R2, respectively V_{1c} and V_{2c} are performed.

Calculation of bridge arm resistors occurs in two phases. During the first phase, the set of ratiometric simultaneous equations is solved, yielding values of resistors with an accuracy limited by the ratiometric accuracy of the voltmeter operating in several volt range. Ratiometric arrangement of equations improves the accuracy over a direct voltage approach. The set of simultaneous equations is very straightforward and easily converging:

$$R1/R4 = V_1/V_4$$

$$R2/R3 = V_2/V_3$$

$$R1/R2 = \text{Abs}(V_1 - V_{1c})/\text{Abs}(V_2 - V_{2c})$$

$$(V_1 + V_4)(R1 + R2 + R3 + R4) = I1(R1 + R4)(R2 + R3)$$

As can be seen, the value of the crosscurrent I2 is not used directly in the equations, so the requirement for the accuracy of this current is drastically reduced. It should be stable from the time of V_{1c} measurement to the time of V_{2c} measurement, typically a 20-ms time frame.

Absolute voltmeter accuracy, for example, at the 10-V range, where arm voltages V1 to V4 are being measured, is typically lower then at the 100-mV range, where the offset V_o is measured. To take advantage of this higher accuracy, the second phase of calculations is implemented. There is only a single measurement made with a better accuracy: offset V_o, creating a singe degree of freedom. There are four previously calculated resistors that possibly require a correction. The correction can be incorporated into one of them or all of them. From a statistical point of view, the most appealing approach is a symmetrical redistribution of values of all resistors to equalize the calculated (based on calculated resistor values R1 through R4) and measured offset values. One possible redistribution possibility using a variable x follows:

$$R1 = R1(1 - x)$$

$$R2 = R2(1 + x)$$

$$R3 = R3(1 - x)$$

$$R4 = R4(1 + x)$$

Using simple algebraic manipulation, the correction value of x may be derived as:

$$x = (-T + \sqrt{Y})/(2Z)$$

where $T = 2\,I1\,(R2\,R4 + R1\,R3) + V_o\,(R1 - R2 + R3 - R4)$
$Z = (R2\,R4 - R1\,R3)\,I1$
$Y = T^2 - 4Z\,[Z - (R1 + R2 + R3 + R4)\,V_o]$

Implementing the modeling approach we have just outlined with an HP3456 voltmeter demonstrated a 10-ppm differential measurement accuracy of resistors between a direct four-wire technique and the modeling technique introduced here. In typical applications with 5 V across the pressure sensor, 10 ppm is equivalent to a

25-UV of offset error during the first modeling phase of calculation. The second phase of calculations corrects this error to below a 1-UV level, providing thermal voltages are not present in the system.

Using the previously outlined model, a system simultaneously testing 200 sensors was built based on a total of only 406 analog input channels with the effective modeling speed of 5 sensors per second (35 measurements/second) per each test point.

BASIC CONSTANT CURRENT COMPENSATION

Basic constant current compensation was pioneered in the 1970s by Foxboro/ICT and Kulite.[33] This technique depends on a proper selection of strain gage doping levels to achieve a condition when the TCR > abs (TCS) and then uses a combination of temperature-sensitive bridge resistors and temperature-stable external resistors to deliver temperature compensation. This technique is briefly outlined in the following sections.

Compensation of Offset Voltage

Compensation of offset voltage involves both initial (e.g., room temperature) bridge unbalance (V_o) and its temperature coefficient (TCV_o).

Correction of an initial offset V_o (often at 25°C) may be easily achieved by the insertion of a series resistance in one arm of the bridge (R3 or R4 in Fig. 70.7). If the offset is positive (the potential at pin 4 is higher than at pin 10) then insertion of a resistor R4 will bring the offset to zero and a resistor R3 should be shorted. When the offset is negative, the reverse is true. Typically, the value of a balancing resistor is smaller than 1 percent of the bridge resistance and thus, practically, it does not change the temperature coefficient of offset in a constant current mode.

Compensation of temperature coefficient of offset (TCV_o) is then achieved by altering the effective TCR of one arm of the bridge with respect to the three remaining arms by shunting it with a temperature-stable resistor (R1 or R2).

When a TCV_o is positive (the potential at pin 4 is increasing faster than potential at pin 10), a decrease of this TC may be achieved by a decrease of the effective TCR of the strain gage connected between pins 12 and 10. This may be accomplished by a parallel connection of a temperature-stable resistor R1 (Fig. 70.8). With a negative coefficient of offset voltage, the decrease of the TCR of the other arm will be accomplished by resistor R2. Only one of these resistors is used for a given sensor. Each of them affect the initial offset, and thus the value of a resistor R3 or R4 has to compensate for this change.

Typical production testing must use at least two test temperatures to gather necessary information. Based on measured data, the computerized sensor model can be developed and a set of simultaneous equations solved to provide the values of compensating resistors. When three

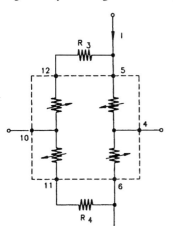

FIGURE 70.7 Offset compensation.

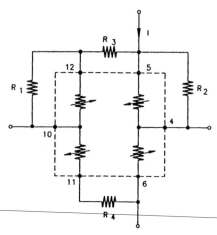

FIGURE 70.8 Offset TC compensation.

test temperatures are used, the offset voltage can be brought to zero at a reference temperature T_r and equal temperature errors between temperatures T_c (cold) and T_h (hot) can be obtained (Fig. 70.9). The temperature error at temperatures T_c and T_h is then a function of the temperature nonlinearity of offset. For a linear offset error with temperature, the errors at T_c and T_h would be equal to zero.

Typically, for OEM pressure sensors, the test temperatures are set as follows:

$$T_c = 0°C$$

$$T_r = 25°C$$

$$T_h = 70°C$$

Over this temperature range, nonlinearities are on the order of 0.3 percent FSO. In practical applications, the inaccuracy of resistors used for compensation may contribute the same amount of error as a sensor itself.

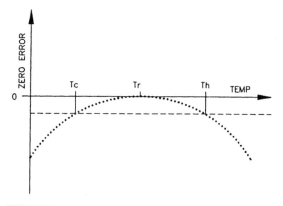

FIGURE 70.9 Typical zero curve after compensation.

It should be noted that the offset voltage of a silicon sensor bridge is not perfectly proportional (ratiometric) to the excitation current. Both voltage sensitivity and self-heating effects may result in a nonlinear change of offset voltage with applied supply current or voltage. In this case, the additional offset error may be expected if a bridge current during a test differs from the value after compensation (when, for example, the R5 resistor is connected).

Compensation of Full-Scale Output

As was discussed previously, a constant current excitation allows for a superposition of the negative temperature coefficient of pressure sensitivity and positive temperature coefficient of bridge resistance to achieve a compensation of a full-scale output. Selection of a current excitation is significantly influenced by a compensation efficiency measured in terms of the lost output after compensation. For many sensor processes, the mean value of the temperature-compensating resistor R5 necessary to compensate FSO is on the order of five times the bridge voltage. The loss of output (as compared to an uncompensated FSO at the same excitation) associated with a current compensation (parallel resistor) is thus about 16 percent. In a voltage mode (resistor in series with the bridge), a signal loss would be about 84 percent. This explains why constant current excitation is recommended for silicon sensors in low-cost applications.

The mean TCR of a sensor bridge is usually designed to be larger than the absolute value of TCS over compensated temperature ranges. This assures that compensation can be achieved for a production variation of sensor parameters. By decreasing the output resistance of the constant current source (Fig. 70.10) with a resistor R5 in parallel to the bridge (or by increasing the output resistance of a constant voltage source with a resistor R5 in series with the bridge), the temperature compensation condition may be achieved.

A single resistor FSO temperature compensation delivers a two-point temperature compensation, where FSO has the same value at two selected temperatures.

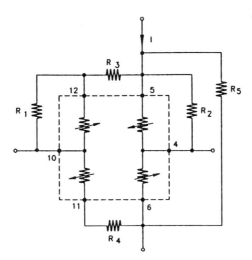

FIGURE 70.10 FSO TC compensation.

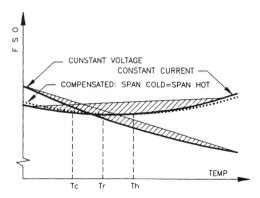

FIGURE 70.11 FSO vs. temperature. Shaded area indicates temperature nonlinearity.

Due to the temperature nonlinearity of both TCS and TCR, as shown in Fig. 70.11, a linear compensated response cannot be achieved with a passive circuitry.

The magnitude of the residual FSO errors after a single resistor compensation for typical processes is typically better than ± 0.5 percent in the compensated range 0 to 700°C rising to about $+3$ percent at 120°C and $+5$ percent at -40°C.

The distribution and repeatability of pressure-sensitivity temperature errors from unit to unit is significantly better than the distribution of offset temperature errors. For less-demanding applications, a statistical temperature compensation of the FSO (no pressure testing at a second temperature) can be easily implemented.

To achieve a flatter temperature response over a wider temperature range, either a different sensor process or a three-point temperature compensation can be implemented. The latter one can be achieved through the use of a resistor/thermistor network (discussed later). This type of compensation can lower a total error band to below 1 percent over a -40 to 125°C temperature range.

Calculation of Compensating Resistor Values

Values of compensating resistors can be calculated based on measurement results of the pressure temperature testing of a sensor. The tests should include measurements of the bridge output voltage and bridge excitation voltage at two test temperatures (T_c and T_h) and two test pressures (P1 and P2) with a constant current (I) excitation.

The basic analog model is used; however, a further simplification was incorporated. Since bridge resistance changes were not needed, the same pressure sensitivity of compressed and tensed resistors was assumed ($p = q$). Testing can thus be simplified and limited to testing a bridge voltage at just one pressure. Test results needed for compensation can be summarized as follows:

	$T = T_c$	$T = T_h$
$p = p_1$	V_{oc}, E_h	V_{oh}, E_h
$p = p_2$	V_{1c}	V_{1h}

where T_c, T_h = test temperatures, cold and hot, respectively
 V, V_h = zero pressure output voltages, cold and hot, respectively
 $V1, V1_{cp} V_{lh}$ = full-scale pressure output voltages, cold and hot, respectively
 E = bridge voltages, cold and hot, respectively
 p_1/p_2 = input pressures, zero and full-scale, respectively

Zero Compensating Resistors. Following are the variables used to calculate zero compensating resistors:

$$A = (V_{oc} + E_c)/I$$

$$B = A - 4 V_{oc} (V_{oc} + E_c)/I/(E_c + 2 V_{oc})$$

$$C = (V_{oh} + E_h)/I$$

$$D = A - 4 V_{oh} (V_{oh} + V_h)/I(E_h + 2 V_{oh})$$

The value of an offset compensating resistor R_s that includes the correction for offset change due to the bridge arm resistive loading by resistor R1 or R2 may be calculated now as follows:

$$R_s = 0.5 \{A + C - \sqrt{\{(A + C)2 - 4[B(D - C) - CD (B - A)]\}}/(D - B)\}$$

The calculated value of resistor R_s may be either positive or negative. The polarity of this value is utilized to define the position of the resistor. As was previously discussed, balancing of the offset voltage can be realized by either R3 or R4 resistor (Fig. 70.6). The truth table for these resistors can be constructed as follows:

when $R_s \geq 0$ then: R4 = R_s, R3 = 0 (shorted)

$R_s < 0$ then: R3 = R_s, R4 + 0 (shorted)

The offset T_c compensating resistor R_p may then be calculated as follows:

$$R_p = (A B - B R_s)/(B - A + R_s)$$

As before, there are two possible positions of the R_p resistor

when $R_p \geq 0$ then: R1 = R_p, R1 = (open)

$R_p < 0$ then: R1 = R_p, R2 = (open)

FSO Compensating Resistor. Temperature compensation of the FSO requires one resistor only. The value of FSO can be calculated as a difference between bridge output at full scale and zero pressure. If we define the FSO, cold (low temperature) as S_c and hot (high temperature) as S_h and the bridge resistance respectively R_c and R_h, the following calculations have to be made:

$$S_c = V_{lc} - V_{oc}$$

$$R_c = E_c/I$$

$$S_h = V_{lh} - V_{oh}$$

$$R_h = E_h/I$$

The value of FSO compensating resistor RS can be derived using the following formula:

$$R_S = (R_h S_c - R_c S_h)/(S_h - S_c)$$

It should be noted that the procedure outlined here does not include the loading effect of zero compensating resistors on bridge resistance change. Usually this effect is small.

Required Performance of Compensating Resistors

For a typical pressure sensor with a 5000-ohm bridge resistance at 25°C, with a TCR of 0.25 percent/°C, TCS of –0.21 percent/°C, and FSO of 15 mV/V, the expected compensating resistor ranges covering a broad production sensor parameters spread can be summarized as follows:

R1, R2:   100 kohms to 10 Mohms, typically 300 kohms to 1.0 Mohms
R3, R4:   0 to 300 ohms, typically 1 to 100 ohms
R5:   10 to 300 kohms, typically 15 to 30 kohms

For the majority of ranges, a thin film 1 percent, 100 ppm/°C resistor such as RN55D or similar is sufficient. Sensitivity of sensor compensation to the accuracy and temperature coefficient of the compensating resistor is graphically presented in Figs. 70.12 to 70.14.

As an example, let's assume that the compensation procedure call is for the following resistors:

R1 = 0.5M
R2 = Open
R3 = 90 ohms
R4 = Shorted
R5 = 20K

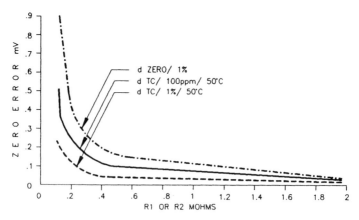

FIGURE 70.12   R1 or R2 resistor tolerance effect.

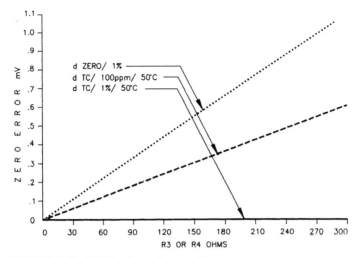

FIGURE 70.13 R3/R4 resistor tolerance effort

The effect of a 1 percent tolerance for resistor R1 (0.5M) can be estimated from Fig. 70.12. A 0.19-mV offset change would be created and a 0.06 mV/50°C offset temperature coefficient would be added. A temperature coefficient of 100 ppm/°C for the resistor would contribute an additional offset temperature error of 0.12 mV/50°C.

The effect of resistor R3 (90 ohms) can be estimated from Fig. 70.13. The offset would change 0.33 mV for a 1 percent resistance deviation and 0.17 mV/50°C due to the of 100 ppm/°C temperature coefficient. Offset temperature coefficient is not affected by the tolerance of this resistor.

Both of these resistors (parallel: R1 or R2 and series: R3 or R4) are affecting the compensated FSO value. Assuming that all the strain gages have the same pressure sensitivity, the change of a bridge arm resistance by 1 percent due to the insertion of zero compensation resistors changes the FSO by 0.25 percent.

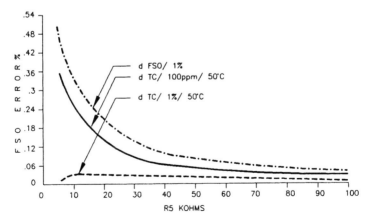

FIGURE 70.14 R5 resistor tolerance effect.

Resistor R5 (20 kOhms) does not effect zero compensation. FSO error (Fig. 70.14) introduced by a 1 percent deviation from the calculated value will be equivalent to a 0.19 percent FSO change and 0.02 percent/°C of the additional FSO temperature coefficient. A temperature coefficient of 100 ppm/°C for resistor R5 would introduce an additional FSO error of 0.15 percent/50°C.

When a known error can be tolerated, trimming ranges for the compensating resistors can be reduced. Assume that a 5-mV offset voltage due to accuracy of R1 or R2 resistor can be tolerated. If 0.5 MOhms (R1) is the starting point, with a 0.19 mV/1 percent offset sensitivity, a 5-mV limit will be reached after 26 increments of 1 percent (26) (0.29 mV). Raising 1.01 to the 26th power gives a factor of 1.295 which translates to 648 kOhms. At this value of R1 or R2, resistance value sensitivity of offset to change in R1 is about 0.16 mV/1 percent. Raising 1.01 to the 31st power gives a factor of 1.361 which translates to 882 kOhms. These two values of resistors could be stocked with the 499 kOhms resistor for ±5-mV offset correction.

This same approach can be applied to all resistors over the entire range and to all related specifications including temperature error. In the preceding example, the worst case assumption was made using the highest error for a given resistance range. Using the average error for a given range would be more realistic (0.18 mV/1 percent over 500 k to 698 kOhms range), but then no room is left for variations of sensor performance due to processing tolerances.

CONSTANT VOLTAGE FSO COMPENSATION

The operation in a constant voltage mode is advantageous in some applications. One of the reasons is a simplicity of the signal conditioning through an elimination of an operational amplifier creating the current source.

As was discussed previously, a basic sensor exhibits a negative temperature coefficient of stress (or pressure or acceleration) sensitivity. To compensate for this change, either a bridge voltage or an amplifier gain has to increase with temperature.

The increase of bridge voltage with temperature with a fixed supply voltage requires insertion of a resistor or elements with a negative temperature coefficient of resistance or voltage in series with the sensor bridge. The negative temperature coefficient of resistance can be achieved using thermistors or polysilicon resistors. Negative temperature coefficient of a voltage can be achieved using either direct or indirect (via a feedback loop) semiconductor junctions. All these techniques are briefly outlined here.

Resistor Compensation

Insertion of a series resistor with a temperature-sensitive bridge resistance increases the bridge voltage with a temperature increase, providing the basis for temperature compensation. For typical silicon sensors, however, this configuration yields a significant FSO (−80 percent) loss which is nevertheless acceptable in many applications, such as disposable blood pressure sensors. By changing sensor process, it is possible to achieve a much lower drop of the output in a voltage excitation mode, lowering the losses to a level of about 20 percent.

A single resistor in series with the bridge causes the common mode voltage (shorted bridge output level to ground) to change with temperature change. For lower-grade instrumentation amplifiers, this change may be excessive and then the calibration resistor has to be split into two halves mounted on both sides of the bridge.

Combination of FSO temperature compensation and normalization combined with offset normalization and compensation is shown in Fig. 70.15. Resistors R3 and R4 provide offset balancing. One resistor R, provides offset TC compensation; however, its position can be changed from the left to the right side of the bridge to accommodate different polarity of the temperature error. TC FSO is compensated by an equivalent resistor created by parallel connection of resistors R5 and R6. Resistor R6 decreases bridge voltage to provide calibration of FSO.

Thermistor Compensation

Thermistors exhibit a large nonlinear temperature coefficient of resistance that may reach a −1 percent/°C level. Large temperature coefficient is advantageous for a compensation efficiency, but the nonlinearity is not quite desired. Using a combination of resistors, thermistor, and a bridge allows not only elimination of an effect of the nonlinearity of a thermistor response, but simultaneously compensates the internal FSO temperature nonlinearity of a sensor. Thermistor compensation provides potential for delivery of a lower loss of output as compared to a simple resistor compensation.

The configuration shown in Fig. 70.16 employs three resistors R5, R6, R7 to provide a three-point temperature compensation. Three-point temperature compensation forces error to be zero at three temperatures and thus requires three degrees of freedom that are provided by resistors. Thermistors are difficult to trim and usually are used as manufactured without any trimming. All compensation parameters can be achieved by trimming resistors only. Values of these resistors are calculated after measurement of sensor and thermistor characteristics, though a solution of three simultaneous equations each zeroing error at one temperature.

Temperature nonlinearity of FSO for silicon sensors can be approximated by a second order polynomial. Upgrading a compensation from a two-point to a three-point decreases the residual temperature error by a factor of about four times, as illustrated in Fig. 70.17.

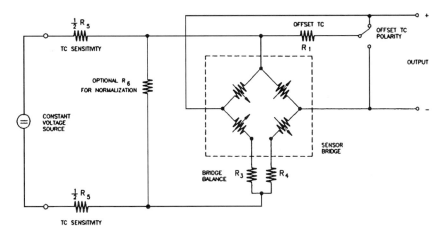

FIGURE 70.15 Calibration compensation circuit used for constant voltage excitation.

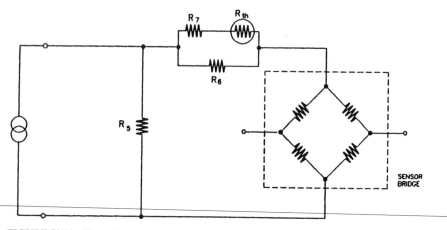

FIGURE 70.16 Thermistor-based temperature compensation unit suitable for a three-point temperature compensation.

Diode Compensation

The circuit shown in Fig. 70.18 provides a simple temperature compensation of pressure sensitivity by inserting forward-biased silicon diodes in series with the bridge. The diodes have a negative temperature coefficient of approximately −2 mV/°C. When placed in series with the sensor bridge powered by a constant voltage, it generates an increase of the bridge voltage with temperature.

Since a forward-bias voltage of the diode and its temperature coefficient are not ratiometric (do not scale up or down with current), this temperature compensation

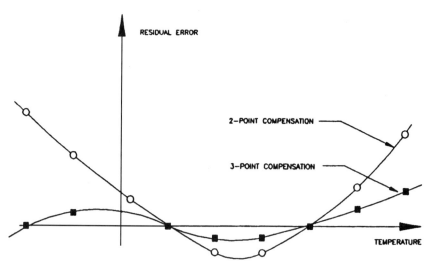

FIGURE 70.17 Three-point temperature compensation reduces errors available in a two-point compensation.

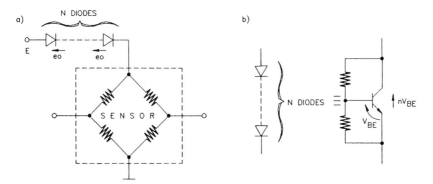

FIGURE 70.18 Compensation of FSO with diodes and transmitters.

technique is sensitive to the value of a supply voltage. To achieve proper temperature compensation of the FSO, the following relationship has to be met:

$$N = E\,\text{TCS}/(A + eo\,\text{TCS})$$

where E = supply voltage (V)
N = number of diodes used
eo = forward voltage of the diode at a reference temperature (V)
A = diode's temperature coefficient (V/°C)
TCS = temperature coefficient of sensitivity (1/°C)

Table 70.1 summarizes the required supply voltage in order to provide temperature compensation of pressure sensitivity equal to −0.21 percent/°C, which is typical for many transducers with 5000-ohm bridge impedance. For four diodes, the power supply voltage providing temperature compensation is approximately 6.2 V.

The following table shows the required supply voltage E for different number of series diodes for $eo = 0.6$ V,

$$A = 0.002\text{ V/°C}, \text{TCS} = -0.0021/\text{°C}$$

A series of diodes can be replaced by the $nVbe$ circuit, shown in Fig. 70.18(b) with a transistor and two feedback resistors forming an effective diode with a voltage equal to $nVbe$ and the overall temperature coefficient multiplied by the same ratio n. This circuit effectively allows the use of any supply voltage.

TABLE 70.1 Required Supply Voltage for Temperature Compensation

Number of diodes	E (V)
1	1.552
2	3.105
3	4.657
4	6.210
5	7.762
6	9.314

GAIN PROGRAMMING FOR NORMALIZATION

Another approach to normalization of sensor output was introduced by Bryzek.[33] This approach provided a resistor in the sensor package that, when used with an external amplifier, provided interchangeability of a full-scale output, Fig. 70.19. The resistor was laser-trimmed to program the gain of the amplifier to compensate for sensor sensitivity variations. This technique thus allowed the output of an amplifier to be independent of the sensor used, providing low-cost interchangeability at a 2-V FSO output signal level when a reference bridge current of 0.996 Ma is used.

Basic Circuit

The effective electrical model of the transducer, together with a basic signal conditioning circuit, is shown in Fig. 70.20. The sensor incorporates temperature-compensated, using thick film, laser-trimmed resistors. The excitation to the bridge is a constant current supplied through pins 5 and 6. Bridge output is at pins 4 and 10, and the gain programming resistor r is connected between pins 7 and 9.

Resistor r is laser-trimmed for each unit using the following algorithm:

$$r = 200\ Si/(2 - Si)$$

where: Si = sensor FSO (V) at bridge current 0.996 mA
r = resistance (kOhms)

The amplifier output FSO, SO, at the differential output of amplifiers A1–A2 is set as follows:

$$SO = ASi\ (r + 2R)/r = 2A[R/100 + Si\ (100 - R)/200]$$

where: $A = I/I_o$, ratio of actual excitation current I to reference current I_o
R = feedback resistors (kOhms)
Si = sensor's FSO at the amplifier input

If 100-kOhm feedback resistors R are used, the expression for the FSO is simplified to:

$$S_o = 2A$$

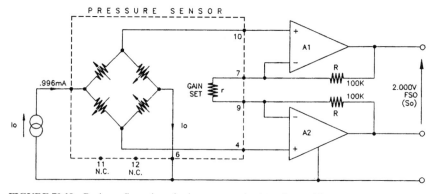

FIGURE 70.19 Basic configuration of gain-programming interchangeable sensor.

and is constant for all sensors, independent of the pressure range. For other values of the feedback resistor, the amplifier output FSO S_o varies with sensor level FSO S_i. Assuming $I = I_o$, we can calculate the S_o variations.

R (K)	S_o (S_i = 40 mV)	S_o (S_i = 90 mv)	S_o max/min
50	1.0200	1.0450	−1.25
75	1.5100	1.5225	−0.41
99	1.9804	1.9809	−0.01
100	2.0000	2.0000	0.00
101	2.0196	2.0191	0.01
200	3.9600	3.9100	0.47
500	9.8400	9.6400	1.02

As can be seen, a large deviation from the optimum feedback resistance of 100 kOhms can be tolerated while still maintaining a reasonable transducer interchangeability.

For the optimum feedback resistance (100 kOhms), calibration accuracy is a function of the accuracy excitation current, feedback resistors, and sensor trimming. The inaccuracy caused by the excitation current feedback resistors can be made negligible by the use of precision components. Since a sensor can be trimmed to a 1 percent interchangeability, without the additional pressure testing, a 1 percent system accuracy can be achieved.

MEASUREMENT OF DIFFERENTIAL PRESSURE USING TWO PRESSURE SENSORS

Measurement of a differential pressure superimposed on a static (common-mode) pressure can be accomplished by using dedicated differential pressure transducers. The disadvantage of using this approach is the high cost of the transducer as well as its large size due to a built-in overload protection for a single static pressure overload.

Differential pressure applications that do not require a high overload protection can be served by the two low-cost gage pressure sensors with a simple circuitry that compensates for the difference in a static pressure sensitivity of these two components. Static pressure sensitivity results as an effect of a mismatched pressure sensitivity and pressure nonlinearity between both units.

Measurement of differential pressure with two sensors requires a summing network with a phase invertion to cancel out the gross static pressure effect. A simple configuration performing this function is shown in Fig. 70.20. Two sensors, sensor "+" and sensor "−," are inverted and connected in parallel. Inverted configuration cancels out static pressure signal if the pressure at sensor "+" is higher then at sensor "−," the output will be proportional to the differential pressure. Due to a parallel connection of the sensors the full-scale output will be approximately half of the value of individual sensors.

With the same pressure applied to both sensors, the output should be constant and independent of the value of the applied static pressure. Due to a mismatch between the sensors, however, the output will change as a function of a static pressure. The circuit shown in Fig. 70.20 allows for a two-point cancellation of the static pressure effect, for example, at the extreme values of a differential pressure range.

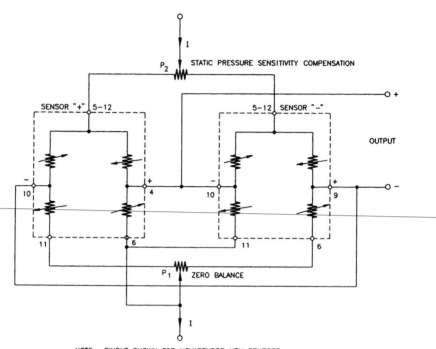

FIGURE 70.20 Two gage sensors allow differential voltage measurement with cancellation of a static pressure.

Potentiometer P1 allows a zero pressure offset balance at the beginning of the operating differential pressure range. Zero balance is simply accomplished by inserting a differential resistor into one side of both bridges. The adjustment of this potentiometer at the beginning of measurement span should occur with the same pressure applied to both sensors, Sensor "+" and Sensor "−."

The other adjustment is accomplished with a potentiometer P2 to compensate offset at the other static pressure. This potentiometer adjusts a current distribution between both bridges to equalize the output of each sensor. This adjustment should be done at the other end of the measurement span to cancel out the effect of static pressure. The error between the two compensated pressure points (with potentiometers P1 and P2) will be a function of the output tracking between both sensors. For the best performance, both units should be matched very closely, ideally they should be from the same wafer. Matching should be done (before temperature compensation) for offset voltage, pressure sensitivity, bridge resistance, pressure nonlinearity, and temperature error for both zero and FSO. Also, the operating temperature of both sensors should be identical to eliminate a potential temperature gradient inducing the measurement error.

Temperature compensation can be performed either individually on both sensors or, preferably, in a common parallel configuration which would take care of temperature errors in the final working configuration.

The static pressure error is a function of not only sensor matching, but also selected pressure compensation points. Sample results shown in Fig. 70.21 were gen-

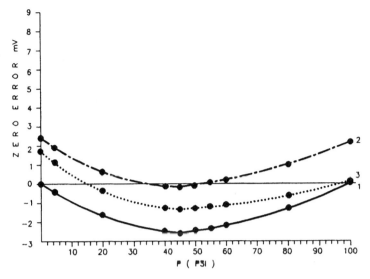

FIGURE 70.21 Static pressure error measured two 5-psi sensors operating over 100 psi static pressure.

erated using three different end points for both zero and pressure sensitivity compensation. In this test, two 5-psi sensors were used and a 100-psi static pressure range was selected. Three boundary conditions were selected for zero and static pressure compensation represented by the curves shown. Compensation of static pressure over 0 to 100 psi, 50 to 55 psi, and 95 to 100 psi pressure resulted in the same magnitude of static pressure error. Distribution, however, was different as a function of the selected compensation points:

Curve	1	2	3
P1 set at (psi)	0	50	95
P2 set at (psi)	100	55	100
Error/5 psi (mV)	0.2	0.2	0.2
Error/0–100 psi (mV)	2.5	2.5	2.5

Since in this example the static pressure range was 20 times larger than the differential pressure range, the static sensitivity adjustments will be magnified at the same ratio. Thus, very stable and high-resolution potentiometers will be required to achieve good results.

The static pressure error is inversely proportional to the selected pressure range for a given static pressure range. Thus, if in our example, a 15-psi sensor would be selected, the errors on that basis related to the FSO would be three times smaller.

While this presented technique allows for a very basic and simple compensation of the static pressure effect, best results can only be achieved using the same sensor. In the example given, the rejection of a static pressure on the order of 50 dB (ratio of 400 times, calculated as 20:1 ratio of static to differential pressure and 50 mVFSO to 2.5 mV static pressure error ratio). A single differential sensor chip with an electronic static error compensation allows for achieving a rejection of the static pressure in excess of 100 dB (100,000 to 1).

DIGITAL COMPENSATION AND NORMALIZATION

Low-cost digital processing electronics allows implementation of digital compensation and calibration techniques. This technique relies on the measurements of pressure- and temperature-related outputs from the sensor. In a sensor configuration shown in Fig. 70.22, the constant current excitation allows measurement of the temperature by the pressure sensor by monitoring the bridge voltage changes related primarily to the temperature change.

The accuracy of this modeling can be very high. It is limited only by the stability of the sensor and accuracy and resolution of the signal-conditioning electronics.

To assure accuracy of the curve-fitting algorithm, there is a need to use a least square technique in order to smooth out the possible oscillation of the higher-order polynomial fitting (if used) due to inaccuracy of measurements.

Several examples of the curve-fitting algorithms are presented in the following sections, offering trade-offs between simplicity and accuracy.

Linear Approximation of Pressure and Temperature Characteristics

The simplest approximation of a pressure P measured by the sensor is given by the following transfer function:

$$P = A_0 + A_1 V$$

where A_0 = constant representing initial offset for a given sensor
A_1 = constant representing sensitivity for a given sensor
V = a bridge output voltage proportional to pressure

Both constants A_0 and A_1 can be made temperature-sensitive to provide a temperature compensation of the sensor transfer function:

$$A_0 = B_{00} + B_{01} E$$

$$A_1 = B_{10} + B_{11} E$$

where E = a bridge excitation voltage of the sensor in a constant-current mode
B_{ij} = constants individually selected for each sensor

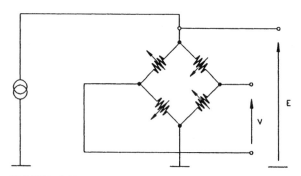

FIGURE 70.22 Basic configuration of piezoresistive sensor for digital compensation.

We can now express pressure as a function of two sensor outputs V and E, and four constants selected individually for each sensor:

$$P = B_{00} + B_{01}E + (B_{10} + B_{11}E)\,V$$

Since the equation for the pressure P represents a linear curve, it fits into both pressure and temperature error surfaces with four degrees of freedom. To define the constants, at least four measurement points are needed. These measurements must include two pressure and two temperature points.

To optimize the overall error, a least mean square fit technique can be implemented with a corresponding increase of the quantity of test points. To illustrate this important technique to digital compensation, a simple example is considered, with three pressure and three temperature points. Selecting input pressures at zero, midpoint, and full scale, denoted as P_0, P_1, P_2, respectively, and temperatures respectively, T_0, T_1, T_2, we get a set of nine measured bridge output voltage V_{pt} and nine bridge excitation voltages E_{pt}, with $p = 0, 1, 2, t = 0, 1, 2$, respectively. With this notation, for example, V12 would indicate a measured bridge output voltage at midpressure and at temperature T1.

We can now formulate a condition for compensation/calibration using a least square fit as a set of nine simultaneous equations with four unknowns B_{00}, B_{01}, B_{10}, B_{11}; if the small change of pressure nonlinearity with temperature can be neglected, then the equations containing results at $P = P_1$ can be eliminated, reducing the system to just six equations with four unknowns.

$$B_{00} + B_{01}E_{00} + B_{10}V_{00} + B_{11}V_{00}E_{00} = P_0$$

$$B_{00} + B_{01}E_{10} + B_{10}V_{10} + B_{11}V_{10}E_{10} = P_1 \qquad T = T_0$$

$$B_{00} + B_{01}E_{20} + B_{10}V_{20} + B_{11}V_{20}E_{20} = P_2$$

$$B_{00} + B_{01}E_{01} + B_{10}V_{01} + B_{11}V_{01}E_{01} = P_0$$

$$B_{00} + B_{01}E_{11} + B_{10}V_{11} + B_{11}V_{11}E_{11} = P_1$$

Where $T = T1$, the following is true:

$$B_{00} + B_{01}E_{21} + B_{10}V_{21} + B_{11}V_{21}E_{21} = P_2$$

$$B_{00} + B_{01}E_{02} + B_{10}V_{02} + B_{11}V_{02}E_{02} = P_0$$

$$B_{00} + B_{01}E_{12} + B_{10}V_{12} + B_{11}V_{12}E_{12} = P_1 \qquad T = T_3$$

$$B_{00} + B_{01}E_{22} + B_{10}V_{22} + B_{11}V_{22}E_{22} = P_2$$

The solution of this system of simultaneous equations can be accomplished by first setting up the four normal equations required by the method of least squares and then solving four simultaneous equations with four unknowns.[32] The first normal equation will be derived by multiplying all equations by the coefficient of B_{00} (in our case, 1) and vertically adding all respective coefficients:

$$9\,B_{00}\,(E_{00} + E_{10} + E_{20} + E_{01} + E_{11} + E_{21} + E_{02} + E_{12} + E_{22})\,B_{01}$$
$$+ (V_{00} + V_{10} + V_{20} + V_{01}V_{11} + V_{02} + V_{12} + V_2)\,B_{10}$$
$$+ (V_{00}E_{00} + V_{10}E_{10} + V_{20}E_{20} + V_{01}E_{01} + V_{11}E_{11})\,V_{21}E_{11}$$
$$+ V_{02}E_{02} + V_{12}\,V_{22}E_{22})\,B_{11} = 3P_0 + 3P_1 + 3P_2$$

Using the following notation:

$$X_1 = (1+1+1+1+1+1+1+1+1) = 9$$

$$Y_1 = (E_{00} + E_{10} + E_{20} + E_{01} + E_{11} + E_{21} + E_{02} + E_{12} + E_{22})$$

$$Z_1 = (V_{00} + V_{10} + V_{20} + V_{01} + V_{11} + V_{21} + V_{02} + V_{12} + V_{22})$$

$$U_1 = (V_{00}E_{00} + V_{10}E_{10} + V_{20}E_{20} + V_{01}E_{01} + V_{11}E_{11}) V_{21}E_{21} + V_{02}E_{02} + V_{12}E_{12} + E_{22}$$

$$T_1 = 3_{P0} + 3_{P1} + 3_{P2}$$

we can now rewrite the main equation as:

$$X_1 B_{00} + Y_1 B_{01} + Z_1 B_{10} + U_1 B_{11} = T_1$$

To set up the second normal equation, we multiply each equation in our simultaneous equations set by the coefficient of B_{01} in that equation. Using analog notation as previously, we can get the following X_2, Y_2, Z_2, U, values:

$$x2 = E_{00} + E_{10} + E_{20} + E_{01} + E_{11} + E_{21} + E_{02} + E_{12} + E_{22}$$

$$y2 = E^2_{00} + E^2_{10} + E^2_{20} + E^2_{01} + E^2_{11} + E^2_{21} + E^2_{02} + E^2_{12} + E^2_{22}$$

$$Z_2 = E_{00}V_{00} + E_1V_1 + E_{20}V_{20} + E_{01}V_{01} + E_{11}V_{11} + E_{21}V_{21} + E_{02}V_{02} + E_{12}V_{12} + E_{22}V_{22}$$

$$U_2 = E^2_{00}V_{00} + E^2_{10}V_{10} + E^2_{20}V_{20} + E^2_{01}V_{01} + E^2_{11}V_{11} + E^2_{21}V_{21} + E^2_{02}V_{02} + E^2_{12}V_{12} + E^2_{22}V_{22}$$

$$T_2 = E_{00}P_0 + E_1P_1 + E_{20}P_2 + E_{01}P_0 + E_{11}P_1 + E_{21}P_2 + E_{02}P_0 + E_{12}P_1 + E_{22}P_2$$

The second normal equation will thus be as follows:

$$X_2 B_{00} + Y_2 B_{01} + Z_2 B_{10} + U_3 B_{11} = T_2$$

In the same way, we can set up the third normal equation with all system equations respectively multiplied by the coefficient of B10 in each equation:

$$X_3 B_{00} + Y_3 B_0 + ZB_{10} + U_3 B_{11} = T_3$$

with the constants as follows:

$$X_3 = V_{00} + V_{10} + V_{20} + V_{01} + V_{11} + V_{21} + V_{02} + V_{12} + V_{22}$$

$$Y_3 = E_{00}V_{00} + E_{10}V_{10} + E_{20}V_{20} + E_{01}V_{01} + E_{11}V_{11} + E_{21}V_{21} + E_{02}V_{02} + E_{12}V_{12} + E_{22}V_{22}$$

$$Z_3 = V^2_{00} + V^2_{10} + V^2_{20} + V^2_{01} + V^2_{11} + V^2_{21} + V^2_{02} + V^2_{12} + V^2_{22}$$

$$U_3 = E_{00}V_{00} + E_{10}V_{10} + E_{20}V_{20} + E_{01}V_{01} + E_{11}V_{11} + E_{21}V_{21} + E_{02}V_{02} + E_{12}V_{12} + E_{22}V_{22}$$

$$T_3 = V_{00}P_0 + V_{10}P_1 + V_{20}P_2 + V_{01}P_0 + V_{11}P_1 + V_{21}P_2 + V_{02}P_0 + V_{12}P_1 + V_{22}P_2$$

Multiplying all equations by the coefficient of the B11 equation and vertically adding equations as before, we set up the last (fourth) normal equation:

$$X_4 B_{00} + Y_4 B_{01} + Z_4 B_{10} + U_4 B_1 = T_4$$

with the constants as follows:

$$X_4 = E_{00}V_{00} + E_{10}V_{10} + E_{20}V_{20} + E_{01}V_{01} + E_{11}V_{11} + E_{21}V_{21} + E_{02}V_{02}$$
$$+ E_{12}V_{12} + E_{22}V_{22}$$

$$Y_4 = 2_{00}V_{00} + E_{21}V_{10} + E_{220}V_{21} + E_{201}V_{01} + E_{211}V_{11} + E_{221}V_{21} + E_{202}V_{02}$$
$$+ E_{212}V_{12} + E_{222}V_{22}$$

$$Z_4 = E_{00}V_{200} + E_{10}V_{210} + E_{20}V_{220} + E_{01}V_{20} + E_{11}V_{211} + E_{21}V_{221} + E_{02}V_{202}$$
$$+ E_{12}V_{212} + E_{22}V_{222}$$

$$Z_4 = E^2_{00}V^2_{00} + E^2_{10}V^2_{10} + E^2_{20}V^2_{20} + E^2_{01}V^2_{01} + E^2_{11}V^2_{11} + E^2_{21}V^2_{21}$$
$$+ E^2_{02}V^2_{02} + E^2_{12}V^2_{12} + E^2_{22}V^2_{22}$$

$$T_3 = E_0V_{00}P_0 + E_{10}V_{10}P_1 + E_{20}V_{20}P_2 + E_{01}V_{01}P_0 + E_{11}V_{11}P_1 + E_{21}V_{21}P_2$$
$$+ E_{02}V_{02}P_0 + E_{12}V_{12}P_1 + E_{22}V_{22}P_2$$

Solution of the four simultaneous normal equations with four unknowns is a simple matter. This solution brings all coefficients B_{00} through B_{11}, allowing calculation of a measured pressure based only on two real-time measurements: bridge excitation voltage E and output voltage V. In most applications, the change of sensor temperature occurs much slower than pressure. The measurement of a bridge voltage E thus can be performed less frequently than pressure output V without a loss of accuracy.

To calibrate a specific sensor, the four constants B_{00}, B_{01}, B_{10}, and B_{11} have to be stored individually for each sensor.

The accuracy of the linear model is limited by the pressure and temperature nonlinearities of the sensor transfer function plus the accuracy and resolution of electronics.

Because the sensor operates at the same current during the compensation procedure test and consecutive normal operation, the power sensitivity and voltage sensitivity errors (change of parameters as a function of applied excitation power or voltage) will be included in compensation.

The additional error will be created by a bridge resistance change with applied pressure, as the effect of strain gages' tension and compression sensitivity mismatch. Bridge resistance change with a full-scale pressure is typically on the order of less than 0.1 percent. This creates temperature measurement error of about 0.4°C (TCR ~ .25 percent/°C). The resulting pressure measurement error is then defined as a function of a compensated sensor temperature error.

Using this approach, about 0.75 percent total error band over a 0 to 70°C temperature range can be achieved.

Linear Approximation of Pressure Characteristics and Parabolic Approximation of Temperature Characteristics

For silicon sensors, a major component of a total error band is temperature errors. Their basic shape is nonlinear in nature with a dominating parabolic term. To improve the compensated error band, we can thus expand the linear model to include a parabolic fit for both zero and sensitivity temperature dependence. We keep, however, a linear fit through a pressure response as before:

$$P = A_0 + A_1 V$$

The coefficients A_0 and A_1, will now be defined as follows:

$$A_0 = B_{00} + B_{01}E + B_{02}E_2$$

$$A_1 = B_{10} + B_{11}E + B_{12}E_2$$

and the measured value of pressure will be defined by the following expression:

$$P = B_{00} + B_{01}E + B_{02}E_2 + B_{10}V + B_{11}E_V + B_{12}E^2{}_V$$

To provide a sufficient number of degrees of freedom for the least square fit of pressure and temperature, we need at least three test pressures P_{01}, P_1, P_2 and four test temperatures T_0, T_1, T_2, T_3 for a total of twelve equations defining six unknown coefficients B_{00} thru B_{12}. If variation of pressure nonlinearity with temperature can be neglected, than only two test pressure points will be sufficient (eliminating P_1 pressure) and the total number of starting equations can be reduced accordingly.

The temperature error band can be reduced about three to four times as compared to a previous model. The solution of the simultaneous equations can be performed in a similar mode to the example given previously.

Parabolic Approximation of Both Pressure and Temperature Characteristics

Significant reduction of sensor pressure nonlinearity can be accomplished by implementing a second-order fit of pressure nonlinearity, utilizing the following model:

$$P = A_0 + A_1V + A_2V^2$$

with respective coefficients as given previously:

$A_0 = B_{00} + B_{01}E + B_{02}E^2$
$A_1 = B_{10} + B_{11}E + B_{12}E^2$
$A_2 = B_{20} + B_{21}E + B_{22}E^2$

and with the final pressure calculating equation as follows:

$$P = B_{00} + B_{01}E + B_{02}E^2 + (B_{10} + B_{11}E + B_{12}E^2)V + (B_{20} + B_{21}E + B_{22}E^2)V^2$$

The last equation contains nine coefficients B00 through B22 that require at least three pressure and three temperature tests for a perfect fit and respectively at least four pressure and four temperature tests for a least mean square fit. The accuracy improvement is significant, delivering total error under 0.1 percent over a 0 to 70°C temperature range.

Third-Order Polynomial Distribution

For a higher-performance application, especially with the overdriven sensor operating with a high output and higher nonlinearity (e.g., 5-psi sensor operating over a 15-psi range), a third-order polynomial (cube) compensation could be used to improve the overall performance. In this case, the basic equations are as follows:

$$P = A_0 + A_1V + A_2V_2 + A_3V^3$$

with the temperature dependence of coefficients defined as follows:

$A_0 = B_{00} + B_{01}E + B_{02}E^2 + B_{03}E^3$
$A_1 = B_{10} + B_{11}E + B_{12}E^2 + B_{03}E^3$
$A_2 = B_{20} + B_{21}E + B_{22}E^2 + B_{23}E^3$
$A_3 = B_{30} + B_{31}E + B_{32}E^2 + B_{33}E^3$

The solution of this equation with 16 unknowns can be done in a similar fashion to the technique outlined previously, with a minimum of five test pressures and five test temperatures for a least mean square fit. The total accuracy of better than 0.04 percent can be achieved with this technique.

For a higher-performance application, even higher-order polynomials are being used. The aerospace pressure transducers, including a vibrating cylinder, Solartron transducer, and piezoresistive Honeywell transducer,[35] both used in air data applications, use a sixth-order polynomial to fit into both pressure and temperature characteristics with 49 coefficients, delivering curve-fitting accuracy on the order of 1 ppm (0.0001 percent). Design of the higher-order polynomial fit follows the same technique as previously outlined. Solution of the simultaneous equations, however, may become a problem by itself and should be delegated to dedicated software.

CURRENT SOURCES FOR SENSOR EXCITATION

As illustrated in the section on basic compensation, one convenient way to achieve first-order full-scale output temperature compensation is to provide a constant-current excitation for the silicon sensor. Constant-current excitation is also useful for other types of sensors such as platinum RTDs and thin film strain gages and it has been used in these applications for many years.

Output current of the constant-current source does not change when bridge resistance changes. In order to minimize the error introduced by a current source, its stability with time and temperature should be significantly better than allowed sensor error. The most efficient technique of building a stable current source is using operational amplifiers driven either by temperature-compensated reference voltage diodes or power supply dividers.

A single op-amp can provide a variety of excellent current source configurations. There are two basic configurations of current sources:

- Fixed
- Ratiometric

Fixed current sources deliver a current value independent from power supply voltage variations while a *ratiometric* one changes current proportionally to the power supply voltage changes.

Each of the examples shown here can provide constant current at a fixed reference value, or a current ratiometric to the supply voltage determined respectively either by a precision voltage reference diode or a resistive divider.

An operational amplifier adjusts the output voltage in such a way that a zero voltage is always maintained between its inputs, no matter what the resistance of the bridge is, thus forming the constant-current source. The amplifier's supply voltage has to provide enough dynamic range so even at the highest operating temperature (highest bridge resistance), the amplifier is not saturated.

The configuration of the current source shown in Fig. 70.23 connects the sensor bridge between the output and inverting input of the amplifier. Reference voltage at noninverting input (divider for a ratiometric configuration, Fig. 70.23, and diode for a fixed configuration, Fig. 70.24) creates the voltage that is transferred on a resistor R1. The inaccuracy of the voltage transfer is set by the amplifier's offset voltage. The value of current is thus set by a noninverting voltage and resistor R1 value.

Some of the applications require a grounding of the sensor, which may be advantageous in specific applications (lower common mode voltage, better noise rejection, etc.). Grounded sensor or floating current source configuration is shown in Fig.

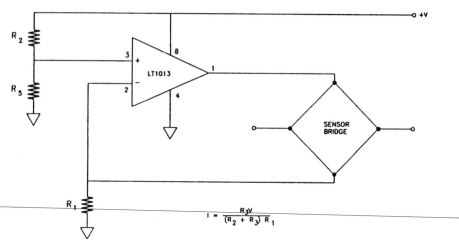

FIGURE 70.23 Basic ratiometric current source configuration.

70.25. Reference voltage in this configuration is connected between an inverting amplifier input and a positive supply voltage. The output transistor creates a floating current source allowing control of the current through a resistor R1 via the feedback loop. When a transistor's base current is very small, then the bridge current will be defined by the resistor R1 current and constant.

Stability of the current is defined by the combined stabilities of input voltage, reference feedback resistor R1, and amplifier's offset voltage. For example, using thin film resistors with TCR of 100 ppm/°C creates a 0.01 percent/°C pressure sensitivity error contribution.

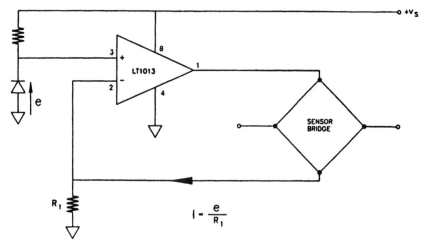

FIGURE 70.24 Basic current source with a fixed current value.

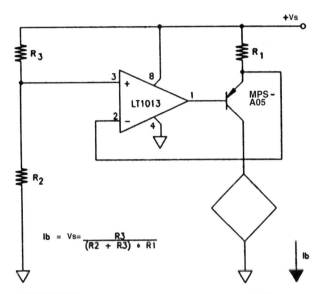

FIGURE 70.25 Ratiometric current source for a grounded sensor.

Another configuration of a grounded current source is possible using a switched capacitor technique. If this technique is used with a differential amplifier, it could create a floating or semifloating current source capable of providing constant current into a grounded or floating sensor. The circuit in Fig. 70.24 shows a simple current source for a grounded sensor.

INSTRUMENTATION AMPLIFIERS

The important step beyond the temperature compensation discussed previously is signal amplification. Because bridge-type sensors inherently produce a fairly low level differential output voltage superimposed on a common mode voltage, the amplifications requires some kind of a differential instrumentation amplifier. The function of the instrumentation amplifier is to amplify this small differential signal (on the order of 1 to 10 percent of bridge voltage) and convert it to a single-ended signal referenced to a single ground. While amplifying the differential signal, the amplifier must also sufficiently reject the much larger (half of bridge voltage) common mode signal present on the output terminals of the sensor.

Other functions of the amplifier might include adjustment of the output offset or output bias (setting the output voltage level at zero sensor offset), rejection of power supply voltage changes, and possibly a correction of second-order nonlinearity through the use of positive or negative feedback to the sensor excitation circuit. For simplicity, all of the examples presented here are shown using a single power supply voltage with negative common.

There are available instrumentation amplifier packages, both hybrid and monolithic, that incorporate internal thin film resistors usually combined with a external gain-setting resistors to implement a highly versatile gain block with very high

CMRR. For example, PMI AMP-01 provides a gain of a 1000 with CMRR of more than 130 dB. These types of amplifiers have the advantage of being factory trimmed for CMRR. Unfortunately, most commercial implementations are relatively expensive and require dual symmetrical power supplies, which may not be available for a self-contained low-cost sensor or transducer. Taking this into account, many applications rely on inhouse-built signal-conditioning circuitry. Broad selection of diverse sensor signal-conditioning applications is included in the references at the end of this chapter.

Amplifier Performance Requirements

Errors introduced by the amplification stage affect the accuracy of both offset voltage and full-scale output. Selection of the proper configuration of the amplifier is thus critical to the overall accuracy of the transducers, as well as to the cost of signal-conditioning implementation.

In general, there are two available options to treating errors introduced by electronics:

- Reduction of electronic errors significantly (at least three times) below the level of the sensor errors
- Compensation of electronic errors together with sensor errors

Reduction of electronic errors requires understanding of their sources. There are four major sources of errors introduced by input amplifiers that are related to the offset voltage. They span quite a wide range, as a function of the amplifier type and cost:

- *Input offset voltage,* ranging from 5 µV to 15 mV. This voltage adds directly to the sensor offset voltage. If the additional bias voltage is created to support the live zero (e.g., 1 V at zero pressure), the stability and accuracy of this bias will directly affect the accuracy of the measurements.
- *Temperature coefficient of input offset voltage,* ranging from 0.01 to 30 µV/°C. Over a 100°C range this error can create up to 30 mV of offset error.
- *Common mode rejection ratio (CMRR),* translating changes of a common mode voltage (the average voltage at the input of the amplifier, typically 50 percent of the bridge excitation voltage) into an input offset voltage change. The range of the CMRR is 30 to 140 Db. This is a logarithmic measure. Each 10:1 ratio adds 20 dB, thus 30 dB is equivalent to about 30:1 reduction ratio. If the bridge is excited with 7 V at 25°C, its resistance temperature sensitivity may increase bridge voltage by 21 percent in compensated mode at 125°C to 8.47 V, or by 1.47 V. Common mode voltage at the amplifier input would be half of this change or 0.735 V. Equivalent input offset voltage change would be reduced by CMRR; if a 30:1 number is used, the input offset would change by 24.5 mV, a very significant and visible error. For a 100-dB (100,000:1) ratio, this error would be reduced to 7.4 microV.

  All input amplifier configurations depending on resistor matching (e.g., a two-op-amp configuration) will have poor CMRR.
- *Power supply rejection ration (PSRR),* translating changes of power supply voltage into an equivalent input offset change. The range span from 60 to 140 dB, creating the worst case of about 1 mV/V offset change.

As far as the FSO, the only practical contribution can be seen from a feedback resistor temperature coefficient. Since most of the input amplifier configuration

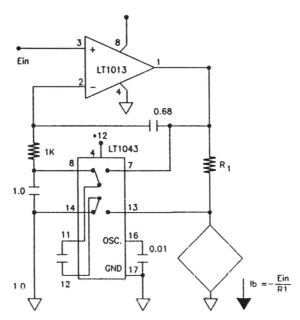

FIGURE 70.26 Switched capacitor current source for a grounded load application.

relies on resistor ratio to set the gain, control of this parameter becomes a relatively easy task.

The interesting and efficient implementation of signal conditioning is to compensate the errors of the electronics simultaneously with the sensor errors. This technique requires the operation of the electronics in the linear mode (outside of a saturation level) during the compensation cycle and then calculation of the apparent sensor errors based on the electronics output change using a well-defined software model including both sensor and electronics. The drawback of this technique is the requirement for compensation know-how by a sensor user, if the procedure is implemented outside of the sensor manufacturer.

Three-Amplifier Configuration

The classic three-amplifier configuration is shown in Fig. 70.27. It provides excellent performance. This configuration consists of two stages. The first stage includes a pair of amplifiers that provides a differential voltage amplification again set by a ratio of Ro and R1 resistors and common mode voltage amplification of one. The first-stage pair offers an excellent common mode rejection (CMR) limited practically by only the CMR of operational amplifiers (80 to 140 dB).

The third amplifier converts the amplified differential output to a single-ended output with a gain set by a ratio of R3 and R4 resistors. Common mode rejection of this stage is set by the feedback resistor ratio matching, where a 1 percent effective match would translate only to a 40-dB (100:1) CMR. Since common mode voltage gain of the first stage is one, it is advantageous to move all amplification to the first stage and leave the second stage with a gain of one to optimize the overall system CMR.

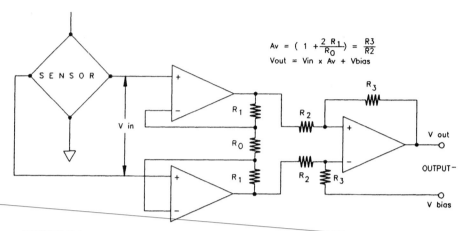

FIGURE 70.27 Three-amplifier instrumentation amplifier with ground-referenced output signal.

Independent output voltage bias adjustment can be easily provided in this stage by a voltage V_{bias}. This configuration can be used to create a live zero, differentiating the damaged transducer from zero signal. Examples of output formats with live zero include 1-6-VDC and 4-20-ma output signals. The 1-V or 4-ma signal allows a system processing sensor output to distinguish a real signal and a fault condition such as an open or shorted signal wire, which would produce a 0-volt or 0-milliamp signal, respectively.

Two-Amplifier Configuration

A simpler version of the instrumentation amplifier is shown in Fig. 70.28. The two-amplifier configuration requires two pairs of matched resistors and one gain-setting resistor. As in the three-amplifier configuration, the output may be biased up from zero by applying the desired voltage to the V_{bias} node.

Assuming for simplification that resistor R_0 is open, the bottom amplifier operates with the gain $1 + R1/R2$ for the negative output of the sensor, amplifying both differential and common mode voltage. The top amplifier provides the same amplification for the positive output of the sensor plus it provides a negative amplification of $R2/R1$ for the bottom amplifier's output, effectively canceling out the common mode voltage. The degree of the cancellation is defined practically by a ratio matching of feedback resistors. Assuming a 40-dB CMR (1 percent effective ratio resistor matching), the input error created by a 0.5-V common mode voltage change would result in a 5-mV input offset change.

Taking this into account, the matching of feedback resistors should be considered carefully, and possibly a trimming of a CMRR should be implemented. This configuration is thus more suitable for a constant-voltage operation where the common mode voltage (half of supply voltage) may be temperature independent.

Switched Capacitor Instrumentation Amplifier

The introduction of high-performance switched capacitor CMOS devices for analog applications has brought a new option for signal conditioning: instrumentation

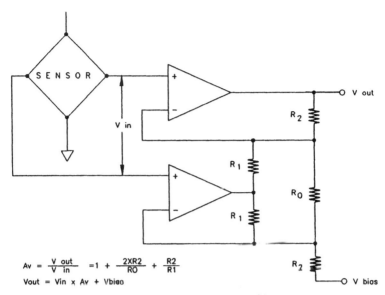

FIGURE 70.28 Two-amplifier instrumentation amplifier with ground-referenced output.

amplifier configurations using only a single operational amplifier, a few inexpensive analog CMOS switches, and a few capacitors.

While not truly a continuous system, the switched capacitor approach can provide very good performance and several unique performance advantages. Special care must be taken, however, when using this approach to avoid errors introduced by charge injection during transitions of the analog switches, increasing noise floor level. It should also be mentioned that the output signal will have a sampled nature. Since sampling can occur at a frequency significant higher than useful input variable range, filtering the output carrier frequency is not a problem in most of the applications.

A commercial building block already highly optimized for these applications is the Linear Technology LT1043 dual-switched capacitor block with on-chip clock oscillator. This IC provides four low-resistance single-pole double-throw (SPDT) CMOS analog switches clocked by an on-chip oscillator. The oscillator can be overdriven by an external clock if synchronization of more than four switches is required. The switches in this IC are symmetrical and have been compensated for charge injection errors to minimize offset voltages that might be introduced by uncompensated switches, delivering outstanding performance in low-level applications. The configuration shown in Fig. 70.29 uses LT0143 as a differential switch, operating in two phases. During the first phase, these switches connect the capacitor between pins 11 and 12 to the sensor differential output, charging it fully. During the second phase, this capacitor is connected to the input of a single-ended amplifier, transferring stored charge to the input capacitor which later maintains the voltage during a repetition of the first phase.

Since the voltage on the floating capacitor driven by switches can be grounded during the second phase, no common mode voltage is transferred to the input of the amplifier. The switched capacitor instrumentation amplifier is thus capable of providing complete isolation of the sensor and the amplifier within the voltage capabilities of the switches used and fairly easily achieves the CMRR of the best monolithic and hybrid amplifiers.

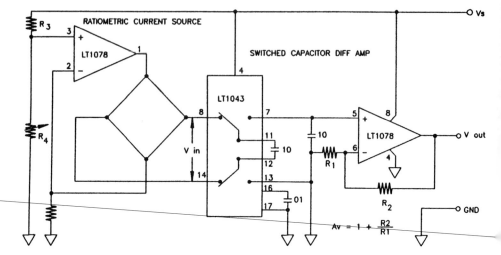

FIGURE 70.29 Ratiometric voltage output transducer with switched capacitor instrumentation amplifier. Components not shown.

Additional features such as amplifier offset cancellation (auto-zero) and output bias point shifting may also be provided entirely through the switched capacitor techniques resulting in very high performance capabilities. Gain adjustment can be realized by changing the switching clock frequency, making it compatible with digital microprocessor programmability. Other configurations of switched capacitor amplifiers are discussed in Ref. 34.

AUTOZEROING CIRCUIT WITH EIGHT-BIT RESOLUTION

Since mixed analog-digital signal processing becomes increasingly important for new applications, we discuss here an example of such an application.

There are some analog applications in which a very long holding period of time is required for the sensor autozeroing circuit. An analog sample-and-hold approach is usually limited to about a one-minute time frame. Mixed analog-digital processing can yield unlimited sample and hold.

The autozeroing circuit presented in Fig. 70.30 consists of a summing amplifier A1, output comparator A2, ramp voltage generator (capacitor C), eight-bit A/D converter ADC, eight-bit D/A converter DAC, and control circuitry. The autozero cycle is initiated by a microswitch SW, starting charging capacitor C with a constant current. The linearly increasing capacitor C's voltage is converted into a digital word by the ADC and later by the DAC into an analog zeroing voltage V_z. Voltage V_z is summed with an input voltage V_{In} in by an amplifier A1. When output voltage V_6 reaches the level set by a reference voltage V_1 connected to a noninverting input of a comparator A2, transistor Q1 stops the conversion, thus freezing zeroing voltage V_z. Switch SW activates transistor Q2 discharging capacitor C and initiating a new autozero cycle.

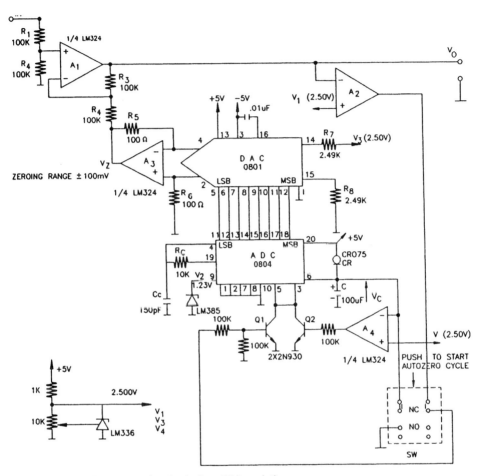

FIGURE 70.30 Autozeroing circuit allows 8-bit resolution.

Since a representation of the value of required zeroing voltage V_z is stored now in a digital form at the output of the ADC converter, the holding time is unlimited. Shown implementation does not require microprocessor control and thus can be implemented quickly.

SMART SENSORS

The explosion of microprocessor-based applications involves the sensor field. While the term *smart sensor* is not yet defined by any standard, there is active work in this area. The term *smart sensor* or *transducer* is applied to devices incorporating a microprocessor or microcontroller for some type of processing. A built-in microcontroller opens up new options, such as the following:

1. Digital calibration and compensation allows all repeatable errors to be electronically calibrated out. This includes the calibration errors of the internal signal-conditioning circuits as well as the errors of the sensor. Compensation for linearity can be extended to any order necessary to correct the linearity errors of a particular sensor. Temperature errors can be compensated at least two orders of magnitude better than with passive analog compensation.
2. Field recalibration without removal from service is possible through the use of portable pressure standards and a simple service module used to store new correction coefficients in the nonvolatile storage area of the microcontroller. There is no need to open the sensor and risk the integrity of the enclosure during the recalibration procedure.
3. Self-diagnostics can be provided to insure that all components of the system are performing up to specifications. The sensor can be made virtually fail safe for critical applications. The degree of self-test capability must be weighed against sensor cost and performance criteria.
4. User programming can be integrated in the sensor to provide such functions as units conversion, scaling, compensation for nonlinear input functions, counting, totalizing, integration, differentiation, and PID closed-loop control.
5. Systems simplification can result from the use of smart sensors distributing functions normally provided by the central computer in large systems to the sensor level. In large systems, the use of sensors with multiplexed communications can reduce wiring cost (often more than the sensor cost) by sharing one twisted pair or coax for tens to thousands of sensors. All major automotive companies are actively pursuing multiplexed wiring as a possibility for the cars of the year 2000.
6. Digital compensation provides not only a lower-cost solution due to a reduction in test and calibration time, but also makes a high-accuracy application requiring high degree of temperature compensation feasible. Applications such as process control can benefit from reduced number of models by using the improved accuracy to increase the turndown ratio of a sensor, thereby reducing the required number of pressure ranges.
7. Remote communication and unattended operation become an easy implementation feature.
8. Addressability of the sensor allows efficient communication.

An example of a smart sensor application taken to the ultimate level might be the air data computer used on military and commercial jet aircraft. The sensor measures pressure and temperature to perform a critical altitude measurement. A state-of-the-art system based on a piezoresistive pressure sensor achieves a total error band of 0.02 percent FSO over −55 to 100°C temperature range and over a period of three years without recalibration for any pressure within the calibrated range.[35]

The block diagram shown in Fig. 70.31 outlines functionally the components of a smart sensor for a 4- to 20-ma current loop output transmitter. In this all-digital open-loop design, the sensor's measurand (e.g., pressure) and temperature outputs are measured and the true value of the measurand is calculated either from a look-up table or an algorithmic calculation or a combination of the two. The implementation technique must be selected based on a number of factors such as response time, accuracy, and power budget for the finished sensor.

Once the true value of measurand is determined, it is converted back to analog form with the appropriate scaling factor and offset and is delivered to the output circuit which buffers the signal for transmission to the control system. In a very high

SIGNAL CONDITIONING FOR SENSORS   70.43

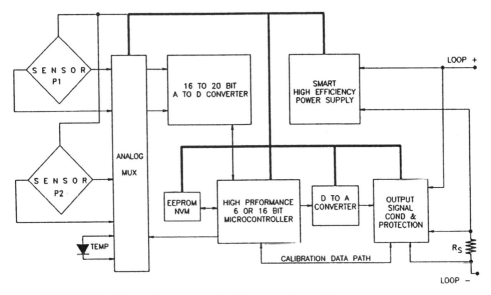

FIGURE 70.31 High-performance two-wire 4-20-ms smart transmitter.

accuracy application, the error in the D/A converter and output circuit could also be included in the algorithms used to correct the signal.

For lower-performance smart sensors with high bandwidth requirements, such as an automotive MAP sensor, an alternate approach to compensation can be used that modifies the uncorrected sensor's output in a feedforward approach with temperature as the only measured parameter. The calibration coefficients for the feedforward correction circuit must be measured and stored along with the corrections for the sensor element. Fig. 70.32.

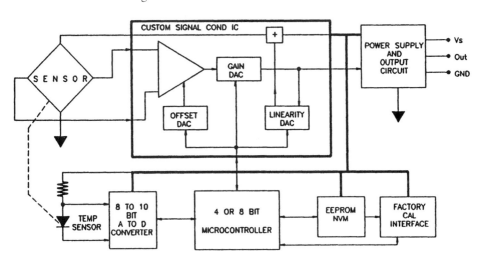

FIGURE 70.32 Feedforward corrected smart sensor for high-speed, medium-accuracy applications.

For high-speed applications, this approach has the advantage that only temperature needs to be measured, and depending on the initial error slopes of the uncorrected sensor, the temperature resolution requirements can be quite modest. The accuracy of the feedforward correction chip must be quite good and also very stable, but the dynamic range (number of bits) of the D/A converters on the chip is quite small.

Typically the integrated on-chip A/D available in most microcontrollers is adequate for achieving 10- to 12-bit sensor accuracy with this approach, but at the expense of a highly customized ASIC chip for the feedforward correction circuit. Additionally, a low-cost, slow microcontroller can be used since temperature does not change very quickly, and therefore the required calculations can be done quite slowly. While there is no clear definition of smart sensor, there is an active movement toward some kind of standardization. The sample of standard and implementation activity in the emerging field of smart sensors include the following:

- ISA SP-50, covering interchangeability of transducers with a 4-20-mA and 0-10-V outputs, primarily for process control
- IEEE P1118 underdevelopment with participation of Intel and Foxboro, primarily for process control
- University of Michigan effort within IEEE framework to develop a standard and architecture for highly on-chip integrated sensors
- Controller Area Network (CAN) developed by a joint effort of Intel and Robert Bosch Gmbh., primarily for automotive applications
- HART protocol implemented by Rosemont, utilizing a modem chip set developed by NCR
- 1553B standard by ERA Consortium, developing a military serial standard
- FIP Club in France, developing a trial standard
- ProfiBus in Germany underwritten by Siemens, developing another trial standard
- IEC TC 65C/Working Group 6 International Field Bus standard for sensor-controller interoperative communication
- NEMA activity in development of a standard for smart switches
- ASHVAC activity in defining a standard for heating, ventilating, and air conditioning industry.

REFERENCES

1. J. Bryzek, "Modeling Performance of Piezoresistive Pressure Sensors," *Proceedings of 1985 IEEE Conference Solid State Sensors and Actuators,* Hilton Head, Philadelphia, June 1985.
2. J. Bryzek, "Approaching Performance Limits in Silicon Piezoresistive Pressure Sensors," *Proceedings of 1983 IEEE Conference Solid State Sensors and Actuators,* Delft, The Netherlands, June 1983.
3. J. Bryzek, "A New Generation of High Accuracy Pressure Transmitters Employing a Novel Temperature Compensation Technique," *Proceedings of Wescon Conference,* San Francisco, 1981.
4. J. Bryzek, "Understanding Design and Performance of Advanced IC Pressure Sensors," *OEM Design Conference,* Philadelphia, September 1985.
5. J. Williams, "Circuits Allow Direct Digitization of Low-Level Transducer Outputs," *EDN,* November 29, 1984.

6. J. Williams, "Digitize Transducer Outputs Directly at the Source," *EDN*, January 10, 1985.
7. D. R. Harrison and J. Dimeff, "A Diode-Quad Bridge Circuit for Use with Capacitance Transducers," Ames Research Center, NASA, Moffett Field, California, June 1, 1973.
8. K. Yamada, M. Nishihara, S. Shimada, M. Tanabe, M. Shimazoe, and Y. Matsuoka, "Nonlinearity of the Piezoresistance Effect of P-Type Silicon Diffused Layers," *IEEE Transactions on Electron Devices*, **ED-29**, January 1982.
9. J. Bryzek, "Pressure Transmitter Employing Non-Linear Temperature Compensation," U.S. Patent 4,414,853.
10. S. Kordic and P. C. M. Van Der Jagt, "Theory and Practice of Electronic Implementation of the Sensitivity-Variation Offset-Reduction Method," Department of Electrical Engineering, Delft University of Technology; Delft, The Netherlands, November 29, 1985.
11. E. B. Merrick, and T. P. Stephens, "Dome and Transducer with Compensating Temperature Coefficient," U.S. Patent 4,581,940.
12. Da Hong Le, "Wheatstone Bridge-Type Transducers with Reduced Thermal Shift," U.S. Patent 4,658,651.
13. R. E. Hickox, "Apparatus and Method for a Pressure-Sensitive Device," U.S. Patent 4,672,853.
14. W. Schulz, "Circuit Arrangement for Compensating for the Temperature Dependence of the Sensitivity and the Null Point of a Piezoresistive Pressure Sensor," U.S. Patent 4,667,516.
15. T. B. Bozarth, Jr., A. M. Olsen, R. H. Rowlands, "Method of Digital Temperature Compensation and a Digital Data Handling System Utilizing the Same," U.S. Patent 4,592,002.
16. F. J. Houvig, "Sensor Output Correction Circuit," U.S. Patent 4,303,984.
17. L. A. Rehn, and R. W. Tarpley, "Thermally Compensated Silicon Pressure Sensor," U.S. Patent 4,320,664.
18. A. R. Zias and D. R. Tandeske, "Pressure Transducer Auto Reference," U.S. Patent 4,051,712.
19. J. E. Solomon and A. R. Zias, "Semiconductor Pressure Transducer Employing Novel Temperature Compensation Means," U.S. Patent 3,899,695.
20. J. E. Solomon and A. R. Zias, "Semiconductor Pressure Transducer Employing Novel Temperature Compensation Means," U.S. Patent 3,836,796.
21. R. J. Billette and J. Vennard, "Semiconductor Pressure Transducer Employing Temperature Compensation Circuits and Novel Heater Circuitry," U.S. Patent 3,886,799.
22. W. P. Mason, J. J. Forst, and L. M. Tornillo, "Recent Developments in Semiconductor Strain Transducers," *Semiconductor and Conventional Strain Gages*, Academic Press, New York, 1962.
23. A. D. Kurtz, "Adjusting Crystal Characteristics to Minimize Temperature Dependency," *Semiconductor and Conventional Strain Gages*, Academic Press, New York, 1962.
24. C. S. Smith, "Piezoresistance Effect in Germanium and Silicon," *Physical Review*, April 1, 1954, **94** (1).
25. F. J. Morin, T. H. Geballe, and C. Herring, "Temperature Dependence of the Piezoresistance of High Purity and Germanium," *Physical Review*, January 15, 1957, **105**(2).
26. R. Lee, "Current-Source IC Compensates for Temperature Errors," *EDN*, October 3, 1985.
27. J. H. Huijsing, "Signal Conditioning on the Sensor Chip," *Sensors and Actuators*, 1986, **10**.
28. K. Najafi and K. D. Wise, "An Implantable Multielectrode Array with On-Chip Signal Processing," *IEEE Journal of Solid State Circuits*, December 1986, **SC21**(6).
29. T. Ishihara, K. Suzuki, S. Suwazono, M. Hirata, and H. Tanigawa, "CMOS Integrated Silicon Pressure Sensor," *IEEE Journal of Solid State Circuits*, April 1987. **SC22**(2).
30. G. Kowalski, "Miniature Pressure Sensors and Their Temperature Compensation," *Sensors and Actuators* 1987, **11**.

31. A.F.P. Van Putten, "A Constant Voltage Constant Current Wheatstone Configuration," Eindhoven University of Technology, Eindhoven, Netherlands, June 18, 1987.
32. I.C.R. Wylie, *Advanced Engineering Mathematics,* McGraw-Hill, New York, 1984.
33. J. Bryzek, "Temperature Compensation—IC Pressure Sensors," *Application Note TN-002, IC Sensors,* March 1985.
34. J. Williams, "Linear Applications Handbook," Linear Technology, 1986.
35. P. DuPuis, "A Novel Primary Data Quality Pressure Transducer," *Proceedings Air Data Systems Conference,* 1985.

CHAPTER 71
SENSOR PACKAGING TECHNOLOGY

INTRODUCTION

The Design Process

As the basic silicon sensor chip progresses through the complex packaging processes of glass bonding, saw, die down, wire bond, and so on through assembly, it gathers various articles of packaging that allow it to interface harmoniously with the surrounding environment, while continuing to do its job of sensing and providing data.

This suiting-up is by no means a random piece of tailoring, but rather a complex matrix of solutions, designed to meet the needs and requirements of the end user, whether those requirements are generic or specific. The more rigorous the requirements, the greater the number of design considerations that contribute to the matrix.

All packaging requirements from the simplest to the most complex require the careful pooling and blending of multiple disciplines, including physicists, scientists, electronics engineers, mechanical engineers, metallurgical and environmental engineers, welding specialists, glassing specialists, fluid specialists, bonding specialists, micromachining specialists, and manufacturing experts.

Functions of the Sensor Package

An integrated silicon piezoresistive pressure sensor die is conceptually a completed sensor requiring no additional packaging. The sensor die incorporates the essential elements of a transducer including the piezoresistive sensing elements, a mechanical flexure to convert force, pressure or acceleration to resistance change, metal thin film interconnections suitable for interfacing the device to a power source, and a readout instrument or electronics for the purposes of measuring and/or controlling mechanical variables such as pressure. Such a device is typically passivated and suitable for use in laboratory environments.

Practically, however, a mechanical package is required to make a useful device. The mechanical package, connector, and their associated manufacturing technologies must fulfill the following functions.

Mechanical Protection. The bare silicon die is too small and too fragile for most applications. The package must allow the device to be easily handled by the user and/or automated assembly equipment. It must provide adequate mechanical protection for day-to-day use in the service intended.

Media protection. Unlike integrated circuits, silicon sensors must operate in harsh environments such as humidity, saltwater, body fluids, fuels, hydraulic fluids, difficult process industry fluids, and gases. While the classic IC package seeks to isolate and hermetically separate the sensing die from the environment, the sensor package must appropriately protect the die while transmitting the mechanical force to the sensor. Thus, the prime goal of the packaging technique is to provide good transfer of the mechanical variable of interest (measurand) from the environment to the sensor while eliminating the effects of environmental contaminants and corrosives.

Mechanical interface. Typically, the sensor is interfaced to a force or a pressure through the use of an appropriate mechanical fitting, tubulation, or other connection. The sensor package must incorporate a convenient low cost means for such interface while isolating undesirable mechanical stresses from the very sensitive silicon die.

Stress isolation. Silicon piezoresistive and capacitive sensors are sensitive to mechanical stress. Indeed, stress is the basic intermediate transduction mechanism for converting, for instance, from pressure to resistance change. The sensor package must be carefully designed to allow efficient transfer of the mechanical variable of intended measure such as pressure, while rejecting extraneous mechanical stress inputs from unwanted variables such as acceleration or vibration.

Since silicon is a relatively low expansion material while most packaging materials such as metals and ceramic exhibit a considerably higher expansion, the package must allow for an appropriate means of reducing the undesirable effects of thermal expansion created stress.

Electrical interface. The die level electrical interface is typically an aluminum or a noble metal thin layer. Typically, the required user interface is a connector such as a ruggedized, metal shell, aerospace connector or the typical pins of an IC. The package must provide an appropriate method of transition from the die to the required user connector.

Compensation and calibration. While modern silicon sensors often incorporate calibration and compensation on the silicon die itself, often analog or active compensation employing hybrid or component resistors is used. The package must provide appropriate means of incorporating and protecting these additional components.

Signal conditioning. The trend is to include more electronics within the sensor. The package must be suitable for housing and protecting the associated signal conditioning electronics.

In addition to the functional requirements which the package must accomplish, a suitable packaging scheme must have a number of characteristics including the following:

- Low cost
- Reliability
- Compatibility with subsequent assembly techniques
- High-volume production capability

The goal of the package designer is to address the functional requirements, adapting the best available technologies to produce a sensor which is manufacturable, serviceable in the intended environment, and reliable at the lowest possible cost.

EVOLUTION OF SILICON SENSOR PACKAGING

The packaging techniques used in silicon sensors come from two distinct roots. A piezoresistive silicon sensor die is essentially a low-density bipolar integrated circuit. One of the great strengths of and an important reason for the success of this technology is that it derives its technological base from the IC industry. This is very apparent in the design and the production of the silicon die itself.

The IC industry has also successfully developed packaging techniques which allow high-volume reliable packages to be manufactured at very low cost. Silicon sensors, from the beginning have employed many of the packaging techniques of the IC industry. Adaptation is sometimes difficult because of conflicting requirements but when successfully employed, these packaging techniques offer tremendous advantage over conventional packages. Figure 71.1 shows a silicon sensor package directly derivative in design from a conventional IC minidip package.

The second major root of sensor packaging technology comes from the conventional mechanical technology of the aerospace, industrial, and process control industries, as adapted primarily to the strain gage transducers which began to be employed from the late 1940s onward. These excellent packaging technologies involving the more traditional mechanical arts of machining of steels, welding, casting of alloys are well developed and can provide elegantly and reliably for all the required functions of a sensor package. Their disadvantage is the relative higher cost and unsuitability for very high volume (million per year) production. Figure 71.2 shows an excellent modern silicon sensor package of this class.

The modern silicon sensor package is a blend and amelioration of these two arts. The silicon sensor packaging engineer seeks to employ the best from each to produce a package which is rugged and reliable while being manufactured at a low cost employing highly automated manufacturing techniques.

The concept of micropackaging is an area that has great future potential in the development of the repertoire of the silicon micromachinist and resulting application to the packaging of practical sensing devices. The incorporation of functions,

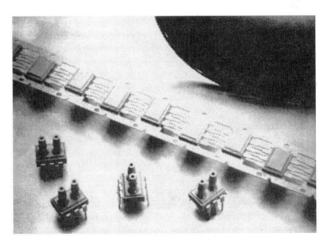

FIGURE 71.1 High-volume low-cost silicon sensor package based on IC minidip leadframe technology.

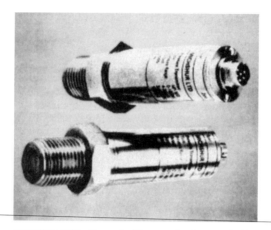

FIGURE 71.2 Rugged silicon packaging employing mechanical stainless steel welded package. (*Courtesy of Minchshur Ltd.*)

normally fulfilled by the package at the wafer level, results in great cost efficiency, improved reliability, and better control of parameters depending upon critical mechanical dimensions.

Figure 71.3 is an example of this concept. If a very narrow gap is formed behind a silicon deflecting diaphragm, the adjacent surface will effectively stop the diaphragm if flexed by pressure beyond a predetermined level, resulting in an increase in overpressure capability from normal overpressure of typically 3 to 10 times full scale to an overpressure capability of 100 to 1000.

This overpressure protection function is normally incorporated in the package through the use of a precisely machined mechanical stop such as a stop behind a stainless steel corrugated diaphragm in an oil media isolated package. Obviously, the difficulty of controlling such a precise mechanical stop is substantial and results in a relatively expensive large and bulky package.

A tremendous area for technological development in the next decade is the development and incorporation of more advanced packaging functions at the sensor level. Media protection, particle filtering, overpressure and overrange protection, and the integration of electronics and first-level advanced readout and interconnection schemes such as bumps for tab bonding are all areas that the micromachinist can be expected to explore in the next decade.

THE APPLICATION-DRIVEN NATURE OF THE SILICON PACKAGE

Silicon sensors are the eyes, ears, and fingers of the microprocessor. Microprocessor-based control systems are machines; they are expected to perform in a wide variety of environments, much wider for instance, than the sensing and controlling systems of a human being which are limited to rather small changes around our ideal environment of dry 72°F and dry air. Silicon sensors on the other hand must perform in environments as diverse as the harsh compartment of an automobile, the interior of a human body, the cold vacuum of outer space, and the hot hydrocarbon fluids of a petrochemical plant.

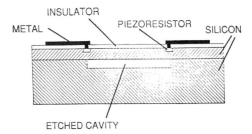

FIGURE 71.3 Silicon die with integral overpressure stop.

Silicon sensors must be incorporated into a wide variety of mechanical systems which have their roots in the industrial development of the last century. Silicon sensor packages must be appropriately packaged to be easily incorporated into systems as diverse as a modern electronic computer, a hydraulically driven XY table in a milling machine, the cockpit of a supersonic aircraft, or the compact package of an artificial heart. Each of these different applications has a set of mechanical and system design criteria based on years or decades of prior development.

The package is also performance-driven, the need to stress isolate the silicon sensor imposes a set of design constraints on the package designer.

Finally, and certainly not least important, is the requirement of cost. Silicon sensors have been particularly successful because of their perceived and demonstrated ability to be manufactured in very high quantities at relatively low cost. There exists a two-decade backlog of known applications whose realization was initially dependent upon the evolution of low-cost computing power but more recently has become very dependent upon the availability of inexpensive, reliable silicon sensors.

The preceding criteria combine in various ways to impose a given set of constraints on sensor package for each individual application. Following are some typical examples of silicon sensor applications which illustrate this point.

Figure 71.4 shows a disposable blood pressure transducer used in clinical care for patients undergoing postcardiac operation, intensive care. The transducer continuously monitors patient blood pressure providing information regarding the patient's health and need for drug therapy.

The application relative factors driving this package design are first of all cost (the product is disposable and must be manufactured in quantities of 10 million annually at $12 each at the hospital), performance (the transducer must provide the accuracy of an aerospace transducer), media compatibility (the transducer must operate in direct contact with bodylike fluids—saline solution—for periods of several days or more), high-voltage isolation (the transducer must withstand defibrillation voltages of 10,000 volts or more while maintaining patient's safety).

Hybrid packaging is the technology of choice for this application. This approach allows low-cost and batch testing in snapstrate form while achieving the excellent quality and reliability needed for the application.

An additional application-related packaging requirement is imposed by the package of Fig. 71.5 which is an intrauterine pressure sensor for use in measuring contractile pressures during child birth. In addition to all the requirements of the previous application, the transducer must be interfaced directly with body tissues and fluids to measure birthing pressures. The hybrid approach is also employed here with the direct overmolding of a plastic package.

The requirements of the research and aerospace industry emphasize an alternate set of price/performance factors. Figure 71.6 shows a thin-line flat-pack-style silicon

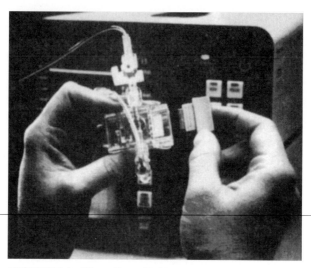

FIGURE 71.4 Silicon disposable blood pressure transducer. (*Courtesy of Abbot Laboratory.*)

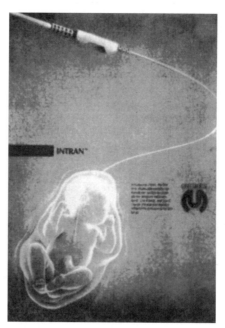

FIGURE 71.5 Silicon intrauterine pressure transducer for measuring child birthing pressure.

pressure sensor used for measuring surface pressures on airplanes and airplane models for wind tunnel and flight testing. For this application, cost is less of a factor; accuracy, reliability, and package customization become more important. The overall prime driver for the package, however, is miniaturization. This type of testing typically requires customization of sensors in relatively small lots, the transducer manufacturer must be able to adapt the basic product line to the individual test requirements in production runs of as little as several hundred pieces.

Certain aerospace/military testing requires somewhat more ruggedized packages as shown in Fig. 71.7. The transducer shown is intended for air blast pressure mea-

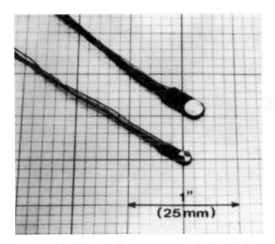

FIGURE 71.6 Thin silicon sensors suitable for testing surface pressures in aerospace applications. (*Courtesy of Entran Devices.*)

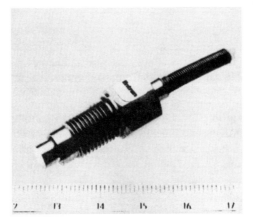

FIGURE 71.7 Ruggedized threaded silicon sensors for military/aerospace applications such as air-blast measurement. (*Courtesy of Entran Devices.*)

surement applications, typically high explosive blasts simulating an atomic explosion for the purposes of nuclear testing. This transducer must survive in a very high acceleration (10–20,000 G), high-pressure shock (10–50,000 PSI) environment. In addition, the blast is normally accompanied with very high velocity dust particles and an initial fireball with high-intensity light from which the transducer must be protected and an instantaneous temperature rise over several milliseconds to 5000°F. The fact that the relatively simple stainless package can provide these requirements very satisfactorily, indicates the tremendous performance and inherent reliability and ruggedness of the silicon piezoresistive die.

Consumer applications such as the bicycle altimeter shown in Fig. 71.8 impose a different set of design criteria. Here, the primary driver is reliability when manufactured in very high volumes at the lowest possible cost. A tire pressure gage, for instance, in order to be commercially successful must be manufacturable, typically offshore, at a target price on the order of $5 in order to sell at a retail price of $10 to $15. The solution to this type of application is the very low cost leadframe-based packages such as shown in Fig. 63.1 or chip-on-board technology, where the silicon sensing die is mounted directly on the PC board containing other electronic components.

FIGURE 71.8 Silicon-based bicycle altimeter. (*Courtesy of Avocet.*)

The pressure sensor shown in Fig. 71.9 is an industrial pressure sensor with 4-20 milliamp output suitable for measuring, for instance, hydraulic pressures at 6000 psi. The driving factor in this case is very high reliability because of safety and cost of replacement issues in the intended industrial application. The cost targets are intermediate and the volumes are in the tens of thousands; consequently, the packaging must be innovative to achieve production economies to serve this market.

The final application is shown in Fig. 71.10. This is a very innovative product which combines three drug infusion pumps into a single miniature package. Drug infusion pumps, widely used in hospitals, are largely replacing the drip bottles previously employed. These pumps are used to inject body fluids such as saline solution and precisely controlled drug dosages to a patient undergoing hospital care.

It is necessary to monitor the pressure in the tubing connected between the pump and the patient in order to detect whether there is a pump malfunction, a clog in the tubing, or a disconnection. For this application, the transducer must be designed to be indirect contact with a disposable plastic membrane cartridge through which the fluid flows. Since the transducer needs to be used many times, clearly a stainless steel diaphragm is the material of choice. The challenge to the package designer was to make the device at a relatively low price incorporating stainless steel isolation/oil-filled technology into a package significantly smaller then typical for this class of sensors. Figure 71.11 shows the package developed for this application.

The previously discussed applications just touch the surface of the many exciting applications employing silicon integrated pressure sensors. In fact, the range of current and future applications is as diverse as the world's manufacturing enterprise

FIGURE 71.9 Industrial 4-20 ma silicon pressure sensor. (*Courtesy of Siemens.*)

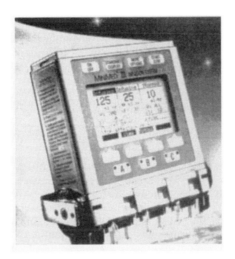

FIGURE 71.10 Medical infusion pump with three channels incorporating silicon pressure sensor for measurement of fluid pressure through a membrane. (*Courtesy of Siemens-Pacesetter.*)

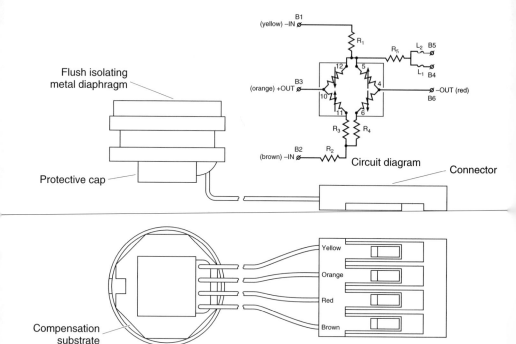

FIGURE 71.11 Miniature stainless steel oil isolation sensor designed for the application of Fig. 71.9.

since most product applications will include in the future the requirements for a closed loop control based on sensing mechanical parameters. Each new application will create its own unique set of packaging requirements.

WAFER-LEVEL OPERATIONS

The Concept of the Micropackage

The greatest strength of the silicon sensor technology is in the wafer-level operations. Incorporation at this level of as many of the functional elements of the package as possible will result in the most cost-efficient and reliable device. The wafer-level operations can be viewed as micropackaging.

The essential elements of the silicon sensor are the silicon piezoresistive elements, the mechanical flexure or pressure-sensing diaphragm and the thin-film electrical interconnections. These elements are incorporated into the silicon sensing die. Such a sensor, however, if unprotected, would be disastrously sensitive to environmental contaminants and lack sufficient stress isolation to accurately increase pressure.

A silicon device relies on the introduction of dopants or impurities with concentrations on the order of 10 ppm. Fortunately, silicon sensors are majority carrier devices and thus less sensitive than active devices such as transistors.

Nevertheless, the performance of the device would be severely affected by environmental contaminants such as water, salts, acids, and heavy metals.

Typically, first-level environmental protection is achieved by the use of dielectric materials such as silicon nitride, silicon oxide, and boro- and phosphorosilicate glasses. These dielectrics, which are incorporated as part of the standard IC process flow, provide a high degree of protection or passivation for the silicon elements.

The micropackaging function typically employed at the wafer level is to adequately constrain the flexure diaphragm. Typically, a stiff peripheral supporting area for the membranelike flexure is etched from an integral silicon wafer. Typical diaphragm thickness are on the order of 20 microns, while wafer thickness is on the order of 500 microns.

This first-level mechanical rigidizing provides for clamping of the silicon diaphragm and additionally provides a first level of stress isolation. The need for the stress isolation is apparent if one considers that typical full-scale stress for a pressure sensor at the piezoresistive elements is approximately 6000 psi which results in a 200 microinch elongation or shortening of the piezoresistive elements. On the other hand, the material of the package is typically steel with an expansion coefficient of 17 ppm/lOC or plastic with a much higher expansion coefficient. Thus a 100°C temperature variation (not untypical for a sensor application) would result in a deflection of 500 microinches or 8 times full scale.

Of course, first-level common-mode rejection is accomplished by the use of a Wheatstone bridge and appropriate symmetrical cancellation of stresses. Still, if this level of stress was transmitted unattenuated to the silicon diaphragm, the result would be an unreliable, nonrepeatable, highly temperature and mechanically sensitive device. Thus, the entire packaging scheme from wafer level onward is geared toward appropriately reducing this stress.

Wafer Lamination Techniques

The second level of packaging/stress reduction which is accomplished at the wafer level is the attachment of an appropriate constraint wafer through the use of wafer lamination techniques.

The conventional wafer lamination technique employed in the manufacture of silicon sensors was developed at Mallory in the late sixties and was subsequently adapted to piezoresistive sensing by many investigators including especially Mallon et al.[1,2] at Kulite who did very early work in the area.

If a silicon die is attached at its bottom surface, the stress at the top surface of the die is greatly reduced due to the geometry. If the lateral dimensions are reduced relative to the die thickness, this effect increases dramatically for aspect ratios (height divided by width) of greater than 1 to 1; stress reductions of 100 to 1 or better are achievable. Unfortunately, it is difficult to fabricate standard silicon wafers with height to width ratio's of greater than 1:1. Thus, the approach taken by the sensor industry has been to laminate a standard thickness (0.020 in) silicon wafer to a thick (0.030 to 1.25 in) pyrex wafer. The technique employed is the use of electrostatic or Mallory bonding shown conceptually in Fig. 71.12.

The sealing technique was discovered serendipitously by Daniel Pomerantz and Gordon Walliss[3] at Mallory while doing semiconductor research. If a flat polished silicon wafer is placed in contact with a similarly polished glass wafer, heated to approximately 400°C, and if an electrostatic bias on the order of 1000 volts DC is applied to the interface, a hermetic seal is formed. The bond is at most several monolayers thick and the mechanism appears to be a silicon-oxygen-silicon bond.

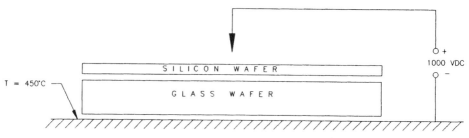

FIGURE 71.12 Electrostatic bonding concept.

This seal is hermetic (absolute silicon sensors have been tested and found to be stable over a decade) and imposes to a first approximation very little sealing stress since borosilicate glass is an excellent thermal expansion match to silicon. The bond is as strong as the glass itself.

Using this technique, it is possible to achieve aspect ratios of 3:1 or better. The electrostatic stress isolation technique has been widely and successfully employed in the sensor industry with 20 million sensors manufactured annually employing the technique.

Recently, however, NovaSensor has began exploiting a new and exciting technique called *silicon fusion bonding* for wafer lamination. Silicon fusion bonding is attractive because the glass is replaced by silicon which is compatible with high-temperature processing opening up a new range of geometrical possibilities to the designer. Furthermore, silicon has three times the elastic modulus of glass and is ten times stronger. Finally, silicon exactly matches the thermal expansion coefficient of the piezoresistive sensing diaphragm, eliminating second-order thermal expansion–related problems in the design of the sensor.

The technique is based on placing flat polished silicon wafers in contact at temperatures in excess of 900°C. Researchers in the United States and Japan had noticed as early as the early 1960s that silicon wafers would fuse together under these conditions. Recently, however, the technique was investigated and perfected by researchers at IBM[4] and Toshiba[5] and employed in the manufacture of dielectrically isolated silicon integrated circuits.

NovaSensor has pioneered[6] in adapting the technique to silicon sensors and has used this wafer lamination technique to provide all-silicon constraints allowing the manufacture of die 1 mm and smaller with aspect ratios of 1 to 1. It is anticipated that over the next decade the silicon fusion bonding technique will largely displace the silicon to glass sealing method.

Both silicon fusion bonding and electrostatic bonding offer the option of producing a gage or differential pressure (reference side of diaphragm exposed to atmospheric or absolute pressure) or an absolute (reference side of diaphragm exposed to vacuum) sensor. Die mounted on various thicknesses of glass are shown in Fig. 71.13.

GENERIC DIE OPERATIONS COMMON TO ALL PACKAGES

As can be seen from the previous section, the market demands can be met by a variety of different packages; certain operations, however, are generic and common to most packages. These operations include

FIGURE 71.13 Individual silicon piezoresistive die electrostatically bonded to glass constraint.

- Wafer sawing
- Die down
- Bonding
- Die protection

Wafer Sawing

After the wafer is probed, the devices are sawn into individual sensor die. The wafer is first mounted to sticky tape for sawing. The saw blade is a diamond/resin or diamond/nickel composite. A number of cutting procedures are utilized depending on the sensor system. Cutting through a thin silicon wafer is easier than cutting a thick silicon-glass bonded combination. In general, larger grit blades, which would incorporate microstresses in the silicon, are suitable for cutting the glass but not the silicon. Except for the added thickness and the glass constraint, standard IC (integrated circuit) cutting practices apply regarding surfactants, cleanliness, and blade width/depth ratio. A sawed wafer is shown in Fig. 71.14.

Die Characterization

The die is now in a form that can easily be tested and characterized on an individual basis (depending on the end use). Samples from each lot are often taken at this point to fully characterize the lot. Die mounted in the final package can be used in accurately determining the lot performance including sensitivity, hysteresis, and temperature effects.

Die Down

The next step is die attach. This is the mechanical and pneumatic connection of the sensor to the package; the method chosen must accomplish both. Since the thermal expansion coefficient of the package material to which the die is attached is typically

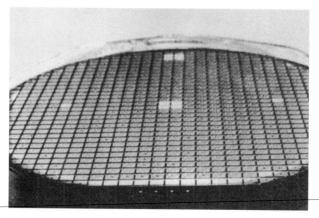

FIGURE 71.14 Sawed silicon wafer.

substantially different than that of silicon and glass, the die attachment must compensate for this. The three generally accepted methods of die attach are eutectic (gold-silicon alloy), epoxy, and silicone rubber, in order of highest to lowest stress. During eutectic die attach, the silicon is brought into intimate contact with gold metallization. With the aid of a gold-silicon alloy preform and the application of heat, pressure, and mechanical scrubbing, a strong bond is formed. The melting point of the gold-silicon system is 370°C. After the alloy freezes and cools to room temperature, the expansion mismatches (the thermal coefficient for silicon is 2.5×10^{-6}, for ceramic 6.5×10^{-6}) causes a significant residual stress. Although the mechanical seal is reliable, the stress generated at the die surface with this technique can cause considerable performance degradation.

Epoxy resin may be used to provide a strong bond with less residual stress while providing a reliable mechanical seal. Temperature cures for this system can vary from 65 to 180°C or higher depending on cure time and whether a catalyst is employed. Since this is a more flexible mounting technique, employing lower temperatures, less stress is transmitted to the die than with eutectic bonding. As in all die attach methods, care must be taken to insure that the backside pressure port is not compromised; the seal must be continuous but not plug the port or interfere with the diaphragm. Epoxy systems that can be screen-printed, b-staged (partially cured), film, transfer-printed, or syringe-dispensed are readily available for die mounting.

When the least amount of stress to the die surface is desired, a silicone rubber or RTV is the system most often used. The flexibility and low-temperature cure (25°C) contribute to the stress-free properties of this system.

The system does have some drawbacks. Since the material skins almost immediately upon exposure to air, it must be dispensed very carefully and die-attached without hesitation. As before, the pressure port must not be comprised. The material is generally quite thick but by carefully controlling the dispenser, the thickness of the die attach material can be tailored accordingly to minimize stresses as necessary. Due to the lengthy cure cycle necessary before subsequent operations, cycle times are increased. Fluoroelastomeric RTV-like materials are employed when the device must be used in the presence of hydrocarbon-based solvents. Figure 71.15 shows a handling tray with a number of silicon die RTV mounted to TO 8 packages.

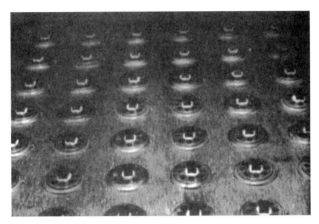

FIGURE 71.15 Silicon die after RTV die, down TO 8 headers.

Wire Bond

The next step in the process is to make the electrical connection to the silicon die. This is typically accomplished by bonding gold or aluminum wire by one of several techniques including thermosonic, thermocompression, or wedge-wedge ultrasonic wire bonding. The various techniques each have their particular process variations (temperature, force, ultrasonic power, etc.). In general, these are not detrimental to most sensors. Only in the case of very fragile structures do the ultrasonics damage the device. Other interconnect methods are usually acceptable in this case. Figure 71.16 shows a TO 8 package with the fine gold wire bonds visible.

Die Protection

The last assembly step is the appropriate protection of the silicon die and the electrical interconnects, depending on the end use of the product. Various methods include the following:

FIGURE 71.16 Silicon die after gold thermosonic wire bonding.

- Vapor-deposited organics (i.e., paralene)
- Silicon gel coating over the die
- A plastic or ceramic cap for particle and handling protection
- A welded-on nickel cap with a pressure port
- Oil-filled stainless steel diaphragm for corrosive environments

It is important to note that any method of protection has some effect on the die, care is taken to minimize these effects.

PACKAGING OPTIONS FOR SILICON SENSORS

The Design Process

In the following sections, various packaging options are discussed. The packages range from the simple to complex. Each package discussion is based on the following design requirements:

- The need for silicon sensing die protection
- The need for silicon all-media compatibility
- The electrical and mechanical interfacing
- Will the silicon sensing die or sensor react benignly with the environment to which it must interface?
- Will the silicon sensing die or sensor function within specification at the required temperature?
- Will costs associated with testing be within target?
- Will the sensor design be expandable into a family or a range of sensors (i.e., modular)?

Each requirement has more than one answer and many novel technical solutions are possible. Design requirements overlap and often a proposal will interact adversely elsewhere in the design. After a thorough analysis of the preceding, eventually the best design recipe emerges, the design is frozen, and prototypes are made and tested. The qualified design then becomes a production unit, where carefully controlled processes ensure that the initial design concepts are realized in each and built into every unit produced.

Unpackaged Die

The first and most basic packaging (and by far the least expensive) is the bare silicon sensor. The most economical method of purchasing a tested sensor requires that the customer take the responsibility of packaging and protecting the sensor. In this case the wafer is supplied fully electrically tested, inked, sawn, and characterized. The sensor die is typically supplied packaged still on wafer tape in a protective carton. Alternatively, the individual sensor chips can be supplied in wafflepack-type carriers. Electrical interconnects can also be bonded to the individual sensors, if desired.

TO Series Packages

A very cost-effective package is TO series headers used for miliary ICs. A cross section of this type of package may be seen in Fig. 71.17.

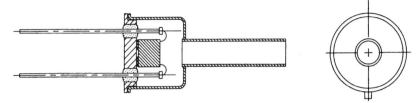

FIGURE 71.17 Cross section of sensor die in TO 8 header.

As is typical with IC technology, the most delicate parts of the sensor are the gold wire bonds, the points of attachment, and the metallized front face of the silicon. In order to provide mechanical protection, the silicon chip is protected by a metal can that is hermetically resistance-welded to the header. Gold-plated pins are hermetically sealed to the header base with glass eyelets which provide a matched hermetic feed through of both high electrical and mechanical integrity. Pressure is interfaced to this assembly by means of a integrated drawn tube, 0.187 inch in diameter.

For a typical service environment such as moist air, protection is required. The recommended protection is a gel coat over the die assembly. For some applications an additional layer over the gel coat over the pins is required.

Because the media is interfaced directly to the surface of the die on which there are voltages and currents, this type of device should not be interfaced to volatile, explosive, or combustible media that may be adversely affected by a mechanical or electrical failure in the die.

Compensation and calibration is by means of hybrid resistor module which can be seen on the package on the left in Fig. 71.18. Electrical interfacing to the unit may take the form of soldering or crimping to the leads or soldering directly into a printed circuit board. This is shown on the right-hand sensor of Fig. 71.18.

A very simple and highly useful variation of this package is the addition of a gage hole and a tube, thereby creating a differential unit. The TO mounted series of solid-state sensor packages is an interesting hybrid; a state-of-the-art 3D micromachined chip married to a relic of the integrated circuit industry form the TO 8 packages.

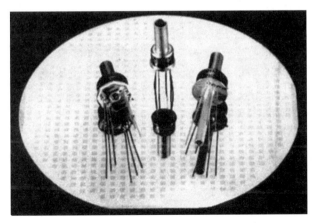

FIGURE 71.18 TO 8 style pressure sensor showing hybrid compensation-calibration substrate.

Because their modularity and simplicity lend themselves to high-volume production, coupling this with efficient testing and relatively low-cost package parts in high volume, NovaSensor has fabricated a low-cost, low-level, high-performance package that is finding more and more areas of application.

Metal Diaphragm, Oil-Isolated, All-Media Package

Tough media applications require a stainless steel interface to separate the silicon die from the corrosive media. The isolated style of package is suitable for pressures from 1 to 10,000 psi.

Figure 71.19 shows a typical cross section of such a transducer. A thin metal stainless steel diaphragm is welded to a stainless steel header. The silicon sensor die is mounted within the chamber thus formed which is filled with silicon oil (see Fig. 71.20). Since oil expands about 6 percent per 100°F, the stiffness of the flexible diaphragm must be a factor of 100 more flexible than the silicon diaphragm. Thus, the deflection of the oil is taken up by flexing of this diaphragm exerting a minimum of back pressure on the silicon sensor. Care must be taken to minimize the oil volume, provide hermetic seals, and completely bake out and vacuum outgas the silicone oil.

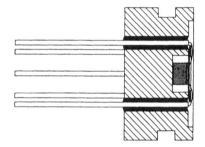
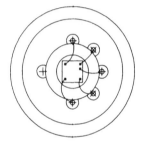

FIGURE 71.19 Cross section of oil-filled, all-media, isolation diaphragm sensor.

FIGURE 71.20 Partially assembled oil-filled pressure sensor.

The advantage of this package is that when the outside world looks at the transducer, it sees an inert metal diaphragm with excellent electrical isolation from the sensor body.

Two areas of application that take advantage of these characteristics are the food processing and medical industries. The pressure module of Fig. 71.21 is designed to be readily adaptable to a number of mechanical configurations. It may be readily TIG welded to a variety of pressure ports. A typical sensor is shown in Fig. 71.22. Thus, an efficient high-volume production testing of the basic module can allow for very low costs while the ability to weld various port configurations can provide the customer with desirable package flexibility. The variety of configurations readily obtainable is shown in Fig. 71.23.

The final key to lower cost is to minimize testing costs. Designing units to be interchangeable in testing equipment is fundamental to speeding up testing and to improving throughput. It also reduces the costs of test manifolds.

Hybrid Substrate Packages. Hybrid interconnection technology is a well-proven, reliable, relatively low tooling cost method of interfacing IC die. Hybrid circuit technology developed in the 1960s and 1970s is an excellent match to the problems of the sensor designer. Highly automated flexible manufacturing machinery is readily available. Low-cost, printed, thick-film resistors can be readily incorporated into the package.

The alumina substrate has a thermal expansion coefficient which, while not a perfect match to silicon, is relatively low compared to other packaging materials such as plastic and stainless steel. The alumina substrate, moreover, has a high elastic modulus and is very dimensionally stable platform upon which to mount the silicon die.

In a sense, the hybrid circuit approach offers many of the advantages of silicon wafer processing in that multiple die can be processed simultaneously in an array. A very successful application has been in disposable blood pressure transducers, where typically 30 to 130 piezoresistive sensors are processed simultaneously.

One of the most significant advantages in the technology is the fact that the sensors can be readily temperature-tested and laser-trimmed while still in snapstrate form. Multiple-head test fixtures with pneumatic interfaces and heating blocks allow for efficient temperature and pressure testing. Data is acquired through a multiplexed data acquisition system. Finally, the device is trimmed using a laser trimmer which trims the compensation resistors according to data transferred via LAN from the data acquisition system. This method of sensor production allows for production volumes on the order of one-quarter million a month or more at costs in the $2.50 to $3.50 range. It has proven to be the method of choice for the manufacturer of disposable blood pressure sensors which are currently manufactured in the United States in quantities of approximately 8 million per year. A typical disposable blood pressure sensor produced by this technology is shown in Fig. 71.23.

Ultra Low Cost Package

Another type of sensor packaging is the plastic leadframe-based package as shown in Fig. 71.24. Although this type of package has been used in the IC market for many years, it is relatively new in the sensor market, driven by the customer demands for a very low cost packaged sensor. In this case, the silicon chip is placed in a premolded (on the leadframe) plastic package; electrical leads and pressure ports are provided. Due to the mismatch of thermal expansion coefficients, the die will often have a glass constraint and be mounted with a soft RTV.

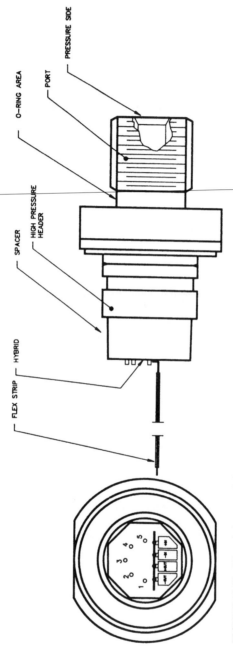

FIGURE 71.21 Threaded port-style iso-sensor.

FIGURE 71.22 Variety of iso-sensors based on modular design.

FIGURE 71.23 Silicon sensor or hybrid substrate.

Because no plastic is truly hermetic when molded over a copper-based leadframe, the silicon gel provides both die protection and a means of sealing the cavity in which the die is site. The external electrical connections are leads on 0.100-inch centers, compatible with standard integrated circuit sockets. They are also available in the gull-wing style for surface mounting to a circuit board. Both pressure ports are provided on the same side of the package in the form of a barbed hose fitting (for $1/16$ tubing), pressure ranges limited by the hose/barb strength, the die attach connection, and the hydrostatic strength of plastic. Rated pressure ranges up to 700 Kpa with 3X overpressure protection. The temperature range for this product is limited by the glass transition temperature of the plastic material used (200°C for Ultem).

Packaging and assembly economies are borrowed directly from the integrated circuit industry, where the equipment has experienced major development. For example, automatic die attach machines are continuously being upgraded to use the best die attach material, such as RTV, which will have its own unique set of handling and

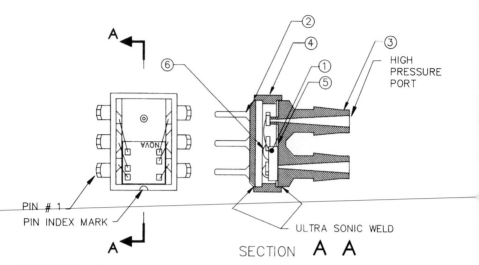

FIGURE 71.24 Cross-section of leadframe–style package.

dispensing problems. Testing is facilitated by the use of walking beam–type custom-designed equipment, which tests the device while still in the leadframe configuration. To achieve the lowest possible cost in this type of package, significant up-front hard-tooling costs and very high volumes of product are required. Compensation of the device can be provided by the customer, the chip can be compensated for, or a ceramic substrate can be provided in some package configurations.

REFERENCES

1. J. R. Mallon and D. G. Germanton, "Advances in High Temperature Ultraminiature Solid State Pressure Transducers," presented at *ISA Silver Jubilee Conference,* October 1970, Philadelphia, Pennsylvania.
2. J. R. Mallon, A. D. Kurtz, and H. Bernstein, "A Solid State Bonding and Packaging Technique for Integrated Sensor Transducers," presented at *The ISA Aerospace Instrumentation Symposium,* May 1973, Las Vegas, Nevada.
3. G. Wallis, and D. I. Pomerantz, "Field Assisted Glass-Metal Sealing," *J.A.P.* October 1969, **40**(10), p. 3946.
4. J. Lasky, S. Stiffler, F. White, and J. Abernathy, "Silicon on Insulator by Bonding and Etch Back," *IEEE Electron Device Meeting,* December 1985, p. 684.
5. H. Ohashi, K. Furakawa, M. Atsuta, A. Nakagawa, and K. Imanura, "Study of Si Wafer Directly Bonded Interface Effect on Power Device Characteristics," *IEEE Electron Device Meeting,* December 1987, p. 678.
6. K. Petersen, P. Barth, J. Poydock, and Mallon J. Brown, "Silicon Fusion Bonding for Pressure Sensors," *Tech Digest, IEEE Solid State Sensor Workshop,* Hilton Head Island, South Carolina, June 6–9, 1988, pp. 144–147.

CHAPTER 72
ADVANCES IN SURFACE MICROMACHINED FORCE SENSORS

INTRODUCTION

Surface micromachining of polysilicon devices on a single crystal silicon substrate is maturing as evidenced by, for instance, the commercial exploitation of polysilicon pressure transducers. This pressure transducer technology is based in part on the batch fabrication of vacuum-sealed pill boxes. Pill boxes, which contain mechanically resonant structures, produce devices for which axial applied loads cause resonant frequency changes. This leads to a quasi-digital force transducer which requires typically 10^{-14} watts of input power ID to maintain the resonance.

The low input power requirement lends itself to optical excitation and interrogation. Photon flux to voltage conversion is achieved with a photodiode. The open circuit photodiode voltage is used to drive the mechanical resonator capacitively. Resonances with modulated and unmodulated input fluxes have been achieved for transducers without connecting wires or critical optical interconnects.

Surface micromachining in its most basic form is used to produce micromechanical structures such as clamped-clamped beams and cantilevers which are only locally attached to the substrate. Structures of this type are formed by starting with a substrate and furnishing it with a patterned sacrificial layer. This layer, typically a thin film, is used to support the unsupported sections of the micromechanical structures temporarily. The construction material is deposited over the patterned substrate and patterned with a second mask. Lateral etching with an etchant which does not attack the substrate or the deposited construction layer produces structures which are locally attached to the substrate.

The surface micromachining system which has received the most attention is the silicon-silicon dioxide-polysilicon system. The starting material is a single crystal substrate with orientation and doping configurations which are not restricted by surface micromachining requirements. The substrate is normally oxidized thermally to form the patterned sacrificial layer. Local oxidations with patterned silicon nitride layers are acceptable. The construction material is an optimized low-pressure chemical vapor deposited polycrystalline silicon film. This film is patterned by reactive ion etching. Removal of the oxide occurs via etching in liquid hydrofluoric acid which will not attack silicon or polysilicon under most processing conditions.

The Si-SiO$_2$-Si system is restricted in achievable geometries. Oxide layers are typically in the micrometer range. Deposited silicon layers which are above four micrometers or so are not very practical because of low growth rates. These restrictions can be overcome in high aspect ratio processing where electroplated metals can be used to produce surface micromachined structures with heights as large as several centimeters. The restrictions to which the SiSiO$_2$-Si system is subjected are beneficial from the IC-compatibility point of view. The limited structural heights produce minor nonplanar processing problems; this is certainly not true for high aspect ratio processing. Furthermore, if one considers that sensors should ideally be noninvasive, which implies small physical dimensions, it is not surprising that the Si-SiO$_2$-Si system is playing a major role in this technology sector.

The Si-SiO$_2$-Si surface micromachining system produces two types of micromechanical structures: open devices and vacuum-sealed sensors. In the second category are absolute pressure transducers and, as of a few months ago, vacuum-sealed resonant force transducers. Since vacuum-sealed sensors appear to have a very large application potential, the following discussion focuses on them.

SURFACE MICROMACHINED ABSOLUTE PRESSURE TRANSDUCERS

Pressure transducers typically use a thin plate or membrane which deflects with applied pressure. This deflection which depends on the pressure difference to which the two sides of the plate are subjected is monitored electronically. The device becomes an absolute transducer if one side of the membrane is connected to a vacuum reference. An absolute pressure transducer is therefore a vacuum-sealed pill box which deforms with applied pressure in a repeatable, elastic manner. The deformation can be sensed by several techniques. Piezoresistive sensing, which requires the decoration by the pill box with carefully placed strain-sensitive resistors, is the transduction mechanism of choice here.[1]

The construction of an open pill box via surface micromachining is, in concept, not very difficult. The silicon substrate is normally quite thick, say 400 μm or so, and can therefore be used to form the rigid support for the box. The polysilicon film, typically 2 μm in thickness, acts as the deflecting membrane and is attached at its periphery to the substrate. The interior of the cavity is defined through the patterned oxide post, which is removed via sacrificial, lateral etching. In order for the etchant to come in contact with the post membrane, attachment to the silicon substrate must occur over a patterned thin oxide which allows etchant entry to the main cavity. The thickness of this surrounding oxide is typically 500 Å with etch channel widths near 1 μm.

Immersion of the pill box structure into HF results in the complete dissolution of all exposed SiO$_2$ in a self-limiting fashion. Subsequent cleaning is followed by vacuum sealing.

The most direct method of producing a vacuum-sealed cavity is that of thermal oxidation under atmospheric pressure conditions. The argument for this hinges on the volume expansion which occurs when silicon is oxidized. Thus, one volume of silicon produces two volumes of SiO$_2$.

Etch channels can therefore be designed in such a way that channel closure occurs before interior cavity dimensions have changed significantly. The etch channel closure interrupts the oxygen supply to the cavity interior. However, trapped oxygen continues to react with the cavity wall and thereby produces a hard vacuum.

There are variations to this technique which use silane and hydrogen outdiffusion. In any case, reliable, batch-fabricated seals can be produced for very low cost.

Sealed devices can be processed further by adding a dielectric isolation layer which is then used to support polysilicon piezoresistors. The details are summarized in Fig. 72.1.[2,3]

In order to understand the performance of these devices, one has to consider first of all the mechanical properties of LPCVD polysilicon. Practical depositions extend from 1000 to 40,000 Å with uniformities of ±1 percent over 3-inch wafers. Film strain levels vary from −0.7 to +0.3 percent and are adjustable via annealing to ±5 percent of a design value. Young's modulus is 1.62×10^{12} dyne/cm^2. The Poisson ratio is assumed to be 0.228. Fracture occurs for strain levels above 1.7 percent.

With this information, mathematical expressions for touchdown pressures as functions of diaphragm size and pill box geometry can be evaluated. This has, in fact,

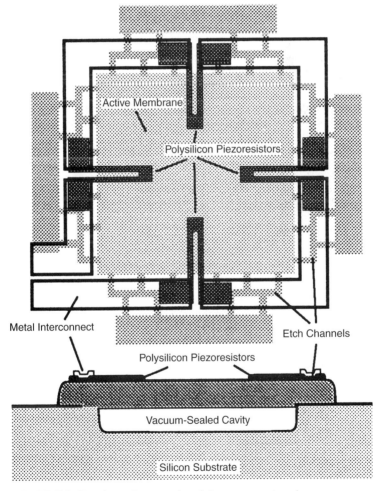

FIGURE 72.1 Top view and cross section of planar pressure transducer.

been done and yields the conclusion that pressure sensors with touchdown pressure to 10,000 psi can be fabricated via this technology.

The lower detection limit for this technology comes from the sensitivity of the polysilicon piezoresistor. These devices have a parallel gage factor near 22 and a perpendicular gage factor of –8. Based on these values, one can calculate the expected output voltage for half-active bridges in millivolt per volt of bridge excitation. Figure 72.2 contains this information.

Since a sensitivity of 10mV/volt is already quite low, a minimum absolute pressure sensitivity of 5 psia is a consequence of the use of piezoresistors in this technology. If this is unacceptable, then the transduction mechanism must be changed.

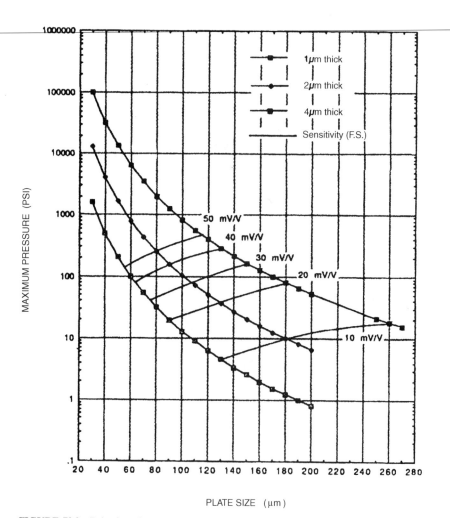

FIGURE 72.2 Behavior of square diaphragm at touchdown with 0.7-μm deflection for plate thicknesses of 1, 2, and 4 μm.

RESONANT INTEGRATED MICROSENSOR (RIM)

Transduction mechanisms which improve pressure transducer performance at the low end can be obtained from capacitive and optical sensing. However, there are practical problems. Capacitive sensing requires the cofabrication of at least some microelectronics as the sensor size is decreased. Optical deflection sensing is difficult to implement in real sensor environments and involves difficult and expensive packaging techniques. A third technique: applied force transduction to resonant frequency changes fits well into this technology and, additionally, produces a quasi-digital device which is monitored in frequency or time and not in amplitude.[4]

The basic concept for a resonating microstructure involves first of all the vacuum-sealed pill box technology which is being used to produce surface micromachined pressure transducers. This cavity can be used to fabricate a clamped-clamped beam in the vacuum environment. The beam can be excited into resonance by applying forces electronically at the modal maxima. The vacuum environment prevents air damping and, since acoustic losses in silicon are low, a high-quality factor resonator results. Figure 72.3 illustrates the basic construction concepts.[5]

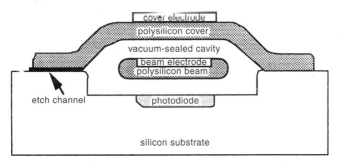

FIGURE 72.3 Polysilicon resonator cross section.

The excitation scheme in Fig. 72.3 uses a reverse biased PN junction at the beam center to couple to the beam capacitively. The fundamental resonance of a clamped-clamped beam is given by

$$\omega^2 = (42\ Eh^2/\rho\ L^4)[1 + (2L^2/7h^2)\varepsilon] \tag{72.1}$$

where E is Young's modulus, ρ the density of the beam length with L and thickness h. The variable ε is the total axial strain to which the beam is subjected. Evaluation of Eq. 72.1 for $L = 200$ μm, $h = 2.0$ μm, and a beam width of 40 μm yields the conclusion that the fundamental will be near 500 kHz if $e = 0$. Axial loads of 1 dyne, roughly a milligram of weight, raise the resonant frequency by 250 Hz. This change in frequency can be measured to 0.001 Hz, which hints at the large potential sensitivity of the device. Whether this large sensitivity can be exploited depends to a large degree on the mechanical stability of the beam material. There are indications that drift of less than 1 ppm/month occurs for polysilicon beams in long-term tests.

Experimental devices have been fabricated and tested. They confirm the behavior predicted by Eq. 72.1. They were also used to measure the quality factor; energy stored to energy dissipated per cycle; of the resonator. Q values above 100,000 are typical and result in the conclusion that the energy supplied per cycle is roughly equal to 1 electron volt. The power dissipation is therefore in the 10- to 14-watt range. These numbers reflect themselves into capacitive driving voltages at resonance of a few microvolts. This type of driving requirement is easily met by optical excitation which is based on the optical transparency of silicon for red wavelengths and photon to voltage conversion of an open-circuited photodiode which is allowed to be quite inefficient. Modulation of the light source at the resonance excites the mechanical vibration. Measurements of the reflected signal via Doppler techniques monitors the beam displacements and may, of course, be used in feedback schemes to produce a locked oscillator. Figure 72.4 illustrates the concept and also indicates that on wafer testing of resonators it is feasible.[6]

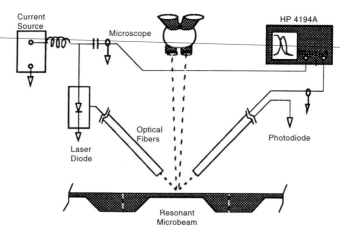

FIGURE 72.4 Optical procedure for resonator testing.

Preliminary optical test data has been used to verify the power dissipation argument. It has also been used to study the possibility of resonance excitation with unmodulated light. For this purpose, the mechanical structure of Fig. 72.3 is considered to be an optical filter for the photodiode. The filter is designed to produce large optical intensities at the diode for an undeflected beam. The resulting photovoltage attracts the beam toward the diode. This movement reduces the light intensity at the diode which causes the beam to reverse direction, overshoot the undeflected beam position, and return to the undeflected position. The cycle repeats and resonance is achieved.

The previously outlined concept has been verified experimentally and theoretically. Source to resonator spacing of several centimeters are acceptable. Beam monitoring via Doppler techniques is not very position sensitive and produces a clean signal output for which only the frequency and not the amplitude are important.

CONCLUSIONS

Surface micromachined silicon structures are starting to produce commercially viable devices. Thus, open structures are being used to produce accelerometers and sealed devices have found a niche in pressure transducers.

The ability to produce batch-fabricated vacuum-sealed cavities has been exploited by filling these pill boxes with mechanically resonant structures. These devices which transduce axial loads to frequency changes are fundamentally replacements for piezoresistors and deformable capacitors. In this sense, resonators find application in precision pressure transducers and accelerometers. These applications are more or less standard and must rely on performance and cost advantages.[7]

There are two unusual attributes to this type of transducer: a unit which requires no physical contact via wires or critical optical connections, and a device which uses silicon only and, in particular, requires no metallization. The first attribute together with the ability to excite and monitor with what essentially is a flash light leads to a variety of infrastructure sensors which cannot be addressed easily with existing technologies. On the other hand, all silicon structures without metals extend the temperature range in which the sensor performs well. This is important in automotive electronics where increasing engine temperatures are the rule.

REFERENCES

1. H. Guckel and D. W. Burns, "Planar Processed Polysilicon Sealed Cavities for Micromechanical Sensors," *IEDM*, San Francisco, 1984, p. 146.
2. H. Guckel, *Sensors and Actuators*, A 28, 1991, p. 133.
3. D. W. Burns, "Micromechanics of Integrated Sensors and the Planar Pressure Transducer," Ph.D. thesis, University of Wisconsin—Madison, May 1988.
4. R. T. Howe and R. S. Muller, "Resonant-Microbridge Vapor Sensor," *IEEE Trans. Electron Devices*, ED-33, 1986, pp. 499–506.
5. J. J. Sniegowski, "Design and Fabrication of the Polysilicon Resonating Beam Force Transducer," Ph.D. thesis, University of Wisconsin—Madison, June 1991.
6. H. Guckel, M. Nesnidal, J. D. Zook, D. Burns, "Optical Drive/Sense for High 0 Resonant Microbeams," *Transducers '93*, Yokohama, Japan, 1993, p. 686.
7. D. W. Burns, J. D. Zook, R. D. Horning, W. R. Herb, H. Guckel, "A Digital Pressure Sensor Based on Microbeams," *Transducers '94*, Hilton Head, South Carolina, 1994, p. 221.

CHAPTER 73
PEER-TO-PEER DISTRIBUTED CONTROL FOR DISCRETE AND ANALOG SYSTEMS

INTRODUCTION: WHAT IS PEER-TO-PEER DISTRIBUTED CONTROL?

In a manner replicating developments in the computer industry, sense and control systems are moving from centralized to decentralized architectures. Large centralized controllers scanning hundreds or thousands of points are being replaced by local controllers, interconnected by a communication path—a bus or network.

Central control schemes came about because processing resources were historically centralized and expensive. Everything in a system tied to the central point of intelligence (Fig. 73.1(a)). But the centralized controller is only an artifact, not a necessity in a control system. The control loop is always closed between the sensor(s) and actuator(s) themselves in a system. Because they are widely distributed across machinery, factories, or huge processing plants, it makes sense that the intelligence to perform control tasks should be similarly distributed (Fig. 73.1(b)).

Large and small users of centralized architectures have come to this realization, whether in factories, process plants, buildings, transportation systems, or other applications where control systems exist. Many have predicted that centralized architectures will be obsolete for new installations in three to seven years.

More importantly, based upon actual bids and installations, users are finding cost savings of up to 40 percent from distributed systems compared to centralized systems. These savings come from reduced costs of wiring, cabinetry, power supplies, and so forth.

The terms *smart sensor, smart transducer,* and *smart actuator* are used in proliferation today, promising benefits of increased accuracy, higher reliability, and more precise sense and control of the physical world. Such *smartness* is the result of embedded microcontrollers. They can be used to calibrate a pressure sensor, watch for signs of improper operation or wear out, and perform a sanity check on its reading before reporting it across a communication network. A motor's smart controller can decide when to operate, report if the motor starts successfully or if a fault exists, monitor current flow, rpm, torque, and even determine if the proper cool-down interval has been observed before restarting.

Each generation of integrated circuit technology allows increased performance and functionality of these controllers at lower costs. (Note the world of PCs!)

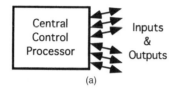

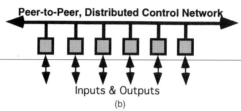

FIGURE 73.1 (a) Inputs and outputs central control process. (b) Peer-to-peer inputs and outputs.

Today's technology already allows enough computing power to do control on the same silicon along with sensing or actuation, at IC costs comparable to the stand-alone microcontroller of the last generation. The tenfold increase in the number of transistors on a single IC seen each decade will correspondingly increase the capability of intelligent control devices at the same rate seen in CPUs and memory chips.

Users have recognized this progress and now prefer distributed system designs which can rapidly follow the price and performance improvements of semiconductor technology.

Ultimately, it is possible to push all decision-making functions down among the devices themselves, communicating as peers across a control network, eliminating the need for central control equipment.

WHY IS PEER-TO-PEER DISTRIBUTED CONTROL USEFUL?

Reliability

Picture a conveyor system in a factory where a central processor or PLC controls all the motors, sensors, bar-code readers, and solenoids. Upon failure of that controller, the entire line either loses control of its I/O devices or stops immediately, perhaps causing product or equipment damage. With a system built from intelligent, distributed control nodes, the shutdown could be accomplished gracefully as everything but the failed device would still be able to make control decisions. Depending upon the site of the failure, perhaps portions of the line could even continue to operate normally.

Systems that tie large numbers of I/O devices to a central controller create the potential for total outage if that controller fails. A well-designed, peer-oriented, distributed control system minimizes the impact of individual failures. It allows the rest of the system to continue to operate, using either default values or the last-read settings from the failed equipment until repairs can be made.

Flexibility

A distributed approach allows for modularity at the lowest level in the control architecture, making the configuration a more exact fit for the application and potentially less expensive. For example, temperature, humidity, and filtered airflow must be tightly controlled on an IC wafer processing floor, but not in the cafeteria, hallways, or rest rooms.

Control system component products each have their individual strengths and weaknesses. The purchaser of a control system trades off many attributes of cost, availability, reliability, accuracy, physical size, the manufacturer's reputation, and ease of service or replacement of the components bought. If the system is built around a flexible, distributed topology, integrators and users have the option to choose from many manufacturer's devices that could be successfully interconnected. They would be free to make the best-fit choice for each function, based upon their own priority of attributes. A $200 sensor doesn't have to be used when a $20 version will do—just because a single control system's manufacturer only offers the higher-cost, high-precision model.

Expandability

Expanding a centralized control system with more field devices degrades its performance as the main processor must input, process, and output information for each attached point. (See Fig. 73.1a.) Even when I/O devices are added to a simple distributed I/O bus—but still mastered by a central controller—more I/O message bandwidth is consumed and now the I/O bus and message processing can become the bottleneck to the control's performance. (See Fig. 73.2.)

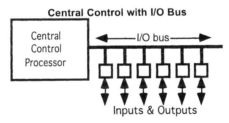

FIGURE 73.2 Central control with I/O bus.

Picture an automated conveyor system, originally designed to move one product line, that now must handle a second or third line and coordinate its movement with the first. If a centralized control is used, either the original system must be greatly oversized in its capability (if the expansion was anticipated) or, more likely, replicated a second or third time.

Distributed control systems using peer communications over a network allow greater expansion potential. Processing is divided up among intelligent devices working together. (See Fig. 73.1b.) Network bandwidth is used more efficiently by communicating only when necessary, no longer requiring every action in the system to require one or more bus transactions. In the conveyor example, each line could be implemented on its own network channel with communication among channels used to coordinate their actions. (See Fig. 73.6 which follows.)

Interoperability

Manufacturers often try to force fit some level of interoperability between their products and others with only partial success. Usually these efforts require a customized design effort for each instance, adding delay and hardware costs in order to settle for a limited scope of interoperable control.

Nodes must be able to work together in a distributed architecture. Not only must they share a common communication protocol in order to exchange data, they also need to be part of a system architecture. It needs to include definitions of what types of node devices will be connected, what data types must be exchanged, how devices are addressed, how action will be taken upon message receipt. The architecture must also include an approach for installation of nodes and the diagnosis and resolution of faults.

An open architecture assures that a wide variety of interoperable devices is available for users' choice in the marketplace. The computer industry saw the development of a huge variety of high-quality hardware and software products to run on PCs. Similarly, emerging standards for the design of control networks will cause increased choices and quality of control network products.

Also, many control problems include the interaction of several related systems working together. While each subsystem of the control is involved with its own devices, subsystems also interact as needed without special hardware interfaces or protocol conversions. (See Fig. 73.3.)

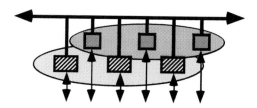

FIGURE 73.3 Interactive systems.

As an example, semiconductor processing involves the interaction of many functions including electronic instrumentation, material handling, building environment (HVAC, lighting, and access control), chemical and heat processes, distribution of chemical gases, and waste water treatment. Traditional control approaches required each function to be controlled by its own system. Coordination of operation was done either manually or by only limited communication allowed between systems. A distributed control approach can integrate these functions into a single architecture, allowing natural interaction as required without the use of expensive communication gateways or 10 independent operator consoles in the plant's control room.

WHAT CHARACTERISTICS ARE NEEDED OF A TECHNOLOGY USED TO IMPLEMENT PEER-TO-PEER DISTRIBUTED CONTROL?

Any technology chosen to implement a peer-to-peer, distributed control architecture must include the following:

- A low-cost yet powerful processor to implement control intelligence at each node site while it shares information with others

- A network architecture designed for the needs of control and communication among intelligent nodes
- A means to integrate this new approach with existing systems allowing smooth upgrades

A Low-Cost, Standardized Controller Element

First, the intelligent nodes themselves are required as processing elements married to digital or analog inputs or outputs. When used in sensors, they allow internal decision-making power. Not only must they sense their intended digital or analog phenomenon, but they can self-calibrate, diagnose, and report their own faults or adjust their own range settings. Furthermore, they might also decide *when* to report their readings to the outside world. On a timed basis? Only when the sensed value changes by 2 percent?

An intelligent actuator can be energized by a specific START signal, but it will also operate based upon its own computation. It will use status information received from sensors or other actuators input to its internal control program from across the network.

This suggests the need for a microprocessor-based element with a flexible I/O architecture to interconnect with various sensor elements and actuator drivers. It must provide reasonable computational power, be easy to program, and must be highly reliable.

Without a central control element, a distributed system needs a way to load and run a distributed control application among its large numbers of nodes. This suggests that each node has nonvolatile memory area which can hold network-downloaded code. Further, it offers the flexibility of making system changes or additions possible without visiting each node site to change program ROMs.

Because intelligent nodes will be used in great numbers within a peer-oriented system, their cost must also be low, with as much functionality provided on-chip as possible.

A Fully Featured Network, Appropriate for Control

Most people intuitively understand that a network allows devices to exchange data in a way that they all understand. To effectively do distributed control, a fully featured network communication scheme is needed, offering services beyond just sending and receiving message packets. The ISO's OSI seven-layer Reference Model provides data communication systems with a template for network services. Most of these services are needed in a well-designed control network as well.

Analog and digital data, device configuration, and diagnostic data types should all be supported with variable-length message packets. Short packets are desired for discrete information to save bandwidth, longer packets are needed for analog data or configuration and diagnostic information to avoid transmitting and reassembling multiple messages.

These data types must also be well defined and agreed upon by all connected equipment types. Communication errors must be identifiable and corrected easily. This means use of error checking at both the OSI link level (like a CRC—cyclical redundancy check—of each packet) and at the transport level (an acknowledgment response from the receiver(s) of the message).

Different types of communication media may exist (wire, power lines, radio, fiber optic, etc.) and should be able to be interconnected to create a single control network. Besides the physical connection, nodes tied to this network need to be logically connected to determine which specific inputs and outputs effect one another. Finally, the network must be built with a flexible technology to allow for changes and additions of more devices or more communication channels without prohibitive costs.

An essential aspect of peer-to-peer, intelligent distributed control is the ability for any node in the system to directly communicate with any other node. This is critical to the mechanism that allows control to take place among the nodes themselves. This requires an addressing scheme that can define node locality as well as unique node addresses. Much like the telephone system uses area codes, exchanges, and extension numbers to route calls, a control network technology must allow for a routing scheme if it is to grow beyond a single communication channel.

Many controls require the use of broadcast messages to coordinate startup or to report a failure to many devices at once. This suggests multicast messages must be supported, to be sent to all or portions of the network, without requiring hundreds of individual, point-to-point transactions.

Other practical questions that deserve consideration about a peer-to-peer network choice include the following:

- How well defined is it? Is it complete?
- Has it been proven to work reliably?
- Is it implemented via low-cost, low power-consuming ICs?
- What features does it have to help diagnose communication problems?
- What development tools exist to help implement its technology?

A Migration Path from Existing Products

While use of new technology always has great personal appeal to the technical community, offering opportunity to work with new products and new approaches, most of us are faced with the reality of control equipment that is still in use, designed around older, centralized control architectures, that cannot be immediately replaced.

Any technology considered as a solution to implement peer-to-peer, distributed control must be able to allow the interoperation of existing products along with the new to the greatest degree possible if the new approach is to be accepted. This means that existing sensors, controllers, or actuator devices should be able to easily join the network of peer-oriented devices without large development effort. (See Fig. 73.4.)

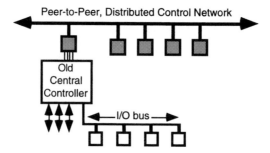

FIGURE 73.4 Peer-to-peer distributed network.

WHAT DOES LONWORKS* INCLUDE THAT MAKES IT A FIT FOR PEER-TO-PEER DISTRIBUTED CONTROL SYSTEMS?

LONWORKS technology, developed by Echelon and largely implemented in hardware and firmware on the NEURON IC manufactured by Motorola Semiconductor (second-sourced by Toshiba Semiconductor), provides a platform with which intelligent distributed control systems can be built. A NEURON chip contains memory, I/O, and processors and can run a user-defined control program while embedded in sensor or actuator devices. It performs input, output, or control operations, programmed for each specific installation. It also implements a complete network communication protocol called LONTALK, allowing for reliable control interaction among NEURON-containing nodes over various communication media. Further, it is complemented by all the diagnostic services, installation schemes, and configuration tools needed to build successful control networks.

Range of Applications

LONWORKS has been widely accepted across a broad range of industries over the past three years with 950 companies involved in development of products and 200 announced products available in the marketplace today. Applications include distributed control systems found in factory automation, process control, building automation, transportation, instrumentation, homes, power distribution, asset management, and others.

NEURON IC—Node Controller

Central to LONWORKS technology is the NEURON IC, integrating all the major features of a distributed control node onto one chip. (See Fig. 73.5.) It includes an 8-bit, 3-processor CPU section, with each processor performing the individual tasks of node application code, network messaging, and media access. This multiprocessor design improves the NEURON's performance by isolating its control tasks from network communication load better than single-processor designs.

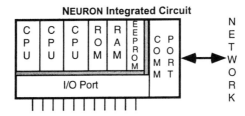

FIGURE 73.5 Neuron integrated circuit.

Implemented in CMOS technology, at 0.8-micron geometry, the NEURON consumes little operating power and shares the low price of single-chip microcontrollers

---

*LONWORKS and NEURON are registered trademarks of Echelon Corporation. LONTALK and LONMARK are trademarks of Echelon Corporation.

that include far less capability in networked architectures. It can also be made to sleep when not active, allowing use in very low power applications such as battery-operated or communication link–powered equipment.

Its variety of memory types allows for storage of protocol and application support libraries in ROM; message buffers and program workspace in RAM; and nonvolatile application programs, data, and network configuration in EEPROM. Furthermore, its EEPROM memory can be programmed over the network after installation.

I/O Structure

The NEURON's I/O structure can be easily tied to digital or analog devices. Its 11 I/O pins are configurable, driven by a choice of 29 different software models that allow easy interface to transducers or off-the-shelf A/D or D/A converters without the need to do low-level programming.

Many digital devices can be directly connected to the NEURON in single-bit, 4-bit, or 8-bit input or output modes. Four I/O pins support output current drive up to 20 milliamps at 5 volts DC, allowing direct connection to visible LEDs, optocouplers, optotriacs, or logic-level power FETs. Four other pins offer on-chip pullup resistors for easy switch or relay contact closure inputs without external components.

Analog inputs can be done in several ways. External A/D controller ICs can be connected in either a parallel or serial fashion. Additionally, the NEURON's on-chip timer structures can be used to measure variable-pulse width or frequency from a voltage-to-frequency converter. Even a dual-slope integrating A/D converter can be built with a handful of external devices, using the NEURON's Dual-Slope I/O model.

Similarly, analog outputs can be generated with an external D/A converter, also connected in either parallel or serial fashion. As an alternative, the NEURON can generate variable-pulse width or frequency outputs which can be integrated to create analog levels.

Another powerful capability of the NEURON's I/O structure is its ability to interface to an external host processor. If the NEURON's processing power is insufficient to do a node's task, a more powerful CPU can be used, tied to the NEURON in order to join the network. In this configuration, the NEURON behaves as a communication coprocessor to the host. This approach can also be used to build a communication gateway to either a non-LONWORKS control bus or to a computer data network like Ethernet.

Further, if an existing piece of equipment offers either a serial control port or some parallel status or control lines, the NEURON can intelligently interface that equipment to the LONWORKS network, allowing it to easily join the peer-to-peer architecture. (See Fig. 73.4.)

A Full OSI Seven-layer Protocol Definition

As previously mentioned, a communication protocol must do more than simply move data messages. It must allow for a wide range of options needed by contemporary control schemes.

While wired networks are most commonly used in control networks today, the installation of communication wiring is often expensive, inappropriate, or sometimes nearly impossible in retrofit situations. LONWORKS offers its users a choice of media options, including twisted-pair wire, the AC power line, radio, multichannel coax, optical fiber, infrared, or others. The LONTALK protocol makes no assumptions about the characteristics of the communication medium, allowing any type to be used by connecting the NEURON to the network with the appropriate transceiver.

Critical to any peer-to-peer network scheme is the approach taken to gain access to the communication medium. How do intelligent devices know *when* they can transmit without a central device in control? LONWORKS's media access scheme uses a patented carrier-sense method with collision avoidance (plus optional collision-detection and priority access). This allows for peer access to the network by large numbers of nodes without the limitations, cost, or complexity of alternative schemes such as time slicing or token passing.

Time-sliced access requires a master time-slot generator (a potential single point of failure) and does not work well for large networks or those that use a mix of transmission media, each with its own data rate and modulation characteristics. Token-passing eliminates the single point of failure and allows for multiple media, but introduces complexity and performance delay in recovery of lost tokens and reassignment of the token-rotation schedule upon adding or removing devices.

LONWORKS's media access approach operates over any physical medium. Nodes can be added or removed from the network with no impact upon how the communication channel is managed.

LONTALK message packets allow for variable data length messages: brief ones for discrete point status and control; longer messages (up to 228 data bytes) for analog information or configuration and diagnostic data.

The maximum bit rate is 1.25 Mbps for highest performance. Lower rates may be selected depending upon the characteristics of the communication medium selected.

Messages can be addressed to any of over 32,000 nodes in a system, even if they exist on different pieces of physical media. Logical groups of nodes can be created, allowing a single data update (from a temperature sensor, perhaps) to be seen by all the devices that need to know this information. Broadcast messages can be sent to the entire network or across more localized subnets.

LONWORKS routers handle the interconnection of different types of data channels on the network. For example, a collection of sensors communicating on a LONTALK twisted-pair wire channel can communicate via a router with a hand-held operator panel using a radio frequency channel or to a motor control running on the AC power line. The router examines each message packet for addressing information and determines whether or not the packet needs to be passed across itself from the wired channel side to the RF or power line side. For example, in Fig. 73.6, nodes 1 through 4 most often communicate with one another. But level sensor node 1 needs to tell pump motor node 11 to stop when the tank is full.

This ability to filter message traffic is important in single-medium networks as well. It allows busy portions of the network to be isolated from one another, controlling the bandwidth needed to reasonable levels, even across very large systems. In the Fig. 73.6 example, nodes 5 through 8 may run a conveyor system with node 6 sending item count information to node 4 for display along with tank level information.

Another LONWORKS mechanism used to streamline the design of peer-to-peer, intelligent distributed control systems involves use of standard shared data definitions called *standard network variable types* (*SNVTs*). SNVTs each define a unique data type, size, range, increment, and engineering unit for the data values that devices will exchange with one another. Manufacturers who produce products with SNVTs make it easy for users to interconnect products which can share matching SNVTs.

A further step to aid interoperation includes definition of objects; that is, a standard definition for entire sensor devices, controller devices, and actuator devices.

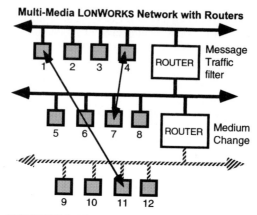

FIGURE 73.6 Multimedia LONWORKS network with router.

This definition includes standardization of the SNVTs that a device will have, plus a description of how its options or special abilities can be controlled.

The recommended use of these application objects, SNVTs, and other network characteristics are documented in the LONMARK Interoperability Guidelines, available to help assure proper device interaction.

All of the protocol features described are embedded in the firmware included with the NEURON IC. Router functions are included in router modules available from Echelon in either packaged or OEM versions.

Operating System Integral to Distributed Control

The integration of a distributed control system requires more than processor-controlled nodes and a communication protocol. It also requires management of input or output actions, calculation, messaging, and decision-making *among* intelligent nodes so that they work together. Every NEURON IC contains an event-driven executive included within its firmware. This firmware governs the order and timing of user tasks within the NEURON, allowing it to make control decisions based upon events.

A NEURON-based intelligent node can react to an I/O event ("My pressure sensor's reading just went down"), by a timer expiration event ("It's been 10 seconds since I last checked. Try again"), or by the test of one or more local program variables ("My pressure reading is now below my local setpoint") This allows a node to recognize and control changes in its own environment.

Message events extend this same control capability to nodes networkwide. This helps users to easily create multiple node applications that work as if they were running on a single processor. A message event occurs when a new piece of data arrives ("Pump 9 out at the pump house cannot maintain pressure. I'll start my own pump here"). This lets the application program running in a node work with external data just as easily as if it were local data.

Other message events help track transmission success or failure so nodes can take appropriate action, perhaps notifying an operator that a repair is needed.

USERS MOVE TOWARD PEER-TO-PEER, DISTRIBUTED CONTROL NETWORKS

The literature on control technology is filled with information on distributed approaches. LONWORKS exists as a solution to make implementation possible and cost ready.

There is a huge installed base of conventional control systems, installed over the past 30 years. Or is that 50 years? . . . or 100 years? Many of these systems are still in operation and, while not optimized, they are slow to be removed from service. Equipment manufacturers, integrators, and users understand this old technology and are often comfortable with its use.

LONWORKS is being used in conjunction with these older, centralized architectures. It can be expanded in modular fashion. Users can start by connecting LONWORKS networks to the existing infrastructure and then expand these networks as more sensors, actuators, or controls need to be added.

Users already understand the benefits of distributed systems. In the computing world, distributed client/server architectures are pervasive today, providing huge benefits in performance at lower cost that nearly everyone recognizes. This precedent will make the transition in the controls world easier.

In particular, visionary users see the benefits of scalability, reliability, modularity, diagnostic capability, and lower costs of distributed architectures. They realize that any other scheme will put them at a disadvantage, especially when they consider that the decision they make today will need to serve them for the next 20 to 30 years—the potential life span of a new control system.

In summary, the move is now underway. *Purchasers* of control systems and related integration services are asking for the capabilities of distributed, peer-connected products with greater intensity as they now understand their benefits of better control, lower cost, and greater flexibility.

A solution set is available today using LONWORKS, uniquely incorporating all the necessary elements needed to do peer-to-peer, intelligent distributed control architectures with cost-effective products.

CHAPTER 74
DISTRIBUTED, INTELLIGENT I/O FOR INDUSTRIAL CONTROL AND DATA ACQUISITION: THE SERIPLEX SENSOR/ACTUATOR BUS

Ever since the introduction of programmable controllers, control manufacturers promised that new developments in automation control system technologies would reduce the cost of control systems. Most factory managers, however, are still waiting for this promise to be kept.

With the advent of PLCs, the continuous stream of product enhancements and major new technologies have resulted in new processors, PO devices, MMIS, and so forth, providing industry with major improvements in manufacturing control, plus a wealth of new process data that managers previously could only dream about.

However, there has been a price to pay, the price of ever more expensive control systems. Within the last few years, products such as small shoe box PLCs have brought the cost of processors down, but, generally, the cost of I/O continues to remain high. Existing efforts to develop new fieldbus technologies, most of which use very sophisticated communication protocols, may continue the upward price pressure on control system costs.

So far, industry in general appears frustrated and impatient with the lack of progress toward usable and affordable solutions to reduce control system costs. However, two developments appear to hold some promise for future cost control. The recent acceptance of the PC/AT platform as a control system CPU and the introduction of low-level PO networking systems that promote the distribution of the I/O control close to the point of use.

INTRODUCTION

There are several effective sensor and actuator networking approaches, some of which have been available for quite some time, that offer simplicity and the control system cost savings many companies are looking for. These approaches, often referred to as *distributed PO systems,* provide for distributing sensors, actuators and pilot devices such as prox and photosensors, push buttons, pilot lights, contactors, solenoids, analog input and output devices in a cost-effective manner, and often over very large distances.

Distributed I/O provides significant installation cost savings (wire, conduit, labor, and start-up), costs which have continued to rise steadily over the years. With the reduction of local PO and point-to-point parallel wiring, control panel enclosures are also significantly reduced in size and cost. It is not uncommon to find that the costs of installation and start-up comprise as much as half the total cost of completed electrical control systems. Some very large companies are already using distributed PO networks on an extensive basis because (1) there is a substantial reduction in the amount of wire used; (2) the cost per device connection is a fraction of that in conventional installations; (3) time to install and debug is usually cut almost in half; (4) there is improved maintenance and fault determination; and (5) reliability and uptime are increased.

Some of these cost-effective approaches include Battebus and ASI (Actuator Sensor Interface), both developed in Europe, SDS (Micro Switch CANbus), Device-Net (Allen-Bradley CANbus), Sensorplex (Turck), VariNet-3 (Pepperl + Fuchs), Photobus (Banner Engineering), ProBlox (NAMCO), and SERIPLEX Sensor/Actuatorbus. A brief comparison of several of these systems is shown in Tables 74.1 and 74.2.

Batibus™ has generally been used in building automation applications and has not seen much use in industrial control applications. It has, however, been used to monitor industrial controls. It was left out of the comparison because sufficient technical data could not be obtained.

The communication media used by these systems varies from twisted-pair to coaxial cable to simple parallel pairs. Almost all systems have a recommended configuration for their media in the form of cables that combine communication and power in one cable for further economies of installation.

The incentive for companies to use distributed control systems is strong. One survey* estimating the use of sensors in the future forecasts over 67 million prox switches alone will be used yearly by 1995. It projects a growth to over 89 million annual usage of prox switches by the year 2000. Add to that sensors for measurement of liquids, rigid materials, chemicals, thermal sensors, acoustic and other binary sensors, plus all the pilot devices and actuators that can connect to a network, and one can see that there is a huge potential savings to be had from distributing as much I/O as practical.

In selecting a distributed I/O system, there are a number of factors to consider: (1) are there real devices in the form of I/O modules or sensors and actuators readily available in the marketplace; (2) speed or throughput required; (3) length of the network; (4) CPU platform that can be used or does the system stand alone and have its own logic capability; (5) is there an installed base of qualified users; (6) does the system interface to other communication buses, LANs, or WANS; and, finally, (7) do the numbers—how does this system compare on an *installed cost basis* to using conventional controls or competing networks?

Installed cost is really the cost benchmark to be considered. The base cost of the hardware devices will be about the same, give or take a bit, as conventional hardware. The cost benefit received from using distributed I/O is really in the material and labor saved for installation and start-up. Depending on the job, start-up labor can be significant. Identifying and sorting out miswired connections when you have

---

* European Prognos Sensor Market Survey 1992.

TABLE 74.1 System Comparison.

Feature	ASI™	SERIPLEX™	SDS™ CANbus™	DeviceNet™ CANbus™
Basic design	One chip for each sensor or actuator	One chip for each sensor or actuator	Several chips, depending on application plus CPU support	Several chips depending on application plus CPU support
Operating voltage	24 VDc	12 VDc—1st generation; 12 or 24 VDc—2d generation	11–24 VDc	24 VDc
Logic capability on chip	Cyclic on-off control	32 boolean logic functions	CPU-based, programmable	CPU-based, programmable
Error checking	Yes	Yes—diagnostic output using logic function	Yes—15-bit CRC (cyclic redundancy check)	Yes—15 bit CRC (cyclic redundancy check)
Method of operation	Master-slave	Master-slave and Peer-to-peer	Peer-to-peer	Peer-to-peer and multimaster
Number of devices per network	31 slaves (nodes) 124 binary devices	>7000 binary/480 analog or combination	128 binary/analog	64 nodes 2,048 devices
System throughput time	5 msec. for 31 slaves	.7 msec. for 32 sensors plus 32 actuators —<5 Ms for 510 I/O	26 msec. min. for 32 devices	Unknown
Deterministic	No	Yes	No—arbitration system	No—arbitration system
Addressing method	External DIP switches	Programmable EE element on chip	Software dependent on network	Software dependent on network
Network structure	Trunkline–dropline only	Open—any combination of tree, loop, multidrop, etc.	Trunkline—dropline only	Trunkline—dropline only
Network length without repeaters	Max. 100 meters including drops	5000 ft.; longer with fiber optics	80, 400, 800, or 1600 ft. with 50% in trunkline	300, 600, or 1600 ft.—single drop max. 10 ft.—cumulative drop 128, 256, 512 ft.
Medium of transfer	Unshielded two-wire power and signal	Four-wire cable—two shielded for data, two unshielded for power	Four-wire cable	Four-wire cable
Signal capability	Digital (binary) only	Binary, analog, ASCII	Binary, analog, ASCII	Binary, analog, ASCII
Known CPU interfaces	Seimens PLC	PC/AT, GE Fanuc 90-30 PLC, VME, EXM, PC/104 STDbus, A-B/GE/Modicon Seimens via VME	GE Fanuc 90-30	Allen-Bradley PLC
Additional cost per switch	<$30, –DM	<$7 USA added cost to devices in moderate quantities	≈$13 USA in moderate quantities	≈$13 USA in moderate quantities
Market introduction	Fall 1993	1990	Early 1994	Mid 1994
Working product	Fall 1994	1990	Early 1995	Early 1995

TABLE 74.2 System Comparison.

Feature	TURCK SENSORPLEX™	P + F VariNet-3,™ uses Seriplex Technology	NAMCO ProxBlox,™ uses Seriplex Technology	BANNER PHOTOBUS,™ uses Seriplex Technology
Basic design	Master/Slave stations	One chip for each sensor or actuator	Intelligent junction boxes for sensors	One chip for each sensor or actuator
Operating voltage	24 VDc	12 VDc—1st generation 12 or 24 VDc—2d generation	12 VDc—1st generation 12 or 24 VDc—2d generation	12 VDc—1st generation 12 or 24 VDc—2nd generation
Logic capability on chip	Unknown	32 boolean logic functions	32 boolean logic functions	32 Boolean logic functions
Error checking	FSK encoding/filtering, no error detection	Yes—diagnostic output using logic function	Yes—diagnostic output using logic function	Yes—diagnostic output, using logic function
Method of operation	Master-slave	Master-slave and peer-to-peer	Master-slave and peer-to-peer	Master-slave and peer-to-peer
Number of devices per network	32 stations, 192 total	>7000 binary	>7000 binary	>7000 binary
System throughput time	<5 msec. for 192 I/O	Adjustable—.7 msec. for 32 sensors plus 32 actuators <5 Ms for 510 devices	Adjustable—.7 msec. for 32 sensors plus 32 actuators <5 Ms for 510 devices	Adjustable—.7 msec. for 32 sensors plus 510 devices
Deterministic	Yes	Yes	Yes	Yes
Addressing method	Physical on	Programmable via EE element on chip	Programmable via EE element on chip	Programmable via EE element on chip
Network structure	Master with daisy chain	Open—any combination of tree, loop, multidrop	Open—any combination of tree, loop, multi-drop, etc.	Open—any combination of tree, loop, multidrop, etc.
Network length	2500 meters with amplifiers	5000 ft., longer with fiber optics	5000 ft., longer with fiber optics	5000 ft., longer with fiber optics
Medium of transfer	Coax and IP67 connectors	Four-wire cable—two shielded for data, two unshielded for power	Four-wire cable—two shielded for data, two unshielded for power	Four-wire cable—two shielded for data, two unshielded for power
Signal capability	Binary	Binary, analog, ASCII	Binary, analog, ASCII	Binary, analog, ASCII
Known CPU interfaces	Seimens PLC, A-B PLC, others unknown	PC/AT, GE Fanuc 90-30 PLC, VME, EXM, PC/104 STDbus, A-B/GE/Modicon Seimens via VME	PC/AT, GE Fanuc 90-30 PLC, VME, EXM, PC/104 STDbus, A-B/GE/Modicon Seimens via VME	PC/AT, VME, GE Fanuc 90-30 PLC, VME, EXM, PC/104 STDbus, A-B/GE/Modicon Seimens via VME
Additional cost per switch	Not applicable	Not known	Not known	Not known
Market introduction	USA 1992	1992	1994	1992
Working product	USA 1992	1992	1994	1992

a couple of hundred pairs in your control system can be an extremely time-consuming chore, to say nothing of the damage that can be done to the control system components from crossed wires and shorted circuits.

Distributed I/O systems are suitable for many types of applications. Some existing applications include machine tool control, material handling equipment controls, press monitoring and control, assembly line monitoring, tank farm monitoring, automated inventory control, automated mining equipment remote controls, and the NASA Mars robot, to name a few.

While several systems are compared in Tables 74.1 and 74.2, the Seriplex system is discussed in detail in this chapter. It was specifically developed for industrial control applications. It can accommodate several thousand I/O devices on a single network, extending over seven-thousand feet in length. Both analog and digital (binary) devices can be controlled and monitored on the network. Small Seriplex chips can be embedded in sensor and actuator devices to provide for direct connection to the bus.

The Seriplex Sensor/Actuator bus technology was developed by Automated Process Control (APC), a Jackson, Mississippi, company. Robert Riley, vice president of engineering and product development, received the first patent award on the system in 1987.

Over the past few years, the technology has demonstrated its ability to provide major control system cost savings. According to users, these savings reduced installation and start-up costs by as much as 50 percent or more, compared to the usual conventional methods.

SYSTEM DESCRIPTION

The Seriplex Sensor/Actuatorbus is a deterministic, serial multiplexed, intelligent, distributed I/O system, providing both master/slave and peer-to-peer I/O control and logic.

The network cable of the Seriplex system connects directly to I/O devices, containing an embedded Seriplex ASIC, or it interconnects to I/O devices via modules containing the Seriplex ASIC. Communication is facilitated by embedding an ASIC (application-specific integrated circuit) directly into sensor and actuator devices. The ASIC is also embedded in modules that connect to sensors and actuators.

The ASIC provides the communication capability, addressablity, and the intelligence to execute logic in virtually any sensor or actuator.

The system supports binary and analog device communication and has the added capability of serial digital communication, such as RS232/485.

System and installation cost savings are achieved by placing I/O modules at the point of I/O device use or as close as possible. Communication takes place over a four-wire, low-voltage cable. The system eliminates the thousands of parallel wires usually run through conduits from local control cabinets to I/O devices, such as thermocouples, push buttons, prox switches, photosensors, valves, solenoids, contactors, thermal sensors, and so forth.

In addition to the material cost savings, there is a significant labor savings, as well. The labor cost to install a small, four-wire cable the size of a little finger that does not require conduit is a mere fraction of the cost of installing conduits, pulling a multitude of control wires through them, and then trying to identify the individual circuits.

The Seriplex system transmits both digital and analog I/O signals in real time for both control and data acquisition applications. It combines distributed and local I/O capability on the same bus.

The system is designed to compliment rather than compete with the higher-level, more sophisticated protocol, fieldbus-type communication systems that are best suited to transmitting large information data packets. It resides primarily at the physical device level providing deterministic, process-critical, real-time PO updates needed by most control systems and leaves the higher-level communication not requiring fast throughput rates to others.

A basic master-slave system includes a host CPU, an interface card that provides the system communication signal source, a power supply, and some I/O interconnect modules or I/O devices with the Seriplex ASIC embedded in them (see Fig. 74.1). Connecting these components is a four-wire cable with two conductors for communication and two for the network power.

A peer-to-peer system (Fig. 74.2) does not use a CPU at all. It simply requires a clock module for the system communication signal source, a power supply, I/O devices, and the cable.

The power of the system lies in its capacity to control over 7000 binary I/O points, or 480 analog channels (240 input plus 240 output), or some combination of discrete and analog I/O over one small four-wire bus. This eliminates the hundreds of wire pairs often seen in a control system running parallel, point to point from a control cabinet through conduit.

The Seriplex Sensor/Actuatorbus may be configured ring, star, multidrop, loop back, or in any topology combination desired. See Fig. 74.3.

Interconnect modules and I/O devices with embedded ASICs are placed as close to their point of use as possible. The Seriplex system differs substantially from other multiplexed systems because addressing, communication capability, and logic functions are programmed directly into (EE elements in ASIC) the module or I/O device. This means that communication capability is inherent in the module or device, rather than relying on a microprocessor for this function. There is no microprocessor in the Seriplex ASIC.

In the case of input devices, such as sensors, with an embedded Seriplex ASIC, the only connections needed are directly to the Seriplex bus (see prox switch in Figs. 74.1, 74.2, and 74.4). There is no additional wiring necessary. Power for the sensor can usually be furnished by the two bus conductors supplying DC power. Actuators with the embedded ASIC also connect directly to the Seriplex bus. However, good engi-

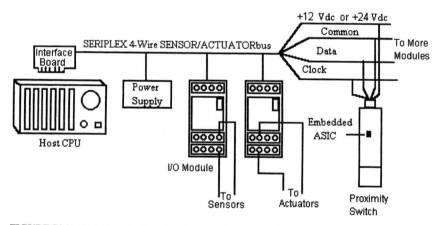

FIGURE 74.1 Typical master/slave CPU operated system configuration.

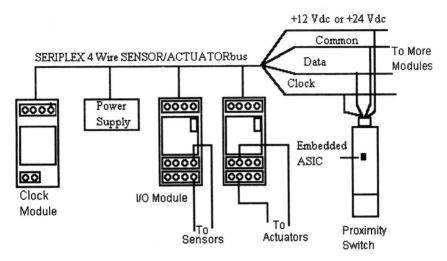

FIGURE 74.2 Typical peer-to-peer system configuration.

neering practice dictates that actuator devices be powered from a separate power source to avoid defeating the devices' optical isolation.

Two of the four conductors in the four-wire bus deliver +12 VDC power or +24 VDC power (user's choice) to the bus. Bus voltage for systems using devices or modules with first generation ASICs are restricted to +12 VDC power only. The other two conductors provide communication to and from the modules and devices. One of the communication lines is for a data signal and the other for a clock signal that controls network timing.

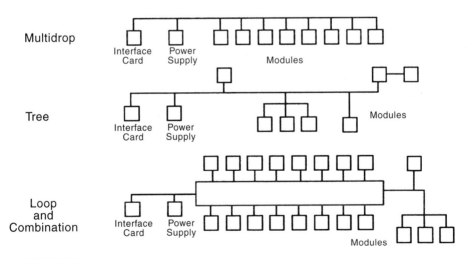

FIGURE 74.3 Sample bus configurations.

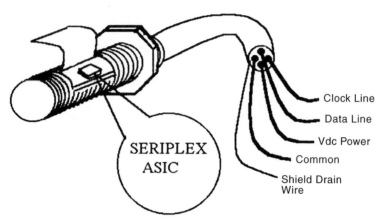

FIGURE 74.4 Proximity switch with ASIC built in.

HOW THE SYSTEM WORKS

The heart of the system is the Seriplex ASIC. It is an application-specific integrated circuit that is embedded in I/O devices or in I/O modules that support the communication between I/O devices and the host CPU over the Seriplex communication bus.

In the master/slave mode (sometimes referred to as *mode 2*) the ASIC provides for the communication of organized field event, occurring at the I/O devices, to a host CPU, to be acted upon by the application program resident in the host.

In the peer-to-peer mode (sometimes referred to as *mode 1*) there is no host CPU. The events occurring at each device are communicated directly between modules or devices containing an embedded ASIC. These events are acted upon at the device level based on the logic functions programmed into the ASICs embedded in each device.

The system is simple and basic in nature. It utilizes large signal swings (12 Vpp), low-pass filtering, and large hysteresis to maintain data integrity. There is also a provision for redundancy at the inputs to provide even further data protection.

The cable used for the Seriplex communication bus is a special design, but available from several manufacturers. The basic design consists of two AWG 22 communication wires for clock and data signals. There are two additional AWG 16 wires for the power and common conductors.

The wires have an overall shield with a drain wire and are covered with an overall jacket. The cable is designed for low capacitance to achieve maximum communication distance and speed. The capacitance in the cable has a direct effect on the distance of communication and the speed of communication. Using one of the standard cables rated at 16 picoFarads per foot, maximum communication distance at 100-kHz clock rate is approximately 500 feet. Maximum communication distance with the same cable at 16 kHz is about 4800 feet. A cable with 20 picoFarads per foot limits communication to about 350 feet at 100 kHz, and 3900 feet at 16 kHz.

Variations of the basic cable exist, such as special application ratings or additional pairs of conductors to provide separate power to actuators. Data sheets on each of the several cables are available from APC. The cables are available from multiple manufacturers.

ASIC GENERAL DESCRIPTION

There are two Seriplex ASICs. The two are similar in function and the second-generation SPXSP256-2A is downward-compatible with the first-generation SPX-SP256 ASIC. They can both reside on the same system provided the bus voltage is +12 VDC. The SPX-SP256-2A also operates at +24 VDC.

The ASICs are programmed differently. The first-generation ASIC requires a programming port on the I/O module or I/O device into which it is embedded. The SPX-SP256-2A is programmed by merely connecting the programmer to the four bus conductors. This is an economical connection that does not require additional real estate for the programming port.

ASICs are generally programmed with a handheld programmer for address, logic function, and mode of operation. Addresses are entered as integer numbers at the programmer address prompt. Each ASIC has two input address channels (A and B) and three output channels (A, B, and C). Logic is programmed by setting each of the input and output points in the ASIC to a normally inverted or noninverted state by means of a control word.

The C output is available as a logical function of A and B outputs. There is generally a choice of using either the C or B output on an I/O interconnect module. This selection is usually made via a jumper located on the module. The choice of B or C output on an I/O device with an embedded ASIC is usually set at the design level within the device.

Additional logic functions can be programmed into the second-generation ASIC. One of these enhancements allows for sampling data once, twice, or three times on consecutive bus scans. The data must be identical on each of the selected number of scans before it is released to the bus.

Both ASICs can operate in master/slave or peer-to-peer mode. This is determined at the time of programming. This selection determines how the ASIC will handle data. The following Figs. 74.5 and 74.6 show a very simplified block diagram of the ASIC's functions in the two modes.

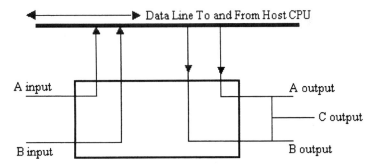

FIGURE 74.5 Master/slave ASIC block diagram.

In the master/slave mode, data is passed from the inputs to the data line and delivered to the host CPU application program so it is acted upon based on the logic in that program. The result of that logic is transmitted over the data line to turn output devices on or off as appropriate.

In master/slave mode, the status of the inputs at various addresses determines the action of outputs at various addresses based on the logic in the application program residing in the host CPU. Inputs and outputs are complimentarily addressed receiving the same numerical address.

There is no direct relationship between inputs and outputs on the same ASIC. This is controlled strictly by the application program logic.

The ASIC functions quite differently in the peer-to-peer mode. In mode 1, inputs at a particular address have a direct effect on outputs with the same numerical address. In Fig. 74.6 the solid lines going from the input to the output represent this direct logical effect between inputs and outputs at the same address.

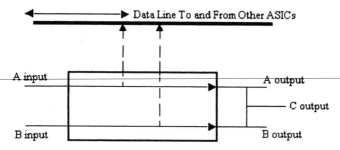

FIGURE 74.6 Peer-to-peer (ASIC to ASIC logic) block diagram.

As previously noted, the status of the A and B inputs, and the A, B, and C outputs can be programmed to be either inverted or noninverted in their normal (inactive) state. Each is programmed individually. This provides 32 boolean logic combinations.

The inputs and the outputs with the same address do not have to be in the same location and, in fact, seldom are. The output can be in a device several hundred feet away from the input with the same address. As long as they have the same address, any action at an input will have a direct effect on all outputs in the system with the same address as the input.

The broken lines with arrows shown in Fig. 74.6 represent the status of the inputs and outputs at each address reflected to the data line. This is useful when running in peer-to-peer mode using a CPU interface card. Seriplex CPU interface cards can be configured at setup for either mode 1 or mode 2 operation. The host CPU can follow the status of each of the I/O points even though the logic is being executed at the I/O level.

COMMUNICATION SYSTEM—MASTER/SLAVE MODE

- A system reset or synchronizing signal is generated on the clock line by one of the various types of interface cards available. It is eight clock cycles long and synchronizes all devices in the system (see Fig. 74.7). On the first negative going pulse edge of the zero clock pulse, the system becomes active. All ASICs in the system are reset and waiting for the first positive clock transition to open the data window.

- The reset pulse is followed by 512 clock pulses. Each pulse corresponds to either an input or output address on the bus. While there are 256 numerical addresses on the bus, only 255 are available to the user. Address 0 is reserved for use by the sys-

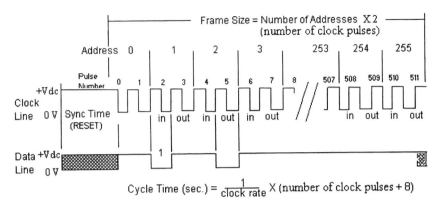

FIGURE 74.7 Computer-controlled (master/slave) mode operations.

tem, leaving addresses 1 through 255 for the user. During the zero address cycle, certain bus housekeeping activities take place, such as checking to see that the data line is operational. If not operational, it will activate the watchdog timer in each ASIC on the system and all outputs will revert to their inactive states.

Complimentary addressing provides an input and output at each numerical address. The pulses are counted by the ASIC embedded in each module or device connected to the bus.

- When the number of pulses received by a module or device equals the address it was programmed for, that module or device is given access to the data line. The status of input is communicated to the host CPU via the data line and the condition of the output is set according to the application program in the host CPU.
- When all the address clock pulses have been sent, a total of 512, the synchronizing signal is generated again and the system repeats itself.

Clock frequencies from 16 to 100 kHz are selectable on host interface cards and clock modules. A clock rate up to 200 kHz is usable on second-generation ASICS, part number SPX-SP256-2A. Future interface cards will have clock rate selections ranging from 16 to 200 kHz.

If all 255 available addresses are scanned on a master/slave system, and a clock rate of 100 kHz is selected, the throughput time of the system to scan all inputs and set all outputs would be 5.2 msec. If faster scan rates are required, a selection is made on the interface card or clock module to scan less than the full 255 addresses. This selection is made at 16 address intervals (16, 32, 48, and so on).

Throughput Time for Master/Slave System

For all 255 addresses with a 100-kHz clock rate: Cycle time (seconds) = $1/100$ kHz $\times (512 + 8) = 5.20$ ms.

For only 16 addresses: Cycle time (seconds) = $1/100$ kHz $\times (32 + 8) = 0.40$ ms.

For all 255 addresses with a 200-kHz clock rate: Cycle time (seconds) = 1/200 kHz × (512 + 8) = 2.60 ms.

For only 16 addresses: Cycle time (seconds) = 1/200 kHz × (32 + 8) = 0.20 ms.

COMMUNICATION SYSTEM—PEER-TO-PEER MODE

A system reset or synchronizing signal is generated on the clock line by an interface card or clock module. It is eight clock cycles long and synchronizes all devices in the system (see Fig. 74.8).

The reset pulse is followed by 256 clock pulses. Each pulse corresponds to an address on the bus. While there are 256 addresses on the bus, only 255 are available to the user. Address 0 is reserved for housekeeping use by the system, leaving addresses 1 through 255 for the user. Complimentary addressing provides an input and output at each address. The pulses are counted by each module or device connected to the bus.

- When the number of pulses received by a module or device equals the address it was programmed for, that module or device is given access to the data line. The status of inputs is communicated to outputs with the same but complimentary address via the data line. The condition of the output is set according to the boolean logic programmed into the module or device. In peer-to-peer mode, inputs have a direct effect on outputs with the same numerical address.

- When all the address clock pulses have been sent (a total of 256), the synchronizing signal is generated again and the system repeats itself.

Throughput Time for Peer-to-Peer System

For all 255 addresses with a 100-kHz clock rate: Cycle time (seconds) = 1/100 kHz × (256 + 8) = 2.6 ms

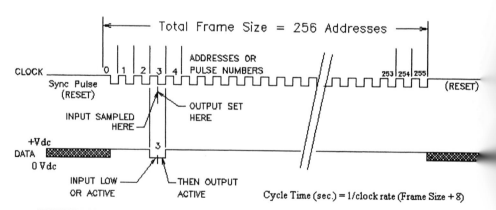

FIGURE 74.8 Peer-to-peer (mode to mode logic) mode.

For only 16 addresses: Cycle time (seconds) = 1/100 kHz × (16 + 8) = 0.20 ms
For all 255 addresses with a 200-kHz clock rate: Cycle time (seconds) = 1/200 kHz × (512 + 8) = 1.30 ms
For only 16 addresses: Cycle time (seconds) = 1/200 kHz × (32 + 8) = 0.20 ms

In both master/slave and peer-to-peer modes of operation, the system updates itself at a rate equal to the number of clock pulses on the bus plus eight divided by the clock frequency.

THE CPU INTERFACES

The host CPU (PC XT/AT, VME, STD, Multibus I and II, PC/104, EXM, or any custom system) is linked to the bus by a host interface card (Fig. 74.9). Each interface card can support 3848 digital input and 3848 digital output addresses or 240 analog input plus 240 output analog channels or some combination of digital and analog devices. Multiple cards can reside in the system for expanded I/O requirements.

Interfacing to a custom CPU is not difficult because of the simple design APC uses. It is readily adaptable to almost any microprocessor, even board-level types. Regardless which host is used, APC maintains a design concept common to all interfaces that makes it simple—the use of dual-port RAM.

Figure 74.10 shows a block diagram for current design interface cards. The host CPU merely looks at bits in dual-port RAM to see the status of I/O or tweaks them to turn I/O on or off. Each bit represents one digital I/O point.

The interface card is a coprocessor card and has an onboard processor that manages the Seriplex network, keeping track of the addresses in use, logic functions, status, and the protocol housekeeping functions. This means there is no protocol for the engineer to contend with! One only need be concerned with the status of bits in memory. A *protocol-less system!* A C programming library is provided for developing the link between the host processor bus and the interface card for those developing custom programs.

There are a number of third-party software suppliers that provide application program software packages with built-in drivers for the Seriplex interface boards. Some third-party software developers are FloPro, Omega Controlware, Red Mountain Technology, 86Ladders, Wonderware, Genesis, Gello, and SoftPLC. A current list of third-party software suppliers is available from APC.

Figure 74.11 shows additional detail on how the Seriplex interface cards interface to the host CPU on one side and to the Seriplex communication bus on the other. There are four main areas on each Seriplex interface card: the host CPU interface circuitry; the dual-port RAM; the onboard CPU; and the interface circuitry to the Seriplex communication bus.

The dual-port RAM on the interface card appears to the CPU as a 2K block of memory. The engineer assigns the location of that memory to the host CPU at initial installation of the card and it is accessed just like any other memory. The memory is partitioned for various uses to provide extensive administrative control of the system.

Dual-port RAM is divided into three segments. The data segment contains the NON-MUX (non-multiplexed) INPUTS data and OUTPUTS data areas, the SAFE STATE buffer area, and MUX (multiplexed) INPUT and OUTPUT BUFFER areas. See Fig. 74.12.

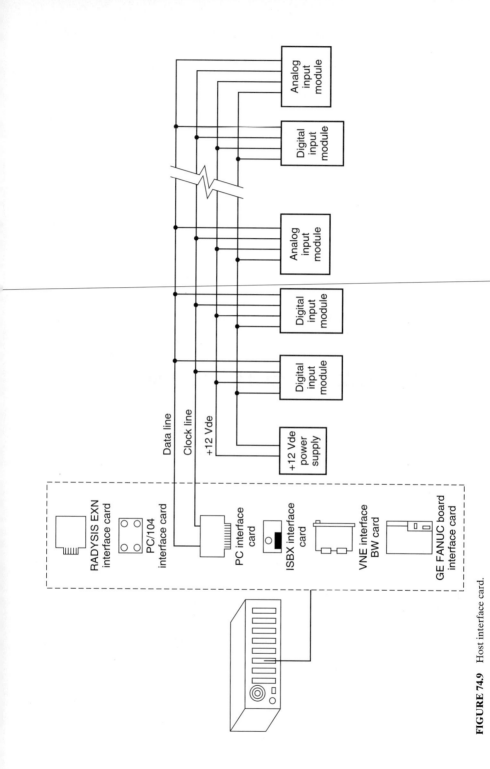

FIGURE 74.9 Host interface card.

THE SERIPLEX SENSOR/ACTUATOR BUS   74.15

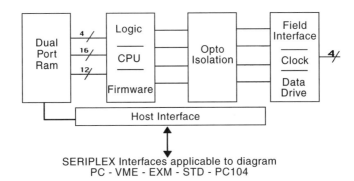

FIGURE 74.10   Seriplex-applicable interfaces.

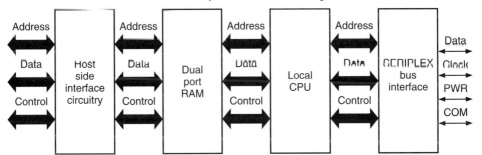

FIGURE 74.11   Seriplex-computer interface.

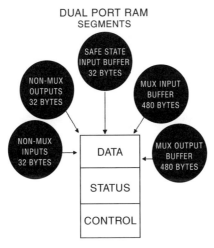

FIGURE 74.12   Dual-port RAM segments.

A unique feature of the dual-port RAM system is that both binary and analog signals can be multiplexed on the bus. This feature is usually used to provide up to 240 analog inputs plus 240 analog outputs on the same bus, or some combination of analog and binary I/O. Ibis data is stored in the dual-port RAM on the host interface card.

It is possible, however, to extend the multiplexing capability of the system to accommodate 7696 binary I/O points.

Consistent with the complimentary addressing scheme, this would provide 3848 inputs and 3848 outputs. Of course, analog values could easily be mixed with binary I/O. They would each consume 8, 12, or 16 binary bits, depending on the resolution desired.

Each of these dual-port RAM areas are defined in terms of a bit table. Each bit in an area represents a Seriplex address and is stored within a byte quantity of memory. Multibit values such as an analog or counter value can also reside in this area. They are registered as a continuous group of address bits.

The status segment of the dual-port RAM has two basic areas (Fig. 74.13). The CONFIGURATION ID status area contains product and firmware revision status information.

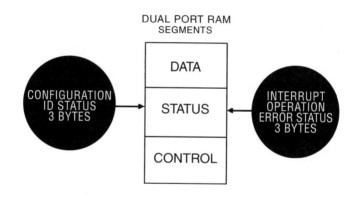

FIGURE 74.13 Dual-port RAM interrupt.

The INTERRUPT, OPERATIONAL, and ERROR STATUS area contains information regarding the status of interrupts, operation modes and parameters, and error indications.

The status information contained in the STATUS BYTES is simply read out to obtain the desired information. The information available would be the error status of the Seriplex bus (current or bus faults), product ID codes, and firmware revision codes. (Figure 74.16 shows examples of C library routines to obtain this information.)

The control segment of the dual-port RAM also has several areas for specific functions. See Fig. 74.14.

The MUX INPUTS ENABLE is used to indicate which of the input blocks of Seriplex bus addresses are to be used in the multiplex mode of operation. This allows the engineer to define up to 15 multiplex blocks on a single bus. (See later Fig. 74.17 also.)

The MUX OUTPUTS ENABLE is used to indicate which of the OUTPUT blocks of Seriplex addresses are to be used in the multiplex mode of operation.

The NON_MUX PASS_THRU is used to indicate which of the non-multiplexed Seriplex address INPUTS will be passed thru directly to the Seriplex address OUT-

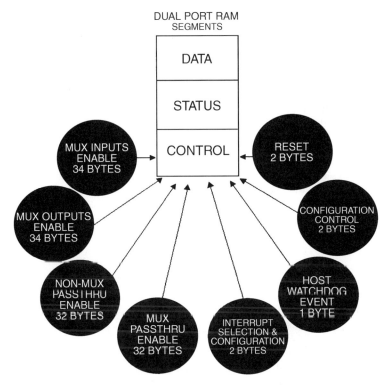

FIGURE 74.14 Dual-port RAM segments reset.

PUTS. This is a feature in master/slave mode that allows the action at an input device to cause an action at an output device having the same address without the data going through the logic in the application program in the host CPU. This provides a faster throughput (at mode 1 speed) for update of output devices.

The MUX PASS_THRU is used to indicate which of the multiplexed Seriplex address INPUTS will be "passed thru" directly to the Seriplex address OUTPUTS.

This is a feature that allows the action at an input device to cause an action at an output device with the same address without the logic going through the application program in the host CPU. This provides a faster throughput (at mode 1 speed) for update of output devices.

The INTERRUPT SELECTION AND CONFIGURATION area is used to select which level of interrupt request is to be used for interrupting the HOST CPU and to identify what event will cause the interruption.

The HOST WATCHDOG EVENT is used to indicate that there is not a host CPU to Seriplex interface card communication fault. A byte is written to this area of dual-port RAM from the host CPU each scan to indicate that the host is communicating.

The CONFIGURATION CONTROL area is used for setting up system parameters such as bus clock rate, multiplex channel size, multiplex priority channel, and so forth.

The RESET area is for using the hard reset feature to reset the Seriplex interface card.

The operating manual, furnished with each of the interface cards, provides complete instructions on set-up and use, including sample C code where applicable.

Figure 74.15 further illustrates how multiplexed addressing is set up and appears on the bus. As was previously noted, address 0 is used by the Seriplex interface card for housekeeping duties. It follows the sync signal used to reset all the ASICs in the system each scan. The following bits, 1, 2, 3, and 4, are used to determine the multiplex channel address, by means of a binary value, that each channel in a block of multiplexed addresses will have.

Figure 74.15 shows the selection of the multiplexed address block as being address 16 through 31.

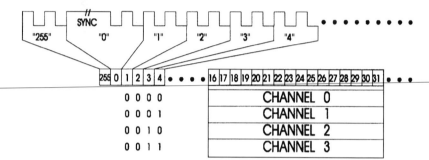

FIGURE 74.15 Multiplexed address block.

The next address block available would be 32 through 47, and so on at each 16-bit address boundary. See Fig. 74.16. The channel address is selected via the binary value at *output address bits 1, 2, 3, and 4*. The Seriplex system allows the multiplexed number of channels to be set at 2, 4, 8, or 16 for each interface card. The number of channels shown in Fig. 74.17 is 4, channels 0 through 3. Figure 74.16 shows 16 channels, channel 0 through 15. This design feature allows for up to 15 blocks of 16 channels on each Seriplex network.

Standard analog modules currently offered do take full advantage of the available multiplex addressing capability on the Seriplex bus. A total of 480 analog addresses (240 input plus 240 output) are available using the multiplex addressing, in addition to 15 binary input and 11 binary output addresses (1 through 4 are used to select the channel address).

If all available multiplexed addresses are used, a total of 7696 binary I/O addresses or some mix of binary and analog addresses is available. Specially designed discrete, binary I/O modules are required to achieve this I/O density. The maximum number of usable discrete I/O addresses per network is 510 without multiplexing.

It is also possible to select any one of the multiplexed channels as a priority channel. This selection would force the system to look at that channel every other scan so the updated data would be available to the host CPU every other scan.

If priority channel is not selected, then each channel will be scanned on a successive scan basis. If four channels were set up for multiplexing, it would take four scans to read the data in all four of the channels. Binary values that were not multiplexed would still be updated every scan.

It should be noted that reference has always been made to the number of addresses available and not the number of modules that can reside on the SENSOR/ACTUATORbus. Multiple modules can be used at all addresses if the I/O

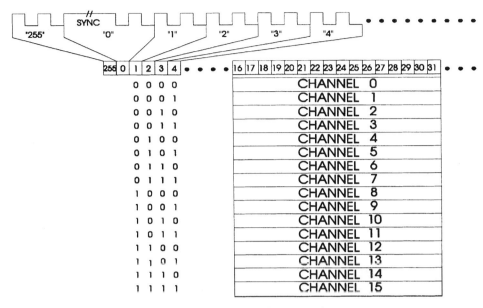

FIGURE 74.16 Fifteen blocks of 16 channels on a Seriplex network.

action is required in multiple locations every time a device at that address is given a command.

If more I/O is required than one host CPU interface card can provide, the solution is to add additional host interface cards. Any number of additional cards may be added. The only restriction is how much memory the host CPU can support.

SERIPLEX Analog to Digital Conversion

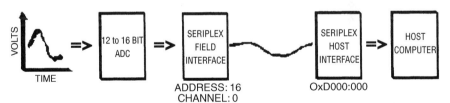

FIGURE 74.17 Seriplex analog-to-digital conversion.

I/O DEVICES

There are few limitations as to what kind of modules can be designed for use with the system. While there are standard catalog devices available, some OEMs design devices specific to their own applications. The ASIC is very simple to adapt to custom I/O modules and directly into sensors and actuators, as well.

Several companies, in addition to APC, manufacture devices or have products in the planning stage incorporating the Seriplex ASIC, such as Banner Engineering (photosensors), Pepper1 + Fuchs (proximity switches), Eaton Corporation, NAMCO and The Square D Company, Turck, and so on. Applications include embedding the ASIC in valve manifolds and bodies, contactors, grippers, process instrumentation, and other actuator and pilot devices.

The second-generation ASIC was released for production in June 1996.

For applications requiring local I/O, high-density analog and digital modules are available. These provide a large number of I/O. Up to 48 digital and 32 analog I/O are available in a small package with only one Seriplex bus connection required for each group of I/O. They connect to the same bus as the more highly distributed block-style devices and can be mixed with the block-style devices.

Standard block-style I/O modules are available from multiple sources in a variety of configurations. They are small, compact, and facilitate installation close to the I/O devices to which they are connected.

Both AC and DC modules are available with as few as two input channels or two output channels per module, or as high as eight input channels per module or eight outputs per module. There are also combination modules with two inputs and two outputs per module. The standard block-style digital modules have break-away connectors on both the network side and the field side, and the circuitry is completely potted. They are generally panel- or din-rail-mounted, but some are available with stand-off mounting. All I/O modules are rated from −20 to +65°C.

Eight-bit and twelve-bit analog input and output modules are available in a variety of designs. Conditioning modules, which plug into them, handle various types of signals.

Analog-to-digital (ADC) and digital-to-analog (DAC) conversions made at the module enhance the integrity of the transmitted signal.

Analog input modules convert the analog signal to a signal value for transmission over the Seriplex bus. Resolutions available are 8, 12, or 16 bits.

If used in the master/slave mode, the digital value is received at the host CPU by the Seriplex interface card, where the value is written to its defined address location in the dual-port RAM. The value is then usable by the application program in the host CPU by accessing the appropriate bits in the dual-port memory.

Analog output modules reverse the process placing a digital value stored in the dual-port RAM of the Seriplex interface card on the Seriplex bus. This value is accessed by the Seriplex module with the associated address, converted within the module to a corresponding analog value and output to the field device controlled by this signal.

OPEN ARCHITECTURE

The Seriplex protocol is offered as an open architecture system through a licensing program. It is very easy and inexpensive to implement the Seriplex ASIC, and does not require an expensive, sophisticated development system.

APC offers a development system that includes discrete and analog modules and a Seriplex interface card. A complimentary quantity of Seriplex ASICs and an ASIC application manual are available upon signing a nondisclosure agreement. ASICs can be preprogrammed to avoid the expense of purchasing a programmer.

This Strategic Partner Program is open to those who desire to develop products that can communicate over the Seriplex Sensor/Actuatorbus. Technical information and support is available to those wanting to develop interfaces to custom CPUs. The Seriplex ASIC is available to companies wishing to embed the ASIC into their sensor or actuator devices from APC or directly from the IC foundry.

As a result of open architecture philosophy, the Seriplex system can be combined with others at the IC level, either as silicon or as a multichip configuration. The interfaces to various field buses are simplified through the use of dual-port RAM.

The Seriplex Sensor/Actuatorbus system is one of the least expensive, easiest, and simplest to implement systems for providing both distributed and local analog and digital I/O available today. It, by far, offers greater installation cost reduction for distributed I/O than any other system currently available.

CHAPTER 75
THIN/THICK FILM CERAMIC SENSORS

INTRODUCTION

In today's automotive applications, there are many sensors such as flow sensors, accelerometers, gas sensors, pressure sensors, and thermal sensors that can be based on thin and thick film ceramic sensor technology. We review various thin and thick film–processing techniques which can be important to the automotive ceramic sensor technology. Process issues which are unique to the thin and thick film ceramic sensors such as tape casting, screen printing, ohmic and nonohmic electrode-ceramic-junction contacts are discussed. Several examples are given to demonstrate that the thin and thick film–processing technology not only can produce economic ceramic sensors but also can improve and enhance the sensing performances.

With the successful applications of silicon microprocessors in automotive vehicles, the demand for sensors is ever increasing. For every conceivable function of the vehicle, there are more and more sensors required for better control, for adding new features to the use of the car. To mass-produce the sensors economically, there is no doubt that thin and thick film technology offers an advantage in terms of production rate and cost.[1,2]

Ceramics, in contrast to other materials such as metal, glass, and polymers, are unique in that it covers a wide ranges of linear and nonlinear properties.[3-9] It can be insulator, capacitor, resistor, semiconductor, superconductor, solid electrolyte, or mixed conductor. It can be catalytic, dielectric, electrooptic, paraelectric, ferroelectric, piezoelectric, ferromagnetic, or ferrimagnetic. Other than bulk (lattice) properties, it possesses many interesting nonlinear properties at the grain boundaries such as varistor, PTCR effect, which can be found only in ceramics.

Therefore, it is no surprise to find people are applying thin and thick film technology on ceramics to make flow sensors, accelerometers, gas sensors, pressure sensors, and thermal sensors. In many cases, the thick and thin film of ceramics are teamed up with silicon technology in making hybrid sensors.

However, today's thin and thick film ceramic sensor technology is still in its primitive stage. Future sensor devices are expected to be more sophisticated, more multifunctional, and more silicon-technology-compatible, and it is our obligation as well as a challenge to explore this possibility.

In this chapter, we review important thin and thick film methods and discuss issues related to the processing of ceramic thin/thick film. We also use several examples to show that the thin/thick film technology not only produces ceramic sensors economically but also improves the performance of ceramic sensors.

THIN FILM PROCESS

Three major methods are suitable for producing thin film ceramic sensors: physical deposition method, chemical deposition method, and chemical-physical deposition method.[10,11]

In physical deposition method, thin film *is* produced from vapor, or liquid. Depending on the energy sources, it can be thermal evaporation with various heat sources such as resistance, electron-beam, rf, or laser. Other than vapor evaporation, it can be liquid-phase epitaxy, glow discharge sputtering, various magnetron sputtering (cylindrical, S-gun, planar), and ion beam deposition.

The chemical deposition method includes solution depositions from chemical reaction, chemical vapor deposition (CVD), and electrochemical vapor deposition.

The physical-chemical method is mainly of plasma film deposition.

Among various thin film deposition methods, those which are potentially low cost (such as sputtering methods) draw people's attention. Because of the refractive nature of ceramics, various physical thermal evaporation methods that have been quite successful with metals and alloys are not particularly suitable for ceramic thin film processing.

The film growth rate is always a concern in developing thin film ceramic sensors. A recently introduced thin film technology promised high film-growth rate, ease of processing, and matching to silicon processing technology—the metalorganic deposition (MOD) method.[12] To start with, a suitable organic precursor, dissolved in solution, is spin-dispensed onto a substrate (could be ceramic or silicon wafer) to remove the excess fluid and solvent and create a uniformly organic film only a few microns thick. The film is then pyrolyxed in a suitable atmosphere to convert the metalorganic precursors to their final ceramic form. Filtering of the metalorganics in solvents to 0.2 μm is typically done before the solution is used. Metalorganics are often highly sensitive to energetic beams of focused particles or light applied prior to pyrolysis. Utilizing this property, various shapes of the film can be developed by the patterning-and-etching techniques borrowed from regular silicon processing steps. It is a liquid-based, nonvacuum, spin-on, thin film method, which should find many applications in future thin film sensor development work.

THICK FILM PROCESS

Traditional thick film deposition methods include: tape casting, roll compacting, extrusion, calendering, screen printing, spraying (electrostatic spray, powder flame spraying, electric harmonic spraying, plasma flame spraying, wire flame spraying), and coating (spin coating, dip coating, brush coating, roller coating, fluidized-bed coating).[13-18]

Among them, there is no doubt that the tape-casting method is the most interesting one. It is a flexible method in which a wide range of film thickness as well as a precise control of the film thickness are possible. In conjunction with the screen-printing technique (which is introduced in the following), the sophisticated multilayer structure of ceramic devices such as multilayer capacitors and multilayer ceramic substrates (modules) has been made. The number of the ceramic layers in such devices can be as high as 50, not including electrode layers, and the thickness of the corresponding ceramic layer can be as thin as 10 microns.

The multilayer approach based on the tape-casting and screen-printing technology can build sophisticated, multifunctional ceramic sensors and its future importance should never be overlooked. In today's automobiles, we already have such multilayer sensor devices in use, the flatplate zirconia-based solid-oxide-electrolyte oxygen sensors for engine exhaust control applications are an example; the piezoelectric displacement modulator for vehicle height applications is another example.

Before the tapes are cast, ceramic slurry has to be made first. Ceramic powder mixed with solvent, dispersant, binder, and plasticizer are ball-milled to the specification. Typical solvent for the slurry can be methylethylketone and toluene (they have to be compatible with powder and be able to dissolve dispersant, binder, and plasticizer), plasticizer can be dioctyl phthalate (to maintain a stable suspension), and binder can be polyvinyl butyl resin (to delay sedimentation, control viscosity, and improve wetting). Their ratio is determined by the ceramic particle size, surface area, and the slurry viscosity specified.

The tapes are produced by using a doctor blade which lets the slurry pass through it and deposit on a tape carrier. The tape thickness can be controlled through the doctor blade gap, the speed of the casting, the slurry viscosity, and the amount of slurry behind the doctor blade. Usually, the relationship between the tape thickness (after drying) and the doctor blade setting is linear and not a strong function of the casting speed. Before the tape cast, the milled slurry is de-aired to avoid air bulbs trapped in the tape. After the drying, the tape is either stored away or is ready for the next step of processing. The control of tape thickness is important to the following process steps, such as screen printing and sintering. To avoid the unisotropic shrinkage effect of the green tape (cast direction versus transverse direction), tapes, after screen printing and hole or via punching, are rotated during the stacking and before the thermal-lamination step. The firing of the thermal-laminated parts includes the cycles for binder-burn-out and sintering. After firing, final inspection as well as electric lead soldering and packaging will be the last steps to complete the thick film ceramic sensors. The furnace used for this part of processing can be a continuous-belt type (gives more throughput) or a batch type (better process of control).

As we have introduced, the screen-printing technology is complementary to the tape cast technology in producing multilayer ceramic sensors. Even standing alone, it is an important thick film deposition technology because of its simplicity in depositing thick film with controlled shapes and thickness.[15–18]

The paste used in the screen printing is made of ceramic powders, fritted glass materials (to enhance the binding of the film to the substrate), binder such as polyvinyl butyl, and solvent such as butyl carbitol. Surfactant can be added to avoid the powder aggregation. The mixture is mixed to form a paste. The paste should be thixotropic and pseudoplastic such that during the printing the paste will flow, but after the printing the pattern spread is negligible. The metal loading factor and the viscosity of the paste usually are controlled by the ratio between the various amount of materials involved.

Many factors can influence the thickness of the printed results. The opening of a screen depends on the number of wires and their diameter in a unit area. Increasing mesh number of the screen decreases the screen opening area but not in a linear relationship. The amount of ink passing through the screen increases as opening area increases but not in a linear relationship. There is an upper and a lower limit for the screen opening areas, beyond which there is no increase in the throughput or no ink will flow. Squeegee force and motion direction can influence the print results through the variation of screen deformation. Thicker emulsion increases the throughput of ink but there is an upper limit beyond which no increase will be observed. The snap-off rate of the screen is controlled by the screen tension and the space between the screen and the substrate. The ratio of ink left in the screen to that on the substrate is determined by the snap-off rate. At low squeegee velocity, the throughput of the ink is fairly constant. After reaching a maximum value, further increasing velocity will decrease the throughput rapidly. The condition of the ceramic substrate surface, such as surface finish, degree of warp, camber, and parallelism can affect the printing results by varying the screen-substrate spacing, which causes a spread of the results. Most of the spread of printing results can be traced to variation of substrate surface and substrate thickness, small component (compare with the mesh size of the screen), screen wear out, and changes in squeegee pressure and speed. The chemical interactions among metal, ceramic powders, binders, substrate, and residual carbons from the organic solvent should be considered if repeated results are needed.[15–18]

Drying and firing are needed to break down and burn out the binders and solvent. Similarly, the furnace used can be a continuous-belt type (more throughput) or a batch type (better control of process).

For protecting the environment, there is a trend to use water as the solvent for tape-cast slurry and screen-print ink paste. For this reason, a tape-forming method without the use of a solvent, such as roll compacting, has advantages over others.[19]

PROCESS FOR ELECTRODE CONTACTS OF THIN/THICK CERAMIC SENSORS

Although it has usually been neglected in the literature, the electrode contacts for thin and thick film ceramic sensors are important subjects for study. Without the development of reliable methods for ohmic contacts and Schottky-barrier junctions, silicon devices cannot have enjoyed their current success. The same argument applies to the future thin/thick film ceramic sensors. This is because many nonlinear electrical phenomena of ceramic materials can be utilized for sensor applications (see introduction section) and these ceramics typically have semiconducting properties.

It has been known for a long time that a rectifying contact is formed by placing a metal in contact with a semiconductor. The rectification is a result of a potential barrier at the electrode-semiconductor junction. Many semiconducting ceramic sensors will require ohmic contacts or controlled nonohmic contacts.[20, 21]

Good ohmic contacts are characterized by linearity and low voltage drop; they are achieved when the device impedance is dominated by the bulk ceramic. The physical reason for the linearity is that the total impedance of the contact is dominated by the series resistance of the ceramic film. Perfect ohmic contact without junction barrier is seldom found. Even for well-developed silicon technology, one of the lowest junction barriers observed on p-type silicon is formed by metallic platinum silicide (PtSi) which has a barrier height of 0.25 eV. At room temperature, this barrier yields a contact resistance of 1.4×10^{-4} ohm-cm^2 (in zero bias).[20, 21]

In the literature, two methods have been proposed to create ohmic contacts: (1) by matching the electrode material to that of the semiconductor to lower the junction barrier for large enough thermal excited current, or (2) by heavily doping the semiconductor junction side to allow quantum mechanical tunneling to take place.

It has been found that (1) a covalent semiconductor has a junction barrier not strongly dependent on the electrode materials, and (2) an ionic semiconductor has a barrier height which is a function of both the work function of the metal and of the particular semiconductor. Therefore, depending on the types of the semiconductors, one of the previously suggested methods will be chosen to make ohmic contacts.[20, 21]

For ferroelectric-type ceramic semiconductors, the screen effect of the spontaneous polarization can be utilized to create ohmic contacts. This phenomenon is similar to the grain boundary PTCR effect in which most of the trapped charges at the grain boundary are compensated by the spontaneous polarization which yields a low barrier at the interface. Ohmic contact based on this effect, potentially, can yield a junction resistance of 10^{-4} ohm-cm^{-2} (for a barrier of 0.3 eV) for ceramic such as n-type barium titanate. It has been found that in order for this type of charge-compensation effect to work, the interface distance between the electrode and the ceramic semiconductors has to be small. A large separation can induce the *intrinsic field* effect of the spontaneous polarization, which bends the band as high as the band gap, as a result of charge compensation by the density states near the top of the valence band. Film deposition followed by a thermal treatment is found to be effective in softening such junctions; this is believed to reduce the interfacial gap and promote the forming of ohmic contact. In principle, this type of ohmic contact is independent of the electrode materials used.[22,23]

To create Schottky barrier junctions, the high dielectric constants of ferroelectric semiconductors makes worse the barrier-height-lowering effect in creating a large nonidea diode phenomenon. The nonlinear dielectric feature plus the possibility of surface piezoelectric effect demand a solid binding between the ceramic and the electrode material.[22,23]

Non-well-controlled electrode process can create mixed-phase-type electrode junctions, in which only part of the electrode is ohmic. Uneven distribution of electrode current can create a reliability problem for the sensor devices.[24-26]

There is another type of electrode contact which is also unique to ceramic sensors: the electrode solid-electrolyte contact.[27-32] An example of such contacts is the platinum electrode of zirconia oxygen sensors. On the platinum side, the electron is the main charge carrier; on the zirconia side, oxide ion is the main charge carrier (or oxygen vacancy), and at the junction, the charges are exchanged and oxygen is pumped in or out of the gas-metal-electrolyte junction. The junction current-voltage characteristic is similar to that of the forward-bias diode behavior in the semiconductor if the electrode polarization is controlled by the charge-transfer process. When the mass transport of gaseous oxygen is limited (blocking electrode), limiting current behavior can be obtained. In such case, unwanted electron current can be directly injected into the electrolyte depending on the degree of oxygen blocking and the applied voltage. The sensor uses two different types of sensing principles: Nernst principle or polarographic principle. In either case, different electrode-processing specifications are involved. Various thin/thick film methods discussed so far can be applicable here for electrode deposition. Mixed-phase electrode (partially ionic, partial electronic conducting junction) should be avoided through good control of the electrode deposition process.[27-32]

All the future ceramic thin/thick film sensors will have electrode contacts to bring out the sensing signals. Whether the sensing principle involves the junction or not, unless the electrode contact processes are understood and well controlled, the success of these devices is in doubt.

WHY THIN/THICK FILMS FOR CERAMIC SENSORS

As mentioned in the introduction, thin/thick film technology potentially represents a cost-effective production method. It has no problem with mass production. The technology has no difficulty in matching the packaging technology (although packaging usually represents more than 70 percent of the cost of an automotive thin/thick ceramic sensor, it is another technology ignored in the literature).

Other than the issues of processing cost and ease of mass production and packaging, we like to stress that there are other compelling reasons to choose thin/thick technology to develop automotive ceramic sensors. That is, the thin/thick film approach can produce multifunctional devices and can possibly enhance the sensing effect of the ceramic film device. We believe this will be one of the main driving forces for the future development of film types of ceramic sensors and allows ceramic sensors to compete head to head with silicon or silicon-based hybrid sensor technologies.

Here are two examples to show the functional enhanced effect of film technology on the ceramic sensors.

Titanium oxide ceramic is a wide-band semiconductor. At room temperature, its high resistance can qualify it as an insulator. At elevated temperature, the oxygen ions in the titania can leave the bulk and form oxygen molecular, leaving behind the oxygen vacancies and/or titanium interstitials for charge compensation. Through the use of mass action law, the resistivity of titania r can be given as

$$r = A\, P_{O_2}^{1/n} \exp(E/kT) \tag{75.1}$$

in which A is a constant, E is the activation energy, and k is the Boltzmann constant, T is the temperature, and P_{O_2} is the partial pressure of oxygen of the gas. The value of n is approximately 4 for the most part of oxygen concentration ranges. Because the P_{O_2} of the engine exhaust gas decreases by many orders of magnitude at the stoichiometric air-to-fuel ratio (A/F), the resistance of the titania ceramic exhibits several orders of magnitude change as the A/F changes from lean to rich values. Therefore, it can be an oxygen sensor for engine A/F stoichiometric control applications.[33-37]

Comparison between the bulk and film titania sensors show that film-type titania sensors have far superior performance in terms of sensor response time and the characteristics of sensor resistance change. This has been attributed to the fact that the film offers more surface areas for gas-solid interaction and offers more room for the Pt catalyst impregnation. In a way, without the approach of thin or thick film, it is not possible to realize the application of titania as an oxygen sensor for a three-way catalyst converter application.

Another wide-band semiconducting ceramic is the ferroelectric barium titanate.[38,39] By proper doping, it behaves as a n-type semiconductor. Like other polycrystalline semiconductors, the grain boundaries can trap charges to create large barriers and become highly resistive. However, barium titanate poses spontaneous polarization that can compensate the charges trapped at the grain boundary and bring down the interfacial potential barrier. As a result of this, at temperatures below the Curie point, it has low resistance. At temperatures higher than the Curie point, the disappearance of spontaneous polarization brings back the barriers at grain boundaries and has the resistance jump several orders of magnitude. Details of this positive temperature coefficient resistor effect (PTCR effect) can be found in the literature as well as in this chapter.

Usually, the device operates as a switch device. A surge of high current can heat up the device beyond its Curie point with the resistance jumped several orders of magnitude, a good effect for sensing and actuating. This nonlinear thermister behavior has been put into many automobile applications. The only problem which limits its wide application is its low-temperature current-carrying capability which is less than 0.5 A-cm^{-2} for a single-layer device.

However, if a multilayer approach is used, one can push up this limitation. Assuming in a normal switch-on condition, the current is I_{on} and the voltage is V_{on} (at temperatures less than the Curie point) and the corresponding current and voltage at off condition (at temperatures higher than the Curie point) are I_{off} and V_{off}, we can have

$$\rho_{\text{off}}/\rho_{\text{on}} = R_{\text{off}}/R_{\text{on}} = (I_{\text{on}}V_{\text{off}})^2/W_{\text{on}}W_{\text{off}} \quad (75.2)$$

where

$$W_{\text{on}} = I_{\text{on}}^2 \times R_{\text{on}} = I_{\text{on}} \times V_{\text{on}} \quad (75.3)$$

and

$$W_{\text{off}} = I_{\text{off}}^2 \times R_{\text{off}} = I_{\text{off}} \times V_{\text{off}} \quad (75.4)$$

Here, the wattage dissipation W_{on} and W_{off} are constants and are dictated by the ambient environment of the device. So, giving a PTCR material and processing, we can decide right away that the maximum current-carrying capacity at the on stage is determined by the resistance ratio between the off and on stages, which should give a number higher than 0.5 A-cm^{-2}. To achieve this potential current-carrying capability, the R_{on} has to satisfy the following equation:

$$R_{\text{on}} = \rho_{\text{on}} \times (L/A/N) - (W_{\text{on}}/I_{\text{on}}^2) \quad (75.5)$$

where N is the number of the layers, A is the nominal electrode area, L is the thickness of the PTCR ceramic between two adjacent electrodes, which we want to be as thin as possible but strong enough for electric breakdown; therefore:

$$L = V_{\text{off}}/E_D \quad (75.6)$$

where E_D is the dielectric strength of the ceramic. Therefore, for a given operation condition where the parameters of ($I_{\text{on}}, W_{\text{on}}, W_{\text{off}}, E_D, V_{\text{off}}$) have been predetermined, the best design of the device is to choose a PTCR formulation and process which give the best value of $\rho_{\text{off}}/\rho_{\text{on}}$ and let N be determined from the preceding equations. Using available PTCR materials and processing, a device of seven layers can easily increase the current capability from 0.5 to 2 A-cm^{-2}, a factor of 4 increase. The device will be able to hold a switch-off voltage of 45 V, which is suitable for automotive applications.

REFERENCES

1. F. Heintz and E. Zabler, "Motor Vehicle Sensors Based on Film-Technology: An Interesting Alternative to Semiconductor Sensors," *SAE* 870477.
2. R. Dell'Acque, "A Bright Future for Thick Film Sensor," *SAE* 890480.
3. W. D. Kingery, H. K. Bowen, and D. R. Uhlmann, *Introduction to Ceramics,* John Wiley & Sons, New York, 1976.
4. L. L. Hench and J. K. West, *Principles of Electronic Ceramics,* John Wiley & Sons, New York, 1990.
5. M. E. Lines and A. M. Glass, *Principles and Applications of Ferroelectrics and Related Materials,* Clarendon Press, New York, 1979.
6. J. M. Herbert, *Ferroelectric Transducers and Sensors,* Gordon and Breach Science Publishers, New York, 1982.
7. *Ceramic Materials for Electronics,* ed. R. C. Buchanan, Marcel Dekker Inc., New York, 1986.

8. *Chemical Sensing with Solid State Devices,* M. J. Madou and S. R. Morrison Academic Press, New York, 1989.
9. *Fine Ceramics,* ed., S. Saito, Elsevier, New York, 1985.
10. *Deposition Technologies for Films and Coatings,* ed., R. F. Bunshah, Noyes Publications, Park Ridge, New Jersey, 1982.
11. *Thin Film Processes,* eds., J. L. Vossen and W. Kern, Academic Press, Orlando, Florida, 1978.
12. J. V. Mantese, A. L. Micheli, A. H. Hamdi, and R. W. Vest, "Metalorganic Deposition (MOD)," *MRS Bulletin,* 48–53, October 1989.
13. *Advanced Ceramic Processing and Technology,* ed., J. G. P. Binner, Noyes Publication, Park Ridge, New Jersey, 1990.
14. W. S. Young, "Multilayer Ceramic Technology," in *Ceramic Materials for Electronics,* ed., R. C. Buchanan, Marcel Dekker Inc., New York, 1986.
15. A. Kosloff, *Ceramic Screen Printing,* Signs of the Times Publishing, Cincinnati, Ohio, 1984.
16. E. A. Apps, *Ink Technology for Printers and Students,* Chemical Publishing, New York, 1964.
17. P. J. Holmes and R. G. Loasby, *Handbook of Thick Film Technology,* Electrochemcial Publications, Ayr, Scotland, 1976.
18. D. W. Hamer and J. V. Biggers, *Thick Film Hybrid Microcircuit Technology,* Wiley-Interscience, New York, 1972.
19. *Surface Phenomena and Additives in Water-Based Coatina and Printing Technology,* ed., M. K. Sharma, Plenum Press, New York, 1991.
20. E. H. Rhoderick and R. H. Williams, *Metal-Semiconductor Contacts,* Clarendon Press, Oxford, United Kingdom 1988.
21. C. A. Mead, "Physics of Interfaces," in *Ohmic Contacts to Semiconductors,* ed., B. Schwartz, *Electrochem. Soc.* 1969, pp. 3–16.
22. S. H. Wemple, "Electrical Contact to N and P-Type Ferroelecttic Oxides," in *Ohmic Contacts to Semiconductors,* ed., B. Schwartz, *Electrochem. Soc.* 1969, pp. 128–137.
23. Da Yu Wang, "Electric and Dielectric Properties of Barium Titanate Schottky Barrier Diodes," *J. Am. Ceram. Soc.* 1994, **77** (4) pp. 897–910.
24. I. Ohdomari and K. N. Tu, "Parallel Silicide Contacts," *J. Appl. Phys.,* 1980, **51** (7) pp. 3735–3739.
25. J. L. Freeouf, T. N. Jackson, S. E. Laux, and J. M. Woodall, "Size Dependence of Effective Barrier Heights of Mixed-Phase Contacts," *J. Vac. Sci. Technol.,* 1982, **21** (2) pp. 570–73.
26. J. L. Freeouf, T. N. Jackson, S. E. Laux, and J. M. Woodall, "Effective Barrier Heights of Mixed-Phased Contacts," *J. Appl. Phys.,* 1982, **40** (7) 634–636.
27. D. S. Eddy, "Physical Principle of the Zirconia Exhaust Gas Sensor," *IEEE Trans. Vehicle Tech.,* 1974, **23** (4), pp. 125–128.
28. L. Heyne, in *Measurement of Oxygen: Preceedings of an Interdiscipilinary Symposium Held at Odense University Denmark, 26–27 September 1974,* p. 65, Elsevier Scientific Pub, New York, 1976.
29. Da Yu Wang, "Electrode Reaction at the Surface of Oxide Ionic Conductors," *Solid State Ionics 40/41,* 849–856, 1990.
30. J. E. Anderson and Y. B. Graves, "Steady-State Characteristics of Oxygen Concentration Cell Sensors Subjected to Nonequilibrium Gas Mixtures," *J. Electrochem. Soc.,* 1981, **128**, pp. 294–300.
31. W. J. Fleming, "Physical Principles Governing Nonideal Behavior of the Zirconia Oxygen Sensor," *J. Electrochem. Soc.,* 1977, **124**, pp. 21–28.
32. T. H. Etsell and S. N. Flengas, "Overpotential Behavior of Stabilized Zirconia Solid Electrolyte Fuel Cells," *J. Electrochem. Soc.,* 1971, **118**, pp. 1890–1900.

33. T. Y. Tien, H. L. Stadler, E. F. Gibbons, and P. J. Zacmanidis, "TiO2 as an Air-to-Fuel Ratio Sensor for Automotive Exhaust," *J. Amer. Ceram. Society,* 1975, **54** (3), pp. 280–285.
34. A. Takami, T. Matsuura, S. Miyata, K. Furusaki, and Y. Watanabe, "Effect of Precious Metal Catalyst on TiO2 Thick Film HEGO Sensor with Multi-Layer Alumina Substrate," *SAE* 870290.
35. D. Howarth and A. L. Micheii, "A Simple Titania Thick Film Exhaust Gas Oxygen Sensor," *SAE* 840140.
36. J. L. Pfeifer, T. A. Libsch, and H. P. Wertheimer, "Heated Thick-Film Titania Exhaust Gas Oxygen Sensors," *SAE* 840142.
37. W. J. Kaiser and E. M. Logothetis, "Exhaust Gas Oxygen Sensors Based on TiO2 Film," *SAE* 830167.
38. W. Heywang, "Barium Titanate as a Barrier Layer Semiconductors," *Solid State Electron.,* 1961, **3**, pp. 51–58.
39. B. M. Kulwicki, "PTC Materials Technology, 1955–80" in *Advances in Ceramics, Vol. 1, Grain Boundary Phenomena in Electronic Ceramics,* eds., L. M. Levinson and D. C. Hill, American Ceramic Society, Columbus, Ohio, 1981, pp. 138–154.

CHAPTER 76
LOW-NOISE CABLE TESTING AND QUALIFICATION FOR SENSOR APPLICATIONS

INTRODUCTION

The objective of this chapter is to raise the awareness of issues such as when a low-noise coaxial cable may be required for an application, what the electrical and mechanical trade-offs of specific designs may be, and how to test and characterize cable performance.

The testing issues are particularly important. The measurement of high-frequency (AC) attributes such as impedance, time delay, and edge and pulse amplitude degradation typically require a high-speed time domain reflectometer (TDR). Inductance, resistance, and capacitance measurements usually involve an LCR meter. Attenuation and shielding effectiveness can be measured using a scalar or vector network analyzer. The test techniques used to measure these attributes are generally well known, and as such, are covered briefly. Not as well understood to system designers are the measurement techniques and values associated with low-level, low-frequency (DC) measurements. Low-current, low-voltage, and high-impedance measurements, particularly for sensor applications, require specialized equipment and techniques. Issues such as leakage, insulation resistance, dielectric withstand, and breakdown are not well understood as they relate to cable performance.

In order to better understand these issues and how they relate to one another, we introduce two important tools: a low-noise transmission line model and the concept of the measurement universe. Using these techniques, we examine low-noise performance as it relates to the noise effects in cables (such as triboelectric, capacitive, and piezoelectric effects) with mechanical stimuli such as drop, flexure, and the Bow-string test. Cable comparisons and low-noise system considerations are included.

OVERVIEW

The self generation of currents within coaxial cables has been recognized for some time.[1] This phenomenon, generally referred to as *triboelectric noise*, has been defined as that component of electrical noise generated in a coaxial cable by mechanical stress. It has also recognized that the noise component may contain other elements, such as capacitive and piezoelectric effects.[2] There has been an evolution of the design of low-noise cables for low-level signal transmission applications such as sensors used for data acquisition. Indeed, traditional MIL-C-17 style RG cables such as RG-58 and RG-174, which utilize polyethylene dielectric, generally exhibit about 30 picoamps of current when excited using the Bowstring test (described later). In order to minimize this effect, manufacturers added a conductive layer over the dielectric and under the shield to create a path for the charge to dissipate. This was usually a graphite dip coating or wrap, which required special handling and cleaning techniques prior to attaching the connector. These cables also suffer high temperature limitations due to their insulation materials (60–85°C maximum).[3,4]

As improvements were made in dielectrics, materials such as polytetrafluoroethylene (PTFE) began to be used for RG-style coaxial cables. While giving these cables better temperature performance (−55 to + 200°C),[4] cables such as RG-178 and RG-316 offered much poorer basic low-noise performance. Indeed, Bowstring testing indicates the performance of these cables was several orders of magnitude worse than their polyethylene counterparts. Again, manufacturers addressed this problem by adding a semiconductive low-noise layer over the dielectric and under the shield. This layer is generally a carbon-filled PTFE tape, which does a good job of minimizing triboelectric effects. This layer is easily peeled out prior to terminating with connectors.

As performance requirements advance, recent innovations in materials technology now allow for new levels of low-noise coaxial cable performance. The reliability of the measurement system is no better than the interconnect cable, since triboelectric noise often limits the system's dynamic range,[5] lower-noise and lower-capacitance cables offer engineers new design freedom. The early cable design example cited previously employed full-density dielectric materials used in conjunction with some form of semiconductive layer. SILENT-LINES low signal-level cables offer lower-noise performance coupled with reduced capacitance. This is accomplished by the use of proprietary low-density expanded PTFE tapes developed by W. L. Gore & Associates that prevent the separation of the layers.[6] This dielectric material, used in conjunction with a conformable, semiconductive expanded PTFE layer, yields patent-pending designs, typically offering one picoamp of current with the Bowstring test. An additional benefit of this new design technique is reduced capacitance per unit length, generally one-third less (19 pf/ft versus 30–32 pf/ft) than that of conventional designs.

Among other topics, this chapter discusses performance considerations for cables used to connect sensors with instrumentation. It includes an electrical model for the sensor/interconnect system and describes tests for determining or verifying the cable performance parameters included in the model.

Basically, a *sensor* can be described as an element that converts nonelectrical information about a system parameter (such as temperature, light level, mechanical stress, vibration, etc.) to another form (such as an electrical signal) so that it may be observed or recorded. For its intended purpose, a sensor will typically have a well-characterized transfer function.

When choosing cables to connect a sensor with other equipment, be aware that the cables used may distort electrical signals or generate unwanted signals in response to environmental conditions. Electrical performance requirements for the interconnect system depend on the signal levels produced by the sensor.

A model such as the Thevenin equivalent circuit shown in Fig. 76.1 may be useful to represent a sensor for this discussion. The cable is represented as a voltage source in series with a source impedance. The Thevenin equivalent source voltage (V_S) is the output voltage that would appear with an open circuit between the sensor output terminals. The short-circuit current that would flow between these terminals would be V_S/R_S, from which the Thevenin equivalent source impedance (R_S) can be calculated. Of course, in some situations, the sensor cannot actually be subjected to an open- or

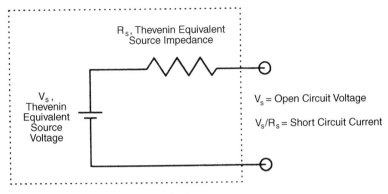

FIGURE 76.1 Thevenin equivalent circuit.

short-circuit condition without damaging it. From any two operating points, one could alternately determine model parameters valid over that limited range.

To compare a wide variety of signal levels, it is useful to plot the Thevenin equivalent model parameters on the measurement universe grid shown in Fig. 76.2. Note that the diagonal lines on this log log plot are lines of constant impedance. Figure

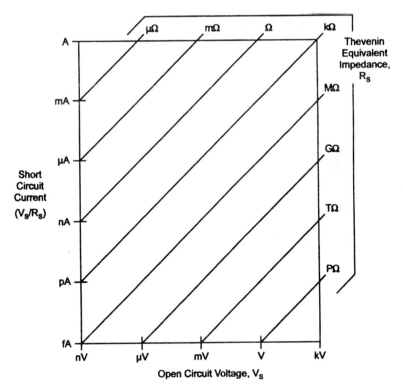

FIGURE 76.2 Thevenin equivalent independence.

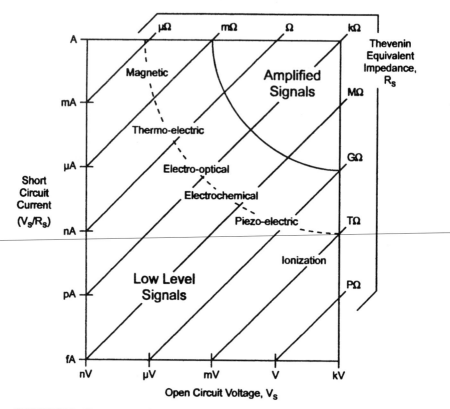

FIGURE 76.3 Energy conversion sensors.

76.3 shows relative signal levels commonly produced by sensors used to convert various forms of energy into electrical signals.

Fig. 76.4 illustrates how to locate a specific sensor on the measurement universe grid by using a simple thought experiment or by taking a few measurements. For example, a sensor such as an electrochemical cell with a high output impedance (R_S) may be susceptible to loading and would require a cable that offers high insulation resistance and low capacitance and that generates very little current when moved or vibrated (i.e., minimal triboelectric effects). Superimposing the information in Fig. 76.5 on Fig. 76.4 indicates that a low-noise coaxial cable is likely the best choice for use with an electrochemical cell to avoid the signal degradation that could be caused by a coaxial cable (e.g., RG-58).

Figure 76.6 represents the coaxial interconnect (cable, connectors) model used throughout this chapter. Insulation resistance is represented by R_{IR}. Dielectric absorption is represented by a lumped RC circuit between the shield and the center conductor. Cable capacitance is shown as C, and any charge trapped in the cable dielectric during manufacture or use is modeled as VST in series with C. As the cable is deformed, cable capacitance changes and a displacement current ($I = VST \times dC/dt$) will flow. Current generated in response to mechanical disturbance is represented as the current source I_{Tribo}.

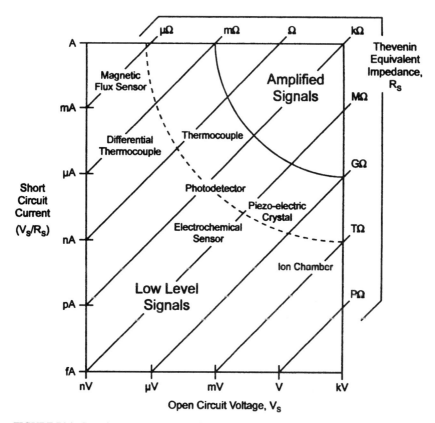

FIGURE 76.4 Locating a sensor on the universal measurement grid.

R_{IR} = Insulation resistance

DA = Dielectric absorption model

C = Cable capacitance

VST = Equivalent voltage to stored charge in the dielectric

I_{Tribo} = Current due to triboelectric generation during movement

R_{Series} = Series resistance

The discussion that follows begins with high-frequency (AC) measurements and testing, continues into the low-frequency (DC) domain, considers the mechanical effects of drop, flexure, and Bowstring excitation, and concludes with some various cable comparisons using these measurement techniques.

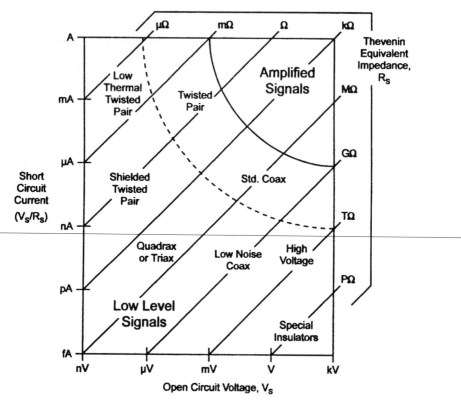

FIGURE 76.5 The use of two-voice coaxial cable in the electromechanical cell.

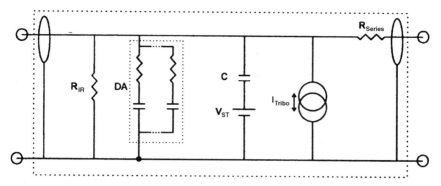

FIGURE 76.6 Coaxial interconnect model.

HIGH-FREQUENCY (AC) CABLE MEASUREMENTS

Measuring High-Frequency Cable Characteristics

Many applications where coaxial cables are used require control of the high-frequency parameters of the cable. These parameters are properties controlled by the cable's cross-sectional dimensions and the materials used. The measurement of these properties may be made in either the time domain or the frequency domain. Although the results from time domain and frequency domain measurements can be mathematically correlated through Fourier approximations, the measurement technique used is usually the one that most closely simulates the intended use of the cable.

Time Domain Measurements

Time domain measurements are quite often the technique of choice for digital applications and applications where a particular position along the cable is of interest. A *time domain reflectometer* (TDR) is the measurement equipment used to make time domain measurements such as characteristic impedance, electrical length, and time delay. The TDR sends a pulse down the length of the cable while monitoring reflected signals coming from variations inside the cable. As the pulse reaches the open end of the cable, the pulse is reflected back through the cable toward the TDR. The display on the TDR provides a visual representation of the pulse traveling through the length of the cable (see Fig. 76.7).

As the pulse enters the coaxial cable, the 50-ohm output impedance of the pulse generator forms a voltage divider with the characteristic impedance of the cable. If the impedance of the cable is 50 ohms, the amplitude will be one-half of the amplitude of the pulse. As the pulse moves through the cable, any changes in the cable's impedance will cause the voltage to increase or decrease from the original value. These changes are called reflections and show up on the oscilloscope trace at the time where the change in impedance occurred. When the signal reaches the end of the cable, the open circuit causes the pulse to increase to the value of the original amplitude at the end of the cable. This pulse edge then moves back through the cable toward the oscilloscope, completing the waveform shown in Fig. 76.8.

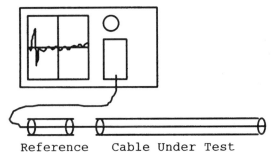

FIGURE 76.7 Visual representation of the pulse through the length of the cable.

Characteristic Impedance

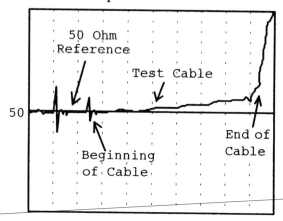

FIGURE 76.8 Sample TDR trace from Fig. 76.7.

The setup for a TDR measurement consists of a precision 50-ohm reference cable connected to the TDR sampling head through a high-quality coaxial cable. The reference cable may be a metal tube with air as the dielectric material to provide a consistent 50-ohm impedance. The cable to be tested is attached to the reference with the opposite end not connected. With this setup, the pulse from the TDR will travel through an interconnect cable and precision reference into the cable being tested. The pulse will be reflected back toward the TDR through the interconnect system in the reverse order.

The display on the screen of the TDR, shown in Fig. 76.8, shows the progress of the pulse through the cable on the horizontal axis. The position on the trace along the horizontal axis will be related to a position along the physical length of the cables attached to the TDR. The vertical scale of the TDR is usually the proportion of the pulse voltage reflected, called the *reflection coefficient*. These pulse reflections are caused by changes in the characteristic impedance in the interconnect system, which includes the cable and connectors.

Characteristic Impedance. The characteristic impedance of a coaxial cable is a function of the resistance, capacitance, and inductance of the cable and is controlled by the cross-sectional dimensions. The impedance is calculated as the square root of the ratio of the inductance to the capacitance. It is because of the ratio of inductance and capacitance, which are both a function of length, that the characteristic impedance is independent of the cable length.

The measurement of the characteristic impedance of a cable is made by comparing the vertical position of the waveform to the position of the precision reference along the TDR trace. The difference in the vertical position is represented in terms of the reflection coefficient (ρ). From the reflection coefficient, the impedance is calculated using the following formula:

$$Z_O = \frac{1+\rho}{1-\rho} \cdot Z_{Ref}$$

The trace in Fig. 76.8 shows the characteristic impedance of the cable as a function of cable length on the display. As the pulse travels through connectors in the interconnect system, short impedance discontinuities can be observed in the trace. These discontinuities cause ringing that distorts the impedance measurement near the connector. The impedance appears to become higher down the length of the cable due to the DC resistance of the center conductor. It is best to make the impedance measurement as close to the reference as possible, where the ringing effects have dampened out.

Electrical length. Another measurement that is made using a TDR is the *electrical length* of a coaxial cable or interconnect. This is the amount of time for a signal to pass through the length of cable. The electrical length of a cable is a function of its physical length and how fast the signal travels through the cable.

In a typical application, a 10-foot sample of cable will be connected to the TDR. The amount of time between the end of the reference cable and the end of the cable under test represents the time for the TDR pulse to travel through the cable and back. Therefore, the electrical length is one-half of this time.

The *propagation delay* is the amount of time required for a signal to travel through a unit length of cable. In theory, electrical signals will travel at the speed of light in a coaxial cable with a dielectric of air between the signal wire and the shield. When dielectric materials are used in flexible coaxial cables, the velocity that the signal travels is slower than the speed of light. The propagation delay is usually reported in nanoseconds per foot.

Another way to express this is the *velocity of propagation,* which is the propagation delay represented as a percentage of the speed of light. This is a common way of specifying the signal-velocity characteristics of a coaxial cable.

Crosstalk. Crosstalk is the measure of the noise from a driven line, generated from a pulse changing states on a nearby quiet line due to electrostatic coupling. One common example of crosstalk is when a person on one telephone may hear a part of another conversation being conducted on a different line. In digital systems, crosstalk is measured as the peak amplitude of a signal in a nearby line caused by a pulse switching states in a driven line. The crosstalk gets worse when the digital pulse switches states faster. Crosstalk can be minimized by using cables with high-quality shielding around the center conductor. (See Fig. 76.9.)

In the crosstalk test, a pulse, monitored on an oscilloscope, is sent down the driven line, which is terminated with a resistor equal to the characteristic impedance of the cable to prevent reflections. A nearby line of interest, called the *quiet line,* is monitored by another channel of the oscilloscope. The peak voltage measured on the quiet line during a pulse transition is the crosstalk. The crosstalk is usually reported as a percentage of the input value. (See Fig. 76.10.)

Crosstalk measurements can be made at either end of the quiet line, depending on the area of interest in a given application. The near-end crosstalk measure-

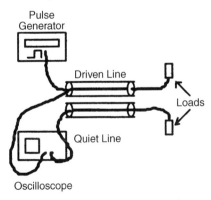

FIGURE 76.9 Quick line load (resistor).

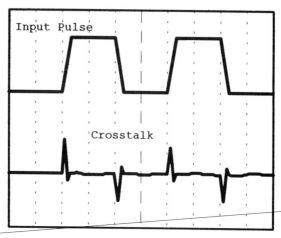

FIGURE 76.10 Crosstalk as percentage of input pulse.

ment is made on the same side that the pulse generator is connected to the driven line. When the measurement is made on the opposite end of the driven line, the measurement is called *far-end crosstalk*. To simulate conditions in an application more closely, the lines on both sides of the quiet line may be driven, which leads to a slightly higher crosstalk value.

Frequency Domain Measurements

Frequency domain measurements are often chosen for radio frequency and microwave frequency applications. A network analyzer is used to make measurements such as attenuation and shielding effectiveness. The network analyzer uses one channel to sweep a reference signal waveform of constant amplitude through a frequency range of interest and a second channel to measure the amplitude of the signal returning through the cable being measured. A comparison of the reference signal and the measured signal is made to determine the frequency response characteristics for the cable of interest (Fig. 76.11).

Attenuation. The *signal attenuation* is a measure of the decrease in signal amplitude as a function of frequency when traveling through a cable. In the attenuation measurement procedure, the reference signal is sent through one end of the cable while the response signal is measured at the other end of the cable. The signal attenuation is the ratio of the measured signal amplitude (V_{OUT}) to the reference signal amplitude (V_{IN}). The attenuation is usually expressed in decibels as shown in the following equation.

$$\text{Attenuation (dB)} = 20 \log \left(\frac{V_{out}}{V_{in}} \right)$$

In coaxial cables, the signal attenuation increases linearly with the length of the cable so the attenuation is expressed in dB/unit length (Fig. 76.11). As shown in the attenuation plot, the attenuation is proportional to the square root of the frequency.

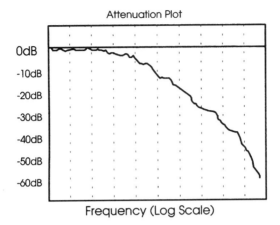

FIGURE 76.11 Attenuation-frequency plot.

The attenuation in a coaxial cable is inversely proportional to the center conductor diameter. Often, cable specifications will specify a maximum cable attenuation at a specific frequency as determined for the application.

Shielding effectiveness. The *shielding effectiveness* is the ratio of electric or magnetic field strength at a point before and after placement of a shield. It is an important property of shielded cables because it is a measure of how well the shield is protecting the signal conductor from external electromagnetic fields. These external fields appear on the desired signal as noise and, in low-level signal applications, can be of equal or greater magnitude than the signal.

The setup for the shielding effectiveness test, shown in Fig. 76.12, is a coaxial cable to be tested next to an injection wire carrying a reference signal. The injection wire is connected to the output channel of the network analyzer and terminated in

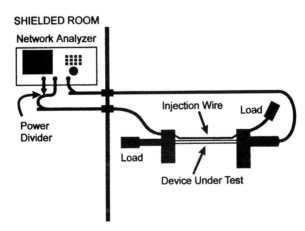

FIGURE 76.12 The shielding effect.

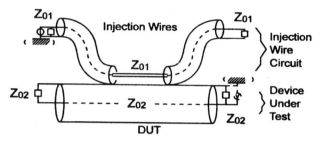

FIGURE 76.13 Shielding effectiveness in injection wires.

its characteristic impedance (50 ohms). One end of the coaxial cable is connected to the input of the network analyzer with the opposite end terminated in the characteristic impedance of the cable.

Shielding effectiveness measurements are made using the network analyzer to inject a swept frequency signal into the injection wire while measuring the signal transferred into the signal wire of the cable under test shown in Fig. 76.13. The shielding effectiveness is determined by the amount of the reference signal that is measured on the cable under test. The results are displayed as the shielding effectiveness as a function of frequency.

The results of shielding effectiveness are reported as the ratio of the measured signal to the reference signal in decibels. A sample waveform in Fig. 76.14 shows that the shielding effectiveness decreases as the frequency of the noise source increases.

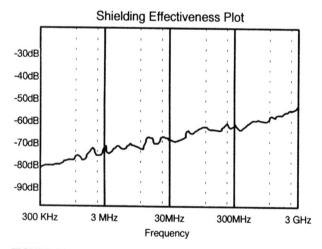

FIGURE 76.14 Shielding effectiveness decreases as the noise source increases.

LOW-FREQUENCY (DC) CABLE MEASUREMENTS

Cable Capacitance Testing

There are many applications where the user needs some knowledge of the cable's capacitance (C). For example, when a relatively high impedance source must be connected through a long cable, the cable's capacitance may degrade the signal-to-noise ratio of the measurement or it may attenuate the signal by forming a capacitive divider. The measurement speed may be adversely affected because the signal must charge the cable capacitance. The signal source may not be able to drive the capacitance without distortion, which would introduce errors into the measurement.

The capacitance of coaxial cable is typically specified in picofarads (pF) per meter or per foot. A cable's capacitance is determined by its length, the dielectric constant of the insulator between the center conductor and the outer shield, and the ratio of the diameter of the outer shield to the diameter of the center conductor. The capacitance can be calculated thus:

$$C = 24.2e/\log D/d \text{ pF/m} \quad \text{or} \quad C = 7.38e/\log D/d \text{ pF/ft}$$

where e is the dielectric constant, D is the diameter of the outer shield, and d is the diameter of the center conductor.

To measure cable capacitance, prepare a length of cable (at least three meters long) by exposing the center conductor and the shield. Connect these two terminals to a four-terminal capacitance meter (see Fig. 76.15). The measured capacitance should then be divided by the cable length to calculate the capacitance per unit length.

To obtain the best results, make certain that the HI current and HI potential meter terminals are connected to the shield and that the LO current and LO potential terminals are connected to the center conductor. Also, be sure the center conductor and the shield are not shorted at the other end of the cable. It's advisable to perform an open/short correction to compensate for any errors contributed by the test fixture and leads. Check the capacitance meter's operation manual for information on making this correction.

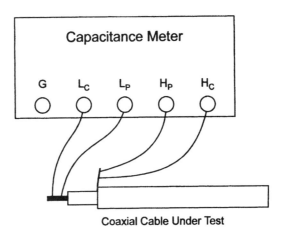

FIGURE 76.15 Measuring cable capacitance.

Since the cable capacitance may vary with the test frequency, the measurement should be made at several frequencies, particularly if the cable is to be used over a wide range of frequencies. Always specify the frequency or frequencies at which a test was performed to simplify future capacitance comparisons.

Typically, a 5O-Ω impedance cable will have about 100 pF/m of capacitance, while 75 Ω cable has about 67 pF/m. The capacitance of expanded dielectric cables may be as much as 30 percent less than these values.

Insulation Resistance and Leakage Current Measurements

In simple terms, a cable's *insulation resistance* (R_{IR}) is the ratio of the voltage applied to the conductors of the cable to the total current between them. This current is known as the *leakage current*. For a coaxial cable, the R_{IR} would be the resistance between the shield and the center conductor. Insulation resistance measurements are often performed on cables with multiple conductors. In these instances, the insulation resistance is measured from one conductor to all other conductors and to the shield.

The insulation resistance between the conductors within a cable should be very high, usually >10 GΩ. Insufficient insulation resistance may indicate a cable failure, such as a soft short, a mechanical failure, insulator defects, or moisture penetration and contamination. High insulation resistance is especially important in applications where the source resistance is very high. If the insulation resistance is of the same magnitude as the source resistance, system errors are inevitable.

Figure 76.16 illustrates a test configuration for measuring the insulation resistance of a coaxial cable. The R_{IR} is measured by sourcing a known voltage (V), measuring the resulting current (I), and calculating the insulation resistance ($R_{IR} = V/I$). These measurements are often performed at a specified cable length. For example, the R_{IR} per 1000 feet of cable would be written as $R_{IR\ 1000}$.

In the test configuration in Fig. 76.16, the HI terminal of the ammeter is connected to the center conductor of the cable, the LO terminal of the ammeter is connected to the LO terminal of the voltage source, and the HI terminal of the voltage source is

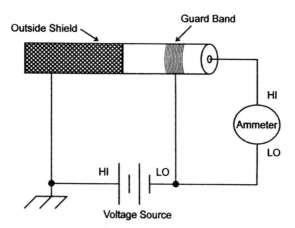

FIGURE 76.16 Measuring the insulation resistance of a coaxial cable.

grounded. When using this configuration, the ammeter must be capable of floating to the test voltage. The outer conductor or shield is connected to earth ground to avoid shock hazards. A guard band should be placed around the outside of the cable to prevent surface leakage from being added to the signal being measured.

After the connections are made, a test voltage is applied to soak the dielectric. The voltage and soak times should be specified in the test criteria and the same values should be used consistently. After the specified time, the resulting current is measured and recorded. The insulation resistance is then calculated, using $R_{IR} = V/I$. The common convention is to calculate the effective insulation resistance as $R_{IR1000} = (R_{IR} \times L)/$ 1000 ft, where cable length (L) is expressed in feet.

Insulation resistance measurement values are typically very high. To ensure their accuracy, some special measurement considerations should be made. The leakage currents being measured are usually in the nanoamp to picoamp ranges, so an electrometer or picoammeter would be required because of its excellent low-current sensitivity. To avoid errors due to leakage current, the resistance of the cables and fixtures used to make the measurement should be as high as possible. If the readings are unstable, electrostatic interference may be responsible.

To eliminate noisy readings, the test sample and all connections must be surrounded with a metal enclosure connected to voltage source ground.

Leakage and insulation resistance can be highly dependent on ambient humidity, particularly in environments with relative humidity readings above 40 percent. To allow for more accurate comparisons of readings taken at various times, the humidity should be measured and recorded during each test.

To prevent overloading the ammeter, which could result in instrument damage, it is advisable to insert a 1-MΩ resistor in series with the ammeter input. This limits current if there is a short across the insulator.

Dielectric absorption. Dielectric absorption in an insulator can occur when a voltage across that insulator causes positive and negatives charges within the insulator to polarize. When the voltage is removed, and the separated charges recombine, they generate a decaying current through circuits connected to the insulator, because various polar molecules relax at different rates.

Dielectric absorption can seriously degrade circuit accuracy in timing and integrating applications. It is a form of loss that dissipates signal energy and causes phase delay.

Dielectric absorption can be defined as the cable's discharge current at a designated time following the initiation of a discharge cycle. The cable is typically charged up to a specified test voltage. The measurement of the discharge current is usually made at a discharge time interval that will be used in the application of the cable.

A diagram of a test to measure dielectric absorption is shown in Fig. 76.16. Note that this is exactly the same test schematic that was used to illustrate measuring the insulation resistance of a coaxial cable. The HI input of the ammeter is connected to the center conductor of the cable and LO input of the ammeter to the LO of the voltage source. The ammeter must be able to float off ground to a voltage equivalent to the magnitude of the applied test voltage. The HI of the voltage source is connected to the shield of the cable. The shield is connected to ground. All connections must be isolated from the work surface and all objects.

To begin the measurement, the test voltage is applied for a precise period of time. This voltage should be specified in the test criteria and used consistently. After the soak period, the test voltage is returned to zero volts, keeping the voltage source connected. The current is plotted as a function of time, starting at the time the source was at zero volts.

Another way to measure the dielectric absorption is to source a known voltage, discharge the cable, and then measure the recovery voltage over a specified time period. A diagram of the test setup is shown in Fig. 76.17. An electrometer voltmeter is particularly useful in measuring dielectric absorption because it draws virtually no charge from the cable during the measurement, nor does it add charge to the cable being measured. Initially, the cable (C) is charged through R1 for the required soak time. This soak period depends on the capacitance of the cable. Next, the voltage source is turned off, and the cable is discharged through R2. The cable is allowed to sit for a few minutes with S2 and S1 open, and the residual voltage is then measured with the electrometer voltmeter. Fig. 76.18 illustrates the voltage waveform produced.

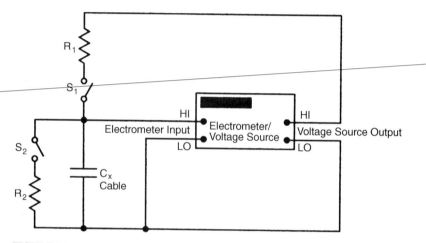

FIGURE 76.17 A dielectric absorption to source.

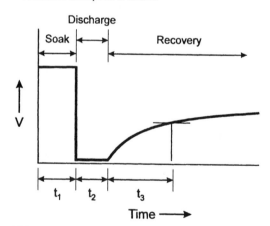

FIGURE 76.18 Produced voltage waveform.

Series Resistance and Continuity Testing

Series resistance (R_{SERIES}) and continuity measurements provide very important information about cable quality. Measuring series resistance gages the integrity of the circuit. If series resistance is higher than expected, the signal is reduced by the R_{IR} losses in the conductors. Continuity checks assure the operator that each connection is not an open circuit.

Each conductor has a finite amount of resistance per unit length. This resistance has been characterized for most types of conductors, wires, and cables used in the test and measurement industry. Coaxial cables, the standard of the industry, have two conductors—the center conductor and the shield. The typical resistance of the center conductor is tens of milliohms per foot.

The *four-wire* or *Kelvin* technique shown in Fig. 76.19 is the standard method for measuring series resistance in cables. This method forces current across the test sample (conductor) and the voltage is measured. With the current known and the voltage measured, the resistance can be calculated.

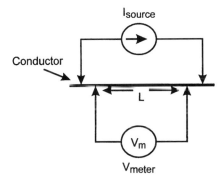

FIGURE 76.19 Kelvin technique (the four-wire).

The measured voltage is the result of the test current flowing through the conductor. The voltmeter used must offer sufficient sensitivity and accuracy to obtain reliable data. In general, the voltage drop will be on the order of millivolts, so a very sensitive voltmeter is required to make the measurement.

It is important to consider the length of the conductor, because the longer the conductor is, the greater the series resistance will be. In turn, the higher the level of series resistance, the greater the effect on the signal will be.

Conductor resistance is normally measured in ohms per foot. However, it's more practical to express conductor resistance in ohms per 1000 foot, simply because the level of resistance per foot is very low (e.g., 16 mΩ/ft.). Conductor resistance for 1000-foot lengths is typically tens or hundreds of ohms, rather than milliohms.

To measure conductor resistance, first, obtain a 100- to 1000-foot length L of the conductor. The longer the sample is, the higher the voltage drop that will be generated. The higher the voltage drop, the easier it will be to measure accurately.

Next, as the test schematic indicates, the current source terminals should be connected near the ends of the conductor. The voltage sense terminals should be connected to the conductor a few inches closer to the cable's midpoint than the current source terminals. Once the current is forced, then the voltage is measured. With the current and voltage known, the resistance can be calculated using the formula $R_{Series} = V/I$.

Thermal EMFs can cause errors in the voltage measurement. They arise when the temperature across a junction changes. These potentials normally range from a few microvolts to tens of millivolts. The degree of error they can introduce into a voltage measurement depends on the magnitude of the voltage being measured and the magnitude of the thermal EMFs themselves.

The two-measurement method is a standard technique for canceling the effects of thermal EMFs. In this technique, the current is forced in one polarity and the voltage is measured. Then, the current is forced in the opposite polarity and another voltage measurement is taken. With this data, the effects of thermal EMFs may be canceled by using the following formula:

$$R = [(V+/I+) + (V-/I-)]/2$$

Handling the cable conductors before or during the test can produce erroneous data for a short time. Body heat from the test operator's fingers can raise the temperature across connections temporarily, generating thermal EMFs that can skew readings.

Dry-circuit test methods are often used to evaluate connections within the cable assembly. These methods evaluate discontinuities, such as insulating films on conductive surfaces, by applying a very low level test voltage in a way designed to limit the open-circuit potential across the current source. In general, the open-circuit voltage is limited to 20 mV. This limited test voltage won't puncture the insulating films, allowing a true measurement of the device to be obtained. With the test voltage limit, the insulating films can be measured, rather than the actual conductors.

The ambient temperature is an important factor because the resistance of conductors varies significantly with temperature. Always record the ambient temperature to simplify correlating readings taken at various times.

Dielectric Withstand Voltage

Dielectric withstand tests are performed on cables to confirm that the cable can carry the maximum specified signal voltage level. This test determines if the dielectric between the center conductor and shield of a coax cable can withstand a specified voltage. This is important to applications in which high voltages are used. If the cable breaks down during use, devices and instrumentation can be destroyed.

Dielectric withstand voltage is measured by applying a specified test voltage and measuring the resulting current. The test voltage is either AC or DC, as required by the test specification.

The test voltage may be applied in any of several ways. A short-time test uses a ramp with a rate of 100 to 5000 V/second. A slow rate-of-rise test is done with a ramp rate of 1 to 100 V/second. A step-by-step requires a staircase waveform with steps of 100 V to several kilovolts.

In all cases, the test voltage is allowed to increase until dielectric breakdown occurs. *Breakdown* is defined as the point at which the current through the sample increases abruptly.

A diagram of the test setup is shown in Fig. 76.20. Connect the ammeter HI to the inner conductor of the cable and the voltage source HI to the outside shield of the cable. The LO of the ammeter and the LO of the voltage source are connected together.

Once the current is measured, it needs to be compared to a specified threshold level. A commonly used threshold level is 0.25 mA.

Because test voltages are usually very high (500 V or greater), follow all appropriate precautions to ensure operator safety.

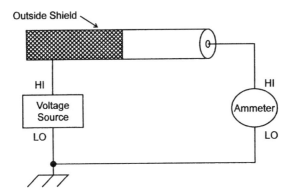

FIGURE 76.20 Dielectric withstand voltage circuit.

Low-Level Systems Considerations

Whenever practical, it's preferable to configure an automated system to control cable testing. Manual testing is typically time-consuming, tedious, and subject to human error. An automatic system will save the operator lengthy setup time and the need to record results manually. Test data can also be easily saved to disk for later statistical analysis.

The most efficient way to control the testing is by using a computer-based system with IEEE-488 bus-controlled scanners and measuring instruments. Using programmable scanners greatly reduces measurement time by switching one set of test instruments to multiple conductors. The scanner usually consists of a mainframe and plug-in switch cards. The mainframe controls the relays on the switch cards. Most manufacturers of scanners offer a variety of switching cards, which are designed for specific test signals such as low current, high voltage, low resistance, or low voltage. Detailed information on switching techniques is provided in Keithley's *Switching Handbook*.[6]

There are some basic steps common to configuring an automated DC test and measurement system for any application, including cable testing. These steps are as follows:

Step 1: Identify the tests to be performed.

Step 2: Select appropriate source and measure instrumentation.

Step 3: Determine the setup for each test.

Step 4: Select a switching topology that might be used to switch the various test points for the cables to be tested.

Step 5: Select the switch specifications.

Step 6: Select the actual switching hardware.

Step 7: Connect the system components.

When connecting a variety of test equipment, there are some system considerations one must understand to make accurate and repeatable measurements. Issues such as system grounding, shielding, guarding, and environmental effects can all cause measurement problems if not properly handled. Timing considerations for the test must also be accounted for. Incorrect test timing can result in readings taken at the wrong time. Connecting the system also requires an understanding of the cable requirements of the system.

Figure 76.21 shows a test configuration for making both insulation and series resistance measurements. This example illustrates many of the system considerations involved in configuring a low-level test setup. The schematic shows how the insulation resistance and path resistance tests can be combined to allow four-wire low-resistance measurements of 10 conductors and insulation resistance between each conductor. This test system shows a switching configuration that uses two-pole switches rated for 1100 volts.

A microohmmeter is used to make four-terminal resistance measurements of the conductors. To measure leakage current, a source-measure unit (SMU) capable of sourcing high voltages and measuring very low currents is used. With the SMU set to source 1000 volts, an insulation resistance of 10^9 Ω will allow 1 μA of current to flow.

Just two 2-pole high-voltage scanner cards in a 10-slot scanner mainframe are needed to make the necessary connections to 10 conductors. Two additional cards are used to switch the source and sense lines from the microohmmeter. The specifications for these should be sufficient to cover the needs of both tests in this configuration. It's important to choose appropriate switches that can be used to switch multiple signal

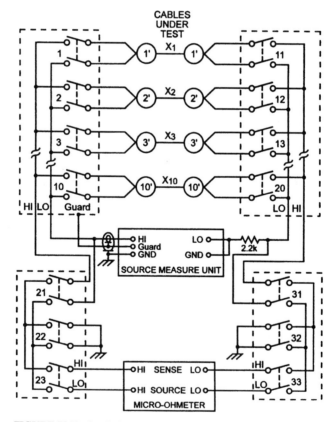

FIGURE 76.21 Insulation resistance and path resistance test setup.

types: low current, high voltage, low resistance for two-wire insulation resistance tests, and four-wire continuity or series resistance measurements, and so on.

As system complexity increases, even more care must be exercised to ensure proper operation of the test system. In this example, the SMU used in this example is capable of sourcing 1100 V at 10 mA of compliance current. The microohmmeter used is protected up to 10 volts. Because of this, relays 22 and 32 are closed to protect the microohmmeter when performing the leakage test.

In these cases, it is advisable to reduce voltages or current to nondamaging levels during development and testing of a new system. While the output data will not be accurate, proper switching can be verified.

MECHANICAL TESTING FOR LOW-NOISE COAXIAL CABLES

In low-signal-level applications, the electrical response to mechanical movement of coaxial cables becomes an important consideration. This electrical response is caused by changes in the cable capacitance, relative motion to fixed charges in the cable, and triboelectric effects that appear as noise on the desired signal. To characterize these effects in cables, standard and repeatable test methods are required to measure the cable noise performance accurately. In addition to the flex life of a cable, it is important to consider the noise performance after a cable is flexed.

Many low-noise cable test methods used today are drop tests where a weight, attached to a cable supported at both ends, is dropped while an electrical response is measured. These test methods are dependent upon test fixturing and are sometimes destructive to the cable. In an effort to characterize the differences in low-noise cable performance, a Bowstring-type of test may be used to separate different noise sources into different electrical waveforms for measurement and analysis.

Drop Tests

A typical drop test fixture is shown in Fig. 76.22. A test cable is supported at two locations a specified distance apart. The amount of slack cable is specified by either an absolute length of cable between the supports or a relative distance from the supports to the cable held in tension at the midspan point. A weight, with mass deter-

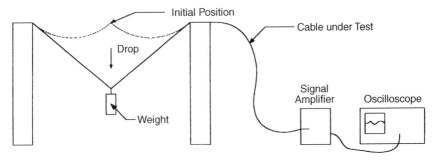

FIGURE 76.22 Drop test fixture.

mined as a function of the cable geometry or mass, is attached to the cable midway between the two supports. To perform the test, the weight is dropped, causing the mechanical excitation that is responsible for the electrical response. One end of the test cable is attached to measurement equipment, such as an electrometer or differential amplifier. The opposite end may be either open-circuited or connected through a known resistance. In either case, this end should be shielded from external electrical fields.

The electrical response can be measured in the form of voltage, current, or charge using an electrometer or similar device. The peak-to-peak value is typically recorded; however, the shape of the waveform also contains useful information. In order to measure the waveform, the electrometer must be capable of rapid sampling or must utilize an external analog voltage output attached to an oscilloscope to record the waveform.

In order to minimize the variability of any type of drop testing, it is important to pay attention to the test fixturing and test procedures. Test results may be greatly influenced by the dimensions of the weight, the method used to attach the weight to the cable and whether the weight is dropped on center each time. Additionally, the supports must be rigid with a firm grip on the cable to prevent slippage during the test. Drop tests can sometimes deform or damage the cable at the location of the weight and cable supports, which adds to the variability of the test results.

There are a few standard test procedures for triboelectric testing of cables; however, these test methods are often vague in the setup and fixturing requirements. Test procedures must specify the dimensions of the weight and the method of attachment to the cable in order to provide consistent results. Another difficulty with developing a standard method is that absolute measurement values for specifications are difficult to specify due to temperature and humidity effects in the cable materials. However, with identical test fixtures and methods, test results can be surprisingly repeatable using these drop test methods.

Bowstring Excitation Test Method

The Bowstring test method has been developed as an engineering tool to measure the noise response generated inside a coaxial cable with no signal applied. The mode of physical excitation is such that the electrical response of different sources inside a coax cable may be identified within the waveform.

The diagram in Fig. 76.23 shows the bowstring cable excitation test fixture used to measure noise generated in coaxial cables due to physical movement. The test consists of a cable, clamped at one end and held under tension by a weight. Located in the center of the fixture is a pneumatically controlled support post, which holds and releases the cable in a consistent manner. As the cable is released, it is free to vibrate back and forth until it stops. Electrical measurement equipment such as an electrometer, differential amplifier, or charge amplifier can be attached to the fixed end of the cable to measure different electrical responses to the motion.

Electrical Response to the Bowstring Test Method

A typical electrical response to the bowstring excitation is shown in Fig. 76.24. There are two dominant components that make up the electrical response displayed with the bowstring test, the sinusoidal response and the exponential response. These signal components originate from different sources inside the cable.

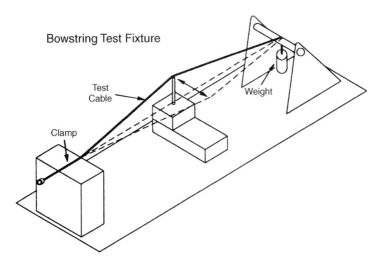

FIGURE 76.23 Bowstring test system.

The exponential response is produced by the cable components returning to a steady-state condition after the impulse of being released. This noise is low frequency in nature and will greatly affect low-level DC signals.

The sinusoidal response is a function of the vibration of the cable. The frequency of oscillation is related to the oscillation of the cable, indicating this response is generated by the vibratory motion. One noise source is from the cable conductors moving through the fields surrounding fixed charges in the cable insulators. Also, triboelectric charges are generated by the dissimilar materials inside the cable rubbing together and discharging through the conductors in the cable.

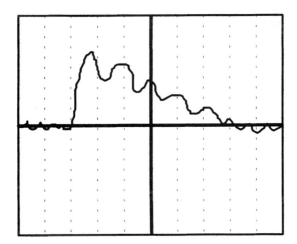

FIGURE 76.24 Electric response to bowstring test.

This measurement method, as with other noise measurements, is greatly affected by changes to the test method. The magnitude of the peak response increases with amount of weight at the end of the cable. The sinusoidal frequency also increases with weight similarly to the way a guitar string is tuned. The distance that the cable is displaced also will increase the magnitude of the response.

Flex Degradation/Flex Life

In low-level signal applications, the performance of a coaxial cable after being subjected to flexing is as important as all of the other cable properties discussed earlier. One common method of testing cables for flex life is the Tick Tock test, where the cable is continuously flexed back and fourth until an electrical failure occurs on one of the cable conductors. Another test method uses a rolling flex fixture to flex the cable a fixed number of times to measure whether or not the electrical properties have degraded after flexing.

Tick Tock Testing

The setup for the Tick Tock test method is shown Fig. 76.25. The test cable is supported by a clamp on the flexing machine arm with a weight hanging from the other end of the cable. The cable passes through two sets of mandrels to control the flexing motion of the cable. The upper mandrel is the diameter of the specified bend radius for the test and the bottom rollers are to keep the lower half of the cable in a vertical position.

The Tick Tock test flexes the cable back and forth 90° from the center line of the lower half of the cable. One cycle consists of the cable being flexed from the center through both directions back to center. During the test, the conductors of the cable are monitored for continuity and the test is finished when there is an open conductor in the cable.

The results of the Tick Tock test are reported in the number of cycles until failure. The number of cycles a cable will withstand is influenced by the minimum bend radius in the test, cable diameter, and materials used. In severe flex applications, cable jacket materials may have special processes that will protect the cable conductors and increase their flex life.

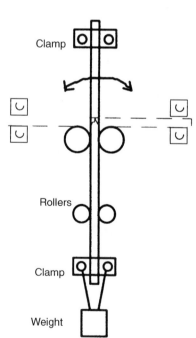

FIGURE 76.25 Tick Tock test setup.

Rolling Flex Test

The rolling flex tester is used to characterize the degradation of the electrical properties and noise performance of a coaxial cable as the cable is flexed in a controlled fashion. The cable under test is supported

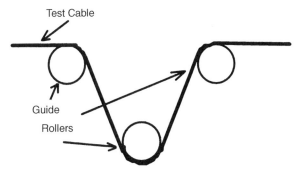

FIGURE 76.26 Rolling test setup.

at one fixed end and a weight holds the cable under tension in the horizontal plane. The cable is routed through a three roller fixture as shown in Fig. 76.26. The roller fixture moves back and forth along the length of the cable to flex the entire length of cable in a consistent manner.

The analysis of the rolling flex test begins with an initial measurement of the cable for the properties of interest. The cable is then installed in the rolling flex machine and cycled a specified number of times as determined by the specific test procedure. The measurements made initially are repeated after the cycling and recorded. This procedure may be repeated several times until a significant degradation in the cable's performance is measured.

The results of the rolling flex test may be reported as individual measurements after a number of flexes or in graphical format where the cable measurement is plotted as a function of flex cycles. The graphical format is useful in relating the cable test results to the performance a cable will see in the application.

CABLE COMPARISONS

The reliability of a measurement system can be no better than that of the input cables. Special thought should be given to the operating environment as well as the electrical requirements of an application. Mechanical and environmental issues include cable diameter and weight, flexibility, flex life, and temperature range. Electrical parameters may include attenuation, shielding effectiveness, capacitance, and low-noise performance.

Coaxial cable design options generally begin with the center conductor. Issues such as attenuation, resistance, flexibility, and flex life requirements will lead to the selection of conductor type and stranding. Generally speaking, the higher the number of strands, the better the cable's flexibility will be, but higher noise levels are likely to accompany a high strand count.

The insulation or dielectric layer will dictate the cable's velocity of propagation, capacitance, and sometimes usable temperature range. Solid polyethylene dielectric cables offer a limited temperature range (–40 to +85°C), moderate flexibility, and attractive prices. Full-density polytetrafluoroethylene (PTFE) has higher insulation resistance and a broader temperature range (–55 to +200°C) when combined with a fluoropolymer jacket. Expanded PTFE dielectric designs offer faster signal speed (85 percent typical versus 69 percent), lower capacitance, and improved flexibility, as well as smaller size and lighter weight.

The shield type will influence cable flexibility, noise performance, and electromagnetic susceptibility. Finally, the jacket chosen will affect environmental resistance, operating temperature range, and flex life.

Just as there is no single ideal sensor type, neither can there be one type of cable that satisfies all applications. Cables should be designed and engineered with the intended application in mind.

REFERENCES

1. Thomas A. Perl, "Electrical Noise from Instrument Cables Subjected to Shock and Vibration," *Journal of Applied Physics,* **23** (6), June 1952.
2. E. P. Fowler, "Microphony of Coaxial Cables," *Proceedings IEEE,* **123** (10), October 1976
3. *Belden Wire and Cable Catalog,* rev. 1, p. 260, 1990.
4. MIL = C = 17.
5. Alfred G. Ratz, "Triboelectric Noise," *ISA Transactions,* **9** (2).
6. Keithley Instruments, Inc., *Switching Handbook: A Guide to Signal Switching in Automated Test Systems,* 3d ed., Keithley Instruments, Inc., Cleveland, Ohio, 1994.

CHAPTER 77
RESONANT MICROBEAM TECHNOLOGY FOR PRECISION PRESSURE TRANSDUCER APPLICATIONS

INTRODUCTION

Resonant sensors have historically provided the finest output for precision sensor applications. Solid-state piezoresistive pressure sensors now meet the stringent performance requirements of air data systems and jet engine controls. Resonant silicon devices are again emerging with increasing levels of performance. Recent developments in sealed-cavity polysilicon resonant microbeam pressure sensors are showing promise for next-generation applications. Differential pressure sensors have been demonstrated with base frequencies of 220,000 Hz and sensitivities of nearly 4000 Hz/psi.

Precision absolute pressure transducers can be divided into two categories: resonant, or vibratory, and nonresonant. *Resonant* sensors employ a sensing element that resonates or vibrates with pressure. Such sensors include the vibrating cylinder sensor, the vibrating diaphragm sensor, vibrating quartz sensor, and resonant microbeam sensors. Resonant sensors rely on a high-Q resonant element to determine the frequency of oscillation. The key advantage of this approach is that the frequency of oscillation is determined almost uniquely by the pressure-sensing element. As a result, the gain circuit employed becomes noncritical, a key advantage for this type of sensor. When the resonant element is made of a highly stable material with good mechanical properties, the result is a highly stable and repeatable sensor. The key disadvantages associated with most resonant sensors are high cost and sometimes size and weight. The main performance deficiency of the sensors is their sensitivity to variations in density with temperature and pressure of the medium, which is especially pronounced for sensors with a large vibratory surface such as the vibrating cylinder sensor.

One of the most successful resonant sensors is Honeywell's vibrating diaphragm sensor. In production since 1970, almost 100,000 sensors have been produced to date. This sensor is used exclusively in commercial and military air data applications. Made from beryllium copper, this sensor uses a thin diaphragm whose resonant frequency changes as a function of pressure loading and temperature. The diaphragm is isolated from package-induced stresses by special packaging techniques. Figure 77.1 illustrates the construction of the vibrating diaphragm sensor. Another resonant sensor developed by Honeywell is the quartz resonant sensor,[1] illustrated in Fig. 77.2.

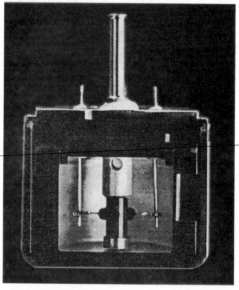

FIGURE 77.1 Cutaway view of BeCu vibrating diaphragm sensor. External pressure is applied to the inlet at the top of the sensor, which introduces gas into the interior of the contoured cylinder. The region between the cylinder and the casing contains a vacuum to provide a pressure reference and maintain a high mechanical Q. The applied pressure stretches the diaphragm at the lower end of the cylinder, causing shifts in the resonant frequency. The diaphragm is driven into resonance by a coil mounted on the medial post, and detection is done through the same coil.

Unlike resonant sensors, *piezoresistive* sensors rely on a change in resistance for sensing pressure. Typically, four piezoresistors are connected in a Wheatstone bridge configuration, which offers maximum sensitivity and good linearity. The sensor conversion circuitry associated with this type of sensor typically dominates the error budget. Piezoresistive silicon sensors use piezoresistors diffused or implanted into a silicon diaphragm. This type of sensor is among those most commonly used in industry due to its small size, low cost, and IC fabrication techniques, but has been rarely used in precision avionics applications. Honeywell has refined this type of sensor to a performance level that is equivalent to or better than that of more expensive resonant sensors. A piezoresistive sensor in an avionics header is shown in Fig. 77.3. In addition, Honeywell has developed proprietary A/D circuitry for nearly error-free conversion and algorithmic computation techniques that eliminate repeatable errors. This type of sensor is now used by Honeywell in precision air data applications for the McDonnell-Douglas DC-10, MD-11, MD-80, and MD-90; Boeing 727,

FIGURE 77.2 Quartz resonant sensors in flat packs. Two quartz resonant pressure sensors are shown with signal-processing electronics for air data applications. The quartz sensors are formed from three bonded wafers of single-crystal quartz. They contain two diaphragms, one forming a reference diaphragm and the other exposed on one side to the pressure-sensing medium. Differences in the diaphragm resonant frequencies provide a measure of applied pressure.

FIGURE 77.3 Piezoresistive air data sensor. Solid-state sensors provide appreciable size and power reductions. The reference vacuum chamber is formed by the welded lid. External pressure is applied through the inlet tube.

737, 747, and 777; Airbus A320, A330, and A340. Typical accuracy of these systems is better than 0.02 percent of full-scale pressure over −55 to +100°C and 0 to 40 in. Hg. Stability is within 0.012 percent of FSO over 20 years. A generic listing of key performance parameters for precision pressure sensors is found in Table 77.1.

TABLE 77.1 Characteristics of High-Performance Pressure Sensors

Stability	<0.02% FSO, 20 years
Resolution	<1 ppm
Noncompensatable errors	<100 ppm
MTBF	>1,000,000 hr
Mechanical hysteresis	<10 ppm
Thermal hysteresis	<50 ppm

RESONANT MICROBEAM TECHNOLOGY

The last few years have seen the emergence of resonant silicon sensors, which adds the inherent advantages of silicon sensor technology.[2-6] Such sensors are now available from several manufacturers for various applications.

Resonant microbeam technology is based on strain-sensitive resonant microbeams fabricated from thin films of polycrystalline silicon. The resonant microbeam elements are stimulated into vibration at their fundamental resonant frequency. The resonant frequency is a strong function of axial loading, resulting in large frequency shifts with small mechanical strains applied to the ends of the microbeam. The microbeams are enclosed by integral vacuum enclosures, allowing the microbeams to vibrate freely in a vacuum environment, unencumbered by undesirable damping from gaseous or liquid media. The resonant microbeams are fabricated on silicon substrates, which are precisely micromachined to produce a pressure-sensitive diaphragm. The microbeams are monolithic with the silicon diaphragm, resulting in a high-sensitivity, low-hysteresis pressure sensor with digital-level frequency output.

Principle of Operation

Microbeam development has focused on electrostatic excitation and piezoresistive detection of the microbeam motion. The microbeams, typically 2.0 μm thick, 40 μm wide, and 100 to 400 μm long, are attached at each end and are free to vibrate in and out of the plane of the wafer. The microbeams contain a drive electrode at the center of the beam and a polysilicon piezoresistor at one end. The microbeams are excited by applying a bias voltage to the shell and substrate and a small AC voltage to the drive electrode. Several millivolts are needed to excite the microbeam into oscillation. The piezoresistor provides a means for externally determining the flexing of the microbeam.

Vacuum enclosures are formed integrally with the microbeam, providing a locus for microbeam movement. A sacrificial layer between the microbeam and the substrate defines the lower cavity. Removal of the sacrificial layer is achieved via small etch channels and a liquid etchant. The microbeams and cavity sides are defined by the patterning and etching of long, thin slits in the deposited microbeam polysilicon. The upper cavity is formed by removal of a second sacrificial layer upon which the shell polysilicon has been deposited and patterned, resulting in a completed silicon enclosure in close proximity to the microbeam. The vacuum environment is created after the sacrificial etching process by reactive sealing, using a thin film deposited at low pressure to seal off the narrow etch channels and seal in a good vacuum.

The microbeams are strategically placed on the silicon diaphragm. Due to wafer-handling considerations, the microbeams are formed prior to diaphragm formation on a thin, epitaxial layer of silicon. The silicon diaphragms are formed using isotropic or anisotropic etchants. An electrochemical etch stop between the epi layer and the substrate produces well-controlled diaphragm thicknesses for accurate pressure sensitivity. Two microbeams are arranged on the diaphragm to provide a differential or push-pull output, where one microbeam frequency increases with applied pressure and the other decreases with applied pressure. The subtraction of one microbeam frequency from the other doubles the pressure sensitivity while canceling first-order temperature and mechanical effects. A packaged sensor die is shown in Fig. 77.4, and an enlarged view of the die is shown in Fig. 77.5.

Differential pressure is measured by applying the pressures of interest to each side of the diaphragm, causing diaphragm flexure, which in turn applies bending strain directly to the microbeams. The integral vacuum enclosure allows the microbeams to operate freely, even with the diaphragm in contact with liquids such as fill fluids. Absolute pressure sensors require formation of a reference vacuum on one side of the diaphragm.

External drive/sense electronics contain amplification stages and automatic gain control to lock the microbeam into oscillation and maintain it there as the resonant frequency shifts. The output is buffered to provide a digital-level frequency that is readily interfaced with system electronics.

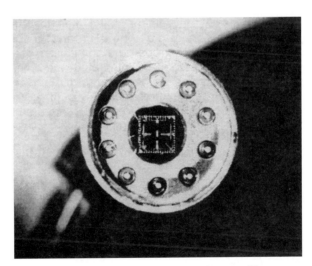

FIGURE 77.4 Resonant microbeam pressure sensor. The pressure sensor die is mounted on a chip tube in an avionics-grade header. One of the microbeams on the diaphragm is wire-bonded out for operation with external drive/sense electronics for testing purposes.

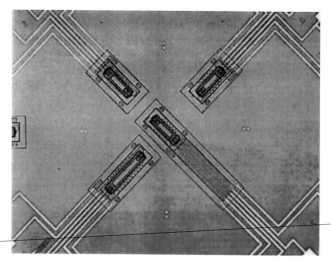

FIGURE 77.5 Resonant sensor die. A multiplicity of resonant microbeams are shown on a micromachined silicon diaphragm. Each microbeam can be independently driven and sensed.

Advantages and Features

Resonant sensors benefit from the direct conversion of physical stimuli to frequency output. Intervening conversions, such as capacitance or resistance shifts to voltage that are then digitized, are not necessary, and pressure information resident in the medium is not occluded. Omission of digitizing electronics consequently reduces system cost and complexity. Resonant microbeam sensors are integrable with cofabricated electronics for further cost reductions and potential performance increases, although current interface electronics use conventional surface-mount technology.

Further increases in long-term stability are driving developments in pressure sensor technology. Resonant microbeam technology offers the potential for higher stability since stability is now determined by precision time-based measurements rather than voltage, charge, or current measurements. Furthermore, the stability of the microbeam is determined by the mechanical properties of the materials composing the microbeam, microstructure, and mounting rather than the electrical properties. The resonant microbeams and microstructures are fabricated from single-crystal silicon and polysilicon, both of which have excellent mechanical properties.

Increases in performance are possible with resonant-microbeam-based sensor technology. Resonant sensors offer a large dynamic range. With resolution better than 0.01 Hz, base frequencies on the order of 500,000 Hz, and 10 percent or higher frequency shifts, dynamic ranges of 5×10^6 and higher can be obtained. The high base frequency and large frequency output allows higher cutoff frequencies and increased fidelity in the output signal. Figure 77.6 shows the frequency output of a single microbeam fabricated on a diaphragm with a sensitivity of 1.0 kHz per psi. Pressure sensitivities up to 4000 Hz/psi have been achieved.

The inherent ability to sense strain with resonant microbeams is strongly dependent on the microbeam geometry. Analysis shows the strain sensitivity or gage factor GF is related to the square of the microbeam aspect ratio.

$$GF = \Delta fr/fr/\text{strain} \cong (1/7) \cdot (L/h)^2$$

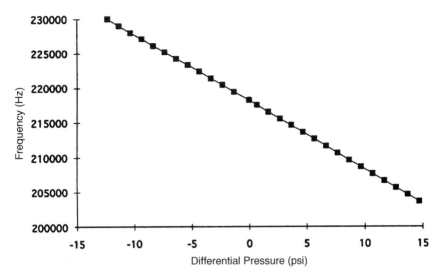

FIGURE 77.6 Frequency output. Running at a base frequency of 215 kHz, this sensor has a sensitivity on the order of 1 kHz per psi and is capable of differential or absolute pressure measurements.

where L is the microbeam length and h is the microbeam thickness. A 200-μm-long, 2-μm-thick microbeam has a gage factor of over 1400. Small corrections are required for internal or residual strain in the microbeam material. Consequently, an applied strain of 100 microstrain (100×10^{-6} strain) results in a frequency shift of over 70,000 Hz or nearly 15 percent for a base frequency of 500 kHz. With the fracture strain of silicon near 10,000 microstrain, significant signals can therefore be obtained with minute applied strains. Proper design of the pressure sensors thus allow high-overrange or burst-pressure specifications. For comparison, the gage factor ($\Delta R/R$/strain) of a single-crystal silicon piezoresistor is typically slightly less than 100. Consequently, much more strain is needed in the piezoresistive sensors to achieve a similar shift in output.

The attainable resolution is set by the noise limit of the microbeam and associated electronics. Measurements of the noise floor below 0.01 ppm of the base frequency have been made on polysilicon resonant microbeams. High-resolution counters are used for measurements and provide 22 to 24 bits of data with short sample times. Fast sample times are allowed by the high base frequencies. A summary of the advantages of resonant microbeam sensors can be found in Table 77.2.

INTERFACE CONSIDERATIONS

Resonant sensors have an advantage over more conventional bridge-type sensors as they are subject to fewer analog error sources. The analog electronics required to close the loop and excite the oscillations are not required to have the precise DC characteristics and stability required for an analog bridge-based sensor. Though most resonant sensors have the capability for extremely precise and repeatable measurements once a calibration algorithm is applied, their full perfor-

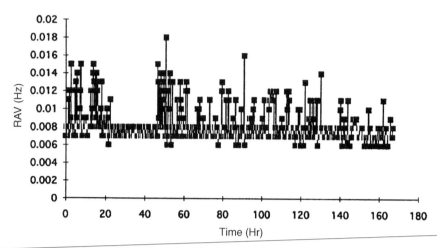

FIGURE 77.7 Root Allan variance data. Data on resonant microbeams taken over a period of one week is shown. Each data point represents 100 frequency measurements taken consecutively, and repeated every 6 min. A noise floor of 0.007 Hz is measured. The device under test operates at 631,594 Hz, indicating attainable short-term stability near 0.01 ppm of the base frequency.

mance may be exploited only if both the sensor and the system in which they are incorporated have been systematically designed to eliminate or mitigate error sources.

The output of a resonant sensor, being a variable frequency, must be compared to a reference frequency or time base in order to convert it into a digital quantity meaningful to a microprocessor. In the past, the most common method for converting such resonant sensor outputs into digital format has either been to gate the sensor output into a counter for a finite period of time or to accumulate a certain number of output cycles from the sensor and then measure the time required to accumulate those cycles. The gating scheme has the advantage of having a finite conversion period and resolution, which is easy to express in system transfer function notation. The finite cycle count method has the disadvantage of having a variable

TABLE 77.2 Advantages of Resonant Sensors

Digital output	Complexity of readout electronics reduced
High stability	Time-base measurements
	Based on mechanical performance, not electrical properties
Increased performance	Large dynamic range
	Large gage factor
	High strain sensitivity
	High resolution
	High resonant frequency
Silicon-based technology	Low cost, batch processing
	On-chip vacuum encapsulation
Self-enclosed resonators	Operation in liquid or gaseous media
	Differential configuration
	High-Q operation
	Environmental protection

conversion time, but the advantage of having constantly improving resolution as the frequency decreases (and hence the integration time interval increases). Both methods are equally simple to implement but have a major drawback—the thermal and time stability of the reference against which the sensor is measured becomes a major limitation to ultimate sensor performance.

A more appropriate method of ensuring long-term stability is to eliminate the need for a precise reference frequency. The preferred approach is to output several frequencies from the resonant sensor. The pressure-sensitive elements are primarily sensitive to pressure but typically have thermal and acceleration sensitivities as well. The other elements, which are constructed similarly, are sensitive only to the thermal and acceleration effects. By measuring these frequencies in terms of ratios, that is, creating a pressure-sensitive function (PSF) where

$$\text{PSF} = F_{p(\text{ressure})}/F_{\text{ref}(\text{erence})}$$

the sensor's perceived output becomes a dimensionless ratio. All measurement error sources are then moved on-chip, where the long-term drift and thermal sensitivity of individual elements are effectively canceled by the nearly identical percentage drift and thermal effect of the measurements in the numerator and denominator. This identical aging and thermal characteristics are achieved by incorporating the elements in the same silicon structure whose elements are similar and simultaneously cofabricated and undergo identical thermal cycles over life.

This mechanical structure requires the raw pressure indications to be adjusted for thermal effects, typically using a polynomial approach, to get precise pressure measurements over the desired operating temperature range. Sensing the temperature for this correction again should follow the approach used for the pressure channel except that a more temperature-sensitive structure than the reference frequency structure should be used to get the dimensionless temperature-sensitive function (TSF), where

$$\text{TSF} = F_{\text{temp}(\text{erature})}/F_{\text{ref}(\text{erence})}$$

Alternatively, since the thermal sensitivity of the PSF is very low compared to piezoresistive analog sensors, such a precise reference is not required to attain the same system-level accuracy. In that case, the F_{temp} may be substituted for the F_{ref} in the PSF, and the external system crystal clock may be substituted for the F_{ref} in the TSF.

The key to attaining the required system-level performance hinges on mitigation of both mechanical and electrical error sources. A factor affecting many resonant sensors having vibrating surfaces in direct contact with the operating medium is the effect of contamination, which can effectively mass-load the resonant element, thereby causing an undesirable change in its frequency. Similarly, such resonant sensors have their resonances changed or even totally damped as the density of the pressure medium changes. For that reason, many applications of that type transducer in liquid-media systems require the use of a bubbler to decouple the sensor from the media and allow it to work with its given calibration gas. The vacuum-encapsulated resonant microbeam, used in place of analog piezoresistors, allows the sensor diaphragm to be in direct contact with the pressure media. This technology enables true differential pressure sensors to be fabricated since both sides of the die (except for the bonding pads) may now be effectively passivated due to the buried resonators. The requirements on the electronics desensitizes the pressure sensor to media-induced electrostatic effects and sensitivity to excitation voltage that are inherent in most analog piezoresistive bridge-based sensors.

CONCLUSIONS AND SUMMARY

Vibratory sensors based on metal and quartz diaphragms are being supplanted by precision solid-state piezoresistive sensors for air data applications requiring absolute (vacuum reference) sensors. Resonant microbeam pressure sensors for differential-pressure-sensing applications have been demonstrated and show capability for precision absolute or differential measurements. Resonant microbeam technology has potential for avionics, residential, and industrial pressure sensor applications.

REFERENCES

1. R. Frische, "Vibratory Pressure Sensors," *Scientific Honeyweller,* Honeywell Inc., 1987, pp. 79–84.
2. K. Saito, T. Nishikawa, H. Kuwayama, K. Yoshioka, and S. Zager, "Resonant Monolithic Silicon Sensor for Intelligent DP Transmitters," *Proceedings of the Conference on Advances in Instrumentation and Control,* 1992, **47** (2), pp. 1019–1029.
3. K. Petersen, F. Pourahmadi, J. Brown, P. Parsons, M. Skinner, and J. Tudor, "Resonant Beam Pressure Sensor Fabricated with Silicon Fusion Bonding," *Proceedings of the 1991 International Conference on Solid-State Sensors and Actuators,* pp. 664–667.
4. P. Parsons, A. Glendinning, and D. Angalidis, "Resonant Sensor for High-Accuracy Pressure Measurement Using Silicon Technology," *IEEE Aerospace and Electronic Systems Magazine,* July 1992, **7** (7), pp. 45–48.
5. Druck Inc., 4 Dunham Drive, New Fairfield, CT 06812.
6. H. Guckel, C. Rypstat, M. Nasnidal, J. D. Zook, D. W. Burns, and D. K. Arch, "Polysilicon Resonant Microbeam Technology for High-Performance Sensor Applications," *IEEE Solid-State Sensors and Actuators Proceedings,* Hilton Head Island, South Carolina, June 22–25, 1992, pp. 153–156.

CHAPTER 78
TWO-CHIP SMART ACCELEROMETER

INTRODUCTION

In this chapter, a signal-conditioned accelerometer is described which offers many advantages to the user. The sensor element, manufactured using silicon micromachining, has proven reliability. The signal conditioning does not require external components and thick or thin film technology but is done entirely within a monolithic IC. The sensor and signal-conditioning chips are hermetically packaged together in a ceramic leadless chip carrier. The output parameters are trimmed electrically after packaging is completed. The chip carrier can be mounted in several orientations to allow measurement of acceleration either perpendicular to or in plane with the mounting surface.

Currently the majority of signal-conditioned accelerometers is packaged using hybrid technology, where thick or thin film resistors are used to set parameters such as offset and sensitivity to the desired values. This approach results in relatively bulky designs with nonuniform mounting configurations. The user is often required to do additional mechanical design, such as, using a mounting bracket. The accelerometer design presented in this chapter is intended to lower not only the cost of the accelerometer itself but the total cost to the user. This is accomplished by mating a silicon micromachined sensor die to a signal-conditioning IC in a ceramic leadless chip carrier. The two-chip approach allows both the sensor and the signal conditioning chips to be optimized and avoids the yield losses associated with complicated single-chip designs. The accelerometer is compatible with automated PC board assembly while offering multiple mounting options.

SENSOR ELEMENT

The accelerometer structure, which measures 3.4 mm square, is shown in Fig. 78.1. A seismic mass and four flexures are formed using bulk micromachining processes. Bulk micromachining technology was chosen over surface micromachining because the entire thickness of the silicon wafer can be used for the seismic mass, resulting in a higher sensor output. Each of the four beams contains two implanted resistors which are interconnected to form a Wheatstone bridge. When the device undergoes an acceleration, the mass moves up or down, causing four of the resistors to increase and the

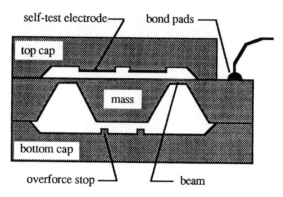

FIGURE 78.1 Cross-sectional view of accelerometer die.

other four to decrease in value. This results in an output voltage change proportional to the applied acceleration. The eight resistors are interconnected such that the effects of any motion other than that caused by an acceleration in the primary direction are canceled out. Piezoresistive transduction provides a relatively high output level with low impedance and good linearity. Therefore, it is not necessary to include signal-conditioning electronics on the same chip as the sensor to obtain good performance.

Silicon top and bottom caps are attached to the section containing the seismic mass and the beams. The silicon caps serve several purposes. Precision gaps are etched into the caps to provide air damping to suppress the resonance peak of the structure. Because the part is critically damped, the response is flat up to several kHz, independent of temperature. Small elevated stops on the top and bottom caps limit the motion of the mass to a fraction of the deflection at which fracture occurs. The mechanical structure does not wear and mechanical latch-up cannot occur. The top and bottom cap form an enclosed cavity around the seismic mass, protecting it against contamination which may obstruct its motion. Because the three sections are bonded together at the wafer level in the clean room, the cavity is free of particles and is protected from particulate contamination during the final chip dicing and assembly operations.

Lastly, the top cap is used to enable testing of the accelerometer in the absence of acceleration.[1,2] The overforce stops on the top cap have been enlarged and a metal electrode has been deposited on them. This electrode is connected to a bond pad. When a voltage is applied between the electrode and the silicon of the seismic mass, an electrostatic force moves the mass toward the top cap. This results in a change in output voltage proportional to the sensitivity and to the square of the applied voltage. It is thus possible to generate an acceleration using an external voltage and check the functionality of the mechanical structure as well as the electronics. The accelerometer has been qualified for and used in air bag crash detection systems and proven to be very reliable.

SIGNAL-CONDITIONING IC

The signal-conditioning IC is made in 1.5-μm CMOS technology. Signals are processed by differential amplifiers throughout most of the circuit in order to minimize common mode effects and noise. Switched-capacitor circuitry is used to save space and because high-accuracy gain stages can be made easily. The −3-dB bandwidth of the signal-conditioning electronics is about 3 kHz. The accelerometer is intended for 5-V operation with an output voltage in the 0.5- to 4.5-V range.

Signal Processing

The accelerometer sensor has a differential output with source impedance of around 4 kΩ and full-scale output voltage of about ± 50 mV. The offset voltage (output at zero applied acceleration) may vary a few millivolts over the temperature range of −40 to 85°C. Also, the full-scale output will decrease over temperature by about −1900 ppm/°C. The signal-conditioning IC converts the differential signal into a single-ended signal in the 0.5- to 4.5-V range while compensating the temperature-related signal variations. As a result, the compensated accelerometers are interchangeable with a total error of less than 5 percent.

The signal path is shown in the block diagram in Fig. 78.2. The output signal of the accelerometer die is processed by the following stages:

- The first stage provides a high-impedance load for the sensor and amplifies the signal to maximize the dynamic range during subsequent processing.
- The offset of the sensor is eliminated by adding a voltage generated by a DAC which is controlled by a digital word representing the programmed offset value.
- The temperature coefficient of offset (TCO) of the sensor is compensated by adding a voltage which is controlled by digital words representing the temperature and the programmed TCO value. Both the offset and TCO voltages are derived from the supply to ensure that the signal remains ratiometric with the supply voltage.
- The signal gain can be varied by changing a capacitor ratio using a digital word. The gain can be varied in a 5 to 1 range to allow for different full-scale specifications.
- The temperature coefficient of sensitivity (TCS) of the sensor is compensated in the next stage. The sensitivity decrease over temperature is compensated by increasing

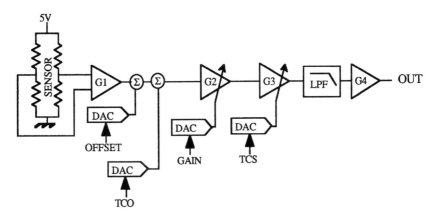

FIGURE 78.2 Block diagram of the accelerometer.

the signal gain linearly with temperature. This method for sensitivity temperature compensation was chosen over a circuit using constant current excitation of the sensor because of the required voltage overhead of the current source. The sensor is now powered with the entire available supply voltage, maximizing its signal.

- The output bias voltage can be set to either 0.5 or 2.5 V by connecting an input pad on the chip to ground during assembly of the part. This allows signals to be processed with either a bipolar or unipolar range.
- A two-pole passive filter removes signals generated by the internal oscillator and switched capacitor networks. Switching noise is further minimized by having separate digital and analog internal supply lines and by the differential signal processing.
- The final stage provides a low-impedance output for driving resistive and capacitive loads without influencing the signal. The output will go in a high-impedance tri-state mode if the part is not addressed.

Error Detection Functions

Because the accelerometer is intended to be used in safety-critical applications such as air bag deployment, several features have been incorporated to detect a failure of the accelerometer or circuitry. It is important to prevent floating signals because the resulting output voltage might look like a crash signal and activate the air bag. Such signals could be caused by a discontinuity between the sensor and the circuit or by a malfunction of the sensor itself. Small current sources have been added between each of the signal inputs and the positive supply. In case one or both of the inputs are open, the output voltage is forced to the positive supply. In addition, two window comparators monitor the voltage at both inputs. If the voltage at one or both inputs exceeds the allowed range, an alarm output pin is made high. This output can be monitored by a microprocessor to alert the user to a malfunction of the sensor.

In addition, the sensor has a built-in self-test function which allows the seismic mass to be moved by means of an externally applied voltage. This allows the entire device to be tested, including the mechanical structure of the sensor and the signal-conditioning electronics. By applying a voltage to the bond pad that is connected to the self-test electrode, the output will exhibit a voltage change which is proportional to the full-scale output, in contrast to other self-test schemes where the output change is fixed. It is thus possible to verify not only complete malfunction of the device but also a parametric error, giving a better indication of a partial or a developing failure.

Addressing

Addressing capabilities have been incorporated in the signal-conditioning electronics in the form of row select and column select digital inputs. Both input lines must be high for the accelerometer to be selected. If one or both of the select lines are in the low state, the signal and alarm outputs are in a high-impedance tri-state mode. This allows the outputs of multiple accelerometers to be connected together as shown in Fig. 78.3. This eliminates the need for analog multiplexers and reduces wiring. The reduced number of wires is an advantage if four or more devices are needed in a system. The Table 78.1 shows the number of lines (including supply and ground) required in a measurement system with sensors used in nonmultiplexed and

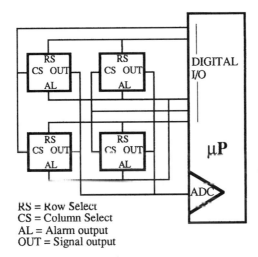

RS = Row Select
CS = Column Select
AL = Alarm output
OUT = Signal output

FIGURE 78.3 Connection of four multiplexed accelerometers.

multiplexed mode. The digital control lines can be driven by a customer-designed logic circuit, a card that plugs into a computer, or the I/O port of a microprocessor.

The addressing capability can also by used during manufacturing of the part. The digital inputs and outputs used during testing and trimming are disabled if the device is not selected, and can therefore be bused together. This greatly simplifies the test hardware if the accelerometers are characterized and trimmed in an array configuration. In case of single-sensor operation, or if multiplexing is not desired, the row select and column select inputs can be left open. Internal pull-up current sources ensure that the accelerometer is selected when these inputs are not connected.

TABLE 78.1 Number of Required Lines in a Measurement System

	Required number of lines	
Number of sensors	Nonmultiplexed	Multiplexed
2	4	6
4	10	8
9	20	10
16	34	12
25	52	14

Electrical Trimming

The optimal trim values for offset, TCO, gain, and TCS are different for each sensor. Often a network of thick or thin film resistors is used to set these coefficients. In that case, the desired resistor values are set by laser trimming after characterization of the untrimmed sensor. This requires a separate trim operation using expensive equipment. Any additional packaging steps done after trimming, such as to seal the substrate in a housing, could change the characteristics of the sensor resulting in sensitivity or offset errors. Furthermore, trimmable resistors and the conductive traces connecting them to the electronics take up space and limit the available packaging options. To avoid these disadvantages the trimming is done internal to the signal-conditioning IC. The trim coefficients for offset, TCO, gain, and TCS are stored in binary registers which are connected to DACs that manipulate the signal. In contrast to some designs that require an additional EEPROM IC which contains the coefficients, the storage registers are on the same chip as the signal-conditioning electronics. The storage registers are made in fuse technology to assure data retention in safety-critical applications.

Before the trim data is permanently programmed into the fused registers, the accelerometer can be operated using data stored in volatile (RAM) registers. This allows for the characterization of the sensor and electronics during manufacturing in order to extract the required coefficients for offset, TCO, gain, and TCS. The actual fuse trimming is handled by circuitry inside the signal-conditioning IC and requires no external equipment. The digital I/O used for characterization and trim consists of a serial input and a serial output line and a clock input for synchronizing the data entry, which uses a 16-bit protocol. All digital I/O lines are available after final packaging. This allows the accelerometer to be trimmed as the last manufacturing step. Because the data transfer is serial rather than parallel, the pin count is not the limiting factor for the package size.

PACKAGE

The package is a leadless chip carrier measuring 0.530 by 0.300 in and is 0.150 in thick. It is manufactured by screening tungsten interconnect traces onto ceramic layers which are then stacked together and fired. The accelerometer die and signal-conditioning IC are mounted into the package cavity and connections are made from the die to the package with gold wire bonds. A gold-plated Kovar lid is then soldered to the package using a Au/Sn preform. This provides a hermetic seal which will withstand the rigorous environmental requirements of the automotive and military industries. The reliability is increased with respect to many other designs because of the reduced number of components. No external components such as capacitors are needed for operation. The stiffness and low mass of the package helps to keep its resonant frequency high. The inputs and outputs needed for operation of the accelerometer and for characterization and trim are brought out to contact pads on the side and on the bottom of the package as shown in Fig. 78.4. Mounting surface 2 is on the opposite side from the metal lid. Because electrical contact can be made on two surfaces and because of the aspect ratio of the package, it is possible to mount the package either flush with or perpendicular to the board. This is particularly useful in air bag applications where the frontal impact crash signal to be detected may be parallel with the board that contains the electronic control unit, while for side impact it is often perpendicular to the board. In many cases, the accelerometer currently needs to be mounted at a 90° with respect to the PC board. This requires addi-

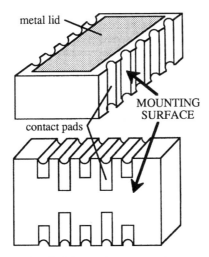

FIGURE 78.4 Accelerometer package.

tional brackets and is not compatible with automated manufacturing. The ceramic package allows the accelerometer to be mounted on the PC board using automatic placement equipment, reducing manufacturing cost and saving space. Another possible application is to make a triaxial accelerometer by mounting two accelerometers perpendicular to the board and one in parallel (see Fig. 78.5). The dimensions of this fully signal conditioned triaxial accelerometer is only 0.73 × 0.53 × 0.30 in. The package will also be used as a building block in products that incorporate more elaborate mounting and connection options.

CUSTOMIZATION

The accelerometer described in this chapter is available in several g-ranges to cover many applications such as ride control, air bag deployment (both frontal and side impact), fusing and arming, vibration monitoring, and general instrumentation. In addition, it will be possible to adapt the device to specific customer needs. Addressing capabilities that are more elaborate than the matrix selection can be added by expanding the logic interface section of the signal-conditioning IC. The signal output can be made digital in the form of a serial word, PWM, or a frequency proportional to the acceleration. Additional filtering can be added to limit the signal bandwidth.

The same signal conditioning is used for pressure transducers. The output bias voltage can be switched between 0.5 V

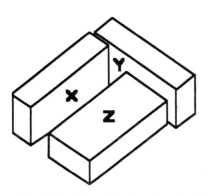

FIGURE 78.5 Orientation of packages in triaxial accelerometer.

for absolute and gage pressure measurements and 2.5 V for differential pressure measurements. Array addressing may even be more useful in pressure applications, where input signals usually vary slowly and multiplexing can be used without losing information.

REFERENCES

1. Henry V. Allen, Stephen C. Terry and Diederik W. de Bruin, "Self-testable Accelerometer Systems," *Proceedings of IEEE Micro Electro Mechanical Systems,* IEEE catalog no. 89TH0249-3, February 1989, pp. 113–115.
2. Henry V. Allen, Stephen C. Terry, and Diederik W. de Bruin, "Accelerometer Systems with Built-in Testing," *Abstracts of Transducers '89: The 5th International Conference on Solid-State Sensors and Actuators,* June 1989, pp. 148–149.

CHAPTER 79
QUARTZ RESONATOR FLUID MONITORS FOR VEHICLE APPLICATIONS*

INTRODUCTION

Thickness shear mode (TSM) quartz resonators operating in a new Lever oscillator circuit are used as monitors for critical automotive fluids. These monitors respond to the density and viscosity of liquids contacting the quartz surface. Sensors have been developed for determining the viscosity characteristics of engine lubricating oil, the state-of-charge of lead-acid storage batteries, and the concentration variations in engine coolant.

Fluids play an extremely important role in today's automotive systems. They serve as fuels, lubricants, coolants, cleaners, hydraulic agents, and charge media. When a fluid depletes, whether it is through consumption or loss, through degradation to the point of nonfunctionality, or through contamination with foreign matter, critical vehicle systems can cease operation or, worse yet, suffer catastrophic damage. It is imperative that automotive fluids be maintained. However, few vehicles have diagnostic sensors, aside from fuel and oil pressure gages, to assist the driver/operator in assessing fluid capacity. One must still visually inspect fluid levels using dipsticks and filler marks and provide fluid replacement at intervals dictated by the odometer or calendar.

This chapter describes a quartz resonator technology that has the capability to sense physical parameters of a contacting medium. The monitor responds to the density and viscosity of a liquid, measures accumulated surface mass, senses phase changes in the contacting material, and can determine viscoelastic properties of the medium. It can also simply respond to the presence or absence of a fluid. Sensors can be configured for in situ use in most environments, and the device response is provided in real time. These characteristics make the quartz resonator monitor useful not only as a vehicle sensor but also as a diagnostic tool for component or system evaluations and qualifications in the laboratory.

Quartz resonators have been partially characterized in three automotive fluid environments: engine lubricating oil, lead-acid battery electrolyte, and engine coolant

---

* This work was performed at Sandia National Laboratories supported by the U.S. Department of Energy under contract No. DE-AS04-94AL85000.

(ethylene glycol). Details of these evaluations are provided in the following sections. The diagnostic ability for the sensor in these fluids is preliminary yet promising. Because of the large viscosities encountered in some of the target liquids, a new oscillator electronic circuit was designed to operate with the quartz resonators. Brief descriptions of this circuit are also provided in the chapter.

QUARTZ RESONATOR SENSORS

The typical thickness shear mode (TSM) quartz resonator is shown in Fig. 79.1. Metal electrodes (few hundred manometers thick) are deposited on each face of a thin wafer of AT-cut quartz. Applying an RF voltage to the electrodes induces a strain in the piezoelectric quartz. Because of the orientation of the crystal axes and the direction of the applied field, shear displacement occurs (see Fig. 79.2).

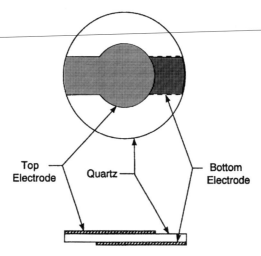

FIGURE 79.1 Top and edge views of a TSM quartz resonator.

The amplitude of the planar displacement in the x direction u_x as a function of time t and position along the quartz thickness y is given by Ref. 1.

$$u_x(y,t) = (Ae^{j\beta qy} + Be^{j\beta qy})e^{j\omega t} \quad (79.1)$$

where A and B are constants, $\omega = 2\pi f$ is the angular excitation frequency, βq is the wavenumber describing the shear wave propagation in the quartz, and $j = (-1)^{½}$ Solving Eq. (79.1) subject to the boundary conditions at the upper and lower surfaces gives sinusoidal standing waves of mode number: $\beta q = N\pi/h, N = 1, 3, 5, \ldots$, where h the quartz thickness. In the unperturbed state (no surface loading), the resonant frequency, f_s, is

$$f_s = \frac{Nv}{2h} = \frac{N}{2h}\sqrt{\frac{\mu_q}{\rho_q}} \quad (79.2)$$

where v is the shear wave propagation (phase) velocity in the quartz, μ_q is the quartz shear stiffness, and ρ_q is the quartz density.

$$\gamma = \frac{1+j}{\delta}$$

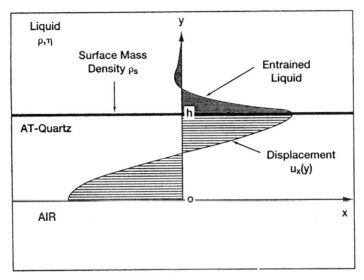

FIGURE 79.2 Cross-sectional view of a smooth TSM resonator with the upper surface contacted by an ideal mass layer and a viscously entrained Newtonian liquid.

Any material in contact with one or both surfaces of the quartz will perturb the resonator. The acoustic wave extends into the contacting material. An ideal mass layer, as shown in Fig. 79.2, acts as a rigid medium and moves synchronously with the oscillating quartz surface. The resonant frequency of the device is reduced from f_s due to added material thickness and a change in density. No loss of resonant magnitude is experienced due to an ideal mass layer.

Liquid in contact with the resonator surface is viscously entrained. The liquid velocity field $v_x(y,t)$ is determined from the one-dimensional Navier-Stokes equation:[2]

$$\eta \frac{\partial^2 v_x}{\partial y^2} = \rho \frac{\partial v_x}{\partial t} \tag{79.3}$$

where ρ and η are the fluid density and viscosity. The solution to Eq. 79.3 for a shear driving force at the solid-liquid interface is[2]

$$V_x(y,t) = v_{xo} e^{-\gamma y} e^{j\omega t} \tag{79.4}$$

where v_{xo} is the particle velocity at the quartz surface and γ is the propagation constant for the damped shear wave radiated into the fluid. For a Newtonian fluid (viscosity independent of shear rate), the propagation constant, found by substituting Eq. 79.4 into Eq. 79.3, is given by

$$\gamma = \frac{1+j}{\delta} \tag{79.5}$$

with the decay length δ defined by[3]

$$\delta \equiv \sqrt{\frac{2\eta}{\omega\rho}} \qquad (79.6)$$

The complex propagation constant creates both oscillatory and dissipative factors in the velocity field of Eq. 79.4, giving rise to a frequency shift and magnitude damping of the acoustic wave.

An equivalent-circuit model (Fig. 79.3) can be used to describe the electrical response of both the unperturbed and loaded TSM resonator.[4] The model consists of a static branch containing C_o, the capacitance created by the electrodes across the insulating quartz, and C_p, the parasitic capacitance of the mounting fixture, in parallel with a motional branch produced by the mechanically vibrating quartz. For the unperturbed resonator, the motional impedance Z_m is given by

$$Z_m \text{ (unperturbed)} = R_1 + j\omega L_1 + \frac{1}{j\omega C_1} \qquad (79.7)$$

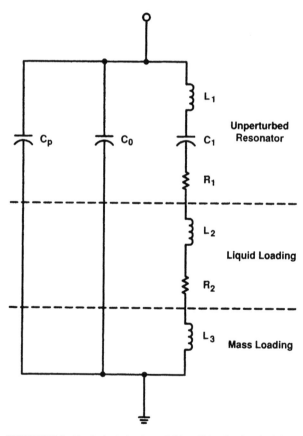

FIGURE 79.3 Equivalent-circuit model describing the electrical characteristics of a TSM resonator having both mass and liquid loading.

with[4]

$$C_1 = \frac{8K^2 C_o}{(N\pi)^2} \tag{79.8}$$

$$R_1 = \frac{\eta_q}{\mu_q C_1} \left(\frac{\omega}{\omega_s}\right)^2 \tag{79.9}$$

and

$$L_1 = \frac{1}{\omega_s^2 C_1} \tag{79.10}$$

where η_q is the quartz viscosity, $\omega_s = 2\pi f_s$ is angular series resonant frequency, and K^2 is the quartz electromechanical coupling factor. The series resonant frequency is defined by

$$f_s \equiv \frac{\omega_s}{2\pi} = \frac{1}{2\pi L_1 C_1} \tag{79.11}$$

The minimum resonant impedance is limited by the motional resistance R_1 and the capacitive reactance in the static branch. Quality factors Q for unperturbed quartz resonators often exceed 10^5.

An ideal surface mass layer contributes only to energy storage in the resonator system. The motional impedance for this term is[4]

$$Z_m(\text{mass}) = j\omega L_3 = j\omega \frac{2n\omega_s L_1 \rho_s}{N\pi \sqrt{\mu_q \rho_q}} \tag{79.12}$$

where n is the number of sides (1 or 2) containing the mass layer and ρ_s is the surface mass density (mass/area).

Liquid loading adds both an energy storage and a power dissipation element to the motional branch. The motional impedance due to the liquid is given by

$$Z_m(\text{liquid}) = R_2 + j\omega L_2 \tag{79.13}$$

with[4]

$$R_2 = \frac{n\omega_s L_1}{N\pi} \sqrt{\frac{2\omega\rho\eta}{\mu_q \rho_q}} \tag{79.14}$$

and

$$L_2 = \frac{n\omega L_1}{N\pi} \sqrt{\frac{2\rho\eta}{\omega\mu_q \rho_q}} \tag{79.15}$$

Note that $R_2 \equiv \omega_s L_2$ for a Newtonian fluid. The liquid shear propagation constant γ can be related to the motional impedance elements as:

$$\gamma = \frac{4K^2 \omega_s C_o \sqrt{\mu_q \rho_q}}{nN\pi\eta} (R_2 + j\omega_s L_2) \tag{79.16}$$

Thus, the equivalent circuit model for liquid-loaded resonators provides a relevant link to the energy storage and power dissipation in the radiated acoustic wave.

The total motional impedance for the mass and liquid-loaded quartz resonator is

$$Z_m = (R_1 + R_2) + j\omega(L_1 + L_2 + L_3) + \frac{1}{j\omega C_1} \tag{79.17}$$

Combining this impedance with the capacitive reactances in the static branch leads to a description of the equivalent admittance (current/applied voltage) for the entire resonator system:

$$Y = j\omega(C_o + C_p) + \frac{1}{Z_m} \tag{79.18}$$

Typical values of the static capacitances are $C_o \approx 4.2$ pF and $C_p \approx 5$ to 10 pF, the latter depending on the mounting fixture.

The motional inductance contributed by a mass layer causes only a shift in resonant frequency. Using Eq. 79.12 in the equivalent circuit model gives

$$\Delta f_s \cong -\frac{L_2 f_s}{2L_1} = -\frac{2 f_s^2 n}{N\sqrt{(\mu_q \rho_q)}} \tag{79.19}$$

which is equivalent to the Sauerbrey equation.[5] Mass sensitivities of less than 1 ng/cm^2 are possible with a 5-MHz resonator in a stable oscillator circuit (frequency fluctuations less than 1 Hz). The mass measurement capability of the resonator gives rise to the often-used name *quartz crystal microbalance* or *QCM*.

Liquid loading produces both a frequency shift and a decrease in the admittance magnitude. Admittances for four fluid-loaded resonators near the fundamental resonance are shown in Fig. 79.4. Peak admittance magnitude for air in contact with the resonator is ~ 100 mmhos, or 30 times the displayed range. The three liquids represent those encountered in automotive systems and cover the spectrum of density-viscosity products: $\rho\eta = 0.01$ to 0.882 at $20°$C. As the value of $\rho\eta$ increases, the peak frequency is shifted by

$$\Delta f_s = -\frac{f_s}{2}\left[1 + \frac{N}{2n}\sqrt{\frac{\pi \mu_q \rho_q}{n f_s \rho \eta}}\right]^{-1} \tag{79.20}$$

while the peak admittance magnitude is reduced:

$$|Y_{max}| = \sqrt{4\pi^2 f_s^2 (C_o + C_p)^2 + \frac{1}{(R_1 + R_2)}} \tag{79.21}$$

where

$$R_1 + R_2 = \frac{(N\pi)^2}{8K^2 C_o}\left[\frac{\eta_q}{\mu_q} + \frac{n}{N}\sqrt{\frac{\rho\eta}{\pi 3 f_s \mu_q \rho_q}}\right] \tag{79.22}$$

When liquid loading is not severe (R_2 is much less than the reactance produced by the static capacitances), then peak admittance magnitudes are approximated by $(R_1 + R_2)^{-1}$.

The density-viscosity can be determined by observing the response of the quartz resonator while contacting the fluid of interest. Contact can be only a few drops of a liquid on one side of the resonator or complete immersion. The acoustic wave decay length for water (Eq. 79.6) is only 0.25 µm at 5 MHz; thus, a thin layer of a liquid is all that is required to make a measurement. An automatic network analyzer (ANA) can be used to scan the frequencies near resonance, a computational fit made to the

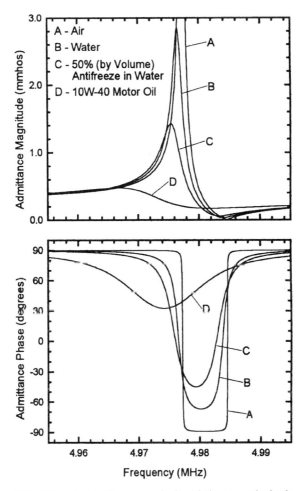

FIGURE 79.4 Admittance magnitude and phase near the fundamental resonance showing the effects of increasing fluid density-viscosity: (*a*) air, $\rho\eta = 2 \times 10^{-7}$, (*b*) water, $\rho\eta = 0.010$; (*c*) 50 percent by volume antifreeze in water, $\rho\eta = 0.044$; and (*d*) 10W-40 motor oil, $\rho\eta = 0.882$. All $\rho\eta$ at 20°C.

equivalent-circuit-model admittance, Eq. (79.18), and the value of $\rho\eta$ extracted from either R_2 or L_2. Equivalently, the frequency shift and the damping of the peak admittance magnitude can be determined and $\rho\eta$ extracted using Eqs. 79.20 or 79.21. An alternate method for measuring $\rho\eta$ requires operating the quartz resonator as a control element in an oscillator circuit. Oscillator frequency approximates f_s when operated in air, and any damping of the oscillator signal near f_s relates to Y_{max} or the motional impedance, $\sim |Y_{max}|^{-1}$. Shifts in the measured parameters while in contact with the fluid can then be related to $\rho\eta$.

A single quartz resonator responds only to the density-viscosity product as described by Eqs. 79.14 and 79.15, and separating these two quantities is impossible.

If each parameter must be known, an independent measurement of either ρ or η is required. This procedure, however, removes the advantages of in situ or real-time determination. It is possible to use two resonators, one with a smooth surface and other with a textured (randomly rough or regularly corrugated) surface to extract both density and viscosity separately.[6,7] The smooth device acts as described previously to measure $\rho\eta$. The textured device traps an additional quantity of the liquid in the crevices so that it moves synchronously with the quartz surface and is weighed. It becomes equivalent to a mass layer, which causes an additional frequency shift according to Eq. 79.19 where $\rho_s = \rho h_o$, h_o being the effective thickness of the trapped fluid. Density is determined from the difference in frequency responses between the textured and smooth devices. Knowing ρ, viscosity can be extracted from $\rho\eta$ of the smooth resonator response.

In this study of automotive fluid monitoring, little use of the dual resonators to measure density and viscosity separately has been attempted. All data presented in this chapter is the result of $\rho\eta$ extraction from a single smooth device.

OSCILLATOR ELECTRONICS

Quartz resonator liquid sensors require the use of specially designed oscillator circuits. It is necessary for the oscillator to accommodate the high resistances due to liquid damping yet provide a predictable output over the wide dynamic range of resonator resistance. Standard oscillator designs, such as the Pierce or Colpitts type, do not function well because of their inherent phase and gain sensitivity to the resonator loss. The Lever oscillator described here uses negative feedback in a differential amplifier configuration to actively and variably divide (lever) the resonator impedance as to maintain the oscillator phase and loop gain.[8] The oscillator servos the resonator impedance to a constant, nearly zero phase over a wide dynamic range of resonator resistances (10 Ω to 5 kΩ). When using a 5-MHz resonator (fundamental mode), this dynamic range will allow single-sided measurements up to $(\rho\eta)^{1/2} \approx 1.6$ g/(cm^2-s$^{1/2}$).

A schematic of the Lever oscillator circuit is shown in Fig. 79.5. A single 5-volt supply is used to power the circuit; source drain is typically 20 mA. There are two signal outputs: a 200 mV$_{p-p}$ RF signal with a frequency that corresponds approximately to the series resonance of the quartz-liquid system f_s; and a dc voltage proportional to the system motional resistance, $R_1 + R_2$.

When operating at f, the resonator impedance is near minimum and is given by

$$Z_{res} = [\,j\omega_s\,(C_0 + C_p) + 1/(R_1 + R_2)]^{-1} \qquad (79.23)$$

If the motional resistance is small compared to the capacitive reactance, then the resonator impedance is essentially real and f is very close to the frequency at which the impedance phase is zero. The Lever oscillator attempts to maintain this condition so that open loop voltage gain, A_v is independent of transistor 0 and is approximated by

$$A_v = \frac{RC}{2h + \dfrac{R_c\,R_m}{R_c + R_f + R_m}} \geq 1 \qquad (79.24)$$

where R_C, R_m, and R_f are the collector resistance, resonator motional resistance ($R_1 + R_2$), and feedback resistance shown in Fig. 79.5, and h is the intrinsic emitter resistance

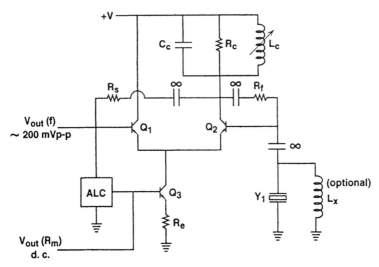

FIGURE 79.5 The Lever oscillator circuit design with automatic level control (ALC).

of the amplifier transistors, Q_1 and Q_2 (at 25°C, $h = 26$ mV/I_e, where I_e is the emitter current). The second term in the denominator of Eq. 79.24 is the lever term since it effectively divides the resonator impedance by the circuit elements R_c and R_f. Wide dynamic range, good sensitivity, and transistor independence is realized when $R_f/R_C \ll \beta$.

Resonator impedance phase is kept small by minimizing the effects of transistor capacitance in Q_2. The parallel L-C tank circuit across R_C makes the collector to ground impedance real at the desired operating frequency. This also prevents the circuit from jumping to a parasitic (capacitance-controlled) frequency when resonator loss becomes large. The feedback impedance Z_f is also forced to be essentially real by selecting R_f to be much less than the collector-to-base capacitive reactance of transistor Q_2. Then, if $R_c \ll Z_f$, the resonator impedance phase follows approximately as the phase of Z_f. And with R_c and Z_f real, the resonator impedance must also be real to satisfy Eq. 79.24. Figure 79.6 shows the measured resonator impedance phase as a function of the magnitude of the resonator impedance for the Lever oscillator operating with a 6-MHz quartz resonator. Above 200 ohms the impedance phase is steady near 6°, which is approximately that determined by the phase of Z_f. The phase shift at low resonator impedances is caused by increased oscillator sensitivity to circuit parameters not included in the approximations used to determine component selection. Fortunately, for small resonator loss, the resonator Q is extremely large and oscillator performance is not dependent on resonator impedance phase.

The automatic level control (ALC) circuit shown in Fig. 79.5 is designed to keep the oscillator gain constant by controlling the current source Q_3. Indirectly, a measure of the control voltage indicates the resonator motional resistance. A plot of ALC output voltage versus resonator resistance is shown in Fig. 79.7. The response is approximately linear with resonator damping up to 1700 ohms. Gain of the ALC can be adjusted to increase the range of resistance measurements. For highly viscous liquids such as lubricants, the resonator motional resistance can exceed 5000 ohms (see Fig. 79.10). The upper resistance limit for oscillator operation is increased dramatically by using the inductor L_x, shown in Fig. 79.5, which effectively tunes out the static capacitance $C_o + C_p$ of the resonator and its fixture.

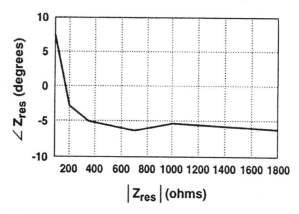

FIGURE 79.6 Measured resonator impedance phase versus the magnitude of resonator impedance for a 6-MHz test device.

LUBRICATING OIL MONITOR

The ability of the quartz resonator monitor to indicate density-viscosity ($\rho\eta$) of fluids makes it a candidate for an in situ, real-time diagnostic for engine lubricating oil. Determination of lubricant viscosity is a standard analytical procedure[9] both in the formulation of products and in the characterization of lubricating quality during use. It is important to protect against excess wear in engines and machines, yet reduce costly maintenance and waste disposal. Several studies note changes in oil viscosity as a function of vehicle use.[10–12] Viscosity shifts result from oil thickening, emulsification, and fuel/water dilution.

Quartz resonator response in contact with lubricating oil is illustrated in Fig. 79.8. The admittance magnitude at three temperatures is plotted versus frequency for the resonator operating in 10W-40 motor oil. In Fig. 79.9, the extracted density-viscosity for temperatures between 0 to 100°C is shown. As observed in both figures, lubricating oil exhibits a particularly strong dependence on temperature. Much of this vari-

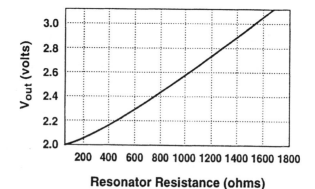

FIGURE 79.7 Oscillator output voltage versus resonator resistance. Gain of the ALC can be adjusted for response to a large range of resonator resistances.

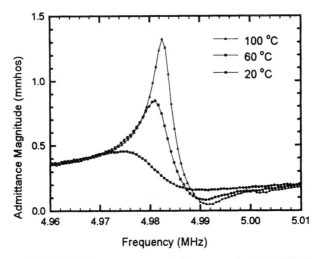

FIGURE 79.8 Quartz resonator admittance magnitude for 10W-40 lubricating oil at three temperatures.

ation is due to the large viscosity increases at the low temperatures. In Fig. 79.8, severe damping of the resonator signal occurs even at room temperature. From Fig. 79.9, $\rho\eta = 0.88$ g^2/cm^4–s at 20°C. A separate density measurement on the 10W-40 oil at 20°C gave 0.88 g/cm^3, indicating $\eta = 100$ cP. This value represents the approximate viscosity upper limit for Lever oscillator operation in conjunction with a resonator having single-sided liquid contact. Thus, high-precision characterization of oil properties is best performed at elevated temperatures where resonator Q and measurement signal-to-noise are large.

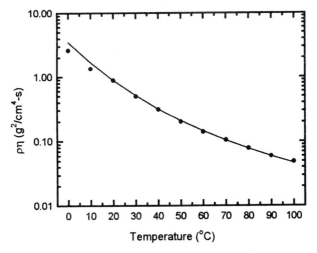

FIGURE 79.9 The liquid density-viscosity product (●) extracted from resonator measurements in 10W-40 motor oil. The solid curve is a theoretical prediction for temperature variations based on ASTM Standard D341-93.

Standard laboratory analysis of lubricating oils extracts the kinematic viscosity v instead of the dynamic viscosity η. The relationship between the two quantities is $v = \eta/\rho$. Quartz resonator response is then related to the kinematic viscosity by $\rho\eta = v\rho^2$. At 20°C, the kinematic viscosity of the previous 10W-40 oil sample is approximately 114 cSt (centistokes or mm²/s). In order to characterize the temperature dependence of the viscosity (often referred to as viscosity index), a second density was measured for the 10W-40 oil at 64°C. Interpolating the $\rho\eta$ data in Fig. 79.9, values for η and v were then determined for this lubricating oil at this second temperature. Variation in the viscosity for petroleum products is described in the ASTM standard handbook[13,14] and is given by the linearized expression

$$\log \log (z) = A - B \cdot \log (T) \tag{79.25}$$

where $z = v + $ (constant), T is the absolute temperature, and A and B are constants. The constant added to v is 0.7 for $v \geq 1.5$ cSt, which is characteristic of lubricants up to ~200°C.[14] Using the two values of kinematic viscosity at two different temperatures, A and B are determined from Eq. 79.25: A = 9.80, B = 3.48.

Temperature-dependent density variations in the 10W-40 oil are assumed due to simple dilation, that is, linear expansion in each direction:

$$\rho = \frac{\rho_o}{[1 + C(T - T_o)]^3} \tag{79.26}$$

where ρ_o is the density at T_o, and C is the coefficient of linear expansion for the liquid. Values for the constants ρ_o, T_o, and C are determined from the two density measurements at 20 and 64°C. Combining Eqs. 79.25 and 79.26 and substituting the appropriate fit parameters produces a theoretical expression for the temperature variations of $\rho\eta$ in the 10W-40 lubricating oil. This curve is plotted as a solid line in Fig. 79.9. Above 20°C, the fit is excellent, having an average error of 1.6 percent. Below 20°C, the measured values show increasing deviation from the curve-fit model as the temperature drops; the relative error is 35 percent at 0°C. This discrepancy is almost entirely due to errors in the resonator determination of viscosity over this range; the viscosity at 0°C is approximately 400 cP.

The response of a single quartz resonator appears adequate for characterizing lubricating oils where the density-viscosity product is a relevant indicator of fluid parameter variations. More precise characterization requires extraction of just the viscosity. As discussed previously, two resonators, one smooth and the other having a textured surface in contact with the liquid, can separate density and viscosity. Such a sensor system has yet to be tried with lubricants.

Many lubricating oils exhibit non-Newtonian characteristics, especially under high shear conditions or when operating at extreme temperatures. Initial developments of quartz resonator response assumed that liquids were strictly Newtonian. Then, energy-storage and power-dissipative components of the motional impedance are simple functions of $\rho\eta$. Resonator theory can be extended to include shear rate dependence in the viscosity:[15]

$$\eta = \frac{\eta_o}{1 + j\omega\tau} \tag{79.27}$$

where η_o is the low-frequency viscosity and τ is a shear relaxation time. For a Maxwell fluid, the relaxation time depends on the viscosity, with $\tau = \eta_o/\mu$ where μ is the high-frequency shear modulus for the liquid. For $\omega t \ll 1$, the liquid is purely Newtonian in behavior. But at high frequencies or for large viscosities, the non-Newtonian character is manifested in a separation of the real and imaginary parts of the liquid-induced motional impedance: $R_2 \neq \omega_s L_2$. This is in contrast to values extracted from Eqs. 79.14 and 79.15.

The non-Newtonian aspects of a liquid can be observed by plotting R_2 and $\omega_s L_2$ for increasing viscosity, or, equivalently, decreasing temperature. In Fig. 79.10, these two parameters are shown for two lubricants: a 10W-40 motor oil and a 50W racing oil. Both sets of curves show some separation in the impedance parameters. The 10W-40 oil is Newtonian above 40°C, exhibiting some non-Newtonian behavior at and below room temperature. The 50W oil shows Maxwell fluid characteristics over the entire measurement range. Only well above 140°C is Newtonian response expected. Based on this limited data set, it appears that quartz resonator monitors have the ability to determine important non-Newtonian behavior in lubricants.

BATTERY STATE-OF-CHARGE MONITOR

The active materials in a lead-acid battery are lead oxide (PbO_2) on the positive plates, porous lead (Pb) on the negative plates, and dilute sulfuric acid (H_2SO_4) as the electrolyte.[16] During the discharging and charging of the battery, the chemical reaction in each cell is described by

$$PbO_2 \text{ (Anode)} + 2\ H_2SO_4 \text{ (Electrolyte)} + Pb \text{ (Cathode)}$$

discharge ↑↓ charge

$$PbSO_4 \text{ (Anode)} + 2\ H_2O \text{ (Electrolyte)} + PbSO_4 \text{ (Cathode)}.$$

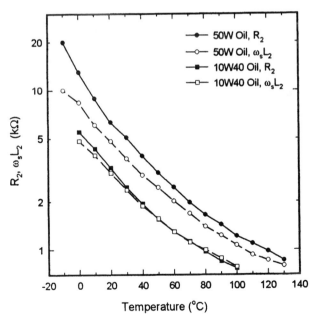

FIGURE 79.10 The variation in motional impedance parameters versus temperature for two lubricating oils: 10W-40 and 50W.

The concentration of sulfuric acid in the electrolyte changes significantly with the battery state-of-charge. Approximate concentration ranges for fully charged and discharged states are shown along the abscissa in Fig. 79.11.[16,17]

Both the density and viscosity of the battery electrolyte vary with the H_2SO_4 concentration. Literature values of these two parameters are plotted (solid lines) in Fig. 79.11.[18] Traditionally, electrolyte density has been used to indicate state-of-charge in a lead-acid battery, although viscosity exhibits a much stronger dependence on acid concentration (5.5 times the sensitivity at 40 percent H_2SO_4). The quartz resonator response, proportional to $(\rho\eta)^{1/2}$, is also plotted. Figure 79.11 shows that density-viscosity values measured with the quartz monitor (open circles) agree with literature values (solid line). Because of the conductivity of the sulfuric acid, only one surface of the quartz resonator can be exposed to the liquid to avoid shorting between electrodes.

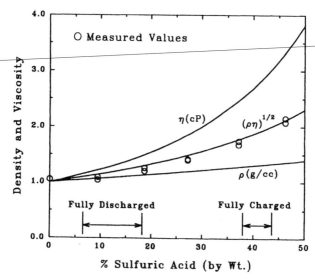

FIGURE 79.11 Parameter variations for the electrolyte in a lead-acid cell as a function of the sulfuric acid concentration.

In a laboratory demonstration, a small 12-V lead-acid battery was discharged and recharged several times in succession. The electrolyte in one cell was continuously pumped across the surface of the quartz resonator. The Lever oscillator circuit was used in conjunction with the resonator and provided the electrical signals for monitoring the frequency shift and the damping magnitude. Periodically, the electrolyte specific gravity was measured using a float hydrometer. Since density and viscosity are temperature-dependent, a separate measurement of temperature was also made.

Four parameters extracted from one discharge cycle in the test series are plotted in Fig. 79.12. The quartz oscillator voltage tracks quite precisely both the electrolyte specific gravity and the estimated charge (battery discharge current × time). Estimating the charge alone could be used to indicate state-of-charge, except this parameter can be highly charge/discharge-rate-dependent and initial charge conditions must be known *a priori*. The battery voltage is also plotted in Fig. 79.12. Battery voltage is not a good measure of the state-of-charge, showing little deviation for most of the discharge cycle and then falling catastrophically as the charge depletes. The cor-

QUARTZ RESONATOR FLUID MONITORS FOR VEHICLE APPLICATIONS 79.15

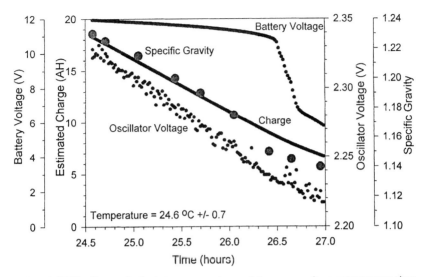

FIGURE 79.12 Changes in the battery parameters and the sensor voltage output versus time for one deep-discharge cycle in a test series.

relation among the estimated charge capacity of the battery, the electrolyte density, and the measured quartz monitor voltage is quite high. Coefficients for pairings of these parameters are given in Table 79.1. These high correlation coefficients are an indication this monitor has the potential for accurately measuring state-of-charge.

TABLE 79.1 Correlation Coefficients Among Three of the Battery and Sensor Parameters Plotted in Fig. 79.12

Correlating parameters	Coefficient
Specific gravity ↔ Estimated charge	0.982
Oscillator voltage ↔ Estimated charge	0.984
Specific gravity ↔ Oscillator voltage	0.993

The shift in resonant frequency of the monitor proved to be an unreliable measurement for determining electrolyte density-viscosity and, consequently, battery state-of-charge. Small quantities of lead oxides and salts from the electrolyte intermittently settled onto the resonator surface. This produced a frequency shift due to the accumulated surface mass. Oscillator voltage, which depends only on the viscous damping contributed by the liquid, does not respond to these changes in surface mass. Thus, a simple oscillator damping measurement is all that is required for this monitor.

The laboratory tests demonstrated that a TSM quartz resonator can be used as a real-time, in situ monitor of the state-of-charge of lead-acid batteries. Quartz resonators are small in size, compatible with the corrosive electrolyte environment, and can be easily configured to fit into each of the cells. This type of monitor is more precise than sampling hydrometers since it responds to changes in both the density and viscosity of the sulfuric acid electrolyte. It is proposed that quartz resonator monitors can provide real-time metering of state-of-charge in present vehicle storage batteries and serve as the sensor for a fuel gage in electric vehicle power sources.

COOLANT CAPACITY MONITOR

Typical vehicle cooling systems contain solutions of a primary agent (mostly ethylene glycol) and water. The capacity of this coolant to perform properly yet resist freezing under nonoperating conditions depends on the component concentrations in solution. Figure 79.13 shows several properties of ethylene glycol at varying concentrations in water.[18] Recommended antifreeze/coolant mixtures for most vehicles is 50 percent by volume (52 percent by weight) in water, giving freezing-point protection to −40°C.

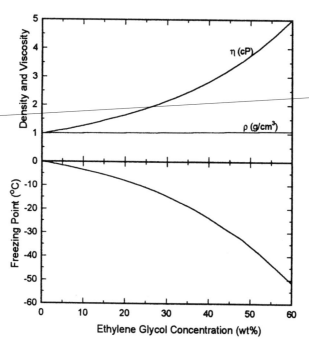

FIGURE 79.13 Parameter variations for aqueous solutions of ethylene glycol as a function of concentration. Literature values are density, viscosity, and freezing point at 20°C.

Present coolant-monitoring systems check capacity or freezing-point protection by measuring the solution density. Like in the battery system discussions in the previous section, density monitoring requires drawing a sample using a hydrometer (floating ball or float level). Density measurements can be performed quite accurately and provide a good indicator of capacity. As observed in Fig. 79.13, however, the viscosity of ethylene glycol solutions is a much stronger function of concentration than is the density, being approximately 100 times more sensitive to changes near 50 percent concentration. Since the quartz resonator monitor responds directly to $(\rho\eta)^{1/2}$, it can take advantage of this increased sensitivity. Ability to operate in situ and to provide information in real time are added advantages.

Figure 79.14 displays plots of the measured and literature values for ρ and $(\rho\eta)^{1/2}$. The quartz resonator response (open circles) determined for a commercial antifreeze/coolant are in good agreement with the literature values for pure ethylene glycol. Some discrepancies in the curves are expected since commercial

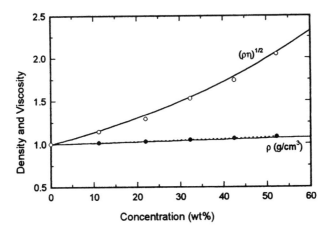

FIGURE 79.14 Quartz resonator measurements (○) of $(\rho\eta)^{1/2}$ for commercial antifreeze solutions at 20°C. Measured densities (●) are determined separately. Solid lines are literature values from Ref. 18.

antifreeze contains small amounts of chemicals other than ethylene glycol. In fact, the measured densities for the antifreeze mixtures (closed circles in Fig. 79.14) were consistently 1 to 2 percent higher than literature values for pure ethylene glycol in water. It is obvious that coolant monitors must account for a range of responses since proprietary additives in commercial products can affect liquid properties.

A 5-MHz quartz resonator was used with the Lever oscillator to monitor characteristics of typical antifreeze mixtures. Concentration was varied from 0 to 50 percent by volume in water; temperature was varied between 0 and 100°C, the exact range depending on the solution concentration. In Fig. 79.15, the oscillator frequency and voltage shifts are plotted versus concentration for three distinct temperatures for each mixture. Maximum shifts occur in both parameters at 10°C (open circles) since densities and viscosities are greatest at that temperature. For the 50 percent solution at 10°C, the liquid viscosity is large enough (estimated to be 7.5 cP) that oscillation at the fundamental frequency of the quartz resonator could not be maintained. A reconfiguration of the oscillator circuit components is required for continued operation at this higher viscosity (or higher motional resistance) range.

It is possible to measure either the voltage shift or the frequency shift from the monitor in vehicle antifreeze/coolant to provide an indication of the system capacity. As in the battery electrolyte monitor system, voltage response is expected to be more accurate since it is directly proportional to $(\rho\eta)^{1/2}$, while frequency response has an added contribution due to mass deposition. Rust and corrosion particulates that are likely to form in vehicle cooling systems could settle on the resonator surface and skew density-viscosity measurements made with frequency shift only.

The quartz resonator monitor also has the ability to determine phase changes in materials contacting the surface. Changes in the properties of a liquid as it transforms to the solid phase are captured by dramatic shifts in both resonator operating frequency and resonance damping. An icing monitor is constructed by spreading a thin layer of water (or modified water solution) over the quartz surface. Immediate response to solidification is evident. By strategically placing a quartz monitor in a vehicle cooling system, it would be capable of predicting fluid freezing prior to bulk ice formation that could seriously damage an engine.

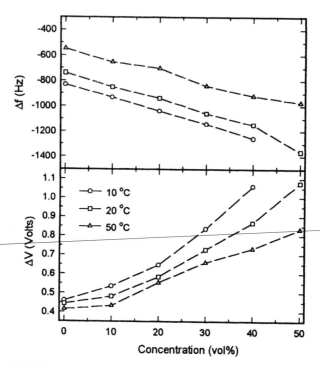

FIGURE 79.15 Oscillator circuit frequency and voltage shifts measured at three temperatures (10, 20, and 50°C) for varying volume concentrations of antifreeze in water.

CONCLUSION

The capability of TSM quartz resonators has been investigated as a diagnostic monitor for three fluids used in vehicle systems: engine lubricating oil, battery electrolyte, and engine coolant. In each case, the sensor responds to changes in the liquid density-viscosity product due to temperature variation, concentration shifts, or a combination of these. Mass deposition on the quartz surface and phase changes in the contacting liquid also produce shifts in the sensor response, although these parameters have not yet been quantified as part of this study. The quartz resonator monitor can track viscosity changes in lubricants and potentially can separate non-Newtonian from Newtonian fluid characteristics. Real-time monitoring of lead-acid battery state-of-charge was demonstrated during charge/discharge cycling of an actual battery. Predictable resonator responses to concentration and temperature changes in antifreeze solutions verified its capability as an in situ coolant monitor.

The quartz resonator technology can be extended to monitor other vehicle fluids—transmission and brake fluids, for example—as well as to explore new properties and characteristics of the three fluids presently under investigation. The technology can also be utilized in other operating modes (i.e., thin-film-coated resonators) to explore liquid or gas chemistry. A wide range of applications in the automotive industry appear possible.

REFERENCES

1. C. E. Reed, K. K. Kanazawa, and J. H. Kaufman, "Physical Description of a Viscoelastically Loaded AT-Cut Quartz Resonator," *J. AppL Phys*, 1990, **68**, pp. 1993–2001.
2. F. M. White, "Fluid Oscillation above an Infinite Plate," Sec. 3–5.1 in *Viscous Fluid Flow*, McGraw-Hill, New York, 1991.
3. K. K. Kanazawa and J. G. Gordon, "Frequency of a Quartz Microbalance in Contact with Liquid," *Anal. Chem.*, 1988, **60**, pp. 1770–1771.
4. S. J. Martin, V. E. Granstaff, and G. C. Frye, "Characterization of a Quartz Crystal Microbalance with Simultaneous Mass and Liquid Loading," *Anal. Chem.*, 1991, **63**, pp. 2272–2281.
5. G. Sauerbrey, "Verwendung von Schwingquarzen zur Wigung dünner Schichten and zur Mikrowdgung (Application of Vibrating Quartz to the Weighing of Thin Films and Use as a Microbalance)," *Z Phys.*, 1959, **155**, pp. 206–222.
6. S. J. Martin, G. C. Frye, A. J. Ricco, and S. D. Senturia, "Effect of Surface Roughness on the Response of Thickness-Shear Mode Resonators in Liquids," *Anal. Chem.*, 1993, **65**, pp. 2910–2922.
7. S. J. Martin, G. C. Frye, R. W. Cemosek, and S. D. Senturia, "Microtextured Resonators for Measuring Liquid Properties," *Proc. Sensors and Actuators Workshop*, Hilton Head Island, South Carolina, 1994.
8. K. O. Wessendorf, "The Lever Oscillator for Use in High Resistance Resonator Applications," *Proc. 1993 IEEE Intl. Freq. Control Symp.*, 1993, pp. 711–717.
9. American Society for Testing and Materials (ASTM), "Kinematic Viscosity of Transparent and Opaque Liquids (and the Calculation of Dynamic Viscosity)," D445-83, *Annual Book of ASTM Standards*, Philadelphia, 1994.
10. N. E. Gallopoulos, "Engine Oil Thickening in High-Speed Passenger Car Service," *SAE Transactions*, 1970, **79** (SAE # 700506).

CHAPTER 80
OVERVIEW OF THE EMERGING CONTROL AND COMMUNICATION ALGORITHMS SUITABLE FOR EMBEDDING INTO SMART SENSORS

INTRODUCTION

Alternative approaches to data processing (e.g., fuzzy set theory, rough set theory, and neural networks, known collectively as *soft computing*) are easy to use and become increasingly popular in engineering applications. In this chapter, we review how these emerging computation schemes may lead to the development of highly flexible sensor components capable of plug-and-play operation. Smart sensors that use these computation schemes integrate signal processing with networking into a simple and compact system architecture. These systems can be easily programmed, and their use results in a substantial reduction of development cost. Using an inverted pendulum problem, the chapter describes a plug-and-play software development cycle that can be applied to a number of practical applications.

The emerging new generation of intelligent sensors for automotive, domestic, and medical applications integrates sophisticated hardware technologies with the flexibility of modern digital communication networks. These modern control applications require highly flexible real-time computation and control algorithms. To develop such systems, progress in hardware must be matched by substantial improvements in adaptive real-time algorithms and in effective programming techniques.

In a typical control system, for example, an automobile, as the driver tries to reach a destination in the shortest time, the decision to accelerate is based on information provided by a clock and a speedometer. In this control system, the driver is a controller, the clock and speedometer are the sensors providing input parameters, and the driver's right foot is an actuator executing control decisions. This system exemplifies a well-known feedback configuration pioneered by James Watts's steam engine controller. Technical progress allowed for a steady replacement of human operators by mechanisms that determine the inputs to the process from process parameters acquired through sensors, Figure 80.1.

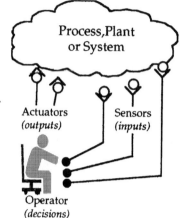

FIGURE 80.1 Feedback control loop links operator, sensors, and actuators in a simple control system.

In a feedback-based system, analog sensors are connected to a processing element that performs simple processing functions using inexpensive analog components. The main advantages of feedback-based controllers are design simplicity and low cost. As a result, they are widely used to control a number of commercial systems. Their main disadvantages are lack of flexibility and inability to handle nonlinear and large-scale systems.

GENERIC MODEL OF A CONTROL SYSTEM

From an operator's point of view, the controlled plant is characterized by the following:

- *Control goals space,* defined by the variables characterizing control goals
- *Observation space,* defined by measurable and observable variables
- *Control space,* characterized by measurable variables and controllable variables

Observation space coordinates are defined as those parameters of the plant which the operator observes and evaluates because the operator is convinced that they are directly related to the control goal.

Determination and analysis of the current situation is the starting point for the operator's evaluation, whether the control goal has been reached or not. This is implied by the fact that the control goal is expressed by the operator with an appropriate configuration of measurable and observable variables.

The coordinates of the control goals space are the notions used by the operator to describe the degree to which the control goal is attained. The values of the notions are derived from the configuration of the coordinates of the current situation. Hence a close correspondence between the observation space and control goals space exists.

The current state of the controlled plant in the observation space may be mapped onto the goal space. At the same time, the control goals from the goal space may be mapped onto the observation space by partitioning or covering it. Regions of the observation space determined by the partition or covering are called *characteristic states* of the plant.

The values of the control space coordinates are determined by the operator in the process of decision-making with the control goal in mind. The operator determines certain typical configurations of the values of the measurable and controllable variables in the control space. For technologically imposed control conditions, this results from the commonsense rules or individual preferences of the operator.

Appropriate sets of measurable and controllable variables are called *characteristic controls.*

The operator's inference model consists of the following:

- Decomposition of the space of observation into areas called *characteristic states* of the plant

- Decomposition of the space of control into areas called *characteristic controls*
- Assignment of a proper characteristic control to every characteristic state

The previously defined model designates the phases of the operator's decision process:

- Evaluate the situation within the space of observations
- Assign the situation to the proper characteristic state of the plant
- Select and execute the proper characteristic control within the space of control

In this chapter, we concentrate on how the previously defined operator's inference model can be used as a foundation of smart sensors.

COMPUTERS AND COMMUNICATION IN CONTROL

The introduction of computers increased the complexity of algorithms that can be used to control processes. Furthermore, as processors executing algorithms become a part of sensors and are placed close to the controlled processes, a notion of a truly distributed control system takes on a new meaning. This migration of intelligence toward sensors lead to new opportunities in the design of control systems. At the same time, the proliferation of new applications has created a demand for technologies to lower development costs. In this section, we describe the structural evolution of control systems and the changing role of sensors.

Hub-Based Control Configuration

Initially, due to the high cost of electronic components, computers served as hubs of control systems. In these systems, Fig. 80.2, sensors and actuators are directly connected to the central computer. Analog sensors are linked to computers using A/D

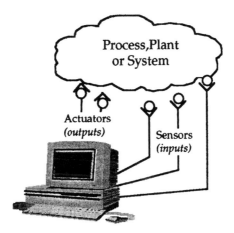

FIGURE 80.2 A centrally located computer eliminates the need for a human operator.

converters that are often located in modules placed at or near the computer chassis. Since only a small number of devices can be connected directly to the backplane of the centrally located computer, designers introduced special I/O subsystems that were combined with multilayered interrupt structures. In these I/O systems, the devices are often hierarchically clustered to ease signal multiplexing.

The development in hardware is matched by the development of special real-time operating systems needed to support the ever increasing complexity of control systems demanded by users.

Bus-Based Control Configuration

Users quickly discovered the low reliability and high weight of thousands of point-to-point links used in a hub-based control system. To avoid these problems, individual links were replaced by a shared communication infrastructure, a network, Fig. 80.3.

The introduction of networks led to the development of networking standards.[1] These standards address two issues:

- *Interchangeability* between the same class of components even if produced by different vendors

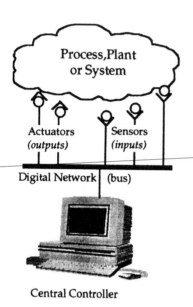

FIGURE 80.3 Networking simplifies the wiring complexity of systems.

- *Interoperability* between subsystems (hardware and software) from different vendors

Communication standards allow devices to talk with each other. To collaborate in a distributed environment, sensors must be able to emulate each other. Therefore, from the information point of view, a smart network component must be capable of self-referencing, that is, it must be able to describe itself when interrogated by the other components. Self-referencing is a part of the protocol establishment process and part of determining a sensor's role in the system. This becomes especially critical when sensors become objects that are autonomous and plug-and-play ready.

Avionics for modern airplanes and electronics for automobiles are the most successful applications of networks in control. They proved that modern low-cost microprocessors are able to provide the necessary processing power to handle signal processing, networks, and computation of decisions.

One of the most popular network topologies is the a bus. In bus-based systems, data is broadcast over a shared data link. Although more sophisticated network topologies are theoretically more powerful, they are less frequently used since they require more processing power to handle the data traffic.

Distributed Control Configuration

As low-cost microprocessors are increasingly placed in the same package with sensors connected by a network, a central computer is no longer needed to execute the control algorithm, Fig. 80.4. In a distributed system, since sensors are equipped with pro-

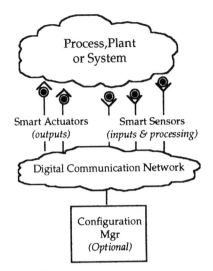

FIGURE 80.4 Distributed control capabilities provided by smart sensors eliminate the need for a central controller.

cessors, control algorithms can be executed in each sensor and the system does not require a single central machine. In such systems, sensors are the autonomous objects capable of processing and adapting to the environment.

Aided by microprocessors, smart sensors are capable of performing a wide range of computation tasks required by digital communication, digital signal processing, and derivation of control rules. One can imagine a truly distributed system in which a sensor plugged into the network determines its role in the system through interaction with other components. Such systems are not only geographically distributed but are also capable of benefiting the most from plug-and-play capabilities offered by the components.

In distributed systems, one of the biggest challenges is to develop a knowledge representation that allows for highly flexible adaptation to changes in the system. Self-referencing, described in the previous section, is a critical part of such systems. In the section entitled "Modern Computation Techniques for Smart Sensors," we discuss two knowledge representations that support learning from examples and provide a framework to develop control systems with highly adaptable plug-and-play capabilities.

Smart Sensor Model

In this section, we describe a simple model of a smart sensor that supports interchangeability and interoperability. The core of the sensor's intelligence is a microprocessor. To support these functions, a sensor contains a read-only memory (ROM) with a sensor model, that is, a complete physical characterization of the sensor, Fig. 80.5. The information contained in the sensor model enables a microprocessor to perform a number of functions, such as calibration, compensation of nonlinearities, and conversion from analog to digital output with automatic scaling within a prescribed range of output values. To assure some level of back-compatibility, the sensor contains an optional analog output.

Using built-in information and processing capabilities, an intelligent sensor can emulate a prescribed class of standard devices, that is, it is interchangeable. To interoperate with other sensors over a network, it must use the same protocols as other sensors in the network.

PLUG-AND-PLAY COMMUNICATION REQUIREMENTS

Control system architectures described in a previous section share a great degree of commonality in terms of information flow. In all these examples, to control a process,

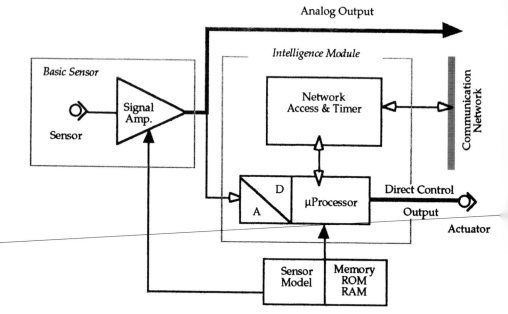

FIGURE 80.5 Smart sensor integrates networking, processing, and interfaces with the external world into a single functional unit.

algorithms reconstruct a global state of the system. In this approach, sensors contribute their measurements to a shared memory. This observation can be a source of functional requirements, linking the complexity of the controlled system with the performance parameters of the controller. In this section, we describe an information flow control and show how it impacts system performance. Let us consider information flow within a control system containing six sensors, s = {a, b, c, d, e, f}. Each sensor collects measurements of five state variables and these measurements are denoted as S_{ij}. Thus, the system state Ω is a sum of all the states of sensors and contains 30 state variables. Since the sensors are interconnected (Fig. 80.6), a shared system state Ω can be reconstructed in each sensor.

In a hub- or network-based system, the system state is available at the central host. In a system with a distributed architecture, the same state must be reconstructed by making each sensor's state visible to other sensors. This observation puts a simple condition on the required speed of network as

sampling rate × number of points × number of measurements per point
= aggregated bandwidth of the system

The total aggregated bandwidth of a distributed system grows linearly with the number of points. This requirement can be reduced by clustering sensors and eliminating the waste associated with the transmission of measurements that will not be used. This analysis can be performed early in the system design specification stage.

CONTROL AND COMMUNICATION ALGORITHMS SUITABLE FOR EMBEDDING

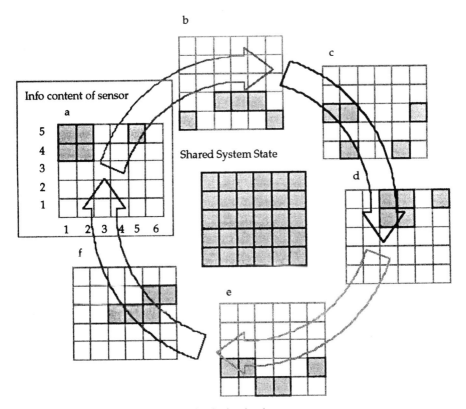

FIGURE 80.6 Communication allows for sharing data between sensors.

MODERN COMPUTATION TECHNIQUES FOR SMART SENSORS

We give the basic concepts associated with fuzzy and rough set theories in this section. These techniques capture knowledge about the control of an object and offer computationally effective approaches to affect the control. Detailed descriptions of those theories can be found in Refs. 2 through 5.

Fuzzy Representation

Classical set theory deals with crisp sets, that is, an item either is a part of a set or not. Fuzzy set theory is a generalization of classical set theory. It allows partial set membership, which may be specified by the *membership function*.

Fuzzy reasoning systems use linguistic variables in place of numerical variables. A linguistic variable assumes words or sentences as its values. The meaning of a linguistic variable is characterized by an associated membership function.

Fuzzy logic is an extension of classical logic. As there is a correspondence between classical set theory and classical logic, there is a correspondence between fuzzy set theory and fuzzy logic. Fuzzy logic allows partial truth and partial falseness, just as fuzzy set theory allows partial set membership. In the broad sense, *fuzzy logic* is almost synonymous with *fuzzy set theory*.[6]

Fuzzy inferences are calculated using fuzzy relationships R which express fuzzy prepositions:

$$Y = X \cdot O \, R_{xy}$$

where X = fuzzy set relating to x
  Y = fuzzy set relating to y
  R_{xy} = fuzzy relation from x to y
  O = fuzzy composition operation

In practice, rules are used such as IF x is A_1 THEN y is B_l or IF x_1 is A_{1k} and ... x_n is A_{nk} THEN y_1 is B_{lk} and y_m is B_{mk}

Fuzzy logic may serve as a basis for the design of fuzzy logic controllers. The configuration of such a controller comprises four major components (see Fig. 80.7):

1. Fuzzifier
2. Inference engine
3. Knowledge base
4. Defuzzifier

The *fuzzifier* and *defuzzifier* perform input/output conversions between crisp real-world signals and fuzzy linguistic values used by the inference engine.

The *knowledge base* includes a rule base modeling the response of the controller and a membership function base for each linguistic variable used in the rules.

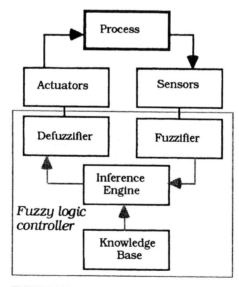

FIGURE 80.7 Block diagram of a fuzzy logic control system (FLC).

The *inference engine* transforms the fuzzified inputs into outputs using the rules and membership functions from the knowledge base. The inputs and truth values serve as conditions for the rules making up the rule base.

Figure 80.8 presents a simple knowledge base for a fuzzy logic controller with two inputs, one output and a triangular membership function typical of such controllers. This knowledge base is used for the inverted pendulum problem discussed in a later section. This scheme of knowledge has been derived from a nondeterministic decision table containing 49 decision rules.

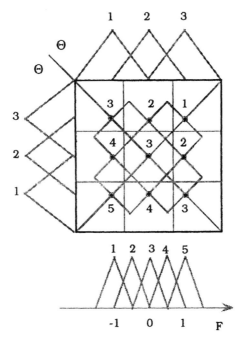

FIGURE 80.8 A scheme of a knowledge base for a fuzzy controller.

Rough Representation

A plant operator's inference model defines sets of conditions which should be satisfied before initiating a set of actions. This imposes the use of a model which checks what conditions hold and denotes what actions are attached to them. Such an interpretation of a plant operator's knowledge is typical of a large class of information systems.

A natural way to represent the plant operator's knowledge is to construct a set of *conditional productions,* each of them having the form IF {set of conditions} THEN {set of actions}.

Assuming that a conditional production is a basic knowledge entity, the problem of plant operator's knowledge acquisition is reduced to the selection of such productions.

The *observation space* is explicitly determined by a finite set of coordinates that we call *condition attributes* and denote $c_1, c_2 \ldots c_n$.

The *control space* is explicitly determined by a finite set of coordinates that we call *decision attributes* and denote $d_1, d_2 \ldots d_n$.

We attach to every condition attribute $c_i, i = 1, 2 \ldots n$, a finite set of its values Vc_i and we attach to every decision attribute $d_j, j = 1, 2, \ldots k$, its domain Vd_j. The operator's decisions within a finite time horizon may now be presented in the form of an observation table (a record of operator's decisions/actions) containing a finite set of conditional productions.

During knowledge acquisition, it is important to conduct the following analysis of the data contained in the observation table:

- Selection of all distinct conditional productions
- Checking whether they are noncontradictory productions (i.e., having the same conditions but different decisions)
- Removal of nondeterminism from the observation table when contradictory productions exist

In the rough set approach, the knowledge base assumes the form of a decision table,[7] shown in Fig. 80.9.

	condition attributes		decision attributes	
rule 1	...		...	
rule 2	...		...	
rule n	...		...	

FIGURE 80.9 Rough set decision table.

The synthesis of a rough controller consists of implementing decision tables and devising an inference engine conditional production processing. The structure of a rough controller is shown in Fig. 80.10. It consists of the following elements:

- Rough analog-to-digital converter (rough classifier)
- Rough inference engine (decision table evaluation engine)
- Decision table–oriented knowledge base
- Rough digital-to-analog converter

Two basic methods can be used to create the knowledge base of a rough controller. The rules can be derived from

- A precise mathematical model of a plant (if such a model exists)
- Analyzing the actions of a human operator controlling a plant

Both methods have advantages and disadvantages. The first one can yield precise rules that are optimal in some respect (time, energy, etc.). Besides, the knowledge base can be guaranteed to be complete.

The second approach has one basic advantage over the first: there is no need for the model. It is a well-known fact that in many practical cases it is extremely difficult to create a quantitative model of the controlled plant. The standard AI approach in such cases is to interview a domain expert in order to create a rule-oriented knowledge base containing an imprecise model (the fuzzy approach) of

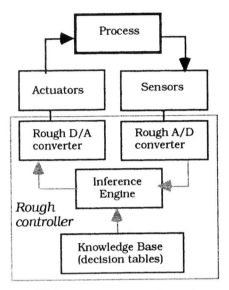

FIGURE 80.10 Block diagram of a rough control system.

the system under control. In another approach, an actual record of control (observation table) performed by the human operator is used. Actions of the operator (decisions) along with the state of the plant that prompted the operator to perform those actions (conditions) are analyzed and used to train an automatic control system. This method is referred to as *learning from example* and is the basis of rough set and neural networks.

When the knowledge base is derived from the observation table, there arises a problem of *completeness* and consistency. *Completeness* is a property of the knowledge base ensuring that every situation in the controlled plant has its counterpart in the condition part of the productions. *Consistency* means that the decision table is deterministic.[7] The rough controller should have an adaptation mechanism ensuring completeness and consistency.

Sample Application

An inverted pendulum problem poses serious problems for qualitative modeling methods, so it is a good benchmark to test their performance. For this reason, it is used in this work to illustrate the application of fuzzy and rough set theories to control and highlight the differences between the two approaches.

The problem consists of balancing a pole hinged to a vehicle by pushing the vehicle to the left or right (see Fig. 80.11). The inverted pendulum system is modeled by the differential equations, Eqs. 80.1 and 80.2.

$$\Theta'' = \{[g \sin \Theta + \cos \Theta [(-F - m1 \, \Theta'^2 \sin \Theta + \mu c \sin \chi)/(M+m)]\}/1(4/3 - m \cos^2 \Theta/M + m)\} - (\mu \rho \, \Theta'/m1) \quad (80.1)$$

$$\chi'' = [F - m1 \, (\Theta'^2 \sin Q - \Theta'' \cos \Theta) - \mu_c \sin \chi]/(M+m) \quad (80.2)$$

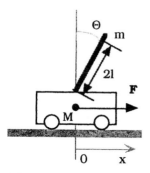

where Θ = deflection angle
x = vehicle position
g = acceleration due to gravity
M = mass of the vehicle
m = mass of the pole
l = half-length of the pole
μ_c = coefficient of friction of vehicle on track
μ_p = coefficient of friction of pole on vehicle
F = force on the vehicle

FIGURE 80.11 A dynamic stabilization of an inverted pendulum is based on measurements of x and Θ to control a force F applied to a vehicle.

Inverted Pendulum Controlled Using Fuzzy Logic. A nine-rule knowledge base described in Ref. 8 was used for the design of the controller. It was created using an ordinary fuzzy partition of input space where each coordinate of the input space was evenly divided into three parts (Fig. 80.8).

The behavior of a dynamic system can be examined using a phase portrait, showing two state variables of a system as a function of time. In the case of the inverted pendulum system, Θ and $\dot{\Theta}$ are the state variables. Figure 80.12 presents a phase portrait for the nine-rule knowledge base fuzzy controller.

Inverted Pendulum Controlled Using Rough Set. The rough set–based approach led to the development of the rules for automatic control of the inverted pendulum system. Phase portrait of the system is shown in Fig. 80.13.

Simulation results show that the controller does achieve its goal. The system eventually approaches state (0,0), though, due to the roughness of the rules, the equilibrium point cannot be reached asymptotically.

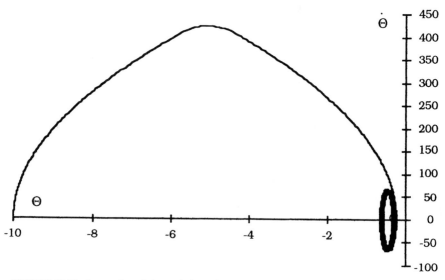

FIGURE 80.12 Inverted pendulum solution using fuzzy logic.

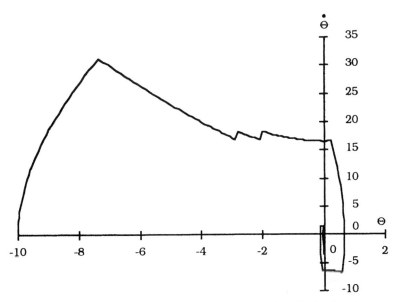

FIGURE 80.13 Phase portrait of the rough control system (initially the pendulum was slightly deflected from the vertical).

Plug-and-Play Approach to Software Development

The learning from example approach and rough set theory were used in our work to obtain a rule-oriented knowledge base for the inverted pendulum controller, as the following describes.

First, a computer simulator of the inverted pendulum system was built using Eqs. 80.1 and 80.2. The simulator has the following functions:

- It displays the vehicle and the pole attached to it in a graphical window.
- It reacts to the operator pressing the arrow keys (left/right) on the keyboard by applying force F (positive or negative) on the vehicle.
- It recomputes $\Theta, \dot{\Theta}, x$, and $\dot{x}$ every time interval (typically set to 0.01 s) using the Runge-Kutta method and moves the vehicle accordingly.
- It stores tuples $(\Theta, \dot{\Theta}, F)$ in a file for later processing.

Pressing an arrow key is equivalent to applying force pushing the vehicle to the left or right for a single time interval. After updating $(\Theta, \dot{\Theta}, x, \dot{x})$, $\dot{x}$ is set to zero (the vehicle is stopped). This is somewhat similar to using an idealized step motor because, at first, the vehicle is accelerated and then is stopped in zero time. This behavior was chosen because it was rather difficult to control the system manually when the vehicle was allowed to continue its inertial motion.

In the experiments that followed, the vehicle was controlled by a human operator whose goal was to balance the pole attached to the vehicle in an upright position. Various initial positions of the pole were used, such as pendent or slightly deflected from vertical. A record of control was collected in each experiment in the form of an observation table. It contained tuples $(\Theta, \dot{\Theta}, F)$ in its rows. A fragment of an example

TABLE 80.1 Fragment of an Observation Table for the Inverted Pendulum Problem

Θ	Θ'	x
1.8145	10.9504	POSITIVE
1.7561	12.4141	POSITIVE
0.0333	3.9719	ZERO
0.0134	3.9702	ZERO
−0.0064	3.9699	ZERO
−0.0223	2.3721	NEGATIVE
−0.03011	0.7751	NEGATIVE

observation table is shown in Table 80.1. The data collected were later analyzed with the DataLogic rough set theory-based tool (DataLogic) and decision rules were created. Rows of the observation table were treated as objects; Θ and Θ' were condition attributes and F was a decision attribute.

The experiment consisted of erecting the pole from a position slightly deflected from vertical (−10°) and stabilizing it in the upright position. Rules generated from an observation table created during this experiment are shown in Fig. 80.14.

These rules model the behavior of the human operator controlling the system during training. They are surprisingly simple and only one presentation by the human operator producing a few hundred examples was needed to obtain them.

The process of deriving rules from examples can be viewed as a game in which the human expert first generates the actions. These actions are analyzed to extract distinct and noncontradictory conditional productions. Next, the productions are further compressed by finding a minimal decision algorithm. The derived algorithm can be loaded into a controller. The whole process can be referred to as play-and-plug (Fig. 80.15).

Once the algorithm is downloaded into a sensor, it can be plugged and played in an application. The overall scheme is simple and quite effective.

---

if {$\Theta > 0.15$ and $-3.70 < \Theta'$ and $\Theta' \leq 11.25$) or
  $\Theta > 0.60$ or
  $\Theta > 0.40$ and $-4.85 < \Theta'$ and $\Theta' \leq 11.25$ or
  $\Theta > 0.2$ and $-3.70 < \Theta'$ and $\Theta' \leq 3.20$ or
  $\Theta > 0.15$ and $11.25 < \Theta'$ and $\Theta' \leq 14.70$ or
  $\Theta > 0.20$ and ($\Theta' \leq 23.25$ or $\Theta' > 11.25$}
then {F = POSITIVE}

if {$\Theta \leq -7.50$ or
  $\Theta \leq -0.02$ and $-3.70 < \Theta'$ and $\Theta' \leq 0.90$ or
  $\Theta \leq -1.60$ or $-3.70 < \Theta'$ and $\Theta' \leq 11.25$ or
  $-4.70 < \Theta$ and $\Theta \leq -0.06$ and ($\Theta' \leq 0.90$ or
  $\Theta' > 23.90$) or
  $\Theta \leq -1.60$ and $11.25 < \Theta'$ and $\Theta' \leq 17.00$}
then {F = NEGATIVE}

Otherwise F = ZERO

FIGURE 80.14 Rules to control an inverted pendulum derived using the rough set approach.

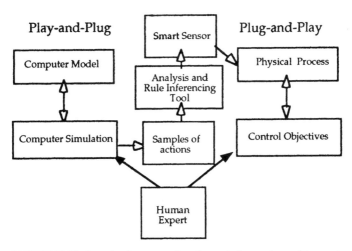

FIGURE 80.15 Learning from examples, plug-and-play can be used to generate rules used by a sensor to process sensory input.

In a distributed system, a complete system can be represented as a set of interconnected decision tables. Partitioning the state variables is done during the preliminary design stage. Next, the rules in the decision tables are augmented to include system invariants that, in case of failures, are used to estimate system state variables.[9] The whole plug-and-play development cycle supports all aspects of development and provides elements of *algorithmic reliability*.

The rough set–based approach to derive rules is similar to neural networks. The main difference between these methods is the symbolic formulation of rules in the rough set approach offers an advantage of explicit symbolic formulation that can be updated. In neural networks, the structure of the rules is hidden within weights associated with synapses.[10,11] These weights are hard to interpret and modify.

FLEXIBLE ARCHITECTURE FOR SMART SENSORS

The approach based on decision tables not only provides a technique to specify and to analyze control problems, but can be used as a basis for a powerful computer architecture.

The reduced set of rules derived from plug-and-play has the form of boolean expressions. The value of a boolean variable can be provided by a predicate in which a value measured by a sensor is thresholded and labeled using one of the predefined *rough values,* that is, small, medium, and so on.

There are two approaches to generate control rules from the decision tables:

- *Direct compilation.* A decision table is used directly to generate actions. An easy-to-implement look-up function serves as an execution engine.
- *Reduced form.* A decision table can be minimized to derive a set of simple boolean expressions that captures the invariants within a decision table. All the minimized expressions (reducts) can be executed simultaneously either within a single clock cycle using standard field programmable array (FPLA) or the parallel execution can be emulated within a microprocessor.

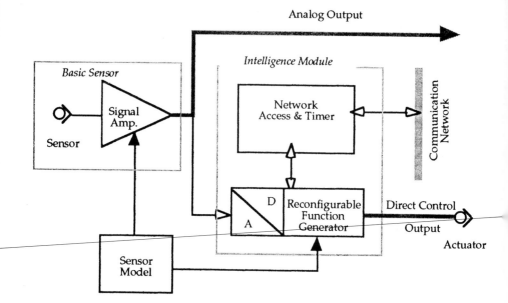

FIGURE 80.16 Rough set–based smart sensor.

Figure 80.16 depicts a structure of a rough set–based smart sensor. The engine of the sensor is a reconfigurable function generator. It contains an array of comparators and can perform simultaneously a number of boolean functions. This module can be emulated by a microprocessor sequentially executing all boolean functions associated with a decision table.

The rough set approach can also be applied to a better partitioning of the system. Given pieces of local information that are augmented by invariants linking their values with global data, a rough set–based information system can infer a global state of the system.[9]

SUMMARY AND CONCLUSIONS

Plug-and-play smart sensors are a combination of flexible communication and algorithms. In this chapter, we have evaluated control algorithms needed by the future sensors. We have reviewed four main class of algorithms: (1) classical control, (2) fuzzy logic, (3) rough set, and (4) neural networks.

Table 80.2 summarizes the findings. Floating-point arithmetic and the complexity of the formalisms of classical control theory are the main bottleneck of the classical control theory approach.

The approaches presented in this chapter are often superior to alternative methods in the following general application areas:

- Complex systems where an adequate model is difficult or impossible to define
- Systems that use human observations as control rules or are controlled by a human operator, and/or have to be trained by example

TABLE 80.2 Comparison of Emergency Adaptive Algorithms for Smart Sensors

Smart sensor algorithm	Model complexity	Computational complexity	Engineering complexity	Hardware implementation	Advantages	Disadvantages
Classic control theory	• Analytical formulation • Explicit treatment of nonlinearities	• Floating point arithmetics • Vector operations • Matrix inversion in real-time	• Derivation of model • Identification • Real-time programming	General purpose computers (MIPS!!)	Mature technology	• Stability • Completeness • Safety • Reliability • Complex theory and methodology
Fuzzy logic	Low	• Floating point arithmetics	Multilevel translation of rules (expert—fuzzyfication—defuzzyfication)	• Microprocessor • Special architectures	Commercial development systems available	• Stability • Completeness • Safety • Reliability
Rough set	Low	• Pattern matching • Integer arithmetics	• Explicit formulation of rules • Expert defines a decision table • Tools to extract symbolic rules	• Microprocessor • Programmable gate arrays	• Fast and low cost • Specifications = program • Parallel or sequential code • Simplicity and high-level semantics • Adaptability	• Stability • Completeness
Neural networks	Low	• Floating point • High complexity of interconnections	• Complex learning schemes • Lack of validation methodology	• Specialized architectures (analog)	• Adaptability • An all-analog system that is well integrated with sensors	• Stability • Completeness • Safety • Reliability • Unknown decision basis

- Systems where vagueness is common
- Systems where performance or controller simplicity issues are vital

The results presented in this chapter are difficult to compare quantitatively because fuzzy controller synthesis utilizes the designer's (expert's) knowledge expressed in terms of linguistic variables and membership functions, and reflects the designer's understanding of the problem. Rough controller synthesis, however, is based on the recorded behavior of the operator (in the form of observation tables) and depends on that person's experience and skill. This leads to a natural style of software development cycle that can be characterized as plug-and-play.

In our opinion, fuzzy control systems suffer from several drawbacks:

- It is difficult to construct the decision rules, since they must be provided by an expert.
- Fuzzification and defuzzification procedures are computation-intensive.
- The control algorithms are impossible to analyze because they are obscured by the knowledge base and the fuzzy inference methods (different variations of modus ponens).

On the other hand, rough controllers

- Utilize rules coming from control experience, that is, derived from observation tables and optimized automatically
- Are faster than fuzzy logic controllers and provide higher accuracy of control
- Have simple hardware implementations

An additional advantage of rough set theory, important from the software engineering point of view, is that both the decision rules and the inference process (trivial, in fact) are fully *transparent* and therefore can be validated more easily and the knowledge base can be tested more thoroughly than in the case of fuzzy or neural models. Rough set methodology relies on a consistent analysis of system. This is important when system safety is a vital factor.

Finally, decision tables used to derive rules are simple to understand and to edit. They serve both as system specification and execution code. System analysis is automated and development tools aid system partitioning that is needed in the design of system networking infrastructure.

REFERENCES

1. R. S. Raji, "Smart Networks for Control," *IEEE Spectrum*, June 1994, pp. 49–55.
2. K. Hirota, ed., *Industrial Applications of Fuzzy Technology*, Springer-Verlag, Tokyo, 1993.
3. E. H. Mamdani, "Application of Fuzzy Algorithms for Control of Simple Dynamic Plant," *Proc. IEEE*, 1974, **121** pp. 1585–1588.
4. R. M. Tong, "A Control Engineering Review of Fuzzy Systems," *Automatica*, 1977, **13**, pp. 559–569.
5. Z. Pawlak, "Rough Sets," *Int. J. Information and Computer Science*, 1982, **11**.
6. L. A. Zadeh, "Fuzzy Logic, Neural Networks, and Soft Computing," *Communications of the ACM*, 1994, **37**.

7. Z. Pawlak, *Rough Sets: Theoretical Aspects of Reasoning About Data,* Kluwer Academic Publishers, Dordrecht, Germany, 1991.
8. E. Czogala, A. Mrozek, and Z. Pawlak, "The Idea of a Rough Fuzzy Controller and Its Application to the Stabilization of a Pendulum-Car System," *Fuzzy Sets and Systems* (to appear), DataLogic reference.
9. J. Maitan, Z. W. Ras, and M. Zemankova, "Query Handling and Learning in a Distributed Intelligent System," in *Methodologies for Intelligent Systems, IV,* ed. Z. W. Ras, North Holland, 1989, pp. 118–127.
10. W. S. McCulloch, and W. Pitts, "A Logical Calculus of Ideas Immanent in Nervous Activity," *Bulletin of Mathematical Biophysics,* 1943, **5**, pp. 115–133.
11. T. M. Nabhan, and A. Y. Zomaya, "Toward generating Neural Network Structures for Function Approximation," *Neural Networks,* 1994, **7**, pp. 89–99.

CHAPTER 81
AUTOMOTIVE APPLICATIONS OF CONDUCTIVE POLYMER-BASED CHEMICAL SENSORS

INTRODUCTION

Conductive polymers have been reported as the active transducer in a number of chemical sensors. Most of the examples, however, demonstrate the detection of materials in the vapor state,[1-3] or in aqueous solution.[4,5] Little has been reported about the use of these materials in sensors that would monitor the chemical changes or compositional information in nonaqueous and nonpolar liquid media. Information about the condition or quality of fluids used for cooking, industrial machining, vehicle lubrication, alternate fuels, or specialty coolants is vital to their efficient use, as well as determining when their properties degrade to the point of needing to replace or replenish them. Currently there is no effective way of obtaining these measurements of such nonpolar, nonaqueous media in real time. We have found, however, that sensors based on overcoating thin gold interdigitated electrodes with a film of conductive polymer[6] of an appropriate thickness to bridge the gap between the digits shows promise as an in situ device for measuring chemical changes in such nonpolar media. In particular, this paper reports the implementation of sensors using polythiophene (and its derivatives) in determining the concentration of methanol in hexane as a model for determining the alcohol content in alternate fuels and the use of polyaniline in determining the useful lifetime of lubricating oil in automotive applications.

EXPERIMENTAL

Gold interdigitated electrodes were deposited on glass slides using standard semiconductor fabrication techniques. In some cases commercially available interdigitated microsensor electrodes (AAI Model IME 1550-CD, AAI-Abtech, Yardley, Pennsylvania) were used. Polythiophene and poly(3-alkylthiophene) were electrochemically polymerized at a constant current density of about 20 mA/cm^2 according to published methods[7] in nitrobenzene using tetrabutylammonium hexafluorophosphate as the electrolyte under argon at 5°C. Polyaniline was deposited electrochemically under cyclic voltammetry conditions[8] from a solution of distilled aniline in a sulfuric acid/sodium hydrogen sulfate electrolyte solution, where the thickness was determined by the number of cycles. Typically, a four-cycle run (8 sweeps between 0 and 900 mV versus a standard Calomel reference electrode) yielded a PAn film about 4.5-μm thick. Deposition was made directly onto the substrate with both sets of digits being the anode. A small DC voltage was applied to the coated sensor and the change in current flow, which ranged from the nA range to mA range, was measured by an electrometer. It was determined that the applied voltage was insufficient to cause any electrochemical deterioration to the conductive polymer films.

RESULTS AND DISCUSSION

Methanol Content in Hexane

With the implementation of the Clean Air Act of 1990, petroleum suppliers as well as automotive manufacturers are experimenting with new, alternative, low-polluting fuels. Among the number of candidates, methanol in gasoline is being actively investigated. The most common composition is 85 percent methanol in gasoline, which is generally referred to as M85. The combustion characteristics of this mixture are significantly different from pure gasoline and are greatly affected by even small changes in the concentration of methanol. Thus, for smooth operation of the engine, it is important to know, in real time, the absolute composition of the alternative fuel. Experimental devices, which measure the dielectric properties of the alternative fuel are currently being investigated, but they are generally large and expensive.

A model sensor was prepared by electrochemically depositing poly(3-methylthiophene) onto gold interdigitated electrodes. This device was immersed in hexane and methanol, in known amounts, was added while the resistance of the device was being monitored. Results of this measurement are shown in Fig. 81.1. The resistance changes by a factor of 9.4 as the concentration of methanol changes from 0 to 100 percent.

Bartlett and coworkers[2] have reported the detection and quantification of methanol vapor in air with model sensors using polypyrrole as the active transducer. A sorption model, in which the methanol interacts with sites either on the surface of, or within, the polymer film was presented to explain the mechanism of conductivity modulation in the polypyrrole. Further, their data appears to be in accord with those obtained by Josowicz and Janata,[9] who studied the response of a suspended-gate field-effect transistor in which the gate was coated with polypyrrole. They observed a positive shift in the threshold voltage on exposure to methanol vapor, which they interpret as meaning that the electrons in the polymer are more tightly bound in the presence of methanol. It seems likely that, even though, these results were obtained from the exposure to methanol in the vapor state, that the interaction with the polypyrrole at room temperature involved condensed methanol in the liquid state. Therefore, we would expect that the sorption model 2 would be appropriate for poly(3-methylthiophene) in the liquid phase detection of methanol as well.

Since the conductivity in conductive polymers is known to be a three-dimensional phenomena, it is important that the individual chains of the polymer remain in a particular orientation with respect to their nearest neighbors. If this orientation should become disrupted, by swelling for instance, caused by the uptake of methanol, then the conductivity would change as a function of the amount of methanol being absorbed. In this case, methanol could have a more pronounced effect on this reorientation than hexane. This physical reorganization of the polymer chains in poly(3-methylthiophene) could also account for the observed modulation in the conductivity seen in the presence of methanol.

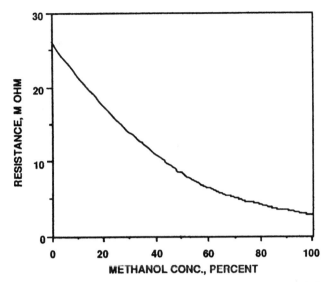

FIGURE 81.1 Change in resistance of a model sensor composed of poly(3-methylthiophene) as a function of methanol concentration in hexane.

Similar experiments were carried out with polypyrrole, but no modulation of the conductivity in liquid solutions of methanol in hexane were observed.

Acid and Water Detection in Nonpolar Media

The conductivity of polyaniline is known to be affected by the degree of protonation caused by mineral acids. The level of conductivity has been studied as a function of pH in aqueous solutions.[10] Nonpolar, hydrocarbon-based fluids often degrade with the formation of acids, which are usually formed by thermal oxidation. The condition of the fluids has been monitored by change in its dielectric constant,[11,12] increase in corrosivity,[13] increase in optical absorption,[14] and an increase in acidity as measured by extraction electrochemical techniques.[15] The use of conductive polymers for the detection of acids in lubricating oil have been reported.[16,17] Our studies were directed to determine whether polyaniline could act as an acid detector in nonpolar media, in which acidic hydrogen atoms are strongly associated (nonionized) with their conjugate base,[6] both in the presence and absence of additional water.

Polyaniline was prepared electrochemically across gold interdigitated electrodes as the emeraldine salt which was neutralized in 10 percent ammonium hydroxide to the insulating emeraldine base. This model sensor was immersed in a nonpolar, hydrocarbon-based oil, to which was added known amounts of water. These results are shown in Fig. 81.2. To an oil mixture containing about 6 percent of added water was added known quantities of acetic acid. Hydrocarbon oils will thermally oxidize to carboxylic acids so acetic acid was chosen as a representative oxidation acid and it has a pKa value similar in magnitude to higher analog carboxylic acids. The modulation in the current flow is shown in Fig. 81.3.

Clearly this sensor is quite sensitive to acid, even in such a nonpolar medium. Modulation of the conductivity of more than four orders of magnitude are observed.

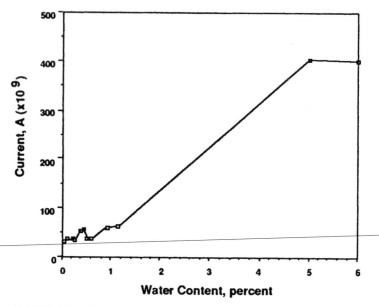

FIGURE 81.2 Change in current flow of a polyaniline-based sensor in lubricating oil as a function of added water at about 100°C.

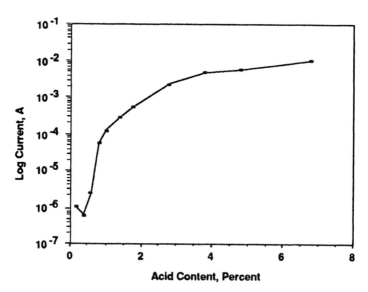

FIGURE 81.3 Change in current flow of a polyaniline-based sensor as a function of added acetic acid in lubricating oil containing about 6 percent water at about 100°C.

Since the sensitivity of the device is diffusion-controlled, the response times are generally in the tens of minutes until maximum readings were obtained. Evidence of reversibility is shown in Fig. 81.4.

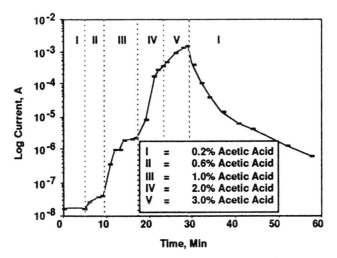

FIGURE 81.4 Change in current flow of a polyaniline-based sensor as a function of acetic acid content in lubricating oil at about 100°C. The sensor was first immersed in I, then II, and so forth.

A polyaniline-based sensor was first immersed into sample I, which contained the lowest concentration of acetic acid. After a certain amount of time the sensor was removed and then immersed into sample II, and so forth. After immersion in sample V, it was then exposed to sample I again and the trend in the conductivity approached the original starting value. Although the abscissa is in units of time, this particular plot does not necessarily indicate a response time since the sensor was moved from one sample to another arbitrarily.

The device shows a direct temperature-dependence to the response. As the temperature increases, so does the current flow. Over the temperature range of 0 to 100°C, this amounts to an increase in conductivity by a factor of about 200. Table 81.1 summarizes the response of our model sensor under varying conditions.

Degradation of Automatic Transmission Fluid

Automatic transmission fluid (ATF) is also a hydrocarbon-based lubricant and therefore it is expected that its degradation should be similar to that of lubricating

TABLE 81.1 Change (D) in Polyaniline-Based Sensor Response Under Varying Conditions

Temperature	Water	Acetic Acid	Δ
30–100°C	—	—	+200X
100°C	0–6%	—	+10X
100°C	6%	0–6%	>1000X
100°C	6%	0–0.5%	+20X

oil. It is known that carboxylic acids are, in fact, formed as one of the degradation products. Samples of degraded ATF were tested with our polyaniline sensors. Correlation of the degree of degradation were observed with a modulation in the conductivity by a factor of nearly 20,000 observed at 100°C between the least- and most-degraded samples. There was a significant time dependency of the readings, but the trend in this dependency appeared to be predictable, and therefore data obtained over a short interval could be extrapolated to a later time. This extrapolated value could then be compared to a known threshold value for degraded ATF and provide an indication of the condition of the ATF being evaluated.

SUMMARY

These results indicate that polyaniline as the active transducer is a highly promising material for sensing the condition of hydrocarbon-based fluids, both in the presence and the absence of water. Water can be an interfering impurity in measuring the acid content of these fluids, particularly when trying to measure the increase of electroactive polar species. Water tends to dissolve and coagulate these species, which effectively removes them from the medium and hence their build-up is not detected. The polyaniline devices described here do in fact detect and measure water in the presence of hydrocarbon-based fluids, and also effectively sense the build-up of acid species, even in the presence of water. Calibration of the device for acid content can yield a device for quantitative measurement in both lubricating oils and transmission fluids. Similarly, poly(3-methylthiophene) appears to be a promising candidate as the transducer material for sensors measuring the alcohol content in hydrocarbon-based solutions.

REFERENCES

1. P. N. Bartlett, P. B. M. Archer, and S. K. LingChung, *Sensors and Actuators,* 1989, **19**, (125).
2. P. N. Bartlett and S. K. Ling-Chung, *Sensors and Actuators,* 1989, **19**, (141).
3. P. N. Bartlett and S. K. Ling-Chung, *Sensors and Actuators,* 1989, **20**, (287).
4. M. E. Boyle and R. F. Cozzens, *Final Report NRL Memorandum Report 6482,* June **12**, 1989.
5. L. D. Couves and S. J. Porter, *Synth. Met.,* 1989, **28**, (C761).
6. F. G. Yamagishi, C. I. van Ast, and L. J. Miller, *patent pending.*
7. S. Hoffa, S. D. D. V. Rughooputh, A. J. Heeger, and F. Wudi, *Macromolecules,* 1987, **20**, (212).
8. A. F. Diaz and J. A. Logan, *J. Electroanal. Chem.,* 1980, **111**, (111).
9. M. Josowicz and J. Janata, *Anal Chem.,* 1986, **58**, (514).
10. J.-C. Chiang and A. G. MacDiarmid, *Synth. Met.,* 1986, **13**, (193).
11. *Sensor Technology,* **May 1991.**
12. C. A. Megerle, U.S. *Patent no. 5,089,780,* **February 18, 1992.**
13. D. Cipris, A. Walsh, and T. Palanisamy, *Proc. Electrochem. Soc.,* 1987, **87–9**, (401).
14. N. Morlya and Y. Kusakabe, *Kokai Tokkyo Koho,* JP 61294337 A2, Japan, **December 25, 1986.**
15. J. Joseph, H.-O. Kim, and S. Oh, *J. Electrochem. Soc.,* 1993, **140**, (L33).
16. R. Yodice and R. E. Gapinski, U.S. *Patent no. 5,023,133,* **June 11, 1991.**
17. Y. Maeda, European *Patent Application no. 0 442 314 A2,* **August 21, 1991.**

CHAPTER 82
MODELING SENSOR PERFORMANCE FOR SMART TRANSDUCERS

INTRODUCTION

Piezoresistive silicon devices dominate the pressure and acceleration sensors market. They represent about 85 percent of the 50-million-unit 1994 pressure sensor market, and 96 percent of the 6-million-unit 1994 acceleration sensor market.

The emerging control networking[1] applications require *smart sensors* (defined as devices with built-in intelligence)[2] with a digital output. Until competitively priced sensing elements with a direct digital output are available (such as resonating sensors),[3] most smart transducers will be based on piezoresistive sensors. The majority of these devices currently deliver an analog output signal. Such sensors require substantial signal processing to enable high-performance digital output. The selection of modeling techniques plays an important factor in cost-effective transducer manufacturing.

COMPENSATING SENSOR ERRORS

Silicon piezoresistive sensors provide high sensitivity to measured variables, almost two orders of magnitude larger than the sensitivity of metal strain gage transducers. This high output comes at the expense of significant temperature sensitivity and relatively high unit-to-unit variation. Traditional sensor compensation and calibration were based on the analog resistor circuit configuration. While temperature errors of silicon sensors are very repeatable, they are somewhat nonlinear. This temperature nonlinearity restricts the sensor performance available with the analog (resistive) compensation.

Two options were available to select values of required compensation resistors: either an iterative slow real-time active trimming or modeling followed by a fast trimming based on a derived numerical model. The first sensor models were in response to production compensation and calibration needs.[4]

Smart sensors redefined a paradigm related to the high-performance piezoresistive sensor accuracy. Using digital compensation and calibration, smart sensors enabled the dramatic reduction of all repetitive sensor errors, achieving two orders of magnitude lower errors than was possible with the analog compensation technology.

Three basic approaches for reducing sensor errors in smart transducer technology evolved (see Fig. 82.1):

- Statistical
- Analog sensor signal processor (ASSP)
- Digital sensor signal processor (DSSP)

Statistical approach is defined as a compensation of errors' mean value. It is usually performed in the analog domain. To make this compensation effective, the statistical distribution of errors must be known, and standard deviation (or range) has to be much smaller than the mean. Statistical compensation is useful for lower-end smart transducers, as the semiconductor manufacturing process is not capable of delivering distribution tight enough for higher-performance applications.

The analog sensor signal processor relies on two signals from the sensor: one proportional to pressure (P in Fig. 82.1) and the other proportional to temperature (T in Fig. 82.1). The temperature signal is used to generate two analog corrections: zero and FSO. The correction is performed in a parallel path to the flow of the pressure signal by adjusting the offset and gain of the instrumentation amplifier. There is no loss of pressure sensor resolution nor response time due to the digital correction. There may be, however, a temperature-related quantization; when temperature changes, zero or FSO signal may change in steps.

Some of the ASSP implementations can directly deliver a digital output in parallel to the analog output (e.g., a pulse-width-modulated signal interfaces with micro-

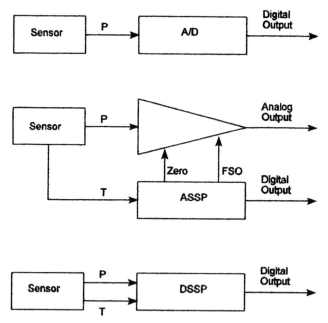

FIGURE 82.1 Three basic approaches to correction of sensor errors in smart transducers: (*a*) statistical compensation; (*b*) analog sensor signal processor (ASSP); and (*c*) digital sensor signal processor (DSSP).

processors, but after filtering creates an analog signal). If another approach is used, an A/D converter must be used to create a digital signal. While the ASSP seems to be somewhat unnecessarily complex, it provides a digitally compensated analog output, thus addressing the currently largest sensor market.

While there are several variations of ASSP, the major difference between them is in the approach to FSO correction. One approach achieves FSO compensation by controlling the gain of an amplifier; the other controls the bridge voltage. While theoretically both approaches are equivalent, the gain control delivers better accuracy, it eliminates the inaccuracy resulting from the implanted or diffused strain gages' voltage sensitivity: resistors' geometry (and thus resulting resistance) changes when bridge voltage varies. It is possible to compensate voltage sensitivity using a proper model, but this requires a longer testing procedure.

The digital sensor signal processor digitizes sensor signals (sometimes after an initial rough normalization) and performs temperature compensation and calibration in the digital domain. It utilizes both pressure and temperature signals from the sensor. No analog output is available. The response time is restricted by the processor's speed. DSSP processing resolution establishes the quantization limits for both pressure measurement resolution and temperature compensation error. The most accurate smart transducers use 24-bit A/D converters.[5]

STATISTICAL COMPENSATION

Statistical compensation for smart sensors is used to reduce the manufacturing cost by eliminating the need for sensor temperature testing. Errors are minimized by removing the mean value and accepting the distribution of sensor parameters around the mean. From a device physics aspect, two almost orthogonal (independent) parameters and their temperature dependencies have to be compensated: zero and FSO. In a constant voltage excitation mode, zero and FSO are orthogonal. In a constant current excitation mode, however, bridge voltage changes resulting from the bridge TCR induce a temperature-dependent zero change when the bridge offset is different from zero. This creates a zero component that correlates to FSO, reducing zero and FSO orthogonality.

Zero

Zero represents the bridge output voltage without applied pressure. It is usually expressed in ratiometric terms of mV (of offset) per V (of bridge voltage). It is affected by the electrical mismatch of sensing resistors and mechanically induced stresses from either the process or sensor packaging. Zero distribution for piezoresistive sensors yields lot-to-lot mean variations typically within ±0.5 to ±2 mV/V. The standard deviation range is typically within 0.5 to 4 mV/V. Distribution variations are a function of the sensor's design and specific process capabilities. One of the key factors limiting zero distribution is the printing capability of the sensor process (photolithography). If, e.g., the sensing resistors line width is 5 μm, 1 mV/V matching requires 0.1 percent average matching of four resistors, equivalent to a 5-nm line width.

For a 3-σ (99.7 percent) distribution, the range of zero can be estimated at ±2 to ±11 mV/V for different products on the market. Transducer errors are typically expressed in %FSO. Full-scale output (FSO) for piezoresistive sensors typically falls in the range of 5 to 50 mV/V. Without any offset trimming, transducer statistical calibration would yield the zero errors as shown in Table 82.1.

TABLE 82.1 Statistical Calibration of Zero

FSO	Zero range
±2 mV/V	±14 mV/V
5 mV/V ±40% FSO	±280% FSO
50 mV/V ±4% FSO	±28% FSO

It is visible that the higher the sensor output, the lower is the resulting zero error. For most of the higher-performance applications, statistical zero distribution based on an untrimmed bridge is not satisfactory, even for the best sensor designs and processes. The use of wafer-level offset trimming, however, may bring zero range to ±0.1 mV/V, reducing zero errors more than an order of magnitude. This makes statistical compensation of zero feasible for many applications.

Full-Scale Output

Full-scale output is defined as the output change in response to a span change of the input variable. It is calculated as a product of pressure sensitivity S and bridge voltage V_b:

$$FSO = S \cdot V_b$$

In a constant-current excitation mode, bridge voltage is a product of bridge resistance R_b and bridge current I_b. In this mode, bridge resistance thus affects the statistical compensation's performance:

$$FSO = S \cdot R_b \cdot I_b$$

Variations of pressure sensitivity in production volume are on the order of ±5 to ±30 percent and result from the nature of processes used to fabricate silicon sensors. Parameters affecting sensitivity include diaphragm thickness, strain gages placement, resistors' doping levels, and junction thickness.

The temperature coefficient of sensitivity (TCS) results from the temperature dependence of piezoresistivity. The value of TCS can be process-adjusted over a range −0.05 to −0.25 percent/°C. For a selected process, the repeatability of TCS is typically on the order of 5 to 15 percent of the mean. This creates an opportunity for a statistical compensation to reduce TC FSO by an order of magnitude without the need for temperature testing.

Bridge resistance variations are typically on the order of 5 to 30 percent. They result from imperfections of the semiconductor manufacturing process. Temperature coefficient of bridge resistance (TCR) can be process-adjusted within a range of +0.03 to 0.35 percent/°C. There is a strong correlation between TCR and TCS,[6] enabling improved statistical compensation: at two doping levels, TCR = −TCS, providing a potential for self-compensation of TC FSO in a constant-current mode. Depending on the sensor design and process, TC FSO can be controlled to about ±1 to ±3 percent over a temperature range of 0 to 70°C. (See Fig. 82.2.)

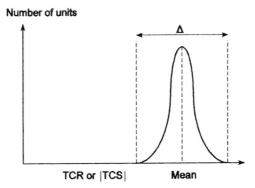

FIGURE 82.2 Statistical compensation of TCS and TCR is feasible, as their distribution range Δ is significantly smaller than the mean value. When the mean is compensated, remaining errors are reduced by about an order of magnitude.

DIGITAL COMPENSATION AND NORMALIZATION

Two basic approaches to sensor output modeling for smart sensors and transducers evolved: look-up tables and compensating algorithms.

Look-up Tables

This approach is based on a memory (look-up table) that stores the output signal values. Based on the sensed pressure and temperature signals, a unique number representing the compensated and calibrated pressure is selected by the processing electronics. The quantity of stored numbers depends on the correctable error band and required accuracy. Two extreme estimates of the required storage can be outlined as follows:

High estimate. Assuming a need to cover all possible cases of sensors characterized by +20 percent FSO zero variations and +30 percent FSO full-scale output variations, for a total of ±50 percent FSO adjustability range at room temperature, and an additional ±20 percent FSO temperature error. This requires the following storage capacity:
- For 2 percent error band: 25 room-temperature curves (spaced every 4 percent FSO), each with 25 pressure values (every 4 percent FSO), repeated for 10 temperatures (every 4 percent FSO), for a total 6250 numbers. If average eight-bit numbers are needed, a memory demand is 50 kbits
- For 0.2 percent error band: Assuming 10-bit numbers, memory demand increases to 200 Mbits

Low estimate. To optimize the required memory, a more sophisticated algorithm must be utilized. One option is to split DSSP into several functions: zero and FSO normalization, and then correction of temperature errors on the normalized output. Storage requirements for the previous example would change to the following:

- For 2 percent accuracy: 10 zero values, 15 FSO values, and 10 temperature error values, for a total of 35 eight-bit numbers, or 280 bits.
- For 0.2 percent accuracy: 100 zero values, 150 FSO values, and 100 temperature error values, for a total of 350 ten-bit numbers, or 3.5 kbits

The actual implementations of sensor look-up tables usually fit somewhere in between these two extreme examples.

Both ASSP and DSSP smart sensor configurations can use the look-up table. It is clear that some kind of compromise will be needed to balance the memory size and performance. Some of the smart sensor configurations use look-up tables only for temperature compensation, after performing initial calibration in the analog domain.

The major advantage of the look-up table approach is its fast frequency response. (See Table 82.2.)

TABLE 82.2 Example of Look-up Table Correction

P Output, mV		Pressure (psi)				
2.0 to 3.0	4.00	4.03	4.04	4.05	4.06
1.0 to 2.0	3.00	3.02	3.03	3.04	3.05
0.0 to 1.0	2.00	2.02	2.03	2.04	2.05
−1.0 to 0.0	0.00	0.01	0.02	0.03	0.04
T Output, mV	1–3	3–5	5–7	7–9	>9

Compensating Algorithm

In this approach, the sensor output is approximated with a set of simultaneous equations. This significantly reduces data storage requirements. Algorithmic approach is a more suitable approach for higher-performance smart transducers.

To develop a sensor model, the sensors pressure and temperature outputs are measured at several pressures, and the measurements are repeated at several temperatures. The number of test pressures and temperatures depends on the desired modeling accuracy. The higher the accuracy, the larger the number of required test points.

Temperature sensing can be accomplished either in a constant-current mode, where a bridge voltage is a good temperature indicator (with a positive slope), or in a constant-voltage mode, where a bridge current changes with temperature (with a negative slope). Since the temperature signal is generated by the pressure sensor itself, the accuracy of modeling can be very high, as it eliminates temperature gradients between pressure and temperature sensors. Modeling accuracy is limited only by the stability and hysteresis of the sensor, accuracy and resolution of the signal conditioning electronics, and accuracy of the test system.

The pressure output at each test temperature can be defined by a polynomial. To find coefficients for a given fitting algorithm, a set of simultaneous equations must be solved. Each of them defines the compensation or calibration objective, such as, the transducer output at all test temperatures should be 0.500 V at zero pressure, and full scale should be 4.500 V.

For higher-order curve or surface fitting, a least squares fitting algorithm should be used (Fig. 82.3) to reduce a possibility of no converging modeling curve or surface or their oscillations between test points.

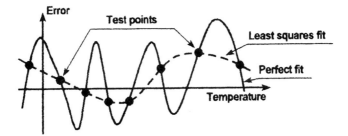

FIGURE 82.3 Higher-order perfect fit through real test data is affected by unavailable errors; this creates oscillations of the fitting algorithm. Use of least squares fit eliminates this problem, at the expense of increased number of test points.

Two basic options exist for sensor models. The first option takes advantage of the orthogonality of zero and FSO, solving two sets of equations that compensate zero and FSO. The second option includes a single function modeling of measured pressure as a function of pressure and temperature sensor outputs.

Modeling of Zero and FSO

In the basic configuration, the processor controls the offset and gain of the instrumentation amplifier to correct the zero and FSO of the sensor, respectively. Sensor output is a combination of temperature-dependent offset $V(T)$ and temperature-dependent pressure sensitivity $S(T)$. The ASSP generates temperature-dependent correction signals: offset E1 (T) or E2 (T), and variable amplifier's gain $K(T)$. ASSP's offset can be fed either to the input (input-related offset) or output (output-related offset) of the amplifier. In the first case, the correction voltage is dependent on the set gain $K(T)$ of the amplifier. In the other case, it is not. The transducer output voltage $U(P,T)$ is thus defined as:

$$U(P,T) = [V(T) + E1(T) + S(T) \times P] \times K(T) + E2(T)$$

This voltage can be divided into FSO and zero (Z):

$$FSO = S(T) \times K(T) \times P$$

$$Z = [V(T) + E1(T)] \times K(T) + E2(T)$$

Since the desired output FSO (e.g., 4 V) and pressure P are known, for each measured pressure sensitivity $S(T_i)$ at given temperature T_i, the required corrective gain $K(T_i)$ can be easily calculated as:

$$K(T_i) = FSO/(S(T_i)/P$$

Once the compensating gain is known, a required corrective offset can also be calculated as either:

$$E1\ (T_i) = Z/K(T_i) - V(T_i), \text{ when } E2 = 0, \text{ or}$$

$$E2(T_i) = Z - V(T_i) \times K(T_i), \text{ when } E1 = 0$$

where Z is the required output zero level (e.g., .5 V). (See Fig. 82.4.)

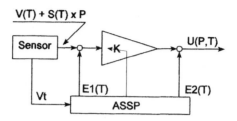

FIGURE 82.4 Compensation of zero and FSO.

Using the outlined procedure, a set of required compensating offset and gain values can be generated. In the next step, a polynomial curve fitting through the calculated data can be performed using the temperature signal V_t:

$$K(T) = \alpha_0 + \alpha_1 \times V_t + \alpha_2 \times V_t^2 + \cdots$$

$$E1(T) = \delta_0 + \delta_1 \times V_t + \delta_2 \times V_t^2 + \cdots$$

For a parabolic fit, only six sensor-dependent constants $\alpha_0, \alpha_1, \alpha_2, \delta_0, \delta_1, \delta_2$ must be stored. Using an average 10-bit data format, the storage requirement is only 60 bits, significantly less than for a look-up table.

Pressure nonlinearity correction is one of the parameters required for higher-performance transducers. It typically falls in the 0.05 to 1 percent range and changes with the temperature. There are two basic options for the nonlinearity correction in ASSP (see Fig. 82.5):

- Analog feedback from a transducer output to the bridge excitation.
- Software feedback from a transducer output to the analog feedback is fast, thus it can be used to correct the dynamic signals. Software feedback frequency response is limited by the speed of the analog signal processor. Thus, it may not work with

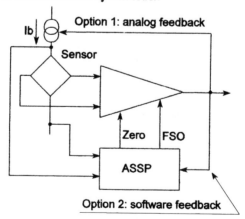

FIGURE 82.5 Two options for correction of pressure nonlinearity: analog or software feedback loop.

all the applications. One of the possible implementations is a mixed analog-software feedback mode, in which software changes the feedback depth (to correct temperature dependence of nonlinearity), and an analog feedback corrects the actual nonlinearity.

Nonlinearity feedback adjusts the system's gain as a function of the transducer output. To correct a negative (concave) pressure nonlinearity, amplifier gain or bridge voltage should decrease with increasing output. This correction somewhat complicates the procedure for calculation of curve-fitting coefficients.

Single-Function Model

The pressure P measured by a sensor can be modeled by a two-dimensional surface:

$$P = f(V_P, V_T)$$

where V_P and V_T are respectively pressure and temperature sensor outputs. Generally, a surface fitting equation can be derived from a set of individual polynomial equations defining pressure output for a set of temperature outputs. This procedure will be illustrated on a parabolic surface. Most silicon pressure sensors can be characterized by a dominating parabolic offset and sensitivity temperature dependence, parabolic pressure nonlinearity, and the linear temperature coefficient of pressure nonlinearity. In Fig. 82.6, all these nonlinearities were highlighted with crosshatched areas. There are three test pressures (P0, P1, P2) and three test temperatures (T1, T2, T3), generating nine test points at which sensor pressure (V_p) and temperature (V_t) outputs are recorded. Surface fitting through all these test points, if does exist, would eliminate or minimize all the nonlinearities.

Using a polynomial curve fitting reflecting physical behavior of the sensor, the following algorithm can be developed:

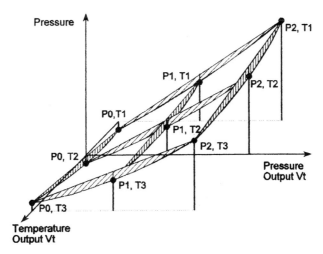

FIGURE 82.6 Surface defining pressure as a function of pressure and temperature outputs.

$$P = A_0 + A_1 \cdot V_t^2 + A_2 \cdot V_t^2 + [A_3 + A_4 \cdot V_t + A_5 \cdot V_t^2] \cdot V_p + [A_6 + A_7 \cdot V_t] \cdot V_p^2$$

Sensor-specific constants $A_0 \ldots A_7$ have easily understood physical representation:

- A_0: Initial offset (zero)
- A_1: Linear temperature coefficient of offset
- A_2: Parabolic temperature coefficient of offset
- A_3: Initial pressure sensitivity
- A_4: Linear temperature coefficient of pressure sensitivity
- A_5: Parabolic temperature coefficient of pressure sensitivity
- A_6: Parabolic pressure nonlinearity
- A_7: Linear temperature coefficient of parabolic pressure nonlinearity

The solution of the expression for pressure is not very straightforward, as it includes eight variables. Based on the test data, nine linear equations can be formed:

$$P_i = f(V_{tj})$$

where $i = 1, 2, 3,$ and $j = 1, 2, 3$.

Since there are eight unknown A constants, there is one redundant test point. This redundant test point would allow for a parabolic fit into the parabolic pressure nonlinearity changes. If it is not used, one of the test points at pressure P2 should be dropped.

Measurement accuracy and a higher nonlinearites order than parabolic may create a lack of convergence of the parabolic surface. If such conditions occur, a least squares fit procedure must be implemented. This procedure requires more test points than are required for a perfect fit; thus, four test pressures and four test temperatures should be used in the preceding example. Procedures to perform the least square fitting are more complex than the outline above.[7,8]

As a function of sensor design, models based on a parabolic surface fit can deliver 0.05 to 0.5 percent total error band over a 0 to 70°C temperature range, and 0.2 to 2 percent over a −40 to 125°C range.

CONCLUSIONS

Modeling sensor performance for smart transducers gains momentum. Sensor model implementation on the silicon chip attracted the attention of not only the established linear IC community, but also several new start-up companies. Growth of the smart sensor market, especially the market for low-cost networkable transducers, is expected to bring many new developments in this area.

While most of the current implementations of sensor models is done outside of the silicon sensor chip, the technology already supports an integration of the smart sensor electronics including the sensor model on the sensor chip.[9] Lack of the sufficiently big market slows down production reality of such devices now, but they are expected to emerge before the turn of the century.

REFERENCES

1. *Proceedings of First IEEE/NIST Smart Sensor Communication Standard Workshop,* Gaithersburg, Maryland, March 31–April 1, 1994.
2. *Draft IEEE TC-9 Standard: Microfabricated Pressure and Acceleration Terminology.*
3. K. Petersen et al., "Resonant Beam Pressure Sensor Fabricated with Silicon Fusion Bonding," *Digest of Technical Papers, Transducers'91,* San Francisco, California, June 1991.
4. J. Bryzek, "Modeling Performance of Piezoresistive Pressure Sensors," *Proceedings of 1985 IEEE Conference Solid State Sensors and Actuators,* Hilton Head, June 1985.
5. P. DuPuis, "Implementing Control Loops Using a Networked Smart Sensor," *Proceedings of Sensors Expo West,* San Jose, California, March 2–4, 1993.
6. J. Bryzek, "Signal Conditioning for Smart Sensors and Transducers," *Proceedings of Sensors Expo 93 Conference,* Philadelphia, Pennsylvania, September 1993.
7. I.C.R. Wylie, *Advanced Engineering Mathematics,* McGraw-Hill, New York, 1984.
8. J. Neter et al., *Applied Linear Statistical Models.* Irwin, 1990.
9. E. Obermeier et al., "Smart Pressure Sensor with On-Chip Calibration and Compensation Capability," *Proceedings of Sensors Expo West,* Anaheim, California, February 8–10, 1994.

CHAPTER 83
INFRARED GAS AND LIQUID ANALYZERS: A REVIEW OF THEORY AND APPLICATIONS

INTRODUCTION

Infrared spectroscopy is one of a number of methods used for the measurement of components in gas and liquid streams. Although other techniques, such as gas or liquid chromatography, mass spectrometry, or atomic absorption, are more sensitive or more selective, the advantages of infrared spectroscopy frequently outweigh its weaknesses. Infrared spectroscopy provides ease of application and maintenance as well as the advantage of continuous analysis. Infrared instrumentation has been applied to such applications as

- Ambient air monitoring for toxic vapors
- Chemical reaction completion
- Dissolved gases in liquids
- Stream purity measurements

The infrared portion of the electromagnetic spectrum lies just beyond the visible portion of the spectrum, starting at approximately 0.8 micrometers (µm). The infrared region is commonly divided into three areas:

Wavelength	Range
0.8–2.5 µm	near-IR
2.5–15 µm	mid-IR
15–1000 µ	far-IR

All organic and many inorganic compounds absorb energy in the infrared portion of the electromagnetic spectrum. This absorption is due to the radiation absorbed by the compound which is converted into kinetic energy by various changes of state of the molecular bonding structures; since many compounds have similar bonding structures, families of compounds can be found to have similar spectra with minor variations allowing for selective analysis in most applications. The *fingerprint* region of the spectrum, from 8 to 15 µm, for many compounds, provides unique bands which can differentiate them from others.

A number of inorganic compounds also absorb infrared energy; these include CO_2, CO, NO, NO_2, N_2O, NH_4, just to list a few.

Various specialized methods have evolved in infrared spectroscopy which usually move from the laboratory to the process line. These include *nondispersive infrared* (*NDIR*), *Fourier transform infrared* (FTIR) as well as grating and other dispersive instruments. The NDIR instrument is most commonly used in process applications, although over the past several years the FTIR instrument has been finding its way into this market as well as into the ambient air market. The FTIR instrument has traditionally been a laboratory spectrometer and when applied to process applications has been unacceptable because of its inability to operate adequately in areas of high vibration; a number of ruggedized process FTIR instruments are now on the market and are operating successfully. The FTIR method has the advantage of scanning its entire spectral range very rapidly and providing information at any or all wavelengths in its range. With all data manipulated in software the FTIR can be a very versatile and flexible instrument.

The traditional NDIR instrument in itself has evolved into a number of configurations which have included single- or dual-beam configurations, single-, dual-, or multiple-wavelength arrangements. Detectors include solid-state or pneumatic (gas-filled) detector, cooled, or uncooled (Fig. 83.1).

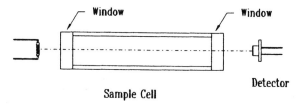

FIGURE 83.1 Sample cell arrangement.

The infrared analyzer, whatever the technique used, incorporates four basic components:

- IR source
- Sample cell
- Optical filter
- IR detector

THE SOURCE

The source is heated to several hundred degrees centigrade and emits infrared energy. Source materials are usually tungsten or nichrome in the form of a ribbon or filament or ceramic element which is heated to several hundred degrees centigrade. The source temperature must be well regulated since its emitted energy varies as the fourth power of its temperature as expressed by the *Stefan-Boltzmann law*:

$$W = \sigma T^4$$

where W = the total energy radiated
T = the absolute temperature in K
σ = a constant
$(5.67 \times 10^{-12}$ watts cm^{-2} deg$^{-4})$

THE SAMPLE CELL

The sample cell is used to contain a fixed length of sample between the source and detector. The length of the sample cell is governed by the concentration of the gas or liquid to be measured as well as its degree of absorbance or *absorbtivity*. The path length can be calculated using *the Beer-Lambert law* which may be expressed as:

$$A = \alpha L C$$

where A = the absorbance
σ = the absorbtivity, a constant for a particular wavelength of the material being measured
L – sample cell effective path length
C = the concentration of the gas or liquid being measured

The proper path length selection is for best linearity and may rely on past experience. Sample cell path lengths may range from 1 mm, for liquid samples or high-concentration gas samples, to as long as 20 m or longer using folded path-length cells, for low ppm gas samples. If the chosen pathlength is too short, the instrument will lack sensitivity, and if too long, the instrument may be very nonlinear or will saturate with high concentrations.

As can be seen from the Beer-Lambert law, if one were to double the sample cell path length, one would get twice the absorbance or sensitivity for a given sample concentration; and if halved, the sample would produce half the absorbance.

As a matter of reference, absorbance can be related to percent transmission as:

$$A = \log \frac{1}{T}$$

where A = the absorbance
T = percent transmission

Absorbance is primarily used in instrumentation because it is linear and increases with concentration, where transmission is logarithmic and decreases with concentration.

SAMPLE CELL WINDOW MATERIALS

Window and lens materials selected must be chosen for their spectral transmission in the region of interest. Some commonly used window and lens materials are shown in Table 83.1.

Other considerations in window material selection include percent transmission as well as tolerance to the sample stream. Some materials are hygroscopic, others are etched by acids, and some can be abraded by particles in the stream; all of these factors must be considered in the selection of window materials.

TABLE 83.1

Material	Symbol	λ Range (μm)	% transmission (uncoated) 1 mm
Sapphire	Al_2O_3	0.17–5.6	88%
Calcium fluoride	CaF_2	0.1–8.8	94%
Barium fluoride	BaF_2	0.15–11.9	94%
Germanium	Ge	1.8–16.7	48%
Silicon	Si	1.2–15	44%
Irtran 4	ZnSe	0.5–22	66%

OPTICAL FILTER(S)

As stated earlier, all organic materials as well as some inorganic materials absorb infrared energy at specific wavelengths. Through the use of optical filters or other methods such as FTIR or gratings, one or several wavelengths can be chosen where the compound of interest absorbs and other materials in the stream do not. The optical filter will allow the detector to only measure transmitted energy at the chosen wavelength(s). For example, if one wishes to measure the concentration of carbon monoxide in a sample stream, a filter transmitting at 4.7 μm would be chosen since carbon monoxide has an absorbance band at that wavelength (Fig. 83.2).

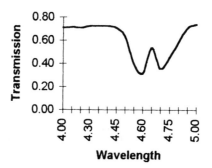

FIGURE 83.2 CO spectrum.

In the preceding example, the degree of IR energy absorbed at 4.7 μm would be proportional to the concentration of carbon monoxide in the stream (Fig. 83.3).

As stated previously, filter selection would be at 4.7 μm for the analysis. This filter would only allow energy at the chosen wavelength to fall on the detector.

For added stability and interference rejection, two or more filters could be used where at least one wavelength could be used as a reference. The reference filter wavelength would **be** chosen were no other components in the stream exhibit absorbance bands. Ideally, the reference wavelength would be selected at a wavelength close to the analytical wavelength to minimize errors due to differences in IR energy between the two wavelengths.

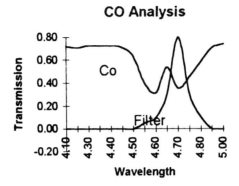

FIGURE 83.3 CO analysis.

DETECTORS

A number of detectors are available for use in the 0.8- to 15-μm region of the spectrum. The four major parameters in selecting an infrared detector for an analyzer are as follows:

- Spectral response
- Sensitivity (D* and Responsivity)
- Ease of support electronics (i.e., power supplies, cooling, etc.)
- Cost

A few commonly used detectors are listed in Table 83.2. Other more exotic detectors are available, such as indium antimonide (InSb) and mercury-cadmium-telluride (HgCdTe), but these are mostly used in laboratory spectrometers and radiometers and are seldom found in production process instruments since they are costly and usually require liquid nitrogen cooling.

APPLICATIONS

Infrared spectroscopy has been used successfully in many applications including the following:

TABLE 83.2

Detector	Symbol	Spectral region	Detectivity (D*)	Remarks
Lead sulfide	PbS	1–3.5 μm	1×10^{11}	Requires bias
Lead selenide	PbSe	1–4.5 μm	2×10^9	Requires bias
Pyroelectric	LiTaO$_3$ also others	1–20 μm	1×10^8	AC operation; no bias required
Thermopile	Bi/Sb also others	1–20+ μm	2.5×10^8	AC/DC operation; no bias required
Luft (gas-filled)		1–20+ μm	Wavelength of interest	Filled with gas to be measured

- Industrial stack analysis for CO and CO_2
- CO_2 and CO levels in ambient air
- Hydrocarbon and carbon monoxide in carbon bed scrubbers
- CO_2 in air separation
- NH_4 in ammonia production
- CH_4 in Syngas production
- Toxic organic vapors in ambient air
- Ethylene oxide in ambient air or autoclaves in the pharmaceutical industry
- Leak detection of hydrocarbons as well as CFCs, HFCs, HCFCs in the plastics industry

CHAPTER 84
INFRARED NONCONTACT TEMPERATURE MEASUREMENT: AN OVERVIEW

INTRODUCTION

Along with pressure and flow, temperature is one of the fundamental parameters in any process or laboratory analysis. The most commonly used devices in industry today for the monitoring of temperature are contact sensors; these include thermocouples, RTDs thermistors, and liquid bulb or other types of contact thermometers. The alternative to contact temperature measurement is noncontact infrared temperature thermometry. In most applications, however, the contact devices will suffice and in many cases they will be superior to or less costly than the noncontact sensors. There are specific applications, however, were the noncontact infrared thermometer is the only device that can successfully perform the measurement or where it is the most practical device.

All bodies above absolute zero emit infrared radiation. This radiation, like X rays, radio waves, and visible and ultraviolet light is electromagnetic in nature and travels at the speed of light. The visible portion of the electromagnetic spectrum, violet through red, lies between the ultraviolet and infrared. (See Fig. 84.1.)

The higher the temperature of an object, the shorter the wavelength of the peak infrared radiation it emits. For example, an object at 500°C will emit infrared energy peaking at 3.75 micrometers (μm) while an object at 1000°C will radiate energy peaking at 2.27 μm. (See Fig. 84.2.)

The wavelength of the peak energy for any temperature may be calculated by using the *Wien Displacement law,*

$$\lambda_{max} = \frac{b}{T} \qquad (84.1)$$

where λ_{max} = wavelength of the peak energy in μm
 b = Wien displacement constant, 2897.9 μm K
 T = temperature of the body in kelvins

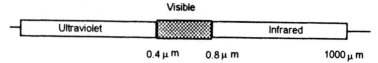

FIGURE 84.1 Electromagnetic spectrum.

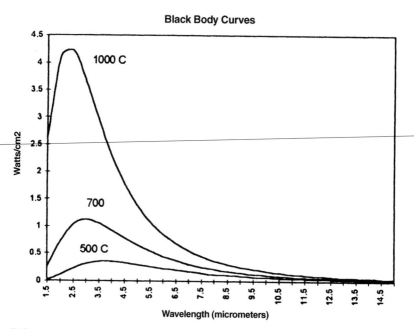

FIGURE 84.2 Blackbody curves.

As the temperature of an object or *body* is increased, the total amount of infrared radiation emitted by it increases as the fourth power of its temperature:

$$W \approx T^4 \tag{84.2}$$

where W = the total energy emitted
T = absolute temperature (K)

By measuring the intensity of this radiation, one can infer the temperature of the object. Unfortunately, things aren't always that simple. No object is an ideal radiator or an ideal absorber and the preceding relationship assumes that the object will radiate 100 percent of its infrared energy.

The theoretical *blackbody* is said to be an ideal absorber and the ideal radiator of energy. That is to say, a blackbody will absorb all the energy that falls upon it while it neither reflects nor transmits any radiation. Since an ideal blackbody emits 100 percent of its radiation, we say it has an *emissivity* (ε) of 1.0; all real objects, known as *graybodies,* are compared to this ideal radiator, and their emissivities are ratioed to the ideal total emissivity of 1.0. Therefore, all graybodies have emissivities of less than 1.0. For example, some typical total emissivities are listed as follows:

Material	Total emissivity
Aluminum	
polished	0.096
oxidized	0.11
Graphite	0.70–0.80
Concrete	0.94
Glass	0.96
Human skin	0.985
Black lacquer	0.96–0.98
Paper	0.94
Plastics	0.94

The total emissivity of the surface of the object governs the amount of energy radiated. The relationship in Eq. 84.2 may be refined to the following:

$$W \approx \varepsilon T^4 \qquad (84.3)$$

where ε = the total surface emissivity of the object

In addition to total emissivity, there are two special cases of emissivity; these are *spectral emissivity*, defined at a particular wavelength, and *specular emissivity*, defined at a particular angle to the surface. For a precision measurement, these should be taken into account.

To calculate the actual infrared energy emitted by an object, the preceding relationship put into the form of the *Stefan-Boltzmann law* becomes

$$W = \varepsilon \sigma T^4 \qquad (84.4)$$

where σ = the Stefan-Boltzmann Constant: $.67 \times 10^{-12}$ W cm^{-2} (K)$^{-4}$

Now we can calculate the *total* energy emitted by an object at a give temperature; for example, to calculate the number of watts/cm^2 emitted by an opaque sheet of paper ($\varepsilon = 0.94$) at 100°C, it would be:

$$W = (0.94)(5.6 \times 10^{-12})(100° + 273°)^4$$

$$W = 0.102 \text{ Watts / cm}^2$$

Using Eq. 84.1, one can calculate the wavelength of the peak energy emitted by the sheet of paper:

$$\lambda_{max} = \frac{2897.9}{373°}$$

$$\lambda_{max} = 7.77 \text{ μm}$$

From the preceding calculations, one has determined that the sheet of opaque paper with an emissivity of 0.94 at a temperature of 100°C will emit 0.102 watts/cm^{-2} total energy and the peak wavelength of the emitted energy will be at 7.77 μm. The energy emitted at the peak can be calculated using *Planck's equation*:

$$W_\lambda = \varepsilon C_1 \lambda^{-5} (eC_2/\lambda T - 1)^{-1}$$

where $C_1 = 3.7413 \times 10^{-12}$ Watts cm^2
$C_2 = 1.4388$ cm degrees
λ = peak wavelength as calculated previously
ε = emissivity at peak λ

(Note: the values chosen for C_1 and C_2 are for watts/cm^2 of energy. For other units C_1 and C_2 will be different)

HARDWARE REQUIREMENTS

The measurement has been defined; the next step is to select the hardware. There are four key elements in an infrared noncontact sensor; these are as follows:

- A target emitting IR energy
- An infrared detector sensitive in the region of interest
- Optical filter
- Collecting optics to focus the energy onto the detector

TARGET

In the preceding example, the target is the sheet of opaque paper. The word *opaque* is a key factor; if the sheet was transparent, energy from objects behind the paper could be transmitted through the paper and detected by the sensor, introducing an error in the measurement. In this example of paper at 100°C, one would need an infrared detector that is sensitive in the region of 7 µm and beyond. We will need optical materials that will transmit this region or one could use reflective optics (Fig. 84.3).

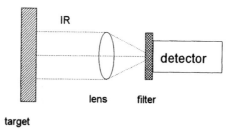

FIGURE 84.3   Detector apparatus.

DETECTORS

There are a variety of infrared detectors to choose from. The detector is usually selected for its wavelength region and then by its *detectivity* (D^*); a few of the typical infrared detectors used for temperature monitoring are as follows:

Material	λ Range (μm)	D*
Silicon	0.5–1.1	12×10^{13}
Germanium	0.7–1.8	5×10^{13}
Lead sulfite	0.7–3.0	1×10^{11}
Thermophile	0.5–20	1×10^{8}

where detectivity is a complex value made up of signal-to-noise ratio, blackbody temperature, chopping frequency, electrical bandwidth, and detector area.

In the preceding application, the ideal detector would be the thermopile with its long-wavelength sensitivity.

OPTICAL MATERIALS

Window and lens materials are chosen for transmission in the region of interest and secondly for their percentage of transmission. Materials commonly used in temperature monitoring are as follows:

Material	Symbol	λ Range (μm)
Quartz	SiO_2	0.2–3.7
Sapphire	Al_2O_3	0.17–5.6
Calcium fluoride	CaF_2	0.1–8.8
Germanium	Ge	1.8–16.7
Silicon	Si	1.2–15
Irtran 4	ZnSe	0.5–22

In this example, one can choose one of a number of materials including germanium, silicon, or Irtran that transmit about 50 percent in this region. With antireflection coatings, this can be improved to over 80 percent.

OPTICAL FILTERS

The broader the accepted wavelength of the measurement the greater the energy gathered. For this application, one could use an optical filter transmitting in the region of 7 to 14 μm.

TWO-COLOR ANALYSIS

One assumption that is made when using infrared temperature sensors is that the emissivity of the target is constant throughout the measurement; this fact is not always the case. In some applications the target changes state during the heating or cooling; in other materials, the emissivity will change dramatically with temperature. In these types of applications, a *two-color* or *dual-wavelength* analysis will greatly reduce the errors due to varying emissivity. In ratio thermometry, the target temperature is inferred from the ratio of the signals received by the sensor at the two different wavelengths. By ratioing Planck's equation for the same temperature at adjacent wavelengths $\lambda 1, \lambda 2$, one arrives at the following relationship:

$$\frac{1}{T} = \frac{1}{T_{AR}} + \frac{\lambda_1 \lambda_2 \ln(\varepsilon_1/\varepsilon_2)}{C_2(\lambda_2 - \lambda_1)} \qquad (84.6)$$

where T_{AR} is the apparent ratio temperature.

If the two emissivities ε_1 and ε_2 are identical, the second term on the right-hand side goes to zero and T_{AR} equals the true temperature T. If the emissivities are not equal, the second term represents the error in $1/T_{AR}$.

Two-color sensors are frequently used in applications where the target is smaller than the field of view of the sensor, such as in filament temperature measurements or in glass fiber pulling. Here the target does not fill the field of view of the sensor and the sensor sees other targets in the background. If the other targets are lower in temperature than the target of interest and uniform in temperature, the two-color sensor will greatly reduce any errors due to the background target.

The two-color sensor will also compensate for water mist, smoke, or dust between the sensor and the target, assuming that the attenuating medium is neutral attenuating.

APPLICATIONS

Applications where infrared thermometry will excel include applications where

- The target is of low mass and a contact device will influence the temperature measurement.
- The target is moving.
- The measurement is in a high-induction field and a contact device could not be used.
- High temperatures where probes would be destroyed.
- Fast response is required; IR sensors can have response times of less than 10 milliseconds.

Typical applications may be found in many industries including the following:

- Agriculture
- Aircraft
- Aluminum
- Automotive
- Food
- Glass
- Engine
- Lightbulbs
- Medical
- Paper
- Plastics
- Paint
- Rubber
- Steel

CHAPTER 85
QUALITY CONTROL CONSIDERATIONS

Semiconductor/silicon wafer processing technology continues to advance an already mature industry to new heights. The silicon sensor industry has adopted the proven technology of silicon wafer fabrication to produce inherently reliable devices that are already mature from a reliability point of view even before the product is made available on the market. The back end operations, which encompass the die attach, wire bond, test, and packaging processes, are also transferred from stable and mature standards developed in the semiconductor industry.

Failure modes identified over the years have been routinely designed out, to the point where even the most complex devices deliver reliability ratings deemed achievable only in theory just a few years ago. The challenge to the reliability of devices being pushed to their limits through shrinking geometries, thinner metals and dielectrics, and higher integration level is being successfully met by the use of the process management tools of Statistical Quality Control, Statistical Process Control (SQC/SPC), and wafer-level reliability, promising even greater breakthroughs in device complexity and reliability. *Statistical Quality Control (SQC)* involves the use of specialized techniques to improve the overall quality of sensors produced by manufacturing. *Statistical Process Control (SPC)* is used to efficiently identify excessive variability of critical process parameters which adversely effect the sensor being processed.

Reliability activities are designed to support quality goals by providing data to R&D as a design tool and to manufacturing as a performance measurement tool. Data is collected for use in the projection of operating life, and engineering analysis is performed to help to discover, understand, and correct any and all failure mechanisms. Products and manufacturing processes are subject to the scrutiny of reliability monitors, used to continuously validate the materials and process controls used in the manufacture of the product.

SPC is defined as the conversion of statistical data to process information for use in documenting, correcting, and improving process performance. Reducing process variability in the manufacture of sensors aids predictability, enhances yields, and increases the achieved sensor reliability.

SPC does more than help to improve the quality of products shipped to customers. The traditional quality control function is to reject nonconforming parts, sorting them from the conforming. Semiconductor manufacturing delivers large volumes of products in short periods of time. When it is learned that products are rejects, large product volume translates into a large financial loss to a manufacturer. SPC allows manufacturers to minimize their loss by monitoring critical process parameters and taking early corrective action when a trend indicates a process is moving toward the control limits.

In practice, statistical quality controls have helped in the manufacture of highly reliable sensors in volume production quantities. Statistical process control is a proven and invaluable tool when used in the pursuit of timely and systematic quality improvements and, when used, has helped displace the need for end of line inspection, helping to reduce manufacturing costs.

DESIGN ASSURANCE

Poorly designed transducer and accelerometer products can be affected by a host of afflictions. Complete device characterization includes chip and packaged part qualification testing for short- and long-term stability, overpressure tolerance, temperature and pressure hysteresis, pressure life-cycling, thermal repeatability, and percent FSO/g acceleration as applicable.

Instrumentation and test setup requirements can be quite demanding as measurement repeatability must be maintained in the microvolt range to assure device performance to standards under all conditions. Industry standards that offer complete guidelines on acceptable device nomenclature, characteristics, and test requirements for transducers and accelerometers are as follows:

- ISA-S37.1, Electric Transducer Nomenclature and Terminology
- ISA-S37.3, Specifications and Tests for Strain Gage Pressure Transducers
- ISA-S37.5, Specifications and Tests for Strain Gage Linear Acceleration Transducers
- ISA-S37.8, Specifications and Tests for Strain Gage Force Transducers

A properly structured life-cycle test can yield valuable insights into the probable lifetime of the product over a range of applications from moderate to the most stressful conditions.

Mechanical and environmental characterization includes some or all of the following tests.

- Temperature cycling
- Thermal shock
- Temperature/humidity
- Constant acceleration
- Vibration
- Mechanical shock
- Salt atmosphere
- Internal visual and mechanical
- Wire bond integrity
- Package lead integrity
- Solderability
- Fine and gross leak
- Resistance to solvents
- Internal water vapor content

CHAPTER 86
MICROSYSTEM TECHNOLOGIES

INTRODUCTION

The use of silicon technologies for sensor fabrication allows in a variety of applications the integration of electronic circuits together with the sensing element on one chip, resulting in complex microsystems. Besides lowering manufacturing costs for high-volume production, the silicon integration of both sensors and electronics offers several significant advantages, such as small size and low weight, close physical proximity of sensors and readout, and the possibility of integrating multisensor arrays in combination with highly complex signal processing. For signal conditioning, CMOS circuit design seems to be the most important technique for the monolithic approach. Compared to bipolar circuits, power consumption is much lower. Compared to BICMOS, the process is much less complex and has a much lower number of masks. This is very important, because, in most cases, monolithic integration means additional process steps and by this a higher process complexity.

Surface Micromachined Pressure

The sensor presented here has been developed for the measurement of absolute pressure in the fields of automotive and mechanical engineering. The integration of the system consists of a pressure sensor, a signal-conditioning circuit with an on-chip clock. The fabrication technique to form the sensor is surface micromachining implemented in a standard 2-μm CMOS process. This technique allows the fabrication of small pressure sensor elements with very low temperature dependence (< 200 ppm/°C) without any compensation. So, the system can be produced very small with the advantages of an easier assembly, a high resistance to electromagnetic interference, and low production costs.

The capacitive pressure sensor consists of an array of single, circular, pressure-sensitive elements switched in parallel, and reference elements. A cross section of such an element with the corresponding reference element is shown in Fig. 86.1. The capacitor is formed by a fixed electrode in the substrate and a movable membrane of polycristalline silicon above. The cavity under the membrane is obtained by etching and later vacuum sealing. Besides building the pressure sensitive elements, pressure insensitive elements are fabricated, too. These reference elements are constructed in an identical manner, with the only difference being the thickness of the membrane. The output voltage signal is proportional to the quotient of the capacitance of the sensor and reference sensor. So, parasitic effects are suppressed.

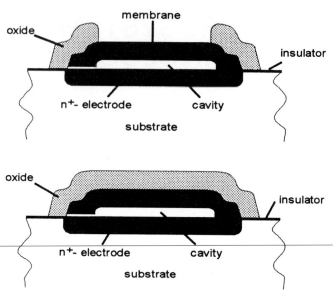

FIGURE 86.1 Cross section of a sensor and a reference sensor element.

The membrane diameters have been optimized for pressure ranges of 1.6, 9, and 30 bars and amount to 120 μm, 70 μm, and 50 μm, respectively. In Fig. 86.1(b), a photograph of a 9-bar sensor chip with the dimensions of 1.8 mm × 3 mm is shown. Because of the small distance between the membrane and the substrate of only 1 μm, the sensor shows an outstanding overpressure resistance.

In Fig. 86.2, the behavior of a pressure sensor by load with 100,000 pressure cycles is shown. No fatigue of the membrane or hysteresis has been observed.

Due to the concept of sensor/reference measurement, the pressure sensor shows a small dependence in temperature. The error is described in Fig. 86.3 for a temperature difference of 100° (−20–80°C). Without any additional temperature compensation, the error is lower than ± 200 ppm/°C FS.

Because of its near $1/x$-behavior, the output signal is nonlinear. By forming the quotient C_{ref}/C_{sen}, a simple linearization is obtained. So, a fit with a function of the second order is sufficient for an accuracy of 1 percent. The calibration data can be saved in an EEPROM on chip.

For signal conditioning and amplification of the capacitive pressure signal, a switched capacitor circuitry in differential path technique is used to convert the signal in a voltage. An oscillator has been integrated to provide the clock for the switched capacitor circuitry. The power supply amounts to 5 V. A photograph of the chip can be seen in Fig. 86.4.

MONOLITHIC MAGNETIC FIELD-SENSOR WITH ADAPTIVE OFFSET REDUCTION

CMOS-compatible magnetic field sensors such as magnetotransistors and Hall plates are well-suited for the integration of the sensor with the analog and digital circuitry on

MICROSYSTEM TECHNOLOGIES

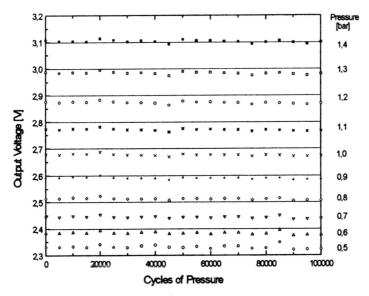

FIGURE 86.2 Reproduction by cycles of pressure.

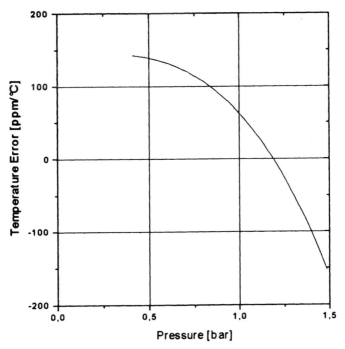

FIGURE 86.3 Temperature error (−20°C to 80°C).

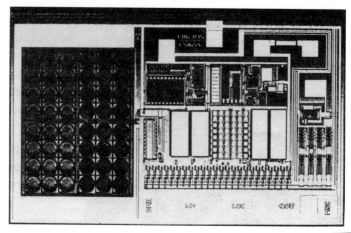

FIGURE 86.4 Photograph of the integrated pressure sensor.

the same chip. Their advantages are well-controlled geometry, high sensitivity, and their low-cost fabrication in a standard CMOS-process. Unfortunately, a irreproducible offset voltage with a magnitude comparable to the sensor signal itself (up to several millivolts) prevents sufficiently high resolution. This offset voltage cannot be compensated by applying a calibration step at the end of the production process due to its dependence on the time-varying mechanical stress (e.g., created by different thermal expansion coefficients of the sensor chip and the carrier). Therefore, an adaptive offset reduction technique is mandatory to achieve a sensor resolution better than lmT. Munter[2] describes an offset reduction method by using the spinning current principle, which benefits from the periodicity of the piezoresistive coefficient in the chip plane. In this chapter, a spinning current Hall plate with eight contacts integrated together with the circuitry for sensor biasing, offset reduction, and signal amplification is presented. A MOS transistor serves as the Hall element. The signal-processing circuit has been realized using fully differential switched capacitor design techniques.

Sensing Element

The sensor signal V_{Hall} of a Hall plate is proportional to its sensitivity constant S_A and the magnetic field component B, which is perpendicular to the surface of the chip:

$$V_{Hall} = S_A B + V_{on} + V_{off} \qquad (86.1)$$

The smallest magnetic field that can be detected is limited by two additional voltage components, the noise voltage V_n and the offset voltage V_{off}. A method of flicker noise suppression in symmetric Hall plates by orthogonal switching of the supply current and subsequent addition of the Hall voltages is described in Ref. 3. The effect of the thermal noise voltage can be reduced by low-pass filtering or averaging. The most significant error is introduced by the offset voltage V_{off}, which is caused by lithographic misalignments and by the piezoresistive effect, where the former has to be minimized by careful layout. Referring to the spinning current principle, we use a NMOS transistor with 8 contacts, shaped like a regular star (see Fig. 86.4), as a sensing element. Cyclic interchanging at a step of 45° of the supply current direction and

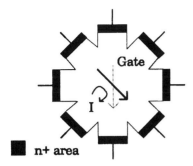

FIGURE 86.5 Schematic view of the Hall sensor.

summing up the Hall voltages at the perpendicular oriented contacts yields an offset attenuation of a factor of 100–1000.[4,5] Measurements show clearly that leakage currents due to resistive loading of the Hall sensor during the voltage measurement must be avoided (Fig. 86.5).

Signal Processing Electronics

The circuit consists of a sensor biasing block, two multiplexers, the signal amplification and integration stage, and the control logic. One measuring cycle consists of one autozero step and eight integration steps, during which the sensors supply current of about 200 µA, and the gate voltages of the Hall plate are held constant. The amplification stage provides fully differential amplification of the sensor signal using switched capacitor techniques, thus avoiding resistive loading of the sensor. Each stage employs correlated double sampling to reduce op-amp offset and 1/f noise. Autozero is carried out at the beginning of a integration cycle. A sensitivity of 10 V/T and an output offset of 5 mV have been measured. The clock frequency (up to 100–200 kHz) is not critical.

A PLANAR FLUXGATE-SENSOR WITH CMOS-READOUT CIRCUITRY

The next example is a miniaturized magnetic field sensor system consisting of a fluxgate-magnetic field sensor fabricated in CMOS-compatible planar technology, and a CMOS-ASIC for sensor supply, readout, and signal processing. The electronic circuit analyzes the sensor output signal at the second harmonic of the excitation frequency.

The fluxgate-sensor as shown in Fig. 86.6 consists of two parallel ferromagnetic cores. Each core is about 1300 µm long, 100 µm wide, and 0.5 µm thick. The sensor

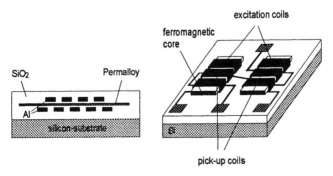

FIGURE 86.6 Planar double core fluxgate magnetic field sensor. Left: cross section of one solenoid with ferromagnetic core. Right: principal sensor design (compensation coils not shown).

coils are fabricated by using the two aluminum layers and the vias of a conventional 1.5 μm double-metal CMOS process. The pickup coil (40 turns) is located at the center of the core, followed by the excitation coil (100 turns) and the compensation coil (100 turns), both divided into two parts. By connecting the pickup coils in a difference circuit the non-field-depending odd-frequency components of the sensor signal are suppressed. The even, magnetic-field-related harmonics are added, resulting in a higher sensitivity in comparison to a single core sensor.

The ASIC consists of three main parts: an oscillator, generating the triangular driving current for the sensor; the readout circuit; and the control unit. The block diagram of the CMOS-ASIC is shown in Fig. 86.7.

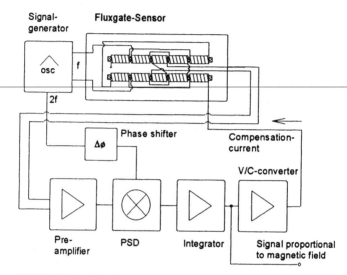

FIGURE 86.7 Circuit arrangement of the CMOS-ASIC with connected sensor.

The internal oscillator generates a symmetrical triangular voltage in a frequency range of 200 kHz–2 MHz. By the following voltage-current converter the premagnetization current in the range 1–3 mA is driven through both excitation coils with antiparallel current direction. The sensor output voltage is proportional to the excitation frequency. A value of 0.15 $V/(kHz*T)$ is measured for the second harmonic. The sensor signal is amplified and then multiplied by a signal of double excitation frequency using the mixer. For achieving highest performance, the phase is adjustable. The DC part of the signal, which is proportional to the external magnetic field, is integrated over a predefined time (0.5–5 ms) by the integrator. After completing integration, the signal at the integrator output can be read out by a sample-and-hold circuit.

A feedback control loop improves the linearity of the sensor system. The integrator signal is converted by a voltage-controlled amplifier to a current driving the compensation coils. The external magnetic field is compensated by the field of compensation coils. So the fluxgate-sensor is operating nearly in a zero field. The compensation current is proportional to the external field. In compensation mode, a sensitivity of 25 µÅ/µT is measured. The dependency of the compensation current on magnetic field (in the range of ±T) is shown in Fig. 86.8. More results will be given at the conference.

Typical applications might be magnetic field measurement, noncontact current measurement, and electronic compass systems.

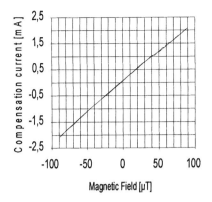

FIGURE 86.8 Magnetic field dependency of compensation current.

A THERMOELECTRIC INFRARED RADIATION SENSOR

In a last example, fabrication and characterization of a monolithic integrated infrared radiation (IR) sensor system is described. The sensor system consists of a thermoelectric IR sensor and a sensor measuring the temperature of the sensor chip and integrated amplifiers for both signals. The fabrication of this system with a chip area of 3.5×3.5 mm^2 is done with a standard CMOS process on SIN40X substrates. A cross sectional view is shown in Fig. 86.9.

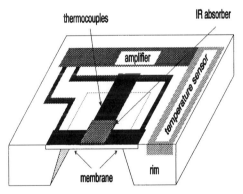

FIGURE 86.9 Cross sectional view of the IR sensor system.

The SIMOX substrates are fabricated by a high dose oxygen implantation (SIMOX) and thermal annealing. They enable the fabrication of crystalline silicon leads on thin, low thermal conductivity membranes of silicon oxide and silicon nitride. So we take advantage of the large, dopant-depending Seebeck coefficient of crystalline silicon and the good thermal insulation of silicon oxide/silicon nitride membranes for highly sensitive thermopiles. In order to achieve a large sensor responsivity, thermocouples made of p^+-crystalline silicon and n-polycrystalline silicon are fabricated in the CMOS process. The thermopile consists of 60 thermocouples connected in series. The area of the thermal insulating membrane between the hot junctions at the center of the membrane and the cold junctions located at the rim of the sensor chip is 1.6 × 1.6 mm². The IR absorber, which converts the incoming radiation to heat, is placed at the center of the membrane and has an area of 0.6 · 0.6 mm². This absorber consists of patterned aluminum and silicon oxide and is fabricated without any additional process steps. An emissivity between 0.6 and 0.9 is measured in the wavelength range from 2.5 to 15 μm. A thermopile responsively R_V = 209 V/W, and an internal resistance R_i = 520 kW are measured.

Assuming thermal noise only we calculate a noise equivalent power NEP= 0.45 nW/Hz 0.5 and a specific detectivity D= 1.3 × 108 cm Hz 0.5/W. The temperature coefficient of the responsively is about –0,1 %/K. In typical applications, thermopile voltages are less then some millivolts. Therefore, amplification is necessary for trouble voltage suppression and getting a usable signal level. This is done by the integrated amplifiers located at the rim of the thermopile. These are designed in switched capacitor technique. The gain is programmable in four steps between 10 and 1250. A temperature drift of the offset voltage smaller than 2.2 pV/K is measured in the temperature range between 10°C and 90°C. An equivalent input noise voltage of about 1 pV/Hz 0.5 is measured in the frequency range below 200 Hz. The knowledge of the thermopile temperature is necessary for the compensation of temperature-dependent sensor parameters and for the contactless temperature measurement. The integrated temperature sensor is based on the temperature dependence of the current voltage characteristic of silicon diodes. The use of a differential arrangement shows a sensor voltage that depends linearly on temperature. A temperature coefficient of 1mV/K is measured.

CONCLUSION

These examples have shown that the monolithic approach using CMOS technologies is possible on a variety of different principles. Other examples during this conference have shown this, too. Institutes like the Fraunhofer Institute of Microelectronic Circuits and Systems that have these technologies available can act as service centers in this field. By this, even small and medium enterprises can have access to new technologies and improve their products or get innovative new ones.

REFERENCES

1. H. Dudaicevs, M. Kandler, Y. Manoli, W. Mokwa, and E. Spiegel, "Surface Micromachined Pressure Sensors with Integrated CMOS Readout Electronics," *Sensors and Actuators*, A43, 1994, p. 151–163.
2. P. Munter, "A Low Offset Spinning Current Hall Plate," *Sensor and Actuators* A21–23, 1990, p. 743–746.
3. Z. Stoessel and M. Resch, "Flicker Noise and Offset Suppression in Symmetric Hall Plates," *Eurosensors*, 1992.
4. R. Gottfiied-G., "Thermal Behaviour of CMOS Hall Sensors for Different Operating Modes," *Eurosensors*, 7, 1993, p. 219.
5. R. Gottfried-G., "Ein CMOS Hall Sensor Mit 8 Kontakten Zur Anwendung Des Prinzips der StromrichtungsmicroMeterkehr," *Proceedings Sensor 1993*, Niimberg, vol. 5, p. 47–54.

INDEX

AbNet, 3.27–3.28
Absolute Sensors, 66.6–66.8
Accelerometers, 11.12, 11.6, 11.8, 11.9, 52.1
  addressing, 78.4–78.5
  customization, 78.7–78.8
  electrical trimming, 78.6
  error detection, 78.4
  packaging for, 78.6–78.7
  signal-condition, 78.1–78.8
  signal processing, 78.3–78.4
  silicon, 66.29–66.31, 67.13–67.17
Acoustic emission (AE)
  laser, 36.9
  monitoring system, 36.1
  strain, 36.1
  strain-gage monitoring, 36.2, 36.4, 36.5, 36.6
  strain-gage testing, 36.8–36.9
  testing, 36.1, 36.2, 36.3, 36.7
Acoustic pyrometers, 33.8, 33.11
  acoustic path, 33.5, 33.12
  boilers, 33.17
  burner control and optimization, 33.14
  depth, 33.11
  distance, 33.11
  emission reduction, 33.16, 33.17
  flight time, 33.11, 33.13
  slagging measurements, 33.16
  sound, 33.11, 33.12, 33.13
  thermal probe, 33.14
Acoustic sensors, 33.1
  flow measurement, 33.1
  gas, 33.6
  temperature, 33.6
Acronym appendix, 29.14–29.15
Addressing:
  in signal-conditioning electronics, 78.4–78.5

Aerospace applications, 71.5, 71.7–71.8
  piezoresistive sensors for, 77.2–77.4
Aluminum extruders, 18.1
  accuracy, 18.1
Aluminum oxide, 39.7–39.8
AMC sensors:
  permalloy sensors, 19.10
  (*See also* Magnetoresistance)
Ammeter, 2.50
  (*See also* Fiber optics)
Amplifiers:
  for current sources, 70.33
  instrumentation, 70.35–70.40
  and switched capacitor, 70.38–70.40
  three-amplifier configuration, 70.37–70.38
  two-amplifier configuration, 70.38
Analog closed-loop system, 52.6
Analog electrical models, 70.4–70.13
Analog links, 2.26
  video links, 2.26–2.27
Analytical modeling, 69.3–69.4
Angular rate, 52.1
Angular rate gyroscope, 52.4
  applications, 52.9
Angular rate sensors, 52.1
ANSYS program, 67.16, 69.6
Applications:
  capacitive proximity, 1.34, 1.35
  industrial (*see* Industrial applications)
  linear indexing, 9.32–9.35
  photo detectors, 1.11
  silicon sensors, 71.4–71.10
Applied force transduction, 72.5
ASI (Actuator Sensor Interface), 74.2, 74.3
Atomic force microscopes (AFMs), 68.20
Attenuation, 76.10–76.11
Attenuators, 2.40
Automated Process Control (APC), 74.5

I.1

Automatic Level Control (ALC) circuits, 79.9
Automation, 6.2
Automotive, 16.1–16.7
  (*See also* MEMS)
Automotive applications:
  all-silicon structures for, 72.7
  battery charges, monitoring, 79.13–79.15
  coolant monitoring, 79.16–79.18
  engine oil monitoring, 79.10–79.13
  quartz resonators for, 79.1–79.18
  silicon accelerometers for, 67.13–67.17
  thin/thick film ceramic sensors for, 75.1–75.7
Automotive Balance Control (ABC), 39.21
Automotive safety, 16.3
  monitoring gas, 16.3
  U.S. vehicles, 16.3
Autozeroing circuits, 70.40–70.41
Aviation applications, silicon accelerometers for, 67.13

Barium titanate, 75.6–75.7
Batibus, 74.2
Battebus, 74.2
Batteries, monitoring charge, 79.13–79.15
Beams, clamped-clamped, 72.1
Belt scales, 47.22
Belt speed, 47.22
Biological materials, 5.33
Biomedical, 5.32–5.33
Blackbody, 18.4, 18.6
  FOV (field of view), 18.5
  infrared filtering, 18.3
Blood pressure sensors, 60.1, 67.2
Boiler performance, 33.17–33.18
  efficiency, 33.18
  heat rate, 33.17, 33.20
  time, 33.17, 33.18, 33.19, 33.20
  volumetric flow rate, 33.18
Bonding methods, 71.13–71.15
Bosses, 66.10–66.11, 67.18
Bowstring testing, 76.2, 76.22
  electrical response to, 76.22–76.24
Breakdown, 76.18
Bridge, 21.1
  composite material, 21.10
  crack, 21.7, 21.9
  infrastructure monitoring, 21.1
  monitoring, 21.1, 21.2, 21.6
  prestressed, 21.9
  problems, 21.1

Bridge (*Cont.*):
  SQUID, 21.10
  structural health monitoring systems, 21.1
  tension, 21.10
  wires, 21.9–21.10
Bridge arm resistance, 70.7
Bridges, 21.2–21.10
Bridges and buildings:
  bridge project monitoring samples, 21.6, 21.8
  embedment applications, 21.9–21.10
  ferromagnetic response, 21.2
  health, 21.1
  martensitic transformation, 21.2
  safety concerns, 21.5
  strains, 21.2, 21.5, 21.6, 21.9, 21.10
Bridge voltage:
  closed sensor testing and modeling, 70.7–70.13
  increase in, 70.20–70.21
  offset voltage of, 70.15
BTU analysis, gas chromatographs, 68.4
Bulk material, 19.1
Bulk micromachined:
  accelerometer, 11.6
  bond metal layer, 11.9
  electrostatic force, 11.8
  forced feedback, 11.6
  position detector, 11.8
  quadrant plates, 11.9
  tip die, 11.9–11.10
Bulk-Polymer-Humidity sensor, 39.3
  (*See also* Humidity)
Burner, 33.14
  control, 33.14, 33.16
  optimization, 33.14
Business management, 4.13
  chart model, 4.13
Bus survey:
  cost, 29.9
  interoperability, 29.9
  loaded network performance, 29.8
  memory, 29.10
  message-passing capability, 20.11
  ownership, 29.10–29.11
  outlook, 29.10
  peer-to-peer capability, 29.10
  product availability, 29.9
  reliability, 29.10
  remaining work, 29.11
  response time, 29.7, 29.8
  selection, 29.7

Bus survey (*Cont.*):
  suppliers, 29.7
  technology, 29.9
  user-friendliness/tools, 29.10

Cables:
  for connecting sensors to equipment, 76.1–76.26
  (*See also* Coaxial cable)
Calibration, 39.26
  errors, 70.2–70.3
  function, of sensor package, 71.2
  National Calibration Laboratories, 39.26
  two-flow method, 39.28
  two-pressure method, 39.28
  two-temperature method, 39.28
Cameras:
  color CCD cameras, 25.6–25.7
  composite, 25.6
  consumer, 25.6
  encode, 25.6
  images, 25.6, 25.7
  non-TV-format in machine vision, 25.7
  pixels, 25.6, 25.7
  scanning, 25.6
Cantilever accelerometer, 67.14–67.16
  formation of, 72.1
Capacitance:
  of coaxial cable, 76.13–76.14
  versus piezoresistance, 66.28–66.29
  variable, SFB for, 67.12–67.13
Capacitive Polymer Sensor, 39.5–39.6
  (*See also* Humidity)
Capacitive pressure sensors, 66.26–66.29
Capacitive proximity, 1.33
Capacitive sensors, 1.33–1.37, 1.38–1.39, 23.9, 52.5, 72.5
  principles, 23.9
  temperature sensitivity of, 66.28–66.29
Catheter-tip sensors, 67.6–67.8
CCM, 39.15
CCM sensor, 39.17
CCM technique, 39.22
Cell devices, 12.5
  biocompatibility, 12.6
  cardiac cells, 12.5
  military toxins, 12.5
  neural repair, 12.6
  pharmaceutical research, 12.5
Cell manipulation, micromechanical techniques for, 68.19–68.20

Ceramic sensors, 54.1, 54.2
  automotive engine testing, 54.3–54.5
  gas sensors, 54.2
  high temperatures, 54.1, 54.2
  immittance spectroscopy, 54.5
  pollution control, 54.1
  thermistors, 54.6–54.9
  thick film, 54.2
  thin film, 56.1, 56.2
  thin/thick films, benefits of, 75.5–75.7
  thin/thick film technology for, 75.1–75.7
Chemical and biochemical sensors, 26.7–26.9
  conductivity sensor, 26.7
  two-dimensional measurement of concentrations matrix, 26.10
Chemical and gas sensors:
  calorimetric sensors, 55.4–55.5
  microfabrication, 55.1
micromachining, 55.1
  Schottky-diode-type sensor, 55.2
  solid electrolyte electrochemical sensors, 55.3
  tin-oxide-based sensors, 55.2
Chemical deposition method, 75.2
Chilled Mirror Hygrometer, 39.9–39.14
  (*See also* Humidity)
Chips, cantilever design, 67.14
CIM (Computer Integrated Manufacturing), 1.2, 1.3, 1.4, 3.1, 4.1
  architecture diagram, 4.29
CIM DB (database), 4.20
  communications, 4.30
  components of, 4.17
  components chart, 4.17
  design, 4.14
  FMS (Flexible Manufacturing System), 4.16
  strategy, 1.4, 4.8, 5.29
CIM Plan, 4.2–4.3
Circuits, digital signal simulators for, 69.12–69.13
Clamped-clamped beams, 72.5
Clinical diagnostics, 12.7
  capillary electrophoresis, 12.8
  medical MEMS, 12.7
  muscle tissue, 12.7
CMOS, 11.1, 11.2, 14.1, 26.1–26.2, 52.4
  circuit technology, 52.4–52.7
  magnetic field sensors, 86.2
  readout circuitry, 86.5

INDEX

Coaxial cable:
  attenuation of, 76.10–76.11
  capacitance testing, 76.13–76.14
  characteristic impedance, 76.8–76.9
  comparison of, 76.25–76.26
  continuity testing, 76.16–76.18
  crosstalk, 76.9–76.10
  dielectric absorption, 76.15–76.16
  dielectric withstand voltages, 76.18–76.19
  electrical length, 76.9
  electrical response, 76.21–76.25
  flex degradation/flex life, 76.24
  frequency domain measurements on, 76.10–76.12
  high-frequency measurements, 76.7–76.12
  insulation resistance, 76.14–76.15
  leakage current, 76.14–76.15
  low-frequency measurements, 76.13–76.21
  series resistance, 76.16–76.18
  shielding effectiveness, 76.11–76.12
  testing, automated, 76.19–76.21
  testing, mechanical, 76.21–76.25
  triboelectric noise in, 76.1–76.2
Color, 5.8, 22.1
  ammunition inspection and sorting, 22.8
  chroma, 22.7
  definition, 22.1–22.2
  food industry, 22.8
  hue, 22.7
  medical closures, assembly of, 22.8
  online, 22.1
  sorting of, automotive parts, 22.8
  sorting of, lumber, 22.8
  theory, 5.8–5.9, 22.7
  valve, 22.7
Color-based objects:
  3D color, 25.10
  classes of, 25.7–25.9
  classification, 25.7–25.9, 25.11
  color distribution, 25.8, 25.9
  cube, 25.7
  HSI space, 25.8
  monochrome, 25.1, 25.7
  reference, 25.12
  RGB, 25.7
  shadows, 25.10
  table lookup, 25.8, 25.11
  test, 25.8
  thresholding, 25.11
  traditional, 25.11
  vector, 25.7, 25.8, 25.9
Color measurement, 5.10, 22.4

Color sensing:
  applications, examples of, 22.8
  AVIRIS (Advanced Visible and Infrared Imaging Spectrometer), 25.3–25.4
  earth, 25.3
  imaging, 25.3
  moon, 25.3
  operation, principle of, 22.7
  principles, 25.2–25.5
  sensing instruments, 22.5
  surface, 5.10
  spectrometer, 5.10, 25.3
  spectrophotometers, 5.10, 22.5, 25.3
  spectroscopy, 25.3
  wavelengths, 5.8, 22.3, 25.2
Color vision:
  applications of, 25.2
  CCD cameras, 25.6–25.7
  gray-scale, 25.7
  human, 22.5, 25.1, 25.4
  monochrome, 25.1, 25.7
Commercial applications:
  vibrating diaphragm sensors for, 77.1–77.2
Common mode rejection ratio (CMRR), 70.36
Communication, 1.7, 3.14, 4.28, 4.30, 9.1
  application layer, 3.15, 3.16
  data link layer, 3.17
  LAN (local area network), 4.15
  network layer, 3.17
  networks, 1.5, 1.7
  physical layer, 3.17–3.18
  presentation layer, 3.16–3.17
  protocol, 3.14, 4.28
  session layer, 3.17
  transport layer, 3.17
Compensating resistors, 70.16–70.20
Compensation:
  constant current, 70.13–70.20
  constant voltage FSO, 70.20–70.23
  digital, 70.28–70.33
  diode, 70.22–70.23
  in package, 71.2
  thermistor, 70.21–70.22
  third-order polynomial, 70.32–70.33
Computer-aided design (CAD), 69.1–69.13
Computer-aided layout (CAL), 69.11
Computer-aided software engineering (CASE), 69.13
Computer communications, 3.14–3.18

Computer design tools, 69.1–69.13
  for sensor and microstructure layout,
    69.10–69.11
Computer Integrated Manufacturing (*see*
    CIM)
Computer modeling, 69.2–69.9
Computer schemes for smart sensors (*see*
    Control systems, smart sensors)
Conductive Polymer-Based sensors, 81.1
Constant current compensation,
    70.13–70.20
Constant current excitation, 70.15, 70.33
Constant voltage, FSO compensation,
    70.20–70.23
Consumer applications, 71.8
Contact, ohmic, 75.4–75.5
Continuity, measurements of, 76.16–76.18
Control loop, 6.1
Control processes, 1.8
  closed-loop control system, 1.9–1.10
  engineering integrated system, 1.8
  open-loop control system, 1.8–1.10
Control schemes, 73.1
  peer-to-peer distributed, 73.1–73.11
Control systems, 1.1–1.3, 1.4, 1.7, 80.1
  bus-based control configuration, 80.4
  computers, use of, 80.3
  control technology, 1.7
  cost of, 74.1
  distributed control configuration,
    80.4–80.5
  fuzzy representation, 80.7–80.8
  fuzzy and rough set theories, sample
    application, 80.11–80.12
  generic model of, 80.2
  hub-based control configuration,
    80.3–80.4
  rough representation, 80.9–80.11
  smart sensor model of, 80.5
  VSATs (very small aperture terminals),
    1.1
Coolants, monitoring, 79.16–79.18
Costs:
  of control systems, 74.1
  of distributed versus central control
    systems, 73.1
  installed, 74.2, 74.5
  low-cost packages, 71.19–71.22
  of silicon sensors, 71.5
Crosstalk, 76.9–76.10
Cryogenic Manufacturing, 7.29
  high temperatures, 7.30

Current:
  constant versus constant voltage excita-
    tion, 70.4
  leakage current, 76.14–76.15
  sources, for sensor excitation, 70.33–70.35
Curve-fitting algorithms, 70.28–70.33

Daisy chain data bus, 2.44
Damping (of oscillation):
  gas and liquid, 67.15
DAQ (*see* Data acquisition)
Data acquisition (DAQ), 34.1–34.7
  intelligent I/O, 34.1
  parallel-port devices, 34.4–34.5
  PCMCIA cards, 34.5–34.6
  plug-ins, alternatives to, 34.2
  portable applications, 34.1–34.2
  power considerations, 34.6
  serial-port devices, 34.3–34.4
Database, 8.26, 8.27
Data Bus networks, 2.27
Datalink layer, 3.17
Debuggers, source-level, 69.13
Density, quartz resonators for monitoring,
    79.12
Deposition methods, 75.2–75.4
  for thin film, 75.2
Design process, of silicon sensor packaging,
    71.16
Detection methods:
  proximity detection, 1.18–1.20
  reflex detection, 1.18, 1.19
DeviceNET (Allen-Bradley CANbus), 74.2,
    74.3
Dew Point hygrometers, 39.23–39.24
Dew Point transmitters, 39.20–39.21
Diaphragms, 66.5–66.10
  metal, 71.18–71.19
  pressure-sensitive, 66.6
Dicing and bonding:
  actuators, 15.3, 15.4, 15.5
  array, 15.3, 15.4, 15.5
  fragile micromachined structure, 15.3
  positioning, 15.5
  testing, 15.3
Die:
  packaging of, 71.1–71.2
  unpackaged, 71.16
  (*See also* Packaging technology)
Die attach, 71.13–71.16
Die characterization, 71.13
Die down, 71.13–71.15

Dielectric absorption, 76.15–76.16
Dielectric withstand voltages, 76.18–76.19
Die protection, 71.15–71.16
Differential capacitive sensing, 52.5
Differential pressure, measurement of, 70.25–70.27
Differential pressure sensors, 66.8–66.10
Digital applications, for cables, time domain measurements, 76.7–76.10
Digital compensation, 70.28–70.33
Digital encoder, 5.4–5.5
   position encoder, 5.5–5.7
Digital links, 2.24–2.26
Digital simulators, 69.12–69.13
Digital trimming, 67.2–67.3
Diodes, compensation with, 70.22–70.23
Displacement sensor, 39.7
   (See also Humidity)
Distributed control:
   LONWORKS technology for, 73.7
   operating systems for, 73.10
   peer-to-peer, 73.1–73.11
Distributed I/O systems, 74.1–74.21
Distributed measurement nodes, 27.2
   properties, measurement-related, 27.3
   properties, system of application-related, 27.3
   properties, transducer-related, 27.2
Distributed measurements, 27.2
   calibration, 27.2
   communication protocol, 27.4
   control protocol and Real-Time issues, 27.5
   data management issues, 27.4
   identity, 27.2, 27.3
   node, 27.2, 27.3
   physical variable, 27.2
   thermocouple, 27.2
   use of, 27.2
Distributed PO systems, 74.1–74.2
Doping level:
   and piezoresistive sensors, 66.12–66.14
   versus piezoresistive temperature coefficients, 66.14–66.16
Drop tests, 76.21–76.22
Dry bulk solids:
   closed loop, 46.1
   control of, 46.1
   flow rate, 46.4
   multi-ingredient proportioning, 47.31
   structure, 46.3
   weigh belt, 46.3

Dual-Mirror Twin-Beam sensor, 39.14–39.15
   (See also Humidity)
Dual-wavelength, 18.8, 18.9, 84.5
   atmosphere, contaminated, 18.8
   emissivity, 18.8
   graybody condition, 18.8
   ratio, 18.8
   spectral response, 18.8
   wire, 18.9
Dynamic errors, 70.3

EBC microbattery, 53.2
   (See also Secondary batteries)
Echelon, LONWORKS technology, 73.7–73.10
Eddy currents, 17.8, 23.8
EDP, for electrical switch fabrication, 68.11–68.15
Electrical contacts, for thin/thick film ceramic sensors, 75.4–75.5
Electrical interface, in sensors, 71.2
Electrical length, 76.9
Electrical modeling, 69.11–69.13
Electrical noise, triboelectric, 76.1–76.2
Electrical switches, 68.11–68.15
Electrical trimming, 78.6
Electrode solid-electrolyte contacts, 75.5
Electrolytic hygrometer, 39.8–39.9
Electromagnetic identification, 5.3
Electrooptic links networking:
   active star, 2.44
   daisy chain, 2.44
   hybrid fibers, 2.44
   hybrid wire, 2.43
Electrostatic bonding, 71.12
Emerging Market Consumer, 64.25
Emerging markets, 64.23–64.29
   aerospace, 64.26
   automotive, 64.23
   HVAC, 64.26
   medical, 64.24
   micromachining, 64.27
   process/industrial, 64.25
Emissivity, 18.6, 33.8
   coefficient, 18.7
   emissivity factor, 33.11
   graybody, 18.6, 33.9, 33.11
   Stefan Boltzmann formula, 33.8
   thermal radiation, 33.8
   wavelength, 33.10, 33.11

Emissivity error, 33.9
    filter, 33.8
    infrared, 33.9
    reflective, 33.8
    temperature accuracy, 33.8
End Effector Camera:
    multiple objects, 7.12
    position and orientation of, 7.12
    shape and size of, 7.11–7.12
Endoscope, 6.27
    angioplasty, 6.28
    bronchoscopy, 6.28
    cardioscopy, 6.28
    cystoscopy, 6.28
    gastroscopy, 6.28
    image transmission, 6.29
    laser surgery, 6.28
Endurance, 19.6
    antiferromagnetic, 19.3, 19.7
    copper, 19.7
    temperature, 19.6
Engines (see Automotive applications)
Enterprise Model:
    engineering management, 4.8
    engineering and research of, 4.6
    facilities engineering, 4.7
    marketing, 4.5–4.6
    process development, 4.6
    product development, 4.6
    production planning, 4.8–4.9
    release control, 4.8
Environmental protection, first-level, 71.11
Epoxy resin, for die attach, 71.14
Errors:
    from amplifiers, 70.36
    compensating for, 70.4
    detection of, 78.4
    in sensors, sources of, 70.2
Ethernet, 3.21–3.22, 3.30
    (See also Manufacturing, computer networks)
Excess gain, 2.12–2.13
    curves, chart of, 2.13
Exponential response, 76.22–76.23

Fabrication processes, conventional versus SFB, 67.6–67.9
FDM (finite difference modeling), 69.8–69.9
Federal Highway Administration (FHWA), 38.1
    bridge management, 38.1, 38.5
    NDE, 38.1

Federal Highway Administration (Cont.):
    steel bridges, 38.2, 38.3, 38.4
    (See also Nondestructive evaluation)
FEM (finite element modeling), 69.2–69.8
Ferromagnetic material, 17.1, 17.3–17.4, 17.5
    coils, 17.5, 17.6
    current, 17.5, 17.6
    domain, 17.4
    earth's field, 17.6, 17.7
    electron types, 19.2
    ferrous, 17.4
    Helmholtz, 17.6
Fiber-Optic power
    advantages of, 28.3–28.4
    interface system, 28.2–28.3
Fiber Optic Rate Gyro:
    gyro design, 50.2–50.4
    performance, 50.5–50.7
    ring laser gyroscope, 50.1
Fiber optics, 2.9–2.10, 2.21, 2.23, 18.5
    ammeter, 2.50
    bundle, design configuration, 2.30–2.32
    communication system, 2.22–2.23
    displacement gages, 20.4
    electrooptic links, 2.42–2.46
    extrinsic, 6.4
    fiber pairs, configurations, 2.33
    industrial applications, 2.47–2.49
    intrinsic and extrinsic sensors, 20.3
    sensors, 23.16–23.18
    strain gages, 20.4
    testing, 2.36–2.42
    vibration sensor, 20.4
Fiber-Optic sensing, 20.3, 23.16–23.18
Fiber Pairs, 2.33
Fields, 8.27, 17.1, 17.5, 17.6
    AC, 17.4, 17.6, 17.8
    DC, 17.5, 17.6, 17.8
    eddy currents, 17.8, 23.8
Filaments, heated, 68.9, 68.10
Finite difference modeling (FDM), 69.8–69.9
Finite element modeling (FEM), 69.4–69.8
Fired boiler (See Boiler performance)
Flex degradation, 76.24
Flexibility, 7.1
Flex life, 76.24
Flow rate, 6.15
Flow restrictors, 68.3
Flow sensor, 6.15
Fluid flow:
    microplumbing for, 68.2–68.7
    valves for, micromachined, 68.5–68.7

Fluidic-amplifier-on-a-chip, 68.3–68.5
Fluids, quartz resonators for maintenance of, 79.1–79.18
Fluorescence derivation, 41.9, 44.5–44.6
FMS (Flexible Manufacturing System), 4.16, 7.1
Food processing applications, 71.19
Force Plate Die, 11.11
    oxide layer, 11.11
    tip wafer, 11.11
Force sensors, surface micromachined, 72.1–72.7
Four-wire technique, 76.17
Frequency domain measurements, on coaxial cable, 76.10–76.12
Full-scale output (FSO):
    compensating resistors, 70.17–70.18
    compensation of, 70.15–70.16
Full-scale output (FSO) voltage, 70.3
Fusion-bonded sensors, 67.5–67.18
Fuzzy Logic, 5.7, 5.13–5.14, 80.8, 80.12
Fuzzy representation:
    computation techniques for smart sensors, 80.7–80.8
    reasoning of, 5.7

Gage sensors, 66.8
Gain programming, 70.24–70.25
Gas-chromatograph-on-a-chip, 68.3–68.4
Gas detection system, 6.22
Gas flow, 33.18
    differential, 33.19, 33.20
    instruments for measuring of, 33.19
    limitations, 33.19, 33.20
    thermal dispersion, 33.20
    ultrasonic, 33.20
    ultrasonic transducers, 33.18
Gas sensors, 55.1–55.5, 6.20–6.22
    with suspended structures, 68.10
Geometrical modeling, 69.2
Giant Magnetoresistance Ratio (*see* GMR)
GMR, 17.10
    coercivity layers, 19.3
    coupling field, 19.4
    electrons, free path of, 19.2
    ferromagnetic material, 19.2
    high temperature endurance, 19.6
    hysteresis and linearity, 19.4–19.5
    integrated, 19.10, 19.11
    magnetic field sensors, 19.7
    magnetic layer, 19.3, 19.4, 19.5
    magnetization, 19.2–19.3, 19.5

GMR (*Cont.*):
    materials, 19.1–19.7
    noise, 19.7
    physics, 19.1
    resistivity, 19.5–19.6
    structures, 19.2, 19.3
    surface scattering, 19.2, 19.3
    TCR (Temperature Coefficient of Resistivity), 19.6
    (*See also* Magnetoresistance)
GMR materials, 19.1–19.7
GMR sensor, 19.7
    antiferromagnetic-coupled multilayer, 19.7
    BICMOS, 19.10
    bridge sensor chip, 19.9
    films of, 19.11, 19.12
    flux, 19.9, 19.11
    IC, 19.10
    polysilicon, 19.11
    silicon wafer, 19.11
    structures, 19.3–19.4, 19.11, 19.13
    technology, potential of, 19.12–19.16
    transistors, 19.10–19.11
    Wheatstone bridge, 19.7, 19.10, 19.11
GMR structures, 19.3
    saturation fields, 19.3–19.4, 19.11, 19.13
    tunneling devices, 19.3, 19.4
Gold-silicon alloy, for die attach, 71.14

Hall sensors, 17.9, 21.3
    gallium arsenide, 17.9
    Hall effect, 17.9, 19.10, 21.2
    indium antimonide, 17.9
    indium arsenide, 17.9
    transducer, 17.9
Heart vibration:
    genetic analysis, 12.4
    monitoring, 12.4
    nerve cells, 12.4
High-frequency parameters, of cables, 76.7–76.12
High-temperature devices, piezoresistive sensors, 66.16–66.18
High-temperature sensors, 11.2
    ZMR (zone-melt-recrystallized), 11.2, 11.3, 11.5–11.6
High-volume applications, SFB fabrication for, 67.5–67.6
Honeywell:
    air data-class sensor of, 66.23
    piezoresistive sensors, 77.2

Honeywell (*Cont.*):
  quartz resonant sensor, 77.2, 77.3
  vibrating diaphragm sensor, 77.1–77.2
Hooke, Robert, 69.4
Human color vision, 25.6
Humidity, 24.5
  aluminum oxide instruments, 39.7
  contaminants, 39.11
  instrumentation, 24.5
  measuring of, applications for, 39.29–39.31
  relative humidity, 39.2–39.3
  sensor types, 39.2–39.7
Humidity generators, 39.27
HVAC (heating, ventilating, and air-conditioning), 10.3
Hybrid substrate packages, 71.19
Hydrazine, 41.1, 41.2
  chemiluminescence, 41.2, 41.3, 41.4
  detection of, 40.1, 41.2, 41.4, 41.10
  fluorescence, 41.2
  fluorescence derivatization, 41.9
  oxidation, 41.4
  ppb, 41.1, 41.3
  vapor monitor, 41.5
Hydrazine sensing:
  area monitor, 44.5
  colorimetric detectors, 44.2–44.3
  colorimetric dosimetry, 44.3–44.4
  conductive polymer, 44.6
  electrochemical sensor, 44.2
  fluorescence detection, 44.5
  infrared spectrometry, 44.2
  human carcinogens, detection of, 44.1
  ion mobility, 44.4–44.5
Hygrometer, 39.8, 39.9, 39.23, 39.24
Hysteresis, 19.4, 62.5, 63.6
  digital applications, 19.5
  linear, 19.5
  material, 19.4
  pinned layer, 19.5
Hysteresis and linearity, 19.4

IC (integrated circuits):
  cooling technique, 68.3
  packaging of, 71.3
  packaging and assembly methods from, 71.21–71.22
  signal-conditioning, 78.2–78.6
IGEFET (interdigitated gate electrode field-effect transistor), 42.1, 42.2, 42.4, 42.5
  sensor concept, 42.4
  sensor fabrication, 42.9

IGEFET (*Cont.*):
  sensor operation, 42.10
  sensor performance, 42.14
Imaging, 40.1, 40.2
  scanning, 40.2
  thermal, 40.1
Immittance Spectroscopy (IS), 54.5
Impedance, characteristics, 76.8–76.9
Inductive methods, 1.23, 1.42
  communication, 23.4
  principles of, 23.5, 23.8
  reed relay, 23.3–23.4
Inductive proximity, 1.21
  oscillator circuit, 1.23
  sensing distance, 1.26–1.29
  surrounding conditions, 1.31–1.33
  target shape, 1.29–1.30
Inductive proximity sensors, 1.21, 1.37–1.39, 2.1, 2.15
  applications of, 1.21–1.23
  basic elements of, 1.23
  sensing range, 1.24–1.26
Inductive sensors, 1.1, 1.58, 23.2, 23.3, 23.8
Industrial applications, 9.1
Industrial control applications, distributed I/O for, 74.1–74.21
Infrared (IR), 1.56
  emissivity coefficient, 18.7, 18.8
  frozen foods, 18.8
  fundamentals of, 18.3
  gas, 18.8, 83.1
  instrumentation applications, 83.1
  liquid, 83.1
  noncontact, 84.1
  quartz infrared lamps, 18.8
  source materials, 83.2–83.3
  scrape rates, 18.12
  spectral response, 18.8
  temperature, 18.1, 18.2, 18.5
  thermometry, 18.1, 18.2, 18.3
Infrared analyzer, 83.2
  IR detector, 83.5
  IR source, 83.2
  optical filter, 83.4
  sample cell, 83.3
Infrared camera, 40.1
  (*See also* Thermal imaging)
Infrared filtering, 18.3
Infrared noncontact sensors, 84.1
  detectors, 84.4–84.5
  optical materials, 84.5

Infrared noncontact sensors (*Cont.*):
  target, 84.4
  two-color analysis, 84.5–84.6
Infrared spectroscopy, 83.1
  applications of, 83.5–83.6
Infrared Spin Gyro:
  cursor control applications, 48.2
  navigation, 48.3, 48.5
  pendulum, 48.4
Infrared thermometry, 84.6
  applications of, 84.6
Infusion, 12.3
  blockage, 12.3
  pumps, 12.3
Infusion pumps, 12.3
Input offset voltage, 70.36
Instrumentation, 24.1
  darkroom, 24.2
  design challenges, 24.4
  design of, 24.4, 24.9
  electromagnetic interference, 24.7
  environment, under extreme, 24.1, 24.5
  FAA, 24.2
  lightning and static discharge, 24.7
  meteorology, 24.1
  mountain top, 24.1, 24.2, 24.4
  observatory's, 24.1–24.2
  power disturbances, 24.6
  reliability and maintenance of, 24.8
  snowfall, 24.1–24.2
  summit, 24.1–24.2
Instrument Society of America (ISA), 63.1
Insulation resistance, 76.14–76.15
Insulators, dielectric absorption, 76.15–76.16
Integrated sensors, geometries of, 66.5–66.10
Integration, 4.33–4.38
Intelligent nodes, 73.5
Intelligent transducer:
  hardware of, 32.7
  implementing of, 32.7
  interoperability of, 32.8
  problems and diagnosis of, 32.12
  software for, 32.10
  systems integration and maintenance of, 32.10–32.11
Interchangeability of units, and testing costs, 71.19
Interconnection methods, 71.15
  hybrids, 71.19
Internal parameters, 8.17–8.18
Intrinsic field effect, 75.5

Intrinsic safety, 35.1, 35.2
  acetylene as hazardous vapor, 35.2
  barrier types, 35.6
  ethylene, as hazardous vapor, 35.2
  fibers, 35.1
  flammable gases, 35.1
  hazardous areas, 35.2–35.4
  hydrogen, as hazardous vapor, 35.2
  internal resistance, 35.7
  propane, as hazardous vapor, 35.2
  proper functioning of, 35.5
  safe apparatus, 35.4
I/O (input/output) devices, 34.1
Ion implantation, 66.5
I/O structures, NEURON, 73.8
I/O systems, distributed, 74.1–74.21
ISO (International Standards Organization), 3.14
ISO 9000:
  certification of, 18.12
  environmental, 18.13
  heat, 18.13
  material, 18.12
  temperature, 18.12
ISO sensors, 60.8

Keithley Instruments, *Switching Handbook*, 76.19
Kelvin technique, 76.17
Kidney dialysis, 12.4
  diaphragm, 12.4
  kidney, 12.4
  silicon sensing, use of, 12.4

Lamination techniques, 71.11–71.12
Laser sensors, 1.63, 1.64
  industrial applications, 1.68–1.80
Leadframe-based packages, 71.19–71.22
Leakage current, 76.14–76.15
Least mean square fit technique, 70.29–70.31
LEDs (light-emitting diode), 1.11–1.12, 1.44, 2.2–2.3, 2.18, 5.8, 5.13, 6.4
Lever oscillator circuits, 79.1, 79.8–79.9
Light, 1.68
  energy, 1.65, 2.3
  definition of, 22.2
  diffused, 22.3
  distribution of, 22.3
  metamerism, 22.5
  reflection of, 22.3, 22.4
  sensing of, 22.1
  spectrum of, 1.64, 1.65, 18.4, 22.2, 22.3

Light (*Cont.*):
  specular reflection, 22.3
  transmission of, 22.3–22.4
Light deflectors, 68.15–68.17
Light modulators, 68.15–68.17
Limit switches, 1.37, 2.1, 2.16, 23.1, 23.2
Linearity, of piezoresistive sensors, 66.22
Linear technology, LT1043 dual-switched
  capacitor block, 70.39–70.40
Link blowing technology, 67.3
Liquid level, 6.17
Liquid level sensors, 6.17
  chemical plants, used in, 6.17
  petroleum, 6.18
  prism, 6.17, 6.18
Liquid loading, 79.5–79.6
Liquids, sensors for, 79.1
Load cells, 57.1
  bolt-on weight sensing, 57.2
  calibration for, 57.7
  Load Stand transducer, 57.6
  Microcell sensor, 75.1, 57.2, 57.6–57.7
  temperature-induced stress, solutions for, 57.5–57.6
  two-axis strain sensors, 57.3
Load Stand transducer, 57.6
  electrical characterization, 57.6–57.7
LONTALK, 73.7, 73.9
LONWORKs technology, 73.7–73.10
  media access approach, 73.8–73.10
  for peer-to-peer distributed control implementation, 73.11
Low-frequency cable measurements, 76.13–76.21
Low-G accelerometers:
  angular gyroscope, 52.4
  angular rate, 52.1, 52.4
  applications of, 52.1, 52.7–52.9
  automotive, 52.1, 52.4
Low pressure solution, 10.4
  HVAC, 10.5
Lung tumors, 5.33

Machine Vision:
  DC regulated, 25.6
  illumination, 25.5–25.6
  incandescent, 25.5–25.6
  lighting, 25.5
Magnet-actuated objects:
  containers, identification of, 23.6
  dirt, in environment, 23.6
  metal, sensing, 23.6

Magnet-actuated objects (*Cont.*):
  safety, 23.5
  switch applications, 23.5–23.6
  switch characteristics, 23.6
  switches, 23.5
Magnetic field sensors:
  sensing, 19.1
Magnetism, 17.2
  flux, 17.3, 17.5
  poles, 17.1, 17.3, 17.6
Magnetometers, 17.1, 17.8
  ABC (antilock braking sensors), 17.8
  EMF, 17.8
Magnetoresistance, 17.9
  AMR (magnetoresistive ratio), 19.1
  AMR sensors, 19.4
  GMR, 17.10
  magnetic field, 17.9
Manufactured components, 5.1
Manufacturing, 1.1–1.7, 1.37–1.39, 2.1
  cells, 1.5
  computer networks, understanding of, 3.19–3.22
  conversion process, example of, 1.6
  corporate strategy benefits, 1.7
  fault detection systems, 3.5–3.13
  functional parameters, 3.2, 3.3–3.5
  integrated system of, 1.7
  major functional areas, chart, 4.4
  networking sensors and control systems in, 3.1
  organizations, 1.1
  sensors, use of, 6.3–6.4
  systems, 1.4, 3.1–3.2
MAP (manifold absolute pressure) (*see* MEMs)
MAP (manufacturing automation protocol), 3.23–3.27, 4.4
Masks, 69.11
Master-slave systems, in seriplex sensor/actuatorbus, 74.6, 74.10–74.12
Maxwell fluids, 79.12–79.13
Mechanical interface, of sensors, 71.2
Mechanical stress, and triboelectric noise, 76.1–76.2
Media protection, on sensors, 71.2
Medical applications, 71.8
  blood pressure transducers, 71.5, 71.6
  flow restrictors for, 68.3
  intrauterine pressure sensor, 71.5, 71.6
  metal diaphragm packages for, 71.19

MEMS (microelectromechanical systems), 10.1, 11.1, 12.1, 13.1, 14.1, 16.1
   airbag crash sensor, 16.1
   automotive, use in, 16.2, 16.3, 16.4, 16.5, 16.7
   definition, 13.1
   diagnosis, 12.7–12.8
   flat panel display, 13.1
   future, use in, 12.4, 13.1, 16.1
   manifold absolute pressure (MAP), 16.1, 16.5
   obstacles, in, 12.9, 13.3
   semiconductor, 13.3
   vehicle, 16.1, 16.7
MEMS improved efficiency, 10.1–10.7
   fuzzy logic, 10.4
MEMS, in the:
   integrated circuits, 13.1
   medical applications of, 12.3, 12.4, 12.8
   medical field, 12.1, 12.3, 12.4
MEMS research:
   NASA/JPL, 11.1
   ONR, 11.1
Message event, 73.10
Metalorganic deposition (MOD) method, 75.2
Microbeams, resonant, 77.4–77.7
Microbiology applications, micromachining for, 68.19–68.20
Microcontrollers, 69.13
Microelectrode array sensors:
   design, 59.1, 59.3
   environmental issues, 59.1
   thin film, 59.2
Microelectromechanical systems (*see* MEMS)
Microfabricated sensors, 45.1
   batteries, solid-state secondary, 53.1
   biosensors, 45.1
Microinstrumentation, micromachined detector, 15.3, 15.5
Micromachining, 11.2, 15.5, 55.1, 66.1, 66.5
   bulk, 78.1–78.2
   diaphragms, 66.10
   sensor technology, 52.2
   sacrificial, 66.18
   surface, 72.1–72.7
   for very low pressure sensors, 67.18–67.19
   (*See also* MEMS; Microstructures)
Micromotors, 68.17–68.18
Micropackages, wafer-level, operations, 71.10–71.12

Micropackaging (*see* Packaging technology)
Microplumbing, 68.2–68.7
Microprocessors, 68.1
Microscopes, micromachining for, 68.20
Microstructures, 68.1–68.21
   microbiology applications, 68.19–68.20
   micromotors, 68.17–68.18
   microplumbing, 68.2–68.7
   resonating, 72.5–72.6
   thermally isolated, 68.7–68.11
Microwave power, 1.54, 1.55
Microwave power sensing, 68.11
Microwave sensing, 1.53
   direction, 1.53, 1.55, 1.59, 1.61
   motion, 1.53, 1.55, 1.56
   presence, 1.53, 1.58–1.59
   range, 1.53, 1.60, 1.62
Microwave sensors, 1.53, 1.54, 1.58, 1.59
Military applications:
   silicon accelerometers for, 67.13
   vibrating diaphragm sensors for, 77.1–77.2
Modeling:
   electrical, 69.11–69.13
   finite difference, 69.8–69.9
   finite element, 69.2–69.8
   piezoresistive pressure sensors, 70.4–70.13
   process, 69.9–69.10
   of sensor performance, 70.5–70.6
Molecular relaxation, 43.1
   interferometer, 43.1, 43.3, 43.4, 43.5
   molecules, 43.1, 43.2, 43.5
   rate, 43.1
   spectrometer, 43.1
Monolithic Magnetic Field sensor, 86.2–86.3
MOSFET, 26.9, 42.4–42.5
Motorola Semiconductor, 66.9
   NEURON IC, 73.7–73.8

Navier-Stokes equation, 79.3
NBC standard hygrometer, 39.26
NC controller:
   absolute control, 9.16–9.17
   geometric commands, 9.22, 9.23
   machining program, 9.12–9.15
   manufacturing procedure, 9.11
   motion commands, 9.25
   NC software, 9.18–9.19
   NC system, 9.19
NDR (negative differential resistance), 56.4–56.5
   (*See also* Oxygen sensor)
Neural interface, 12.4–12.7

Neural networks, 8.29–8.30
NEURON IC, 73.7–73.8
 governing firmware, 73.10
Newtonian fluids, 79.3–79.5
 lubricating oils as, 79.12–79.13
NMR (*see* Nuclear Magnetic Resonance)
Node architecture, 27.7
 data path, 27.8
 digital representation, 27.8
 script, 27.8
Node controller, NEURON IC as, 73.7–73.8
Noise:
 in coaxial cable, 76.23–76.24
 crosstalk, 76.9–76.10
 low testing for, 76.21–76.25
 triboelectric, 76.1–76.2
Nondestructive evaluation (NDE), 38.1
 background, 38.1–38.5
 bridges, 38.1–38.5
 research and development objectives, 38.1
 research programs, 38.6, 38.7
 strain-fatigue cracks, 38.1, 38.6, 38.7
Nonlinearity, 70.3
Nova sensor, 66.6
 high-pressure sensors of, 66.11
 low-pressure sensor chip of, 67.18
 SenStable sensor, 66.23, 70.3
 silicon fusion bonding, 71.12
Nozzles, 68.2
Nuclear Magnetic Resonance (NMR), 17.9
 atoms, 17.9
Numerical modeling, 69.2–69.3

Offset voltage, 70.3
 compensation of, 70.13–70.15
On-chip:
 capacitance, 15.2
 digital and analog, 15.2
 microinstrumentation, 15.2
Optical applications:
 light modulators and deflectors, 68.15–68.17
Optical character recognition, 5.3
Optical deflection sensing, 72.5
Optical filters, 83.4, 84.5
Optical pyrometers, 33.8
 blackbody, 33.8, 33.9
 hot gases, 33.8, 33.10
 nonintrusive, 33.8
 radiation, 33.8, 33.9
 wavelengths, 33.8, 33.9, 33.10
Oscillation damping, 67.15

Oscillators, in quartz resonators, 79.8–79.9
OSI (open system interconnect), 3.14, 3.18
Output, sensor, normalization of, 70.24–70.25
Overpressure capabilities:
 of piezoresistive sensors, 66.20–66.22
 protection functions, 71.4
Overrange stops, at wafer-level, 67.14–67.15
Oxygen sensor, 56.1
 device characteristics, 56.4–56.6
 device structure, 56.2
 electrical measurements, 56.3
 MIM (metal-insulator-metal) devices, 56.1, 56.2
 NDR, 56.1, 56.4
 operation of, 56.6–56.7

PACER (Programmable Automatic Contaminant Error Reduction), 39.11–39.15
 circuit, 39.11
 cycle, 39.14, 39.22
 techniques, 39.12–39.14
Packaging technology, 9.31, 71.1–71.22
 application-driven nature, 71.4–71.10
 die operations, 71.12–71.16
 evolution, 71.3–71.4
 functions of, 71.1–71.2
 hybrid, 71.19
 leadframe-based, 71.19–71.22
 oil-isolated, 71.18–71.19
 for signal-conditioning accelerometers, 78.6–78.7
 for silicon sensors, 71.16–71.22
 wafer-level operations, 71.10–71.12
PCMCIA DAQ cards, 34.5–34.6
 (*See also* Data acquisition)
Peer-to-peer distributed control, 73.1–73.11
 expandability, 73.3
 flexibility, 73.3
 interoperability, 73.4
 LONWORKS technology, 73.7–73.10
 moves toward, 73.11
 reliability of, 73.2
 requirements of, 73.4–73.6
Peer-to-peer transducers, 32.1, 32.2
 architecture, 32.2
 communication, 32.1, 32.2
 control loop, 32.2
Personal computer, 9.6
Petersen, Robert, *IEEE Transaction Electron Devices*, 64.1
Photobus (Banner Engineering), 74.2, 74.4

Photo detectors:
  manufacturing, 1.12
  photoelectric control, 1.11
Photoelectric principles, 23.10
  diffuse, 23.10
  principles of, 23.10
  sensing, types of, 23.10, 23.11
  terminology of, 23.10
  thru-beam scanning, 23.11–23.12
Photoelectric sensors, 1.11–1.56, 1.58, 2.1–2.2
  applications of, 2.8
  detection methods, 1.17, 1.18, 1.19, 1.20, 2.1
  detectors, 1.12, 1.13 1.14, 2.2, 2.3. (*See also* photo detectors)
  DPDT (double pole, double throw), 1.12
  light, 1.12, 1.15, 1.16
  operation, principles of, 1.11
  SPDT (singular pole, double throw), 1.12
Photolithography, 69.11
Photomasks, 69.11
Physical-chemical deposition method, 75.2
Physical deposition method, 75.2
Physical distribution, 4.12–4.14
Physical sensors, 26.2
  flow sensor principle, 26.6
  integrated flow sensor, 26.5–26.9
  surface machining pressure elements, advantages of, 26.4–26.5
  surfaced-micromachined capacitive pressure sensor, 26.2, 26.5
Piezoelectric crystals, 6.11
  (*See also* Temperature sensors)
Piezoelectric hygrometer, 39.24
Piezoresistance, 52.2
  versus capacitance, 66.28–66.29
  in silicon, 66.1–66.5
  single crystalline silicon, 11.6
  SOI wafers, 11.5
  of ZMR, 11.5
Piezoresistive pressure sensors, 65.10
  modeling, 70.4–70.13
Piezoresistive sensors, 77.2–77.4
  and dopant concentration, 66.12–66.14
  for extended temperature operation, 66.16
  high-stability, 66.22–66.24
  for high temperatures, 66.16–66.18
  linearity performance, 66.22
  overpressure capabilities, 66.20–66.21
  performance, dynamic, 66.25–66.26
  voltage sensitivity of, 66.24–66.25
Pill boxes, 72.1
  vacuum-sealed, 72.2, 72.5, 72.7

Planar Fluxgate sensor, 86.5
Plant operations. 4.9
  chart, example of, 4.10
PLC (programmable logic controller), 9.7, 18.13, 74.1
  input, 9.7, 23.23
Plug-and-play communication, 80.5–80.6
  software development, 80.13–80.15
Plug-and-play techniques, 8.8–8.9
Plug-in DAQ (*see* DAQ, Plug-and-play communication)
POEM (portable object-oriented environment model), 30.1
  compliance, 30.1
  convert, 30.1
  synchronization, 30.1
Polarization, 2.14–2.15
Polarized Reflex Detection, 2.5–2.7
Polytetrafluorethylene (PTFE), 76.2
Pomerantz, Daniel, 71.11
Position sensors, 2.1, 2.17
  certification, 2.18
  cost, 2.17
  environment, 2.7
  intangibles, 2.18
  sensing distance, 2.17
  speed, 2.17
Positive temperature coefficient resistor effect (PTCR effect), 75.6–75.7
Power supply rejection ratio (PSRR), 70.36
ppb level, 41.1
Presence detection, 23.1
  (*See also* Presence sensors)
  general types of, 23.1
Presence sensors, 23.1
  noncontact, 23.1
  noncontact, sensing technologies, 23.2
  noncontact sensor, advantages, 23.2
  physical contact, 23.1
  reed relay, 23.3, 23.4, 23.5
  sensor die, 60.2, 60.6
  switches, 23.1–23.2
  ultrasonic, 23.2
Pressure:
  linear approximation of, 70.28–70.32
  parabolic approximation of, 70.32
Pressure hysteresis, 70.3
Pressure nonlinearity, 70.3
Pressure scales, 66.6
Pressure sensors, 6.10, 12.1, 12.3, 26.2, 31.1
  advanced designs, 67.1–67.19
  automotive engine cylinders, 31.2

Pressure sensors (*Cont.*):
  capacitive absolute, 67.12–67.13
  characteristics of, 70.2–70.3
  communication networking, 60.10–60.11
  disposable sensor, 60.1
  head design, 31.3–31.4
  high temperature, 31.1, 67.11–67.12
  low cost, 60.2, 60.6
  measurement configurations of, 66.6
  medical disposable sensors, 60.1
  miniature sensors, 60.5
  on-chip compensated, calibrated, 67.2–67.5
  transducer, 12.1
  very low pressure, 67.18–67.19
  with SI/SI bonding, 67.5–67.18
  surfaced micromachined, 72.2–72.4
Principles of operations, sensing range, 1.11–1.12
ProBlox (NAMCO), 74.2, 74.4
Process diagram, 9.1
Processes, 11.16–11.17
Process modeling, 69.9 69.10
Production planning, 4.8–4.9
Proof Mass die, 11.10
  bond pads, 11.11
  crab legs, 11.11
  front surface, 11.11
Propagation delay, 76.9
Prototype system, 27.6
  application example, 27.6–27.7
  Laboratory Ambient Condition monitoring, 27.11
  miscellaneous systems, 27.11
  observations, 27.11–27.12
  Printed Circuit Board manufacturing, 27.10–27.11
  smart nodes, appendix of, 27.13–27.17
Proximity detection, 1.21, 2.7
Prox switches, use of, 74.2
Psychrometer, 39.24

Quality control, 85.1
  considerations, 85.1–85.2
  design assurance, 85.2
  SPC (statistical process control), 85.1
  SQC (statistical quality control), 85.1
Quantitative analysis, 8.27–8.28
Quartz crystal microbalance (QCM), 79.6
Quartz resonators, 79.1–79.18
  for battery charge monitoring, 79.13–79.15
  for coolant monitoring, 79.16–79.18

Quartz resonators (*Cont.*):
  for engine oil monitoring, 79.10–79.13
  oscillators for, 79.8–79.9
Quartz Rotation Rate:
  applications, 49.4–49.6
  construction, 49.3
  control, 49.6–49.9
  instrumentation, 49.4–49.6
  theory, 49.1–49.3
Quiet lines, 76.9

Radiation detectors, 68.9
Radiation pyrometers, 33.8–33.11
Radio, 23.24
  frequency, 23.24
Reflective Strip Imaging, 5.30
Relaxation, 69.12
Relays, 1.39
Repeatability, 23.8, 62.5, 70.3
  capacitive, 23.8
  inductive, 23.8
  precision, rating for, 23.8
  relay(reed), 23.8
  switches, limit, 23.8
Research applications, 71.5, 71.7
Resistive Polymer sensor, 39.3
  (*See also* Humidity)
Resistors:
  compensating, 70.16–70.20
  series, 70.20–70.21
Resonance, 17.9
Resonant integrated microsensor (RMI), 72.5–72.6
Resonant microbeam technology, 77.4–77.7
Resonant sensors, 77.1–77.10
  interfacing, 77.7–77.9
  quartz, 79.1–79.18
Resonant structures, silicon, 66.18–66.19
Response, 1.49
Response time, 1.49, 1.50, 1.51, 23.25, 23.27
  inertia, 23.25
RG-58 cable, 76.2
RG-174 cable, 76.2
RG-178 cable, 76.2
RG-316 cable, 76.2
RGB (red,green,blue), 5.10, 5.11
  color sensing, 5.10, 5.12, 22.7
  GO/NO GO, 22.6
  halogen light source, 22.6
  web movement, 22.6
Riley, Robert, 74.5

Robot assembly:
  accuracy, 7.38
  gripper, 5.26
  handling, 7.7, 7.36
  robot, programming of, 7.35–7.36
Robot Guidance with Vision, 7.9
  inspection of, 7.4, 7.5, 7.9, 7.31
  robot, components of, 7.10
Robotic gripper, 5.26
Rolling reflex testing, 76.24–76.25

Saturated salt (lithium salt) sensor, 39.25
Saturation, 5.8, 5.9
Saving current scan, 8.12
Scanners, for cable testing, 76.19
Scanning, 8.14, 8.15, 8.19, 8.20
  proximity(diffuse), 23.10, 23.14, 23.15
  reflector, 8.5
  retroreflective, 23.13
  retroreflective, polarized, 23.14
Scanning Tunneling Microscope (STM), 68.20
Schottky barrier junctions, 75.4–75.5
Screen printing, 75.2–75.4
SDS (Micro Switch CANbus), 74.2, 74.3
Sealed gage sensors, 66.8
Secondary batteries:
  anode, 53.5
  cathode, 53.1, 53.2, 53.3–53.4
  DC magnetron, 53.2
  electrolyte, 53.2, 53.4
  electrolyte layer, 53.1
  Lil layer, 53.1, 53.2, 53.5
  lithium anode, 53.1
  microbattery, 53.1, 53.5
  results and discussion, 53.3
  RF magnetron, 53.2
  sputtered film, 53.3
  vacuum evaporation, 53.2
Seismic mass, 67.14
Semiconductor, 1.67, 5.10, 6.4, 6.5, 6.8
Semiconductor pressure transducer, 62.1
  details of, 62.1
  electrical, 62.1, 62.4
  performance, 62.1, 62.4–62.8
  physical/mechanical, 62.2–62.3
  selection of, 62.2
Semiconductors:
  covalent and ionic, 75.4
  device fabrication, 6.8
  photoreceiver, 5.10
Semiconductor strain gage, 66.2–66.3

Sensing distance, terminology for, 23.6–23.7
Sensing technologies:
  differential, 23.7–23.8
  repeatability, 23.8
  temperature, 24.3, 24.5
Sensor Actuator Bus (SAB):
  candidates, 29.3
  evaluation efforts, 29.2–29.3
  evaluation, process of, 29.4–29.6
  interchangeability, 29.1
Sensor die (*see* Pressure sensors)
Sensor errors, compensating for, 70.4
Sensor packaging, technology of, 15.3
  (*See also* Packaging technology)
Sensor performance, 15.5
Sensorplex (Turck), 74.2, 74.4
Sensors, 1.4, 1.7, 6.1, 76.2
  bridge-based versus resonant, 77.7–77.8
  ceramic, 75.1–75.7
  chemical and biochemical, 26.7–26.9
  chemical and gas, 55.1–55.5
  excitation current sources, 70.33–70.35
  force, 72.1–72.2
  gas, 6.20–6.22
  GMR, 19.7–19.16
  Hall, 17.9, 21.3
  high-pressure, 66.11
  high-pressure, SFB, 67.8–67.10
  high-temperature, 11.2
  inductive, 1.1, 1.58, 23.2, 23.3, 23.8
  inductive proximity (*see* Inductive proximity sensors)
  infrared noncontact, 84.1
  interchangeable, 70.24
  ISO, 60.8
  laser, 1.63, 1.64, 1.68–1.80
  liquid level, 6.17, 6.18
  low-pressure, 66.10–66.11
  low pressure, SFB, 67.10–67.11
  magnetic field, 19.1
  microfabricated, 45.1
  microwave, 1.53, 1.54, 1.58, 1.59
  Monolithic Magnetic Field, 86.2–86.3
  oxygen, 56.1–56.7
  packaging technology, 71.1–71.22
  photoelectric (*see* Photoelectric sensors)
  piezoresistive, 77.2–77.4 (*See also* Piezoresistive sensors)
  piezoresistive pressure (*see* Piezoresistive pressure sensors)
  presence, 23.1, 23.2, 60.2, 60.6
  pressure (*see* Pressure sensors)

Sensors (*Cont.*):
    quartz resonators, 79.1–79.18
    resonant, 77.1–77.10
    resonant integrated microsensor, 72.5–72.6
    saturated salt, 39.25
    signal conditioning for, 70.1–70.44
    silicon piezoresistive pressure, 66.1–66.26
    smart, 70.41–70.44
    thin/thick film ceramic, 75.1–75.7
    vacuum-sealed, 72.2–72.4
Series resistance, 76.16–76.18
SERIPLEX sensor/actuatorbus, 74.2, 74.3, 74.5
    ASIC of, 74.9–74.10
    CPU interfaces, 74.13–74.19
    description of, 74.5–74.8
    I/O devices of, 74.19–74.20
    master/slave mode, 74.10–74.12
    open architecture of, 74.20–74.21
    peer-to-peer mode, 74.12–74.13
SFB (silicon fusion bonding), advantages of, 67.17–67.18
Shape function, 69.6
Shear geometry, 66.4–66.5
Shielding effectiveness, 76.11–76.12
Signal conditioning, 4.24, 78.2–78.6
    in sensor package, 71.2
    for sensors, 70.1–70.44
Signal domain, 64.3
    chemical, 64.4, 74.7
    electrical, 64.4
    magnetic, 64.4, 64.6
    mechanical, 64.3, 64.5
    radiant, 64.3, 64.5
    thermal, 64.4, 64.6
Signal processing. autozeroing circuit for, 70.40–70.41
Single-board computer, 9.1
    RAM, 9.1, 9.4
    ROM, 9.1
Single wavelength, 18.7
    total energy, 18.7
    webs of paper, 18.8
Silicon, 15.1, 15.6, 16.2, 16.3, 17.9, 19.9, 26.1, 63.1, 64.1
    definition of, 65.1
    microfabrication in chemical and gas sensors, 55.1
    micromachining, 26.2
    as optical material, 68.16–68.17

Silicon (*Cont.*):
    p-doped, 66.3
    piezoresistance in, 66.1–66.5
    for resonant structures, 66.18–66.19
    single-crystal, 65.1
Silicon accelerometers, 66.29–66.31
Silicone rubber, for die attach, 71.14
Silicon fusion bonding, 64.16
    for wafer lamination, 71.12
Silicon market:
    emerging markets, 64.23–64.29
    market trends, 64.32–64.33
    size and growth, 64.19–64.23
Silicon micromechanics:
    advantages/obstacles, 64.13–64.18
    chip stress isolation, 64.15
    DEEP Silicon Etch, 64.15
    silicon fusion bonding (SFB), 64.16
    silicon wafers, 64.15, 64.17
Silicon microstructures, 65.4, 65.10, 68.1–68.21
Silicon piezoresistive pressure sensors, 65.10, 82.1
Silicon pressure sensor chip, 65.13
    etching, 65.9–65.10, 65.12
Silicon pressure sensors, 65.10
Silicon processing:
    additive processes, 65.8
    diffusion, 65.5–65.6
    electrical properties, 65.2
    insulators, 65.7–65.8
    ion implantation, 65.5–65.6
    mechanical performance, 65.3
    mechanical properties of silicon, table, 65.3
    metal layers, 65.7
    micromachining, 65.11, 65.12
    photolithography, 65.5, 65.6
    subtractive processes, 65.9–65.10
Silicon resistors, 64.17
Silicon sensor, 15.1, 15.2, 15.6
    accuracy, 63.1
    aerospace/military applications, 64.28
    anodic bonding, 63.2
    boron, 63.2
    burst pressure, 63.2
    capacitive pressure sensors, 66.26–66.29
    chronology of technology, 64.11
    compensation, 63.2
    consumer applications, 64.27
    cross-axis sensitivity, 63.2, 66.4
    CVD, 63.2

Silicon sensor (*Cont.*):
  dead volume, 63.2
  definitions, 63.1–63.5
  diffusion, 63.2
  evolution and growth in technology, 64.8–64.13
  excitation, 63.2
  frequency, 63.2
  on-axis sensitivity, 66.4
  packaging options for, 71.16–71.22
  performance of, 66.18–66.20
  piezoresistive, 66.1–66.26
  resonance, 63.8
  temperature sensitivity, 66.12
  thermal zero shift, 63.9
  transverse sensitivity, 63.9
Silicon sensors and microstructures, 64.1
  characteristics, 64.2–64.3
  characterization of effects on design, 64.5–64.7
  classification, 64.3–64.4
  market trends, 64.32–64.33
  micromachining as foundation for, 64.1, 64.2
  sensor market definition, 64.18–64.18
  technology trends, 64.29
Silicon sensor terminology:
  general definitions, 63.1–63.5
  performance-related definitions, 63.5–63.9
Silicon-silicon bonding, 67.5–67.18
Silicon-silicon dioxide-polysilicon micromachining, 72.1–72.2
Sinusoidal response, 76.22–76.23
Small air bubbles, 6.16
Smart design, 73.1
Smart Force Transducer (SFT), 47.16
Smart nodes, 27.7
  actuators, 27.8
  appendix of various models and interfaces, 27.13–27.17
  network interface, 27.10
  node monitors, 27.9
  operational aspects, 27.9
  transducer, 27.8
  transducer, interface, 27.10
Smart sensing:
  accelerometer, 61.3
  discrete component, 61.1–61.2
  pressure sensor, 61.4
  single-chip, 61.2
  two-chip, 61.2–61.3

Smart Sensor, 15.3
  actuators, 27.1
  computer control system model, 80.5
  distributed measurement, 27.1–27.3
  flexible architecture, 80.15–80.16
  networked, 27.1
  reactive object classes, 30.13–30.14
  technology of, 60.9–60.10
Smart sensors, 15.5, 27.1, 27.2, 70.41–70.44
  active object classes, 30.11–30.12
  active objects, 30.8, 30.15
  key attributes, 30.2
  object model, 30.8
  OO technology sample, 30.6–30.7
  programming model, 30.7
  reactive objects, 30.10, 30.16
  software engineering issues, 30.5
  system integration, 30.2
  technology, 30.6
Smart structure, 20.1, 20.2
  civil, 20.1, 20.2
  intelligence, 20.2
Smart transducers, 32.1
  digital compensation and normalization, 82.5
  errors, reducing, 82.2
  Single-Function model, 82.9–82.10
  statistical compensation, 82.3–82.4
Smith, C. S., 66.1
Snapstrate, 60.4
Software development, 4.28
  aids, 69.13
  compatible, 3.9
Software feedback loop, 82.8
Software structure, 7.34–7.35
SOI (silicon-on-insulator), 11.1–11.3, 11.5, 11.18, 66.17–66.18
  process of, 67.12
SOI sensor, 11.1
Solid state, 23.20
Solid-state sensor, 23.7, 52.1
  differential, 23.7–23.8, 52.2
  genealogy table in the U.S., 64.12
  microbatteries, secondary batteries for, 53.1
  technologies, 23.18–23.19
  TO mounted, 71.17
Sound Vision:
  positioning, 7.24
  sensitivity, 7.23
  surface measurements, 7.22, 7.23
  standoff, 7.21

Spectrometer, 6.20, 8.4, 8.6, 43.1
Spectroscopy, gas monitoring, 6.20
Spectrum, 8.10, 8.14–8.16, 8.18–8.19, 8.21
SpectRx Alignment Procedure, 8.9
  optimum distance, 8.10
  production parameters, 8.11
  production run, 8.11
  reference, 8.10
  standard scan, 8.10–8.11
SPICE circuit simulator, 69.12
SQUID (superconducting quantum interference device), 17.10, 19.7
Stability errors, 70.3
STACKWATCH, 33.26–33.28
  (*See also* Volumetric flow)
Standard network variable types (SNVTs), 73.9–73.10
Static errors, 70.3
Static pressure compensation, 70.26–70.27
Static pressure sensitivity, 66.8
Strain gages, semiconductor, 6.3
Strain, sensing, with resonant microbeams, 77.6–77.7
Strain sensor, 57.1–57.2
Strategic Partner Program, 74.20
Stress isolation:
  first-level, 71.11
  second-level, 71.11–71.12
  in sensors, 71.2
Stress measurement, structure, 5.27–5.28
Stress/strain relationship, FEM for, 69.5
Structural problems, FEM for, 69.5–69.6
SUPREM (Stanford University Process Engineering Model), 69.10
Surface acoustic waves, 5.3
Surface machining:
  accelerometers, 52.3
  capacitive pressure sensors, 26.2–26.5
  pressure sensors, 86.1
Surface micromachining, 72.1–72.7
  micromachining, 11.12
  substrate, 11.14
Switches, electrical, 68.11–68.15
Switching Handbook (Keithley), 76.19
Synchronous production, 9.35
Synthetic aperture radar, 1.1

Tape casting, 75.2–75.3
Temperature, 6.4, 6.5, 6.8, 18.1, 24.3
  linear approximation of, 70.28–70.31
  parabolic approximation of, 70.31–70.32
  and signal conditioning, 70.2

Temperature coefficient of input offset voltage, 70.36
Temperature coefficient of pressure sensitivity (TCS), 70.4
Temperature coefficients, 66.12
  piezoresistive versus doping level, 66.14–66.16
Temperature compensation: 70.23–70.26
  applications, 70.7–70.10
  with series resistors, 70.20–70.21
Temperature effects, modeling, 69.8
Temperature errors, 70.3
Temperature sensors, 6.4, 6.5, 66.4
  photoluminescence, 6.6–6.8
  piezoelectric crystals, 6.11
  pyrometers, 6.8–6.10
  semiconductor absorption sensors, 6.5–6.6
  strain gages, 6.11
Thermal compensation, 66.12–66.14
Thermal conductivity, sensors for, 68.7–68.9
Thermal dispersion, 33.20
Thermal imaging, 40.1
  camera cooling systems, 40.2–40.12
  defect detection system, 40.4
  focal plane array camera, 40.2
  infrared camera evaluation, 40.1
  processors, 40.3–40.4
  scanning camera, 40.2
Thermistors, 54.1, 54.6
  ceramic sensors, 54.1, 54.6
  compensation for, 70.21–70.22
Thermocouples, 33.6
Thermoelectric infrared radiation sensors, 86.7
Thevenin equivalent circuit, 76.2–76.4
Thick film process, 75.2–75.4
  for ceramic sensors, 75.5–75.7
Thick film sensors, applications of, 75.1
Thickness shear mode (TSM) quartz resonators, 79.1–79.18
Thin film process, 75.2
  for ceramic sensors, 75.5–75.7
Thin film sensors, applications of, 75.1
Thin/thick film for ceramic sensors, electrode contacts, 75.4–75.5
Third-order polynomial compensation, 70.32–70.33
Thru-Beam:
  detection, 2.3–2.5, 23.12
  scanning, 23.11
Tick Tock testing, 76.24

Time domain measurements, on coaxial cable, 76.7–76.10
Time domain reflectometer (TDR), 37.1, 76.7
  cable installation, 37.4–37.7
  cable selection, 37.4
  definition, 37.2
  high-speed, 76.1
  technology, 37.1–37.2
Titanium oxide ceramic, 75.6
Torsional mirror silicon, 68.15–68.16
TO series packages, 71.16–71.18
Transducers, 12.1, 12.2, 27.2, 27.15, 30.1
  disposable blood pressure, 71.5, 71.6
  metal diaphragm, oil-isolated, 71.18–71.19
  pressure, for differential pressure applications, 70.25
  resonant, 72.5–72.7
  selection factors, 62.2
Transduction, applied force, 72.5
Transistor, 1.41, 1.42
  sinking, 23.20
  sourcing, 23.20
  switching, 1.41–1.42, 23.20
Transponders, 5.3
Triac devices, 1.39–1.40
Triboelectric noise, 76.1–76.2, 76.23
Trimming, 70.4
  electrical, 78.6
Tubes, microminiature, 68.2–68.5
Tuning fork gyroscope, 51.1
  electronics, 51.5–51.6
  fabrication, 51.3–51.5
  micromachining, 51.1
  results, 51.6–51.8
  theory, 51.2
  tuning fork, 51.1
Two-axis strain sensor, 57.3

Ultrasonic end effector, 7.19

Vacuum-sealed sensors, 72.2–72.4
Vacuum sensors, 66.8, 68.9
Valves, micromachined, 68.5–68.7
VariNet-3 (Pepper1 + Fuchs), 74.2, 74.4
Velocity of propagation, 76.9
Vibrating sensors, 77.1
  (*See also* Resonant sensors)
Vibration measurement, 5.17

Viscosity, determination of, 79.10–79.13
VLSI, and piezoresistive sensors, 66.22
Voltage, dialectric withstand, 76.18, 76.19
Voltage excitation, constant versus constant current, 70.4
Volumetric flow, 33.26
  sound waves, 33.26
  STACKWATCH, 33.26–33.28

Wafer-level operations, 71.10–71.12
  lamination techniques, 71.11–71.12
Wafers, 15.5
  bonding, 15.5
  lamination of, 64.18
  sawing, 71.12
  self fine alignment, 15.6
Walliss, Gordon, 71.11
Weigh belt feeders, 47.1
  applications of, 47.20
  basics, 47.4
  belt scales, 47.22
  calibration, 47.22
  feeding principles, 47.5
  gravimetric, 47.1, 47.2
  mass flow, 47.4
  mechanical design strategies, 47.6
  sensors and controls, 47.13–47.17
  triangle, 47.3–47.4
  weigh belt, 47.9
Weighing instrumentation, 58.1
  communication networks, 58.6
  digital filtering, 58.2–58.3
  MODBUS, 58.6–58.7
  model, 58.6–58.7
  sigma delta A/D conversion, 58.1–58.2
  synchronous A/D control, 58.3
  system diagnostics, 58.3–58.6
  technologies, example of, 58.1
Weld field, 23.24–23.25
Wheatstone bridges, 72.2
  for common-mode rejection, 60.2, 71.11
Wire bonding, 71.15
Wire technology:
  AC/DC, 23.23
  three-wire, 23.21
  two-wire, 23.22
W.L. Gore & Associates, 76.2
Workstations, 1.3–1.5
  cell, 1.4, 1.5, 6.2
  center, 1.4, 1.5

ABOUT THE AUTHOR

Sabrie Soloman (Ridgewood, NJ) is chairman and CEO of American SensoR, Inc., and professor in the Columbia University–founded Advanced Manufacturing Technology Program. He is author of McGraw-Hill's *Sensors and Control Systems in Manufacturing, Affordable Automation, Introduction to Electro-mechanical Engineering,* and *Modern Welding Technology.*